Biology of the Laboratory Mouse

By
THE STAFF
of
THE JACKSON LABORATORY

EARL L. GREEN, *Editor*

Douglas L. Coleman
Charles P. Dagg
John L. Fuller
Margaret C. Green
Nathan Kaliss
Elizabeth S. Russell
Joan Staats

Section Editors
Eunice U. Fahey, *Assistant Editor*

SECOND EDITION

DOVER PUBLICATIONS, INC., NEW YORK

Published in Canada by General Publishing Company, Ltd., 30 Lesmill Road, Don Mills, Toronto, Ontario.

Published in the United Kingdom by Constable and Company, Ltd., 10 Orange Street, London WC 2.

This Dover edition, first published in 1975, is an unabridged and unaltered republication of the second (1966) edition, as originally published by the McGraw-Hill Book Company.

International Standard Book Number: 0-486-63185-0
Library of Congress Catalog Card Number: 75-2614

Manufactured in the United States of America
Dover Publications, Inc.
180 Varick Street
New York, N.Y. 10014

Dedicated to

CLARENCE COOK LITTLE

founder of The Jackson Laboratory
and diligent student of the
biology of the laboratory mouse
in the cause of human welfare

Preface to the Second Edition

Since the publication of the first edition of the *Biology of the Laboratory Mouse* in 1941, a large amount of new information appropriate for inclusion in such a book has accumulated. Mice have been used in increasing numbers as materials for investigations of basic issues in biology and as model systems for studying diseases and processes related to human health. In particular, there has been a great increase in the use of inbred strains, an increase in the number and use of stocks carrying named mutant genes, and an increase in awareness on the part of most biologists of the importance of hereditary factors in determining the characteristics of organisms. The second edition has been written to summarize the studies with mice published in the years since the first edition and to record much new information not previously published. In keeping with the trend toward use of genetically controlled stocks, the emphasis throughout the book is upon genetic variations of mice and their exploitation for the solution of biological problems.

The second edition follows somewhat the same plan as the first but is much enlarged. The passages on histology and early embryology have been changed little, but nearly all the other parts have been completely rewritten, most of them by different authors. The thirty-three chapters may be grouped into seven sections. Chapters 1 to 6 introduce the reader to the techniques of husbandry and to genetic nomenclature. Chapters 7 to 10 give information about the genetics of mice, including an alphabetical catalogue of the named mutations. Chapters 11 to 15 deal with the anatomy, development, and reproductive physiology of mice; 16 to 21 with physiology and biochemistry; 22 to 25 with the responses of mice to radiation, drugs, and foreign tissues; and 26 to 31 with pathological conditions and immune functions. The final two chapters, 32 and 33, deal with behavioral traits. In every chapter the authors' objective has been to provide the reader with a summary of the known facts and an introduction to the earlier literature. In most cases the references are not intended to be exhaustive, but are extensive enough to allow the reader to find his way into the older literature if he so desires.

Many topics had to be omitted entirely. These include experimental embryology, organogenesis, late responses to radiation, responses to stresses other than radiation and drugs, and induced tumors. A list of characteristics of inbred strains of mice is not included, but publications in which lists may be found are given in Chapter 1.

The authors were supported by a variety of sources of funds while writing the manuscript. These sources are individually acknowledged in a footnote on the first page of each chapter.

An undertaking of this magnitude is impossible without the devoted and conscientious assistance of many people with many different talents. The Section Editors, whose names appear on the title page, served as advisers to the Editor while the outlines and arrangements of all chapters were being developed, as readers and critics of the chapters in their respective sections, and as critical reviewers of the entire manuscript before final revisions. In addition to the Section Editors, Dr. Katharine P. Hummel and Dr. Edwin D. Murphy also helped me, especially on questions about the contents of chapters and about the illustrations. Dr. Gunther Schlager devised a method of constructing the index by use of punched cards. Mrs. Eunice U. Fahey handled the successive outlines and drafts of all the chapters and scrutinized hundreds of pages of text for accuracy of notations, symbols, spelling, and the like.

I am happy to record my indebtedness to Isabelle Stover, Lillian Runstuk, Josephine Foley, Florence Smith, Dencie Anthony, Rita Richard, Jennie Jenkins, Dorothy Killam, Jeannette Cleaves, and Bernice Sylvester who collectively typed successive versions of the chapters and to Ann Schlager who prepared the index cards. I appreciate the diligent efforts of George C. McKay, Jr., and of Ruth M. Soper in preparing the photographs and the drawings for the figures.

Finally, I am grateful to a number of people whose contribution to the book is less specific but not less significant. This group includes Dale J. Foley, Alan P. Russell, George T. Vose, Austin C. Carter, and Robert E. Theriault. They made it possible for the Editor, the Section Editors, and the authors to complete their tasks.

Bar Harbor, *March, 1966* Earl L. Green, *Editor*

Preface to the First Edition

Of all the laboratory mammals, probably none has contributed more to the advancement of knowledge than the common mouse. Certainly among all the mammals it is the most widely used, for not less than one million mice are raised each year in this country for research in bacteriology, cancer and genetics.

A result of this extensive use of the mouse is that a large body of information has grown up concerning it. This, however, is so widely scattered through the literature that it is often a major undertaking for the research worker who wishes to use it to locate and gather the particular facts that he needs. Much of this information is assembled in this book. In a number of cases, where there are important gaps in the literature, these have been filled in by special research projects. In general, controversial material has been avoided or given only brief mention. The emphasis is placed on established facts useful to the research worker.

Certain fields, for example anatomy and endocrinology, have of necessity been largely omitted. In most cases material omitted is adequately covered in other recent books.

Because it deals with the mouse alone, this book presents a vertical cross-section of biological knowledge rather than the more usual horizontal cross-section. It contains information about one animal drawn from various branches of zoology, rather than information about one branch of zoology drawn from observation of a variety of animals. There is, I believe, one notable virtue in this vertical method of presentation, namely, that it makes the synthesis of biological knowledge somewhat easier. There is a widespread feeling among biologists that progress will depend increasingly on the synthesis of the specialized techniques which have been developed within the individual cubby-holes into which science is somewhat arbitrarily divided. The departmentalization of biology is a convenience not to say an absolute necessity, but within the organism the tissues, the genes, the endocrines, the diseases and the processes of development are all intimately related, and the biologist frequently finds that research in his own specialty is leading him straight into another field of knowledge. At the present time there are, for example, increasingly well beaten paths between genetics and embryology, between endocrinology and cancer research, between cancer research and bacteriology, between bacteriology and genetics. It is a major purpose of this book, by gathering together the fundamental knowledge about the mouse from several fields of study, to make it easier for the research worker using mice as his experimental material to traverse these interconnecting paths of science.

The preparation of the book has been financed by a grant from the John and Mary R. Markle Foundation. This generous support has made possible the conduct of several pertinent research projects and the preparation of many original photographs and drawings. The embryological studies described in Chapter 1 have also been aided by a grant from the Alexander Dallas Bache Fund of the National Academy of Sciences. In the preparation of their material the authors have been ably assisted by the following persons: Miss Olive Bartholomew, preparation of embryological and histological sections; Miss Bernette Bohen, drawings; Mr. Joshua Burnett, tabulation of linkage data; Dr. Elizabeth Chase, histological sections; Dr. Katharine P. Hummel, photography; Mr. Arthur Lieberman, bibliography; Mr. John Mowat, photography and construction of apparatus; Mr. William Payne, photography; Miss Ella Rowe, preparation of sections; Miss Elizabeth Keucher, assistance in preparation of the index. Prof. C. H. Danforth has made valuable suggestions in regard to several parts of the text.

In conclusion, the editor would like to express his appreciation to the other members of the Laboratory Staff for their continued cooperation and for many valuable suggestions, and to Dr. C. C. Little for his hearty support and, in a broader sense, for the wise direction in a large measure responsible for the friendly atmosphere so essential for successful collaboration.

GEORGE D. SNELL, *Editor*
Roscoe B. Jackson Memorial Laboratory
Bar Harbor Maine

Contents

1

The Laboratory Mouse[1]

Joan Staats

Mus musculus Linn., the common house mouse, has been a member of man's immediate environment for many centuries. Along with other members of the order Rodentia, rats and mice constituting the family Muridae spread with man and his commerce from their origin in Asia to all parts of the world. A thorough account of the antiquity of the fancy mouse and its interactions with the human species is given by Clyde E. Keeler (1931).

The conversion of the mouse from pest to pet to productive element of the scientific community took place slowly. During the 19th century a number of European zoologists bred fancy mice to investigate varietal characters and tried to interpret the results by Galton's law of ancestral inheritance. The valuable information thus acquired, however, could not be correctly interpreted until the rediscovery of Mendel's work in 1900. The problems of inbreeding, selection, decrease in fertility, appearance of abnormalities, and increased susceptibility to disease noted by earlier workers were attacked from a new viewpoint in the light of Mendel's findings. Cuénot's 1902 papers in *Archives de Zoologie Expérimentale et Générale* seem to be the first to apply Mendelian principles to animals (Bateson, 1903). William E. Castle working with *Drosophila,* Sewall Wright with guinea pigs, and S. Hatai and Helen Dean King with rats provided early examples of the new scientific breeding of animal forms (Castle et al., 1906; Wright, 1922; King, 1911).

ORIGINS OF COMMON INBRED STRAINS

Clarence Cook Little, a Harvard undergraduate, began studying the inheritance of coat color in mice under Castle's tutelage in 1907. Two years later he obtained a pair of mice carrying the recessive genes for dilution, brown, and nonagouti. During the next few years he inbred the descendants of this pair brother to sister for more than 20 generations, with selection for vigorous animals, thus creating the first inbred strain of mice, which he named dbr. It was later called dba after the three recessive genes, and since about 1950 has been written DBA.

Little was interested in the study of neoplastic diseases and recognized that difficulties were bound to arise in dealing with a condition which appears relatively late in life and is subject to much environmental influence. He reasoned that elimination of the great genetic diversity in unrelated animals would facilitate that study.

Within 15 years after the origin of the first inbred strain, most of the others used in cancer research had been established. Relationships between strains and families of strains are shown in Fig. 1-1.

In 1913, Halsey J. Bagg obtained some albino mice from a dealer in Ohio, maintained them as a closed colony, and used them in behavioral experiments. In 1921, Leonell C. Strong mated a mouse of the Bagg albino stock with one from an albino stock Little maintained at that time at

[1] The writing of this chapter was supported in part by National Science Foundation grant G-18485, The Louise H. and David S. Ingalls Foundation, and the Edwin S. Webster Foundation.

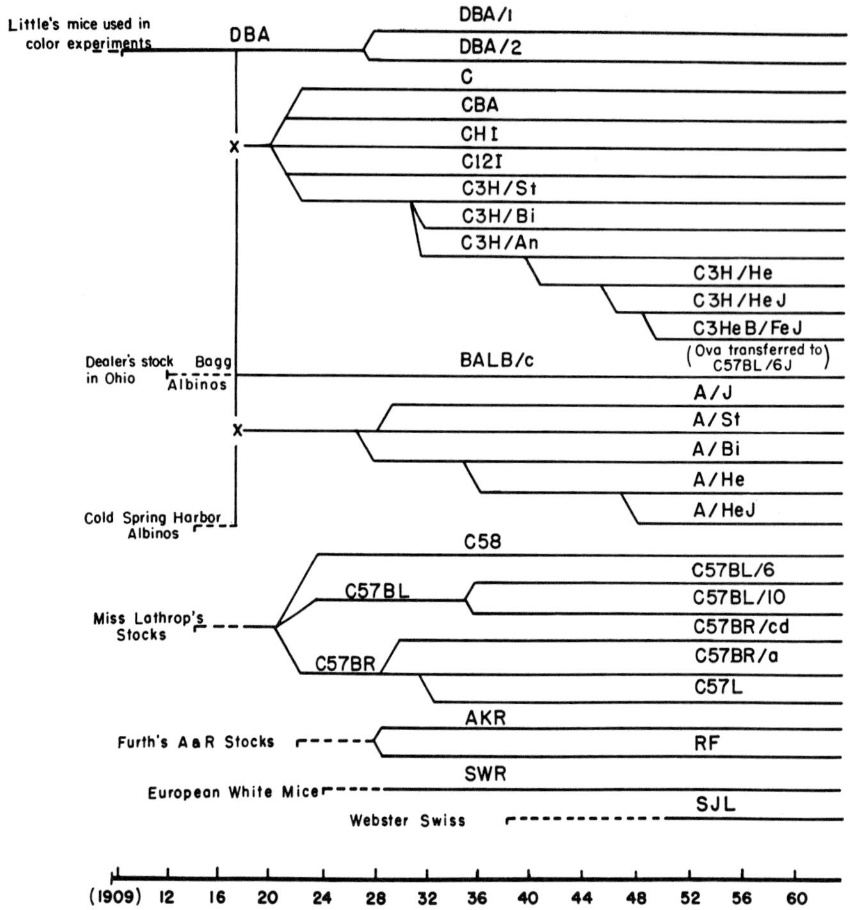

Fig. 1-1. Origins of the older inbred strains of mice.

Cold Spring Harbor. From this cross Strong started the A strain, a high mammary and lung tumor strain (Heston, 1949). In 1920, Strong made a series of crosses between the Bagg albinos and strain DBA, and from the hybrids developed a number of inbred lines: C3H, CBA, C, CHI, and C12I. Of these the C3H has been the most widely used and has been split into several sublines with well-defined differences between them (Staats, 1964).

Another well-known and widely used family of strains also dates from 1921. While he was at Cold Spring Harbor, Little obtained mice from Miss A. E. C. Lathrop, a fancier in Granby, Massachusetts, and mated littermates ♀57 and ♂52. Progeny of this black pair segregated as black and brown; inbreeding then led to the C57BL and C57BR strains. C57L was developed by J. M. Murray from a color mutant in a C57BR subline.

Also at Cold Spring Harbor, E. Carleton MacDowell received from Little the descendants of Miss Lathrop's ♂52, the progenitor of the C57 lines, and ♀58. He inbred these mice, forming the C58 strain, and by selection was able to establish an incidence of leukemia of about 90 per cent. MacDowell also inbred the Bagg albinos and sent some to George D. Snell about 1932. Snell used the letter "c" in his laboratory records as a convenient indication that the animals were white. The letter became attached to "Bagg alb," and the designation evolved to BALB/c, a widely used strain.

While at the Henry Phipps Institute in Philadelphia in 1928, Jacob Furth purchased three different stocks of mice, designated A, R, and S. Stock A "was claimed to yield many cancers," and stock R was stated to be cancer-free. He and his collaborators inbred a number of families in each stock, from which were derived the AK and RF strains (Furth et al., 1933).

The widely distributed Swiss albino mice, largely noninbred, are mainly derived from two males and seven females which Clara J. Lynch of the Rockefeller Institute obtained from A. de Coulon of Lausanne in 1926 (Lynch, 1961, personal communication). The original stock probably came from Paris. Descendants of these mice were distributed to other laboratories and to commercial breeders. Some lines, such as the SWR/J and SJL/J, have since been inbred. Origins of these and many other inbred strains are listed in Table 1–1. References given are not in all cases the original ones, for an attempt has been made to list papers best describing the strain origin. In most cases multiple references are supplementary, offering additional information or enabling the searcher to go from a reference in which the strain is clearly named to one in which the real origin is given but the mice are not so named. In some cases, no really satisfactory reference can be found. The many fostered and congenic strains have not been included. For these interesting formulations the latest issue of *Inbred Strains of Mice* should be consulted. Strains known to be extinct have not been listed.

USE OF MICE IN RESEARCH

The majority of inbred strains, from the most recent back to the DBA, were developed for use in cancer research, to prove or disprove the existence of genetic factors influencing the incidence of cancer and the independence of inheritance of different types of cancers. By selection during inbreeding, various types of malignancies in predictable frequencies were established in the several genotypes. Or, conversely, resistance to all forms of neoplasia was established (Chap. 27, 28). As inbred strains became available and information about them began appearing in the scientific literature, investigators recognized that these animals could contribute greatly to medical research. It became possible to use biological material in experiments with confidence that the only variables were those the investigator chose to include in the experimental design. The greater the uniformity among animals, the fewer are needed to attain a given standard of accuracy or repeatability.

A large proportion of cancer research has been built upon inbred strains of mice. Many types of projects were made possible only by the development of the strains and the tumors the mice produce or tolerate, and a large part of the remainder is dependent on the strains for suitable material.

Investigators in many fields have come to realize the value of F_1 hybrids from crosses between inbred strains. Such mice are genetically homogeneous although heterozygous for those gene pairs by which the parent strains differ. Hybrids have been found to be as predictable in response as the parent strains, though not necessarily like either one. The greatest general advantage of F_1 hybrids is their increased vigor and, in certain types of terminal experiments, they are preferred over inbred mice. Such mice cannot be used for propagating their own characteristics, however, since genetic segregation will occur in F_2 generations.

As an extra dividend of fixing coat-color genes and cancer-affecting genes by inbreeding, it was found that other constitutional diseases also became established in the various genotypes (Chap. 29). Many of these conditions parallel pathological states in man, thus providing unique material for studies on causation as well as on the march of the disease process.

Other differences between strains have been found and exploited in many fields and situations, as following chapters in this book attest. These include differences in disease susceptibility, nature of disease produced by a given pathogen, and survival time of infected individuals; nature and severity of radiation response, length of reproductive life, litter size, number of litters, and maternal care; sensitivity to and production of various hormones, and reaction to implantation or extirpation of endocrine organs; cold tolerance, growth performance on varying dietary formulas, and capacity for antibody production; blood constituents including normal blood-cell values, and enzyme levels in various organs.

SOURCES OF INFORMATION

Information about mice is available from a variety of institutions and publications. Among the standard scientific periodicals, the *Journal of Heredity* (United States), *Genetical Research* and *Heredity* (England), *Bulletin of the Experimental Animals* (Japan), and *Zeitschrift für Vererbungslehre* and *Zeitschrift für Versuchstierkunde* (Germany) are important vehicles for reporting new mutations, linkage tests, and methods of husbandry.

Table 1-1. REFERENCES TO ORIGINS OF INBRED STRAINS

Strain	*Reference*	*Strain*	*Reference*
A	Strong (1936)	DB	Staats (1964)
AB	*Inbred Strains of Mice* (1965)	DBA	Murray (1934)
AG	Staats (1964)	DBA/1_0	Staats (1964)
AH	Staats (1964)	DBA/2eB	Deringer (1962)
AK	Furth et al. (1933), Lynch (1954)	DBR	Diadkova and Lotosh (1962)
AKR	Lynch (1954)	DD	Heston et al. (1964)
AL	Staats (1964)	DDM	*Inbred Strains of Mice* (1965)
AU	Staats (1964)	DE	*Inbred Strains of Mice* (1963)
A2G	Lab. Animals Centre (1958), Staats (1964)	DL	Rauch and Yost (1963)
		DLK	*Inbred Strains of Mice* (1965)
BALB	Heston (1945), MacDowell et al. (1927)	DM	Staats (1964)
BAMA	Staats (1964)	DP	*Inbred Strains of Mice* (1963)
BAN	*Inbred Strains of Mice* (1965)	DS	*Inbred Strains of Mice* (1965)
BC	Gardner (1954), Staats (1964)	DW	*Inbred Strains of Mice* (1965)
BD	Staats (1964)	D103	Staats (1964)
BDP	Gates (1927)	E	Eaton (1941), Staats (1964)
BL	Deringer (1959*b*)	F	Strong (1942)
BN/a	Dux (1957)	FB	*Inbred Strains of Mice* (1965)
BN/b	Dux (1957)	FL	Staats (1964)
BNT	Staats (1964)	FU	Staats (1964)
BRS	Bagshaw and Strong (1950)	GFF	Lab. Animals Centre (1958)
BRSUNT	Bagshaw and Strong (1950)	GM	Runner (1957)
BRVR	Webster (1933), Webster (1937)	GR	*Inbred Strains of Mice* (1965)
BSL	Staats (1964)	GRS	*Inbred Strains of Mice* (1965)
BSVS	Webster (1933), Webster (1937)	H	Staats (1964)
BRO	Foulds (1956)	HA	Hummel (1957)
BR6	Foulds (1956)	HALB	Staats (1964)
BT	*Inbred Strains of Mice* (1965)	HC	Staats (1964)
BUA	Staats (1964)	HG	Staats (1964)
BUB	Staats (1964)	HM	*Inbred Strains of Mice* (1965)
BUC	Staats (1964)	HR/De	Deringer (1951)
BUE	Staats (1964)	HR/Rl	Staats (1964)
C	Strong (1942)	HRS	*Inbred Strains of Mice* (1965)
CBA	Strong (1942)	I	Strong (1942)
CC57BR	Medvedev (1957)	IBA	*Inbred Strains of Mice* (1965)
CC57W	Medvedev (1958)	ICR/Kp	Staats (1964)
CD	*Inbred Strains of Mice* (1965)	ICRC	Ranadive et al. (1961)
CE	Detlefsen (1921), Woolley and Little (1945)	IF	Bonser (1938)
		IHB	Blount and Blount (1961)
CFCW	Staats (1964)	IPBR	Strong (1951)
CFW	Snell (1941), Staats (1964)	JB	Staats (1964)
CHI	Strong (1942)	JK	Strong (1942)
CM	Tibón (1953)	JU	Bowman and Falconer (1960)
CT	Staats (1964)	KE	*Inbred Strains of Mice* (1965)
C12I	Strong (1942)	KI	Staats (1964)
C3H	Strong (1942)	KK	Staats (1964)
C3HeB/De	Deringer (1959*a*)	KL	Chase et al. (1948)
C3HeB/Fe	*Inbred Strains of Mice* (1963)	KP	*Inbred Strains of Mice* (1965)
C3HA	Maliugina (1957)	KSA	Staats (1964)
C3HBL	Taylor and McKenna (1961)	L	Strong (1942)
C57BL	Heston (1949)	LAA	*Inbred Strains of Mice* (1965)
C57BR	Heston (1949)	LCSa	Crispens (1963)
C57L	Heston (1949), Murray (1933)	LG	Goodale (1938)
C58	MacDowell (1935), MacDowell (1936)	LIS	*Inbred Strains of Mice* (1965)
DA	Hummel (1958)	LP	Staats (1964)

Table 1-1. **REFERENCES TO ORIGINS OF INBRED STRAINS** (*Continued*)

Strain	*Reference*	*Strain*	*Reference*
L(C)	Ranadive (1958)	RIIIeB	*Inbred Strains of Mice* (1963)
L(P)	Ranadive (1958)	S	Schott (1932)
LT/A	Staats (1964)	SD	Staats (1964)
LT/Ch	MacDowell (1950)	SEA	Green and Green (1946)
LTS	*Inbred Strains of Mice* (1965)	SEAC	Staats (1964)
MA	Murray (1938)	SEAC/c	Staats (1964)
MA/My	Murray (1963)	SEC/1	Staats (1964)
MABA	Staats (1964)	SHC/a	Staats (1964)
MAS	*Inbred Strains of Mice* (1965)	SIM	Staats (1964)
MO	Kobozieff (1959)	SJL	Murphy (1963)
MOB	Crispens (1963)	SL	Staats (1964)
MY	Staats (1964)	SLT	Staats (1964)
MYA	Staats (1964)	SM	MacArthur (1944)
N	Strong (1942)	SMA	Staats (1964)
NAW	Staats (1964)	SP	Staats (1964)
NB	Green (1941), Law (1948)	ST	Stamer (1943)
NBL	Staats (1964)	STL	Staats (1964)
NCN/a	*Inbred Strains of Mice* (1965)	STOLI	Heston (1949), MacDowell et al. (1927)
NCN/b	*Inbred Strains of Mice* (1965)	STR	*Inbred Strains of Mice* (1963)
NH	Strong (1940)	STS	*Inbred Strains of Mice* (1965)
NHO	Strong (1940)	STR/1	Sokoloff and Barile (1962)
2NHO	Strong (1940)	SWR	Staats (1964)
NLC	Rudali et al. (1956)	T	Staats (1964)
NS	Fraser et al. (1953)	TA	Staats (1964)
NU	Staats (1964)	TH	Wolfe (1961)
NZB	Helyer and Howie (1958)	TL	Wolfe (1961)
NZC	Bielschowsky et al. (1956)	TM	Casas (1963)
NZO	Bielschowsky et al. (1956)	TP	*Inbred Strains of Mice* (1965)
NZX	Staats (1964)	TS	*Inbred Strains of Mice* (1965)
NZY	Bielschowsky et al. (1956)	TIII	Staats (1964)
O20	van Gulick and Korteweg (1940)	T6	Staats (1964)
P/A	Staats (1964)	TXII	Staats (1964)
P/J	Cooper (1939)	TXXIX	*Inbred Strains of Mice* (1965)
PBR	Bagshaw and Strong (1950)	WB	Russell and Lawson (1959)
PE	Staats (1964)	WC	Russell and Lawson (1959)
PET	Nichols et al. (1960)	WD	*Inbred Strains of Mice* (1965)
PHH	Weir (1953)	WH	Russell and Lawson (1959)
PHL	Weir (1953)	WHT	Staats (1964)
PL	Nelson (1948)	WK	Russell and Lawson (1959)
PLA	Staats (1964)	WLH	Staats (1964)
POLY	*Inbred Strains of Mice* (1965)	WLL	Kreyberg (1952)
PS	Mouriquand et al. (1960)	WLO	Kreyberg (1952)
PT	*Inbred Strains of Mice* (1965)	WV	Staats (1964)
P20	Mühlbock (1947)	X	*Inbred Strains of Mice* (1965)
QV	*Inbred Strains of Mice* (1965)	Y/Wi	Staats (1964)
RAP	Kobozieff (1959)	Y (YBR)	Heston (1945)
RB	Lehmann and Busnel (1962)	YS	Staats (1964)
RF	Furth et al. (1933)	Z	Staats (1964)
R.I.L.	Lynch (1954)	IV/B	Dobrovolskaïa-Zavadskaïa (1930*b*)
RU	Staats (1964)	XVII	Dobrovolskaïa-Zavadskaïa (1937)
RUS	Russell (1957)	19	Staats (1964)
RIII	Dobrovolskaïa-Zavadskaïa (1929) and (1930*a*)	101	Dunn (1936)
		129	Russell and Hurst (1945)

Hans Grüneberg's *The Genetics of the Mouse* (1952) is the standard work in this field. It gives detailed descriptions of the genetics and pathology of all mouse mutants known up to 1950. Chapter 8 of this volume contains a check list of all mouse mutants known to early 1965 with short descriptions of each. The number of recorded mutants is more than three times as large as it was in 1950.

Mouse News Letter (MNL) is a continuing source of information on the location of mutants, discovery of new ones, and research news. It is not a publication, but an informal document circulated privately. It is issued semiannually by the Laboratory Animals Centre, Carshalton, Surrey, England, and provides an information exchange among those working with mutant mice. Contributing laboratories provide lists of the mutant stocks they maintain.

Inbred Strains of Mice (ISM) is issued biennially by The Jackson Laboratory, Bar Harbor, Maine, as a companion to *Mouse News Letter.* Contributing laboratories send lists of their inbred strains only. Both MNL and ISM are arranged alphabetically by reporting laboratories.

Standardized Nomenclature for Inbred Strains of Mice (Staats, 1964) is compiled mainly from the contributions to ISM and is arranged by strains rather than by laboratories, thus being essentially a locator list. It gives information on the origin of strains, their particular characteristics, and the institutions maintaining them. It contains also the rules of nomenclature (Chap. 6) and a list of abbreviations of names of institutions or persons to be used in identifying substrains.

More general information, not restricted to genetics, is gathered and disseminated by many national centers. The Universities Federation for Animal Welfare (UFAW) in London has published an excellent handbook (Worden and Lane-Petter, 1957) covering all of the common and some of the uncommon animals. The topics to which UFAW has given particular attention during its history include the regulation of wild populations (especially inhumane trapping and poisoning), the treatment of animals, the techniques of euthanasia, and humane education. The Institute of Laboratory Animal Resources (ILAR) in Washington, D.C., issues a mimeographed quarterly publication, *Information on Laboratory Animals for Research,* containing news of meetings and publications, with occasional information-exchange sections or husbandry hints. Distribution is limited to persons active in biological research or in the production of laboratory animals. The ILAR has published minimum standards for the commercial production of various laboratory animals. Whereas UFAW and ILAR act as informational bodies, promote legislation, and work toward accreditation and improved standards, they do not generate scientific research or raise animals. The Laboratory Animals Centre, set up near London in 1947 by the Medical Research Council, does generate research and raise animals, as well as promote the use of better animals and the proper training of animal caretakers.

Three important publications in the general field of laboratory animal husbandry are the *Proceedings of the Laboratory Animal Science Association* (and its predecessor the *Laboratory Animals Centre Collected Papers*), the *Journal of the Animal Technicians Association,* and *Laboratory Animal Care.*

SOURCES OF QUALITY MICE

Inbred mice, once maintained in very few places, are now widely available from national or regional stock centers. The Jackson Laboratory raises 2 million mice a year, maintains more than 75 strains and substrains, and in addition has many stocks carrying one or more named mutant genes. Other large colonies in the United States are at the National Institutes of Health in Bethesda, Maryland, and at the Oak Ridge National Laboratory, Oak Ridge, Tennessee. England, France, Germany, Japan, Russia, Hungary, and Scotland all have at least one center where large numbers of genetically controlled mice are raised for the benefit of the scientific community. Many other countries are establishing or planning such centers, either under the auspices of the national health ministry or as private laboratories.

SUMMARY

Inbred mice have been used for cancer research since C. C. Little established the DBA strain in 1909. Other differences in disease incidence between strains, whether fixed in the genotype by selective inbreeding or arising as mutations, have made possible concerted attacks on major diseases afflicting mankind. As more becomes known about subtle biochemical differences between strains, ever more refined tools will be available to medical research workers.

LITERATURE CITED

Bagshaw, M. A., and L. C. Strong. 1950. The occurrence of tumors of the forestomach in mice after parenteral administration of methylcholanthrene: a histopathologic and genetic analysis. J. Nat. Cancer Inst. 11:141–175.

Bateson, W. 1903. The present state of knowledge of colour-heredity in mice and rats. Proc. Zool. Soc. Lond. 2:71–99.

Bielschowsky, M., F. Bielschowsky, and D. Lindsay. 1956. A new strain of mice with a high incidence of mammary cancers and enlargement of the pituitary. Brit. J. Cancer 10:688–699.

Blount, R. F., and I. H. Blount. 1961. Strain differences and the relation of aging to salt susceptibility in mice. Texas Rep. Biol. Med. 19: 739–748.

Bonser, G. M. 1938. The hereditary factor in induced skin tumours in mice: establishment of a strain specially sensitive to carcinogenic agents applied to the skin. J. Pathol. Bacteriol. 46: 581–602.

Bowman, J. C., and D. S. Falconer. 1960. Inbreeding depression and heterosis of litter size in mice. Genet. Res. 1:262–274.

Casas, C. B. 1963. Induction of hepatomas by thiouracil in inbred strains of mice. Proc. Soc. Exp. Biol. Med. 113:493–494.

Castle, W. E., F. W. Carpenter, A. H. Clark, S. O. Mast, and W. M. Barrows. 1906. The effects of inbreeding, cross-breeding, and selection upon the fertility and variability of *Drosophila.* Proc. Amer. Acad. Arts Sci. 41: 731–786.

Chase, H. B., M. S. Gunther, J. Miller, and D. Wolffson. 1948. High insulin tolerance in an inbred strain of mice. Science 107:297–299.

Cooper, C. B. 1939. A linkage between naked and caracul in the house mouse. J. Hered. 30:212.

Crispens, C., Jr. 1963. Factors which influence normal values for serum lactic dehydrogenase in mice. Experientia 19:97–98.

Davenport, C. B. 1900. Review of von Guaita's experiments in breeding mice. Biol. Bull. 2: 121–128.

Davenport, C. B. 1906. Inheritance in poultry. Carnegie Inst. Pub. No. 52. 136 p.

Deringer, M. K. 1951. Spontaneous and induced tumors in haired and hairless strain HR mice. J. Nat. Cancer Inst. 12:437–445.

Deringer, M. K. 1959*a*. Occurrence of tumors, particularly mammary tumors, in agent-free strain C3HeB mice. J. Nat. Cancer Inst. 22: 995–1002.

Deringer, M. K. 1959*b*. Necrotizing arteritis in strain BL/De mice. Lab. Invest. 8:1461–1465.

Deringer, M. K. 1962. Development of tumors, especially mammary tumors, in agent-free strain DBA/2eB mice. J. Nat. Cancer Inst. 28:203–210.

Detlefsen, J. A. 1921. A new mutation in the house mouse. Amer. Natur. 55:469–473.

Diadkova, A. M., and E. A. Lotosh. 1962. Onkologicheskaia kharakteristika novoi linii laboratornykh myshei DBR. [The oncological characteristics of a new line of laboratory mice DBR.] Vop. Onkol. 8(11):46–47.

Dobrovolskaïa-Zavadskaïa, N. 1929. Sur l'hérédité de la prédisposition au cancer spontané chez la souris. Compt. Rend. Soc. Biol. 101:518–520.

Dobrovolskaïa-Zavadskaïa, N. 1930*a*. Sur une lignée de souris, riche en adénocarcinome de la mamelle. Compt. Rend. Soc. Biol. 104:1191–1193.

Dobrovolskaïa-Zavadskaïa, N. 1930*b*. Sur une lignée de souris, pauvre en adénocarcinome de la mamelle. Compt. Rend. Soc. Biol. 104:1193–1195.

Dobrovolskaïa-Zavadskaïa, N. 1937. Efficacité de la sélection en vue de l'élimination des facteurs héréditaires responsables du cancer spontané dans une lignée de souris (lignée XVII nc). Compt. Rend. Soc. Biol. 126:287–289.

Dunn, L. C. 1936. Studies of multiple allelomorphic series in the house-mouse. I. Description of agouti and albino series of allelomorphs. J. Genet. 33:443–453.

Dux, A. 1957. Experimental studies on the transplantability of the mammary cancer in mice [in Polish, English summary]. Nowotwory 7: 67–90.

Eaton, O. N. 1941. Crosses between inbred strains of mice. J. Hered. 32:393–395.

Foulds, L. 1956. The histologic analysis of mammary tumors of mice. I. Scope of investigations and general principles of analysis. J. Nat. Cancer Inst. 17:701–711.

Fraser, F. C., T. D. Fainstat, and H. Kalter. 1953. The experimental production of congenital defects with particular reference to cleft palate. Neo-Natal Stud. 2:43–48.

Furth, J., H. R. Seibold, and R. R. Rathbone. 1933. Experimental studies on lymphomatosis of mice. Amer. J. Cancer 19:521–604.

Gardner, W. U., and J. Rygaard. 1954. Further studies on the incidence of lymphomas in mice exposed to x-rays and given sex hormones. Cancer Res. 14:205–209.

Gates, W. H. 1927. A case of non-disjunction in the mouse. Genetics 12:295–306.

Goodale, H. D. 1938. A study of the inheritance of body weight in the albino mouse by selection. J. Hered. 29:101–112.

Green, E. L. 1941. Genetic and non-genetic factors which influence the type of the skeleton in an inbred strain of mice. Genetics 26:192–222.

Green, E. L., and M. C. Green. 1946. Effect of the short ear gene on number of ribs and presacral vertebrae in the house mouse. Amer. Natur. 80:619–625.

Grüneberg, H. 1952. The genetics of the mouse, 2nd ed. Nijhoff, The Hague. 650 p.

Helyer, B. J., and J. B. Howie. 1958. Spontaneous haemolytic anaemia in NZB/Bl mice. 36th Ann. Rep. Brit. Empire Cancer Campaign, p. 458–459.

Heston, W. E. 1945. Genetics of mammary tumors in mice, p. 55–84. *In* F. R. Moulton [ed.] A symposium on mammary tumors in mice. Amer. Ass. Adv. Sci., Washington, D.C.

Heston, W. E. 1949. Development of inbred strains in the mouse and their use in cancer research, p. 9–31. *In* Lectures on genetics, cancer, growth and social behavior. Roscoe B. Jackson Memorial Laboratory, Bar Harbor, Maine.

Heston, W. E., G. Vlahakis, and Y. Tsubura. 1964. Strain DD, a new high mammary tumor strain, and comparison of DD with strain C3H. J. Nat. Cancer Inst. 32:237–251.

Hummel, K. P. 1957. Mouse News Letter 17:54.

Hummel, K. P. 1958. The inheritance and expression of disorganization, an unusual mutation in the mouse. J. Exp. Zool. 137:389–423.

Inbred Strains of Mice, No. 3. 1963. The Jackson Laboratory, Bar Harbor, Maine.

Inbred Strains of Mice, No. 4. 1965. The Jackson Laboratory, Bar Harbor, Maine.

Keeler, C. E. 1931. The laboratory mouse: its origin, heredity, and culture. Harvard Univ. Press, Cambridge, Mass. 81 p.

King, H. D. 1911. The sex ratio in hybrid rats. Biol. Bull. 21:104–112.

Kobozieff, N. 1959. Mouse News Letter 19:14.

Kreyberg, L. 1952. The origin and the development of the "white label" mouse strain. Brit. J. Cancer 6:140–147.

Laboratory Animals Centre. 1958. Catalogue of uniform strains, 2nd ed. Carshalton, Surrey.

Law, L. W. 1948. Mouse genetic news, No. 2. J. Hered. 39:300–308.

Lehmann, A., and R. G. Busnel. 1962. A new test for detecting MAO-inhibitor effects. Int. J. Neuropharmacol. 1:61–70.

Little, C. C. 1913. Experimental studies of the inheritance of color in mice. Carnegie Inst. Pub. No. 179:17–102.

Lynch, C. J. 1954. The R.I.L. strain of mice: its relation to the leukemic AK strain and AKR substrains. J. Nat. Cancer Inst. 15:161–176.

MacArthur, J. W. 1944. Genetics of body size and related characters. I. Selecting small and large races of the laboratory mouse. Amer. Natur. 78:142–157.

MacDowell, E. C. 1936. Genetic aspects of mouse leukemia. Amer. J. Cancer 26:85–101.

MacDowell, E. C. 1950. "Light"—a new mouse color. J. Hered. 41:35–36.

MacDowell, E. C., E. Allen, and C. G. MacDowell. 1927. The prenatal growth of the mouse. J. Gen. Physiol. 11:57–70.

MacDowell, E. C., and M. N. Richter. 1935. Mouse leukemia. IX. The role of heredity in spontaneous cases. Arch. Pathol. 20:709–724.

Maliugina, L. L., and O. G. Prokof'eva. 1957. Oncological characteristics of mice of strain C3HA [English transl.]. Probl. Oncol. 3:201–207.

Medvedev, N. N. 1957. K onkologicheskoi Kharakteristike nizkorakovykh myshei *CC-57-Korichnevye*. [On the oncological characteristic of the low-tumorous *CC-57-brown* stock mice.] Biull. Moskov. Obshchest. Ispytatelei Prirody 62:63–67.

Medvedev, N. N. 1958. O lineinykh myshakh CC-57-belye. [The CC-57 white mice lines.] Dokl. Akad. nauk SSSR 119:369–371.

Mouriquand, J., C. Mouriquand, and J. Petat. 1960. Premières observations à propos d'une nouvelle souche de souris hautement cancérigene. Compt. Rend. Soc. Biol. 154:632–633.

Mühlbock, O. 1947. On the susceptibility of different inbred strains of mice for oestrone. Acta Brev. Neer. 15:18–20.

Murphy, E. D. 1963. SJL/J, a new inbred strain of mouse with a high, early incidence of reticulum-cell neoplasms. Proc. Amer. Ass. Cancer Res. 4:46. (Abstr.)

Murray, J. M. 1933. "Leaden," a recent color mutation in the house mouse. Amer. Natur. 67:278–283.

Murray, W. S. 1934. The breeding behavior of the dilute brown stock of mice (Little dba). Amer. J. Cancer 20:573–593.

Murray, W. S. 1938. Genetic segregation mammary cancer to no mammary cancer, in the mouse. Amer. J. Cancer 34:434–441.

Murray, W. S. 1963. MA/My strain of the Marsh albino mouse. J. Nat. Cancer Inst. 30:605–610.

Nelson, J. B. 1948. The nasal transmission of pleuropneumonia-like organisms in mice and rats. J. Infect. Dis. 82:169–176.

Nichols, S. E., Jr., and W. M. Reams, Jr. 1960. The occurrence and morphogenesis of melanocytes in the connective tissues of the PET/MCV mouse strain. J. Embryol. Exp. Morphol. 8:24–32.

Ranadive, K. J., and S. A. Hakim. 1958. A biological study of strain L(P) and its response to 20-methylcholanthrene treatment. Brit. J. Cancer 12:44–54.

Ranadive, K. J., K. A. Kamat, T. G. Coutinho, and V. R. Khanolkar. 1961. Incidence of spontaneous mammary carcinoma in the new strain of Indian laboratory mouse. Indian J. Med. Res. 49:562–567.

Rauch, H., and M. T. Yost. 1963. Phenylalanine metabolism in dilute-lethal mice. Genetics 48: 1487–1495.

Rudali, G., N. Yourkovski, L. Juliard, and M. Fautrel. 1956. Sur quelques caractères des souris appartenant à la nouvelle lignée cancéreuse. Lignée NLC de la Fondation Curie. Bull. Ass. Franc. Ét. Cancer 43:364–383.

Runner, M. N. 1957. Mouse News Letter 17:56.

Russell, E. S., and F. A. Lawson. 1959. Selection and inbreeding for longevity of a lethal type. J. Hered. 50:19–25.

Russell, L. B. 1957. Mouse News Letter 17:84.

Russell, W. L., and J. G. Hurst. 1945. Pure strain mice born to hybrid mothers following ovarian transplantation. Proc. Nat. Acad. Sci. 31:267–273.

Schott, R. G. 1932. The inheritance of resistance to *Salmonella aertrycke* in various strains of mice. Genetics 17:203–229.

Snell, G. D. [ed.] 1941. Mouse genetic news No. 1, p. 7. Roscoe B. Jackson Memorial Laboratory, Bar Harbor, Maine. (Mimeo.)

Sokoloff, L., and M. F. Barile. 1962. Obstructive genitourinary disease in male STR/1N mice. Amer. J. Pathol. 41:233–246.

Staats, J. 1964. Standardized nomenclature for inbred strains of mice, Third listing. Cancer Res. 24:147–168.

Stamer, S. 1943. Effect of a carcinogenic hydrocarbon on manifest malignant tumors in mice. Munksgaard, Copenhagen. 158 p.

Strong, L. C. 1936. The establishment of the "A" strain of inbred mice. J. Hered. 27:21–24.

Strong, L. C. 1940. A genetic analysis of the induction of tumors by methylcholanthrene, with a note on the origin of the NH strain of mice. Amer. J. Cancer 39:347–349.

Strong, L. C. 1942. The origin of some inbred mice. Cancer Res. 2:531–539.

Strong, L. C. 1951. Litter seriation phenomena in fibrosarcoma susceptibility. J. Gerontol. 6: 340–357.

Strong, L. C. 1952. Susceptibility to fibrosarcomas in 2NHO mice. Yale J. Biol. Med. 24:109–115.

Taylor, A., and G. F. McKenna. 1961. A new mouse strain susceptible to mammary cancer. Texas Rep. Biol. Med. 19:706–707.

Tibón, G. 1953. Mouse News Letter 9:10.

Van Gulick, P. J., and R. Korteweg. 1940. Susceptibility to follicular hormone and disposition to mammary cancer in female mice. Amer. J. Cancer 38:506–515.

Webster, L. T. 1933. Inherited and acquired factors in resistance to infection. I. Development of resistant and susceptible lines of mice through selective breeding. J. Exp. Med. 57: 793–817.

Webster, L. T. 1937. Inheritance of resistance of mice to enteric bacterial and neurotropic virus infections. J. Exp. Med. 65: 261–286.

Weir, J. A. 1953. Association of blood-pH with sex ratio in mice. J. Hered. 44:133–138.

Wolfe, H. G. 1961. Selection for blood-pH in the house mouse. Genetics 46:55–75.

Woolley, G. W., and C. C. Little. 1945. The incidence of adrenal cortical carcinoma in gonadectomized female mice of the extreme dilution strain. I. Observations on the adrenal cortex. Cancer Res. 5:193–202.

Worden, A. N., and W. Lane-Petter. 1957. The UFAW handbook on the care and management of laboratory animals, 2nd ed. Universities Federation for Animal Welfare, London. 951 p.

Wright, S. 1922. The effects of inbreeding and crossbreeding on guinea pigs. U.S. Dep. Agr. Bull. No. 1090:1–63.

2

Breeding Systems[1]

Earl L. Green

Characteristics of mice, like the characteristics of other things, living or nonliving, may be treated as *discrete* or *continuous* variables. Litter size, sex, and coat colors are examples of discrete variables. Weight, body size, and lifespan are examples of continuous variables. We may describe groups by the proportions of males, of brown mice, or of brown male mice (discrete variables), or by the means and variances of their weights, lengths, or lifespans (continuous variables).

A collection of samples will usually exhibit some variability in the proportions or means of the characteristics under study. If a sufficient number of characteristics are simultaneously considered, no two mice will be alike. The variability, aside from chance differences between samples, may have either *genetic* or *nongenetic* causes. The nongenetic causes of variation in the characteristics under study may be of many kinds, including seasonal fluctuations, differences in kind and amount of food consumed, exposures to noise, light, or humidity, or differences in prior experiences. The genetic causes of variation are simpler, at least in principle, in that they are exclusively the consequences of different genetic contents of zygotes.

Investigators who use mice are concerned with the control of both genetic and nongenetic sources of variation. Some of the nongenetic sources may be controlled by standardizing the management of the mouse colony or, when it is desired, by deliberately imposing two or more different conditions of rearing or treating the mice. The genetic sources of variability may be controlled by the choice of an appropriate breeding system. Various breeding systems useful with mice, their theoretical consequences, and the criteria for choosing one in preference to another are the subjects of this chapter.

More extensive general discussions of breeding systems may be found in the books by Lush (1945), Malécot (1948), Fisher (1949), Mather (1949), Lerner (1950), Li (1955), Kempthorne (1957), Falconer (1960), and Le Roy (1960), and in the papers by Wright (1921*a*, 1963), Bartlett and Haldane (1935), Green and Doolittle (1963), Kimura and Crow (1963), and Robertson (1964), as well as others.

PURPOSES OF BREEDING SYSTEMS

The purpose of all breeding systems is to preserve or control the genetic causes of variability in traits of interest. In theoretical populations of infinite size under specified environmental conditions, random mating in the absence of selection or mutation will keep the means and variances of all quantitative traits constant. Inbreeding will subdivide a population and, in the individual subpopulations, will increase or decrease the means and will reduce the genetic variances. Outcrosses (matings between populations) will usually change the mean and increase the genetic variance of the resulting population. Selective breeding of like with like will increase or decrease or stabilize the mean, depending upon the direction of selection, and will also ultimately decrease the genetic variance, but not necessarily

[1] The writing of this chapter was supported in part by the Aaron E. Norman Fund and by Contract AT(30-1)-1979 with the U.S. Atomic Energy Commission.

eliminate it. If the selected mates are deliberately as unlike as possible, the genetic variance will be kept large.

Combinations of inbreeding and selection systems give geneticists a wide variety of methods for controlling the inherited characteristics of research animals.

Four general types of mice have come into common use: *inbred, hybrid, mutant-bearing,* and *selected.* Breeding systems to produce inbred and hybrid mice and to propagate mutant-bearing mice are described in the following sections of this chapter. Systems of selective breeding are described in Chap. 9.

It is necessary to distinguish between two uses of mice in research. Mice for *breeding experiments* must be individually identified in such a way that the parents and more remote ancestors, the offspring and later descendants, and collateral relatives can, when necessary, also be identified. Methods for keeping breeding records to accomplish this aim are given in Chap. 3. In *terminal experiments,* the mice are used for investigations of physiology, biochemistry, pathology, behavior, etc., but not for further breeding. They need not necessarily be individually identified, but must be identified by lot. That is, the strain, generation, age, weight, and other characteristics will need to be known, but the exact relationships to other mice in the same or different lots may not always be needed.

SYSTEMS OF BREEDING

The genetic consequences of a system of breeding may be grasped by considering how the system affects the probabilities or theoretical relative frequencies of the alleles or genes, the genotypes, and the mating types at any one locus and hence at all loci.

We shall refer to the *a* locus as any locus whose heterozygosity is in question under any given system of breeding. For simplicity we shall suppose that there are two alleles, + and *a*, at the *a* locus. In certain systems we shall need to refer to the locus of a dominant mutation (D) and to the locus of a recessive mutation (r). These will be denoted as the D locus with alleles + and D and as the r locus with alleles + and r. The a, D, and r loci are distinct and may be linked or not. The D and r loci are called loci of interest in any system of breeding for control of a segregating locus.

From three genotypes, such as $+/+$, $a/+$, a/a at a locus with two alleles, there will be nine mating types, which may be grouped into four kinds:

1. Matings of like homozygotes, $+/+ \times +/+$ and $a/a \times a/a$, called *incrosses*
2. Matings of unlike homozygotes $+/+ \times a/a$ and $a/a \times +/+$, called *crosses*
3. Matings of a homozygote and a heterozygote, $+/+ \times a/+$, $a/+ \times +/+$, $a/+ \times a/a$, and $a/a \times a/+$, called *backcrosses*
4. Matings of heterozygotes, $a/+ \times a/+$, called *intercrosses*

The theoretical consequences of any breeding system are easily displayed in terms of the probability or frequency of heterozygotes, denoted by h, that is, $P(a/+) = h$, and of the probability of incrosses, denoted by p, that is, $P(+/+ \times +/+$ or $a/a \times a/a) = p$. The probability of incrosses (p) may increase in a given system of mating only at the expense of the crosses, backcrosses, and intercrosses. The magnitude of p therefore reveals the expected proportion of matings of like homozygotes, and $1 - p$ gives the probabilities of matings of other types.

As h decreases, the probability of homozygotes ($1 - h$) increases. If these probabilities are interpreted as referring to all loci, we may say that the probability of homozygosity is $1 - h$. In the limit when $h = 0$, the strain may be said to be *isogenic.* In practice the term isogenic is used to mean homozygous at nearly all loci as well as homozygous at all loci. If two animals or two groups of animals are isogenic (or nearly so) at all loci except one or two at which they are known to be different, either by being opposite homozygotes ($+/+$ vs. a/a) or by one being homozygous ($+/+$ or a/a) and the other being heterozygous ($a/+$), the two animals or groups are called *coisogenic* (Chovnik and Fox, 1953). If a new mutation arises in and is propagated within an already inbred strain, the mutant and nonmutant mice are called coisogenic. To avoid the risk of a false claim inherent in the use of *coisogenic* when in fact h is not zero, Loosli et al. (1961) proposed the term *congenic.* This term is especially useful when referring to stocks produced by repeated crossing of mutant-bearing animals to animals of inbred strains.

Methods of analyzing breeding systems not given here are amply described in several books and papers. The method of *path analysis,* invented by Sewall Wright, is the most versatile general method of analyzing regular and irregular breeding systems (Wright, 1921*a*, 1921*b*, 1934, 1954, 1963). The *generation matrix* method, first used by Bartlett and Haldane (1935), is useful in analyzing regular breeding systems and was adopted by Fisher (1949) in an extensive general analysis of inbreeding. The introduction of concepts of probability to the analysis of breeding systems is largely due to Malécot (1948). An exposition of the methods and results of the analysis of some regular breeding systems useful with laboratory mice will be found in Green and Doolittle (1963).

In the following sections of this chapter are descriptions of random breeding, avoidance of inbreeding, brother-sister inbreeding, other types of inbreeding, strain crossing, breeding systems for putting mutant genes on standard inbred backgrounds and for decomposing complex genotypes, and inbreeding with forced heterozygosis. Special breeding systems using linked markers are described in Chap. 8. The construction and maintenance of linkage testing stocks are described by Carter and Falconer (1951).

RANDOM BREEDING

Random breeding is easy to define, but difficult to achieve. In principle, random breeding requires that the product rule of probability be satisfied. This means that the chance of choosing any one male out of m males must be $1/m$ and of choosing any one female out of f females be $1/f$ and that the chance of mating any two specific mice be $1/mf$. Tables of random sampling numbers or other equivalent randomizing devices must be used to insure random choices and random matings. Indiscriminate or nonsystematic ways of making up matings are likely to permit unknown biases to occur in the choices of mates.

The effect of random breeding on the genetic structure of a stock of mice has to be considered separately for theoretical and actual populations.

In theoretical or large populations, random breeding will preserve the gene and genotype frequencies generation after generation. Discrete traits usually determined by one or a few pairs of genes will be retained at stable frequencies in the population, assuming there is no selection and no mutation. The means and variances of continuous or metrical traits will likewise remain constant, under the same assumptions, provided the nongenetic components of variances do not change. These assertions about stability of gene frequencies, genotype frequencies, means, and variances follow from the Hardy-Weinberg principle (Li, 1955).

In actual populations of small size the expected results of a random breeding system will be slightly different. First, the effective breeding number (N) will determine the rate of random fixation or loss of alleles at each locus, the rate being approximately equal to $1/2N$ (Wright, 1931). Random breeding in populations of finite size will lead to homozygosity (fixation and loss of alleles), slowly if N is large, rapidly if N is small. The proportions of discrete traits will change to higher or lower values, ultimately to all or none, and the means of continuous traits will increase or decrease. The genetic variance will stochastically decrease to zero. These changes arising from the finite size of the population have been variously called the "inbreeding effect," "random drift," "genetic drift," and the "Sewall Wright effect." Second, it is impossible to raise populations of mice in the laboratory in the absence of selection, mutation, and varying nongenetic or environmental causes of variation. Selection may favor one allele or one set of nonalleles, or one or more genotypes, or one or more phenotypes at the expense of all the other alleles, genotypes, and phenotypes. Mutation may be either "forward" or "reverse." Nongenetic sources of variation may not act uniformly on all genotypes. The consequences of all these forces interacting in the population may impair attempts to preserve the gene and genotype frequencies of a population by random breeding. Nonetheless, the experimenter who wishes to maintain the genetic variability of a population has no recourse but to carry out a random breeding system in as large a population as feasible and to try to minimize the effects of selection and of nongenetic causes of variation. There is nothing he can do about natural mutations.

AVOIDANCE OF INBREEDING

As indicated in the preceding section, random breeding in populations of finite size results in some loss of heterozygosity. Certain systems of breeding minimize the loss of heterozygosity even in finite populations. Wright (1921*a*) showed that regular matings of double first cousins, quadruple second cousins, and octuple third cousins preserve any existing heterozygosity better than random matings in finite populations. Other systems have since been devised that are superior to the regular cousin-mating systems in preserving heterozygosity in later generations, even though the loss of heterozygosity may be greater in early generations. These new systems, called circular, circular pair, and circular subpopulation mating systems, are described and analyzed by Kimura and Crow (1963) and Robertson (1964).

BROTHER-SISTER INBREEDING

The system of brother-sister inbreeding is by far the easiest from the standpoint of the investigator. He needs merely to put a male and one of his sisters, usually from the same litter, in a breeding pen. The record-keeping is minimal (Chap. 3). Repeating this act, generation after generation, will produce an inbred strain of mice. Meanwhile, mice with certain characteristics of interest may be selected in order to fix the character or characters, if possible, within the line, along with the inevitable selection for viability, fertility, and fecundity. The only major difficulty is that mice, like other naturally outbreeding species, exhibit inbreeding depression, that is, loss of reproductive fitness as inbreeding progresses. Astute selection for the more vigorous mice in each generation may enable the line to escape reproductive failure, the most severe consequence of inbreeding.

From the large number of strains of mice, successfully propagated for more than 20 generations and some for more than 100 generations of brother-sister inbreeding, one may gain the impression that mice accept inbreeding easily. Two observations deny this. First, many attempts to inbreed mice have failed. Although the number of failures has never been recorded, it probably exceeds the number of successes about fivefold. Second, crosses between unrelated inbred strains of mice characteristically produce vigorous, healthy progeny which in turn give birth to and rear large families. Hybrid vigor is the opposite of inbreeding depression and can be manifested only if some degree of depression has already occurred.

A strain of mice is called an "inbred strain" when there have been 20 or more consecutive generations of brother-sister matings or 20 or more consecutive generations of parent-offspring matings, provided the offspring was mated to the younger parent (Staats, 1964). For convenience, strains with fewer generations of inbreeding are said to be "partially inbred" or "on the way to being inbred." It is necessary to understand that this definition of an inbred strain of mice is in no way inconsistent with the more general concept of inbreeding as matings between individuals related by descent more frequently than expected by the product principle of probability.

A word of caution is necessary. When inbred mice are used in terminal experiments, it is often necessary to make up the experimental groups by pooling mice from several different litters, usually of about the same age and therefore born of different parents. The parents of such litters should themselves be closely related through a recent common pair of ancestors. If the common ancestors were in, say, the 30th generation and the mice to be used in the experiment are in the 33rd to 38th generation, they may confidently be expected to be genetically similar. They may not be genetically identical for there is no way of arresting the evolutionary forces that create genetic diversity. On the other hand, if the common ancestors were in the second generation and the mice to be used in the experiment are in the 33rd generation, one should not be surprised to find phenotypic diversity due to genetic diversity in the group as a whole, even though each individual litter satisfies the operational definition of an inbred strain.

The genetic consequences of brother-sister inbreeding are easily seen from the probability of heterozygosity (h) and the probability of incrosses (p) at specified generations with respect to any neutral locus, the a locus, whose heterozygosity is in question. In the case of starting to inbreed following a cross ($a/a \times +/+$), h approaches zero so that at 20 generations only about 1 per cent of neutral loci like a are expected to be heterozygous. Meanwhile p approaches unity so that 98 per cent of the matings are expected to be incrosses (Table 2-1).

At the same time that all neutral or unselected loci are being driven to homozygosity by brother-sister inbreeding, one or more loci may deliberately be forced to remain heterozygous by matings such as $D/+ \times +/+$ or $D/+ \times D/+$, and $r/+ \times r/r$ or $r/+ \times r/+$. Brother-sister inbreeding with forced heterozygosis is described in a later section of this chapter.

Table 2-1. THE PROBABILITY OF HETEROZYGOSITY (h) AND THE PROBABILITY OF INCROSSES (p) AT THE a LOCUS AT n GENERATIONS OF BROTHER-SISTER INBREEDING, ASSUMING A CROSS ($a/a \times +/+$) IN GENERATION ZERO

n =	0	1	2	4	8	12	16	20
h (%) =	0	100	50	37.5	16.4	7.0	3.0	1.2
p (%) =	0	0	12.5	41.4	74.8	89.2	95.4	98.0

OTHER SYSTEMS OF INBREEDING

Systems of inbreeding, other than brother-sister, may sometimes be necessary. A sequence of mating offspring to the younger parent is genetically equivalent to brother-sister mating.

Any system of breeding within a closed colony will inevitably produce increases in the probability of homozygosity and of incrosses, the rates of increase being dependent upon the size of the colony and the specific system of breeding.

Pen breeding is a system of mating a number of males, usually one, with a number of females, usually two or more, of rearing all the progeny together in a single pen, and of making no distinction among the progeny when mates are chosen to produce the next generation. The mated mice may thus be related as brother and sisters, as brother and half-sisters, or, if there were two or more males, as cousins of some degree. (It is usually not practical to put more than one male mouse in a pen with females because of fighting between males leading to emasculation or death.) If pen breeding starts with mice from an inbred strain, the genetic uniformity already achieved will not only be retained but improved upon. If it starts with noninbred stock, genetic uniformity will be approached less rapidly than under brother-sister inbreeding. At worst, when a large number of females are mated with a single male in each generation, the rate of approach to homozygosity is still a little more than half the rate of approach under brother-sister inbreeding.

If it is not practical to use a regular system of inbreeding such as brother-sister, parent-offspring, or a fixed number of mates for pen breeding, it is still possible to estimate the probability of homozygosity at the *a* locus by computing the inbreeding coefficient, *F* (Wright, 1931; Malécot, 1948; Emik and Terrill, 1949; Cruden, 1949). *F* may be interpreted as a probability of homozygosity at any locus.

One further type of mice may be produced by matings of mice within an inbred strain, although the males and females are not necessarily related as brother and sister. The progeny of such matings are called "inbred-derived" to distinguish them from inbred mice. Obviously the mated mice should have recent common ancestors within an inbred line if the genetic variance is to be kept low. In terminal experiments requiring large numbers of mice from a single strain, inbred-derived mice may be superior to inbred mice. Any genetic heterogeneity within an inbred strain will be more uniformly spread over the inbred-derived progeny and so their use will help to avoid accidental differences between experimental groups traceable to genetic heterogeneity.

STRAIN CROSSES

Crosses between inbred strains of mice may serve any one of six purposes:

1. To produce uniform F_1 hybrids for use in terminal experiments
2. To produce F_1, F_2, B_1, B_2, and other generations to analyze the genetic basis of a variable trait, either discrete or continuous
3. To produce 2-way, 3-way, 4-way, etc., crossbred populations to be used as the starting point for genetic selection experiments
4. To produce all possible F_1 generations (diallel crosses) between several inbred strains to reveal the nature of a multiple factor genetic system determining variations in a continuous trait or to estimate general and specific combining abilities
5. To transfer a mutant gene from one genetic background to another
6. To decompose a complex genotype into its single-locus constituents and thus lead to the discovery and identification of new genes

The first four of these purposes are briefly described in the following subsections. The last two, because of their special usefulness in experiments with mice, are described in the next two sections. See Chap. 9 for more discussion of the third and fourth purposes.

F_1 hybrids

F_1 hybrids produced by crosses of two inbred strains of mice are usually decidedly more hardy than mice of the inbred strains. They grow faster, survive to maturity in greater proportions, live longer, and in turn reproduce earlier and more abundantly. Hybrid litters are usually, but not always, larger than inbred litters (Forsthoefel, 1954; Butler, 1958; Franks et al., 1962; McCarthy, 1965). Hybrid mice will usually accept grafts of tumors, ovaries, skin, and other tissues from either parental strain. In some traits F_1 hybrids may be less variable than mice of inbred strains (Yoon, 1955).

F_1 hybrids will be heterozygous at all loci for which the inbred strains are homozygous for different alleles. But if the inbred strains are isogenic or nearly so, the F_1 hybrids will be uniformly heterozygous or nearly so and thus be as genetically uniform as the mice of the inbred strains. Therefore F_1 hybrids are especially valuable in terminal experiments when genetic and phenotypic uniformity are requisites, but are useless for propagating their kind.

Crosses between two strains

One may cross two inbred strains of mice to analyze the genetic basis of any trait for which the strains are different. The basic steps are the same irrespective of whether the trait is discrete or continuous, but the methods of analysis of the data are decidedly different.

The basic steps are to cross the strains, denoted as P_1 and P_2, reciprocally if possible, to produce one or two first hybrid or F_1 generations. Then the F_1 mice are mated among themselves (intercrossed) to produce a second hybrid or F_2 generation and are also backcrossed to the parental strains to produce two first-generation backcross generations, B_1 and B_2. The matings may stop at this point, but it is usually desirable to produce some additional generations. The choice of which ones to produce depends upon whether the trait is discrete or continuous and upon the results obtained in the hybrid and backcross generations.

If the trait is discrete and appears in sharply alternative states, such as "all" vs. "none," "black" vs. "brown," or "high" vs. "low" in the parental strains, and if, further, a small number of discretely distinguishable types appear in Mendelian proportions in the segregating generations (F_2, B_1, B_2, etc.), one may infer that one or two or three pairs of genes distinguish the two parental strains with respect to the trait in question. It is rarely possible to discern the discrete effects of more than three pairs of alleles. There may be any degree of interaction between alleles (dominance) and any degree of interaction between nonalleles (epistasis). Even if the results suggest only one pair of alleles with dominance, it is desirable and sometimes of crucial importance to perform either of two breeding tests.

One test is to ascertain that the F_2 mice with the dominant phenotype are in fact of two types, homozygous and heterozygous, and that the two types have occurred in the expected frequencies of $\frac{1}{3}$ and $\frac{2}{3}$. This may be done by mating the F_2 mice with dominant phenotypes to any recessive mice. The other test requires mating B_2 mice with the dominant phenotype to the supposed recessive mice to produce a second backcross B_{22} in which the two types should again occur in equal frequencies. Yet another backcross of B_{22} mice with the dominant phenotype to the supposed recessive mice should be performed in doubtful cases to ascertain that the dominant phenotypes in the B_2 and B_{22} are in fact all heterozygotes.

Examples of breeding experiments of the sort described abound in the literature. Experiments with clear-cut results were reported by Shreffler and Owen (1963), Russell and Coleman (1963), and Ruddle and Roderick (1965).

If the trait is continuous, there should be at least a significant separation of the means of the parental strains to make a breeding experiment worthwhile. The mice of the P_1, P_2, F_1, F_2, B_1, B_2, and such other generations as B_{11}, B_{12}, B_{21}, and B_{22} are individually measured for the trait. The data may then yield estimates of the amount of additive genetic variance, dominance variance, epistatic variance, and the number of segregating loci. The methods of analysis are set forth in Mather's (1949) book and elsewhere. Honeyman (1957) used these methods for analyzing a nutritional difference between two strains of mice. Bruell (1962) has given an exposition of the method of analysis and has analyzed the genetics of spontaneous activity and of exploratory behavior of two strains of mice.

Crossbred populations

The motive for crossing inbred strains of mice may be to produce a high degree of genetic heterogeneity within a population to serve as a starting point for a selection experiment. If strains are designated as A, B, C, . . . , one may start with the progeny of a 2-way cross of A $\times$ B, or a 3-way cross of (A $\times$ B) $\times$ C where females of the F_1 of A $\times$ B are chosen to mate with males of C in order to take advantage of the extra vigor, if any, of the F_1 hybrids. Or 4-way, 6-way, or 8-way crosses may be constructed.

Diallel crosses

A diallel cross is the set of all possible matings between several genotypes, defined as individuals, lines, or inbred strains. If there are n genotypes there are n^2 mating combinations, counting all inbred, crossbred, and reciprocal matings. The results of a diallel cross may be set forth in a diallel table of n^2 measurements corresponding to the n^2 mating combinations (Hayman, 1954). A diallel table may, therefore, be constructed from all first generation data (inbreds and F_1's) or from all second generation data (inbred and F_2's or inbreds and backcrosses) for one sex or for both sexes.

The type of analysis of a diallel table will depend in part on the experimenter's objectives. He may be interested in an analysis that yields estimates of the additive genetic variance, the dominance variance (allelic interactions), the epistatic variance (nonallelic interactions), and the environmental variance. Or he may be interested in judging the specific and general combining ability of each strain. Or his interest may lie in maternal effects or differences between reciprocal hybrids. The data from the F_1 generations and the parental strains may yield all the requisite information. Thus the diallel crossing system has an advantage of simplicity of execution since segregating generations will not be required except for the most exhaustive analyses.

The literature on diallel crosses has grown enormously. For discussions of the method and of the interpretation of results, see Allard (1956), Griffing (1956 and earlier), Jinks (1956 and earlier), Dickinson and Jinks (1956), Kempthorne (1956), Hayman (1960 and earlier), Broadhurst (1960), Kempthorne and Curnow (1961), Falconer and Bloom (1962), Wearden (1964), and Bloom and Falconer (1964).

TRANSFERRING A MUTATION TO AN INBRED BACKGROUND

The purpose of transferring a mutation to the genetic background of an inbred strain of mice is to permit comparisons of the effects of the mutant and nonmutant alleles with the greatest possible precision, that is, as free as possible from the effects of variability due to other unidentified loci. This may be regarded as comparing the results of different doses (0, 1, 2) of the mutant gene in the nonmutant homozygote, the heterozygote, and the mutant homozygote. Further, the relative quantitative effects of different mutations may be ascertained only by comparing them against otherwise uniform genetic backgrounds. For this purpose one of the standard inbred strains, such as C57BL/6J, provides the requisite background.

Three breeding systems have passed into general use for this purpose (Green and Doolittle, 1963). The *backcross* system is most useful if the mutant is a dominant, but may be used if the mutant is a recessive, particularly if the recessive homozygote is inviable or infertile. The *cross-intercross* and the *cross-backcross-intercross* systems are useful if the mutation is a recessive. Other breeding and manipulative techniques for maintaining special genetic stocks are described in Chap. 8 and by Lyon (1963).

Backcross system

Mice bearing a dominant mutation of interest, either as heterozygotes ($D/+$) or, if viable, as homozygotes (D/D), are mated with mice of an inbred strain which are $+/+$ at the D locus. The $D/+$ progeny are backcrossed to $+/+$ mice from the inbred strain, following which the $D/+$ progeny in the next generation are also backcrossed to $+/+$ mice from the inbred strain. This sequence may be continued as long as desired.

The probability of incrosses with respect to the a locus in the nth generation following a mating in the zeroth generation of $D/+ \times +/+$ is $p_n = 1 - (1-c)^{n-1}$, where c is the probability of crossing over between the a and D loci. The probability of heterozygosity is $h_n = (1-c)^{n-1}$. Values of p_n and h_n in generation n for selected values of c are shown in Table 2-2. In eight generations p exceeds 99 per cent if c is near 50 per cent, but 45 generations will be required for p to exceed 99 per cent if c is near 10 per cent.

After seven or eight generations of backcrossing, nearly all the alleles not closely linked with D will have been replaced by alleles from the inbred strain. At this point it may be desirable to cease backcrossing to the inbred strain and to continue with brother-sister matings of the types $D/+ \times D/+$ or $D/+ \times +/+$. Sib matings of these sorts will allow variable loci linked to the D locus to become homozygous faster than will continued backcrossing to the inbred strain (see below). Further, if D/D is viable, a few generations of intercrossing following seven or eight generations of backcrossing will allow all three geno-

Table 2-2. THE PROBABILITY OF INCROSSES p AT THE a LOCUS AFTER n GENERATIONS OF BACKCROSSING, WHERE c IS THE PROBABILITY OF RECOMBINATION BETWEEN THE a AND THE D LOCI. THE PROBABILITY OF HETEROZYGOSITY (h) EQUALS $1 - p$

c	n =	1	2	4	8	12	16	20
0.5	p (%) =	0	50.0	87.5	99.2	99.9	99.9+	99.9+
0.4		0	40.0	78.4	97.2	99.6	99.9+	99.9+
0.3		0	30.0	65.7	91.8	98.0	99.5	99.9
0.2		0	20.0	48.8	79.0	91.4	96.5	98.6
0.1		0	10.0	27.1	52.2	68.6	79.4	86.5

types to be examined against a relatively uniform background. As a practical matter, therefore, seven or eight generations of backcrossing to an inbred strain followed by 10 or 12 generations of backcrossing or intercrossing sibs will usually be adequate for producing genetically similar mice.

The backcross system may also be used if the mutation of interest is recessive. Heterozygotes ($r/+$) are mated with homozygous mice ($+/+$) of an inbred strain and the heterozygous progeny are again mated to $+/+$ mice of the inbred strain. Backcrossing in this way may continue as long as desired. If the heterozygotes ($r/+$) and the homozygotes ($+/+$) produced in each backcross generation are indistinguishable, it will be necessary to identify the heterozygotes by a breeding test. This requires more time, but not more generations, to reach a given probability of incrosses than backcrossing with a dominant. The use of heterozygotes will be obligatory when the recessive homozygotes (r/r) are inviable or infertile.

Cross-intercross system

Mice homozygous for a recessive mutation (r/r) are crossed with mice of an inbred strain which are $+/+$ at the *r* locus. Their progeny ($r/+$) are intercrossed to recover the recessive homozygotes (r/r) which are used to start another cycle of crossing and intercrossing. The number of such cycles, with two generations of matings per cycle, may be made as large as desired.

The probabilities of incrosses (p_m) and of heterozygotes (h_m) at the *a* locus after *m* cycles of the cross-intercross system are shown in Table 2-3 for selected values of *c*. Eight cycles or 16 generations will raise the probability of incrosses above 99 per cent if there is independence or loose linkage, but 46 cycles will be required to raise it to 99 per cent if there is close linkage ($c =$ 10 per cent).

Table 2-3. THE PROBABILITY OF INCROSSES (*p*) AND OF HETEROZYGOSITY (*h*) AT THE *a* LOCUS AFTER *m* CYCLES OF THE CROSS-INTERCROSS SYSTEM WITH TWO GENERATIONS PER CYCLE, WHERE *c* IS THE PROBABILITY OF RECOMBINATION BETWEEN THE *a* AND THE *r* LOCI

c *m*	=	1	2	4	6	8	10
0.5 *p* (%)	=	25.0	59.4	89.6	97.4	99.3	99.8
0.4		16.0	45.3	79.7	92.7	97.4	99.0
0.3		9.0	31.1	65.1	82.8	91.6	95.9
0.2		4.0	18.1	45.7	65.1	77.6	85.7
0.1		1.0	7.3	22.9	37.3	49.1	58.8
0.5 *h* (%)	=	50.0	31.2	8.3	2.1	0.5	0.1
0.4		48.0	37.4	14.7	5.3	1.9	0.7
0.3		42.0	39.7	21.8	10.8	5.3	2.6
0.2		32.0	35.8	26.6	17.4	11.1	7.2
0.1		18.0	23.5	22.9	19.2	15.6	12.7

Table 2-4. THE PROBABILITY OF INCROSSES (*p*) AND THE PROBABILITY OF HETEROZYGOSITY (*h*) AT THE *a* LOCUS AFTER *m* CYCLES OF THE CROSS-BACKCROSS-INTERCROSS SYSTEM WITH THREE GENERATIONS PER CYCLE, TION BETWEEN THE WHERE *c* IS THE PROBABILITY OF RECOMBINA- *a* AND *r* LOCI

c *m*	=	1	2	3	4	6	8
0.5 *p* (%)	=	56.2	89.1	97.3	99.3	99.9+	99.9+
0.4		41.0	78.7	92.3	97.3	99.6	99.9+
0.3		26.0	63.7	82.2	91.3	97.9	99.5
0.2		13.0	44.3	64.3	77.2	90.7	96.2
0.1		3.6	21.9	36.8	48.8	66.4	77.9
0.5 *h* (%)	=	37.5	9.4	2.3	0.6	0.0+	0.0+
0.4		46.1	16.6	6.0	2.2	0.3	0.0+
0.3		50.0	25.0	12.0	6.0	1.4	0.3
0.2		46.1	29.5	18.9	12.1	4.9	2.0
0.1		30.8	24.9	20.2	16.4	10.7	7.0

Cross-backcross-intercross system

This system starts the same as the cross-intercross system by matings of recessive homozygotes with mice of an inbred strain which is $+/+$ at the *r* locus. All the progeny ($r/+$) are backcrossed to mice of the inbred strain, thus producing two indistinguishable kinds of mice ($r/+$ and $+/+$) expected in equal proportions. Among these backcross progeny 12 or more matings should be made up to insure recovery of the homozygotes (r/r). The homozygotes are then crossed with mice of the inbred strain and a new cycle of three generations—cross, backcross, and intercross—is thereby started. The sequence may go on for as many cycles or generations as desired.

The question of how many intercross matings to make up may be treated as follows. The Mendelian probability of a heterozygote in the backcross is $\frac{1}{2}$ and the probability of a mating between two heterozygotes, a "right" mating, is $\frac{1}{4}$. Thus the probability of the "wrong" type of mating is $\frac{3}{4}$. If *k* matings are made up, the probability that all of them are "wrong" is $(\frac{3}{4})^k$. When *k* exceeds

11 this probability is less than 5 per cent and when it exceeds 17, the probability is less than 1 per cent. One should plan therefore to set out about 12 intercrosses, with the understanding that the "right" intercross may be found among the first few attempts.

The chief advantage of this system appears when the recessive homozygotes (r/r) are not easily detectable, but rather require expensive time-consuming tests. The tests are necessary only once in each three generations. This has to be weighed against the extra expense and time of maintaining more breeding pens of mice.

The probabilities of incrosses (p) and of heterozygotes (h) after m cycles of three generations each are shown in Table 2-4. Four cycles or 12 generations raise the probability of incrosses to about 99 per cent if there is loose linkage or independence, but 23 cycles will be required to achieve the same probability if there is close linkage ($c = 10$ per cent).

ISOLATION OF GENES

In the preceding section, three systems for transferring a mutant allele to an inbred background are described under the assumption that a locus with a discrete effect has already been identified by the usual breeding tests. The same three systems of breeding may be used to decompose a complex character into identifiable subunits and thus to isolate single pairs of genes affecting or controlling the character of interest. Even though the breeding systems are identical, the terminology must be changed to avoid confusion. Thus instead of referring to mutant and nonmutant mice, we refer to those with and those without the trait, or to those with high and those with low levels of the trait. The point of the change in terminology is that one does not know, when attempting to decompose a trait into its genetic units, how many loci are responsible for the alternative categories or levels of the trait in any given generation.

The breeding techniques for the isolation of single pairs of genes have been used most effectively by Snell (1958 and earlier; Chap. 24) to isolate histocompatibility loci. The histocompatibility reactions fulfill all of the desiderata for the most successful use of the cross-intercross and cross-backcross-intercross breeding systems in uncovering and isolating the histocompatibility loci: discrete all-or-none differences between inbred strains, reasonably small numbers of differential loci distinguishing any pair of strains, and each mouse readily classifiable as having or not having the trait. It is not likely that other genetic systems of such analytic beauty will readily be found in mice.

In the foregoing discussion I have assumed that the trait of interest exhibits discrete alternatives. If the trait is continuous, the same breeding procedure may be used, but the outcome is likely to be less clear cut because the environmental sources of variability may blur the differences between alleles at a single locus. Nonetheless, Chai (1961) has reported some success in isolating genes affecting body size in mice.

One final word of caution is necessary. These breeding techniques do not actually isolate single loci, but do isolate successively smaller and smaller bits of heterozygous chromosomes which carry at least one pair of alleles affecting the trait of interest. See Haldane's appendix in Snell's (1948) paper.

INBREEDING WITH FORCED HETEROZYGOSIS

Brother-sister inbreeding may be pursued while heterozygosity is forced upon a specific locus, either by backcrossing or by intercrossing. All matings may be backcrosses, such as $r/+ \times r/r$, if r is a viable recessive, or $D/+ \times +/+$, if D is a dominant or semidominant lethal; or they may be intercrosses, such as $r/+ \times r/+$, if r is a recessive lethal, or $D/+ \times D/+$, if D is a dominant or a semidominant lethal.

Inbreeding with forced heterozygosis may be desirable if an inbred strain is not available, making it impossible to try to transfer a mutant allele to the background of a standard inbred strain. Inbreeding with forced heterozygosis may be obligatory if the mutant is incompatible with the genetic backgrounds of available standard inbred strains.

Loci not linked with r (or D) will become homozygous with the same probabilities as under brother-sister inbreeding. Linked loci will become homozygous less rapidly, the rate depending upon the recombination probability c. Further, the probabilities of heterozygosity at linked loci (a) differ in the homozygotes and heterozygotes of the locus of interest (r or D) (Bartlett and Haldane, 1935; Green and Doolittle, 1963). The probability of incrosses (p) for selected values of c are shown in Table 2-5.

With the exception of intercrossing with a recessive, the systems of brother-sister inbreeding with forced heterozygosis have nearly the simplicity of regular brother-sister inbreeding. Mat-

Table 2-5. THE PROBABILITY OF INCROSSES p AT THE a LOCUS AFTER n GENERATIONS OF BROTHER-SISTER INBREEDING WITH HETEROZYGOSIS FORCED BY BACKCROSSING, ASSUMING THE INITIAL MATING IS $+ r/+ r \times a +/a\ r$ OR $a\ r/a\ r \times + +/+ r$, AND BY INTERCROSSING, ASSUMING THE INITIAL MATING IS $+ +/+ r \times a +/a\ r$, WHERE c IS THE PROBABILITY OF RECOMBINATION BETWEEN THE a AND r LOCI

c	n =	1	2	4	8	12	16	20
By backcrossing								
0.5	p (%) =	0	12.5	41.4	74.8	89.2	95.4	98.0
0.4		0	12.0	40.3	73.8	88.5	95.0	97.8
0.3		0	10.5	37.1	70.5	86.2	93.5	97.0
0.2		0	8.0	31.3	63.8	81.0	90.0	94.7
0.1		0	4.5	22.4	50.7	68.3	79.4	86.6
By intercrossing								
0.5	p (%) =	0	12.5	41.4	74.8	89.2	95.4	98.0
0.4		0	12.5	41.4	74.8	89.2	95.4	98.0
0.3		0	12.8	41.1	74.4	88.9	95.2	97.9
0.2		0	14.1	39.9	72.2	87.2	94.1	97.3
0.1		0	17.6	36.6	63.9	79.5	88.4	93.4

ings of the desired types ($r/r \times r/+$, $D/+ \times +/+$, or $D/+ \times D/+$) may usually be made up from littermates. The task of keeping a standard inbred strain on hand for the sole purpose of putting a mutant gene on an inbred background is thus unnecessary.

Intercrossing with a recessive is somewhat bothersome. The desired matings are $r/+ \times r/+$ and these must be selected from a number of matings which are $?/+ \times ?/+$ until an r/r mouse is produced. Consequently, to be 99 per cent confident of getting at least one mating of the desired type, one must make up at least $k = 8$ matings of $?/+ \times ?/+$. This is so because the probability for $r/+$ among $?/+$ progeny is 2/3 and the probability of $r/+ \times r/+$ is 4/9. Hence to reduce the probability of not getting the right mating to less than 1 per cent requires that k be selected so that $(5/9)^k < 0.01$.

From Tables 2-2, 2-3, and 2-4 in comparison with 2-5, one may see that the backcross, cross-intercross, and cross-backcross-intercross systems require fewer generations than the systems with forced heterozygosis to achieve a specified high probability of incrosses at the a locus when the a locus is independent of or loosely linked with the r or D locus. When the a locus is closely linked with the r or D locus, the relationship is reversed. This suggests that the two kinds of systems may be used effectively in sequence. To produce a congenic line with segregation at the D locus, for instance, one may use the backcross system for seven or eight generations to bring about a high percentage of incrosses at loci independent of or loosely linked with the D locus and follow this with 10 or 12 generations of brother-sister inbreeding with forced heterozygosis to increase the percentage of homozygosity at loci closely linked with the D locus, while not only preserving the percentage of incrosses already achieved but improving it at the same time.

PROPAGATING MUTATIONS WITHOUT INBREEDING

Some dominant mutations in mice depress viability and fertility so severely that they can neither be put on standard inbred backgrounds nor inbred with forced heterozygosis. It is therefore necessary, just to preserve the mutation, to propagate it in heterogeneous mice. This may be done either by deliberately avoiding inbreeding within a stock, by routinely backcrossing mutant-bearing mice to a vigorous randombred or noninbred stock, or by routinely backcrossing mutant-bearing mice to vigorous F_1 hybrids between two standard inbred strains.

SUMMARY

An experimenter may manipulate the genetic variability in mice by means of the breeding systems described in this chapter. Inbreeding will reduce it, outcrossing will increase it, and random breeding will preserve it.

Brother-sister inbreeding has been useful in producing genetically uniform mice and has yielded a large number of inbred strains. Various systems of mating for preserving heterozygosity at a specific locus have also come into general use. These produce mice which are as genetically similar as possible except for the different genotypes at a controlled or segregating locus.

Strains may be crossed to analyze the genetic basis of discrete or of continuous traits. A classic Mendelian experiment with a discrete trait requires the examination of parental, first- and second-generation hybrids, backcrosses, and sometimes other generations. Biometrical techniques will be needed to analyze the genetic constitution of groups of strains differing in quantitative characters. The breeding techniques include selection and diallel crosses.

Mutant genes may be transferred to standard inbred backgrounds by the backcross, the cross-intercross, and the cross-backcross-intercross systems of mating. The same systems may be used to decompose a complex trait into its genetic components and thus to isolate new alleles. Inbreeding with forced heterozygosis may be advantageous when a mutant cannot, for some reason, be transferred to a standard inbred background.

LITERATURE CITED

Allard, R. W. 1956. The analysis of genetic-environmental interactions by means of diallel crosses. Genetics 41:305–318.

Bartlett, M. S., and J. B. S. Haldane. 1935. The theory of inbreeding with forced heterozygosis. J. Genet. 31:327–340.

Bloom, J. L., and D. S. Falconer. 1964. A gene with major effect on susceptibility to induced lung tumors in mice. J. Nat. Cancer Inst. 33: 607–618.

Broadhurst, P. L. 1960. Analysis of a diallel cross, p. 71–102. *In* H. J. Eysenck [ed.] Experiments in personality. Vol. 1. The Humanities Press, New York.

Bruell, J. H. 1962. Dominance and segregation in the inheritance of quantitative behavior in mice, p. 48–67. *In* E. L. Bliss [ed.] Roots of behavior. Hoeber, New York.

Butler, L. 1958. The inheritance of litter size, body weight, and variability, in a cross between two inbred strains of mice. Can. J. Zool. 36: 969–983.

Carter, T. C., and D. S. Falconer. 1951. Stocks for detecting linkage in the mouse, and the theory of their design. J. Genet. 50:307–323.

Chai, C. K. 1961. Analysis of quantitative inheritance of body size in mice. IV. An attempt to isolate polygenes. Genet. Res. 2:25–32.

Chovnik, A., and A. S. Fox. 1953. Immunogenetic studies of pseudoallelism in *Drosophila melanogaster.* I. Antigenic effects of the lozenge pseudoalleles. Proc. Nat. Acad. Sci. 39:1035–1043.

Cruden, D. 1949. The computation of inbreeding coefficients for closed populations. J. Hered. 40: 248–251.

Dickinson, A. G., and J. L. Jinks. 1956. A generalized analysis of diallel crosses. Genetics 41: 65–78.

Emik, L. O., and C. E. Terrill. 1949. Systematic procedures for calculating inbreeding coefficients. J. Hered. 40:51–55.

Falconer, D. S. 1960. Introduction to quantitative genetics. Ronald Press, New York. 365 p.

Falconer, D. S., and J. Bloom. 1962. A genetic study of induced lung-tumours in mice. Brit. J. Cancer 16:665–685.

Fisher, R. A. 1949. The theory of inbreeding. Oliver and Boyd, Edinburgh. 120 p.

Forsthoefel, P. F. 1954. A study of the effect of heterozygosis on litter size in the mouse *Mus musculus.* Ohio J. Sci. 54:135–141.

Franks, E., N. S. Fechheimer, and C. Cohen. 1962. An examination of heterosis in crosses of certain inbred strains of mice. Ohio J. Sci. 62:177–184.

Green, E. L., and D. P. Doolittle. 1963. Theoretical consequences of systems of mating used in mammalian genetics, p. 3–41. *In* W. J. Burdette [ed.] Methodology in mammalian genetics. Holden-Day, San Francisco.

Griffing, B. 1956. A generalized treatment of the use of diallel crosses in quantitative inheritance. Heredity 10:31–50.

Hayman, B. I. 1954. The theory and analysis of diallel crosses. Genetics 39:789–809.

Hayman, B. I. 1960. The theory and analysis of diallel crosses. III. Genetics 45:155–172.

Honeyman, M. S. 1957. A quantitative genetic study of a nutritional strain difference in mice. J. Hered. 48:84–87.

Jinks, J. L. 1956. The F_2 and backcross generations from a set of diallel crosses. Heredity 10:1–30.

Kempthorne, O. 1956. The theory of the diallel cross. Genetics 41:451–459.

Kempthorne, O. 1957. An introduction to genetic statistics. Wiley, New York. 545 p.

Kempthorne, O., and R. N. Curnow. 1961. The partial diallel cross. Biometrics 17:229–250.

Kimura, M., and J. F. Crow. 1963. On the maximum avoidance of inbreeding. Genet. Res. 4:399–415.

Lerner, I. M. 1950. Population genetics and animal improvement. Cambridge Univ. Press, London. 342 p.

Le Roy, H. L. 1960. Statistische Methoden der Populationsgenetik. Birkhäuser Verlag, Basel und Stuttgart. 397 p.

Li, C. C. 1955. Population genetics. Univ. Chicago Press, Chicago. 366 p.

Loosli, R., E. Ş. Russell, W. K. Silvers, and J. L. Southard. 1961. Variability of incidence and clinical manifestation of mouse hereditary muscular dystrophy on heterogeneous genetic backgrounds. Genetics 46:347–355.

Lush, J. L. 1945. Animal breeding plans, 3rd ed. Iowa State College Press, Ames. 443 p.

Lyon, M. F. 1963. Genetics of the mouse, p. 199–

234. *In* W. Lane-Petter [ed.] Animals for research. Academic Press, London and New York.

Malécot, G. 1948. Les mathématiques de l'hérédité. Masson, Paris. 63 p.

Mather, K. 1949. Biometrical genetics. Dover, New York. 158 p.

McCarthy, J. C. 1965. The effect on litter size of crossing inbred strains of mice. Genetics 51: 217–222.

Robertson, A. 1964. The effect of non-random mating within inbred lines on the rate of inbreeding. Genet. Res. 5:164–167.

Ruddle, F. H., and T. H. Roderick. 1965. The genetic control of three kidney esterases in C57BL/6J and RF/J mice. Genetics 51:445–454.

Russell, R. L., and D. L. Coleman. 1963. Genetic control of hepatic δ-aminolevulinate dehydratase in mice. Genetics 48:1033–1039.

Shreffler, D. C., and R. D. Owen. 1963. A serologically detected variant in mouse serum: inheritance and association with the histocompatibility-2 locus. Genetics 48:9–25.

Snell, G. D. 1948. Methods for the study of histocompatibility genes. J. Genet. 49:87–108.

Snell, G. D. 1958. Histocompatibility genes of the mouse. II. Production and analysis of isogenic resistant lines. J. Nat. Cancer Inst. 21:843–877.

Staats, J. 1964. Standardized nomenclature for inbred strains of mice, Third listing. Cancer Res. 24:147–168.

Wearden, S. 1964. Alternative analyses of the diallel cross. Heredity 19:669–680.

Wright, S. 1921*a*. Systems of mating. Genetics 6:111–178.

Wright, S. 1921*b*. Correlation and causation. J. Agr. Res. 20:557–585.

Wright, S. 1931. Evolution in Mendelian populations. Genetics 16:97–159.

Wright, S. 1934. The method of path coefficients. Ann. Math. Statist. 5:161–215.

Wright, S. 1954. The interpretation of multivariate systems, p. 11–33. *In* O. Kempthorne, T. A. Bancroft, J. W. Gowen, and J. L. Lush [ed.] Statistics and mathematics in biology. Iowa State College Press, Ames.

Wright, S. 1963. Discussion of paper by Green and Doolittle, p. 42–53. *In* W. J. Burdette [ed.] Methodology in mammalian genetics. Holden-Day, San Francisco.

Yoon, C. H. 1955. Homeostasis associated with heterozygosity in the genetics of time of vaginal opening in the house mouse. Genetics 40:297–309.

3

Keeping Records[1]

Margaret M. Dickie

The history of record keeping in mouse colonies is, in part, the history of the development of inbred strains of mice. The genealogy, family tree, or pedigree of the DBA strains traces back to 1909. The C57 strains owe their origin to the record of matings of two mice at Cold Spring Harbor in 1921 (Chap. 1). The pedigrees of these strains are valuable not only for their length but also for their extensive history of the strains and sublines of the strains in the intervening decades.

A record-keeping system for inbred lines and various experimental colonies must enable the investigator to trace the collateral relatives, ancestors, and progeny of any animal with ease. Many systems have been devised for use with colonies of mice (Dickie, 1963; Porter and Lane-Petter, 1962). Any investigator who wishes to start keeping records on a colony may study the various systems, decide what sorts of information he wishes, what is essential for accurate maintenance of an inbred line and for various types of experimental colonies, and what will be most efficient and least cumbersome. In this chapter we shall review the basic components of the systems, including ledgers, pedigree or record cards, pedigree charts, cage tags, and card files (Fig. 3–1). The first two components are essential in building pedigree charts; the latter two provide efficient tools for managing colonies.

IDENTIFICATION SYSTEM AND LEDGER

The first essential for a record-keeping system is a source of sequence numbers which may be a continuous series, a code, or an evolving-number system. A continuous number series may be entered in a ledger beginning with 01 (or 0001 or 0101 if a punched card system is to be used) and continuing to 99. Thereafter the numbers are 101 to 199, 201 to 299, and so forth. Note that 100, 200, and so forth are omitted, since such numbers would mean unmarked animals in the usual earmarking system. The animals are marked with an ear punch according to the code shown in Fig. 3–2. In some cases a hole punched in the center of the ear may be used to denote zero. The center hole is used primarily so that each ear will be punched whether the number is 07 or 70.

Toe clipping is also a successful means of identification but is not in very widespread use. It is more common to clip toes of animals at birth to signify classification of some condition such as a transitory anemia that will be unrecognizable at the time of weaning. Tail clipping is often used to mark animals treated in some way shortly after birth. This is useful not only to distinguish them from all other litters or animals in adulthood, but also to distinguish treated from untreated members of the same litter. Painting has been used to identify mice of albino strains but this practice is not recommended. Each of the other methods is more permanent.

Ledgers are usually bookkeeping ledgers or spiral-bound notebooks. A ledger may be used for each strain or mutant stock or, in small colonies, for all animals of several stocks. The ledger includes the following information:

1. The sequence number assigned to the mouse

[1] The writing of this chapter was supported in part by Contract AT(30-1)-3249 with the U.S. Atomic Energy Commission.

2. Sex
3. Phenotype, if needed
4. Fate or destiny of the animal
5. Dam and sire (dam number always given first)
6. Strain or genotype of dam and sire (if single master ledger)
7. Generation and subline, if any
8. Number born in litter
9. Date of birth

A line of the ledger is used for each individual in the litter. At the beginning of the record for a litter the ancestry, generation, parentage, and birth date of the litter are recorded. After all individuals in the litter are recorded, a line is drawn across the page under the notations for

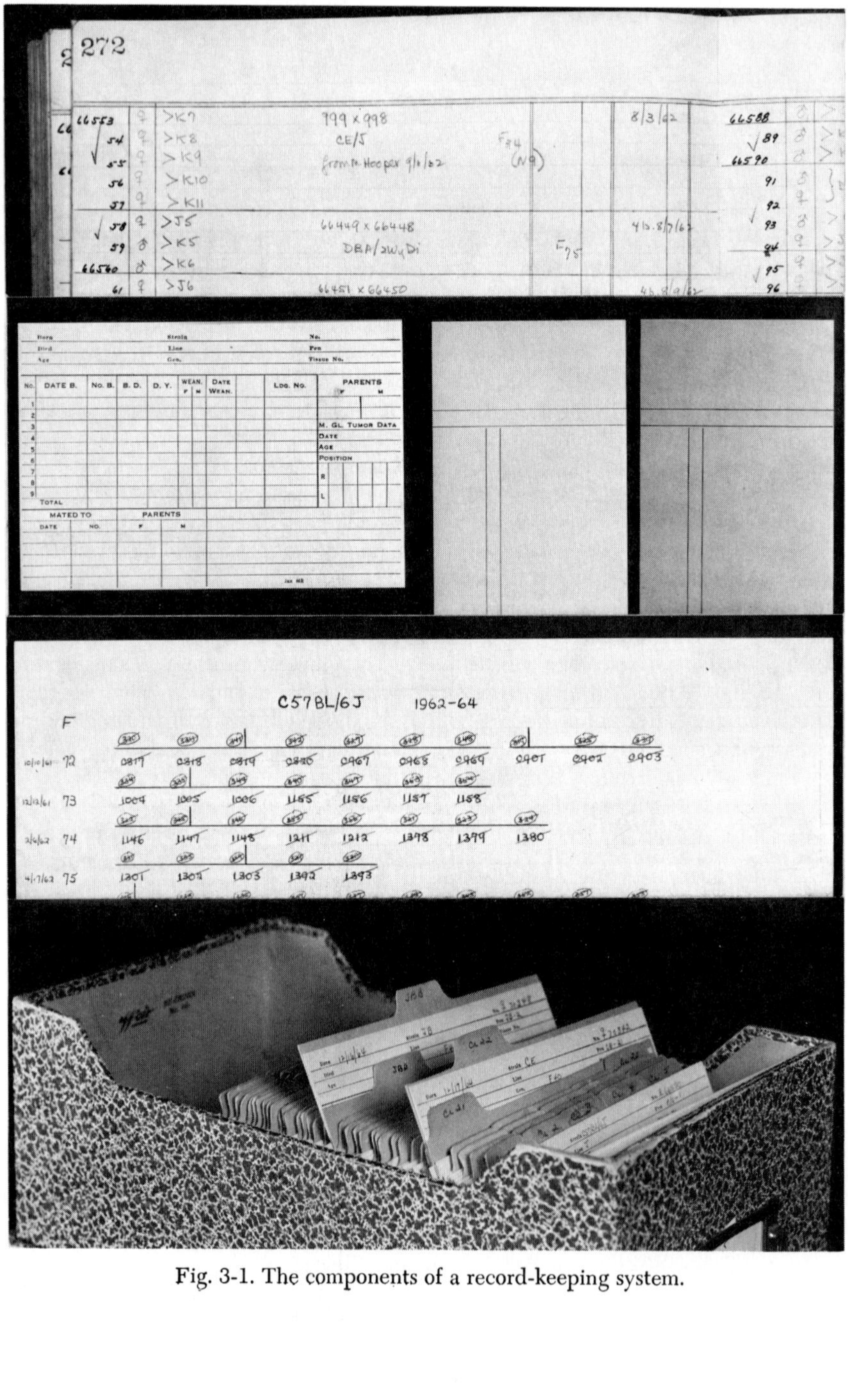

Fig. 3-1. The components of a record-keeping system.

the last animal in the litter. The record for a subsequent litter begins on the next line. Therefore one can see at a glance, by study of the ledger, the disposition of any particular litter.

Some systems employ a ledger for each stock and assign each mating a double-page space for all information concerning offspring of that mating. When this system is used, a modified numbering or code system is used for identification.

It is essential to enter the phenotype or genotype when recording mutant stocks and various experimental animals. When such notations are made, the symbol + is standard for wild-type or nonmutant alleles. This information is noted after the sex of the animals.

At The Jackson Laboratory the production of mice on a large scale takes place in three steps: Foundation Stocks (FS), a small primary source for inbreeding control without subline differentiation; Pedigreed Expansion Stocks (PES), for expansion to large numbers of breeding pairs in a few generations; and Production Stocks (PS), where all progeny are harvested for use in research. An envolving-number system and a master ledger are used to identify the breeding units in all three colonies. In FS, each breeding pair is given a four-digit serial number such as 0352 from a master ledger, the source of numbers for all strains. When weanling pairs are transferred from FS to PES, they are given an additional single-place number separated from the serial number of the parents by a hyphen, such as 0352-1, 0352-2, . . . , which identifies the serial order of the pairs made up from the 0352 pair in FS. The 10th pair is designated as 0352-0; the 11th through 20th pairs are designated by the addition of the same single serial numbers, but each is encircled; the 21st through the 30th pairs may, if necessary, be designated by the addition of the same single serial numbers enclosed in squares.

When the pairs transferred to PES from FS in turn reproduce, the pairs made up from their progeny are designated by serial numbers in the second position, such as 0352-11, 0352-12, . . . , 0352-21, 0352-22, ... The second position is thus used for serial numbers of matings two generations removed from the FS, just as the first position designated serial numbers of matings one generation removed. The third, fourth, . . . positions may thus be used for enumerating matings in the third, fourth, . . . generations removed from FS.

When breeding pairs are transferred from PES to PS, another number is added, separated by a

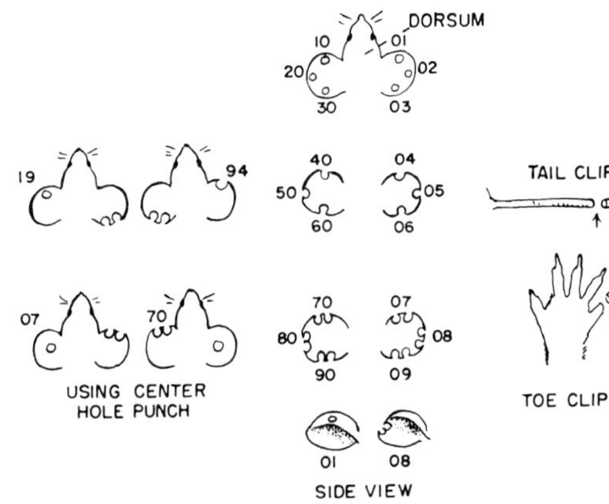

Fig. 3-2 The codes used in identification of animals.

hyphen. Thus 0352-131-6 is the sixth mating in PS, following three generations of matings in PES of descendants of pair 0352 in FS (Fig. 3-3).

This evolving-number system shows (1) the source pair in FS, (2) the location of any particular mating. i.e., whether it is from FS, PES, or PS, and (3) the number of generations the mating has been separated from FS.

PEDIGREE CARD AND CAGE TAG

A pedigree card is designed to suit the needs of the investigator and the type of colony under study. One side of the card may be used for all information about breeding performance of the female and the reverse side of the card may be used for auxiliary data. At The Jackson Laboratory the reverse side of the card is used for comments on gross autopsy findings. It is efficient to have headings on the front of the pedigree card and on a ledger page read across the card and page in the same order to minimize the chance for error when one makes out the card. Space is provided for date of mating, identification number of male, kind of male, and date when he is removed from the pen. The mating date is important to make sure that the litter is the product of the pair in the pen and not a father-daughter mating or a litter from a previous male. Space is also provided for the following information: date of birth, date of death, strain or stock, phenotype or genotype (if desired), generation, individual identification number, location (pen or experiment), date of birth of each litter, number born, number and sex of mice weaned, and their disposition (ledger number or a note if killed without being recorded).

In some systems two cards are used, a mating

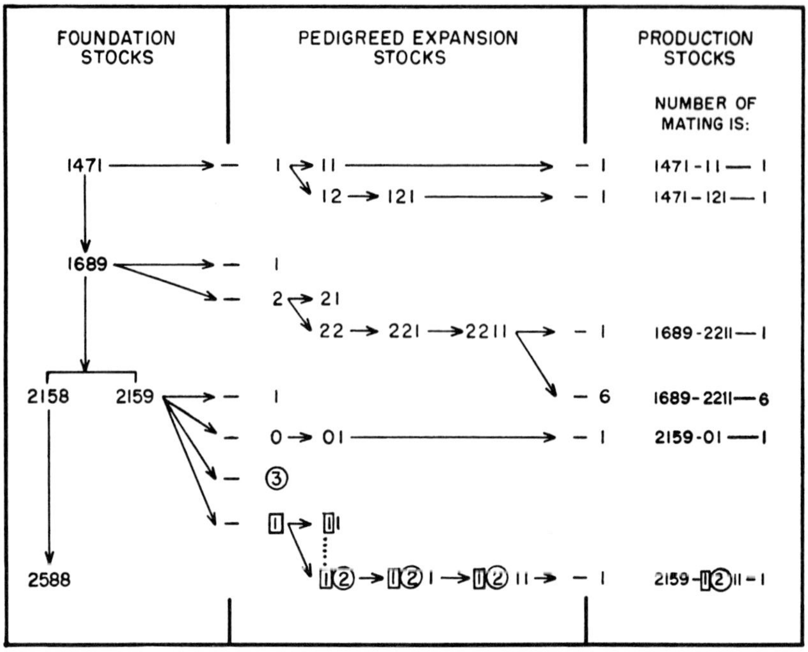

Fig. 3-3. The evolving number system.

card and a litter card. In this instance the use of the ledger is omitted and the identification of animals evolves from the litter cards.

When inbred strains are being maintained (always brother-sister matings), it is unnecessary to provide an individual card for the male since all data for the female are applicable to the male and neither will be mated again. In breeding ex-

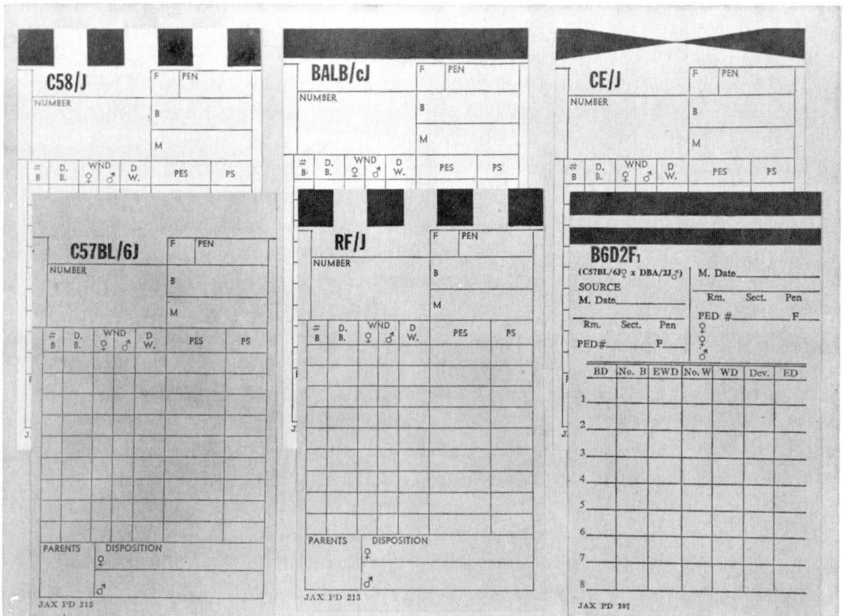

Fig. 3-4. Coded cage tags.

periments it is advisable to have a pedigree card for each mate so that each may be moved from pen to pen, as required, without confusion or loss of information.

Cage tags are mainly of two types: those providing only location on the shelf and those providing not only the location but also information such as the identification numbers of the animals in the pen, or number of mice in the pen, and any other information desired. In large breeding colonies it has proved efficient to have the cage tag or card serve as the pedigree card and thus eliminate one record-keeping step. In this case the basic information desired on the pedigree card is placed on the cage card. When it is necessary to have several strains of similar phenotype (e.g., several albino strains) in one room, cage cards of different colors and with different patterns on the top border of the card provide an extra measure of security against mix-ups. A color and pattern system can be devised so that no two strains identical in appearance will be next to each other, since cards of similar color and pattern would not ordinarily be next to one another (Fig. 3-4).

Keypunch and keysort cards may be used as pedigree cards or as cage cards to eliminate time-consuming transfers of data in large-scale summaries and analyses of breeding records.

CARD FILE

Several systems for maintenance of card files have been used. If all the pedigree cards in one file pertain to one strain, they may be filed in sequence according to pen number, dam or sire number, or parental birth date. A birth date file is especially useful when large numbers of pairs must be replaced at regular intervals.

When the colony is small and there is a variety of strains, stocks, or experimental animals, it is more efficient to use file guide cards which correspond to cage numbers. In some colonies it is useful to divide the cards into three groups, each group marked by an appropriate guide card labeled mated, pregnant, and litters (with oldest litters at front of group). This system provides a continuous check on the status of the colony.

PEDIGREE CHART

Pedigree charts are necessary in maintaining inbred strains. They must be used to insure that a strain is propagated through a single pair in each generation and that sublines are prevented from developing and coexisting for many generations. The charts may contain any additional information desired for selection of the main line such as tumor occurrence, litter size, number of litters, and time between generations. These charts need only contain the female number and should be kept current with the matings in the pens. Pedigree charts are also useful in determining relationships of deviants that may arise in a colony, learning about their inheritance, and eliminating carriers of mutations from the colony.

SUMMARY

The components of record-keeping systems are ledgers, pedigree cards, pedigree charts, and cage tags. Methods of using these components in various record-keeping systems have been described. Use of these components in these systems should provide the basic objective of easy traceability of ancestors, collateral relatives, and progeny of each animal.

LITERATURE CITED

Dickie, M. M. 1963. Methods of keeping records, p. 522–537. *In* W. J. Burdette [ed.] Methodology in mammalian genetics. Holden-Day, San Francisco.

Porter, G., and W. Lane-Petter. 1962. Notes for breeders of common laboratory animals, p. 96. Academic Press, London and New York.

4

Husbandry[1]

Edwin P. Les

Animal husbandry is the applied science of providing an optimal environment for a population of animals. The practice of mouse husbandry encompasses all of the activities necessary to produce mice of high quality, including care and feeding, cleaning of the equipment and physical surroundings, and constant concern for the health of each animal. The implementation of good mouse husbandry begins with the design and construction of the physical facilities and continues through maintenance of an environment conducive to good health, growth, reproduction, and survival of the animals.

FACILITIES FOR MICE

Rearing rooms

The design and construction of a mouse room should meet certain standards regardless of the type of research carried on or of the use of the room solely for breeding or maintenance. If mice are produced internally (not procured from an outside source) the breeding rooms should be separated from rooms in which experimental procedures are carried out. If mice are procured from an outside source, they should be quarantined at least 2 weeks in a room remote from the research area. The isolation of such areas may be accomplished in separate buildings, on different floors of the same building, or perhaps only at opposite ends of the same floor. The objective is to minimize the transfer of diseases between the groups of animals.

The size of a mouse room (when choice is possible) should be determined by striking a balance between isolation and efficiency of operation. Small rooms (100 sq ft of floor space or less) provide greater protection against spread of disease, but they are inefficient in terms of access to the mice (bringing in clean equipment and removing soiled equipment). Small rooms also result in excessive corridor areas in relation to mouse room areas.

In addition to the racks and aisle space, there must be space for temporary storage, a work table, a sink (possibly outside the room), and waste receptacles.

Mouse rooms should have two doors, one opening from a clean supply corridor and the other into an evacuation corridor. The appearance and degree of contamination of the evacuation corridor should not differ appreciably from the clean supply corridor. In a research situation there must be access from the mouse rooms to laboratories. To incorporate the supply-and-evacuation corridor concept into a workable relationship with laboratory rooms, it is necessary to compromise the traffic pattern to permit the people from the laboratory to cross one of the two corridors.

The decision about which of the corridors adjacent to the mouse rooms will function as the clean supply corridor can be made on the basis of the amount of contamination present in the colony. If disease incidence is low, the evaculation corridor can be adjacent to the laboratories; if disease incidence is high or if the experimental

[1] The writing of this chapter was supported in part by Public Health Service Research Grant GM 10236 from the National Institute of General Medical Sciences.

procedure involves communicable disease vectors, the supply corridor should be adjacent to the laboratories.

Materials and methods used in construction of mouse rooms should not only conform to local building codes, but should meet or exceed the recommendations of the Institute of Laboratory Animal Resources (1962).

Other sources of information on laboratory animal facilities and mouse husbandry include Public Health Service (1963), Hill (1963), Hoag and Les (1963), and Lane-Petter (1963).

Racks and shelving

Racks for cages may be either mobile or fixed. Fixed racks can be attached to the floor, the walls, or the ceiling. Regardless of the system used, provision must be made for cleaning beneath and behind the racks. The bottom shelf of a rack should be no less than 6 in. off the floor; the top shelf no higher than the reach of a person of average height. The use of stepladders should be avoided.

The spacing between shelves should allow ample room for insertion and removal of cages, i.e., about 2 in. greater than the over-all height of a cage unit with the water bottle attached.

Shelves may be made of metal pipe or rods, or they may be solid sheet metal like book shelves. The supposed advantage of pipe racks is that the open construction allows greater circulation of air around the cages; however, an undesirable feature is that materials (feces, food, bedding) thrown out of the cages by mice can fall from an upper cage into one on a lower shelf. A combination of solid shelves with open structure in the back wall of the rack (or no back wall in the rack) would probably be most desirable.

Regardless of the design, the racks should be constructed with a minimum of crevices or joints where dirt might accumulate. For this reason, materials such as expanded metal, perforated metal, or structural members with many holes for adjusting shelf height should be avoided. Unless the racks are stainless steel or galvanized metal, they should be painted with coating that prevents corrosion and withstands the action of detergents and disinfectants.

Cages

Materials for construction of cages include galvanized and stainless steel, plastic, wood, glass, and paper. The most common type of mouse cage for a breeding pair is a box approximately $12 \times 6 \times 6$ in. deep. Cages are available commercially with many dimensional variations of this basic design. Among the desirable features of cage design are the following:

1. The cages should be seamless and designed for easy cleaning
2. They should confine the mice adequately without restricting movement; there should be about 8 sq in. of floor space per adult mouse
3. They should be stackable without binding
4. They should be light in weight, yet sufficiently sturdy to withstand repeated washing and rough handling
5. They should be resistant to the corrosive effects of mouse urine.

If transparency is a requisite, the cage must be plastic or glass. If transparency is not a necessity, metal or opaque or translucent plastic may be used. Of the metal cages the stainless steel are by far the most durable; next in order are aluminum and galvanized steel. Plastic cages include polyethylene, polypropylene, polystyrene, polycarbonate, cellulose-acetate butyrate, acrylic resin (Lucite), acrylate and methacrylate resin (Plexiglas), and melamine-formaldehyde filled with glass fibers. Polycarbonate is probably the nearest to an ideal plastic for cages. It is transparent, autoclavable, and virtually indestructible.

For research installations where small numbers of mice are to be kept it might be desirable to use disposable cages, thereby avoiding the expense of facilities for washing them. Several types of disposable plastic cages are available, including a semirigid form requiring a supporting frame and a rigid (but rather brittle) self-supporting type.

Covers

Covers for mouse cages are usually made of metal, either of punched sheets, hardware cloth, rods, or expanded metal. The material can be stainless steel, aluminum, or metal that has been galvanized, tinned, zinc plated, or aluminized. Covers can be designed to stack; however, this is not essential since the cover will generally be in use rather than in storage. The cover can either overlap the cage or be recessed within it. If the cage is made of plastic or wood, the cover should be recessed to prevent the mice from nibbling the rim of the cage. If the cage is metal, the cover may overlap the sides of the cage although this presents some disadvantages. Mice often push fecal matter or bedding material out through the cover, and recessed covers tend to minimize the consequences of this activity. If hardware cloth is used for cover material, the mesh should be no larger than three spaces to the inch. Smaller mesh is more difficult to clean; larger mesh allows mice to escape. Covers should have a tag-holder attachment and some provision for lifting the cover from the cage. The cover should have a support for the water bottle and a hole in the mesh to accommodate the water drinking tube. Cage manufacturers can usually supply appropriate covers.

The food hopper can be incorporated into the cover or attached to the cage separately. A slotted hopper or one of hardware cloth having a mesh of three spaces to the inch is preferred. A hopper which is an integral part of the cover should be so constructed that mice cannot cling to it and soil the food with urine and feces. The design of the hopper should permit the caretaker to add food without removing the cage from the shelf. It should also be possible to position a water bottle without moving the cage. The hopper should extend low enough into the cage to permit small or young mice to feed, but should be high enough so that large or adult mice do not have to lie on their backs to feed. The hopper must be sufficiently large to avoid bridging of the food inside. Hoppers may be covered to prevent dust from getting onto the food and to prevent contaminaion by stray mice.

Water supply

Water may best be supplied from bottles resting on top of the cages. Each bottle is fitted with a rubber stopper and a tube of stainless steel. The tube must be constricted at the end so that the water does not flow out unless the mouse licks the end of the spout. The diameter of the tube should be $\frac{1}{4}$ to $\frac{3}{8}$ in. A large-mouthed bottle, such as a 1-pint milk bottle, facilitates cleaning. Other types and sizes of bottles can also be used; e.g., syrup bottles, 1-pint prescription bottles, and almost any other type which can be fitted with a stopper and tube. The position of the water bottle on the cage depends on the type of cover. If the water bottle rests on top of the cover, the tube must be bent at an angle of about 105° so that the water will drain from the bottle and the tube will extend in a vertical position into the cage. The tip of the tube should reach a point about 2 in. above the floor of the cage; i.e., low enough to permit small mice to reach it and yet high enough so that the bedding material cannot be piled up to the point of touching the tube. If the cover of the cage is designed with a bottle receptacle depressed into the cage cover, a straight tube may be used. The opening on the end of the tube should be free of burrs and sharp edges. Aluminum tubes are too soft for this purpose; mice can nibble and distort the size of the opening. When ordering bottles it is essential to specify the inside diameter of the neck of the bottle, since this is the dimension which must be fitted with the proper size of stopper. Half-pint and 1-pint milk bottles will usually take a No. $6\frac{1}{2}$ rubber stopper. The cage cover should be so designed that the rubber stopper is not touching the surface of the cover. This can be acomplished by providing a metal disc on the cover against which the rubber stopper will rest. If the bottle rests on top of the cage and a bent tube is used, this is not necessary.

Water may also be provided from a central system without the use of bottles by extending pipes or tubes from a central reservoir into each cage. At the end of each of the tubes a small valve allows water to run when the mouse touches the valve stem. The tube from the central reservoir may enter the cage either through the cover or through a hole in the end of the cage.

Rack caging

A second widely used system of caging mice is the suspended cage. Each cage consists of hardware cloth bottom and sides which are either solid sheet metal or hardware cloth. Cages of this type are not usually used for breeding mice. Suspended cages usually have no tops. They slide into a rack on metal guides and require a pan underneath to catch droppings and urine. The pans can be lined with paper which is easily changed or they may be filled with various kinds of absorbent litter. Water bottles similar to those mentioned above may be used on this type of cage, although the method of attaching the bottle to the cage is different. The bottle is held in a vertical position on the cage by a loop of wire or a metal bracket. A bent glass or steel tube protrudes into the cage so that the tip of the tube is about 2 in. from the floor. Feed hoppers for this type of cage consist of a small chute attached to the outside of the cage. Feed placed in the chute sildes down against the wire front of the cage through which the mice are able to nibble the pellets. Suspended cages of this type are usually attached to the shelves of mobile racks.

Bedding and nesting material

Bedding material serves three purposes: to provide material from which the mice can make nests, to absorb the fecal and urine wastes, and to provide insulation. Among the materials in common use are pine, cedar, or other wood shavings, sawdust, ground-up corn cobs, shredded flax, chopped straw, hay, absorbent clay such as fuller's earth, shredded paper, and blotter paper in the form of sheets. Regardless of the type of material used, several requisites must be met. The material should be clean, dry, and free of extraneous objects and waste matter. It should be metabolically neutral, since mice will nibble the bedding regardless of its composition. The bedding should be about $\frac{3}{4}$ in. deep in the case of shavings and other fluffy material, and about $\frac{1}{4}$ in. deep for such things as fuller's earth. In any case, it should be sufficient to absorb all of the urine, wet fecal matter, and water which may drip from the water bottle. The bedding material must fall freely from the inverted cage during cleaning. Some types of bedding tend to stick to the surface of the cage. White pine shavings meet most of the criteria indicated above.

Cage tags and card holders

For record keeping and other information it is convenient to have a card holder of sheet aluminum or other thin metal attached to the cage or cover. If cages and covers are washed with equal frequency, the card holder may be attached to either the cover or the cage. However, if the cages are washed more frequently than the covers, the card holder should be attached to the cover to minimize the number of times the card has to be transferred from the dirty card holder to the clean.

SANITATION PROCEDURES

Floors, walls, ceilings

The floors of the mouse room should be cleaned regularly during periods of active use, usually at the end of each day. A vacuum cleaner or a wet mop should be used to pick up pieces of shavings or other spilled material. Sweeping should be avoided; any essential sweeping should be done in a manner that will raise the least amount of dust. After the floor has been swept or vacuumed, it should be swabbed with a disinfectant containing a detergent. The disinfectant should be allowed to stay on the floor for the length of time necessary to achieve disinfection. The walls and ceilings of a mouse room should be cleaned and disinfected as frequently as is necessary. This can be determined on the basis of bacteriological sampling or visual observation.

Fixtures and equipment

Light fixtures and any flat surface which accumulates dust should be cleaned with a vacuum cleaner and swabbed with a disinfectant. Cage racks should be similarly cleaned at least once a week, preferably when they are empty. Sinks should be disinfected at least once a week. Tables and carts used as work surfaces should be disinfected after each use. In installations where walk-in washers are available, tables and carts may be removed from the mouse room and put through the washing machine.

Forceps and ear punches

Forceps used for handling mice should be disinfected between successive cages. For this purpose it is convenient to have two containers of about $\frac{1}{2}$-pint liquid capacity, each about three-quarters full so the forceps are immersed about 3 or 4 in. Ear punches should also be disinfected between successive cages. The disinfectant should be of a type which acts quickly but is not harmful to the mice, e.g., bound iodine or alcohol.

Disinfectant cleansers

Mouse rooms equipped with sinks should have dispensers for disinfectant hand soap. Caretakers should wash their hands upon entering and before leaving a mouse room. For general cleaning purposes solutions containing phenyl derivatives, bound iodine, or quaternary ammonium compounds can be used. Sodium hypochlorite or bound iodine solutions in low concentrations can be used for disinfecting items which may come in contact with mice. Phenolic and quaternary ammonium type compounds should not be used to disinfect items which will come in contact with mice, since either or both might be carcinogenic or cause skin irritations.

Waste removal

Dirty cages should be taken from the mouse room for disposal of the soiled bedding. Dirty cages may be stored temporarily in the wash area, but in no case should they be allowed to remain uncleaned beyond the end of the working day. Soiled bedding accumulated in the wash area should be incinerated or disposed of in a way to render it incapable of contaminating animal areas or people who come in contact with it. The waste bedding material may be removed simply by inverting the cage or by use of a vacuum hose. After the waste material has been removed, anything clinging to the cage surface should be scraped off with a rubber-tipped scraper.

Cage washing

In small animal colonies cages can be washed by hand in a sink with a washing and rinsing basin. Cages should not be hand dried; they should be allowed to dry in air in a clean area. Rinsing the cages in hot water will hasten the drying.

Various types of machines are available for washing cages. The so-called batch washer is designed so that a number of cages may be stacked in the machine and washed as a group. The cages must be stacked so that the spray jet will reach all surfaces. Another type is the tunnel washer. This is a long tunnel-shaped device through which the cages pass on a conveyor belt over jets of hot water. A tunnel washer is easily adapted to a supply-evacuation corridor and barrier system. Cages enter the machine in the evacuation area and emerge into the supply area, having passed through a wall. In mouse colonies where extreme cleanliness is required cages may be sterilized after they are washed. The autoclave should be built as a pass-through device so that the clean, sterilized cages can be removed from the chamber in the supply area.

For cages mounted on racks, it is necessary to have a walk-in type washer to receive the entire rack of cages. For this type of washer the old food and waste material clinging to the cage should be removed before the rack is put into the machine.

If autoclaving is not practical, it may be possible to achieve a degree of sterility by dipping the clean, washed cages in a chemical disinfectant. Solutions which are relatively harmless to mice, such as iodine dip, may be used. The dipping tank should be installed so that cages can be passed from the evacuation into the supply area through the disinfectant solution.

Bottle washing

In a small mouse colony bottles may be washed with a hand brush. For a large operation various kinds of bottle-washing machines are available. Some of these are of the soak type, others use a system of spray jets. Where machinery can be used, it is desirable to wash the bottles in their handling cases. One type of machine handles bottles under the batch system; that is, a case or more (up to three cases of bottles) may be placed in the machine and be washed and removed from the clean side. In a tunnel-type bottle washer the inverted cases of bottles are placed on a conveyor belt and passed over a series of spray jets for washing, rinsing, and sterilizing, or rinsing with a chemical disinfectant. Bottles may be sterilized in an autoclave and passed into a clean area through a double-door system. If the bottles have been autoclaved, they should not be filled with cold water as this will cause breakage; they should be allowed to cool to 100°F or to room temperature before cold water is added.

Bottles may be filled with water either automatically by a device integral with the washing machine or they may be filled by means of a multiple-jet filling header, the number of filler jets being equal to the number of bottles per case.

Stopper washing

Stoppers should be washed at the same time as the bottles, but in a different machine. The stoppers may be placed in washing water in a wire basket and the water brought to a boil in a steam-heated sink. They may also be rinsed in a wash solution and then sterilized in an autoclave as indicated above for the cages and bottles. If the stoppers are to be inserted in the bottles by hand, they should be immersed in a disinfectant solution just prior to being handled. This will insure that the stoppers will remain clean until they reach the mouse room. Dirty stoppers should be rinsed with a strong jet spray to remove the bits of shavings and other materials clinging to them.

Other equipment

Other equipment used in mouse rooms must be sanitized either in the room by means of swabbing with a disinfectant or in a separate wash area. The equipment should be passed through a suitable washing machine or swabbed with washing compound and passed through a chemical dip from the evacuation area to the supply area. Unwashable materials and equipment used in the mouse room should be of a disposable type.

MOUSE ROOM EQUIPMENT

Tables, carts, dollies

Two kinds of operations are usually carried on in a mouse room: maintenance and research. For either operation it is convenient to have a table mounted on casters. The surface of the table should be smooth and preferably made of stainless steel or galvanized iron. One or more shelves may be fitted beneath the surface. The size of the working table depends on the aisle width and on the size of the cages being used. The width of a table should be at least twice the width of the cages being used, and the length of the table should be approximately three times the length of the cages. It is best to allow a little leeway in both directions. Tables can be obtained which are adjustable in height although the standing-height work surface of 35 in. is usually used. The table may have four swivel casters or two swivel and two rigid casters. In either case the diameter of the caster should be no less than 4 in. The larger the caster the easier it is to move the table. If the mouse room floors are not level, the table should be provided with locking type casters. Although metal or hard fiber wheels roll most easily on a smooth floor, rubber-tired wheels are easier to roll over small obstructions.

If mice must be removed from the mouse room, it is best to have a cart with several shelves that will hold 10 to 20 cages. Since this type of cart will be moved over doorsills and through corridors, it is desirable to have casters with fairly large (6 to 8 in.) rubber-tired wheels.

Cages, covers, water bottles, feed, and various types of containers can be moved most conveniently on dollies designed specifically for the load. The dolly may be made of angle iron with the inside of the angle upward and mounted on rubber-tired casters. Cages can be stacked directly on the dolly; covers and water bottles must be held in trays or cases stacked on the dolly.

Containers

Containers for feed, bedding, or other material used in a mouse room should be covered and kept off the floor. Since all containers must be cleaned frequently, they must be constructed of noncorroding metal or plastic that will withstand high temperatures. If the containers are to be moved, they should be placed on dollies mounted on casters. For dispensing food, it is convenient to have a covered hopper mounted above the work table with an opening at the bottom from which the food may be picked up with a scoop. The advantages of this type of hopper are: (1) Food may be added to the top and removed from the bottom, so that the older food is used first; (2) it prevents the accumulation of ground particles of food; and (3) contamination of food is minimized, since the hopper is located above the work surface.

Devices for killing mice humanely

Every mouse room should be equipped with a device for killing mice humanely. A simple device is a covered can or jar of about 5-quart capacity lined with a plastic bag. If mice are to be killed in the mouse room, ether or chloroform should not be used; instead, carbon dioxide from a small cylinder fastened under the work table or located at some central point in the room may be used. When enough mice have been accumulated in the jar to cover the bottom, a tube from the gas cylinder is inserted and enough gas is released into the jar to kill the mice.

If the mice can be taken out of the room to be killed, chloroform may be used. In this case the animals may be accumulated in a large cage for transporting them. Chloroform as liquid should not be allowed to come in direct contact with the mice. A wad of cotton or paper towel can be moistened with the liquid and placed in the bottom of the jar beneath a piece of paper or plastic sheet. Regardless of the method used the number of mice should not exceed that which would cover the bottom of the container. Mice should never be accumulated to the extent that they suffocate or trample each other before they are killed. After the mice are killed, the plastic bag should be tied with a tight knot and discarded in a container designated for that purpose. Dead mice should be removed from the mouse room and incinerated before the end of each working day. Dead mice accumulated over weekends and holidays should be placed in plastic bags and stored in a freezer until they can be taken to the incinerator.

HANDLING PROCEDURES

Mice

There are usually only two situations in which mice must be handled: during the changing of cages and during experimental procedures. Regardless of circumstances under which the mice are handled, the mice should be touched with the hands as little as possible. At the time of cage changing the mice should be transferred from the dirty cage to the clean cage with a pair of long forceps. The mice may be picked up either by the tail a short distance from its base or by the loose skin at the back of the neck. In either case the forceps should be used so that the mouse is not pinched, although held firmly enough that it cannot break loose from the grip. With practice the caretaker should be able to pick up the mice deftly with a minimum of pursuit in order to avoid undue stress to the mice. The mouse should be lowered into the clean cage to the point where its feet are touching the bedding material. Mice should not be dropped but placed in the cage. Pregnant female mice should be handled with extreme care. In rare instances a female will be found giving birth to a litter at the time the cages are being changed; it is advisable to leave this cage unchanged but marked for changing at a later time. Disturbing the female at this time might cause her to destroy her litter or result in abnormal delivery of those not yet born. Newborn or very young mice may also be handled with the forceps although extreme caution must be used to avoid injury due to squeezing. Very young mice should be placed in a cluster preferably touching each other when they are placed in the clean cage. If the newborn mice are scattered when they are placed in the clean cage, it may be some time before the female has recovered from the stress of being changed and gets around to picking them all up. If they are placed in a cluster, she will be more likely to attend to them sooner. Sick mice found in a breeding colony should be removed from the mouse room and either killed or sent to a diagnostic laboratory.

Under certain circumstances it is necessary to pick up the mice by hand, for example, if the ears are to be notched or the toes clipped for identification. Toe clipping should be done only when it is necessary to identify young mice whose ears have not yet grown to an extent where the ear punch can be used. In special cases it might be necessary to use a combination of ear punching and toe clipping even on adult mice.

The frequency of cage changing depends on several factors. If wood shavings or similar material is used as bedding, the cages should be changed at least once a week and in some cases twice a week. If a specially absorbent material is used, it may be possible to extend the changing period to 2 weeks or more. A balance must be struck between the degree of cleanliness desired for the mice and the stress that mice undergo every time they are changed to a new cage. If the cage covers are constructed in such a way that they do not easily become soiled, it is possible to extend the changing period for covers to as long as 4 to 8 weeks. If the covers become soiled, they should be changed with the same frequency as the cages. The principal reason for less frequent changing of covers is to avoid wasting usable food.

Young mice may be weaned any time after they are 18 days of age. The actual age chosen will depend on the strain of mice being maintained. Some mice mature more rapidly than others. It is essential that the litter be removed from the parents before the young mice become sexually mature. This is usually at 4 to 5 weeks of age. In some cases it is necessary to wean a litter earlier than anticipated because of the presence of a second litter. If mice are to be held for any period of time after weaning, males and females should be kept in separate cages. All mice should be carefully examined at the time of weaning and those not conforming to the desired quality should be culled.

Feed and feeding

Mouse food is commercially available in the form of pellets. Some manufacturers can supply several varieties for specific purposes, e.g., a special formulation for breeding mice, pasteurized feed for mice maintained under pathogen-free conditions, sterilized diets for germ-free mice, diets fortified with extra amounts of heat-labile vitamins (for installations where the feed is autoclaved before it is brought into a mouse room), and diets with specified levels of fat or protein (Chap. 5). The standard formulations are usually sufficient for general use; however, it is advisable to investigate the formulation of the diet, the ingredients, the chemical analysis, methods of shipping, and whether the diet is free of potentially pathogenic organisms. Feed should not be stored for more than 30 days from the date of milling. In warm, humid climates storage time must be even shorter unless air-conditioned or refrigerated storage facilities are available.

New food can be added to hoppers either at the time of cage changing or as a separate operation. Feed should not be touched with the hands; a small metal scoop is much more sanitary and is actually more convenient and efficient to use. Old feed, that is, feed not used up by mice in a cage that is being terminated, should never be reused; it should be discarded and incinerated along with other waste materials from the mouse room.

Cage changing

When cages are being changed it is more practical to change in a vertical sequence; that is, columns of cages, rather than rows, are changed successively. The vertical sequence system requires less moving of the work table.

Cages containing mice should be handled with a smooth, even motion to minimize disturbance of the animals. Whether empty or occupied the cages should be picked up and set down with a minimum of noise. Careful handling will not only reduce the level of noise in the room, it will also extend the useful life of the cage.

Bottles

Bottles may be removed from the cages as each cage is about to be changed or they may be removed by sections, that is, about 50 bottles in a group. When clean bottles are installed on cages each bottle should be checked to ascertain that the stopper is firmly fastened and that the water has entered the water-delivery tube. The latter can be accomplished by holding the bottle downward and shaking it gently until air bubbles come out of the tube indicating that the tube has been filled with water. Clean water should be provided at least once a week. Under certain special circumstances two or three changes per week may be necessary. Bottles should not be refilled, since any contamination accumulated in the bottle will be retained.

ENTRY AND EGRESS

Personnel

Systems of entry into animal quarters depend on the degree of cleanliness to be maintained. In so-called conventional animal colonies, the only requirement is that personnel should wear garments which are worn only in the animal area, such as laboratory gowns or coveralls, and shoes which are worn only in the animal areas. Clothes should be changed outside of the animal room in a location accessible from the outside and also from the animal rooms. Personal belongings should be kept out of the mouse room, since these items usually cannot be sanitized in any way before entering. Food for human consumption should not be brought into the animal quarters.

Mice

All mice brought to an established animal colony should be quarantined for a period of 2 to 4 weeks prior to entry. If possible, castrated mice from the established colony should be exposed to the quarantined mice to ascertain whether a transmissible disease might be carried by the new arrivals. Various bacteriological tests should also be carried out. The mice should be examined for ectoparasites and small numbers should be killed and examined for endoparasites.

When mice are transferred from one room to another within the same building, transfer should be made in small disposable paper or plastic containers or in freshly changed cages. Dirty cages should never be moved from one mouse room to another. Mice to be shipped from the animal colony to an outside destination should be removed from the animal room and assembled in a special holding or shipping room. Preparations for shipment should not be carried out in the breeding rooms.

Feed

Feed for a conventional colony of mice should be delivered to a room accessible both from outside the building and from the animal room. Feed should be stored in a dry room at low temperatures and should not be kept longer than 30 days from milling before it is fed to the mice. Feed manufacturers can stamp the milling date on each bag. Conventional (not pasteurized) feed is usually delivered in paper sacks. The outside of these bags can be disinfected with a sponge when the bags are brought to the animal rooms. It is also possible to have the feed delivered in double wrappers in a manner such that the outside wrapper is removed as the bag of feed is delivered to the storeroom. Thus any contamination accumulated during shipment will be left outside of the storage area.

Bedding

Bedding for conventional colonies is usually safe enough to use as it comes from the vendor provided it has been checked to ascertain that the material is clean and free from obvious contamination. However, it is preferable to sterilize or at least pasteurize the bedding when it is being introduced into the animal quarters. Material such as wood shavings, ground corn cobs, shredded flax, etc., can easily be sterilized or pasteurized. Bedding should never be sterilized with ethylene oxide gas. Such a procedure results in the accumulation of ethylene glycol as a result of the reaction of ethylene oxide with the moisture in the bedding. The glycols are harmful to certain strains of mice and may cause hemorrhagic disease (Chap. 18).

QUALITY CONTROL

Disease recognition

Mice should be closely observed for signs of disease or evidence of incapacity of any type at the time cages are being changed. More frequent observations may be necessary in certain types of experimental procedures, although it must be remembered that cages should not be disturbed so frequently that the mice will become unduly stressed. Gross observation is not sufficient to detect all types of disease so it is advisable to remove representative mice periodically for autopsy by a pathologist. If mice cannot be killed for autopsy, it is possible to detect some diseases by fecal examination. Mice should be placed in a beaker or some type of disposable container; after the fecal pellets have been accumulated, the containers should be covered and sent to a diagnostic laboratory for analysis. Detailed records should be kept of the mice sampled so that if some undesirable condition is detected, the mice can be removed from the colony or medication can be applied.

Culling criteria

Unless precluded by the protocol of the experiment, sick animals should be removed from the colony whenever they are observed. In a breeding colony maintained for perpetuation of a strain the culling can begin with newborn mice; those smaller than normal should be removed from the litter. At any time prior to weaning, mice obviously behind their littermates in growth or apparently ungroomed should also be removed. The most severe culling should be practiced at the time mice are weaned. It may be advisable to weigh the mice at weaning to determine whether they conform to the expectation for that strain. With experience it is possible to detect mice that are underweight by observation. In a breeding colony selection of breeders for the next generation is of utmost importance. The criteria for selection of breeders can be determined only on the basis of experience, although certain signs are indicative of good quality. The mice should have glossy coats, they should be active and vigorous, they should not be under- or overweight, and they should not have any injury that might have been sustained between birth and weaning.

SUMMARY

Implementation of good mouse husbandry includes design and construction of facilities and provision of cages and other equipment for proper care of the mice. Sanitation of facilities and equipment is of utmost importance; clean cages, bedding, and water bottles must be provided on a regular schedule; mice, feed, water, and mouse room equipment should be handled in a manner that results in clean, efficient operation with minimum disturbance of the mice. Entry and egress of personnel, mice, equipment, feed, bedding, and waste must be accomplished so as to minimize the possibility of introducing disease organisms or vectors of disease. Constant vigilance, supplemented with good diagnostic services, must be exercised to promote the health and phenotypic quality of the mice.

LITERATURE CITED

Hill, B. F. [ed.] 1963. Proceedings of the symposium on research animal housing. Lab. Anim. Care 13:221–467.

Hoag, W. G., and E. P. Les. 1963. Husbandry, equipment, and procurement of mice, p. 538–557. *In* W. J. Burdette [ed.] Methodology in mammalian genetics. Holden-Day, San Francisco.

Institute of Laboratory Animal Resources. 1962. Standards for the breeding, care and management of laboratory mice. National Academy of Sciences–National Research Council, Washington, D.C. 29 p.

Lane-Petter, W. 1963. The physical environment of rats and mice, p. 1–20. *In* W. Lane-Petter [ed.] Animals for research. Academic Press, London.

Public Health Service. 1963. Guide for laboratory animal facilities and care. U.S. Government Printing Office, Washington, D.C. 33 p.

5

Nutrition[1]

Warren G. Hoag and Margaret M. Dickie

In this chapter we discuss the basic components of various mouse diets, ancillary factors influencing the effects of diets, and the composition of the diets used at The Jackson Laboratory.

We have learned that certain associated factors should be considered when designing an experiment because they may effect the outcome. If experimental mice are born and reared on one diet, there should be a period of adjustment when a different food is used. The type of feed formulation may be important also. In open-formula commercial feeds the stated nutritional analysis may be unchanging, but the source and, therefore, the quality of the ingredients may vary depending upon the season, availability, and market prices of these ingredients. Though such a diet may appear constant by chemical analysis, the form in which the nutrients are actually available to the animal may be variable. For example, the protein content may have quite different nutritional value, depending upon the type of protein (biological source), certain plant proteins being of much less value than animal proteins as a source of energy. In this chapter we present formulations for four diets which have proved best in terms of reproductive performance and health in the strains for which they are used. The diets and some representative strains are: Diet No. 1 for strains C57BL/6J, A/J, CBA/J; Diet No. 2 (high protein–low fat) for strains DBA/2J, A/HeJ, C57L/J, DBA/1J; Diet No. 3 (low fat) for strains DBA/2J, AKR/J; Diet No. 4 (modified

Table 5-1. DIET FORMULATIONS AT THE JACKSON LABORATORY IN PARTS PER THOUSAND BY WEIGHT

Ingredients	Diet 1	Diet 2	Diet 3	Diet 4
Ground milling wheat (Penn. or Ohio)	515.00	329.50	560.00	602.00
Nonfat skim milk, edible	200.00	120.00	200.00	200.00
50% dehulled soybean meal	112.50	67.50	112.50	25.00
Corn oil, edible grade	102.50	34.50	57.50	102.50
Dried brewer's yeast	40.00	24.00	40.00	40.00
Sodium chloride	13.75	8.75	13.75	13.75
Dicalcium phosphate	10.00	10.00	10.00	10.00
Ferric citrate	1.25	0.75	1.25	1.25
Vitamin premix (calculated from Table 5-2)	5.00	5.00	5.00	5.00
Wheat germ meal, edible		400.00		
Total	1,000.00	1,000.00	1,000.00	1,000.00

[1] The writing of this chapter was supported in part by Public Health Service Research Grant CA 04691 from the National Cancer Institute and by Contract AT(30-1)-3249 with the U.S. Atomic Energy Commission.

Table 5-2. SUMMARY OF VITAMIN NEEDS AND EFFECTS IN THE MOUSE AND AMOUNTS USED IN JACKSON LABORATORY DIETS (1965)

Vitamins	Range of amounts recommended/kg of feed	Daily suggested amount/25-g mouse
A	242 –5,500 IU	1.1– 2.0 IU
E (α-tocopherol)	9.9 – 27.5 mg	1.0 μg
B_{12}	3.9 – 5.5 μg	
Riboflavin	2.4 – 11.0 mg	3.0 μg
Niacin	26.4 – 143.0 mg	45 μg
Pantothenic acid	9.9 – 55.0 mg	40 μg
Choline	495.0 –1,452.0 mg	
Folic acid	0.506–2,750.0 mg	
Menadione (vitamin K activity)		
Pyridoxine (B_6)	0.99 – 5.5 mg	
Biotin	19.8 – 165.0 μg	
D	143 –5,060 IU	0.7 IU
Thiamine	2.2 – 9.9 mg	

from Morris's, 1944, diet) for good reproduction in most strains, though it tends to increase the deposition of body fat in reproductively inactive or older mice (Hoag and Dickie, 1960, 1962; Fenton and Cowgill, 1947; Lee et al., 1953).

NUTRITIVE REQUIREMENTS

Vitamins

In general, vitamins are not synthesized by the animal body. The amounts of these essential ingredients in a diet are dependent upon availability for absorption in the mouse, presence or absence of antagonists, and balance or equilibrium with other ingredients, although little quantitative information is available about this presumed "balance" (Albanese, 1963).

Estimates of vitamin requirements of the mouse vary widely among investigators. Amounts of vitamins in feeds have a great range (Bell, 1962; Cuthbertson, 1957; Lane-Petter, 1963; Porter and Lane-Petter, 1962; Spector, 1956). The ingredients used in the Jackson Laboratory diets are given in Table 5-1 and the recommended amounts of vitamins and the calculated amounts in these diets are given in Table 5-2.

Recommendations as to the quantities of vitamins to be included in mouse diets are unfortunately based on limited evidence. Not only must the diets be formulated to supply the calculated amounts, but also they should be shipped or stored under conditions which minimize deterioration of vitamins. Sometimes additional amounts of vitamins are provided to counteract deterioration. Unfortunately, it is not known what effect excess vitamin consumption will have on mice. Our diets supply an adequate amount of these vitamins if the feed is used within 12 weeks of manufacture, although we prefer to avoid storage beyond 30 days.

Proteins

Proteins act as biocatalysts or as structural elements in every metabolic process. The stated requirements for vitamins, minerals, carbohydrates, calories, and fats are, therefore, of little importance unless expressed in terms of general dietary protein levels. High protein levels can compensate for lower or borderline levels of other dietary ingredients; for example, with low-calorie diets increased amounts of protein serve as a substitute energy source (Bosshardt et al., 1950; Frazer, 1961; Albanese, 1963). Conversely, one may use lower protein levels when higher amounts of other components are supplied. Little information is available about specific amino acid requirements of mice, but the few studies reported indicate that they are similar to those of the rat (Bell, 1962; Morris, 1944; Spector, 1956).

Bell (1962) found no differences in growth rates of mice (strains not specified) at levels of 17, 19, and 20 per cent total dietary protein. We

Major signs of deficiency in mice	*Calculated analysis for Jackson Laboratory diets/lb of feed* Diet 1	Diet 2	Diet 3	Diet 4
Tremors, diarrhea, eye exudates, hemorrhage, sterility	6,200 IU	3,900 IU	8,100 IU	4,150 IU
Muscular lesions, kidney degeneration, reproductive failure	60.5 mg	118.3 mg	60.5 mg	25.05 mg
Retarded growth, renal atrophy	4.1 μg	2.4 μg	4.15 μg	4.09 μg
Dermatosis, posterioı paralysis, spinal cord lesions, diarrhea	4.1 mg	6.7 mg	7.8 mg	8.5 mg
	22.3 mg	21.2 mg	22.3 mg	24.06 mg
Alopecia, posterior paralysis, dermatosis, diarrhea, neurasthenia	8.3 mg	15.1 mg	21.7 mg	23.1 mg
Kidney lesions	500.0 mg	309.63 mg	515.75 mg	416.88 mg
Poor growth, lowered reproduction	2.06 mg	2.00 mg	2.06 mg	2.41 mg
Hemorrhages, failure of blood clotting	5.0 mg	5.0 mg	5.0 mg	5.0 mg
Posterior paralysis, poor growth, alopecia, necrotic degeneration of tail	4.5 mg	4.5 mg	4.5 mg	4.5 mg
Alopecia, achromatrichia, poor growth	100.0 μg	100.0 μg	100.0 μg	100.0 μg
Rickets	700 IU	700 IU	700 IU	700 IU
Testicular degeneration, convulsions, muscle lesions	3.7 mg	6.6 mg	3.7 mg	3.7 mg

compared 17, 19, 21, and 23 per cent protein and found differences in reproductive performance in strains AKR/J, DBA/2J, and C57BL/6J (Hoag and Dickie, 1962). C57BL/6J mice produced more litters when given a diet with 23 per cent protein, whereas strains DBA/2J and AKR/J produced more litters with 19 per cent protein, provided the fat level remained constant. Most available diets contain 17 to 30 per cent protein. Levels as low as 16 per cent have been satisfactory in a few inbred and noninbred mice. Table 5-3 gives the dietary protein levels and other components of the feed in use (1965) at The Jackson Laboratory.

Fats

Unsaturated fatty acids, such as linoleic and linolenic, are dietary requirements of mice (Decker et al., 1950; Bell, 1962; Morris, 1944). Since

Table 5-3. COMPARATIVE ANALYSIS OF VARIOUS DIETS IN PERCENTAGE

Components	Diet 1	Diet 2	Diet 3	Diet 4
Protein	21.2	24.9	21.3	17.9
Fat	10.9	8.2	7.1	11.2
Fiber	1.7	2.1	1.1	1.1
Ash	4.7	4.7	4.8	4.3
Ca	1.4	0.9	1.0	0.9
P	0.9	1.1	0.9	0.8

most animal fats are saturated, the source of the fat component for feeds should be noted. In the rat an interdependence exists even between the unsaturated fats, so that combinations provide a more effective protection against X-irradiation injury than does a large amount of a single fatty acid (Cheng et al., 1954). These studies suggest that the source of dietary fat in radiation experiments should be considered, since radiation sensitivity in rats can be altered by changing the fatty acid levels.

We have found that certain strains of mice such as C57BL/6J exhibit improved reproductive performance on 11 to 12 per cent dietary fat levels (using corn oil as the source), whereas other strains (DBA/2J, C3H/HeJ, and AKR/J) have an improved reproductive performance with only a 6 per cent fat level. Production of $B6D2F_1$ and $C3D2F_1$ hybrids is increased when the high-fat feed is used, even though one or both parental strains have a better within-strain reproductive performance on the low-fat diet.

In our experiments with different levels of dietary fat, using strains DBA/2J and AKR/J, total reproductive performance (numbers of fertile matings, litters, and mice born and weaned) was significantly improved with a low fat level (6 per cent). There was further improvement when wheat-germ meal was added to the low-fat feed.

During their active reproductive period mice can utilize large amounts of unsaturated fats and will eat more of a diet supplying these ingredients. When reproduction ceases such animals may become obese. It has been suggested that the obesity may cause cessation of reproductive activity, but many males retain their reproductive capability. The obesity may be concurrent with reproductive inactivity rather than its cause.

Minerals

Quantitative studies indicate that mice need calcium, phosphorus, manganese, iron, zinc, and potassium (Bell, 1962; Lee et al., 1953). The amounts used in our diets are given in Table 5-3. The mouse undoubtedly requires certain other trace minerals but specific needs have not been reported. Most feeds probably supply levels of such trace elements sufficient to meet these requirements.

ADEQUATE DIETS

The adequacy of a diet is determined by the animal's physiological status, metabolic rate, age, length of time it is to be kept, and its genetic constitution. A diet adequate for maintenance of body weight of an adult confined in a small cage may be completely inadequate for one kept in a larger cage. A diet adequate at one ambient temperature and humidity may be inadequate at another. It has been shown that rats require an increased amount of thiamine and choline when maintained at high temperatures (Mills, 1943).

A diet's adequacy often varies, depending upon the nature of the experiment. Schneider (1956) demonstrated a "salmonella resistance factor" in mouse feeds containing large quantities of wheat; this would be an important consideration when designing experiments on salmonellosis. Other investigations demonstrate the important role of diet in resistance or susceptibility of mammals to radiation, viruses, bacteria, and drugs (Schneider, 1960; Simon, 1960). Adequacy, therefore, must be defined in terms of experimental designs and the goals to be achieved.

There are many reports of deficiency syndromes produced by experimental diets lacking a specific ingredient, but such conditions are rarely encountered in the average colony. The problems usually encountered are malnutrition or dietary imbalances which are often masked by secondary microbial infections and occasionally by inherent anatomical defects. Dietary imbalances may be encountered only after several generations of feeding. We studied an infertility problem in strain DBA/2J manifested only after 2 years (8 to 15 generations) on Diet No. 4 (Table 5-1). When the strain was first put on the diet, weaning weight increased and reproductive performance improved. But after 2 years, although the weights at weaning remained the same as when the dietary change was initiated, infertility increased to over 50 per cent of the matings, the number of litters per female was reduced, and the number of spontaneous abortions increased. After investigation of microbiological, endocrinological, pathological, and toxicological aspects of the problem, it was found that lowering the fat level and adding wheat-germ meal corrected infertility so that 85 to 90 per cent of breeding pairs became productive.

Finally, adequacy depends also on the manner in which the diet is presented to the animal, i.e., form, hardness, and palatability of feed pellets.

NUTRITION STUDIES

In early studies using inbred mice we considered a 12- to 14-week period an adequate test of a diet. The investigations were principally concerned with body weight and reproductive performance. However, experience with strains DBA/2J and AKR/J has shown that we must include not only the reproductive performance of the original experimental and control animals, but also the data through at least three generations. Mice may react to any change in diet with either an increase or decrease in weight or reproductive performance, but this effect may disappear after a few generations. It is also evident that dietary studies are conclusive only for a given strain of mice.

SUMMARY

The investigator and the animal breeder must both be concerned with the nutrition of experimental animals. Reproduction and growth are not the only criteria for evaluation of adequacy of diets; dietary factors and their role in health and in disease should be given consideration in any experiment. Nutritional imbalance can be as important as dietary deficiency, so that an excess or deficiency of a particular essential can either in itself produce pathological signs or interfere with the utilization of other essential components.

LITERATURE CITED

Albanese, A. 1963. Newer methods of nutritional biochemistry. Academic Press, New York. 583 p.

Bell, J. M. 1962. Nutrient requirements of the laboratory mouse, p. 39–49. *In* Nutrient requirements of laboratory animals. National Research Council–National Academy of Sciences, Publ. 990.

Bosshardt, D. K., W. F. Paul, and R. H. Barnes. 1950. The influence of diet composition on vitamin B_{12} activity in mice. J. Nutrit. 40:595–604.

Cheng, A. L. S., M. Ryan, R. Alfin-Slater, and H. J. Deuel, Jr. 1954. The effect of fat level of the diet on general nutrition. XI. The protective effect of varying levels of ethyl linoleate against multiple sublethal doses of X-irradiation in the rat. J. Nutrit. 52:637–643.

Cuthbertson, W. F. J. 1957. The nutritional requirements of rats and mice, p. 27–37. *In* Laboratory Animals Bureau Collected Papers. Vol. 5. Laboratory Animals Bureau, Medical Research Council Lab., London.

Decker, A. B., D. L. Fillerup, and J. F. Mead. 1950. Chronic fatty acid deficiency in mice. J. Nutrit. 41:507–521.

Fenton, P. F., and G. R. Cowgill. 1947. Nutrition of the mouse. I. A difference in the riboflavin requirements of two highly inbred strains. J. Nutrit. 24:273–283.

Frazer, A. C. 1961. Role of lipids in normal metabolism. Fed. Proc. 20 (No. 1, Part 3, Suppl. 7):146–151.

Hoag, W. G., and M. M. Dickie. 1960. A comparison of five commercial diets in two inbred strains of mice. Proc. Anim. Care Panel 10:109–116.

Hoag, W. G., and M. M. Dickie. 1962. The effect of various levels of dietary protein in inbred mice. Proc. Anim. Care Panel 10:7–10.

Lane-Petter, W. [ed.] 1963. Animals for research. Academic Press, New York. 531 p.

Lee, Y. C. P., J. T. King, and M. B. Visscher. 1953. Strain differences in vitamin E and B_{12} and certain mineral trace element requirements for reproduction in A and Z mice. Amer. J. Physiol. 173:456–458.

Mills, C. A. 1943. Heightened thiamine and choline requirements in tropical heat. Proc. Soc. Exp. Biol. Med. 54:265–266.

Morris, H. P. 1944. Review of the nutritional requirements of normal mice for growth, maintenance, reproduction and lactation. J. Nat. Cancer Inst. 5:115–142.

Porter, G., and W. Lane-Petter. [ed.] 1962. Notes for breeders of common laboratory animals. Academic Press, London. 205 p.

Schneider, H. A. 1956. Nutritional and genetic factors in the natural resistance of mice to Salmonella infections. Ann. N.Y. Acad. Sci. 66:337–347.

Schneider, H. A. 1960. Nutritional factors in host resistance. Bacteriol. Rev. 24:186–191.

Simon, H. J. 1960. Attenuated infection. Lippincott, Philadelphia. 349 p.

Spector, W. S. [ed.] 1956. Handbook of biological data (Nat. Res. Counc.–Nat. Acad. Sci.). Saunders, Philadelphia. 584 p.

6

Nomenclature[1]

Joan Staats

This chapter presents a history of the efforts to achieve standardization of genetic nomenclature for mice and gives the 1964 rules for naming strains and the 1963 rules for naming genes. The confusing terminology of the earlier years has gradually given way to increasing orderliness and uniformity; we recognize, however, that complete uniformity may not always be possible.

HISTORY

Several decades ago, Professor H. de Haan (1932) wrote: "It is high time to bring unity into genetic nomenclature. . . . In my opinion it is due to the International Congress of Genetics to appoint a Committee for Genetic Nomenclature. A permanent committee which keeps drawing the attention to the international rules to be drafted by it could do much to promote unity in the nomenclature."

Five years earlier, at the Fifth International Congress of Genetics, A. S. Serebrovsky (1928) had presented an entirely new system of denoting hereditary factors analogous to a decimal bibliographic system. Casual inspection of this system indicates the motivation for de Haan's plaint.

Earlier still, in 1919, the American Society of Naturalists appointed a Committee on Genetic Form and Nomenclature with C. C. Little as chairman and Sewall Wright, G. H. Shull, O. E. White, and A. H. Sturtevant as members. The scheme drawn up by this group was adopted and published (Little, 1921). The hope of the chairman, expressed in his covering letter, that this publication would stimulate discussion of modifications was apparently largely ignored, and scant attention was paid to this succinct report.

It was indeed, as de Haan had suggested, left to future Genetics Congresses, both to appoint committees and to serve as meeting places for them. At the Sixth Congress in Ithaca, N.Y., in 1932, C. C. Little announced that several suggestions had been received that the Congress appoint a standing committee to consider matters of genetical nomenclature. There was no discussion and no action was taken. However, at a later plenary session of the same Congress, the resolutions committee moved that the problem of standardizing genetic symbolism and nomenclature be reconsidered. The motion was carried (Jones, 1932).

As a result of a circular letter sent in the spring of 1939 from The Jackson Laboratory to biologists interested in mouse genetics, a committee was chosen consisting of L. C. Dunn, G. D. Snell, and F. A. E. Crew to prepare recommendations for symbols for mouse genes. Hans Grüneberg later replaced Crew. During the Seventh Congress in Edinburgh, an open meeting on rodent gene symbolism, chaired by A. L. Hagedoorn, considered a set of nomenclature rules drawn up by the committee (Punnett, 1941). Recommendations from this meeting were incorporated into rules concerning only gene symbols, published in the *Journal of Heredity* (Dunn et al., 1940).

In 1949, after consultation with the genetics group at The Jackson Laboratory and correspondence with other geneticists, G. D. Snell drew up

[1] The writing of this chapter was supported in part by National Science Foundation grant G18485 and the William E. and Bertha Schrafft Charitable Trust.

a list of suggested rules of nomenclature for inbred strains of mice (*Mouse News Letter*, 1950). This list, with minor variations, was sent by Snell and T. C. Carter to geneticists in 1951, asking for a vote on two alternative systems of notation. That giving maximum uniformity was adopted. Workers were also asked to select abbreviations of their own names for use in designating substrains.

This Committee on Standardized Nomenclature for Inbred Strains of Mice, as the informal working group came to be known, was responsible for issuing the first standardized nomenclature list in *Cancer Research* in 1952 (Committee, 1952), following the format devised by Snell (1941) and Law (1948). This paper contained the recommended rules for strain names, a list of known inbred strains with their histories and characteristics, and a list of users of mice with abbreviations for their names or institutions.

As time went on, more substrains of existing strains were developed by spontaneous mutations, manipulation, or merely physical separation. New mutations and linkages were discovered, strains became more widely distributed, and the list of workers grew enormously. It became apparent that a reappraisal of the rules of nomenclature was in order.

Following the suggestion of Hans Grüneberg, the old nomenclature committee was reactivated, reconstituted, and renamed. During the 10th International Congress of Genetics in Montreal in 1958, the group met, was named the Committee on Standardized Genetic Nomenclature for Mice, and discussed both strain and gene symbols. This body supersedes and represents a merger of the Mouse Genetics Nomenclature Committee (genes) and the Committee on Standardized Nomenclature for Inbred Strains of Mice (strains) (Staats, 1963).

This body was responsible for issuing the second (Committee, 1960) and third (Staats, 1964) alphabetic listing of inbred strains in *Cancer Research* and the first major revision of the gene nomenclature rules (Committee, 1963). The latter was made necessary by the detection of many presumed recurrences of old mutations, reversions to wild type, long series of alleles at single loci, translocations, and by the increased use of mice generally. In 1965, the members of the Committee were: Margaret C. Green, Hans Grüneberg, Paula Hertwig, W. E. Heston, Mary F. Lyon, N. N. Medvedev, George D. Snell, and Joan Staats.

STRAIN RULES

The rules for symbols to designate inbred strains of mice as recommended by the Committee on Standardized Genetic Nomenclature are as follows (from Staats, 1964, with permission of *Cancer Research*):

1. ***Definition of inbred strain.*** A strain shall be regarded as inbred when it has been mated brother × sister (hereafter called b × s) for twenty or more consecutive generations. Parent × offspring matings may be substituted for b × s matings, provided that in the case of consecutive parent × offspring matings the mating in each case is to the younger of the two parents.

2. ***Symbols for inbred strains.*** Inbred strains shall be designated by a capital letter or letters in Roman type. It is urged that anyone naming a new stock consult [Staats, 1964, Appendix 2 or the latest issue of *Inbred Strains of Mice* (see references)] to avoid duplication. Brief symbols are preferred.

An exception is allowed in the case of stocks already widely used and known by a designation which does not conform.

3. ***Definition of substrain.*** The definition of substrain presents some of the same problems as the definition of species. In practice, the determination of whether two related strains should be treated as substrains, and whether, in published articles, substrain symbols should be added to the strain symbol, must rest with the investigators using them. The following rules, however, may be of help.

Any strains separated after 8 to 19 generations of b × s inbreeding and maintained thereafter in the same laboratory without intercrossing for a further 12 or more generations shall be regarded as substrains. It shall also be considered that substrains have been constituted (*a*) if pairs from the parent strain (or substrain) are transferred to another investigator, or (*b*) if detectable genetic differences become established.

4. ***Designation of substrains.*** A substrain shall be known by the name of the parent strain followed by a slant line and an appropriate substrain symbol. Substrain symbols may be of two types.

a. Abbreviated name as substrain symbol. The symbol for substrains should usually consist of an abbreviation of the name of the person or laboratory maintaining it. The initial letter of this symbol should be set in Roman capitals; all other letters should be in lower case. Abbreviations should be brief, should as far as possible be standardized, and should be checked with published lists [Staats, 1964, Appendix 3, and the latest issue of *Inbred Strains of Mice*] to avoid duplication. Examples: A/He (Heston substrain of strain A), A/Icrc (Indian Cancer Research Centre substrain of strain A).

When a new substrain is created by transfer, the old symbol may be retained and a new one added. Example: YBR/He, on transfer from Heston to Wilson, becomes YBR/HeWi. The accumulation of substrain symbols in this fashion provides a history of the strain. If the substrain symbols are not accumulated, the history of transfers should be recorded in *Inbred Strains of Mice*.

b. Numbers or lower-case letters. Numbers or lower-case letters may be used as substrain symbols in certain circumstances. The position of these relative to other parts, if any, of the substrain symbol should be suggestive of a historical or time sequence. Thus, two substrain branches, separated in and maintained by one laboratory, may be designated by terminal numbers, with or without a preceding slant line. Example: two sublines of A/HeCrgl, separated and maintained by Crgl, become A/HeCrgl/1 (or A/Crgl/1) and A/HeCrgl/2 (or A/Crgl/2). Lower-case letters immediately following the strain symbol, with a slant line only intervening, may be employed when two substrains are separated from a common strain prior to complete inbreeding. Example: C57BR/a and C57BR/cd. (These were separated after nine generations of b × s.) The use of numbers or lower-case letters immediately after the slant line, to designate lines separated after twenty or more generations b × s, is ordinarily not recommended, but may occasionally be justified for sublines widely recognized as different. Example: DBA/1 and DBA/2. Appropriate checks to avoid duplication should be made before this type of symbol is adopted.

5. *Coisogenic stocks.* Coisogenic stocks produced by the occurrence of a single major mutation within an inbred strain, or by the introduction of a gene into an inbred background by a series of crosses, shall be designated by the strain symbol and, where appropriate (see rule 7), substrain symbol, followed by a hyphen and the gene symbol (in italics in printed articles). Example: DBA/Ha-*D*. Where the mutant or introduced gene is maintained in the heterozygous condition, this may be indicated by including a + in the symbol. Examples: A/Fa-+*c*. C3H/N-+W^j.

When a coisogenic strain is produced by inbreeding with forced heterozygosis, indication of the segregating locus is strictly optional. Examples: 129 or 129-$c^{ch}c$ (129 is customary); SEAC-*d*+/+*se* or SEAC/Gn.

In the case of coisogenic stocks produced by repeated crosses of a dominant gene into a standard inbred strain, it may be desirable to indicate the number of backcross generations. Strains shall be regarded as coisogenic when at least seven such crosses have been made. Example: C57BL/6-+W^v(N8). The first hybrid or F_1 generation should be counted as generation 1, the first backcross generation as generation 2, etc.

6. *Substrains developed through foster nursing, ova transfer, or ovary transplant.* Substrains developed by foster nursing shall be indicated by appending an "f" to the strain symbol. Example: C3Hf. The strain used as foster parent may be indicated if desired by the addition of its symbol or an abbreviation for the same. Example: C3HfC57BL or C3HfB (C3H fostered on C57BL). In like manner, strains developed through egg transfer or ovary transplant shall be indicated by adding an "e" or "o," respectively. Example: AeB (A ova transferred to C57BL). Where the symbol for fostering or transfer might be confused with an adjoining substrain symbol, it may be used in a subscript position. Example: A/He_fB (Heston substrain of A fostered on C57BL).

7. *Compound substrain symbols for stocks of complex origin.* When a stock has been produced by manipulation of a standard inbred strain, as, for example, by fostering or by introduction of a foreign gene, compound substrain symbols may be necessary. In general, the elements of such a compound symbol should be arranged in an order indicating a historical or time sequence. Specifically, different positions should be interpreted as follows:

a. Substrain symbol that immediately follows strain symbol. Examples: BALB/cf, DBA/2eB, C3H/Ha-*p*. In this position the substrain symbol (c, 2, or Ha in examples given) designates the substrain which was fostered or otherwise manipulated, or in which a mutation occurred.

b. Substrain symbols following symbol for manipulative process or introduced gene. Substrain symbols in this position refer either to the person performing the fostering or other manipulation, or to the person or laboratory currently maintaining the strain, or to both. The symbol or symbols may or may not be immediately preceded by a slant line. Examples: DBA/2eB/De or DBA/2eBDe (strain derived from ova of DBA/2 transferred by Deringer to C57BL, maintained by Deringer); C3H/He$_f$Ha (C3H/He fostered by Heston, currently maintained by Hauschka); CBA/Ca-*se*/Gn (Carter's substrain of CBA with mutation to *se*, maintained by Green). Since a single symbol in this position (e.g., the De in DBA/2eBDe) may refer either to the person producing or the person maintaining the strain, the intended meaning should be clearly recorded.

8. ***Indication of inbreeding.*** Where it is desired to indicate the number of generations of b × s inbreeding, this shall be done by appending, in parentheses, an F followed by the number of inbred generations. Example: A(F87). If, because of incomplete information, the number given represents only part of the total inbreeding, this should be indicated by preceding it with a question mark and plus sign. Example: YBL(F?+10).

9. ***Priority in strain symbols.*** If two inbred strains are assigned the same symbol, the symbol to be retained shall be determined by priority in publication. For this purpose, listing in *Mouse News Letter* or *Inbred Strains of Mice* shall be regarded as publication.

STRAIN ABBREVIATIONS

Abbreviations for symbols of some older strains are useful for forming hybrid designations and in other places where brevity is desired. However, such abbreviations should be used only after giving the full symbols. The following abbreviations for standard inbred strains are used in this book.

AKR	AK	C57BL/6	B6
BALB/c	C	C57BL/10	B10
DBA/1	D1	C57L	L
DBA/2	D2	C57BR/cd	BR
C3H	C3		

Example: AKD2F$_1$ for (AKR × DBA/2)F$_1$.

GENE RULES

The rules for symbols to designate genes in mice as recommended by the Committee on Standardized Genetic Nomenclature for Mice (1963) are as follows (reprinted with permission of *The Journal of Heredity*):

1. ***The names of genes,*** as distinguished from the symbols, shall be written with a lower case initial letter, regardless of whether the mutant is dominant or recessive, except at the beginning of a sentence or other place in which capitalization would normally be used, or when the word is a proper noun. Examples: albinism, rex, Jay's dominant spotting.

2. ***The symbols for genes*** should typically be abbreviations of the name, e.g., *d* for dilution, *dw* for dwarf, *ac* for absence of corpus callosum. For convenience in alphabetical listings, the initial letters of names and symbols should be the same. Certain exceptions are indicated in subsequent rules.

3. ***Recessive mutations*** shall be indicated by the use of a small initial letter for the symbol of the mutant gene, e.g., *a* for nonagouti.

4. ***Dominant mutations*** (mutations with consistent heterozygous expression) shall be indicated by the use of a capital initial letter for the symbol of the mutant gene, e.g., *Re* for rex.

5. ***The locus symbol*** shall be the symbol of the first named mutant gene or allelic pair, except that any superscript indicative of a specific allele shall be omitted. Examples: *d* for the dilution locus, *Re* for the rex locus, *H-2* for the histocompatibility-2 locus.

6. ***The wild type may be designated by:***

a. The locus symbol with a small initial letter and a plus superscript, e.g., t^+, re^+.

b. A + sign only, when the context leaves no doubt as to the locus represented.

c. The same symbol as the mutant gene but with a capital initial letter for recessive mutants, a small initial letter for dominant mutants. Examples: *D* for the wild type allele of *d*; *re* for the wild type allele of *Re*.

Alternatives (*a*) and (*b*) are recommended, except for teaching purposes or when there is doubt as to which allele is wild type, e.g., biochemical polymorphisms, the agouti locus.

7. ***Mutants of similar phenotype but different location*** (mimics) shall be indicated either by entirely different names and symbols (e.g., *ln* for leaden and *d* for dilution) or by the same name and symbol with the addition of a hyphen and a distinguishing number. Examples: *H-1, H-2, H-3,* etc. for different histocompatibility loci; *wa-1* (waved-1) and *wa-2* (waved-2). The preferred system is to give mimics distinctive names.

8. ***Multiple alleles determining visible or other clearly characterized distinctions*** shall be represented by the locus symbol with an added superscript, typically a small letter or letters suggestive of the name. The initial letter of the symbol shall be capitalized according as the allele behaves as a dominant or a recessive. Examples: c^{ch} (chinchilla allele of *c* or albino), A^y (the yellow allele of *A* or agouti), $H\text{-}2^k$ (the *k* allele of histocompatibility-2).

9. ***Variants which are members of a series*** (e.g., recurrences, reversions to wild type, translocations, some series of multiple alleles) should usually be separated one from another by the use of a series symbol. For this purpose an Arabic numeral corresponding to the serial number of the variant in any given laboratory, plus an abbreviation indicative of the name of the laboratory or the discoverer, shall be used. Where the laboratory or discoverer has already been assigned an abbreviation for the designation of inbred substrains, this shall be the abbreviation used. Where there is no preassigned abbreviation, any appropriate abbreviation may be used, except that it should not duplicate an existing symbol in the standard list of abbreviations (Staats, 1964). In every case the first letter of the abbreviation should be capitalized. Example: *7Rl* (the seventh recurrence of any particular type of variant found by Russell). To avoid the confusion of the numeral one and the letter l, any first-discovered variant may be left unnumbered. The second variant is then numbered 2.

10. ***Indistinguishable alleles of independent origin*** (supposed recurrences) shall be designated by the existing gene symbol with the series symbol (rule 9) appended as a superscript. If the gene symbol already has a superscript, this shall be separated from the appended superscript by a hyphen. Examples: c^{4Rl} (the fourth recurrence of *c* found by Russell); $a^{t\text{-}7J}$ (the seventh recurrence of a^t found at The Jackson Laboratory).

11. ***Reversions to wild type*** shall be designated by the symbol for the wild-type allele (rule 6a) with the series symbol (rule 9) appended as a superscript. Examples: d^{+J}, d^{+2J} (the first and second reversions from *d* to d^+ found at The Jackson Laboratory).

12. ***Translocations,*** when initially discovered, shall be designated by a *T* followed by the series symbol (rule 9). Example: *T138Ca* (the 138th translocation found by Carter). When the linkage groups involved in the translocation have been identified, these shall be indicated by adding the appropriate numbers. These shall be inserted between the *T* and the series symbol, shall be Arabic rather than Roman numerals (except that the symbol for the X chromosome may be either X or 20, and that the Y chromosome shall be indicated by Y), shall be enclosed in parentheses, and shall be separated by a semicolon. Example: after it is found that *T138Ca* involves linkage groups II and IX, the symbol becomes *T(2;9) 138Ca*. If one of the linkage groups involved in the translocation has been identified but the other has not, the unknown linkage group shall be indicated by a question mark. Example: *T(11;?)7Ca*.

In general, the translocation symbol should be reserved for reciprocal translocations. If a translocation is shown to be something other than reciprocal, e.g., a transposition, some other appropriate initial letter or letters should be used.

13. ***In published articles*** in which symbols are used, the symbols should be set in italics.

ADDITIONAL CONVENTIONS

The authors of this edition of the *Biology of the Laboratory Mouse* have adhered to the following conventions wherever possible.

Genotypes are designated with alleles separated by "/," nonalleles by a space. In heterozygotes with the type allele (+) at a given locus, the mutant allele is written first. Genotypes of males carrying sex-linked alleles are designated as shown in the examples using *Ta*. Examples: *a/a*, *a/+*, $H\text{-}2^d/H\text{-}2^k$, and in males only, *Ta*/Y or *Ta*.

Linked genes are written in sets with spaces between nonalleles. The sets are written on one line separated by a "/" or are written on two lines like a fraction. Examples:

$$fr\ c\ p/+c+ \qquad \frac{fr\ c\ p}{+\ c\ +}$$

Matings are denoted by a multiplication or times sign "×." Example: $a/a \times a/+$.

Hybrids between two inbred strains are designated as follows: (C57BL/6 × DBA/2)F_1 or, for greater brevity, B6D2F_1 (see Strain Abbreviations, this chapter). In every instance the female is listed first.

SUMMARY

Since 1940, concerted efforts have been made to promote uniformity in genetic nomenclature for mice. The Committee on Standardized Genetic Nomenclature for Mice has published rules for designating both inbred strains and mutant genes. The rules are given in this chapter.

LITERATURE CITED

Committee on Standardized Genetic Nomenclature for Mice. 1963. A revision of the standardized genetic nomenclature for mice. J. Hered. 54:159–162.

Committee on Standardized Nomenclature for Inbred Strains of Mice. 1952. Standardized nomenclature for inbred strains of mice. Cancer Res. 12:602–613.

Committee on Standardized Nomenclature for Inbred Strains of Mice. 1960. Standardized nomenclature for inbred strains of mice, Second listing. Cancer Res. 20:145–169.

De Haan, H. 1932. The symbolizing of hereditary factors. Genetica 15:1–21.

Dunn, L. C., H. Grüneberg, and G. D. Snell. 1940. Report of the Committee on Mouse Genetics Nomenclature. J. Hered. 31:505–506.

Inbred Strains of Mice. An informal biennial mimeographed document distributed by The Jackson Laboratory and edited by Joan Staats. It carries lists of inbred strains of mice, arranged alphabetically by the cooperating laboratories.

Jones, D. F. [ed.] 1932. Proc. VI Int. Cong. Genet. (Ithaca, 1932) p. 16, 19. Vol. I. Brooklyn Botanic Garden, New York.

Law, L. W. 1948. Mouse genetic news, No. 2. J. Hered. 39:300–308.

Little, C. C. 1921. Report of the Committee on Genetic Form and Nomenclature. Amer. Natur. 55:175–178.

Mouse News Letter, No. 2. 1950. p. 8–9. An informal semiannual mimeographed document distributed by the Laboratory Animals Centre, Carshalton, Surrey. It carries information on mutant genes, research stocks of mice, and research news, and is edited by Mary F. Lyon.

Punnett, R. C. [ed.] 1941. p. 37. Proc. VII Int. Cong. Genet. (Edinburgh, 1939), Cambridge Univ. Press, London.

Serebrovsky, A. S. 1928. Versuch einer allgemeinen Nomenklatur des Genes. Proc. V Int. Cong. Genet. (Berlin, 1927) Z. Indukt. Abstamm. Vererb., Suppl. 2:1320.

Snell, G. D. [ed.] 1941. Mouse genetic news, No. 1. Roscoe B. Jackson Memorial Laboratory, Bar Harbor, Maine. 18 p. (Mimeo.)

Staats, J. 1963. International rules of nomenclature for mice, p. 517–521. *In* W. J. Burdette [ed.] Methodology in mammalian genetics. Holden-Day, San Francisco.

Staats, J. 1964. Standardized nomenclature for inbred strains of mice, Third listing. (Prepared for the Committee on Standardized Genetic Nomenclature for Mice.) Cancer Res. 24:147–168.

7

Nuclear Cytology[1]

Allen B. Griffen

Throughout the history of animal nuclear cytology the consensus among investigators has been that the nuclei in the cells of the "germ line" must be the most dependable sources of chromosome counts which will reveal the true diploid chromosome number for the species. This was, and still is, based upon the logical assumption that the progenitors of the germ cells contain, in as nearly inviolate a state as is possible under the exigencies of early development and differentiation, a reserved portion of the individual's unique inheritance from the past, to be used only as the source of his contributions to the future. These cells are the "gonial" cells: oogonia in the female, and spermatogonia in the male.

GONIAL CHROMOSOMES

Quite in keeping with their being on reserve, the chromosomes of gonial cells at metaphase are invariably highly condensed and individually discrete, providing the ideal condition for absolute chromosome counts. In the mouse, both oogonia and spermatogonia provide an absolute count of 40. Figures 7-1 and 7-2 illustrate oogonial and spermatogonial metaphases, respectively. These

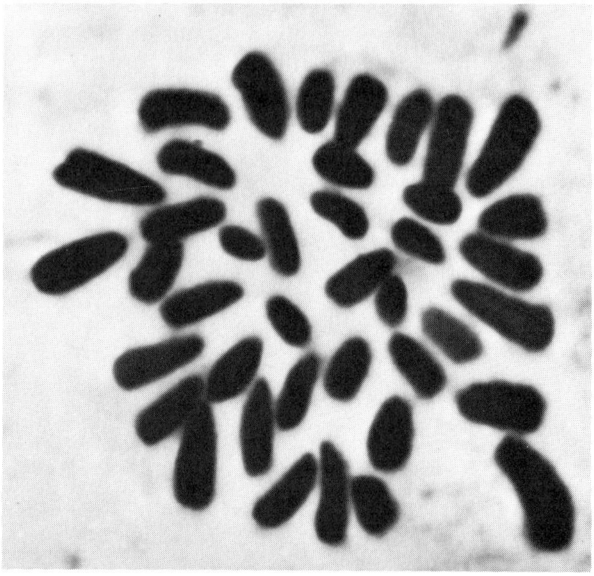

Fig. 7-1. Oogonial metaphase of the mouse Sudan Black squash (×3250).

[1] The writing of this chapter was supported in part by Public Health Service Research Grant CA 04362 from the National Cancer Institute and in part by Contract AT(30-1)-2113 with the U.S. Atomic Energy Commission.

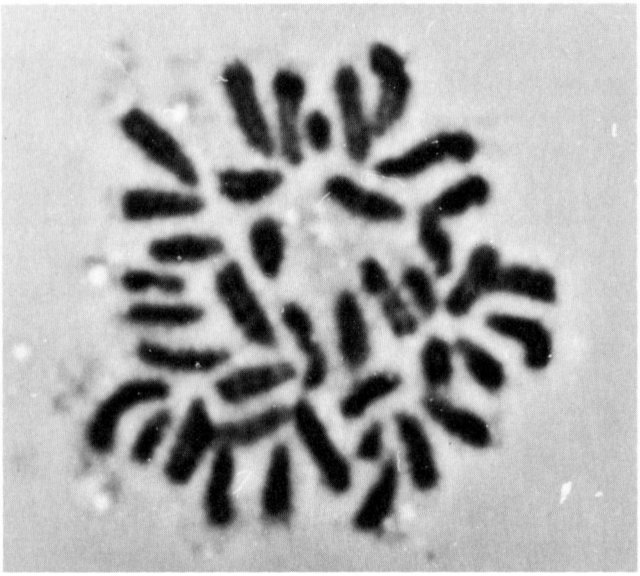

Fig. 7-2. Spermatogonial metaphase of the mouse. Sudan Black squash (×3250).

photographs, taken from Sudan Black squash preparations, clearly show the sharp definition of gonial metaphase chromosomes; they also reveal the prevailing lack of individuality of gonial metaphase chromosomes of the mouse, all of which appear as blunt rod-like elements differing only in length. When gonial metaphase pictures are dismembered and the chromosomes paired in serial order according to length, the resulting idiograms show the presence of two "smallest" elements in the female (Fig. 7-3) and three smallest elements in the male. For one of these elements in the male there is no mate of the same size (Fig. 7-4). Instead there is a long chromosome, left over after all possible pairings have been made, which is the only available partner for the small element.

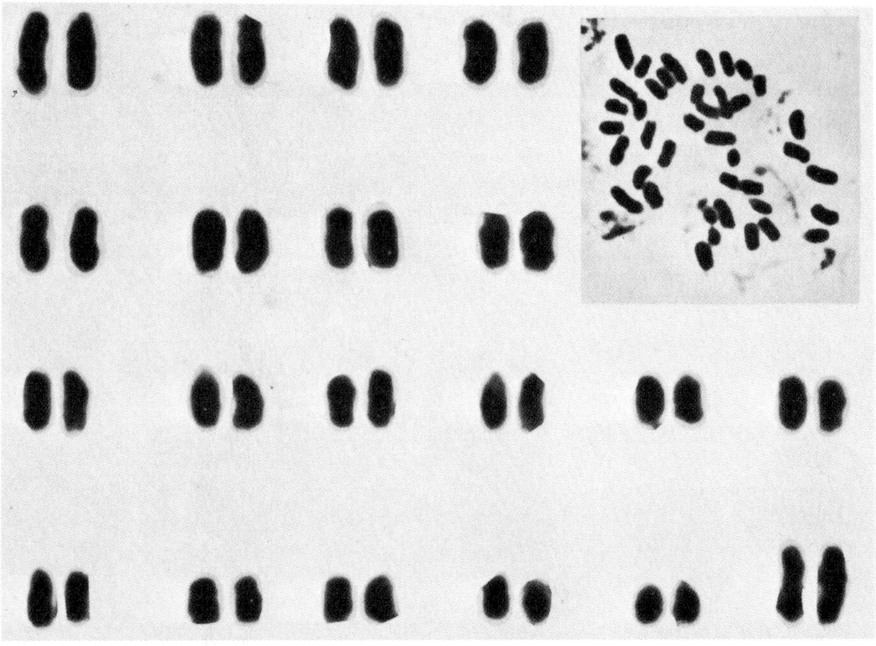

Fig. 7-3. Idiogram prepared from an oogonial metaphase.

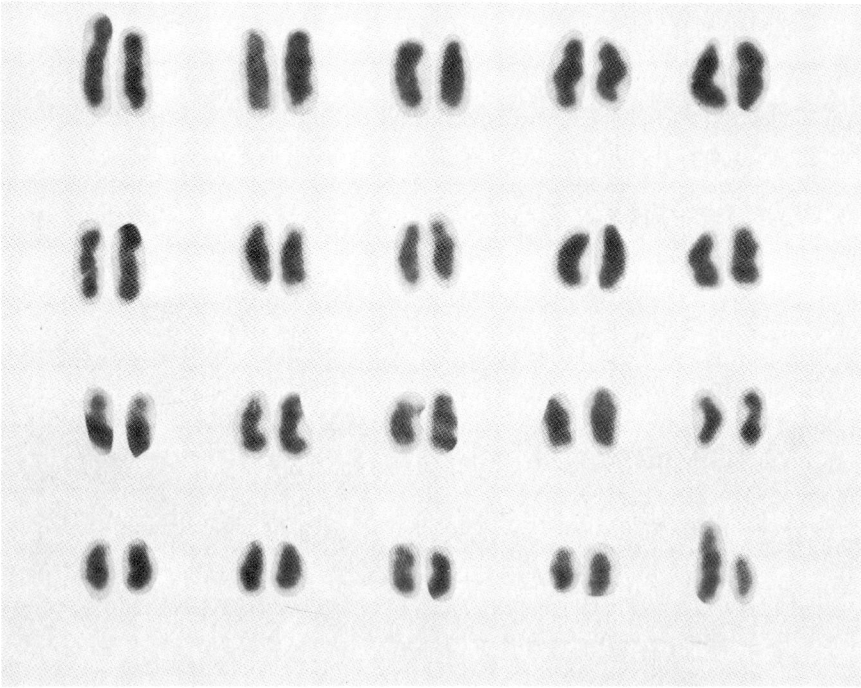

Fig. 7-4. Idiogram prepared from a spermatogonial metaphase.

This pair of unlike bodies in the male is concluded to be the XY pair, the shorter being the Y since it is not present in the female. The first diploid counts from gonial cells were made by Cox (1926) and verified by Painter (1927) and by Cutright (1932).

Aside from revealing the basic chromosome number and, in the male, the sex chromosome pair, gonial nuclei are of little diagnostic use in the mouse. All metaphase elements appear to be telocentric, showing neither constrictions nor accompanying small arms or satellites which might indicate that the centromeres have subterminal positions. The prophase chromosomes are thick and heavy, without marked chromomeric appearance. Anaphase and telophase stages always show compact chromosome clusters in which individual elements are not readily discernible.

In the female, oogonia complete their mitoses before birth. The ovary of the newborn individual shows nearly all gonial cells to be in the final interphase during which occurs the growth in volume immediately preceding meiosis. Borum (1961) and Ohno (1963) stated that oocytes appear on the 14th day of fetal life. Borum added that all are in pachytene by the 18th day; by the fourth or fifth day after birth all oocytes are in the arrested dictyotene or dictyate stage.

In the male, spermatogonia are visible from the last third of fetal life through old age. In the fully formed spermatogenic tubule the spermatogonia appear as a thin layer of cells with compact deeply staining nuclei forming a poorly defined row immediately internal to the wall of the tubule (Fig. 7-5). Most of these cells are type B spermatogonia, the last stage before the transformation into primary spermatocytes. Although occasional metaphases and anaphases are seen in this layer, most of the nuclei are in the resting or interphase stage, characterized by the presence of irregular blocks and "clots" of chromatin. Type A and intermediate spermatogonia are less prominent in the gonial layer, being far outnumbered by the type B spermatogonia. They are occasionally discerned in squash preparations. In the gonial layer, type B spermatogonia become transformed into primary spermatocytes. The inconspicuous spermatocytic interphase appears finely granular and may show a single pale vesicular nucleolus. Being of very short duration, spermatocytic interphases are usually outnumbered by spermatocytic leptotene, zygotene, and earliest pachytene nuclei (Fig. 7-6). By the time of the transformation of all preceding stages into the early pachytene stage, the row of primary spermatocytes is found to have migrated inward, allow-

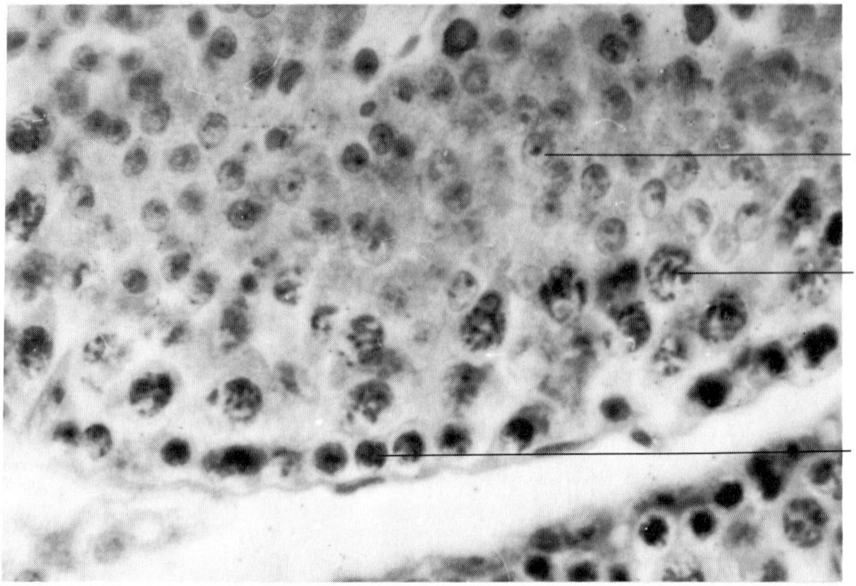

Fig. 7-5. Section of a spermatogenic tubule of the mouse; B, type B spermatogonia; C, primary spermatocytes; ST, spermatids. Bouin fixative, hematoxylin and eosin (×400).

ing for the formation of a new gonial layer from the multiplication of type A spermatogonia.

GONIAL MITOSIS

Interphase nuclei

Interphase nuclei in type A spermatogonia have been described by Regaud (1901) as "dust-like," with reference to the appearance of their fine and evenly dispersed chromatin particles. A single prominent chromatin nucleolus, whose nature is not known, usually appears in an eccentric position. This body does not seem to represent a cluster of centromeres, as is the case for the chromatin nucleolus in *Drosophila* (Warters and Griffen, 1958). Interphase nuclei in intermediate and type B spermatogonia are characterized by deeply staining peripheral spots and patches of chromatin which impart a "crust-like" appearance to the nuclei. A prominent nucleolus somewhat resembling a plasmosome, but surrounded by irregular crust-like chromatin bodies, usually appears in a central rather than a peripheral position in each nucleus. Oakberg (1956*a*, *b*) provided descriptions of the several types of spermatogonia.

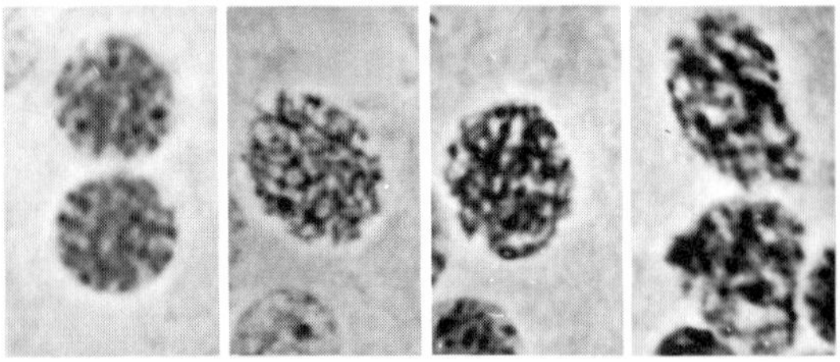

Fig. 7-6. Primary spermatocytic interphase, leptotene, zygotene, and earliest pachytene. Sudan Black squash (×1200).

Prophase

Prophase in type B spermatogonia is characterized by the formation of thick, smooth threads which seem to fill the nucleus completely. As the nuclear membrane disintegrates, the final condensation and shortening of the chromosomes into their metaphase form occurs in a steady progression showing no clearly demarcated steps. Prophase in type A and intermediate cells apparently has not been described.

Metaphase

Metaphase in spermatogonia has the expected equatorial arrangement of the chromosomes, with no morphological indications of spindle attachment points. In side views the spindle is readily visible as a double cone with a broad common base (the equator) and short altitudes; often the spindle may appear rather barrel-shaped. At the poles, centrosomes of very small size often may be observed. Type A and type B spermatogonial metaphases are easily distinguished. In A cells the chromosomes are long and slender, whereas in B cells they are usually short and are rounded into bean-like shapes.

Anaphase

Anaphase is not often observed in sectioned material, but is frequently observed in spermatogonia in squash preparations. Chromatid separations appear to be both clean-cut and quick, with no lagging elements and no precocious passages to the poles. As at metaphase, A cells show long, slender chromosomes, and B cells show rounded, short chromosomes.

Telophase

Telophase in all types of spermatogonia is a fleeting stage seen occasionally in the extreme periphery of the spermatogenic tubules. It has no unusual characteristics other than location and scarcity. As the gonial layer becomes crowded through cell divisions, the daughter nuclei of B spermatogonia tend to remain associated in pairs. This association frequently appears to be carried over into the darkly staining sharp layer of early primary spermatocytes. Figures 7-7 and 7-8 illustrate spermatogonial mitoses.

MITOSIS IN SOMATIC CELLS

In vivo

Aside from routine observations made in the regular course of histological and pathological studies, somatic mitosis in the mouse has received little attention. There have been no indications from sectioned material that somatic chromosome studies might be of any cytogenetic use, aside from the demonstrations of polyploidy, aneuploidy, anaphase bridging, multipolar mitoses, endomitosis, and other phenomena always associated with malignancy and tumorigenesis. Through the use of mechanical means of cell separation, combined with improved techniques for squashing and spreading, Bunker (1961, 1965) has produced diagnostically usable prometaphase and metaphase preparations from adult bone marrow, spleen, and kidney and from several fetal tissues. The chromosomes are regularly separated into component chromatids, held together at subterminal centromere positions. Minute chromosome arms or satellites are regularly visible and make possible the construction of somatic idiograms in which most of the members of the chromosome complement can be characterized and "paired up" with accuracy. Figure 7-9 illus-

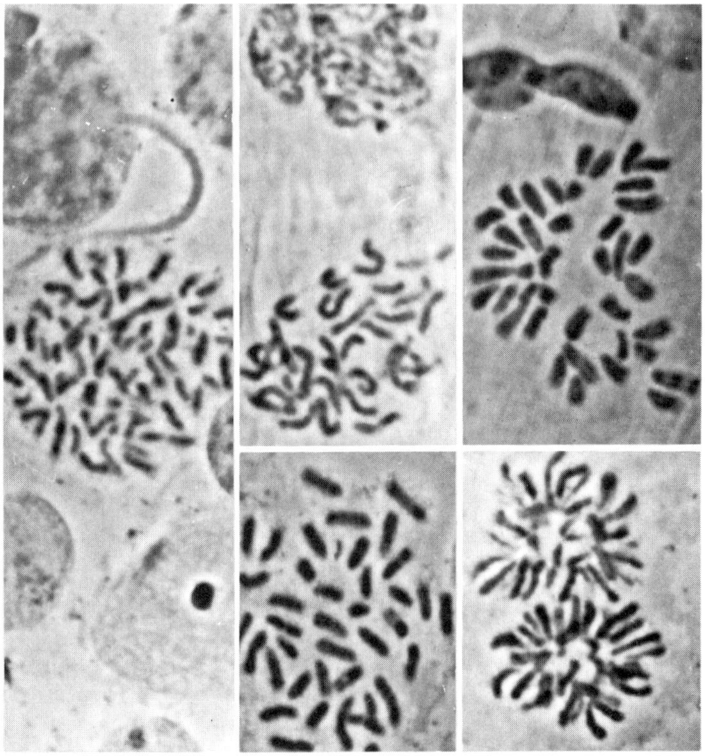

Fig. 7-7. Spermatogonial mitoses in adjacent cells in a Sudan Black squash. Left, top to bottom: type B interphases, type B late prophase, type A interphase with nucleolus; center, top to bottom: type B early prophase, type A late prophase, type A metaphase; right, top to bottom: type B metaphase, flattened type A anaphase ($\times1650$).

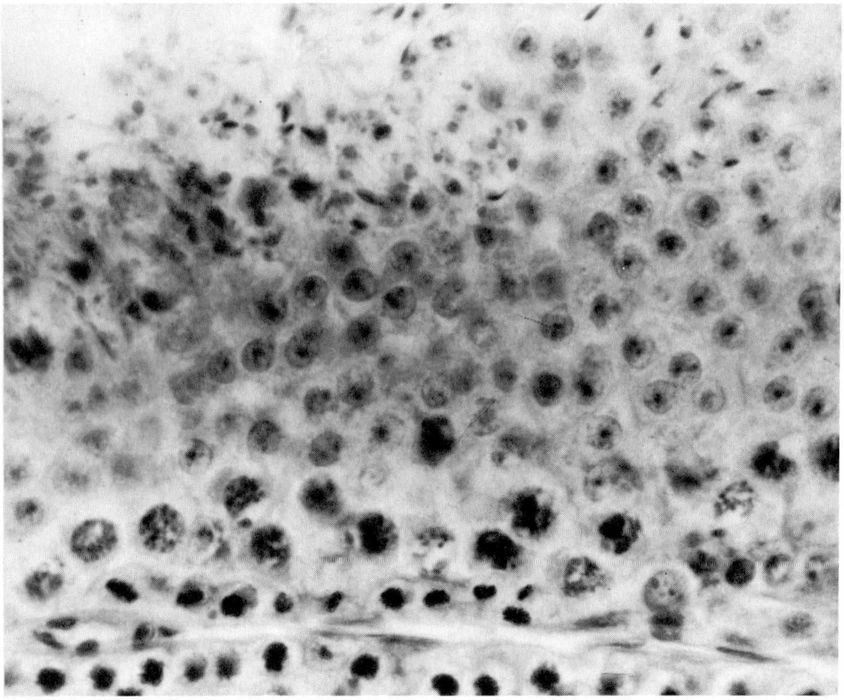

Fig. 7-8. Spermatogonial mitoses in a sectioned spermatogenic tubule. The large primary spermatocytes immediately above the dense gonial metaphases are in the midpachytene stage. Bouin fixative, hematoxylin and eosin (×400).

trates a typical metaphase from the adult spleen. Figure 7-10 is an idiogram of the same cell, showing the small arms and centromere positions in several of the component pairs. A late prophase nucleus from the spleen of an adult female is shown in Fig. 7-11; Fig. 7-12 is an idiogram of the same cell.

In vitro

The advent of somatic mitosis into great prominence as a source of cytogenetic information came with Hughes's (1952) suggestion that the hypotonic pretreatment of cells grown in vitro improved the opportunities for studying the metaphase chromosome complement of these cells. The independent discovery and immediate successful use of hypotonic pretreatment by Hsu and Pomerat (1953) upon cells grown in short-term tissue cultures opened new horizons for the cytology of all mammals, and the technique has been in many respects contributory to the clarification of sex determination in these animals. In the mouse, hypotonic pretreatment of actively dividing cells from cultures of any soft tissue, such as fetal gonads, kidney, and bone marrow, provides material which can be handled directly with squash techniques.

Somatic metaphases, whether derived from direct squashes or from cultures, reveal chromosome individualities not seen in gonial metaphases prepared by the same methods, presumably indicating that the "reserved" state of the gonia causes the occlusion of many fine points of chromosome structure. For example, the metaphase and idiograms of Fig. 7-9 to 7-12 may be compared with those of Fig. 7-1 and 7-2. All received hypotonic pretreatment, yet the somatic chromosomes show constrictions, satellites, and short arms never visible in the gonia. The X and Y seem to retain their gonial type of condensation, appearing more dense, more deeply-staining, and less frequently separated into component chromatids.

Allocycly of the X and Y

Although the X and Y under such conditions may be called "heteropycnotic," a better term is "allocyclic," indicating that these elements are not strictly in phase with other members of the chromosome complement. True allocycly, however, is best seen in XX cells such as those in

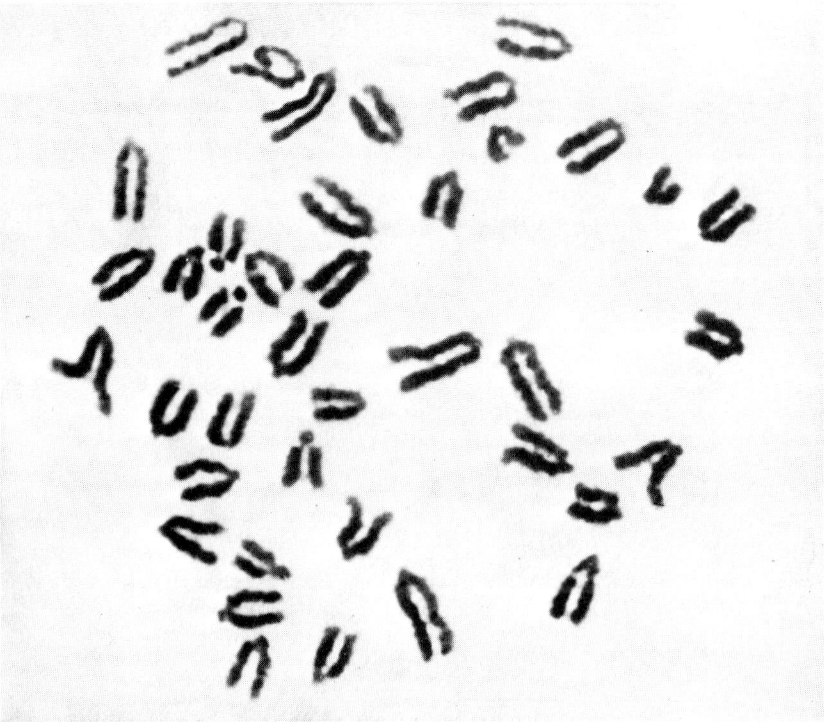

Fig. 7-9. Metaphase from the spleen of an adult female mouse. Note the indications of somatic pairing and the concordance of arm shapes. Sudan Black squash (×3250). (*Courtesy of M. C. Bunker.*)

Fig. 7-10. Idiogram of the metaphase in Fig. 7-9. Small arms and centromeric constrictions visible in several pairs. The presumed X's are the last pair.

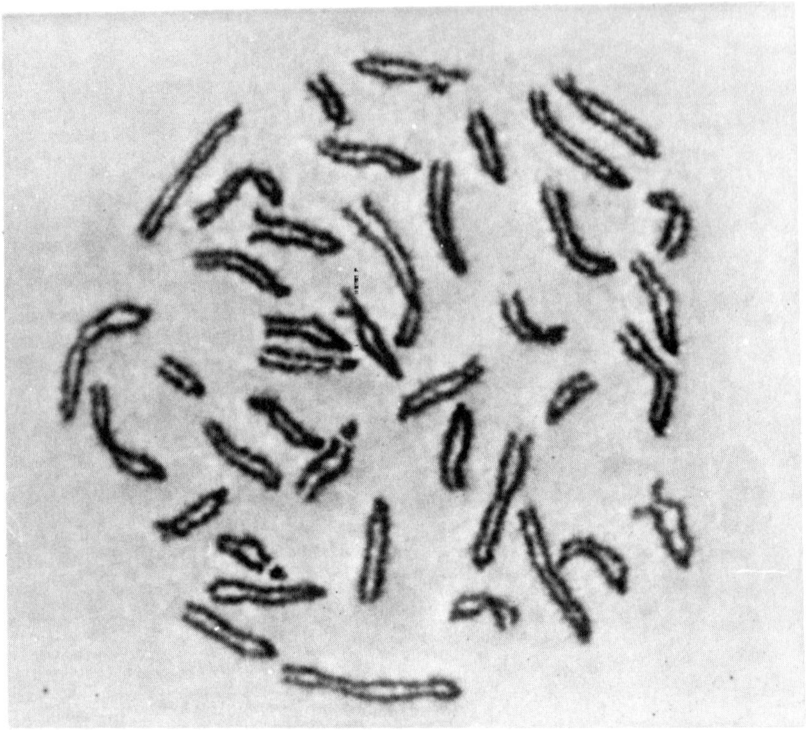

Fig. 7-11. Late prophase chromosomes from the spleen of an adult female mouse. Giemsa stain. ($\times 3300$). (*From M. C. Bunker,* 1965.)

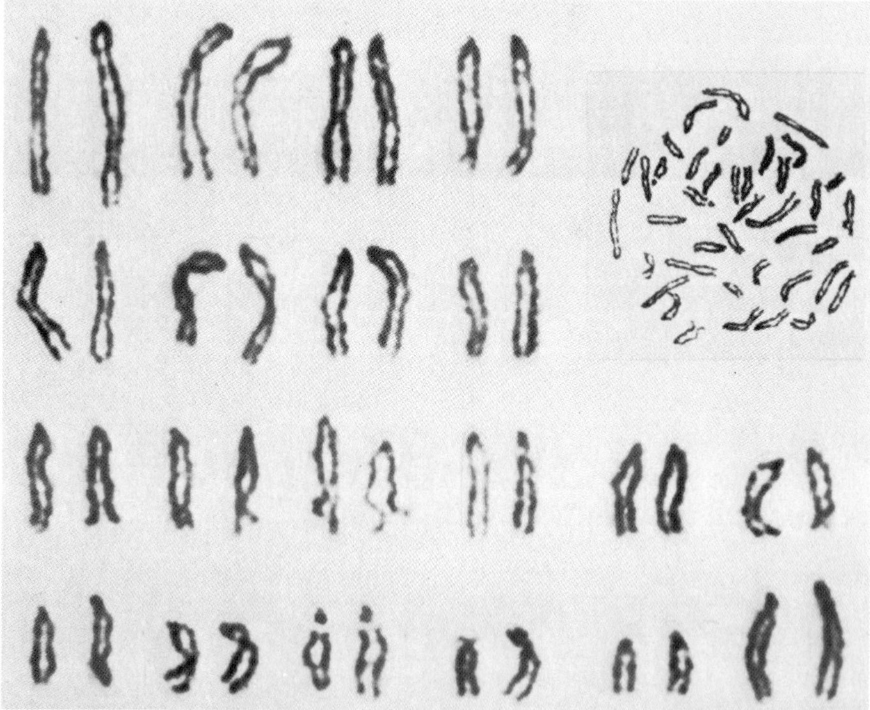

Fig. 7-12. Idiogram of the late prophase shown in Fig. 7-11. The presumed X's are the last pair.

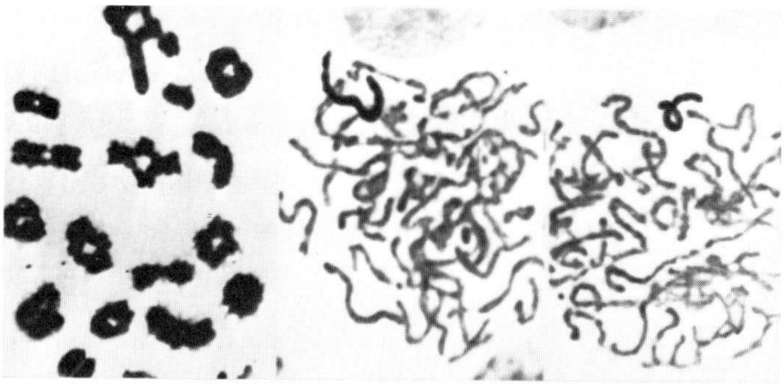

Fig. 7-13. Allocycly of an autosomal interval inserted into an X chromosome. Left: spermatocytic metaphase I showing the X-Y-autosome association at the top of the picture; center: somatic prophase showing allocycly of the X chromosome and of the autosomal segment which is inserted into it; right: somatic prophase showing allocycly of a normal X chromosome (×2100). (*After Ohno and Cattanach,* 1962.)

which Ohno and Hauschka (1960) first pointed out the density of one X, whereas the other was hardly distinguishable from the autosomes in certain tumor samples. That the condensed X in such cases is not in phase with other elements is demonstrable through its late labeling with tritiated thymidine. The demonstration of allocycly was a direct forerunner of the inactive-X hypothesis of Lyon (1961, 1962). Lyon demonstrated genetically that an autosomal segment bearing the dominant allele of a coat color factor, when inserted into an X, is inactivated along with that X to permit a mosaic phenotypic expression of the recessive allele for color borne in the homologous intact autosome. Ohno and Cattanach (1962) provided cytological confirmation of the hypothe-

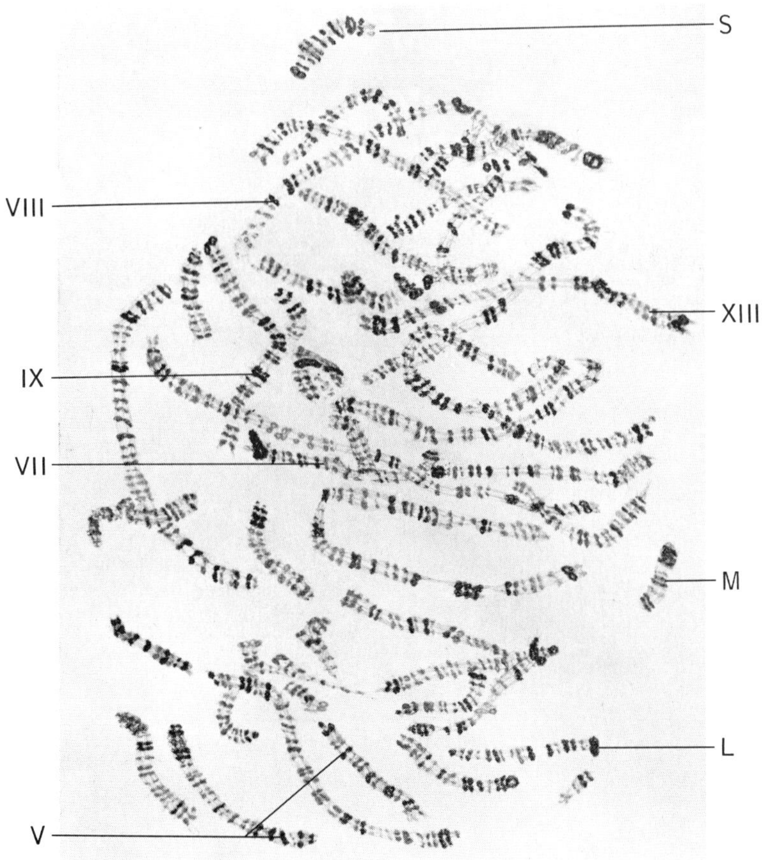

Fig. 7-14. Tentative chromosome identifications in a tumor cell prophase. Letters and numbers refer to the pachytene maps of Slizynski and of Griffen (× ca.4000).

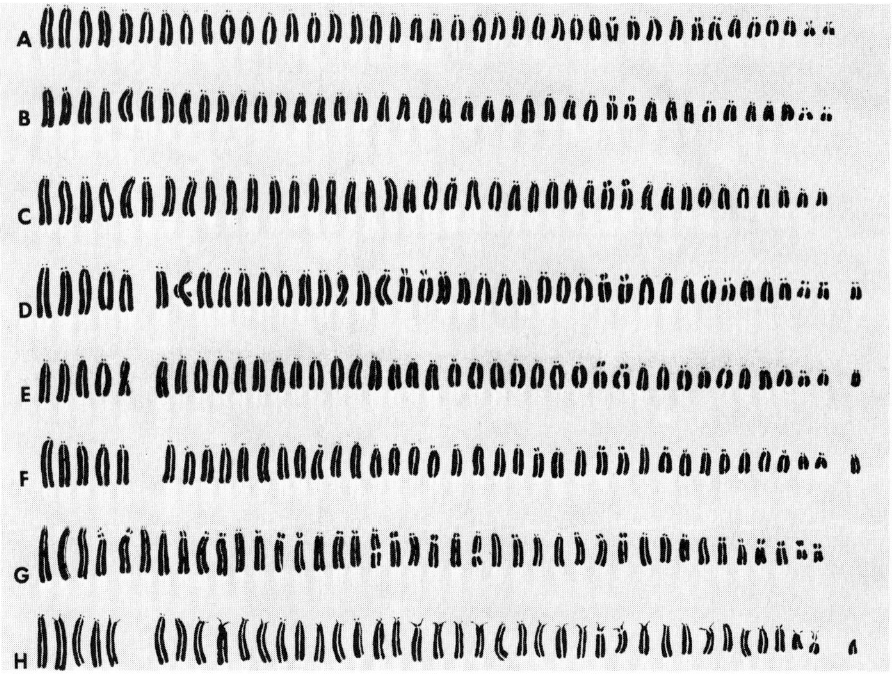

Fig. 7-15. Karyotypes from cultured embryonic mouse cells. A, D, G, Swiss mice; B, E, H, C3H; C, F, C57BL; A, B, C, G, female; D, E, F, H, male (×1100). (*After Levan, Hsu, and Stich,* 1962.)

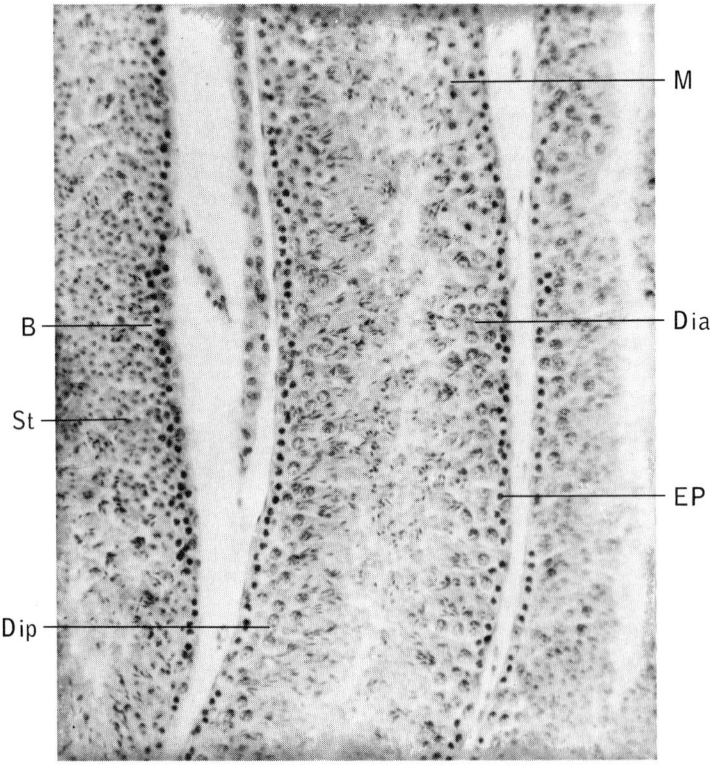

Fig. 7-16. Spermatogenic tubules from an adult male mouse. B, type B spermatogonia; Dia, diakinesis; Dip, diplotene; EP, early pachytene; M, primary spermatocytic metaphase; St, spermatids. Bouin fixative, hematoxylin and eosin (×80).

sis through the study of the chromosomal constitution of the mosaic areas. Figure 7-13 illustrates the active and inactivated autosomal segment.

Somatic chromosome identification

Chromosome identification and chromosome mapping in somatic cells of mice have not been accomplished. However, with the development of techniques for spreading the chromosomes, somatic prophases may prove to be useful in these respects. Griffen (1960) was able to make tentative chromosome identifications in tumor cell prophases in which chromomeric structure was prominent. Figure 7-14 illustrates such a cell and indicates the identified chromosome parts.

Utilizing all information and available techniques, Levan et al. (1962) prepared somatic cell idiograms of the mouse, based upon tissue cultures. These idiograms represent a comprehensive approach to the problem of somatic chromosome identification. They also reveal the inadequacies of mouse metaphase chromosomes in this usage (Fig. 7-15).

MEIOSIS

Spermatogenesis

Spermatogonia. The location and appearance of the type B spermatogonia, which transform directly into primary spermatocytes, has been presented in the discussion of gonial mitosis. The thin peripheral layer of the spermatogenic tubule was shown to consist of these cells. This layer and several stages of the primary spermatocytes are shown in Fig. 7-16,

Primary spermatocytes. The earliest stages of the primary spermatocytic prophase form a sharp, darkly staining row which consists of leptotene-zygotene or of early pachytene stages. The preceding interphase and similarly inconspicuous early leptotene stages seldom appear in this row. Internal to the dark row the later stages of the meiotic prophase are characterized by their large size, all nuclei from midpachytene through diakinesis being more than twice the diameter of the early pachytene nuclei. The growth period in primary spermatocytes is thus indicated to be the pachytene stage.

Interphase. The only well known interphase in meiosis in the male mouse is that of the primary spermatocyte, found in the gonial layer of tubules and distinguished from type B cells principally by the finely granular nucleus with its pale nucleolus. Griffen and McMahon (unpublished) have estimated the time of duration for this stage to be approximately 3 per cent of the total time of the spermatogenic cycle as determined by Oak-

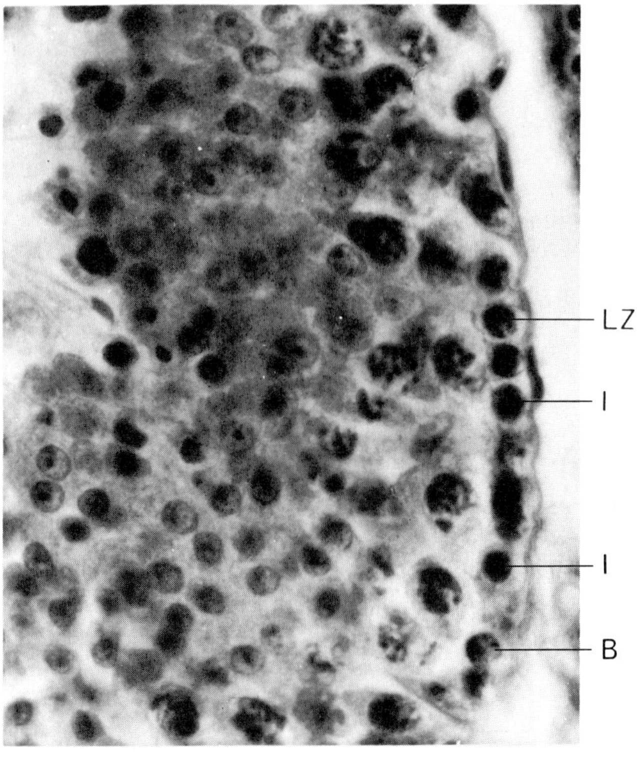

Fig. 7-17. Spermatogenic tubule showing early stages of meiosis. I, primary spermatocytic interphase; LZ, leptotene-zygotene; B, type B spermatogonium. Bouin fixative, hematoxylin and eosin (×400).

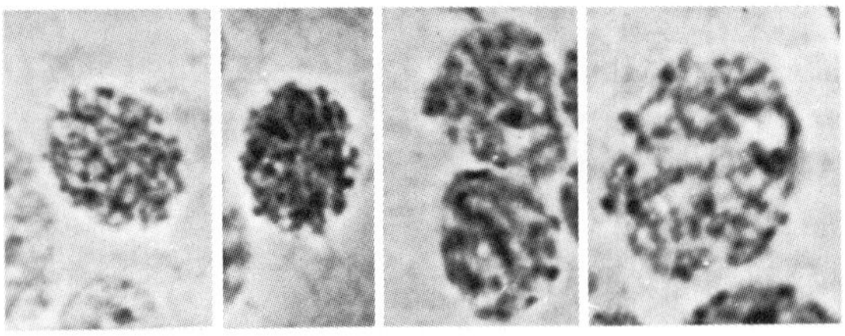

Fig. 7-18. Early primary spermatocytes of the mouse. Left to right: leptotene, zygotene, early pachytene, midpachytene. Sudan Black squash ($\times$1200).

berg (1956*b*). Since Oakberg found the length of the cycle to be 212 hours, the interphase exists for about 6 hours. The interphase and other early stages of meiosis as they appear in a sectioned tubule are illustrated in Fig. 7-17.

Leptotene. The first stage of the meiotic prophase, the leptotene, appears in portions of the dark row of early spermatocytes. Leptotene nuclei are finely granular, with indications of fine strands connecting the chromatin particles (ultimate chromomeres?) for short distances. Eventually the nuclei appear as masses of lightly stainable beaded threads, among which the nucleolus appears as a clear space bounded by a distinct membrane.

Zygotene. In its early stages the zygotene is not distinguishable from the leptotene, except in the fact that thickened regions of the chromosome strands indicate the beginning of pairing. The late zygotene is readily identifiable through its many regions in which the double thickness of the chromosome strands indicates advanced pairing. Figure 7-18 illustrates the early stages of the meiotic prophase in a squash preparation.

Pachytene. The pachytene stage is characterized by complete pairing. With all homologous strands in synapsis, the chromosomes resemble minute railroad tracks winding about within the nuclear membrane with gently spiralled twists. The plasmosome is now sharply distinct and, according to Slizynski (1949) and Griffen (1955), is regularly attached through several heavy chromatic bodies to a particular member of the synapsed chromosome complement. Sachs (1955) and Geyer-Duszynska (1963) challenged this interpretation and stated that the plasmosome is not a part of a stranded XY pair, but is the sex vesicle,

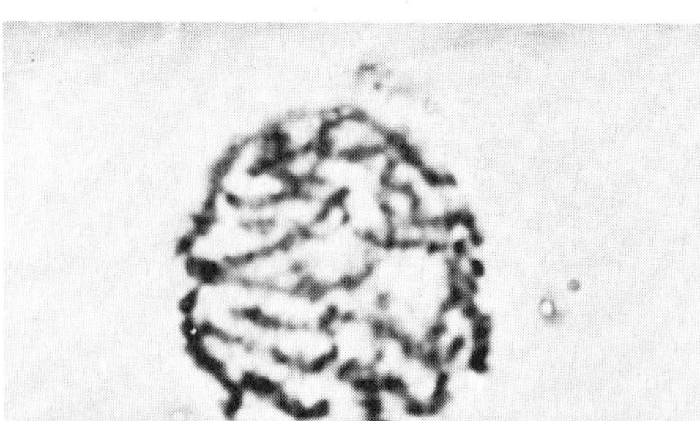

Fig. 7-19. Late pachytene nucleus of the male mouse. Sudan Black squash ($\times$2000).

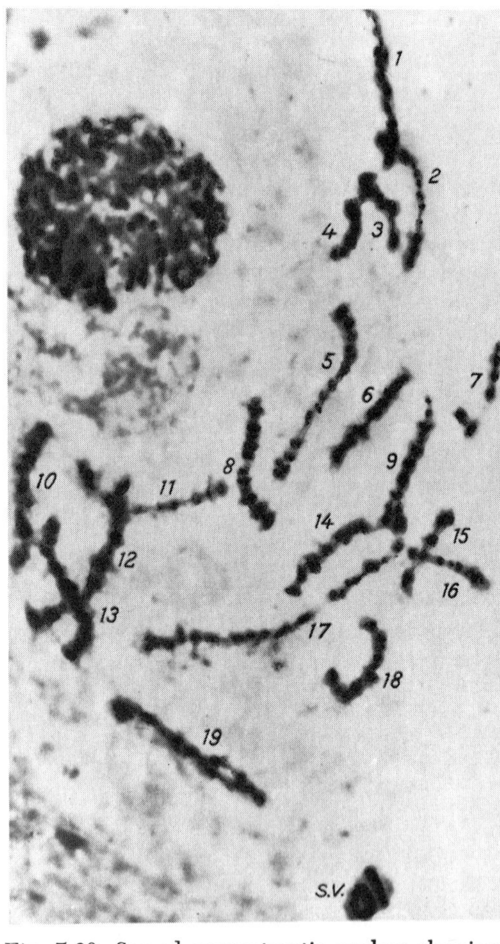

Fig. 7-20. Spread spermatocytic nucleus showing 19 chromomeric chromosomes and the sex vesicle (SV) containing the X and Y ($\times$2250). (*After Geyer-Duszynska,* 1963.)

containing the X and Y in precondensed form. The nature of the vesicle or plasmosome and its true relation to the sex chromosomes is a matter of fundamental importance in chromosome mapping and in sex-chromosome inactivation studies. An early solution to the problem is therefore desirable.

Through the use of hypotonic solutions and squash methods in preparing slides, pachytene nuclei can be ruptured and the chromosomes separated with some success. Figure 7-19 illustrates a ruptured pachytene nucleus similar to those used by Slizynski and by Griffen for map construction. The chromomeric nature of the paired strands is clearly visible. Figure 7-20 shows 19 chromomeric chromosomes and the presumed X and Y in the sex vesicle, such as described by Geyer-Duszynska.

Diplotene. The passage from pachytene to diplotene is characterized by the relaxation of pairing, resulting in the disclosure of a four-stranded composition for each former pachytene pair. So gradual is the change that sharp demarcation of the diplotene is not possible. Early diplotene is almost identical with pachytene in superficial appearance, yet four strands can be discerned at numerous, if not all, levels along each chromosome pair. Chiasmata first appear at diplotene, occasionally visible as interchanges between members of the otherwise parallel pairs of chromatids. However, these exchanges are readily detectable in ordinary practice only at the onset of diakinesis. Figure 7-21 illustrates the diplotene nucleus. Diplotene and all preceding stages of the primary spermatocyte are notoriously difficult subjects for photography.

Diakinesis. At its most easily recognized stage diakinesis shows the tetrads of the diplotene in the process of condensation by means of coiling. Each tetrad has the appearance of a fuzzy, loose loop, ring, or cross, the configuration depending upon the position and the degree of terminalization of one or more chiasmata. Figure 7-22 shows several appearances of diakinesis, progressing from loose coiling to the compact coiling which approaches the condensation seen in metaphase I.

Metaphase I. The primary spermatocytic metaphase is the most easily recognized stage in meiosis. In the male mouse 20 discrete tetrads are visible, each usually contorted by the terminalization of chiasmata. Occasionally tetrads may be found which appear to have undergone complete terminalization, for no evidence of a chiasma is visible. Such tetrads usually appear as closely paired rod-like bodies, usually representing the smaller chromosome pairs. In contrast to those of diakinesis, metaphase I tetrads are highly condensed, especially deep-staining, and sharply smooth in outline. The XY pair is readily identified, the X appearing as a long rod and the Y as

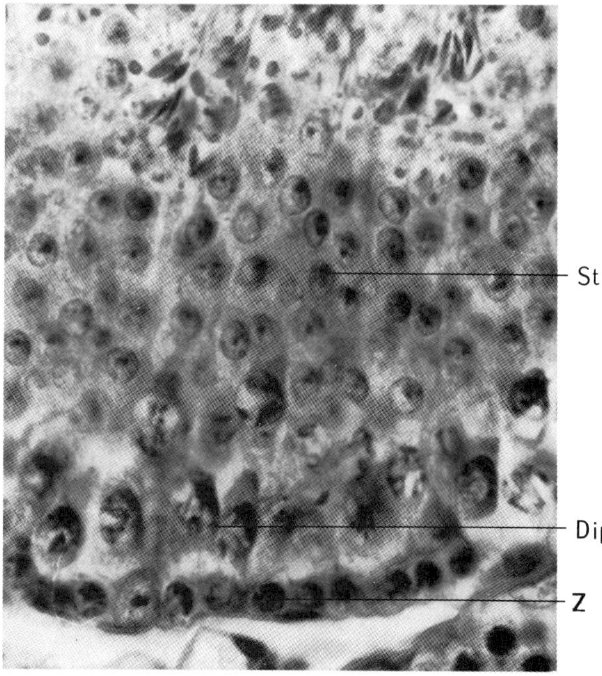

Fig. 7-21. Diplotene nuclei (Dip) in a sectioned spermatogenic tubule; Z, zygotene nuclei; St, spermatids. Bouin fixative, hematoxylin and eosin ($\times 400$).

an elliptical short rod, in end-to-end synapsis. The X and Y seldom appear as dyads, the chromatids remaining closely appressed. In side views the meiotic spindle is clearly visible, having the same short form as that of the mitotic metaphase.

Anaphase I. The first meiotic anaphase is only occasionally seen in sections and squash preparations, being of very short duration. The spindle is usually quite distinct, and polar movement of the dyads is uneventful unless induced aberrations lend distortions in the form of dicentric bridging. The X and Y normally reduce (separate) at this stage, although reports and photographs indicate occasional equational division of these bodies so that one monad of each body passes to each pole. Figure 7-23 illustrates reductional and equational separation as observed by Slizynski (1955*b*). Sachs (1955) and Ohno et al. (1959) have reviewed the significance of the terminal association of the X and Y at meiosis.

Telophase I. After anaphase I the two groups of dyads form tightly clumped aggregates at the poles, representing the first part of telophase I. As the cytoplasm constricts to form the two new cells, nuclear membranes become distinct and the dyads uncoil progressively to form the second meiotic interphase.

Interphase II. In studying the regeneration of the germinal epithelium after depopulation by irradiation, Griffen and Bunker (unpublished) found and described the interphase of the second meiotic division. Interphase II is always intimately associated with metaphase I, metaphase II, or both. In cursory examinations of sectioned testes, interphases II grossly resemble spermatids, with which they regularly have been confused; however, they are fully twice the diameter of spermatids and half as numerous. In the mouse, as in the rat (Tjio and Levan, 1956), X-bearing interphase II nuclei show a large allocyclic element representing the X-chromosome; Y-bearing nuclei have a similar but smaller element.

Prophase II. The second meiotic prophase is of short duration and therefore is never prominent in either sections or squashed preparations. Having the same diameter as interphase II, prophase II is easily distinguished from late pachytene which it otherwise resembles. Second prophases probably have been classified usually as partial sections of pachytene nuclei.

Prometaphase II. The first recognizable stage in the second spermatocytic division is best described as a prometaphase. The lightly stainable dyads of fuzzy outline, often showing clear indications of the major condensation spiral, appear as ragged pairs held together only at the centromeres. Short arms or satellites are regularly found in many of the dyads. Figure 7-24 illustrates interphase II, prophase II, and metaphase II in a section. Prometaphase II is seen in Fig. 7-29.

Metaphase II. The second spermatocytic metaphase is frequently observed and is readily

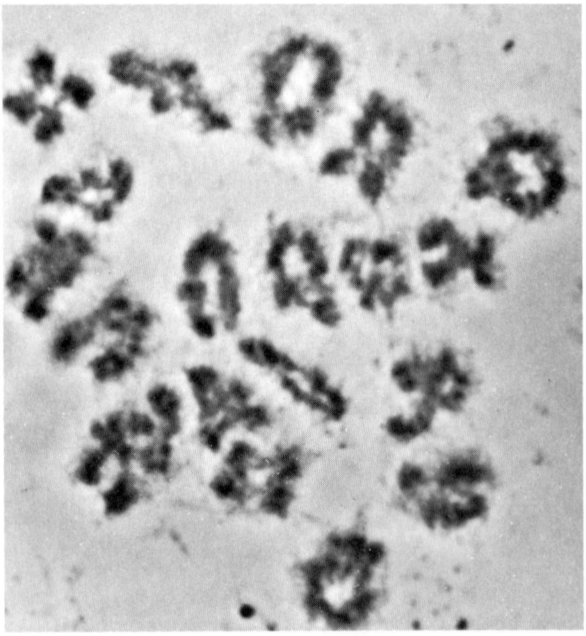

Fig. 7-22. Primary spermatocyte at diakinesis. Sudan Black squash (×3150).

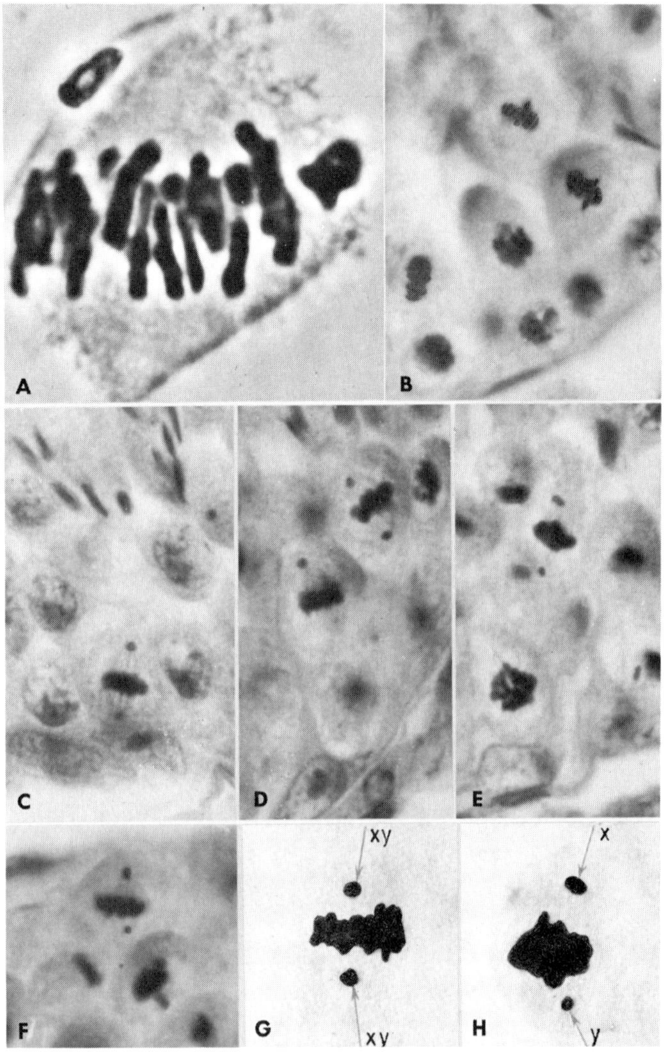

Fig. 7-23. Early anaphases of primary spermatocytes. *A*. Tetrads beginning polar movement, X and Y at center. Sudan Black squash (×2500). *B*. Two cells showing protruding X and Y at onset of precocious passage to the poles; *C, D, E, F*. Precocious arrival of X and Y at the poles; *C, E, F*, reductional; *D*, equational. Sections, hematoxylin and eosin (×500).

G. Equational (symmetrical) anaphase, XY at each pole. *H*. Reductional (asymmetrical) anaphase, X at upper pole, Y at lower. (*After Slizynski,* 1955*b*).

identified through the presence of 20 small dyads not easily confused with the 20 large tetrads of metaphase I. In squash preparations sister cells in metaphase II often lie side by side, one showing the X and the other the small Y. Figure 7-25 illustrates metaphase II in two sister cells.

Anaphase II. The second spermatocytic anaphase is of short duration and therefore scarce in all preparations. In squash preparations the second spindle is distinct and more elongated than the first spindle. The chromosomes in passage to the poles are very small and seem to be approximately the same size.

Telophase II. The second telophase has not been recognized in the mouse, the transition from second spermatocyte to spermatid thus seeming to be inconspicuous.

Spermatids. The spermatids resulting from division II lie next inward beyond the spermatocyte layer in the spermatogenic tubules. Aside from their small size, approaching that of spermatogonia, and the presence of several prominent chromatin bodies within the otherwise finely granular nucleus, the spermatids are of little cytological interest.

Spermatozoa. Oakberg (1956) has comprehensively described the development of spermatozoa from spermatids. Mature spermatozoa appear in sections as deeply stained short rods which form the innermost layer of the spermatogenic tubule, the flagella extending into the lumen. In squash preparations sperm heads usually appear as arrowhead-like bodies with curved points. The flagella are seldom found attached to the heads.

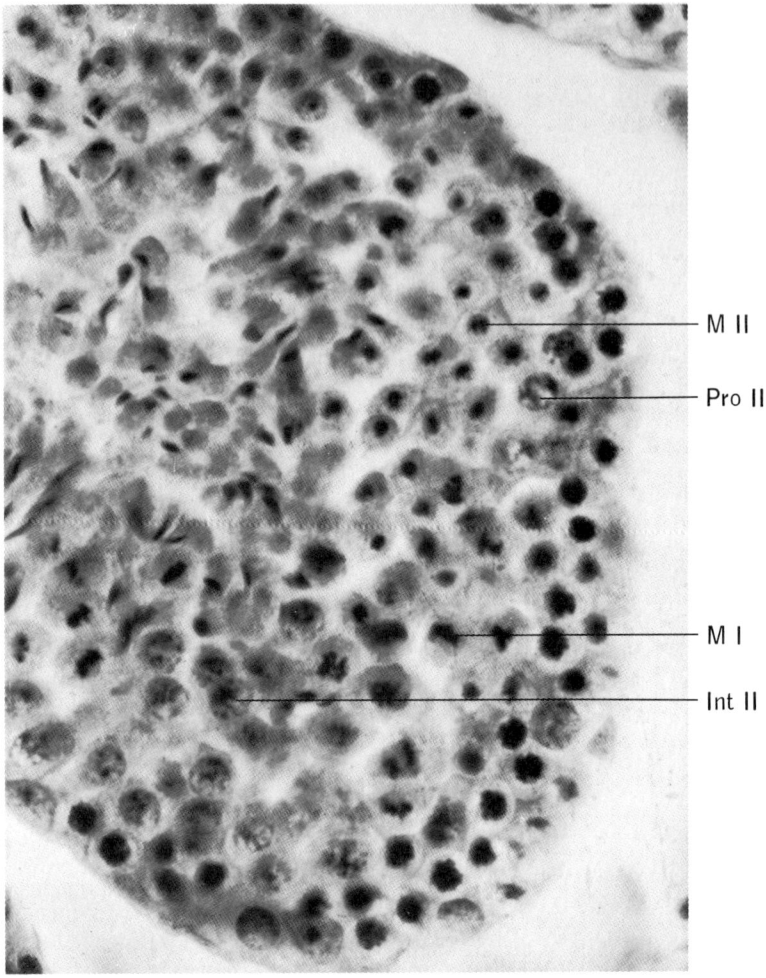

Fig. 7-24. Secondary spermatocytes in a sectioned spermatogenic tubule; Int II, interphase; M II, metaphase; Pro II, prophase; M I, metaphase of primary spermatocyte. Bouin fixative, hematoxylin and eosin (×400).

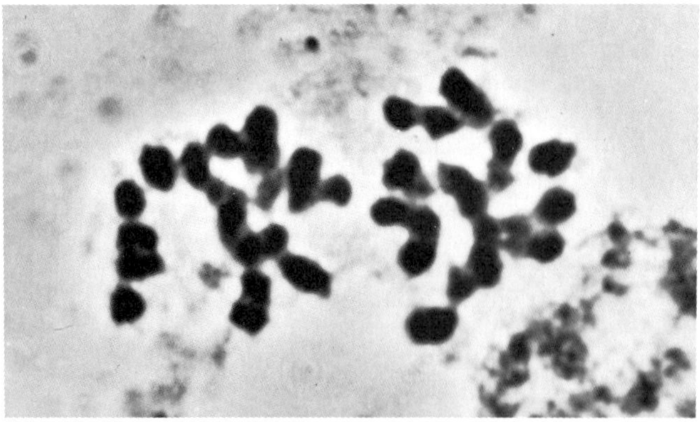

Fig. 7-25. Secondary spermatocytic metaphases. Sudan Black squash (×1725).

Oogenesis

Slizynski (1957*a*), Borum (1961), Ohno (1963), and Kuhlmann (1965, personal communication) have analyzed the nuclear phenomena of oogenesis. According to Borum the mouse gonad is first identified as such in the 10-day fetus. In the 13-day fetus the gonad can be identified as testis or ovary. In the ovary some oogonial divisions are seen (Fig. 7-1), but most cells appear to be in the interphase growth stage of the primary oocyte (Fig. 7-26*A*). On the 14th day most of the cells have entered meiosis and are in leptotene and zygotene; some oogonial divisions are still found. On the 15th day all cells are in meiosis, with zygotene and pachytene predominating. The 16th day shows practically all oocytes in pachytene, with many beginning the rather massive degeneration which will eventually limit the final number of potential ova. Degenerating cells have heavily staining clumped chromosomes which appear to be highly condensed pachytene elements. On the 17th day cells in pachytene and degenerating pachytene continue to be predominant. By the 18th day many of the innermost cortical cells become large and clear, entering the diplotene stage; they become surrounded by flattened stromal cells. The more peripheral cells remain in pachytene. On the 19th day diplotene nuclei are increased in number and vast numbers of degenerating pachytene stages are seen. A few hours after birth many cells are in late diplotene, representing the resting stage, known also as dictyate or dictyotene. At 19 hours after birth, early and late diplotene and degenerating cells predominate. At 40 hours after birth pachytene nuclei become scarce and diplotene nuclei are principally found. By 60 hours after birth many oocytes have enlarged and received their follicle cells. Further nuclear stages of meiosis await the onset of estrus.

Leptotene. Kuhlmann (1965, personal communication) has shown that leptotene nuclei contain 40 paired heterochromatic blocks from which the chromosome threads extend as very thin coiled strands (Fig. 7-26*B*). The blocks are considered to be the centromeric regions of the chromosomes, and the paired condition is interpreted as the beginning of synapsis.

Zygotene. At zygotene the chromosomal threads become progressively paired and thickened. Kuhlmann demonstrated the formation of a distinct "bouquet" stage (synizesis) at zygotene. The bouquet is doubly polarized, showing a cluster of the early staining centromeric ends of the chromosomes and beside it a cluster of noncentromeric ends. The pairing chromosomes extend from one cluster to the other in long loops (Fig. 7-26*C*).

Pachytene. Full synapsis at pachytene is generally similar to that in the male except for the pronounced polarization (Fig. 7-26*D*) and the absence of a large plasmosome (or sex vesicle). According to Slizynski (1957) no plasmosome is found in early oocytes.

Diplotene. At diplotene, pairing relaxes slightly and four strands are detectable in each pair of synapsed elements; otherwise this stage closely resembles pachytene.

Dictyotene. Instead of entering diakinesis, which is a clearly defined stage in the primary spermatocyte, the primary oocyte passes from diplotene into the dictyotene or dictyate stage. This stage is peculiar to oogenesis and constitutes the long stage of arrest in which the oocyte nuclei remain from early infancy until the time of fertilization in the adult. At this stage in the mouse the tetrads become progressively denuded of nucleic acids, the major coiling of late diplotene disappears, and the nuclei assume an "empty" appearance. Chromosome organization becomes indistinct. A large nucleolus appears and persists as the outstanding feature of the nucleus. It is notable that dictyotene persists throughout the growth of the oocyte from the small primordial to the large mature state. Although invisible, the chromosomes remain in tetrad association and chiasmata persist, as indicated by the presence of terminalizing chiasmata at the first meiotic metaphase and anaphase. Figure 7-26*E* illustrates the dictyotene nucleus in a primordial follicle of an adult mouse. Figure 7-26*F* shows the late dictyotene nucleus of an oocyte in a graafian follicle. The conspicuous large nucleolus is accompanied by several micronucleoli and connecting eosinophilic strands.

Prometaphase I. In connection with studies on the radiosensitivities of oocytes at the time of estrus, Kuhlmann has defined and described several stages which augment the classic first and second metaphase, anaphase, and telophase stages. After dictyotene the first stage in the continuation of the meiotic cycle is prometaphase I, characterized by the disappearance of the large and small nucleoli, the dissolution of the nuclear membrane, the formation of chromatin mass I, and the formation of the first meiotic spindle. Figure 7-27*A* illustrates the formation of "chromatin mass I"

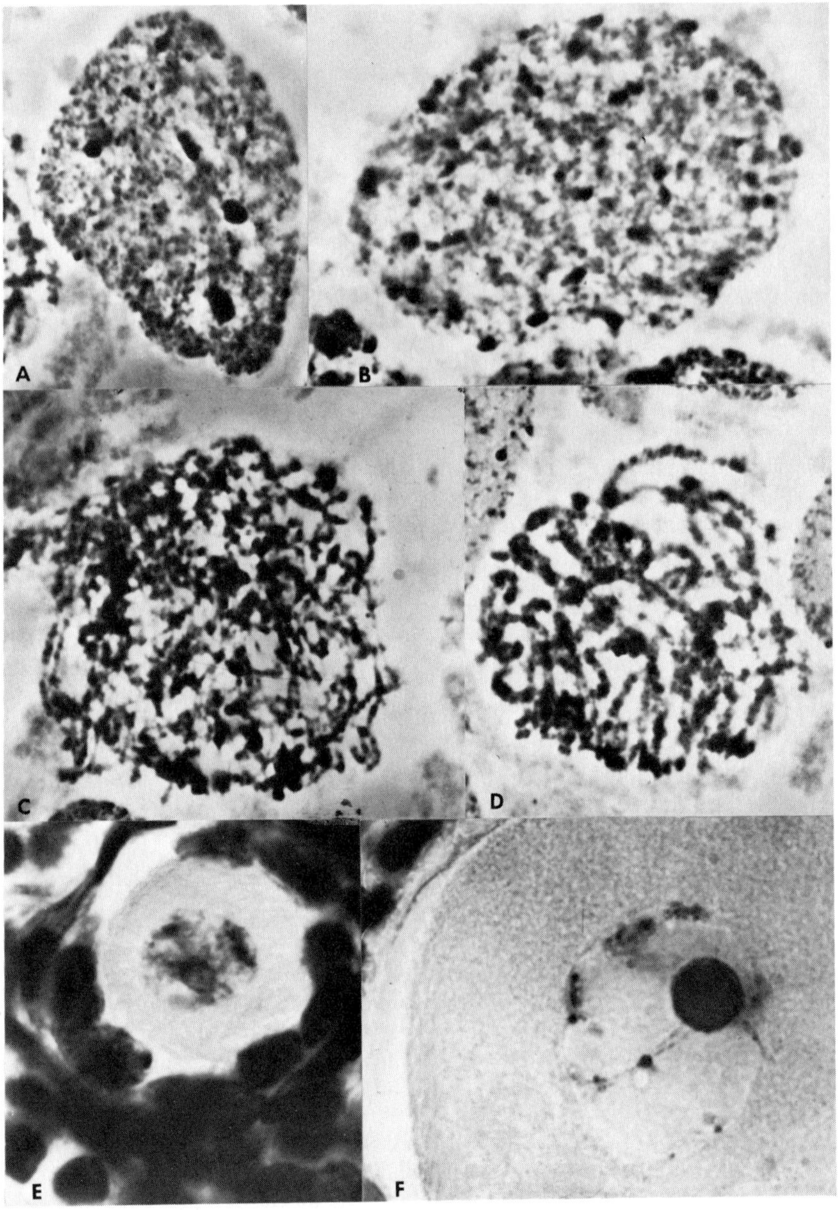

Fig. 7-26. Early meiotic stages in the female mouse. *A*, *B*, *C*, *D*. Sudan Black squash preparation from a 15-day fetus (×3000); *E*, *F*. Sections of adult ovary stained with hematoxylin and eosin (×1000). (*Courtesy of Dr. W. Kuhlmann.*)

A. Interphase nucleus of a primary oocyte; *B*, leptotene nucleus of a primary oocyte. Paired heterochromatic centromeric regions indicate the beginning of synapsis; *C*, zygotene nucleus of a primary oocyte. Polarization of centromeric ends and of free ends at lower periphery to form doubly polarized "bouquet," synapsis well advanced; *D*, pachytene nucleus of a primary oocyte. Synapsis complete, polarization distinct, with free ends grouped at lower left and centromeric ends grouped at lower right; *E*, dictyotene nucleus in a primordial follicle. Conspicuous nucleolus and small heterochromatic bodies; *F*, late dictyotene nucleus of an oocyte in a graafian follicle. Nucleolus and eosinophilic strands.

and the weakening of the nuclear membrane. Figure 7-27*B* shows Kuhlmann's "transitional phase I" in which discrete chromosome bodies (tetrads) emerge from the irregular mass. Figure 7-27*C* is Kuhlmann's "forming spindle I," with the appearance of the spindle fibers and visible stretching of the bivalents.

Metaphase I. In Fig. 7-27*D* the bivalents are

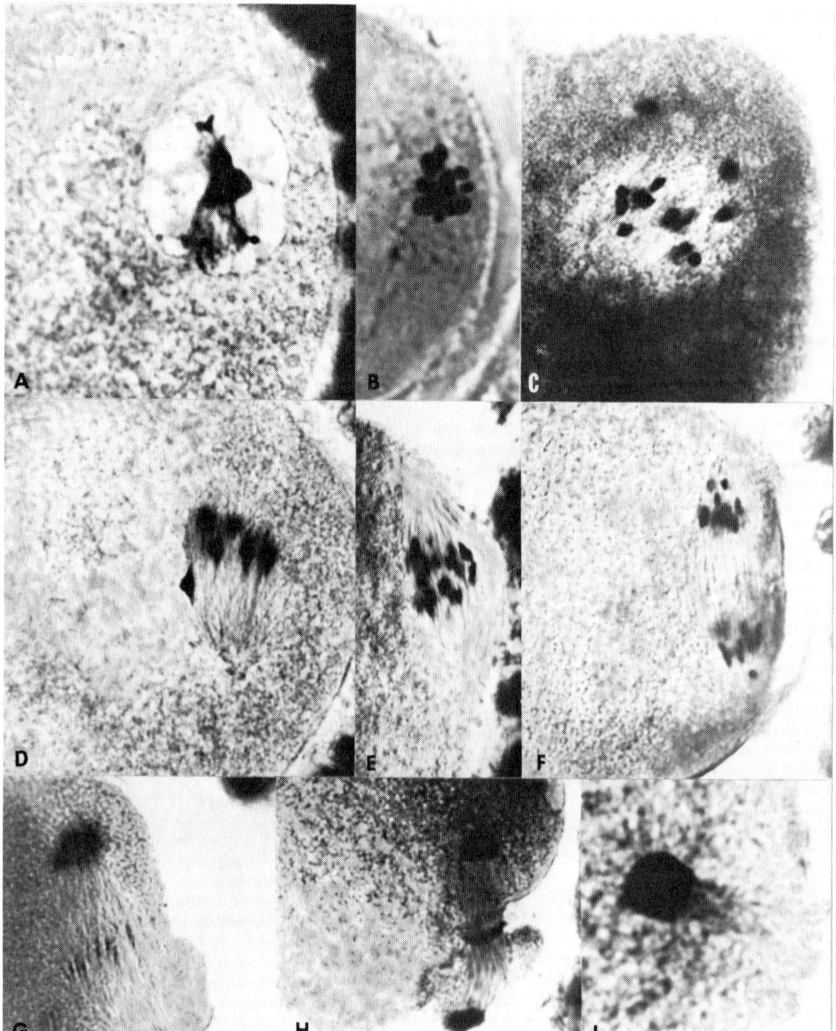

Fig. 7-27. Midmeiotic stages in the female mouse. All from adult ovaries, sectioned and stained with hematoxylin and eosin (×900; *H* ×600). (*Courtesy of Dr. W. Kuhlmann.*)

A. "Chromatin mass I" showing aggregation of chromatic material (condensation of chromosomes) and weakening of nuclear membrane; *B*, "transitional phase I" showing absence of nuclear membrane and relaxation of the chromatin mass into discrete bivalents; *C*, "forming spindle I" with bivalents becoming oriented among the fibers; *D*, metaphase I, side view. Central bulges in the bivalents indicate terminalizing chiasmata; *E*, early anaphase I showing separation of tetrads into component dyads; *F*, late anaphase I. Thickening of spindle fibers at the equator to form rudiments of midbody; *G*, end of anaphase I. "Chromatin mass II" forming from aggregation of chromosomes at inner pole of spindle; polar body I forming at outer pole; thickening of midbody rudiments; rotation of spindle toward perpendicular position; *H*, telophase I. Compression of midbody by the constriction of the limiting membrane which separates polar body I from the secondary oocyte. "Chromatin mass II" at the inner pole of the spindle; *I*, "chromatin mass II." The cup-shaped aggregate of dyads marks the first stage of the secondary oocyte.

becoming oriented in the equatorial plane of metaphase I. The bulging centers of the bivalents indicate the presence of chiasmata in the process of terminalization.

Anaphase I. At early anaphase I (Fig. 7-27*E*) terminalization is completed and the released homologous dyads separate toward the poles. At midanaphase I (Fig. 7-27*F*) the dyads approach the poles preparatory to the formation of "chromatin mass II" and the first polar body. At the equator the spindle fibers are thickened and present the first indication of the midbody which is

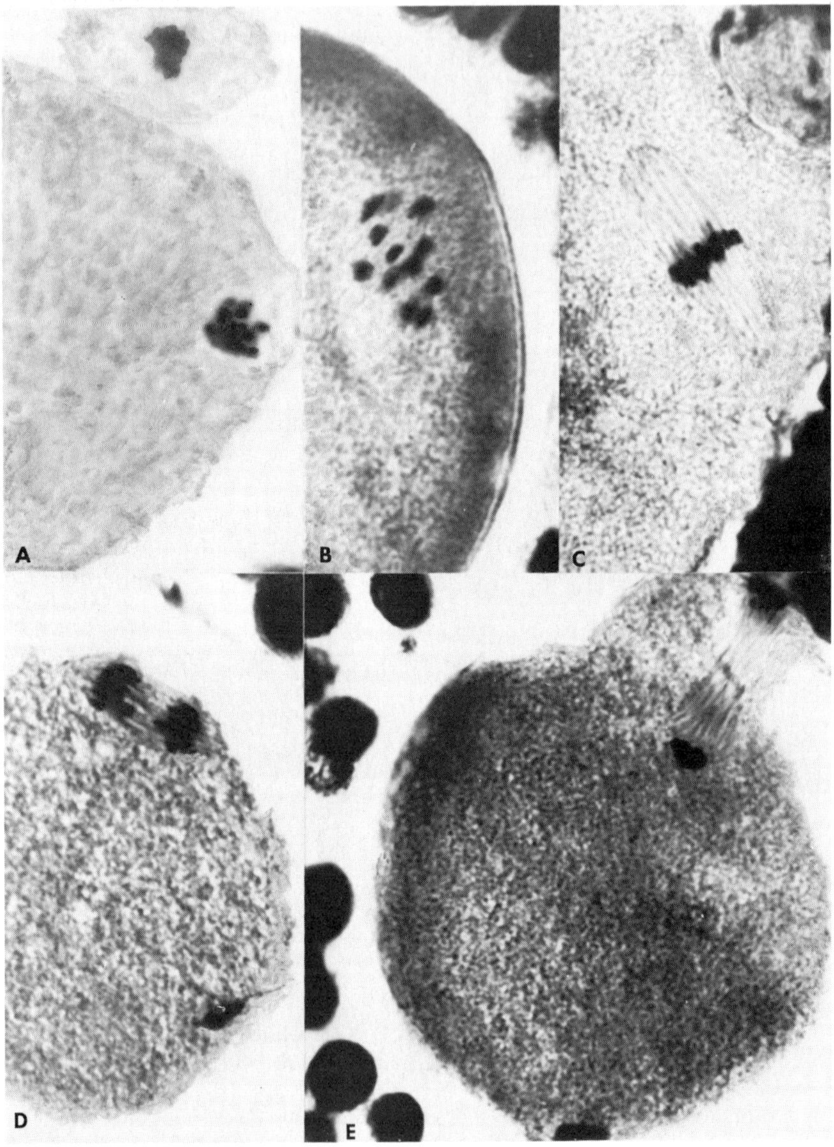

Fig. 7-28. Late meiotic stages in the female mouse. All from adult ovaries, sectioned and stained with hematoxylin and eosin (×600). (*Courtesy of Dr. W. Kuhlmann.*)

A, "Transitional phase II." Dyads of "chromatin mass II" become separated and rounded; *B*, "forming spindle II." Orientation of dyads toward equator of second meiotic spindle; *C*, metaphase II. Orderly arrangement of dyads on the spindle; *D*, anaphase II. Spindle operation proceeding after entry of sperm; sperm head and portion of flagellum at lower right periphery of oocyte; *E*, telophase II. Delimitation of polar body II; second midbody formed at equator of spindle; sperm head at lower periphery of ovum.

a prominent feature of the first meiotic telophase. In Fig. 7-27*G* late anaphase I shows "chromatin mass II," the continued thickening of the spindle elements at the equator, the beginning of polar body I, and the rotation of the spindle from the tangential toward a vertical position with reference to the surface of the oocyte.

Telophase I. The first oocytic telophase is shown in Fig. 7-27*H*. "Chromatin mass II" is highly condensed, the midbody is condensed and is compressed by the constricting cell membranes, and polar body I is becoming well defined. The midbody is destined to be excluded from each cell; this structure constitutes the material which Ohno (1963) described as "non-DNA materials" accumulated during the long dictyotene growth phase and extruded by the condensed chromosomes at the equatorial plate.

Prometaphase II. Kuhlmann's prometaphase II begins with "chromatin mass II," a cup-shaped chromosome aggregation which formed at the inner pole of the first meiotic spindle (Fig. 7-27*I*). No spindle fibers are visible. In "transitional phase II" the dyads of the mass become discrete and somewhat rounded (Fig. 7-28*A*). At "forming spindle II" spindle fibers appear and the dyads become oriented toward the equatorial region (Fig. 7-28*B*).

Metaphase II. The second oocytic metaphase (Fig. 7-28*C*) shows uniform equatorial arrangement of the dyads on the second spindle, in contrast to the irregular or staggered position of the tetrads on the first spindle. Metaphase II persists until sperm entry after ovulation.

Anaphase II. At sperm entry anaphase II begins and proceeds rapidly. Figure 7-28*D* shows late anaphase II and, at the bottom of the photograph, the head of the fertilizing spermatozoon.

Telophase II. At the second meiotic telophase a midbody is again formed at the equator of the spindle and is extruded. At the inner pole the definitive monads of the ovum form a dense aggregate in which individual chromosomes are not distinguishable. The second polar body contains a similar chromatin mass (Fig. 7-28*E*).

CHROMOSOME MECHANICS

Coiling

The condensation of chromosomes through coiling occurs in the prophase of both mitosis and meiosis. In the mouse, as in other animals, three types of coiling are postulated: molecular, primary (or minor), and secondary (or major). All are best exemplified in the meiotic prophase because of the prolongation of this stage and the large size of the nuclei.

Molecular coiling. The initial step in the coiling sequences is classically specified as coiling at the molecular level, providing for the first condensation of individual interphase chromosomes which thereby become visible for the first time. The spermatocytic interphase ordinarily shows prominent chromomeres whose serial alignment can be inferred from the short stretches of such alignment which are visible at many points as the observer focuses upon successive levels through the interphase nucleus. As prophase begins, the chromomeres come closer together in serial fashion and are seen to be constituents of a stainable thread. At full leptotene the individuality of the chromomeric chromosomes is easily observed, yet no visible indication of coiling is detected. Accordingly, molecular coiling is presumed to have been in operation.

At zygotene the chromomeric homologous chromosomes begin the process of pairing, with further condensation and thickening of the involved homologues. Since there is still no visible coiling, this condensation represents a continuation of the initial type of condensation. Figure 7-18 illustrates the spermatocytic interphase, leptotene, and zygotene in squash preparations.

Primary coiling. At diplotene the first indications of primary or minor coiling are seen in the linear contraction of each of the two homologous dyads. Each dyad appears to be thickened and to have a finely serrated outline which, with proper optical resolution, is seen to result from the presence of a fine coil. The coiled state is best observed adjacent to the chiasmata which hold the homologous dyads together; at these points of cross-connection the tension causes stretching of the coils so that their gyres are quite distinct for short distances.

Secondary coiling. At diakinesis the secondary or major coil appears, consisting of gross gyres superimposed upon the primary coil. The major coil drastically shortens the dyads and brings them eventually to their highly condensed metaphase condition. Major coiling is especially clear in certain somatic prometaphases such as those of Fig. 7-9 and 7-11. It is also visible at diakenesis (Fig. 7-22) and in the prometaphase of secondary spermatocytes (Fig. 7-29).

Relational coiling. The loose twining of one chromatid about another, the two being held together at the centromere positions, constitutes relational coiling. This type of spiralization is not readily demonstrable in the mouse, although it is

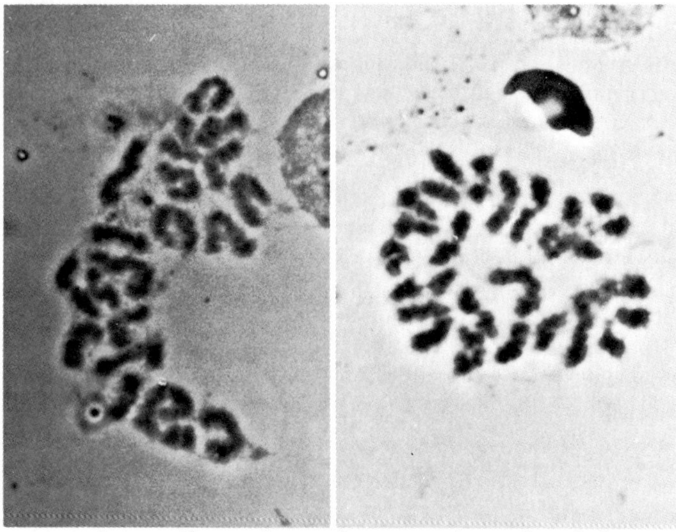

Fig. 7-29. Prometaphases in secondary spermatocytes showing major coiling and the presence of short arms in several dyads. Sudan Black squash (×1600).

suggested in well-spread squash preparations of the spermatocytic diplotene.

Inasmuch as the phenomenon of coiling is the subject of a large portion of nuclear cytology, discussions such as that of Swanson (1957) should be consulted for details and interpretations not obtainable from studies on the mouse.

Synapsis

In the mouse the synapsis of homologous chromosomes at zygotene through pachytene reveals no unusual phenomena. Crew and Koller (1932) have emphasized the nonrandom nature of synapsis, indicating that it regularly begins at the ends of leptotene threads and proceeds along the chromosomes until the pairing is complete. The starting points are presumed to be the centromere regions of the threads.

Somatic pairing

The phenomenon of somatic pairing or somatic association of homologous chromosomes is a classic feature of somatic nuclear cytology in organisms such as the *Drosophila* species. In the mouse there are strong, but not compelling, indications of extensive associations of homologues. The marked chromosome pairs of Fig. 7-9 illustrate this presumed association, found too frequently to be ac-

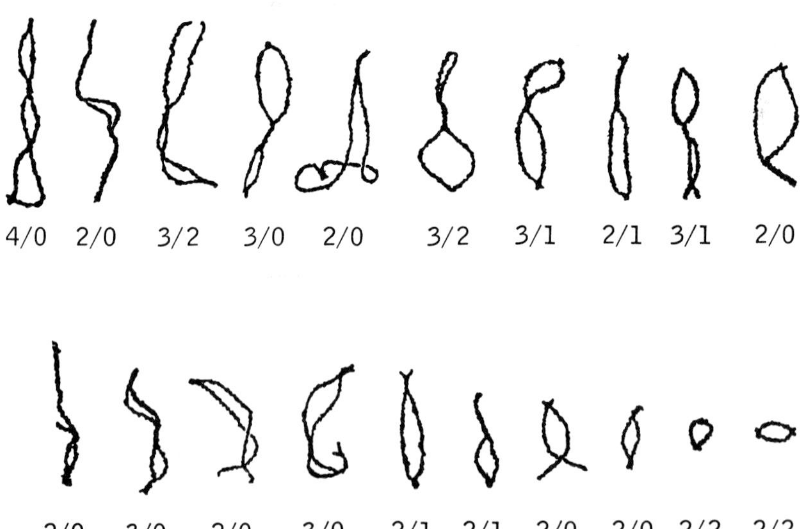

Fig. 7-30. Complete diplotene chromosome complement of a male mouse, showing chiasmata. Numerals indicate the total/terminal chiasmata. (*From Crew and Koller,* 1932.)

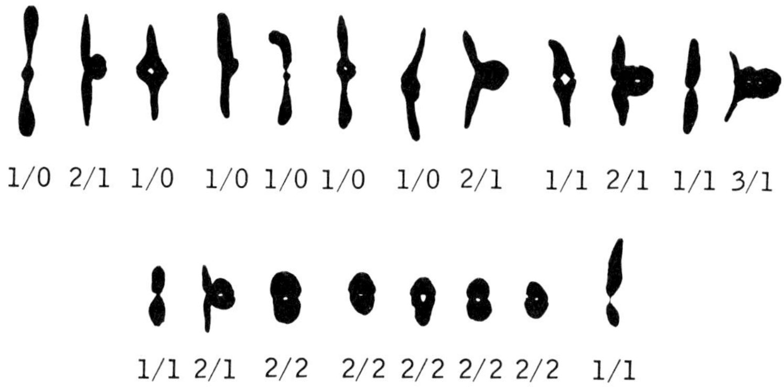

Fig. 7-31. Complete metaphase I chromosome complement of an older male mouse. Numerals indicate the total/terminal chiasmata. The XY bivalent is the last bivalent in the second row. (*From Crew and Koller,* 1932).

credited to chance. In squash preparations of testicular teratomas of the mouse Stevens and Bunker (1964) frequently observed pairs of chromosomes similar in size and with correspondingly curved chromatids. Gropp and Odunjo (1963) reported the same phenomenon in metaphases of cultured human leukocytes. Evidence clearly exists, therefore, that somatic association in mammalian cells may be a chromosomal feature with hitherto unexpected prevalence. An especially interesting aspect of the associations is the concordance of chromatid configurations, curvatures, and positions. There can be little doubt that somatic pairing in the mouse is far more extensive than is revealed in squash preparations.

Chiasma formation

Nature of chiasmata. At the diplotene stage in meiosis, chiasmata are the points of exchange between homologous (nonsister) chromatids and represent the physical evidence that genetical crossovers have occurred (Fig. 7-29). In the mouse the classic study of chiasmata was made by Crew and Koller (1932) who showed that, although diakinesis and metaphase configurations in the mouse have fewer chiasmata than can be counted at diplotene, the decrease in number and alteration of position are brought about by terminalizations of chiasmata. Therefore, at diakinesis and metaphase, chiasma counts which include terminalizations provide a close approximation of genetically determined crossover frequencies in both males and females. Figure 7-30 illustrates Crew and Koller's drawings of the complete diplotene chromosome complement of a male. Below each bivalent are numerals indicating the total number of chiasmata and the number of terminal chiasmata.

Frequency in males. Figure 7-31 is a drawing of the complete male metaphase I complement as presented by Crew and Koller. Numerals indicate total/terminalized chiasmata as in Fig. 7-30. The XY bivalent is the last bivalent in the second row. Comparison of the metaphase and diplotene drawings exemplifies the reduction in chiasma count at metaphase as compared with diplotene. Table 7-1 indicates the chiasma frequency as determined from five such metaphase cells in a

Table 7-1. CHIASMA FREQUENCY IN ADULT MALE CELLS AT METAPHASE.
(*After Crew and Koller, 1932*)

Number of cell	*Number of chiasmata*				Total number of chiasmata	Terminal chiasmata	Mean number of chiasmata per bivalent
	1	2	3	4			
1	7	12	1	0	34	20	1.7
2	11	9	0	0	29	21	1.4
3	10	9	1	0	31	19	1.5
4	13	7	0	0	27	20	1.3
5	14	6	0	0	26	20	1.3
Total	55	43	2	0	147	100	1.47

Table 7-2. CHIASMA FREQUENCY IN YOUNG MALE CELLS (6 WEEKS OLD) AT METAPHASE. (*After Crew and Koller, 1932*)

Number of cell	*Number of chiasmata* 1	2	3	4	Total number of chiasmata	Terminal chiasmata	Mean number of chiasmata per bivalent
1	9	13	1	0	35	21	1.7
2	6	10	1	0	32	22	1.6
3	8	9	2	1	36	20	1.8
4	11	8	1	0	30	19	1.5
5	7	11	2	0	35	21	1.7
6	12	7	1	0	29	19	1.4
Total	53	58	8	1	197	122	1.64

fully mature male. Table 7-2 shows the higher chiasma frequency as determined at metaphase in six cells from a male 6 weeks old. These tables exemplify the decrease in chiasma frequency which occurs as age increases, a finding which is comparable with Dunn's (1920) demonstration of a decrease in crossing over as age increases. Fisher (1949) found a similar decrease in recombination in the agouti-undulated interval of the fifth linkage group in both sexes.

Crew and Koller presented the graph shown in Fig. 7-32 as a comparison of their determinations of the number of chiasmata per bivalent in males at metaphase with information obtained in studies by Cox (1926) and Painter (1927). They attributed Painter's lower chiasma frequency to the possible influence of a major chromosomal deficiency which was present in his material, the v/O of Gates.

Table 7-3 compares Crew and Koller's metaphase chiasma frequency for the male with the diakinesis chiasma frequencies found by Huskins and Hearne (1936) and by Slizynski (1955*a*). Since diakinesis and metaphase I are closely related in form and time, the comparisons are quite valid and compatible. Crew and Koller's value for a male 6 months old corresponds to Huskins and Hearne's and to Slizynski's values for CBA males. Crew and Koller's "old" male value corresponds to that found in males other than CBA by Huskins and Hearne and to that found in unspecified hybrids by Slizynski.

Frequency in females. Figure 7-33 shows Crew and Koller's drawing of a complete female metaphase I complement along with numerical indicators of total/terminalized chiasmata. The XX bivalent is not identifiable. Table 7-4 indicates the chiasma frequency in five metaphases from a

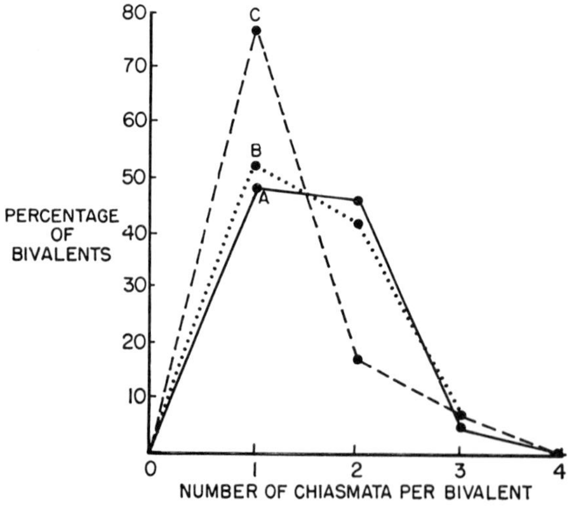

Fig. 7-32. Comparison of chiasma frequency data as obtained by Crew and Koller (A), Cox (B), and Painter (C). (*After Crew and Koller,* 1932.)

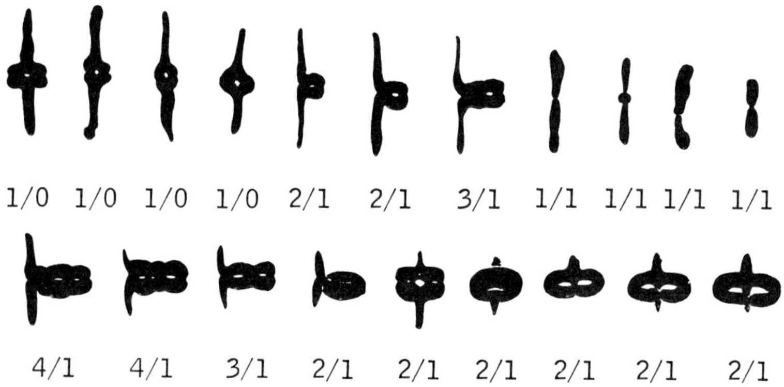

Fig. 7-33. Complete metaphase I chromosome complement of a female mouse. Numerals indicate the total/terminal chiasmata. (*From Crew and Koller,* 1932.)

Table 7-3. COMPARISON OF CHIASMA FREQUENCIES IN MALE MOUSE CELLS AT DIAKINESIS AND METAPHASE

Spermatocyte stage	Strain	Mean number of chiasmata per bivalent	Reference
Metaphase I	(?) "Old"	1.47	Crew and Koller
	(?) "Young"	1.64	(1932)
Diakinesis	CBA	1.61	Huskins and
	Non-CBA	1.43	Hearne (1936)
Diakinesis	CBA	1.58	Slizynski (1955*a*)
	Unspecified hybrids	1.52	

female which was a littermate of the male of Table 7-1.

With these studies Crew and Koller proved a difference in the sex incidence of metaphase chiasma frequency corresponding with the established sex difference in crossover frequency (Chap. 8).

Chiasmata and chromosome map length

Slizynski (1955*a*) counted the chiasmata in 50 diplotene and 56 diakinesis nuclei from male mice of the CBA strain and from hybrid males of unspecified origin. In comparing his diplotene data with those which Crew and Koller obtained from a single cell, Slizynski found that his counts indicated a total map length of 1,916 centi-Morgans, while Crew and Koller's single-cell count indicated 2,450 cM. By combining his diplotene data for CBA and hybrid males, Slizynski found the total map length of the male to be 1,954 cM. Average lengths for short, medium, and long chromosomes were 76, 106, and 127 cM, respectively. Cytologically based calculations for the female have not been made.

Chiasma positions

Most of the mouse metaphase configurations are readily interpreted on the basis of the partial chiasmatype theory and terminalizations. Yet there is a class of configurations requiring clarification: namely, the small ring or doughnut-shaped metaphase bivalents such as those in the second row

Table 7-4. CHIASMA FREQUENCY IN FEMALE MOUSE CELLS AT METAPHASE. (*After Crew and Koller, 1932*)

Number of cell	*Number of chiasmata* 1	2	3	4	Total number of chiasmata	Terminal chiasmata	Mean number of chiasmata per bivalent
1	4	13	1	2	41	14	2.0
2	8	8	2	2	38	16	1.9
3	7	10	2	1	37	19	1.8
4	6	11	1	2	39	17	1.9
5	4	12	2	2	42	20	2.1
Total	29	54	8	9	197	86	1.97

of Fig. 7-31. Crew and Koller (1932) considered these to be produced by the terminalization of two chiasmata. Since the centromere is nearly terminal in all mouse chromosomes, this interpretation requires the assumption that chiasmata form in the extremely minute arms of the chromosomes with remarkably high frequency. An alternative explanation requires that one of the end associations be based upon failure of the homologous centromeres to separate at diakinesis with an even more remarkable high frequency. The first explanation appears to be preferable but must await confirmation.

SEX DETERMINATION

According to the genic-balance principle of sex determination, long accepted for *Drosophila* (Bridges, 1922, 1932), autosomal factors for maleness are balanced against the female-determining capacities of the X chromosome. This system of sex determination came to be accepted for higher animals, including mammals, in which cytogenetic experimentation had not become feasible. The first indication that mammalian sex determination was not of the *Drosophila* type came with the studies by Welshons and Russell (1959) and by Russell et al. (1959). In studying the aberrant features of the inheritance of the sex-linked gene scurfy, these investigators discovered that mice of the XO constitution were female and that the Y chromosome bears factors which determine maleness. Cattanach (1961*a*) found that XXY mice are males and thus firmly established the Y chromosome as the male determiner.

Table 7-5. SEX DETERMINATION IN DROSOPHILA, MELANDRIUM, AND THE MOUSE

Chromosome constitution[a]	*Drosophila*	*Melandrium*	Mouse
XX : 2A	Female	Female	Female
XY : 2A	Male	Male	Male
XO : 2A	Male		Female
XXY : 2A	Female		Male
XXX : 2A	Female		
XXX : 3A	Female	Female	
XXY : 3A	Intersex	Male	
XX : 3A	Intersex	Male	
XO : 3A	Male		
XXY : 4A		Male	
XXXY : 4A		Male	
XXXXY : 4A		Hermaphrodite	

[a] A = one complete haploid complement of autosomes.

A similar system having maleness determined by the Y has long been known for a relatively obscure organism, the dioecious plant *Melandrium dioicum.* Wamke and Blakeslee (1940) showed that Y-bearing plants were male in all cases except those in which a great preponderance of X's existed; even in the XXXXY constitution the maleness factors are expressed, this combination being hermaphrodite. Comparisons of the *Drosophila, Melandrium,* and mouse situations are given in Table 7-5, where each letter A represents a complete haploid complement of autosomes. The eventual finding and cytological verification of sex modifications in polyploid mice can provide further information about the strength of the mouse Y, as well as any possible role of the autosomes, in determining maleness.

CHROMOSOME BREAKAGE AND ANEUPLOIDY

Radiation cytogenetics in the mouse began with Painter's (1928) cytological study of Little and Bagg's (1924) abnormal-eyed mice, discovered in the progeny of animals exposed to X-rays. No abnormal chromosomal conditions were revealed, the diploid number of 40 being present in cells of each of four males studied. No aberrant features were found in the study of separating dyads on the first meiotic spindle, employing the exceedingly difficult procedures previously successful in disclosing for the first time the phenomenon of deletion in Gates' "nondisjunction" *v*/O mice.

After Muller's discovery of the production of mutation in *Drosophila* by X-rays, several investigators accumulated data indicative of chromosome abnormalities in the offspring of irradiated mice. Russell (1962) summarizes most of the studies and provides extensive references. Snell (1932, 1933*a, b, c*) first reported the occurrence of semisterility in the offspring of irradiated males and attributed this effect to the presence of chromosome translocations. Snell et al. (1934), Snell (1935), and Snell and Picken (1935) related the production of abnormal embryos by the semisterile offspring of irradiated mice to the chromosomal imbalance resulting from the several possibilities of segregation in translocation heterozygotes. Figure 7-34 illustrates the production of aneuploid gametes by a generalized translocation heterozygote. This diagram postulates a slightly subterminal position for the centromeres and in-

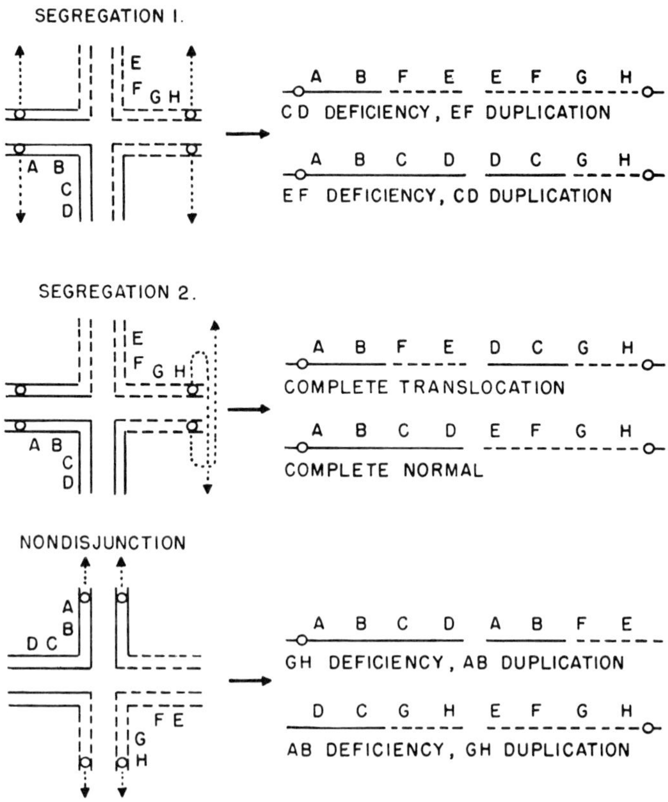

Fig. 7-34. Reciprocal translocation configuration at diplotene, showing the types of gametes produced by several possible types of separation at anaphase I.

cludes the classes of gametes producible through nondisjunction of homologous centromeres.

Koller and Auerbach (1941) conclusively demonstrated that semisterility in the offspring of irradiated mice was caused by reciprocal translocations readily detected in primary spermatocytes of semisterile males. Koller (1944) further explained the mechanical bases for semisterility, indicating that for any particular translocation the degree of fertility is correlated with the frequency of nondisjunctional arrangement of members of the translocation double tetrad or ring-of-four. Slizynski (1949) prepared the first pachytene chromosome map for the mouse and employed it with success (1952) in making a cytological analysis of Snell's T(5:8)*a* translocation at pachytene. He extended the pachytene studies (1957) in the analysis of several translocations produced by Carter et al. (1955). Jaffe (1952) reported segmental interchanges in primary spermatocytes of semisterile males derived from certain of Dunn's tailless series but he did not find evidence for inversions in any of these lines. Griffen (1958, 1964) found translocations, supernumerary fragments, trisomy, and asynapsis to be common aberrations in the offspring of irradiated males. Deletions and inversions were not detected, since pachytene nuclei were not studied.

Table 7-6 lists the gene-tagged translocations and other heritable chromosome abnormalities thought to be available in existing stocks in 1965.

Two-break aberrations induced in haploid cells

Translocations. The cytological investigations thus far mentioned have dealt principally with segmental interchanges, which are often readily visible as rings-of-four or chains-of-four in the first meiotic division of spermatogenesis and are readily detected because of the reduced number of tetrads at diakinesis and at metaphase I. Whereas the normal number of tetrads at metaphase I is 20, the presence of one translocation involving two chromosomes reduces the count to 19 through the formation of a ring or chain "double tetrad." Two independent translocations, each resulting from two breaks and each involving two different chromosomes, reduces the count to 18, as in cases

Table 7-6. TRANSLOCATIONS AND OTHER HERITABLE CHROMOSOMAL ABNORMALITIES

Name	Synonyms and remarks	References
Reciprocal translocations		
T(5;8)Sn	T(5;8)*a*	Snell (1946), Slizynski (1952, 1957)
T(5;8)1Ca		Carter et al. (1955, 1956), Slizynski (1957)
T(5;13)5Ca		Carter et al. (1955, 1956), Slizynski (1957)
T(3;14)6Ca		Carter et al. (1955, 1956), Slizynski (1957), Green (1965)
T(5;11)7Ca		Carter et al. (1955, 1956), Slizynski (1957)
T(2;9)138Ca		Carter et al. (1955, 1956), Slizynski (1957)
T(9;13)190Ca		Carter et al. (1955, 1956), Slizynski (1957)
T(14;17)264Ca		Carter et al. (1955, 1956), Slizynski (1957), Phillips (1961), Lane (1965)
T(X;?)16H	TX2(Searle)	Ford and Evans (1964), Lyon et al. (1964)
T(8;X)1Rl	R1	Russell and Bangham (1961), Russell (1963)
T(1;X)2Rl	R2	Russell and Bangham (1961), Russell (1963)
T(1;X)3Rl	R3	Russell and Bangham (1961), Russell (1963)
T(1;X)4Rl	R4	Russell and Bangham (1961), Russell (1963)
T(1;X)5Rl	R5	Russell and Bangham (1961), Russell (1963)
T(1;X)6Rl	R6	Russell and Bangham (1961), Russell (1963)
T(1;X)7Rl	R7	Russell and Bangham (1961), Russell (1963)
Hydrops	Linkage not known. Offspring of aneuploid gametes may survive to birth or later	Hertwig (1955)
Transposition and deletion		
T(1;X)Cattanach	T(*fd*). Piece of chromosome I transposed to the X. Heterozygous transposition without the deletion is viable. Heterozygous deletion without the transposition is lethal	Cattanach (1961), Ohno and Cattanach (1962)
Deletions		
d se (many), L. G. II[a]	Viable in heterozygotes, diagnosed as deletions by genetic evidence only	Russell and Russell (1960)
t^0, L. G. IX	Viable in heterozygotes, diagnosed as deletion by cytological evidence	Geyer-Duszynska (1964)
t^{w6}, L. G. IX	Viable in heterozygotes, diagnosed as deletion by cytological evidence	Geyer-Duszynska (1964)
t^{w18}, L. G. IX	Viable in heterozygotes, diagnosed as deletion by cytological evidence	Geyer-Duszynska (1964)
Sex-chromosome monosomics		
X/O females	Occur frequently (ca. 1%) and can be detected in stocks with appropriate sex-linked markers. Fertile but produce about ⅓ as many X/O daughters as expected	Russell et al. (1959), Welshons and Russell (1959), Cattanach (1962)

[a] Linkage group.

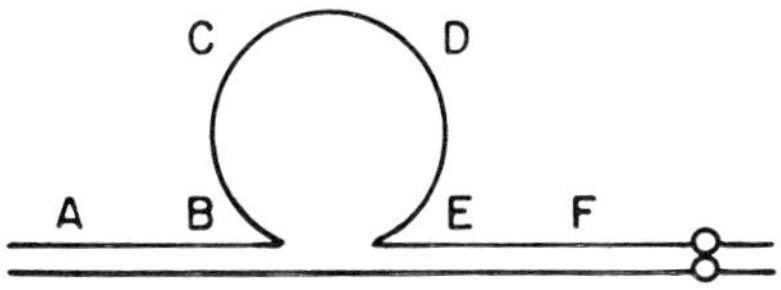

Fig. 7-35. Deficiency configuration at pachytene, showing loop formation in the normal homologue.

such as that of Falconer et al. (1952) and in those of Griffen (1964). As a general rule it may be stated that each translocation involving two chromosomes reduces by one the count of metaphase I bodies.

Inversions. Inversions have not been reported in the mouse, although they are among the most common of all spontaneous chromosome aberrations and also are of very frequent occurrence in experimental organisms exposed to ionizing radiations. We may presume they have been produced in mice in the course of numerous radiation experiments. Inversions are readily detected in only two ways: (1) through the use of breeding tests which may reveal drastic modifications of the recombination frequencies of linked markers and (2) through the discovery of loops between inverted and noninverted homologous chromosomes at meiotic synapsis, particularly at pachytene. Since pachytene is extremely difficult to study in mouse cells, it has not been feasible to search the pachytene stage of individuals sired by irradiated males for the presence of inversions. Partial sterility may indicate the possible presence of inversions, particularly those of greater length, since crossing over within the limits of an inversion may produce dicentric and acentric fragments whose loss results in zygotic death through hypoploidy.

Deficiencies. Deficiencies or deletions have been reported as probable in the offspring of irradiated mice by Russell and Russell (1960) and by Kidwell et al. (1961). Cytological verifications have not been made for these deficiencies, although the genetic evidence for their occurrence is unimpeachable. Eventual pachytene demonstrations of irradiation-induced deletions may be expected to resemble the deficiencies of spontaneous origin studied by Geyer-Duszynska (1964) in a cytological analysis of the *t* series of alleles in chromosome IX. Three *t* "alleles" (t^o, t^{w6}, and t^{w18}) were shown to be represented by small loop formation at pachytene. Figure 7-35 provides a generalized example of synapsed deficient and normal chromosomes at pachytene.

External duplications. External duplications are those portions of the genome carried as supernumerary free fragments, the remnants of more or less grossly deleted two-break chromosomes. Such free bodies are partial trisomics and will be discussed in a later section.

Three-break aberrations induced in haploid cells

Translocations. Translocations involving three chromosomes, with one break in each, form especially prominent rings-of-six at metaphase I of meiosis (Griffen, 1964).

Such aberrations are especially rare and may constitute the class of "progressive" translocations in which the attachments result in a 1-2, 2-3, 3-1 type of union (Fig. 7-36). Each aberration of this type reduces the count of metaphase I bodies by two. It should be noted that a ring-of-six and reduction of the count by two will also appear in the heterozygote formed by crossing two individuals, each carrying a single translocation and the two translocations having one chromosome in common. Thus the heterozygote for T(1:2) and T(2:3) shows a 1-2-3 ring-of-six and a count of 18 at metaphase I. Slizynski (1957*b*) illustrated such configurations in mice heterozygous for Carter, Lyon, and Phillips' translocations 2/281, 5/7, 2/5, and others.

Transpositions. Transpositions result when the interval between two breaks in one chromosome becomes inserted between the ends of a single break in a second chromosome (Fig. 7-37). Ohno and Cattanach (1962) illustrated such an aberration in the "flecked" mouse, showing that a segment of chromosome I bearing the wild-type

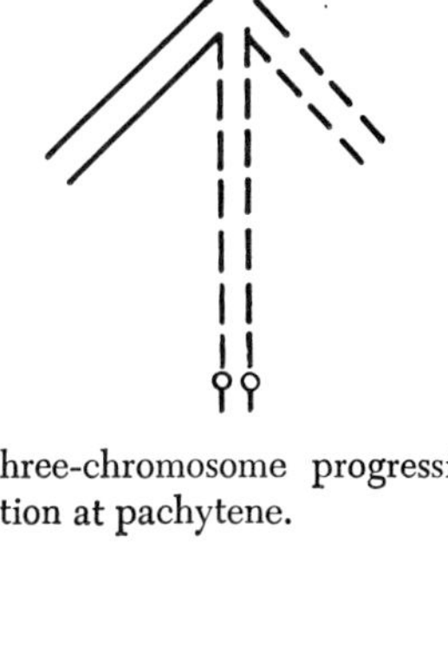

Fig. 7-36. Three-chromosome progressive translocation configuration at pachytene.

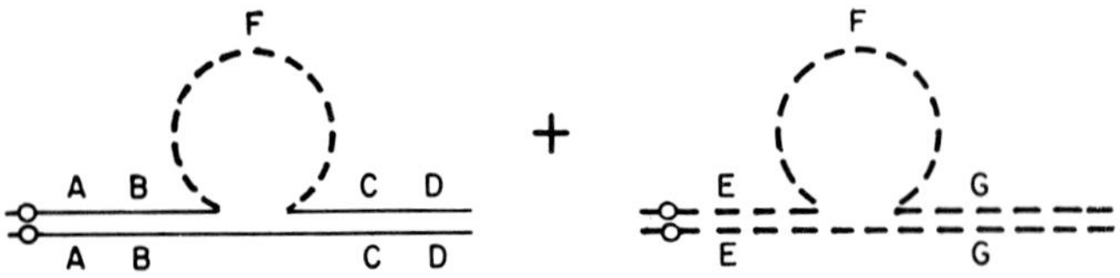

Fig. 7-37. Transposition type of translocation at pachytene.

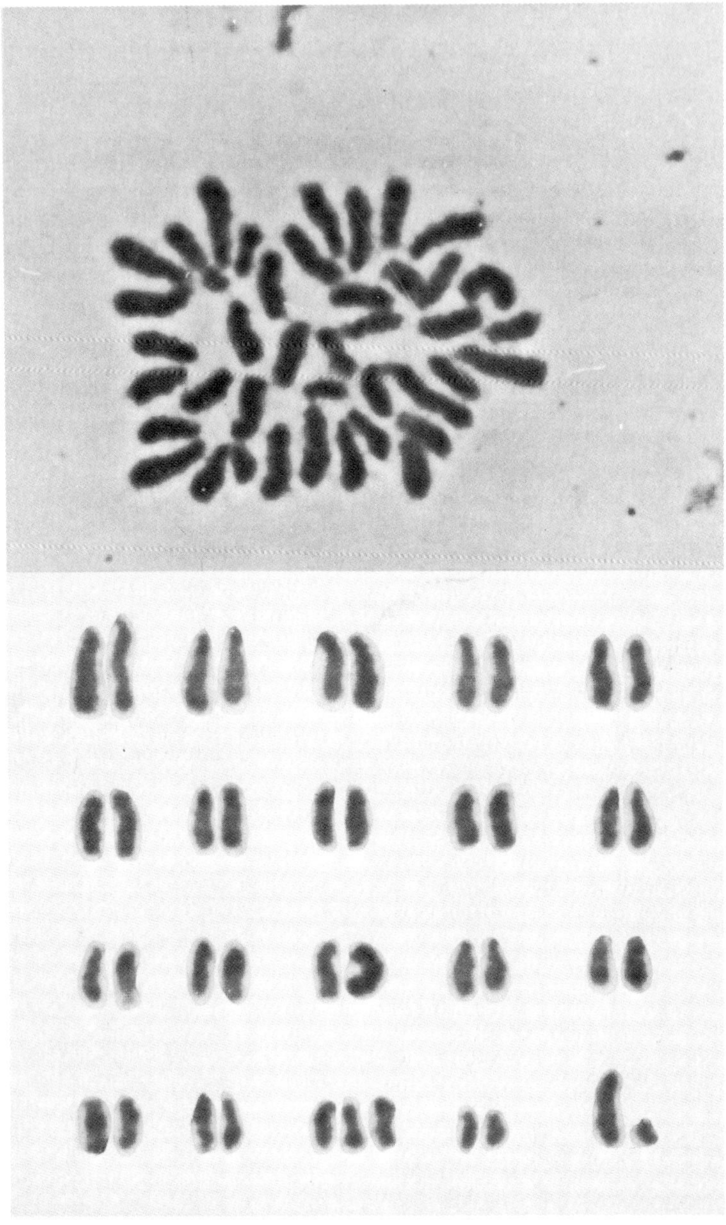

Fig. 7-38. Trisomic spermatogonial metaphase and its idiogram. Sudan Black squash (×3200).

alleles of genes *p* and *c* is inserted into the X chromosome. Even though this transposition was chemically induced (Cattanach, 1961*b*), it may nevertheless exemplify the similar instances of radiation-induced variegated aberrations reported by Russell and Bangham (1959, 1961) and by Russell et al. (1962). Cytological analyses of these cases have not been reported.

Segregational internal duplications. Internal duplications are those contained within some member of the regular chromosome complement and are derived from transpositions such as Cattanach's "flecked" transposition. Segregants having two entire chromosomes I plus an X carrying the inserted segment of I are viable and fertile in both sexes. (Duplications in Russell and Bangham's cases are lethal; legitimate inclusion of these cases must await cytological analyses.) Other types of internal duplication depend upon origination in diploid cells and are discussed in a following section.

Segregational deficiencies. Segregational deficiencies are the complements of segregational internal duplications. The "flecked" transposition, for example, readily segregates so as to produce a class of lethal zygotes presumed to include both males and females and to have one normal chromosome I, one deficient chromosome I, and two normal X's or a normal X and a normal Y.

Autosomal aneuploidy

Radiation-induced deviations from the normal haploid and diploid chromosome numbers usually may be defined as the gain or loss of a centromere, along with all or part of the portion of the genome normally transported by that centromore in cell division. Nondisjunction, the failure of centromeres to separate at the first meiotic division, is usually invoked in accounting for the production of aneuploidy.

Trisomy. Through nondisjunction, gametes may be formed which contain a single whole chromosome as a supernumerary element. Union with a normal haploid gamete results in a $2N + 1$ zygote in which one linkage group is present in a triplicate condition in an otherwise diploid genome. The zygote is said to be trisomic for the triply represented linkage group. Four cases of autosomal trisomy have been reported for the mouse, one by Cattanach (1964) and three by Griffen and Bunker (1964). Cattanach's trisomic was a phenotypically normal, but sterile, male which resulted from chemically-induced nondisjunction of one of the smallest autosomal pairs at spermatogenesis in the male parent. Griffen and Bunker's three trisomics occurred among the offspring of irradiated males, presumably as the result of nondisjunction at spermatogenesis. All were phenotypically normal, but two were sterile and one semisterile. In each case the trisomy involved members of the smaller chromosome classes. A trisomic spermatogonial metaphase and its idiogram are shown in Fig. 7-38.

Monosomy. Gametes lacking one complete member of the haploid chromosome set are produced by nondisjunction as the complements of gametes containing a supernumerary element. Union of a deficient with a normal gamete results in a $2N - 1$ zygote in which one linkage group is present in the haploid state in an otherwise diploid genome. The zygote is said to be monosomic for the singly represented linkage group. No case of viable autosomal monosomy has been reported for the mouse.

Partial trisomy. Supernumerary free fragments, also classifiable as external duplications, result from the gross deletion of single chromosomes whose small remains, carried by centromeres, may become added to an otherwise normal diploid genotype through nondisjunction. Partial trisomics are of especially common occurrence in the offspring of irradiated mice (Griffen, 1964). Partial trisomy may result also from reciprocal translocations whose reunion products consist of a very large plus a very small element. If the small element becomes a supernumerary body through nondisjunction, partial trisomy exists for a minute portion of each of two linkage groups rather than for a single linkage group. Partial trisomics cannot be classified as to origin unless, in the translocation type, the original intact segmental interchange has been recognized in precursors of the trisomic individuals. A spermatogonial metaphase showing a partial trisomic is illustrated in Fig. 7-39.

Aneuploidy of the sex chromosomes

The X and Y chromosomes are the only elements of the mouse genome which, in the monosomic condition, are known to permit survival. The normal XY male is, by definition, monosomic for both the X and the Y. It is known that normal XX females, through the X-inactivation phenomenon, are *functionally* monosomic even though cytologically *disomic*. True cytological X-monosomy has been found by Russell et al. (1959), Welshons and Russell (1959), McLaren (1960), Cattanach (1961*a*), and Kindred (1961) who have described

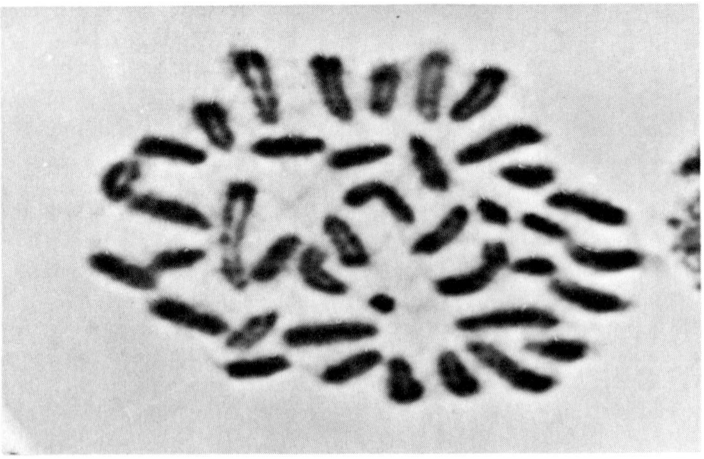

Fig. 7-39. Partial trisomy in a spermatogonial metaphase. Sudan Black squash (×3200).

viable XO females, some of which were derived from irradiated sperm whose X or Y had been destroyed.

Hyperploidy for the sex chromosomes in the mouse is represented by sterile XXY males such as found by Russell and Chu (1961) and by Kindred (1961). XXY individuals cannot be considered as trisomics, since the X and Y cannot be equated cytologically or physiologically. True trisomic XXX female mice have not been reported; such individuals, if comparable to multiple-X human beings, would be functionally monosomic through inactivation of two of the three X's, although cytologically trisomic. Diagnostically useful sex chromatin bodies, such as those of humans and cats, have not been found in somatic cells of the mouse.

Chromosome breakage sensitivities in germinal cells

A comprehensive review of the radiosensitivity of germ cells by Mandl (1964) and a review of chromosome aberrations in experimental animals by Russell (1962) indicate the gross lack of precise information on cytologically demonstrable chromosome aberrations which are produced in germinal cells of precise morphological types and at definite mitotic and meiotic phases. Although the investigations of Oakberg and his colleagues (summarized by Oakberg and Clark, 1964) have made possible the accurate timing and identification of spermatogenic stages in the mouse, only sensitivity determinations have been made, as determined by cell survival. Extensive information on the production of aberrations in irradiated spermatogonia has been obtained by Griffen (1964), who found translocations, fragments, trisomics, and asynaptics among partially fertile individuals derived from treated gonia.

SUMMARY

The nuclear cytology of the mouse generally shows chromosomal features and activities conforming to the classic patterns for both animal and plant cytology. Deviations are seen in the prophase of oogenesis, presumably associated with the long stage of nuclear arrest and cytoplasmic growth.

Chromosome identifications are not feasible in gonial cells, nor in meiocytes except for the XY bivalent in primary spermatocytes. Identifications are possible in prometaphases of somatic cells grown in vivo and in vitro and in prophases of certain large tumor cells. Prophase somatic chromosomes can be identified to some extent with elements of the pachytene chromosome maps.

Allocycly of the X and Y chromosomes permits the location of these elements in somatic and tumor prophases. However, in female cells no sex chromatin body is diagnostically visible as it is in certain other mammals.

Evidence of somatic pairing in somatic cells is strongly but not conclusively indicated, along with an unexplained concordance of position, curvature, and other attributes of the chromatids of homologous chromosomes at prometaphase and metaphase.

Chiasma frequencies at diplotene, diakinesis, and metaphase can be correlated with crossover frequencies which are genetically detected. Terminalizations of chiasmata cause understandable discrepancies between the two phenomena. Like crossover frequencies, chiasma frequencies are lower in males than in females; no cytological explanation for this difference is available.

Sex determination in the mouse resembles the system in *Melandrium,* the Y chromosome being male-determining. There appears to be no male-determining role of the autosomes such as exists in the genic balance system of *Drosophila* and certain other organisms.

Chromosome aberrations in the mouse, principally induced by X-rays, include translocations, transpositions, deficiencies, and segregational internal duplications. Autosomal monosomy has not been found. Trisomy and partial trisomy have been demonstrated for several of the smaller autosomes.

LITERATURE CITED

Borum, K. 1961. Oogenesis in the mouse. Exp. Cell Res. 24:495–507.

Bridges, C. B. 1922. The origin of variations in sexual and sex-limited characters. Amer. Natur. 56:51–63.

Bridges, C. B. 1932. The genetics of sex in *Drosophila*, p. 53–93. *In* Sex and internal secretion. Baillière, London.

Bunker, M. C. 1961. A technique for staining chromosomes of the mouse with Sudan Black B. Can. J. Genet. Cytol. 3:355–360.

Bunker, M. C. 1965. Chromosome preparations from solid tumors of the mouse: a direct method. Can. J. Genet. Cytol. 7:78–83.

Carter, T. C., M. F. Lyon, and R. J. S. Phillips. 1955. Gene-tagged chromosome translocations in eleven stocks of mice. J. Genet. 53:154–166.

Carter, T. C., M. F. Lyon, and R. J. S. Phillips. 1956. Further genetic studies of eleven translocations in the mouse. J. Genet. 54:462–473.

Cattanach, B. M. 1961*a*. XXY mice. Genet. Res. 2:156–158.

Cattanach, B. M. 1961*b*. A chemically-induced variegated-type position effect in the mouse. Z. Vererb. 92:165–182.

Cattanach, B. M. 1962. XO mice. Genet. Res. 3:487–490.

Cattanach, B. M. 1964. Autosomal trisomy in the mouse. Cytogenetics 3:159–166.

Cox, E. K. 1926. The chromosomes of the house mouse. J. Morphol. Physiol. 43:45–54.

Crew, F. A. E., and P. C. Koller. 1932. The sex incidence of chiasma frequency and genetical crossing-over in the mouse. J. Genet. 26:359–383.

Cutright, P. R. 1932. The spermatogenesis of the mouse (*Mus musculus* var. *albula*). J. Morphol. 51:197–219.

Dunn, L. C. 1920. Linkage in mice and rats. Genetics 5:325–343.

Falconer, D. S., B. M. Slizynski, and C. Auerbach. 1952. Genetical effects of nitrogen mustard in the house mouse. J. Genet. 51:81–88.

Fisher, R. A. 1949. A preliminary linkage test with *agouti* and *undulated* mice. Heredity 3:229–241.

Ford, C. E., and E. P. Evans. 1964. A reciprocal translocation in the mouse between the X chromosome and a short autosome. Cytogenetics 3:295–305.

Geyer-Duszynska, I. 1963. On the structure of the XY bivalent in *Mus musculus* L. Chromosoma 13:521–525.

Geyer-Duszynska, I. 1964. Cytological investigation on the T-locus in *Mus musculus* L. Chromosoma 15:478–502.

Green, M. C. 1965. Mouse News Letter 32:46.

Griffen, A. B. 1955. A late pachytene chromosome map of the male mouse. J. Morphol. 96:123–136.

Griffen, A. B. 1958. Occurrence of chromosomal aberrations in prespermatocytic cells of irradiated male mice. Proc. Nat. Acad. Sci. 44:691–694.

Griffen, A. B. 1960. Mammalian pachytene chromosome mapping and somatic chromosome identification. J. Cell. Comp. Physiol. 56 (Suppl. 1):113–121.

Griffen, A. B. 1964. The occurrence of chromosomal aberrations in prespermatocytic cells of irradiated male mice. II. Cytological studies of sterile and semi-sterile F_1 individuals, p. 175–188. *In* W. D. Carlson and F. X. Gassner [ed.] Effects of ionizing radiation in the reproduction system. Pergamon Press, New York.

Griffen, A. B., and M. C. Bunker. 1964. Three cases of trisomy in the mouse. Proc. Nat. Acad. Sci. 52:1194–1198.

Gropp, A., and F. Odunjo. 1963. Beobachtungen zur morphologischen Konkordanz homologer Chromosomen somatischer Zellen. Exp. Cell Res. 30:577–582.

Hertwig, P. 1935. Über sterilitätserscheinungen bei röntgenbestrahlten Mäusen und deren Nachkommenschaft. Z. Indukt. Abstamm. Vererb. 70:512–523.

Hertwig, P. 1938. Untershiede in der Entwicklungsfähigkeit von F_1-Mäusen nach Röntgenbestrahlung von Spermatogonien, fertigen und unfertigen Spermatozoen. Biol. Zentralbl. 58: 273–301.

Hertwig, P. 1940. Vererbbare Semisterilität bei Mäusen nach Röntgenbestrahlung verursacht durch reziproke Chromosomentranslokationen. Z. Indukt. Abstamm. Vererb. 79:1–29.

Hertwig, P. 1955. Der *Hydrops*-Stamm. Züchter 25:194–198.

Hsu, T. C., and C. M. Pomerat. 1953. Mammalian

chromosomes *in vitro*. II. A method for spreading the chromosomes of cells in tissue culture. J. Hered. 44:23–29.

Hughes, A. 1952. Some effects of abnormal tonicity on dividing cells in chick tissue cultures. Quart. J. Microscop. Sci. 93:207–219.

Huskins, C. L., and E. M. Hearne. 1936. Spermatocyte chiasma frequency in strains of mice differing in susceptibility or resistance to the spontaneous occurrence of malignant tumours. Can. J. Res. 14:39–58.

Jaffe, J. 1952. Cytological observations concerning inversion and translocation in the house mouse. Amer. Natur. 86:101–104.

Kidwell, J. F., D. G. Nash, J. Stadler, and J. W. Gowen. 1961. An X-ray induced deficiency in linkage group VI of the mouse. Amer. Zool. 1:365.

Kindred, B. M. 1961. Abnormal inheritance of the sex-linked Tabby gene. Austral. J. Biol. Sci. 14:415–418.

Koller, P. C. 1944. Segmental interchange in mice. Genetics 29:247–263.

Koller, P. C., and C. A. Auerbach. 1941. Chromosome breakage and sterility in the mouse. Nature 148:501–502.

Lane, P. W. 1965. Mouse News Letter 32:42.

Levan, A., T. C. Hsu, and H. F. Stich. 1962. The idiogram of the mouse. Hereditas 48:676–687.

Little, C. C., and H. J. Bagg. 1924. The occurrence of four inheritable morphological variations in the mouse and their possible relation to treatment with X-rays. J. Exp. Zool. 41: 45–92.

Lyon, M. F. 1961. Gene action in the X-chromosome of the mouse (*Mus musculus* L.). Nature 190:372–373.

Lyon, M. F. 1962. Sex chromatin and gene action in the mammalian X-chromosome. Amer. J. Hum. Genet. 14:135–148.

Lyon, M. F., A. G. Searle, C. E. Ford, and S. Ohno. 1964. A mouse translocation suppressing sex-linked variegation. Cytogenetics 3:306–323.

Mandl, A. M. 1964. The radiosensitivity of germ cells. Biol. Rev. 39:288–371.

McLaren, A. 1960. New evidence of unbalanced sex-chromosome constitution in the mouse. Genet. Res. 1:253–261.

Oakberg, E. F. 1956*a*. A description of spermiogenesis in the mouse and its use in analysis of the cycle of the seminiferous epithelium and germ cell renewal. Amer. J. Anat. 99:391–409.

Oakberg, E. F. 1956*b*. Spermatogenesis in the mouse and timing of stages of the cycle of the seminiferous epithelium. Amer. J. Anat. 99: 507–516.

Oakberg, E. F., and E. Clark. 1964. Species comparisons of radiation response of the gonads, p. 11–24. *In* W. D. Carlson and F. X. Gassner [ed.] Effects of ionizing radiation in the reproductive system. Pergamon Press, New York.

Ohno, S. 1963. Life history of female germ cells in mammals, p. 36–39. II Int. Conf. Congen. Malformations, Int. Med. Cong., New York.

Ohno, S., and B. M. Cattanach. 1962. Cytological study of an X-autosome translocation in *Mus musculus*. Cytogenetics 1:129–140.

Ohno, S., and T. S. Hauschka. 1960. Allocycly of the X-chromosome in tumors and normal tissues. Cancer Res. 20:541–545.

Ohno, S., W. D. Kaplan, and R. Kinosita. 1959. On the end-to-end association of the X and Y chromosomes of *Mus musculus*. Exp. Cell Res. 18:282–290.

Painter, T. S. 1927. The chromosome constitution of Gates' 'non-disjunction' (*v–o*) mice. Genetics 12:379–392.

Painter, T. S. 1928. The chromosome constitution of the Little and Bagg abnormal eyed mice. Amer. Natur. 62:284–285.

Phillips, R. J. S. 1961. Mouse News Letter 24:34.

Regaud, C. 1901. Études sur le structure des tubes seminifères et sur la spermatogénèse chez mammifères. Arch. Anat. Microscop. Morphol. Exp. 4:101–156, 231–380.

Russell, L. B. 1963. Mammalian X-chromosome action: inactivation limited in spread and in region of origin. Science 140:976–978.

Russell, L. B., and J. W. Bangham. 1959. Variegated-type position effects in the mouse. Genetics 44:532.

Russell, L. B., and J. W. Bangham. 1960. Further analysis of variegated-type position effects from X-autosome translocations in the mouse. Genetics 45:1008–1009.

Russell, L. B., and J. W. Bangham. 1961. Variegated-type position effects in the mouse. Genetics 46:509–525.

Russell, L. B., J. W. Bangham, and C. L. Saylors. 1962. Delimitation of chromosomal regions involved in V-type position effects from X-autosomal translocations in the mouse. Genetics 47:981–982.

Russell, L. B. and E. H. Y. Chu. 1961. An XXY male in the mouse. Proc. Nat. Acad. Sci. 47: 571–575.

Russell, L. B., and W. L. Russell. 1960. Genetic analysis of induced deletions and of spontaneous nondisjunction involving chromosome two of the mouse. J. Cell. Comp. Physiol. 56:169–188.

Russell, W. L., L. B. Russell, and J. S. Gower. 1959. Exceptional inheritance of a sex-linked gene in the mouse explained on the basis that the X/O sex-chromosome constitution is female. Proc. Nat. Acad. Sci. 45:554–556.

Sachs, L. 1955. The possibilities of crossing-over between the sex chromosomes of the house mouse. Genetica 27:309–322.

Slizynska, H. 1963. Origin of repeats in *Drosophila* chromosomes. Genet. Res. 4:154–157.

Slizynski, B. M. 1949. A preliminary pachytene chromosome map of the house mouse. J. Genet. 49:242–245.

Slizynski, B. M. 1952. Pachytene analysis of Snell's T(5;8)A translocation in the mouse. J. Genet. 50:507–510.

Slizynski, B. M. 1955*a*. Chiasmata in the male mouse. J. Genet. 53:597–605.

Slizynski, B. M. 1955*b*. The sex bivalent of *Mus musculus* L. J. Genet. 53:591–596.

Slizynski, B. M. 1957*a*. Meiotic prophase in female mice. Nature 179:638.

Slizynski, B. M. 1957*b*. Cytological analysis of translocations in the mouse. J. Genet. 55:122–130.

Snell, G. D. 1932. The effects of X-rays on the fertility of the male house mouse. Proc. VI Int. Cong. Genetics (Ithaca, 1932). Vol. 2. Brooklyn Botanic Gardens, New York.

Snell, G. D. 1933*a*. Genetic changes in mice induced by X-rays. Amer. Natur. 67:24.

Snell, G. D. 1933*b*. X-ray sterility in the male house mouse. J. Exp. Zool. 65:421–441.

Snell, G. D. 1933*c*. The production of translocations and mutations in mice by means of X-rays. Rec. Genet. Soc. Amer. 2:70.

Snell, G. D. 1935. The induction by X-rays of heredity changes in mice. Genetics 20:545–567.

Snell, G. D. 1946. An analysis of translocations in the mouse. Genetics 31:157–180.

Snell, G. D., E. Bodemann, and W. Hollander. 1934. A translocation in the house mouse and its effect on development. J. Exp. Zool. 67:93–104.

Snell, G. D., and D. I. Pickens. 1935. Abnormal development in the mouse caused by chromosome unbalance. J. Genet. 31:213–235.

Stevens, L. C., and M. C. Bunker. 1964. Karyotype and sex of primary testicular teratomas in mice. J. Nat. Cancer Inst. 33:65–74.

Swanson, C. P. 1957. Cytology and cytogenetics. Prentice-Hall, Englewood Cliffs, N.J. 596 p.

Tjio, J. H., and A. Levan. 1956. Note on the sex chromosomes of the rat during male meiosis. An. Estac. Exp. Aula Dei 4:173–184.

Warmke, H. E., and A. F. Blakeslee. 1940. The establishment of a 4N dioecious race in *Melandrium*. Amer. J. Bot. 27:751–762.

Warters, M., and A. B. Griffen. 1958. The centromeres of *Drosophila*. Genetica 30:152–167.

Welshons, W. J., and L. B. Russell. 1959. The Y-chromosome as the bearer of male determining factors in the mouse. Proc. Nat. Acad. Sci. 45:560–566.

8

Mutant Genes and Linkages[1]

Margaret C. Green

This chapter presents (1) a list of the named mutant genes of the mouse, (2) a summary of information about known linkages, and (3) a discussion of certain kinds of stocks useful for studying and maintaining mutants. It is an outgrowth of a chapter entitled Gene and Chromosome Mutations by George D. Snell in the first edition of the *Biology of the Laboratory Mouse.*

NAMED MUTANT GENES

Well over 300 mutant genes occupying more than 250 loci are known in the mouse. These numbers are subject to considerable uncertainty because the information about individual mutants varies in reliability. Many have not been adequately tested for allelism with other similar mutants. Many known to be alleles have not been compared sufficiently to determine whether or not their effects are identical by known criteria. Some well described mutants have become extinct. The choice of mutants to be included in the following list has therefore been somewhat arbitrary. I have tried to include all named mutants that seem likely to be useful, with the exception of apparently identical repeat mutations. I have omitted many of the mutants at the complex *T* locus and all of the alleles at the histocompatibility loci. The histocompatibility loci are described in detail in Chap. 24.

The list of references is not intended to be complete. References in addition to those listed here and appearing before 1950 may be found in Grüneberg (1952). Among papers published after that date I have tried to select those giving access to the previous literature. Useful descriptions of mutants affecting the skeleton may be found in Grüneberg (1963) and of mutants affecting the nervous system in Sidman et al. (1965).

Mutants discussed in other chapters in this book are so indicated by cross references to the appropriate chapters.

The linkage group if known is given in Roman numerals.

a locus, V, Chap. 21. The agouti locus controls the relative amount and distribution of yellow pigment (phaeomelanin) and black pigment (eumelanin) in the hairs of the coat. Experimental evidence indicates that the genes at this locus act only in the hair follicles. Pigment produced at sites other than the hair follicles is always eumelanin regardless of the genotype at the *a* locus (Silvers and Russell, 1955; Markert and Silvers, 1956). Mutations from *a* to a^t or A^w are very common, but other mutations within the locus appear to be rare.

A^y, yellow, semidominant, Chap. 19, 29. An old mutant of the mouse fancy. All the hair pigment is yellow. Heterozygotes usually become obese and sterile after the first few months. They are more susceptible to several kinds of tumors than normal mice (Vlahakis and Heston, 1963). Homozygotes die before implantation (Robertson, 1942) or shortly thereafter with failure of trophoblast giant cell development (Eaton and Green, 1963). A^y is dominant to all other members of the series in the sense that heterozygotes with all other members are indistinguishable from each other.

A^{vy}, viable yellow, dominant, Chap. 29. Arose

[1] The writing of this chapter was supported in part by research grants G15826 and G18485 from the National Science Foundation.

spontaneously in the C3H/HeJ inbred strain. A^{vy} resembles A^y except that homozygotes are viable and indistinguishable from heterozygotes with lower members of the series. Both homozygotes and heterozygotes show considerable variation in appearance ranging from clear yellow, through mottling with dark patches, to a complete agouti-like coat (Dickie, 1962*a*). Homozygotes and heterozygotes tend to become obese, and the degree of obesity is correlated with the amount of yellow in the coat (Wolff, 1965; Dickie, 1965, personal communication).

A^w, white-bellied agouti. Hairs of back are black with a subapical yellow band, the typical "agouti" pattern. The belly is white or cream. Dominant to A and all lower alleles.

+ or A, agouti. Like A^w except that the belly is dark. Usually regarded as the wild-type allele.

A^i, intermediate agouti. Arose spontaneously in the C57BL/6J strain. A^i/A^i, A^i/A, and A^i/a^{td} resemble $A^w/-$. A^i/a^t and A^i/a mice have a dark back and light belly with agouti hairs along the sides (Dickie, 1962*b* and personal communication).

a^{td}, tanoid. Arose spontaneously in the C57BL/6J strain. a^{td}/a^{td}, a^{td}/a^t, and a^{td}/a all have a light belly and a very dark agouti back. A/a^{td} resembles $A^w/-$ (Loosli, 1963).

a^t, black and tan. Found by Dunn (1928) in a strain obtained from an English fancier. Black back and cream or yellow belly. Recessive to A on back but dominant on belly. A/a^t resembles $A^w/-$.

a, nonagouti. An old mutant of the mouse fancy. a/a mice are plain black on back and belly, with some yellow hairs on the ears and perineum.

a^e, extreme nonagouti. Found among descendants of an irradiated mouse (Hollander and Gowen, 1956). a^e/a^e mice are very dark all over with no yellow hairs in the ears or around the nipples and perineum. Recessive to all other alleles.

a^x, radiation-induced, lethal when homozygous. Recessive to A^y, A^w, A, a^t, dominant to a. a^x/a has lighter belly than a/a. a^x shows about 0.5 per cent recombination with A^y (L. B. Russell et al., 1963).

ab, asebia, recessive, linkage not known. Arose as a spontaneous mutation in the BALB/cCrglGa strain. Homozygotes may be recognizable at 7 days of age by retarded growth of the coat. Alopecia increases and at adulthood the hair is very sparse. Sebaceous glands are completely absent. The base of the hair follicles sometimes exhibits excessive development, but hair production is faulty. The centers of many follicles are plugged with keratotic material. The follicles extend deep into the fat tissue and single hair-follicle units rather than the usual multiple hair-follicle units are observed. Results of reciprocal transplantation of skin between normal and mutant mice suggest that some diffusible substance synthesized by normal skin can stimulate hair growth and alleviate the hyperkeratosis characteristic of the mutant skin. Homozygotes are viable and fertile, but fertility of females is somewhat reduced (Gates and Karasek, 1965).

ac, absent corpus callosum, recessive, linkage not known. Discovered accidentally in the course of studies of the brain anatomy of mice with rodless retina (King and Keeler, 1932). The corpus callosum is partially or completely absent, but a tract of fibers, the longitudinal callosal bundle, which does not occur in normal mice, is present (King, 1936). A condition similar to that in *ac/ac* mice and probably genetically identical to it occurs in the BALB/cJ and 129/J strains (Wimer, 1965, personal communication).

ad, adipose, recessive, linkage not known, Chap. 19, 29. Arose spontaneously in a strain selected for large size. Homozygotes are recognizably larger than normal between 4 and 5 weeks of age. They may be nearly twice as heavy as normals when adult. Both sexes are probably sterile (Falconer and Isaacson, 1959; Batt and Harrison, 1963).

ag, agitans, recessive, III. Arose spontaneously in a moderately inbred stock. Homozygotes can be recognized from about 10 days onward. They show retarded growth, generalized tremor, and ataxia. They usually die at weaning but may live as long as 3 months. Atrophy of the Purkinje cells has been found in some regions of the cerebellum (Hoecker et al., 1954; Martinez and Sirlin, 1955).

Al, alopecia, dominant, VII. Appeared as a spontaneous mutation in the first generation of an outcross. Hair loss is first noticeable at 25 to 30 days. *Al/Al* mice lose all hair over the body with the exception of some guard hairs. *Al/+* mice have more hair but the distinction between *Al/Al* and *Al/+* is not always clear. After the first moult, hair grows in but is lost again at the next moult. Mice appear more patchy as they grow older (Dickie, 1955).

ald, adrenocortical lipid depletion, recessive, linkage not known. Found in the AKR/O inbred strain by Arnesen (1955, 1956). In homozygotes, spontaneous lipid depletion in the adrenal cortex takes place at the time of puberty and depends on the presence of the gonads in both males and females. In adult males the lipid depletion measured as sudanophilia is almost total. In females the depletion is subtotal and shows considerable variation (Arnesen, 1963).

am, amputated, recessive, linkage not known. Found among the fetuses of dissected pregnant females in an irradiation experiment. Homozygous fetuses lack a tail and at late stages of gestation the limbs appear to have been severed close to the trunk. The head is short and the body edematous dorsally. Vertebrae in the lumbo-sacro-caudal region are absent and there is disorganization of the remainder of the spinal column and some rib fusion. In the forelimbs most of the bones are present but abnormal (Meredith, 1964).

Amy-1 locus, salivary amylase, linkage group not known, but closely linked to *Amy-2*. Two codominant alleles are known. $Amy\text{-}1^a$ is recognized by a slowly migrating component demonstrated with agar-gel electrophoresis. This allele occurs in several laboratory strains including C3H, DBA, and AKA. $Amy\text{-}1^b$ is recognized by a fast migrating component. Found in a single mouse obtained from a fancier. $Amy\text{-}1^a/Amy\text{-}1^b$ mice have both bands (Sick and Nielsen, 1964).

Amy-2 locus, pancreatic amylase, linkage group not known, but closely linked to *Amy-1*. Two codominant alleles are known. $Amy\text{-}2^a$ is recognized by a single band demonstrated with starch-gel electrophoresis. This band migrates faster than either of the bands produced by the *Amy-1* locus. $Amy\text{-}2^b$ is recognized by two bands of equal strength, one in the same position as that produced by the allele $Amy\text{-}2^a$ and another faster migrating band. $Amy\text{-}2^a/Amy\text{-}2^b$ has 2 bands but the fast one is weaker than the slow one (Sick and Nielsen, 1964).

an, anemia, recessive, VIII, Chap. 17, 29. X-ray-induced. Homozygotes are anemic at birth, although there is enough variation in the severity of the condition to make classification a little uncertain. Affected mice are small and usually die young. Survivors are sterile (Hertwig, 1942). The anemia is the normochromic macrocytic type. It can be seen as early as 14 days of gestation and is caused by a disturbance of development of all erythropoietic cells (Thoms, 1952; Kunze, 1954). The sterility is a secondary effect of the anemia (Menner, 1957).

ap, alopecia periodica, recessive, linkage not known. Arose spontaneously in an inbred stock at the National Institute of Genetics, Japan. Homozygotes are recognizable at 3 to 4 days of age by their short whiskers and short, sparse first coat. The first coat is shed almost completely between 13 to 15 days and 20 to 24 days. Thereafter the hair is lost and regenerated in cycles which begin at the anterior end and spread toward the tail. During growth of the first coat the skin may have thick scales, and parts of the tail may be lost. All four kinds of hair are present but they are somewhat abnormal (Tutikawa, 1953). This mutant is not allelic with furless (*fs*) nor linked to it (Tutikawa, 1955).

As, agouti-suppressor, semidominant, V. Found among the offspring of an irradiated (C3H × 101)F_1 male. Very closely linked to the *a* locus. *As* appears to suppress the yellow pigment on the middorsum and belly of agouti mice. Homozygotes are darker than heterozygotes with black hair in the ears (Phillips, 1963*a*).

av, Ames waltzer, recessive, IV. Arose as a mutation in the K strain (Schaible, 1956). Viability and fertility are about normal, but females are poor mothers. Homozygotes show the typical circling, head-tossing, deafness, and hyperactivity of the circling mutants. They have defects of the membranous labyrinth similar to those of *sh-2* (Deol, 1965, personal communication).

ax, ataxia, recessive, XV. Arose as a spontaneous mutation in the CBA/H strain. Viability is low; both sexes are sterile. A mutation at the same locus (at first called "paralytic") occurred independently at The Jackson Laboratory. Homozygotes are hyperactive during the second week of life. From the third week on they show a steadily increasing paralysis up to 3 months of age. The paralysis is characterized by tremor, and loss of coordination and weakness of limbs (Lyon, 1955*b*). In the central nervous system there is degeneration, gliosis, and focal destruction of lower-spinal white matter. The corpus callosum, hippocampal and anterior commissures, and some brain-stem tracts are small (Coggeshall et al., 1961).

b locus, VIII, Chap. 21.

b, brown, recessive. An old mutant of the mouse fancy. The eumelanin of the hair is brown, rather than black. The pigment granules also appear brown rather than black and are spheroid rather than ovoid in shape (Markert and Silvers, 1956). Mutations from + to *b* have been recorded frequently.

B^{lt}, light, semidominant. Arose as a spontaneous mutation in the C58 inbred strain (MacDowell, 1950). B^{lt}/B^{lt} mice have almost white hair except for the tips which are brown. The pigment granules are the same shape as in *b*/*b*, but somewhat darker in color (Markert and Silvers, 1956). The granules are clumped (Quevedo and Chase, 1958; Pierro, 1963*a*). $B^{lt}/+$ mice have darker hair tips and the pigment extends much further down.

b^c, cordovan. Arose as a spontaneous mutation in an F_1 hybrid between C57BL and DBA/1. Recessive to + and dominant to *b*. The hair color is a rich deep brown (Miller and Potas, 1955).

ba, bare, recessive, linkage not known. Arose in a strain of Swiss albino mice in 1953 at the Poona Virus Research Centre (Randelia and Sanghvi. 1951). Homozygotes have no vibrissae at birth. The first coat is much delayed in appearance (13 to 14 days) and the hairs are very thin and short. It is shed at about 30 days. A new hair cycle starts at 45 days but the hairs are thin and are soon shed. Affected mice are entirely naked at 6 months. The number of hair follicles is not reduced, but the follicles contain keratinized globular masses instead of straight hair. The hairs formed are very thin, with no regular internal structure.

bf, buff, recessive, XVII. Arose spontaneously in the C57BL/6J strain. Nonagouti buff mice (*a/a bf/bf*) are khaki colored. Eye color does not appear to be affected (Dickie, 1964*b*).

bg, beige, recessive, XIV. Probably radiation-induced. The eye color is light at birth and varies from ruby to almost black in adults. Ear and tail pigmentation is reduced and the coat color is lighter, particularly at the base of the hairs (Kelly, 1957). There have been at least two recurrences of this mutation or a similar allele. One of them, called slate (*slt*) before its allelism with *bg* was discovered (Lyon, 1965, personal communication), has been described by Pierro and Chase (1963, 1965) and Pierro (1963*b*). The pigment granules are reduced in number but larger in size in the cortex and medulla of the hair and in the melanocytes of the choroid and retina of the eye and of the Harderian gland. The granules are clumped in the medulla of the hair. The smaller number of granules is due both to the synthesis of fewer melanin granules and to the fusion of individual granules into progressively larger clumps. Temporary hair loss during the third week of life occurs in some homozygotes.

bh, brain-hernia, recessive, probably I, Chap. 29. Appeared as a spontaneous mutation in the second generation following an outcross. Viability of homozygotes is reduced but some survive and may breed well. Penetrance is probably variable and dependent on the genetic background. At birth homozygotes may have cerebral hernia, hydrocephaly, anophthalmia, or microphthalmia. Later in life they all develop polycystic kidneys (Bennett, 1959). The polycystic disease is preceded by the appearance of PAS-positive plugs in the tubules of the kidney. Analysis of the urine shows that *bh/bh* mice have a higher than normal level of free amino acids and protein prior to the development of the disease (Bennett, 1961*a*).

Bld, blind, semidominant, linkage not known. Arose spontaneously in the Bagg albino strain. Heterozygotes are born with their eyelids open and the cornea becomes damaged. Homozygotes die at the end of the seventh day of gestation. Primitive streak and mesoderm fail to form. Entoderm appears very abnormal. In the proximal entoderm the cells are greatly enlarged and may be polyploid. The over-all size of the egg cylinder is about one-fifth that of the normal (Vankin, 1956). Penetrance in the heterozygote is incomplete (Watson et al., 1961). The mutation recurred at Harwell, England (Watson, 1962).

Blo, blotchy, semidominant, XX (sex-linked), Chap. 15. Arose spontaneously. Heterozygous females have irregular patches of light-colored fur. Expression is occasionally poor at weaning age but is complete at adulthood. Hemizygous males and homozygous females are light all over with no blotching, are usually small, and occasionally have deformed hind legs. Viability and fertility of heterozygotes is normal. Hemizygotes and homozygotes have reduced viability and many are infertile (L. B. Russell, 1960*b*). *Blo* may be an allele of *Mo*.

Bn, bent-tail, semidominant, XX (sex-linked), Fig. 8-2A, Chap. 15. Appeared as a spontaneous mutation in a crossbred male. Tails of heterozygous females are more or less shortened, with one or several bends. Penetrance is incomplete. Tails of homozygous females and hemizygous males are shorter and more kinked and not always clearly distinguishable from the tails of heterozygotes. Heterozygous females have normal viability and fertility, but homozygotes and hemizygotes may be less viable and fertile than normal (Garber, 1952*b*; Grüneberg, 1955*b*).

bp, brachypodism, recessive, V, Chap. 15. Arose in a stock of Swiss mice at Harvard University. Fully viable and fertile. The phalangeal bones of all four feet are reduced in number, the metacarpals and metatarsals are shortened and all other bones of the feet are small and irregular in shape. The long bones of the legs are significantly shortened (Landauer, 1952). There have been at least two recurrences of this mutation or a similar allele.

Bph, brachyphalangy, semidominant, linkage not known. Found in a low-intensity neutron irradiation experiment. In heterozygotes, digits of the forefeet are splayed out with marked shortening and thickening of phalanges and metacarpals. Hind feet are less severely affected. Homozygotes die before birth, all feet showing brachyphalangy and high grade polydactyly. There are other severe limb defects and frequently exencephaly (Searle, 1965).

bt, belted, recessive, VI, Chap. 21. Arose as a spontaneous mutation in the DBA inbred strain (Murray and Snell, 1945). Homozygotes have a white belt across the back in the midtrunk region and a white belly patch. The two may coalesce to form a complete belt. The white spotting appears to be due to a defect in the hair follicles which prevents pigment cells either from entering the hair follicles or from developing there (Mayer and Maltby, 1964).

c locus, I, Chap. 21. The albino locus affects the amount of tyrosinase in the pigment cells, but does not interfere with the production of pigment cells themselves (Coleman, 1962; Silvers, 1958). It is probably the structural locus for tyrosinase. All the mutant alleles are recessive to wild type in appearance, but heterozygotes with wild type produce intermediate amounts of tyrosinase. Spontaneous mutation at this locus appears to be relatively infrequent.

c^{ch}, chinchilla. First described by Feldman (1922). Agouti chinchilla mice are gray rather than brownish gray. The yellow pigment in the hair is greatly reduced and the black pigment slightly so. Eyes are black. Tyrosinase activity in the skin is about one-third that of normal (Coleman, 1962). Heterozygotes with c^h, c^e, and c are intermediate both in color and in tyrosinase activity.

c^e, extreme dilution. Found in the wild by Detlefsen (1921). Hair is very light gray and eyes are black. Tyrosinase activity in the skin is very low (Coleman, 1962). c^e/c are almost white with black eyes.

c^h, himalayan. Found in the offspring of a cross between the DBA/2 and AKR/J strains. Body hair is light and ears, nose, and tail are dark as in Siamese cats. Body hair may be very light or quite sooty, depending on the genetic background. Eyes are slightly pigmented and appear red. c^h/c and c^h/c^e are intermediate between the homozygotes (Green, 1961*a*). The tyrosinase of c^h/c^h mice is temperature sensitive (Coleman, 1962).

c, albino. A very old mutant, already known in Greek and Roman times. Hair and eyes are completely devoid of pigment. Tyrosinase activity is almost absent (Coleman, 1962).

Ca, caracul, dominant, VI. Arose in a stock of Swiss mice (Dunn, 1937*b*). Homozygotes are indistinguishable from heterozygotes. Both are fully viable and fertile. Vibrissae are curved and hair is wavy from time of first appearance until about 4 weeks of age. After that the waves disappear but the hair has a plush-like look. Several recurrences of mutations at this locus similar to *Ca* have been reported.

Cat, dominant cataract, dominant, linkage not known, Chap. 29. Arose in an albino strain of unknown origin (Paget, 1953). Homozygotes and heterozygotes are fully viable and fertile. They are indistinguishable except that the age of onset may be earlier in homozygotes. Age of onset varies between 10 days and 14 weeks. Liquefaction of the subcapsular zone of the lens leads to cataract of the cortex and the lens nucleus. The lens capsule always remains intact. With the exception of the nucleus the whole lens material may become liquefied and much of it may pass through the intact lens capsule into the vitreous (Paget and Baumgartner-Gamauf, 1961). An allelic mutant, originally called shrivelled (*Svl*) was found in the A/J strain by Fraser and Schabtach (1962) (Verrusio, 1964).

cb, cerebral degeneration, recessive, linkage not known. Found by Deol and Truslove (1963). Homozygotes usually die before reaching maturity. If they live on they are always sterile. The degeneration produces an *ex vacuo* hydrocephalus visible in the living animal and usually recognizable at birth. The most susceptible parts of the brain are the cerebral hemispheres and olfactory lobes. Later the epithelium of the nose and trachea also degenerates.

Cd, crooked, semidominant, linkage not known. Arose spontaneously in the A inbred strain (Morgan, 1954). Heterozygotes have crooked tails with abnormal caudal vertebrae and often have abnormal sacral and lumbar vertebrae as well. Some have no abnormalities of the skeleton at all. Homozygotes are more severely affected and show no normal overlapping. They may die early in development, or later with exencephaly, or, if they survive, are smaller than normal from about 15 days of age and may have crooked tails, microphthalmia, small or nonerupted lower incisors, absence of third molars, irregular tail rings with few hairs, and nervous movement of the head

(Morgan, 1954; Grewal, 1962*a*). The vertebral anomalies can be traced back to the somite stage when the somites are seen to be irregular in size and often partly fused with their neighbors (Matter, 1957).

Ce, low liver catalase, dominant, linkage not known, Chap. 19. This mutant reduces liver catalase activity to about half its normal level but has no effect on kidney catalase. The low allele is fully dominant over the normal allele. The low allele is found in most of the C57 family of strains including C57BL/6, C57BL/10, C57L, C57BR/cd, and C58/Lw, but not in C57BL/He and C57BL/An. The normal allele occurs in C3H/He, YBR/He, and BALB/cDe (Rechcigl and Heston, 1963; Heston and Rechcigl, 1964; Heston et al., 1965).

ch, congenital hydrocephalus, recessive, XIV. Appeared as a spontaneous mutation in descendants of an outcross of the CBA inbred strain (Grüneberg, 1943*a*). Homozygotes die immediately after birth, possibly from inability to inflate their lungs. They are very uniform in appearance, with bulging forehead, and with twin protuberances corresponding to the cerebral hemispheres and filled with hemorrhagic fluid. Development of cartilage is severely affected, beginning at the mesenchymal stage. The mesenchymal condensations are smaller than normal, the reduction being more marked in some parts than in others. Once laid down, cartilages grow at a normal rate but may be abnormal in shape. Many are absent altogether. The hydrocephalus is due to an abnormal shortening of the base of the skull (Grüneberg, 1953*b*).

cl, clubfoot, recessive, linkage not known. Appeared in a stock of unknown ancestry. Homozygotes are much less viable and fertile than normal. The clubbing is a simple calcaneum (dorsiflexed) type, always affecting both hind feet and sometimes one or both forefeet. Muscles of the lower limb are smaller than normal and there is some carpal and tarsal fusion (Robins, 1959).

cn, achondroplasia, recessive, linkage not known, Fig. 8-1*H*, Chap. 29. Arose in strain AKR/J in 1960. Homozygotes show a disproportionate dwarfing which is probably due to abnormal growth of the epiphyses. Viability is considerably reduced and very few of the survivors are fertile (Dickie, 1961).

cr, crinkled, recessive, XIV. Appeared in the progeny of a male treated with nitrogen mustard. Homozygotes are characterized by absence of guard hairs and zigzag hairs in the coat. This produces a bald patch behind the ears, and a bald tail, and also leads to kinks at the tail tip, reduced aperture of the eyelids, respiratory disorder, and a modification of the agouti pattern. The hair follicles that produce guard hairs and zigzags develop at 14 to 17 days of gestation and after birth, respectively. These follicles are absent in *cr/cr*. The viability and breeding performance of homozygotes is somewhat reduced (Falconer et al., 1951).

ct, curly-tail, recessive, linkage not known. Arose spontaneously in the GFF inbred strain. Genetic data indicate that *ct* is probably a recessive with incomplete penetrance (20 to 50 per cent) but do not absolutely rule out the possibility that it is a dominant with reduced penetrance. Presumed homozygotes, if abnormal, usually have some degree of spina bifida and, occasionally, anencephaly. If the mice survive, the spina bifida usually results in a curly or kinky tail (Grüneberg, 1954*a*).

cv, calvino, recessive, linkage not known. Arose in the C58/Hr inbred strain. The first coat of homozygotes is somewhat sparse. Adults have a reduced quantity of all hairs, especially zigzags. Males are more affected than females, and the back is more affected than the belly. There is extensive silvering of the hair. Calvinos are usually underweight and often sterile (Pizarro, 1957).

cw, curly whiskers, recessive, II. Arose in the CBA strain maintained at the Chester Beatty Research Institute, London. Homozygotes have strongly curled whiskers; their coats are not waved but slightly abnormal in appearance. Heterozygotes have slightly bent whiskers but cannot be reliably identified on this basis (Falconer and Isaacson, 1962*b*).

d locus, II, Chap. 21, 23, 32. Mutations at this locus alter coat color, reduce phenylalanine hydroxylase activity (Coleman, 1960), and affect the nervous system. Mutations from + to d^l and from *d* to + have been observed repeatedly.

d, dilute, recessive. An old mutant of the mouse fancy. The blue-gray color of the hair produced by this mutant in nonagouti (*a/a*) mice is caused by clumping of the melanin pigment into a few large masses (E. S. Russell, 1949*a*). The clumping is due to the shape of the melanocytes which have fewer and thinner dendritic processes than wild-type melanocytes, with the result that melanin granules are largely clumped around the nucleus (Markert and Silvers, 1959).

d^l, dilute-lethal, recessive to both + and *d*. Arose in the C57BL/Gr inbred strain. The color of d^l/d^l is identical to that of *d/d*, but the mice develop a severe neuromuscular disorder characterized by convulsions and opisthotonus (arching upward of the head and tail) and usually die at about 3 weeks (Searle, 1952). Myelin degeneration occurs in the central nervous system (Kelton and Rauch, 1962). Phenylalanine metabolism is severely disturbed (Rauch and Yost, 1963).

d^s, slight dilution, recessive. Discovered in a C57BL/6J × DBA/2 hybrid. Homozygotes are darker than d/d mice. d^s/d is intermediate between d^s/d^s and d/d (Dickie, 1965).

d^{15}, dilute-15, recessive to +. Homozygotes have slightly diluted coats, darker than d/d. They develop behavior abnormalities similar to d^l/d^l at about 20 days but some may live, one male surviving to breed. d/d^{15} are similar in color to d^{15}/d^{15} but behave normally (Phillips, 1962).

da, dark, recessive, I. Arose in the CBA/Fa inbred strain. Probably extinct. Homozygotes are smaller than normal. In combination with *A* or A^y the yellow pigment on the back is replaced by black so that both look like a/a except on the flanks. The darkening of the back of *A* lessens as the animals get older (Falconer, 1956).

Dc, dancer, semidominant, linkage not known. Arose as a spontaneous mutation in the C3H/J-*ob* stock (Lane, 1958). Not allelic with *Tw*, which it somewhat resembles. Viability and fertility of heterozygotes are normal. Penetrance is incomplete in heterozygotes. Homozygotes die at birth with cleft lip and cleft palate. Heterozygotes exhibit circling and head-tossing behavior, but are not deaf. They usually have a small patch of white hairs on the forehead. There is complete absence of the macula of the utriculus accompanied by gross defects of the bony and membranous labyrinths in the vestibular region (Deol, 1965, personal communication).

de, droopy-ear, recessive, XVI. Arose spontaneously in the J stock at the Institute of Animal Genetics, Edinburgh. In homozygotes the ears are set low on the sides of the head and the pinnae project laterally. On an a^t background the light belly hair comes up farther around the sides of the body and face. Homozygotes tend to be smaller than their sibs but they breed well, except that some females are poor mothers. The skeleton of the occiput and shoulder girdle shows specific abnormalities, while the rest of the skeleton shows immaturity of form with disproportionate shortening of the limb bones. The skeletal abnormalities can be traced back to disturbed mesenchymal condensations. It is possible that the growth of cartilage itself is also abnormal (Curry, 1959).

df, Ames dwarf, recessive, VII, Chap. 20, 29 Arose spontaneously in descendants of a cross of Goodale giant mice to a pink-eyed stock. Homozygotes resemble *dw/dw* mice. Growth retardation occurs after the first week and weight at 2 months is only about one-half normal. Both sexes are sterile. Treatment with bovine growth hormone for 6 weeks produces nearly normal growth and males become fertile (Schaible and Gowen, 1961). The anterior hypophysis lacks acidophils and has very few thyrotropic hormone-producing cells (Bartke, 1964).

Dh, dominant hemimelia, semidominant, XIII, Chap. 15. Arose spontaneously in a crossbred stock. Heterozygotes show preaxial polydactyly or oligodactyly of the hind limbs, tibial hemimelia, and sometimes reduction of the femur and pubic element of the pelvic girdle. Number of ribs, sternebrae, and presacral vertebrae tend to be reduced. The spleen is entirely lacking; the stomach is somewhat smaller and the alimentary canal shorter than normal. Homozygotes die within a few days after birth. Abnormalities are similar to those of heterozygotes but much more severe and include also abnormalities of the urogenital system (Searle, 1964).

di, duplicate incisors, recessive, linkage not known. Found in the F_2 generation from mice treated in utero with nitrogen mustard. Penetrance is high in mice of the original stock, but low and variable in descendants of outcrosses. Affected mice are characterized by the appearance on one or both sides of supernumerary lower incisors located directly behind, or slightly medial to the corresponding normal teeth. The extra teeth usually erupt a few days later than the normal incisors. Usually they become deflected and are broken off, after which it may be difficult to say whether an extra tooth had ever been present (Danforth, 1958).

dl, downless, recessive, probably IV. Appeared in strain A/H. Homozygotes closely resemble homozygous crinkled (*cr*) mice, but the two mutants are not allelic. Defects of the incisors prevent the mice from eating hard food (Phillips, 1960*b*).

dm, diminutive, recessive, V, Chap. 15, 17. Appeared in descendants of a cross of strains 129 and C57L/J. Viability of homozygotes is low but many surviving to maturity are fertile. Homozygotes are consistently smaller than normal and have a macrocytic anemia which becomes slightly less severe with age. They have short kinked tails and tend to have an additional rib at both the cervical and lumbar ends of the thorax, and an additional presacral vertebra. The vertebrae are malformed and ribs may be fused (Stevens and Mackensen, 1958).

dn, deafness, recessive, linkage not known, Chap. 32. Discovered in a stock at University College, London, in the course of a systematic search for uncomplicated deafness (Deol and Kocher, 1958). Fertility of homozygotes is normal. Homozygotes are deaf their entire life, and a few of them show slight head-tossing. Degeneration of the cochlea begins on the 10th day. The macula of the sacculus may degenerate in both head-tossing and normally behaving mice, but remains histologically normal in many of them.

dr, dreher, recessive, XIII, Chap. 32. Arose in the wild, presumably as a spontaneous mutation, and was obtained in a litter of four, all homozygous, captured in a factory in Detmold, Germany (Falconer and Sierts-Roth, 1951). Homozygotes show the circling, head-tossing, deafness, and hyperactivity characteristic of the circling mutants. They may also have a partial or complete white belt (Lyon, 1961*a*). The bony and membranous labyrinths develop abnormally and there may be hydrocephalus of the hindbrain (Fischer, 1956, 1958; Bierwolf, 1956, 1958). The defects of the ear and hindbrain trace back to an abnormality of the neural tube at 9 days of gestation. The roof of the posterior part of the rhombencephalon, normally very thin, in *dr/dr* embryos is thick like that of the neural tube in the cervical region, as though joining of the lateral plates had extended more anteriorly than normal (Deol, 1964*b*).

Ds, disorganization, semidominant, III. Arose spontaneously in the inbred DA/Hu strain. *Ds* disrupts the orderly processes of development. Homozygotes probably die in utero after implantation and before term. Heterozygotes are normal, deviate slightly, or are monstrous individuals with multiple defects. The defects show great variety, range of severity, and random distribution among derivatives of all germ layers. Penetrance varied between 9 and 89 per cent on different genetic backgrounds observed by Hummel (1958, 1959).

dt, dystonia musculorum, recessive, XIII, Fig. 8-2*F*. Arose spontaneously in mice bred at the Institute of Animal Genetics, Edinburgh. Homozygotes can be first recognized at 7 and 10 days of age by clasping of the hind limbs when the mice are lifted by the tail. There is increasing incoordination with alternating hyperextension and hyperflexion of the limbs. There is no obvious paralysis. Many affected animals die before weaning, but some survive several months. The clinical condition becomes relatively stationary after the first phase of deterioration. Histologically the nervous system shows widespread degeneration and loss of nerve fibers in the peripheral nerves, dorsal root ganglia, and roots, and in the dorsal and ventrolateral columns of the spinal cord. In longer surviving animals there is also evidence of lower motor neurone involvement (Duchen and Strich, 1964). There have been several recurrences of mutation at this locus at The Jackson Laboratory.

du, ducky, recessive, II. Arose as a spontaneous mutation in a noninbred stock (Snell, 1955). Viability is somewhat less than normal. Males living to maturity may be fertile but are poor breeders. Females rarely breed. Homozygotes show a waddling or reeling gait and a tendency to fall to one side. They are slightly smaller than their normal sibs and may occasionally have seizures.

dv, dervish, recessive, linkage not known. Possibly radiation-induced. Homozygotes are viable (L. B. Russell, 1961). No tests for allelism with other waltzing mutants have been reported.

dw, dwarf, recessive, linkage not known, Fig. 8-1*B*, Chap. 20, 29. Arose in a stock of silver mice obtained from an English fancier (Snell, 1929). Homozygotes are about one-fourth to one-third normal size and are sterile. The small size is due to a defective anterior pituitary in which there is a great deficiency of acidophils and also a deficiency of thyrotrop cells (Elftman and Wegelius, 1959). Dwarf mice grow to normal size and may become fertile when given daily implants of whole pituitaries (Smith and MacDowell, 1930).

dy, dystrophia muscularis, recessive, IV, Fig. 8-1*C*, Chap. 19, 29. Arose as a spontaneous mutation in the 129/Re inbred strain. Homozygotes are characterized by progressive weakness and paralysis beginning at about 3½ weeks of age. The hind limbs are affected first, later the axial and forelimb musculature. Death usually occurs before 6 months of age and the mice are usually sterile. There is no demonstrable pathology of the nervous system, but the muscular tissue shows degeneration typical of the muscular dystrophies, with proliferation of sarcolemmal nuclei, increase in amount of interstitial tissue, and size variation among individual fibers (Michelson et al., 1955).

Ea-1 locus, erythrocyte antigen-1, linkage not known. Wild populations of house mice have been found to contain four phenotypes with respect to erythrocyte antigens, designated A, B, AB, and O and appearing to be under the control of a system of three alleles, $Ea\text{-}1^a$, $Ea\text{-}1^b$, $Ea\text{-}1^o$. The locus appears to differ from the *H-2* locus. Erythrocytes of 13 inbred strains lacked both A and B antigens (Singer et al., 1964).

eb, eye-blebs, recessive, IV. Arose in a noninbred stock of hairless mice. It resembles *my* but is not allelic to it. Penetrance is incomplete. Homozygotes may be normal or have defects of the eyes, kidney, or feet including: anophthalmia or microphthalmia; small, missing, or cystic kidneys; and clubbed feet, webbed toes, or extra digits (Hummel and Chapman, 1963).

ec, ectopic, recessive, linkage not known. Found among the descendants of mice exposed to 1,000 R of X-rays. Homozygotes are first distinguishable from normal at 28 to 30 days, when there is a slight increase in the size of the eyeball. The increase disappears after 6 weeks. At 5 weeks an opacity of the lens develops. The nucleus of the lens is gradually displaced through the posterior pole of the ruptured capsule. Histologically the lens epithelium is abnormal at about 4 weeks. In normal mice it is a single layer with no mitotic figures but in *ec/ec* mice the epithelium is irregularly stratified with many mitotic figures. The posterior part of the lens capsule is much thinner than normal (Beasley, 1963). *ec* is similar to *lr* but allelism tests have not been made.

Ee-1 locus, erythrocytic esterase-1, linkage not known. Independent of *Ee-2*. Two condominant alleles are known, *Ee-1*^{a} and *Ee-1*^{b}, which produce esterases with different electrophoretic mobility. Both alternative molecular forms are sensitive to diisopropyl fluorophosphate. *Ee-1*^{a} is found in inbred strains C57BL/6, C57BL/10, and C57BR/cd. *Ee-1*^{b} is found in strains AT, BALB/c, SEC/Re, A/He, C3H/He, CBA/J, and DBA/2J (Pelzer, 1965).

Ee-2 locus, erythrocytic esterase-2, linkage not known. Independent of *Ee-1*. Two alleles at this locus control the presence (*Ee-2*^{a}) or absence (*Ee-2*^{o}) or an erythrocytic esterase inhibited by eserine. *Ee-2*^{a} is found in strains A/He, C3H/He, CBA/J, and DBA/2J. *Ee-2*^{o} is found in strains C57BL/6, C57BL/10, C57BR/cd, AT, BALB/c, and SEC/Re (Pelzer, 1965).

Eo, eye-opacity, dominant, linkage not known. Radiation-induced. Eye-opacity and small eyes, and occasionally open lids at birth (Gower, 1953).

ep, pale-ears, recessive, XII. Arose as a spontaneous mutation in the C3HeB/FeJ inbred strain. The ears and tail are pale in color. The juvenile coat is slightly diluted in color but the adult coat is almost normal. Eyes are pale at birth and darken with age. *ep* is an almost exact mimic of *le* but not allelic to it (Lane and E. L. Green, 1965, personal communication).

Er, repeated epilation, dominant, linkage not known. Probably radiation-induced. Homozygotes probably die prenatally. There is complete penetrance in heterozygotes. They grow a normal appearing but probably somewhat dry coat until the age of about 13 days; then hair loss begins until the fur becomes sparse. Repeated growth and re-epilation follow without any definite pattern. Heterozygotes are slightly reduced in size but completely viable and fertile (Hunsicker, 1960).

Es-1 locus, serum esterase-1, XVIII, Chap. 15. This locus controls a difference in electrophoretic mobility on starch gel of a serum esterase which has a mobility similar to that of serum albumin. Two codominant alleles are known: *Es-1*^{a}, which produces a faster migrating single band and is present in C57BL/Cum and C57L/Rl; and *Es-2*^{a}, which produces a slower double band and is present in many other inbred strains (Popp and Popp, 1962; Popp, 1965*a*).

Es-2 locus, serum esterase-2, linkage not known, Chap. 15. This locus controls variation in the fastest anodally migrating serum esterase as revealed by starch-gel electrophoresis at pH 8.5. Two alleles are known: *Es-2*^{a}, absence of the esterase band, found in a wild mouse from Washtenaw County, Michigan; and *Es-2*^{b}, presence of the esterase band, present in many inbred laboratory strains. *Es-2*^{a}/*Es-2*^{b} produces a lighter band than *Es-2*^{b}/*Es-2*^{b} (Petras, 1963).

Es-3 locus, kidney esterase-3, linkage not known. Independent of *Es-4*. This locus controls variation in a kidney esterase revealed by vertical starch-gel electrophoresis at pH 8.65. Two alleles are known: *Es-3*^{a}, absence of the esterase band, found in inbred strain C57BL/6J; and *Es-3*^{b}, presence of the esterase band, found in inbred strain RF/J. *Es-3*^{a}/*Es-3*^{b} is recognizable by a lighter band than in *Es-3*^{b}/*Es-3*^{b}. There are five zones of banding of kidney esterases, and this esterase is found in the fastest migrating region of zone IV, numbered from the origin. The esterase is sensitive to inhibition by eserine sulfate (Ruddle and Roderick, 1965).

Es-4 locus, kidney esterase-4, linkage not known. Independent of *Es-3*. This locus controls variation in a kidney esterase revealed by vertical starch-gel electrophoresis at pH 8.65. Two alleles are known: *Es-4*^{a}, absence of the esterase band, found in inbred strain C57BL/6J; and *Es-4*^{b}, presence of the esterase band, found in inbred strain RF/J. *Es-4*^{b} is dominant to *Es-4*^{a}. The esterase band occurs in the cathodal region of zone IV (see *Es-3*) (Ruddle and Roderick, 1965).

et, elevated-tail, recessive, linkage not known. Arose spontaneously in the BALB/c inbred strain. In homozygotes the tail emerges high and is about two-thirds normal length and wavy (L. B. Russell, 1951).

ey-1, eyeless-1, *ey-2,* eyeless-2, linkage not known. Chase (1942, 1944) postulated that these two recessive loci were responsible for the microphthalmia and anophthalmia observed in nearly 90 per cent of the mice in certain selected strains. It is not certain that two discrete loci with large effects exist (Beck, 1963*a, b*). If they exist they are not yet distinguishable from each other either by their effects or by known linkage map position. Mice heterozygous at one or both of these loci have a higher incidence of abnormal eyes than homozygous normal mice as a result of trypan blue injected into their mothers during gestation (Barber, 1957; Beck, 1963*c*).

f, flexed-tail, recessive, XIV, Chap. 17, 21. Appeared in a stock maintained by Hunt (Hunt et al., 1933). Homozygotes are born with a transitory siderocytic anemia caused by a disturbance of the hematopoietic function of the fetal liver. Most homozygotes also have flexed tails and a belly spot, but these are not constant manifestations of the mutant. Because of the anemia there is probably slightly greater postnatal mortality among *f/f* than among normal mice. The anemia begins on the 12th day of embryonic life when the liver first starts to produce blood cells (Grüneberg, 1942*a*). Both heterozygotes and homozygotes have low levels of liver δ-aminolevulinate dehydratase (Margolis, 1956). Flexed is closely linked to *Lv* and may be an allele (Coleman and E. S. Russell, 1965, personal communication). The tail abnormalities are first noticeable on the 14th day as abnormal differentiation of the fibers of the annulus fibrosis of the intervertebral discs. Kinking of the tail results, as well as occasional fusions of the other vertebrae (Kamenoff, 1935).

fa, falter, recessive, linkage not known. Arose as a spontaneous mutation in a DBA subline. Homozygotes have an abnormal side-to-side swaying walk beginning at about 10 days of age. The abnormality becomes more pronounced with age and all mice die at about 20 days. *fa* has not been tested for allelism with other similar mutants (Yosida, 1960).

fi, fidget, recessive, V, Chap. 15, 32. Discovered by Grüneberg (1943*b*) in a heterogeneous strain. Viability of homozygous fidgets is usually less than normal and may be as low as 50 per cent on some genetic backgrounds. Fertility is poor but some males are reliable breeders. Homozygotes toss their heads from side to side and tend to run in circles but can hear throughout life. The bony labyrinth is quite defective. There are other abnormalities including small eyes, absence of lachrymal glands, dislocation of the hip, increased incidence of polydactylism, and displaced paraflocculur lobes of the cerebellum (Truslove, 1956).

fl, flipper-arm, recessive, linkage not known. Possibly radiation-induced. Viability of homozygotes is poor. Males are sterile, females only occasionally fertile. Homozygotes are smaller than their normal sibs. The ulna and radius bend sharply outward near the wrist. The hind limbs are normal (Kelly, 1957).

fm, foam-cell reticulosis, recessive, linkage not known. Arose spontaneously in the CBA/H strain. In homozygotes lipid-containing foam-cells replace the lymphoid tissue of the thymus and Peyer's patches and occur in smaller numbers in other tissues. The lipid may be a complex of lysolecithin and cholesterol and the disease is similar to Niemann-Pick disease in man. It is first detectable in mice 3 to 4 months old, and death occurs within a few weeks (Lyon et al., 1965).

fr, frizzy, recessive, I. Arose in a stock of mixed origin at The Jackson Laboratory. Homozygotes have wavy or curly vibrissae at 1 or 2 days after birth. The coat is short and rough at first and then may become normal at 6 to 7 weeks. It is usually short and thin in older animals (Falconer and Snell, 1952). Following some outcrosses the proportion of *fr/fr* mice identifiable in the F_2 is low, an indication that penetrance may be low on some backgrounds.

fs, furless, recessive, XIV. Appeared in an unpedigreed stock maintained in the Department of Zoology at Ohio State University. Homozygotes have short or missing vibrissae 2 days after birth. The first coat grows normally but begins to thin out at 19 days, starting on the head. A new coat grows but persists only a short time. Mature mice are partly devoid of hair at all times. The hairs present are shorter than normal. Furless mice are slightly smaller than their normal sibs but are normally fertile (Green, 1954).

ft, flaky tail, recessive, linkage not known. Discovered in a heterogeneous stock. Homozygotes have flaky tails, often with constrictions resulting in amputation of the tail, and slightly smaller than normal ears. The stratum granulosum is thinner than normal, presumably resulting in slower than normal proliferation of the stratum corneum and keratin (Lane and Green, 1962).

Fu locus, IX.

Fu, fused, semidominant. Arose in stocks at the Bussey Institution prior to 1931. Expression of this mutant shows great variability. Both homozygotes and heterozygotes may have shortened and kinked tails or they may both be normal. Homozygotes tend to be more severely affected than heterozygotes. Expression is not dependent on the residual genotype but offspring of *Fu*/+ or *Fu*/*Fu* mothers are less likely to express the character than offspring of +/+ mothers (Reed, 1937). Homozygotes and heterozygotes occasionally show an abnormal behavior similar to the circling mutants and are deaf (Dunn and Caspari, 1945). Embryos of homozygotes show some overgrowth and duplication of the posterior part of the neural tube (Theiler and Gluecksohn Waelsch, 1956).

Fu^{ki}, kinky, semidominant. Arose in the stocks of a Florida mouse fancier (Caspari and David, 1940). Heterozygotes are very similar to fused in abnormalities of the skeleton and may show the same behavioral abnormalities. Dunn and Caspari (1945) found five probable crossovers between *Fu* and Fu^{ki} among 505 gametes tested, but Dunn and Gluecksohn-Waelsch (1954) found none in 971 and concluded that the previous study was in error and that kinky is an allele of *Fu*. Unlike *Fu*, kinky homozygotes are inviable. They show tissue hyperplasia and twinning at 7 days and die between 8 and 10 days of embryonic life (Gluecksohn-Schoenheimer, 1949).

fy, frowzy, recessive, linkage not known. Arose spontaneously in either the C3H or 101 inbred strain. Homozygotes are viable and usually fertile. The skin is wrinkled and the hair is sparse and frizzy. Vibrissae are curly. Ear tufts are absent, and feet, nose, and tail are white (Major and Hawkins, 1958, as "hair abnormality").

fz, fuzzy, recessive, XIII. Discovered in 1945 in the CFW stock at Carworth Farms. Homozygotes are viable and fertile. They are classifiable at 6 days of age by their thin, wavy vibrissae and uneven coat. In the adult the coat is thin and wavy or curly. Penetrance seems to be complete (Dickie and Woolley, 1950). Follicles of all hair types are present in the skin of newborn mice, but the structure of the hair is altered (Mann, 1964).

g, low glucuronidase, recessive, XVII, Chap. 19. Several sublines of the C3H strain (C3H/He, C3H/Ha) produce a low level of liver β-glucuronidase. The low level behaves as if controlled by a single recessive gene in crosses with strains having normal levels (CBA/St, CBA/An, C57BL, BALB/c, DBA/2, RIII) (Law et al., 1952). In *g*/*g* mice the intracellular distribution of the enzyme as well as the time course of accumulation in the liver is different from that of +/− mice. The enzyme of *g*/*g* mice differs in susceptibility to heat denaturation and to inactivation by alkali from that of +/+ mice and is therefore presumably different in structure (Paigen, 1961*a*, 1961*b*, 1964).

gl, grey-lethal, recessive, IV, Chap. 15, 20, 29. Arose spontaneously in a stock segregating for c^e (Grüneberg, 1935*b*). Homozygotes die between 20 and 30 days of age. Most of the yellow pigment is absent from the fur, resulting in a slate gray color in agouti mice. The color of *a*/*a* mice is not noticeably influenced by *gl*. Homozygotes lack the power of secondary bone resorption. Consequently the bones cannot grow normally, they do not form normal marrow cavities, and the teeth do not erupt. An atlas of the skeletal anatomy of grey-lethal and normal sibs at 3 weeks of age has been prepared by Bateman (1954). Grey-lethals are considerably more resistant to injections of parathyroid hormone than normal mice (Barnicot, 1945, 1948). Evidence favors the hypothesis that failure of secondary bone resorption is due to an accelerated rate of inactivation of parathormone (Hirsch, 1962).

gm, gunmetal, recessive, linkage not known. Arose spontaneously in the C57BL/6J inbred strain. Homozygotes have a diluted coat color somewhat similar to that of dilute (*d*/*d*). Eye color does not appear to be affected. Homozygotes have high mortality and do not breed well (Dickie, 1964*b*).

go, angora, recessive, probably XVII. Arose spontaneously in the BALB/cJ inbred strain. Homozygotes are recognizable at about 18 days by the extra length of the guard hairs. At weaning the guard hairs are more than twice normal length. Vibrissae are also very long (Dickie, 1963).

gp, gaping lids, recessive, linkage not known. Not allelic with *oe* or *lg* (Miller, 1964) but has not been tested for allelism with *lo*. Arose in a stock carrying *ax*. In homozygotes the eyes are open at birth, and become opaque with age. Homozygotes are viable and fertile. The gene appears to have full penetrance. From 15 days of gestation onward the lens is about twice normal size (Kelton, 1964).

gr, grizzled, recessive, X. Homozygotes resemble c^{ch}/c^{ch} mice. The yellow pigment becomes white, but the black pigment is unaffected. In combination with nonagouti the hair on the ears and genitalia is white instead of yellow (Falconer, 1950). Segregation is abnormal with a deficiency of *gr*/*gr* which is due to prenatal mortality from unknown causes (Bloom, 1962).

Gs, greasy, semidominant, XX (sex-linked). Discovered among later descendants of an irradiated male but is probably of spontaneous origin. Possibly an allele of *Ta*. *Gs*/+ resembles *Ta*/+. Homozygous females and hemizygous males have shiny fur like the corresponding *Ta* genotypes but lack the bare patches behind the ears, the dark middorsal stripe in agouti mice, and the characteristic sticky feel of the tail (Larsen, 1964).

Gy, gyro, semidominant, XX (sex-linked). Hemizygous males show circling behavior, abnormal development of the long bones and ribs, and are sterile. Heterozygous females show incomplete penetrance of the circling behavior and no bony abnormalities have been detected (Lyon, 1961*b*).

H-1 locus, histocompatibility-1, I. For this and other histocompatibility loci see Chap. 24.

H-2 locus, histocompatibility-2, IX.

H-3 locus, histocompatibility-3, V.

H-4 locus, histocompatibility-4, I.

ha, hemolytic anemia, recessive, linkage not known, Chap. 17. Arose spontaneously in the DBA/1J inbred strain. Homozygotes show neonatal hypochromic anemia, microcytosis, and jaundice. The abnormality is not the result of either maternal-fetal incompatibility or the presence of abnormal hemoglobin. Many die within a few days after birth but occasional animals survive 6 months or more. All are sterile (Bernstein, 1963).

Hba locus, hemoglobin α-chain, linkage not known, Chap. 17. The several alleles at this locus were first recognized by their effect on the solubility of carbon monoxyhemoglobin, and the locus was called *Sol* (Popp, 1962*a*). It was later shown that alleles at the same locus govern differences in tryptic peptides of the α-chain of the hemoglobin molecule. *Hba* is therefore probably the structural locus for the α-chain (Popp, 1962*b*; Hutton et al., 1962*b*). It is not linked to *Hbb*, the probable structural locus for the β-chain. *Hba* alleles are classified on the basis of solubility, crystal shape, or peptide differences revealed by electrophoresis and chromatography (Popp, (1965*b*; Hutton et al., 1964). At least four different alleles have been described.

Hbb locus, hemoglobin β-chain, I, Chap. 17. Called *Hb* when first discovered. Two codominant alleles are well established, Hbb^s which produces an electrophoretically single hemoglobin, and Hbb^d which produces a diffuse electrophoretic pattern with two distinct hemoglobins (Ranney and Gluecksohn-Waelsch, 1955; Gluecksohn-Waelsch et al., 1957; E. S. Russell and Gerald, 1958; Popp, 1965*b*). The two hemoglobins produced by Hbb^d have different "β"-chains and both of these are somewhat different from the β-chain produced by Hbb^s (Hutton et al., 1962*a*, 1962*b*; Popp, 1962*b*). The *Hbb* locus is thought to be the structural locus for the β-chain and to be structurally complex, possibly duplicated in the Hbb^d allele. Morton (1962) has described a possible third allele, provisionally symbolized Hbb^p, which produces hemoglobin distinguishable from Hbb^d by minor electrophoretic differences.

Hc locus, hemolytic complement, linkage not known, Chap. 15. The locus was independently discovered by immunological methods and the antigen produced by Hc^1 was called MuB1 (Cinader et al., 1964). Two alleles are known, Hc^1, found in strains DBA/1J, BALB/cJ, C57BL/6J, and others, which determines the presence of hemolytic complement in serum, and Hc^0, found in strain DBA/2J, which determines the absence of detectable hemolytic complement in the serum (Herzenberg et al., 1963*b*). Hc^1/Hc^0 mice are indistinguishable from Hc^1/Hc^1. Absence of complement has no apparent effect on viability and does not reduce bacterial phagocytosis (Stiffel et al., 1964).

hf, hepatic fusion, recessive, I. Discovered in the MA/MyJ inbred strain in which the mutant is homozygous. Homozygotes have varying degrees of fusion between the left portion of the central lobe and the adjacent left lateral lobe. Penetrance decreases on outcrossing, suggesting that the residual genotype of the MA strain is favorable for the complete expression of the gene. Similar fusion has not been discovered in other strains routinely autopsied at The Jackson Laboratory (Bunker, 1959).

Hk, hook, semidominant, XVIII. Arose spontaneously in a piebald shaker stock. Homozygotes are viable and fertile, and at least on some genetic backgrounds are more severely affected than heterozygotes. Heterozygotes may be not detectably abnormal, or they may have shortened, crooked, or hook-shaped tails, and an abnormal anus. The anus tends to be slitshaped rather than round, and to be bare rather than surrounded by hair. It may be very much elongated and displaced toward the tail, sometimes running well out onto the tail (Holman, 1951).

hl, hair-loss, recessive, VI. Arose spontaneously. Homozygotes show progressive degeneration of hair follicles over the entire body, loss of hair becoming obvious after about 5 weeks. Some specimens never lose all hair, others become very wrinkled and resemble rhino (Hollander and Gowen, 1959). An interesting peculiarity of this mutant is that *hl*/+ offspring of *hl*/*hl* mothers have extremely fragile skeletons which break easily and lead to excessive mortality until the age of about 2 weeks. The antagonism is probably prenatal. It is not immunological. The suggestion has been made that *hl/hl* mice may maintain and require a smaller pool of calcium than normal mice. Their *hl/hl* offspring are able to obtain enough to satisfy their smaller requirement but their *hl/+* offspring obtain too little (Hollander, 1960).

Hm, hammer-toe, dominant, XVII. Found in a linkage cross. Homozygotes are viable and fertile. Both homozygotes and heterozygotes are characterized by feet in which the second phalanx of digits 2, 3, 4, and 5 is strongly flexed. All four feet are affected. Homozygotes have webbing between digits 2, 3, 4, and 5, extending to the nails in the hind feet and to the base of the distal phalanges in the forefeet. Heterozygotes also show webbing, but it usually does not extend so far on the toes (Green, 1964*b*; Green and Kaufer, unpublished).

Hp, hairpin tail, probably semidominant, linkage not known. Arose in the AKR/J strain. Heterozygotes usually have a short kinky tail and abnormalities of the sacral, lumbar, and thoracic vertebrae, with most severe effects in the sacral region. Viability is reduced and females have small litters. Spermatogenesis is abnormal but males sire litters of about normal size. The fate of homozygotes is unknown (Dickie, 1965).

hr locus, III.

hr, hairless, recessive, Fig. 8-1*G*, Chap. 23. Found in a mouse caught in an aviary in London (Brooke, 1926). Homozygotes develop a normal coat up to the age of about 10 days, at which time they begin to lose their hair. The complete hair is lost from the follicle. After a short time a few thin fuzzy hairs grow in but are soon lost. Waves of similar growth occur at intervals of about a month thereafter but the animals appear essentially hairless (Crew and Mirskaia, 1931). The regenerating hairs come exclusively from guard-hair (tylotrich) follicles, which are, however, very abnormal (Mann and Straile, 1961). The vibrissae, which also grow from tylotrich follicles are repeatedly shed and become more abnormal with age. The toenails are excessively long and curved. Histologically there is a hyperkeratosis of the stratified epithelium and the upper part of the hair canals beginning at about 14 days. Hair club formation is abnormal and the lower part of the follicles tend to separate from the upper part. The isolated lower parts develop into cysts which become quite large and numerous (David, 1932; Fraser, 1946). Some cysts arise from isolated sebaceous glands. Regardless of their origin all cysts at first undergo a sebaceous transformation and later a keratinization (Montagna et al., 1952). Hairless mice are fertile but most females do not nurse their young well.

hr^{rh}, rhino, recessive. Arose as a spontaneous mutation in the descendants of a cross between two inbred lines. Homozygotes are similar to *hr/hr* except that no hair regeneration occurs, and the skin becomes thickened and enormously wrinkled (Howard, 1940). The hyperkeratosis in the hair follicles is very extensive, producing balls of keratin, utriculi, in such large numbers that the skin area is greatly increased. In addition, the subepidermal cysts are much larger than in *hr/hr* (Fraser, 1946). Rhino females do not nurse their young well. hr/hr^{rh} mice resemble *hr/hr* but histologically are somewhat intermediate. The rhino condition is not cured by massive doses of vitamin A (Mauer, 1961). Rhino skin when transplanted to normal hosts grows some hair around the edges, but this is probably due to the nonspecific effect of transplantation (Tsuzi and Yosida, 1960).

hr^{ba}, bald, recessive. Arose spontaneously in a stock of albino mice of unknown origin. This allele is intermediate between *hr* and hr^{rh} in its effects (Garber, 1952*a*).

Ht, hightail, semidominant, VI. Radiation-induced. Homozygotes die before birth. In heterozygotes the tail emerges high, is short and thick at the base, but is never kinked. *Ht/+* mice are usually smaller than their normal littermates (Gower, 1957).

hy-1, hydrocephalus-1, recessive, linkage not known, probably extinct. Discovered by Clark (1932). Penetrance is incomplete. Homozygotes with recognizable hydrocephalus usually die but a few reach maturity and breed. The abnormality is recognized externally as a dome-shaped head which may sometimes be seen at birth and in other cases develops during the first 2 weeks (Clark, 1934). Internally, the whole ventricular system is dilated, but the cause has not been determined (Bonnevie and Brodal, 1946).

hy-2, hydrocephalus-2, recessive, linkage not known, probably extinct. Discovered by Zimmerman (1933) in the offspring of mice caught in the wild. Penetrance was reported to be complete and homozygotes never bred. Not allelic with *hy-1* (Clark, 1935). The anomaly resembles *hy-1* but is more severe.

hy-3, hydrocephalus-3, recessive, linkage not known, Fig. 8-1*D*. Found by inbreeding a heterogeneous stock of laboratory mice (Grüneberg, 1943*b*). It has not been tested for allelism with *hy-1* or *hy-2*. Penetrance is incomplete. Homozy-

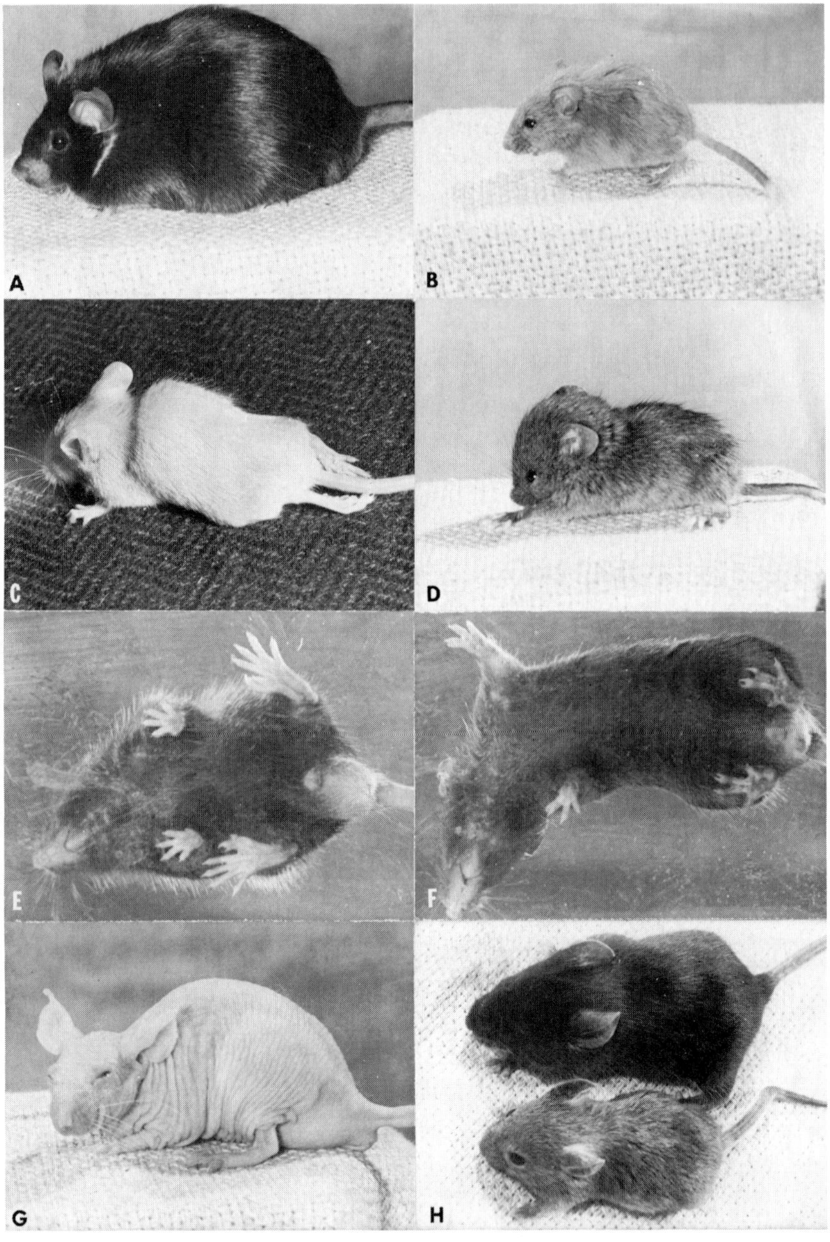

Fig. 8-1. Photographs of some mutant mice, approximately one-half natural size but not all reproduced at the same magnification.

A. Obese (*ob/ob*). *B.* Dwarf (*dw/dw*). *C.* Dystrophia muscularis (*dy/dy*). *D.* Hydrocephalus-3 (*hy*-3/*hy*-3). *E.* Luxoid (*lu*/+), from below through glass. *F.* Oligosyndactylism (*Os*/+), from below through glass. *G.* Hairless (*hr/hr*). *H.* Achondroplasia (+/— above, *cn/cn* below).

gotes with frank hydrocephalus die by 4 or 5 weeks of age. It seems likely that some apparently normal mice which survive and breed are *hy*-3/*hy*-3, but no pure breeding line in which all matings produce some hydrocephalus has been established. The lateral ventricles and the third ventricle are enlarged, the aqueduct of Sylvius and the fourth ventricle are only slightly affected, and there is some dilatation of the ventral subarachnoid cistern. The hydrocephalus seems to be due to a defect in the subarachnoid space caused by an abnormal postnatal differentiation of the arachnoid mater and the pia mater which prevents their separation (Berry, 1961*a*).

hz, haze, recessive, linkage not known. Arose in the DBA/2J strain. In homozygotes the hair tips are brownish and the underfur is white. At birth the eyes are light colored (Dickie, 1965).

ic, ichthyosis, recessive, linkage not known. Arose spontaneously in a sibmated stock. Homozygotes are recognizable at 2 days of age by their shorter vibrissae. The vibrissae become curved a

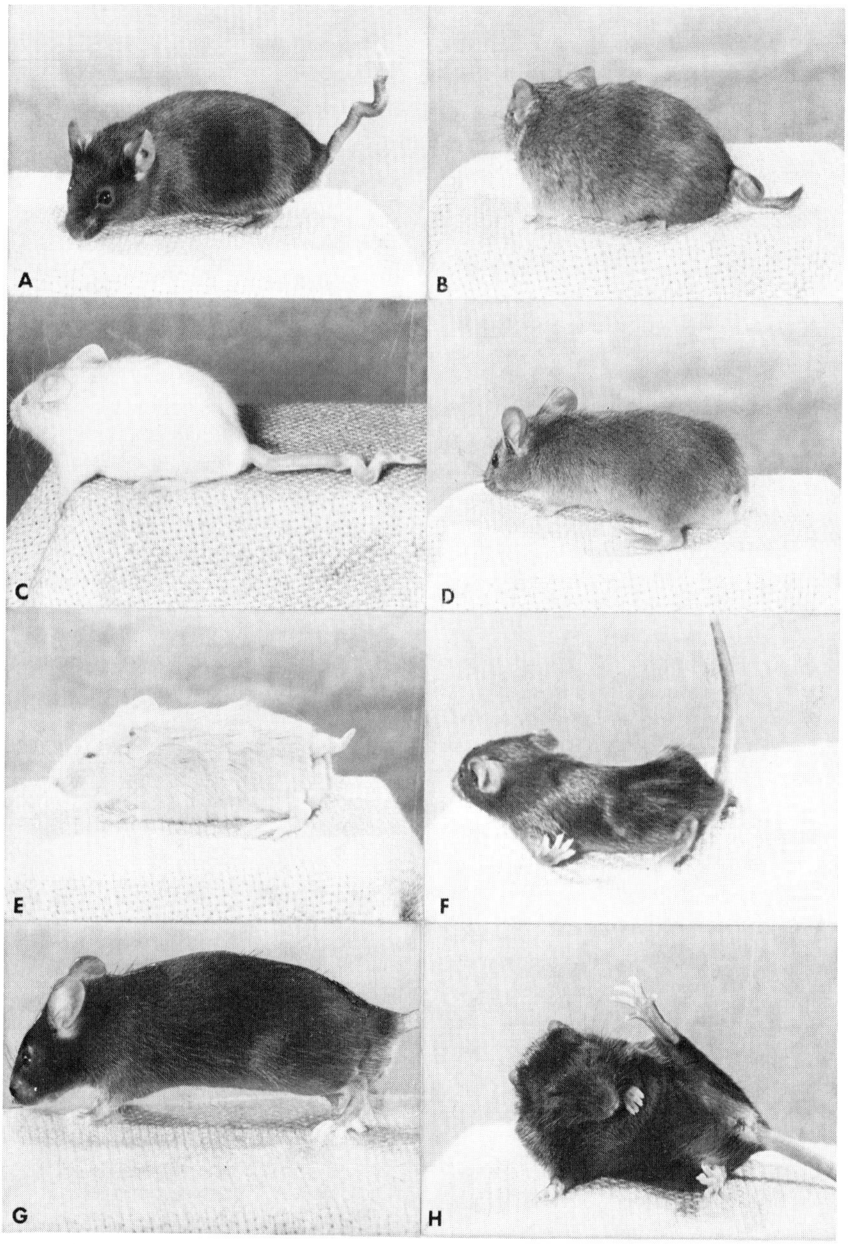

Fig. 8-2. Photographs of some mutant mice, approximately one-half natural size but not all reproduced at the same magnification.

A. Bent-tail (*Bn*/Y). *B.* Loop-tail (*Lp*/+). *C.* Tail-kinks (*tk*/*tk*). *D.* Tailless (T/t^6). *E.* Pudgy (*pu*/*pu*). *F.* Dystonia musculorum (*dt*/*dt*). *G.* Spastic (*spa*/*spa*). *H.* Lethargic (*lh*/*lh*).

day or so later. The first coat is delayed and is short, thin, and somewhat curly. During the third and fourth weeks the skin may develop scales. Sometimes these are large and hard and may reduce viability considerably. Often the tail has constrictions resulting in loss of the tip. Adults may be bare or may have a thin fuzzy coat. Viability and fertility are considerably reduced (Carter and Phillips, 1950). Hairs of the coat are thin and wavy and have an unevenly compressed medulla with irregular distribution of melanin granules. The tail epidermis is thicker and more active than normal (Spearman, 1960).

Id-1, isocitrate dehydrogenase-1 locus, linkage not known. This locus controls an electrophoretic difference in one of the three isocitrate dehydrogenases known to be present in mouse cells, namely the one that is specific for triphosphopyridine nucleotide (NADP) and is located in the soluble fraction of the cell. It migrates anodally on starch gel at pH 6.0. Two alleles are known; $Id\text{-}1^a$, controlling the slow migrating variant and found in strains C3H/HeJ, C57BL/6J, BALB/cJ and others, and $Id\text{-}1^b$, controlling the fast migrating variant and found in strains DBA/1J, DBA/2J, AKR/J, CBA/J and others. $Id\text{-}1^a/Id\text{-}1^b$ mice have three electrophoretic forms, the two parental forms and a more abundant intermediate form. This suggests that the enzyme controlled by this locus is a dimer (Henderson, 1965).

Ig-1 locus, immunoglobulin-1, Chap. 15. Linkage group not known, but closely linked to *Ig-2*. There is no agreement on the symbol to be used for this locus. Synonyms are: γB^A (Kelus and Moor-Jankowski, 1961), *MuA2* (Dubiski and Cinader, 1963) As_a (Dray et al., 1963), *Gg* (Wunderlich and Herzenberg, 1963; Herzenberg et al., 1963*a*), and *Iga* (Mishell and Fahey, 1964). The terminology used here (Herzenberg et al., 1965) is in conformity with that used for the histocompatibility loci. The antigenic activities controlled by this locus were demonstrated by preparing isoprecipitins in strains carrying one allele to γ-globulins from strains carrying another allele. The antigenic activity is found on the H chain of the 7S γ_{2a}-globulin (Mishell and Fahey, 1964; Fahey et al., 1964). At least eight codominant alleles are known (Lieberman and Dray, 1964; Herzenberg et al., 1965). In Herzenberg's classification the alleles and strains they occur in are: $Ig\text{-}1^a$ in C3H/HeJ and BALB/c; $Ig\text{-}1^b$ in C57BL/10J; $Ig\text{-}1^c$ in DBA/2J and DBA/1J; $Ig\text{-}1^d$ in AKR/J; $Ig\text{-}1^e$ in A/J; $Ig\text{-}1^f$ in CE/J; $Ig\text{-}1^g$ in RIII/J; and $Ig\text{-}1^h$ in SEA/Gn and BDP/J.

Ig-2 locus, immunoglobulin-2, linkage not known, but very close to *Ig-1*, Chap. 15. This locus specifies antigens on the H chain of the β_{2A}-globulin. Two alleles have been identified directly, one in C3H/HeSn and the other in DBA/2J. At least one other allele remains to be identified, as a number of strains have neither the C3H or the DBA/2 antigen (Herzenberg, 1964).

The antigenic specificity of a third immunoglobulin has been shown to be controlled by a locus very closely linked to *Ig-1* or identical to it (Lieberman et al., 1965). The specificity is found on the H chain of the 7S globulin called γ_{2b} in the terminology of Fahey et al. (1964) or γG-Be2 in the terminology of the authors. Different alleles are present in the BALB/c and C57BL/6 strains.

iv, situs inversus viscerum, recessive, linkage not known. Arose spontaneously in a noninbred stock (Hummel and Chapman, 1959). Penetrance is incomplete, 71 per cent in the stocks examined. Homozygotes may show left-right transposition of thoracic and abdominal viscera and associated blood vessels, anomalous relationships of postcaval and azygous veins, anomalous position of the hepatic portal vein, abnormalities in spleen position and shape, and abnormalities in liver and lung lobation.

j, jaw-lethal, recessive, linkage not known, probably extinct. Found in the descendants of both irradiated and control mice by Little and Bagg (1924). Homozygotes are stillborn, death resulting from abnormalities of the head including shortening of the head, reduced or absent lower jaw, microphthalmia, anophthalmia or cyclopia, abnormal or missing tongue, lack of opening between oral and nasal passages and the trachea and pharynx, and various abnormalities of the brain (Johnson, 1926).

ja, jaundiced, recessive, linkage not known, Chap. 17. Arose in the 129 inbred strain. Homozygotes are anemic at birth and become jaundiced within a few hours. The anemia is present in 15-day fetuses. Examination of the blood of newborn jaundiced mice shows that they have a severe microcytic anemia with about half the normal number of red cells. There is considerable erythrocyte destruction and a compensating release of immature cells into the blood. Homozygotes die within a day or two of birth (Stevens et al., 1959).

je, jerker, recessive, linkage not known (Wallace, 1958*a*), Chap. 32. Originated as a dancing mouse found by a fancier and given to Grüneberg (Grüneberg et al., 1941). Homozygotes show the typical behavior of the circling mutants, head-tossing, circling, hyperactivity, and deafness. The abnormal behavior is associated with postnatal degeneration of the sensory cells of the cochlea and the sacculus and utriculus (Deol, 1954).

jg, jagged-tail, recessive, XVII. Arose spontaneously in the C3H/HeJ inbred strain. Homozygotes have tails with a normal tapering tip but usually much shortened and kinked. In the most severely affected individuals other parts of the vertebral column may be abnormal; in the less severely affected it appears nearly normal. Testes are much reduced in size and the males are probably sterile. Females have small ovaries. They may breed but litters are few and small. Viability of both sexes is considerably reduced (Green, 1964*a*).

ji, jittery, recessive, X. Discovered by DeOme (1945) as a spontaneous mutation in the Bagg albino strain. Homozygotes usually die by 4 weeks of age. They show muscular incoordination beginning at 10 to 16 days. Epileptiform seizures follow within a few days. As seizures and incoordination become more severe the mice lose ability to obtain food and die apparently from starvation and thirst. Polycystic alterations in the white matter of the brain have been reported (Harman, 1950).

jo, jolting, recessive, linkage not known. Arose in the DBA/2WyDi strain. Homozygotes shiver or quake in the anterior part of the body and show some incoordination of the hind limbs. They often survive to adulthood and some have bred (Dickie, 1965).

jp, jimpy, recessive, XX (sex linked), Chap. 15, 32. Arose spontaneously in an inbred line. Hemizygous males die between 20 and 40 days of age. Homozygous females cannot therefore be produced. Affected males appear normal when sitting quietly, but beginning at about 2 weeks they show a marked tremor when attempting movement. The tremor is most noticeable in the hindquarters. From 3 weeks on convulsions may occur (Phillips, 1954). The central nervous system is very deficient in myelin. Cells of certain white matter tracts contain nonpolar lipids in the form of strongly sudanophilic cytoplasmic granules. Impaired myelin formation appears to be accompanied by formation of an abnormal lipid resulting from abnormal myelin synthesis or from myelin destruction (Sidman et al., 1964).

jt, joined-toes, recessive, linkage not known. Arose spontaneously in the 129 inbred strain. Homozygotes show fusion of the soft tissue of the digits, more commonly on the hind feet than on the forefeet. Fusions are more common between the second and third, and third and fourth digits, but all five may be involved (Stevens, 1955, as "syndactylism").

kd, kidney disease, recessive, linkage not known. Found in the CBA/H strain. Homozygotes develop nephrosis recognizable at the age of 2 to 3 months when the urine becomes pale or colorless. Affected animals can produce two or three litters before becoming ill (Hulse et al., 1965).

kr, kreisler, recessive, V. Found by Hertwig (1942) in the descendants of an irradiated male. Viability and fertility are less than normal. Males are more likely to breed than females. In homozygotes both the bony and membranous labyrinths are very abnormal. The abnormality traces back to faulty segmentation of the neural tube in the region of the rhombencephalon in 8¾-day embryos, followed by degeneration of cells in the fourth rhombomere (Deol, 1964*a*). The abnormalities of the neural tube are thought to lead to an abnormal position of the ear vesicles which in 9-day kreisler embryos are not directly in contact with the neural tube as they are in normal embryos. An imperfect capsule forms around the vesicle and extracapsular cysts result from growth of the vesicle through gaps in the capsule. The cysts are often quite large and located under the brain (Hertwig, 1944; Deol, 1964*a*). The abnormal behavior resulting from these defects is very similar to that of the degenerative group of circling mutants.

Kw, kinky-waltzer, dominant, or semidominant, linkage not known. Probably radiation induced. Heterozygotes are viable and fertile. Penetrance is incomplete in heterozygotes. No description of homozygotes has been published. In heterozygotes the amount of kinking of the tail is variable and may or may not be accompanied by a balance defect. The balance defect has not been seen without the kinky tail (Gower and Cupp, 1958, as "kinky tail balance defect"). This mutant resembles *Q*, *Fu*, and Fu^{ki} but tests for allelism have not been reported.

la, leaner, recessive, linkage not known. Arose spontaneously in the AKR/J strain (Dickie, 1962*b*, as "heeler"). Most homozygotes survive and can breed. They are recognizable at 10 days of age. Adults show instability of trunk and altered muscle tone in trunk and limbs. They have degenerative changes affecting virtually all neuron types in a patchy distribution in the cortex of the vermis of the cerebellum. The distribution of the disorder within the cerebellum accounts for the more severe ataxia in the trunk than in the limbs, in contrast to the ataxia of *rl*, *sg*, and *wv*, which have malformations affecting the cerebellar cortex more diffusely (Sidman, 1965).

Lc, lurcher, semidominant, XI. Arose as a spontaneous mutation in a male homozygous for white (Mi^{wh}). Heterozygotes show a characteristic swaying of the hindquarters during which they fall to one side or the other. There is no trembling but a jerky up and down movement, particularly in older mice. They are identifiable with sureness at 12 to 14 days of age. Homozygotes die shortly after birth but have no visible abnormalities. Heterozygotes are smaller than normal at maturity. Males are fully fertile but the litter size of heterozygous females is reduced about one-quarter (Phillips, 1960*a*).

ld, limb deformity, recessive, V, Chap. 15. Arose at Oak Ridge National Laboratory, possibly radiation induced. Homozygotes have reduced viability but often live and breed. Skeletal deformities appear to be confined to the portions of the limbs below the elbow and knee. The radius and ulna appear fused into a broad flat bone; the tibia and fibula are replaced by a short bone, usually triangular. The fore and hind feet are radically reduced and disorganized (Cupp, 1960). A similar mutation at the same locus occurred at The Jackson Laboratory. The tibia and fibula are replaced by a single long bone in homozygotes of this allele, and the viability appears to be lower. Whether the difference is an effect of the residual genotype or due to a real difference in the alleles is not known (Green, 1962; Cupp, 1962).

le, light ears, recessive, XVII. Arose spontaneously in the C3H/HeJ strain. Closely resembles *ep*. Homozygotes have light ears, tail, and feet, and a slightly diluted juvenile coat which becomes darker in the adult. Eyes are pale at birth and darken with age (Lane and E. L. Green, 1965, personal communication).

lg, lid gap, recessive, linkage not known. Spontaneous. Homozygotes are viable (L. B. Russell, 1961). A repeat mutation at this locus occurred at the University of British Columbia and was studied by Watney and Miller (1964). Penetrance is incomplete. About 70 per cent of homozygotes have one or both eyes open at birth. Cortisone administered to the mother on the 15th day of pregnancy reduces the incidence of the defect in homozygotes to zero (Chap. 14). *lg* is not allelic with *gp* and *oe*, and has not been tested for allelism with *lo* (Miller, 1964).

lh, lethargic, recessive, linkage not known, Fig. 8-2*H*. Arose spontaneously in the BALB/cGn strain. Homozygotes make long pauses between movements, often hunch the back, and may raise first a front foot and then a hind foot. They may sit up with the front feet contracted for several minutes. Both sexes breed and there is little early mortality (Dickie, 1964*b*).

ln, leaden, recessive, XIII, Chap. 21. Arose spontaneously in the C57BR inbred line (Murray, 1933). In its effect on coat color *ln* is indistinguishable from *d*. Like *d* it causes clumping of melanin granules into larger masses, but no change in color of the pigment. The clumping is due to the shape of the melanocytes, which have fewer and thinner dendritic processes than wild-type melanocytes with the result that the pigment granules are largely clumped around the nucleus (Markert and Silvers, 1959).

lo, lids open, recessive, linkage not known. Arose spontaneously. Homozygotes are viable (Saylors, 1961). Tests for allelism with other similar mutants (*gp*, *lg*, *oe*) have not been reported.

Lp, loop-tail, semidominant, XIII, Fig. 8-2*B*. Discovered by Strong and Hollander (1949) as a spontaneous mutation in the A strain. Lethal when homozygous, probably not fully penetrant in heterozygotes. Heterozygotes have looped or crooked tails and often show a wobbling of the head. In females the vagina may be imperforate. Homozygotes have open neural folds in the region of the brain and may or may not have open spinal cords as well. They are usually alive until birth but never survive thereafter (Strong and Hollander, 1949). The axial skeleton and ribs of homozygotes are abnormal, apparently as a secondary result of the abnormalities of the nervous system (Stein and Mackensen, 1957). In *Lp/Lp* embryos of 9 to 9½ days, the shortening or regression of the primitive streak is retarded and the neural plate and notochord are shorter than normal. Failure of the neural tube to close apparently results from its failure to elongate (Smith and Stein, 1962).

lr, lens rupture, recessive, linkage not known, Chap. 29. Arose spontaneously in an inbred albino strain (Fraser and Herer, 1948). Penetrance is complete on some genetic backgrounds, but may be reduced on others. In homozygotes, beginning at about 3 weeks of age there is cataractous degeneration of the lens with rupture of the capsule and expulsion of the lens nucleus into the vitreous chamber. The lens shrinks, tearing the suspensory ligaments, and may pass through the pupil into the anterior chamber (Fraser and Herer, 1950). This mutant resembles *ec* but has not been tested for allelism with it.

ls, lethal spotting, recessive, V, Chap. 15, 21,

29. Arose in a subline of an inbred strain, C57BL-a^t. Homozygotes resemble piebald mice (s/s) in having considerable white spotting. They die usually in the third week of life with megacolon associated with lack of ganglion cells in the lower colon (Lane, 1966). The mutant seems to exert its effect on pigmentation by reducing the number of melanoblasts, possibly through an effect on the neural crest (Mayer and Maltby, 1964).

lst, Strong's luxoid, semidominant, linkage not known, Chap. 15, 29. Arose in a line treated with methylcholanthrene for 22 generations (Strong and Hardy, 1956). Heterozygotes show preaxial abnormalities of the hind feet, including polydactyly and triphalangy of the hallux. Very rarely the forefeet may also be affected. Penetrance is incomplete and dependent on the genetic background. Homozygotes show preaxial polydactyly of all four feet, reductions and duplications of the radius, reductions and rarely duplications of the tibia, reduction of the pubis, modifications of the skull, temporary dorsal alopecia, a posterior shift of the umbilicus (Forsthoefel, 1962), and a temporary postnatal anemia due to bleeding at the umbilicus (Kuharcik and Forsthoefel, 1963). There is no effect on the number of ribs, presacral vertebrae, or sternebrae in either heterozygotes or homozygotes. Males are sterile but females sometimes breed. Forsthoefel (1963) found that embryonic development of parts of the limbs, head, integument, and belly is retarded.

lt, lustrous, recessive, linkage not known. Arose in the DBA/2J strain. The fur of homozygotes has a very satiny appearance and the vibrissae are curly. Not allelic with satin (Dickie, 1965).

lu, luxoid, semidominant, II, Fig. 8-1*E*, Chap. 14, 15. Arose spontaneously in the C3H/He inbred strain (Green, 1955). Heterozygotes show preaxial polydactyly or hyperphalangy of the hind feet. Penetrance is close to zero in the C3H/He strain and about 90 per cent in the C57BL/10 strain. Homozygotes show preaxial polydactyly, hyperphalangy, or oligodactyly of the hind feet, preaxial polydactyly or hyperphalangy of the forefeet, tibial hemimelia, and occasionally radial hemimelia and tail kinks. The gene tends to increase the number of presacral vertebrae, ribs, sternebrae, and total number of vertebrae, the effect being greater in homozygotes than in heterozygotes. Males are sterile but an occasional female may breed (Forsthoefel, 1958). The developmental effects of *lu* in homozygotes can be traced back to the $10\frac{1}{2}$-day stage when the posterior end of the coelom is found to be more caudal than in normal controls. The somites of the tail may be abnormal, and there are more somites than in controls. An increased rate of growth along the longitudinal axis with some loss of orderly control is proposed as a possible explanation of the axial anomalies. How the limb abnormalities are related to the axial abnormalities is not clear (Forsthoefel, 1959). A very similar mutant has been described by Kobozieff and Pomriaskinsky-Kobozieff (1962 and earlier). It has not been tested for allelism with *lu.*

Lv locus, levulinate dehydratase, XIV, Chap. 19. An interstrain difference in the hepatic activity of δ-aminolevulinate dehydratase is controlled by this locus. Two alleles have been described, Lv^a for high activity (in strain AKR/J) and Lv^b for low activity (in strain C57BL/6J). The enzyme effects the condensation of two molecules of δ-aminolevulinate to form porphobilinogen. Lv^b/Lv^b mice have lower levels of enzyme than Lv^a/Lv^a mice at all stages from 8 days before birth to adulthood. Heterozygotes are intermediate (R. L. Russell and Coleman, 1963).

lx, luxate, semidominant, XVII, Chap. 15. Found in descendants of a silver mouse obtained from a fancier (Carter, 1948). A similar mutant had been reported by Rabaud (1914; see Grüneberg, 1952, for description) but is extinct and cannot now be tested for allelism with *lx.* Heterozygous *lx*/+ mice show preaxial polydactyly (including hyperphalangy of the first digit) of the hind feet. Penetrance is incomplete and dependent on the genetic background. Homozygotes show preaxial polydactyly or oligodactyly of the hind feet, reduction of the tibia, loss of part of the femur and pubis, decrease in the number of presacral vertebrae, and anomalies of the urogenital system including horseshoe kidney, hydronephrosis, and hydroureter (Carter, 1951*c*, 1953). The abnormalities of homozygotes can be traced back to the 10-day stage when the posterior end of the coelom is nearly a full segment anterior to its position in normal sibs. A day later the right umbilical artery is highly abnormal and, with the left umbilical artery, forms a ring which is often too small to allow free passage of the developing kidneys as they migrate forward. The posterior limb buds of homozygotes are narrower than normal from a very early stage (Carter, 1954). The hind limb bud, the umbilical arteries, and the posterior end of the coelom are all close together in the 10-day embryo, but the causative relationship between the abnormalities of the three parts is not clear.

m, misty, recessive, VIII. Arose spontaneously in the DBA/J inbred strain. Homozygotes are lighter in color than normal and usually have a white belly spot and tail tip. The color of *m/m* mice is not so diluted as that of *d/d* or *ln/ln*, and the hairs of the coat have more cortical pigment than in *d/d* or *ln/ln* (Woolley, 1941, 1945).

ma, matted, recessive, XVI. Arose spontaneously in the CBA/Gr inbred strain. Homozygotes can first be identified between 2 and 4 weeks of age. The hairs are more erect than normal and are matted in clumps, there is a tendency to baldness depending on areas of friction and resulting hair breakage, and a color change to russet happens toward the end of each hair cycle in black hair. Hairs are normal in morphology, but are brittle and tend to split longitudinally. Viability and fertility are normal (Searle and Spearman, 1957). The defect in matted hairs is thought to be in the cuticle (Jarrett and Spearman, 1957).

mc, marcel, recessive, linkage not known. Arose in the C57BL/10 strain. Homozygotes have pronounced waves in the coat at 14 to 21 days. Female homozygotes are sterile. Not allelic with *we*, *wa-1*, or *wa-2* (Foerster, 1956).

md, mahoganoid, recessive, linkage not known. Arose in the C3H/HeJ inbred strain. This gene is identical to mahogany (*mg*) in its effects on the coat color of *A/—* and *a/a* mice. The two mutants are neither allelic nor linked (Lane, 1960*a*).

mdg, muscular dysgenesis, recessive, linkage not known, Chap. 29. Arose spontaneously in a tailless line. Homozygotes never develop normal skeletal muscle. The first observable effects occur at 13 days of gestation when there is marked edema and at the same time failure of myoblasts to differentiate into striated myotubes. Subsequently all skeletal muscle cells degenerate. At birth the thorax and limbs are almost bare of muscle. Cardiac and smooth muscle as well as nervous tissue are normal. There are numerous skeletal abnormalities, including a short lower jaw and usually a cleft palate, which may be secondary to the abnormalities of muscle, Homozygotes do not survive birth (Gluecksohn-Waelsch, 1963; Pai, 1965).

mg, mahogany, recessive, V. Appeared in an agouti stock of unknown origin. In homozygous condition *mg* darkens the back, ears, and tail of agouti mice and darkens the ears and tail of non-agouti mice. *a/a mg/mg* mice resemble a^e/a^e (Lane and Green, 1960).

mi locus, XI, Chap. 21, 29.

mi, microphthalmia, semidominant. Found among the descendants of an irradiated male (Hertwig, 1942). In heterozygotes the eyes have less iris pigment than normal, both at birth and throughout life (Grüneberg, 1948). Heterozygotes may also show white spotting on the belly, head, and tail. Homozygotes are devoid of pigment in the hair and eyes. The eyes are very small usually with cataracts (Tost, 1958) and the eyelids never open. Most homozygotes die at about weaning. There is a failure of secondary bone absorption, as in grey-lethal but a little less severe (Grüneberg, 1948; Freye, 1956). During development the optic vesicle fails to form a proper cup and the choroid fissure remains permanently open (Müller, 1950, 1951).

mi^{bw}, black-eyed white, recessive to wild type. Arose spontaneously in the C3H inbred strain. Homozygotes have white coats and black eyes. Occasionally a few pigmented hairs may be found on the back (Kreitner, 1957). Melanoblasts cannot be demonstrated in the skin of embryos of these mice by transplantation to the eyes of albino hosts (Markert, 1960). Mi^{wh}/mi^{bw} mice have pigmented eyes and are white with some pale yellow spots which become white in the adult (Schaible, 1963*c*).

mi^{sp}, mi-spotted. Found in a C57BL/6J-Mi^{wh} stock. Homozygotes and $mi^{sp}/+$ are not detectably different from $+/+$ in color but have slightly less tyrosinase activity in the skin; Mi^{wh}/mi^{sp} are light yellow with dorsal and ventral white spots and pigmented eyes. Medullary pigment granules show much clumping and are yellowish brown (Wolfe and Coleman, 1964). mi/mi^{sp} mice are white with some pigmentation in the eyes and some flecks of pigmented hair on the back. All these combinations are viable and fertile (Wolfe and Coleman, 1964).

Mi^{wh}, white, semidominant. Found among offspring of a cross between the DBA and C57BL strains (Grobman and Charles, 1947). Heterozygotes are somewhat lighter than *d/d* mice and have light ears. They also may have a white belly spot and rarely a dorsal spot. Medullary pigment granules in the hairs appear to vary greatly in size and shape, apparently because of clumping (Wolfe and Coleman, 1964). Homozygotes are white with very slightly pigmented eyes. They have no pigment cells except a few in the retina (Markert and Silvers, 1956). They are slightly microphthalmic and are usually smaller and less fertile than normal. Mi^{wh}/mi are also white. They have slightly pigmented eyes, normal skeleton, and are slightly less microphthalmic than Mi^{wh}/Mi^{wh} (Grüneberg, 1952, 1953*c*).

mi^{ws}, white spot, semidominant. Arose in the

C57BL/6 strain. $mi^{ws}/+$ mice have a white diamond on the belly. mi^{ws}/mi^{ws} are all black-eyed white. Mi^{wh}/mi^{ws} are white with some dark patches and dark eyes (Miller, 1963; Hollander, 1964).

mk, microcytic anemia, recessive, linkage not known, Chap. 17. Arose spontaneously in a hybrid stock. Homozygotes are pale at birth and slightly smaller than normal. The red blood cells are extremely microcytic. At birth the number of red blood cells is slightly less than normal but rises so that by 8 weeks of age it may be one and two-thirds times normal. During the first week of life some homozygotes show an extreme scaliness of the skin. Only about 50 per cent of homozygotes survive to weaning (Nash et al., 1964).

mn, miniature, recessive, probably VI. Appeared in the descendants of an outcross of AKR to a stock at Columbia University. Homozygotes are smaller than their normal sibs at birth and have a dorsoventral flattening of the skull. They grow more slowly and are only about one-fourth to one-third normal size at 75 days. They do not seem to differ much from normal in body proportions or in other anatomical features or behavior, but mortality is high at all times up to 2 months when 96 per cent were dead (Bennett, 1961*b*).

Mo locus, XX (sex-linked). Mutations at the *Mo* locus seem to be very common. Numerous mutations similar in phenotype have been reported, and although allelism is difficult to prove in this case, it is likely that most of them are remutations at this locus (Lyon, 1960; Welshons and Russell, 1959).

Mo, mottled, semidominant, Chap. 15. Arose spontaneously in a cross segregating for several color factors (Fraser et al., 1953). Heterozygous females have irregular patches of full colored and very lightly colored fur over the whole coat. Patches with well-defined edges very rarely cross the middorsal or midventral line. Vibrissae are curly but the coat is not waved. Viability is reduced, some heterozygotes dying prenatally and some postnatally. Survivors are usually fertile. Hemizygous males die in utero at about 11 days of gestation with no visible abnormality (Falconer, 1953).

Mo^{br}, brindled, semidominant. Arose spontaneously in the C57BL inbred strain (Fraser et al., 1953). Heterozygous females are very similar to $Mo/+$ females in appearance but have normal viability. Hemizygous males are almost devoid of pigment except in the eyes and ears. The vibrissae are strongly curled and the coat is wavy. Males usually die when 2 weeks old but a few have lived and been fertile. They have a behavioral abnormality consisting of slight tremor, uncoordinated gait, and clasping of the hind feet when held up by the tail. These males have been used to produce Mo^{br}/Mo^{br} females which are identical in phenotype to hemizygous males. Mo/Mo^{br} females are indistinguishable from Mo^{br}/Mo^{br} females and Mo^{br}/Y males and die at the same age (Falconer, 1956).

Mo^{dp}, dappled, semidominant. Arose in a low-dosage γ-irradiation experiment. Heterozygous females are similar in color and curliness of vibrissae to $Mo/+$ females. Some have clubbing of the forefeet at birth or, at weaning, a tendency to walk on the dorsal surfaces of the hind feet. With age, calcified lumps may appear in the region of the periosteum of the thoracic and lumbar vertebrae. Hemizygous males die at about 17 days of gestation. They show bending and thickening of the ribs and distortion of the pectoral and pelvic girdles and limb bones (Phillips, 1961).

mr, maroon, recessive, linkage not known. Arose spontaneously in a "lactation" stock. Eyes are colorless at birth, but darken to a rich maroon. The coat color varies from near normal to extreme pallid, even among littermates and darkens with age. Not allelic with *d*, *ln*, *p*, *pa*, or *ru* and interacts with *si* in the F_1. May be an allele of *si* (Bateman, 1957).

mu, muted, recessive, linkage not known. Arose in a stock of *t*-alleles. Homozygotes have light eyes at birth and fur of a muted brown shade, often with white underfur. Some have a balance defect, similar to that of *pa*, which is due to absence of otoliths in one or both ears. Not allelic with *pa* (Lyon and Meredith, 1965).

Mx, myxovirus resistance, dominant, linkage not known. Strain A2G is resistant to intracerebral inoculations of neurotropic influenza A virus, and strains A/J and C3H/HeJ are susceptible. In crosses between A2G and the other two strains, resistance segregated as if controlled by a dominant gene at a single locus (Lindenmann, 1964).

my, blebs, recessive, linkage not known. Found among the descendants of irradiated mice (Little and Bagg, 1923). Effects are variable and penetrance is incomplete. This mutant has been the subject of a large number of investigations reviewed by Grüneberg (1952). The genetics and development were reinvestigated by Carter (1956, 1959*b*) who concluded that the effects of *my* include: embryonic subepidermal blebs leading to abnormalities of the eyes, skin, and hair, clubbing of the feet, ectopic viscera, and split sternum; pseudencephaly, acrania, renal agenesis, preaxial polydactyly, and syndactyly of the middle digits; and probably also hydronephrosis, ectopia viscerum, and midcerebral lesions.

N, naked, semidominant, VI. Arose as a spontaneous mutation in a stock at the Latvian University of Riga (Lebedinsky and Dauwart, 1927). Heterozygotes grow a nearly normal first coat. At 10 to 14 days the hairs begin to break off because of weakness due to incomplete keratinization. Breaking off and regeneration occurs in cycles and produces animals with irregular bare and haired patches which change as the cycle progresses. Heterozygotes are viable and fertile. Homozygotes lack vibrissae at birth. They often die before 10 days but some live and even breed. The effect on the coat is much more severe than in heterozygotes. Homozygotes never grow a complete coat. They have cycles of hair regeneration and loss, but the new hairs break off almost as soon as they are formed (David, 1932; Fraser, 1946).

Nil, neonatal intestinal lipoidosis, semidominant, I. Discovered in the A/Cam strain. Heterozygotes and homozygotes are classifiable at 0 to 5 days of age by the presence of white lipid in the wall of the small intestine visible through the skin. There is no milky fluid in the peritoneum. Expression is variable. Heterozygotes have variable penetrance, and homozygotes have greater expression and less viability than heterozygotes (Wallace, 1963*a*, 1965).

nu, nude, recessive, VII. Found in mice in the Virus Laboratory, Ruchill Hospital, Glasgow. Homozygotes do not grow even a first coat of hair and most die within a few weeks of weaning (Isaacson and Cattanach, 1962).

ob, obese, recessive, XI, Fig. 8-1*A*, Chap. 19, 20, 23, 29, 33. Arose as a spontaneous mutation in a multiple recessive stock (Ingalls et al., 1950). Homozygotes are first recognizable at about 4 weeks. They increase in weight rapidly and may reach three times the normal weight. Females are always sterile but an occasional male may breed, particularly if maintained on a restricted diet. Obesity is due to a moderate hyperphagia and to marked inactivity. There is moderate hyperglycemia which is not due to lack of insulin but probably to insulin insensitivity. There is increased secretion of insulin and hyperplasia of the islets of Langerhans. Incorporation of acetyl-CoA into fatty acids in adipose tissue is increased and lipolysis of tissue-triglycerides is depressed. Storage of fat and its nonutilization is thus favored (Christophe, 1963; see also Mayer, 1960).

oe, open eyelids, recessive, VII. Arose spontaneously in inbred strain 129/RrSv. Not allelic with *gp* but has not been tested for allelism with *lg* and *lo*. Homozygotes have open eyelids at birth. The cornea usually becomes opaque in adults, and the eyes are usually smaller than normal. A corneal staphyloma may be present. Homozygous embryos can be distinguished at 17 days when the eyelids normally close. Some homozygotes at this stage have a protruding lens, folded retina, and constricted cornea (Mackensen, 1960).

oel, open eyelids with cleft palate, recessive, linkage not known. Appeared in a phocomelic strain. Homozygotes have cleft palates and die soon after birth (Gluecksohn-Waelsch, 1961).

ol, oligodactyly, recessive, I, Chap. 15. Appeared among the descendants of an X-rayed male (Hertwig, 1939, 1942). Homozygotes show reduction of the postaxial digits of all four limbs in varying degrees of severity. There may also be reduction or absence of the ulna and fibula. The tail may be shortened and kinked, the last rib is reduced in size or absent, and the ribs and sternebrae may show fusions. The spleen may be reduced in size and deformed, and there may be horseshoe kidney or the kidneys may be cystic or absent (Freye, 1954). Few homozygotes live beyond 1 month. Survivors are small and do not breed.

or, ocular retardation, recessive, linkage not known. Appeared in a stock segregating for patch (*Ph*) (Truslove, 1962). Homozygotes have small eyes at birth and can be easily classified. The eyes develop normally up to the age of 11 days of gestation. After that they are smaller than normal because of the failure of the central artery and vein to establish a pathway along the choroid fissure, which closes completely. In the adult there is no optic nerve or optic chiasma. The space between the small eyeball and the orbit of normal size is filled with hypertrophied Harderian and lachrymal glands (Truslove, 1962; Konyukhov et al., 1962). Eyes of *or/or* embryos cultured in the anterior chamber of eyes of adult mice develop a more normal lens than when left in situ (Konyukhov et al., 1963).

Os, oligosyndactylism, semidominant, XVIII, Fig. 8-1*F*, Chap. 15, 29. Arose in an irradiation experiment and was probably X-ray-induced. Homozygotes die early in embryonic life but have not been described. Heterozygotes are affected on all four feet. Fusion usually occurs between the second and third digits and occasionally involves the fourth (Grüneberg, 1956). The muscles of the forearms and lower legs as well as of the feet show anomalous arrangements not necessarily correlated with the skeletal changes (Kadam, 1962). At 11 days of gestation the preaxial border of the limbs can be seen to be reduced (Grüneberg, 1961*a*) and histological examination at this time shows that there is cellular degeneration of the preaxial part of the apical ectoderm (Milaire, 1962). *Os*/+ mice have a mild diabetes insipidus present at 5 weeks and increasing with age (Falconer et al., 1964).

p locus, I, Chap. 21.

p, pink-eyed dilution, recessive. This is a very old mutant carried in many varieties of fancy mice. Homozygotes have pink eyes with pigmentation very much reduced but not completely absent. The black pigment of the hair is very much diluted but the yellow pigment is only slightly affected. Pigment granules are irregular and shredlike in shape (E. S. Russell, 1949*a*). Numerous recurrences of this mutation have been noted. One, which appears identical with *p*/*p* when homozygous, undergoes reverse mutation at the rate of about 0.3 per cent (Wolfe, 1963*b*).

p^d, dark pink eye, recessive. Probably X-ray-induced. Eyes of homozygotes are slightly pigmented at birth and darken in the next few days; the coat is only slightly diluted. Eyes of p^d/p mice are colorless at birth but darken during the first 2 weeks; the coat is diluted but darker than that of *p*/*p* (Carter, 1959*a*).

p^r, Japanese ruby, recessive. Discovered in a stock of Japanese waltzers (Sô and Imai, 1926). The eyes of homozygotes are very variable in color and may be different on the two sides of the same animal. Coat color is intermediate between wild type and *p*/*p*. p^r/p mice resemble *p*/*p*. This mutant may be extinct.

p^s, p-sterile, recessive. Probably X-ray-induced. This mutant is like *p* in its effect on pigmentation, but homozygotes are small and males are almost completely sterile. Females may be fertile but are poor mothers. Both sexes show slightly uncoordinated behavior and have incisors that wear abnormally (Hollander et al., 1960).

pa, pallid, recessive, V. Found in a mouse caught in the wild (Roberts, 1931). Eyes of homozygotes are pink and the coat is a little lighter than that of *p*/*p*. Both black and yellow pigment are diluted. Homozygotes have slightly reduced viability and some have slightly abnormal behavior. The head is held tilted to one side or the other, and the mice may be unable to orient themselves if submerged in water. The abnormal behavior is due to the absence of otoliths in the sacculus and utriculus which occurs in many but not all *pa*/*pa* mice (Lyon, 1953, 1955*a*).

pc, phocomelic, recessive, linkage not known, Chap. 15. Found in a tailless stock. Homozygotes show disproportionate dwarfism and die shortly after birth as a result of a large median cleft palate. The skull is narrow and pointed, the upper incisors are small or absent, and the limbs are disproportionately shortened. Both polydactyly and syndactyly may occur (Gluecksohn-Waelsch et al., 1956). Embryos show retardation of precartilage formation in the limbs and head at 12 days of gestation, and chondrification and ossification of the extremities is retarded throughout development (Sisken and Gluecksohn-Waelsch, 1959; Fitch, 1957). Extra pieces of cartilage are often found in the head and limbs. Two abnormal bars of cartilage in the area where palatine closure normally occurs are thought to interfere mechanically with closure.

Pd locus, pyrimidine degrading, linkage not known, Chap. 19. Two alleles are known at this locus: Pd^a produces a low level of pyrimidine degrading enzyme activity (in strains C57BL/6J and BDP/J); Pd^b produces a high level (in strains SJL/J and RF/J). Three enzymes are affected simultaneously; they catalyse the stepwise reactions, uracil → dihydrouracil → β-ureido-propionic acid → $CO_2 + NH_3 + \beta$-alanine. If the rate of reaction is measured by determining the radioactivity of respiratory CO_2 after injection of radioactive uracil, Pd^b is fully dominant to Pd^a. If the same reaction is carried out with liver slices, Pd^a/Pd^b is intermediate between the two homozygotes (Dagg et al., 1964).

pe, pearl, recessive, XIV, Chap. 21. Arose spontaneously in the C3H/He inbred strain. Homozygotes have a smaller amount of pigment in the eyes at birth than normal, but the eye color cannot be distinguished from normal in adults. The yellow and black pigments are diluted. Viability until maturity is good, but females have a tendency to die during pregnancy and lactation and the survivors are poor mothers (Sarvella, 1954). Somatic reverse mutations to wild type are frequent, but the frequency is different on different genetic backgrounds (L. B. Russell and Major, 1956).

pf, pupoid fetus, recessive, VIII. Found among the fetuses of dissected pregnant females in an irradiation experiment. Homozygotes die at birth. The skin of late fetuses is stretched in an anteroposterior direction and lacks hair follicles. Fore and hind limbs and tail are seen as smooth stubs covered and held bound to the body by an abnormal epidermis. The cartilaginous skeleton appears grossly normal (Meredith, 1964).

pg, pigmy, recessive, IV. Appeared as undersized segregants in a strain selected for small size (MacArthur, 1944). Homozygotes are recognizably smaller than normal at birth. Their growth rate is low and they weigh about one-third as much as normal when adult (King, 1950). They are healthy and active but some are sterile. Fertility may be almost normal on genetic backgrounds favoring large size. The dwarfing is not due to a pituitary deficiency, and the thyroid and adrenals appear normal (King, 1955). The gene increased in frequency during five generations of selection for small size, indicating that *pg* probably reduces the size of heterozygotes (Warwick and Lewis, 1954).

Ph, patch, semidominant, XVII, Chap. 21. Arose spontaneously in the C57BL strain. Heterozygotes have sharply defined white spotting in the belt region. The spot may vary from a large belly spot to a wide complete belt which includes the fore and hind legs. The skull is a little wider and shorter than normal and has a large interfrontal bone. Homozygotes die in utero from about 10 days of gestation onward. At 9 days they have an excessive amount of fluid in the circulation, the pericardium, the tissues, and under the epidermis. Those that survive the 10-day period have a large bleb in the middle of the face which interferes with the development of the nose and palate (Grüneberg and Truslove, 1960).

pi, pirouette, recessive, XVII, Chap. 32. Found by Woolley and Dickie (1945) as a spontaneous mutation in the C3H strain. Somewhat less viable and fertile than normal. Males are more reliable breeders than females. Homozygotes show the typical behavior of the circling mutants, circling, head-tossing, hyperactivity, and deafness. Deafness may be present from the beginning or may be delayed until 1 to 4 months of age depending on the genetic background (Kocher, 1960*a*). Behavioral abnormalities are due to degenerative changes in the inner ear including degeneration of Corti's organ, the spiral ganglion, stria vascularis, saccular macula, and (after 9 months) the cristae ampullaris (Deol, 1956*b*; Kocher, 1960*a*).

pk, plucked, recessive, linkage not known. Arose in the DBA/2J strain. Not allelic with *ic* or *hr*. Homozygotes develop a thickened folded skin with no fur at 5 days of age when normal mice have begun to show their first coat. The skin gradually becomes thinner and by 30 days areas of very short fur have developed. The tail lacks fur. Both sexes are fertile (Dickie, 1965).

pl, platino, recessive, linkage not known. Arose in the C57BL/10-$H\text{-}2^d$ strain. Homozygotes are near white with dark eyes. They resemble c^e/c^e but *pl* is not allelic with c^e. Tests with other loci have not been reported. Homozygotes are fully viable (Pizarro, 1957).

pn, pugnose, recessive, III. Arose spontaneously in the S strain at the Rockefeller Institute for Medical Research. Homozygotes are distinguishable at 4 weeks by the relatively shorter, wider head. The forehead protrudes slightly and the skin over the nose is wrinkled. Adults are 2 to 4 g lighter than their normal littermates. The parietal, frontal, and nasal bones and the mandible are shorter and wider than normal. The xiphisternum is bent ventrally. There is reduced survival to weaning. Males are normally fertile but females are poor breeders, often failing to deliver their young or to care for them properly (Kidwell et al., 1961).

Po, postaxial polydactyly, dominant, linkage not known, Chap. 15. Found in mice obtained from a fancier in Japan. This abnormality is characterized by an extra toe on the ulnar side of one or both forefeet. The character depends on a major dominant gene (*Po*) and possibly some minor genes which may modify its expression (Nakamura et al., 1963).

Pre locus, prealbumin component, linkage not known, Chap. 15. Two alleles are known at this locus: Pre^a, presence of a minor prealbumin component of serum protein, detected by starch-gel electrophoresis and found in the AKR/J, BALB/cJ, CBA/J and other strains; and Pre^o, absence of this component found in the A/HeJ, C57BL/6J and other strains. The function of the prealbumin component is not known (Shreffler, 1964*a*).

Ps, polysyndactyly, semidominant, linkage not known. Arose in a low-intensity neutron irradiation experiment. In heterozygotes forefeet show postaxial poly- and syndactyly, as well as preaxial fragmentation of the pollex. In hind feet the hallux is shortened and thickened; other digits may also be shortened or syndactylous. Homozygotes die shortly after birth with malformed feet in which all digits seem to be syndactylous. Claws are vestigial in heterozygotes and absent in homozygotes (Searle, 1965).

Pt, pintail, semidominant, VIII. Arose in a strain protractedly treated with methylcholanthrene. Heterozygotes have tails of variable length usually characterized by kinks near the end and a thin threadlike tip. Homozygotes have tails similar to those of heterozygotes but usually much shorter. Homozygotes are smaller than normal and have a high preweaning mortality rate, but survivors are healthy and fertile (Hollander and Strong, 1951). Examination of the skeletons of *Pt*/+ and *Pt*/*Pt* mice shows that there is a progressive reduction of the nucleus pulposus of the intervertebral discs in a cephalocaudad direction. The reduction is more severe in *Pt*/*Pt* than in *Pt*/+. The defect can be traced back to the 10th day of embryonic development where it appears as a reduced rate of cell division of the notochord. This leads to a smaller than normal notochord, and this in turn to small intervertebral discs (Berry, 1960, 1961*b*).

ptr, pulmonary tumor resistance, recessive, linkage not known. The C57BL/Fa and A/Fa inbred strains differ at a single locus with large effect on resistance to urethane-induced lung tumors. The allele in C57BL/Fa, *ptr*, acts as a recessive and appears also to be recessive to the alleles in several other inbred strains (Bloom and Falconer, 1964).

pu, pudgy, recessive, I, Fig. 8-2*E*. Appeared in descendants of an X-rayed male. Homozygotes are identifiable at birth by their extremely short tails and shortened trunk region. The whole vertebral column is greatly shortened and highly irregular. Ribs and sternum are also abnormal. The rest of the skeleton is normal. Viability is somewhat reduced. Females are very poor breeders but many males, surprisingly, breed well. The axial abnormalities arise from defective segmentation. Somite tissue with an epithelially arranged outer layer is formed, but segmentation is abortive or absent (Grüneberg, 1961*b*). *pu* is very possibly a recurrence of stub (*sb*).

px, postaxial hemimelia, recessive, XI, Chap. 15. Arose spontaneously in a stock carrying *Ra*. A similar allele, called px^r, arose at the Oak Ridge National Laboratory. The forelimbs are regularly affected. There may be absence of digits 5, 4, and sometimes 3, and reduction or absence of the ulna. There is always a large, oval foramen in the scapula. The hind limbs are usually normal, but digit 5 may be absent and occasionally the fibula is reduced. On the dorsal surface of both fore and hind feet there are epidermal papillae, situated a little behind the base of the digits, which sometimes develop a rudimentary claw. Mice with more severely affected limbs tend to have an extra pair of ribs and a slight reduction in number of presacral vertebrae. Both sexes are sterile and show anomalies of the Müllerian ducts including a partly or wholly double vagina and uncoiled oviducts in the female, and persistent Müllerian ducts in the male (Searle, 1964).

py, polydactyly, XIII, Chap. 15. Polydactyly unassociated with other major defects has been described several times in the mouse. In one such case, described by Holt (1945), the polydactyly is on the preaxial side and affects the hind feet almost exclusively. Penetrance is variable and may be increased by selection. In appropriate stocks it has been possible to follow the segregation of this gene sufficiently to determine its linkage (Fisher, 1953). In stocks selected for regular manifestation of *py*/*py*, heterozygotes (*py*/+) were occasionally polydactylous on one foot (Parsons, 1958*a*). *py* is therefore not completely recessive.

Q, quinky, semidominant, linkage not known. Like *Kw* but not tested for allelism with it. Not linked to *T*. Arose as a spontaneous mutation in the Q strain. Heterozygotes are viable and fertile, but penetrance is incomplete. Heterozygotes may have short kinked tails and shaking or circling behavior. Most homozygotes die soon after implantation (Schaible, 1961).

qk, quaking, recessive, linkage not known, Chap. 32. Arose spontaneously in the DBA/2J strain. Homozygotes have a marked rapid tremor which disappears when the mouse is at rest and in contact with bedding, but increases during locomotion. The tremor is most marked in the caudal part of the trunk. It begins at about 10 to 12 days and is fully developed by 3 weeks. Mature mice may have attacks in which a motionless posture is maintained for many seconds. Both sexes are viable and fertile but males seldom sire litters. The entire central nervous system is very deficient in myelin at all ages. Apparently myelin fails to develop. Axons and cells in gray matter appear normal (Sidman et al., 1964).

qv, quivering, recessive, I. Found as a spontaneous mutation in a noninbred stock (Yoon and Les, 1957). Viability at weaning is normal but the lifespan is short, the majority dying before 5 months of age. Males are sterile but females may be fertile and nurse their litters. Homozygotes are characterized by locomotor instability, pronounced quivering, varying degrees of paralysis of the hind legs, clasping of the hind legs when held up by the tail, and priapism in most males (Yoon and Les, 1957; Yoon, 1960). Serial sections of the brain, cord, and nerve roots revealed no abnormalities, and urinary amino acids are normal (McNutt, 1962). Studies of the effect of *qv* on serum proteins and serum cholinesterase have been made, but it is not clear how the results are related to the effects of the gene on behavior (Yoon, 1961*a*; Yoon and Harris, 1962).

r, rodless retina, recessive, IV, Chap. 29, 32. Found by Keeler (1924) in derivatives of the Bagg albino strain. Fully viable and fertile. It may be extinct, although it was found by Keeler (1927*a*) to be widely distributed in various mouse colonies in the United States and to be present in a stock at Berlin-Dahlem. The eyes of homozygotes are devoid of rods and the outer nuclear layer of the retina is reduced. No visual purple is present. The abnormality was not recognized in the first week after birth but was apparent at 13 days (Keeler, 1927*b*). Keeler (1928) found that rodless mice were blind.

Ra locus, V.

Ra, ragged, semidominant. Arose spontaneously in a crossbred stock. In heterozygotes the first coat develops a little more slowly than normal. The coat contains guard hairs and awls but no auchenes and very few zigzags. This gives the coat a thin, ragged appearance. The agouti pattern is modified in *Ra*/+, the entire coat being unusually dark. Heterozygotes are normally viable and fertile. Homozygotes are almost completely naked. Many are edematous at birth and most die before weaning. A few survive and may breed (Carter and Phillips, 1954). Developmental studies have shown that in *Ra*/+ mice growth of the late-differentiating hair follicles which produce auchenes and zigzags is very retarded or arrested. Growth of nearly all follicles in *Ra*/*Ra* mice is arrested (Slee, 1962). A low percentage of *Ra*/+ mice in some stocks have a white chylous fluid in the abdomen from shortly after birth until a week or two of age (Herbertson and Wallace, 1964). The authors have proposed a mutation at a separate closely linked locus as the explanation, but in view of a similar condition in Ra^{op}/+ (see below), it seems more likely that chylous ascites is a pleiotropic effect of *Ra*.

Ra^{op}, opossum, semidominant. Appeared spontaneously in a hybrid between the C57BL/6J and DBA/2J inbred strains (Green and Mann, 1961). Heterozygotes have very sparse fur consisting mostly of guard hairs and awls. The anterior vibrissae are absent. Viability is low, many heterozygotes dying shortly after birth in an edematous condition. Some develop a white fluid in the peritoneal cavity as in *Ra*/+ and usually die before weaning. The surviving females are very poor breeders, but males may be normally fertile. Homozygotes die in utero, probably before 11 days. Ra/Ra^{op} mice also die in utero, probably before 11 days (Mann, 1963).

rd, retinal degeneration, recessive, XVII, Chap. 29, 32. Described by Brückner (1951) and Tansley (1954) in various stocks and later found to be present in many inbred strains including C3H/He, C3H/St, C3H/Ha, C3HeB/Fe, P/J, BDP/J, PL/J, ST/J, SJL/J, and CBA/J. Homozygotes are fully viable and fertile. Eyes develop normally up to about the 12th day of age. At this stage the outer segment of the rod cell has begun to form, and in normal mice it elongates rapidly during the 12th and 13th day. In *rd*/*rd* mice the outer segments, which have begun to form, degenerate during the 12th and 13th day. The rod cells degenerate and by the 21st day have disappeared (Sidman, 1961).

Re, rex, dominant, VII. First described by Crew and Auerbach (1939) who obtained it from a commercial breeder. Both homozygotes and heterozygotes are fully viable and fertile. Homozygotes have a slightly more extreme expression than heterozygotes when young (Carter, 1951*b*). Both have curly whiskers and wavy coats. The waviness of the coat disappears in adults but the vibrissae and guard hairs remain curly.

Rf, rib fusion, semidominant, linkage not known. Arose spontaneously in strain 129/RrSv. Heterozygotes have various skeletal anomalies, the commonest of which is fusion of ribs, usually somewhat distal to the vertebrae. The vertebrae may also be abnormal, and the tail may be kinked. Homozygotes have very abnormal shortened vertebral columns and a short wide thoracic basket with extensive neural arch and rib fusions (Mackensen and Stevens, 1960). Embryos of homozygotes fail to form somites. The rib fusions of heterozygotes are probably the result of abnormalities of the ventrolateral extensions of otherwise normal somites (Theiler and Stevens, 1960).

rl, reeler, recessive, XVII, Chap. 32. Found by Falconer (1951) as a spontaneous mutation in a mildly inbred stock. Homozygotes are unable to keep their hindquarters upright and frequently fall over on their sides when walking or running. Viability is much reduced. Males are sterile and females almost always so. Falconer thought that reelers appeared mentally deficient. Subsequent investigation showed that in reelers the typical organization and lamination of the cerebellar cortex, the cerebral cortex, and the hippocampal cortex are destroyed (Hamburgh, 1963; Meier and Hoag, 1962). The cerebellum is much smaller than normal, but the cholinesterase activity per unit of wet weight is about twice that of normal (Hamburgh, 1963).

ro, rough, recessive, V. Arose spontaneously in the RIII inbred strain. Homozygotes have wavy vibrissae, apparent a few days after birth and still visible, though reduced, in the adult. The hair of the coat is not wavy, but the hairs tend to stick together in bundles as if wet or greasy. The septa between the air spaces of the hair are thicker and the air spaces smaller than normal. Many of the air spaces are filled with fluid which does not dry out because the hairs are greasier than normal (Falconer and Snell, 1952).

Rp, reduced pinna, semidominant, linkage not known. Occurred spontaneously in a randombred stock. Heterozygotes have either or both pinnae affected. Expression varies from a slight reduction along the dorsal edge of one pinna to an almost complete absence of the pinna. Homozygotes do not live past 9 days of gestation. Up to this stage abnormal embryos may be recognized by inversion of the head or tail folds of the ectoderm so that all or part of the embryo lies outside the yolk sac. Penetrance in heterozygotes is incomplete and can be improved by selection (Lyon and Meredith, 1963).

ru, ruby-eye, recessive, XII, Chap. 21. Found by Dunn (1945) in a silver piebald stock of Danforth. Homozygotes at birth have unpigmented eyes which later darken to a dark ruby color. The black pigment of the coat is diluted to a dark slate color, and the yellow pigment is diluted slightly. Ruby in homozygous condition has the same effect on shape of the pigment granules as *b*, i.e., it makes them spheroidal rather than ovoid as in wild type, and it changes the color of the granules to dark brown. It greatly reduces the number of melanocytes in the retina, ear skin, Harderian gland, and nictitans (Markert and Silvers, 1956).

Rw, rump white, semidominant, linkage not known. Arose in a low-intensity neutron irradiation experiment. Heterozygotes show partial depigmentation confined to the posterior end of the body. The tail, hind legs, and area around the base of the tail are usually white. The white area is more extensive ventrally than dorsally. Homozygotes probably die before birth (Searle, 1965).

s locus, III, Chap. 15, 21, 29. Radiation-induced mutation at this locus is very common (W. L. Russell, 1965).

s, piebald, recessive. This is a very old mutant and it is possible that some piebalds in existing stocks may be of independent origin. Homozygotes show irregular white spotting, the amount of which is greatly influenced by minor modifying genes (Dunn and Charles, 1937). They have dark eyes. The white areas are completely lacking in melanocytes and there is a reduction in the number of melanocytes in the choroid layer of the eye (Markert and Silvers, 1956; Billingham and Silvers, 1960). There may also be defects in the structure of the iris, suggesting that pigment cells make some structural or inductive contribution to normal development of the iris (Dunn and Mohr, 1952). Homozygotes may develop megacolon which is always associated with lack of ganglion cells in the distal portion of the colon. The incidence of megacolon is affected by minor modifying genes (Bielschowsky and Schofield, 1962). Mayer (1965) has shown by explantation of embryonic tissues that the defect leading to white spotting is in the neural crest rather than in the skin.

s^l, piebald-lethal. Found in the F_2 of a cross between C3H/HeJ and C57BL/6J. This allele is recessive to wild type and nearly recessive to *s*. Homozygotes are almost completely white with dark eyes and with only an occasional small pigmented spot on the head or rump (Lane, 1962*a*). All homozygotes develop megacolon with lack of ganglion cells in the posterior end of the colon. They usually die at about 2 weeks of age, but some live a year or more and may breed (Lane, 1966).

sa, satin, recessive, XIV. Possibly radiation-induced. Homozygotes are viable and fertile. The hair texture is altered to produce a silky coat with a high sheen. The effect is probably on the cellular arrangement within each type of hair rather than changes in distribution of different types of hair (Major, 1955).

sb, stub, recessive, linkage not known. This mutant is extinct but may possibly have recurred in *pu* which it resembles closely. It arose in the inbred CF #1 stock of Carworth Farms. The embryology of *sb* was not studied, but the description of postnatal homozygotes differs in no way from that of *pu* (Dunn and Gluecksohn-Schoenheimer, 1942).

sc, screw-tail, recessive, linkage not known, extinct, Chap. 15. Arose spontaneously in the inbred Bagg albino strain at Cold Spring Harbor. Homozygotes show widespread abnormalities of the skeleton including shortened kinked tails, defective vertebral centra which may result in sharp kinks in the thoracic or lumbar region, a short broad unsegmented sternum, abnormal teeth and jaws, and abnormalities of the calvarium. The number of presacral vertebrae is increased by about one (MacDowell et al., 1942). The abnormalities of the sternum appear to result from retarded growth of the ribs which fails to bring the bilateral sternal bands into close union (Bryson, 1945). Abnormalities of dentition appear to result from retarded growth of the mandibular condyle, maxillary sutures, and connective tissue of the dental pulp (Bhaskar et al., 1951).

Sd, Danforth's short tail, semidominant, V, Chap. 29. Found as a spontaneous mutation by Danforth (Dunn et al., 1940). Many heterozygotes and all homozygotes die shortly after birth from urogenital abnormalities. The surviving heterozygotes may have good viability and fertility. Heterozygotes have short tails with a reduced number of caudal vertebrae and some kinking. Tails may be absent and the third and fourth sacral vertebrae missing. The bodies of all the vertebrae are reduced (Grüneberg, 1953*a*). One or both kidneys may be reduced in size or absent. In the absence of a kidney its ureter may be short or absent. Homozygotes have similar but much more severe abnormalities. In addition the anus is imperforate and the rectum and sometimes the urethra and bladder are absent (Gluecksohn-Schoenheimer, 1943). The developmental effects can be traced to a structurally abnormal notochord, more severe toward the caudal end, leading to abnormal vertebrae and to reduction of the cloaca and tail gut (Grüneberg, 1958*a*). Organ-culture experiments attempting to determine whether the lack of kidneys in *Sd/Sd* embryos is due to defective kidney mesenchyme or to defective ureters indicated that both are quantitatively defective (Gluecksohn-Waelsch and Rota, 1963; Chap. 25).

se, short-ear, recessive, II, Chap. 29. Arose spontaneously in mice obtained from a commercial breeder (Lynch, 1921). Homozygotes have short, slightly ruffled ears recognizable at about 14 days of age. The defective ears are due to a defective cartilage framework. The whole skeleton is abnormal, being slightly smaller than normal with numerous local defects including a reduced or bifurcated xiphisternum, reduced number of ribs and sternebrae, reduction or absence of the ulnar sesamoid bone of the wrist, the medial sesamoid bone of the knee, and the anterior tubercles of the sixth cervical vertebra, and other similar defects (Green, 1951). The skeletal abnormalities are traceable to defective condensation of mesenchyme in the embryo (Green and Green, 1942). In the adult, proliferation of periosteal cells in the healing of bone fractures is reduced in *se/se* mice (Green, 1958). Homozygotes also show variable frequency of abnormalities of the soft tissues probably secondary to the skeletal defects and including hydroureter and hydronephrosis, multiple lung cysts, medially displaced left ovary, displaced right renal artery, and multiple small giant cell granulomas on the ventral surface of the liver (Green, unpublished). Viability and fertility of homozygotes are normal on some genetic backgrounds but reduced on others.

sf, scurfy, recessive, XX (sex-linked). Arose spontaneously. Hemizygous males can first be recognized at about 11 days of age by a reddening of the genital papilla. They develop scaliness, first of the tail and later of other parts of the body. The skin appears tight, and the eyelids open late. Scurfy males usually die before or shortly after weaning. The survivors are small and sterile. Occasional scurfy females have occurred and have proved to be X/O in sex chromosome type. These females resemble scurfy males in appearance and viability (W. L. Russell et al., 1959).

sg, staggerer, recessive, II. Occurred spontaneously in a stock of obese mice in 1955. Homozygotes usually die during the fourth week. Some survive to adulthood, and one male has bred. Homozygotes show a staggering gait, mild tremor, hypotonia, and small size. The cerebellar cortex is grossly underdeveloped with too few granule cells and unaligned Purkinje cells (Sidman et al., 1962).

sh-1, shaker-1, recessive, I, Chap. 32. Found by Lord and Gates (1929) in a Bagg albino strain. Viability is normal and breeding ability is high for a circling mutant. Hemozygotes show the circling, head-tossing, deafness, and hyperactivity characteristic of mutants of this type. They can sometimes hear and swim well on the surface of water up to 4 weeks or more but lose the ability later. The degenerative changes of the labyrinth may occur a little later than in some of the other waltzing mutants. They consist of degeneration of the organ of Corti, the spiral ganglion, and the stria vascularis in the cochlea, and of the saccular macula and the vestibular ganglion in the vestibular labyrinth (Deol, 1956*b*). The eighth nerve action potential and the cochlear potentials begin to develop but are never normal and have disappeared by 22 days of age (Mikaelian and Ruben, 1964). The double heterozygote, *sh-1*/+ *v*/+, becomes deaf at 10 weeks to 8 months with some degeneration in the organ of Corti.

sh-2, shaker-2, recessive, VII, Chap. 32. Discovered by Dobrovolskaïa-Zavadskaïa (1928) in the descendants of an irradiated male. Viability and fertility are nearly normal. This mutant is very similar in behavior and pathology to *sh-1*, with the exception that the abnormalities are observed a little earlier in *sh-2*. Homozygotes appear to be deaf from the beginning, and the saccular macula is abnormal at birth (Deol, 1954).

Sha, shaven, semidominant, VI. Arose in a stock of the Endocrinology Department, Edinburgh. Heterozygotes have curled whiskers and a "greasy" coat. Homozygotes grow sparse coats of very short hair. The histology of homozygotes is very much like that of *N*, to which *Sha* is closely linked (Isaacson and Cattanach, 1962; Falconer and Flanagan, 1963).

sho, shorthead, recessive, linkage not known, Chap. 15. Arose spontaneously in an inbred line at the Nevis Biological Station of Columbia University. Newborn homozygotes are characterized by small size, disproportionate shortening of the head and limbs, and a large median cleft palate. They die soon after birth. The forelimbs are relatively more shortened than the hind limbs. The small intestine is about one-half normal length, and in females one ovary or oviduct may occur medial to the kidney on that side. At 14 days of gestation the chondrocranium is severely foreshortened and the part of the basal plate that will become the presphenoid is widened. Mutants can be recognized at 12 days of gestation by their small size and short heads, showing that the gene acts before cartilage formation (Fitch, 1961*a*). The cleft palate can be traced to 14-day embryos, when the tongue and lower jaw are seen to be abnormally small and the palatal shelves are prematurely closed at their anterior end. They enclose the tip of the tongue preventing it from descending and thereby preventing the posterior part of the shelves from closing (Fitch, 1961*b*).

si, silver, recessive, IV, Chap. 21. This mutant is widespread in the English fancy. Homozygotes are extremely variable in appearance. Individual hairs of the coat of nonagouti silvers may be all white, all black, black with white tips, or white with gray or black bands. Silvering results from a reduction in number of pigment granules. Nonagouti silvers heterozygous for brown (*si/si a/a b/+*) have very light underfur (Dunn and Thigpen, 1930). The effect of *si* is so variable that it is often difficult to classify in crosses, and its usefulness as a genetic marker is therefore limited.

Sk, scaly, semidominant, linkage not known. Arose spontaneously. Penetrance is probably complete in heterozygotes. In mild cases they show scaliness on shoulders and back of head only; in extreme expression the scaliness may cover the entire dorsal surface and ears, producing almost immobile animals which occasionally die. Scaliness is first noticeable at about 4 days. The coat grows in sparsely at first but may in milder cases become indistinguishable from normal. The fate of the homozygotes is unknown (Kelly and Bangham, 1955).

Sl locus, IV, Chap. 15, 17, 21, 29. Mutants at this locus are very common and have occurred in many laboratories.

Sl, steel, semidominant. Arose spontaneously in the C3H inbred strain (Sarvella and Russell, 1956). Heterozygotes have a slight dilution of coat color, more extreme on the belly than on the back, light ears, feet, and tail, and occasionally a belly spot or blaze. They also have a slight macrocytic anemia but are viable and fertile. Homozygotes are severely anemic in utero and die usually at 15 to 16 days of gestation (Sarvella and Russell, 1956). Primitive germ cells are absent in *Sl/Sl* and deficient in *Sl/+*. There is no pigment-forming ability in the skin of *Sl/Sl* embryos (Bennett, 1956).

Sl^d, steel-Dickie, semidominant. Arose spontaneously in the DBA/2J inbred strain (Bernstein, 1960). This allele is similar to *Sl* in its effect on coat color in heterozygotes, but homozygotes are viable. They are white with black eyes, severely anemic, and sterile. Sl/Sl^d mice are also black-eyed white and anemic and many live as long as a year. Both Sl^d/Sl^d and Sl/Sl^d are sterile with very few germ cells in the gonads. The germ-cell-deficient ovaries develop tumors in old age (Bernstein and Russell, 1965, personal communication).

Sl^{gb}, grizzle-belly, semidominant. Arose spontaneously in a belted white female. Heterozygotes have light bellies. Homozygotes are anemic and die during the first week of life (Schaible, 1961, 1963*b*).

Sl^{so}, sooty, semidominant. Arose in the C57BL/6 strain. $Sl^{so}/+$ mice have a dilute coat and light tail. Sl^{so}/Sl^{so} are black-eyed white and anemic but live 10 to 40 days (Miller, 1963; Hollander, 1964).

sla, sex-linked anemia, XX (sex-linked), Chap. 17. Probably radiation-induced. There is some misclassification. Two heterozygous females were classified as anemic, so it is uncertain whether *sla* is completely recessive (Falconer and Isaacson, 1962*a*). Homozygous females and hemizygous males can be recognized at birth by their pale color and small size. The color becomes normal within a few days, but the size difference persists throughout life. An abnormal blood picture persists virtually unchanged throughout life and consists in a red cell count about three-quarters of normal and a slight reduction in mean corpuscular hemoglobin concentration and cell size. There is reduction of hemopoietic tissue in both liver and bone marrow (Grewal, 1962*b*).

sm, syndactylism, recessive, linkage not known, Chap. 15. Arose spontaneously in the A/Fa inbred strain (Grüneberg, 1956). There is considerable preweaning mortality of homozygotes. Fertility is about normal in surviving males but low in females. Homozygotes have all four feet affected. The third and fourth digits are always syndactylous and the first and second sometimes so. Digits of the forefeet are usually joined by soft tissue only, digits of the hind feet by fusions of cartilage and bone. Many homozygotes have tail kinks. The defect can be traced to hyperplasia of the apical ectodermal ridge of the limbs, of part of the limb epidermis, and of the epidermis in the distal part of the tail. Enlargement and deformation of the foot plates follows. The foot plates are bent over in a palmar direction causing crowding of the middle digits. The enlargement of the foot plates is thought to result from increased stimulation of mesenchymal growth by the hyperplastic ectodermal ridge (Grüneberg, 1960).

sn, scanty, recessive, linkage not known. Found in offspring of a cross. Homozygotes grow a thin coat of uneven hair. The thickness of the coat is variable in an individual and between individuals. The coat starts thinning out after some months, but this process also is variable. By 9 months the coat is usually very ragged, consisting of small tufts of hair. Homozygotes are usually small and difficult to breed (Nolte, 1957).

So, sombre, semidominant, XVIII. Arose spontaneously in the C3H inbred strain (Bateman, 1961). Heterozygotes resemble nonagouti (a/a) except that they have darker ears and nipples and have a stronger tendency to develop yellow hairs on their sides as they mature. Homozygotes are black all over. Except for a few yellowed hairs on the perineum they exactly resemble extreme nonagoutis (a^e/a^e). Viability and fertility are normal in both homozygotes and heterozygotes.

Sp locus, XIII, Chap. 21. Mutations at this locus occur frequently (Dickie, 1964*a*).

Sp, splotch, semidominant. Arose spontaneously in the C57BL inbred strain (W. L. Russell, 1947). Heterozygotes show white spotting on the belly and occasionally on the back, feet, and tail. Homozygotes die at 13 days of gestation with malformations which include rachischisis in the lumbosacral region and frequently in the region of the hindbrain, overgrowth of neural tissue, reduction or absence of spinal ganglia and their derivatives, and abnormal tail morphology. Neural crest and the adjacent regions of *Sp/Sp* embryos fail to develop any pigment when transplanted to chick coelom or to the anterior chamber of mouse eyes (Auerbach, 1954).

Sp^d, delayed splotch, semidominant. Arose spontaneously in the C57BL/6J strain. Heterozygotes have a white belly spot. Homozygotes are similar to *Sp/Sp* but have caudal rachischisis only, and survive to birth. Sp^d/Sp mice resemble Sp^d/Sp^d (Dickie, 1964*a*).

spa, spastic, recessive, linkage not known, Fig. 8-2G, Chap. 23, 32. Found in a hybrid stock by Chai (1961). Preweaning mortality is high and breeding performance is low. Homozygotes can usually be recognized at 14 days of age but sometimes not until 5 or 6 weeks. They show spastic symptoms which sometimes occur spontaneously and can always be induced by handling. The spasms consist of rapid tremor, stiffness of posture, and difficulty of righting when placed on the back (Chai, 1961). The symptoms can be markedly alleviated by intraperitoneal injection of aminooxyacetic acid but not by Dilantin or trimethadione (Chai et al., 1962).

spf, sparse-fur, recessive, XX (sex-linked). Found in a descendant several generations removed from an irradiated male. In affected mice (presumably homozygous females and hemizygous males) development of the fur is late and patchy. By weaning age the fur looks practically normal (Cupp, 1958; L. B. Russell, 1960*a*).

sph, spherocytosis, recessive, linkage not known, Chap. 17. Arose in an unpedigreed C3H stock. Homozygotes are extremely pale at birth and acquire a yellow color during the first few hours thereafter. They die within the first 24 hours. Examination of the blood shows numerous spherocytes accompanied by all the evidence of hemolytic anemia (Joe et al., 1962).

sr, spinner, recessive, linkage not known. Arose spontaneously in the C57BL/How inbred strain. Homozygotes show the typical head-tossing, circling, deafness, and hyperactivity of the shaker-waltzer mutants. The abnormal behavior can be recognized as early as 7 days. The anomalies of the inner ear consist of degeneration of the organ of Corti and spiral ganglion and reduction in size of the stria vascularis of the cochlea, and of lesser degeneration of the vestibular part of the labyrinth limited to the saccular macula. Both sexes are fertile but females do not make good mothers (Deol and Robins, 1962).

Ss locus, serum serological, semidominant, IX, Chap. 24. Two alleles are known which differ in the amount of a serologically detected serum globulin. Ss^h determines a high level of the component and Ss^l determines a low level. No qualitative difference has been found between the proteins determined by the two alleles. The *Ss* locus is either very closely linked to *H-2* or identical with it. Ss^l has been found in all $H\text{-}2^k$ strains and Ss^h in all other strains examined (Shreffler and Owen, 1963; Shreffler, 1964*b*).

st, shaker-short, recessive, linkage not known. Probably extinct. Arose in a stock of hairless mice (Dunn, 1934). Viability is much reduced. Females were sterile but a few males bred. Most newborn homozygotes have one or two cerebral hernias covered by skin near the median point of the parieto-occipital suture of the skull. The blebs later dry to a scab. Tail length varies from absence to three-quarters normal length. From 5 days on there is disturbed equilibrium with erratic circus movements and in adults marked lack of coordination. In the development of the brain no foramen of Magendie or choroid plexus is formed, leading to disturbed circulation of the cerebrospinal fluid (Bonnevie, 1936). By the 16th day of gestation the lumen of the brain is narrow and the walls are abnormally thick. About the 17th day there is a rupture of the roof of the mesencephalon and cerebellum, which gives rise to the cerebral hernias. The ear vesicles, as in kreisler, are very abnormal in morphology, possibly because of a separation of the vesicles from the inductive influence of the myelencephalon.

stb, stubby, recessive, V. Found in a linkage cross. Resembles achondroplasia (*cn*) but is less severe. Homozygotes are recognizable shortly after birth by the stubby appearance of the head. Adults have shorter heads, bodies, and legs than normal but there is some variation in expression. Most females are fertile and bring up their litters but males are probably sterile (Lane, 1965).

Str, striated, semidominant, XX (sex-linked), Chap. 15. Found among the progeny of an X-rayed male. Heterozygous females show a transverse striping similar to *Ta*/+ females. The dark stripes are due to shortening of the hairs and not to a lack of zigzags as in *Ta*/+. Viability appears to be normal, but there is a shortage of *Str*/+ in segregating generations probably due to misclassification. Hemizygous males die at about 11 to 13 days of gestation (Phillips, 1963*b*).

su, surdescens, recessive, linkage not known. Found in mice of the C57BL/Gr inbred strain and in a strain carrying Mi^{wh} (Kocher, 1960*b*). Homozygotes hear normally when young but become deaf or hard of hearing between 2 and 5 months. Behavior is normal. In some *su/su* mice reduction in size of the stria vascularis and degeneration of the organ of Corti and the spiral ganglion are found. In others the changes are more severe and include also a decrease in the size of the endolymphatic space in the scala media and the sacculus, somewhat resembling the changes caused by *sy*. Middle ear anomalies also occur in *su/su* (Kocher, 1960*b*).

sv, Snell's waltzer, recessive, II. Found in a histocompatibility stock of Snell (Green, 1960). Viability is nearly normal. Breeding ability is reduced, and males are more reliable breeders than females. Homozygotes show the typical circling, head-tossing, deafness, and hyperactivity of other mutants of this type. They have defects of the membranous labyrinth similar to those of *Va*, with degeneration occurring in the cristae as well as in other parts of the labyrinth (Deol, 1965, personal communication).

sy, shaker-with-syndactylism, recessive, linkage not known, Chap. 15. Found by Hertwig (1942) among the descendants of an irradiated male. Most homozygotes die within the first month and none have lived to breed. Homozygotes may be syndactylous on all four feet; the forefeet may often be normal and possibly one or both hind feet very occasionally so. The remainder of the skeleton shows many slight anomalies in addition to small size. The shafts of the long bones are considerably thinner than normal, and there are differences in shape of the sacral vertebrae and the scapula (Grüneberg, 1956). The osseous skeleton is less densely constructed than normal (Grüneberg, 1962). Abnormal behavior appears during the first week. It consists of head-tossing and some circling, and the affected mice are always deaf. Abnormalities of the labyrinth can be seen at 13 days of gestation. There is an excessive amount of mesenchymal tissue from which develops an excessive amount of perilymphatic space. Partial collapse of the endolymphatic space follows and eventually extensive degeneration of all parts of the membranous labyrinth occurs. A quantitative abnormality of the mesodermal tissue has been suggested as the basis of the whole syndrome (Deol, 1963).

T locus, IX. At this very complex locus several semidominant (T) and numerous recessive (t^n) alleles are known. In general $T/+$ mice are short tailed, $t^n/+$ mice are normal, T/t^n mice are tailless (Fig. 8-2*D*) and the T/T and t^n/t^n homozygotes are lethal. For some *t* alleles homozygotes are viable and normal tailed. Compounds of two different *t* alleles (t^n/t^m) are usually normal tailed and viable but males are sterile. The *t* alleles greatly reduce crossing over in this region of linkage group IX and are thus probably chromosome abnormalities of some kind. They mutate frequently to new *t* alleles and such mutations are usually accompanied by crossing over (Lyon and Phillips, 1959). Lyon and Meredith (1964), after examining the properties of a number of *t* alleles, have concluded that they probably consist of a functional change in chromatin covering various lengths of chromosome between *T* and *tf* and perhaps beyond. Loss of specific pairing might cause unequal crossing over which gives rise to small duplications or deficiencies, thus producing new alleles.

The segregation ratios in males but not in females of the genotype $t^n/+$ or T/t^n are very abnormal, usually with a large excess of offspring from the *t*-bearing gametes. In wild populations and probably also in some laboratory populations *t* alleles may be very common. Once introduced they are maintained by their high segregation ratios in spite of the lethality of the homozygotes (see Grüneberg, 1952, chap. XV; Dunn et al., 1960, 1962 for a more complete description and references). The sperm and spermatogenesis of T/t^n and $t^n/+$ males appear normal, but the t-bearing sperm fertilize a disproportionately high number of eggs when such males are mated. The proportion becomes more nearly normal if matings are made late in the estrous period of the female so that a shorter interval occurs between ejaculation and fertilization (Yanagisawa et al., 1961). On the other hand, the sterile males that are compounds of two lethal *t* alleles or of a lethal and a viable *t* produce sperm of normal numbers and appearance but unable to effect fertilization (Braden and Gluecksohn-Waelsch, 1958). No mechanism has yet been proposed to account for both of these effects. Some of the alleles and their effects are:

T, brachyury, Chap. 15. Discovered by Dobrovolskaïa-Zavadskaïa (1927). Tail length of heterozygotes varies from nearly normal to absent. A few or nearly all tail vertebrae may be missing. There tends to be one fewer presacral vertebra than in $+/+$ sibs. Studies of development have shown that the notochord in the tail region is abnormal (Chesley, 1935). Grüneberg (1958*b*) later showed that the notochord was present in the whole tail but tended to become incorporated either in the neural tube or the tail gut. T/T embryos never develop a notochord or only a rudimentary one, the neural tube and somites are irregular, and the posterior portion of the body is greatly reduced. They die at 10 days of gestation (Chesley, 1935). In organ culture, T/T neural tube can induce cartilage in normal somites, but cartilage cannot be induced in T/T somites (Bennett, 1958; Chap. 25).

T^H, T-Harwell. Found in the control series of an irradiation experiment. $T^H/+$ are indistinguishable from $T/+$ but T^H/T^H die at an earlier stage than T/T, being already highly abnormal at 8 days (Lyon, 1959).

t^0-t^n, tailless-0 to tailless-n, t^{w1}-t^{wn} tailless wild-1 to tailless wild-n. See Dunn et al. (1962) for review of the genetics, and Bennett et al. (1959) for description of the effects in homozygotes and compounds. Briefly, t^n/t^n or t^{wn}/t^{wn} mice are either normal, or die at a very early stage with abnormalities which are primarily ectodermal, or die at a late fetal stage with striking pycnosis and degeneration of the ventral portion of the neural tube.

t^b, tailless-b (Braden and Gluecksohn-Waelsch, 1958).

t^e, tailless-Edinburgh (Bateman, 1960). See also *Mouse News Letter* for unpublished alleles.

Ta locus, XX (sex-linked), Chap. 14, 15.

Ta, tabby, semidominant. Arose spontaneously in a strain selected for large size (Falconer, 1953). Hemizygous males and homozygous females are identical in phenotype with homozygous crinkled mice (*cr/cr*). They are characterized by absence of guard hairs in the coat, a bald patch behind each ear, bald tail with a few kinks near the tip, reduced aperture of the eyelids, a respiratory disorder, and a modified agouti pattern. Heterozygous females are most easily recognized if they are agouti, in which case they show transverse dark stripes. The dark stripes have no agouti bands on the hairs. Guard hairs are present in the bands, but zigzags are very deficient or absent. Tabby males breed satisfactorily but homozygous females are often sterile. Heterozygous females are fully fertile (Falconer, 1953).

Ta^J, tabby-J, semidominant. Arose spontaneously in the 129/Sv inbred strain. Ta^J resembles *Ta* except that hemizygous males and homozygous females have some hair on the tails, and the hairs are curved. *Ta*/Ta^J females have both hairless areas and areas with curved hairs on their tails (Stevens, 1963).

tb, tumbler, recessive, linkage not known. Found among offspring of a cross. Homozygotes walk in crab-like fashion. They may somersault, fall over, or jump when trying to go forward. They can swim but cannot hold onto a rope. Most homozygotes survive and can breed (Dickie, 1965).

tc, truncate, recessive, XI. Arose spontaneously in a Swiss albino stock at The Jackson Laboratory (Theiler, 1957). Penetrance is incomplete. Homozygotes have short or absent tails and many have missing sacral or lumbar vertebrae. In some cases several vertebrae seem to be "picked out" in an intermediate position. The primary effect is an interruption of the notochord at 9½ to 10 days of gestation. The sclerotomic cells, which normally migrate from the somites to the midline under the neural tube to form the vertebrae, degenerate in the region of the interrupted notochord. These defects lead eventually to interruption of the spine, paralysis of the hind legs, difficult parturition in females, and absence of the median ventral fissure of the spinal cord (Theiler, 1959).

td, torpid, recessive, II. Arose in the DBA/2J strain. Homozygotes move very slowly. They may lie on their backs for minutes at a time, sometimes with the mouth and the feet quivering. When standing one foot often is raised and quivers. Homozygotes are viable. Both sexes are fertile, but males are more productive (Dickie, 1965).

tf, tufted, recessive, IX. Probably arose spontaneously in a multiple recessive stock. Homozygotes show repeated waves of hair loss and regrowth which begin at the nose and pass posteriorly along the body. Viability and penetrance are good (Lyon, 1956).

tg, tottering, recessive, XVIII, Chap. 32. Found in the DBA/2J strain as a spontaneous mutant (Green and Sidman, 1962). Viability is nearly normal. Homozygotes of both sexes are fertile but breeding performance may be low. Affected mice are characterized by intermittent seizures which may begin as early as 2 weeks of age and continue throughout life, and by a wobbly gait affecting particularly the hindquarters.

th, tilted head, recessive, XIII. Possibly radiation-induced (Kelly, 1958). Homozygotes are viable and fertile. The head tilts to one side, the side being constant for an individual. The animals shake when held up by the tail.

ti, tipsy, recessive, VII. Found among the descendants of a cross between the C3H/H and 101/H inbred strains. Penetrance is complete and homozygotes are viable and fertile. They have a rabbit-like gait from the age of 1 week, followed soon by a tendency for the forepart of the body to sway from side to side, leading to a reeling locomotion and a tendency to fall over. There is marked variation in expression and some amelioration in older animals (Searle, 1961).

tk, tail-kinks, recessive, II, Fig. 8-2*C*, Chap. 15. Arose spontaneously in a BALB/c subline at the Chester Beatty Research Institute. Homozygotes are recognized by their shortened kinky tails. The tail vertebrae are very abnormal, and the cervical and upper thoracic vertebrae are also severely affected. There tends to be one more presacral vertebra than normal. The defects can be traced to the 10-day embryonic stage when the cervical sclerotomes in normal mice have differentiated into anterior and posterior halves which differ in density. No such differentiation occurs in *tk/tk* embryos (Grüneberg, 1955*a*). Penetrance is complete and fertility is good.

tl, nonerupted teeth, recessive, linkage not known. Possibly radiation-induced. Homozygotes usually die at weaning but have been raised by special feeding to 3 months. Their teeth do not erupt and there are bony spurs on the long bones and ribs. The tail is slightly shortened (Kelly, 1955).

Tla locus, thymus leukemia antigen, IX, Chap. 24. This locus determines the presence or absence of an antigen, *TL*, in the normal thymus. The antigen is present (Tla^a) in thymus of strains A and C58 and absent (Tla^b) in the thymus of strains C57BL/6, C3H/An, BALB/c, and AKR (Old et al., 1963). The antigen is found in no other normal tissue but is found in a proportion of leukemias of probably all mouse strains. Tla^a/Tla^b mice are intermediate between the two homozygotes in the quantity of antigen present in the thymus (Boyse et al., 1964). Transplantation experiments have shown that the *TL* type of the thymus cells is determined by their own genotype rather than by the genotype of the thymic stroma or of other cells (Schlesinger et al., 1965).

tm, tremulous, recessive, linkage not known. Tremor is the most conspicuous symptom of *tm/tm* mice. They do not have convulsions. Both sexes are sterile. Affected mice have a higher concentration of creatine phosphates and ribose in the supernatant fraction of calcium-precipitated trichloroacetic acid extracts of the brain than their normal sibs (Yoon and Denuccio, 1963). They also have lower serum cholinesterase and serum β-globulins but higher serum albumin (Yoon, 1961*a;* Yoon and Harris, 1962).

tn, teetering, recessive, VII. Arose spontaneously in the C3H/HeJ inbred strain. Homozygotes die between 5 and 6 weeks of age. They are first recognizable at 25 to 30 days by their stiff, slow, unstable movements. They may assume and maintain unusual postures. Just prior to death they look emaciated and lie on their sides with all limbs extended (Lane, 1962*b*).

To, tortoiseshell, semidominant, XX (sex-linked), Chap. 21. Arose spontaneously in an obese stock. Heterozygous females resemble *Mo*/+ females in color. Some have slight skeletal abnormalities of the fore and hind limbs. Hemizygous males die before birth (Dickie, 1954). *To* shows about the same linkage relation with *Bn* as *Mo* (Lane, 1960*b*) and is probably an allele of *Mo.*

tp, taupe, recessive, I, Chap. 21. Arose spontaneously in the C57BL/10 inbred strain. Viability of homozygotes is normal and fertility of males is normal, but females may have difficulty gestating their young and do not nurse them. Their nipples are abnormal. Homozygotes have a diluted coat color, *a/a tp/tp* mice being slate gray (Fielder, 1952).

Tr, trembler, dominant, VII, Chap. 32. Arose as a spontaneous mutation in 1946 in a stock at the Institute of Animal Genetics in Edinburgh (Falconer, 1951). Mortality of heterozygotes is high at 3 to 4 weeks but can be improved if moist food is provided within easy reach during this critical period. Females breed normally but males are frequently sterile. Homozygotes are viable and cannot be distinguished from heterozygotes. Affected mice show spastic paralysis beginning during the third week, and exhibit convulsions when stimulated. Within a week or two the convulsions usually cease and are replaced by a tremor that consists of a very rapid side-to-side movement of the head and neck. The trembling ceases when the mice are at rest. The paralysis persists into adulthood but is less severe (Falconer, 1951). Tremblers have normal electrocorticograms. Histological examination of the brain and spinal cord has revealed no lesions (Braverman, 1953).

Trf locus, transferrin, II. Alleles at this locus control variation in the electrophoretic properties of the iron-binding β-globulin component, transferrin, of the serum (Cohen, 1960; Shreffler, 1960; Cohen and Shreffler, 1961). Ashton and Braden (1961) and Thompson et al. (1954) have independently described variation in the electrophoretic properties of β-globulin which is probably controlled by the *Trf* locus, although this has not been tested genetically. Two codominant alleles are known, Trf^a, found only in the CBA strain, and Trf^b, found in all other strains tested. The serum component of Trf^a is more negatively charged at pH 8.5 and migrates more rapidly than that of Trf^b.

Ts, tail-short, semidominant, linkage not known, Chap. 15, 17. Arose spontaneously in the BALB/c inbred strain (Morgan, 1950). Homozygotes probably die before implantation. Viability and expression of heterozygotes is strongly dependent on the genetic background (Morgan, 1950). Heterozygotes are recognizable by their short kinked tails. They are smaller than normal and have numerous skeletal abnormalities including vertebral fusions and dyssymphyses, bilateral asymmetry of the length of the humerus and the tibia, triphalangy of digit 1 of the forefoot, an additional pair of ribs, and often a shortened and highly abnormal skull. There is a prenatal anemia which disappears before birth and can be traced to a deficiency of the blood islands in the yolk sac of 8-day embryos. The anemia is thought to be the primary effect of *Ts* and to be the cause of the subsequent abnormalities (Deol, 1961).

Tw, twirler, semidominant, XV, Chap. 32. Spontaneous. Heterozygotes show head-shaking and circling behavior but are not deaf. Viability and fertility are normal except that adults tend to become obese and may then become sterile. Penetrance is probably incomplete. There are morphological abnormalities of the inner ear which consist of irregularities in the outline of the semicircular canals, sometimes amounting to branching, and reduction or absence of otoliths. Homozygotes have cleft lip and palate or cleft palate only. They die within 24 hours after birth (Lyon, 1958).

U, umbrous, semidominant, linkage not known. Found among descendants of a stock carrying *wa-1. U* in homozygous condition causes a marked darkening in *A/A* mice. In heterozygous condition the darkening is somewhat less. In *U/U* mice *A/a* is distinguishably darker than *A/A. U* is thus a modifier of dominance of *A* (Mather and North, 1940). No effect of *U* is recognizable in *a/a* mice. A similar mutant has been described by Robinson (1959), but no tests for allelism have been made.

un, undulated, recessive, V. Found in mice obtained from a Cambridge fancier (Wright, 1947). Homozygotes have a shortened and usually kinked tail. The caudal vertebrae are reduced in size but not in number. The kinks can be flattened out with the fingers but return immediately when released. Some homozygotes have marked kyphosis of the lower thoracic and upper lumbar region. The vertebrae are abnormally formed over the whole spine. The acromion process of the scapula is reduced or absent (Grüneberg, 1950). The anomalies can be traced to the 11th day of gestation. The condensations of mesenchyme cranial to the sclerotomic fissure are smaller than normal. Instead of joining with the primitive centra in front of them to form the body of the vertebra, they remain with the material posterior to the sclerotomic fissure and enter into the intervertebral disk. The vertebrae are thus smaller than normal and the disks larger than normal (Grüneberg, 1954*b*).

Up-1 locus, urinary protein, II. This locus controls variation in the major protein of the urine as revealed by electrophoresis. Two codominant alleles are known: *Up-1*^{a} produces the electrophoretic components numbered 1 and 3; *Up-1*^{b} produces components 2 and 3. *Up-1*^{a} is found in strains BALB/cN, DBA/2Lw, C3H/Lw, NH, C57BR/cd, and some others. *Up-1*^{b} is found in C57BL/10Sc, C58, C57L, and some others (Finlayson et al., 1963).

ur, urogenital, recessive, linkage not known, Chap. 15, 29. Discovered in a balanced tailless stock (Dunn and Gluecksohn-Schoenheimer, 1947). Homozygotes have short tails and are small at birth. They usually have cleft palates and die within a day or two. Some have normal palates and may survive for a month or more, but they remain small and are sterile. Older animals are usually found to have hydronephrosis or polycystic kidneys. At birth the kidneys have only slight histological abnormalities but are deficient in alkaline phosphatase and probably do not function normally (Gluecksohn-Waelsch and Kamell, 1955). The skeletal abnormalities consist of a reduction in length but not in width of midline structures, particularly at the anterior and posterior ends. There may be a small extra pair of ribs, and fusions of ribs and sternebrae. The shortening of the skeleton can be seen in the tail at 10½ days of gestation and in the head at 12 days (Fitch, 1957). The origin of the kidney abnormalities is not known.

uw, underwhite, recessive, linkage not known. Arose spontaneously in the C57BL/6J inbred strain. The fur of homozygotes is a light buff color on top with very white underfur. Eyes are unpigmented at birth but darken to a dark reddish color at maturity (Dickie, 1964*b*).

v locus, X, Chap. 32.

v, waltzer, recessive. Probably originated in China many centuries ago (Keeler, 1931). Viability and breeding ability are somewhat less than normal. Homozygotes show the typical circling, head-tossing, deafness, and hyperactivity of the circling mutants. Most of them are deaf from the beginning. Abnormalities of the inner ear include degeneration of the organ of Corti, spiral ganglion, stria vascularis, and saccular macula. Double heterozygotes with shaker-1 (*v*/+ *sh-1*/+) are deaf beginning at 3 to 6 months. They have changes similar to those of the homozygotes in the organ of Corti, stria vascularis, and spiral ganglion, but less severe and with much later onset (Deol, 1956*b*; Kocher, 1960*a*).

v^{df}, deaf, recessive. First noticed in a laboratory stock through a somewhat anomalous position of the ears (Deol, 1956*a*). Homozygotes may be deaf from the beginning or may hear for a few days before weaning but otherwise behave normally. Deafness is caused by degeneration of the organ of Corti, the spiral ganglion, and the stria vascularis. v/v^{df} mice are like v^{df}/v^{df} mice in hearing ability and behavior. Double heterozygotes with shaker-1 (v^{df}/+ *sh-1*/+) become deaf at a late stage (Kocher, 1960*a*).

Va, varitint-waddler, semidominant, XVI, Chap. 21, 32. Found by Cloudman and Bunker (1945) in descendants of a cross of C57BL × C57BR strains. Viability of heterozygotes is nearly normal but fertility is reduced. Mortality is very high in homozygotes and very few of the survivors are fertile. Heterozygotes are deaf and show circling behavior, head-tossing, and hyperactivity. They circle somewhat less than some of the other circling mutants. Their coats are variegated with patches of normal-colored, diluted, and white fur. Homozygotes show more intense behavioral abnormalities than heterozygotes and their coats are white, except for small patches of unaltered color near the ears and base of the tail (Cloudman and Bunker, 1945). The pathological changes in heterozygotes include degeneration of the organ of Corti, stria vascularis, spiral ganglion, saccular macula, cristae ampullares, and vestibular ganglion. In homozygotes the degenerative changes are more severe and include also the utricular macula (Deol, 1954).

vb, vibrator, recessive, probably VII. Arose in the DBA/2J strain. Homozygotes show a constant vibration of the whole body recognizable from about 10 days of age. None live beyond weaning age (Lane, 1965).

vc, vacillans, recessive, VIII. Arose spontaneously in the DBA/1 inbred strain. Homozygotes can be identified with certainty at 14 days of age by a violent tremor when walking and by swaying of the hindquarters. With age the instability lessens and a ducklike gait sets in. The tremor is less marked with age. Vacillans mice are less aggressive than normal. They are smaller, and females do not take good care of their litters. Muscular strength is about half that of normal mice. There is a peak mortality at weaning age, after which survival appears normal. Sexual maturity in males did not occur before 5½ months. The central nervous system shows no gross anatomical abnormality on pathological examination (Sirlin, 1956).

Ve, velvet coat, dominant or semidominant, linkage not known. Appeared in offspring of an irradiated male. Heterozygotes closely resemble *sa/sa* mice. The coat has a velvety sheen and on the ventral surface appears more luxuriant than normal. Heterozygotes are viable and fertile. The fate of the homozygotes is not known (Maddux, 1964).

vi, visceral inversion, recessive, linkage not known. Found in descendants of a 4-way cross. Probably extinct. Penetrance is probably not complete, and there may be some loss of homozygotes in segregating generations because of inviability. The mutant resembles *iv* but was not tested for allelism with it. Homozygotes show complete or partial reversal of the viscera as well as a general sickliness and often hydrocephalus (Tihen et al., 1948).

vt, vestigial-tail, recessive, VII, Chap. 15. Arose spontaneously in the C57BR inbred strain (Heston, 1951). Homozygotes have very short tails, varying from complete absence to about half normal length. They tend to have fewer presacral vertebrae than normal, and the bodies of the lumbar vertebrae often ossify from bilateral twin centers rather than from a single center. At 10 days of gestation when there is a ventral ectodermal ridge at the zone of growth of the tail tip, the ridge is much reduced in *vt/vt* embryos. There is also a reduction in the tail gut. The neural tube in the tail bud is abnormal in the ventral region, tending to split off and round up to form accessory tubes. The abnormality of the ventral ectodermal ridge may be the primary effect of the gene (Grüneberg, 1957).

W locus, XVII, Chap. 15, 17, 21, 29. Mutation at this locus is common. Homozygotes or compounds of the mutant alleles at this locus tend to develop ovarian tumors, a defect thought to be due to the absence of germ cells resulting in underproduction of sex hormone and overproduction of pituitary gonadotrophic hormone with excess stimulation of the gonad (Russell and Fekete, 1958; Murphy and Russell, 1963). Alleles with published descriptions are:

W, dominant spotting, semidominant. This is an old mutant of the mouse fancy. Heterozygotes have variable amounts of white spotting depending on the genetic background. Colored areas may be interspersed with white hairs to produce a roan type of pattern (Dunn, 1937*a*). Heterozygotes have normal blood and are fully viable and fertile (Grüneberg, 1942*b*). Homozygotes are white with black eyes in which the pigment is restricted to the retina. They have a severe macrocytic anemia and a severe shortage of primary germ cells (Mintz and Russell, 1957). They die within the first week after birth.

W^a, Ames dominant spotting, semidominant. Found among offspring of an X-rayed male of the Z strain. Resembles *W* except that heterozygotes have a prominent blaze and more often have belly spotting. Homozygotes are anemic and die within a few days after birth (Schaible, 1963*c*).

W^b, Ballantyne's spotting, semidominant. Arose in the C57/St strain. Heterozygotes are more extensively spotted than $W^v/+$ and the coat color is more diluted. Homozygotes may survive to maturity. They are anemic, black-eyed white, and sterile (Ballantyne et al., 1961).

W^j, Jay's dominant spotting, semidominant. Arose spontaneously in the C3H strain. W^j resembles *W* in its effects in heterozygotes and homozygotes with the exception that it causes more white spotting in heterozygotes (E. S. Russell et al., 1957).

W^v, viable dominant spotting, semidominant. Discovered by Little and Cloudman (1937) in the C57BL inbred strain. Heterozygotes have a variable amount of white spotting and also a slight dilution of the coat color. They have a slight macrocytic anemia (E. S. Russell, 1949*b*) but are fully viable and fertile. Homozygotes resemble *W/W* in color, anemia, and germ cells but many of them survive to maturity. W/W^v mice are also viable. W^v/W^v mice utilize the L form of several amino acids and excrete the D form, whereas normal mice utilize both the D and L forms. $W^v/+$ mice are intermediate in this respect (Goodman, 1956; Chap. 19).

W^x, a mutant indistinguishable from *W*. It occurred in the C3H/HeJ strain in 1952 (E. S. Russell et al., 1957; Murphy and Russell, 1963).

wa-1, waved-1, recessive, XI. Found in a mixed mouse colony (Crew, 1933). Homozygotes are recognizable at 2 or 3 days of age by their curly whiskers. The first coat is strongly waved but the hair becomes straight in later coats. Most of the whiskers also become straight but the guard hairs are curved and shorter than normal. The hair follicles of the first coat are curved but the curvature disappears in later hair generations. The hairs are somewhat less keratinized than normal (David, 1937). Bennett and Gresham (1956) found that many *wa-1/wa-1* mice had eyelids open at birth and ascribed the condition to a closely linked gene, but the possibility that open eyelids is a pleiotropic effect of *wa-1* was not excluded.

wa-2, waved-2, recessive, VII. Found in an "abnormal corpus callosum" stock (Keeler, 1935). This mutant is very similar to *wa-1*. Homozygotes can be recognized at 2 to 3 days by their curly whiskers. The first coat is waved but later coats are not. The vibrissae usually remain curled and the guard hairs curved. Some homozygotes have eyelids open at birth (Butler and Robertson, 1953; Green, unpublished). Butler and Robertson have ascribed the condition to an independent gene, squint (*sq*), but it is probably a pleiotropic effect of *wa-2*.

wd, waddler, recessive, VIII. Arose spontaneously in a stock carrying furless. Homozygotes can be identified at about 14 days of age. The hindquarters sway from side to side in a smooth arc and the mice often fall on their hips. No trembling or paralysis occurs. Homozygotes are smaller than their normal littermates from 3 weeks on, but viability is good and many waddlers are fertile. The abnormalities of behavior do not become more severe with age (Yoon, 1959). Studies have been made of the serum proteins (Yoon, 1961*a*), serum cholinesterase (Yoon and Harris, 1962), and acid-soluble phosphorus (Yoon and Denuccio, 1963) of the brain in *wd/wd* mice, but the results are not easily interpretable.

we, wellhaarig, recessive, V. Arose as a spontaneous mutation in the stocks of Agnes Bluhm. Homozygotes are fully fertile. They have curly whiskers at 2 or 3 days of age and a wavy first coat, most strongly evident between 10 and 21 days. In later coats the waviness is lost. The hairs have a lower average diameter than those of normal mice (Hertwig, 1942).

wh, writher, recessive, linkage not known. This mutant is similar to *dt*, but has not been tested with it. Possibly radiation-induced. Affected mice die before weaning. They are unable to support themselves on their legs. Convulsions of the torso occur in later stages. The condition is first recognizable at 12 days (Kelly, 1953).

wi, whirler, recessive, VIII. Arose as a spontaneous mutation in a multiple recessive stock. Viability may be slightly reduced. Both sexes are fertile but females do not always make good mothers. Homozygotes show the characteristic symptoms of the shaker syndrome, deafness, head-tossing, circling, and hyperactivity (Lane, 1963). They have defects of the membranous labyrinth similar to those of *sh-2* (Deol, 1965, personal communication). Oxygen consumption and adrenal weight are greater than normal (Sackler et al., 1964).

wl, wabbler-lethal, recessive, III. Arose in a stock carrying pirouette (Dickie et al., 1952). Homozygotes are first recognizable at 12 days of age and usually die at about 4 weeks. They have an abnormal wobbly gait and a pronounced tremor when walking. Difficulty in walking increases in severity until the animal dies. Histological examination shows myelin degeneration widely distributed throughout the central nervous system, particularly in the vestibulocerebellar and spinocerebellar systems (Harman, 1954). The onset of demyelinization of individual tracts is in the same order as their myelinization. Axons are normal. The interfascicular glia show extensive atrophy in the tracts where abnormal myelin is present (Anderson and Harman, 1961). Leukopenia is found from 2 weeks of age until death (Mufson and Starr, 1958).

wr, wobbler, recessive, linkage not known. Arose in the C57BL/Fa strain. The main features are tremor and paralysis which affect the forelimbs more than the hind limbs. Penetrance appears to be complete, but both sexes are sterile and much reduced in viability (Duchen et al., 1966).

Wt, waltzer-type, semidominant, linkage not known, Chap. 32. Arose spontaneously at the Oak Ridge National Laboratory. Heterozygotes may show behavior varying from slight nervousness to rapid circling. They are not deaf. The posterior and lateral semicircular canals are abnormal, with defects ranging from slight shortening to complete absence distal to the ampulla. The rest of the labyrinth is normal. All *Wt/+* mice appear to have morphological abnormalities of the labyrinth, but many of them do not show any behavioral abnormality. Circling behavior is associated with defective lateral rather than posterior canals. Both canals seem important for normal swimming, and a larger proportion of mice with defective canals can be detected on this basis. *Wt/Wt* mice die at about the 11th day of gestation (Stein and Huber, 1960).

wv, weaver, recessive, linkage not known. Arose spontaneously in the C57BL/6J inbred strain. Homozygotes resemble staggerer and reeler mice but *wv* is not allelic with *sg* or *rl*. Affected mice do not live to maturity. Gross inspection of the brain shows that the cerebellum is defective (Lane, 1964).

Xt, extra-toes, semidominant, XIV, Chap. 15. Probably X-ray-induced. Heterozygotes have extra toes on the preaxial side of the hind feet. Penetrance is nearly complete. On the front feet one or more preaxial digits may be lacking. Homozygotes die with cranioschisis and have many extra toes on all feet (Lyon et al., 1964*a*).

Bacillus enteritidis resistance. It was shown by Webster (1937) that a single locus controls resistance to this bacterium. Alleles at this locus segregate independently of those at the locus controlling resistance to the Arbor B viruses. The allele for resistance is dominant to that for susceptibility. The BRVR strain is resistant and the BSVS strain is susceptible.

Arbor B virus resistance. A single locus with alleles controlling susceptibility and resistance to the Arbor B viruses has been demonstrated by Webster (1937) and Sabin (1952). Resistance seems to be a property of the macrophages, since macrophages from resistant mice do not support growth of the virus, whereas kidney and lung cultures from resistant mice, and macrophages from susceptible mice do (Goodman and Koprowski, 1962). The allele for resistance is dominant to that for susceptibility. The PRI and BRVR strains are resistant and C3H/He is susceptible. This locus is probably independent of that controlling resistance to mouse hepatitis virus (Kantock et al., 1963).

Mouse hepatitis virus resistance. The C3H/An strain is resistant to mouse hepatitis virus and the PRI strain is susceptible. The difference is due to differences at one or possibly two loci with susceptibility dominant to resistance. Resistance or susceptibility is a property of the macrophages. Cultivation of resistant macrophages in a medium containing an extract of susceptible macrophages changes them to susceptible cells. Preliminary comparisons with Arbor B virus resistance based on tests of survivors of various backcrosses indicates that the two characters are independent (Kantock et al., 1963).

Leukencephalosis. A dystrophy of the white matter of the brain producing an *ex vacuo* hydrocephalus occurred in Hertwig's *sy* strain (Fischer, 1959). It is recessive and appears to have full penetrance. The dystrophy begins in the occipital parts of the telencephalon and spreads nasad and laterad. Other parts of the brain are unaffected. This mutant is similar to cerebral degeneration (*cb*) but has not been tested for allelism with it.

Neonatal jaundice. A recessive mutant causing jaundice in the newborn usually with death before 10 days has been described by Scheufler (1963). In the blood there are few normocytes, many polychromatic erythrocytes, sperocytes, poikilocytes, fragments, and a few erythroblasts. Hemoglobin is very deficient. One surviving homozygous female was fertile but no males were fertile. Not tested for allelism with *ha, ja,* or *sph.*

Biosynthesis of corticosteroids. The in vitro synthesis of cortisol and corticosterone by the adrenal glands is about twice as high in strain A/Cam as in strain CBA/Fa. The difference appears to be controlled by a single locus with the allele in strain A dominant to that in strain CBA (Badr and Spickett, 1965).

LINKAGE

Linkage groups

The mouse has 20 pairs of chromosomes and should therefore have 20 linkage groups. Eighteen autosomal groups and a sex-linked group are known (Fig. 8-3 and Table 8-1). Linkage groups I, II, V, VII, VIII, IX, XI, XIII, XVII (formerly part of III, Lane, 1965), and XX have been shown by Slizynski (1954, 1957) to be on different chromosomes. VI, IX, XII, and XV are very short, and at least one end of X, XI, XVI, and perhaps of some others has not been adequately tested against all other known linkage groups. It is therefore likely that some present assignments of loci will eventually be changed.

In Fig. 8-3 the loci whose order is uncertain are shown in boldface. Brackets indicate that the order within the bracketed group has not been established. Where recombination differs in males and females, the unweighted average is given. The recombinations between adjacent loci have in many cases been determined directly, but in others the values were obtained by subtraction. The map is therefore not to be taken too literally.

Table 8-1 gives the recombination values between adjacent loci when these are known and some values between nonadjacent loci. Other recombination values that have been determined can be found in the references cited. Whenever recombination is different in the two sexes, the values for males and females are given separately. When there is no sex difference or when values were not determined for the sexes separately, only one value is given. In combining estimates from different experiments each estimate was weighted by the reciprocal of its variance.

Centromere position

The position of the centromere is not known with certainty for any mouse linkage groups. Most attempts to map the centromere make use of a phenomenon known as affinity.

Occasionally genes known to be located on different chromosomes behave in crosses as if they were loosely linked. Such a case was first noticed by Gates (1926) in a cross between Japanese waltzers and European laboratory mice. Michie and Wallace (1953) have reinvestigated the phenomenon. They have called it affinity and have offered the explanation that affinity is due to the presence in a cross of different states of homologous centromeres, such that nonhomologous chromosomes with centromeres of the same state tend to move to the same pole at meiosis rather than assort independently. Genes on nonhomologous chromosomes may thus appear to be linked. Wallace has made crosses which demonstrate association between markers in linkage groups V and III and has used the results to map the position of the centromere in linkage group V (Wallace, 1957*a*, 1958*b*, 1959, 1961). She concluded that it lies between *fi* and *Sd*. Parsons (1959) has found a possible case of affinity between linkage groups V and XIII and has suggested that the evidence favors the conclusion that the centromere in XIII is near *Dh* and *ln*. Michie, in his reanalysis of Gates's results (Michie, 1955*b*) concluded that the centromere in linkage group II is near *d*.

These positions are all median or subterminal with respect to the known markers in the three linkage groups. On cytological examination, however, the mitotic chromosomes of the mouse appear to have terminal centromeres (Chap. 7, Levan et al., 1962), an observation in conflict with the results of affinity studies. It is possible that the centromeres are actually subterminal with a very short arm not easily seen with the light microscope. If a chiasma occurred with high frequency in the short arm, the genetic map of the arm would appear to be quite long. However, the observations of Ford and Evans (1964) on the position of chiasmata in ring bivalents at meiosis argue against the probability that chiasmata occur in the short arm with appreciable frequency. It seems safest, therefore, not to regard the position of the centromeres in linkage groups II, V, and XIII as established from affinity studies, but to reserve judgment until there is confirmation by other methods.

Evidence on the position of the centromere in linkage group IX is discussed in Chap. 24.

STOCKS FOR MAINTAINING MUTANT GENES

Since most named mutant genes have consistent and easily recognized effects not obscured by the normal range of variation in environment or genetic background, most of them can be maintained without attention to systematic breeding schemes. Many systematic schemes which are useful for various purposes have been devised, however. Several widely used kinds are mentioned below. Consult Lyon (1963*b*) for a more comprehensive discussion of methods of maintaining mutants.

Inbred stocks

Methods for keeping a mutant and its normal allele segregating on an otherwise homozygous background, either by crossing to a standard inbred strain or by inbreeding with forced heterozygosis, are described in Chap. 2. They will not be further discussed here.

Balanced stocks

Inviable or infertile recessive mutants must be bred from heterozygous carriers. Since the heterozygous offspring of such matings are not distinguishable from homozygous normal mice except by breeding tests, maintenance of such mutants is difficult. If closely linked mutants are known, they can be used to help distinguish carriers of the mutant in question in so-called balanced stocks.

If r is an inviable recessive mutant with wild-type allele +, and a a closely linked recessive with wild-type allele +, matings of the type $r\ +/+\ a \times r\ +/+\ a$ will produce the nonrecombinant offspring $r\ +/r\ +$ (homozygotes for r), $+\ a/+\ a$ (homozygotes for a), and $r\ +/+\ a$ (wild type carrying both r and a). In the absence of recombination, wild-type offspring mated together will again produce the same three types in the next generation. Recombination will cause some wild-type offspring to be $r\ +/++$ or $++/+\ a$, but the closer the linkage the rarer will be such progeny. The stock can be propagated by mating wild-type animals together. If such matings produce both $r\ +/r\ +$ and $+\ a/+\ a$ offspring, they are then proved to be of the expected type and their wild-type offspring can in turn be used to produce the next generation. Figure 8-4 shows the expected proportion of such matings which will be of the proper type for various degrees of linkage between r and a. It was noted in Chap. 2 that in propagating a recessive mutant by intercrosses, the probability that matings between the sibs of homozygous mutants will in fact be matings

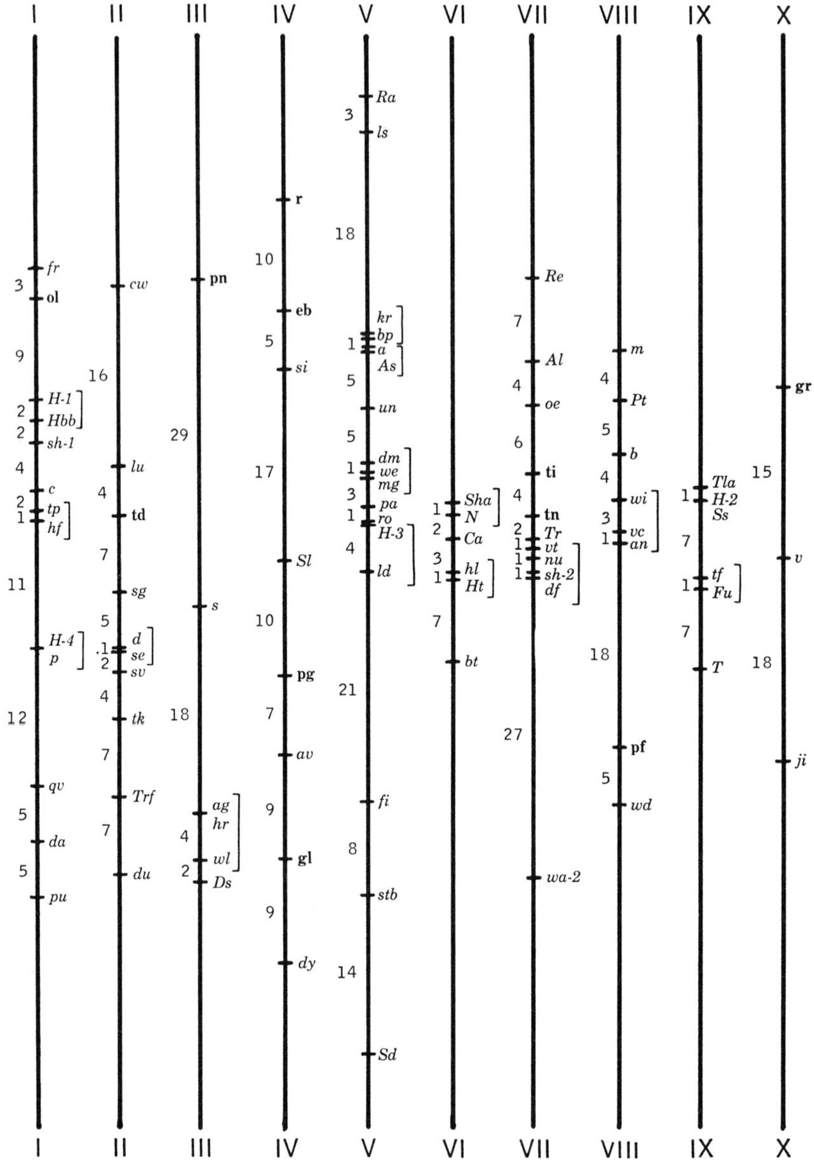

Fig. 8-3. Linkage map of the mouse.

between heterozygotes is ⅔ × ⅔ or 4/9 (44 per cent). From Fig. 8-4 it can be seen that use of a linked recessive marker will increase this probability only if it is closer than 17 map units to the mutant in question. A linked semidominant marker is somewhat more efficient. This is because the heterozygotes for the semidominant can be positively identified, thus eliminating half of the uncertainty in the selection of animals for mating.

D. S. Falconer (1965, personal communication) has pointed out that if small numbers of offspring are raised, the probability that a suitable mating will be recognized as such is somewhat lower for a balanced stock with a linked recessive marker than for a stock with a linked dominant or with no linked marker. This is because both mutants must be recovered in the stock with a linked recessive to prove that a mating is suitable, but only one must be recovered in stocks with a linked dominant or no linked marker. If as many as 12 offspring are obtained, however, the probability that a suitable mating will be discovered is 94 per

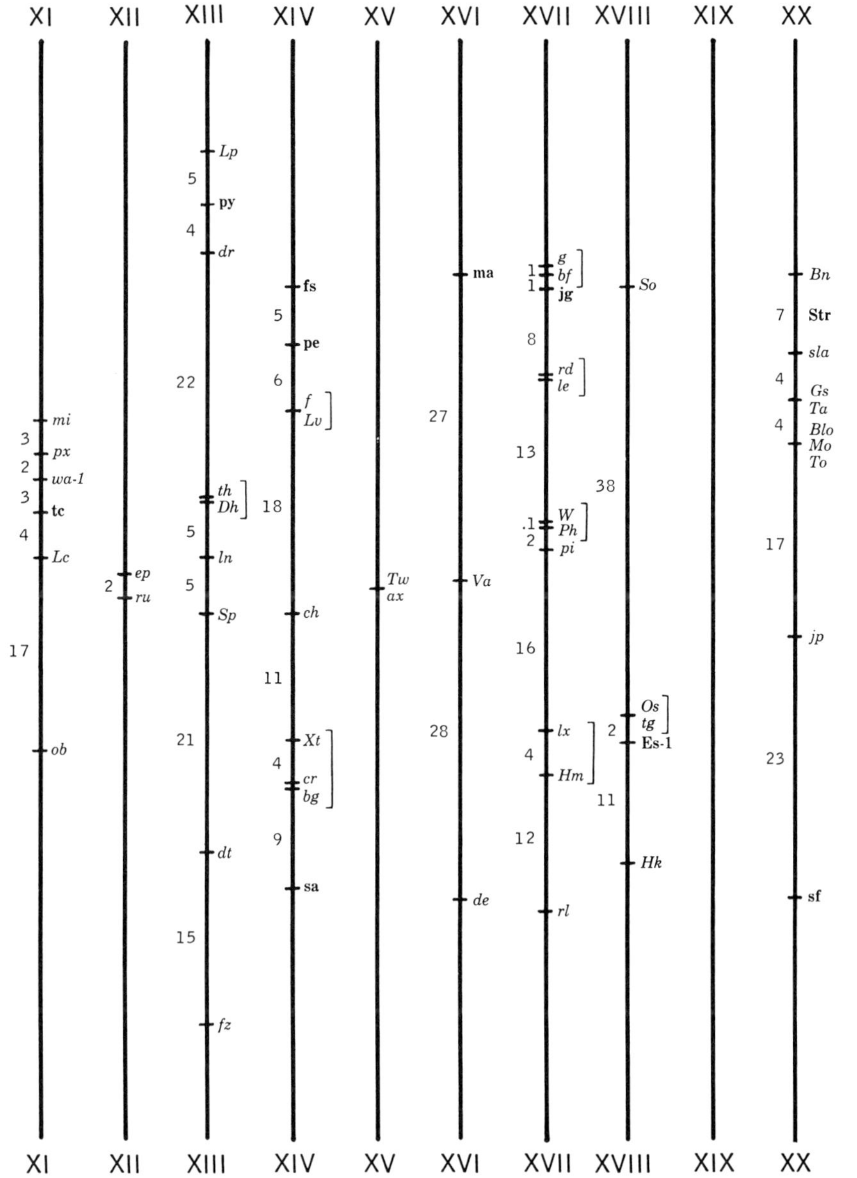

Fig. 8-3. (*continued*)

cent in a stock with a linked recessive and 97 per cent in the other two cases. The dashed lines in Fig. 8-4 show the proportion of matings that will be discovered to be suitable with 12 offspring for the three cases. I am indebted to Dr. Falconer for the method used in calculating these proportions. In practice it is usually easier to raise a second litter if the first one fails to prove a mating than to discard the mating and make up another one. The linked marker should be close enough and enough offspring should be produced to raise the proportion of suitable matings discovered to about 65 to 70 per cent to result in much saving of space.

An example of the use of a linked recessive marker is the case of wabbler lethal (*wl*) balanced against hairless (*hr*). Recombination between the two loci is about 4 per cent. Figure 8-2 shows that about 80 per cent of the matings will produce both *wl*/*wl* and *hr*/*hr* among 12 offspring and will therefore be suitable for propagating the stock. For practical purposes, considering that

Table 8-1. LINKAGE GROUPS AND RECOMBINATION PERCENTAGES

Linkage group	Locus symbol	Interval	Per cent recombination ♀	♂	Average or sex not stated	References
I	*fr*	*fr–sh-1*			16	Falconer and Snell (1952)
	*ol**	*ol–c*			17	Hertwig (1942)
	H-1	*H-1–c*			8	Snell (1958), Snell and Stevens (1961)
	Hbb	*Hbb–sh-1*			2	Popp (1962*a*)
	sh-1	*sh-1–c*	4	3		Falconer and Snell (1952), Gates (1931), Grüneberg (1936)
	c	*c–p*	16	12		Detlefsen (1925), Detlefsen and Clemente (1924), Falconer and Snell (1952), Grüneberg (1935*a*, 1936), Popp and St. Amand (1960)
	tp	*c–tp*	2			L. B. Russell (1963)
	hf	*c–hf*			3	Bunker (1959)
	H-4	*H-4–p*			0	Snell and Stevens (1961)
	p	*p–qv*			12	Yoon and Les (1957)
	qv	*c–qv*			28	Yoon and Les (1957)
	da	*p–da*			17	Falconer (1957)
	pu	*p–pu*			22	St. Amand and Cupp (1957*b*)
	*bh**	*c–bh*			Probably linked	Wallace (1963*b*)
	*Nil**	*c,p–Nil*			Close	Wallace (1965, personal comm.)
II	*cw*	*cw–se*			32	Falconer and Isaacson (1962*b*)
	lu	*lu–d se*			16	Green (1961*b*) (corrected)
	*td**	*td–d*			12	Dickie (1965)
	sg	*sg–d se*			5	Lane (1965, personal comm.)
	d	*d–se*			0.1	Gates (1928), Snell (1928), Goodwins and Vincent (1955)
	se	*d se–tk*			6	Falconer (1961), Green (unpub.)
	sv	*se–sv*	3	1		Green (unpub.)
	tk	*lu–tk*			22	Green (unpub.)
	Trf	*d–Trf*	17	8		Shreffler (1963)
	du	*d se–du*			20	Green (1961*b*), Snell (1955)
	*Up-1**	*Up-1–d*			Linked	Finlayson et al. (1963)
III	*pn**	*pn–s*			29	Kidwell et al. (1961)
	s	*s–hr*			18	Snell (1931), Hummel (1965, personal comm.)
	ag	*ag–hr*			0	Hoecker et al. (1954)
	hr	*hr–Ds*			6	Hummel (1965, personal comm.)
	wl	*hr–wl*			4	Lane and Dickie (1961)
	Ds	*s–Ds*			25	Hummel (1965, personal comm.)
IV	*r**	*r–si*			15	Keeler (1930)
	*eb**	*eb–Sl*			22	Chapman (1965, personal comm.)
	si	*si–Sl*	17			Shaible (1964, personal comm.), Green (unpub.)
	Sl	*Sl–av*			17	Schaible (1963*a*), Nash (1964)
	*pg**	*si–pg*			20	Falconer and Isaacson (1965)
		Sl–pg			14	Falconer and Isaacson (1965)
	av	*si–av*			33	Schaible (1961)
	*gl**	*Sl–gl*			26	Lane (1965, personal comm.)
	dy	*Sl–dy*			32	Wolfe (1963*a*)
		av–dy			18	Southard (1965, personal comm.)
	dl	*dl–av*			Linked	Lane (1965, personal comm.)
V	*Ra*	*Ra–a*			22	Carter and Phillips (1954), Parsons (1958*b*), Lane and Green (1960)
	ls	*Ra–ls*			3	Phillips (1965, personal comm.)
	kr	*kr–a*			1	Hertwig (1942), Lane (1960, personal comm.)
	bp	*bp–a*			0.3	Runner (1959)

Table 8-1. LINKAGE GROUPS AND RECOMBINATION PERCENTAGES (*Continued*)

Linkage group	Locus symbol	Interval	Per cent recombination ♀	♂	Average or sex not stated	References
	a	*a–un*			5	Carter (1947), Fisher and Landauer (1953), Runner (1959)
	As	*a–As*			Closely linked	Phillips (1963*a*)
	un	*un–we*	7	5		Fisher and Landauer (1953), Falconer (1954)
	dm	*a–dm*			10	Mackensen (1962)
	we	*we–pa*	4	2		Fisher and Landauer (1953), Falconer (1954)
	mg	*a–mg*	13	10		Lane and Green (1960)
	pa	*pa–ro*			1	Falconer (1954)
	ro	*a–fi*	36	27		Wallace (1957*b*)
	H-3 †	*we–H-3*			4	Snell and Bunker (1964)
	ld	*mg–ld*			8	Green (1965)
	fi	*pa–fi*			26	Carter (1951*a*), Bodmer (1961)
	stb	*stb–Sd*			14	Lane (1965)
	Sd	*fi–Sd*			22	Wallace (1957*b*)
VI	*Sha*	*Sha–N*			1	Flanagan (1964)
	N	*N–Ca*	1	3		Cooper (1939), Murray and Snell (1945), Mallyon (1951)
	Ca	*Ca–bt*	4	11		Murray and Snell (1945), Mallyon (1951)
	hl	*Ca–hl*	1	5		Hollander (1959 and 1965, personal comm.)
	Ht	*Ca–Ht*	2	7		St. Amand and Cupp (1957*a*)
	bt	*hl–bt*			8	Hollander (1959)
	*mn**	*N–mn*			Probably linked	Wallace (1963*b*)
VII	*Re*	*Re–Al*			7	Lane (1960, personal comm.)
	Al	*Al–sh-2*	23			Dickie (1955), Lane (1960, personal comm.)
	oe	*oe–sh-2*			15	Kelton (1965, personal comm.)
	*ti**	*Re–ti*			20	Searle (1961)
		ti–vt			9	Searle (1961)
	*tn**	*Re–tn*			21	Lane (1965, personal comm.)
	Tr	*Re–Tr*			23	Falconer and Sobey (1953)
		Tr–sh-2			3	Falconer and Sobey (1953)
	vt	*Re–vt*	27	18		Michie (1955*a*)
	‡	*vt–sh-2*			2	Michie (1955*a*)
	nu	*Re–nu*			25	Flanagan (1964)
	sh-2	*Re–sh-2*	28	19		Falconer (1947), Carter and Phillips (1953), Nasrat (1956)
		sh-2–wa-2	24	30		Snell and Law (1939), Heston (1941), Wright (1947), Fisher et al. (1947), Carter and Phillips (1953), Dickie (1955)
	df	*Re–df*	26			Bartke (1965)
	wa-2	*vt–wa-2*			28	Michie (1955*a*)
		df–wa-2			Linked	Schaible (1963*b*)
	vb	*Re–vb*			Linked	Lane (1965, personal comm.)
VIII	*m*	*m–Pt*			4	Lane (1963)
	Pt	*Pt–b*			5	Hollander and Strong (1951), Yoon (1961*b*), Lane (1963)
	b	*m–b*			7	Woolley (1945), Sirlin (1957)
	wi	*b–wi*	6	2		Lane (1963)
	vc	*b–vc*			7	Sirlin (1956, 1957)
	an	*b–an*			8	Hertwig (1942), McFarland (1965, personal comm.)
	*pf**	*b–pf*			26	Meredith (1965)
	wd	*b–wd*	31§			Yoon (1961*b*)

Table 8-1. LINKAGE GROUPS AND RECOMBINATION PERCENTAGES (*Continued*)

Linkage group	Locus symbol	Interval	Per cent recombination ♀	♂	Average or sex not stated	References
IX	*Tla*	*Tla–H-2*			1	Boyse et al. (1964)
	H-2	*H-2–tf*			7	Green and Stimpfling (unpub.)
		H 2–Fu			4	Allen (1955)
	Ss	*H-2–Ss*			0	Shreffler (1964*b*)
	tf ⎤	*tf–T*			8	Lyon and Phillips (1959), Green and Stimpfling (unpub.), Dunn et al. (1962)
	Fu ⎦	*tf–Fu*^{ki}			1	Dunn et al. (1962)
	T	*Fu–T*	8	4		Dunn and Caspari (1945), Dunn and Gluecksohn-Waelsch (1954)
X	*gr**	*gr–v*			15	Falconer and Isaacson (1965)
	v	*v–ji*			18	Snell (1945)
	ji					
XI	*mi*	*Mi*^{wh}*–px*			3	Searle (1964)
	px	*px–wa-1*			2	Searle (1964)
	wa-1	*Mi*^{wh}*–wa-1*			9	Bunker and Snell (1948)
	*tc**	*Mi*^{wh}*–tc*			8	Lane (1961)
	Lc	*wa-1–Lc*			7	Phillips (1960*a*), Searle (1964)
	ob	*wa-1–ob*			24	Lane (1965, personal comm.)
XII	*ep*	*ep–ru*	1	3		Lane and E. L. Green (1965, personal comm.)
	ru					
XIII	*Lp*	*Lp–ln*	38	35		Snell et al. (1954)
	*py**	*py–ln*	38	24		Fisher (1953), Parsons (1958*a*)
	dr	*dr–Dh*			22	Lyon (1961*a*)
	th ⎤	*th–ln*			5	Larsen (1961, 1963)
	Dh ⎦	*Dh–ln*			5	Lyon (1961*a*), Searle (1964)
	ln	*ln–Sp*			5	Snell et al. (1954), Parsons (1958*a*), Dickie (1964*a*)
	Sp	*Sp–fz*	40	33		Snell et al. (1954), Parsons (1958*a*), Dickie (1964*a*)
	dt	*ln–dt*			32	Falconer and Isaacson (1965), Kelton (1965), Lane (1965)
	fz	*dt–fz*			18	Falconer and Isaacson (1965), Kelton (1965), Lane (1965)
XIV	*fs**	*fs–f*			11	E. L. Green (1965, personal comm.)
	*pe**	*pe–Xt*	42	27		Lyon (1965)
		pe–cr			Loose	Lyon (1965)
	f	*f–cr*			33	King (1956), Phillips (1956)
	Lv	*f–Lv*			0	Coleman and E. S. Russell (1965, personal comm.)
	ch	*f–ch*			10	Phillips (1956)
	Xt ⎤	*f–Xt*	32	26		Lyon (1965)
	cr ⎥	*ch–cr*			8	Phillips (1956)
	bg ⎦	*cr–bg*			0	Lane (1965)
	*sa**	*bg–sa*			9	St. Amand and Cupp (1958)
XV	*Tw*	*Tw–ax*			0	Lyon (1958)
	ax					
XVI	*ma**	*ma–Va*			27	Lane (1964)
	Va	*Va–de*			28	Curry (1959)
	de					
XVII	*g* ⎤	*g–rd*	5	15		Paigen and Noell (1961), Sidman and Green (1965)
	bf ⎦	*bf–W*^{v}			22	Dickie (1965)
	*jg**	*jg–W*^{v}			21	Green (unpub.)

Table 8-1. LINKAGE GROUPS AND RECOMBINATION PERCENTAGES (*Continued*)

Linkage group	Locus symbol	Interval	Per cent recombination ♀	♂	Average or sex not stated	References
	rd]	*rd*–W^v			13	Sidman and Green (1965)
	le]	*le*–W^v	18	9		Lane and E. L. Green (1965, personal comm.), Lane and M. C. Green (unpub.)
	W]	W^v–*lx*			18	Carter (1951*d*)
	Ph]	W^v–*Ph*			0.1	Grüneberg and Truslove (1960)
	pi	W^v–*pi*	3	1		Dickie and Woolley (1946), Lane and M. C. Green (unpub.)
	lx]	*pi*–*lx*			11	Lane and M. C. Green (unpub.)
	Hm]	W^v–*Hm*			22	Green and Fox (1965)
	rl	*lx*–*rl*			16	Falconer (1952)
	go	*go*–W^v			Linked	Dickie (1965, personal comm.)
XVIII	*So*	*So*–*Os*			38	Falconer and Isaacson (1965)
	Os]	*Os*–*tg*			0	Green et al. (1963)
	tg]					
	*Es-1**	*Os*–*Es-1*			2	Popp (1965*a*)
	Hk	*Os*–*Hk*	18	8		Green et al. (1963)
XIX						
XX	*Bn*	*Bn*–*Ta*	11			Falconer (1954), Phillips (1954), Auerbach et al. (1962)
	*Str**	*Str*–*Ta*	12			Phillips (1963*b*), Lyon (1963*a*), Lyon et al. (1964*b*)
	sla	*sla*–*Ta*	4			Falconer and Isaacson (1962*a*)
	Gs	*Gs*–*Ta*	0			Larsen (1964)
	Ta	*Ta*–*Mo*	4			Falconer (1953, 1954), Auerbach et al. (1962)
	Blo	*Ta*–*Blo*	3			L. B. Russell (1960*b*)
	Mo					
	To	*Bn*–*To*	20			Lane (1960*b*)
	jp	*Ta*–*jp*	21			Phillips (1954)
	*sf**	*Ta*–*sf*	44			Welshons and Russell (1959)
	*Gy**	*Ta*–*Gy*	Not far			Lyon (1961*b*)
	*spf**		Sex-linked			L. B. Russell (1960*b*)

NOTE: Order within bracket not established.
* Listed order not established.
† Order not established with respect to each other or to *pa* and *ro*.
‡ Order not established with respect to each other or to *Tr* and *sh-2*.
§ Recomb. = 47% in females at 10°C higher than normal room temperature.

extra matings must be maintained to insure against sterility, this is very nearly as good as 100 per cent, and a great increase in efficiency over the 43 per cent of suitable matings obtained in maintaining *wl* by itself.

Another example is the use of the semidominant luxate (*lx*) balanced against reeler (*rl*). The recombination is about 16 per cent. About 65 per cent of matings between heterozygous luxate mice will therefore produce *rl*/*rl* and be suitable for propagating the stock.

Balanced stocks are also useful for determining the effect of a mutant in both single and double dose. For a recessive mutant (r) this means that comparisons must be made between $+/+$, $+/r$, and r/r. A closely linked marker (a) can be useful in distinguishing $+/+$ from $+/r$. Among offspring of matings of $r\ +/+\ a \times r\ +/+\ a$, those

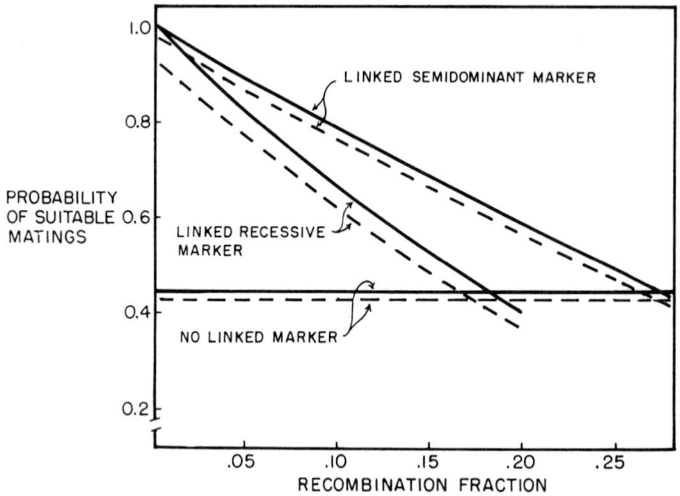

Fig. 8-4. Relation of closeness of linkage to usefulness of a linked marker in maintaining an inviable or infertile mutant. Solid lines: probability of occurrence of suitable matings. Dashed lines: probability of discovery of suitable matings when 12 offspring are raised.

+/+ at the *r* locus will be + *a*/+ *a*, and recognizable as homozygotes for *a*, whereas those heterozygous at the *r* locus will be *r* +/+ *a* and recognizable as wild type for both *r* and *a*.

An example is the use of short ear (*se*) in studies of the effect of dilute lethal (d^l) on phenylalanine metabolism. Matings of d^l +/+ *se* × d^l +/+ *se* produce + *se*/+ *se* (short-eared, homozygous wild type for d^l), d^l +/+ *se* (wild type, heterozygous for d^l) and d^l+/d^l + (homozygous for d^l). This linkage is very tight (recombination = 0.15 per cent) and identification is therefore virtually without error. Looser linkages may be useful if the effect under investigation is large enough. The linked marker must, of course, be known not to influence the effect in question.

Linked markers for early identification of a mutant

In studying the mode of action of a mutant gene during development it is often useful to know whether a mouse is the mutant type before any known effect of the gene is detectable. If the mutant is viable and fertile when homozygous, this can be achieved by mating homozygotes together. But for mutants which must be bred from heterozygotes and which therefore have nonmutant littermates, some means of identifying the mutants early may be very useful. Even in the case of fertile mutants it is often desirable to have nonmutant littermates as controls. If there exists a closely linked recessive or semidominant marker which is recognizable earlier in development than the mutant of interest, the phenotype of the marker can be used to identify the mutant individuals in offspring of appropriate matings. Table 8-2 shows some of the kinds of matings that can be used for various kinds of mutants, where *r* is a recessive mutant of interest, *D* is a dominant or semidominant mutant of interest, *a* is a linked recessive, and *A* is a linked semidominant.

This method has not been widely used in the mouse, in part because suitable markers have only rarely been available. E. S. Russell et al. (1956) were able to use the semidominant luxate (*lx*) in coupling with W^v to identify W^v/W^v and W^v/+ embryos at 12 days of gestation. Even a fairly loosely linked marker may be useful provided enough offspring can be examined to establish a correlation between the presence of the marker and the effect being examined. It may, of course, be necessary to establish by means of appropriate controls that the effect is not due to the marker itself. It should also be pointed out that it is not easy to put two closely linked recessives together in the same chromosome, particularly if one of them is lethal or sterile.

Linked dominant marker for crossing a recessive to an inbred strain

Transferring a recessive mutant to an inbred background requires a two-generation cross-intercross cycle, or, if the mutant is infertile, continued backcrossing with test crossing of the prospective parents in each generation (Chap. 2). If a closely linked dominant marker is available, the intercross generation can be omitted and continued backcrossing used with a high probability that the animals backcrossed will be carrying the mutant of interest. Using the conventions of the previous section, matings of the type *D r*/++ × ++/++ produce *D r*/++ and ++/++. The *D r*/++

Table 8-2. KINDS OF MATINGS WHICH ALLOW USE OF A LINKED MARKER FOR EARLY IDENTIFICATION OF A MUTANT

Mating[a]	Nonrecombinant offspring
r/r or D/D infertile	
$r\ a/+\ + \times r\ a/+\ +$	$r\ a/r\ a,\ r\ a/+\ +,\ +\ +/+\ +$
$D\ a/+\ + \times +\ a/+\ a$	$D\ a/+\ a,\ +\ a/\ +\ +$
$D\ a/+\ + \times D\ a/+\ +$	$D\ a/D\ a,\ D\ a/+\ +,\ +\ +/\ ++$
$r\ A/+\ + \times r\ A/+\ +$	$r\ A/r\ A,\ r\ A/+\ +,\ +\ +/+\ +$
$r\ +/+\ A \times r\ +/+\ A$	$r\ +/r\ +,\ r\ +/+\ A,\ +\ A/+\ A$
$D\ A/+\ + \times +\ +/+\ +$	$D\ A/+\ +,\ +\ +/+\ +$
$D\ A/+\ + \times D\ A/+\ +$	$D\ A/D\ A,\ D\ A/+\ +,\ +\ +/+\ +$
r/r fertile	
$r\ a/+\ + \times r\ a/r\ a$	$r\ a/r\ a,\ r\ a/+\ +$
$r\ +/+\ A \times r\ +/r\ +$	$r\ +/r\ +,\ r\ +/+\ A$

[a] See text for explanation of symbols.

offspring are chosen and backcrossed again to the inbred strain.

This plan has been used to transfer the very deleterious mutant, anemia (*an*), to the C57BL/6J background using the dominant marker, light (B^{lt}). Recombination between the two loci is about 8 per cent, so that 92 per cent of the light offspring from matings of $B^{lt}\ an/{+}{+} \times$ C57BL/6J will be carrying *an* ($B^{lt}\ an/{+}{+}$). It is wise to maintain several lines and to intercross light offspring in every generation or every other generation to make sure the recessive has not been lost by recombination.

Multiple mutant stocks for linkage testing

Multiple linkage group stocks. For efficient testing of a new mutant to locate it on the linkage map it is desirable to have linkage testing stocks so designed that linkage with a maximum length of the known map can be tested with a minimum number of crosses. The first such stocks in the mouse were designed and produced by Snell (Cooper, 1939). With many new mutants available, Carter and Falconer (1951) devised a new set of stocks and described the theory of their design. The theory allows calculation of the length of linkage map tested and so makes possible comparison of the relative efficiency of different sets of stocks. Since 1951, new mutants and an extended linkage map have made various modifications of these stocks possible. Similar though not identical sets of stocks are now maintained in several laboratories. Lists of such stocks and the laboratories which maintain them are carried periodically in *Mouse News Letter*, a mimeographed bulletin produced and distributed by the International Committee on Laboratory Animals and the Laboratory Animals Centre, MRC Laboratories, Woodmansterne Road, Carshalton, Surrey, England.

The linkage testing stocks maintained at The Jackson Laboratory in 1965 are listed in Table 8-3. These stocks leave some large regions untested including the *wa-2* end of VII and the *bg* end of XIV. Markers covering these regions are available in other stocks.

Instructions for use of such stocks and for the statistical treatment of the results can be found in Carter and Falconer (1951). More extensive and general discussions of linkage methods may be found in Mather (1951) and Green (1963).

Single linkage group stocks. Once a linkage has been detected for a new mutant, it is desirable to determine the linear order of the mutant with respect to other loci in the linkage group. If the loci are close together three-point backcrosses are usually necessary. If the loci are marked by recessive mutants, the mutants must be in coupling, i.e., on the same chromosome, to make backcrosses possible. Combining closely linked recessive mu-

Table 8-3. LINKAGE TESTING STOCKS MAINTAINED AT THE JACKSON LABORATORY IN 1965

Name	Markers	Linkage groups
MWT	a^t, Mi^{wh}, W^v, T	V, XI, XVII, IX
ROP	*Ra*, *Os*, *Pt*	V, XVIII, VIII
E1	*Re*, *Sd*, *Va*	VII, V, XVI
CST	*Ca*, *Sl*, *Tw*	VI, IV, XV
V	*a*, *ln*, *fz*, *s*, *v*	V, XIII, III, X
SEC/1	c^{ch}, *se*, *a*, *b*	I, II, V, VIII
JE	*a*, *f*, *ru*, *je*	V, XIV, XII
BP	*a*, *bt*, *p*, *wa-1*	V, VI, I, XI

tants is a tedious process, and it is therefore useful to maintain whatever such combinations have been produced, to have them available for future use. Numerous such stocks are now in existence in various laboratories and can be found listed in *Mouse News Letter.* At the Jackson Laboratory stocks carrying at least two linked recessives in coupling are available for linkage groups I, II, III, IV, V, VII, VIII, XIII, XIV, and XVII.

SUMMARY

The over 300 named mutant genes of the mouse, as well as a few conditions which have been shown to be due to mutants at as yet unnamed loci, are listed and briefly described. The known linkage groups are shown in a table with the pertinent recombination values. Efforts to map the positions of the centromeres by use of affinity are discussed.

Several kinds of stocks for maintaining mutants have been developed. Linked markers may be used to aid in the identification of carriers of recessive inviable or infertile mutants and to aid in the identification of a mutant early in development before any known effect of the mutant is detectable.

Multiple mutant stocks have been constructed for use in testing for linkage.

LITERATURE CITED

Allen, S. L. 1955. Linkage relations of the genes histocompatibility-2 and fused tail, brachyury and kinky tail in the mouse, as determined by tumor transplantation. Genetics 40:627–650.

Anderson, F. D., and P. J. Harman. 1961. Neurohistology of two mammalian mutations. Neurology 11:676–680.

Arnesen, K. 1955. Constitutional difference in lipid content of adrenals in two strains of mice and their hybrids. Acta Endocrinol. 18:396–401.

Arnesen, K. 1956. The adrenothymic constitution and susceptibility to leukemia in mice. Acta Pathol. Microbiol. Scand. Suppl. 109. 95 p.

Arnesen, K. 1963. The cytology of the adrenal cortex in mice with spontaneous adrenocortical lipid depletion. Acta Pathol. Microbiol. Scand. 58:212–218.

Ashton, G. C., and A. W. H. Braden. 1961. Serum β-globulin polymorphism in mice. Austral. J. Biol. Sci. 14:248–253.

Auerbach, C., D. S. Falconer, and J. H. Isaacson. 1962. Test for sex-linked lethals in irradiated mice. Genet. Res. 3:444–447.

Auerbach, R. 1954. Analysis of the developmental effects of a lethal mutation in the house mouse. J. Exp. Zool. 127:305–329.

Badr, F. M., and S. G. Spickett. 1965. Genetic variation in the biosynthesis of corticosteroids in *Mus musculus.* Nature 205:1088–1090.

Ballantyne, J., F. G. Bock, L. C. Strong, and W. C. Quevedo, Jr. 1961. Another allele at the *W* locus of the mouse. J. Hered. 52:200–202.

Barber, A. N. 1957. The effects of maternal hypoxia on inheritance of recessive blindness in mice. Amer. J. Ophthalmol. 44:94–101.

Barnicot, N. A. 1945. Some data on the effect of parathormone on the grey-lethal mouse. J. Anat. 79:83–91.

Barnicot, N. A. 1948. The local action of the parathyroid and other tissues on bone in intracerebral grafts. J. Anat. 82:233–248.

Bartke, A. 1964. Histology of the anterior hypophysis, thyroid and gonads of two types of dwarf mice. Anat. Rec. 149:225–235.

Bartke, A. 1965. Mouse News Letter 32:52–53.

Bateman, N. 1954. Bone growth: a study of the grey-lethal and microphthalmic mutants of the mouse. J. Anat. 88:212–262.

Bateman, N. 1957. Mouse News Letter 16:7.

Bateman, N. 1960. High frequency of a lethal gene (t^e) in a laboratory stock of mice. Genet. Res. 1:214–225.

Bateman, N. 1961. Sombre, a viable dominant mutant in the house mouse. J. Hered. 52:186–190.

Batt, R. A. L., and G. A. Harrison. 1963. The reproductive system of the adipose mouse. J. Hered. 54:135–138.

Beasley, A. B. 1963. Inheritance and development of a lens abnormality in the mouse. J. Morphol. 112:1–11.

Beck, S. L. 1963*a*. The anophthalmic mutant of the mouse. I. Genetic contribution to the anophthalmic phenotype. J. Hered. 54:39–44.

Beck, S. L. 1963*b*. The anophthalmic mutant of the mouse. II. An association of anophthalmia and polydactyly. J. Hered. 54:79–83.

Beck, S. L. 1963*c*. Frequencies of teratologies among homozygous normal mice compared with those heterozygous for anophthalmia. Nature 200:810–811.

Bennett, D. 1956. Developmental analysis of a mutation with pleiotropic effects in the mouse. J. Morphol. 98:199–234.

Bennett, D. 1958. In vitro study of cartilage induction in *T/T* mice. Nature 181:1286.

Bennett, D. 1959. Brain hernia, a new recessive mutation in the mouse. J. Hered. 50:265–268.

Bennett, D. 1961*a*. A chromatographic study of abnormal urinary amino acid excretion in mutant mice. Ann. Hum. Genet. 25:1–6.

Bennett, D. 1961*b*. Miniature, a new gene for small size in the mouse. J. Hered. 52:95–98.

Bennett, D., S. Badenhausen, and L. C. Dunn. 1959. The embryological effects of four late-lethal *t*-alleles in the mouse which affect the neural tube and skeleton. J. Morphol. 105:105–143.

Bennett, J. H., and G. A. Gresham, 1956. A gene for eyelids open at birth in the house mouse. Nature 178:272–273.

Bernstein, S. E. 1960. Mouse News Letter 23:33.

Bernstein, S. E. 1963. Analysis of gene action and characterization of a new hematological abnormality, hemolytic anemia, p. 186. *In* S. J. Geerts [ed.] Proc. XI Int. Congr. Genet. Vol. 1. Pergamon Press, New York. (Abstr.)

Berry, R. J. 1960. Genetical studies on the skeleton of the mouse. XXVI. Pintail. Genet. Res. 1:439–451.

Berry, R. J. 1961*a*. The inheritance and pathogenesis of hydrocephalus-3 in the mouse. J. Pathol. Bacteriol. 81:157–167.

Berry, R. J. 1961*b*. Genetically controlled degeneration of the nucleus pulposus in the mouse. J. Bone Joint Surg. 43B:387–393.

Bhaskar, S. N., I. Schour, E. C. MacDowell, and J. P. Weinmann. 1951. The skull and dentition of screw tail mice. Anat. Rec. 110:199–229.

Bielschowsky, M., and G. C. Schofield. 1962. Studies on megacolon in piebald mice. Austral. J. Exp. Biol. Med. Sci. 40:395–404.

Bierwolf, D. 1956. Kleinhirnmissbildungen durch hereditären Hydrocephalus bei der Hausmaus. Wiss. Z. Martin-Luther-Univ. 5:1237–1282.

Bierwolf, D. 1958. Die Embryogenese des Hydrocephalus und der Kleinhirnmissbildungen beim Dreherstamm der Hausmaus. Morphol. Jahrb. 99:542–612.

Billingham, R. E., and W. K. Silvers. 1960. The melanocytes of mammals. Quart. Rev. Biol. 35:1–40.

Bloom, J. L. 1962. Mouse News Letter 27:30.

Bloom, J. L., and D. S. Falconer. 1964. A gene with major effect on susceptibility to induced lung tumors in mice. J. Nat. Cancer Inst. 33:607–618.

Bodmer, W. F. 1961. Viability effects and recombination differences in a linkage test with pallid and fidget in the house mouse. Heredity 16:485–495.

Bonnevie, K. 1936. Abortive differentiation of the ear vesicles following a hereditary brain-anomaly in the "short-tailed waltzing mice." Genetica 18:105–125.

Bonnevie, K., and A. Brodal. 1946. Hereditary hydrocephalus in the house mouse. IV. The development of the cerebellar anomalies during foetal life with notes on the normal development of the mouse cerebellum. Skr. Norske Vidensk.-Akad. Oslo, I. Mat.-Natur. Kl. 1946. (4). 60 p.

Boyse, E. A., L. J Old, and S. Luell. 1964. Genetic determination of the *TL* (thymus-leukaemia) antigen in the mouse. Nature 201:779.

Braden, A. W. H., and S. Gluecksohn-Waelsch. 1958. Further studies of the effect of the *T* locus in the house mouse on male fertility. J. Exp. Zool. 138:431–452.

Braverman, I. M. 1953. Neurological actions caused by the mutant gene "trembler" in the house mouse (*Mus musculus*, L.): an investigation. J. Neuropathol. Exp. Neurol. 12:64–72.

Brooke, H. C. 1926. Hairless mice. J. Hered. 17:173–174.

Brückner, R. 1951. Spaltlampenmikroskopie und Ophthalmoskopie am Auge von Ratte und Maus. Doc. Ophthalmol. 5–6:452–554.

Bryson, V. 1945. Development of the sternum in screw tail mice. Anat. Rec. 91:119–141.

Bunker, H., and G. D. Snell. 1948. Linkage of white and waved-1. J. Hered. 39:28.

Bunker, L. E. 1959. Hepatic fusion, a new gene in linkage group I of the mouse. J. Hered. 50:40–44.

Butler, L., and D. A. Robertson. 1953. A new eye abnormality in the house mouse. J. Hered. 44:13–16.

Carter, T. C. 1947. A new linkage in the house mouse: undulated and agouti. Heredity 1:367–372.

Carter, T. C. 1948. A new strain of luxate mice. Heredity 2:405–406. (Abstr.)

Carter, T. C. 1951*a*. The position of fidget in linkage group V of the house mouse. J. Genet. 50:264–267.

Carter, T. C. 1951*b*. Wavy-coated mice: phenotypic interactions and linkage tests between rex and (a) waved-1, (b) waved-2. J. Genet. 50:268–276.

Carter, T. C. 1951*c*. The genetics of luxate mice. I. Morphological abnormalities of heterozygotes and homozygotes. J. Genet. 50:277–299.

Carter, T. C. 1951*d*. The genetics of luxate mice. II. Linkage and independence. J. Genet. 50:300–306.

Carter, T. C. 1953. The genetics of luxate mice.

III. Horseshoe kidney, hydronephrosis and lumbar reduction. J. Genet. 51:441–457.

Carter, T. C. 1954. The genetics of luxate mice. IV. Embryology. J. Genet. 52:1–35.

Carter, T. C. 1956. Genetics of the Little and Bagg X-rayed mouse stock. J. Genet. 54:311–326.

Carter, T. C. 1959*a*. Mouse News Letter 21:40.

Carter, T. C. 1959*b*. Embryology of the Little and Bagg X-rayed mouse stock. J. Genet. 56:401–435.

Carter, T. C., and D. S. Falconer. 1951. Stocks for detecting linkage in the mouse, and the theory of their design. J. Genet. 50:307–323.

Carter, T. C., and R. S. Phillips. 1950. Ichthyosis, a new recessive mutant in the house mouse. J. Hered. 41:297–300.

Carter, T. C., and R. J. S. Phillips. 1953. The sex distribution of waved-2, shaker-2, and rex in the house mouse. Z. Indukt. Abstamm. Vererb. 85:564–578.

Carter, T. C., and R. J. S. Phillips. 1954. Ragged, a semidominant coat texture mutant in the house mouse. J. Hered. 45:151–154.

Caspari, E., and P. R. David. 1940. The inheritance of a tail abnormality in the house mouse. J. Hered. 31:427–431.

Chai, C. K. 1961. Hereditary spasticity in mice. J. Hered. 52:241–243.

Chai, C. K., E. Roberts, and R. L. Sidman. 1962. Influence of aminooxyacetic acid, a γ-aminobutyrate transaminase inhibitor, on hereditary spastic defect in the mouse. Proc. Soc. Exp. Biol. Med. 109:491–495.

Chase, H. B. 1942. Studies on an anophthalmic strain of mice. III. Results of crosses with other strains. Genetics 27:339–348.

Chase, H. B. 1944. Studies on an anophthalmic strain of mice. IV. A second major gene for anophthalmia. Genetics 29:264–269.

Chesley, P. 1935. Development of the short-tailed mutant in the house mouse. J. Exp. Zool. 70:429–459.

Christophe, J. 1963. Biochimie des obésités expérimentales. Prob. Actuel. Endocrinol. Nutr. 7:23–76.

Cinader, B., S. Dubiski, and A. C. Wardlaw. 1964. Distribution, inheritance, and properties of an antigen, MuB1, and its relation to hemolytic complement. J. Exp. Med. 120:897–924.

Clark, F. H. 1932. Hydrocephalus, a hereditary character in the house mouse. Proc. Nat. Acad. Sci. 18:654–656.

Clark, F. H. 1934. Anatomical basis of a hereditary hydrocephalus in the house mouse. Anat. Rec. 58:225–233.

Clark, F. H. 1935. Two hereditary types of hydrocephalus in the house mouse (*Mus musculus*). Proc. Nat. Acad. Sci. 21:150–152.

Cloudman, A. M., and L. E. Bunker. 1945. The varitint-waddler mouse. J. Hered. 36:258–263.

Coggeshall, R. E., C. J. D'Amato, M. A. Brodbine, and S. P. Hicks. 1961. Developmental neuropathology of genetic mutant mouse "paralytic." Fed. Proc. 20:330. (Abstr.)

Cohen, B. L. 1960. Genetics of plasma transferrins in the mouse. Genet. Res. 1:431–438.

Cohen, B. L., and D. C. Shreffler. 1961. A revised nomenclature for the mouse transferrin locus. Genet. Res. 2:306–308.

Coleman, D. L. 1960. Phenylalanine hydroxylase activity in dilute and nondilute strains of mice. Arch. Biochem. Biophys. 91:300–306.

Coleman, D. L. 1962. Effect of genic substitution on the incorporation of tyrosine into the melanin of mouse skin. Arch. Biochem. Biophys. 69:562–568.

Cooper, C. B. 1939. A linkage between naked and caracul in the house mouse. J. Hered. 30:212.

Crew, F. A. E. 1933. Waved: an autosomal recessive coat form character in the mouse. J. Genet. 27:95–96.

Crew, F. A. E., and C. Auerbach. 1939. Rex: a dominant autosomal monogenic coat texture character in the mouse. J. Genet. 38:341–344.

Crew, F. A. E., and L. Mirskaia. 1931. The character "hairless" in the mouse. J. Genet. 25:17–24.

Cupp, M. B. 1958. Mouse News Letter 19:37.

Cupp, M. B. 1960. Mouse News Letter 22:50.

Cupp, M. B. 1962. Mouse News Letter 26:51.

Curry, G. A. 1959. Genetical and developmental studies on droopy-eared mice. J. Embryol. Exp. Morphol. 7:39–65.

Dagg, C. P., D. L. Coleman, and G. M. Fraser. 1964. A gene affecting the rate of pyrimidine degradation in mice. Genetics 49:979–989.

Danforth, C. H. 1958. The occurrence and genetic behavior of duplicate lower incisors in the mouse. Genetics 43:139–148.

David, L. T. 1932. The external expression and comparative dermal histology of hereditary hairlessness in mammals. Z. Zellforsch. 14:616–719.

David, L. T. 1937. Die gegenseitige Beeinflussung der Erbfaktoren für Haarlosigkeit und Welligkeit bei der Hausmaus (*Mus musculus*). Rev. Suisse Zool. 44:397–400.

Deol, M. S. 1954. The anomalies of the labyrinth of the mutants varitint-waddler, shaker-2, and jerker in the mouse. J. Genet. 52:562–588.

Deol, M. S. 1956*a*. A gene for uncomplicated deafness in the mouse. J. Embryol. Exp. Morphol. 4:190–195.

Deol, M. S. 1956*b*. The anatomy and development of the mutants pirouette, shaker-1, and waltzer in the mouse. Proc. Roy. Soc. B 145:206–213.

Deol, M. S. 1961. Genetical studies on the skeleton of the mouse. XXVIII. Tail-short. Proc. Roy. Soc. B 155:78–95.

Deol, M. S. 1963. The development of the inner ear in mice homozygous for *shaker-with-syndactylism*. J. Embryol. Exp. Morphol. 11:493–512.

Deol, M. S. 1964*a*. The abnormalities of the inner ear in *kreisler* mice. J. Embryol. Exp. Morphol. 12:475–490.

Deol, M. S. 1964*b*. The origin of the abnormalities of the inner ear in dreher mice. J. Embryol. Exp. Morphol. 12:727–733.

Deol, M. S., and W. Kocher. 1958. A new gene for deafness in the mouse. Heredity 12:463–466.

Deol, M. S., and M. W. Robins. 1962. The spinner mouse. J. Hered. 53:133–136.

Deol, M. S., and G. M. Truslove. 1963. A new gene causing cerebral degeneration in the mouse, p. 183–184. *In* S. J. Geerts [ed.] Proc. XI Int. Congr. Genet. Vol. 1. Pergamon Press, New York. (Abstr.)

DeOme, K. B. 1945. A new recessive lethal mutation in mice. Univ. Calif. Pub. Zool. 53:41–66.

Detlefsen, J. A. 1921. A new mutation in the house mouse. Amer. Natur. 55:469–473.

Detlefsen, J. A. 1925. The linkage of dark-eye and color in mice. Genetics 10:17–32.

Detlefsen, J. A., and L. S. Clemente. 1924. Linkage of a dilute color factor and dark eye in mice. Genetics 9:247–260.

Dickie, M. M. 1954. The tortoise shell house mouse. J. Hered. 45:158–190.

Dickie, M. M. 1955. Alopecia, a dominant mutation in the house mouse. J. Hered. 46:31–34.

Dickie, M. M. 1961. Mouse News Letter 25:36.

Dickie, M. M. 1962*a*. A new viable yellow mutation in the house mouse. J. Hered. 53:84–86.

Dickie, M. M. 1962*b*. Mouse News Letter 27:37.

Dickie, M. M. 1963. Mouse News Letter 29:39.

Dickie, M. M. 1964*a*. New splotch alleles in the mouse. J. Hered. 55:97–101.

Dickie, M. M. 1964*b*. Mouse News Letter 30:30.

Dickie, M. M. 1965. Mouse News Letter 32:43–46.

Dickie, M. M., J. Schneider, and P. J. Harman. 1952. A juvenile wabbler-lethal in the house mouse. J. Hered. 43:283–286.

Dickie, M. M., and G. W. Woolley. 1946. Linkage studies with the pirouette gene in the mouse. J. Hered. 37:335–337.

Dickie, M. M., and G. W. Woolley. 1950. Fuzzy mice. J. Hered. 41:193–196.

Dobrovolskaïa-Zavadskaïa, N. 1927. Sur la mortification spontanée de la queue chez la souris nouveau-née et sur l'existence d'une caracter hereditaire "non-viable." Compt. Rend. Soc. Biol. 97:114–116.

Dobrovolskaïa-Zavadskaïa, N. 1928. L'irradiation des testicules et l'hérédite chez la souris. Arch. Biol. 38:457–501.

Dray, S., R. Lieberman, and H. A. Hoffman. 1963. Two murine γ-globulin allotypic specificities identified by ascitic fluid isoprecipitins and determined by allelic genes. Proc. Soc. Exp. Biol. Med. 113:509–513.

Dubiski, S., and B. Cinader. 1963. A new allotype specificity in the mouse (MuA2). Nature 197:705.

Duchen, L. W., D. S. Falconer, and S. J. Strich. 1966. Hereditary progressive neurogenic muscular atrophy in the mouse. J. Physiol. In press.

Duchen, L. W., and S. J. Strich. 1964. Clinical and pathological studies of an hereditary neuropathy in mice (dystonia musculorum). Brain 87:367–378.

Dunn, L. C. 1928. A fifth allelomorph in the agouti series of the house mouse. Proc. Nat. Acad. Sci. 14:816–819.

Dunn, L. C. 1934. A new gene affecting behavior and skeleton in the house mouse. Proc. Nat. Acad. Sci. 20:230–232.

Dunn, L. C. 1937*a*. Studies on spotting patterns. II. Genetic analysis of variegated spotting in the house mouse. Genetics 22:43–64.

Dunn, L. C. 1937*b*. Caracul, a dominant mutation. J. Hered. 28:334.

Dunn, L. C. 1945. A new eye color mutant in the mouse with asymmetrical expression. Proc. Nat. Acad. Sci. 31:343–346.

Dunn, L. C., A. B. Beasley, and H. Tinker. 1960. Polymorphisms in populations of wild house mice. J. Mammal. 41:220–229.

Dunn, L. C., D. Bennett, and A. B. Beasley. 1962. Mutation and recombination in the vicinity of a complex gene. Genetics 47:285–303.

Dunn, L. C., and E. Caspari. 1945. A case of neighboring loci with similar effects. Genetics 30:543–568.

Dunn, L. C., and D. R. Charles. 1937. Studies on spotting patterns. I. Analysis of quantitative variations in the pied spotting of the house mouse. Genetics 22:14–42.

Dunn, L. C., and S. Gluecksohn-Schoenheimer. 1942. Stub, a new mutation in the mouse. J. Hered. 33:235–239.

Dunn, L. C., and S. Gluecksohn-Schoenheimer. 1947. A new complex of hereditary abnormalities in the house mouse. J. Exp. Zool. 104:25–51.

Dunn, L. C., S. Gluecksohn-Schoenheimer, and V. Bryson. 1940. A new mutation in the mouse affecting spinal column and urogenital system. J. Hered. 31:343–348.

Dunn, L. C., and S. Gluecksohn-Waelsch. 1954. A genetical study of the mutation "fused" in the house mouse with evidence concerning its allelism with a similar mutation "kink." J. Genet. 52:383–391.

Dunn, L. C., and J. Mohr. 1952. An association of hereditary eye defects with white spotting. Proc. Nat. Acad. Sci. 38:872–875.

Dunn, L. C., and L. W. Thigpen. 1930. The silver mouse, a recessive color variation. J. Hered. 21:495–498.

Eaton, G. J., and M. M. Green. 1963. Giant cell differentiation and lethality of homozygous *yellow* mouse embryos. Genetica 34:155–161.

Elftman, H., and O. Wegelius. 1959. Anterior pituitary cytology of the dwarf mouse. Anat. Rec. 135:43–49.

Fahey, J. L., J. Wunderlich, and R. Mishell. 1964. The immunoglobulins of mice. II. Two subclasses of mouse $7S\gamma_2$-globulins: γ_{2a}- and γ_{2b}-globulins. J. Exp. Med. 120:243–251.

Falconer, D. S. 1947. Linkage of rex with shaker-2 in the house mouse. Heredity 1:133–135.

Falconer, D. S. 1950. Mouse News Letter 2:3.

Falconer, D. S. 1951. Two new mutants "trembler" and "reeler," with neurological actions in the house mouse. J. Genet. 50:192–201.

Falconer, D. S. 1952. Location of "reeler" in linkage group III in the mouse. Heredity 6:255–257.

Falconer, D. S. 1953. Total sex-linkage in the house mouse. Z. Indukt. Abstamm. Vererb. 85:210–219.

Falconer, D. S. 1954. Linkage in the mouse: the sex-linked genes and "rough." Z. Indukt. Abstamm. Vererb. 86:263–268.

Falconer, D. S. 1956. Mouse News Letter 15:23.

Falconer, D. S. 1957. Mouse News Letter 17:40.

Falconer, D. S. 1961. Mouse News Letter 25:30.

Falconer, D. S., and S. P. Flanagan. 1963. Mouse News Letter 29:30.

Falconer, D. S., A. S. Fraser, and J. W. B. King. 1951. The genetics and development of "crinkled," a new mutant in the house mouse. J. Genet. 50:324–344.

Falconer, D. S., and J. H. Isaacson. 1959. Adipose, a new inherited obesity of the mouse. J. Hered. 50:290–292.

Falconer, D. S., and J. H. Isaacson. 1962*a*. The genetics of sex-linked anaemia in the mouse. Genet. Res. 3:248–250.

Falconer, D. S., and J. H. Isaacson. 1962*b*. Mouse News Letter 27:30.

Falconer, D. S., and J. H. Isaacson. 1965. Mouse News Letter 32:30–31.

Falconer, D. S., M. Latyszewski, and J. H. Isaacson. 1964. Diabetes insipidus associated with oligosyndactyly in the mouse. Genet. Res. 5:473–488.

Falconer, D. S., and U. Sierts-Roth. 1951. Dreher, ein neues Gen der Tangmausgruppe bei der Hausmaus. Z. Indukt. Abstamm. Vererb. 84:71–73.

Falconer, D. S., and G. D. Snell. 1952. Two new hair mutants, rough and frizzy, in the house mouse. J. Hered. 43:53–57.

Falconer, D. S., and W. R. Sobey. 1953. The location of "trembler" in linkage group VII. J. Hered. 49:159–160.

Feldman, H. W. 1922. A fourth allelomorph in the albino series in mice. Amer. Natur. 56:573–574.

Fielder, J. H. 1952. The taupe mouse. J. Hered. 43:75–76.

Finlayson, J. S., M. Potter, and C. C. Runner. 1963. Electrophoretic variation and sex dimorphism of the major urinary protein complex in inbred mice: a new genetic marker. J. Nat. Cancer Inst. 31:91–107.

Fischer, H. 1956. Morphologische und mikroskopisch-anatomische Untersuchungen am Innenohr eines Stammes spontanmutierter Hausmäuse (dreher). Z. Mikroskop.-Anat. Forsch. 62:348–406.

Fischer, H. 1958. Die Embryogenese der Innenohrmissbildungen bei dem spontanmutierten Dreherstamm der Hausmaus. Z. Mikroskop.-Anat. Forsch. 64:476–497.

Fischer, H. 1959. Mikroscopische Untersuchungen am Gehirn einer neuen Hausmausmutante mit Leukodystrophie. Verhandl. Deut. Zool. Ges. Münster/Westf. 1959:519–524.

Fisher, R. A. 1953. The linkage of *polydactyly* with *leaden* in the house mouse. Heredity 7:91–95.

Fisher, R. A., and W. Landauer. 1953. Sex differences of crossing over in close linkage. Amer. Natur. 87:116.

Fisher, R. A., M. F. Lyon, and A. R. G. Owen. 1947. The sex chromosome in the house mouse. Heredity 1:355–366.

Fitch, N. 1957. An embryological analysis of two mutants in the house mouse, both producing cleft palate. J. Exp. Zool. 136:329–357.

Fitch, N. 1961*a*. A mutation in mice producing dwarfism, brachycephaly, cleft palate and micromelia. J. Morphol. 109:141–149.

Fitch, N. 1961*b*. Development of cleft palate in mice homozygous for the shorthead mutation. J. Morphol. 109:151–157.

Flanagan, S. P. 1964. Mouse News Letter 31:21.

Foerster, E. 1956. Mouse News Letter 14:23.

Ford, C. E., and E. P. Evans. 1964. A reciprocal translocation in the mouse between the X chromosome and a short autosome. Cytogenetics 3:295–305.

Forsthoefel, P. F. 1958. The skeletal effects of the luxoid gene in the mouse, including its interactions with the luxate gene. J. Morphol. 102:247–287.

Forsthoefel, P. F. 1959. The embryological development of the skeletal effects of the luxoid gene in the mouse, including its interactions with the luxate gene. J. Morphol. 104:89–141.

Forsthoefel, P. F. 1962. Genetics and manifold effects of Strong's luxoid gene in the mouse, including its interactions with Green's luxoid and Carter's luxate genes. J. Morphol. 110:391–420.

Forsthoefel, P. F. 1963. The embryological development of the effects of Strong's luxoid gene in the mouse. J. Morphol. 113:427–451.

Fraser, A. S., S. Sobey, and C. C. Spicer. 1953. Mottled, a sex-modified lethal in the house mouse. J. Genet. 51:217–221.

Fraser, F. C. 1946. The expression and interaction of hereditary factors producing hypotrichosis in the mouse: histology and experimental results. Can. J. Res. D, 24:10–25.

Fraser, F. C., and M. L. Herer. 1948. Lens rupture, a new recessive gene in the house mouse. J. Hered. 39:149.

Fraser, F. C., and M. L. Herer. 1950. The inheritance and expression of the "lens rupture" gene in the house mouse. J. Hered. 41:3–7.

Fraser, F. C., and G. Schabtach. 1962. "Shrivelled," a hereditary degeneration of the lens in the house mouse. Genet. Res. 3:383–387.

Freye, H. 1954. Anatomische und entwicklungsgeschlichtliche Untersuchungen am Skelett normaler und oligodactyler Maüse. Wiss. Z. Martin-Luther-Univ. 3:801–824.

Freye, H. 1956. Untersuchungen über die Zahnanomalie des Mikrophthalmus-syndrome der Hausmaus. Z. Mensch. Vererb. Konst. 33:492–504.

Garber, E. D. 1952*a*. Bald, a second allele of hairless in the house mouse. J. Hered. 43:45–46.

Garber, E. D. 1952*b*. "Bent tail," a dominant, sex-linked mutation in the mouse. Proc. Nat. Acad. Sci. 38:876–879.

Gates, A. H., and M. Karasek. 1965. Hereditary absence of sebaceous glands in the mouse. Science 148:1471–1473.

Gates, W. H. 1926. The Japanese waltzing mouse; its origin, heredity, and relation to the genetic characters of other varieties of mice. Pub. Carnegie Inst. Wash. No. 337:83–138.

Gates, W. H. 1928. Linkage of the factors for short-ear and density in the house mouse. Genetics 13:170–179.

Gates, W. H. 1931. Linkage of the factor shaker with albinism and pink-eye in the house mouse. Z. Indukt. Abstamm. Vererb. 59:220–226.

Gluecksohn-Schoenheimer, S. 1943. The morphological manifestations of a dominant mutation in mice affecting tail and urogenital system. Genetics 28:341–348.

Gluecksohn-Schoenheimer, S. 1949. The effects of a lethal mutation responsible for duplications and twinning in mouse embryos. J. Exp. Zool. 110:47–76.

Gluecksohn-Waelsch, S. 1961. Mouse News Letter 25:12.

Gluecksohn-Waelsch, S. 1963. Lethal genes and analysis of differentiation. Science 142:1269–1276.

Gluecksohn-Waelsch, S., D. Hagedorn, and B. F. Sisken. 1956. Genetics and morphology of a recessive mutation in the house mouse affecting head and limb skeleton. J. Morphol. 99:465–479.

Gluecksohn-Waelsch, S., and S. A. Kamell. 1955. Physiological investigations of a mutation in mice with pleiotropic effects. Physiol. Zool. 28:68–73.

Gluecksohn-Waelsch, S., H. M. Ranney, and B. F. Sisken. 1957. The hereditary transmission of hemoglobin differences in mice. J. Clin. Invest. 36:753–756.

Gluecksohn-Waelsch, S., and T. R. Rota. 1963. Development in organ tissue culture of kidney rudiments from mutant mouse embryos. Develop. Biol. 7:432–444.

Goodman, G., and H. Koprowski. 1962. Macrophages as a cellular expression of inherited natural resistance. Proc. Nat. Acad. Sci. 48:160–165.

Goodman, R. M. 1956. The effect of the W^v allele in the mouse on the differential excretion of the optical isomers of several amino acids. J. Exp. Zool. 132:189–217.

Goodwins, I. R., and M. A. C. Vincent. 1955. Further data on linkage between short-ear and Maltese dilution in the house mouse. Heredity 9:413–414.

Gower, J. S. 1953. Mouse News Letter 8, Suppl. 14.

Gower, J. S. 1957. Mouse News Letter 16:37.

Gower, J. S., and M. B. Cupp. 1958. Mouse News Letter 19:37.

Green, E. L. 1954. The genetics of a new hair deficiency, furless, in the house mouse. J. Hered. 45:115–118.

Green, E. L., and M. C. Green. 1942. The development of three manifestations of the short ear gene in the mouse. J. Morphol. 70:1–19.

Green, E. L., and S. J. Mann. 1961. Opossum, a semi-dominant lethal mutation affecting hair and other characteristics of mice. J. Hered. 52:223–227.

Green, M. C. 1951. Further morphological effects of the short ear gene in the house mouse. J. Morphol. 88:1–22.

Green, M. C. 1955. Luxoid, a new hereditary leg and foot abnormality in the house mouse. J. Hered. 46:91–99.

Green, M. C. 1958. Effects of the short ear gene in the mouse on cartilage formation in healing bone fractures. J. Exp. Zool. 137:75–88.

Green, M. C. 1960. Mouse News Letter 23:34.

Green, M. C. 1961*a*. Himalayan, a new allele of albino in the mouse. J. Hered. 52:73–75.

Green, M. C. 1961*b*. The position of luxoid in linkage group II of the mouse. J. Hered. 52:297–300.

Green, M. C. 1962. Mouse News Letter 26:34.

Green, M. C. 1963. Methods for testing linkage, p. 56–82. *In* W. J. Burdette [ed.] Methodology in mammalian genetics. Holden-Day, San Francisco.

Green, M. C. 1964*a*. Mouse News Letter 30:32.

Green, M. C. 1964*b*. Mouse News Letter 31:27.

Green, M. C. 1965. Mouse News Letter 32:46.

Green, M. C., and S. C. Fox. 1965. Mouse News Letter 32:46.

Green, M. C., and R. L. Sidman. 1962. Tottering, a neuromuscular mutation in the mouse and its linkage with oligosyndactylism. J. Hered. 53:233–237.

Green, M. C., G. D. Snell, and P. W. Lane. 1963. Linkage group XVIII of the mouse. J. Hered. 54:245–247.

Grewal, M. S. 1962*a*. The development of an inherited tooth defect in the mouse. J. Embryol. Exp. Morphol. 10:202–211.

Grewal, M. S. 1962*b*. A sex-linked anaemia in the mouse. Genet. Res. 3:238–247.

Grobman, A. B., and D. R. Charles. 1947. Mutant white mice. A new dominant autosomal mutant affecting coat color in *Mus musculus*. J. Hered. 38:381–384.

Grüneberg, H. 1935*a*. A three-factor linkage experiment in the mouse. J. Genet. 31:157–162.

Grüneberg, H. 1935*b*. A new sub-lethal colour mutation in the house mouse. Proc. Roy. Soc. B 118:321–342.

Grüneberg, H. 1936. Further linkage data on the albino chromosome of the house mouse. J. Genet. 33:255–265.

Grüneberg, H. 1942*a*. The anaemia of flexed-tail mice. II. Siderocytes. J. Genet. 44:246–271.

Grüneberg, H. 1942*b*. Inherited macrocytic anaemias in the house mouse. II. Dominance relationships. J. Genet. 43:285–293.

Grüneberg, H. 1943*a*. Congenital hydrocephalus in the mouse, a case of spurious pleiotropism. J. Genet. 45:1–21.

Grüneberg, H. 1943*b*. Two new mutant genes in the house mouse. J. Genet. 45:22–28.

Grüneberg, H. 1948. Some observations on the microphthalmia gene in the mouse. J. Genet. 49:1–13.

Grüneberg, H. 1950. Genetical studies on the skeleton of the mouse. II. Undulated and its "modifiers." J. Genet. 50:142–173.

Grüneberg, H. 1952. The genetics of the mouse, 2nd ed. Nijhoff, The Hague. 650 p.

Grüneberg, H. 1953*a*. Genetical studies on the skeleton of the mouse. VI. Danforth's short-tail. J. Genet. 51:317–326.

Grüneberg, H. 1953*b*. Genetical studies on the skeleton of the mouse. VII. Congenital hydrocephalus. J. Genet. 51:327–358.

Grüneberg, H. 1953*c*. The relations of microphthalmia and white in the mouse. J. Genet. 51:359–362.

Grüneberg, H. 1954*a*. Genetical studies on the skeleton of the mouse. VIII. Curly-tail. J. Genet. 52:52–67.

Grüneberg, H. 1954*b*. Genetical studies on the skeleton of the mouse. XII. The development of undulated. J. Genet. 52:441–455.

Grüneberg, H. 1955*a*. Genetical studies on the skeleton of the mouse. XVI. Tail-kinks. J. Genet. 53:536–550.

Grüneberg, H. 1955*b*. Genetical studies on the skeleton of the mouse. XVII. Bent-tail. J. Genet. 53:551–562.

Grüneberg, H. 1956. Genetical studies on the skeleton of the mouse. XVIII. Three genes for syndactylism. J. Genet. 54:113–145.

Grüneberg, H. 1957. Genetical studies on the skeleton of the mouse. XIX. Vestigial tail. J. Genet. 55:181–194.

Grüneberg, H. 1958*a*. Genetical studies on the skeleton of the mouse. XXII. The development of Danforth's short-tail. J. Embryol. Exp. Morphol. 6:124–148.

Grüneberg, H. 1958*b*. Genetical studies on the skeleton of the mouse. XXIII. The development of brachyury and anury. J. Embryol. Exp. Morphol. 6:424–443.

Grüneberg, H. 1960. Genetical studies on the skeleton of the mouse. XXV. The development of syndactylism. Genet. Res. 1:196–213.

Grüneberg, H. 1961*a*. Genetical studies on the skeleton of the mouse. XXVII. The development of oligosyndactylism. Genet. Res. 2:33–42.

Grüneberg, H. 1961*b*. Genetical studies on the skeleton of the mouse. XXIX. Pudgy. Genet. Res. 2:384–393.

Grüneberg, H. 1962. Genetical studies on the skeleton of the mouse. XXXII. The development of shaker with syndactylism. Genet. Res. 3:157–166.

Grüneberg, H. 1963. The pathology of development. Wiley, New York. 309 p.

Grüneberg, H., J. B. Burnett, and G. D. Snell. 1941. The origin of jerker, a new gene mutation of the house mouse, and linkage studies made with it. Proc. Nat. Acad. Sci. 27:562–565.

Grüneberg, H., and G. M. Truslove. 1960. Two closely linked genes in the mouse. Genet. Res. 1:69–90.

Hamburgh, M. 1963. Analysis of the postnatal development of "reeler," a neurological mutation in mice. A study in developmental genetics. Develop. Biol. 8:165–185.

Harman, P. J. 1950. Polycystic alterations in the white matter of the brain of the "jittery" mouse. Anat. Rec. 106:304. (Abstr.)

Harman, P. J. 1954. Genetically controlled demyelination in the mammalian central nervous system. Ann. N.Y. Acad. Sci. 58:546–550.

Henderson, N. S. 1965. Isozymes of isocitrate dehydrogenase: subunit structure and intracellular location. J. Exp. Zool. 158:263–273.

Herbertson, B. M., and M. E. Wallace. 1964. Chylous ascites in newborn mice. J. Med. Genet. 1:10–23.

Hertwig, P. 1939. Zwei subletale recessive Mutationen in der Nachkommenschaft von röntgenbestrahlten Mäusen. Erbarzt 6:41–43.

Hertwig, P. 1942. Neue Mutationen und Koppelungsgruppen bei der Hausmaus. Z. Indukt. Abstamm. Vererb. 80:220–246.

Hertwig, P. 1944. Die Genese der Hirn- und Gehöroganmissbildungen bei röntgenmutierten Kreisler-Mäusen. Z. Mensch. Vererb. Konst. 28:327–354.

Herzenberg, L. A. 1964. A chromosome region for gamma$_{2a}$ and beta$_{2a}$ globulin H chain isoantigens in the mouse. Cold Spring Harbor Symp. Quant. Biol. 29:455–462.

Herzenberg, L. A., R. I. Mishell, and L. A. Herzenberg. 1963*a*. Gamma-globulin isoantigens (allotypes) in the house mouse, p. 196. *In* S. J. Geerts [ed.] Proc. XI Int. Congr. Genet. Vol. 1. Pergamon Press, New York. (Abstr.)

Herzenberg, L. A., D. K. Tachibana, L. A. Herzenberg, and L. T. Rosenberg. 1963*b*. A gene locus concerned with hemolytic complement in *Mus musculus*. Genetics 48:711–714.

Herzenberg, L. A., N. L. Warner, and L. A. Herzenberg. 1965. Immunoglobulin isoantigens (allotypes) in the mouse. I. Genetics and cross-reactions of the 7S γ_{2a}-isoantigens controlled by alleles at the Ig-1 locus. J. Exp. Med. 121:415–438.

Heston, W. E. 1941. Relationship between susceptibility to induced pulmonary tumors and certain known genes in mice. J. Nat. Cancer Inst. 2:127–132.

Heston, W. E. 1951. The "vestigial tail" mouse. J. Hered. 42:71–74.

Heston, W. E., H. A. Hoffman, and M. Rechcigl. 1965. Genetic analysis of liver catalase activity in two substrains of C57BL mice. Genet. Res. 6:387–397.

Heston, W. E., and M. Rechcigl. 1964. Genetic analysis of liver catalase activity in C57BL mice. Proc. Amer. Ass. Cancer Res. 5:26 (Abstr.)

Hirsch, M. S. 1962. Studies on the response of osteopetrotic bone explants to parathyroid explants in vitro. Bull. Johns Hopkins Hosp. 110: 257–264.

Hoecker, G., A. Martinez, S. Markovic, and O. Pizzaro. 1954. Agitans, a new mutation in the house mouse with neurological effects. J. Hered. 45:10–14.

Hollander, W. F. 1959. Mouse News Letter 20: 34–35.

Hollander, W. F. 1960. Genetics in relation to reproductive physiology in mammals. J. Cell. Comp. Physiol. 56:61–72.

Hollander, W. F. 1964. Mouse News Letter 30:29.

Hollander, W. F., J. H. D. Bryan, and J. W. Gowen. 1960. Pleiotropic effects of a mutant at the p locus from X-irradiated mice. Genetics 45:413–418.

Hollander, W. F., and J. W. Gowen. 1956. An extreme non-agouti mutant in the mouse. J. Hered. 47:221–224.

Hollander, W. F., and J. W. Gowen. 1959. A single-gene antagonism between mother and fetus in the mouse. Proc. Soc. Exp. Biol. Med. 101:425–428.

Hollander, W. F., and L. C. Strong. 1951. Pintail, a dominant mutation linked with brown in the house mouse. J. Hered. 42:179–182.

Holman, S. P. 1951. The hook-tailed mouse. J. Hered. 42:305–306.

Holt, S. B. 1945. A polydactyl gene in mice capable of nearly regular manifestation. Ann. Eugen. 12:220–249.

Howard, A. 1940. "Rhino," an allele of hairless in the house mouse. J. Hered. 31:467–470.

Hulse, E. V., M. F. Lyon, and R. Meredith. 1965. Mouse News Letter 32:38.

Hummel, K. P. 1958. The inheritance and expression of disorganization, an unusual mutation in the mouse. J. Exp. Zool. 137:389–423.

Hummel, K. P. 1959. Developmental anomalies in mice resulting from action of the gene, disorganization, a semi-dominant lethal. Pediatrics 23:212–221.

Hummel, K. P., and D. B. Chapman. 1959. Visceral inversion and associated anomalies in the mouse. J. Hered. 50:9–13.

Hummel, K. P., and D. B. Chapman. 1963. Mouse News Letter 28:32.

Hunsicker, P. R. 1960. Mouse News Letter 23:58.

Hunt, H. R., R. Mixter, and D. Permar. 1933. Flexed tail in the mouse, *Mus musculus.* Genetics 18:335–366.

Hutton, J. J., J. Bishop, R. Schweet, and E. S. Russell. 1962*a*. Hemoglobin inheritance in inbred mouse strains. I. Structural differences. Proc. Nat. Acad. Sci. 48:1505–1513.

Hutton, J. J., J. Bishop, R. Schweet, and E. S. Russell. 1962*b*. Hemoglobin inheritance in inbred mouse strains. II. Genetic studies. Proc. Nat. Acad. Sci. 48:1718–1724.

Hutton, J. J., R. S. Schweet, H. G. Wolfe, and E. S. Russell. 1964. Hemoglobin solubility and α-chain structure in crosses between two inbred mouse strains. Science 143:252–253.

Ingalls, A. M., M. M. Dickie, and G. D. Snell. 1950. Obese, a new mutation in the house mouse. J. Hered. 41:317–318.

Isaacson, J. H., and B. M. Cattanach. 1962. Mouse News Letter 27:31.

Jarrett, A., and R. I. Spearman. 1957. The keratin defect and hair-cycle of a new mutant (matted) in the house-mouse. J. Embryol. Exp. Morphol. 5:103–110.

Joe, M., J. M. Teasdale, and J. R. Miller. 1962. A new mutation (*sph*) causing neonatal jaundice in the house mouse. Can. J. Genet. Cytol. 4: 219–225.

Johnson, P. L. 1926. An anatomical study of abnormal jaws in the progeny of X-rayed mice. Amer. J. Anat. 38:281–317.

Kadam, K. M. 1962. Genetical studies on the skeleton of the mouse. XXXI. The muscular anatomy of syndactylism and oligosyndactylism. Genet. Res. 3:139–156.

Kamenoff, R. J. 1935. Effects of the flexed-tailed gene on the development of the house mouse. J. Morphol. 58:117–155.

Kantoch, M., A. Warwick, and F. B. Bang. 1963. The cellular nature of genetic susceptibility to a virus. J. Exp. Med. 117:781–798.

Keeler, C. E. 1924. The inheritance of a retinal abnormality in white mice. Proc. Nat. Acad. Sci. 10:329–333.

Keeler, C. E. 1927*a*. Sur l'origine de caractère "sans bâtonnets" chez la souris domestique. Bull. Soc. Zool. France 52:520–521.

Keeler, C. E. 1927*b*. Rodless retina, an ophthalmic mutation in the house mouse, *Mus musculus.* J. Exp. Zool. 46:355–407.

Keeler, C. E. 1928. Blind mice. J. Exp. Zool. 51: 495–508.

Keeler, C. E. 1930. Hereditary blindness in the house mouse with special reference to its linkage relationships. Howe Lab. Ophthalmol. Bull. No. 3. 11 p.

Keeler, C. E. 1931. The laboratory mouse. Harvard Univ. Press, Cambridge, Mass. 81 p.

Keeler, C. E. 1935. A second rexoid coat character in the house mouse. J. Hered. 26:189–191.

Kelly, E. M. 1953. Mouse News Letter 8 (Suppl.): 15.

Kelly, E. M. 1955. Mouse News Letter 12:48.

Kelly, E. M. 1957. Mouse News Letter 16:36.

Kelly, E. M. 1958. Mouse News Letter 19:37.

Kelly, E. M., and J. W. Bangham. 1955. Mouse News Letter 12:47.

Kelton, D. E. 1965. Mouse News Letter 32:60.

Kelton, D. E., and H. Rauch. 1962. Myelination and myelin degeneration in the central nervous system of dilute-lethal mice. Exp. Neurol. 6: 252–262.

Kelton, D. E., and V. Smith. 1964. Gaping, a new open-eyelid mutation in the house mouse. Genetics 50:261–262. (Abstr.)

Kelus, A., and J. K. Moor-Jankowski. 1961. An iso-antigen (γB^A) of mouse γ-globulin present in inbred strains. Nature 191:1405–1406.

Kidwell, J. F., J. W. Gowen, and J. Stadler. 1961. Pugnose—a recessive mutation in linkage group 3 of mice. J. Hered. 52:145–148.

King, J. W. B. 1950. Pygmy, a dwarfing gene in the house mouse. J. Hered. 41:249–252.

King, J. W. B. 1955. Observations on the mutant "*pygmy*" in the house mouse. J. Genet. 53:487–497.

King, J. W. B. 1956. Linkage group XIV of the house mouse. Nature 178:1126–1127.

King, L. S. 1936. Hereditary defects of the corpus callosum in the mouse, *Mus musculus*. J. Comp. Neurol. 64:337–363.

King, L. S., and C. E. Keeler. 1932. Absence of corpus callosum, a hereditary brain anomaly of the house mouse. Preliminary report. Proc. Nat. Acad. Sci. 18:525–528.

Kobozieff, N., and N. A. Pomriaskinsky-Kobozieff. 1962. Hémimélie chez la souris. Rec. Méd. Vét. 138:671–686.

Kocher, W. 1960*a*. Untersuchungen zur Genetik und Pathologie der Entwicklung von 8 Labyrinthmutanten (*deaf-waltzer-shaker*-Mutanten) der Maus (*Mus musculus*). Z. Vererb. 91:114–140.

Kocher, W. 1960*b*. Untersuchungen zur Genetik und Pathologie der Entwicklung späteinsetzender hereditärer Taubheit bei der Maus (*Mus musculus*). Arch. Ohr. Nas. Kehlkopfheilk. 177: 108–145.

Konyukhov, B. V., and D. M. Glukharev. 1962. The genetic and morphological characteristics of microphthalmic mice (blind mutants) [English transl.]. Bull. Exp. Biol. Med. 52:1437–1440.

Konyukhov, B. V., O. G. Stroeva, M. V. Sazhina, and T. A. Lipgart. 1963. Retinal injury as a cause of microphthalmia in mice of the ocular retardation mutant line [in Russian, English summary]. Arkh. Anat. Gistol. Embriol. 44: 36–43.

Kreitner, P. C. 1957. Linkage studies in a new black-eyed white mutation in the house mouse (not *W*). J. Hered. 48:300–304.

Kuharcik, A. M., and P. F. Forsthoefel. 1963. A study of the anemia in Strong's luxoid mutant. J. Morphol. 112:13–21.

Kunze, H. G. 1954. Die Erythropoese bei einer erblichen Anämie röntgenmutierter Mäuse. Folia Haematol. 72:391–436.

Landauer, W. 1952. Brachypodism, a recessive mutation of house-mice. J. Hered. 43:293–298.

Lane, P. W. 1958. Mouse News Letter 19:25.

Lane, P. W. 1960*a*. Mouse News Letter 22:35.

Lane, P. W. 1960*b*. Mouse News Letter 23:36.

Lane, P. W. 1961. Mouse News Letter 25:38.

Lane, P. W. 1962*a*. Mouse News Letter 26:35.

Lane, P. W. 1962*b*. Mouse News Letter 27:38.

Lane, P. W. 1963. Whirler mice, a recessive behavior mutation in linkage group VIII. J. Hered. 54:263–266.

Lane, P. W. 1964. Mouse News Letter 30:32.

Lane, P. W. 1965. Mouse News Letter 32:47.

Lane, P. W. 1966. Association of megacolon with two recessive spiting genes in the mouse. J. Hered. 57:29–31.

Lane, P. W., and M. M. Dickie. 1961. Linkage of wabbler-lethal and hairless in the mouse. J. Hered. 52:159–160.

Lane, P. W., and M. C. Green. 1960. Mahogany, a recessive color mutation in linkage group V of the mouse. J. Hered. 51:228–230.

Lane, P. W., and M. C. Green. 1962. Mouse News Letter 27:38.

Larsen, M. M. 1961. Mouse News Letter 24:60.

Larsen, M. M. 1963. Mouse News Letter 29:73.

Larsen, M. M. 1964. Mouse News Letter 30:47.

Law, L. W., A. G. Morrow, and E. M. Greenspan. 1952. Inheritance of low liver glucuronidase activity in the mouse. J. Nat. Cancer Inst. 12: 909–916.

Lebedinsky, N. G., and A. Dauwart. 1927. Atrichosis und ihre Vererbung bei der albinotischen Hausmaus. Biol. Zentralbl. 47:748–752.

Levan, A., T. C. Hsu, and H. F. Stich. 1962. The idiogram of the mouse. Hereditas 48:677–687.

Lieberman, R., and S. Dray. 1964. Five allelic genes at the Asa locus which control γ-globulin allotypic specificities in mice. J. Immunol. 93: 584–594.

Lindenmann, J. 1964. Inheritance of resistance to

influenza virus in mice. Proc. Soc. Exp. Biol. Med. 116:506–509.

Little, C. C., and H. J. Bagg. 1923. The occurrence of two heritable types of abnormality among the descendants of X-rayed mice. Amer. J. Roentgenol. 10:975–989.

Little, C. C., and H. J. Bagg. 1924. The occurrence of four inheritable morphological variations in mice and their possible relation to treatment with X-rays. J. Exp. Zool. 41:45–91.

Little, C. C., and A. M. Cloudman. 1937. The occurrence of a dominant spotting mutation in the house mouse. Proc. Nat. Acad. Sci. 23: 535–537.

Loosli, R. 1963. Tanoid—a new agouti mutant in the mouse. J. Hered. 54:26–29.

Lord, E. M., and W. H. Gates. 1929. Shaker, a new mutation of the house mouse. Amer. Natur. 63:435–442.

Lynch, C. J. 1921. Short ears, an autosomal mutation in the house mouse. Amer. Natur. 55: 421–426.

Lyon, M. F. 1953. Absence of otoliths in the mouse: an effect of the pallid mutant. J. Genet. 51:638–650.

Lyon, M. F. 1955*a*. The developmental origin of hereditary absence of otoliths in mice. J. Embryol. Exp. Morphol. 3:230–241.

Lyon, M. F. 1955*b*. Ataxia—a new recessive mutant of the house mouse. J. Hered. 46:77–80.

Lyon, M. F. 1956. Hereditary hair loss in the tufted mutant of the house mouse. J. Hered. 47:101–103.

Lyon, M. F. 1958. Twirler: a mutant affecting the inner ear of the house mouse. J. Embryol. Exp. Morphol. 6:105–116.

Lyon, M. F. 1959. A new dominant *T*-allele in the house mouse. J. Hered. 50:140–142.

Lyon, M. F. 1960. A further mutation of the mottled type in the house mouse. J. Hered. 51: 116–121.

Lyon, M. F. 1961*a*. Linkage relations and some pleiotropic effects of the dreher mutant of the house mouse. Genet. Res. 2:92–95.

Lyon, M. F. 1961*b*. Mouse News Letter 24:34.

Lyon, M. F. 1963*a*. Attempts to test the inactive-X theory of dosage compensation in mammals. Genet. Res. 4:93–103.

Lyon, M. F. 1963*b*. Genetics of the mouse, p. 199–234. *In* W. Lane-Petter [ed.] Animals for research. Academic Press, London and New York.

Lyon, M. F. 1965. Mouse News Letter 32:39.

Lyon, M. F., E. V. Hulse, and C. E. Rowe. 1965. Foam-cell reticulosis of mice: an inherited condition resembling Gaucher's and Niemann-Pick diseases. J. Med. Genet. 2:99–106.

Lyon, M. F., and R. Meredith. 1963. Mouse News Letter 28:29.

Lyon, M. F., and R. Meredith. 1964. The nature of *t*-alleles in the mouse. Heredity 19:301–330.

Lyon, M. F., and R. Meredith. 1965. Mouse News letter 32:38.

Lyon, M. F., and R. J. S. Phillips. 1959. Crossing-over in mice heterozygous for *t*-alleles. Heredity 13:23–32.

Lyon, M. F., R. J. S. Phillips, and A. G. Searle. 1964*a*. The overall rates of dominant and recessive lethal and visible mutation induced by spermatogonial X-irradiation of mice. Genet. Res. 5:448–467.

Lyon, M. F., A. G. Searle, C. E. Ford, and S. Ohno. 1964*b*. A mouse translocation suppressing sex-linked variegation. Cytogenetics 3:306–323.

MacArthur, J. W. 1944. Genetics of body size and related characters. Amer. Natur. 78:142–157, 224–237.

MacDowell, E. C. 1950. "Light"—a new mouse color. J. Hered. 41:35–36.

MacDowell, E. C., J. S. Potter, T. Laanes, and E. N. Ward. 1942. The manifold effects of the screw-tail mouse mutation. J. Hered. 33:439–449.

Mackensen, J. A. 1960. "Open eyelids" in newborn mice. J. Hered. 51:188–190.

Mackensen, J. A. 1962. Mouse News Letter 27:38.

Mackensen, J. A., and L. C. Stevens. 1960. Rib fusions, a new mutation in the mouse. J. Hered. 51:264–268.

Maddux, S. C. 1964. Mouse News Letter 31:41.

Major, M. H. 1955. Mouse News Letter 12:47.

Major, M. H., and M. S. Hawkins. 1958. Mouse News Letter 19:37.

Mallyon, S. A. 1951. A pronounced sex difference in recombination values in the sixth chromosome of the house mouse. Nature 168:118–119.

Mann, S. J. 1963. The phenogenetics of hair mutants in the house mouse: opossum and ragged. Genet. Res. 4:1–11.

Mann, S. J. 1964. The hair of the fuzzy mouse. J. Hered. 55:121–123.

Mann, S. J., and W. E. Straile. 1961. New observations on hair loss in the hairless mouse. Anat. Rec. 140:97–102.

Margolis, F. L. 1965. Reduced δ-aminolevulinate dehydratase (ALD) activity in a mutant mouse anemia. Fed. Proc. 24:469. (Abstr.)

Markert, C. L. 1960. Biochemical embryology and genetics. *In* N. Kaliss [ed.] Symposium on normal and abnormal differentiation and development. Nat. Cancer Inst. Monogr. 2:3–17.

Markert, C. L., and W. K. Silvers. 1956. The effect of genotype and cell environment on melanoblast differentiation in the house mouse. Genetics 41:429–450.

Markert, C. L., and W. K. Silvers. 1959. Effects of genotype and cellular environment on melanocyte morphology, p. 241–248. *In* M. Gordon [ed.] Pigment cell biology. Academic Press, New York.

Martinez, A., and J. L. Sirlin. 1955. Neurohistology of the agitans mouse. J. Comp. Neurol. 103:131–137.

Mather, K. 1951. The measurement of linkage in heredity, 2nd ed. Wiley, New York. 149 p.

Mather, K., and S. B. North. 1940. Umbrous: a case of dominance modification in mice. J. Genet. 40:229–241.

Matter, H. 1957. Die formale Genese einer vererbten Wirbelsäulenmissbildung am Beispiel der Mutante Crooked-tail der Maus. Rev. Suisse Zool. 64:1–38.

Mauer, T. 1961. The effect of vitamin A in hyperkeratotic mouse mutants. J. Exp. Zool. 146: 181–207.

Mayer, J. 1960. Genetic factors in obesity. Bull. N.Y. Acad. Med. 36:323–343.

Mayer, T. C. 1965. The development of piebald spotting in mice. Develop. Biol. 11:319–334.

Mayer, T. C., and E. Maltby. 1964. An experimental investigation of pattern development in lethal spotting and belted mouse embryos. Develop. Biol. 9:269–286.

McNutt, W. 1962. Urinary amino acid excretion in quivering mice (*qvqv*) Anat. Rec. 142:257. (Abstr.)

Meier, H., and W. G. Hoag. 1962. The neuropathology of "reeler," a neuromuscular mutation in mice. J. Neuropathol. Exp. Neurol. 21:649–654.

Menner, K. 1957. Die postnatale Gonadenentwicklung bei Mäusen, die an einer angeborenen Anämie leiden. Wiss. Z. Martin-Luther-Univ. 6:335–344.

Meredith, R. 1964. Mouse News Letter 31:25.

Meredith, R. 1965. Mouse News Letter 32:39.

Michelson, A. M., E. S. Russell, and P. J. Harman. 1955. Dystrophia muscularis: a hereditary primary myopathy in the house mouse. Proc. Nat. Acad. Sci. 41:1079–1084.

Michie, D. 1955*a*. Genetical studies with "vestigial tail" mice. I and II. J. Genet. 53:270–284.

Michie, D. 1955*b*. "Affinity." Proc. Roy. Soc. B 144:241–259.

Michie, D., and M. E. Wallace. 1953. Affinity: a new genetic phenomenon in the house mouse. Nature 171:26.

Mikaelian, D. O., and R. J. Ruben. 1964. Hearing degeneration in shaker-1 mouse. Arch. Otolaryngol. 80:418–430.

Milaire, J. 1962. Détection histochimique de modifications des ébauches dans les membres en formation chez la souris oligosyndactyl. Bull. Acad. Roy. Belgique 1962. p. 505–528.

Miller, D. S. 1963. Coat color and behavior mutations in inbred mice under chronic low-level γ-irradiation. Radiat. Res. 19:184–185.

Miller, D. S., and M. Z. Potas. 1955. Cordovan, a new allele of black and brown color in the mouse. J. Hered. 46:293–296.

Miller, J. R. 1964. Mouse News Letter 30:16.

Mintz, B., and E. S. Russell. 1957. Gene-induced embryological modifications of primordial germ cells in the mouse. J. Exp. Zool. 134:207–237.

Mishell, R. K., and J. L. Fahey. 1964. Molecular and submolecular localization of two isoantigens of mouse immunoglobulins. Science 143: 1440–1442.

Montagna, W., H. B. Chase, and H. P. Melaragno. 1952. The skin of hairless mice. I. The formation of cysts and the distribution of lipids. J. Invest. Dermatol. 19:83–94.

Morgan, W. C. 1950. A new tail-short mutation in the mouse. J. Hered. 41:208–215.

Morgan, W. C. 1954. A new crooked tail mutation involving distinctive pleiotropism. J. Genet. 52:354–373.

Morton, J. R. 1962. Starch-gel electrophoresis of mouse haemoglobins. Nature 194:383–384.

Mufson, M. A., and A. Starr. 1958. Abnormal myelin and leukopenia in the wabbler-lethal mouse. J. Hered. 49:233–237.

Müller, G. 1950. Eine entwicklungsgeschichtliche Untersuchung über das erbliche Kolobom mit Mikrophthalmus bei der Hausmaus. Z. Mikroskop.-Anat. Forsch. 56:520–558.

Müller, G. 1951. Die embryonale Entwicklung eines sich rezessiv vererbenden Merkmals. (Kolobom bei der Hausmaus). Wiss. Z. Martin-Luther-Univ. 1:27–43.

Murphy, E. D., and E. S. Russell. 1963. Ovarian tumorigenesis following genic deletion of germ cells in hybrid mice. Acta Un. Int. Contra Cancrum 19:779–782.

Murray, J. M. 1933. "Leaden," a recent color mutation in the house mouse. Amer. Natur. 67: 278–283.

Murray, J. M., and G. D. Snell. 1945. Belted, a

new sixth chromosome mutation in the mouse. J. Hered. 36:266–268.

Nakamura, A., H. Sakamoto, and K. Moriwaki. 1963. Genetical studies of post-axial polydactyly in the house mouse. Ann. Rep. Nat. Inst. Genet. Jap. (1962) 13:31.

Nash, D. J. 1964. Mouse News Letter 30:53–54.

Nash, D. J., E. Kent, M. M. Dickie, and E. S. Russell. 1964. The inheritance of "mick" a new anemia in the house mouse. Amer. Zool. 4:404–405. (Abstr.)

Nasrat, G. E. 1956. Estimation of the recombination fraction between the two linked genes *Re* and *sh*-2 in the house mouse when the female is the heterozygous parent. Proc. Zool. Soc. (Bengal) 9:85–87.

Nolte, D. J. 1957. Mouse News Letter 17:100.

Old, L. J., E. A. Boyse, and E. Stockert. 1963. Antigenic properties of experimental leukemias. I. Serological studies *in vitro* with spontaneous and radiation-induced leukemias. J. Nat. Cancer Inst. 31:977–995.

Paget, O. E. 1953. *Cataracta hereditaria subcapsularis:* ein neues, dominantes Allel bei der Hausmaus. Z. Indukt. Abstamm. Vererb. 85: 238–244.

Paget, O. E., and M. Baumgartner-Gamauf. 1961. Histologische Untersuchungen an einer dominant erblichen Form einer Cataract bei der Hausmaus. Zool. Anz. 166:55–69.

Pai, A. C. 1965. Developmental genetics of a lethal mutation, muscular dysgenesis (*mdg*) in the mouse. I and II. Develop. Biol. 11:82–109.

Paigen, K. 1961*a*. The genetic control of enzyme activity during differentiation. Proc. Nat. Acad. Sci. 47:1641–1649.

Paigen, K. 1961*b*. The effect of mutation on the intracellular location of beta-glucuronidase. Exp. Cell Res. 25: 286–301.

Paigen, K. 1964. The genetic control of enzyme realization during differentiation, p. 181–190. *In* Second Int. Conf. Congen. Malformations. Int. Med. Congr. Ltd., New York.

Paigen, K., and W. K. Noell. 1961. Two linked genes showing a similar timing of expression in mice. Nature 190:148–150.

Parsons, P. A. 1958*a*. A balanced four-point linkage experiment for linkage group XIII of the house mouse. Heredity 12:77–95.

Parsons, P. A. 1958*b*. Additional three-point data for linkage group V of the mouse. Heredity 12:357–362.

Parsons, P. A. 1959. Possible affinity between linkage groups V and XIII of the house mouse. Genetica 29:304–311.

Pelzer, C. F. 1965. Genetic control of erythrocytic esterase forms in *Mus Musculus*. Genetics 52: 819–828.

Petras, M. L. 1963. Genetic control of a serum esterase component in *Mus musculus*. Proc. Nat. Acad. Sci. 50:112–116.

Phillips, R. J. S. 1954. *Jimpy,* a new totally sex-linked gene in the house mouse. Z. Indukt. Abstamm. Vererb. 86:322–326.

Phillips, R. J. S. 1956. The linkages of congenital hydrocephalus in the house mouse. J. Hered. 47:302–304.

Phillips, R. J. S. 1960*a*. "Lurcher," a new gene in linkage group XI of the house mouse. J. Genet. 57:35–42.

Phillips, R. J. S. 1960*b*. Mouse News Letter 23: 29–30.

Phillips, R. J. S. 1961. "Dappled," a new allele at the mottled locus in the house mouse. Genet. Res. 2:290–295.

Phillips, R. J. S. 1962. Mouse News Letter 27:34.

Phillips, R. J. S. 1963*a*. Mouse News Letter 29: 37–38.

Phillips, R. J. S. 1963*b*. *Striated*, a new sex-linked gene in the house mouse. Genet. Res. 4:151–153.

Pierro, L. J. 1963*a*. Effects of the *Light* mutation of mouse coat color on eye pigmentation. J. Exp. Zool. 153:81–87.

Pierro, L. J. 1963*b*. Pigment granule formation in slate, a coat color mutant in the mouse. Anat. Rec. 146:365–372.

Pierro, L. J., and H. B. Chase. 1963. Slate—a new coat color mutant in the mouse. J. Hered. 54:47–50.

Pierro, L. J., and H. B. Chase. 1965. Temporary hair loss associated with the slate mutation of coat colour in the mouse. Nature 205:579–580.

Pizarro, O. 1957. Mouse News Letter 17:96.

Popp, R. A. 1962*a*. Studies on the mouse hemoglobin loci. II, III, IV. J. Hered. 53:73–80.

Popp, R. A. 1962*a*. Studies on the mouse hemoglobin loci. V, VI, VII. J. Hered. 53:142–151.

Popp, R. A. 1965*a*. Loci linkage of serum esterase patterns and oligosyndactylism. J. Hered. 56:107–108.

Popp, R. A. 1965*b*. Hemoglobin variants in mice. Fed. Proc. 24:1,252–1,257.

Popp, R. A., and D. M. Popp. 1962. Inheritance of serum esterases having different electrophoretic patterns among inbred strains of mice. J. Hered. 53:111–114.

Popp, R. A., and W. St. Amand. 1960. Studies on the mouse hemoglobin locus. I. Identification of hemoglobin types and linkage of hemoglobin with albinism. J. Hered. 51:141–144.

Quevedo, W. C., Jr., and H. B. Chase. 1958. An analysis of the light mutation of coat color in mice. J. Morphol. 102:329–345.

Rabaud, E. 1914. Sur une anomalie héréditarie des membres postérieurs chez la souris. Compt. Rend. Soc. Biol. 77:411–412.

Randelia, H. P., and L. D. Sanghvi. 1961. "Bare," a new hairless mutant in the mouse—genetics and histology. Genet. Res. 2:283–289.

Ranney, H. M., and S. Gluecksohn-Waelsch. 1955. Filter-paper electrophoresis of mouse haemoglobin: preliminary note. Ann. Hum. Genet. 19:269–272.

Rauch, H., and M. T. Yost. 1963. Phenylalanine metabolism in dilute-lethal mice. Genetics 48: 1487–1495.

Rechcigl, M., and W. E. Heston. 1963. Tissue catalase activity in several C57BL substrains and in other strains of inbred mice. J. Nat. Cancer Inst. 30:855–864.

Reed, S. C. 1937. The inheritance and expression of fused, a new mutation in the house mouse. Genetics 22:1–13.

Roberts, E. 1931. A new mutation in the house mouse (*Mus musculus*). Science 74:569.

Robertson, G. G. 1942. An analysis of the development of homozygous yellow mouse embryos. J. Exp. Zool. 89:197–231.

Robins, M. W. 1959. A mutation causing congenital clubfoot in the house mouse. J. Hered. 50: 188–192.

Robinson, R. 1959. Sable and umbrous mice. Genetica 29:319–326.

Ruddle, F. H., and T. H. Roderick. 1965. The genetic control of three kidney esterases in C57BL/6J and RF/J mice. Genetics 51:445–454.

Runner, M. N. 1959. Linkage of brachypodism. A new member of linkage group V of the house mouse. J. Hered. 50:81–84.

Russell, E. S. 1949*a*. A quantitative histological study of the pigment found in the coat color mutants of the house mouse. IV. The nature of the effects of genic substitution in five major allelic series. Genetics 34:146–166.

Russell, E. S. 1949*b*. Analysis of pleiotropism at the *W*-locus in the mouse: Relationship between the effects of *W* and W^v substitution on hair pigmentation and on erythrocytes. Genetics 34:708–723.

Russell, E. S., and E. Fekete. 1958. Analysis of *W*-series pleiotropism in the mouse: Effect of W^vW^v substitution on definitive germ cells and on ovarian tumorigenesis. J. Nat. Cancer Inst. 21:365–381.

Russell, E. S., and P. S. Gerald. 1958. Inherited electrophoretic hemoglobin patterns among 20 inbred strains of mice. Science 128:1569–1570.

Russell, E. S., F. Lawson, and G. Schabtach. 1957. Evidence for a new allele at the *W*-locus of the mouse. J. Hered. 48:119–123.

Russell, E. S., L. M. Murray, E. M. Small, and W. K. Silvers. 1956. Development of embryonic mouse gonads transferred to the spleen: effects of transplantation combined with genotypic autonomy. J. Embryol. Exp. Morphol. 4:347–357.

Russell, L. B. 1951. Mouse News Letter 5:50.

Russell, L. B. 1960*a*. Mouse News Letter 22:50.

Russell, L. B. 1960*b*. Mouse News Letter 23:58.

Russell, L. B. 1961. Mouse News Letter 25:64.

Russell, L. B. 1963. Mouse News Letter 29:73.

Russell, L. B., and M. H. Major. 1956. A high rate of somatic reversion in the mouse. Genetics 41:658. (Abstr.)

Russell, L. B., M. N. C. McDaniel, and F. N. Woodiel. 1963. Crossing-over within the A "locus" of the mouse. Genetics 48:907. (Abstr.)

Russell, R. L., and D. L. Coleman. 1963. Genetic control of hepatic δ-aminolevulinate dehydratase in mice. Genetics 48:1033–1039.

Russell, W. L. 1947. Splotch, a new mutation in the house mouse, *Mus musculus*. Genetics 32: 102. (Abstr.)

Russell, W. L. 1965. Evidence from mice concerning the nature of the mutation process. p. 257–264. *In* S. J. Geerts [ed.] Proc. XI Int. Congr. Genet. (The Hague, 1963) Vol. 2. Pergamon Press, New York.

Russell, W. L., L. B. Russell, and J. S. Gower. 1959. Exceptional inheritance of a sex-linked gene in the mouse explained on the basis that the X/O sex-chromosome constitution is female. Proc. Nat. Acad. Sci. 45:554–560.

Sabin, A. B. 1952. Nature of inherited resistance to viruses affecting the nervous system. Proc. Nat. Acad. Sci. 38:540–546.

Sackler, A. M., A. S. Weltman, P. Steinglass, and S. D. Kraus. 1964. Endocrine differences between whirler and normal strains of female mice. Fed. Proc. 23:252. (Abstr.)

Sarvella, P. A. 1954. Pearl, a new spontaneous coat and eye color mutation in the house mouse. J. Hered. 45:19–20.

Sarvella, P. A., and L. B. Russell. 1956. Steel, a

new dominant gene in the house mouse. J. Hered. 47:123–128.

Saylors, C. L. 1961. Mouse News Letter 25:64.

Schaible, R. H. 1956. Mouse News Letter 15:29.

Schaible, R. H. 1961. Mouse News Letter 24:38.

Schaible, R. H. 1963*a*. Mouse News Letter 28:39.

Schaible, R. H. 1963*b*. Mouse News Letter 29: 48–49.

Schaible, R. H. 1963*c*. Developmental genetics of spotting patterns in the mouse. Ph.D. Dissertation. Iowa State Univ., Ames. 225 p.

Schaible, R. H., and J. W. Gowen. 1961. A new dwarf mouse. Genetics 46:896. (Abstr.)

Scheufler, H. 1963. Erbliche Neugeborenengelbsucht—eine neue Mutante der Hausmaus. Z. Versuchstierk. 3:27–29.

Schlesinger, M., E. A. Boyse, and L. J. Old. 1965. Thymus cells of radiation-chimeras: *TL* phenotype, sensitivity to guinea-pig serum, and origin from donor cells. Nature 206:1119–1121.

Searle, A. G. 1952. A lethal allele of dilute in the house mouse. Heredity 6:395–401.

Searle, A. G. 1961. Tipsy, a new mutant in linkage group VII of the mouse. Genet. Res. 2:122–126.

Searle, A. G. 1964. The genetics and morphology of two "luxoid" mutants in the house mouse. Genet. Res. 5:171–197.

Searle, A. G. 1965. Mouse News Letter 32:39.

Searle, A. G., and R. I. Spearman. 1957. "Matted," a new hair-mutant in the house mouse: genetics and morphology. J. Embryol. Exp. Morphol. 5:93–102.

Shreffler, D. C. 1960. Genetic control of serum transferrin type in mice. Proc. Nat. Acad. Sci. 46:1378–1384.

Shreffler, D. C. 1963. Linkage of the mouse transferrin locus. J. Hered. 54:127–129.

Shreffler, D. C. 1964*a*. Inheritance of a serum pre-albumen variant in the mouse. Genetics 49: 629–634.

Shreffler, D. C. 1964*b*. A serologically detected variant in mouse serum: further evidence for genetic control by the histocompatibility-2 locus. Genetics 49:973–978.

Shreffler, D. C., and R. D. Owen. 1963. A serologically detected variant in mouse serum: inheritance and association with the histocompatibility-2 locus. Genetics 48:9–25.

Sick, K., and J. T. Nielsen. 1964. Genetics of amylase isozymes in the mouse. Hereditas 51: 291–296.

Sidman, R. L. 1961. Tissue culture studies of inherited retinal dystrophy. Dis. Nerv. Syst. 22: 1–7.

Sidman, R. L. 1965. Mouse News Letter 32:37.

Sidman, R. L., M. M. Dickie, and S. H. Appel. 1964. Mutant mice (*quaking* and *jimpy*) with deficient myelination in the central nervous system. Science 144:309–311.

Sidman, R. L., and M. C. Green. 1965. Retinal degeneration in the mouse. Location of the *rd* locus in linkage group XVII. J. Hered. 56: 23–29.

Sidman, R. L., M. C. Green, and S. H. Appel. 1965. Catalog of the neurological mutants of the mouse. Harvard Univ. Press, Cambridge. 82 p.

Sidman, R. L., P. W. Lane, and M. M. Dickie. 1962. Staggerer, a new mutation in the mouse affecting the cerebellum. Science 137:610–612.

Silvers, W. K. 1958. Origin and identity of clear cells found in hair bulbs of albino mice. Anat. Rec. 130:135–144.

Silvers, W. K., and E. S. Russell. 1955. An experimental approach to action of genes at the agouti locus in the mouse. J. Exp. Zool. 130: 199–220.

Singer, M. F., M. Foster, M. L. Petras, P. Tomlin, and R. W. Sloane. 1964. A new case of blood group inheritance in the house mouse, *Mus musculus*. Genetics 50:285–286. (Abstr.)

Sirlin, J. L. 1956. Vacillans, a neurological mutant in the house mouse linked with brown. J. Genet. 54:42–48.

Sirlin, J. L. 1957. Location of vacillans in linkage group VIII of the house mouse. Heredity 11: 259–260.

Sisken, B. F., and S. Gluecksohn-Waelsch. 1959. A developmental study of the mutation "phocomelia" in the mouse. J. Exp. Zool. 142:623–642.

Slee, J. 1962. Developmental morphology of the skin and hair follicles in normal and in ragged mice. J. Embryol. Exp. Morphol. 10:507–529.

Slizynski, B. M. 1954. Partial sex linkage in the mouse. Nature 174:309–310.

Slizynski, B. M. 1957. Cytological analysis of translocations in the mouse. J. Genet. 55:122–130.

Smith, L. J., and K. F. Stein. 1962. Axial elongation in the mouse and its retardation in homozygous looptail mice. J. Embryol. Exp. Morphol. 10:73–87.

Smith, P. E., and E. C. MacDowell. 1930. An hereditary anterior-pituitary deficiency in the mouse. Anat. Rec. 46:249–257.

Snell, G. D. 1928. A cross-over between the genes for short-ear and density in the house mouse. Proc. Nat. Acad. Sci. 14:926–928.

Snell, G. D. 1929. "Dwarf," a new mendelian

recessive character of the house mouse. Proc. Nat. Acad. Sci. 15:733–734.

Snell, G. D. 1931. Inheritance in the house mouse, the linkage relations of short ear, hairless, and naked. Genetics 16:42–74.

Snell, G. D. 1945. Linkage of jittery and waltzing in the mouse. J. Hered. 36:279–280.

Snell, G. D. 1955. Ducky, a new second chromosome mutation in the mouse. J. Hered. 46:27–29.

Snell, G. D. 1958. Histocompatibility genes of the mouse. II. Production and analysis of isogenic resistant lines. J. Nat. Cancer Inst. 21:843–877.

Snell, G. D., and H. P. Bunker. 1964. Histocompatibility genes of mice. IV. The position of H-3 in the fifth linkage group. Transplantation 2:743–751.

Snell, G. D., M. M. Dickie, P. Smith, and D. E. Kelton. 1954. Linkage of loop-tail, leaden, splotch and fuzzy in the mouse. Heredity 8: 271–273.

Snell, G. D., and L. W. Law. 1939. A linkage between shaker-2 and wavy-2 in the house mouse. J. Hered. 30:447.

Snell, G. D., and L. C. Stevens. 1961. Histocompatibility genes of mice. III. H-1 and H-4, two histocompatibility loci in the first linkage group. Immunology 4:366–379.

Sô, M., and Y. Imai. 1926. On the inheritance of ruby eye in mice. Jap. J. Genet. 4:1–9.

Spearman, R. I. 1960. The skin abnormality of "ichthyosis," a mutant of the house mouse. J. Embryol. Exp. Morphol. 8:387–395.

St. Amand, W., and M. B. Cupp. 1957*a*. Mouse News Letter 16:37.

St. Amand, W., and M. B. Cupp. 1957*b*. Mouse News Letter 17:88.

St. Amand, W., and M. B. Cupp. 1958. Mouse News Letter 19:38.

Stein, K. F., and S. A. Huber. 1960. Morphology and behavior of waltzer-type mice. J. Morphol. 106:197–203.

Stein, K. F., and J. A. Mackensen. 1957. Abnormal development of the thoracic skeleton in mice homozygous for the gene for looped-tail. Amer. J. Anat. 100:205–223.

Stevens, L. C. 1955. Mouse News Letter 13:41.

Stevens, L. C. 1963. Mouse News Letter 29:40.

Stevens, L. C., and J. A. Mackensen. 1958. The inheritance and expression of a mutation in the mouse affecting blood formation, the axial skeleton, and body size. J. Hered. 49:153–160.

Stevens, L. C., J. A. Mackensen, and S. E. Bernstein. 1959. A mutation causing neonatal jaundice in the house mouse. J. Hered. 50:35–39.

Stiffel, C., G. Biozzi, D. Mouton, Y. Bouthillier, and C. Decreusefond. 1964. Studies on phagocytosis of bacteria by the reticulo-endothelial system in a strain of mice lacking hemolytic complement. J. Immunol. 93:246–249.

Strong, L. C., and L. B. Hardy. 1956. A new "luxoid" mutant in mice. J. Hered. 47:277–284.

Strong, L. C., and W. F. Hollander. 1949. Hereditary loop-tail in the house mouse, accompanied by imperforate vagina and with lethal craniorachischisis when homozygous. J. Hered. 40: 329–334.

Tansley, K. 1954. An inherited retinal degeneration in the mouse. J. Hered. 45:123–127.

Theiler, K. 1957. Boneless tail, ein recessive autosomales Gen der Hausmaus. Arch. Julius Klaus-Stift. Vererb. 32:474–481.

Theiler, K. 1959. Anatomy and development of the "truncate" (boneless) mutation in the mouse. Amer. J. Anat. 104:319–343.

Theiler, K., and S. Gluecksohn-Waelsch. 1956. The morphological effects and the development of the fused mutation in the mouse. Anat. Rec. 125:83–104.

Theiler, K., and L. C. Stevens. 1960. The development of rib fusions, a mutation in the house mouse. Amer. J. Anat. 106:171–183.

Thompson, S., J. F. Foster, J. W. Gowen, and O. E. Tauber. 1954. Hereditary differences in serum proteins of normal mice. Proc. Soc. Exp. Biol. Med. 87:315–317.

Thoms, G. 1952. Das histologische Verhalten von Milz, Leber, und Knockenmark bei der erblichen Anaemie röntgenmutierter Mäuse. Wiss. Z. Martin-Luther-Univ. 1:13–26.

Tihen, J. A., D. R. Charles, and T. O. Sippel. 1948. Inherited visceral inversion in mice. J. Hered. 39:29–31.

Tost, M. 1958. Cataracta hereditaria mit Mikrophthalmus bei der Hausmaus. Z. Mensch. Vererb. Konst. 34:593–600.

Truslove, G. M. 1956. The anatomy and development of the fidget mouse. J. Genet. 54:64–86.

Truslove, G. M. 1962. A gene causing ocular retardation in the mouse. J. Embryol. Exp. Morphol. 10:652–660.

Tsuzi, S., and T. H. Yosida. 1960. Reciprocal transplantation of skin between rhino and nonrhino mice. Ann. Rep. Nat. Inst. Genet. Jap. (1959) 10:32.

Tutikawa, K. 1953. Studies on an apparently new mutant, "alopecia periodica" found in the mouse. Ann. Rep. Nat. Inst. Genet. Jap. (1952) 3:9–10.

Tutikawa, K. 1955. Test for allelism of "alopecia

periodica" and "furless" in the house mouse. Ann. Rep. Nat. Inst. Genet. Jap. (1954) 5:16.

Vankin, L. 1956. The embryonic effects of "blind," a new early lethal mutation in mice. Anat. Rec. 125:648. (Abstr.)

Verrusio, C. 1964. Mouse News Letter 31:35.

Vlahakis, G., and W. E. Heston. 1963. Increase of induced skin tumors in the mouse by the lethal yellow gene (A^y). J. Nat. Cancer Inst. 31:189–195.

Wallace, M. E. 1957*a*. The use of affinity in chromosome mapping. Biometrics 13:98–110.

Wallace, M. E. 1957*b*. A balanced three-point experiment for linkage group V of the house mouse. Heredity 11:223–258.

Wallace, M. E. 1958*a*. New linkage and independence data for ruby and jerker in the mouse. Heredity 12:453–462.

Wallace, M. E. 1958*b*. Experimental evidence for a new genetic phenomenon. Phil. Trans. Roy. Soc. 241:211–254.

Wallace, M. E. 1959. An experimental test of the hypothesis of affinity. Genetica 29:243–255.

Wallace, M. E. 1961. Affinity: evidence from crossing inbred lines of mice. Heredity 16:1–23.

Wallace, M. E. 1963*a*. Mouse News Letter 28:22.

Wallace, M. E. 1963*b*. Mouse News Letter 29:22.

Wallace, M. E. 1965. Mouse News Letter 32:20.

Warwick, E. J., and W. L. Lewis. 1954. Increase in frequency of a deleterious recessive gene in mice. J. Hered. 45:143–145.

Watney, M. J., and J. R. Miller. 1964. Prevention of a genetically determined congenital eye anomaly in the mouse by the administration of cortisone during pregnancy. Nature 202:1029–1031.

Watson, M. L. 1962. A test for the identity of "dysoptic" with "blind" in mice. Proc. Iowa Acad. Sci. 69:591–593.

Watson, M. L., A. Orr, and T. D. McClure. 1961. A study of blindness in the house mouse. Proc. Iowa Acad. Sci. 68:558–561.

Webster, L. T. 1937. Inheritance of resistance of mice to enteric bacterial and neurotropic virus infections. J. Exp. Med. 65:261–286.

Welshons, W. J., and L. B. Russell. 1959. The Y-chromosome as the bearer of male sex determining factors in the mouse. Proc. Nat. Acad. Sci. 45:560–566.

Wolfe, H. G. 1963*a*. Mouse News Letter 29:40.

Wolfe, H. G. 1963*b*. Two unusual mutations affecting pigmentation in the mouse, p. 251. *In* S. J. Geerts [ed.] Proc. XI Int. Congr. Genet. Vol. 1. Pergamon Press, New York. (Abstr.)

Wolfe, H. G., and D. L. Coleman. 1964. Mi-spotted: a mutation in the mouse. Genet. Res. 5:432–440.

Wolff, G. L. 1965. Body composition and coat color correlation in different phenotypes of "viable yellow" mice. Science 147:1145–1147.

Woolley, G. W. 1941. "Misty," a new coat color dilution in the mouse, *Mus musculus*. Amer. Natur. 75:501–508.

Woolley, G. W. 1945. Misty dilution in the mouse. J. Hered. 36:269–270.

Woolley, G. W., and M. M. Dickie. 1945. Pirouetting mice. J. Hered. 36:281–284.

Wright, M. E. 1947. Undulated: a new genetic factor in *Mus musculus* affecting the spine and tail. Heredity 1:137–141.

Wunderlich, J., and L. A. Herzenberg. 1963. Genetics of a gamma globulin isoantigen (allotype) in the mouse. Proc. Nat. Acad. Sci. 49: 592–598.

Yanagisawa, K., L. C. Dunn, and D. Bennett. 1961. On the mechanism of abnormal transmission ratios at *t* locus in the house mouse. Genetics 46:1635–1644.

Yoon, C. H. 1959. Waddler, a new mutation and its interaction with quivering. J. Hered. 50: 238–244.

Yoon, C. H. 1960. Genetic and non-genetic factors affecting the quivering condition in mice. Amer. Natur. 94:435–440.

Yoon, C. H. 1961*a*. Electrophoretic analysis of the serum proteins of neurological mutants in mice. Science 134:1009–1010.

Yoon, C. H. 1961*b*. Linkage relationship of the waddler gene in mice with evidence for temperature effect on crossing-over. J. Hered. 52: 279–281.

Yoon, C. H., and D. J. Denuccio. 1963. Acid-soluble phosphorus in brains of neurological mice. J. Hered. 54:202–205.

Yoon, C. H., and S. R. Harris. 1962. Cholinesterase studies of neurologic mutants in mice. I. Alterations in serum cholinesterase levels. Neurology 12:423–426.

Yoon, C. H., and E. P. Les. 1957. Quivering, a new first chromosome mutation in mice. J. Hered. 48:176–180.

Yosida, T. H. 1960. Study on the new mutant "falter" found in the house mouse. Ann. Rep. Nat. Inst. Genet. Jap. (1959) 10:27–28.

Zimmerman, K. 1933. Eine neue Mutation der Hausmaus: "hydrocephalus." Z. Indukt. Abstamm. Vererb. 64:176–180.

9

Multiple Factor Inheritance[1]

Thomas H. Roderick and Gunther Schlager

Genes may have large effects and be called "major genes," or they may have individually small effects and be called "minor genes." The distinction between major and minor genes is somewhat arbitrary, because the effect of individual allelic differences on the phenotypic variation may range from small to very large. Nevertheless, it is useful to determine whether the genotypic variance of a trait is caused by allelic differences at one or two loci or by allelic differences at several loci. For example, body weight of mice is a characteristic in which the genetic variation is usually due to segregation and assortment of alleles with minor effects at many loci. However, in addition to the genetic causes of normal phenotypic variation, there are mutations with drastic major effects on body weight such as the obese gene (*ob*) and the pituitary dwarf gene (*dw*).

Characteristics are said to show *multiple factor inheritance* if their genetic variation depends on allelic differences at more than just a few loci. Multiple factor inheritance is usually caused by the action of many "minor genes." Falconer (1963) defined the problem in terms of the character itself rather than in terms of the mode of inheritance. He defined a *quantitative character* as "any attribute for which individual differences do not divide the individuals into qualitatively distinct classes." In this definition it does not matter whether there are few or many loci concerned with the genetic variation of the trait so long as the segregation of single alleles does not produce discrete phenotypic discontinuity. In general it is convenient to regard the genetic variation as of two types (one or a few loci vs. many loci) and the phenotypic variation as of two types (discrete vs. continuous). All four combinations of genetic with phenotypic variations except that of one or a few loci affecting variation in a discrete trait require analytical methods appropriate for quantitative characters. In this chapter we deal with these other three combinations and chiefly with the combination of many loci affecting the variation in a continuous trait.

Two treatises of Falconer (1960*a*, 1963) explain the details of the methods used in multiple factor inheritance. For further introduction to the subject consult Mather (1949), the Cold Spring Harbor Symposium on Quantitative Biology (1955), Li (1955), Kempthorne (1957), Lerner (1958), and Le Roy (1960). These references give details of the historical development of the field of multiple factor inheritance and of the great contributions of R. A. Fisher, Sewall Wright, and J. B. S. Haldane.

GENERAL PRINCIPLES

Continuously varying traits present many analytical challenges. In this section we shall consider the methods of analyzing quantitative characteristics, physiological mechanisms underlying them, inbreeding depression, heterosis, and subline divergence within inbred strains.

[1] The writing of this chapter was supported in part by contracts AT(30-1)-1979 and AT(30-1)-3249 with the U.S. Atomic Energy Commission and in part by Public Health Service Research Grant GM 07249 from the National Institute of General Medical Sciences.

Methods

We assume that the phenotypic variation of a quantitative trait is caused, at least in part, by many allelic differences having small individual effects on the trait. One cannot separate out and study the mode of inheritance of each of the alleles contributing to the phenotypic variation, but one can determine the relative importance of genetic and environmental factors responsible for the phenotypic variation. Although resemblance between relatives indicates some genetic determination of the characteristic, there may also be factors in the environment shared by relatives which cause their resemblance. Maternal effects and cage effects fall in this category. By using appropriate experimental designs and analyses, the genetic and environmental factors can be estimated and the relative importance of each determined. Inbred strains of mice are very useful in these studies.

The terms "genetic determination" and "heritability" are often used interchangeably, but they have different meanings in the following discussion. Genetic determination will mean that portion of the total phenotypic variation attributable to any genetic causes, and heritability, denoted as h^2, will mean that portion of the total phenotypic variance attributable to additive genetic effects alone. Heritability can therefore be a part of genetic determination.

Comparing two strains. The degree of genetic determination can be estimated by comparing two inbred strains, their F_1, F_2, and backcross generations. Under the most simplifying assumptions, the phenotypic variance (σ_P^2) may be considered as the sum of the genetic variance component (σ_G^2) and the environmental variance component (σ_E^2): $\sigma_P^2 = \sigma_G^2 + \sigma_E^2$. If the parental strains are highly inbred, there will be only environmental and no detectable genetic variance within the two parental strains and the F_1 hybrids. In the F_2 animals the phenotypic variance will be ascribable only in part to environmental variance. In addition there will be a genetic component of variance due to segregation of alleles and assortment of nonalleles. Therefore, one can estimate genetic determination by subtracting the F_1 computed phenotypic variance (V_{F_1}, environmental) from the F_2 computed phenotypic variance (V_{F_2}, genetic and environmental) and dividing by the F_2 computed phenotypic variance:

$$\text{Genetic determination} = \frac{V_{F_2} - V_{F_1}}{V_{F_2}} = \frac{(\sigma_G^2 + \sigma_E^2) - \sigma_E^2}{(\sigma_G^2 + \sigma_E^2)} = \frac{\sigma_G^2}{\sigma_P^2}.$$

Grahn (1958) used this method in estimating the genetic determination of radiation resistance in mice.

An approximation of the minimum number of loci affecting a character is often of interest. This can be estimated from the equation:

$$n = \frac{(m_1 - m_2)^2}{8\,(V_{F_2} - V_{F_1})}$$

where m_1 and m_2 are the means of the parental strains and V_{F_2} and V_{F_1} are the computed variances of the F_2 and F_1 generations. This is considered as a minimum estimate because the equation is based on a number of assumptions. Violation of these assumptions will tend to underestimate *n*. A discussion of these assumptions and the consequences of their nonfulfillment is given by Wright (1952). Estimating the minimum number of loci affecting a quantitative character, Falconer (1960*a*) found that 35 loci affect 6-week body weight of mice.

Comparing many strains. Differences between strains have been found for so many characters that genetic influences on the variation of any character may almost be taken for granted. The relative importance of genetic and environmental factors can be estimated by comparing many strains, the reliability of the estimate increasing as the number of unrelated inbred strains used in the analysis increases. The procedure is to measure the character in several highly inbred strains and to obtain estimates of the between-strain and within-strain variance components by the standard analysis-of-variance techniques. The expected mean square between strains is equal to $\sigma_E^2 + k(1+F)\sigma_G^2$, where F is the inbreeding coefficient, σ_E^2 and σ_G^2 are the environmental and genetic components of variance respectively, and k is the number of individuals measured per strain (or the harmonic mean of the numbers in the strains if the sample sizes are not equal). F is effectively equal to 1 in highly inbred strains, so the genetic variance component is doubled. Using strain comparisons, Roderick (1963*b*) estimated the genetic determination of radioresistance of 27 strains of mice to be as high as 53 per cent.

Comparing inbred and randombred strains. The methods above describe how nongenetic variance can be estimated from the phenotypic variance within inbred strains and within their F_1 hybrids, and then used to determine the degree of genetic influence on a character by comparing the estimate with the variance of the segregating F_2 generation. This method can also be used to estimate the degree of genetic determination of a character in a comparison of inbred and randombred strains. Falconer and Bloom (1962) used six inbred strains and their 15 F_1 hybrids to estimate the nongenetic variance in the incidence of induced lung tumors in mice. Using this estimate, they partitioned the variance in two randombred lines (JC and LX) into genetic and environmental components and showed that individual genetic differences account for 82 per cent of the variation of susceptibility in the JC strain and 88 per cent in the LX strain.

Diallel cross. The diallel cross technique is useful for estimating the additive and nonadditive components of the genetic variance, but it has not been extensively used in studies of quantitative traits in mice. Essentially the method involves making all possible crosses between a number of inbred strains and estimating the general combining ability of each line and specific combining ability of each pair of strains. The average performance of a strain represents the general combining ability, while the deviation in performance from the average combining ability of the two strains is the specific combining ability of a cross. The specific combining ability component is analogous to an interaction term in a two-way analysis of variance and is attributable to the nonadditive genetic variance. Differences in general combining ability are attributable to the additive genetic variance. References to the theory and methods of the diallel cross are given in Chap. 2.

Bruell (1963) used this technique with five inbred strains of mice and the 20 F_1 hybrids including the reciprocals to show that alleles affecting serum cholesterol have additive effects. Eaton et al. (1950) used nine inbred lines of mice and all possible F_1 hybrids in an investigation of body weight. Fuller (1964) also used this method for investigating the inheritance of alcohol preference in mice.

Selection. Lerner (1958) has succinctly defined selection as "the nonrandom differential reproduction of genotypes." *Natural* selection produces a change in gene frequencies through natural causes which can often be complex and obscure. The complexities and obscurities are inherent in the nature of fitness, the characteristic upon which natural selection acts. *Artificial* selection denotes a conscious human effort to bring about a differential propagation of genotypes. "Artificial" does not imply that selection by human choice is not *genuine* selection. Studies of natural selection in mice have been made principally to observe its effect on a particular mutant introduced into a wild population, as in the study of Anderson et al. (1964) who introduced a *t* allele into an island population of mice. Artificial selection in mice has been used principally for two purposes: to explore the genetic theory of selection, as in Falconer's (1954*b*) selection for tail length, and to produce a large divergence in a particular trait in order to study the physiological genetics of the trait, as in Wolfe's (1961) selection for blood pH.

Selection can alter the phenotypic mean of a population only if there is some additive genetic variance affecting the trait in question. Because there is no detectable genetic variability within inbred strains, they are unsuitable for a selection experiment. However, genetically heterogeneous foundation populations can be synthesized by crossing two, three, four, or more inbred strains of genetically different origins. The use of inbred strains crossed in a specific way to produce a foundation population for selection has the additional advantage of enabling one to replicate the foundation population at another time and place.

The higher the proportion of additive genetic variance relative to the phenotypic variance (that is, the greater the heritability, h^2) the greater will be the gains under selection. Falconer (1963) gave ways of estimating h^2 prior to selection by using the regression of the means of offspring on the means of parents and by paternal halfsib analysis. Under most conditions, with a reasonably high heritability, selection on the basis of the individual's phenotype is the most economical. Falconer (1960*a*) gave an example of a method of selection based on individual phenotypes. Ten mated pairs per generation are used and selection is made within each of the 10 litters for the two most extreme individuals for the next generation. These 20 individuals are then mated at random among themselves, sib matings being avoided. This system is good in that a reasonably small number of mice are used in each generation, and inbreeding is kept reasonably low for several generations. Because it is within-litter selection, this method avoids selecting for environmental variation due to common litter or cage effects, but does not permit selecting for maternal effects under genetic control which may affect the trait.

Not many selection experiments have been attempted in mice for characteristics measured by destroying the animals. Two methods can be used, both necessitating the raising of more mice than when choice of parents for the subsequent generation can be made before mating: (1) One can estimate the genetic value of particular individuals by measuring the characteristic on close relatives of the individual, or (2) one can breed the animals first, save all offspring, measure the parents, and then select the offspring on the basis of results on the parents. The first method makes it possible to mate assortatively, i.e., to mate like extremes together to produce a greater selection differential (the value of the difference between the mean of the selected parents and the mean of the generation as a whole from which the selected parents come), but it also has the drawback of being a poorer estimate of the true breeding value of the animals. The second method has the disadvantage of random mating and therefore a lowered selection differential, but the animal's breeding value is more accurately measured. An example of the second method is given by Roderick (1963*a*).

Physiological mechanisms underlying quantitative traits

Sometimes it is possible to gain insight into physiological mechanisms by using the methods of quantitative genetics.

Strain correlations. It is common to find that animals which rank high in one characteristic tend to rank high (or low) in another characteristic. The characteristics then are said to exhibit a positive or negative phenotypic correlation depending on whether the relationship is direct or inverse. Under the simplest assumptions, characteristics may be correlated for two major reasons: (1) because they are both influenced by one or more common factors in the environment, and (2) because they are both influenced by the same genes. The first is an environmental correlation and the second a genetic correlation. In mice, body weight and disease resistance may be correlated through environmental factors if one mouse is raised in a good environment and another mouse is raised in a poor environment. Or the characeristics may be correlated through genetic factors if it is shown that heavier strains of mice are generally healthier than lighter strains raised in the same environment. Because there is no detectable genetic variation between animals within an inbred strain, any correlation of characteristics found between animals within a strain must be due to environmental influences. Environmental influences can be any causative factor other than the genetic material itself, such as differences in cytoplasmic factors, prenatal environments, postnatal environments, and daily environmental fluctuations.

If a correlation between traits is found among inbred strains, one must ascertain that this correlation is greater than expected from random samples of mice within a strain to be sure it is to some extent a genetic correlation. If the correlation within strains is zero, as appeared in a study of lifespan and litter size (Roderick and Storer, 1961), then all the correlation between strains must be due to genetic factors. Estimates of the genetic and environmental correlations are methodologically analogous to the estimates of genetic and nongenetic determination of variation as described earlier. The correlation analysis is an analysis of covariance in the same way that the genetic determination analysis is an analysis of variance. Environmental and genetic correlations may be present simultaneously, and they need not be equal nor in the same direction.

One word of caution should be mentioned in such analyses. Inbred strains are unusual genetic entities because they are almost completely homozygous and because they have often been subjected to artificial selection during the course of inbreeding. Thus any genetic correlation found from an analysis of two variables among inbred strains should probably be checked in other ways, possibly by a similar analysis of a group of F_1 hybrids between the various strains. Also one should be cautioned about estimating genetic correlations based on related sublines of inbred strains. For instance, correlation found among strains C57BL/6J, C57BL/10J, C57L/J, C57BR/cdJ, C3H/HeJ, and C3HeB/FeJ could be due to associations of the traits in the original C57 group and obverse associations in the original C3H group. The characters may be associated because of genetic drift and not because of genetic pleiotropy. In the example given, the degrees of freedom (N-2) for estimating the correlation are zero, based on the number of actual genetically independently derived strains, and not four, based on the number of strains observed.

Significant genetic correlations from studies of independently derived inbred strains do not of themselves give any indication about the physiological mechanism for the correlation. Further exploration is necessary to determine if the association is due to linkage or pleiotropy, although linkage is not a likely cause of significant genetic correlations found among independently derived strains. If the association is due to pleiotropy, one must also explore further to ascertain whether the characteristics are related in a direct cause and effect manner, or whether they are both influenced independently by a third factor acting earlier in development.

Correlated responses to selection. Usually the response to selection will be accompanied by changes in unselected characteristics as well as in the trait directly selected. These associated responses are called correlated responses and the traits are referred to as correlated characters. Correlated characters are those which change because they are truly genetically associated with the trait under selection. The correlation may be caused by linkage of the genes for the two traits, but is more probably the result of pleiotropy of the many genes affecting both traits. In pleiotropy the association could be either through independent physiological pathways between the genes and the traits, or it could be through a direct causative pathway where one trait is a necessary developmental intermediate between the genes and the other trait. If the genetic variation of the associated trait is dependent on only one or a few loci, it is inappropriate to assume that the trait is a correlated character on the basis of evidence from only one selection experiment. Such traits could be genetically entirely independent of the selected trait and could manifest chance associations in the selected lines because of genetic drift. If it is a quantitative trait and varies gradually with each generation as the selected trait varies with selection, the trait can be assumed to be a true correlated character.

A study of the correlated responses to selection may afford insight into the physiology of the selected characteristic. For example, by examining the physiological differences of animals selected in two ways for disease resistance, one may get some understanding of the natural mechanisms protecting animals against disease. Selection for radiation resistance in mice was undertaken chiefly to study the correlated responses in body weight, litter size, and ability to survive other regimens and doses of radiation (Roderick, 1963*a*). One of the more interesting responses was an apparent susceptibility to a disease in the radiation-sensitive line. The disease was not identified, but it is possible that, through the mechanism of disease susceptibility, the animals have a lowered resistance to irradiation.

Inbreeding depression

The phenomenon of inbreeding depression has been witnessed in laboratory mammals for decades as evidenced by the extensive investigations with guinea pigs by Wright (1922). The mean value of characteristics associated with fitness in animals tends to decrease as inbreeding increases. The fitness characters affected by inbreeding in mice include: litter size (Falconer, 1960*b* and earlier), body weight (Falconer, 1960*a*), disease resistance (Weir, 1960*a*), and lifespan (Chai, 1959). For example, Falconer found that the average size of first litters decreased at a rate of 0.56 animals for each 10 per cent increase in the inbreeding coefficient. A decline was also found in lines where selection was practiced for an increase in litter size. The study demonstrated that part of the decline in litter size was due to inbreeding in the litters and part to the inbreeding of the dam. Subsequent work showed a difference between inbred and noninbred dams in preimplantation losses, but ovulation rate measured as number of corpora lutea was not affected by inbreeding (Falconer and Roberts, 1960). McCarthy (1965) studied the physiological basis of heterosis of litter size in a study of four inbred strains and their crosses. He varied the inbreeding of the offspring and kept the dams inbred. The only significant difference in prenatal characteristics between incrosses and outcrosses was a significantly greater number of early postimplantation deaths (moles) in the incrosses. Inbreeding in this case, therefore, reduces litter size by acting through the dam to increase preimplantation loss and by acting through the embyro to increase early postimplantation mortality.

Decrease of litter size does not always result from increase in inbreeding. Falconer (1960*b*) found a few examples where inbreeding did not depress litter size. Weir and Schlager (1962) found no perceptible decrease in litter size in 25 generations of fullsib matings in a study of selection for leukocyte counts. That inbreeding depression is a real phenomenon, however, is emphatically demonstrated by the difficulties associated with the establishment and maintenance of inbred lines. Falconer found that only one line of an original 20 started could be maintained beyond an inbreeding coefficient of 99 per cent; 17 were lost at 76 per cent and two more after 80 per cent. On mathematical grounds Falconer (1960*a*) showed that the changes in gene frequencies during inbreeding are in the direction of increasing the number of recessive alleles being fixed. Generally the most probable cause of inbreeding depression is the fixation of deleterious recessive alleles normally masked and rendered ineffective by dominant alleles in a genetically heterogeneous population. For further discussion see Lerner (1954) and Mather (1955).

Heterosis

Heterosis or hybrid vigor is the converse of inbreeding depression and is said to occur when the mean of the character in the F_1 falls outside the range of the means of the two parental strains, lines, or populations. As inbreeding depression is generally considered to be the result of the fixation of harmful recessive alleles, heterosis is considered to be the result of the dominance of favorable alleles not held in common by the two parental populations. All types of genetic interaction may also affect expression of the character in the F_1 hybrid. Discussions of the mechanisms of heterosis can be found in the papers of Rendel (1953), Crow (1952), and Dobzhansky (1952), and in the book by Lerner (1954). The mathematical treatment is given by Falconer (1960a). Several reviews are available, including a summary of the early work by Shull (1952) and a later review by Bowman (1959). The mathematical model extending the techniques of the diallel cross in an analysis of variance are given by Kidwell et al. (1960) who studied heterosis of body weights in rats. Litter size in the mouse is a fitness characteristic that can exhibit heterosis (Roberts, 1960; Barnett and Coleman, 1960; McCarthy, 1965).

Inbred strains and subline differentiation

The inbred lines of mice today so common in biological research have long been thought to be homozygous at almost all of their loci. The genetic theory of inbreeding devised by Wright (1921) showed that animals bred brother to sister for at least 20 generations (the criterion by which geneticists presently define inbred strains of mice) were homozygous at more than 99 per cent of their loci. In addition the descendants of highly inbred animals should all be genetically uniform with respect to the fixed alleles at each locus. In other words the animals are homozygous *and* as genetically alike as identical twins except for the necessary dimorphism caused by the X and Y chromosomes. General observations of mice within strains on size, specific morphology, coat color genes, and behavior support the theory of uniformity of inbred strains. Work on tissue transplantation within strains showed that mice within a single strain were as alike as identical twins at the histocompatibility loci (Chap. 24). Mouse biologists have not observed segregation of alleles within inbred strains other than those occurring at such low frequency that they could be considered as new mutations. However, the recent work of Deol et al. (1960) has put the assessment of homozygosity of inbred strains on a firmer and less intuitive footing. In an array of different skeletal variants they looked for parent-offspring correlations within inbred strains which would indicate the presence of genetic variance. They found no evidence for such correlations and concluded that the inbred strains of mice show no phenotypic variation within strains ascribable to an additive genetic component. Therefore, since there is no evidence for other mechanisms, such as balanced lethal systems retaining hidden genetic variation between alleles in these strains, the inbred strains of mice can be considered homozygous. Deol et al. point out that this high percentage of homozygosity in the mouse may not necessarily be possible in all other species, as there is considerable evidence from other studies for retention of heterozygosity far beyond the time predicted on the basis of the breeding systems used. Selection may in some circumstances favor heterozygotes or it may hold back inbreeding by selecting against unfavorable homozygous alleles. The work of Falconer (1960*b*) showed that, even for mice, inbreeding achievements probably did not come easily.

The fact still remains that sublines of established inbred lines, if separated long enough, eventually do develop differences. However, they never seem to lose their similarity (in the 40 years or more that some sublines have existed) to the extent that they are as different as independently derived inbred strains (Roderick and Storer, 1961; Roderick, 1963*b*). It is the opinion of most authors who have studied the problem that subline differentiation is not the result of fixation of genes which were in heterozygous state at the time of the separation of the sublines, but rather is due to fixation of different mutations which have occurred in the separate sublines (Deol et al., 1957; Carpenter et al., 1957; Bailey, 1959; Kindred, 1963).

Green (1953) presented evidence showing considerable heterogeneity in vertebral characteristics among the sublines of strain C3H. This variation was confirmed in a study by McLaren and Michie (1954, 1955), who pointed out the tremendous effect that different environments have on the vertebral column. Environmental variation probably does not explain the subline differences in Green's work because all except one subline were raised in the same laboratory and examined at approximately the same time. The variation Green found, however, seemed much too great to be explained on the basis of mutation alone. He presented several alternative explanations of which the most likely, we think, is the possible inadvertent genetic contamination from a phenotypically similar mouse at some time in the history of one of the major offshoots of the strain. These sublines had a history of frequent transfers between several investigators and laboratories. Grewal (1962) studied the skeletons of sublines of the C57BL strain and noted divergence in excess of that expected from mutational origin alone. He suggested that residual heterozygosity in the original inbred strain might be a factor. He also considered the possibility that breeding practices in the early history of the line may not have been as uniform as they were later. It is also possible, when sublines are raised in different laboratories, that different environments present different selection forces (both natural selection and subconscious artificial selection) which act on newly arisen mutations to widen the genetic gap between the sublines.

In order to estimate experimentally the rate at which parallel sublines diverge genetically, the most important requirement is to reduce the environmental variations and the possibility of inadvertent genetic contamination as much as possible. Separation and subsequent maintenance of the incipient sublines should be carried on in one laboratory, and there should be a common uniform environment for both sublines, rigid genetic control, and the supervision of one investigator. Most of these requirements were satisfied by the study of Bailey (1959), who estimated that six to nine generations would be required to show significant differences between sublines in particular skeletal traits.

The Committee on Standardized Genetic Nomenclature for Mice (Staats, 1964) recommended that independent inbred sublines from a single inbred line be considered different genetic entities, i.e., different sublines, (1) when they have been separated for at least eight generations, (2) when the lines are under the care of different investigators, and (3) when any detectable genetic differences between the sublines arise. If subline differentiation is due to residual heterozygosity in the original line alone, "crossing of sublines reduces the coefficient of inbreeding to that of the generation at which the sublines were separated" (Dinsley, 1963). However, with the likelihood of mutation and the problem of genetic contamination, the homozygosity of an F_1 between sublines which meet any of the three criteria above should remain in doubt.

INHERITANCE OF SPECIFIC CHARACTERISTICS

In this section we present some examples of studies in multiple factor inheritance in the mouse which have particular importance either because the characteristics appear to be important in many fields of biology or because the traits have been used to test or verify the theory of multiple factor inheritance.

Physiological characteristics

***Blood* pH.** Selection for differences in the level of blood pH has been successful on two occasions. Weir and Clark (1955) selected for high and low levels of blood pH, and Wolfe (1961) replicated the experiment using the same outbred control stock. In the first experiment various degrees of inbreeding were practiced as the breeding systems used included fullsib, halfsib, and "unrelated" (animals less closely related than first cousins) matings in the second and third selected generations. Parents of the first generation were mass-selected from an expanded colony derived from MacArthur's randombred control stock and were mated assortatively. There was an immediate response to selection. The mean pH in the third generation was 7.476 ± 0.004 in the high line, 7.444 ± 0.007 in the control (outbred) line, and 7.426 ± 0.004 in the low line.

Wolfe's replication of the experiment was continued for seven generations with minimal inbreeding for the first five generations and fullsib mating thereafter. Selection by individual merit was practiced, with 11 to 13 single-pair matings used to maintain the selected lines in each generation. Immediate response to selection was again evident with no additional response after the fourth generation. The mean pH in the fourth generation was 7.582 ± 0.006 in the high line, 7.548 ± 0.007 in the outbred control line, and 7.513 ± 0.005 in the low line. Estimates of heritability were 14.0 per cent from the ratio of summed response to summed selection differential, 16.8 per cent from the ratio of the regression coefficients for cumulative response and selection differential, and 14.5 per cent from the intraclass correlation of polygamous matings of the outbred stock using the combined sire and dam variances. The immediate response and the early limits to selection in both experiments indicated that few loci determine the variation of blood pH.

Response to hormone treatments. Endocrine systems are essential to the growth of an organism and to maintenance of physiological homeostasis during life. A great portion of the natural variation in response to hormone treatments present in mouse populations can be attributed to a genetic component of variance based on allelic differences at many loci. Further discussion of the genetics of endocrine variation is given in Chap. 20.

Morphological characteristics

Skeletal traits. Many studies have been made on the skeleton of the mouse (Chap. 13; Grüneberg, 1963). Particularly for the axial skeleton consult the studies of Searle (1954), McLaren and Michie (1955), and Green (1962). These and earlier studies of the components of variation of axial skeletal traits show that a large part of the phenotypic variance can be attributed to genetic variation but that the number of gene differences may not be very great. The inheritance of number of thoracic and lumbar vertebrae can best be explained by a model of multiple factor inheritance underlying a threshold characteristic. The environment also has much influence on these traits, since the phenotypic variation within strains is large. Differences exist between colonies of the same substrain maintained in different places. Diet has a strong influence. Very important, too, is a maternal effect which is expressed both as an effect of age of the mother and as an effect of the particular uterine environment of the mother. Environmental differences do exist within litters, but these effects are not attributable to differences between uterine horns, or differences of position and presentation within the uterine horns. The environmental causes of the phenotypic variation of individuals within a single litter have not been ascertained.

Body weight. Grüneberg (1952) presented a good summary and bibliography of the studies on the inheritance of body weight published up to that time. Body weight is an example of a trait whose phenotypic variation is controlled by a large number of loci although many examples exist where body weight is greatly affected by single loci. There is a maternal effect on weaning weight, but this influence is almost negligible for adult body weight. Heterosis in body weight is often observed in crosses of inbred strains of mice, although some cases have been reported where intermediate body weight was found in hybrids (Franks et al., 1962; McCarthy, 1965). The reasons for the presence of heterosis in some cases and its absence in others are not clearly understood.

The most thorough studies on the genetics of body weight are those of Falconer (1955) and his colleagues. Two-way selection for 6-week body weight produced differences between lines 16 times larger than the original genetic standard deviation. The response was gradual with an apparent plateau at about the 20th generation. Many genes appear to affect the variability of the trait. Six-week weight is a complicated combination of the influence of the maternal environment (also genetically controlled) and the animal's own endowment. Selection for 6-week weight produced genetic changes in both components. Selection success in the low line was greater than in the high line; this asymmetry could be attributed to the asymmetrical response of the maternal effect and not to the effect of the individual's own genotype. MacArthur (1949 and earlier) found no asymmetry in response to selection for 60-day body weight, but the weights at later ages did not seem to be influenced very much by maternal effects. Fowler (1958) studied the correlated responses of Falconer's selection experiment as well as of another independent 2-way selection by Falconer for 3- to 6-week weight gain. Differences between the two experiments in the correlated responses of proportion of fat and protein and associated water in the carcass could be attributed to the genetic differences in the foundation populations or to the slight difference in the selection criterion. These findings are important because they suggest the possible variety of perhaps independent physiological mechanisms through which selection may alter a single trait.

Vibrissa number. Vibrissa number in the house mouse is a very stable character with rare deviations from the count of 19 in normal mice. In a survey of several inbred lines and randombred stocks, Dun and Fraser (1959 and earlier) found no variation in the number of postorbital vibrissae and little variation in the other four groups: supra-orbitals, postorals, inter-ramals, and ulnar-carpals. On the other hand, stocks carrying the sex-linked semidominant gene tabby (*Ta*) showed a comparatively wide range of variation in vibrissa number. Dun and Fraser formed a foundation population from crosses of three tabby stocks and two inbred lines. The wild-type mice in this population showed no variation in vibrissa number. Mice were then se-

Table 9-1. RESULTS OF AN EXPERIMENT FOR SELECTION OF VIBRISSA NUMBER.
(*Fraser and Kindred, 1962*)

	Mean vibrissa number			
	Generation 2		*Generation 19*	
Genotype	High	Low	High	Low
+/+ (♀), +/Y (♂)	18.9	18.8	19.5	16.7
Ta/+ (♀)	15.3	15.0	18.0	10.0
Ta/Y (♂)	8.8	8.6	12.0	5.4

lected for high and low vibrissa number by a combination of mass selection of *Ta*/+ females and sib-testing of *Ta*/Y males. A self-continuing mating scheme was initiated in the third generation that produced wild-type mice in each subsequent generation. The effects of selection were recorded for the three phenotypes +/+, *Ta*/+, and *Ta*/Y. The results of selection through the 19th generation were given by Fraser and Kindred (1962 and earlier). The results are summarized in Table 9-1; the values for the vibrissa number in generation 19 are from Fraser and Kindred (1962, Fig. 1). This experiment demonstrated that the variability of vibrissa number in the tabby mice is largely genetic, and continued selection for changes in the vibrissa number in the tabby mice brings about concomitant changes in the wild-type mice where variation in number is rare.

Tail length. Falconer (1953) found that a change in tail length was a correlated response to selection for 6-week body weight. He then selected for tail length (Falconer, 1954*b*) to see if the correlated response of 6-week body weight would be that expected from the first experiment on the theory of genetic correlation. The theory is that the response of a correlated character is fully explained by its genetic correlation (r_G) with the selected trait. If the two estimates of r_G proved to be the same, then the responses could be accurately predicted from either experiment and the theory of genetic correlation would be supported. If ΔG and $\Delta G'$ represent the genetic changes in the selected and correlated characters respectively, σ_G and $\sigma_{G'}$ the respective genetic standard deviations, and r_G the genetic correlation, then two estimates of r_G, one from each selection experiment, are obtainable from the equation:

$$\frac{\Delta G'}{\sigma_{G'}} = r_G \frac{\Delta G}{\sigma_G}$$

where σ_G and $\sigma_{G'}$ must be equated to the measurable parameters $h\sigma_P$ and $h'\sigma_{P'}$, respectively. His estimates of r_G, 0.62 and 0.57, were in close agreement, so no evidence against the theory of genetic correlation was found.

Behavioral characteristics

Perhaps the greatest reason for the study of multiple factor inheritance is that many of the intrinsically interesting traits of living organisms have multiple factor inheritance. Behavioral characteristics and other components of fitness are intriguing to investigators because their phenotypic variation is so vitally important to the well-being of the individual, the family group, the population, and the species. Examples of quantitative behavioral characteristics are audiogenic seizures and alcohol preference. Other examples of quantitative traits with perhaps a more complex physiology are emotionality and exploratory behavior (Chap. 32, 33).

Fitness characteristics

Disease resistance. A review of the role of inheritance in the immunity of mice and other animals was given by Gowen (1948). There is evidence from earlier papers that some of the resistance to various diseases is controlled by single dominant autosomal genes (Chap. 8). However, other studies indicate multifactorial inheritance. Nadel et al. (1955) demonstrated differences between 12 inbred strains, F_1 hybrids, F_2 hybrids, and backcrosses in survival after experimental infection with *Plasmodium berghei.* There was evidence of heterosis as the hybrids of the C57BL/6 and C57L parents exhibited the longest mean survival time. Greater variance was found in the F_2 and in the backcross animals than in the parents and F_1. Additional evidence indicating that multiple genetic factors were involved came from the backcross data. The study was complicated by the wide range of survival times exhibited by the mice in all series and by the presence of a bimodal curve of time until death.

Gowen (1963) maintained strains of mice at fixed levels of mortality from infection with 200,000 *Salmonella typhimurium* organisms by inbreeding and selection. The most resistant S strain had 83.9 per cent survival and the most susceptible Ba strain (= BALB/Gw) had 0.2 per cent survival. The observed survival indicated that a number of genes affect different physiological functions responsible for resistance or susceptibility. The survival of hybrids between three resistant and two susceptible strains suggested dominance or partial dominance for some of the genes controlling resistance. Crosses of two highly susceptible strains resulted in hybrids with greater resistances than either parent. Crosses between resistant lines produced hybrids somewhat less resistant.

Gowen and Calhoun (1943) found a positive strain correlation between resistance to *Salmonella* and total leukocyte count. This finding stimulated Weir and Schlager (1962) to select for total leukocyte count and examine the correlated response of disease resistance. Total leukocyte count responded to 2-way selection for 13 generations, increasing from 9,100 in an outbred line (T strain) to 17,500 in the high line and decreasing to 4,000 in the low line. Selection was continued for seven more generations with no apparent response. The selection was based on family and individual merit of 30-day-old mice and fullsib matings were employed during most of the program. The response was asymmetrical, being more rapid in the high line than in the low line. Expected heritability from sire and dam intraclass correlations in the T strain was 22.4 ± 3.4 per cent from the sire component. The realized heritability was 18.7 ± 1.8 per cent for the first 13 generations. In studying the correlated responses the authors found *no* significant differences between strains in resistance to a dose of 200,000 live organisms of S. *typhimurium,* in the toxic effects of massive doses of heat-killed organisms, or to reinoculation with live organisms after immunization. Therefore differences in *Salmonella* resistance cannot be explained by differences in total leukocyte count alone.

In a related study Chai (1957) compared the variances of leukocyte counts of F_1 and F_2 mice from the Small (SM) and Large (LG) strains of mice. He estimated that genetic determination of the character "leukopenia" was approximately 50 per cent.

Heston (1942) compared the F_1 and F_2 of crosses of strains A and L and estimated that approximately 86 per cent of the total variance of susceptibility to induced pulmonary tumors was due to genetic factors. Falconer and Bloom (1962) also studied the genetics of susceptibility to induced lung tumors in mice. The degree of genetic determination of susceptibility estimated by comparing the variances of six inbred strains and their 15 F_1's with the variances of two randombred populations was 82 per cent in one randombred line and 88 per cent in the other. The genetic variability was further partitioned into additive and nonadditive components by parent-offspring regression to give heritability estimates of 23 per cent in one line and 49 per cent in the other.

Litter size. The optimal litter size for the greatest fitness both for dam and offspring is difficult to ascertain. If too few animals are born, the species risks extinction. If too many are born, the offspring may get a poor start, the mother's fitness may drop, a longer period between litters may come about, and a reduced number of grandchildren may be born. Falconer (1960*b*) studied the effects of inbreeding and selection on litter size, as mentioned above. In the selection experiment no obvious progress was made until the 15th generation when the high, control, and low lines took their respective predicted positions. Results after 31 generations led him to conclude tentatively that the upward response was due to an increased ovulation rate of the dams and the downward response was due to lowered viability of the embryos. This conclusion is important because it suggests that, although 2-way selection is being applied on one characteristic, the actual physiological mechanisms underlying the characteristic are vastly different in the two lines.

Sex ratio. Genes acting at different times in embryogenesis can influence the X-Y system of sex determination in the mouse. Any number of genes could be involved in the processes of gametogenesis and fertilization which affect the differential survival of X- and Y-bearing sperm. Also genes may influence implantation and prenatal development to bring about differential survival of the sexes. Changes in the frequencies of these alleles will then bring about changes in the sex ratio as recorded by the investigator. Indirect evidence for the existence of genes that modify the sex ratio comes from the differences in sex ratio reported in the literature for different inbred strains (Weir, 1960*b;* Cook and Vlček, 1961).

Falconer (1954*a*) attempted to select for differences in sex ratio. After four generations both the high line and the low line had sex ratios of 53 per cent males. Calculations of the maximum amount of genetic variation compatible with the observed data yielded heritability estimates of 6.9 per cent, but this was judged to be too high since there was no evidence for genetic variance in an outbred stock.

Changes of sex ratio have occurred as a correlated response in a selection experiment for high and low levels of blood pH (Weir, 1960*b*). The high-pH line had a sex ratio of 52.8 ± 1.0 per cent males and the low line had 41.8 ± 0.9 per cent. The selection for blood pH was repeated by Wolfe (1961) with a correlated response in sex ratio opposite to the direction observed by Weir. The pooled data of eight generations showed that the high line had a sex ratio of 45.7 per cent males and the low line had 52.5 per cent males.

Radiation resistance. Differences between inbred strains in susceptibility to irradiation have been known since Henshaw (1944) compared the resistance of two strains. Extensive studies of the radiation resistance of strains reported by Grahn (1960 and earlier), Reinhard et al. (1954), Kohn and Kallman (1957 and earlier), Stadler and Gowen (1957), Luchnik (1959), Frölén et al. (1961), Roderick (1963*b*), and Ehling (1964) indicate that strain differences are extremely large but that the various strains do not necessarily bring about their resistance to radiation through precisely the same physiological mechanisms. That at least some of the variation between strains has a multiple factor basis was demonstrated by Roderick (1963*a*), who found that selection for survival under 100 R of X-rays per day was still effective after the eighth generation of selection. The upper and lower limits of selection at eight generations did not exceed the extremes of the inbred strains studied. Vogel et al. (1962) showed that these selected lines were also different in the same direction in their resistance to fission neutron irradiation, where the criterion again was days survival. With strains available differing so greatly in their radiation susceptibility, the physiological nature of the natural mechanisms underlying susceptibility can be investigated. Radiation resistance is discussed more fully in Chap. 22.

Lifespan. Chai (1959) reported greater lifespan in a hybrid than in inbred strains and attributed the difference to the reduction of early mortality among the hybrids. Some of the results of E. S. Russell's studies on the difference in lifespan between inbred strains of mice were reported by Roderick and Storer (1961), who found a significant genetic correlation between lifespan and litter size. Roberts (1961) found a significant difference in lifespan of strains selected for 6-week body weight, but observed a negative genetic correlation between lifespan and body weight. The hybrids of a cross between his long- and short-lived strains nearly equalled the long-lived strains. However, the cross exhibited heterosis in reproductive capacity. Probably many genes are involved in the genetic variation of lifespan, although no study has been attempted to estimate the number. Selection of lifespan might be possible by the method of freezing spermatozoa for later use. Or it might be possible by selecting for traits, such as high litter size and low body weight, genetically correlated with lifespan, but measurable in younger animals. It is interesting to note that high litter size and low body weight, although both appear to be positively genetically correlated with lifespan, are themselves negatively genetically correlated. For a further discussion of lifespan see Chap. 26.

SUMMARY

We have outlined some of the main problems in studying multiple factor inheritance and noted specific studies where mice have been used effectively in solving these problems. The problems discussed are methods for studying multiple factor inheritance, physiological mechanisms underlying quantitative traits, inbreeding depression, heterosis, and subline differentiation. Because of their small size and the availability of a great variety of inbred strains, mice are highly valuable in studies of multiple factor inheritance.

Studies of multiple factor inheritance in mice are too numerous to list here. As an introduction to this large literature we have discussed a few studies including response to hormone treatment, skeletal traits, body weight, vibrissa number, tail length, disease resistance, blood pH, litter size, sex ratio, radiation resistance, and lifespan.

LITERATURE CITED

Anderson, P. K., L. C. Dunn, and A. B. Beasley. 1964. Introduction of a lethal allele into a feral house mouse population. Amer. Natur. 98:57–64.

Bailey, D. W. 1959. Rates of subline divergence in highly inbred strains of mice. J. Hered. 50: 26–30.

Barnett, S. A., and E. M. Coleman. 1960. 'Heterosis' in F_1 mice in a cold environment. Genet. Res. 1:25–38.

Bowman, J. C. 1959. Selection for heterosis. Anim. Breeding Abstr. 27:261–273.

Bruell, J. H. 1963. Additive inheritance of serum cholesterol level in mice. Science 142:1664–1666.

Carpenter, J. R., H. Grüneberg, and E. S. Russell. 1957. Genetical differentiation involving morphological characters in an inbred strain of mice. II. American branches of the C57BL and C57BR strains. J. Morphol. 100:377–388.

Chai, C. K. 1957. Leukopenia: an inherited character in mice. Science 126:125.

Chai, C. K. 1959. Life span in inbred and hybrid mice. J. Hered. 50:203–208.

Cold Spring Harbor Symposium on Quantitative Biology. 1955. Population genetics: the nature and causes of genetic variability in populations. Vol. 20.

Cook, M. J., and A. Vlček. 1961. Sex ratio in mice. Nature 191:89.

Crow, J. F. 1952. Dominance and overdominance, p. 282–297. *In* J. W. Gowen [ed.] Heterosis. Iowa State College Press, Ames.

Deol, M. S., H. Grüneberg, A. G. Searle, and G. M. Truslove. 1957. Genetical differentiation involving morphological characters in an inbred strain of mice. I. A British branch of the C57BL strain. J. Morphol. 100:345–375.

Deol, M. S., H. Grüneberg, A. G. Searle, and G. M. Truslove. 1960. How pure are our inbred strains of mice? Genet. Res. 1:50–58.

Dinsley, M. 1963. Inbreeding and selection, p. 235–259. *In* W. Lane-Petter [ed.] Animals for research. Academic Press, London and New York.

Dobzhansky, Th. 1952. Nature and origin of heterosis, p. 218–223. *In* J. W. Gowen [ed.] Heterosis. Iowa State College Press, Ames.

Dun, R. B., and A. S. Fraser, 1959. Selection for an invariant character, vibrissa number, in the house mouse. Austral. J. Biol. Sci. 12:506–523.

Eaton, N. E., W. E. Neville, and G. E. Dickerson. 1950. General and specific combining abilities in mouse crosses. J. Anim. Sci. 9:636–637. (Abstr.)

Ehling, U. H. 1964. Strain variation in reproductive capacity and radiation response of female mice. Radiat. Res. 23:603–610.

Falconer, D. S. 1953. Selection for large and small size in mice. J. Genet. 51:470–501.

Falconer, D. S. 1954*a*. Selection for sex ratio in mice and *Drosophila*. Amer. Natur. 88:385–397.

Falconer, D. S. 1954*b*. Validity of the theory of genetic correlation. J. Hered. 45:42–44.

Falconer, D. S. 1955. Patterns of response in selection experiments with mice. Cold Spring Harbor Symp. Quant. Biol. 20:178–196.

Falconer, D. S. 1960*a*. Introduction to quantitative genetics. Ronald Press, New York. 365 p.

Falconer, D. S. 1960*b*. The genetics of litter size in mice. J. Cell Comp. Physiol. 56 (Suppl. 1): 153–167.

Falconer, D. S. 1963. Quantitative inheritance, p. 193–216. *In* W. J. Burdette [ed.] Methodology in mammalian genetics. Holden-Day, San Francisco.

Falconer, D. S., and J. L. Bloom. 1962. A genetic study of induced lung-tumours in mice. Brit. J. Cancer 16:665–685.

Falconer, D. S., and R. C. Roberts. 1960. Effect of inbreeding on ovulation rate and foetal mortality in mice. Genet. Res. 1:422–430.

Fowler, R. E. 1958. The growth and carcass composition of strains of mice selected for large and small body size. J. Agric. Sci. 51:137–148.

Franks, E., N. S. Fechheimer, and C. Cohen. 1962. An examination of heterosis in crosses of certain inbred strains of mice. Ohio J. Sci. 62: 177–184.

Fraser, A. S., and B. M. Kindred. 1962. Selection for an invariant character, vibrissae number, in the house mouse. III. Correlated responses. Austral. J. Biol. Sci. 15:188–206.

Frölén, H., K. G. Lüning, and C. Rönnbäck. 1961. The effect of X-irradiation on various mouse strains due to their genetic background. I. Lethality after acute irradiation. Radiat. Res. 14: 381–393.

Fuller, J. L. 1964. Measurement of alcohol preference in genetic experiments. J. Comp. Physiol. Psych. 57:85–88.

Gowen, J. W. 1948. Inheritance of immunity in animals. Annu. Rev. Microbiol. 2:215–254.

Gowen, J. W. 1963. Genetics of infectious diseases, p. 383–404. *In* W. J. Burdette [ed.] Methodology in mammalian genetics. Holden-Day, San Francisco.

Gowen, J. W., and M. L. Calhoun. 1943. Factors affecting genetic resistance of mice to mouse typhoid. J. Infect. Diseases 73:40–56.

Grahn, D. 1958. The genetic factor in acute and chronic radiation toxicity. Proc. 2nd United Nations Int. Conf. Peaceful Uses of Atomic Energy, Geneva 22:394–399.

Grahn, D. 1960. Genetic control of physiological processes: the genetics of radiation toxicity in animals, p. 181–200. *In* R. S. Caldecott and L. A. Snyder [ed.] Radioisotopes in the biosphere. Center for Continuation Study, Univ. Minnesota, Minneapolis.

Green, E. L. 1953. A skeletal difference between sublines of the C3H strain of mice. Science 117: 81–82.

Green, E. L. 1962. Quantitative genetics of skeletal variations in the mouse. II. Crosses between four inbred strains (C3H, DBA, C57BL, and BALB/c). Genetics 47:1085–1096.

Grewal, M. S. 1962. The rate of genetic divergence of sublines in the C57BL strain of mice. Genet. Res. 3:226–237.

Grüneberg, H. 1952. The genetics of the mouse, 2nd ed. Nijhoff, The Hague. 650 p.

Grüneberg, H. 1963. The pathology of development; a study of inherited skeletal disorders in animals. Wiley, New York. 309 p.

Henshaw, P. S. 1944. Experimental roentgen injury. II. Changes produced with intermediate-range doses and a comparison of the relative susceptibility of different kinds of animals. J. Nat. Cancer Inst. 4:485–501.

Heston, W. E. 1942. Genetic analysis of susceptibility to induced pulmonary tumors in mice. J. Nat. Cancer Inst. 3:69–78.

Kempthorne, O. 1957. An introduction to genetic statistics. Wiley, New York. 545 p.

Kidwell, J. F., H. J. Weeth, W. R. Harvey, L. H. Haverland, C. E. Shelby, and R. T. Clark. 1960. Heterosis in crosses of inbred lines of rats. Genetics 45:225–231.

Kindred, B. 1963. Skin grafting between sub-lines of inbred strains of mice. Austral. J. Biol. Sci. 16:863–868.

Kohn, H. I., and R. F. Kallman. 1957. The influence of strain on acute X-ray lethality in the mouse. II. Recovery rate studies. Radiat. Res. 6:329–338.

Lerner, I. M. 1954. Genetic homeostasis. Wiley, New York. 134 p.

Lerner, I. M. 1958. The genetic basis of selection. Wiley, New York. 298 p.

Le Roy, H. L. 1960. Statistische Methoden der Populationsgenetik. Birkhäuser, Basel und Stuttgart. 397 p.

Li, C. C. 1955. Population genetics. University of Chicago Press, Chicago. 366 p.

Luchnik, N. V. 1959. Radiation injuries and means of affecting them. II. The relationship of the mortality rate of irradiated mice and rats to their strains, sex, weight, dose of irradiation and the distribution of this mortality rate with respect to time. U.S. Dep. Commerce, Office of Technical Service, OTS 59-13409:97–149.

MacArthur, J. W. 1949. Selection for small and large body size in the house mouse. Genetics 34:194–209.

Mather, K. 1949. Biometrical genetics. Dover Publications, New York. 158 p.

Mather, K. 1955. The genetic basis of heterosis. Proc. Roy. Soc. B 144:143–150.

McCarthy, J. C. 1965. The effect on litter size of crossing inbred strains of mice. Genetics 51: 217–222.

McLaren, A., and D. Michie. 1954. Factors affecting vertebral variation in mice. I. Variation within an inbred strain. J. Embryol. Exp. Morphol. 2:149–160.

McLaren, A., and D. Michie. 1955. Factors affecting vertebral variation in mice. II. Further evidence on intra-strain variation. J. Embryol. Exp. Morphol. 3:366–375.

Nadel, E. M., J. Greenberg, G. E. Jay, and G. R. Coatney. 1955. Backcross studies on the genetics of resistance to malaria in mice. Genetics 40:620–626.

Reinhard, M. C., E. A. Mirand, H. L. Goltz, and J. G. Hoffman. 1954. Mouse-strain differences in response to radiation. Proc. Soc. Exp. Biol. Med. 85:367–370.

Rendel, J. M. 1953. Heterosis. Amer. Natur. 87: 129–138.

Roberts, R. C. 1960. The effects on litter size of crossing lines of mice inbred without selection. Genet. Res. 1:239–252.

Roberts, R. C. 1961. The lifetime growth and reproduction of selected strains of mice. Heredity 16:369–381.

Roderick, T. H. 1963*a*. Selection for radiation resistance in mice. Genetics 48:205–216.

Roderick, T. H. 1963*b*. The response of twenty-seven inbred strains of mice to daily doses of whole-body X-irradiation. Radiat. Res. 20:631–639.

Roderick, T. H., and J. B. Storer. 1961. Correlation between mean litter size and mean life span among 12 inbred strains of mice. Science 134:48–49.

Searle, A. G. 1954. Genetical studies on the skeleton of the mouse. XI. The influence of diet on variation within pure lines. J. Genet. 52:413–424.

Shull, G. H. 1952. Beginnings of the heterosis concept, p. 14–48. *In* J. W. Gowen [ed.] Heterosis. Iowa State College Press, Ames.

Staats, J. 1964. Standardized nomenclature for inbred strains of mice, Third listing. Cancer Res. 24:147–168.

Stadler, J., and J. W. Gowen. 1957. Contributions to survival made by body cells of genetically differentiated strains of mice following X-irradiations. Biol. Bull. 112:400–421.

Vogel, H. H., Jr., D. L. Jordan, and T. H. Roderick. 1962. Variation of radiosensitivity to daily neutron exposures in mouse strains selected for survival under X-irradiation. Radiat. Res. 16: 577. (Abstr.)

Weir, J. A. 1960*a*. Genetics and laboratory animal diseases. Proc. Anim. Care Panel 10:177–188.

Weir, J. A. 1960*b*. A sex ratio factor in the house mouse that is transmitted by the male. Genetics 45:1539–1552.

Weir, J. A., and R. D. Clark. 1955. Production of high and low blood-pH lines of mice by selection with inbreeding. J. Hered. 46:125–132.

Weir, J. A., and G. Schlager. 1962. Selection for leucocyte count in the house mouse and some physiological effects. Genetics 47:1199–1217.

Wolfe, H. G. 1961. Selection for blood-pH in the house mouse. Genetics 46:55–75.

Wright, S. 1921. Systems of matings. Genetics 6:111–178.

Wright, S. 1922. The effects of inbreeding and crossbreeding on guinea pigs. I. Decline in vigor. II. Differentiation among inbred families. U.S. Dep. Agric. Bull. 1090. Washington. 63 p.

Wright, S. 1952. The genetics of quantitative variability, p. 5–41. *In* E. C. R. Reeve and C. H. Waddington [ed.] Quantitative genetics. H. M. Stat. Office, London.

10

Radiation Genetics[1]

Earl L. Green and Thomas H. Roderick

The laboratory mouse has been and probably will continue to be the mammal of choice for experiments on the genetic effects of radiation. This is so because the existing inbred strains, the stocks bearing named mutations, and the already known genetic linkage groups and chromosome maps provide a desirable degree of experimental sophistication. In addition, the large numbers of mice required for mutation studies may be reared in relatively small space with relatively little expense. More importantly, as mammals, mice may help to bridge the gap between *Drosophila* and other small organisms, on the one hand, and human beings and other large organisms, on the other, in attempts to predict the effects of high-energy radiation on genetic constitutions not amenable to direct experimental analysis.

The genetic effects of radiation are classifiable into three categories: the induction of mutations, the breakage of chromosomes, and the effects of these changes on populations. We use the term "mutation" to mean any heritable change in the genetic constitution detectable by breeding methods. The breakage of chromosomes by radiation and the related effects are discussed in Chap. 7. The other two aspects are discussed in this chapter.

MUTATION RATES

Little and Bagg (1923) pioneered in attempting to induce heritable changes in the germ plasm of the mouse. Although it is believed that the variations they observed in the descendants of X-rayed mice may not have been induced by X-rays, still this study called attention to the need for such investigations and to the use of laboratory animals for the purpose. Within a few years their work was followed by studies by Hertwig and Snell and later by larger studies by Charles, Russell, Carter, Searle, Lyon, Lüning, and others. Falconer, Wright, Haldane, and Woolf, among others, have contributed important concepts of theory and method.

The most extensive investigations of radiation-induced mutations and mutation rates in mice have been carried out at the Oak Ridge National Laboratory, Oak Ridge, Tennessee, and at the MRC Radiobiological Research Unit at Edinburgh, Scotland, and Harwell, England. W. L. Russell at Oak Ridge and T. C. Carter at Edinburgh each started large-scale studies of mice exposed to radiation in 1947 with related but not identical major objectives and with practically the same genetic techniques. A few years later, the mouse stocks at Edinburgh were transferred to Harwell and in 1953 were supplemented by sublines from Oak Ridge to make the studies even more comparable. This section will be based mostly on these two studies.

Russell (1954) has thoroughly reviewed the studies up to 1952. Since then several reviews of the rapidly expanding literature on radiation mutagenesis of mice have appeared, including a later paper by Russell (1963), opening the way

[1] The writing of this chapter was supported in part by Contract AT(30-1)-1979 with the U.S. Atomic Energy Commission and in part by Public Health Service Research Grant GM 07249 from the National Institute of General Medical Sciences.

to the preceding papers. We will refer to a few specific experiments in this section. References to studies not in the list of Literature Cited may be found in either the 1954 or 1963 papers of Russell.

Types of mutation

Mutations may be classified with respect to type of cell (somatic vs. germinal), chromosomal location (autosomal vs. sex-linked), nature of effect (dominant vs. recessive, visible vs. lethal), and size of effect (major vs. minor, oligogenic vs. polygenic). The studies reviewed in this section deal predominantly with recessive autosomal visible mutations in spermatogonia because these are easier to detect than others in large routine breeding experiments.

Cell stages

Male mice, exposed to 300 R or more of high-energy radiation, are usually fertile for about 3 weeks and then enter a sterile period which lasts 4 weeks or more depending on the dose. Doses less than 300 R may depress fertility without a clear-cut sterile period. After the sterile period the males remain fertile throughout the rest of their lives.

Matings in the presterile period make use of germ cells which were in postmeiotic stages (sperm, spermatids, spermatocytes) at the time of irradiation. Matings after the sterile period make use of germ cells descended from irradiated spermatogonia. As found by Hertwig, the types of genetic changes induced in the gonial stages are characteristically different from those in the post-gonial stages. Irradiated sperm tend to transmit dominant mutations, whereas irradiated gonia tend to transmit recessive mutations. By selecting the interval of time between irradiation and copulation one may thus determine whether irradiated postmeiotic or premeiotic cells produce the sperm.

Young adult female mice exposed to 50 R or more of high-energy radiation may produce one, sometimes two, litters and thereafter may be permanently sterile. Their ovaries contain a quite uniform population of oocytes, nearly all in a resting stage prior to the first meiotic metaphase and a few about to be extruded as fertilizable ova without yet having completed the second meiotic division. As a consequence, it is not possible to effect a clear-cut separation of germ cell stages in irradiated adult females. Females must be irradiated as fetuses to expose oogonia.

The type of cell—somatic or germinal—in which a mutation occurs affects its detectability and, more importantly, one's judgment about the seriousness of the consequences. In this section we are concerned chiefly with the problem of detecting and measuring the frequency of mutations in germ cells exposed to high-energy radiation.

Breeding tests

The objective of breeding tests of irradiated and control mice is to analyze the genetic constitution of single germ cells by examining the zygotes they produce or of somatic cells by examining the cell clusters they produce. This may require one, two, or three generations of breeding depending on the nature of the genetic change under investigation.

There are four breeding systems each with a large number of variants useful in radiation genetic experiments with mice. We propose that the systems be called: parental outcross or P-cross, filial outcross or F_1-cross, outcross-backcross or cross-backcross, and outcross-intercross or cross-intercross. These are operational terms referring to what one does with the mice and not to what type of genetic change he expects to find. Each of the systems may be used to detect several types of mutations.

P-cross method. Each parental mouse, irradiated or control, to be tested for mutations is crossed to an unrelated mouse and the progeny are scrutinized for phenotypic deviations or presumptive mutations (Fig. 10-1A). Each viable deviant should be tested by further matings to establish whether a mutation has occurred. The P-cross method may be used to search for all types of dominant mutations—autosomal or sex-linked, visible or lethal—and, with appropriate choices of the mates in the cross, for specified types of recessive mutations as well. The rationale may best be explained by considering a few types of mutations. We shall denote alleles at a given locus as dominant D, recessive r, or wild type $+$, and newly induced mutations as D^* or r^*.

If the mating type is $+/+ \times +/+$ (autosomal) or $+/+ \times +/Y$ (sex-linked), the F_1 progeny will be $+/+$ or $+/+$ and $+/Y$, unless there is a new mutation. Either sex may be irradiated. Dominant visible mutations are indicated by the occurrence of some progeny with the D-phenotype. Since each F_1 mouse tests one treated or control gamete in combination with an untreated gamete, the relative frequency of the D-phenotype in the treated group, corrected for the relative frequency of the D-phenotype in the control group, is an

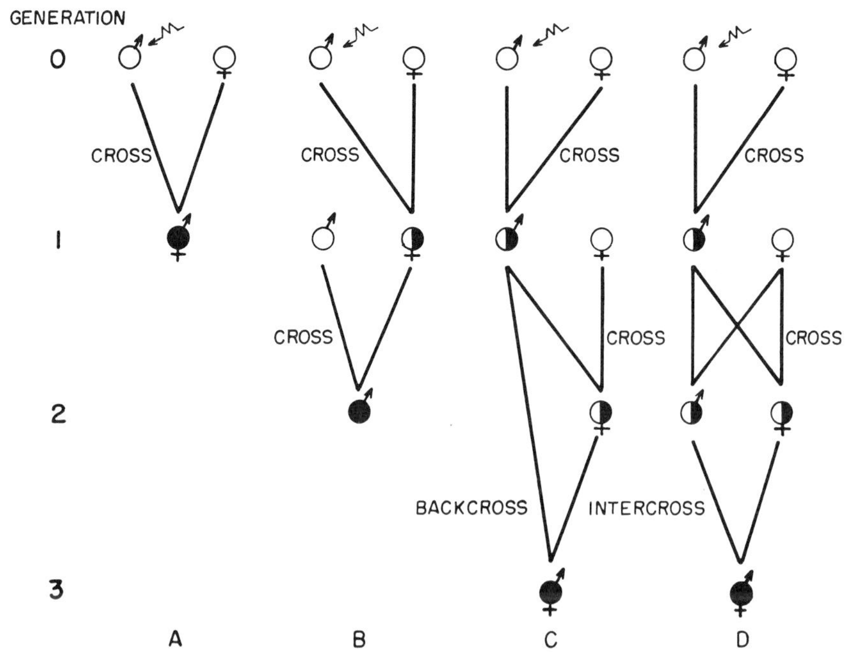

Fig. 10-1. Breeding techniques for detecting dominant and recessive autosomal and sex-linked mutations. *A*. P-cross. *B*. F_1-cross. *C*. Cross-backcross. *D*. Cross-intercross. Explanation in text. ○ = nonirradiated or, with arrow, irradiated mice, ◑ = carriers of recessive mutations, ● = carriers of expressed mutations.

estimate of the radiation-induced mutation rate at this one locus. If n_1 and n_2 are the numbers of F_1 mice tested in the control and irradiated groups, respectively, and if v_1 and v_2 are the numbers of mice with the *D*-phenotype in the respective groups, then $p_2 = v_2/n_2$ is the relative frequency or fraction of mice with spontaneous (D) and radiation-induced (D^*) mutations in the irradiated group and $p_1 = v_1/n_1$ is the fraction with spontaneous mutations (D) in the control group. The ratio $(p_2 - p_1)/(1 - p_1)$ is an estimate of the radiation-induced rate of mutation at the D locus. If p_1 is very small, the difference $(p_2 - p_1)$ is a sufficiently accurate estimate of the induced mutation rate. If $q_1 = 1 - p_1$ and $q_2 = 1 - p_2$, the induced rate is also estimated by $1 - (q_2/q_1)$.

It may not often be practical to confine one's attention to a single locus. Therefore the progeny are examined for any departure from the expected type and, either supposing the departures to be due to dominant mutations or testing them by breeding to establish them as such, the resulting frequency relative to the total examined gives an estimate of the rate of mutation to dominant visible alleles in the genome as a whole. Computing the ratio given above gives the radiation-induced mutation rate per gamete.

The irradiated and control mice should be mated with unrelated or distantly related mice to minimize the possibility of mistaking a recessive allele, present in the stock in a low frequency, for an induced dominant mutation. That is, if the mating is in fact $r/+ \times r/+$, the r/r progeny may be erroneously scored as $D^*/+$.

If females are irradiated, recessive sex-linked mutations may appear in their male progeny. The mating $+/+ \times +/Y$ may produce r^*/Y sons.

One variant of the P-cross, the "specific locus" method, has been extensively used for detecting recessive mutations. If the irradiated and control mice are dominant homozygotes for one or more loci, say $+/+$, $+/+$, . . . , and if they are mated with recessive homozygotes a/a, b/b, . . . , mutations from $+$ to a, $+$ to b, . . . , will be detected by the appearance of the recessive phenotypes among the progeny. The number of proved mutations recovered may be used directly to estimate the mutation rate at each locus tested, the treated rate being adjusted for the control rate to give an induced rate. Obviously the rates from several loci may be averaged.

The P-cross method may be used to detect dominant autosomal lethals by a significant reduction in litter size, particularly if fathers are irradiated. If mothers are irradiated, litter size cannot be used to measure the incidence of dominant lethals, as Russell and Russell (1956) showed. This must be done by using the ratio of the number of living embryos $(n - v)$ to the number of corpora lutea (n). The induced rate is then estimated as $1 - (q_2/q_1)$.

If males are irradiated, dominant sex-linked lethals may be detected by a deficiency of daughters. If females are irradiated, recessive sex-linked lethals may be detected by a deficiency of sons. Both litter size and sex ratio are so highly variable that changes in their magnitude are difficult to detect without enormous samples, except in cases of dominant semisterility (due to reciprocal translocations) which can be detected by the half-size of a few litters.

The P-cross method or one of its variants has been widely used. In addition to its being used in the early studies by Hertwig and by Snell on the effects of X-rays and by Snell and Aebersold on the effects of neutrons, it has been used by most later investigators as well. Russell and Carter each devised a multiple recessive tester stock (a/a b/b c^{ch} p/c^{ch} p d se/d se s/s) to mate with irradiated mice normally carrying wild-type alleles at all seven loci. By far the largest number of mice in radiation genetic experiments have been bred under this system. Lyon (1960) used a balanced lethal line of genotype $T\ tf/t^6 +$ to detect mutations from T to t^m which restore the tail length to normal. Russell and Major (1957) adapted the specific locus technique to the detection of somatic mutations.

F_1-cross method. This method requires two generations and may be used to detect sex-linked lethal mutations (Fig. 10-1*B*). It makes use of dominant sex-linked markers such as *Bn*, Mo^{br}, and *Ta*. Tabby males, irradiated or control, are mated to females heterozygous for bent-tail and brindle in coupling:

$$\frac{+\ Ta\ +}{\text{Y}}\ ♂ \times \frac{Bn + Mo^{br}}{+\ +\ +}\ ♀$$

F_1 daughters, heterozygous for *Ta*, *Bn*, and Mo^{br} and possibly carrying an induced sex-linked mutation in the chromosome marked by *Ta*, are backcrossed to wild-type males:

$$\frac{+\ +\ +}{\text{Y}}\ ♂ \times \frac{Bn + Mo^{br}}{+\ Ta\ +}\ ♀$$

The absence of tabby males $(+\ Ta + /\text{Y})$ among the backcross progeny of any one family of adequate size indicates that a lethal mutation is located on the *Ta* chromosome between the *Bn* and *Mo* loci. Auerbach et al. (1962), in proposing and first using this method with mice, pointed out the difficulties of using these specific markers. As more sex-linked alleles are discovered in mice, new improved stocks may be synthesized for detecting sex-linked lethal mutations.

Cross-backcross method. This method, requiring three generations, is useful for detecting recessive visible and lethal mutations. Each exposed or control mouse in the zeroth generation (G0) is mated to a stock mouse with wild-type or specific mutant alleles. If a recessive mutation has occurred in some of the gametes, the progeny, each representing one gamete from an irradiated or control parent, will be heterozygous for an induced or natural mutation ($r^*/+$ or $r/+$). These mutations are to be uncovered by two more generations of breeding: an outcross and a backcross. Accordingly, the G1 mice are outcrossed to unrelated wild-type mice to produce prospective heterozygous carriers in G2. The G2 mice are, in turn, backcrossed to their parents to produce G3. Homozygotes (r^*/r^* or r/r) in G3 can be recovered if + mutated to r^* or r, if the new phenotype is viable, and if a sufficient number of progeny have been reared to reduce the chance of nonrecovery to a low level. If r^*/r^* or r/r is lethal, its existence must be inferred from a reduction in litter size in G3 (Fig. 10-1*C*).

Using this method, Carter and Phillips (1952) found two new recessive visible mutations in the progeny of nonexposed mice, and Carter (1957*a*) found evidence for the induction of recessive lethal mutations, manifested by reductions in litter sizes.

The strength of this method may be markedly increased by the use of single markers of selected linkage groups as Snell (1935) proposed and, still further, by the use of two or more linked markers as Carter (1957*b*) proposed. Each marker having normal viability in homozygotes should appear in one-half of the progeny in G3. The absence of a marker class is evidence for a nearby induced recessive lethal mutation. The statistical problems of estimating mutation rates from the collected data may be quite formidable. They arise from the uncertainty of detecting all recessive mutations, present or induced in the G0 parents, through their failure to appear in litters of finite sizes three generations later. Falconer (1949) has dealt with this problem by introducing the idea of reducing the number of incompletely tested gametes to the equivalent smaller number of completely tested gametes.

Cross-intercross method. This method, the first

used for detecting recessive mutations in mice, also requires three generations. It is like the preceding method except that the outcross progeny in G2 are intercrossed instead of backcrossed, but its efficiency is less (Fig. 10-1*D*). For this method also Falconer (1949) solved the problem of scoring the results to estimate the equivalent number of fully tested gametes.

Haldane (1956, 1957) proposed a breeding design using the cross-intercross method. Males or females or both of a stock of mice homozygous for several known recessives are exposed to radiation for one or more generations, ending with a generation (G1) not itself exposed. Recessive lethal mutations may have accumulated in these mice. The objective of further breeding is to detect the presence of such lethals by observing deficiencies in the recessive marker classes of an F_2 (= G3) generation, indicating that a lethal mutation has been induced at or near each known locus whose recessive class is deficient. Accordingly the G1 mice are outcrossed and their progeny are intercrossed to produce a multiply-segregating F_2 generation. The method has been used by Carter (1959) and by Sugahara and his colleagues (Sugahara, 1964 and earlier).

A related method is adapted to estimating recessive mutation rates within inbred lines being propagated for other purposes. Following a cross and one generation of intercrossing, brother-sister matings are continued indefinitely. The articles by Snell (1945) and Woolf (1954) set forth the theory of this method.

Estimates of mutation rates

Charles (1950) was the first to estimate the rate of induction of mutations by radiation, using fairly large samples of mice. His estimate of 7×10^{-5} per roentgen per gamete was for all types of mutations, chiefly dominants, induced in sperm under repeated small doses of penetrating radiation. The very large scale studies at Oak Ridge and Harwell in the ensuing years have established the importance of specifying the sex, cell stage, test system, radiation quality, dose rate, and other variables in estimating mutation rates.

The results of numerous experiments are summarized in Tables 10-1 to 10-9. The data for male cells exposed to single doses of radiation at high and low rates are given first. Data on female cells are in the last two tables. The breeding methods and other distinctive features of each experiment are also given in the tables. This arrangement does not reveal the sequence of completion of the various experiments and so masks the excitement of the successive stages of discovery. This side of the story may be found in Russell's (1963) review paper.

The mutation rates in Tables 10-1 to 10-9 are given in two ways. First, if mutations of a designated type were detectable at all loci, the rate per gamete is given for the radiation dose employed. For example, in row 3 of Table 10-2, five dominant visible mutations in 31,253 gametes gives an overall rate of 16.0×10^{-5} per gamete exposed to 200 R of acute X-rays. Dividing by 200 gives 80.0×10^{-8}/R/gamete. This is called an overall dominant visible rate because all loci were presumably at risk and dominant visible mutations at any loci were detectable. In this example the rate also includes any spontaneous or natural dominant visible mutations and so must be adjusted to give an induced rate. Second, if mutations only at certain specific loci were counted, the rate per locus is given for the radiation dose employed. For example, in row 4 of Table 10-2, nine recessive visible and lethal mutations in 31,253 gametes tested gives a rate of 28.8×10^{-5} for seven loci exposed to 200 R. Dividing by seven gives 4.1×10^{-5}/locus. A further division by 200 gives 20.6×10^{-8}/R/locus. If the spontaneous rate is independently estimated as 6.2×10^{-5} for seven loci, the adjusted or induced rate comes out 16.2×10^{-8}/R/locus. Leaving the rates on a per gamete or a per locus basis, rather than a per roentgen per gamete or per roentgen per locus basis, provides numbers for easy comparison as in row 20 of Table 10-4. In the text a few rates are given on a per roentgen basis.

The average mutation rate of seven specific loci to recessive visible and lethal alleles in postmeiotic male germ cells irradiated with 300 R of X-rays was estimated as 13.6×10^{-5} per locus or 45.3×10^{-8} per roentgen per locus (Table 10-1, row 1). No extensive data are available for sex-linked lethals (Table 10-1, row 2). Irradiated fetal spermatogonia gave a rate of 80.0×10^{-5}/R/gamete for dominant visible mutations at all loci (Table 10-2, row 3) and a rate of 20.5×10^{-8}/R/locus for recessive visibles and lethals at seven specific loci, roughly one-half that of the postmeiotic rate (Table 10-2, row 4). An earlier small-scale experiment had given a much smaller rate (Table 10-2, row 5). A variety of experiments with all spermatogonial stages exposed to low dose rates and relatively low total doses of γ-rays have yielded too few data to provide good estimates of mutation rates (Table 10-3, rows 6 to 10).

In the first large experiment, 48,007 progeny

Table 10-1. MUTATION RATES IN POSTMEIOTIC MALE GERM CELLS EXPOSED TO SINGLE ACUTE DOSES OF X-IRRADIATION

Row	Type of mutation	Total dose	Dose rate	Number of gametes tested	Number of mutants	Mutation rate $\times 10^5$	Source
(C3H × 101)F_1 ♂♂ X-rayed (250 kvp), mated to test stock ♀♀ (P-cross):							
1	Recessive visibles and lethals at 7 loci	300 R	[a]	33,874	[b]	13.6/locus[c]	[d]
Ta ♂♂ X-rayed, mated with *Bn Mo*^{br} ♀♀; *Bn Ta Mo*^{br} F_1 ♂♂ backcrossed to + + + ♀♀ (F_1-cross):							
2	Sex-linked lethals, all loci	500 R		154	0		[e]

[a] Dose rate not given; probably 90 R/min.
[b] Not given for 33,874 gametes. There were 16 in 16,813 gametes (Russell, 1956).
[c] Not given for 33,874 gametes; said to be about twice the rate in spermatogonia (Table 10-4, line 20). Rate quoted is from 16,813 gametes (Russell, 1956).
[d] Russell et al. (1958).
[e] Auerbach et al. (1962).

of males exposed to 600 R of acute X-rays and bred after the sterile period gave an induced mutation rate of 25.0×10^{-8}/R/locus for recessive visibles and lethals at seven specific loci. By 1962, with a much larger sample, the improved estimate was very nearly the same: 22.1×10^{-8}/R/locus (Table 10-4, row 20, 600 R). This is about the same as the rate for fetal spermatogonia and about one-half the rate for postmeiotic stages. The spermatogonial mutation rates at 0, 300, and 600 R were satisfactorily fitted by a straight line, but the point for 1,000 R fell significantly below the line (Table 10-4, row 20). There is no obvious explanation of this result. It can be accounted for by supposing that spermatogonia are differentially susceptible to the action of large acute doses of radiation so that the observed drop in mutation rate at 1,000 R is possibly due to cell selection. Estimates of other rates for acute exposures of spermatogonia are also given in Table 10-4, rows 11 to 18 and 21.

While the results of exposures to acute doses were being accumulated, the results of exposure of males to chronic radiation suggested that radiation at low dose rates yielded fewer mutations than radiation at high dose rates (Table 10-5, rows 22 to 24). Further, mutation rates in oocytes also turned out to depend upon dose rate. Since nearly all oocytes in young adult females, unlike spermatogonia in young adult males, appear to be in a single premeiotic stage, the idea of cell selection as an explanation of the dose-rate effect was less satisfactory than a hypothesis of repair of a premutational state induced by radiation at low dose rates.

To investigate further the low mutation rate at 1,000 R, Russell (1963) exposed samples of mice to total doses of 1,000 R of acute radiation in various fractions (Table 10-6, row 25). The mutation rates obtained with some of the fractions fell near the line extrapolated from the 0, 300, and 600 R points. Although the explanation is not

Table 10-2. MUTATION RATES IN FETAL SPERMATOGONIA EXPOSED TO SINGLE ACUTE DOSES OF X-IRRADIATION

Row	Type of mutation	Total dose	Dose rate	Number of gametes tested	Number of mutants	Mutation rate $\times 10^5$	Source
(C3H × 101)F_1 ♂♂ X-rayed *in utero* (250 kvp), mated as adults with test stock females (P-cross):							
3	Dominant visibles, all loci	200 R	72 R/min	31,253	5	16.0/gamete	[a]
4	Recessive visibles and lethals at 7 loci	200 R	72 R/min	31,253	9	4.1/locus	[a]
5	Recessive visibles and lethals at 7 loci	300 rads	70 rads/min	10,155	1	1.4/locus	[b]

[a] Carter et al. (1960).
[b] Carter (1958).

Table 10-3. MUTATION RATES IN ALL SPERMATOGENETIC STAGES EXPOSED TO VARYING DOSES OF γ-RAYS AT LOW DOSE RATES FROM A 400 MILLICURIE RADIUM SOURCE

Row	Type of mutation	Total dose	Dose rate	Number of gametes tested	Number of mutants	Mutation rate × 10^5	Source
CBA ♂♂ controls, mated with CBA ♀♀, F_1 ♂♂ mated with sisters[a] and progeny intercrossed (Cross-intercross):							
6	Recessive visibles, all loci	0		193.7[b]	1	Included with row 7	[c]
Same as 6 except that F_1 ♂♂ mated with sisters[a] and daughters backcrossed to F_1 fathers (Cross-backcross):							
7	Recessive visibles, all loci	0		145.8[b]	2	883.7/gamete	[c]
Same as 7 except CBA ♂♂ γ-rayed:							
8	Recessive visibles, all loci	30 R	1.2 R/wk	72.4[b]	0		[c]
		150	6.0				
		450	30.0				
CBA ♂♂ γ-rayed, mated with test stock ♀♀ (P-cross):							
9	Recessive visibles and lethals at 7 loci	0		10,795	0		[d]
		40 R		4,986	1		
(C3H × 101)F_1 ♂♂ γ-rayed, mated with test stock ♀♀ (P-cross):							
10	Recessive visibles and lethals at 7 loci	0		7,560	2		[d]
		40 R		5,038	0		

[a] Reference says sisters, but context allows that females were unrelated.
[b] Fully-tested equivalent (Falconer, 1949).
[c] Carter and Phillips (1952).
[d] Carter et al. (1956).

clear, it appeared that some factor other than, or in addition to, cell selection must be operative. Finally, two doses of 500 R of acute radiation given one day apart yielded the highest mutation rate so far found in spermatogonia. This result suggests that the first dose imposes a synchronization and heightened mutability upon the cells and that the second dose thus yields more mutations than otherwise expected. Obviously this phenomenon requires further study by variation of the total doses, the sizes and numbers of the fractional doses, the intervals between the fractions, and the cell stage and type irradiated.

Snell and Aebersold (1937) first attempted to induce mutations in mice with neutrons and got evidence that neutrons are five to six times as effective as X-rays in inducing dominant lethals in spermatozoa. Snell (1939) used the cross-backcross method of searching for recessive visible mutations but found none. Later studies by Batchelor et al. (1964), Searle and Phillips (1964), and Russell and Kelly (1964*a, b*) have confirmed the evidence that neutrons are five to six times more effective than X-rays for inducing mutations in premeiotic as well as postmeiotic stages. Batchelor et al. (1964) found a mutation rate of 25.0×10^{-5}/locus or about 110×10^{-8}/rad/locus for spermatogonia exposed to 215 rads of neutrons with 100 R of γ-contamination at a rate of 0.002 rads/min or less (Table 10-7, row 26). These authors suggested that high linear-energy-transfer radiation (neutrons) may have a dose-rate effect opposite in direction from that of low-linear-energy-transfer radiation (X-rays). Russell and Kelly (1964*b*) did not find such an effect, however.

Investigations of mutation rates in female germ cells have been less extensive. The rates for recessive visibles and lethals at seven specific loci in acutely irradiated fetal oogonia were about 10 to 14×10^{-8}/R/locus, slightly more than one-half the spermatogonial rate (Table 10-8, rows 28 and 29). The mutation rates in oogonia also turned out to be dependent upon the dose rate. When 400 R of X-rays were given at a rate of 0.009 R/min, the specific locus rate was about 1.9×10^{-8}/R/locus, whereas at 0.8 R/min it was 12×10^{-8}/R/locus, or about six times higher (Table 10-9, row 31).

The estimation of somatic mutation rates presents special technical problems, chief of which is that breeding tests of the presumed mutant cells

Table 10-4. MUTATION RATES OF SPERMATOGONIA EXPOSED TO ONE OR TWO ACUTE DOSES OF RADIATION

Row	Type of mutation	Total dose	Dose rate	Number of gametes tested	Number of mutants or dead fetuses	Mutation rate $\times 10^5$	Source
(C3H $\times$ 101)F_1 ♂♂ X-rayed (250 kvp), mated with test stock ♀♀ (P-cross):							
11	Dominant visibles, all loci	0		37,868	0		a
		600 R	90 R/min	48,007	5	10.4/gamete	
(C3H $\times$ 101)F_1 ♂♂ X-rayed (250 kvp), mated with (C3H $\times$ 101)F_1 ♀♀ (P-cross):							
12	Dominant visibles, all loci	0		854	0		b
		600 R	67.6 R/min	838	0		
(C3H $\times$ 101)F_1 ♂♂ X-rayed (250 kvp), mated with CBA ♀♀ (P-cross):							
13	Dominant visibles, all loci	0		4,290	0		d
		1,200 R[c]	217 R/min	3,612	2	55.4/gamete	
Same as 12, pregnant females dissected before birth of litters:							
14	Dominant lethals, all loci	0		216[e]	29	13.4 $\times 10^3$/gamete[f]	b
		600 R	67.6 R/min	220[e]	39	17.7	
Same as 13, pregnant females dissected before birth of litters:							
15	Dominant lethals, all loci	0		1,470[e]	384	26.1 $\times 10^3$/gamete[f]	d
		1,200 R[c]	217 R/min	1,418[e]	488	34.4	
(C3H $\times$ 101)F_1 ♂♂ X-rayed (250 kvp), mated with (C3H $\times$ 101)F_1 ♀♀, sons crossed with 101 ♀♀, their daughters backcrossed to the sons (their fathers) (Cross-backcross):							
16	Recessive visibles	0		34.2[g]	0		b
		600 R	67.6 R/min	34.0[g]	0		
(C3H $\times$ 101)F_1 ♂♂ X-rayed (250 kvp), mated with CBA ♀♀, sons crossed with C57BL ♀♀, their daughters backcrossed to the sons (their fathers) (Cross-backcross):							
17	Recessive visibles	0		142[g]	1	704/gamete	d
		1,200 R[c]	217 R/min	142[g]	4	2,817	
Same as 16 except pregnant daughters dissected before birth of litters:							
18	Recessive lethals	0		1,423[e]	384	27.0 $\times 10^3$/gamete[f]	b
		600 R	67.6 R/min	1,321[e]	326	24.7	
Same as 17, pregnant females dissected before birth of litter:							
19	Recessive lethals	0		7,310[e]	1,266	17.3 $\times 10^3$/gamete[f]	d
		1,200 R[c]	217 R/min	7,335[e]	1,501	20.5	
Same as 11:							
20	Recessive visibles and lethals at 7 loci	0		531,500	28	0.8/locus	h
		300 R	90 R/min	65,548	40	8.7	
		600	90	119,326	111	13.3	
		1,000	90	44,649	29	9.3	
(C3H $\times$ 101)F_1 ♂♂ γ-rayed (Co^{60}), mated with test stock ♀♀ (P-cross):							
21	Recessive visibles and lethals at 7 loci	600 R	24 R/min	28,916	21	10.4/locus	i

[a] Russell (1951).
[b] Carter and Lyon (1961).
[c] Dose split; 600 R at 6 to 7 weeks, 600 R at 14 to 15 weeks of age.
[d] Lyon et al. (1964).
[e] Number of corpora lutea.
[f] Read as per cent per gamete. For example: $26.1 \times 10^3 \times 10^{-5} = 26.1\%$.
[g] Fully tested equivalent (Falconer, 1949).
[h] W. L. Russell (1962).
[i] Russell et al. (1960).

Table 10-5. MUTATION RATES OF SPERMATOGONIA EXPOSED TO SINGLE CHRONIC DOSES OF RADIATION

Row	Type of mutation	Total dose	Dose rate	Number of gametes tested	Number of mutants	Mutation rate $\times 10^5$	Source
(C3H $\times$ 101)F_1 ♂♂ γ-rayed (Cs^{137}), mated with test stock ♀♀ (P-cross):							
22	Recessive visibles and lethals at 7 loci	86 R	0.001 R/min	59,810	6	1.4/locus	[a]
		300	0.001	49,569	15	4.3	
		600	0.001	31,652	13	5.9	
		300	0.009	58,457	10	2.4	
		516	0.009	26,325	5	2.7	
		861	0.009	24,281	12	7.1	
		600	0.8	28,059[c]	10	5.1	
Same as 21:							
23	Recessive visibles and lethals at 7 loci	650 R	0.005 R/min	22,375	5	3.2/locus	[b]
Same as 22 except males were X-rayed (250 kvp):							
24	Recessive visibles and lethals at 7 loci	600 R	9 R/min	40,326[c]	23	8.2/locus	[a]

[a] Russell (1963).
[b] Batchelor et al. (1964).
[c] Incomplete data.

Table 10-6. MUTATION RATES OF SPERMATOGONIA EXPOSED TO 1,000 R OF X-RAYS AT 90 R/MIN IN FRACTIONATED DOSES [a]

Row	Type of mutation	Number of fractions	Dose each fraction	Time between doses	Number of gametes	Number of mutants	Mutation rate $\times 10^5$
(C3H $\times$ 101)F_1 ♂♂ X-rayed (250 kvp), mated with test stock ♀♀ (P-cross):							
25	Recessive visibles and lethals at 7 loci	2	500 R	2 hr	14,879	12	11.5/locus
		5	200	1 wk	10,968	15	19.5
		5	200	1 day	8,588	16	26.6
		2	600 + 400	15 days	4,904[b]	10	29.1
		2	500	1 day	11,164	39	49.9

[a] Russell (1963).
[b] C3H ♂♂ rather than F_1 ♂♂ X-rayed.

Table 10-7. MUTATION RATES OF SPERMATOGONIA EXPOSED TO CHRONIC NEUTRON RADIATION

Row	Type of mutation	Total dose	Dose rate	Number of gametes tested	Number of mutants	Mutation rate $\times 10^5$	Source
(C3H $\times$ 101)F_1 ♂♂ irradiated with fast neutrons (mean energy 0.7 Mev) with γ-ray contamination, mated with test stock ♀♀ (P-cross):							
26	Recessive visibles and lethals at 7 loci	215 rads neutrons, 100 R γ-rays	0.002 rads/min maximum	17,169	30	25.0/locus	[a]

[a] Batchelor et al. (1964).

Table 10-8. MUTATION RATES IN FETAL OOGONIA EXPOSED TO SINGLE ACUTE OR CHRONIC DOSES OF RADIATION

Row	Type of mutation	Total dose	Dose rate	Number of gametes tested	Number of mutants	Mutation rate × 10^5	Source
(C3H × 101)F_1 ♀♀ X-rayed in utero (250 kvp), mated as adults with test stock ♂♂ (P-cross):							
27	Dominant visibles	200 R	72 R/min	30,289	1	3.3/gamete	[a]
28	Recessive visibles and lethals at 7 loci	200	72	30,289	3	1.4/locus	
Same as 27 except that F_1 ♀♀ were γ-rayed:							
29	Recessive visibles and lethals at 7 loci	300 rads	50 rads/16 hrs	18,753	4	3.0/locus	[b]

[a] Carter et al. (1960).
[b] Carter (1960).

are ruled out. Russell and Major (1957) used the specific locus method (P-cross) to estimate the average mutation rate of four loci. C57BL females, mated with NB males, were irradiated on the 10¼ day of pregnancy. The zygotes should be a/a $b/+$ c^{ch} $p/++$ d $se/++$ and the offspring should be self-colored black. Spots of color other than black anywhere in the coat may be taken to indicate a mutation at the *b*, *c*, *d*, or *p* locus. This method gave a somatic rate of 70×10^{-8}/R/locus compared with a germinal rate of 24×10^{-8}/R/locus for the same four loci exposed to the same amount of acute X-irradiation.

The preceding discussion has referred only to estimates of rates of mutation to recessive visible and lethal mutations in spermatogonia, postspermatogonial cells, and oogonia, obtained in experiments using the specific locus method. The chief value of the specific locus method lies in its usefulness in comparing mutation rates in different cell stages under exposures to various types of radiation given at various rates and at various total doses. The method has relatively little value in predicting overall mutation rates, nor is any other method fully satisfactory for doing so. Estimates of mutation rates to dominant visibles and dominant lethals, as steps toward estimating overall mutation rates, are given for a variety of types of cells and of radiation in Table 10-2 (row 3), Table 10-4 (rows 11 to 13), and Table 10-8 (row 23). The estimates work out to 17.3×10^{-8}/R/gamete for dominant visible mutations

Table 10-9. MUTATION RATES IN OOCYTES EXPOSED TO SINGLE ACUTE AND CHRONIC DOSES OF RADIATION

Row	Type of mutation	Total dose	Dose rate	Number of gametes tested	Number of mutants	Mutation rate × 10^5	Source
(C3H × 101)F_1 ♀♀ X-rayed (250 kvp), mated with test stock ♂♂ (P-cross):							
30	Recessive visibles and	0		47,612	0		[a]
	lethals at 7 loci	400 R	96 R/min	1,729	1	17.8/locus	
		400	92	11,124	15		
Same as 30 except that F_1 ♀♀ were γ-rayed (Cs^{137}):							
31	Recessive visibles and	258 R	0.009 R/min	27,174	1	0.5/locus	[b]
	lethals at 7 loci	400	0.009	37,049	2	0.8	
		400	0.8	20,827	7	4.8	
Same as 30 except that F_1 ♀♀ were γ-rayed (Co^{60}) and mated with test stock ♂♂:							
32	Recessive visibles and lethals at 7 loci	600 rads	0.05 rads/min	10,117	1	1.4/locus	[c]

[a] Russell et al. (1960).
[b] Russell (1963).
[c] Carter (1958).

in spermatogonia exposed to acute X-rays and 16.7×10^{-8}/R/gamete in oogonia. In contrast Sugahara (1964) found a recessive lethal mutation rate of about 4.5×10^{-3}/R/autosomal set, and Searle (1964*a*) found a rate of 1.8×10^{-4}/R/gamete, both using chronic γ-irradiation. It is apparent that many more studies need to be done before we shall have a clear understanding of the mutability of mouse genes exposed to radiation.

Mutations at specific loci

There are several notable features of mutations induced at specific loci which are briefly mentioned here. Further information will be found in papers by Carter et al. (1958), Russell and Russell (1959*b*), Phillips (1961), L. B. Russell (1962), and W. L. Russell (1965).

First, in spermatogonia the induced mutation rates show a 27-fold variation among the seven loci studied by the specific locus method. In postspermatogonial stages, however, the rates are much more uniform (Table 10-10). Second, mutations recovered from irradiated spermatogonia are mostly point mutations, whereas those from later stages contain a high proportion of chromosomal deficiencies. Specifically, only one deficiency at the closely linked *d* and *se* loci was found in irradiated spermatogonia, whereas five were found in later spermatogenic stages and six in oocytes. Third, about 75 per cent of the induced mutations are lethal when homozygous, lethals have occurred at all loci, and the time of death is frequently after birth. Fourth, in heterozygotes the induced mutations, although recessive by the usual criteria of their effects on pigmentation, frequently affect body size and viability.

One question of considerable interest is not yet answerable. How do radiation-induced mutations in general compare, in the spectrum of phenotypic effects, with natural mutations? An answer to this question will require the same careful search for and analysis of spontaneous mutations that has been applied to radiation-induced mutations (Green et al., 1965). The fact that lethal mutations at the seven loci used in the specific locus studies were not known at the time the radiation studies were started may mean that natural mutations are less likely to be lethal when homozygous or it may mean that the existing mutations are the viable survivors out of a very much larger group containing many lethals. Until better information is available, we must withhold judgment about the comparability or noncomparability of induced and natural mutations in mice.

EFFECTS IN POPULATIONS

The occurrence of radiation-induced mutations and chromosome aberrations implies that populations exposed to radiation will suffer deleterious effects. The magnitude of the effect following specified radiation exposures can only be predicted after we have more knowledge about the viability, fertility, and fecundity of nonirradiated descendants of irradiated ancestors. Several studies in which populations of mice or selected members thereof have been irradiated for one or more generations will be summarized in this section. For an introduction to the literature on other laboratory populations of mammals see Roderick (1964).

To understand the effects of ancestral irradiation on the genetic constitution of populations, it is necessary to control or deliberately vary the important variables: dose level and dose rate of irradiation, stage of germ cells irradiated, the sex irradiated, size of the breeding population, and genetic background of the population. The mouse is the only mammal for which a wide variety of genetically independent inbred strains and mutant-bearing stocks are available. The foundation populations of long-term experiments can be of wide variety, either hybrid or inbred, and each variety can be almost perfectly replicated genetically at another time or place.

The pioneer work of Snell and Hertwig (see review by Russell, 1954) showed that the reduced number of animals born from irradiated post-

Table 10-10. PERCENTAGE DISTRIBUTION OF MUTATIONS AT SEVEN SPECIFIC LOCI INDUCED IN SPERMATOGONIAL AND POSTSPERMATOGONIAL CELLS BY γ- AND X-IRRADIATION. (*From Russell, 1965, with permission of the author and the Pergamon Press.*)

Cell stage	*a*	*b*	*c*	*p*	*d*	*se*	*d se*	*s*	Total
Spermatogonia	1.9	16.7	10.0	14.1	15.6	1.1	0.4	40.2	269
Postspermatogonia	5.3	21.4	5.3	13.3	8.0	4.0	9.3	33.4	75

meiotic germ cells of the sire was primarily due to dominant mutations of the chromosomal type. From these and subsequent studies it is apparent that dominant lethals and serious postnatal deleterious effects are easily detected among the zygotes formed from irradiated sperm and oocytes, but are comparatively very uncommon in the zygotes formed from irradiated spermatogonia. In female mammals premeiotic germ cells are not present after birth. In population studies, therefore, it is important to know the extent of the male's contribution to successive generations by irradiated sperm and by irradiated spermatogonia.

The following discussion of experiments is arranged according to the germinal stage irradiated.

Irradiation of postgonial stages

In an early study Kalmus et al. (1952) paired mice and allowed each pair to produce two litters. Then they X-rayed the gonads of males with 150 to 300 R and reconstituted the pairs. They found a reduction in litter size in the subsequent third litter compared to the second preirradiation litter. Also the sex ratio changed in the direction of expectation (fewer females among the offspring of irradiated males) in the third litter. The results are not clearly the consequence of irradiation because parity has effects on litter size and sex ratio.

Spalding et al. (1964 and earlier) have reported an irradiation experiment of many generations. All the animals originated from a single pair of inbred RF mice. Two sublines were begun from the pair, one of which received no irradiation and functioned as a control line. The males of the other line were given 200 rads of whole-body X-rays at 24 to 28 days of age and mated to unexposed sisters in each generation. Throughout the experiment mice were conceived as early as 25 days and as late as 120 days after irradiation. The earliest animals conceived were probably used for the parents of following generations, since the sublines average about four generations each year. Therefore, the postspermatogonial germinal stages at the time of irradiation probably constitute most of the sires' genetic contribution to subsequent generations. Differences between the lines were observed in the sixth generation. The irradiated line was found to be significantly lower than the control in weaning weight, litter size at birth and weaning, and length of survival under protracted γ-ray and fractionated X-ray exposures. The irradiated line also exhibited a greater number of litters eaten by their mothers, but no consistent differences were observed in sex ratio, weaned-to-born ratio, or lifespan.

In their 1964 study they also showed values of some of these characteristics for a third line, an irradiated subline derived from the irradiated line but propagated in parallel for a few generations without irradiation. In those traits showing significant differences between irradiated and control lines, the values for the irradiated subline were not intermediate; sometimes the values fell beyond those of the control and sometimes beyond those of the irradiated line. One would expect the irradiated subline to be intermediate or at least consistently nearer one line or the other if the differences between the irradiated and control lines were due solely to the amount of ancestral irradiation.

Russell (1957) investigated the lifespan of offspring of $(101 \times C3H)F_1$ males exposed to neutrons from a nuclear detonation. The range of exposure was 0 to 186 rep, and the offspring were conceived in the presterile period. A general life shortening of 0.61 days for each rep was calculated from the data over the entire range of doses. There was no evidence of nonlinearity of this mortality over the dose range.

In a similar study Spalding (1964) investigated the lifespan of male RF mice exposed to both fission neutrons and γ-rays and the lifespan of their first and second generation offspring. He exposed five groups of males to various doses of whole-body fission neutrons ranging from 32 to 117 rads and another group to whole-body Co^{60} γ-rays over a range of 60 to 300 rads. A third major group, a control population of males, received no irradiation. To obtain offspring from both pre- and postspermatogonial stages during the time of irradiation, he mated the males immediately after irradiation and also 16 weeks after irradiation to two different samples of RF females. He was unable to demonstrate any shortening of life or any effect on the sex ratio.

Proshina (1961) studied the fitness of nonpedigreed white mice over five generations after a series of graded doses, 50 to 300 R of X-rays, were given to 400 males of the initial generation. We do not know the time between irradiation and mating of the animals nor the germinal stage irradiated. She found more deaths in the first month of life in the descendants of the irradiated males than in the control group. This effect was apparent in all five generations. The weights of 1-month-old mice were also less than in the controls in all generations, and developmental anomalies were noted in the descendants of the irradiated males in the first two generations.

From the summary of the experiment of Lindop and Rotblat (1963), it is not clear what stages of spermatogenesis at the time of irradiation contributed to subsequent generations. Offspring from male mice irradiated with a whole-body dose of 350 R of X-rays at 4 weeks of age were compared with offspring of nonirradiated males. They found no effect of three generations of such parental exposure on lifespan or radiation sensitivity.

Continuous irradiation exposure

In some studies, populations of mice were either left continuously in a radiation field or males were periodically removed for irradiation and then replaced with their nonirradiated mates. Fertilizing gametes, therefore, could come from all possible stages of gametogenesis at the time of irradiation.

The early studies of Charles et al. (reported in 1960 and 1961) were on males of strain DBA exposed to radiation of the gonads and mated to nonirradiated females of strain C57BL. The gonadal doses were 0.1, 0.5, 1, and 10 R/day of X-rays given in a period of 1 to 1½ hours each day except Sunday. The males remained with the females throughout the period except for the short time each day necessary for the irradiation. Therefore, part and probably most of the fertilizing sperm were irradiated in postspermatogonial stages. Charles et al. found increased juvenile mortality in the offspring of the males receiving 10 R/day or those receiving a total dose of 100 R or more. The effect was significantly expressed in larger litters but only suggestive in smaller litters. These investigators thoroughly autopsied all of the male offspring of the irradiated and control sires. Examining mice in such detail in a radiation-genetics experiment has never been attempted elsewhere. They observed a small but significant increase in rare external and internal morphological abnormalities among the offspring of the irradiated males. The female offspring were mated to an albino tester stock for the analysis of partial sterility. There were a greater number of partially sterile females whose fathers had been irradiated. About one-fifth of these females were tested for another generation and proved to be semisterile; that is, their partial sterility could be explained as a consequence of a reciprocal translocation induced by irradiation (Chap. 7).

Charles et al. were unable to find any differences between the groups in sex ratio at weaning. In the male offspring of the irradiated and control sires they found no differences in body weight, tail length, body length, strength, and number of Peyer's patches on the small intestines. In the female offspring they found no difference in age at time of their fourth litters nor in total offspring born in their first four litters.

In a long-term investigation Gowen and Stadler (1964 and earlier) studied the fitness of 10 inbred strains exposed to continuous Co^{60} γ-rays over many generations. They exposed mice to doses ranging from 0.06 to 0.10 R/hour given during 22-hour days. Throughout the study the strains were propagated by brother-sister matings. Mice receiving 3.0 R or more per day were sterile. Five strains did not survive five generations, and the other five survived 12 generations. At least one strain reached the 24th generation. Gowen and Stadler believed that irradiation probably was not a significant factor in extinguishing the five lines because these strains were less vigorous at the outset of the experiment. The surviving strains appeared healthy and did not appear to differ from their respective nonirradiated controls in first-litter productivity or in sex ratio. Lifespan data were collected on all the surviving strains over all the generations and analyzed through the sixth generation. Although there was a significant decline in lifespan with succeeding generations, the authors said the decline could not be explained as a result of the increased number of generations of ancestral irradiation. The average correlation of lifespan and ancestral irradiation within strains and within generations was −0.17. These correlations differed between strains but not between generations.

In another long-term experiment Touchberry and Verley (1964) studied the effects of whole-body X-rays given for successive generations to several hybrid populations derived from $B6D2F_1/J$ and several inbred populations derived from C57BL/6J. Males of the initial populations were divided into five groups and given, 0, 50, 100, 150, and 200 R, respectively. Females were similarly divided but into groups given 0, 10, 20, 30, and 40 R, respectively. In the next generation, matings were made at random between the irradiated groups within the inbred or within the hybrid populations. Mice were between 60 and 85 days old at the time of irradiation and were mated 14 days after irradiation. They were permitted to raise two litters, from which parents of the subsequent generation were chosen randomly.

Data on six generations of the hybrid populations and four generations of the inbred populations were used for the analysis. In both the inbred and hybrid populations they found a decrease in litter size with an increase in parental irradiation. Surprisingly they also found an increase in 32-day body weight and in growth rate from 2 to 38 days with an increase in parental irradiation. Touchberry and Verley believed the decrease in litter size to be evidence of induced dominant mutations and the increase in 32-day body weight and growth rate to be evidence of heterotic effects of the induced mutations. They pointed out that maternal irradiation produced both genetic and somatic effects, whereas paternal irradiation produced genetic effects only. The supposition of a heterotic effect seems less probable than the possibility that 32-day weight and growth rate are negatively correlated with and partially dependent upon litter size.

Sugahara (1964) and his colleagues raised and bred two populations of a tester stock of mice for three generations in a field of Co^{60} γ-rays. In the first experiment the mice were given about 5 to 8 R/22-hour day and in the second experiment 0.43 R/22-hour day. In a third experiment they gave acute doses of 600 R of X-rays to the gonads of males of another population of the same stock. In a fourth experiment they irradiated the adult males of one generation with 597.3 R of Co^{60} γ-rays given at a dose rate of 8.3 R/22-hour day. Nonirradiated but genetically similar mice were raised and maintained in a similar manner to serve as control populations. The fourth generation in the first two experiments and the second generation in the third and fourth experiments, all nonirradiated, were crossed to strain CBA, according to Haldane's (1956, 1957) plan. From the F_1 population, siblings were intercrossed to produce an F_2 generation. In addition to the data on suspected lethals induced in these populations (mentioned in the first part of this chapter), they collected data on mean litter size, incidence of sterile pairs, sex ratio, and juvenile death (death before weaning). Only for juvenile death was there a clear effect of ancestral irradiation in the second, third, and fourth experiments. This effect was found in the F_1 and F_2 generations following the outcross.

To determine whether irradiation would produce subline differentiation Searle (1964*a*) studied the effect of continuous Co^{60} γ-rays on 28 pairs of sublines of the inbred strain C3H/HeH. One subline of each pair was exposed to 1 R/night with a total dose of about 80 R/generation, whereas the other subline was set aside as a nonirradiated control. All sublines were propagated by brother-sister matings. Data on fitness characteristics were analyzed through 24 generations. Searle found a significant reduction in litter size in the irradiated sublines particularly apparent with greater parity, but he attributed the effect to damage of irradiated somatic tissue and not to mutations in irradiated ancestral germ plasm. No differences were found between the irradiated and control lines in complete sterility. Searle studied the genetic divergence of two pairs of sublines by comparing the lines with respect to 23 quasi-continuous characteristics and four metrical characters of the skeletal system. The unexpected result was that the rate of divergence was somewhat greater for the nonirradiated lines than for the irradiated lines.

Irradiation of spermatogonia

Dominant mutations or chromosomal aberrations can be induced by irradiating spermatogonia, and occasionally such changes do reach zygotes and are passed on to subsequent generations (Griffen, 1964; Lyon et al., 1964). However, they are far fewer than can be found among zygotes descended from irradiated postspermatogonial stages. There is apparently no period after irradiation when the germ line can be assumed to be free from any kind of genetic damage. Irradiation will apparently have some effect on the fitness of populations, no matter what stage of gametogenesis is irradiated. The problem still remains, however, to show clearly in what ways *population fitness* is damaged by irradiating spermatogonia. The effect of spermatogonial irradiation on population fitness is important, too, because it is this type of irradiation damage which may be most important to a human population.

Ehling and Randolph (1962) reported an experiment in which strain 101 males were exposed to different sources and conditions of radiation and then crossed to C3H females so that both postspermatogonial and spermatogonial stages could be studied independently with respect to their effects on skeletal abnormalities in the F_1 animals. The progeny were conceived during the presterile period after their fathers were irradiated with 600 rads of X-rays or 80 to 200 rads of neutrons with varying intensities. Males mated after the poststerile period had received 600 rads. The litters from the exposed sires were killed for examination between 26 and 28 days of age. Skeletons of 618 offspring from nonirradiated sires and 515 offspring of irradiated sires were examined. The authors classified the abnormalities as either of two types depending upon their frequency: class I abnormalities were those occurring only once in all the animals studied, and class II abnormalities were those occurring more than once. No differences between irradiated and control groups were found in frequencies of class II abnormalities, nor in class I abnormalities when progeny were derived from irradiated spermatogonial cells. A significant difference, however, was discovered between the irradiated and control groups in class I abnormalities when the sires' contribution was derived from irradiated postgonial cells. The authors concluded that class I abnormalities were probably due to genetic effects and class II to nongenetic effects.

In studies of general fitness Kohn (1960) irradiated CAF_1 males with X-rays over a range of 0 to 720 rads. During the poststerile period surviving males were mated to nonirradiated BALB/cCrgl females and the offspring were studied for the characteristics: males per litter, females per litter, progeny per litter, sex ratio at birth, and sex ratio at weaning. Regression lines were fitted to these characteristics plotted against the sires' irradiation doses, but the slopes were not significant. Thus no effect of irradiation was demonstrated. Samples of the progeny were also studied for lifespan and tolerance to 500 rads of X-rays per week, but no significant effect of paternal irradiation was observed (Kohn et al., 1965).

However, Russell and Russell (1959*a*) reported a reduction of mean litter size at 3 weeks of age among offspring derived from spermatogonia given 300 R of X-rays in a single generation. The lower mean litter size at 3 weeks was consistently found in all 10 groups of animals studied over different periods of time.

In further studies of fertility Carter and Lyon (1961) gave 600 R of whole-body X-irradiation to $(101 \times C3H)F_1$ males. The males were mated 6 weeks after irradiation to females of the same F_1. The females were allowed to bear four litters and were then dissected during their fifth pregnancies for counts of corpora lutea, moles, and live and dead fetuses. The males from the liveborn litters were crossbred to detect a possible reduction of their fertility, and their mates were dissected during their pregnancies for similar counts. The same males were then mated to females of strain 101 to produce offspring to be backcrossed to the males to get an indication of the induction of recessive lethals and visibles. There was no evidence of reduced fertility in the F_1 males. There was also no difference found in litter size at birth when the F_1 males were crossed to strain 101 females. Nor was there any difference in the counts of corpora lutea, number of implantations, moles, or live and dead embryos. But from backcross matings the granddaughters of the irradiated males had a significant reduction of corpora lutea per pregnancy over that in the control granddaughters, *but* no differences were found in number born per litter. The authors concluded that there is a compensating mechanism tending to equalize the litter sizes despite differences in the number of eggs ovulated.

Later experiments at Harwell have considerably clarified the mechanisms involved in reduced fertility following spermatogonial irradiation. Lyon et al. (1964) irradiated the males of a cross of strains C3H/HeH and 101/H with an acute gonadal dose of 1,200 R of X-rays in two fractions 8 weeks apart. Twelve weeks after the latter dose, the F_1 males were outcrossed to CBA/H females. Controls were propagated and manipulated in the same manner but with no irradiation. The sizes of the litters of the irradiated sires were reduced 15 per cent at birth. The decrease in litter size was mostly due to increased embryonic death just after implantation. Part of this lethality was due to semisterility caused by the induction of reciprocal translocations at a rate of 3.3 per cent per gamete. Subsequent crosses also revealed a higher incidence of recessive lethal mutations induced in the irradiated group. No differences were found in sex ratios.

In the same experiment Searle (1964*b*) outcrossed the F_1 daughters and recorded the sizes of their first four litters. Their fifth litters were observed at the 14-day fetal stage. The mean litter size was 5.5 per cent lower in the daughters of the irradiated males. This reduction in fertility was almost wholly attributable to semisterility, present in 6.7 per cent of the daughters of the irradiated males. Searle estimated the rate of induction of translocations in spermatogonia to be 6.4×10^{-5}/R/genome. This rate is about one-sixth that of spermatozoa; he suggested that spermatogonial stages are less sensitive to irradiation.

In a related study, Phillips and Searle (1964) looked for a dose-rate effect by spreading the 1,200 R over 12 weeks at 100 R/week. The lethal effect was reduced to a 2 per cent decrease in litter size, and the rate of induction of translocations to about 1 per cent. Neither of these effects was significant when compared with the control values, but the fact that they were far less than the effects observed after an acute dose of 1,200 R is evidence for a dose-rate effect.

In long-term studies Lüning and Sheridan (1964 and earlier) irradiated the spermatogonia of males of a randomly mated population from strain CBA with 276 R of X-rays each generation and compared their descendants with those of a parallel nonirradiated control population. The radiation was delivered to the gonads of the males with the rest of the body shielded. In studying the reproductive performance of the fourth to sixth generations under irradiation, they found a consistent reduction of litter size in the irradiated group independent of the number of litters the females produced. The females of the irradiated group had a slightly greater frequency of losses of complete litters. Lüning and Sheridan also found a decrease in the number of corpora lutea in the first litters of the fourth generation in the irradiated group. These findings suggested that damage to this population is through a lowered maternal fitness. There was no change in the sex ratio because of irradiation. Further studies of offspring of sib and nonsib matings of these lines indicated that the major dominant effects of irradiation were on the immediate offspring of the irradiated males and not on future generations. In other words, there was no evidence that dominant detrimental effects accumulated with increasing generations of irradiation. However recessive lethals apparently did accumulate with successive generations of irradiation.

In other long-term studies Green and his colleagues (Green et al., 1964; Roderick and Schlager, 1965, personal communication) propagated both genetically heterogeneous and genetically homogeneous populations of mice under different levels of irradiation for several generations. The genetically heterogeneous populations were all descended from a 4-way cross of genetically independent inbred strains. The genetically homogeneous populations were all descended from a single pair of mice from strain C57BL/10Gn. The two major populations were subdivided at the outset of the experiment into several different subpopulations and propagated in parallel with different amounts of inbreeding in order to study the possible effect of inbreeding in combination with the effect of irradiation. All groups were further subdivided so that different levels of irradiation could be delivered to populations of each breeding size. Irradiation doses were 0, 50, and 100 R per generation. In the data collected from the mice in the first and second litters used to propagate the main lines, no differences were found in age of dam at parturition, number born, number weaned, sex ratio, weaned-to-born ratio, or frequency of phenotypic deviations attributable to differences in irradiation dose or breeding size of the population.

Samples of nonirradiated males from the 11th generation were mated to siblings as well as to females from strain C57BL/6J for a study of prenatal mortality. The females were dissected before parturition, and data were taken on number of corpora lutea, moles, dead fetuses, and living fetuses. No differences in prenatal mortality between the populations were found which could be attributed to ancestral radiation or inbreeding level.

Every three generations some nonirradiated animals of each dose and breeding size group were set aside to determine their lifetime reproductive performance. In these studies a significant reduction in number of litters was found in the genetically homogeneous populations after nine generations, and this difference could be attributed to the ancestral radiation exposure. The mean litter sizes, however, did not differ. These findings suggested that the animals with ancestral irradiation perform just as well through most of their lives, but that they deteriorate sooner physiologically. The authors point out the value of studies of lifetime performance, because these differences would not have been observed on data of first and second litters only. The effect was *not* observed in the heterogeneous populations after nine generations, and no effects could be attributed to the inbreeding levels of the subpopulations.

Green and Les (1964) tried to bring about genetic extinction in a small hybrid mouse population by irradiating the spermatogonia (body shielded) with a dose of 900 R of X-rays in each generation. After nine generations they found no differences in fertility, number of litters, number of offspring born and weaned, and average litter size between the irradiated and nonirradiated control populations. It is not clear why this study and that of Kohn (1960) failed to show the reduced litter size reported in other studies with comparable radiation doses.

Conclusions

Table 10-11 gives an abridged summary of the methods and results of the papers discussed in this section of the chapter. Characteristics such as viability and fertility are the principal components of population fitness. It appears that reduction in fitness of mouse populations as measured in these characteristics can be shown to be a result of irradiation of doses as low as 100 to 300 R per generation. Furthermore, the data suggest that decreased fitness, like chromosomal abnormalities, can be more easily brought about and the damage more easily passed on to subsequent generations as a result of postspermatogonial irradiation than of spermatogonial irradiation. Genetically heterogeneous populations of mice may be more resistant than inbred populations to radiation-induced damage to fitness. Selection probably eliminates many mutations in each generation and thus reduces the damage to population fitness. There may be mechanisms of physiological compensation, such as that reported by Carter and Lyon (1961), ameliorating the outward phenotypic manifestations of a fitness decrement. Certain fitness traits seem to express the genetic damage

Table 10-11. EXPERIMENTS DESIGNED TO ASSESS THE EFFECTS OF IRRADIATION ON THE HEREDITARY FITNESS OF MOUSE POPULATIONS

Investigator[a]	*Irradiation*		Exposure time	Germinal stage irradiated	Effects attributable to ancestral irradiation
	Amount	Kind			
Kalmus et al. (1952)	150 to 300 R	X-rays	Brief	Postspermatogonial stages	Reduced litter size
Spalding et al. (1964)	200 rads/gen	X-rays	Brief	All stages of males	Inconclusive
Russell (1957)	0 to 186 rep	Neutrons	Brief	Postspermatogonial stages	Reduced lifespan
Spalding (1964)	32 to 117 rads 60 to 300 rads	Neutrons X-rays	Brief Brief	Spermatogonial and postspermatogonial stages separately	No effect on lifespan
Proshina (1961)	50 to 300 R	X-rays	Brief	Unknown stages of males	Increased preweaning mortality Reduced weight at age 1 month
Lindop and Rotblat (1963)	350 R	X-rays	Brief	Unknown stages of males	No effect on lifespan
Charles et al. (1960, 1961)	0.1 to 10 R/day	X-rays	Continuous	All stages of males	Increased phenotypic abnormalities Increased partially sterile daughters
Gowen and Stadler (1964)	0.06 to 0.10 R/hr	γ-rays	Continuous	All stages of males and females	Slightly reduced lifespan
Touchberry and Verley (1964)	0 to 200 R/gen 0 to 40 R/gen	X-rays X-rays	Brief Brief	All stages of males } Oocytes }	Decreased litter size Increased 32-day weight and growth rate
Sugahara (1964)	5 to 8 R/day 0.43 R/day	γ-rays γ-rays	Continuous } Continuous }	All stages of males and females	Increased juvenile death
	600 R	X-rays	Brief	Probably all male stages	Increased juvenile death
	597.3 R	γ-rays	8.3 R/day continuous	All male stages	Increased juvenile death
Searle (1964*a*)	80 R/gen	γ-rays	Continuous	All stages of males and females	No effect observed
Ehling and Randolph (1962)	600 rad 80 to 200 rad	X-rays Neutrons	Brief } Brief }	Spermatogonial and postspermatogonial stages separately	Increased skeletal abnormalities from irradiated postspermatogonial stages only
Kohn et al. (1965)	0 to 720 rads	X-rays	Brief	Spermatogonial	No effect observed
Russell and Russell (1959*a*)	300 R	X-rays	Brief	Spermatogonial	Reduced litter size

Table 10-11. EXPERIMENTS DESIGNED TO ASSESS THE EFFECTS OF IRRADIATION ON THE HEREDITARY FITNESS OF MOUSE POPULATIONS (*Continued*)

Investigator[a]	*Irradiation*		Exposure time	Germinal stage irradiated	Effects attributable to ancestral irradiation
	Amount	Kind			
Carter and Lyon (1961)	600 R	X-rays	Brief	Spermatogonial	Reduced numbers of corpora lutea but no change in litter size
Lyon et al. (1964)	1,200 R in two fractions	X-rays	Brief	Spermatogonial	Reduced litter size
Searle (1964*b*)	1,200 R in two fractions	X-rays	Brief	Spermatogonial	Reduced litter size attributable to semisterility
Lüning and Sheridan (1964)	276 R/gen	X-rays	Brief	Spermatogonial	Decreased corpora lutea Lowered maternal fitness Accumulation of recessive lethals
Green et al. (1964)	0 to 100 R/gen	X-rays	Brief	Spermatogonial	Reduced reproductive lifetime

[a] Results given are in this and earlier papers by the same author and, in some cases, by his colleagues.

from irradiation better than others in different experimental designs and under different conditions. Sex ratios apparently do not change in mouse populations under the irradiation and conditions studied, even though it is the only characteristic which showed an apparent change after irradiation of a human population (Neel, 1963).

SUMMARY

Several experiments, chiefly at the Oak Ridge National Laboratory and at the MRC Radiobiological Research Unit at Harwell, have yielded estimates of induced mutation rates in various types of germinal cells of mice exposed to X-rays, γ-rays, and neutrons. These experiments have revealed that mutation rates depend upon many factors including: sex and cell stage irradiated, type of mutation scored, genotype or strain, and kind, amount, and dose rate of radiation.

A number of other experiments have been initiated by the desire to predict the effect of irradiating one or more generations on the fitness of a population. Measures of fitness include fertility, numbers and sizes of litters, reproductive life, and lifespan. No clear generalization emerges from the experiments so far reported, except that irradiation of postgonial stages appears to be more detrimental to fitness than irradiation of gonial stages. This does not necessarily mean that gonial stages are less sensitive to radiation damage; possibly the damage is compensated for and overcome by selection in favor of nonmutated cells.

LITERATURE CITED

Auerbach, C., D. S. Falconer, and J. H. Isaacson, 1962. Test for sex-linked lethals in irradiated mice. Genet. Res. 3:444–447.

Batchelor, A. L., R. J. S. Phillips, and A. G. Searle. 1964. High effectiveness of chronic neutron exposures for the induction of specific locus mutations in mice. Nature 201:207–208.

Carter, T. C. 1957*a*. Recessive lethal mutation induced in the mouse by chronic γ-irradiation. Proc. Roy. Soc. B 147:402–411.

Carter, T. C. 1957*b*. The use of linked marker genes for detecting recessive autosomal lethals in the mouse. J. Genet. 55:585–595.

Carter, T. C. 1958. Radiation-induced gene mutation in adult female and foetal male mice. Brit. J. Radiol. 31:407–411.

Carter, T. C. 1959. A pilot experiment with mice, using Haldane's method for detecting induced autosomal recessive lethal genes. J. Genet. 56:353–362.

Carter, T. C. 1960. Mutation induced in germ

cells of the foetal female mouse. Genet. Res. 1:59–61.

Carter, T. C., and M. F. Lyon. 1961. An attempt to estimate the induction by X-rays of recessive lethal and visible mutations in mice. Genet. Res. 2:296–305.

Carter, T. C., M. F. Lyon, and R. J. S. Phillips. 1956. Induction of mutations in mice by chronic gamma irradiation; interim report. Brit. J. Radiol. 29:106–108.

Carter, T. C., M. F. Lyon, and R. J. S. Phillips. 1958. Genetic hazard of ionizing radiation. Nature 182:409.

Carter, T. C., M. F. Lyon, and R. J. S. Phillips. 1960. The genetic sensitivity to X-rays of mouse foetal gonads. Genet. Res. 1:351–355.

Carter, T. C., and R. J. S. Phillips. 1952. An experimental attempt to investigate the induction of visible mutations in mice by chronic gamma irradiation, p. 73–81. *In* A. Haddow [ed.] Biological hazards of atomic energy. Clarendon Press, Oxford.

Charles, D. R. 1950. Radiation-induced mutations in mammals. Radiology 55:579–581.

Charles, D. R., J. A. Tihen, E. M. Otis, and A. B. Grobman. 1960. Genetic effects of chronic X-irradiation exposure in mice. Univ. of Rochester, Atomic Energy Project Rep., UR-565. 354 p.

Charles, D. R., J. A. Tihen, E. M. Otis, and A. B. Grobman. 1961. Genetic effects of chronic X-irradiation exposure in mice. Genetics 46:5–8.

Ehling, U. H., and M. L. Randolph. 1962. Skeletal abnormalities in the F_1 generation of mice exposed to ionizing radiations. Genetics 47: 1543–1555.

Falconer, D. S. 1949. The estimation of mutation rates from incompletely tested gametes, and the detection of mutations in mammals. J. Genet. 49:226–234.

Gowen, J. W., and J. Stadler. 1964. Lifespans of mice as affected by continuing irradiation from Cobalt-60 accumulated ancestrally and under direct irradiation. Genetics 50:1115–1142.

Green, E. L., and E. P. Les. 1964. Persistence of populations of mice exposed to substerilizing doses of spermatogonial X-irradiation. Genetics 50:497–507.

Green, E. L., T. H. Roderick, and G. Schlager. 1964. Embryonic mortality in mouse populations after eleven generations of spermatogonial irradiation. Genetics 50:1053–1063.

Green, E. L., G. Schlager, and M. M. Dickie. 1965. Natural mutation rates in the house mouse: plan of study and preliminary estimates. Mutation Res. 2:457–465.

Griffen, A. B. 1964. The occurrence of chromosomal aberrations in pre-spermatocytic cells of irradiated male mice. II. Cytological studies of sterile and semi-sterile F_1 individuals, p. 175–188. *In* W. D. Carlson and F. X. Gassner [ed.] Proc. Int. Symp. Effects of ionizing radiation on the reproductive system. Pergamon Press, New York.

Haldane, J. B. S. 1956. The detection of autosomal lethals in mice induced by mutagenic agents. J. Genet. 54:327–342.

Haldane, J. B. S. 1957. Methods for the detection and enumeration of mutations produced by irradiation in mice, p. 35–39. Proc. Int. Genet. Symp. 1956. Science Council of Japan, Ueno Park, Tokyo.

Kalmus, H., J. D. Metrakos, and M. Silverberg. 1952. Sex ratio of offspring from irradiated male mice. Science 116:274–275.

Kohn, H. I. 1960. The effect of paternal X-ray exposure on the secondary sex ratio in mice (F_1 generation). Genetics 45:771–778.

Kohn, H. I., M. L. Epling, P. H. Guttman, and D. W. Bailey. 1965. Effect of paternal (spermatogonial) X-ray exposure in the mouse: Lifespan, X-ray tolerance, and tumor incidence of the progeny. Radiat. Res. 25:423–434.

Lindop, P. J., and J. Rotblat. 1963. Study of the offspring of irradiated mice. Summarized by E. C. Amoroso, p. 408–429. *In* M. Ebert and A. Howard [ed.] Radiation effects in physics, chemistry and biology. North-Holland, Amsterdam.

Little, C. C., and H. J. Bagg. 1923. The occurrence of two heritable types of abnormality among the descendants of X-rayed mice. Amer. J. Roentgenol. Radiat. Therap. 10:975–989.

Lüning, K. G., and W. Sheridan. 1964. Dominant effects on productivity in offspring of irradiated mouse populations. Genetics 50:1043–1052.

Lyon, M. F. 1960. Effect of X-rays on the mutation of *t*-alleles in the mouse. Heredity 14:247–252.

Lyon, M. F., R. J. S. Phillips, and A. G. Searle. 1964. The overall rates of dominant and recessive lethal and visible mutation induced by spermatogonial X-irradiation of mice. Genet. Res. 5:448–467.

Neel, J. V. 1963. Changing perspectives on the genetic effects of radiation. Charles C Thomas, Springfield, Ill. 97 p.

Phillips, R. J. S. 1961. A comparison of mutation induced by acute X and chronic gamma irradiation in mice. Brit. J. Radiol. 34:261–264.

Phillips, R. J. S., and A. G. Searle. 1964. The effect of dose-rate on the yield of translocations and dominant lethals following spermatogonial irradiation of mice. Genet. Res. 5:468–472.

Proshina, A. D. 1961. Comparative characteristics of the fertility of five generations of mice from irradiated males. [Transl. from Russian] Med. Radiol. 6:41–48.

Roderick, T. H. [ed.] 1964. Proceedings of the symposium: The effects of radiation on the hereditary fitness of mammalian populations. Genetics 50:1019–1217.

Russell, L. B. 1962. Chromosome aberrations in experimental mammals. Prog. in Med. Genetics 2:230–294.

Russell, L. B., and M. H. Major. 1957. Radiation-induced presumed somatic mutations in the house mouse. Genetics 42:161–175.

Russell, L. B., and W. L. Russell. 1956. The sensitivity of different stages in oogenesis to the radiation induction of dominant lethals and other changes in the mouse, p. 187–192. *In* J. S. Mitchell, B. E. Holmes, and C. L. Smith [ed.] Progress in radiobiology. Proc. 4th Int. Conf. Radiobiol., Cambridge, August, 1955. Oliver and Boyd, London.

Russell, W. L. 1951. X-ray-induced mutations in mice. Cold Spring Harbor Symp. Quant. Biol. 16:327–336.

Russell, W. L. 1954. Genetic effects of radiation in mammals, p. 825–859. *In* A. Hollaender [ed.] Radiation biology. Vol. 1, Part 2. McGraw-Hill, New York.

Russell, W. L. 1956. Radiation-induced mutations in the mouse, p. 25. US AEC Report ORNL-2155, Oak Ridge National Laboratory.

Russell, W. L. 1957. Shortening of life in the offspring of male mice exposed to neutron radiation from an atomic bomb. Proc. Nat. Acad. Sci. 43:324–329.

Russell, W. L. 1962. An augmenting effect of dose fractionation on radiation-induced mutation rate in mice. Proc. Nat. Acad. Sci. 48:1724–1727.

Russell, W. L. 1963. The effect of radiation dose rate and fractionation on mutation in mice, p. 205–217. *In* F. Sobels [ed.] Repair from genetic radiation damage. Pergamon Press, New York.

Russell, W. L. 1965. Evidence from mice concerning the nature of the mutation process, p. 257–264. *In* S. J. Geerts [ed.] Proc. XI Int. Congr. Genet. (The Hague, 1963) Vol. 2. Pergamon Press, New York.

Russell, W. L., J. W. Bangham, and J. S. Gower. 1958. Comparison between mutations induced in spermatogonial and postspermatogonial stages in the mouse. Proc. X Int. Congr. Genet. 2:245–246. (Abstr.)

Russell, W. L., and E. M. Kelly. 1964*a*. Relative biological effectiveness of neutron to X-ray induction of gene mutations in mouse postspermatogonial stages. Semiannual Prog. Rep., Biol. Div., Oak Ridge Nat. Lab. 3601:83–85.

Russell, W. L., and E. M. Kelly. 1964*b*. Neutron-induced mutation in mouse spermatogonia. Lack of effect of dose rate. Relative biological effectiveness of neutrons. Semiannual Prog. Rep., Biol. Div., Oak Ridge Nat. Lab. 3700:83–85.

Russell, W. L., and L. B. Russell. 1959*a*. Radiation-induced genetic damage in mice, p. 179–188. *In* J. G. Bugher, J. Coursaget, and J. F. Loutit [ed.] Prog. in Nuclear Energy, Ser. 6, Biol. Sci. Pergamon Press, New York.

Russell, W. L., and L. B. Russell. 1959*b*. The genetic and phenotypic characteristics of radiation-induced mutations in mice. Radiat. Res., Suppl. 1:296–305.

Russell, W. L., L. B. Russell, and E. M. Kelly. 1960. Dependence of mutation rate on radiation intensity, p. 311–320. *In* Immediate and low level effects of ionizing radiations: Conference held in Venice, June, 1959. Int. J. Radiat. Biol. Spec. Suppl.

Searle, A. G. 1964*a*. Effects of low-level irradiation on fitness and skeletal variation in an inbred mouse strain. Genetics 50:1159–1178.

Searle, A. G. 1964*b*. Genetic effects of spermatogonial X-irradiation on productivity of F_1 female mice. Mutation Res. 1:99–108.

Searle, A. G., and R. J. S. Phillips. 1964. Genetic effects of neutron irradiation in mice, p. 361–370. *In* Biological effects of neutron and proton irradiation. Vol. 1. Int. Atomic Energy Agency, Vienna.

Snell, G. D. 1935. The induction by X-rays of hereditary changes in mice. Genetics 20:545–567.

Snell, G. D. 1939. The induction by irradiation with neutrons of hereditary changes in mice. Proc. Nat. Acad. Sci. 25:11–14.

Snell, G. D. 1945. The detection of mutations. Relative efficiency of various systems of brother-sister inbreeding of mice. J. Hered. 36:275–278.

Snell, G. D., and P. C. Aebersold. 1937. The production of sterility in male mice by irradiation with neutrons. Proc. Nat. Acad. Sci. 23:374–378.

Spalding, J. F. 1964. Longevity of first and second generation offspring from male mice exposed to fission neutrons and gamma rays, p. 147–152. *In* W. D. Carlson and F. X. Gassner [ed.] Proc. Int. Symp. Effects of ionizing radiation on the reproductive system. Pergamon Press, New York.

Spalding, J. F., M. Brooks, and P. McWilliams. 1964. Observations on lifespan, radioresistance, and productivity in offspring from 5 to 25 generations of X-irradiated male mice. Genetics 50:1179–1186.

Sugahara, T. 1964. Genetic effects of chronic irradiation given to mice through three successive generations. Genetics 50:1143–1158.

Touchberry, R. W., and F. A. Verley. 1964. Some effects of X irradiation in successive generations of an inbred and a hybrid population of mice. Genetics 50:1187–1209.

Woolf, B. 1954. Estimation of mutation rates. I. Visible recessive characters detected in inbred lines maintained by single sib-matings. J. Genet. 52:332–353.

11

Reproduction[1]

Franklin H. Bronson, Charles P. Dagg, and George D. Snell

This chapter is a partial revision of a chapter in *Biology of the Laboratory Mouse* (Snell, 1941). A few of the previously published sections—sexual maturity, estrous cycle, mating, and gestation—have been rewritten. The remaining sections have been altered only slightly. Only references actually dealing with laboratory mice have been included and, particularly in the rewritten sections, emphasis has been placed on the descriptive details of normal reproduction (e.g., time and magnitude characteristics). More comprehensive texts (such as Young, 1961) should be consulted for the general principles of reproductive physiology.

SEXUAL MATURITY

Reproduction in the mature female mouse consists of a series of related and dependent hormonal and neural events which function to insure the successful production of new members of the species. Basic in this process is the interaction of anterior pituitary, placental, and gonadal hormones. Also basic is the role of the central nervous system, particularly the hypothalamus, in regulating hormone release by the anterior pituitary and, in turn, the effects of gonadal hormones on the hypothalamus itself. Follicle stimulating hormone (FSH), a gonadotropin released by the anterior pituitary, acts primarily to promote gametogenesis in both sexes. Luteinizing hormone (LH), another gonadotropin, has a primary function of promoting secretion of the gonadal hormones; estrogens and progesterone in the female and androgen in the male. A third anterior pituitary hormone, prolactin, functions in lactation and the development of the ovary during pregnancy. The gonadal hormones have the responsibilities of maintenance of secondary sexual characteristics and of the proper condition of the reproductive tract as well as acting on the central nervous system to insure successful mating. Full sexual maturity depends upon a delicate and as yet incompletely understood interplay among all of these components and the maturation of each proceeds at a somewhat different rate.

At birth the mouse hypophysis is physiologically undifferentiated from the gonadotropic viewpoint (Vivien, 1950) and, in addition, the ovaries are unresponsive to gonadotropin (Francke, 1948). Sex differentiation of the hypophysis is usually realized by day 6 in males, and before day 12 in females. Ovarian weight becomes responsive to exogenous FSH by days 6 to 9, but follicle size cannot be altered by FSH injections much before 12 to 15 days of age (Ben-or, 1963). Antrum formation normally begins in some follicles by days 12 to 14.

The threshold of ovarian (follicle) responsiveness decreases following initiation of competence to a low at about 3 weeks of age and, thereafter, is elevated to a relatively mature level by 30 days of age. Zarrow and Wilson (1961), after injections of pregnant mare serum (PMS), a hormone preparation having activity similar to FSH, and human chorionic gonadatropin (HCG; LH-like in

[1] The writing of this chapter was supported in part by Public Health Service Research Grants HD 00767 and HD 00473 from the National Institute of Child Health and Human Development.

activity), found the follicles competent to respond by about 14 days of age. Ovulation occurred in 40 per cent of the females at 13 to 14 days of age; 69 per cent ovulated by 15 to 18 days; 80 per cent by 19 to 20 days; and about 86 to 100 per cent thereafter. Maximum ova counts ran from 3.6 at 13 to 14 days to 60 at 23 to 26 days, and then dropped to 29 by 30 days of age. Gates and Runner (1957) reported the beginnings of follicle responsiveness at 15 days of age, the maximum superovulatory response at 3 weeks, and a decrease in responsiveness thereafter. Ovarian weight has been reported to be most responsive to PMS on day 20 (Soper, 1963). References to the techniques for induction of superovulation are given in the Bibliography of Techniques.

Sexual maturity normally occurs coincidentally with rising titers of circulating gonadotropin at some time after 4 weeks of age. Precisely when maturity occurs is highly variable and, in addition, such a statement must be interpreted within the framework of the measurement used to determine when "sexual maturity" has been reached. The first observable signs of puberty in females are estrogen-dependent: vaginal introitus and a cornified vaginal smear. Other indices of sexual maturity such as willingness to mate, the ability to conceive and carry a litter to term, and conceivably sexual maturity as measured by the ability to produce weanling-age young have more complex hormonal bases.

The vagina may open as early as day 24 and is often reported open by 4 weeks of age. The first vaginal cornification often occurs 24 to 120 hours after establishment of the vaginal opening, but this also is subject to considerable variability. In addition, estrus, in the sense of willingness to mate, does not always occur on schedule. Mirskaia and Crew (1930) reported that mating occurred at the first (vaginal) estrus in 75 and 85 per cent of the cases in two strains; percentages of pregnancy resulting from these first matings, however, were a low 48 and 57 per cent compared to 80 to 90 per cent for more mature animals. The difficulties inherent in attempts to determine when an animal is fully sexually mature should be obvious.

Attainment of sexual maturity, regardless of the measurement of such development, is a highly variable process. For example, utilizing time of vaginal opening, Parkes (1925) stated that his albino mice were usually mature by 7 weeks of age, whereas Engle and Rosasco (1927) reported a median age of 35 days and a range of 28 to 49 days also in an albino strain. Vaginal opening in C57BL/6J mice in this laboratory has been observed, in rare instances, as early as 24 days. Two factors contributing heavily to such variation are genetic background and season. Many reports establish that some degree of interstrain variability may occur (MacDowell and Lord, 1927; Heinecke and Grimm, 1958; Rudali et al., 1957). With respect to seasonal effects, Yoon (1955) reported that vaginal introitus occurred at an earlier age in summer than it did during the winter months. He accounted for much of the associated variance by fitting a sine curve to the data, thereby establishing strain differences after accounting for seasonal effects. The effect of season is relatively large. Time of vaginal opening averaged 29 days at the trough of the curve (winter) and 41 days at the peak (summer) for C57BL/10 mice. Vaginal opening, first cornified smear, and first typical estrus are all delayed by experimental exposure to cold (Barnett and Coleman, 1959).

Considerable work has been done on pituitary-ovarian function in immature mice, much of it directed toward inducing precocious maturity. Smith and Engle (1927) induced vaginal opening by day 15 after five daily transplants of mature pituitaries. Three daily transplants resulted in ovulation and mating by day 20. Pfeiffer and Hooker (1942) obtained a cornified vaginal smear by day 15, following PMS injection. These studies again indicate the inability of the ovary to respond to pituitary hormones before days 14 or 15. Utilizing estradiol injections, however, Takasugi and Bern (1962) were able to induce vaginal opening as early as 5 to 7 days of age, indicating the readiness of the vagina to respond to ovarian hormones at a much earlier age.

A series of papers by Runner, Gates, and Smithberg probably best illustrates induction of (relatively complete) sexual maturity in immature females. Females 30 to 35 days of age were caused to ovulate, mate, and conceive by priming with PMS 60 hours before the desired ovulation time and administering an ovulatory dose of HCG 13 hours before the desired time. Implantation would not occur (Runner and Gates, 1954), but the eggs developed with normal viability if transplanted to mature recipients (Gates, 1956). Implantation failure apparently occurred because of incomplete development of corpora lutea, since implantation was accomplished by daily injections of progesterone. Pregnancy was maintained with larger doses of progesterone and parturition occurred normally if relaxin was given (Smithberg and Runner, 1956).

Only a few of the other reports dealing with sexual maturity in mice will be mentioned. Strong and Fuller (1958) examined vaginal opening within the framework of selection experiments. Backman (1939) utilized a mathematical approach to characterize attainment of sexual maturity. Time of attainment of sexual maturity has been experimentally altered by many factors, including neonatal castration (Reynaud and Reynaud, 1947) and caloric restriction (Ball et al., 1947). Histological and histochemical characteristics of developing mouse ovaries and uteri are described by Ben-or (1963), Barraclough (1955), and Crelin and Levin (1955), and in many older works summarized by Snell (1941). Little is known about the details of sexual maturation in males. The process, however, is thought to proceed somewhat more slowly than in females (Parkes, 1925; Bishop and Leathem, 1946, 1948).

GAMETOGENESIS

The mitotic and meiotic phases of gametogenesis show marked chronological differences in females and males. The last of the mitotic divisions of the gonocytes in oogenesis occurs on day 12 or 13 of gestation. The existing cells then enter meiotic prophase and no new cells are formed thereafter. In spermatogenesis, the gonocytes are transformed into spermatogonia during early neonatal life and these continue to divide by mitosis throughout the reproductive life of the male. The discussion of oogenesis, then, is more properly delegated to Chap. 12, while the remainder of this section will describe spermatogenesis. For a complete description of the chromosomal aspects of meiosis, the reader is referred to Chap. 7.

The primordial germ cells, the gonocytes, divide mitotically during prenatal and neonatal life. During early postnatal life, the gonocytes are transformed into spermatogonia. In the mouse, three types of spermatogonia can be recognized: Type A, Intermediate, and Type B (Oakberg, 1956*a*). Type A spermatogonia divide four times, usually being transformed in the process into Intermediate type spermatogonia. A few of the resulting cells remain as Type A, enter a period of dormancy of about 207 hours, and then act as stem cells for a new cycle of multiplication (Oakberg, 1956*a;* Monesi, 1962). The Intermediate types divide by mitosis again and their progeny become transformed into Type B spermatogonia. The Type B cells undergo mitosis and change into spermatocytes in the preleptotene stage of meiotic prophase. These cells have progressed to the diplotene stage approximately 12½ days later and then rapidly go through the meiotic divisions. The duration of the complete process for the development of Type A spermatogonia into spermatozoa takes about 34½ days (Oakberg, 1956*b*). The entire sequence of events for spermatogenesis in the mouse, as well as other mammals, has been presented diagrammatically by Roosen-Runge (1962).

Not all of the cells survive spermatogenesis. A few spermatogonial cells, particularly Type A, break down in mitosis and thereafter degenerate. A second loss of cells, approximately 13 per cent, occurs in the period from early primary spermatocyte to spermatid. Degeneration has been observed at the first and second meiotic division (Oakberg, 1956*a*).

ESTROUS CYCLE

Knowledge of the estrous cycle in the mouse dates from the classic study by Allen (1922). Unless interrupted by pregnancy, by pseudopregnancy, or by other events to be discussed later, mice normally tend to display signs of estrus, including mating behavior, every 4 or 5 days. The mouse therefore is a polyestrous mammal. Individual cycles are actually complexes of related secretory, anatomical, and behavioral cycles in which the rhythmic interaction of pituitary and ovarian hormones is fundamental and which, in their totality, have the function of insuring fertilization. The periodicity of estrus observed in mature females is a direct result of cyclic changes that occur in the ovary which, in turn, reflect altered hypothalamic activity and changes in gonadotropin secretion. The problem of the basic responsibility for such rhythmicity is still not thoroughly understood and relatively little of the research into this problem has utilized the mouse. The key to cyclic reproductive activity apparently lies in the hypothalamus, which communicates with the anterior pituitary by way of a portal system. FSH, a protein, is released by the anterior pituitary and acts to promote follicle growth. Another anterior pituitary protein, LH, aids in the final development of the mature follicle and facilitates production of estrogens by the theca interna cells of the FSH-primed follicle. Further release of LH by the hypophysis results in rupture of the follicle and ovulation. Increasing titers of estrogen during the later phases of follicular growth are thought to act, by way of the hypothalamus, both to suppress further release of FSH and to favor release of more LH. Progesterone, another gonadal steroid, is also produced in the ovary in small quantities during the follicular growth phases. Progesterone, in small doses, promotes ovulation by enhancing LH release. Thus the gonadal hormones produced during follicular growth act on the hypothalamus to suppress further release of FSH while promoting release of LH and, hence, ovulation. Functional development of the corpus luteum in the mouse is induced by mating and, when this does not occur, gonadal hormone titers decrease allowing succession of a new cycle. The prime factor allowing for the periodicity of estrous phenomena is thought to be cyclic activity in the hypothalamus which is reflected in LH release.

While the basic responsibility of the rhythmicity remains somewhat in doubt, considerable work has been devoted to descriptions of the individual anatomical and secretory cycles which, together, constitute the estrous cycle in mice; these may be discussed in some detail.

Divisions of the estrous cycle

The mouse estrous cycle has been divided into as few as four phases: diestrus, proestrus, estrus, and metestrus; or as many as 13 (Thung et al., 1956). The latter system consists of a single stage of diestrus, four of proestrus, two of estrus, and six of metestrus. Since the cycle is continuous and a division into so many stages of the cycle would rarely be used, this section will rely on a five-stage description of the cycle (Allen, 1922). The first two stages (proestrus and estrus) are anabolic stages during which active growth is in progress in various parts of the genital tract. They

Table 11-1. SCHEMATIC OUTLINE OF CHANGES IN THE REPRODUCTIVE ORGANS OF THE MOUSE DURING THE ESTROUS CYCLE

Stage	Smear[a]	Histology of the vaginal epithelium	Uterus	Ovary and oviduct
Proestrus	E to EC or ECL to EC	Many cell layers (10 to 13). Outer 4 or 5 nucleated, stain lightly with eosin. Under these, granulosa layer showing increasing cornification. Active mitoses. Few leukocytes.	Increasing hyperemia and distension. Active mitoses in epithelium, few leukocytes.	Follicles large and distended with considerable liquor folliculi. Few mitoses in germinal epithelium and in follicular cells.
Estrus	EC to C+	Superficial nucleated layer lost. Cornified layer now superficial. About 12 layers of nucleated cells under this. Mitoses decreasing. Leukocytes absent.	Distension and mitotic activity reach maximum during estrus, and then decrease. No leukocytes.	Ovulation occurs followed by distension of the upper end of oviduct. Active mitoses in germinal epithelium and in follicular cells.
Metestrus-1	C++	Cornified layer delaminated. Leukocytes begin to appear under epithelium.	Distension decreased. Leukocytes begin to penetrate epithelium.	Early corpora lutea present. Eggs in oviduct. Many follicles undergoing atresia.
Metestrus-2	C++ EL++	4 to 7 layers of epithelial cells, with very many leukocytes in outer layers.	Walls collapsed. Epithelium shows degeneration. Mitoses rare. Leukocytes numerous.	Growing corpora lutea. Eggs in oviduct. Few mitoses in germinal epithelium and in follicular cells.
Diestrus	EL, more or less mucus	4 to 7 layers of epithelial cells, with leukocytes in outer layers. Growth commences towards end of diestrus.	Anemic, walls collapsed. Epithelium healthy but contains many leukocytes. Some secretion by uterine glands.	Follicles begin rapid growth toward end of period.

[a] E = epithelial cells, C = cornified cells, L = leukocytes, + indicates many cells, ++ indicates very many cells. The smears given are typical; there is considerable variation.

culminate in ovulation and, where mating occurs, in fertilization. The third and fourth stages, metestrus-1 and metestrus-2, are catabolic stages characterized by degenerative changes in the genital tract. The last (or first) stage, diestrus, is a period of quiescence or slow growth.

The cycle may be conveniently charted by examining vaginal smears. Cellular characteristics of vaginal smears reflect changes in the structure of the vaginal epithelium which, in turn, are dependent upon estrogen and follow a regular and predictable course during the cycle. Three types of cells are found in vaginal smears: leukocytes, cornified epithelial cells, and nucleated epithelial cells. Relative abundance of the various types of cells during the different stages of the cycle are given in Table 11-1. A number of variations from this basic pattern may occur. As estrus approaches, the smear may contain epithelial cells with dark-staining cytoplasm and karyolytic nuclei. Cells intermediate between cornified and nucleated epithelial cells also occur. It should be noted that the smear marking the termination of estrus is often denoted by the presence of clumps or sheets of cornified cells (Snell, 1941). A more detailed description of the changes in cell type and stainability in vaginal smears of mice may be found in Thung et al. (1956).

Ovary

The mouse ovary reveals a marked cycle of ovum and hormone production but little change in weight (except in pregnancy). Mice are spontaneous ovulators with ovulation normally occurring sometime in estrus. Newly formed (nonfunctional) corpora lutea, which persist for two to four cycles, are present in metestrus-1. Counts of the number of mitoses occurring in the germinal epithelium reveal mitotic activity to be least in diestrus, rising slowly through preovulatory estrus, and are characterized by a sudden high spurt of activity in postovulatory estrus. Much of the sequence of oogenesis takes place in this same short postovulatory spurt of activity (Bullough, 1942). By proestrus the follicles that will ovulate at estrus are definitely set apart by their larger size; follicles average only 380 μ in diameter at the beginning of the estrus cycle and reach a maximum size of about 550 μ immediately before rupturing. Most of this change occurs in the last 48 hours.

Not all of the follicles which begin the rapid preovulatory growth are destined to ovulate. In mouse ovaries collected in late diestrus, Nakamura (1957) found an average of 8.0 follicles per ovary, which appeared to be
in readiness for the next ovul
had decreased to an average
and to 4.0 at estrus. The numb
licles (e.g., second maturation division
gross chromosomal aberrations, etc.) averaged
in late diestrus, 5.3 in proestrus and had decreased to 1.1 at estrus.

"Free" progestin in the plasma has been measured during the various phases of the cycle by Guttenberg (1961). An average of 0.5 progesterone equivalents per milliliter was found at diestrus, 3.5 at proestrus, 4.0 at estrus, 2.1 at metestrus-1 and 0.8 at metestrus-2. Plasma progestin levels, therefore, rise well in advance of corpora lutea formation. Estrogen levels, on the other hand, are apparently highest at proestrus as shown by a variety of indicators (e.g., vaginal alkaline phosphatase studies by Vaczy et al., 1955). Uterine and vaginal changes during the estrous cycle are the responsibility of these two ovarian hormones, particularly estrogen.

The effect of the ratio of estrogen to progesterone on the vaginal smear has been examined experimentally by Jones and Astwood (1942). An adequate dose of progesterone will convert an estrous smear to diestrus regardless of the amount of estrogen given. A relatively large dose of progesterone is needed to inhibit the vaginal response to estrogen, but only a small amount will change an estrous smear to diestrus. Endocrine ramifications may sometimes be complex, however, as illustrated by the finding that adrenocortical neoplasms are responsible for some estruslike cycling in gonadectomized females (Christy et al., 1950). Green (1957) found no change in ovarian sensitivity to injected PMS during the estrous cycle.

Vagina

There is a tendency for the vaginal orifice to gape and the vulva to swell at proestrus and estrus. These signs are variable, however, and of less precision in detecting estrus than the characteristics of vaginal smears. Snell et al. (1944) suggested that the condition of the cervix (as determined by observation during artificial insemination) may be of more value than vaginal smears in predicting estrus. An absence of moisture in the cervix, tension of the cervix muscles, and absence of leakage following artificial insemination are all indicators of estrus.

The successive stages within the vagina itself are briefly summarized in Table 11-1. In proestrus the epithelium consists of three layers. The outer yer is composed of epithelial cells sometimes ıore or less filled with mucus and with nuclei showing signs of pyknosis. Below this is the stratum granulosum which, with the approach of estrus, becomes the stratum corneum. Third is the stratum germinativum, some seven cell layers in thickness. During proestrus and early estrus the cells of the outer layer are sloughed, producing the characteristic nucleated cell smear. The degree of delamination is not uniform in all parts of the vagina; prior to the onset of estrus the cornified layer may be fully exposed in some regions, not in others. Cells are sloughed from the cornified layer during late proestrus and throughout estrus. The onset of metestrus-1 is characterized by the peeling off of the whole layer, with an accompanying rise in the cornified cell count in the smear. During metestrus-2 there is a rise in the nucleated cell count, indicating that in the last stages of the delamination process some of the superficial layers of the stratum germinativum are included. The superficial layers of the stratum germinativum, meantime, have become heavily infiltrated with leukocytes which also appear abundantly in the smear. As a result of the loss of the superficial layers, the vaginal epithelium at diestrus contains only one layer, the stratum germinativum, some three to seven cell layers in thickness. Active growth begins in the stratum germinativum late in diestrus, and by proestrus a stratum granulosum has formed several cell layers below the surface, thus completing the cycle.

Weight of the vagina, either wet or oven-dried, is lowest at diestrus and greatest at proestrus. A comparison of dry-to-wet weight ratios reveals that the vagina is dehydrated at proestrus and estrus, relative to the other stages of the cycle (Balmain et al., 1956). Several studies have examined biochemical correlates of the process of keratinization in the vaginal epithelium: glycogen content (Balmain et al., 1956; Biggers, 1953), alkaline phosphatase levels (Vaczy et al., 1955), sulphydryl and disulphide groups (Asscher and Turner, 1955), and nucleic acids (Thiery, 1960).

Uterus

The uterus, like the vagina, undergoes a series of anabolic and catabolic changes during the estrous cycle, but they are somewhat less striking (Table 11-1). In general appearance, the uterus is distended because of the activity of the uterine glands in proestrus and estrus. The distension starts to diminish in late estrus, and in diestrus the uterine wall is collapsed and anemic. The wet and dry uterine weights are lowest at diestrus and heaviest at proestrus; the uterus is relatively hydrated at proestrus and estrus (Balmain et al., 1956). Glycogen content is greatest at proestrus. The uterine epithelium is composed of low columnar cells in estrus (Clauberg, 1931). In metestrus-1, degenerative processes become apparent. The basement membrane fades into a pink-staining band which includes the basal sides of the epithelial cells and the superficial stroma. The epithelium loses its definite organization and shows vacuolar degeneration. Leukocytes appear in the region of the basement membrane. In metestrus-2 the degeneration of the epithelium is further advanced, so that almost all the epithelial cells are lost (Rietschel, 1929). Cell walls at this stage are no longer recognizable and leukocytes are numerous. The uterine glands show minimum activity. The onset of diestrus is marked by the beginning of regenerative processes. Fuxe and Nilsson (1963) have examined the ultrastructure of the mouse uterine epithelium and have confirmed that maximum secretory activity occurs at estrus.

Mammary glands

The mammary glands show cyclic growth and regression, though the changes are slight compared with those occurring during pregnancy (Cole, 1933). In proestrus, buds appear on the ducts, particularly around the periphery of each gland, and large, blunt projections appear on the main ducts near the nipples. In estrus the mammary ducts become dilated, and the buds formed during proestrus elongate. Metestrus-1 introduces regressive changes and by the end of metestrus-2 the ducts are decreased in width and the duct endings collapsed. In diestrus the mammary gland consists of a very open network of narrow, thread-like ducts with comparatively few branches.

Other correlates of the estrous cycle

Slight changes in body weight have been reported for mice at estrus (Dewar, 1957). Henin (1941) described changes in the oviduct correlated with the estrous cycle. Kliman and Salhanick (1952) reported a slight relaxation of the pubic symphysis at estrus. Bullough (1942, 1946) has described a correlation between mitotic activity in the skin and the estrous cycle.

Time relations of the estrous cycle

Two aspects of the timing of the mouse estrous cycle bear discussion: the occurrence of cycles in relation to lifespan and the length of the entire cycle and of its arbitrary divisions. A third aspect of timing, the relationship between the estrous and light-dark cycles, will be discussed in the sections on mating and ovulation.

The onset of estrous cycling has been discussed previously. Several workers have examined the relationship between advanced age and occurrence of cycles. Such a relationship is, apparently, quite variable and dependent upon the particular strain under study. Caschera (1959) found increasing irregularity in cycles with advanced age, including some rather prolonged periods of diestrus. Thung et al. (1956) also found decreasing numbers of cycles per unit of time in advanced age. Cycles were characterized by somewhat lengthened metestrus and diestrus in older mice, interspersed with some cycles in which full estrus was not obtained, irrespective of breeding history. Strain differences occurred in both the pattern of irregularity which developed with age and in the age of cessation of cycles, DBA mice showing a prolonged anestrus at the end of life due to ovarian degeneration whereas some other strains continued cycling until death. Boot and Mühlbock (1957) reported some ovulation after cessation of fertility. Suntzeff et al. (1938) reported that the average depth of the epithelial processes extending into the underlying connective tissue in the mouse vagina increases with age. Jones and Krohn (1961) could find no evidence (except in one strain, CBA) that the decline in fertility at older ages was a result of decreased numbers of oocytes. They postulated that such losses in fertility are probably traceable to the hormonal control of the ovary or in the uterine environment.

The average length of individual cycles is usually considered to be 4 to 5 days but this is highly variable and, apparently, easily influenced. Parkes (1928), for example, found the following distribution of lengths for 1,000 cycles in unmated albino mice: 2 days, 4 per cent; 3 days, 3 per cent; 4 days, 16 per cent; 5 days, 29 per cent; 6 days, 22 per cent; 7 days, 12 per cent; 8 days, 6 per cent; 9 days, 3 per cent; and 10 to 28 days, 8 per cent. Estrus itself has been variously reported to last from ½ to 9 days (Allen, 1922; Snell, 1941). Probably, considering the environmental effects which will be discussed shortly, the best set of data relative to length of the various phases of the "average" cycle comes from Boot (in Van Ebbenhorst Tengbergen, 1955): early proestrus, 11 hours; late proestrus, 21.4 hours; estrus, 20.7 hours; metestrus, 21.8 hours; and diestrus, 21.8 hours for a total of 104 hours or 4.3 days.

The causes of such variation in the length of the cycle can be, initially, divided into genetic and environmental (realizing that these are not necessarily independent). Strain differences in cycle characteristics have been the subject of numerous studies which will not be cited here. In general, it may be said that various strains do differ in length of the cycle and, sometimes, of its phases. In addition, some of this type of variation has been traced to different thresholds for the vaginal response to estrogen (Mühlbock, 1947). Worthy of note, with respect to genetics and estrous cycles, is a paper by Mizuhara et al. (1958), who developed a graphical method for more accurate comparison of estrous cycles between strains.

Environmental variation may be classified into that stemming from the social and that from the nonsocial environment. Noteworthy in the second category are effects of such factors as season and diet. Seasonal effects and their interaction with genetic background have been established. Mice of strains A and C57 averaged cycles of 5.2 and 5.4 days in length in February, whereas both averaged 5.8 days in May. The longer cycles in May were due to a somewhat prolonged proestrus and, particularly, to a much longer metestrus. Strain A mice were characterized by longer estrus and shorter diestrus with metestrus and proestrus about the same length as occurred in C57 females (Laguchev, 1959). The relationship between diet and estrous characteristics has been examined many times (e.g., Halberg and Visscher, 1952; Lugo, 1959). Worthy of note are the effects of estrogenic plant substances (e.g., Cranston, 1945; East, 1955).

A series of findings implicating the social environment as a prime source of variation in estrous cycles casts doubt on the validity of many earlier reports on cycle characteristics and, actually, poses the problem: What is a "normal" estrous cycle? Estrus is somewhat suppressed (due to pseudopregnancy or prolonged diestrum) in mice crowded in all-female groups (van der Lee and Boot, 1956; Whitten, 1959; Marsden and Bronson, 1965). This phenomenon has, in part, been traced to olfactory-mediated stimuli. In addition to the effects of the all-female environment, olfactory stimuli originating from a male are known to, in a sense, override the effects of the female stimuli by regularizing some of the abnormal cycles found under crowded conditions (Whitten, 1957). The effects of a male (i.e., to regularize and actually accelerate estrus) have been mimicked utilizing male urine alone (Marsden and Bronson, 1964). The relationship between olfactory stimuli and mammalian reproduction has been reviewed by Parkes and Bruce (1961). In addition to olfactory stimuli, neurotropic stimuli, loosely categorized as "social stress" and stemming from the increase in behavioral competition accompanying crowding, affect estrous characteristics. Crowding, particularly when accompanied by some degree of social competition between males, is known to alter some aspects of central nervous system activity and increase synthesis and release of ACTH (Christian, 1960) and it is known that increased ACTH production may, directly or indirectly, alter ovarian function (Christian, 1964).

Postpartum estrus

The interval between parturition and the following ovulation in mice has been reported to be about 14 to 28 hours (Long and Mark, 1911). Runner and Ladman (1950) have examined this relationship more closely in light of the fact that they found parturition occurring at random across a 24-hour period whereas, as will be shown in following sections, ovulation and mating are strongly influenced by the light cycle. Ovulation during the postpartum estrus is, apparently, not timed exclusively by either the light cycle or the time of parturition. They found good predictability in ovulation times only when both factors were considered, the tendency to ovulate nocturnally being incompletely countered by a tendency to ovulate 12 to 18 hours after parturition. Cornification of the vagina is not complete at postpartum estrus, and the cornified cell content of the smear normally never reaches 100 per cent. Fertile matings are less often obtained during this period than during the course of the normal cycle. In addition, there is less fluid in the uterus than during a normal estrus (Merton, 1939).

OVULATION

Ovulation occurs spontaneously during estrus in mice whether or not mating has occurred. However, ovulation may not occur at every estrus, particularly in young virgin females (Togari, 1927), and estrus may not always accompany ovulation (Young, 1941). It should be remembered that the two phenomena, estrus and ovulation, have different endocrinological bases, estrus being dependent upon gonadal hormones whereas ovulation occurs in response to gonadotropins. Following initial stimulation of follicular growth by FSH, LH titers begin to rise, resulting in drastically increased secretion of follicular liquor and, finally, ovulation. Experimental production of ovulation by injections of PMS and HCG is discussed in the section on sexual maturity.

Strain differences in the state of ovarian development reached at birth are known (Jones and Krohn, 1961). However, by 3 to 6 days postpartum the oocytes usually have a follicular epithelium (Brambell, 1927). The primordial follicles in the adult consist of a single thin layer of cells surrounding the germ cell and lie near the ovarian epithelium. The oocytes measure approximately 13 μ in diameter, and the entire follicle about 16.5 μ (Brambell, 1928). As the oocyte and follicle increase in size they gradually move towards more central regions of the ovary. As the egg and follicle approach maturity they, in turn, resume their peripheral position on the ovary.

The epithelial cells of the follicle divide and transform into cuboidal and thence columnar cells as the oocyte begins to enlarge. When the egg has reached maximum growth, the multilayered follicular epithelium develops a fluid-filled cavity, the antrum. In the mouse, the growth of the follicle relative to the growth of the oocyte can be divided into two phases. During the first phase both the oocyte and the follicle grow rapidly, in a correlated fashion. At the end of this phase the oocyte measures 70 μ in diameter as compared to 125 μ for the follicle. During the second phase, there is no significant increase in size of the oocyte, but the diameter of the follicle increases to about 550 μ (under the influence of LH) just before ovulation. The follicular growth during the second phase is due largely to distention of the antrum with fluid.

Just prior to ovulation the egg undergoes the final stages of maturation. The centrosome divides into two centrioles, the asters are formed, and the rest of the spindle apparatus appears. In the mouse the spindle lies tangential to the surface (Long and Mark, 1911). The first meiotic division is completed and the first polar body is cut off and comes to lie beneath the zona pellucida. The second meiotic division begins in the oocyte nucleus but does not proceed beyond metaphase. The nucleus will remain in this condition, normally, until the sperm has entered the egg cytoplasm (Snell et al., 1940).

The period of estrus and the time of ovulation are normally controlled by the diurnal rhythm of light and darkness. Reversing the time of light and darkness reverses the time of estrus and ovulation (Snell et al., 1940; Snell et al., 1944; Braden, 1957). Published accounts differ considerably as to the time of ovulation in relation to the onset of estrus. Ovulation has been reported to occur at or near the beginning of estrus (Brambell and Parkes, 1927; Lewis and Wright, 1935; Snell et al., 1940), and at or near the end of estrus (Allen, 1922; Togari, 1927). In MacDowell-Bagg albino mice, ovulation usually took place between midnight and 3:00 AM, while mating most commonly occurred between 10:00 PM and 1:00 AM. Thus, in this strain, mating preceded ovulation by an average of only 2 hours. The interval varied considerably, ranging from 1 to 3¾ hours (Snell et al., 1940). Ovulation within 15 minutes of mating has been noted (Lewis and Wright, 1935).

Investigations in which the light-dark cycles were artificially controlled have supported the conclusion that the midpoint of the ovulation period is determined more by the midpoint of the dark phase than by its onset or completion. Averaging the results of three stocks, Braden (1957) found the midpoint of the ovulation period to occur around 3½ hours after the midpoint of the dark period when animals were kept on a 10-hours-dark to 14-hours-light cycle. The delay averaged 4¾ hours after the midpoint of the dark period in animals on a 4-hours-dark to 20-hours-light cycle. For two inbred strains, 129/Rr and BALB/cGn, the midpoint of ovulation was 4.5 to 5.5 hours after the midpoint of the dark period in a 7-hours-dark to 17-hours-light cycle (Whitten and Dagg, 1961).

Some of the apparent discrepancies in earlier reports may have been due to using mating time rather than the light-dark cycle as a reference point. Undoubtedly, some of the reported differences are due to interstrain variation. Three noninbred stocks (L, C, and PCT) were found to differ in the midpoint of the ovulation period, indicating that mechanisms controlling the time of ovulation are modified according to the genetic constitution of the stock (Braden, 1957). Braden also noted that the interval between mating and ovulation was approximately 5 hours, somewhat longer than the 2-hour interval found by Snell et al. (1940). The time required for ovulation of 75 per cent of the eggs released by any one mouse has been estimated as ½ to 1 hour (Braden, 1957).

MATING

Pertinent literature on sex behavior, including some description, is covered in Chap. 33. Mating in house mice is normally detected by the occurrence of vaginal plugs. Such plugs are formed by a mixture of the secretions of the vesicular and coagulating glands of the male and usually fill the vagina from cervix to vulva. Occasionally, smaller, less conspicuous plugs are formed, a condition more common in postpartum matings. Plugs persist for 16 to 24 hours and may last as long as 48 hours (Parkes, 1926).

The efficiency of utilizing plugs to predict pregnancy is usually high. For example, Snell (1941) reported that, among two strains of mice, 80 to 90 per cent of the mature females showing a plug became pregnant. However, like so many biological parameters in laboratory mice, predictability of pregnancy from vaginal plugs is strain-dependent. Fainstat (1951) reported that only 14 per cent of his strain A females with plugs became pregnant, whereas comparative figures ranged from 29 to 100 per cent in four other strains. Housing conditions during the postinsemination period were not described. Conceivably this could account for some of the large differences, but that strains differ in this respect is certainly true. Eaton (1941) suggests that the proportion of successful matings of hybrids might be a factor amenable to genetic manipulation.

Receptivity among laboratory mice is not limited to estrus but may take place during proestrus or metestrus-1 (as indicated by vaginal smears). Fowler and Edwards (1957) reported good success in predicting whether or not mating would occur by the examination of a single vaginal smear and, in addition, presented the following frequency distribution of inseminations according to the type of vaginal smear observed in conjunction with the plugs: proestrus or early estrus (moderate numbers of nucleated epithelial cells, few or no leukocytes), 57 per cent (of 7); estrus (moderate numbers of cornified and some nucleated epithelial cells), 83 per cent (of 12); late estrus (moderate numbers of cornified cells), 36 per cent (of 11); metestrus-1 (many clumped cornified cells), 22 per cent (of 9); and none (of 9 and 22, respectively) in metestrus-2 or diestrus.

The length of time between pairing and insemination is strongly dependent upon housing conditions of the female prior to pairing. Whitten (1956*a*, 1957) first described the phenomenon of postpairing synchrony of estrus among females previously housed together. Such synchrony manifests itself as a nonrandom frequency distribution of inseminations (with respect to days) during a 4- to 5-day postpairing period. Specifically he found a third-day peak in inseminations with correspondingly lower levels on days 1, 2, and 4. Exteroceptive factors arising both from the all-female grouping prior to pairing and from the male at the time of pairing contribute to such synchrony (Marsden and Bronson, 1965). As mentioned previously, stimuli which may or may not be olfactory-mediated tend to suppress the estrous cycle among crowded females. Release from such stimuli by removing the female for individual pairing with a male, as well as olfactory stimuli associated with male urine, acts to recommence her cycle and leads to most females attaining estrus on the third day after pairing.

Matings may occur out of phase with the light-dark cycle on the first day after pairing depending upon the time of pairing. Although it is normal for behavioral estrus, mating, and ovulation all to take place somewhere during the dark cycle, mating may take place throughout much of the light cycle if pairing with a male is done just after the light phase begins. Such out-of-phase mating is probably restricted to animals that would normally have mated during the preceding dark phase. For example, in a recent series of C57BL/6J females paired with a male at 9:00 AM (light phase 5:00 AM to 7:00 PM) in this laboratory, 51 per cent of 99 plugs formed during the first 24 hours after pairing were formed before 4:00 PM. The corresponding figures for the second and third days after pairing were, respectively, 6 per cent of 54 and 4 per cent of 50. It is probable that only insemination itself is out of phase in these cases and that some of these matings would be infertile.

The interactions between the environment, central nervous system activity, and the gonodal hormones in producing behavioral symptoms of estrus (spontaneous activity, sexual receptivity) in females or mating behavior in males cannot be denied. Unfortunately the mouse, as opposed to such mammals as the rat and guinea pig, has been little utilized in studies designed to elucidate the physiological causes of mating behavior (Young, 1961). Engle (1942) reported somewhat ambivalent sex-behavior changes in males after sex-steroid injections. Induction of estrus and mating in immature females has been accomplished as described in a previous section on sexual maturity. Several early investigators reported induction of mating by estrogen treatment alone, e.g., Allen and Doisy (1923). Ring (1944) re-examined this problem and found that, regardless of dose, not all females treated with estrogen would mate. When both progesterone and estrogen were given, 100 per cent of the females mated. Fowler and Edwards (1957) injected 1 IU of PMS followed in 40 hours by 2 IU of HCG or 3 IU of PMS followed by 3 IU of HCG into mature mice selected at random with respect to estrus and found that mating occurred in 75 per cent and ovulation in 99 per cent of the animals within 20 hours. Three-fourths of the injected animals became pregnant as opposed to 90 per cent for controls. Partial dependence of dose requirements of PMS and HCG on genetic background has been demonstrated (Edwards et al., 1963).

Artificial insemination procedures, consisting of insemination via a hypodermic needle inserted through the cervix, followed by copulation with a castrated male have been described by Snell et al. (1944). This procedure yielded, in general, 70 per cent or fewer pregnancies. Kile (1951) injected sperm through the uterine wall 1½ to 4½ hours after copulation with a vasectomized male and obtained 72 per cent fertility.

Many cases of mating at various times during pregnancy have been reported; for example, Mirskaia and Crew (1930), Bilewicz and Mikiewiczowa (1954), and Bloch (1958).

FERTILIZATION

Fertilization usually occurs in the ampulla or the upper end of the oviduct, and sperm may be present in this area for several hours before ovulation occurs. Although the main mass never reach the ovarian end of the oviduct, travel may be rapid by those that do attain this distance (as little as 15 minutes after ejaculation; Lewis and Wright, 1935). A relatively small number of sperm (fewer than 100) have been found near the eggs at the time of fertilization (Braden and Austin, 1954*a*).

The mechanisms for transport of sperm through the uterus and Fallopian tubes are not completely understood. Motility of the sperm appears to play only a small part; contractions of the uterus and possibly the oviducts probably hasten transport (Parkes, 1960). Sperm that reach the Fallopian tubes retain their fertilizing ability for about 6 hours, even though their motility has been reported to persist for 13½ hours (Merton, 1939). Sperm remaining in the uterus are phagocytosed by polymorphonuclear leukocytes within 20 hours after mating (Austin, 1957). The egg has usually completed the first maturation division by the time ovulation occurs. The first polar body is present and the second maturation spindle has formed. If the egg has not completed the first maturation division it does so very quickly after ovulation. The polar body degenerates as early as the one-cell stage or as late as the morula (Lewis and Wright, 1935). Each egg is surrounded by a zona pellucida and, external to this, a mass of cumulus cells, the cumulus oophorus. The cumulus cells are adhesive; consequently, all the eggs from one ovary are frequently gathered into a clump by attachment of the masses of cumulus cells. Sperm penetrate the cumulus cells and reach the outer egg membrane in about 2 hours (Lewis and Wright, 1935).

More than one spermatozoon can enter the perivitelline space. However, usually only one sperm penetrates the vitellus and fertilizes the egg. As the egg ages, its capacity to prevent polyspermy diminishes. In one study the local application of heat to the Fallopian tube raised the frequency of polyspermy from 0.3 to 3.8 per cent, but delayed coitus did not alter the frequency (Braden and Austin, 1954*c*). As many as four supplementary sperm have been observed in the perivitelline space. The frequency distribution of eggs with extra sperm differed from the expected distribution, indicating that fewer sperm penetrated into the zona pellucida than expected by chance. The entry of the first or fertilizing sperm seems to restrict the admission of additional sperm into the zona pellucida (Braden et al., 1954).

The egg of the mature mouse is capable of being fertilized and producing normal embryos for about 10 to 12 hours after ovulation (Long, 1912; Braden and Austin, 1954*b*). Eggs undergo a gradual aging in which functions successively disappear. An early manifestation is the production of abnormal or inviable embryos. Further deterioration leads to abnormalities of fertilization and syngamy and even to loss of protection against polyspermy. Unfertilized eggs undergo degeneration in the Fallopian tubes and uterus. Frequently these cells fragment, producing clusters of two or more cells resembling normal division stages. Closer examination, however, reveals that cell sizes are usually atypically unequal and the nuclear material is not distributed normally. Some cells resulting from the spontaneous fragmentation may be without a nucleus, whereas others appear to have more than one. See Chap. 12 for additional comments on fertilization.

GESTATION

While estrogens assume primary (but not complete) responsibility during the estrous cycle, progestagens may be thought of as assuming a primary role for gestational success. Neither of these statements is of course completely true since progesterone, among other things, is required for successful mating and plays a prime role in ovulation, while it is known that estrogens act synergistically with progestagens to insure a successful pregnancy.

Functional corpora lutea are not found in mice unless mating (or cervical stimulation) has occurred. The act of mating produces widespread activation of the nervous system including the hypothalamus, thereby promoting release of luteotrophic hormone (LTH) by the anterior pituitary. LH administration can cause luteinization in the mouse ovary even without ovulation, but the development of a functional corpus luteum is under the control of LTH (prolactin in mice).

There is probably a neural block to LTH release which prevents functional corpus luteum formation and, consequently, high titers of progesterone during infertile estrous cycles. If mating occurs, functional corpora lutea develop, the uterus comes under progestational dominance, and ovulation and succeeding estrous cycles are blocked. Implantation normally occurs about 5 days after insemination coincident with rising progesterone levels (Forbes and Hooker, 1957). Placental hormone secretion is apparently of less importance for the maintenance of pregnancy in mice than it is in many other species, since gonadectomy at any time during gestation is followed by termination of pregnancy.

Sterile matings or cervical stimulation with a glass rod may induce pseudopregnancy. The average interval between a sterile mating and the next estrus in mice has been reported to be about 11 days (Deanesly, 1930), but later reports have shown this to be quite variable. The initial stages of pseudopregnancy and pregnancy are similar in hormone responses (Atkinson and Hooker, 1945). Following a sterile mating the corpora lutea become functional, sensitizing the uterine epithelium which can then respond to trauma. In the pseudopregnant mouse, the maximum capacity for deciduoma formation following local injury of the uterus occurs about 3 days after mating. By 5 days after mating the sensitivity is almost lost (Parkes, 1929). Contrary to findings in some other mammals, electrical stimulation is relatively ineffective as a means of producing pseudopregnancy in mice (Shelesnyak and Davies, 1953).

Gestation in the nonsuckled mouse normally lasts 19 to 21 days. Fekete (1954) has calculated that a female bearing a litter of 5 or 10 gains, respectively, 2.5 or 3.5 times as much weight during pregnancy as a female with a litter of one. Length of gestation, like most other aspects of mouse reproduction is strain-dependent. For example, in both DBA and C57BL strains, the modal length of gestation has been reported to be 20 days, but a larger proportion of C57BL than DBA females delivered at 19 days and, in contrast, a larger proportion of DBA than C57BL females gave birth on the 21st day (Fekete, 1941). Hybrid mice may have shorter gestation periods than inbred mice. Superfetation has been reported for the mouse (Stowell, 1941); however, as in most other mammals, this is apparently a rare phenomenon. Superfetation has been induced experimentally in mice (Edwards and Fowler, 1958).

Time of parturition is probably subject to both genetic and environmental influences. Runner and Ladman (1950), examining the relationship between the light-dark cycle and time of parturition, reported that in no 4-hour period did fewer than 12 per cent or more than 20 per cent of the births occur. In an older study using a different stock of mice it was found that two-thirds of the births occurred between 4:00 PM and 4:00 AM (dark phase) (Merton, 1938).

Gestation and lactation often proceed simultaneously but under these conditions the gestation period is lengthened because of a delay in implantation. If only one or two young are being suckled, the prolongation of the gestation period does not exceed 7 days. With three or more young suckling, prolongations up to 12 or 13 days are not uncommon and may go as high as 16 days (Brambell, 1937; Fekete, 1940).

The reproductive potential of mice is high. Some females in some strains may consistently ovulate 10 or more ova and the reproductive lifespan of some strains approaches 2 years. When one considers the relatively high number of ova that may be shed, the fact that pregnancy and lactation may occur simultaneously, and the fact that some strains may reproduce until quite late in life, the total production of young which could occur is impressive. For many reasons such production is almost never realized. Some females never produce young, others show inconsistent production, and, even among those reproducing more or less consistently, there is normally some degree of loss between ovulation and parturition.

The number of ova shed, as indicated by corpus luteum counts, is usually lower for the first pregnancy but remains fairly constant thereafter. It is normal, however, to expect litter size to decrease in older age because of an increasing level of postovulation loss of ova or intrauterine problems (Hollander and Strong, 1950) and reproduction usually ceases entirely, or at least becomes very sporadic, in older animals. The usual pattern in litter size is a rise for the first few litters followed by a gradual decline until sterility or death. The number of young born in any given litter is usually between 1 and 10 but is too variable, due to genetic and environmental influences, to justify speaking about "an average litter size" of mice.

The degree of embryonic loss is highest in the earlier stages of pregnancy. Hollander and Strong (1950) found an average level of postimplantation mortality of 15 per cent in their stocks. Mortality was noted in all stages of embryonic life but 72 per cent of the observed loss occurred in the first 3 days after implantation. Loss of ova before implantation would further decrease the realized reproduction. One factor not correlated with intrauterine loss was crowding of embryos in the uterus. Bowman and Roberts (1958) also reported little correlation between the number of implanted blastocysts and the amount of intrauterine mortality but did find that the loss of eggs before implantation was correlated with the number shed.

The possible causes for decreases in the reproductive potential are many and may, initially, be categorized as genetic or environmental. Most laboratory mice are inbred to some extent and inbreeding is known often to be accompanied by some degree of reproductive inefficiency. For example, a decrease of 0.56 young per litter for each 10 per cent increase in the inbreeding coefficient was reported by Bowman and Falconer (1960). Starting with 20 lines of mice, only three lines were surviving by the time the inbreeding coefficient had reached 76 per cent. The causes for losses of reproductive efficiency which accompany selection or inbreeding experiments are probably manifold. Examples are: increased preimplantation loss of eggs (Falconer and Roberts, 1960), low libido of the male, or hypofunctioning of the adenohypophysis of the female (Fowler and Edwards, 1960).

The increased demands placed on a female during pregnancy make her susceptible to many environmental influences. Biggers et al. (1958) have reported increased pre- or postimplantational mortality in environments that were either too hot or too cold. Barnett (1962) found that mice maintained in a cold environment generally began breeding later with a longer time period between litters and, as a result, produced about half as many young as controls. Mice of one strain, however, produced about the same number of young as controls because of a twofold increase in their reproductive life span. As another example of the possible interactions between genetic background and the environment in producing reproductive inefficiency, Hoag and Dickie (1962) reported that strains may differ with respect to the optimum level of protein and fat required for most successful reproduction (Chap. 5). An example of a common environmental stress placed on pregnant females in the laboratory is the report by Runner (1959) who found that gestational success in 129/J mice was decreased by daily handling. Experimentally Weir and DeFries (1963) have demonstrated that enforced swimming and activity decrease the number of births.

Pregnancy and pseudopregnancy, like some other aspects of mouse reproduction, are susceptible to alteration by the social environment. Olfactory stimuli are known to induce true pseudopregnancy (as shown by deciduoma formation) in crowded all-female groups (van der Lee and Boot, 1955, 1956). Biancifiori and Caschera (1963) reported that 46 per cent of the 35 cycles examined in females which had mated with a vasectomized male were pseudopregnancies. Comparable figures for unmated females housed five per cage and anosmic females maintained one per cage were, respectively, 14 per cent of 150 and none of 54. In addition, a series of papers by Bruce has disclosed that olfactory stimuli originating from a strange male may prevent implantation of the blastocysts or pseudopregnancy. Following exposure to a strange male the female returns to an estrous condition within 7 days after insemination. Any male other than the particular stud male will accomplish this and, if the strange male is of a different strain than the stud male, the proportion of females showing blocked pregnancies is increased still further (Bruce, 1960; Bruce and Parrott, 1960; Parkes and Bruce, 1961). Experimental evidence suggests that the mechanism involved in this phenomenon includes a failure of the normal luteotrophic activity of the pituitary (Parkes, 1961). This phenomenon may not occur in many of the highly inbred strains. Testing of adequate numbers of C57BL/6J, SWR/J, SJL/J, and CBA/J strains at The Jackson Laboratory revealed no effect of postinsemination exposure to a nonstud male.

The role of olfaction in mouse reproduction may be more important than previously thought. An intimate relationship has been demonstrated by the large degree of inhibition imposed on reproduction by ablation of the olfactory lobes (Whitten, 1956*b*; Lamond, 1958). This may be compared to a study by Chase (1941), who found normal development of sexual maturity in a stock of mice suffering from congenital eyelessness.

Stimuli, other than olfactory, originating in the social environment are also known to affect gestational success. Crowding of mice, particularly when accompanied by some degree of fighting between males, is known to have profound effects on several endocrine systems. That such a situation is accompanied by increased pre- and postimplantational mortality is well documented (Christian and LeMunyan, 1958). Bruce (1954) has confirmed that monogamous pairs show the best production but, considering the savings of space and numbers of mice, has concluded that one male and two females per cage is the most efficient housing method with respect to maximum production of mice. There can be little doubt that the relationship between the number of mice per cage on the one hand and genetic and (nonsocial) environmental factors on the other is complex and highly variable.

LACTATION

The suckling period is a critical period, in terms of development, in the life of a mouse. It has been estimated that maternal influence accounts for about 72 per cent of the variance associated with body weights taken at 12 days of age (Cox et al., 1959). The normal (average) duration of lactation in mice is about 4 weeks but this is somewhat variable undoubtedly, at least in part, because of strain differences. Milk production is not constant throughout lactation. Production rises for about 10 days and then declines until weaning.

Nandi (1959) has presented extensive work on the hormonal control of mammogenesis and lactogenesis and has included a literature review. A series of three papers by Munford (1963) deals extensively with histological and biochemical characteristics of mammary glands during pregnancy, lactation, and involution.

SUMMARY

Characteristics of the process of reproduction in laboratory mice, while different in detail, bear general similarity to those found in many other mammals. As an attempt to classify their reproduction, mice may be said to be polyestrous; they are spontaneous ovulators in which the formation of luteal tissue is induced by mating; and, in addition, like many other rodents, they possess a high reproductive potential. For a variety of reasons such a high capacity for production of young is almost never fully realized. One characteristic encountered time and again while reviewing the literature for this chapter is the variability found among reports dealing with reproduction in "the mouse"; so much so that difficulty is encountered when one attempts to discuss normal parameters of reproduction. Strain differences have been found in almost every instance where they have been sought and, in addition, most aspects of mouse reproduction are amenable to alteration by one or many environmental factors. These range from physical aspects such as light or temperature to the presence or absence of other animals or even their odors. An important category of causes of variability resides in the interaction between genetic background and environment. Strains differ in their reproductive responses to the environment; for example, in their optimum dietary needs or in their capacity to respond to their social environment. Because of such variability it should be emphasized that all exact measurements given in this chapter (e.g., time of ovulation with respect to the light-dark cycle) should be suspect unless the same stocks of animals are used and the same environmental conditions prevail.

LITERATURE CITED

Allen, E. 1922. The oestrous cycle in the mouse. Amer. J. Anat. 30:297–371.

Allen, E., and E. A. Doisy. 1923. An ovarian hormone: preliminary report on its localization, extraction and partial purification and action in test animals. J. Amer. Med. Ass. 81:819–821.

Asscher, A. W., and C. J. Turner. 1955. Vaginal sulphhydryl and disulphide groups during the oestrous cycle of the mouse. Nature 175:900–901.

Atkinson, W. B., and C. W. Hooker. 1945. The day to day level of estrogen and progestin throughout pregnancy and pseudopregnancy in the mouse. Anat. Rec. 93:75–89.

Austin, C. R. 1957. Fate of spermatozoa in the uterus of the mouse and rat. J. Endocrinol. 14:335–342.

Backman, G. 1939. Das Wachstum der weissen Maus. Lunds Univ. Arsskr. Ard. Z. 35(12):1–26.

Ball, Z. B., R. H. Barnes, and M. B. Visscher. 1947. The effects of dietary caloric restriction on maturity and senescence, with particular reference to fertility and longevity. Amer. J. Physiol. 150:511–519.

Balmain, J. H., J. D. Biggers, and P. J. Claringbold. 1956. Glycogen, wet weight and dry weight changes in the vagina of the mouse. Austral. J. Biol. Sci. 9:147–158.

Barnett, S. A. 1962. Total breeding capacity of mice at two temperatures. J. Reprod. Fertil. 4:327–325.

Barnett, S. A., and E. M. Coleman. 1959. The effect of low environmental temperature on the reproductive cycle of female mice. J. Endocrinol. 19: 232–240.

Barraclough, C. A. 1955. Influence of age on the response of preweaning female mice to testosterone propionate. Amer. J. Anat. 97:493–521.

Ben-or, S. 1963. Morphological and functional development of the ovary of the mouse. I. Morphology and histochemistry of the developing ovary in normal conditions and after FSH treatment. J. Embryol. Exp. Morphol. 11:1–12.

Biancifiori, C., and F. Caschera. 1963. The effect of olfactory lobectomy and induced pseudopregnancy on the incidence of methylcholanthrene-induced mammary and ovarian tumors in C3Hb mice. Brit. J. Cancer 17:116–118.

Biggers, J. D. 1953. The carbohydrate components of the vagina of the normal and ovariectomized mouse during oestrogenic stimulation. J. Anat. 87:327–336.

Biggers, J. D., M. R. Ashoub, A. McLaren, and D. Michie. 1958. The growth and development of mice in three climatic environments. J. Exp. Biol. 35:144–155.

Bilewicz, S., and Z. Mikiewiczowa. 1954. Czteryprzypadki nieprawidłowego rozrodu u hiełych [in Polish, English summary]. Folia Biol. (Warsaw) 2:113–121.

Bishop, D. H., and J. H. Leathem. 1946. Response of prepuberal male mice to equine gonadotropin. Anat. Rec. 95:313–319.

Bishop, D. H., and J. H. Leathem. 1948. Effect of equine gonadotrophin on prepuberal male mice. Exp. Med. Surg. 6:28–30.

Bloch, S. 1958. Beobachtungen über Falle von früh zeitiger Trachtigheit bei der Albino-Maus. Experientia 14:141–142.

Boot, L. M., and O. Mühlbock. 1957. The ovarian function in old mice. Acta Physiol. Pharmacol. Neer. 3:463.

Bowman, J. C., and D. S. Falconer. 1960. Inbreeding depression and heterosis of litter size in mice. Genet. Res. 1:262–274.

Bowman, J. C., and R. C. Roberts. 1958. Embryonic mortality in relation to ovulation rate in the house mouse. J. Exp. Biol. 35:138–143.

Braden, A. W. H. 1957. The relationship between the diurnal light cycle and the time of ovulation in mice. J. Exp. Biol. 34:177–188.

Braden, A. W. H., and C. R. Austin. 1954*a*. The number of sperms about the eggs in mammals and its significance for normal fertilization. Austral. J. Biol. Sci. 7:543.

Braden, A. W. H., and C. R. Austin. 1954*b*. The fertile life of mouse and rat eggs. Science 120:610–611.

Braden, A. W. H., and C. R. Austin. 1954*c*. Fertilization of the mouse egg and the effect of delayed coitus and of hot-shock treatment. Austral. J. Biol. Sci. 7:552–565.

Braden, A. W. H., C. R. Austin, and H. A. David. 1954. The reaction of the zona pellucida to sperm penetration. Austral. J. Biol. Sci. 7:391.

Brambell, F. W. 1927. The development and morphology of the gonads of the mouse. I. The morphogenesis of the indifferent gonad and of the ovary. Proc. Roy. Soc. B 101:391.

Brambell, F. W. 1928. The development and morphology of the gonads of the mouse. III. The growth of the follicles. Proc. Roy. Soc. B 103:258.

Brambell, F. W. 1937. The influence of lactation on the implantation of the mammalian embryo. Amer. J. Obstet. Gynecol. 33:942–953.

Brambell, F. W., and A. S. Parkes. 1927. The normal ovarian cycle in relation to oestrus production. Quart. J. Exp. Physiol. 18:185–198.

Bruce, H. M. 1954. Feeding and breeding of laboratory animals. XIV. Size of breeding group and production of mice. J. Hyg. 52:60–66.

Bruce, H. M. 1960. A block to pregnancy in the mouse caused by proximity of strange males. J. Reprod. Fertil. 1:96–103.

Bruce, H. M., and J. East. 1956. Number and viability of young from pregnancies concurrent with lactation in the mouse. J. Endocrinol. 14:19–27.

Bruce, H. M., and D. M. V. Parrott. 1960. Role of olfactory sense in pregnancy block by strange males. Science 131:1526.

Bullough, W. S. 1942. Oogenesis and its relation to the oestrous cycle in the adult mouse. J. Endocrinol. 3:141–149.

Bullough, W. S. 1946. Mitotic activity in the adult female mouse, *Mus musculus* L. A study of its relation to the oestrous cycle in normal and abnormal conditions. Phil. Trans. Roy. Soc. B 231:453–517.

Caschera, F. 1959. La "menopause" nei topi femmine vergini (RIII/DmSe, C3Hb/Se, A/He/Se substrains). Lav. Anat. Patol. Perugia 19:13–20.

Chase, E. B. 1941. Studies on an anophthalmic strain of mice. II. Effect of congenital eyelessness on reproductive phenomena. Anat. Rec. 80:33–36.

Christian, J. J. 1960. Endocrine adaptive mechanisms and the physiologic regulation of population growth. Nav. Med. Res. Inst. Lect. Rev. Series No. 60–2, 49–150.

Christian, J. J. 1964. Effect of chronic ACTH treatment on maturation of intact female mice. Endocrinology 74:669–679.

Christian, J. J., and C. D. LeMunyan. 1958. Adverse effects of crowding on reproduction and lactation of mice and two generations of their progeny. Endocrinology 63:517–529.

Christy, N. P., M. M. Dickie, and G. W. Woolley. 1950. Estrus and mating in gonadectomized female mice with adrenal cortical abnormalities. Endocrinology 47:129–130.

Clauberg, C. 1931. Genitalcyclus und Schwangerschaft bei der weissen Maus. (Anatomische studien am Ovarium, Uterus and Scheide.) Dauer des Genitalcyclus. Arch. Gynakol. 147: 549–596.

Cole, H. A. 1933. The mammary gland of the mouse, during the estrous cycle, pregnancy and lactation. Proc. Roy. Soc. B 114:136–160.

Cox, D. F., J. E. Legates, and C. C. Cockerham. 1959. Maternal influence on body weight. J. Anim. Sci. 18:519–527.

Cranston, E. M. 1945. The effect of *Lithospermum ruderale* on the estrous cycle of mice. J. Pharmacol. Exp. Therap. 83:130–142.

Crelin, E. S., and J. Levin. 1955. The prepuberal pubic symphysis and uterus in the mouse: their response to estrogen and relaxin. Endocrinology 57:730–747.

Deanesly, R. 1930. The corpora lutea of the mouse, with special reference to fat accumulation during the oestrus cycle. Proc. Roy. Soc. B 106:578–595.

Dewar, A. D. 1957. Body weight changes in the mouse during the oestrous cycle and pseudopregnancy. J. Endocrinol. 15:230–233.

East, J. 1955. The effect of certain plant preparations on the fertility of laboratory mammals. 1. *Polygonum hydropiper* L. J. Endocrinol. 12: 252–260.

Eaton, O. N. 1941. Crosses between inbred strains of mice. J. Hered. 32:393–395.

Edwards, R. G., and R. E. Fowler. 1958. The experimental induction of superfoetation in the mouse. J. Endocrinol. 17:223–236.

Edwards, R. G., E. D. Wilson, and R. E. Fowler. 1963. Genetic and hormonal influences on ovulation and implantation in adult mice treated with gonadotrophins. J. Endocrinol. 26:389–399.

Engel, P. 1942. Female mating behavior shown by male mice after treatment with different substances. Endocrinology 30:623.

Engle, E. T., and J. Rosasco. 1927. The age of the albino mouse at normal sexual maturity. Anat. Rec. 36:383–388.

Fainstat, T. D. 1951. Hereditary differences in ability to conceive following coitus in mice. Science 114:524.

Falconer, D. S., and R. C. Roberts. 1960. Effect of inbreeding on ovulation rate and foetal mortality in mice. Genet. Res. 1:422–430.

Fekete, E. 1940. Observations on three functional tests in a high-tumor and a low-tumor strain of mice. Amer. J. Cancer 38:234–238.

Fekete, E. 1941. Data presented p. 56, *In* G. D. Snell [ed.] Biology of the laboratory mouse. Blakiston, Philadelphia.

Fekete, E. 1954. Gain in weight of pregnant mice in relation to litter size. J. Hered. 45:88–98.

Forbes, T. R., and C. W. Hooker. 1957. Plasma levels of progestin during pregnancy in the mouse. Endocrinology 61:281–286.

Fowler, R. E., and R. G. Edwards. 1957. Induction of super-ovulation and pregnancy in mature mice by gonadotrophins. J. Endocrinol. 15:374–384.

Fowler, R. E., and R. G. Edwards. 1960. The fertility of mice selected for large or small body size. Genet. Res. 1:393–407.

Francke, C. 1948. Some observations on the morphological structure of explanted immature ovaries and on "ovariohypophyses." Acta Neer. Morphol. Pathol. 6:129–140.

Fuxe, K., and O. Nilsson. 1963. The mouse uterine surface epithelium during the estrous cycle. Anat. Rec. 145:541–548.

Gates, A. H. 1956. Viability and developmental capacity of eggs from immature mice treated with gonadotrophins. Nature 177:754–755.

Gates, A. H., and M. N. Runner. 1957. Influence of prepuberal age on number of ova that can be superovulated in the mouse. Anat. Rec. 128:554. (Abstr.)

Green, J. A. 1957. Ovarian weight and responsiveness to gonadotrophin throughout the estrous cycle of mice. Proc. Soc. Exp. Biol. Med. 95:504–506.

Guttenberg, I. 1961. Plasma levels of "free" progestin during the estrous cycle in the mouse. Endocrinology 68:1006–1009.

Halberg, F., and M. B. Visscher. 1952. A difference between the effects of dietary calorie restriction on the estrous cycle and on the 24-hour adrenal cortical cycle in rodents. Endocrinology 51:329–335.

Heinecke, H., and H. Grimm. 1958. Untersuchungen zur Öffnungszeit der Vaginal-membran bei verschiedenen Mäusestammen. Endokrinologie 35:205–213.

Henin, A. 1941. Étude des modification de l'oviducte au cours du cycle oestral (souris). Arch. Biol. 52:97–115.

Hoag, W. G., and M. M. Dickie. 1962. Studies of the effect of various dietary protein fat levels on inbred laboratory mice. Proc. Anim. Care Panel 12:7–10.

Hollander, W. F., and L. C. Strong. 1950. Intrauterine mortality and placental fusions in the mouse. J. Exp. Zool. 115:131–149.

Jones, G. E. S., and E. B. Astwood. 1942. The physiological significance of the estrogen:progesterone ratio on vaginal cornification in the rat. Endocrinology 30:295–300.

Jones, E. C., and P. L. Krohn. 1961. The relationships between age, numbers of oocytes and fertility in virgin and multiparous mice. J. Endocrinol. 21:469–496.

Kile, J. C., Jr. 1951. An improved method for the artificial insemination of mice. Anat. Rec. 109: 109–117.

Kliman, B., and H. A. Salhanick. 1952. Relaxation of pubic symphysis of the mouse during the estrous cycle and pseudopregnancy. Proc. Soc. Exp. Biol. Med. 81:201–202.

Laguchev, S. S. 1959. Comparison of the estrous cycles in mice of high and low cancer lines [English transl.]. Bull. Exp. Biol. Med. 48: 1149–1152.

Lamond, D. R. 1958. Infertility associated with extirpation of the olfactory bulbs in female albino mice. Austral. J. Exp. Biol. Med. Sci. 36:103–108.

Lewis, W. H., and E. S. Wright. 1935. On the early development of the mouse egg. Carnegie Inst. Washington Pub. 459:113–144.

Long, J. A. 1912. Studies on early stages of development in rats and mice. III. The living eggs of rats and mice, with a description of apparatus for obtaining and observing them. Univ. Calif. Pub. Zool. 9:105–136.

Long, J. A., and E. L. Mark. 1911. The maturation of the egg of the mouse. Carnegie Inst. Washington Pub. 142:1–72.

Lugo, F. P. 1959. The effect of a diet supplemented with hens' eggs upon the estrus cycle of mice. Anat. Rec. 133:408. (Abstr.)

MacDowell, E. C., and E. M. Lord. 1927. Reproduction in alcoholic mice. Arch. Entwickl. Org. 109:549–583. (Quoted by Strong and Fuller, 1958.)

Marsden, H. M., and F. H. Bronson. 1964. Estrous synchrony in mice: alteration by exposure to male urine. Science 144:1469.

Marsden, H. M., and F. H. Bronson. 1965. Oestrous synchrony in mice: relative roles of male and female environments. J. Endocrinol. 32:313–319.

Merton, H. 1939. Studies on reproduction in the albino mouse. I. The period of gestation and the time of parturition. Proc. Roy. Soc. Edinb. B 58:80–96.

Mirskaia, L., and F. A. E. Crew. 1930. On the genetic nature of the time of attainment of puberty in the female mouse. Quart. J. Exp. Physiol. 20:299–304.

Mizuhara, F., T. Kamei, and T. Fujii. 1958. Studies on the estrous cycle in uniform strains of mice. An application of the dimension method. Endocrinol. Jap. 5:243–246.

Monesi, V. 1962. Relation between X-ray sensitivity and stages of the cell cycle in spermatogonia of the mouse. Radiation Res. 17:809–838.

Mühlbock, O. 1947. On the susceptibility of different strains of mice for oestrone. Acta Brev. Neer. 15: 18–20.

Munford, R. E. 1963. Changes in the mammary glands of rats and mice during pregnancy, lactation and involution. 3. Relation of structural and biochemical changes. J. Endocrinol. 28:35–44.

Nakamura, T. 1957. Cytological studies on abnormal ova in mature ovaries of mice observed at different phases of oestrous cycle. J. Fac. Fish Anim. Husbandry, Hiroshima Univ. 1:343–362.

Nandi, S. 1959. Hormonal control of mammogenesis and lactogenesis in the C3H/HeCrgl mouse. Univ. Calif. Pub. Zool. 65:1–128.

Oakberg, E. F. 1956*a*. A description of spermiogenesis in the mouse and its use in analysis of the cycle of the seminiferous epithelium and germ cell renewal. Amer. J. Anat. 99:391–413.

Oakberg, E. F. 1956*b*. Duration of spermatogenesis in the mouse and timing of stages of the cycle of the seminiferous epithelium. Amer. J. Anat. 99:507–516.

Parkes, A. S. 1925. The age of attainment of sexual maturity of the albino mouse. J. Roy. Microscop. Soc. 315–319. (Quoted by Engle and Rosasco, 1927.)

Parkes, A. S. 1926. Observations on the oestrous cycle of the albino mouse. Proc. Roy. Soc. B 100:151–170.

Parkes, A. S. 1928. The length of the oestrous cycle in the unmated normal mouse: records of 1000 cycles. Brit. J. Exp. Biol. 5:371–377.

Parkes, A. S. 1929. The internal secretions of the ovary. Longmans, Green and Co., London.

Parkes, A. S. 1960. The biology of spermatozoa and artificial insemination, p. 161–263. *In* A. S. Parkes [ed.] Marshall's physiology of reproduction, 3rd ed. Vol. I, Part 2. Longmans, Green, London.

Parkes, A. S. 1961. An olfactory block to pregnancy in mice. Part 2: Hormonal factors involved. Proc. 4th Int. Cong. Anim. Reprod., The Hague: 163–166.

Parkes, A. S., and H. M. Bruce. 1961. Olfactory stimuli in mammalian reproduction. Science 134:1049–1054.

Pfeiffer, C. A., and C. W. Hooker. 1942. Early and late effects of daily treatment with pregnant mare serum upon the ovary of mice of the A strain. Anat. Rec. 84:311–330.

Reynaud, J., and A. Reynaud. 1947. Observations sur la structure du tractus genital des souris femelles castrées à la naissance. Ann. Endocrinol. 8:81–86.

Rietschel, P. E. 1929. Zur Morphologie der Genitalausführungsgänge im Individualcyclus der weissen Maus. Z. Wiss. Zool. 135:428–494.

Ring, J. R. 1944. The estrogen-progesterone induction of sexual receptivity in the spayed female mouse. Endocrinology 34:269–275.

Roosen-Runge, E. C. 1962. The process of spermatogenesis in mammals. Biol. Rev. 37:343–377.

Rudali, G., L. Juliard, N. Yourkovski, and M. Fautrel. 1957. Sur l'âge de maturité sexuelle dans différentes lignées de souris (avec étude particulière de la lignée NLC). J. Physiol. (Paris) 49:657–666.

Runner, M. N. 1959. Embryocidal effect of handling pregnant mice and its prevention with progesterone. Anat. Rec. 133:330–331. (Abstr.)

Runner, M. N., and A. Gates. 1954. Conception in prepuberal mice following artificially induced ovulation and mating. Nature 174:222–223.

Runner, M. N., and A. J. Ladman. 1950. The time

of ovulation and its diurnal regulation in the post-parturitional mouse. Anat. Rec. 108: 343–361.

Shelesnyak, M. C., and A. M. Davies. 1953. Relative ineffectiveness of electrical stimulation of the cervix for inducing pseudopregnancies in the mouse. Endocrinology 52:362–363.

Smith, P. E., and E. T. Engle. 1927. Induction of precocious sexual maturity in the mouse by daily homeo- and heterotransplants. Proc. Soc. Exp. Biol. Med. 24:561–562.

Smithberg, M., and M. N. Runner. 1956. The induction and maintenance of pregnancy in prepuberal mice. J. Exp. Zool. 133:441–457.

Snell, G. D. [ed.] 1941. Biology of the laboratory mouse. Blakiston, Philadelphia. 497 pp.

Snell, G. D., E. Fekete, K. P. Hummel, and L. W. Law. 1940. The relation of mating, ovulation and the estrous smear in the house mouse to time of day. Anat. Rec. 76:39–54.

Snell, G. D., K. P. Hummel, and W. H. Abelmann. 1944. A technique for the artificial insemination of mice. Anat. Rec. 90:243–253.

Soper, E. H. 1963. Ovarian and uterine responses to gonadotrophin in immature mice as related to age. Anat. Rec. 145:352–353. (Abstr.)

Stowell, R. E. 1941. A case of probable superfetation in a mouse. Anat. Rec. 81:215–220.

Strong, L. C., and C. A. Fuller. 1958. Maternal age at time of first litters in mice. J. Gerontol. 13:236–240.

Suntzeff, V., E. L. Burns, M. Moskop, and L. Loeb. 1938. On the proliferative changes taking place in the epithelium of vagina and cervix of mice with advancing age and under the influence of experimentally administered estrogenic hormones. Amer. J. Cancer 32:256–289.

Takasugi, N., and H. A. Bern. 1962. Crystals and concretions in the vaginae of persistent-estrous mice. Proc. Soc. Exp. Biol. Med. 109:622–624.

Thiery, M. 1960. Les variations de la teneur en acide désoxyribonucléique (DNA) des noyaux de l'épithélium vaginal de la souris au cours du cycle oestral. Arch. Biol. 71:389–406.

Thung, P. J., L. M. Boot, and O. Mühlbock. 1956. Senile changes in the oestrous cycle and in ovarian structure in some inbred strains of mice. Acta Endocrinol. 23:8–32.

Togari, C. 1927. On the ovulation of the mouse. Nagoya J. Med. Sci. 2:17–50.

Váczy, L., T. Sándor, and D. Juhos. 1955. Die histochemische Untersuchung des Östruszyklus. Acta Endocrinol. 18:87–98.

Van der Lee, S., and L. M. Boot. 1955. Spontaneous pseudopregnancy in mice. Acta Physiol. Pharmacol. Neer. 4:442–444.

Van der Lee, S., and L. M. Boot. 1956. Spontaneous pseudopregnancy in mice. II. Acta Physiol. Pharmacol. Neer. 5:213–215. (Abstr.)

Van Ebbenhorst Tengbergen, W. J. P. R. 1955. The morphology of the mouse anterior pituitary during the oestrous cycle. Acta Endocrinol. 18:213–218.

Vivien, J. H. 1950. Époque de la différenciation hypophysaire chez la souris albinos. Compt. Rend. Soc. Biol. 144:284–287.

Weir, M. W., and J. C. DeFries. 1963. Blocking of pregnancy in mice as a function of stress. Psychol. Rep. 13:365–366.

Whitten, W. K. 1956*a*. Modification of the oestrous cycle of the mouse by external stimuli associated with the male. J. Endocrinol. 13: 399–404.

Whitten, W. K. 1956*b*. The effect of removal of the olfactory bulbs on the gonads of mice. J. Endocrinol. 14:160–163.

Whitten, W. K. 1957. Effect of exteroceptive factors on the oestrous cycle of mice. Nature 180: 1436.

Whitten, W. K. 1959. Occurrence of anoestrus in mice caged in groups. J. Endocrinol. 18: 102–107.

Whitten, W. K., and C. P. Dagg. 1961. Influence of spermatozoa on the cleavage rate of mouse eggs. J. Exp. Zool. 148:173–183.

Yoon, C. H. 1955. Homeostasis associated with heterozygosity in the genetics of time of vaginal opening in the house mouse. Genetics 40:297–309.

Young, W. C. 1941. Observations and experiments on mating behavior in female mammals. Quart. Rev. Biol. 16:135–156.

Young, W. C. [ed.] 1961. Sex and internal secretions. Vol. I, II. Williams and Wilkins, Baltimore.

Zarrow, M. X., and E. D. Wilson. 1961. The influence of age on superovulation in the immature rat and mouse. Endocrinology 69:851–855.

12

Early Embryology[1]

George D. Snell and Leroy C. Stevens

This description of the early embryology of the mouse is a revision of Snell's (1941) description in the first edition of the *Biology of the Laboratory Mouse*. Most of the new information was obtained from a search of the literature published since then, and some of this was based on investigations on the embryology of the rat. In general there is much similarity in the development of Muridae, but we will make it clear when we discuss information obtained from the rat. Since 1941 there have been many investigations devoted to certain aspects of mouse embryology, but only a few have attempted a comprehensive study. For other descriptions the reader is referred to Boyd and Hamilton (1952), Bonnevie (1950), and Rugh (1964). Graves (1945) described early development of the golden hamster. Grüneberg (1943) has provided a very useful description of mouse development from 9⅓ to 18⅓ days, and Otis and Brent (1954) have compared human and mouse developmental stages. Grüneberg (1963) surveyed the field of developmental pathology of the skeleton, including many inherited malformations of the mouse.

The material used in this study is the same as that used in 1941 and consists of sections of embryos spaced at 6-hour intervals from 4 to 9 days. In some cases 10 or more embryos of a single stage have been sectioned. The sections were prepared by Olive Bartholomew, Elizabeth Fekete, and G. D. Snell, using a technique described by Fekete et al. (1940). In most cases the females used as mothers were hybrids between two strains and fathers were from a third strain, thus giving both embryos and mothers a maximum of hybrid vigor. Because of this the stages described here are usually earlier, often by as much as a day, than comparable stages described by other authors. Although this procedure gave embryos which developed rapidly and were usually normal, it did not eliminate variability. We have not attempted to describe the variations noted in the rate of development of embryos or in the rate of development or form of separate parts. It should be emphasized, however, that the range of variation in these respects is considerable.

PREIMPLANTATION

Eggs

In several mammalian species, including the mouse, mitotic activity of oogonia ceases and the prophase of meiosis is known to be initiated about 5 days before birth (Mandl, 1963), and primary oocytes may progress to the dictyate condition by birth (Mintz, 1960). It is these cells which transform into the definitive germ cells, and there is no somatic contribution. In adult mice the interval between the dictyate (resting) stage and metaphase of the second meiotic division which occurs just before ovulation is about 12 hours (Edwards and Gates, 1959).

According to Edwards and Gates (1959), oocytes apparently remain in the dictyate stage until stimulated by luteinizing hormone to complete their maturation. They found that 2 hours

[1] The writing of this chapter was supported in part by Public Health Service Research Grant CA 02662 from the National Cancer Institute, and in part by a grant from The John A. Hartford Foundation, Inc.

after injection of luteinizing hormone oocytes were in late prophase, and from 5 to 8 hours all were in metaphase of the first meiotic division. At 14 hours all eggs were ovulated and were in the metaphase of the second maturation division.

The ovarian oocyte of the rabbit, mouse, and mole has been shown to be bilaterally symmetrical, and this symmetry is traceable throughout early development to the primitive streak stage (Jones-Seaton, 1950*a*). Mouse and rat eggs show animal and vegetal poles which are indicated by the eccentric germinal vesicle or by the polar bodies (Jones-Seaton, 1950*b*; Dalcq and Van Egmond, 1953; Dalcq, 1956).

The mouse ovary is almost completely enveloped by a bursa, and it is very unlikely that there can be any internal escape of the eggs into the abdominal cavity. No case of naturally occurring extrauterine pregnancy has been reported in the mouse.

The freshly ovulated mouse egg is enveloped by the vitelline membrane, a product of the egg; the zona pellucida, a product of the surrounding follicular cells; and by the cumulus oophorus composed of follicular cells and an intercellular matrix consisting of acidic mucopolysaccharide, hyaluronic acid, and protein (Braden, 1962).

The zona pellucida is a mucoprotein. It serves to maintain the normal cleavage pattern and to prevent fusion of eggs (Mintz, 1962). Eggs will adhere to one another when the zona is artificially removed with enzymes. This has served as the basis of methods for synthesizing genetic mosaics (Mintz, 1964, 1965). The zona is not necessary for development in mice (Mintz, 1962). Nicholas and Hall (1942) separated the first two blastomeres of the rat egg, transplanted them to the uterus, and obtained implantation and egg cylinders of early embryos. Gwatkin (1964) removed the zona enzymatically and observed fusions of

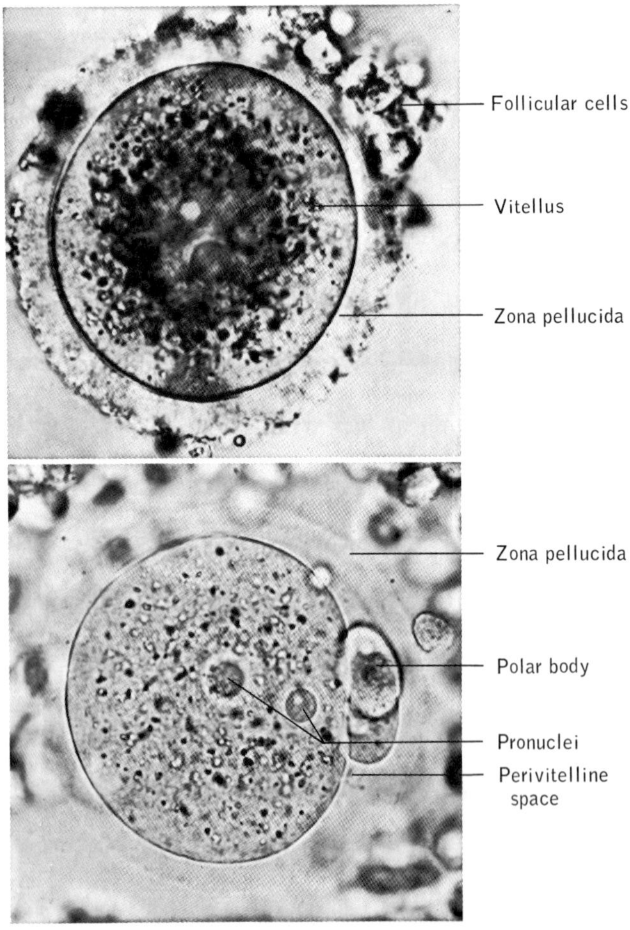

Fig. 12-1. Photographs of mouse eggs (×600). *A.* Egg removed from ovary. *B.* Fertilized egg from oviduct 20 hours after copulation. Two polar bodies and pronuclei. (*A from Lewis and Wright,* 1935.)

eggs. At the four-cell stage blastomeres separated easily and sometimes gave rise to miniature blastocysts. He concluded that the importance of the zona was in preventing loss of blastomeres and fusion of eggs. He also found that the zona in the mouse was not a complete barrier to virus infection of the ovum.

The innermost cells of the cumulus oophorus are tightly packed and oriented radially to form the corona radiata. Chang et al. (1962) have shown that the corona radiata plays a major role in the fertilization of rabbit ova or in the maintenance of their fertilizability. Sperm penetration may be dependent upon the corona. More probably the fertilizability of the ova is protected by the corona and would disappear shortly after its removal. The outside diameter of the zona pellucida of mature ovarian mouse eggs averages about 95 μ. Following fertilization the zona pellucida expands until its outer diameter becomes about 113 μ. This is just within the limits of visibility for the unaided eye (Lewis and Wright, 1935). Within the vitellus is the egg nucleus, not clearly visible in living eggs such as the one shown in Fig. 12-1, but easily seen in fixed and stained material.

Sperm

The sperm head contains one complete set of chromosomes, and the middle piece contributes mitochondria. There is some evidence that Golgi material is also carried by the sperm (Gresson, 1940; Huber, 1915; Lams and Doorme, 1907).

The acrosome is a caplike structure on the anterior part of the sperm. Its function is thought to be to carry hyaluronidase, which probably enables the sperm head to pass through the cumulus oophorus (Austin and Bishop, 1958). The perforatorium lies under the acrosome and is thought to be a modification of the sperm nucleus. Austin and Bishop concluded that the acrosome becomes modified in sperm passing through the female reproductive tract and is detached before the sperm penetrates the zona pellucida. The perforatorium may carry a lysin capable of altering the zona in such a way as to permit penetration into the perivitelline space.

Dziuk and Runner (1960) found that sperm motility apparently influences fertilization. The percentage of blastocysts increases as the motility score of the sperm suspension increases. Cooled mouse sperm retains fertilizing capacity for at least 24 hours.

Numerous morphological and behavioral differences have been found in the sperm from eight inbred strains of mice (Beatty and Sharma, 1960). Sharma (1960) investigated the phenotype of mouse spermatozoa in four inbred strains and their F_1 crosses and found heterosis in F_1 crosses.

Insemination

The mean number of sperm ejaculated into rats is 58,060,000 (Blandau and Money, 1944; Blandau and Odor, 1949). Within a few minutes after ejaculation the sperm of the rat are disseminated throughout the uterine cornua. This rapid dispersion is brought about by the presence of large quantities of uterine fluid and marked muscular activity of the cornua. In some rats, within 15 minutes after the sperm have entered the cornua, some may be observed in the ampullae of the oviducts. In all animals examined 1 hour after ejaculation, sperm had reached the ampullae. These observations are similar to those of Lewis and Wright (1935) for the mouse.

Fertilization

Fertilization in the mouse occurs in the ampullae at the upper end of each oviduct where the eggs are found, usually in clumps, about 2 hours after ovulation. Sperm are usually present in the ampullae at the time of ovulation, but there is a delay of several hours before the sperm penetrates the egg (Braden, 1954, 1960; Edwards, 1955). Apparently the egg membranes, particularly the cumulus oophorus, need some form of maturation before sperm penetration can occur (Braden, 1960, 1962). Braden (1958) suggested that cells of the cumulus and the corona radiata are less dense and loosen earlier in some strains of mice than in others. The delay between ovulation and penetration of eggs by sperm is considerably shorter (1 to 3 hours) in artificially ovulated mice than in mice ovulating naturally (3 to 5 hours).

Sperm penetration of the mouse egg normally takes place before dissolution of the cumulus (Lewis and Wright, 1935). The number of sperm found at the site of fertilization is small, and it is generally considered that the hyaluronidase carried by the individual sperm enables it to make its own way through the cumulus to the zona pellucida (Leonard, 1947; Braden, 1962). In the mouse the first sperm to arrive at the egg usually penetrates and fertilizes it.

Bishop and Tyler (1956) have demonstrated a sperm agglutinin of mammalian eggs, including eggs of mice, acting like fertilizins of invertebrate eggs. They tentatively indicated the zona pellucida as the source of mammalian "fertilizin."

According to Edwards and Gates (1959) mouse sperm penetrates the cumulus and zona in approximately 1 hour and remains in the perivitelline space for 50 minutes. Pronuclei are formed 4¼ hours later, and the first cleavage division occurs at about 25 hours after penetration of sperm. Usually only one sperm enters each egg. Almost immediately after entry, which may occur through any part of the egg's surface, the vitellus shrinks slightly in size and the zona pellucida expands, so that a perivitelline space forms between them (Lewis and Wright, 1935; Pincus, 1936).

The zona pellucida increases in thickness and changes after the entry of the first sperm so that the chance of subsequent penetration is reduced (Braden, 1958). This "zona reaction" is apparently stimulated in the vitelline cortex by the sperm head. Upon stimulation, the vitelline cortex may release a substance which traverses the perivitelline space and causes an alteration in the penetrability of the zona pellucida. But Austin and Braden (1956) suggested that the "zona reaction" is initiated in the zona itself at the point of entry of the first sperm and then is propagated through the zona.

In the rat and mouse the sperm passes rapidly through the thick zona and the head apparently adheres almost immediately to the surface of the vitellus (Austin and Braden, 1956). Shortly after this the vitellus loses its adhesiveness and supplementary sperm remain free in the perivitelline space. Thus the attachment of the sperm head to the vitellus probably elicits both the "zona reaction" and the block to polyspermy at the surface of the vitellus. The head lies flat on the vitelline surface for about ½ hour. It then becomes engulfed by the vitellus and becomes the male pronucleus which moves toward the center of the egg. The female pronucleus is larger than the male and is close to the second polar body. After fertilization, the pronuclei come together and fuse in approximately the center of the zygote to form the zygote nucleus. At the first cleavage division the nuclear walls break down, the chromosomes split longitudinally, and one-half of each split chromosome is carried to each daughter cell. Hence at this division, as at all future somatic divisions, each cell receives a full complement of chromosomes from each parent.

Within 2 to 3 hours after fertilization the second polar body is formed and is everted into the perivitelline space. The first polar body has already been formed in the ovarian egg. Fertilization is discussed in Chap. 11.

Cleavage

Preimplantation development in many mammals may vary in some details, but in general a similar pattern exists throughout cleavage and the formation of the morula. Edwards (1957) has published a table of time intervals between cleavages up to the sixth division for the mouse.

Jones-Seaton (1950*a*) has shown that the dorsal regions of the rat, rabbit, mouse, and mole ova are marked by a cortical zone of ribonucleic acid and the ventral by a vacuolate cortex. The plane of the first cleavage in the rat has no set relation to the dorsoventral axis. When the early cleavages are unequal, the larger blastomeres divide more rapidly.

The first two cleavages in the mouse occur while the eggs are still in the oviduct. The first cleavage occurs about 24 hours after copulation and results in two cells almost equal in size (Fig. 12-2*A*). Following divisions occur somewhat more rapidly, giving rise to four-cell, eight-cell stages, etc., and are usually nearly synchronous. Occasionally, however, eggs are found with some divisions completed and others still incomplete, hence showing an odd number of cells (Fig. 12-2*B*). The actual act of division requires only 5 or 10 minutes; the interval between divisions lasts about 12 hours. Eggs of 16 cells or more, in which no cavity has appeared, are called morulae. Eggs usually reach this stage about 60 hours after fertilization and pass from the oviduct, through which they have been gradually moving, into the uterus, some 6 to 12 hours later (Lewis and Wright, 1935). This is subject to considerable variation, however, and in one study passage into the uterus at 4 days was found to be the rule (Burckhard, 1901).

Whitten and Dagg (1961) demonstrated that some factor(s) contributed by the spermatozoa can influence the intermitotic times during early cleavages of the eggs of BALB/c mice. The nature of this influence is unknown.

Smith (1956) showed that the mutant gene t^{12} acts prior to blastocoel formation. Homozygous animals cease development just prior to blastocyst formation. This is the earliest action of any gene or chromosomal deficiency (Geyer-Duszynska, 1964) known in the mouse. Mintz (1964) showed that blastocyst formation does not depend upon an early morphogenetic prepattern set before or during cleavage. Any normal cell, rather than certain ones, may be involved at *first* in regions

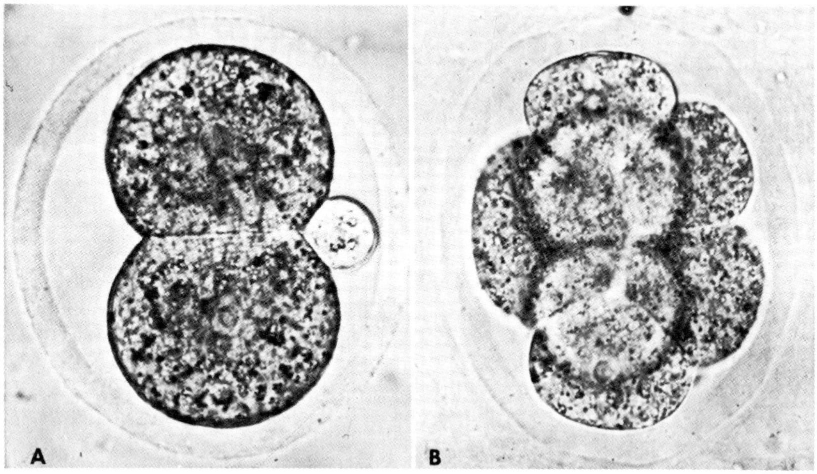

Fig. 12-2 Photographs of mouse eggs (×600). *A.* Two-cell egg from oviduct 24 hours after copulation. Large second polar body and disintegrating first (on opposite side). *B.* Seven-cell egg from oviduct 48 hours after copulation. Note one cell on left larger than the rest. Division of this cell would give the eight-cell stage. (*From Lewis and Wright,* 1935.)

which will become inner cell mass or trophoblast by other, epigenetic rather than prepatterned mechanisms. At the eight-cell stage and even beyond, the egg is still entirely labile, and the decision for further regional differentiation is made much closer to the blastocyst stage. In spite of the regulative capacity of the egg, all of its cells continue to undergo progressive change in increased adhesiveness between cells and increased cell motility during cleavage. Pronounced modifications in ribonucleic acid and protein synthesis are also occurring (Mintz, 1964).

Mintz (1965) has obtained genetic mosaic mice by aggregating, during cleavage stages, the blastomeres of two embryos of different genotype into a single cluster, and by transferring the developing aggregates to the uterus of a surrogate mother. These remarkable animals include hermaphrodites and mosaics for alleles at the *H-2* and *a* loci.

Blastocysts

Shortly after the egg has entered the uterus and usually sometime after it has reached the 32-cell stage, an eccentrically located fluid-filled cavity appears among the cells of the morula. This enlarges rapidly to produce the blastocoel (Fig. 12-3). The cavity is bounded by a single layer of cells except on one side where most of the cells are grouped to form a structure called the inner cell mass. Eggs in this stage are known as blastocysts. Blastocysts settle into the uterine crypts between 4⅓ and 4½ days (Eaton and Green, 1963).

It is not quite clear what part the uterine environment and what part the blastocyst itself plays in shedding the zona pellucida. In normal development these two factors probably contribute jointly (Tarkowski, 1962). It is known that the uterine environment is more acid than the fallopian tubes and that lowering the pH of the medium dissolves the zona. However, the blastocysts liberate themselves from their zonae when transferred to extrauterine sites.

Oviducal segmented eggs of the mouse, transplanted beneath the kidney capsule, gave rise to trophoblast and extraembryonic membranes, but not to embryonic derivatives (Kirby, 1962*a*). Uterine blastocysts (but not blastocysts locked in the fallopian tubes) similarly transplanted gave rise to embryonic structures. Kirby concluded that possibly the denial of a "uterine factor" renders mouse eggs incapable of developing structures other than extraembryonic ones. One of us (L. C. S.) transplated segmenting ova of strain 129 mice to the adult testis and obtained disorganized teratoid growths composed of a wide variety of mature tissues indicating that a "uterine factor" is not necessary for development. Kirby (1962*b*) found a higher proportion of successful transplants of uterine blastocysts into the testis than for any other extrauterine organ tried.

In a study on the fine structure of the 5-day rat blastocyst, Schlafke and Enders (1963) found that the cells of the trophoblast form a single continuous layer of low cuboidal cells. Microvillae are irregularly distributed on the apical sur-

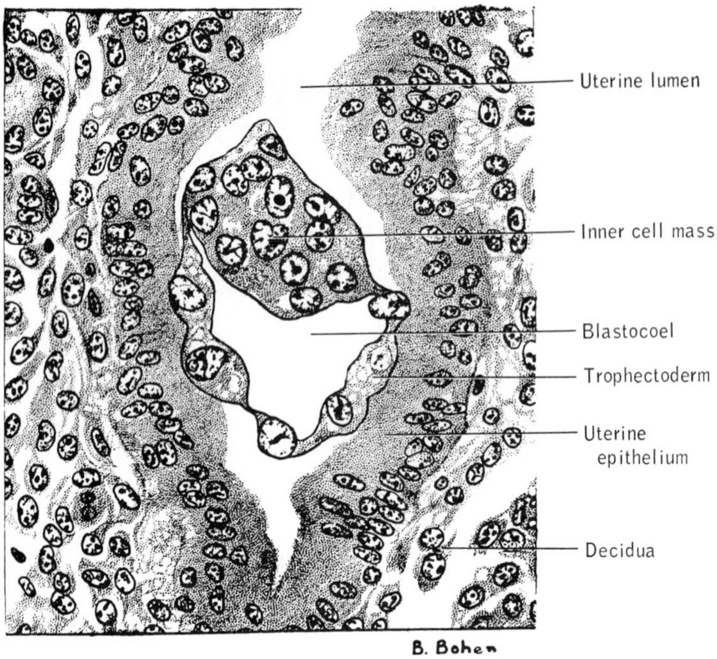

Fig. 12-3. Blastocyst in uterine crypt 4 days after copulation. Projection drawing (×600).

faces projecting toward the zona. At 7 days the microvillae arc less numerous and less regular than previously. There is no evidence of differentiation within the inner cell mass.

IMPLANTATION AND EARLY GROWTH

The uterus in the mouse is duplex, consisting of two horns uniting just anterior to their junction with the vagina, and each attached to the dorsal body wall by a mesentery, the mesometrium (Fig. 12-4). There are two layers of muscle in each horn, an outer longitudinal layer and an inner circular layer. The uterine lumen is lined with epithelium. Between the epithelium and the muscle layers is the mucosa, a tissue forming the bulk of the uterine wall. The epithelium is indented by numerous small crypts.

Very shortly after entering the uterus the eggs become spaced more or less evenly throughout its length, and each egg finds its way into a uterine crypt on the ventral or antimesometrial side of the lumen, thereby coming into close contact with the uterine epithelium (Fig. 12-3). The presence of the blastocysts quickly sets up changes at the implantation site. Within a few hours the epithelium begins to loosen, and its nuclei show degenerative changes (Fig. 12-5). Within 15 hours it is sloughed entirely (Fig. 12-6). At the same time active growth commences in the mucosa, so that by 1 day after implantation (5 days after mating) there is an appreciable swelling in the uterus at the implantation site. The swollen mucosa at the implantation site is known as decidua.

Hollander and Strong (1950) observed that in the mouse the external appearance of pregnant uteri and the frequency of placental fusion suggest that blastocysts implant at random after being scattered throughout each uterine horn by its churning. Krehbiel (1962) studied the distribution of ova in the rat uterus and reached a similar conclusion.

Dickerman and Noyes (1960) showed that on the fifth day of pregnancy in rats both the blastocyst and the endometrium have reached a specific stage of development favorable for implantation. The ovum must have reached the stage of development of the 5-day blastocyst, and the endometrium must be "ripe" but not overripe for implantation to begin. The uterine environment is unsuitable for implantation on both the fourth and sixth days of pseudopregnancy.

Up to the time of implantation there has been no growth in size in the egg. Cleavage has resulted in a division of the egg, originally one large cell, into numerous smaller cells, but little if any new protoplasm has been formed in the process. Beginning with implantation, however, rapid growth commences. At first the blastocoel enlarges while the inner cell mass assumes a flattened cup shape with the concave face towards the cavity (Fig. 12-5). In the living condition the blastocoel is probably distended with fluid, and its walls are tightly pressed against the uterine epithelium, but in fixed material at this stage there is always some collapse. This initial expansion of the blastocoel requires only a few hours and is quickly followed by a growth of the inner cell mass down into the enlarged cavity (Fig. 12-6). The blastocoel and inner cell mass are called thereafter the yolk cavity and the egg cylinder. A comparison of Figs. 12-7, 12-8, 12-10, and 12-12 will show the rapid growth of the egg cylinder occurring during the next $2\frac{1}{2}$ or 3 days.

Formation of endoderm

As the blastocoel begins to enlarge, the inner cell mass is composed of two types of cells (Fig. 12-5). Adjacent to the blastocoel is a single layer of darkly staining cells, the endoderm, one of the three primary germ layers. The rest of the blastocyst is composed of ectoderm, divided into the ectoderm of the inner cell mass, and the trophectoderm, a single-celled layer bounding the blastocoel ventrally and laterally. The trophecto-

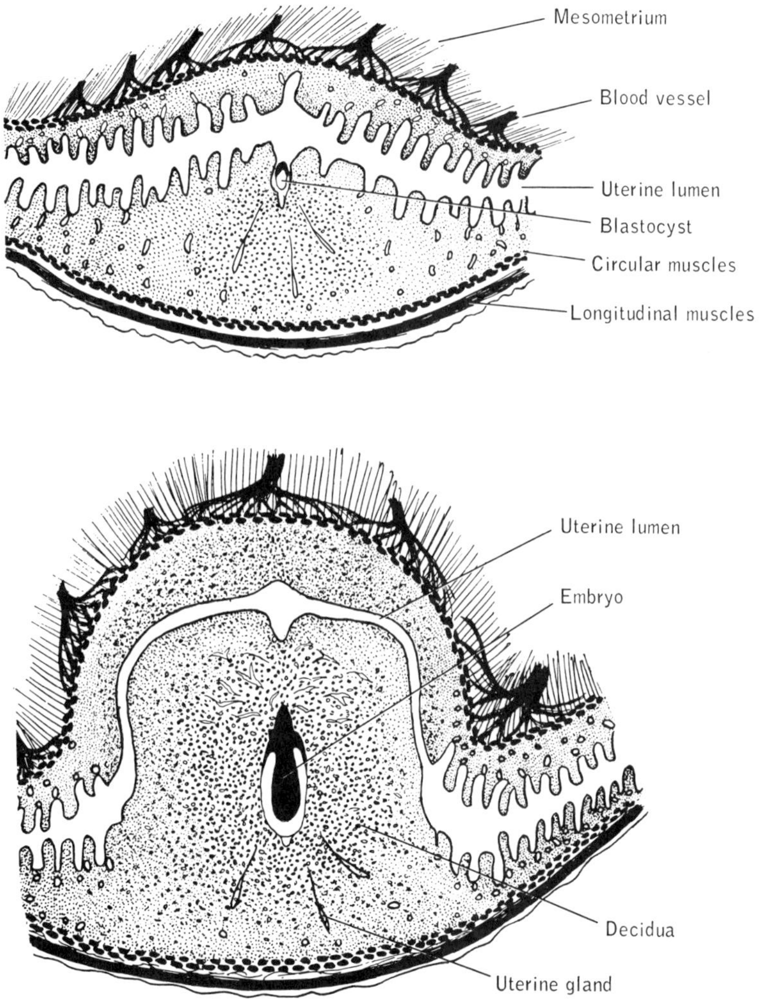

Fig. 12-4. Diagrams showing implantation (×45). *A*. Longitudinal section through horn of uterus about 5 days after mating. An ovum has recently become implanted in one of the uterine crypts. *B*. Longitudinal section through implantation site about 7 days after mating. (*After Burckhard*, 1901.)

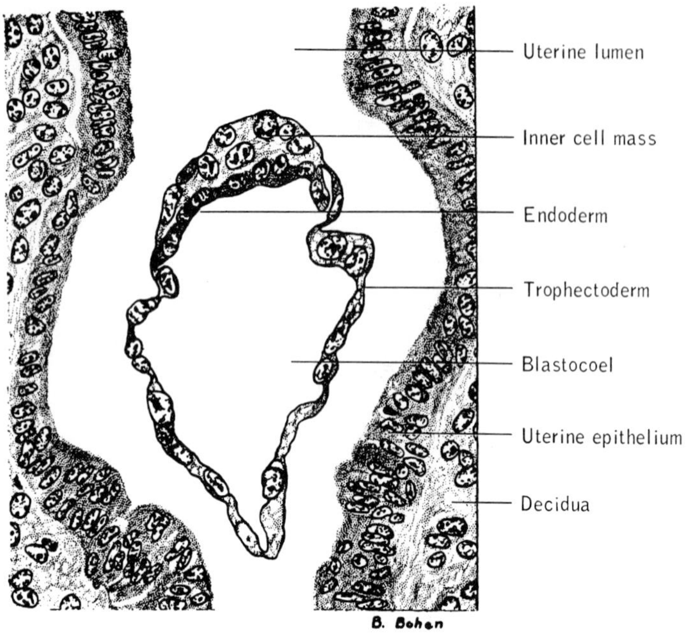

Fig. 12-5. Section of implanting blastocyst 4 days 5 hours after mating. Projection drawing (×400).

derm (troph from the Greek word for nourishment) derives its name from its probable role in the nourishment of the young embryo. The mesoderm has not yet appeared.

Shortly after the first appearance of the endoderm, single cells or strands of cells grow out from its margin along the inner surface of the trophectoderm. At first these cells are few and widely separated (Fig. 12-7, 12-8), but by 6½ days they lie evenly spaced and quite close to-

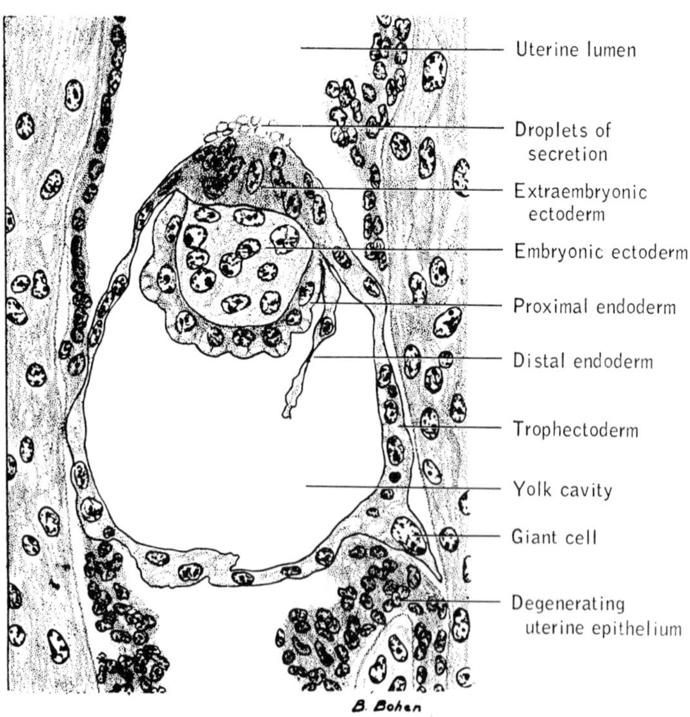

Fig. 12-6. Longitudinal section of early egg cylinder stage at 4 days 15 hours after mating. Projection drawing (×400).

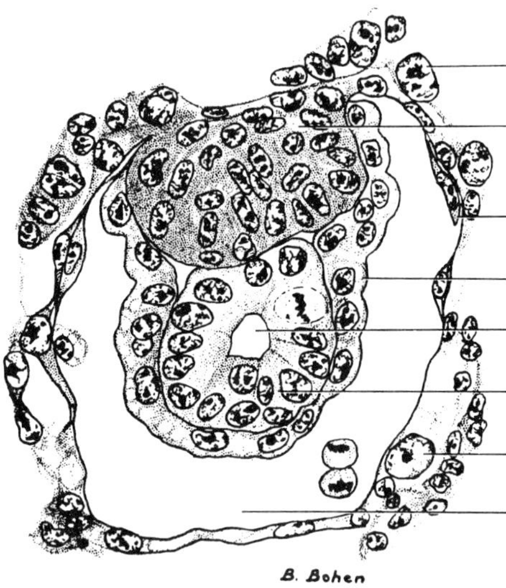

Fig. 12-7. Longitudinal section of 5- or 6-day egg cylinder. Projection drawing (×400).

gether over the trophectoderm's entire inner surface (Fig. 12-10). The layer of cells thus formed is known as the distal (or parietal) endoderm. Meanwhile the inner cell mass has grown down into the yolk cavity to form the egg cylinder. This is composed of an inner mass of ectoderm cells, which represents considerably more than the ectoderm of the future embryo, and an outer layer of endoderm cells (Fig. 12-8). This layer of endoderm cells bounding the egg cylinder is known as the proximal (or visceral) endoderm. The endoderm is thus divided into two distinct parts, distal and proximal, lining the distal and proximal walls of the yolk cavity.

Embryonic and extraembryonic ectoderm

At about 4½ days, the egg cylinder ectoderm, first beginning to form, is divided into two parts: a dorsal, more darkly staining (when counterstained with congo red) region with elongated nuclei, and a ventral, more lightly staining portion with round nuclei (Fig. 12-6). The dorsal part gives rise to various extraembryonic structures and is called the extraembryonic ectoderm; the ventral part gives rise to the ectoderm of the embryo proper and is called the embryonic ectoderm. Although the difference in staining reaction and in the shape of the nuclei has disappeared by 5½ days, the division between the two regions is still quite distinct (Fig. 12-8). Strictly speaking the trophectoderm is also extraembryonic ectoderm, but as a matter of convenience the term will be used only for the extraembryonic ectoderm of the egg cylinder.

At about 5 days a cleft or cavity, the proamniotic cavity, appears in the embryonic ectoderm (Fig. 12-7). At this stage, according to Bonnevie (1950), there is evidence of bilateral symmetry. The proamniotic cavity is covered on what may be the anterior side of the embryo by a single layer of cells, whereas on the posterior side the cell layer is thicker. Bonnevie suggested that this thickening may represent a localized growth center within the posterior-median line of the inner cell mass. The formation of the proamniotic cavity in the embryonic ectoderm is followed very shortly by the appearance of a similar cleft in the extraembryonic ectoderm and by the fusion of these two, so that by 5½ days the egg cylinder contains a narrow lumen (Fig. 12-8).

Ectoplacental cone

Beginning after 5 or 5½ days of active growth, the dorsal end of the extraembryonic ectoderm gives rise to a new structure, the ectoplacental cone, which joins the egg cylinder ventrally and extends dorsally towards the lumen of the uterus (Fig. 12-8). This develops rapidly, its cells showing numerous mitoses, and by 6½ days it composes almost one-half the total length of the embryo. Its structure, particularly at the upper extremity, is porous, and the interstices between the strands of cells that compose it soon become infiltrated with maternal blood (Fig. 12-10). In the rat the outermost cells of the cone become true giant cells on the seventh day. This suggests that the cone cells in the 6-day embryo represent undifferentiated trophoblast (Bridgman, 1948). In later stages it becomes part of the placenta.

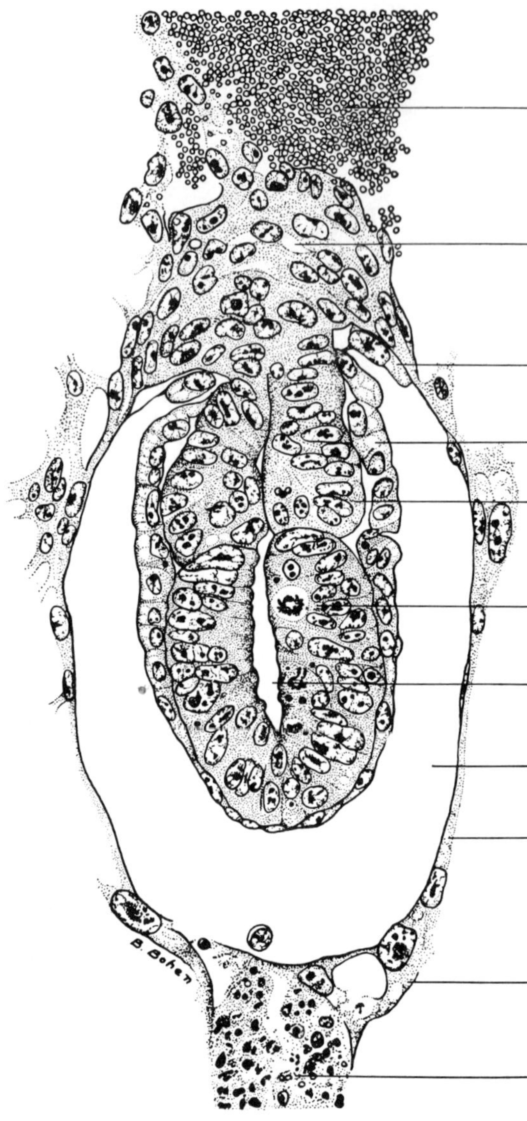

Fig. 12-8. Longitudinal section at 5 days 12 hours after mating. Projection drawing (×300).

Inversion of germ layers

At 5½ days (Fig. 12-8) the egg cylinder consists of a double wall enclosing a narrow lumen, the inner layer of which is composed of ectoderm, the outer of endoderm. This relation of ectoderm and endoderm, found in the mouse, rat, rabbit, guinea pig, and their close relatives, proved very puzzling to early embryologists, being the reverse of that found in all other chordates. It has been called the inversion of the germ layers. Although at first sight it seems to indicate a drastic alteration in early development, actually there is no very fundamental change in the relations of the important structures. Those changes that are involved are easily understood from a comparison of the early development of the mouse with that of a primitive rodent, the thirteen-striped ground squirrel. Three comparable stages for each species are shown diagrammatically in Fig. 12-9. Beginning students of embryology will want to refer again to this figure after completing the study of later stages in the mouse.

In primitive rodents, as represented by the thirteen-striped ground squirrel, the embryonic area (embryonic ectoderm and underlying endoderm) forms a disc that overlies an almost spherical yolk cavity. In the mouse, the embryonic area forms a deep cup pushed far down into the yolk cavity, which thereby is greatly reduced in size. During some period in the evolution of the mouse the embryonic area invaginated into the yolk cavity, the curvature of the embryonic area thereby being reversed and the relation of ectoderm and endoderm inverted. The change is comparable to that produced when a rubber ball has one side pushed in, being altered thereby from a sphere to a cup.

In Fig. 12-9*D* the lumen of the egg cylinder is shown extending through the ectoplacental cone to the outside. This condition may be the excep-

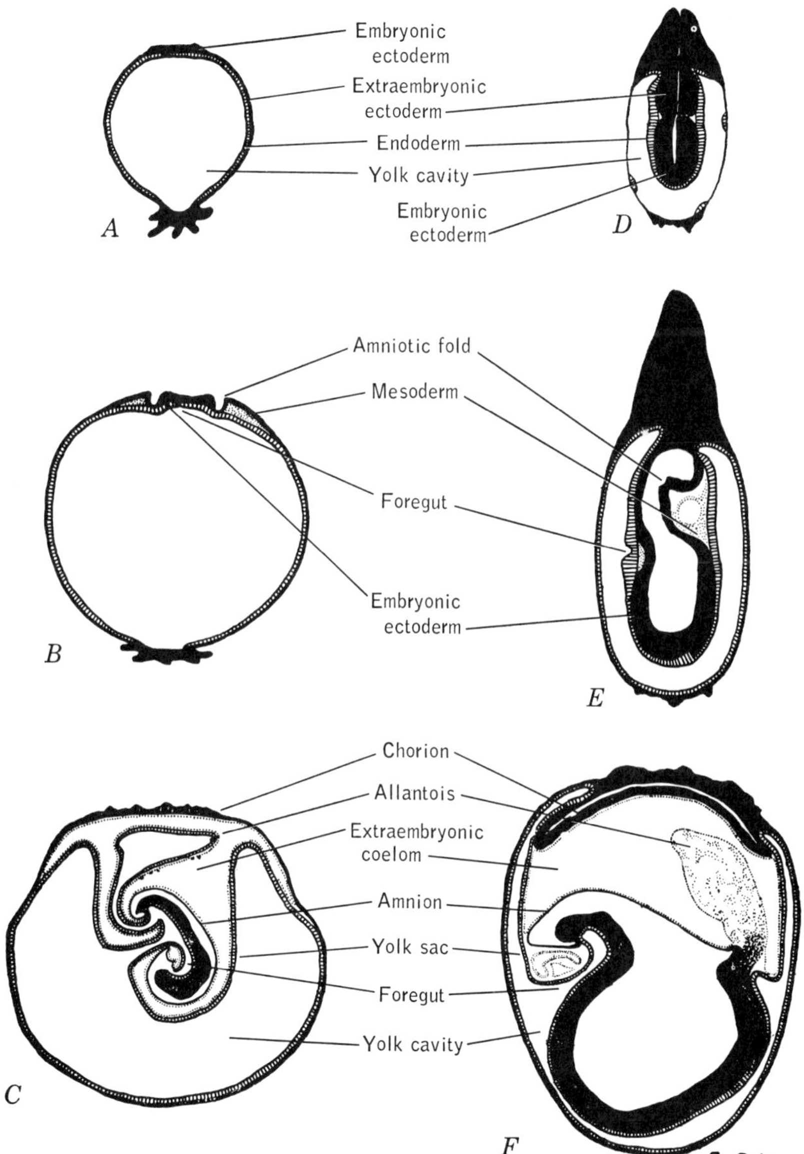

Fig. 12-9. Diagram comparing the early stages of development in a primitive rodent, the thirteen-striped ground squirrel, and in the mouse. *A, B,* and *C*. The thirteen-striped ground squirrel. (*After Mossman and Weisfeldt,* 1935). *D, E,* and *F*. The mouse (*Mus musculus*). (*A* and *B*, ×24; *C*, ×8; *D* and *E*, ×100; *F*, ×50.)

tion rather than the rule, but it has been described by Sobotta (1911) and Melissinos (1907), and we have found it in a few cases in our material. It is significant evidence for the theory that the inversion of the germ layers is due to an invagination of the embryonic area. Bonnevie (1950) described a sloughing of cells from the lining of the proamniotic cavity and suggested that they may be expelled into the ectoplacenta through this lumen.

Further evidence is provided by the later development of the thirteen-striped ground squirrel (Fig. 12-9*C*). In this species the whole embryo sinks down into the yolk cavity, carrying the splanchnopleure with it. The splanchnopleure is thereby inverted, but no inversion of embryonic ectoderm and endoderm occurs because of the advanced development of the embryo at the time. However, if the sinking or invagination of the embryonic area were pushed back to an earlier period of development, the condition found in the mouse would result.

One interesting consequence of the inversion of the germ layers is a very compact form of early development. Much seemingly waste space in the yolk cavity is eliminated. Note that the drawings of the mouse embryos in Fig. 12-9 are at a higher scale of magnification than those of the ground squirrel embryos. Actually, at comparable stages of early development, the total volume of a mouse embryo is perhaps one-fiftieth that of the total volume of an embryo of the ground squirrel. This reduction in total volume involves little if any reduction in the volume of the embryonic area proper.

Primitive streak and mesoderm formation

At 6½ days the middle germ layer or mesoderm makes its appearance (Fig. 12-10). The first mesoderm cells are formed at the posterior end of the embryo by delamination from a narrow strip of embryonic ectoderm extending from the line of junction of the embryonic and extraembryonic ectoderm anteriorly to about halfway to the tip of the egg cylinder. This strip of ectoderm is known as the primitive streak, and since it lies at the posterior end of the embryo an anteroposterior axis is at once established with its appearance.

In our material we have noted that from 5 to 5½ days the egg cylinder and more particularly the proamniotic cavity, instead of being round in cross-section, are slightly flattened along an axis perpendicular to the mesometrium. This is the same as the future anteroposterior axis. However, it cannot be determined until the appearance of the mesoderm which end of the axis is anterior and which end posterior. With the appearance of the mesoderm the flattening of the egg cylinder, if any, is along the opposite axis.

Mesoderm cells form a loose tissue of very characteristic appearance. They multiply rapidly, wedging their way laterally between ectoderm and endoderm toward the anterior margin of the egg cylinder (Fig. 12-14*A*). The forward growth is particularly rapid along the line marking the junction between embryonic and extraembryonic ectoderm, and in this line mesoderm may be found at the anterior margin of the egg cylinder about 12 hours after the first mesoderm cells appeared (Fig. 12-12). Elsewhere the two lateral wings of mesoderm do not penetrate to the midsagittal region until much later. Some mesoderm cells also push between the extraembryonic ectoderm and the adjacent endoderm, thus leaving the region of the embryo proper. These mesoderm cells, for the most part, are destined to take part in the formation of the yolk sac, an extraembryonic membrane which later envelops the embryo and which is discarded at birth.

Bennett and Dunn (1960) described the effects of the homozygous t^{w18}/t^{w18} genotype on mouse development. During the seventh day the main abnormality was a thickening of the wall of the egg cylinder in the region of the primitive streak, just at the time when the primitive streak begins to form. On the eighth day the overgrowth of the primitive streak produced a median bulge projecting dorsally into the proamniotic cavity. Since the swollen area was still covered with the neural material normally overlying the primitive streak, it produced, in effect, two neural grooves separated by the median mass of overgrown primitive-streak material.

Orientation of embryo in uterus

Since the primitive streak is at the posterior margin of the egg cylinder, its formation, heralded by the appearance of the mesoderm, establishes an anteroposterior axis in the embryo. It is appropriate at this point to consider how this axis and the other axes of the embryo are oriented in relation to the uterus.

At the time of implantation the embryo settles to the ventral or antimesometrial side of the uterus. When it first implants, the inner cell mass is up or towards the mesometrium, the blastocoel is down or away from the mesometrium (Fig.

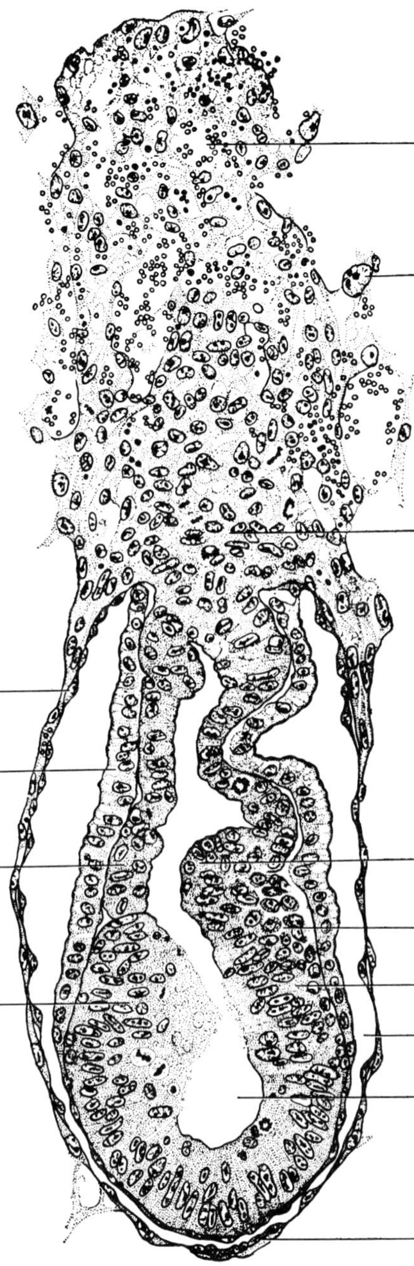

Fig. 12-10. Sagittal section of mouse embryo of 6 days 13 hours, showing early stage of mesoderm formation. Projection drawing (×300).

12-4*A*). In terms of an older embryo, the ectoplacental cone is up and the embryonic portion of the egg cylinder is down (Fig. 12-4*B*). The dorsoventral axis of the embryo is thus parallel to the mesometrium and perpendicular to the long axis of the uterus (Fig. 12-11). The anteroposterior axis of the embryo likewise has a definite orientation with respect to the uterus, being as a rule perpendicular to the mesometrium. Departures from this orientation by as much as 45°

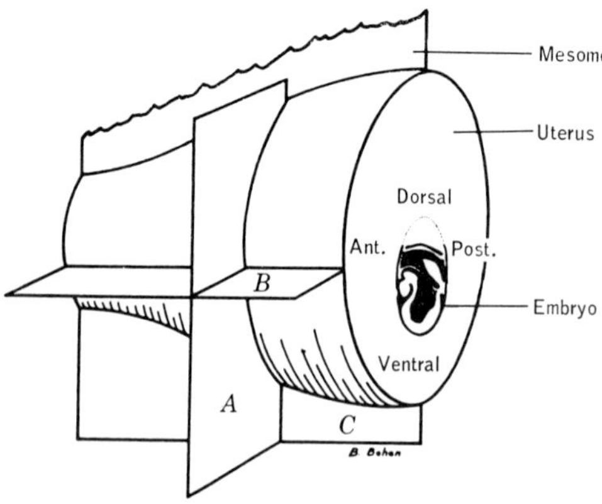

Fig. 12-11. Diagram showing the orientation of an 8-day embryo in the uterus and of the planes in which sections are cut. Plane *A*: transverse to uterus, sagittal to embryo. In the early egg cylinder stages this may be referred to also as a longitudinal section of the egg cylinder. The orientation of the embryo is not always consistent and may sometimes depart by as much as 45° from this plane. Plane *B*: transverse section of embryo. Note that in embryos past the egg cylinder stage this plane, though transverse to head and tail regions, is frontal with respect to the midtrunk region. Plane *C*: frontal section of embryo. Note that in embryos past the egg cylinder stage this plane though frontal to head and tail regions is transverse with respect to the midtrunk region. In early egg cylinder stages this may be referred to also as a longitudinal section of the egg cylinder.

may, however, occur. This orientation persists until about 8 or 8½ days when the embryo begins to shift its position in the uterus.

Amnion, chorion, and exocoelom

When mesoderm cells first appear between the ectoderm and endoderm at the posterior margin of the egg cylinder, they cause the ectoderm at the line of junction between its embryonic and extraembryonic portions to bulge into the proamniotic cavity. This bulge is the beginning of the posterior amniotic fold (Fig. 12-10). In like manner the lateral wings of mesoderm, progressing around the egg cylinder towards its anterior margin, give rise to folds along the sides of the cylinder. These are the lateral amniotic folds. Finally, when the mesoderm reaches the anterior margin of the egg cylinder, a small anterior fold is produced (Fig. 12-12). The posterior, lateral, and anterior folds should be thought of not as separate structures, but as a continuous constriction about the middle of the egg cylinder drawn tighter and tighter as the folds develop. Because of the very precocious development of the posterior as compared with the anterior amniotic fold, the constriction is eccentric, the point of final closure being far towards the anterior margin of the egg cylinder. In the rat, the anterior amniotic fold is much better developed than in the mouse, and the constriction, therefore, less eccentric (Jolly and Férester-Tadié, 1936).

In Fig. 12-10 a second fold may be seen pushing into the proamniotic cavity just dorsal to the posterior amniotic fold. Sobotta (1911) showed this in his Fig. 5 but interpreted it as an artifact. Our material indicates that it is regularly though briefly present. Its significance is unknown, but it is perhaps indicative of the very rapid growth that occurs in the whole posterior wall of the egg cylinder at the time of mesoderm formation.

Before the anterior fold forms, small cavities begin to appear in the mesoderm of the posterior and lateral folds. (None of the embryos in our collection show this early stage in the formation of the exocoelom. This description is based on the observation of Jolly and Férester-Tadié (1936). Bonnevie (1950) also described the formation of the exocoelom in the mouse.) These soon coalesce to form a single large cavity, the extraembryonic coelom, or, more concisely, the exocoelom (Fig. 12-12). The exocoelom at this stage and at all future stages is lined by mesoderm. The extraembryonic mesoderm lying between the posterior amniotic fold and the columnar endoderm appears infiltrated with fluid, and small lumina are formed. A large number of small lumina gradually unite into larger ones and the proamnion wall is reduced in thickness. For a short time a second cavity is present in the posterior amniotic fold between the mesoderm and ectoderm (Fig. 12-12), but this is a transitory structure of no particular significance.

In less than a day after the first appearance of the amniotic folds the girdle which they form has closed. The resulting condition is shown in the sagittal section reproduced in Fig. 12-13. As the amniotic folds coalesce, they become thin and delicate as a result of continued delamination of mesodermal cells from them (Bonnevie, 1950).

Three cavities are now present in the egg cylinder in place of the single proamniotic cavity it formerly contained. The most ventral of these is the amniotic cavity, lined with embryonic ectoderm. At this stage it is cupshaped, as can be seen from Fig. 12-13 and 12-14A, which show it in sagittal and cross section respectively. In the middle is the exocoelom, lined with mesoderm.

At the dorsal extremity of the egg cylinder is the ectoplacental cavity lined with extraembryonic ectoderm. The ectoplacental cavity, the smallest of the three, gradually becomes narrower and finally disappears.

The membrane separating the amniotic cavity from the exocoelom is called the amnion. It is composed of two thin cellular layers, one of ectoderm, the other of mesoderm. Separating the exocoelom from the ectoplacental cavity is another membrane, the chorion, likewise composed of ectoderm and mesoderm.

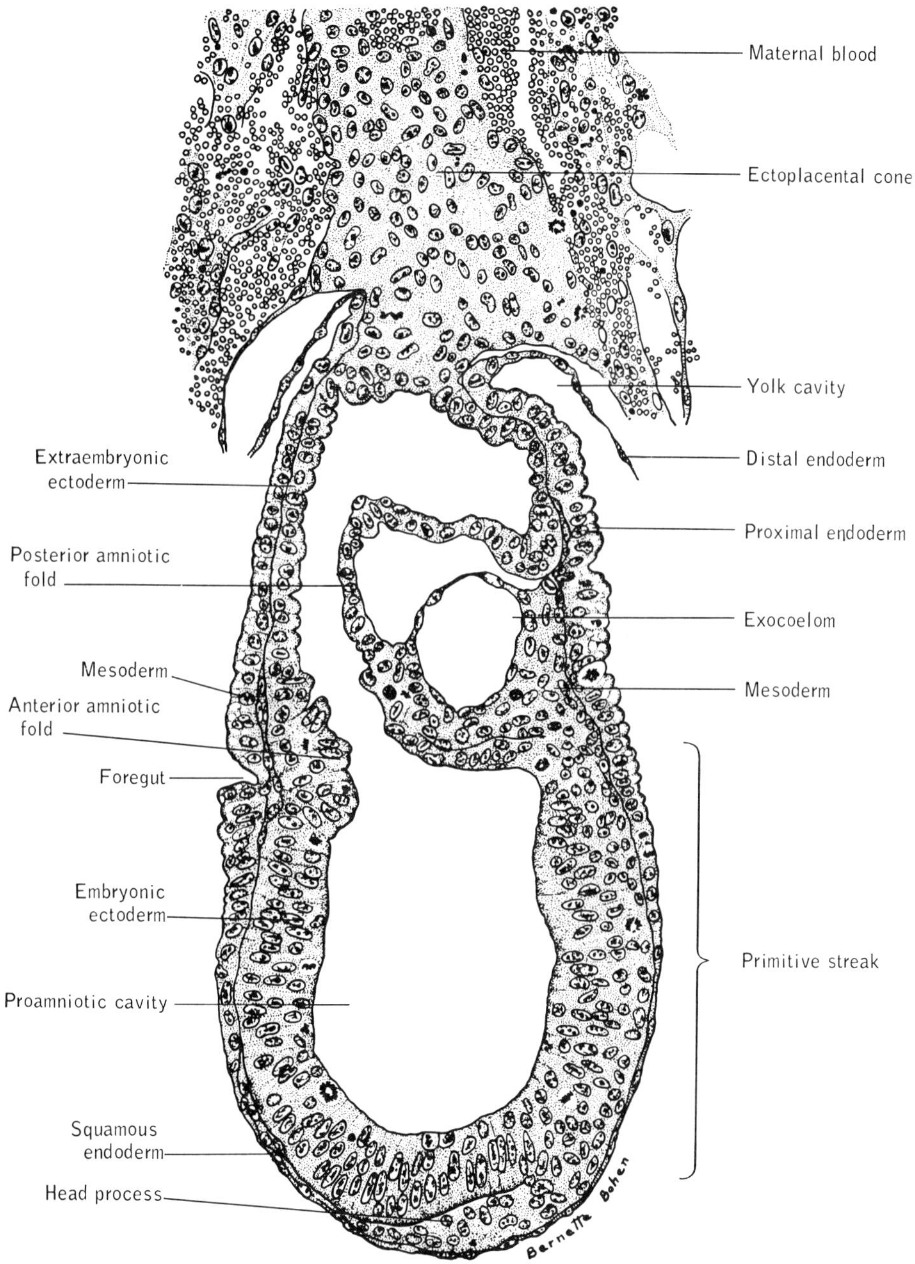

Fig. 12-12. Sagittal section of embryo of 7 days 1 hour showing the amniotic folds and the head process. Reichert's membrane omitted. Projection drawing (×300).

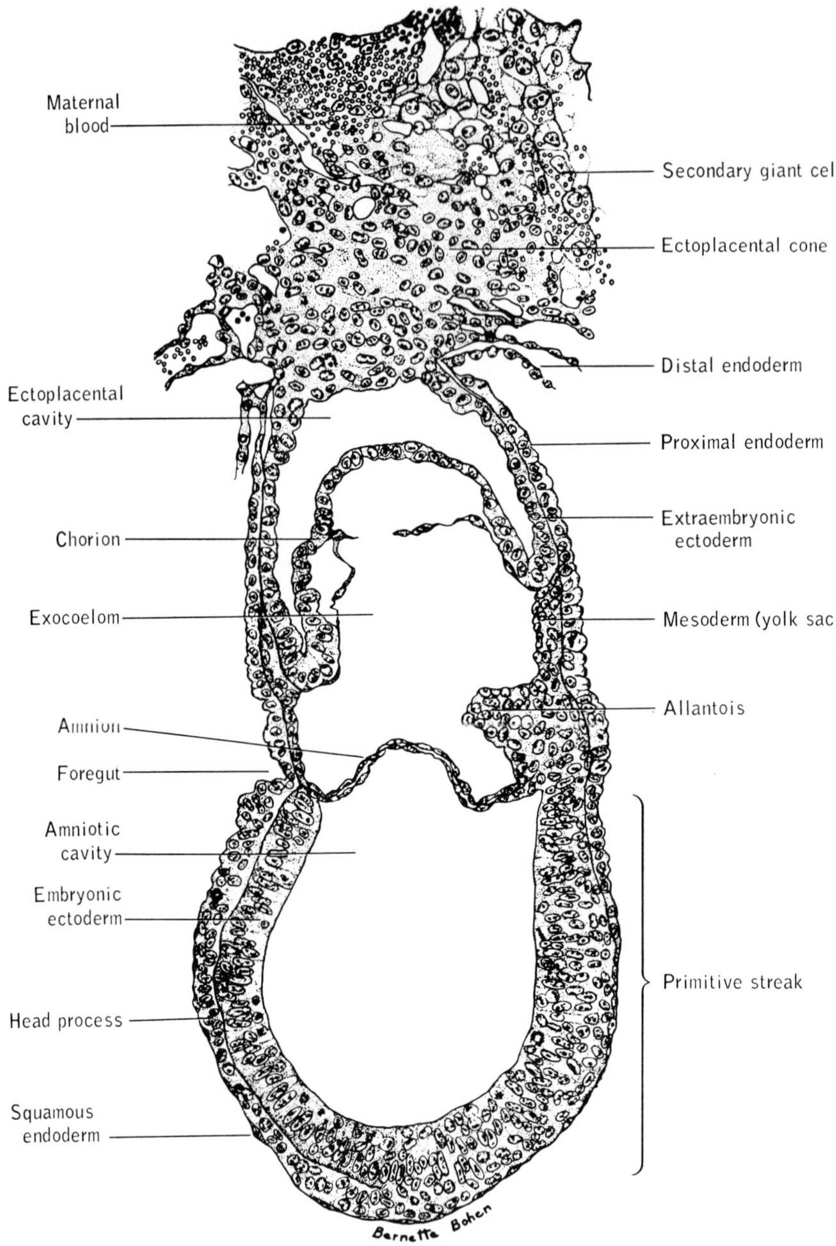

Fig. 12-13. Sagittal section of embryo of 7 days 6 hours showing completion of amnion formation. Reichert's membrane omitted. Projection drawing (×200).

Head process

Jolly and Férester-Tadié (1936) first described the head process for the mouse and rat; our observations are in accord with theirs. Bonnevie (1950) also described the formation of the head process in detail. In cross-sections the head process is found exactly opposite the primitive streak. It will be remembered that mesoderm is first proliferated by the primitive streak in embryos about 6½ days old. The growth is entirely from the lateral and caudal margins of the primitive streak; no mesoderm is proliferated from its cephalic extremity. Beginning at about 7 days, however, growth does occur in this region, but the structure formed shows greater affinity to the endoderm than to the mesoderm. It is known as the head process. The head process takes its origin from Hensen's node, a globular mass of mesoderm cells appearing at the cranial end of the primitive streak. In sagittal sections the head process first appears as a wedge-shaped group of cells between the ectoderm and endoderm at the ventral extremity of the egg cylinder (Fig. 12-12). The base of the wedge is attached to the cranial end of the primitive streak from which it takes its origin; the tip of the wedge points forward towards the anterior margin of the egg cylinder. Cells grow out rapidly from the margins of the wedge, forming a thin spreading sheet between ectoderm and endoderm.

The endoderm and the margins of the head process are so thin and close together at this stage that favorable conditions are necessary to distinguish them. In the section shown in Fig. 12-12 there are several cells at the anterior limit of the head process that cannot be distinguished as either head process or endoderm. The division in the drawing in this region is somewhat arbitrary. When head process and mesoderm come into contact there is also possibility for confusion. However, in well-fixed preparations cut at a favorable angle, the division can almost always be precisely determined.

As shown in Fig. 12-10 the endoderm over the ventral extremity of the egg cylinder is stretched and the cells flattened, but near the upper margin of the embryonic portion of the cylinder there is a sudden change to a higher type of cell. The transition is particularly abrupt at the anterior margin of the cylinder. The thin or flattened endoderm we shall refer to as squamous endoderm, the thick endoderm as columnar endoderm, the line of junction between the two as the transition line. According to Bonnevie (1950) the nutritive function of the columnar endoderm is evident from the small or large fluid drops within the cells. Chiquoine (1958) found that the apical cytoplasm of the embryonic endoderm of 5- to 8-day rat embryos contains striking amounts of periodic acid–Schiff material. The amount of this material varies with the height of the cells. The tall columnar cells contain the largest amounts. Note the precise location of the transition line in Fig 12-10 and 12-12. The flat squamous endodermal cells are characteristic of the embryonic part of the egg cylinder during further development. Bonnevie (1950) pointed out that the varying height of the columnar endodermal cells and their fluid contents are useful for indicating the seat of the most intensive nutritive activity.

The limits of the head process are as follows. Caudad, it begins at the anterior extremity of the primitive streak, that is, just a little above and caudad to the ventral tip of the egg cylinder. Cephalad, it extends to the transition line. Laterad, at its broadest point it may extend almost around the anterior half of the circumference of the egg cylinder (Fig. 12-14*A*), but mostly it is narrower than this, filling perhaps the anterior fifth of the egg cylinder's circumference.

When its forward growth brings it to the transition line, the head process fuses with the columnar endoderm with which it has thus newly come in contact (Fig. 12-13). The fusion is so complete that in later stages the line of junction is completely lost. Laterally, its outer margins fuse with the squamous endoderm. Meantime the squamous endoderm underlying the head process, already very thin, becomes increasingly attenuated, its nuclei become widely separated and very flat, and the cytoplasm almost disappears (Fig. 12-14*A*, *B*). At 7½ days no trace of it remains.

In the course of the upward and laterad growth of the head process and the forward growth of the mesoderm the two cell layers come in contact and overlap (Fig. 12-14*A*). In the regions of overlapping, the head process stays adjacent to and advances over the surface of the endoderm, whereas the mesoderm remains next to the ectoderm. At 7½ days the development of the mesoderm has brought it between ectoderm and head process everywhere except for a strip along the midsagittal plane of the embryo. As we shall see later, the head process of this midsagittal strip gives rise to notochord, whereas the remainder of the head process contributes to the lining of the gut.

Neural groove

It can be seen from Fig. 12-14*B*, a cross-section of the anterior part of an egg cylinder of a 7¼-day embryo, that the ectoderm in the midsagittal plane forms a definite trough or truncated V. This trough extends forward in the midsagittal plane from the cephalic end of the primitive streak well towards the cephalic limit of the embryonic ectoderm. Developed between the 7 and the 7¼-day stages (Fig. 12-14*A*, *B*), it is the beginning of the neural groove which later gives rise to the central nervous system. The further development

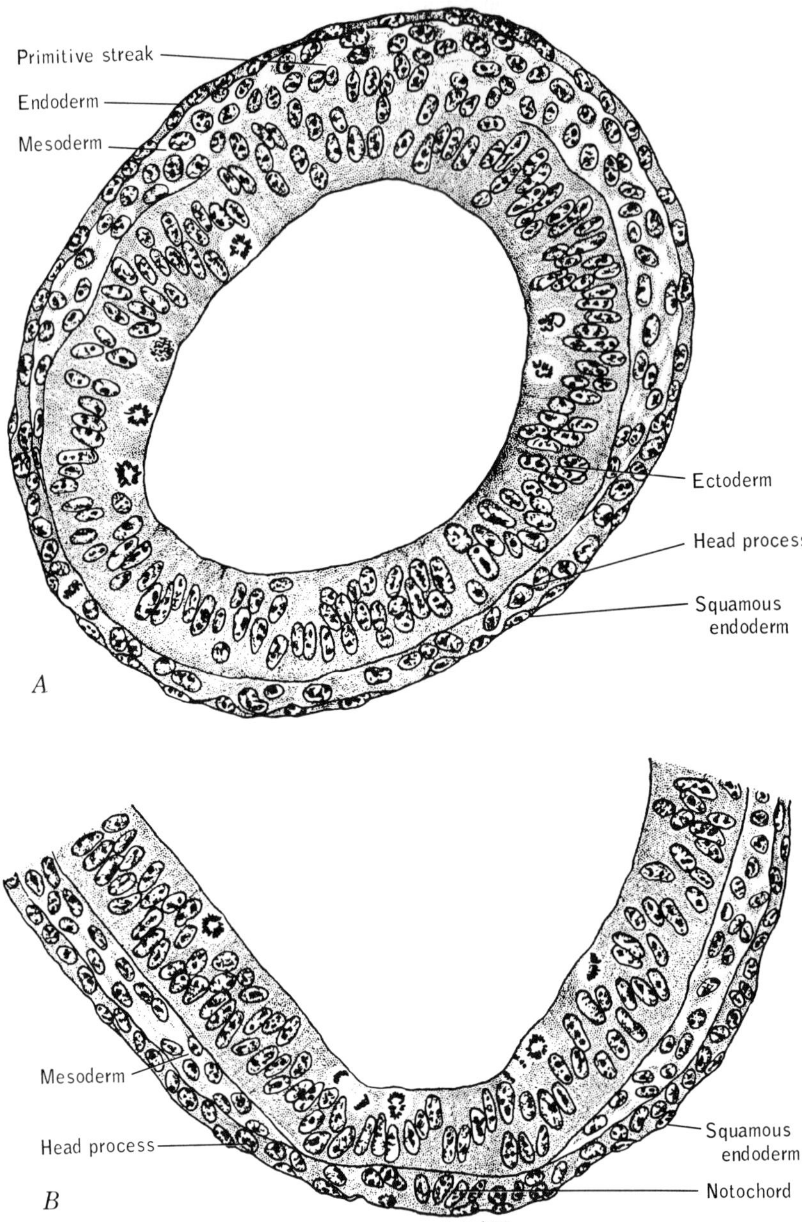

Fig. 12-14. Transverse sections of head process. The location of sections is indicated on the small key diagrams in Fig. 12-15. *A*. 7-day 1 hour embryo. *B*. 7-day 6-hour embryo.

of the neural groove will be discussed later; look ahead to Fig. 12-20 and 12-21 showing the way in which it deepens and narrows and finally closes at the top to form the neural tube. The point to be emphasized is that the appearance of the neural groove establishes a clear anteroposterior axis throughout the length of the embryo. The neural groove anteriorly and the primitive streak posteriorly lie in the midplane.

Notochord

At the same time that the neural groove is differentiating in the midsagittal area of the ectoderm, changes are also going on in the midsagittal region of the head process which immediately underlies it (Fig. 12-15). In this region the head process thickens and the oval nuclei become oriented in general perpendicular to the ectoderm. Elsewhere it forms a thin membrane with the nuclei oriented parallel to the plane of the membrane. The structure thus differentiated ventral to and in contact with the ectoderm of the neural groove is the notochord. It is the axis about which the vertebral column is later laid down. The remainder of the head process, together with a part of the endoderm to which it is fused, becomes the lining of the gut. This part of the head process will hereafter be referred to as gut endoderm. It seems likely that most or all of the midgut is lined by head process. Whether or not any of it enters into the formation of the fore- and hindguts is not clear. For a considerable period notochord and gut endoderm remain joined. Eventually, however, the two halves of the gut endoderm grow across the ventral surface of the notochord and unite in the midventral line, leaving the notochord as an axial, rodlike structure between ectoderm and endoderm.

Huber (1918) describes the head process in the guinea pig as giving rise to notochord only. Our material, however, confirms completely the contention of Jolly and Férester-Tadié (1936) that in the mouse at least some gut endoderm is also derived from the head process. As shown in Fig. 12-14*B*, the head process extends laterally considerably beyond the limits of the differentiating notochord.

It is questionable whether the notochord should be classed as ectoderm, endoderm, or mesoderm (Kingsbury, 1920). Since it is formed

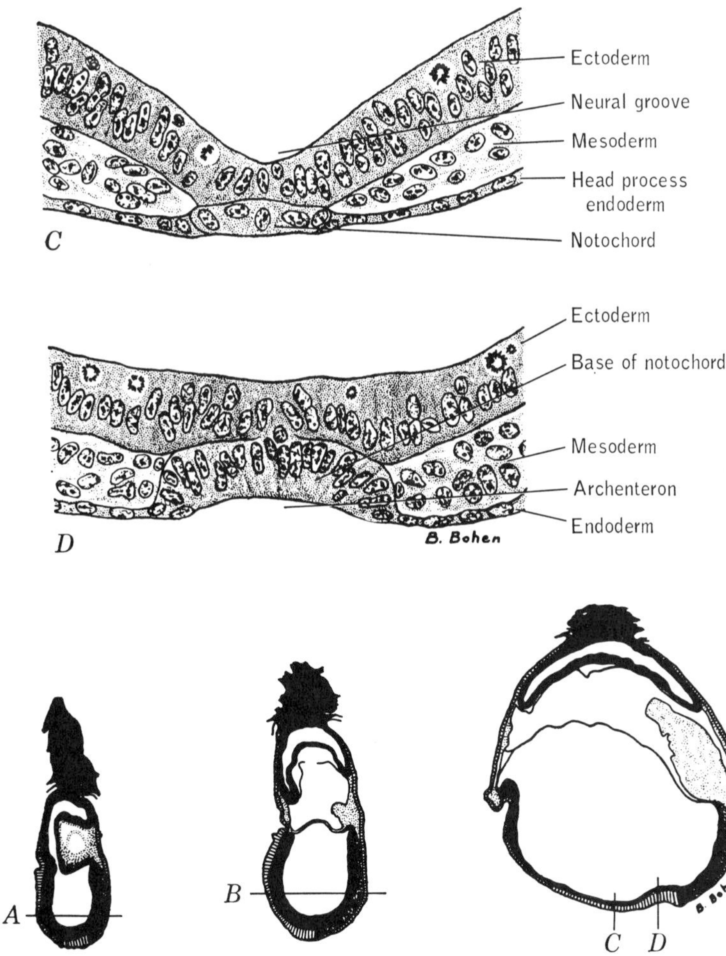

Fig. 12-15. The location of sections is indicated on the small key diagrams. *C.* and *D.* 7-day 10-hour embryo. Projection drawings (×400).

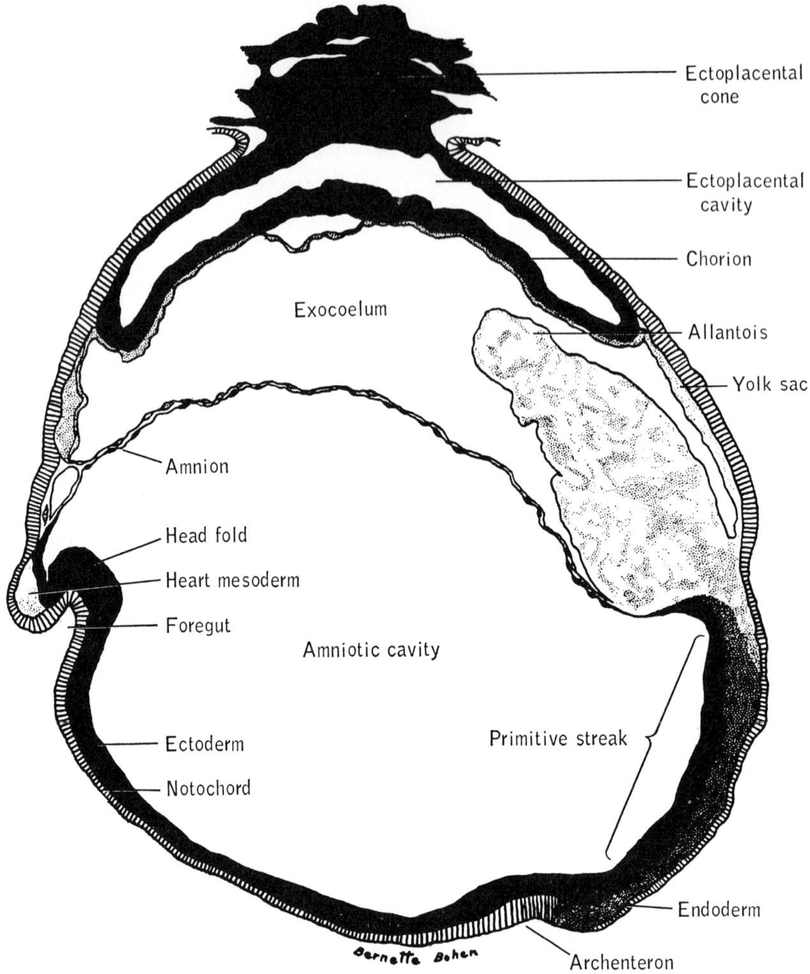

Fig. 12-16. Sagittal section of mouse embryo of 7 days 15 hours. Reichert's membrane omitted. Projection drawing ($\times 150$).

from the head process and since the very complete fusion of the margins of the head process with the endoderm indicates a close affinity between the two tissues, classification as endoderm would seem logical. If, however, head process is classed as endoderm, it must be remembered that its origin in time is quite different from that of all the other endoderm, and two separate stages of endodermal proliferation must be recognized. As to the place of origin, there is a certain similarity between the two tissues, one forming at the ventral margin of the inner cell mass, the other near the ventral tip of the egg cylinder, which is, so to speak, simply the inner cell mass grown up. Cell lineage studies might reveal a closer similarity in origin than is superficially apparent.

Archenteron

At 7½ days there is a broad depression in the rather thick base of the notochord adjacent to its junction with the primitive streak (Fig. 12-15*D*, 12-16). The depression is a conspicuous landmark at this stage, but it is a transitory structure, the first signs of it appearing at 7¼ days and disappearance being complete about 12 hours later. It probably is homologous to the archenteron of the lower chordates.

Allantois

Soon after the exocoelomic cavity becomes well established, the allantois begins to grow into it from the mesoderm at the caudal end of the primitive streak. The allantois (Fig. 12-13) is an extraembryonic mesodermal structure whose function is to convey blood vessels from the embryo to the placenta where they establish contact with the maternal circulation. In many vertebrates the allantois contains a cavity lined with endoderm and connected with the gut. There is no endoderm-lined cavity in mice, but there are numerous small cavities in the mesoderm giving the organ a porous structure.

After its first appearance at 7¼ days the allantois grows rapidly across the exocoelom in the direction of the ectoplacental cone (Fig. 12-16, 12-17). Meantime the chorion becomes flattened against the base of the cone, constricting the ectoplacental cavity and finally eliminating it altogether. When the allantois makes contact with the chorion at about 8 days, a continuous structure is established connecting the posterior end of the primitive streak with the ectoplacental cone. In due course embryonic blood vessels will find their way along this pathway to make contact with the maternal blood supply. In the rat the allantoic vessels reach the chorion and the chorioallantoic placenta is established by the latter half of the ninth day (Bridgman, 1948).

Foregut and hindgut

In the early stages of its formation the digestive tract consists of three parts; foregut, hindgut, and midgut. These appear in the order named. The foregut can be traced back to the 7-day stage when it consists merely of a deep notch in the thick or columnar endoderm at the anterior margin of the egg cylinder (Fig. 12-12). Six hours later there is little change (Fig. 12-13), but by 7½ days (Fig. 12-16) the notch has been replaced by a definite pocket in the endoderm, and the endoderm surrounding the pocket together with the overlying ectoderm form a bulge which projects into the amniotic cavity. From this stage on, growth of the foregut is exceedingly rapid, the pocket changing in a few hours into a deep pouch (Fig. 12-17). The process is due to a progressive drawing together in the midventral line of the folds of endoderm that bound the anterior intestinal portal (Fig. 12-26*C*, 12-28) the portal thus being shifted toward the posterior end

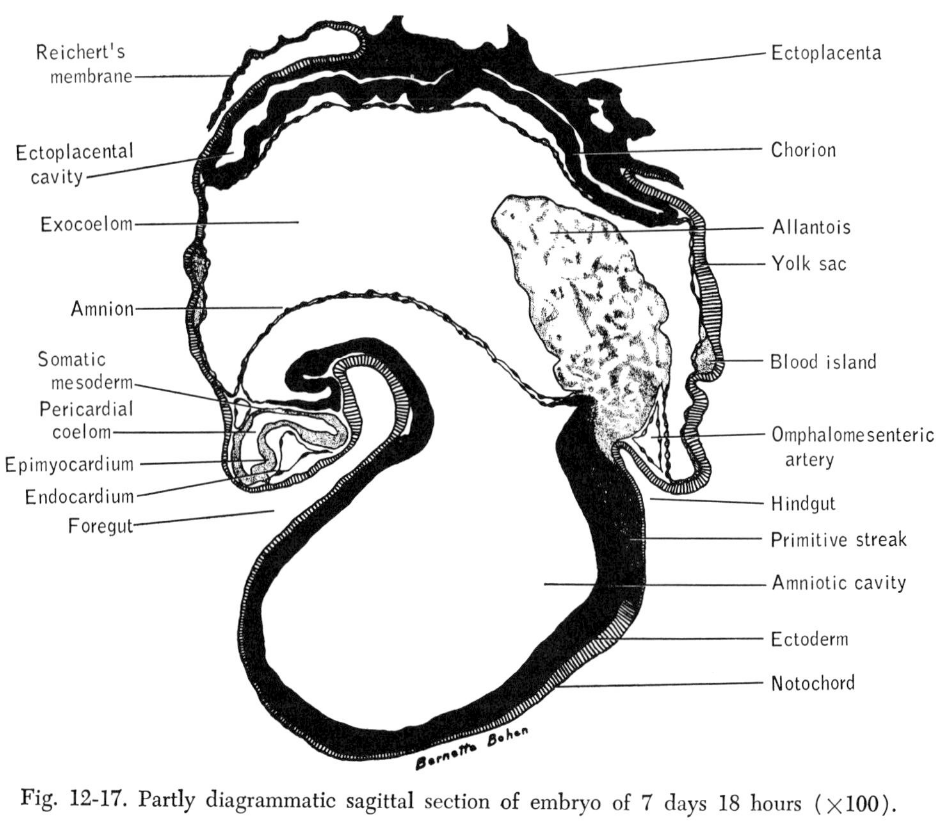

Fig. 12-17. Partly diagrammatic sagittal section of embryo of 7 days 18 hours (×100).

of the embryo. The process has been aptly described as a "zipper action."

The foregut is lined by endoderm which is surrounded by mesoderm and ectoderm. Thus in this region the process of invagination has reversed the inversion of the germ layers found in the early egg cylinder. This is the first of the steps by which the germ layers in mice are brought into the relation characteristic of the adult, i.e., endoderm on the inside, ectoderm on the outside, mesoderm in between.

The hindgut, less precocious than the foregut, appears at about 7¾ days as an invagination in the endoderm and overlying layers at the posterior end of the primitive streak (Fig. 12-17).

The open ends of the fore- and hindguts are eventually joined by the midgut, whose formation will be described in a later section. The blind ends of the two guts break through to the outside and give rise to mouth and anus. An early stage in the development of the mouth may be seen in 8-day embryos (Fig. 12-23). In the ectoderm of the head there is an invagination directed toward the anterior end of the foregut. This is the stomodeum. The wall between the stomodeum and the foregut is the oral plate. In the course of time this ruptures and the mouth opening is formed. The anus develops in a similar manner at a somewhat later stage.

Head fold

The invagination of the foregut involves a pushing or folding of the adjacent tissues into the amniotic cavity. The structure thus produced is the head fold (Fig. 12-16). First appearing at about 7½ days, it becomes a large and conspicuous structure within less than 24 hours (Fig. 12-23, 12-26A). The growth of the neural folds in this region is more rapid than elsewhere, presaging the formation of the brain. The heart, just ventral to the head fold, is also conspicuous by its rapid growth. In 8 or 8½-day embryos the difference in size between the head region and the middle of the trunk is striking. The head fold is a region of particularly rapid growth (Hyman, 1927).

Somites

Since the somites are mesodermal structures, it will be useful before discussing their development to review the distribution of the mesoderm at the 7½-day stage just as somite formation begins. In the extraembryonic region, the entire exocoelom is lined with mesoderm. Also, the exocoelom contains the allantois, a wholly mesodermal structure. In the embryo proper there is little mesoderm in the midsagittal region. One small mass which will later contribute to the formation of the heart occurs anterior to the foregut (Fig. 12-16). The primitive streak in the midsagittal plane consists of a tissue which joins and, in structure, is intermediate between ectoderm and mesoderm. Whether or not this should be called mesoderm is a matter of definition. At the caudal end of the embryo its structure is essentially that of true mesoderm and there is mesoderm in the midsagittal plane in this region.

Even though the notochord blocks the entrance of mesoderm into most of the midsagittal area, there are well-developed sheets of mesoderm on each side. These lie between ectoderm and endoderm and are continuous laterally with the extraembryonic mesoderm (Fig. 12-18). Anterior to the primitive streak, there are two distinct areas in these mesodermal sheets: an area of paraxial mesoderm adjacent to the notochord and an area

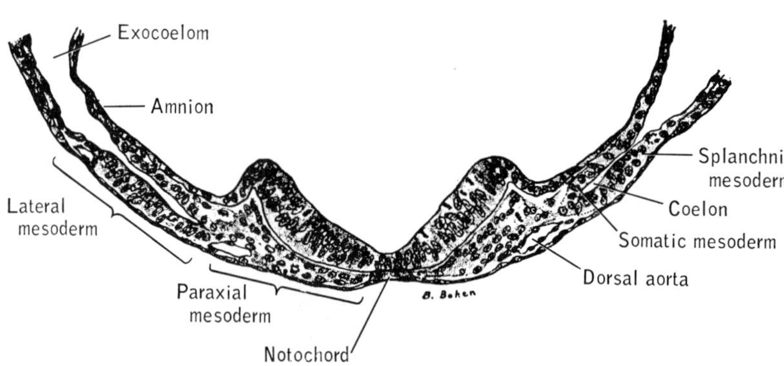

Fig. 12-18. Section transverse to midtrunk region of 8-day, 4-somite embryo. Through second somite. Projection drawing (×150).

of lateral mesoderm adjacent to the extraembryonic coelom. The former gives rise to the somites, the latter to the mesoderm of the embryonic coelom. At 7½ days there is no visible division between the two areas (Fig. 12-20*B*), but beginning at about 7¾ days, coincident with the development of the somites, they are separated by a longitudinal cleft that becomes increasingly pronounced as the differentiation of the somites progresses (Fig. 12-18).

The somites are paired, segmental structures arising in the paraxial mesoderm (Fig. 12-19, 12-26*D*). They are the first indication of metamerism in the developing embryo. The first pair forms a little anterior to the anterior end of the primitive streak. Each member of the pair appears as a localized denser area which grades off anteriorly into loose mesoderm, and which posteriorly is separated by a cleft from the undifferentiated mesoderm. The second pair forms posterior to the first and is likewise separated by a cleft from the undifferentiated caudal mesoderm. Additional pairs of somites form at more or less regular intervals, each new pair differentiating just posterior to the pair last formed until in the rat a total of 65 pairs has appeared (Butcher, 1929). The number of somites may be slightly different for the mouse, and there is some individual variation. Butcher states in his excellent paper on the somites of the rat that the first pair dedifferentiates and disappears at about the 7-somite stage. We found no evidence of such a dedifferentiation in our material. As may be seen from Fig. 12-19*A*, *B*, and *C*, the first somite can be traced clearly to the 11-somite stage. Continued and rapid proliferation of the mesoderm on each side of the primitive streak maintains a supply of undifferentiated cells. These push forward to about the level of the base of the notochord where the new somites are segmented off in regular succession. The anterior somites are older and, at any given stage, more highly differentiated than the posterior somites.

Figure 12-19 shows how the "zipper action" by which the foregut is formed moves the location of the anterior intestinal portal steadily caudad. At the 2-somite stage the opening of the shallow foregut lies anterior to the first somite. At the 7-somite stage the anterior portal has moved posteriorly until it is about at the level of the first somite. At the 11-somite stage it has reached approximately the level of the sixth somite. (The portion of the gut shown in Fig. 12-19*C* is midgut just caudal to the anterior intestinal portal.)

The number of somites in an embryo is a convenient means of stating its stage of development.

Primitive streak as a growth center

The primitive streak is remarkable as being a region in which the three germ layers meet (Fig. 12-16). It is continuous dorsally with the ectoderm, laterally and posteriorly with the mesoderm, and anteriorly with the head process (which is endodermal in nature and indistinguishably fused with the original proliferation of endoderm). Of these three, it gave rise to two: the mesoderm and the head process. It may be added that the proliferation of mesoderm cells that produced the allantois occurred at its posterior end. Its own cells are undifferentiated and cannot be classified as either ectoderm, endoderm, or mesoderm. The only structure in primitive chordates possessing these characteristics is the dorsal lip of the blastopore, and the primitive streak and the dorsal lip of the blastopore may be homologous. Besides being a point of origin for new tissues, it is the center of a region of rapid growth. The adjacent mesoderm is full of dividing cells which are continually pushed forward to give rise to somites anterior to the primitive streak. Much of the increase in length of the embryo is due to growth in this region. There are thus two regions of particularly active growth in the developing embryo: the primitive streak and the head fold (Hyman, 1927). There is one point of contrast between these two: The tissues in the head region are well advanced in differentiation, whereas the tissues of the primitive streak region remain relatively undifferentiated.

Coelom

Coincident with the formation of the somites in the paraxial mesoderm, the coelom or body cavity develops in the lateral mesoderm. It is formed by a division of this mesoderm into two layers, a dorsal or somatic layer adjacent to the ectoderm, and a ventral or splanchnic layer adjacent to the endoderm. The coelom is the space between the two (Fig. 12-18). Because the somatic mesoderm and the ectoderm are closely associated and undergo many foldings in common, it is convenient to designate the two layers together by the term somatopleure. For the same reasons splanchnic mesoderm and endoderm together are designated as splanchnopleure. The mesoderm and ectoderm of the somatopleure dorsal to the coelom are continuous with the similar layers in the amnion. The amnion, therefore, may

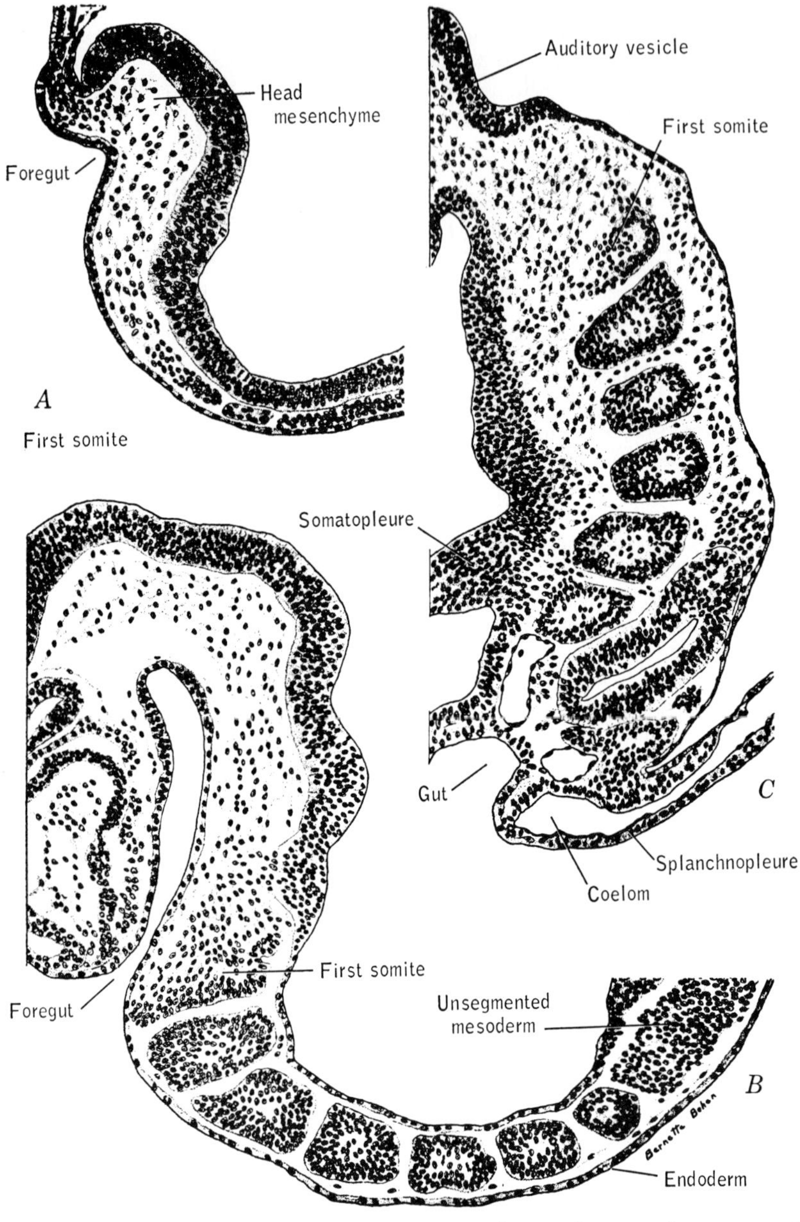

Fig. 12-19. Sagittal sections through somites. *A*. Embryo of 7 days 18 hours with 2 somites formed. *B*. Embryo of 8 days 1 hour, 7 somites. *C*. Embryo of 8 days 11 hours, 11 somites. Because the embryo begins turning at about the 7-somite stage, the plane at which this 11-somite embryo is cut, although sagittal to the first 6 somites, is transverse to the midtrunk region. Projection drawings (×150).

also be classed as somatopleure. Similarly the mesoderm and endoderm of the splanchnopleure ventral to the coelom are continuous with the similar layers in a tissue which bounds the extra-embryonic coelom laterally. This tissue, therefore, may also be classed as splanchnopleure.

It has been previously stated that there is a mass of mesoderm in the midsagittal plane anterior to the foregut. This extends to right and left across the front of the foregut and is continuous laterally with the lateral sheets of mesoderm. It thus forms the base of a U of which the lateral mesoderm forms the sides. By about the 4-somite stage or slightly later the coelom extends not only throughout the lateral mesoderm but also as a passage through this anterior mesoderm (Fig. 12-17, 12-30). The coelom also is thus U-shaped. The whole posterior portion of the coelom opens laterally into the extraembryonic coelom (exocoel) (Fig. 12-18, 12-29). The anterior part of the coleom on the other hand, forming the base of the U and extending as far posteriorly as the second somite, is separated from the extraembryonic coelom by a partition of mesoderm. Much of this anterior portion of the coelom becomes the pericardial coelom enclosing the heart. The connection between the anterior and the lateral parts of the coelom is called the pericardial-peritoneal canal (Fig. 12-29).

The relations of coelom and extraembryonic coelom can be studied from the series of sections of a seven-somite embryo shown in Fig. 12-23*A* to *G*.

Reichert's membrane

The mouse embryo is protected during its development by three extraembryonic membranes: Reichert's membrane, the amnion, and the yolk sac. There is no essential difference between the amnion of rodents and of other mammals. The yolk sac in Rodentia has come to have rather unusual relations to other structures. The chorion, an important fetal membrane in most mammals, is present in the mouse but remains small and unimportant as a protective structure.

To follow the development of Reichert's membrane we must go back to the 5½-day stage (Fig. 12-8). Except in the region of the ectoplacental cone, the embryo is bounded by the trophectoderm. This is continuous with the margins of the cone and is separated from the egg cylinder by the yolk cavity. Laterally its cells are in close contact with the maternal decidua, a contact so intimate in fact that in some cases it is impossible to tell whether a given cell is of embryonic or maternal origin. Ventrally it stretches across the remains of the uterine lumen, now filled with a degenerating mass of uterine epithelium. On the inner surface of the trophectoderm are a few widely separated endoderm cells.

A day later (Fig. 12-10) these endoderm cells have increased in number and form a uniform though not quite continuous layer over the inner surface of the trophectoderm. Between the two cell layers there soon begins to appear a thin, noncellular, pink-straining membrane called Reichert's membrane. The first signs of it are often visible at the ventral extremity of the egg where there is apt to be a cluster of endoderm cells. It soon becomes continuous over the entire inner surface of the trophectoderm. The fully developed membrane is of uniform thickness and, as can be demonstrated by dissection, surprisingly tough for so delicate a structure. Pierce et al. (1962) and Midgley and Pierce (1963) have shown that parietal (distal) endoderm cells of the mouse embryo secrete Reichert's membrane. This membrane has the ultrastructural and immunochemical attributes of a basement membrane. Late on the 16th day in the rat, Reichert's membrane and the decidua capsularis rupture (Bridgman, 1948). This permits the placenta to shorten in diameter and become thicker.

Amnion

The early stages of the development of the amnion have been described. Owing to the inversion of the germ layers, the amniotic folds have only a short distance to grow, and amnion formation is consequently precocious in the mouse as compared with most other mammals (Fig. 12-9*B*, *E*). For the same reason the area of the amnion at first is small. It expands rapidly, however, to accommodate the growing embryo and by 8 days it forms a large sac over the embryo's entire dorsal surface (Fig. 12-23, 12-26*C*). In the later stages of development the embryo floats free in the amniotic cavity attached only by the umbilical cord.

Yolk sac

The mammalian ovum contains virtually no yolk. The mammals are, however, descended from reptilian ancestors in whose eggs yolk was abundant, and this long period in their evolutionary history has left an indelible imprint on mammalian development. Most striking, perhaps, is the development of a yolk sac so similar in many details to the reptilian yolk sac as to be unmistakably homologous. In the small rodents the yolk sac constitutes an important accessory nutritive organ during the whole period of gestation (Böe, 1951). Yolk sac cells probably exercise both absorptive and secretory activities simultaneously (Wislocki et al., 1946).

The yolk cavity of the mouse may be defined as the cavity derived from the original segmentation cavity or blastocoel and lying between the egg cylinder and the distal endoderm with its secretion, Reichert's membrane (Fig. 12-20*A*). The yolk sac is only a part of the boundary of this cavity; namely, that middle portion of the egg cylinder wall composed of mesoderm and endoderm or, in other words, of extraembryonic splanchnopleure. In many mammals, e.g., the pig, the allantois as well as the yolk sac are derived from splanchnopleure. This is not the case in the mouse. In this species the yolk sac, as we are using the term, and the extraembryonic splanchnopleure are identical.

At 7¼ days the extraembryonic splanchnopleure or yolk sac is a structure of limited area forming the central or ectoderm-free portion of the egg cylinder wall (Fig. 12-20*A*). Although small at first, it is an area of rapid growth and by 8 or 8½ days forms an extensive membrane enveloping the amnion and a greatly enlarged exocoelomic cavity (Fig. 12-20*B, C*). The whole embryo changes its shape in the process, the egg cylinder becoming an ovoid and the ovoid a sphere. At 8 days the yolk sac is still attached to the embryo along a band that runs anterior to the opening of the foregut and posterior to the opening of the hindgut so that most of the

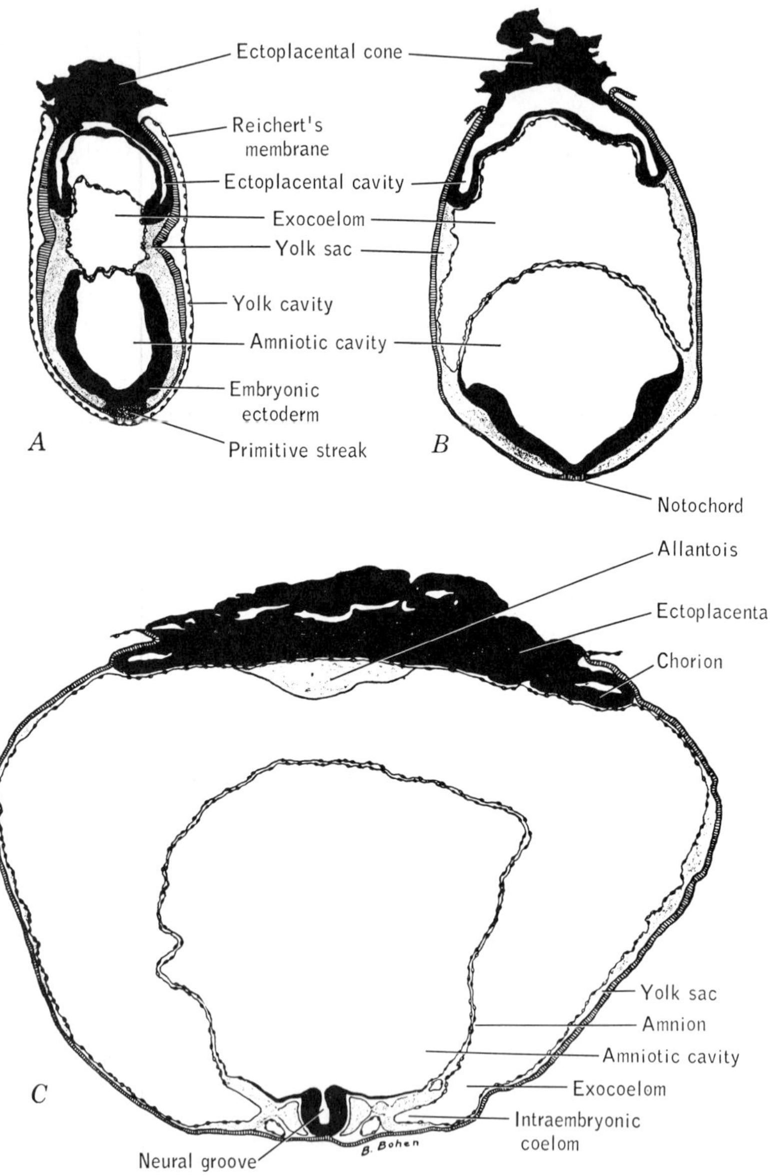

Fig. 12-20. Frontal sections (×75) showing development of the yolk sac. *A*. 7 days 6 hours. *B*. 7 days 10 hours. *C*. 8 days 10 hours, 9 somites, through eighth somite. Reichert's membrane omitted except in *A*.

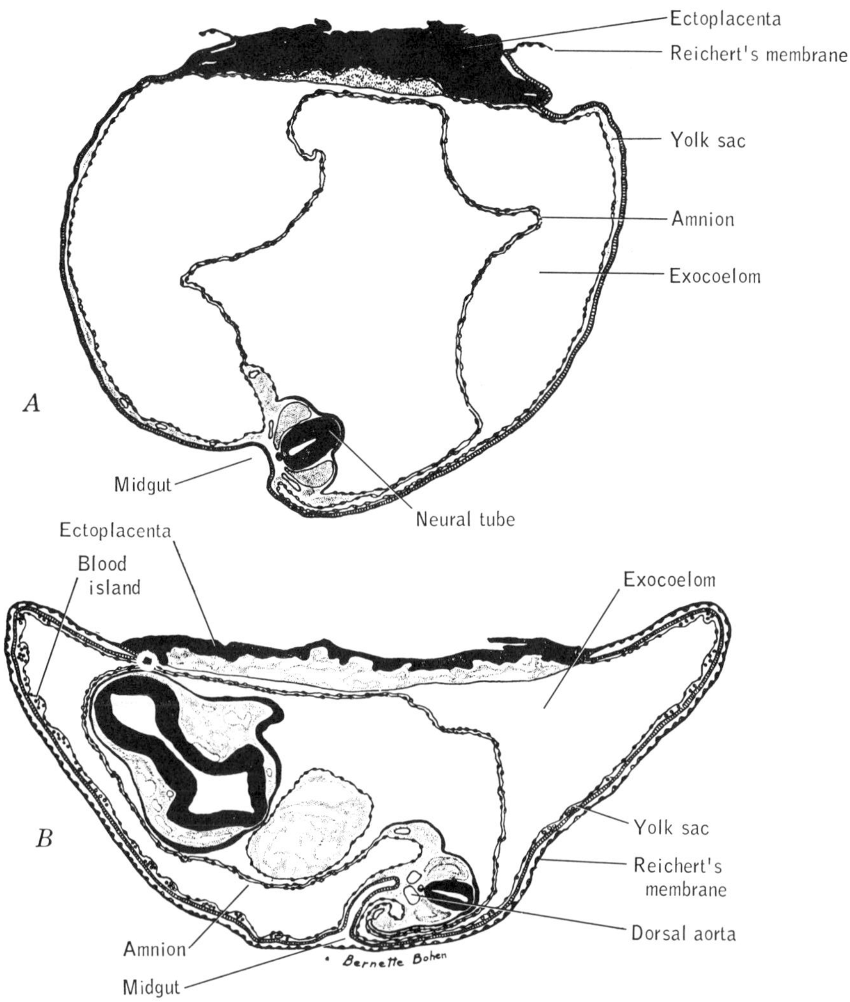

Fig. 12-21. Frontal sections showing development of extraembryonic membranes and formation of midgut. The sections are viewed from the head towards the tail, so that the right side of the embryo is on the left side of the drawing, and vice versa. *A.* 8 days 11 hours, 11 somites, through ninth somite ($\times 60$). *B.* 8 days 18 hours, 16 somites, through ninth somite ($\times 45$).

ventral surface of the embryo is outside it (Fig. 12-23). After the midgut has formed, however, this portion of the embryo, too, is enveloped by the yolk sac (Fig. 12-21). The details of this process will be described later.

Blood islands

Associated with the yolk sac splanchnopleure are structures known as the blood islands. These appear in the mouse at 7½ days as thickenings in the inner or mesodermal layer of the yolk sac about which they form an irregular girdle (Fig. 12-17). As the name implies, the blood islands give rise to part of the circulatory system. The peripheral cells differentiate to form the endothelium of a system of blood vessels encircling the yolk sac while the inner cells become primitive blood corpuscles that circulate in the embryonic blood stream. Block (1946) performed an experimental analysis of hematopoiesis in the rat yolk sac. Böe (1951) described the vascularization of the rat yolk sac.

Germ cells

The germ cells of the mouse make their appearance early in embryonic life (Chiquoine, 1954; Mintz and Russell, 1957), and mitotic descendants of the same cell lineage persist to maturity in the adult in both males and females (Mintz, 1960). All definitive germ cells are derived from fewer than 100 cells first seen in the endodermal yolk sac epithelium localized near the base of the allantois at 8 days of gestation. At 9 to 12 days, as the paired genital ridges make their appearance, the germ cells migrate—apparently actively and probably selectively—up the dorsal gut mesentery and into these ridges. The original few cells multiply to 5,000 or more when the migratory period terminates (Mintz and Russell, 1957). At 11 days of development the distribution of the germ cells in the genital ridges is peripheral in females and central in males.

All female germ cells enter meiosis before birth (Mintz, 1960). Leptotene chromosomes are already present in some germ cells at 13 days. Pachytene oocytes are numerous at 16 to 17 days. They may continue through diplotene and enter the dictyate stage by birth. The dictyate stage persists until just before ovulation when the chromosome organization is again visible and anaphase and telephase of meiosis I are completed.

In at least two inbred strains of mice (strain 129 and A/HeJ) the male primordial germ cells may be experimentally induced to initiate development and form testicular teratomas (Stevens, 1964). When genital ridges from 12½-day (but not older) male fetuses are grafted to the adult testis, primordial germ cells in most of them begin to develop germ layers and subsequently many kinds of differentiated, but not organized, tissues.

Changes in the uterus

Implantation in the mouse is accompanied by a rapid growth of the uterine mucosa adjacent to the implantation site to produce a definite swelling, the decidual swelling. For a while the uterine crypt containing the embryo maintains its connection with the uterine cavity, but by about 7½ days the growth of the decidua has blocked this off so that the cavity containing the embryo is separated from the main lumen (Fig. 12-4*B*). The bridge of tissue thus formed dorsal to the ectoplacental cone will later become part of the placenta. Further growth of the decidua constricts and finally, by about 8 days, completely closes the uterine lumen dorsal to the embryo except for one or more small isolated chambers (Fig. 12-22). On each side of the decidual swelling the uterine lumen remains open, but at this period in development there is no continuous passage throughout the length of the uterus. A little later a continuous lumen is reestablished, but the new lumen is on the opposite side of the decidual swelling from the old, passing ventral instead of

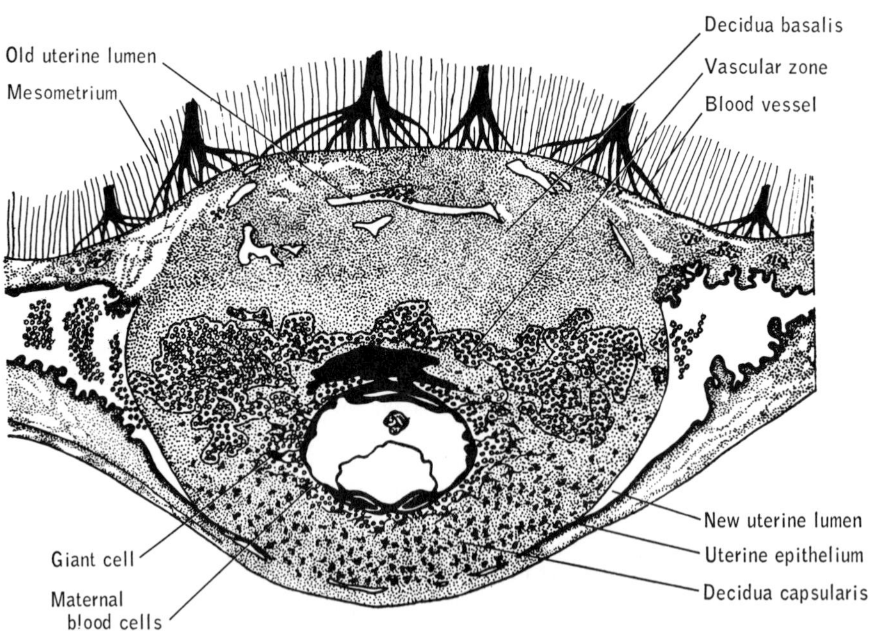

Fig. 12-22. Longitudinal section (partly diagrammatic) of uterus at site of implantation of 8 day 6 hours, 5-somite embryo. Cut parallel to mesometrium.

dorsal to the embryo. An early stage in this reestablishment of the lumen may be seen at about 8 days (Fig. 12-22). The epithelium lining the lumen on each side of the decidual swelling has grown in between the muscle layers and the decidua ventral to the embryo. The extreme limits of this growth consist of a double but unsplit layer of epithelium. In the slightly older epithelium nearer the lumen the two layers have split so that two wedge-shaped spaces extend from the lumen between decidua and muscles on each side of the decidual swelling. In course of time the wedges penetrating from the two sides meet ventral to the embryo, thus completing the formation of the new lumen.

Besides the changes in the uterine lumen there are interesting changes in the histology of the decidua. Starting as a relatively homogeneous tissue, it differentiates into several different zones, each with its characteristic structure. As many as six zones can be distinguished (Krehbiel, 1937), but we will note no more than three (Fig. 12-22). Ventrally there is an antimesometrial zone or decidua capsularis characterized by large bi-, tri-, or tetranucleate cells. In the rat this region is characterized by binucleate cells (Krehbiel, 1937). The individual nuclei in this zone as well as the cells are larger than elsewhere in the decidua, and this together with the grouping of the nuclei gives the zone a very characteristic appearance. It will be noted that it lies between the embryo and the new uterine lumen. With the growth of the embryo it becomes stretched until, in the later part of the gestation period, it is hardly more than a thin membrane separating embryo and lumen. Dorsally there is a mesometrial zone, or decidua basalis, whose cells at 8 days still closely resemble those of the unaltered mucosa. It later contributes to the formation of the placenta. Between the antimesometrial and mesometrial zones is an intermediate or vascular zone characterized by the presence of numerous irregular endothelial-lined blood spaces or sinusoids. Its cells tend to be multinucleate like those of the decidua capsularis.

Nourishment of the embryo

Prior to the attachment of trophoblast cells the nutrient requirements of the blastocyst are thought to be supplied by the uterine milk (Eaton and Green, 1963). The degenerating cells of the uterine epithelium that originally lined the implantation chamber may serve as a source of food. The epithelium is sloughed off and begins to undergo degenerative changes at just about the same time that the first real increase in size of the embryo occurs. At the mesometrial pole of the embryo at 4½ days may be seen droplets of secretion that may contain an enzyme concerned with the digestion of the epithelial cells (Fig. 12-6). This stage in the nourishment of the embryo is brief; by 5½ days only a remnant of the epithelial cells remains (Fig. 12-8).

At the same time a new source of nourishment makes its appearance. At 5½ days the blood-filled sinusoids of the intermediate zone of the decidua begin to rupture, pouring their contents into the lumen surrounding the embryo. In a very short time the embryo is completely bathed in maternal blood. This blood is not stagnant as was once supposed but remains a part of the maternal circulation. In the rat Everett (1935) found that there is a complete replacement every 20 minutes.

The maternal blood is separated from the embryo by Reichert's membrane, by the yolk cavity, and in later stages by the yolk sac. Reichert's membrane probably plays an entirely passive role in the transportation of nutrient substances from the maternal blood to the embryo, acting simply as a semipermeable membrane. The yolk sac, on the other hand, probably actively absorbs nutrients. This is particularly true after the blood islands girdling the yolk sac have developed into a capillary network and after the embryonic circulation is established. By the time this occurs the yolk sac has become pressed against and partly fused with Reichert's membrane, obliterating the yolk cavity (Fig. 12-21*B*). The embryonic yolk sac circulation is thus brought very close to the maternal circulation, and the yolk sac is established as "an organ of exchange whose importance is not secondary to that of the allantoic placenta" (Everett, 1935). It is interesting to observe that in rodents the yolk sac has thus recovered a role as an organ of absorption it possessed in the reptiles with, however, the important difference that the material absorbed comes from the maternal blood instead of from yolk deposited within the egg.

The sinusoids in the intermediate zone of the decidua extend from the decidual cavity containing the embryo to the periphery of the decidual swelling where this borders on the uterine lumen. Beginning at about 7½ or 8 days there is bleeding into the uterus from these peripheral sinusoids (Stafford, 1930; Venable, 1939). At about 10 days some of this blood finds its way into the vagina, persisting there for 3 or 4 days (Sato, 1936). It is a convenient early sign of pregnancy.

In the later stages of development the decidua

basalis, the ectoplacental cone, the chorion, and parts of the allantois fuse to give rise to a true placenta which thereupon assumes a major role in transferring nutritive material to the embryo. Bridgman (1948) has described the histology and cytology of the developing chorioallantoic placenta of the rat. Wislocki et al. (1946) performed a histochemical investigation of the placentas of several rodents, including the mouse.

Trophoblastic giant cells

A conspicuous feature in sections of mouse embryos of 6 to 14 days is the presence of certain remarkably large cells lying between Reichert's membrane and the decidua. These are the trophoblastic giant cells (Fig. 12-22). Fawcett et al. (1947) observed these giant cells in grafts of mouse ova to the anterior chamber of the eye and concluded that they were derived from trophectoderm and not from uterine elements. Transformation of cells to trophoblast giant cells is retarded in blastocysts undergoing delay of implantation so as to be coincident to implantation (Dickson, 1963).

It is convenient to distinguish three types of giant cells. The first large and unmistakable giant cells to appear are at the ventral extremity of the embryo (Fig. 12-6, 12-8). Already quite large at 5½ days they have become enormous by 7 days, at which time they have penetrated for some distance into the remains of the implantation cavity ventral to the embryo. These are primary giant cells. The trophectoderm cells lateral to the egg cylinder also give rise to similar though somewhat smaller primary giant cells. Although mitotic figures are seen in the earliest stages, cell division is not characteristic of these enlarged cells (Alden, 1948).

A second and much more numerous group of trophoblastic giant cells is derived from the ectoplacental cone. At 5 days cells may be seen growing outside the trophectoderm from the region of the future cone (Fig. 12-7). Later, when the embryo is surrounded by maternal blood, these become long strands of cells extending down, within the blood or along the inner surface of the decidua, from the cone toward the ventral extremity of the egg cylinder. At first small, these cells increase in size and at 8 days form a loose meshwork of large cells whose long protoplasmic processes extend across the blood-filled space between Reichert's membrane and the decidua (Fig. 12-22). Other similar cells may be seen adjacent to the ectoplacental cone. These are the secondary giant cells. At 8 days their continuity with the cells of the ectoplacental cone is still quite obvious. It should be pointed out that the division between primary and secondary giant cells is partly arbitrary; the trophectoderm and the ectoplacental cone are continuous structures, and cells from near the line of junction might be said to give rise to either type. One obvious function of the giant cells is to anchor Reichert's membrane to the decidua.

The third class of giant cells appear multinucleate. They first appear in the decidua adjacent to the embryo at 7 or 7½ days. The number of nuclei per cell appears to be extraordinary, mounting into the dozens by 8 days, but Kirby and Malhotra (1964) reported that all stages of mouse trophoblast showed distinct delimiting cell membranes between nuclei.

The precise roles played by the trophoblastic giant cells and the uterus during implantation are not clear. Alden (1948) tentatively suggested that both may play an active role. The denudation of the mucosa at the very beginning of implantation may be attributed to the action of trophoblast cells or to edematous changes in the underlying mucosa or to both (Boyd and Hamilton, 1952). Blandau (1949) showed that the decidual reaction in the rat starts before any apparent damage to uterine epithelium. Finn and Hinchliffe (1964) suggested that epithelial degeneration is an inherent property of the uterus during the period of implantation and is not directly caused by the activities of the giant cells of the trophoblast.

The uterine epithelial cells first struck by the trophoblastic giant cells are apparently completely ingested and, it seems, under this stimulus the primary giant cells become decidedly larger (Alden, 1948). Once a breach in the epithelium has been established, the succeeding stages of cell removal seem to be as follows: The cytoplasm of the giant cell commonly exhibits an advancing border of weakly staining hyaline material often seen in other motile cells. This ectoplasmic tongue flows around a group of epithelial cells, engulfing them.

Eaton and Green (1963) found that the lethal action of the A^y/A^y genotype in mice is due to the fact that the trophoblastic giant cells fail to differentiate normally. These giant cells differentiate at the equatorial plane of the blastocyst and at the ventral pole. In normal animals the giant cells form filamentous pseudopodia which penetrate between cells of the intact uterine epithelium. Completion of penetration is followed by extension of giant cell pseudopodia between the uterine epithelium and basement membrane. The sloughed uterine epithelium is then phagocytized by giant cells making attachment of the trophoblast to the endometrium possible. Penetration and attachment of giant cells begins at 4 days and 20 hours after copulation. In homozygous yellow (A^y/A^y) animals, the giant cells do not differentiate from the trophectoderm and attachment fails to take place. Kirby and Malhotra (1964) reported that the mouse probably has the most invasive trophoblast.

Wilson (1963) described "primary invasive" cells in the early mouse embryo. When the blastocysts reach their implantation sites, certain cells within the inner cell mass apparently migrate around the inner surface of the ventral trophectoderm. Approximately 100 hours after copulation cells of this type can be found between the trophoblast cells but protruding from its surface and penetrating the uterine epithelium. The uterine epithelium starts degenerating some time after the first of these "primary invasive" cells have penetrated it. According to Wilson (1963) these primary invasive cells are characteristic and had not been described previously.

Seven-somite embryo

In embryos from genetically vigorous stock, the 7-somite stage is reached at about 8 days. Thereafter the embryo begins a series of important changes, and it will be useful to review here the development attained at this point (Fig. 12-23, 12-24). In sagittal section the embryo is seen to form a letter S (facing to the left in Fig. 12-23) with the head region convex, the trunk region concave towards the dorsal surface. In transverse section, whereas the embryo was formerly conspicuously cupshaped with the ectoderm on the inside, it has now flattened out: in the regions of the fore- and hindgut the endoderm has become the inner layer. The neural groove, deep and well developed, is still open dorsally, though in the midtrunk region the walls are quite close together. Anteriorly, precocious growth of

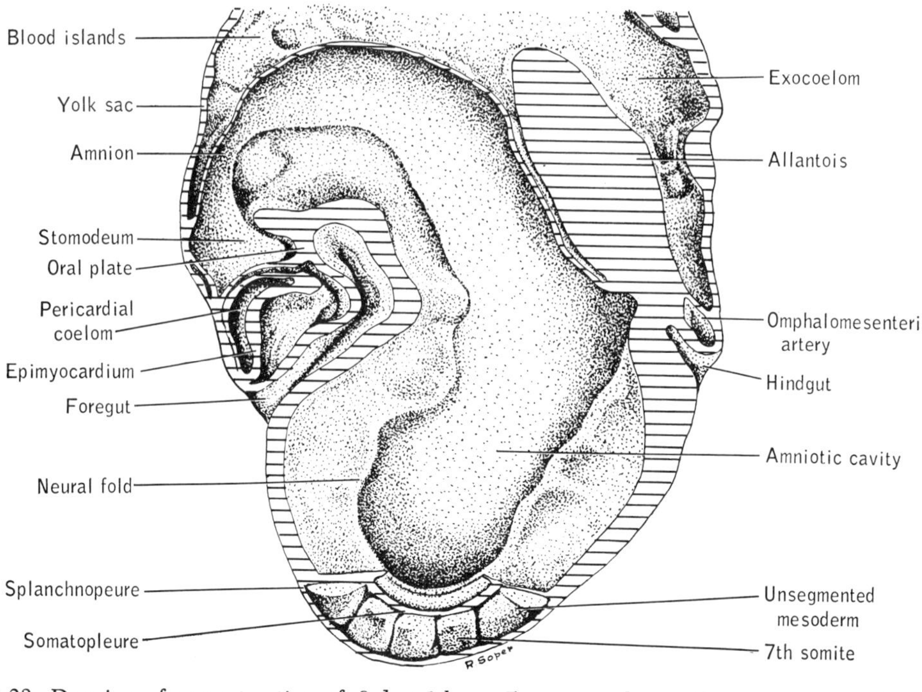

Fig. 12-23. Drawing of reconstruction of 8-day 1-hour, 7-somite embryo. The reconstruction is cut in the midsagittal plane and only the right half shown except at the ventral extremity where the last 4 somites and part of the undivided mesoderm of the left side are included. Cut areas are shown by horizontal shading (×75).

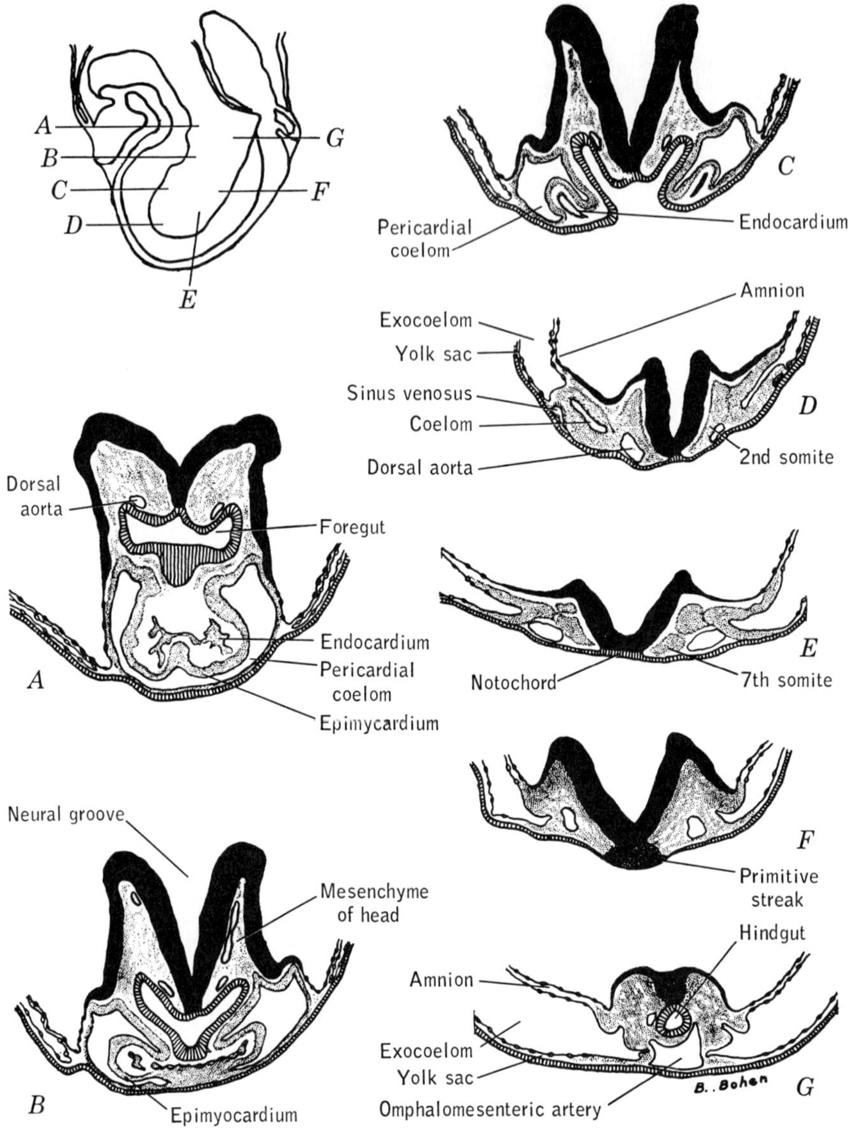

Fig. 12-24. Sections transverse to neural groove of 8-day 1-hour, 7-somite embryo. All except *E* are from the embryo shown in Fig. 12-22. The location of each section is indicated on the key diagram (×90).

certain parts of the neural groove ectoderm indicates early differentiation of the brain. The hindgut is small, but there is a deep foregut, and the heart, just anterior to the foregut, is a conspicuous structure. No midgut has formed. The allantois has almost reached the chorion; in fact in some embryos at this stage the allantois has already reached and fused with the chorion. The amnion and the yolk sac plus the chorion form a double arched roof over the whole dorsal surface of the embryo. The blood islands appear as a conspicuous hummocky band around the inner surface of the yolk sac. Blood vessels have begun to form within the embryo.

Tail fold

The hindgut appears much later than the foregut, but soon overtakes it in development. In 10-somite embryos the two are of approximately equal size (Fig. 12-30). A necessary concomitant of hindgut growth is the appearance of a tail fold; the gut endoderm pushes the overlying ectoderm and mesoderm ahead of it away from the yolk sac wall. A beginning of this process can be seen in 6-somite embryos (Fig. 12-17), and in 10-somite embryos the tail fold is well developed (Fig. 12-27*A*). The process is comparable to the formation of the head fold except for one interesting difference; whereas the head fold lies entirely within the amniotic cavity, the tail fold lies only partly within it. The ventral surface of the tail fold is in the exocoelom. This is because in its growth away from the yolk sac it pushes the base of the allantois and the adjacent margin of the amnion ahead of it. The amnion remains attached to its caudal and lateral walls, and only its dorsal surface is within the amniotic cavity (Fig. 12-25).

At about 9 days of gestation a transitory structure, the ventral ectodermal ridge of the tail (Grüneberg, 1963 and earlier), appears on the ventral aspect of the tail tip. The ectoderm is thickened and columnar and covers unsegmented paraxial mesoderm. Histologically the ventral ectodermal ridge resembles the apical ectodermal ridge of the limb buds, which is widely believed to be closely associated with the outgrowth of the limbs. The ventral ectodermal ridge of the tail may bear a relation to the outgrowth of the tail similar to that of the apical ectodermal ridge to the outgrowth of the limb buds.

Turning of the embryo

Almost immediately after the 7-somite stage the embryo begins a process of turning which results in a reversal of the curvature of the whole trunk region. Thus instead of being S-shaped the embryo becomes C-shaped with the ventral surface everywhere on the inside of the C. The turning begins in the head and tail folds and consists of a rotation of each along its long axis or, in other words, on axes parallel to the fore- and hindguts (Fig. 12-25 to 12-27). Viewing each fold from its cephalic toward its caudal end, the direction of rotation is clockwise in each case. Of course, both folds cannot be viewed in this direction from any one point because of the curvature of the embryo. Viewed from the mesometrial pole, in sections the turning of the head fold appears to be clockwise, of the tail fold counterclockwise (Fig. 12-25).

At first the turning is confined to the head and tail folds; the midtrunk region, still firmly attached to the yolk sac, remains in its original position. At about 8½ days and at about the 11- or 12-somite stage, the midtrunk region suddenly turns also. Transverse sections of the trunk region at about this period show it to be either turned or not turned (Fig. 12-20*C*, 12-21*A*). It is quite possible that after the growth of the head and tail folds reduces sufficiently the attachment of the trunk region to the yolk sac, this region snaps over like a spring whose tension has come to exceed the forces holding it. Some time elapses after the turning of the midtrunk before the head and tail regions complete their rotation, which eventually amounts to a full 180°. Essentially, however, by about 9 days the embryo has become concave towards the ventral surface (Fig. 12-27*B*, C).

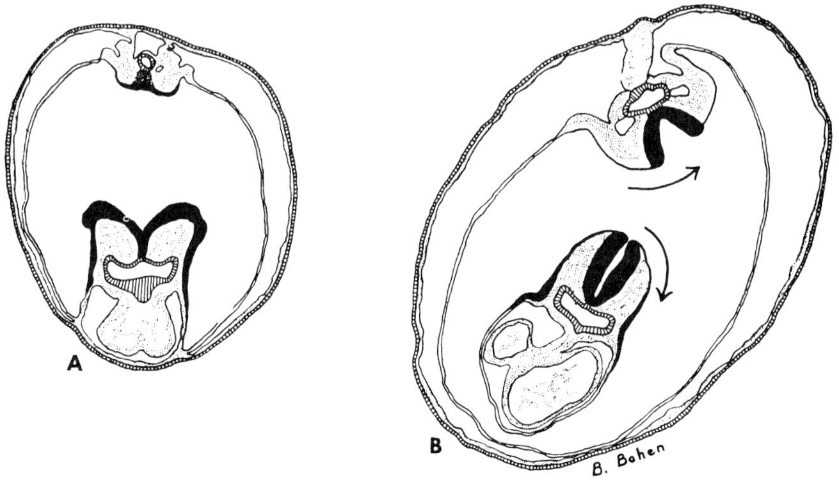

Fig. 12-25. Transverse sections showing the turning of the embryo. *A*. 8 days 1 hour, 7 somites. *B*. 8 days 10 hours, 10 somites (×50).

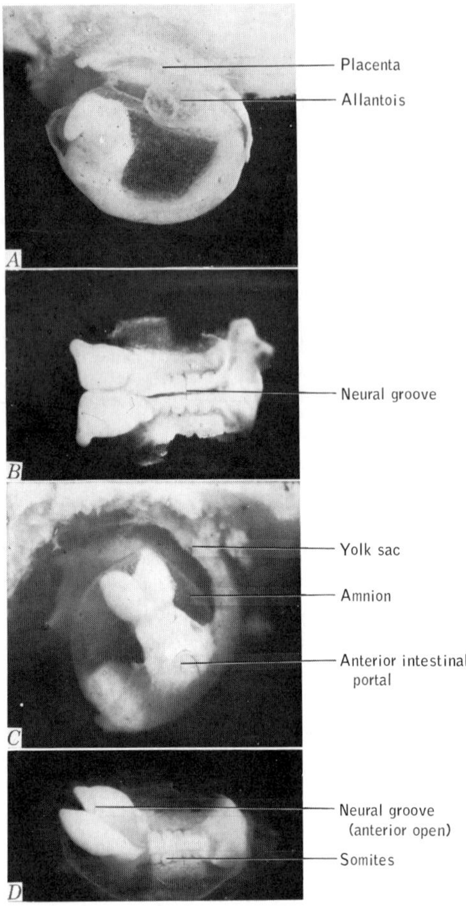

Fig. 12-26. Photographs (×22.5) of mouse embryos. *A*. Lateral view of 7-day 18-hour, 6-somite embryo, with decidua and most of yolk sac dissected. *B*. Dorsal view of same embryo, amnion also dissected. *C*. 10-somite embryo, age 9 to 9½ days. Slightly retouched. Embryo from inbred stock. *D*. Same embryo as *C*, dorsal view, amnion removed.

Midgut

The turning of the midtrunk region automatically results in the formation of the midgut. Prior to turning, the two sheets of embryonic splanchnopleure in the midtrunk region extend straight out from the sides of the embryo, forming a virtually plane surface (Fig. 12-20*C*). There is thus no indication of a midgut. When the midtrunk region turns suddenly towards its left side, the two sheets are pulled after it, forming between them a groove which is continuous anteriorly and posteriorly with the fore- and hindguts. This groove is the midgut (Fig. 12-21*A*). The two sheets of splanchnopleure rapidly draw closer together (Fig. 12-21*B*), and at the 19-somite stage, which may be reached as early as 8¾ days, have fused distally to form a closed tube.

Heart

In 7½-day embryos a small region of mesoderm anterior to the foregut (Fig. 12-16) forms the base of a U of which the two lateral sheets of mesoderm form the sides. Within this U the coelom develops and is, therefore, itself U-shaped. The base of the U and the two sides approximately as far caudad as the second pair of somites, contain that portion of the coelom which ulti-

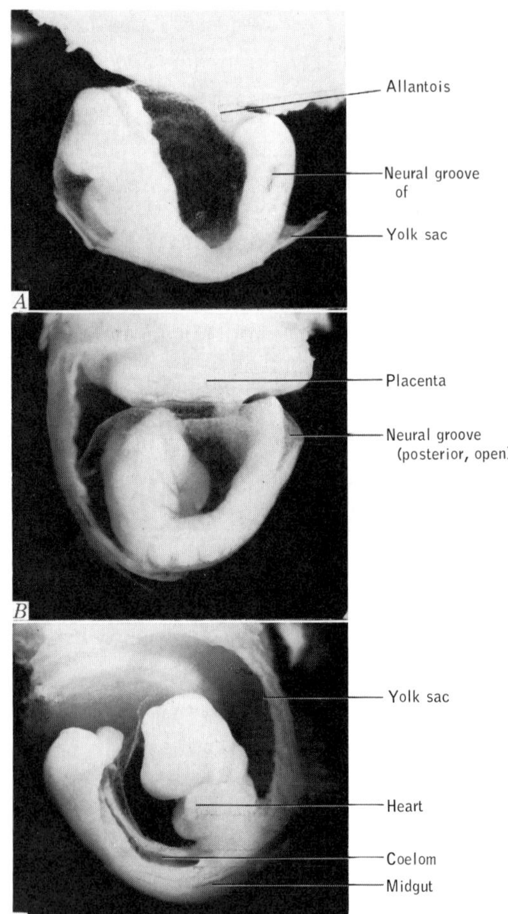

Fig. 12-27. Photographs (×22.5) of mouse embryos. *A*. 13-somite embryo age 9¼ to 9¾ days, from inbred stock. *B* and *C*. 14-somite embryo, age 8 days 22 hours. Note the greater degree of turning of this embryo, particularly in the midtrunk region, as compared with the one in *A*.

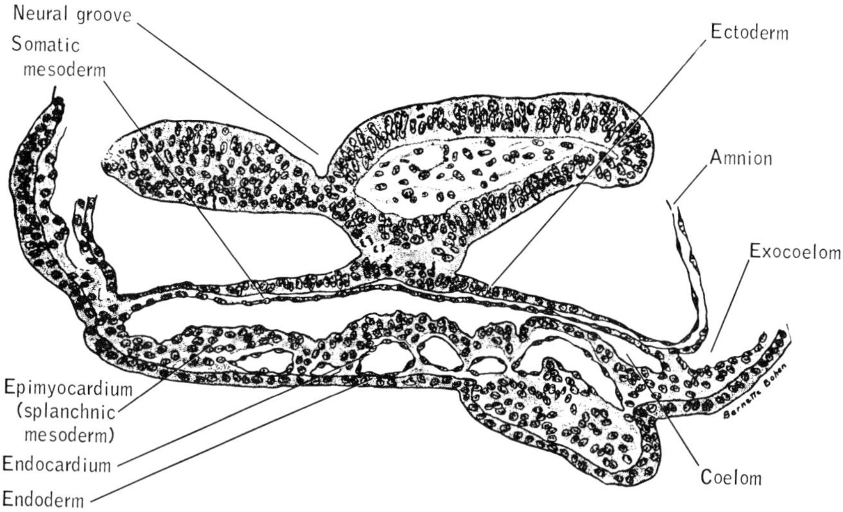

Fig. 12-28. Slightly diagonal transverse section through median endocardial primordium (see Fig. 12-29), just cephalad to foregut. Embryo of 8 days 6 hours, 5 somites. Projection drawing (×150).

mately encloses the heart and which, therefore, is known as the pericardial coelom (Fig. 12-23). The curved shape of the pericardial coelom in cross section in Fig. 12-23 should not be confused with the U-shape of the paricardial coelom as a whole.

The heart is derived from the splanchnic mesoderm forming the ventral wall of the pericardial coelom (Fig. 12-28). In 5-somite embryos this mesoderm has differentiated into two layers. Adjacent to the pericardial coelom is a thick continuous layer known as the epimyocardium because it will give rise both to the heavy muscular layer of the heart wall (myocardium) and to its outer covering (epicardium). Between the epimyocardium and the underlying endoderm are a number of irregular cavities which later fuse to form the cavity of the heart. The lining of these cavities is the endocardium.

Because of its relation to the U-shaped pericardial coelom, the heart is itself a U-shaped structure at this stage with the base of the U lying just cephalad to the anterior intestinal portal (Fig. 12-29). In many vertebrates the heart originates as two entirely distinct primordia which later fuse. As has been shown by Goss (1940) and by Burlingame and Long (1939), this is not the case in the rat. Our observations indicate that the condition in the mouse corresponds closely to that in the rat. As the intestinal portal moves caudad due to the "zipper action" causing the progressive folding together and fusion in the midventral line of the bounding endoderm, the sides of the U are likewise brought into approximation and fused together in the midventral line. The endocardium is thus transformed from a U-shaped structure into a single tube. At the 3-somite stage (in the rat) the different regions of

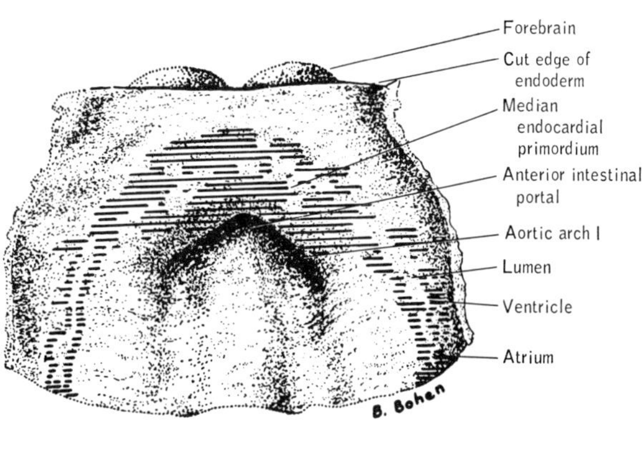

Fig. 12-29. Diagram of the foregut region viewed from the ventral surface, showing distribution of the endocardium. Endocardial tissue is represented by horizontal lines. Rat embryo of 9 days 16 hours, 3 somites. (*Modified after Goss,* 1940.)

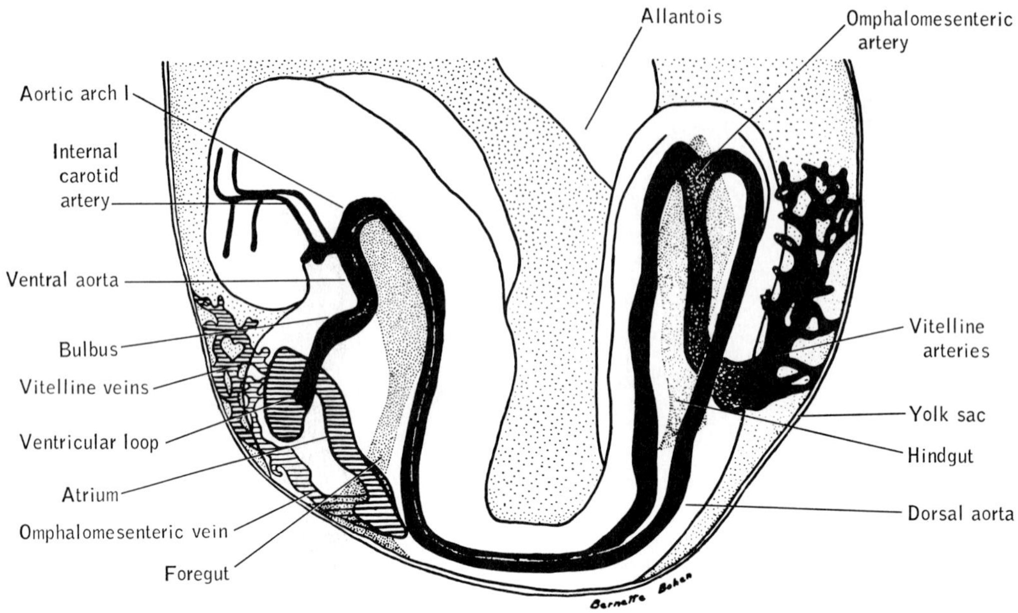

Fig. 12-30. Diagram of the circulatory system in an 8-day 10-hour, 10-somite embryo. The head and tail folds of this embryo have begun to turn but there is as yet no turning in the midtrunk region. Traces of the allantoic veins are present but are not shown as they do not yet form a continuous channel (×64).

the heart are not clearly set apart, though a slight constriction serves to mark the boundary between the atrium and the ventricle. As a result of subsequent foldings of the endocardial tube the different regions of the heart are clearly differentiated (Fig. 12-30).

Blood vessels

In 10-somite embryos, still in the process of turning, a number of blood vessels have become established (Fig. 12-30). The dorsal aorta at this stage is a paired vessel running the length of the trunk. It connects anteriorly with the heart by way of the aortic arches and the ventral aorta. Posteriorly its two halves fuse at the caudal extremity of the hindgut to form the single median omphalomesenteric artery. This runs cephalad for a short distance ventral to the hindgut and then turns away from the embryo towards the inner surface of the yolk sac on which it spreads out into a network of capillaries. These capillaries are derived from the blood islands. At this stage actual blood channels have not appeared in most of the blood islands, but when these are established, a capillary network is formed encircling the yolk sac. Blood is collected from this network anteriorly by the paired, omphalomesenteric veins which convey it back to the heart. When the heart starts beating, this system of blood vessels provides a generous circulation through the yolk sac serving at this time as the principal organ for the procurement of food from the mother.

Change in shape of yolk sac

When the embryo starts turning, the yolk sac and ectoplacental cone form a slightly flattened sphere (Fig. 12-20*C*). When turning has been completed, these bounding structures of the embryo shortly assume the form of a slightly saucered-out hemisphere (Fig. 12-21*B*). The ectoplacental cone becomes flattened and then dorsally concave, and the yolk sac adjacent to the cone pushes outward into the porous blood-filled vascular zone of the decidua. The embryo meanwhile, still attached to the yolk sac by the walls of the midgut, tips over so that it lies with its left side adjacent to the yolk sac, its right side facing the placenta.

SUMMARY

After fertilization the mouse egg undergoes cleavage in the oviduct and forms a morula. During cleavage, the cells are labile and have not yet been determined. The morula enters the uterus and becomes a blastocyst: a fluid-filled sphere formed dorsally by the inner cell mass and laterally and ventrally by the trophectoderm. About 5 days after fertilization the blastocyst implants in the uterine mucosa. Endoderm delaminates from the ventral part of the inner cell mass and migrates around the inner surface of the trophectoderm to form the distal (parietal) endoderm. The inner cell mass is composed of ectoderm covered laterally and ventrally by a sheet of proximal endoderm. It grows down into the blastocoel or yolk cavity. The ectoderm of the inner cell mass becomes divided into a ventral region which will form the embryo proper and a dorsal region which is extraembryonic ectoderm. The dorsal pole of the extraembryonic ectoderm becomes the ectoplacental cone which will form trophoblastic giant cells and will become part of the placenta. Because the inner cell mass, composed of ectoderm covered with endoderm, grows down into the yolk cavity, the germ layers are temporarily inverted. The embryonic and extraembryonic ectoderm line the proamniotic cavity. At about 6½ days the mesoderm is proliferated from the posterior part of the embryonic ectoderm, called the primitive streak. Mesodermal cells migrate between the embryonic and extraembryonic ectoderm and the endoderm. At the posterior end of the primitive streak the proliferation of cells causes a bulge into the proamniotic cavity. This bulge of ectodermal and mesodermal cells is the primordium of the amnion separating the amniotic cavity from the dorsal cavity lined by extraembryonic ectoderm. The amniotic cavity closes by fusion of the amniotic folds and separates the embryo from the extraembryonic structures dorsal to it. The amnion is the floor of the exocoelom which forms in the mesoderm underlying the extraembryonic ectoderm. The chorion forms the roof of the exocoelom. The chorion fuses with the ectoplacental cone and becomes part of the placenta.

The head process is formed at the anterior end of the primitive streak, near the ventral pole of the embryo. It migrates anteriorly between the endoderm and ectoderm and gives rise to the notochord and part of the lining of the gut. The neural groove forms dorsal to the notochord. Soon after the exocoelom is formed, a mesodermal fingerlike process, the allantois, grows into it from the posterior end of the primitive streak. The allantois elongates and fuses with the chorion and ectoplacental cone. Embryonic blood vessels will form within the allantois and will connect the embryo to the maternal blood supply.

The foregut and hindgut form as pockets in the endoderm at the anterior and posterior ends of the embryo. There is a progressive drawing together in the midventral line, a "zipper-like" action, of the endodermal lining of the foregut and hindgut which contributes to the formation of the midgut. The invagination of the foregut pushes the endoderm and overlying mesoderm and ectoderm as a bulge into the amniotic cavity. This bulge is the head fold. There is rapid growth of neural folds in this region which will form the brain. The heart will develop from the mesoderm of the head fold.

At about 7½ days the somites begin to form as paired segmental structures in the paraxial mesoderm a little anterior to the primitive streak. Approximately 65 pairs are formed.

The primitive streak is a region of rapid growth of undifferentiated cells giving rise anteriorly to the head process and posteriorly to the mesoderm and allantois. Coincident with the formation of the somites in the paraxial mesoderm, the lateral mesoderm splits into two layers forming the coelom. The coelom extends anteriorly around the foregut where it will become the pericardial coelom.

The embryo is enveloped by three membranes: Reichert's membrane, the amnion, and the yolk sac. Reichert's membrane is a protective structure secreted by the distal endoderm and is anchored to the maternal decidua by the trophoblastic giant cells. The amnion is small at first since it merely covers the cavity formed by the U-shaped embryo. It expands rapidly as the embryo straightens and completely encloses it. The yolk sac is formed from extraembryonic endoderm and its underlying mesoderm. It grows rapidly and after the midgut is formed it envelops the entire embryo. Blood islands develop in the mesodermal layer of the yolk sac. The primordial germ cells are located in the yolk sac near the allantois of 8½-day embryos. These cells are the progenitors of all the definitive germ cells of both sexes.

Implantation occurs on the antimesometrial wall of the uterus with the ectoplacental cone projecting dorsally into the uterine lumen. The uterine mucosa (decidua) grows rapidly at the point of implantation so that it completely surrounds the embryo and extraembryonic structures. The decidua grows and contacts the mesometrial wall of the uterus and fuses with it to form part of the placenta. At the same time, it breaks away from the antimesometrial wall so that the embryo is now attached to the wall opposite to the implantation site.

The preimplanted embryo is thought to be nourished by the uterine milk. Early implantation stages may utilize digested uterine epithelial cells. After implantation the embryo is bathed in maternal blood from which the yolk sac probably actively absorbs nutrients. Later, the decidua basalis, the ectoplacental cone, the chorion, and parts of the allantois fuse to give rise to a true placenta which assumes a major role in transferring nutrients to the embryo. Trophoblastic giant cells are derived from the trophectoderm and the ectoplacental cone. They may play an active role in implantation and they anchor Reichert's membrane to the decidua.

At about 8 days, the 7-somite embryo becomes S-shaped, the foregut and hindgut are forming, the neural groove is well developed, and the heart develops in the mesoderm anterior to the foregut. The allantois has almost reached the chorion, the blood islands appear, and blood vessels have begun to form within the embryo.

In the 10-somite embryo the hindgut endoderm pushes the overlying mesoderm and ectoderm into the base of the allantois to form the tail fold.

Shortly after the 7-somite stage the embryo rotates 180° along its longitudinal axis and the embryo changes from being S-shaped to being C-shaped with the ventral surface on the inside of the C. This turning brings the walls of the midgut parallel to each other to form a groove. The walls then fuse to form a closed tube, the midgut.

In 10-somite embryos the dorsal aorta is paired and connects with the heart by way of the aortic arches and the ventral aorta. Posteriorly its two halves fuse to form the single median omphalomesenteric artery which spreads out over the yolk sac as capillaries. The omphalomesenteric veins convey blood from the yolk sac back to the heart.

LITERATURE CITED

Alden, R. H. 1948. Implantation of the rat egg. III. Origin and development of primary trophoblast giant cells. Amer. J. Anat. 83:143–181.

Austin, C. R., and M. W. H. Bishop. 1958. Role of the rodent acrosome and perforatorium in fertilization. Proc. Roy. Soc. B 148:241–248.

Austin, C. R., and A. W. H. Braden. 1956. Early reactions of the rodent egg to spermatozoan penetration. J. Exp. Biol. 33:358–365.

Beatty, R. A., and K. N. Sharma. 1960. Genetics of gametes. III. Strain differences in spermatozoa from eight inbred strains of mice. Proc. Roy. Soc. Edinb. B 68:25–53.

Bennett, D., and L. C. Dunn. 1960. A lethal mutant (t^{w18}) in the house mouse showing partial duplications. J. Exp. Zool. 143:203–219.

Bishop, D. W., and A. Tyler. 1956. Fertilizin of mammalian eggs. J. Exp. Zool. 132:575–602.

Blandau, R. J. 1949. Embryo-endometrial interrelationship in the rat and guinea pig. Anat. Rec. 104:331–360.

Blandau, R. J., and W. L. Money. 1944. Observations on the rate of transport of spermatozoa in the female genital tract of the rat. Anat. Rec. 90:255–260.

Blandau, R. J., and D. L. Odor. 1949. The total number of spermatozoa reaching various segments of the reproductive tract in the female albino rat at intervals after insemination. Anat. Rec. 103:93–109.

Block, M. 1946. An experimental analysis of hematopoiesis in the rat yolk sac. Anat. Rec. 96:289–304.

Böe, F. 1951. Studies on placental circulation in rats. III. Vascularization of the yolk sac. Acta Endocrinol. 7:42–53.

Bonnevie, K. 1950. New facts on mesoderm formation and proamnion derivatives in the normal mouse embryo. J. Morphol. 86:495–546.

Boyd, J. D., and W. J. Hamilton. 1952. Cleavage, early development and implantation of the egg, p. 1-126. *In* A. S. Parkes [ed.] Marshall's physiology of reproduction, 3rd ed. Vol. 2. Longmans, Green, London.

Braden, A. W. H. 1958. Variation between strains of mice in phenomena associated with sperm penetration and fertilization. J. Genet. 56:37–47.

Braden, A. W. H. 1960. Genetic influences on the morphology and function of the gametes. J. Cell. Comp. Physiol. 56 (Suppl. 1):17–29.

Braden, A. W. H. 1962. Spermatozoon penetration and fertilization in the mouse. Symp. Genet. Biol. Ital. 9:1–8.

Braden, A. W. H., and C. R. Austin. 1954. The fertile life of mouse and rat eggs. Science 120: 610–611.

Bridgman, J. 1948. A morphological study of the development of the placenta of the rat. II. An histological and cytological study of the development of the chorioallantoic placenta of the white rat. J. Morphol. 83:195–224.

Burckhard, G. 1901. Die implantation des Eies der Maus in die Uterusschleimhaut und die Umbildung derselben zur Decidua. Arch. Mikroskop. Anat. 57:528–569.

Burlingame, P. L., and J. A. Long. 1939. The development of the heart in the rat. Univ. Calif. Pub. Zool. 43:249–320.

Butcher, E. O. 1929. The development of the somites in the white rat (*Mus norvegicus albinus*) and the fate of the myotomes, neural tube, and the gut in the tail. Amer. J. Anat. 44:381–439.

Chang, M. C. 1962. Fertilizability of rabbit ova after removal of the corona radiata. Fertil. Steril. 13:421–425.

Chiquoine, A. D. 1954. The identification, origin, and migration of the primordial germ cells in the mouse embryo. Anat. Rec. 118:135–146.

Chiquoine, A. D. 1958. The distribution of polysaccharides during gastrulation and embryogenesis in the mouse embryo. Anat. Rec. 129: 495–516.

Dalcq, A. M. 1956. Effets du réactif de Schiff sur les oeufs en segmentation du rat et de la souris. Étude qualitative. Exp. Cell Res. 10:99–119.

Dalcq, A., and M. Van Egmond. 1953. Effets de la centrifugation sur l'oocyte de trois Mammifères (Rat, Hamster, Taupe). Arch. Biol. 66: 312–397.

Dickerman, Z., and R. W. Noyes. 1960. The fate of ova transferred into the uterus of the rat. J. Reprod. Fertil. 1:197–212.

Dickson, A. D. 1963. Trophoblastic giant cell transformation of mouse blastocyst. J. Reprod. Fertil. 6:465–466.

Dziuk, P. J., and M. N. Runner. 1960. Recovery of blastocysts and induction of implantation following artificial insemination of immature mice. J. Reprod. Fertil. 1:321–331.

Eaton, G. J., and M. M. Green. 1963. Giant cell differentiation and lethality of homozygous *Yellow* mouse embryos. Genetica 34:155–161.

Edwards, R. G. 1957. The experimental induction of gynogenesis in the mouse. I. Irradiation of the sperm by X-rays. Proc. Roy. Soc. B 146: 469–487.

Edwards, R. G., and A. H. Gates. 1959. Timing of the stages of the maturation divisions, ovulation, fertilization and the first cleavage of eggs of adult mice treated with gonadotrophins. J. Endocrinol. 18:292–304.

Everett, J. W. 1935. Morphological and physiological studies of the placenta in the albino rat. J. Exp. Zool. 70:243–287.

Fawcett, D. W., G. B. Wislocki, and C. M. Waldo. 1947. The development of mouse ova in the anterior chamber of the eye and in the abdominal cavity. Amer. J. Anat. 81:413–443.

Fekete, E., O. Bartholomew, and G. D. Snell. 1940. A technique for the preparation of sections of early mouse embryos. Anat. Rec. 76: 441–447.

Finn, C. A., and J. R. Hinchliffe. 1964. Reaction of the mouse uterus during implantation and deciduoma formation as demonstrated by changes in the distribution of alkaline phosphatase. J. Reprod. Fertil. 8:331–338.

Geyer-Duszyńska, I. 1964. Cytological investigations on the T-locus in *Mus musculus L.* Chromosoma 15:478–502.

Goss, C. M. 1940. First contractions of the heart without cytological differentiation. Anat. Rec. 76:19–27.

Graves, A. P. 1945. Development of the golden hamster, *Cricetus auratus* Waterhouse during the first nine days. Amer. J. Anat. 77:219–251.

Gresson, R. A. R. 1940. Presence of the sperm middle-piece in the fertilized egg of the mouse (*Mus musculus*). Nature 145:425.

Grüneberg, H. 1943. The development of some external features in mouse embryos. J. Hered. 34:89–92.

Grüneberg, H. 1963. The pathology of development. A study of inherited skeletal disorders in animals. Wiley, New York. 309 p.

Gwatkin, R. B. I. 1964. Effect of enzymes and acidity on the zona pellucida of the mouse egg before and after fertilization. J. Reprod. Fertil. 7:99–105.

Hollander, W. F., and L. C. Strong. 1950. Intrauterine mortality and placental fusions in the mouse. J. Exp. Zool. 115:131–150.

Huber, G. C. 1915. The development of the albino rat, *Mus norvegicus albinus*. I. From the pronuclear stage to the stage of mesoderm anlage; end of the first to the end of the ninth day. J. Morphol. 26:1–114.

Huber, G. C. 1918. On the anlage and morphogenesis of the chorda dorsalis in mammalia, in particular the guinea pig (*Cavia cobaya*). Anat. Rec. 14:217–263.

Hyman, L. H. 1927. The metabolic gradients of vertebrate embryos. III. The chick. Biol. Bull. 52:1–38.

Jolly, J., and M. Férester-Tadié. 1936. Recherches sur l'oeuf du rat et de la souris. Arch. Anat. Microscop. 32:323–390.

Jones-Seaton, A. 1950*a*. A study of cytoplasmic basophily in the egg of the rat and some other mammals. Ann. Soc. Roy. Zool. (Belgium) 80:76–86.

Jones-Seaton, A. 1950*b*. Étude de l'organization cytoplasmique de l'oeuf des rongeurs principalment quant à la basophilie ribonucléique. Arch. Biol. 61:291–444.

Kingsbury, B. F. 1920. The developmental origin of the notochord. Science 51:190–193.

Kirby, D. R. S. 1962*a*. Influence of uterine environment on the development of mouse eggs. J. Embryol. Exp. Morphol. 10:496–506.

Kirby, D. R. S. 1962*b*. The development of mouse blastocysts transplanted to the scrotal and cryptorchid testis. J. Anat. 97:119–130.

Kirby, D. R. S., and S. K. Malhotra. 1964. Cellular nature of the invasive mouse trophoblast. Nature 201:520.

Krehbiel, R. H. 1937. Cytological studies of the decidual reaction in the rat during early pregnancy and the production of deciduomata. Physiol. Zool. 10:212–234.

Krehbiel, R. H. 1962. Distribution of ova in the rat uterus. Anat. Rec. 143:239–241.

Lams, H., and J. Doorme. 1907. Nouvelles recherches sur la maturation et la fécondation de l'oeuf des Mammifères. Arch. Biol. 23:259–365.

Leonard, S. L., P. L. Perlman, and R. Kurzrok. 1947. Relation between time of fertilization and follicle cell dispersal in rat ova. Proc. Soc. Exp. Biol. Med. 66:517–518.

Lewis, W. H., and E. S. Wright. 1935. On the early development of the mouse egg. Carnegie Inst. Wash. Pub. No. 459:133–144.

Mandl, A. M. 1963. Pre-ovulatory changes in the oocyte of the adult rat. Proc. Roy. Soc. B 158: 105–118.

Melissinos, K. 1907. Die Entwicklung des Eis der Mäuse von der ersten Furchungs-Phänomenen bis zur Festsetzung der Allantois an der Ectoplacentarplatte. Arch. Mikroskop. Anat. 70: 577–628.

Midgley, A. R., and G. B. Pierce. 1963. Immunohistochemical analysis of basement membranes of the mouse. Amer. J. Anat. 63:929–944.

Mintz, B. 1960. Embryological phases of mammalian gametogenesis. J. Cell. Comp. Physiol. 56 (Suppl. 1):31–47.

Mintz, B. 1962. Experimental study of the developing mammalian egg: removal of the zona pellucida. Science 138:594–595.

Mintz, B. 1964. Formation of genetically mosaic mouse embryos, and early development of "lethal (t^{12}/t^{12})-normal" mosaics. J. Exp. Zool. 157:273–292.

Mintz, B. 1965. Genetic mosaicism in adult mice of quadriparental lineage. Science 148:1232–1233.

Mintz, B., and E. S. Russell. 1957. Gene-induced embryological modifications of primordial germ cells in the mouse. J. Exp. Zool. 134:207–238.

Mossman, H. W., and L. A. Weisfeldt. 1939. The fetal membranes of a primitive rodent, the thirteen-striped ground squirrel. Amer. J. Anat. 64:59–109.

Nicholas, J. S., and B. V. Hall. 1942. Experiments on developing rats. II. The development of isolated blastomeres and fused eggs. J. Exp. Zool. 90:441–460.

Otis, E. M., and R. Brent, 1954. Equivalent ages in mouse and human embryos. Anat. Rec. 120: 33–64.

Pierce, G. B., A. R. Midgley, J. Sri Ram, and J. D. Feldman. 1962. Parietal yolk sac carcinoma: clue to the histogenesis of Reichert's membrane of the mouse embryo. Amer. J. Pathol. 41:549–566.

Pincus, G. 1936. The eggs of mammals. Macmillan, New York. 161 p.

Rugh, R. 1964. The mouse: a discoid placentate, p. 236–303. *In* R. Rugh, Vertebrate embryology; the dynamics of development. Harcourt, New York.

Sato, K. 1936. Über die Entwicklungsgeschichte des Mäuseeies I. Die intratubare Entwicklung deselben. Okayama-Igakkai-Zasshi 48:423–441.

Schlafke, S., and A. C. Enders. 1963. Observations on the fine structure of the rat blastocyst. J. Anat. 97:353–360.

Sharma, K. N. 1960. Genetics of gametes. IV. The phenotype of mouse spermatozoa in four inbred strains and their F_1 crosses. Proc. Roy. Soc. Edinb. B 68:54–71.

Smith, L. J. 1956. A morphological and histochemical investigation of a preimplantation lethal (t^{12}) in the house mouse. J. Exp. Zool. 132:51–84.

Snell, G. D. 1941. The early embryology of the mouse, p. 1–54. *In* G. D. Snell [ed.] Biology of the laboratory mouse. Blakiston, Philadelphia.

Sobotta, J. 1911. Die Entwicklung des Eies der Maus vom ersten Auftreten des Mesoderms an bis zen Ausbildung der Embryonalanlage und dem Auftreten der Allantois. Arch. Mikroskop. Anat. 78:271–352.

Stafford, E. S. 1930. The origin of the blood of the "placental sign." Anat. Rec. 47:43–57.

Stevens, L. C. 1964. Experimental production of testicular teratomas in mice. Proc. Nat. Acad. Sci. 52:645–661.

Tarkowski, A. K. 1962. Interspecific transfers of eggs between rat and mouse. J. Embryol. Exp. Morphol. 10:476–495.

Venable, J. W. 1939. Intra-uterine bleeding in the pregnant albino rat. The "placental sign." Anat. Rec. 74:273–293.

Whitten, W. K., and C. P. Dagg. 1961. Influence of spermatozoa on the cleavage rate of mouse eggs. J. Exp. Zool. 148:173–183.
Wilson, I. B. 1963. New factor associated with the implantation of the mouse egg. J. Reprod. Fertil. 5:281–282.
Wislocki, G. B., H. W. Deane, and E. W. Dempsey. 1946. The histochemistry of the rodent's placenta. Amer. J. Anat. 78:281–345.

13

Anatomy[1]

Katharine P. Hummel, Flavia L. Richardson, and Elizabeth Fekete

This chapter on gross and microscopic anatomy is based in part on Chap. 3, Histology, by Elizabeth Fekete (1941) in the first edition of the *Biology of the Laboratory Mouse*. We presuppose a general knowledge of mammalian anatomy and histology. We have added observations on the gross anatomy of skeletal, circulatory, and organ systems; on development of organs; and on sex, age, and strain differences in morphology. Except in rare instances, we have not included fine structure nor functional attributes as revealed by electron microscopy and histochemistry.

We have made no comprehensive detailed study of gross anatomy, referring rather to atlases on anatomy of the rat (Greene, 1955) and of the mouse (Cook, 1965). Some morphological differences between rat and mouse have been recorded. No descriptions of muscular or nervous systems are included; organs of special sense are described in Chap. 32 and blood and hematopoietic tissues in Chap. 17.

We have adhered to the terminology of comparative vertebrate anatomy rather than that of human anatomy in the following respects: anterior and cephalic for head end, posterior and caudal for tail end, dorsal for the back or vertebral side, and ventral for the under or belly side.

SKELETON

The axial skeleton is composed of skull, vertebral column, sternum, and ribs; the appendicular skeleton of the paired girdles and appendages. In addition, there are 43 sesamoid bones of three general types (Wirtschafter and Tsujimura, 1961). For a detailed description of the anatomy and growth of the bones of the normal skeleton, see Bateman (1954).

Axial Skeleton

This is the main support of the body and also serves to protect the brain within the cranial cavity, the spinal cord within the vertebral column, and the heart and lungs within the thoracic basket formed by the vertebrae, ribs, and sternum.

Skull. The skull closely resembles that of the rat but has a relatively wider and more rounded cranium (Kutuzov and Sicher, 1953). From the dorsal aspect, the following bones can be identified: the unpaired occipital and interparietal; and the paired parietals, frontals, nasals, premaxillae, maxillae, zygomatics, and squamosals (Fig. 13-1). The occipital forms the posterior wall of the cranial cavity, the interparietal a portion of the roof, and the parietals most of the roof and a portion of the sides of the cranial cavity. Frontals and nasals cover the anterior portions of the brain and with premaxillae, maxillae, zygomatics, and squamosals form the face and upper jaw.

Vertebral column. The vertebral column (spine) articulates anteriorly with the occipital bone of the skull, forms the middorsal support of the thoracic and abdominal cavities, and constitutes the skeleton of the tail. It is composed of the following groups of vertebrae: seven cervical, the most anterior being the atlas and axis; 12, 13, or 14 thoracic, each articulating with a pair of ribs; five or six lumbar; four sacral, some articu-

[1] The writing of this chapter was supported in part by Public Health Service Research Grant HD 00468 from the National Institute of Child Health and Human Development and in part by the Elsa U. Pardee Foundation.

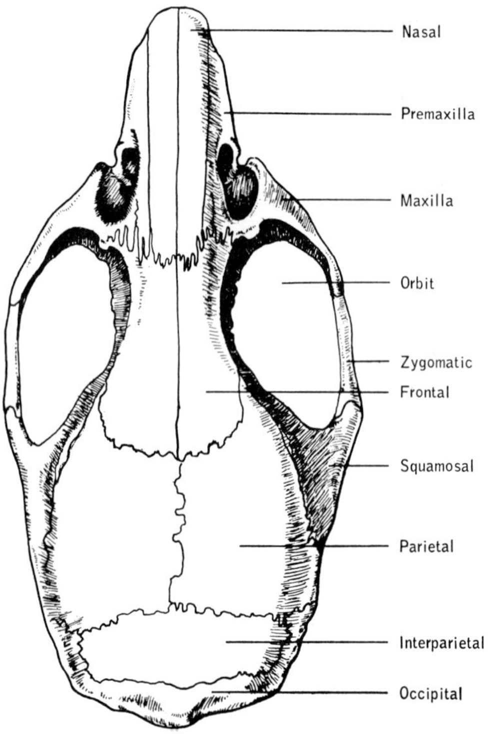

Fig. 13-1. Drawing of the skull, dorsal aspect.

lating with the innominate bones of the pelvic girdle; and 27 to 30 caudal. Of the 25 to 27 presacral vertebrae, seven are invariably cervical, but the numbers of thoracic and lumbar vertebrae differ with strain, sex, and environmental factors (Grüneberg, 1952, 1963; Green, 1962).

Sternum. The sternum forms the ventral support of the thoracic basket and consists of six connected parts: the manubrium articulating laterally with the paired clavicles and first ribs, four sternebrae articulating with more posterior ribs, and the xiphisternum, which is a slender segment extending posteriorly beyond the rib attachments and terminating in a cartilaginous xiphoid process. Bifurcation of the xiphisternum is not uncommon and is characteristic of short ear (*se/se*) mice (Green and McNutt, 1941). The sternum is first visible in the 12-day embryo as a mesenchymal condensation on either side of the midventral line in the thorax. Elongation, chondrification, and medial movement of the sternal bands result in a single continuous cartilage bar by the 16th embryonic day, and segmentation and ossification follow, producing the six bony segments present at birth (Chen, 1952).

Ribs. These are 12, 13, or 14 pairs of slender curved bones, all articulating dorsally with thoracic vertebrae and some ventrally with the sternum, thus forming the thoracic basket. The dorsal segments are ossified and the ventral segments calcified. Usually the ventral segments of the first seven pairs of ribs attach to the sternum, whereas more posterior pairs join the ventral segment of the seventh rib, and the most posterior three or four have no ventral attachments.

Appendicular skeleton

This includes the paired pectoral and pelvic girdles and the bones of the limbs. There is considerable variation in size and structural detail of girdles and long bones of limbs in mice of different strains (Stein, 1957).

Girdles. Each pectoral girdle consists of dorsal scapula and ventral clavicle. The scapula is a flat trapezoid bone resting against the anterior ribs with its base toward the vertebral column and its pointed end articulating laterally with the humerus and clavicle. The clavicle is a slender curved bone articulating ventrally with the manubrium and laterally with the scapula.

The pelvic girdle of the adult consists of right and left innominate bones attached dorsally to one or more sacral vertebrae and joined ventrally at the pubic symphysis. Three separate bones (ilium, ischium, and pubis) fuse to form the innominate bone, although fusion between the ischium and pubis may be delayed or incomplete in mice of certain strains (Stein, 1957). The pelves of adults exhibit sexual dimorphism. The presence of the testis is necessary for the development of the male type bony pelvis, the influence of the male sex hormone becoming irreversible by 21 days of age (Crelin, 1960). The pelvis of the female is wider and is located farther ventrally and caudally, the pubic symphysis being opposite the second caudal vertebra rather than opposite the first as in the male.

Limbs. The bones of the forelimb are humerus, radius and ulna, eight carpals, five metacarpals, five first phalanges, four second phalanges, and five third phalanges with claws at their tips. The preaxial digit (thumb or pollex) is much shorter than the other four, lacks a middle phalanx, and has a very short distal one. The bones of the hind limb are femur, tibia and fibula, seven tarsals, five metatarsals, and the same number of phalanges as in the forepaw. The first or great toe (hallux) lacks the middle phalanx but is not so short as the pollex. The tibia and fibula are separate for the proximal two-thirds of their length, are fused along the distal third, but again free at the distal extremities. For the articulations of the bones of the hind limb and foot, see Carter (1951) and for their development, see Carter (1954); for development of fore and hind limbs, see Forsthoefel (1959, 1963).

CIRCULATORY SYSTEM

This brief description includes only heart and principal blood vessels. For details, see Greene (1955) and Cook (1965).

Heart

Gross anatomy. The heart lies in the pericardial cavity, a division of the thoracic cavity, and consists of four muscular-walled chambers, the left and right atria and the left and right ventricles (Fig. 13-2). As in other mammals there is no direct communication between left and right sides after birth. Blood from the body is carried in veins to the right atrium, then into the right ventricle, and through the pulmonary artery to the lungs. Oxygenated blood from the lungs enters the left atrium, passes into the left ventricle, and then into the aorta to be distributed to all parts of the body.

Valves at the atrioventricular orifices and at the bases of the large arteries prevent reverse flow of blood. The right atrioventricular orifice is guarded by the tricuspid valve, three leaflets extending into the ventricular cavity and held in place by tendinous cords. These lead from the edges of the leaflets to muscular pillars (papillary muscles) projecting from the wall of the ventricle. A similar, but bicuspid, valve (the mitral) guards the left atrioventricular orifice. The openings of right and left ventricles into pulmonary artery and aorta respectively are guarded by trios of semilunar valves. These pocketlike extensions of the ventricular lining fill with blood and close the openings during relaxation of the ventricle walls.

Microscopic anatomy. Three main layers can be identified in heart wall: the endocardium, a thin layer lining the cavities and covering the valves; the myocardium, a muscle layer that is thin in the atria and considerably thicker in the left than in the right ventricle; and the epicardium, a thin covering layer. The endocardium consists of endothelial cells plus loose connective tissue binding them to the underlying muscle. The valves are folds of endocardium in which the connective tissue is fibrous. In old mice cartilage cells often can be observed within the fibrous tissue at the bases of the valve cusps. The myocardium consists of spiral sheets of cardiac muscle bound together by connective tissue supporting blood vessels and nerves. The epicardium consists of a thin layer of connective tissue and a mesothelial layer or pericardium continuous with the lining of the pericardial cavity surrounding the heart.

Arteries

Gross anatomy. The principal vessels carrying blood from the heart are the pulmonary artery from the right ventricle and the aorta from the left ventricle (Fig. 13-2). Branches of these arteries supply all parts of the body, their origins and pathways being somewhat variable in individuals as well as in strains (Froud, 1959).

The pulmonary artery arises from the base of the right ventricle ventral to the base of the left ventricle and aorta. It is a short vessel, soon dividing into right and left branches to the right and left lung lobes.

The aorta arises from the left ventricle dorsal to the pulmonary artery, extends anteriorly (ascends) a short distance, arches to the left, then passes posteriorly (descends) through the thoracic and abdominal cavities and into the tail (Fig. 13-3). The right and left coronary vessels supplying the heart are the only branches from

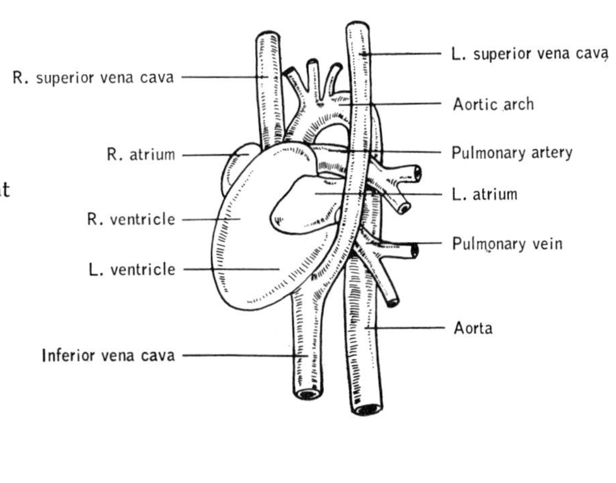

Fig. 13-2. Drawing of the heart reflected to right to show principal arteries and veins.

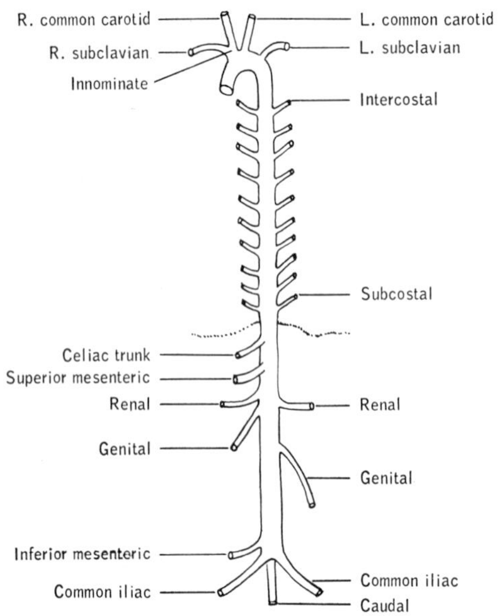

Fig. 13-3. Drawing of the main arteries branching from the aorta.

the ascending limb. The innominate, left common carotid, and left subclavian arise from the arch. A short distance from its origin, the innominate divides into right subclavian and right common carotid. The subclavians and their branches supply the pectoral girdles, forelimbs, and thorax, and the carotids supply head and neck.

The thoracic portion of the descending aorta extends from the aortic arch to the diaphragm, lying slightly left of the vertebral column anteriorly and approaching the midline near the diaphragm. Its main branches supply the thoracic viscera (trachea, bronchi, esophagus, etc.) and the intercostal muscles.

The abdominal portion of the descending aorta extends along the midline of the vertebral column from the diaphragm to the lumbosacral region where it divides into right and left common iliac arteries supplying pelvic and hind limb regions. Visceral branches of the abdominal aorta supply the organs; parietal branches supply the body wall. The first large branch is the celiac trunk, an unpaired artery supplying liver, stomach, spleen, duodenum, and pancreas. The next branch is the superior mesenteric, a large unpaired vessel branching through the mesentery and supplying all of the small intestine, cecum, and ascending and transverse colons. Next are the paired renal arteries to the adrenals and kidneys. These are asymmetrical, the right being more anterior than the left. The right renal may arise anterior to the superior mesenteric and usually is dorsal to the inferior vena cava and to the renal vein and slightly anterior to the latter. The paired genital arteries (spermatic or ovarian) usually arise from the aorta posterior to the renals but their pattern is extremely variable; they may branch from the renals or from the renal aortic junction and they are seldom bilaterally symmetrical. The unpaired inferior mesenteric artery arises from the ventral surface of the aorta close to its division into common iliacs and supplies the descending colon and rectum. The lumbar arteries are segmental pairs arising from the dorsal surface of the aorta and supplying the dorsal musculature. The common iliacs supply the pelvis and hind limbs, and the middle caudal continues the course of the aorta into the tail.

Microscopic anatomy. Three layers, intima, media, and adventitia (externa), can be distinguished in the walls of most arteries. The intima consists of endothelial cells on a connective tissue bed which in large arteries contains a network of elastic fibers. The media is composed of alternating layers of smooth muscle fibers and elastic fibers, the number and relative width of the layers varying with vessel caliber. In the aorta there are 6 to 10 layers intermingled with fine collagenous fibers; in smaller arteries there are more muscle than elastic fibers; and in arterioles few muscle and no elastic fibers. The adventitia is a layer of loose connective tissue merging with that of surrounding structures.

Capillaries and sinusoids

Extensive networks of capillaries connect terminal arterioles and beginning veins in most organs. The capillary walls are composed of single layers of endothelial cells accompanied by reticular tissue. In some organs irregular spaces or sinusoids connect arteries and veins. The thin walls are formed of irregularly scattered phagocytic and nonphagocytic reticulum cells accompanied by networks of reticular fibers.

Veins

Gross anatomy. The principal veins entering the atria of the heart are the pulmonary veins from the lungs; the superior venae cavae from the head, neck, chest, and forelimbs; and the inferior vena cava from regions of the body posterior to the diaphragm (Fig. 13-2).

The left and right pulmonary veins—short vessels formed by the union of smaller veins from the lobules and lobes of the lung—enter the left atrium of the heart. The right pulmonary vein passes dorsal to the right superior vena cava, the left pulmonary dorsal to the left superior vena cava.

The right and left superior venae cavae, carrying blood from regions of the body anterior to the diaphragm, enter the right atrium. The right vein is short and opens directly into the atrium on its anterior border. The left vein extends farther posteriorly, passes ventral to the pulmonary arteries and veins and dorsal to the base of the ventricles, joins the unpaired inferior vena cava, and enters the right atrium on its posterior border. Each superior vena cava is formed by the confluence of the jugular vein from head and neck and the subclavian vein from forelimb and chest. Coronary veins from the heart join the left superior vena cava and also enter the right atrium directly. The unpaired azygos vein from the intercostal spaces lies to the left of the aorta and vertebral column and joins the left superior vena cava.

The inferior vena cava is an unpaired vessel originating in the union of left and right iliac veins from the pelvic and hind limb regions. It lies to the right of the vertebral column and aorta, is dorsal to the aorta in the lumbar region, and ventral to it in the renal region. Veins from the dorsal musculature and gonads join it posterior to the large renal veins from the kidneys and adrenals. Like the renal arteries, the renal veins are asymmetrical and vary in position relative to nearby arteries, veins, and organs. The inferior vena cava enters the liver at the base of the posterior subdivision of the right lateral lobe, passes through the liver receiving therein numerous hepatic branches, and emerges from the anterior surface of the median lobe still to the right of the midline. It then pierces the diaphragm, traverses the thoracic cavity, and joins the left superior vena cava where it enters the right atrium.

An important tributary of the inferior vena cava is the hepatic portal vein, formed by the union of veins from the stomach, intestine, pancreas, and spleen. It passes through the mesentery dorsal to the duodenum to enter the median lobe of the liver close to the gall bladder. In the liver, branches of the hepatic portal vein communicate with the liver sinusoids which drain into the hepatic veins and thence into the inferior vena cava. In anomalous conditions, the hepatic portal vessel may pass ventral to the duodenum, enter one of the lateral lobes, and branch before entering the liver.

Microscopic anatomy. Veins are similar to arteries but have thinner, softer, and less elastic walls in which the three layers are frequently indistinct. In small veins the smooth muscle of the media is replaced by connective tissue, and in pulmonary veins by cardiac muscle. The walls of many veins contain paired semilunar valves formed of folds of the intima.

LYMPHATIC SYSTEM

This includes the vessels through which lymph is transported from tissue spaces to the blood circulation, the nodes that lie in the course of the vessels, and the peripheral nodules present at the beginnings of lymph channels in the digestive tube. There are no palatine and no pharyngeal tonsils.

Vessels

Gross anatomy. Lymph flows from lymph capillaries into successively larger vessels, thence into ducts that empty into veins in the neck. In the rat, lymph vessels unite to form two main ducts one on each side of the vertebral column (Job, 1915). The small right duct opens into the right superior vena cava at the junction of right jugular and subclavian veins. The thoracic duct is the larger duct on the left that opens into the junction of left subclavian and jugular veins. A pouchlike dilatation, the cisterna chyli, on the dorsal abdominal wall slightly anterior to the kidneys, identifies the beginning of the thoracic duct. In the mouse only certain subsidiary pathways such as that from the tail have been traced (Engeset and Tjötta, 1960).

Microscopic anatomy. Lymph capillaries are thin-walled vessels with irregular lumina and frequent dilatations and constrictions. Their walls are single layers of endothelial cells. In the larger vessels, collagenous, elastic, and smooth muscle fibers surround the endothelial layer. In still larger vessels, intima, media, and adventitia can be distinguished and folds of the intima form paired valves. The intima consists of endothelium and a layer of elastic fibers, the media of circularly arranged smooth muscle fibers, and the adventitia of longitudinally oriented elastic, collagenous, and smooth muscle fibers.

Nodes

Gross anatomy. The nodes are bean-shaped structures of varied size located in the course of lymph vessels, interrupting their continuity. Nodes are found in connective tissue subcutaneously, between muscles, and near viscera in body cavities. The number and size of visible nodes varies

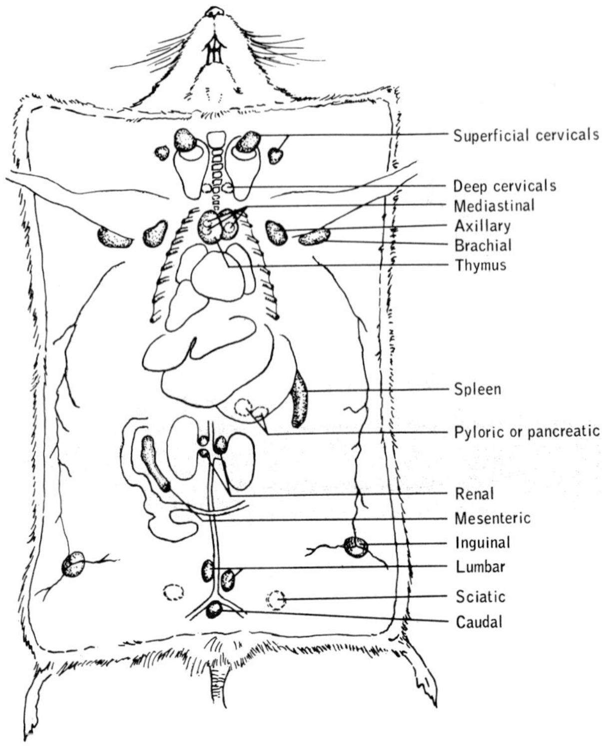

Fig. 13-4. Diagram to show the locations of the principal lymph nodes, the spleen, and the thymus. (*From Dunn,* 1954, *courtesy of the author.*)

with individuals and environmental conditions. Strain differences in size have been reported, nodes of C3H mice being larger at all ages than those of mice of other strains (Albert and Johnson, 1960). The locations (Fig. 13-4) and the names and descriptions of nodes visible under normal conditions are based on the observations of Dunn (1954).

All nodes beneath the skin and between muscles are bilateral. The principal ones are two superficial cervicals on the anterior ventral margin of the submandibular gland, one or two very small deep cervicals, buried in the connective tissue alongside the trachea, a large axillary in the axillary fossa, a large brachial on the belly of the biceps muscles, a large inguinal adherent to the skin of the groin, and a small sciatic buried between bundles of the gluteal muscles near the emergence of the sciatic nerve.

Most of the nodes of the viscera are not bilateral, although some are on or near the midline. The principal nodes are three or four mediastinal near the thymus and bifurcation of the trachea; two pyloric (pancreatic) attached to the anterior margin of the pancreas near the pyloric end of the stomach; left and right renals between aorta and kidneys, the left being anterior to the renal vessels and the right partly obscured by the renal vessels passing ventral to it; a mesenteric, usually single and elongated, in the mesentery of the ascending colon; two small lumbars anterior to the bifurcation of the abdominal aorta, the left being smaller and more posterior than the right; and a single small caudal just posterior to the bifurcation of the abdominal aorta.

Microscopic anatomy. A fibrous capsule, a marginal subcapsular sinus, a cortex of dense lymphatic tissue, and a medulla of diffuse lymphatic tissue with large intercommunicating sinusoids are constant features of all nodes. Reticular fibers and cells form the framework in which the lymphocytes are massed, and flattened reticulum (littoral) cells line the sinusoids. Indistinctly outlined nodules of lymphatic tissue, occasionally with pale centers (secondary or germinal centers), are transitory structures in the cortex. The boundary between cortex and medulla is often obscured by an intermediate zone of diffuse tissue which in old mice may be composed largely of plasma cells (Dunn, 1954). Afferent lymph vessels open through the capsule into the marginal sinus; efferent vessels leave the node at the hilus where blood vessels enter and leave. The lining of postcapillary veins is unusual in that the cells are cuboidal and have distinctive histochemical features (Smith and Hénon, 1959).

The nodes have been classified into three types according to relative amounts and arrangements of cortex and medulla (Dunn, 1954). In nodes such as the inguinal, the cortex surrounds the medulla, except at the hilus, and the intermediate zone is wide. The cortex and medulla are at opposite poles in nodes of a second type, represented by the lumbar node. In the mesenteric node, the third type, the cortex and intermediate zone are narrow and eccentric and the medulla wide with large sinusoids.

Peripheral nodules

Gross anatomy. Although scattered foci of lymphatic tissue may occur anywhere in the walls of the respiratory, urinary, reproductive, and digestive tracts, aggregates of lymphatic nodules are found only in the small intestine, cecum, colon, and rectum (Fig. 13-36*B*). The conspicuous nodules on the antimesenteric wall of the small intestine are the Peyer's patches or intestinal tonsils. They vary in size with environmental conditions and in number with strain (Kelsall, 1946; Hummel, unpublished, Table 13-1).

Microscopic anatomy. The peripheral nodules differ from nodes in having no capsules, marginal sinuses, afferent lymph vessels, nor definite cortical and medullary areas; otherwise they are structurally similar accumulations of lymphatic tissue.

SPLEEN

The spleen, although composed chiefly of lymphatic tissue, is not classed with lymph nodes and peripheral nodules because it is neither in the course of lymph vessels nor at their beginnings. Instead, the spleen is in the course of blood vessels and has no lymph vessel connections.

Table 13-1. PEYER'S PATCHES IN MICE OF SEVEN STRAINS

Strain[a]	Number of mice	Average number of patches
C58	286	7.9
C3H	77	8.7
C57BL/6	132	7.4
DBA/1	69	9.5
A	54	7.7
BALB/c	169	8.3
RIII	160	8.0

[a] No sex differences observed.

Gross anatomy

The spleen is a slightly curved, elongated oval organ lying diagonally in the left anterior quadrant of the abdominal cavity along the greater curvature of the stomach (Fig. 13-4). The splenic artery enters through several branches along a longitudinal ridge on the dorsal aspect. The ventral aspect is very slightly convex; a cross-section through the central portion of the spleen is roughly triangular, the dorsal ridge forming the apex (Fig. 13-5). The spleen varies in appearance, shape, and size depending on age, strain, sex, and especially environmental conditions. The healthy spleen is deep red in color and has a smooth glistening surface. The right extremity is usually rounded but the left may be rounded, pointed, or bifurcated (Table 13-2). Variation in size is so great that spleen weight is not a useful indicator of response of lymphatic tissue to experimental manipulations (Dunn, 1954; Santisteban, 1960). One or more nodules of accessory splenic tissue are often embedded in nearby pancreas. Strains of mice have been found to differ with respect to the percentage of individuals having accessory nodules (Hummel, unpublished, Table 13-2).

Microscopic anatomy

The surface layer is peritoneum, beneath which is a thin but tough elastic capsule of dense connective tissue containing some elastic fibers. Irregularly arranged thin trabeculae containing smooth muscle cells project inward, and with a network of reticular fibers and cells form the framework for the white and red splenic pulp.

Table 13-2. ACCESSORY SPLEENS AND SPLEENS WITH FORKED TIPS IN MICE OF NINE STRAINS

Strain[a]	Number of mice	Percentage with accessory spleens	Percentage with forked left lateral tip
C57L	146	38	6
C57BL/6	352	32	45
C58	134	21	33
BALB/c	262	20.5	4
DBA/1	372	17	1
A	185	14	8
RIII	115	4	12
CE	137	0.9	7
C3H	261	0.2	1

[a] No sex differences were observed.

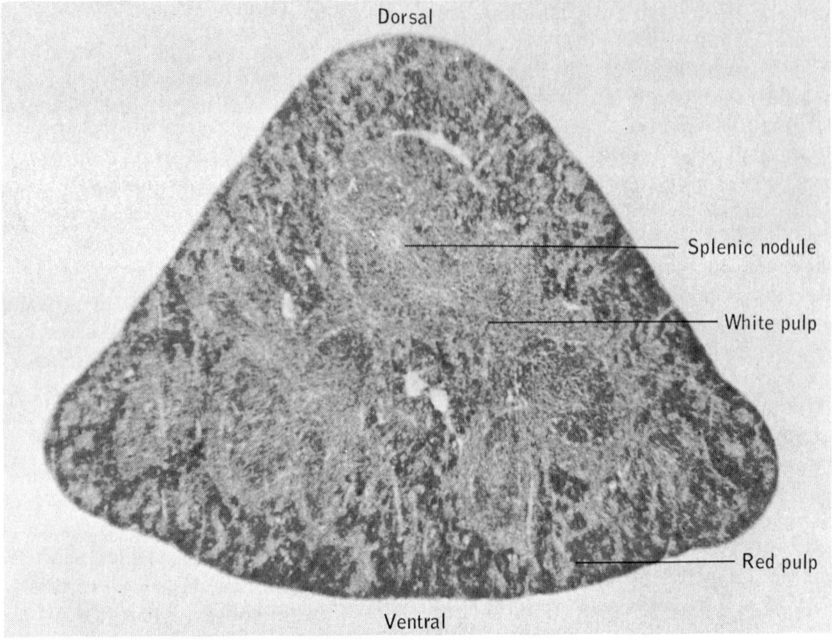

Fig. 13-5. Spleen, cross section (×30).

White pulp, lymphatic tissue sheathing small arteries, makes up the bulk of the organ. In white pulp, lymphocytes are massed into indistinctly outlined nodules, the splenic nodules or Malpighian corpuscles. They are similar to the cortical nodules of lymph nodes and like them are transitory structures occasionally exhibiting pale secondary or germinal centers. Marginal zones of diffuse lymphatic tissue surround the nodules, separating them from the red pulp which contains erythrocytes in its meshes. In mice of some strains, anastomosing clefts between the nodules and marginal zones have been described (Snook, 1950). In addition to erythrocytes, red pulp contains granular leukocytes, lymphocytes, blood-forming elements, mast cells, plasma cells, and megakaryocytes. Extramedullary hematopoiesis is found regularly in the red pulp, erythropoiesis being especially evident in young healthy mice (Dunn, 1954).

The complex vascular arrangement has been the object of much study and controversy. Transillumination of living spleens resulted in divergent opinions with respect to a closed (through sinuses) or open (through tissue spaces) circulation. Snook (1950), after studying graphic reconstructions of spleens of many mammals, concluded that both types of circulation exist. Whereas man and the rat are in the "sinusal" group, the mouse belongs in the "nonsinusal" group. The red pulp of spleens of rat, dog, man, and many other mammals contains an elaborate plexus of sinuses; that of mouse, cat, pig, and others has no true sinuses and relatively few veins leading from the meshes into collecting veins.

After entering the spleen, arteries branch repeatedly, decrease in size, and become ensheathed by splenic nodules. The small branched arteries extend from the white pulp into the red pulp where they divide into short straight penicilli. The penicilli of the mouse (and rat) do not have ellipsoid sheaths characteristic of their structure in spleens of man. Zones of red pulp with many erythrocytes in their meshes intervene between the ends of the penicilli and the capillary venules leading to collecting veins (Snook, 1950).

There are strain differences in the pattern of the reticulum and its arrangement within the nodule, the relative amounts of red and white pulp, amount of hematopoiesis, numbers and distribution of hemosiderin-containing macrophages, and numbers of such cells as mast cells and megakaryocytes (Dunn, 1954).

THYMUS

The thymus, long regarded as a vestigial organ with little or no function in the adult animal, has been classed with lymph nodes and spleen as an organ of reticular tissue and with other pharyngeal derivatives as an endocrine gland. The thymus differs from lymph nodes and spleen in being epithelial in origin and character and in having no sinusoids, and from endocrine glands in having no proved hormone-secreting activity. Recognition of the important role of the thymus in leukemogenesis (in the mouse) and in immunobiology has kindled interest in its form as well as its function. For studies on morphology the mouse thymus is ideal, having few lobules and a relatively simple vascular pattern (Smith, 1964).

Gross anatomy

The thymus consists of two separate lobes located close together on either side of the midline in the anterior portion of the thoracic cavity ventral to the base of the heart and aortic arch (Fig. 13-4). In young mice the lobes are white and firm with smooth surfaces and few lobular indications. The two lobes differ in shape and the right more oval slightly overlaps the left more triangular one. The hilus where blood vessels enter and leave is a narrow inconspicuous cleft, variable in location but usually on the dorsolateral surface (Smith et al., 1952). The thymus is larger in young than old mice, reaching the maximum absolute size at about the time of sexual maturity and thereafter decreasing (Smith and Ireland, 1941). Decrease in weight is rapid between 35 and 80 days, but becomes gradual later. Although complete involution does not occur in old age, there is a decrease in all dimensions and the lobes become thin and leaflike (Smith et al., 1952).

The thymic lobes are epithelial thickenings in the region of third and fourth pharyngeal pouches in 11-day embryos. During the 15th day the epithelial vesicles separate from the pharyngeal epithelium and come to lie anterolateral to the heart and during the subsequent 4 days they grow, migrate posteromedially, and become lymphoidal (Auerbach, 1960).

Microscopic anatomy

Each lobe is composed of a dense cortex surrounding a pale irregularly arranged medulla (Fig. 13-6). The connective tissue capsule is thin and fibrous, and the septa which extend into the cortex at irregular intervals are inconspicuous and usually fail to reach the medulla. The stroma is a network of reticular fibers closely associated with blood vessels, and the parenchyma consists of a mass of cells classified on the basis of their appearance in the electron microscope into "lymphoid" cells, macrophages, and epithelial cells (Clark, 1963). The latter, derived presumably from either endoderm or ectoderm of the pharynx, are distinguishable from the macrophages (mesenchymal reticular cells of Smith, 1964) by their ultrastructure. The "lymphoid" cells or thymocytes are indistinguishable from the small, medium, and large lymphocytes of nodes and spleen.

The epithelial cells and reticular fibers of the cortex are concealed by the densely packed masses of small lymphocytes. These cells, shown by tissue culture and transplantation techniques to be of epithelial origin, may be the precursors of all lymphocytes in nodes and spleen (Auerbach, 1964). As there are fewer lymphocytes in the medulla, the large pale-staining indistinctly outlined epithelial cells with their pale spherical nuclei are visible, arranged in cords and small groups. Small cysts lined with cuboidal or ciliated

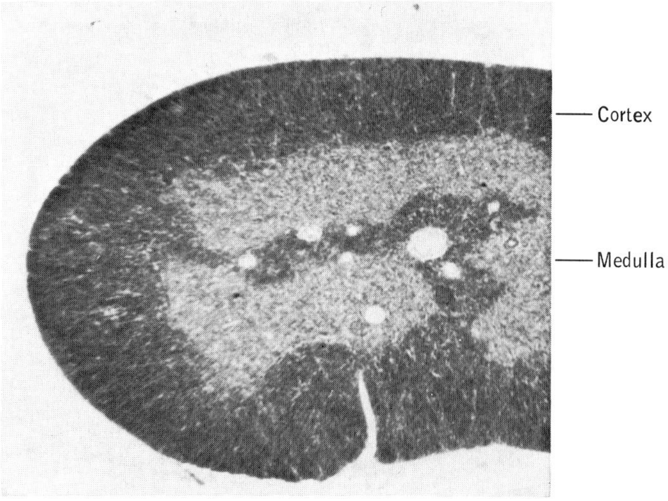

Fig. 13-6. Thymus, section (×30).

mucus-secreting cells are frequent components of the medulla; they have been interpreted as secretory units (Arnesen, 1958). The large macrophages with inclusions of phagocytosed debris are components of medulla, cortex, and surrounding connective tissue. Postcapillary veins in the cortex are lined, as in lymph nodes, by tall endothelial cells (Clark, 1963). The presence of sheathlike intrathymic lymph vessels accompanying large veins and arteries of the medulla (Smith, 1964) has been denied by Clark, who observed large spaces often packed with lymphocytes but saw no endothelial lining. Medullary structures composed of one or two hypertrophied epithelial cells resemble Hassall's corpuscles of the thymus of man but do not form keratin.

With normal aging and after irradiation, the cortex becomes narrow and less cellular and large foamy chromolipoid cells appear near blood vessels and beneath the capsule, often accompanied by accumulations of mast cells. With age the vascular pattern changes, collagenous and argyrophil fibers increase around blood vessels, thymic cysts increase in number, plasma cells accumulate, and in old age, the thymus becomes almost entirely replaced by lobules of fat (Smith, 1964).

MISCELLANEOUS ORGANS

These structures belong to no particular anatomical system, but each is sufficiently prominent to warrant description. Included are fat organs (aggregations of white and brown fat), the Harderian gland, and the lacrimal glands.

Fat organs

Both white and brown fat are found as discrete bodies with predilections for specific sites, and with the functional attributes of organs.

White fat bodies. Although fat cells may develop and accumulate anywhere within areolar connective tissue at any time during life, certain depots of fat are of constant appearance in specific sites. Among these are aggregations along blood vessels in the mesenteries, around kidneys and adrenals, attached to the gonads and excretory ducts, and in subcutaneous tissue of inguinal and axillary regions. There is compelling evidence that these aggregations of fat cells are specific embryonically determined structures and that genetic and sex influences may have important roles in their physiology (Liebelt, 1959).

White adipose tissue is modified connective tissue appearing in fixed sections as a lacework of spherical or polyhedral cells in which the cytoplasm has been almost entirely replaced by fat and the nuclei are pressed against the cell membranes. Interspersed among the fat-laden cells are others with light-staining cytoplasm and centrally located oval nuclei. A delicate reticular framework surrounds the cells and supports the capillary network.

Brown fat organs. This type of adipose tissue is morphologically, histogenetically, and physiologically distinct. The compact light brown bodies develop from specific embryonic anlagen, and no new areas appear postnatally (Fawcett, 1952). Among the most conspicuous are lobes between the scapulae, in the axillae, in the cervical region along jugular veins, adjacent to the thymus, along the thoracic aorta, at the kidney hilus, and alongside the urethra. The relatively large paired lobes in the interscapular depression have been called hibernating glands and have been classified as endocrine organs. The brown fat is a form of adipose tissue physiologically more active than white fat, but there is no evidence that it secretes hormones and it does not function as an organ of hibernation in the mouse (Fawcett, 1952).

Brown adipose tissue is composed of groups of polygonal cells containing lipid droplets in a coarsely granular cytoplasm. The fat droplets are not coalescent and the nuclei are central. The cells are surrounded by prominent, relatively coarse reticular fibers and numerous capillaries. The cells are unusually rich in phospholipids, and contain glycogen deposits under certain conditions (Sidman and Fawcett, 1954).

Harderian gland

Gross anatomy. The Harderian is a large horseshoe-shaped gland located deep within the orbit. A small superior arm is connected to a large inferior arm by a narrow band medial to the optic nerve. A single excretory duct opens at the base of the nictitating membrane. The color of the gland varies from pink to dark grey depending upon the abundance and characteristics of melanocytes in the capsule and interlobar septa (Markert and Silvers, 1956). In addition the gland is speckled with a pigment identified as a porphyrin that fluoresces under ultraviolet light. There is evidence that this pigment is produced as well as excreted by the gland (Cohn, 1955), and the amount is said to vary with strain, sex, and age (Figge and Davidheiser, 1957).

Microscopic anatomy. The tubuloalveolar gland is covered by a delicate connective tissue capsule bound loosely to the orbital fascia. Strands from the capsule divide the gland into lobes and lobules. The epithelial cells are pyramidal, their

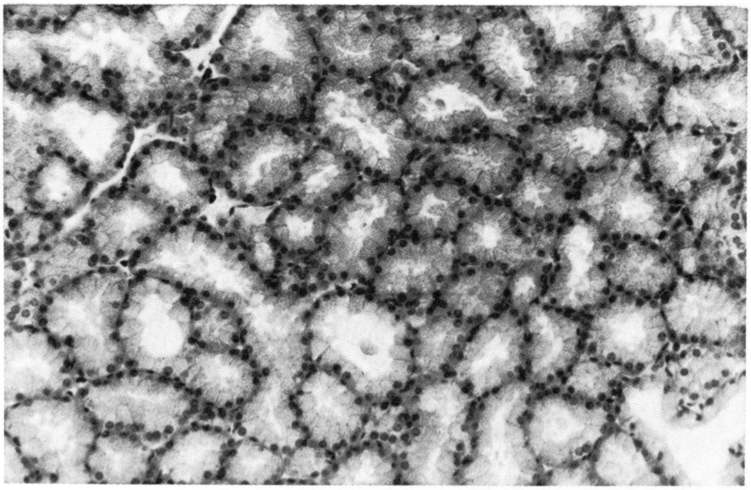

Fig. 13-7. Harderian gland, section (×150).

height depending upon the secretory phase. A round nucleus containing two or three nucleoli is located near the base, and the cytoplasm, packed with lipid droplets, appears vacuolated (Fig. 13-7). Myoepithelial cells are present between the epithelial cells and the prominent basement membrane (Chiquoine, 1958). The lamina propria is delicate fibrous connective tissue and among its cells are the melanocytes with nonfluorescent pigment. The secretion in the lumina of the tubules is oily and occasionally a yellow or reddish-brown color. Pigmented secretion may be present also in the lumina of the ducts which are lined with cuboidal epithelium. Alveolar, lobular, and lobar ducts join to form a single excretory duct.

Lacrimal glands

There are two pairs of lacrimal glands, the exorbital located subcutaneously ventral and anterior to the ear, and the intraorbital at the outer canthus where the joint excretory duct opens. The glands are tubuloalveolar and structurally identical. Each is enclosed in a connective tissue capsule and divided by connective tissue septa into lobes and lobules. The alveoli are larger and more loosely arranged than those of the parotid gland which they resemble (Fig. 13-8). The secretory

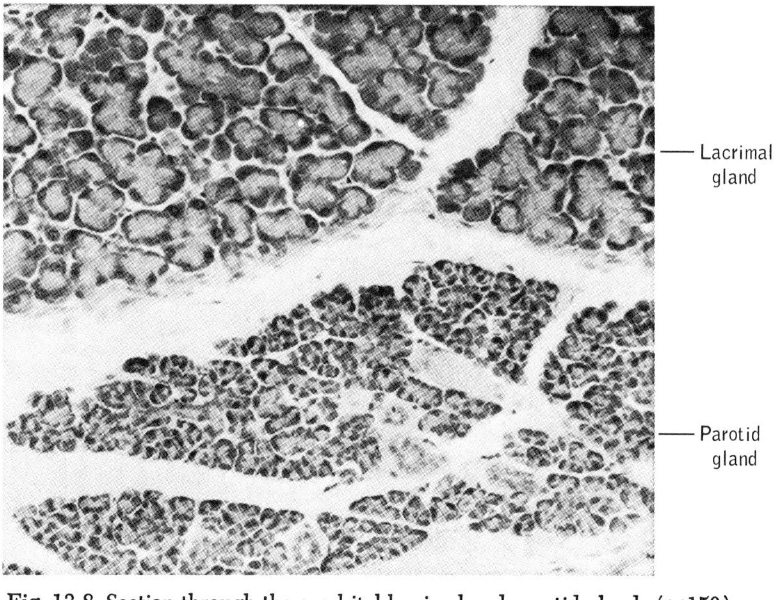

Fig. 13-8. Section through the exorbital lacrimal and parotid glands (×150).

cells are pyramidal with granular basophilic cytoplasm that stains intensely near the round basally located nuclei and much less intensely between the nuclei and the narrow lumen. Myoepithelial cells are found between the epithelium and the basement membrane. The intralobular ducts are lined by cuboidal cells without basal striations, and the excretory duct by stratified columnar epithelium.

ENDOCRINE GLANDS

The endocrine or ductless glands are but one part of a complex neurosecretory apparatus that regulates countless body functions. Only the glands or parts thereof that are discrete endocrine units are described in this chapter. These include the thyroid, parathyroids, adrenals, pituitary (hypophysis), pineal (epiphysis), and the islets of Langerhans. The testis and ovary, which have exocrine as well as endocrine functions, are described under male and female genital systems. For discussions of endocrine interactions and concomitant histological variations see Chap. 20.

Thyroid gland

Gross anatomy. The thyroid gland consists of two elongated oval lobes, one on either side of the trachea, joined near their posterior poles by a thin isthmus crossing the trachea ventrally. The lobes, buried under the muscles of the neck region, are richly vascularized and made up of groups of hollow spheres often visible macroscopically. The lobes extend anteriorly as far as the cricoid cartilage of the larynx and posteriorly over the first three or four tracheal rings. Variations in size, extent, and position are common.

Microscopic anatomy. The gland is made up of hollow spheres or follicles of varying size surrounded by a fibrous connective tissue capsule and supported by richly vascular interfollicular connective tissue (Fig. 13-9). The follicles are lined by simple cuboidal cells having distinct outlines, large spherical nuclei, and clear cytoplasm. The central cavities contain varying amounts of colloid. The height of the epithelial cells and the amount and staining quality of the colloid are indications of secretory activity.

In young mice the follicles are uniform in size, are lined by tall cuboidal cells, and contain homogeneous slightly acidophilic colloid. During the early postnatal months (1 to 3 months in mice of strain C3H and 1 to 5 months in those of strain C57), there is a rapid decrease in cell height and an increase in follicle diameter (Jacobs, 1958). With increasing age follicles become larger and more variable in size, the interfollicular tissue decreases, and the colloid becomes more eosinophilic. Senile changes, which occur as early as 12 months in mice of some strains and are more marked in females, include loss of stainable colloid, increase in fibrous interfollicular connective tissue, and great variation in follicle size with coalescence of contiguous large follicles to form bilocular and

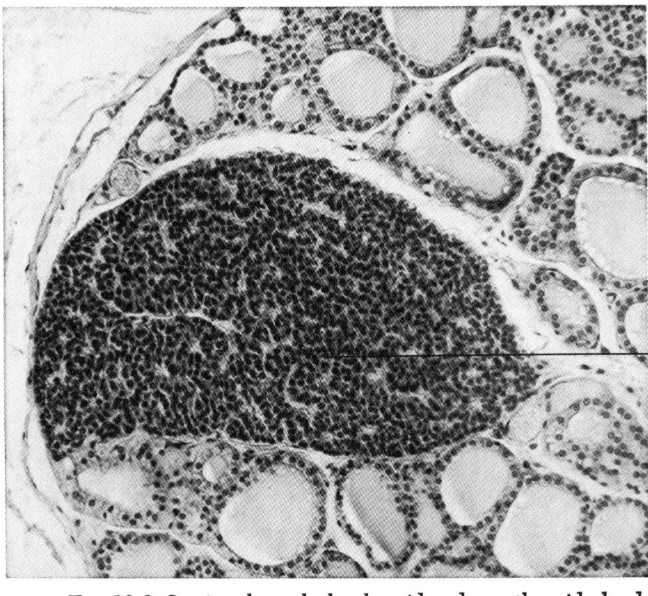

Fig. 13-9. Section through the thyroid and parathyroid glands (×150).

trilocular cysts with flattened epithelium (Andrew and Andrew, 1942; Blumenthal, 1955).

The thyroid gland develops from a medial epithelial mass growing ventrally at the level of first and second pharyngeal pouches. Thus, it is not surprising that aberrant thyroid follicles are occasionally found in regions both anterior and posterior to that described above (Hunt, 1963). Ultimobranchial tissue from pouches IV and V normally becomes closely integrated with the median mass and may form structures that persist in the adult thyroid, some of which are physiologically as well as morphologically distinguishable from the medially derived thyroid (Gorbman, 1947). Ultimobranchial follicles are recognized by the presence of ciliated epithelial cells. Cysts with ciliated epithelium are particularly conspicuous in strain C3H where they have been observed in newborn mice (Dunn, 1944).

Thyroid function is initiated in 15- to 17-day fetuses and observations indicate that colloid secretion precedes follicle formation (Van Heyningen, 1961).

Parathyroid glands

Gross anatomy. The position as well as the number of parathyroid lobes is variable, although usually a single lobe lies just under the capsule near the dorsolateral border of each lobe of the thyroid. Two members of a pair are seldom at the same anteroposterior level; sometimes one or both may be posterior to the thyroid (Dunn, 1949*b*); they may be deeply embedded in the thyroid tissue; and there may be more than two.

Microscopic anatomy. Each parathyroid gland is separated from the thyroid by a connective tissue capsule and consists of sheetlike masses and anastomosing cords of polygonal cells separated by a network of capillaries or sinusoids (Fig. 13-9). Two or three cell types can be recognized, their relative abundance varying with age (Foster, 1943; Blumenthal, 1955). The principal cells have large vesicular nuclei and scanty basophilic cytoplasm. Ovoid-to-fusiform–shaped cells with smaller more hyperchromatic nuclei and more abundant granular eosinophilic cytoplasm occur in small groups in the interstitial connective tissue. These increase in number with age. Very large cells with large vesicular nuclei and prominent nucleoli become conspicuous only in old age (Blumenthal, 1955). Pigmented dendritic cells may occur in the parathyroid stroma of pigmented mice and have been seen most frequently in mice of strain C58 (Dunn, 1949*b*).

The parathyroid glands develop from the third and fourth pharyngeal pouches in close proximity to the developing thymus, ultimobranchial bodies, and thyroid, with any or all of which they may remain in contact in the adult. Parathyroid rests, distinguishable histochemically, have been found consistently in the thymus septa or surface connective tissue (Smith and Clifford, 1962), and parathyroid, thyroid, and thymus are sometimes connected by a ciliated cyst (Gorbman and Bern, 1962).

Adrenal glands

Gross anatomy. The adrenal glands are a pair of small ovoid structures situated one on either side of the midline near the anterior pole of the kidney. The right and left adrenals differ in respect to closeness to kidney, renal vessels, and inferior vena cava; strain differences in these respects have been observed (Hummel, 1958). The glands of males and females differ in size and appearance, those of females being consistently larger and more opaque due to the presence of more lipid.

The adrenal gland consists of two parts, cortex and medulla, each with a separate origin, structure, and endocrine function. The cortical anlagen develop in the mesodermal coelomic epithelium near the genital ridges at about the 12th day, the medullary anlagen from sympathetic nervous system ganglia at the 13th day, and the first signs of union of the two occur at about the 14th day (McPhail and Read, 1942). Accessory units of cortical and medullary (chromaffin) tissues may be present in the adult, the chromaffin tissue in scattered small groups near the left renal vein (Coupland, 1960) and cortical nodules near the kidneys and adrenals on both sides (Hummel, 1958).

Microscopic anatomy. The cortex is surrounded by a fibrous connective tissue capsule often laden with adipose cells. In man and most mammals three zones are visible in the adrenal cortex, but in the mouse only two zones are clearly defined (Fig. 13-10). The outer glomerulosa is a narrow zone consisting of small cells arranged in arches. The cells have relatively large nuclei, basophilic cytoplasm, and a rich capillary blood supply. Internal to the zona glomerulosa is the wide zona fasciculata, composed of radial columns of cells separated by fine connective tissue septa bearing capillaries. The nuclei are vesicular and the cytoplasm acidophilic and foamy due to the presence of finely distributed lipid droplets. Although some investigators describe an inconspicuous third zone, the reticularis, others ques-

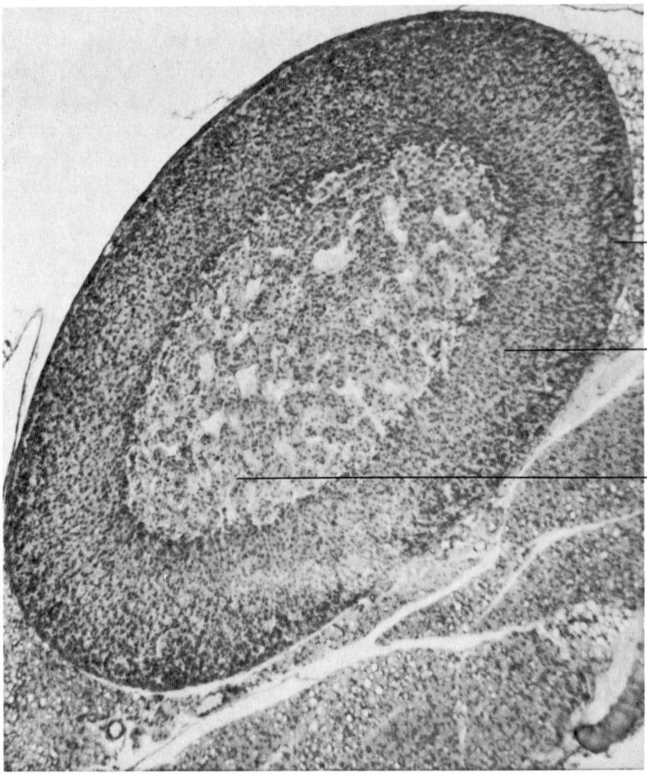

Fig. 13-10. Adrenal gland section showing cortex and medulla (×75).

tion its existence (Jones, 1950; Miller, 1950). In young nulliparous females and in males prior to sexual maturity, there is a zone of variable width between cortex and medulla. This juxtamedullary zone, the X-zone, disappears with sexual maturity in the male and with first pregnancy in the female. It persists in castrated males and in virgin females for periods varying with strain (Delost and Chirvan-Nia, 1958).

Certain conspicuous changes occur with age. A proliferation of small spindle-shaped cells starts between capsule and zona glomerulosa and spreads laterally and centrally to the juxtamedullary zone. Large multinucleate foamy cells bearing brown pigment appear in the juxtamedullary zone and central portions of the fasciculata zone. This "brown degeneration" is believed identical to that in aged ovaries; the pigment resembles ceroid (Jones, 1950; Deane and Fawcett, 1952).

The medulla consists of homogeneous polyhedral cells arranged in irregular groups separated by sinusoids (Fig. 13-10). Nuclei are large and centrally located, the cytoplasm lightly basophilic and finely granular. The medullary cells have an affinity for chromium and are thought to be a special type of neurosecretory cell. Erythropoietic foci are common in adrenals in newborn mice as well as in adult mice under certain pathological conditions (Borghese, 1952).

Pituitary gland

Gross anatomy. The pituitary gland or hypophysis rests on the dorsal surface of the basisphenoid bone of the skull and is attached to the floor of the brain by a fragile stalk (Fig. 13-11). The gland is slightly flattened dorsoventrally and has an elongated oval shape with its long axis perpendicular to that of the head. The ventral surface is well vascularized and homogeneous and the dorsal surface is demarcated into three distinct regions (Fig. 13-12). The central and most opaque is the pars nervosa (neural lobe), which is derived embryologically from the floor of the third ventricle of the brain. Bordering the pars nervosa is the slightly less opaque narrow pars intermedia (intermediate or proximal lobe) and, outside of this, extending far laterally and covering the other lobes ventrally, is the well-vascularized pars distalis (anterior or distal lobe). The latter two lobes are derived from Rathke's pouch,

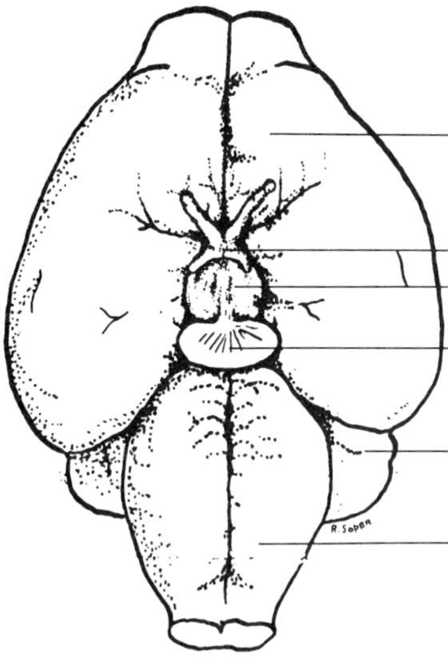

Fig. 13-11. Drawing of the ventral aspect of the brain to show location of the pituitary gland.

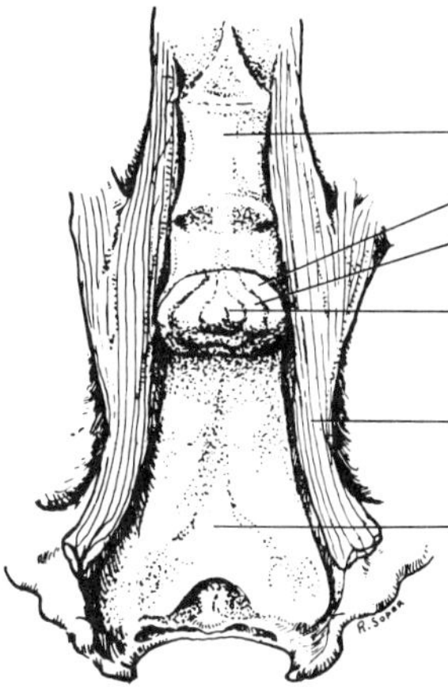

Fig. 13-12. Drawing of the dorsal aspect of the pituitary gland, brain removed.

a thickening and evagination of the ectoderm of the roof of the mouth cavity where this is in close proximity to the developing brain. The pars nervosa loses its central cavity but remains attached to the hypothalamus of the brain by a thin stalk. The other lobes lose their connection with the oral cavity, the only remains of Rathke's pouch being a residual cleft separating pars intermedia and distalis (Fig. 13-13, 13-14). The inconspicuous tuberalis is the small part of the anterior lobe partially wrapped around the stalk. The development of the pituitary has been studied and described in detail (Kerr, 1946).

The pituitary gland is consistently heavier in females than in males and there are also strain differences in size. The average weight of the pituitary of the adult female exceeds that of the male of the same age and strain by 0.5 to 1.5 mg (Chap. 20).

Microscopic anatomy. The pars nervosa is made up of endings of neurons, glial cells, ependymal cells, and connective tissue supporting capillaries and sinusoids. Cell bodies of neurons are seldom seen. The lining cells of the sinusoids are phagocytic.

The border between neural and intermediate lobes is uneven and indistinct (Fig. 13-14). The cells of the pars intermedia are arranged in groups separated by fibrous connective tissue strands with scarce vascular elements. Most of the cells are polygonal with pale-staining nongranular baso-

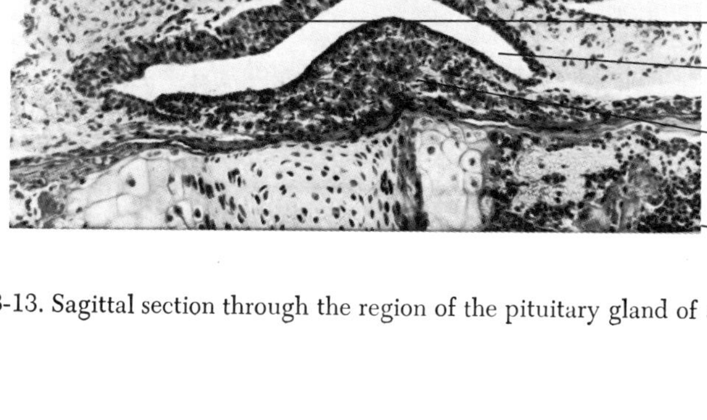

Fig. 13-13. Sagittal section through the region of the pituitary gland of a newborn mouse (×100).

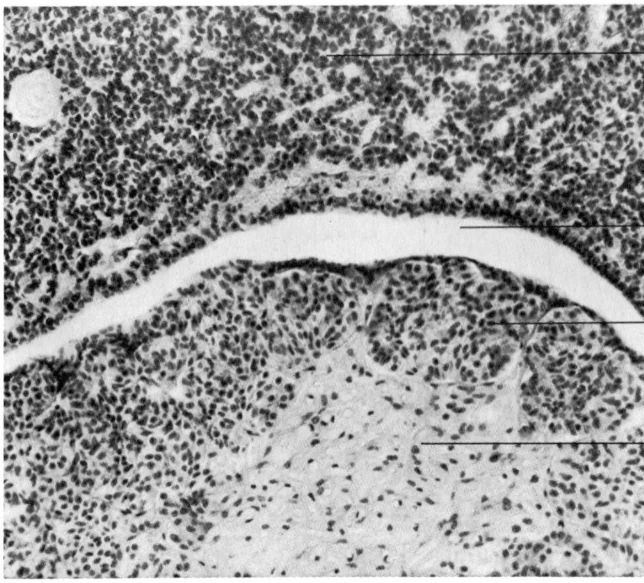

Fig. 13-14. Pituitary gland, section through pars distalis, residual cleft, pars intermedia, and pars nervosa ($\times$150).

philic cytoplasm and densely staining oval nuclei. A few stellate cells with long processes are interspersed. The intermediate lobe of the mouse is wide compared to that of some mammals; it is especially wide and conspicuous in mice of strain 129.

The pars intermedia and pars distalis are separated by the residual cleft, lined throughout by low cuboidal epithelium (Fig. 13-14). In the pars distalis, the largest and most vascular of the lobes, the cells are arranged in branching cords separated by sinusoids and are of several cytological types. The cells do not stain differentially as readily as in some other mammals, special techniques being required to bring out cytological details. Chromophobe cells with large light-staining nuclei surrounded by small amounts of nongranular cytoplasm make up about 50 per cent of the population, the rest being chromophils of three types (Halmi and Gude, 1954). The most common chromophil is the acidophil, a small round or oval cell with a centrally located nucleus and eosinophilic cytoplasm, which makes up about 40 per cent of the whole. The smallest group, making up less than 11 per cent, is composed of beta and delta cells, formerly classed together as basophils. The beta cells are angular or crescentic, have coarse aldehyde-fuchsin–positive granules, are most numerous in the central portions of the lobe, and are more numerous in males than in females. The delta cells are small with aldehyde-fuchsin–negative granules and are more numerous in females. Others have classified the cells of the pars distalis into acidophils and amphophils (cells that show affinity for aniline blue) and described a transformation from one to the other (Van Ebbenhorst Tengbergen, 1955). Cysts of varying size, often lined with ciliated epithelium and containing eosinophilic secretion, occur frequently.

Pineal gland

Gross anatomy. The pineal gland or epiphysis is a very small cone-shaped body situated on the

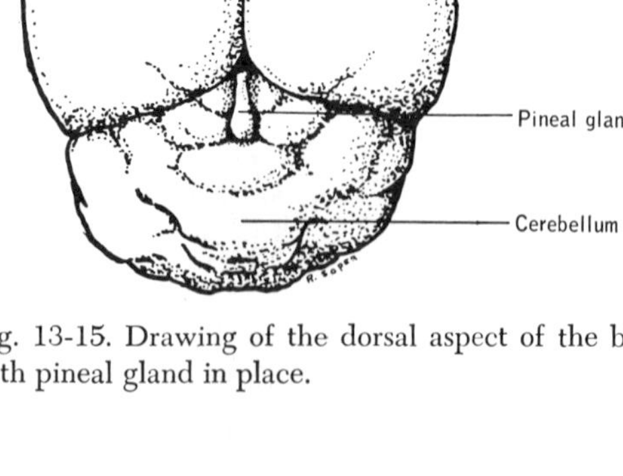

Fig. 13-15. Drawing of the dorsal aspect of the brain with pineal gland in place.

dorsal surface of the brain between cerebrum and cerebellum, with the peak of the cone directed anteriorly (Fig. 13-15). The attachment to the brain stem is long, extremely delicate, and filmy; if the roofing bones of the skull are removed, the epiphysis remains attached to them at the suture between parietal and interparietal bones. The pineal develops from a medial evagination of the roof of the third ventricle of the brain into a well-vascularized area and in the adult retains an association with the richly vascular cells of the choroid plexus. The blood supply, however, is independent of that of choroid plexus and brain (Von Bartheld and Moll, 1954).

Microscopic anatomy. The thin fibrous capsule surrounding the gland merges with cells of the choroid plexus. The gland is complex histologically and contains ependymal, glial, neuronal, and other elements in a richly vascular framework (Fig. 13-16). Two types of cells can be identified: large indistinctly outlined spherical cells with pale, slightly granular, basophilic cytoplasm and large vesicular nuclei with prominent nucleoli; and elongated stellate cells with deeply staining basophilic cytoplasm and oval nuclei with fine chromatin granules.

Islets of Langerhans

Gross anatomy. The islets, distributed among the exocrine secretory alveoli of the pancreas, are not visible macroscopically under normal conditions. The solid spherical groups of cells are always closely associated with pancreatic ducts and blood vessels of the connective tissue septa (Fig. 13-40).

Microscopic anatomy. The islets vary considerably in size and are distinguishable from pancreatic alveoli by their pale-staining quality. The cells, grouped in irregular cords about sinusoids, are round, polyhedral, and cuboidal and have round, faintly staining nuclei. Special staining techniques allow identification of as many as four types of cells based on differences in cytoplasmic granules. Cellular differentiation from 13 days of gestation to the adult has been described (Munger, 1958).

SKIN AND DERIVATIVES

Derivatives of the skin are pelage hairs, sensory or tactile hairs, sebaceous glands, and mammary glands. Another skin derivative, the sudoriferous gland, is reportedly absent in the mouse (Hardy, 1949) although rudimentary and transitory tubules resembling these glands have been observed (Gibbs, 1941).

Skin

The skin covers the entire outer surface of the body and, except in certain regions and in mice of certain genotypes, bears hair over the greater part. Hairless skin surrounds and extends a variable distance into all external openings (nipples, nostrils, mouth, urethra, vagina, and anus). The

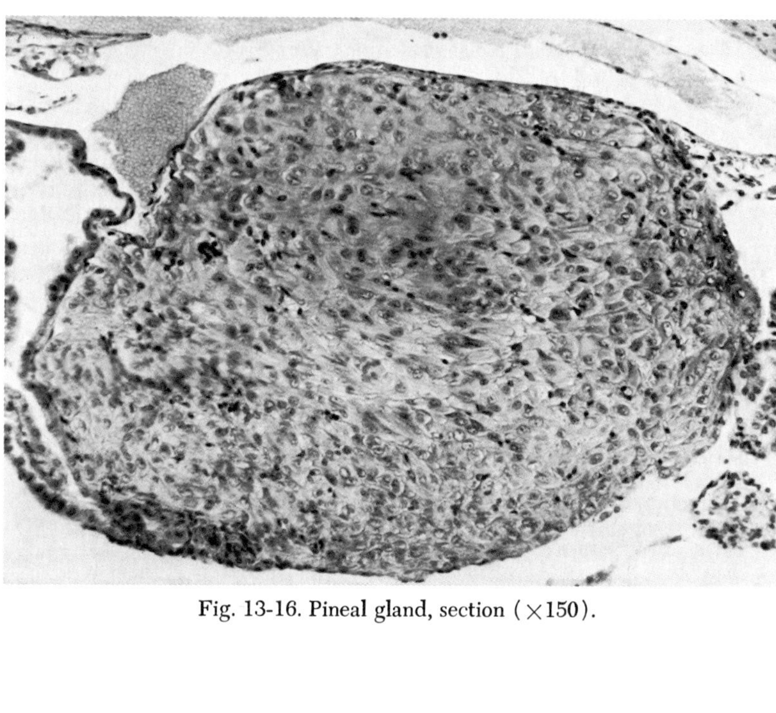

Fig. 13-16. Pineal gland, section (×150).

skin consists of two parts, an outer epidermis of stratified squamous epithelium and an inner dermis or corium of dense connective tissue, continuous with adipose and loose connective tissue of the subcutaneous areas.

Epidermis. The epidermis is thin in haired areas and considerably thicker on the hairless or relatively hairless portions such as feet, tail, snout, nipples, and genital and anal areas. In these thick areas there are three or more strata, each of several cell layers. The basal layer, stratum germinativum, rests on a basement membrane and consists of vertically compressed cells with indistinct cell outlines and clear oval nuclei, plus several layers of polyhedral cells connected across intercellular spaces by tonofibrils. The next four or five layers of cells, which are compressed horizontally and may contain coarse keratohyalin granules, make up the stratum granulosum. The outermost stratum corneum is composed of several layers of dead cornified cells that are shed at the surface and replaced from deeper layers.

In haired areas the epidermis rarely exceeds six cell layers and the strata are hard to define. The basal cells of the germinativum are cuboidal, the stratum granulosum is represented by a few scattered cells, and there are only one or two layers of cells in the stratum corneum. The epidermis is well developed at birth, becomes thicker during the first 4 or 5 days after birth, and then decreases in conjunction with hair-follicle development (Gibbs, 1941). There are no blood vessels or nerves in the epidermis, and, although melanocytes potentially capable of producing pigment are scattered among the basal cells, pigment is usually not detectable in epidermal cells (Billingham and Silvers, 1960).

Dermis. The connective tissue of the dermis contains collagenous and elastic fibers, blood vessels, nerves, fat cells, and strands of smooth muscle (arrector pili). In head, neck, and trunk regions thin sheets of striated muscle (the panniculus carnosus) insert on the fibers of the dermis at its boundary with the subcutis.

Branched melanocytes containing pigment are cellular components of the dermis in pigmented areas such as muzzle, ears, soles of feet, tail, genital papilla, and scrotum; but are not demonstrable in the dermis of haired areas (Billingham and Silvers, 1960). Where the epidermis is thick, the epidermodermal boundary is very uneven, the dermis with its blood vessels and nerves being extended into the epidermis in tall elevations or papillae. In haired areas dermal papillae are inconspicuous and the boundary between dermis and epidermis is only slightly uneven. The loose connective tissue on which the dermis rests becomes transformed soon after birth into an adipose layer of packed fat cells.

Hair and sebaceous glands

Hair follicles are invaginations of epidermis into the dermis giving rise to hair, both pelage and tactile, and to sebaceous glands.

Pelage hair. Mouse hairs and hair follicles are similar structurally to those of other mammals. The hair projecting above the skin surface is a three-layered column or shaft of cornified cells consisting of an outer cuticle of translucent scales, a cortex of elongated cells or septa containing small amounts of pigment, and an inner core or medulla of irregularly shaped septa containing pigment granules. Below the skin surface the hair shaft is enclosed in a double sheath or follicle at the base of which the hair matrix is formed by proliferation of epidermal cells. In development of the hair follicle, the basal layer of the epidermis thickens, grows into the dermis, and forms a bulb over a core or papilla of dermal cells. The external root sheath of the follicle is derived from the downgrowing epidermal cells; the internal root sheath, hair matrix, and hair shaft arise from actively dividing cells of the bulb area. The follicles are not perpendicular to the skin surface but slant obliquely toward the posterior. A single sebaceous gland develops in the obtuse angle between follicle and epidermis by proliferation and evagination of epidermal cells of the external sheath. A strand of smooth muscle fibers (arrector pili muscle) forms in the dermis, extending from the follicle just below the sebaceous gland to the epidermodermal junction. As the basal epidermal cells push into the dermis they carry melanocytes which produce pigment in this environment. The melanocytes become concentrated in the hair bulb where pigment granules are transferred to the growing hair shaft by some undetermined process. The number, size, shape, clumping pattern, color, and color intensity of the pigment granules in the medullary septa all contribute to the definitive coat color (Russell, 1946; Chap. 21).

There are several types of hair in the coat and for each there is a hair follicle that produces the one type only. Hairs were first classified by Dry (1926) into overhairs (monotrichs, awls, and auchenes) making up about 16 per cent and underfur (zigzags) about 84 per cent of the coat, on the basis of such characteristics as length, number of bends or constrictions in the shaft, and number of rows of cells or septa in the medulla.

The monotrichs, also called guard hairs, are the long straight tylotrichs, classified and described below as tactile hairs. The awls are straight hairs, auchenes have a single bend, and zigzags have several bends. Awls may have as many as four, usually have three, and always have at least two rows of medullary septa. Auchenes have two septal rows, seldom more, and zigzags only one row.

Initiation of follicle formation has been reported to occur in the 14-day embryo and to continue until 9 days after birth (see Chase, 1954). However, Mann (1962) observed first pelage follicles at 16¼ days and no new follicle initiation after birth. Initiation of all follicle types occurs first in the shoulder region; spreads in a wave anteriorly, caudally, ventrally, and dorsally; and is completed everywhere before birth. First initiation of awl follicles occurs in 16¼-day embryos, and of auchene and zigzag follicles in 18¼-day embryos. The duration of development is approximately the same for all follicle types; hairs emerge through the skin surface 8 to 9 days after initiation of the follicle (Mann, 1962). Cycles of hair growth subsequent to first coat emergence occur throughout life with periods of active growth of 17 to 19 days alternating with resting phases of variable duration, during which old hairs are retained in the follicles as dead clubs (see Chase et al., 1951, for growth stages). Individual follicles do not enter the growth phase independently of neighboring follicles. Instead, all follicles within a particular area enter the growth phase of the cycle synchronously resulting in orderly waves or spreads of hair growth and hair follicle activity. Chase and Eaton (1959) studied the wave patterns over five generations of hair in male and female mice of four inbred strains and found that patterns varied with strain, sex, and age, as well as in individual mice.

During cycles of hair growth, changes take place in the skin and sebaceous glands (Chase, 1954; Borodach and Montagna, 1956). During active hair proliferation the capillaries around the follicle become enlarged, the dermis and adipose layers thicken, the epidermis becomes thin, and the sebaceous glands small. Although dermal papillae are necessary in the production of hair, they contain few if any capillaries at any stage.

Tactile hair. Sensory or sinus hairs differ from pelage hairs in several respects: they are larger and longer; their tips contain little pigment, and in agouti mice they lack the yellow band. The follicles are large and contain blood sinuses and abundant nerve endings (Fig. 13-17); and dead club hairs are not retained in the follicles. There are two types of tactile hairs: the vibrissae located almost exclusively on the face, and the tylotrichs (monotrichs or guard hairs) scattered among the pelage hairs.

Patterns of major and minor groups of vibrissae have been plotted for mice of different genotypes (Davidson and Hardy, 1952; Dun, 1958). Those in the major group are in five horizontal rows and one vertical row on the snout and three rows on the lower lip, but they vary in number and are difficult to count. There are usually 19 in the minor group which includes three interramals, paired supraorbitals (two in one tubercle), post-

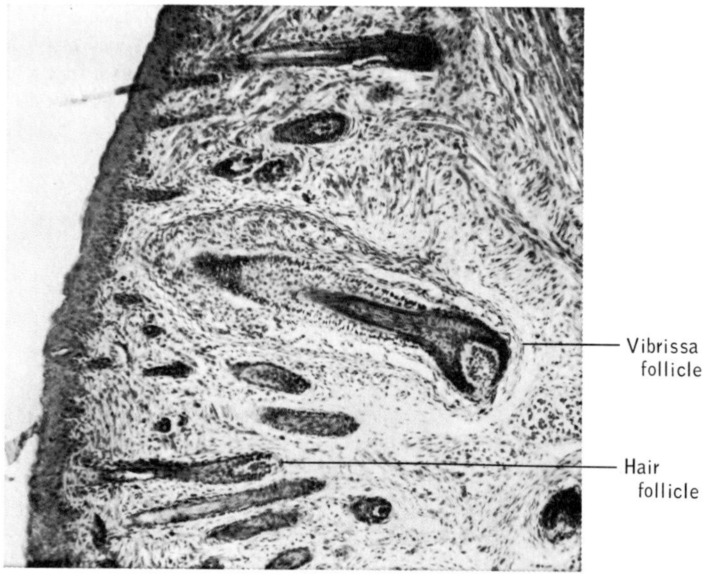

Fig. 13-17. Section of skin of the head if a newborn mouse to show differences between vibrissa and pelage follicles (×100).

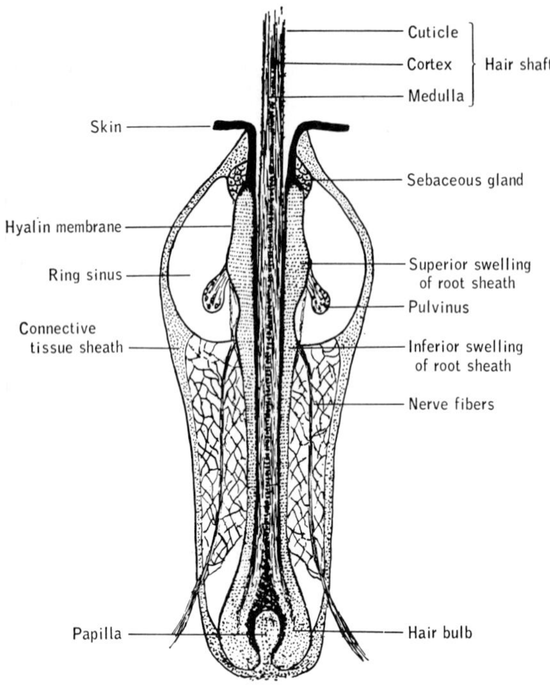

Fig. 13-18. Diagram of a vibrissa follicle (×112.5).

orbitals (one), postorals (two), and ulnar carpals (three in a large tubercle). This pattern is fairly constant, postorbitals being invariably present and others rarely absent (Dun, 1958).

The structure of the vibrissa hair is similar to that of the pelage hair except in size, length, and distribution of pigment. The vibrissa follicle is large and contains venous blood sinuses and nerve fibers within a heavy connective tissue (dermal) sheath (Fig. 13-18). Large capillaries are present in the papilla, and striated muscle fibers replace smooth muscle bundles. There is a single sebaceous gland for each as in pelage follicles. Vibrissa follicles are the first to develop in the embryo and all vibrissae have emerged by birth. Initiation of vibrissa follicles is first observable during the 13th embryonic day and the hairs emerge 5 to 6 days later. For detailed descriptions of the structure and development of vibrissa follicles and vibrissae see Melaragno and Montagna (1953) and Davidson and Hardy (1952).

Tylotrichs are the monotrichs of Dry (1926) and the guard hairs of others. They are scattered singly among the shorter awls, auchenes, and zigzags of the trunk, and small hairlets are associated with some of them. Although recognized as tactile hairs, they were classified as pelage hair until their similarity to vibrissae and to tylotrichs of other mammals made reclassification desirable (Straile, 1960). Tylotrich hairs are long and straight with a graduated tip that projects above the other coat hairs. There is but one hair in each follicle, dead club hairs not being retained, and there are two sebaceous glands for each follicle. Embryonic development of a tylotrich follicle starts as a raised area on the epidermal surface, in contrast to the basal-cell downgrowth of pelage follicles (Mann, 1962). Part of this epidermal thickening becomes the *Haarschiebe,* an acentric thickening of the epidermis of the follicle orifice and a distinguishing characteristic of the tylotrich.

Initiation of tylotrich follicles starts as does that of pelage follicles in the shoulder region and spreads similarly in a wave. Initiation starts in 13¼-day embryos and is completed 2 to 3 days later at about the time initiation of awl follicles starts (Mann, 1962). The hairs erupt 8 to 9 days after follicle initiation (about 3 days after birth).

Sebaceous glands. These glands, one to each pelage and vibrissa follicle and two to each tylotrich, develop as focal thickenings, cellular hypertrophies, and evaginations of the external root sheath into the surrounding dermis. Excretory ducts formed by canalization of the thickened area of the root sheath open into the space between hair shaft and follicle just below the orifice. The gland is a pear-shaped structure surrounded

Fig. 13-19. Diagram of the maximum extent of the mammary system, ventral and lateral aspects. The large black dots represent the nipples and the stippled areas the mammary glands.

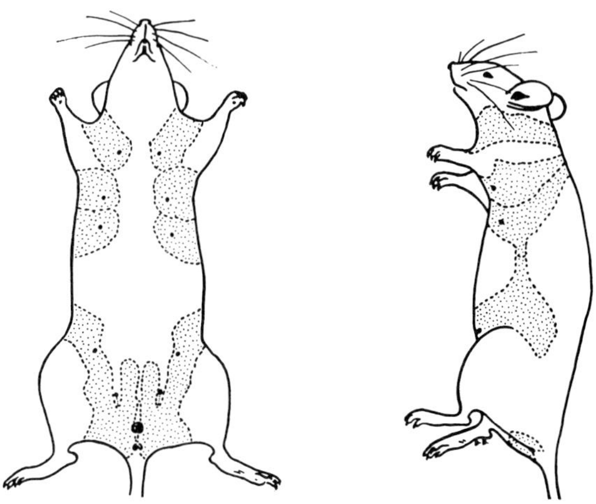

by a basement membrane and dermal connective tissue. There are two layers of cells, an outer basal layer of thin flat cells, and an inner layer of large rounded secretory cells. These latter accumulate secretion, die, disintegrate, and are replaced from the basal layer. If destroyed, sebaceous glands will redifferentiate from cells of the external root sheath but only while the hair is in the active growing phase (Montagna and Chase, 1950).

Mammary glands

For extensive reviews of the literature and of studies on the prenatal, prepuberal and postpuberal development and morphology of the mammary glands of mice, male and female, virgin and parous, see Raynaud (1961) and Cowie and Folley (1961).

Gross anatomy. Female mice normally have five pairs of nipples and mammary glands, three in the thoracic and two in the abdominal region (Fig. 13-19). (There is an additional abdominal pair in female rats.) When fully developed, the glands consist of extensive duct systems and lobules of secretory alveoli embedded in subcutaneous fat pads. Each gland is a separate unit branching from a primary duct that opens to the exterior at the tip of the nipple. The nipples are slightly depressed and are surrounded by circular folds of thickened hairless and sometimes pigmented skin; they are inconspicuous and obscured by hair except in infant mice and during late pregnancy and lactation. Variation in number and arrangement of nipples is frequent in some strains. For example, fewer than five pairs are common in strain A females; more than five pairs in BALB/c females (Little and MacDonald, 1945). Male mice have no nipples and usually only four pairs of rudimentary glands consisting of branching ducts with no alveoli, primary ducts, or openings to the exterior.

Mammary glands develop from lines of thick-

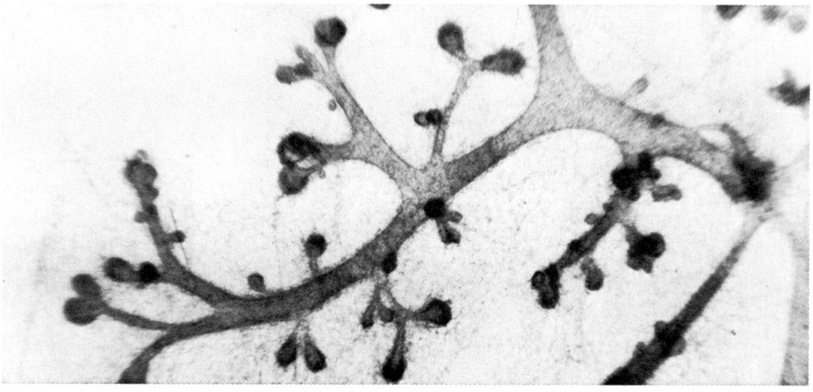

Fig. 13-20. Mammary gland showing rapidly growing end buds, whole mount (×25).

ened epithelium overlying mesenchymal condensations on either side of the ventral midline. At about the 13th embryonic day, five pairs of anlagen or buds are formed on this line by proliferation of basal epidermal cells. During the 15th day, sex differences begin to appear coincident with differentiation of the fetal testis. The mammary bud lengthens in the female and grows into the mesenchyme as an epithelial cord that remains connected to the epidermis. Canalization of the cord takes place in the 16th and 17th days, resulting in a primary duct opening to the exterior. In the meantime nipples have become demarcated by invagination of epidermal cells encircling the mammary buds. At birth, the female gland consists of a primary duct opening through a nipple and branching distally into several secondary ducts.

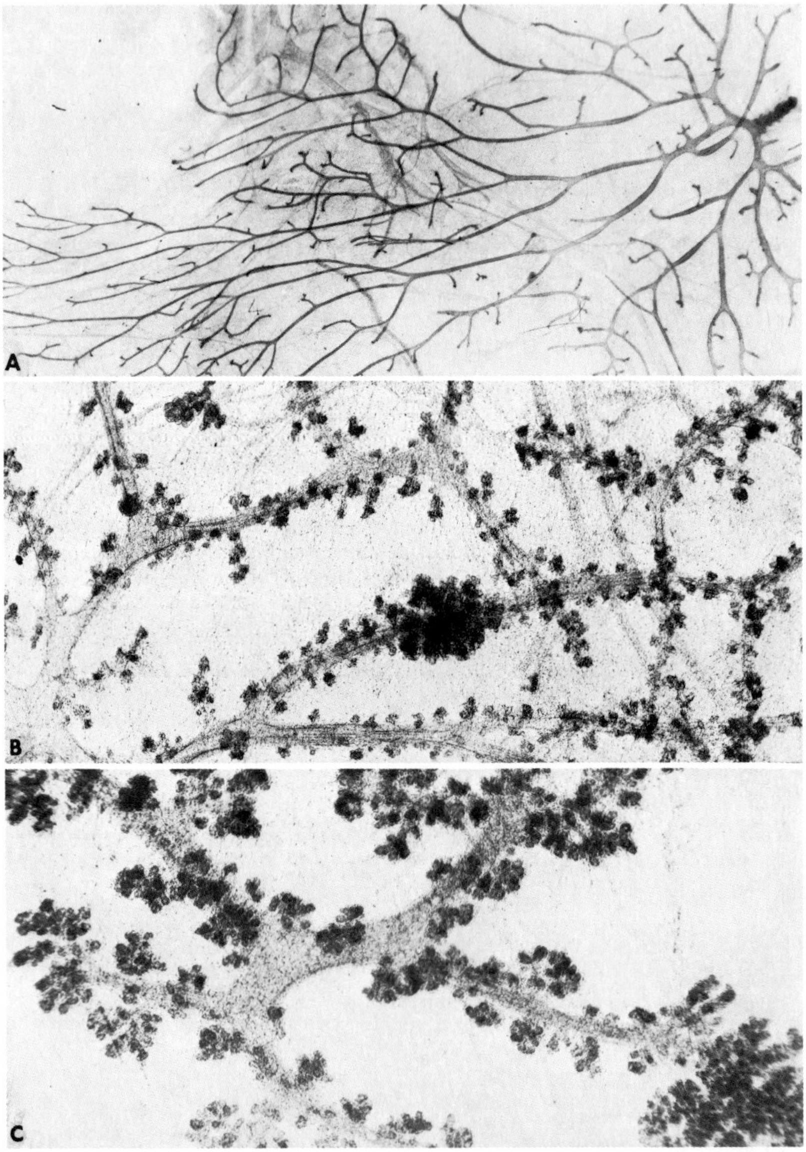

Fig. 13-21. Third mammary glands of virgin female mice. *A*. Gland of RIII/J female 6 months old showing duct system with no alveoli, whole mount ($\times 6$). *B*. Gland of (C3H $\times$ RIII)F_1 female 7 months old showing small clusters of alveoli, whole mount ($\times 18$). *C*. Gland of (C3H $\times$ RIII)F_1 female 7 months old showing many alveoli, whole mount ($\times 18$).

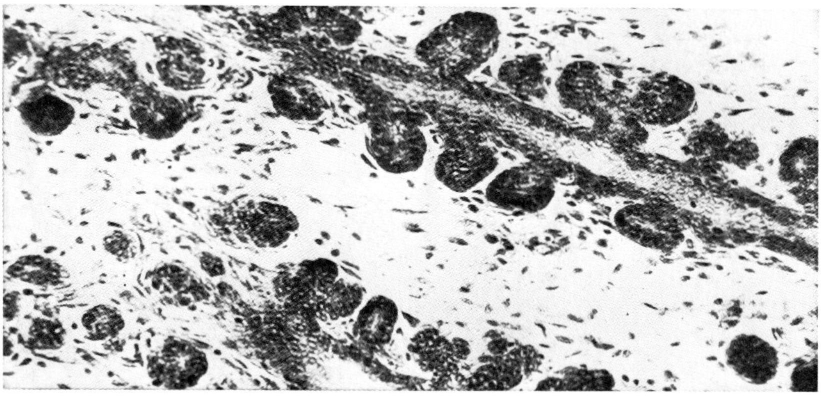

Fig. 13-22. Developing mammary gland on the 11th day of pregnancy, section (×200).

In males, the mammary buds become separated from the epidermis by condensation of mesenchyme around the neck constricting it and are completely isolated from the epidermis by the 16th day. Further development is usually limited to formation of small epithelial cords and simple branching. Some buds such as the inguinal pair may degenerate completely. No invagination of epidermal cells occurs around the buds and thus no nipples are formed. The fetal testis is responsible for the separation of mammary buds from the epidermis and for the inhibition of development of the nipple (Raynaud, 1961).

Microscopic anatomy. Mammary glands of prepuberal mice consist of branching ducts embedded in adipose tissue. The ducts are lined by low columnar or cuboidal cells with dark staining oval nuclei and small amounts of cytoplasm. There is often a layer or scattering of myoepithelial cells between the epithelium and basement membrane which show an intense alkaline phosphatase reaction (Richardson and Pearson, 1954). Circularly arranged connective tissue fibers, thicker in the large main ducts than in the smaller terminal ones, complete the duct walls. Just before puberty, at 3 to 4 weeks of age, there is a period of rapid growth with elongation of ducts, development of side branches, and formation of end buds on the branches (Fig. 13-20). Postpuberal development reaches the maximum in virgin females of 4 to 7 months, and consists of further duct proliferation and the formation of a few isolated alveoli (Nandi, 1958). There are strain differences in the extent of alveolar development in the glands of virgin females (Richardson and Hummel, 1959). The glands of strain RIII virgins have few alveoli (Fig. 13-21*A*), whereas many glands of strain C3H and of hybrids between C3H and RIII contain lateral buds and clusters of alveoli (Fig. 13-21*B*, *C*) (Richardson and Hall, 1960). Although hypertrophy and ductal dilatation take place during first estrous cycles, biopsies of mammary glands of 3- to 7-month-old virgin females showed no cyclic changes (Ferguson, 1956).

Complete lobuloalveolar development occurs

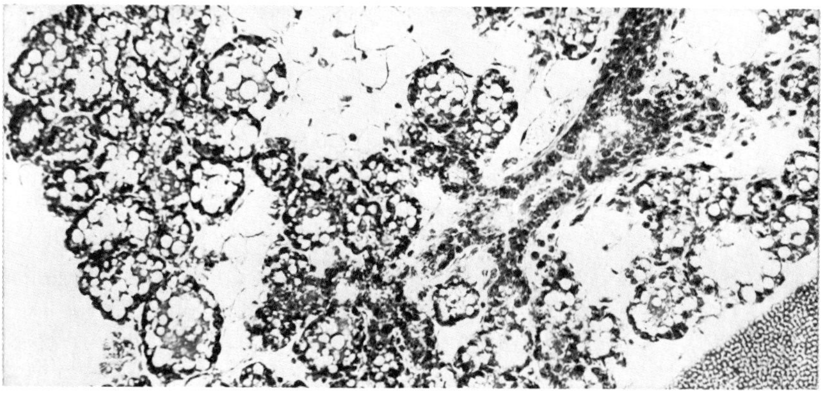

Fig. 13-23. Mammary gland near term showing secretory activity, section (×200).

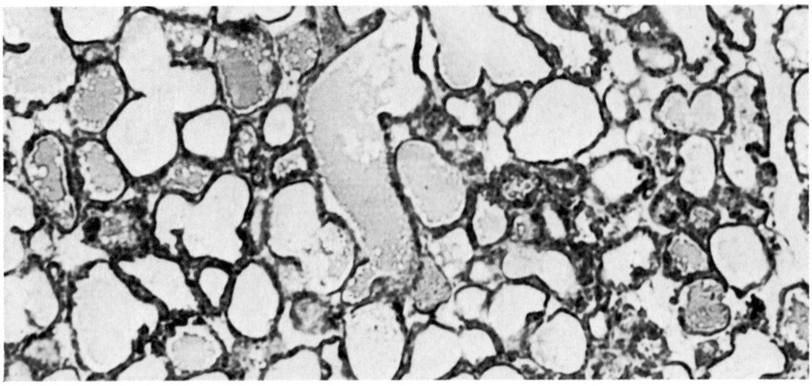

Fig. 13-24. Mammary gland on the seventh day of lactation (×100).

only during pregnancy and lactation. During the first week of pregnancy many new end buds and alveoli appear, connective tissue increases around the ducts, and there are numerous mitoses in the epithelial cells of the alveoli (Fig. 13-22). Growth reaches a peak by the end of the second week when there are numerous lobules made up of alveoli lined with single layers of low columnar and cuboidal cells (Wellings et al., 1960). Further development consists of enlargement of alveoli with increase in lumen size, hypertrophy of epithelial cells, and beginning secretory activity. Proliferation and enlargement of alveoli is accompanied by decrease in adipose tissue of the fat pad and increase in vessels of the capillary network. The secretion of milk starts in the alveoli near the nipple and is usually well established distally at term (Fig. 13-23).

By the fourth day of lactation, lobules of closely packed alveoli have replaced most of the adipose tissue, and alveoli and ducts are distended with milk (Fig. 13-24). The epithelial cells are low cuboidal and squamous with flattened nuclei. Although secretory activity with increased distension of alveoli and ducts is apparent throughout lactation, growth appears to have ceased. However, studies using DNA as an index of growth have shown that proliferation continues throughout pregnancy and at least to day 14 of lactation (Brookreson and Turner, 1959).

Following the cessation of suckling, the glands remain congested with milk for 24 to 48 hours, and during this time alveolar epithelial cells have started to degenerate and are found detached in the lumina (Fig. 13-25). By the fifth postweaning day the glands are devoid of milk and the alveoli reduced to masses of cells without lumina (Nandi, 1958). In the completely regressed and resting gland, capillaries are inconspicuous, adipose tissue fills the spaces between the irregular clumps of cells from collapsed alveoli, and the duct lumina are narrow. In old multiparous females, glands undergo gradual involution, distal duct branches becoming atrophic, leaving only the main ducts and a few secondary branches.

No lateral buds or alveoli ever develop on the

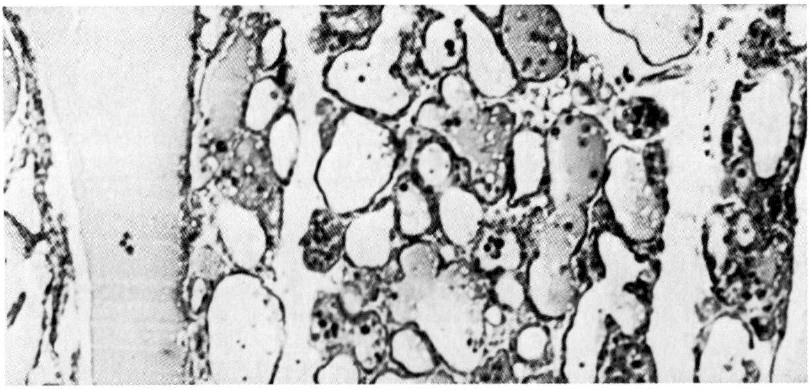

Fig. 13-25. Mammary gland after cessation of lactation that had continued for 22 days (×100).

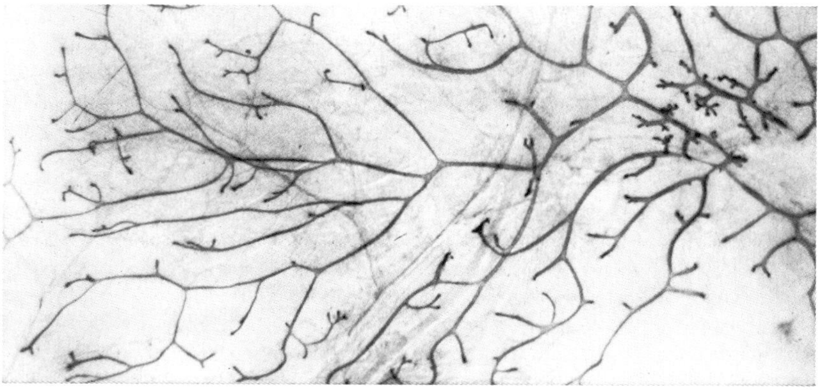

Fig. 13-26. Third mammary gland of (MA/My $\times$C57BL)F_1 male mouse 9 weeks old, whole mount ($\times$7). (*From Richardson,* 1953, *with permission of the Wistar Institute of Anatomy and Biology, publisher of the Anatomical Record.*)

ducts of normal male mice although the duct system may be extensive (Fig. 13-26). Glands of an individual seldom develop uniformly and strain differences in the extent and architecture of the duct systems have been noted (Richardson, 1951, 1953). Although the glands of most males are rudimentary or absent, some glands in males of a few strains have well-developed duct systems resembling those of virgin females except for absence of primary ducts.

DIGESTIVE SYSTEM

Oral cavity and pharynx

The oral cavity is bounded anteriorly by the lips, laterally by the cheeks, and posteriorly by the epiglottis and rim of the soft palate. The roof is formed by the hard and soft palates, and the floor by the tongue. The cavity is widest in the region of the molar teeth, narrowing anteriorly and posteriorly. The pharynx, a small chamber posterior to the oral cavity, is common to digestive and respiratory tracts. It has four openings: anteroventral into the oral cavity, anterodorsal into the nasopharynx, posteroventral into the larynx, and posterodorsal into the esophagus. The epithelium of the oral cavity is stratified squamous cornified throughout, as is that of most of the pharynx including extensions into the nasopharynx and larynx.

Lips. The lips are covered externally with skin bearing short hairs and internally with hairless skin (mucous membrane). The upper lip is cleft, exposing the two upper incisor teeth; posterior to them the lip curls inward on each side, forming lobes with their hairy surfaces in contact with the dorsal surface of the tongue and their mucous surfaces with the palate. The lower lip partly conceals the lower incisor teeth and, at the angles of the mouth, turns upward along the cheeks, forming folds bearing hairs on the surfaces facing the cheeks and palate. The folds of upper and lower lips occupy the spaces (diastemata) between incisor and molar teeth. Large sebaceous glands are located at the angles of the mouth, their short ducts opening directly onto the mucous surface between lip fold and cheek.

Teeth. The dental formula is incisor 1/1, cuspid 0/0, premolar 0/0, and molar 3/3. The incisors are long and bow-shaped and have roots extending posteriorly, dorsal and ventral to the roots of the upper and lower molars. The pointed tips of the lower incisors lie slightly posterior to the blunted ends of the uppers when the mouth is closed. The incisors grow and are worn down continuously, the apical foramina remaining open. The crown on the outer convex surface is covered with enamel, whereas enamel is lacking on the inner concave surface and the dentine is covered with cementum.

The molars are similar structurally to those of man, and as in man the third molars are small and poorly developed. In one strain (CBA) the third molars are especially small and one or more are lacking in 18 per cent of the mice, the lower molars being more affected than the uppers (Grüneberg, 1952). In mice with absent molars the germ develops normally until about the sixth postpartum day when it begins to regress (Grewal, 1962). The cusp and root patterns of upper and lower molars differ, the uppers being more uniform with three cusps and three roots per molar (Cohn, 1957). Upper and lower molars also differ in the direction of tilt, the uppers being tilted toward the cheek, the lowers toward the tongue (Gaunt, 1961). The development of molar teeth, from initiation of tooth bud to eruption into the oral cavity and functional occlusion, has been described by Cohn (1957) and the development of enamel and dentine and their distribution on the crowns of the developing teeth described by Gaunt (1956). At the time of eruption the cusps are sharp and pointed and are linked by transverse ridges with deep clefts between them. Areas on the margins of the cusps and in the clefts (apical pits) are enamel-free. As the exposed dentine and surrounding enamel are worn down and the crown eroded by mastication, there is a compensatory deposition of cementum at the apical end of each root. In young mice before grinding has worn the surface, distinctive strain-limited cusp patterns can be identified (Bader, 1962, personal communication). Strain differences have been observed also in the trabecular pattern of the alveolar bone and in the thickness of the periodontal membrane (Baer and Lieberman, 1959).

Tongue. The tongue extends from the epiglottis to the lower incisors. The distal portion anterior to the molars is not attached to the floor of the mouth; the proximal portion posterior to the molar teeth is attached at the sides and forms the floor of the mouth cavity. Except for a small area immediately anterior to the epiglottis, the dorsal surface is roughened by the presence of many horny papillae. The ventral and lateral surfaces are smooth. Distinguishing features are the median dorsal groove at the tip, the abruptly rising median intermolar eminence, and the postmolar solitary vallate papilla.

The epithelium is thick stratified squamous cornified extended into papillae on the dorsal surface (Fig. 13-27). The most numerous of these are the conical papillae anterior to the intermolar eminence and the structurally identical filiform papillae posterior to it (Kutuzov and Sicher, 1953). Their horny tips, formed by overlapping layers of cornified cells, point posteriorly. Giant filiform papillae are located in the depression just anterior to the intermolar eminence, their very horny tips being directed toward the center of the depression (Fig. 13-27). Less numerous mound-shaped fungiform papillae, formed of epithelium with cores of connective tissue, are scattered among the conical papillae. The vallate papilla with its connective tissue core is bounded laterally by deep curved troughs. Four or five rows of low foliate papillae, separated by shallow oblique furrows, are located on the sides of the tongue opposite the molar teeth.

Clusters of taste buds are numerous in the epithelium on both sides of the deep troughs of the vallate papilla, on the dorsal surface of the vallate papilla (Fig. 13-28*A*), and on the surfaces of the foliate papillae (Fig. 13-28*B*). The solitary taste buds of the fungiform papillae extend into the connective tissue cores (Fig. 13-28*C*).

The lamina propria, a narrow band of fibrous connective tissue between the epithelium and underlying muscle bundles, forms numerous interepithelial papillae producing an uneven waved epithelial border. The muscles, all striated, are both extrinsic and intrinsic, the former attaching to the hyoid bone, mandibles, and cartilages of the larynx. The muscle bundles are separated by

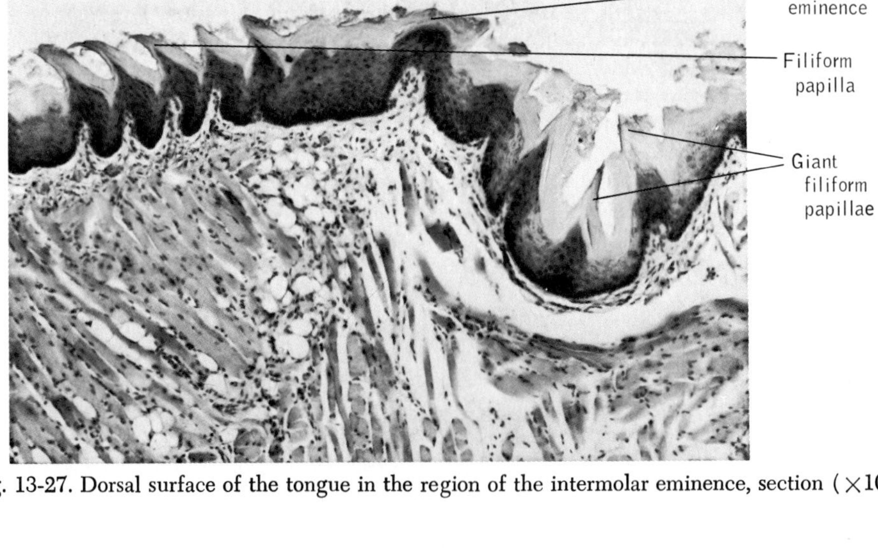

Fig. 13-27. Dorsal surface of the tongue in the region of the intermolar eminence, section (×100).

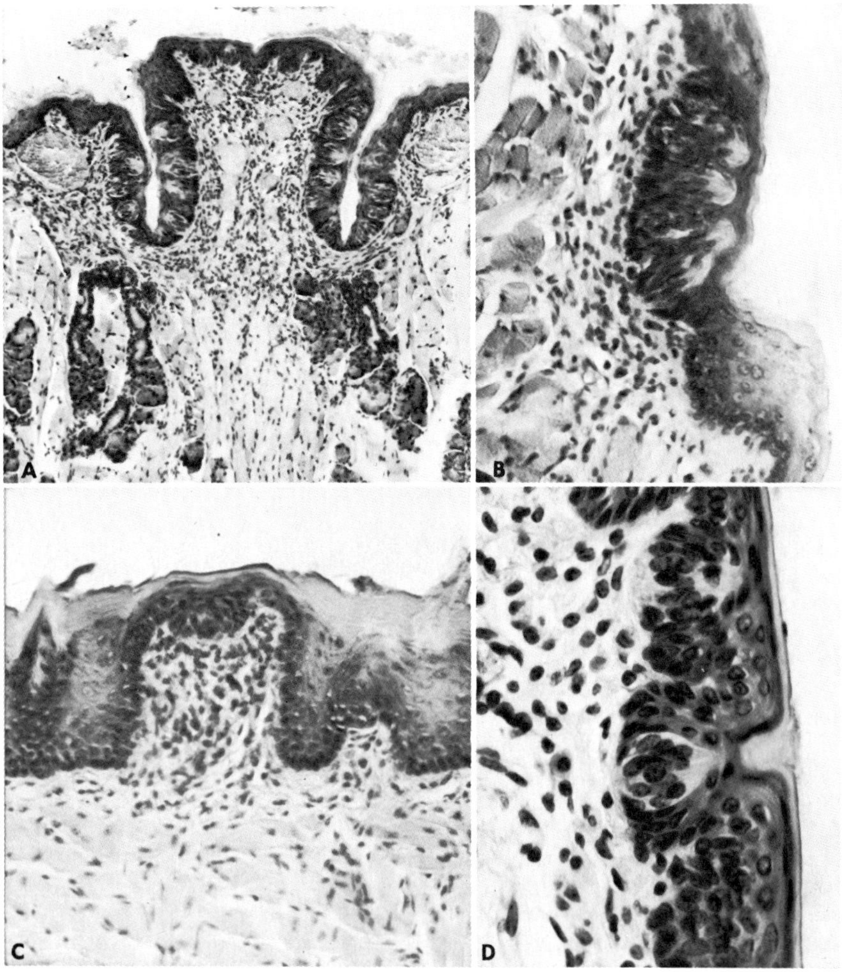

Fig. 13-28. Taste buds. *A*. In epithelium of vallate papilla (×100). *B*. In epithelium of foliate papilla (×200). *C*. In connective tissue core of fungiform papilla (×200). *D*. In epithelium of postrugal hard palate (×400).

thin sheets of fibroelastic and adipose connective tissue, supporting blood and lymph vessels, nerve fibers, and ganglia. Mast cells are a common constituent of the connective tissue.

Serous glands (von Ebner's) are embedded in the muscles in the postmolar region (Fig. 13-28*A*). Their ducts open into the troughs of the vallate papilla. Numerous mucous glands open directly onto the dorsal surface posterior to the vallate papilla, and deeply embedded mucous glands open by long ducts into the shallow furrows of the foliate papillae.

Palate. The anterior hard palate, extending from the incisors to beyond the third molars, is firmly attached to the palatine processes of the premaxillary, maxillary, and palatine bones of the cranium. The hard palate bears eight rows of membranous ridges (rugae) formed by condensations of dense connective tissue continuous with the periosteal connective tissue. A short anterugal region posterior to the incisor teeth is covered with a many-layered, heavily cornified, stratified squamous epithelium. On either side of the midline just anterior to the first ridge are small openings guarded by folds of epithelium. These lead into narrow channels extending dorsally and anteriorly to the nasal cavities. The anterior three ridges are transverse and unpaired; the five posterior pairs are V-shaped and irregular but tend to meet in the midline. A bar of cartilage is contained within the connective tissue of the first ridge. Although the epithelium over the ridges is thick and cornified it does not project into spinous processes as in the rat. The postrugal region is short and bears

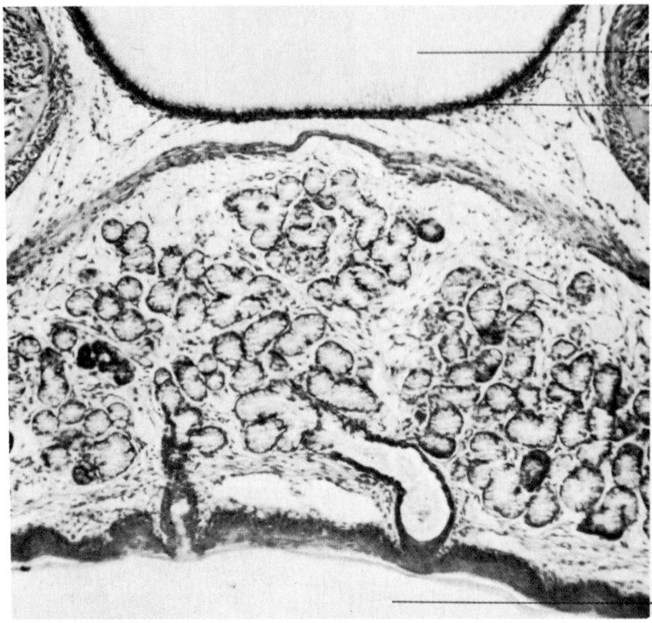

Fig. 13-29. Soft palate, cross-section (×100).

numerous solitary taste buds in its epithelium (Fig. 13-28*D*) except in the median area opposite the vallate papilla.

The palate originates as a vertical shelf of tissue on each side of the tongue. Closure is accomplished in 14- to 15-day embryos by rapid movement from sagittal to transverse planes and meeting and fusion of the palatine shelves (Walker and Fraser, 1956). There are strain differences in the developmental age of palate closure according to these authors.

The soft palate is boneless, glandular, and flexible, forming not only the roof of the posterior oral cavity but also the floor of the nasopharynx (Fig. 13-29). The posterior border forms a semicircular arch (glossopalatine) around the opening of the oral cavity into the pharynx. Fibroelastic connective tissue underlies the epithelium of both surfaces, and striated muscle fibers lie beneath the nasopharyngeal surface in the posterior third of the palate. Numerous mucous alveolar glands surrounded by loose vascular connective tissue open through short ducts directly onto the oral surface.

Pharynx. The posterodorsal opening of the pharynx into the esophagus is depressed and continuous with deep channels on either side of the slightly elevated larynx. The epithelium of this region is thick stratified squamous with many layers of cornified cells. Solitary taste buds are scattered in the thinner stratified squamous epithelium of other regions of the pharynx. The lamina propria of the pharynx is dense and fibrous and merges with the connective tissue of striated muscle bundles of the neck and head regions.

Salivary glands

There are three pairs of salivary glands (parotid, submandibular, and sublingual) located in the subcutaneous tissue of the face and neck. Each gland retains a connection with the oral cavity through a single excretory duct. None is a mixed gland, each being limited to one type of secretory cell, serous in the parotid and submandibular and mucous in the sublingual gland.

Gross anatomy. The parotids are diffuse lobulated glands that extend over a considerable area from the ears to the clavicles. Each gland is divided into lobules by delicate connective tissue septa that merge with the surrounding loose connective tissue. Anteriorly some of the lobules overlie the leaflike exorbital lacrimal gland (Fig. 13-8) and posteriorly others are in contact with the submandibular and sublingual glands (Fig. 13-30). A single excretory duct, formed by the union of interlobular ducts, extends anteriorly over the muscles of the jaw and opens into the vestibule of the oral cavity opposite the molar teeth of the lower jaw.

The large compact submandibular glands are located in the ventral neck region. They extend posteriorly to the sternum and clavicles, anteriorly to overlie the hyoid bone, and medially to meet or overlap slightly in the midventral line. The glands of males and females differ, those of males being larger and more opaque, but both are lobulated and well vascularized. A single excretory duct from the anterior dorsal surface of each gland extends anteromedially to open on the floor of the mouth just posterior to the incisor teeth.

The small compact flattened sublingual glands are pressed against the ventral anterolateral surfaces of the submandibular glands. Each is a single lobe subdivided into lobules and each has a single excretory duct that follows the submandibular duct, opening close to it, but separately, into the oral cavity.

Microscopic anatomy. The three glands have structural features in common. All are compound tubuloalveolar glands separated into lobules by connective tissue septa. The secretory alveoli are lined with tall pyramidal cells resting on a delicate basement membrane, with stellate myoepithelial (basket) cells scattered between epithelium and basement membrane. The very small alveolar lumina are continuous with intercalated (terminal) tubules that are tributaries of larger intralobular ducts. The intralobular ducts or striated tubules are lined by rodded epithelium composed of low columnar cells with large round centrally located nuclei and characteristic striations in the basal cytoplasm. The interlobar and main excretory ducts are lined with columnar or stratified columnar epithelium except near their openings where the epithelium is stratified squamous, continuous with that of the oral cavity.

The secretory alveoli of the parotid glands are very small (Fig. 13-8, 13-30), each consisting of three or four tall pyramidal serous cells arranged eccentrically about the lumen. The relatively large spherical nucleus is near the base of the cell. Below the nucleus the cytoplasm is deeply basophilic; above the nucleus it is more lightly staining and more coarsely granular. The intercalated ducts are short and narrow and are lined by low cuboidal cells with large central nuclei. The septa are delicate and their fibers merge with the loose connective tissue surrounding the lobules.

The secretory alveoli of the submandibular glands are larger, the septa are fibrous, and a fibrous connective tissue capsule surrounds and encloses the lobules. The fibrous septa supporting the large interlobar ducts, blood and lymph vessels, and nerves contain smooth muscle fibers. The alveoli are lined by tall pyramidal serous cells with centrally located nuclei and granular, lightly basophilic, somewhat vacuolated cytoplasm (Fig. 13-30). The intercalated tubules and some of the alveoli have a different structure in adult males from that in females and young males. In adult males the cells are tall columnar and have basally

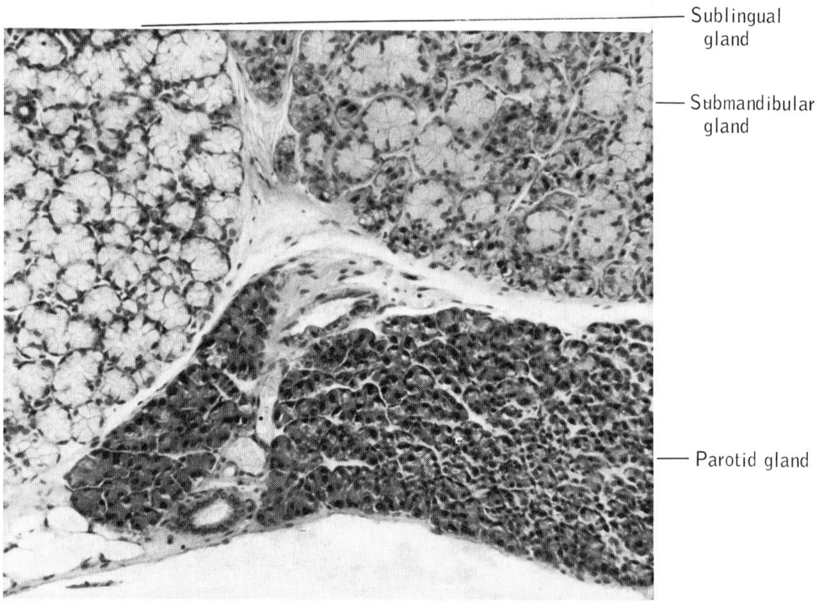

Fig. 13-30. Section through the sublingual, submandibular, and parotid salivary glands of a male mouse (×150).

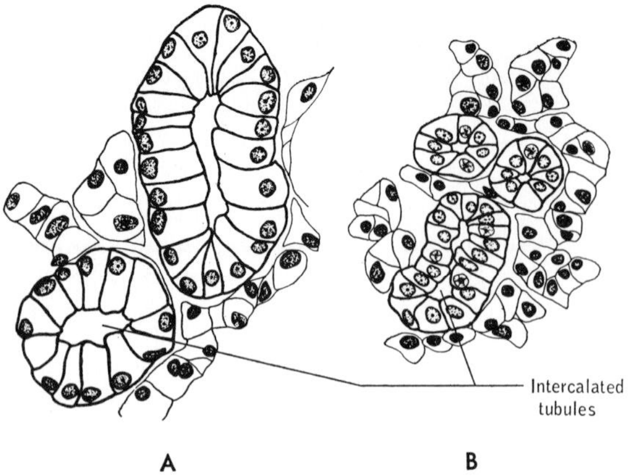

Fig. 13-31. Camera lucida drawings (×450) to show sexual dimorphism of the intercalated tubules of the submandibular gland. *A.* male. *B.* female.

located nuclei and granular vacuolated eosinophilic cytoplasm (Fig. 13-31*A*). In females and young males the columnar cells are not so tall and the nuclei are centrally located (Fig. 13-31*B*).

Each sublingual gland is surrounded by a connective tissue capsule and is divided into lobules by projections of the capsule. The secretory mucous cells are tall pyramidal with pale-staining basophilic cytoplasm and basal nuclei (Fig. 13-30). The intercalated ducts are short and narrow and are lined by low cuboidal epithelium. The epithelium of the main excretory duct is stratified columnar.

Digestive tube

The digestive tube extends from the pharynx to the anus and includes esophagus, stomach, small intestine, and large intestine. The portions within the body cavity are supported by dorsal mesenteries continuous with the lining of the body cavity, in which blood and lymph vessels and nerves are carried.

Gross anatomy. The esophagus is a straight tube from the pharynx to the stomach. In the neck region it is dorsal to the larynx and anterior end of the trachea, but as it enters the thoracic cavity it is slightly to the left of the trachea. It traverses the thoracic cavity in a channel between pleural cavities, pierces the diaphragm to the left of center, and enters the stomach in the middle of the lesser curvature.

The stomach is a large dorsoventrally flattened sac located in the left anterior quadrant of the abdominal cavity, partly concealed by the left lateral lobe of the liver. The lesser curvature of the anterior border is short and concave; the greater curvature of the posterior border is long and convex. A division into thin-walled left and thick-walled right portions is visible grossly (Fig. 13-32). The stomach narrows at the right of the lesser curvature where a constriction, the pylorus, marks the boundary between stomach and small intestine.

The small intestine is a coiled tube about 18 inches long, extending from pylorus to cecum and divided into three functionally but not morphologically delimited units, duodenum, jejunum, and ileum. The duodenum consists of a U-shaped loop extending from the pylorus posteriorly to the level of the umbilicus and a transverse portion dorsal to the posterior margin of the stomach. The ileum ends at the cecum, a blind sac in the lower abdominal cavity, and the jejunum is the portion between duodenum and ileum. Elevated and sometimes prominent opaque nodules are

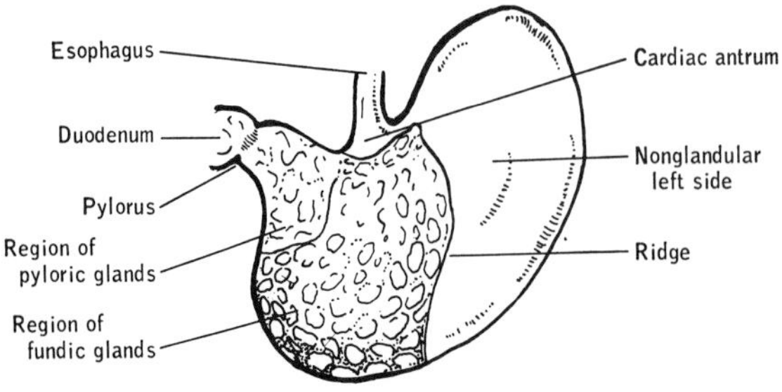

Fig. 13-32. Drawing of the stomach, ventral aspect.

scattered in the intestinal wall opposite the mesenteric attachment. These nodules, Peyer's patches, are aggregates of lymphatic tissue and are described with other organs of the lymphatic system.

The large intestine consists of cecum, colon, and rectum. The cecum is an elongated blind sac in the lower abdominal cavity. Its two openings, one from the ileum and one into the colon, are close together. Accumulations of lymphatic tissue are usually visible in the wall near the narrow blind tip. The mouse has no vermiform appendix. The colon is not coiled and consists of an ascending section from cecum to the level of the pylorus, a short transverse section, and a descending section extending posteriorly to the border of the body cavity. The descending colon often appears beaded because of fecal pellets distending its walls. One or more nodules of lymphatic tissue are usually present although not antimesenteric as in the small intestine. The rectum is outside of the body cavity and is a short, thick passageway from the colon to the anus.

Microscopic anatomy. The digestive tube has a uniform histological organization throughout its length. The wall is made up of four coats: mucosa, submucosa, muscularis, and either adventitia or serosa. The innermost layer or mucosa consists of the epithelium and associated glands and the lamina propria, composed of reticular and fibroelastic connective tissue supporting blood and lymph vessels and nerves. In some regions the outer boundary of the mucosa is marked by a layer of smooth muscle, the muscularis mucosae. The mucosa is structurally the most varied of the four coats, the epithelial lining being modified for different functions in different portions of the tube. The submucosa is a layer of loose fibrous connective tissue supporting larger vessels and sympathetic neurons and nerve fibers. Glands from the epithelial layer may extend into it. The muscularis is usually arranged as inner circular and outer longitudinal layers of smooth muscle, or striated muscle at the extremities. A sympathetic nerve plexus lies in the connective tissue between the muscle layers. The outermost layer of portions of the tube not within body cavities is the adventitia, a loose fibroelastic connective tissue layer merging with that of adjacent tissues. The covering of portions of the tube within the body cavity is the serosa, loose fibroelastic connective tissue covered by mesothelium (visceral peritoneum) and continuous with the supporting mesenteries and the lining of the body cavity (parietal peritoneum).

The esophagus is lined by stratified squamous cornified epithelium, and the mucosa is in longitudinal folds. There are no glands and no muscularis mucosae. The external muscle coat is striated muscle throughout (Fig. 13-33). The thin-walled left side of the stomach has an epithelium similar to and continuous with that of the esophagus (Fig. 13-34). The thick-walled right side has a much folded glandular epithelium of columnar cells and a muscularis mucosae. The muscularis is in three layers, a thin inner oblique, a wide circular, and a thin outer longitudinal. The circular layer is especially wide at the pylorus.

The two regions of the stomach are separated by a U-shaped ridge around the entrance of the esophagus (the cardiac antrum) with the open ends of the U extending around the body of the stomach (Fig. 13-32). The ridge is formed by thickened lamina propria of the nonglandular stomach. The stomach glands are simple branched tubules, appearing as parallel deep pits perpendicular to the stomach wall. Several short straight tubular glands open into the bottom of each pit. There are three types of gland: cardiac, pyloric, and fundic, the latter being the most numerous.

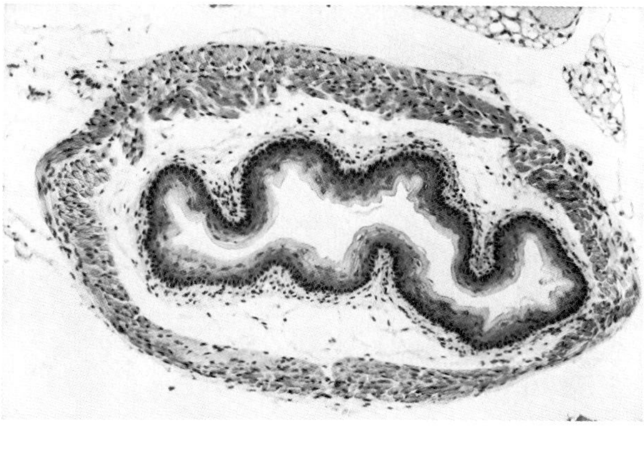

Fig. 13-33. Esophagus, cross section (×100).

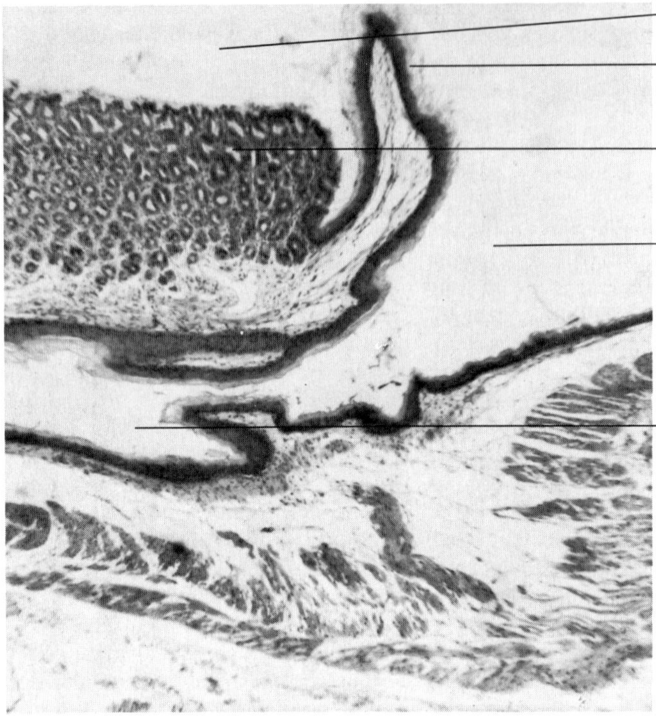

Fig. 13-34. Stomach, section through the ridge separating glandular and nonglandular portions (×60).

The cardiac glands are adjacent to the ridge and consist of a few tubules made up of columnar cells without secretory granules. The fundic glands are made up of mucous cells, chief or zymogenic cells, and acid-secreting parietal cells. The pyloric glands have deep pits and short gland tubules with mucus-secreting cells. The transition from fundic to pyloric glands is gradual and marked by disappearance of chief and parietal cells.

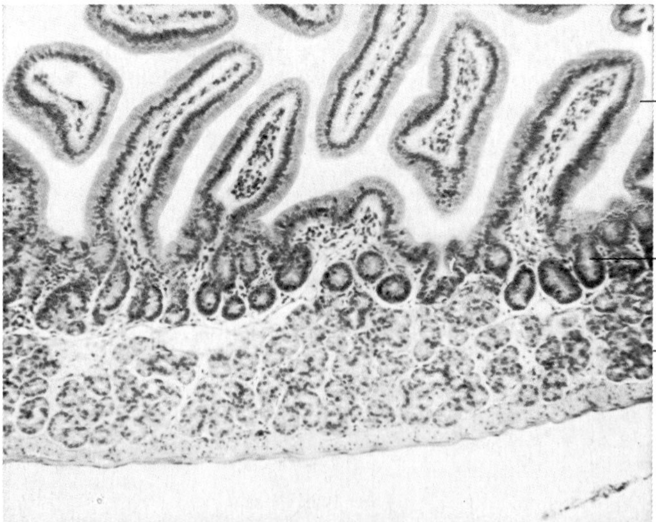

Fig. 13-35. Duodenum, longitudinal section near the pylorus (×100).

The surface area of the small intestine is increased by fingerlike projections, the villi. Those of the duodenum are tall and leaf-shaped (Figs. 13-35, 13-37); those of the jejunum, tall and cylindrical; those of the ileum, short and cylindrical. The short tubular intestinal glands (crypts of Lieberkühn) open between adjacent villi. Near the pylorus, groups of coiled tubuloalveolar duodenal glands (of Brunner) extend into the submucosa and open into the bases of the intestinal glands (Fig. 13-35). The epithelial cells covering the villi are columnar with oval basal nuclei and striated cuticular borders. Oval goblet cells are scattered among the columnar cells and are especially numerous in the ileum. Other cells, Paneth cells, with acidophilic cytoplasm are found at the bases of the villi, especially in the jejunum. They become conspicuous after several hours of fasting and probably do not represent a unique cell type (Dunn and Kessel, 1945). The intestinal glands are lined by low columnar and cuboidal cells and the duodenal glands by cuboidal cells.

Vascular lamina propria forms the core of each villus and fills the spaces between glands. Each villus contains a central lymph vessel, the lacteal. Lymphocytes and granular leukocytes are numerous both in the connective tissue and in the epithelium through which they migrate. The solitary lymphatic nodules and aggregated nodules (Peyer's patches) are in the lamina propria and submucosa, interrupting the muscularis mucosae by their presence.

The mucosa of the cecum and of the ascending and transverse colons is in transverse folds, that of the descending colon and rectum in longitudinal folds. The epithelium is composed of columnar cells and many goblet cells (Fig. 13-36*A*, *B*). The simple tubular glands become shorter and disappear near the anal opening where the lining of the rectum is cornified stratified squamous epithelium. The muscularis mucosae is poorly developed in the colon and well developed in the rectum (Fig. 13-36*A*, *B*). The inner circular layer of the muscularis is very thick and the outer longitudinal layer thin in the rectum, and smooth muscle is replaced by striated muscle near the anal opening. Numerous anal glands of the sebaceous type open into the anal cavity.

With aging the connective tissue of the lamina propria becomes more fibrous and less cellular especially in the duodenum, the villi become shorter and reduced in number, and the mucosa

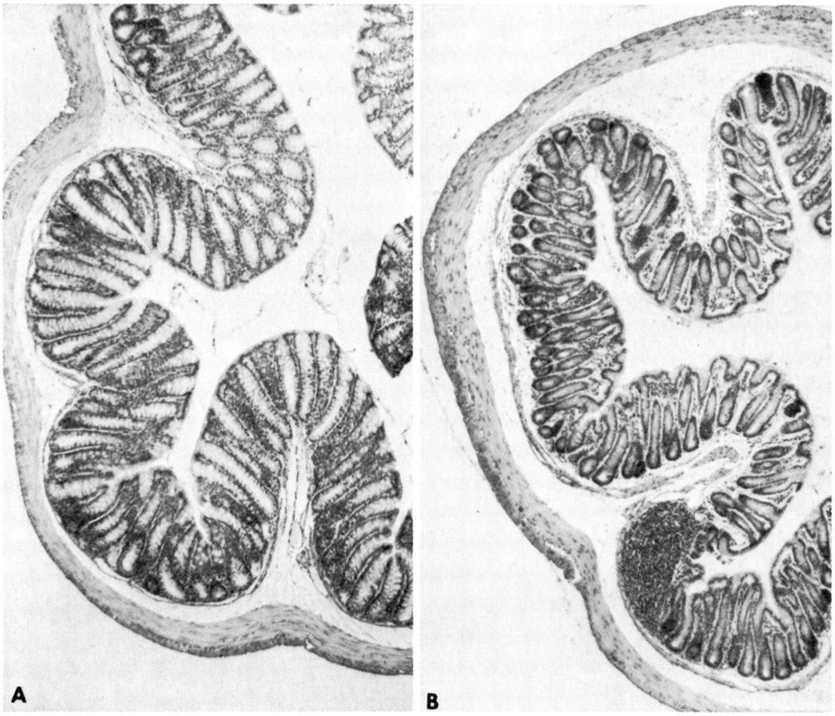

Fig. 13-36. *A*. Colon, cross-section (×50). *B*. Rectum, cross-section showing a lymphatic nodule (×50).

of the large intestine shows signs of atrophy (Andrew and Andrew, 1957; Suntzeff and Angeletti, 1961).

Liver

Gross anatomy. The liver is a large gland occupying the anterior third of the abdominal cavity. The anterior convex surface is pressed against the arch of the diaphragm, and the posterior concave surface fits over and partially conceals the stomach and duodenum. There are four main lobes joined dorsally: the large median subdivided into right and left portions by a deep bifurcation, the undivided left lateral, the right lateral divided horizontally into anterior and posterior portions, and a caudal consisting of two leaf-shaped lobes dorsal and ventral to the esophagus at the lesser curvature of the stomach. Although this is the most frequent pattern of lobation, at least 13 different patterns have been described, the trend being to fission rather than to fusion, and sex and strain differences have been reported (Rauch, 1952). A characteristic abnormal pattern involving degrees of fusion has been shown to be inherited as a recessive trait (Bunker, 1959). The adult pattern of lobation takes shape in 15- to 16-day embryos and is greatly influenced by the growth patterns of the developing gonads (Danforth and Center, 1953). The liver of the female consistently weighs more than that of the male (Webster and Liljegren, 1955).

The mouse differs from the rat in having a gall bladder; this is located at the base of the deep bifurcation of the median lobe near the point of origin of the falciform ligament, a membrane continuous with the covering of the liver and with the lining of the body cavity at the median ventral line. The hepatic duct from the liver and the cystic duct from the gall bladder unite to form the common bile duct. This narrow duct extends posteriorly to the duodenum traversing a portion of the pancreas before passing through the intestinal wall to open on a large papilla (Fig. 13-37).

Microscopic anatomy. The surface of the liver is covered by a thin serosa from which fine strands of reticular connective tissue project inward to form the supporting framework for hepatic cells, blood vessels, and bile ducts. Separation into lobules or hepatic units is very indistinct and septa are visible only around interlobular branches of the hepatic artery, hepatic portal vein, and bile duct (Fig. 13-38). Circulation of blood is through interlobular branches of the hepatic artery and hepatic portal vein into sinusoids separating cords of liver cells and converging on a central vein (Fig. 13-38). The sinusoids communicate not only with central veins (tributaries of hepatic veins) but also with hepatic veins (Lee et al., 1960). The lining of the sinusoids is endothelium, containing specialized phagocytic cells, the Kupffer cells, which possibly function to insure an even distribution of blood to liver cells.

The hepatic parenchymal cells are large and polygonal with large central nuclei (sometimes

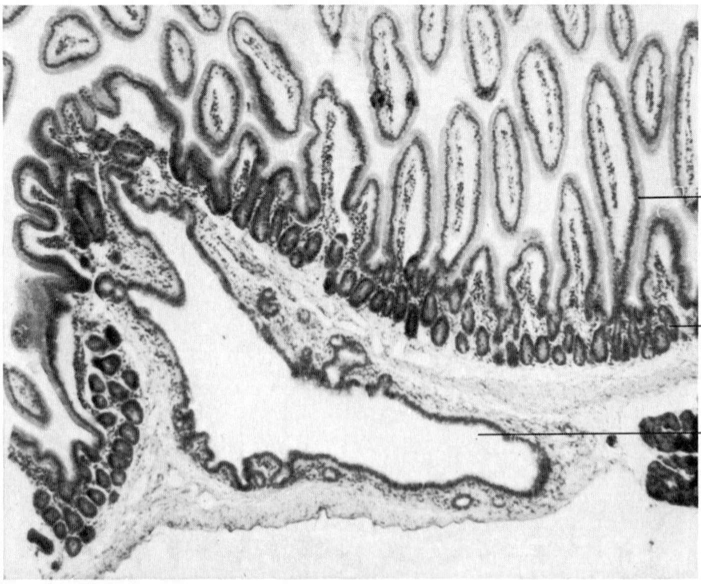

Fig. 13-37. Section showing the opening of the bile duct into the duodenum ($\times 60$).

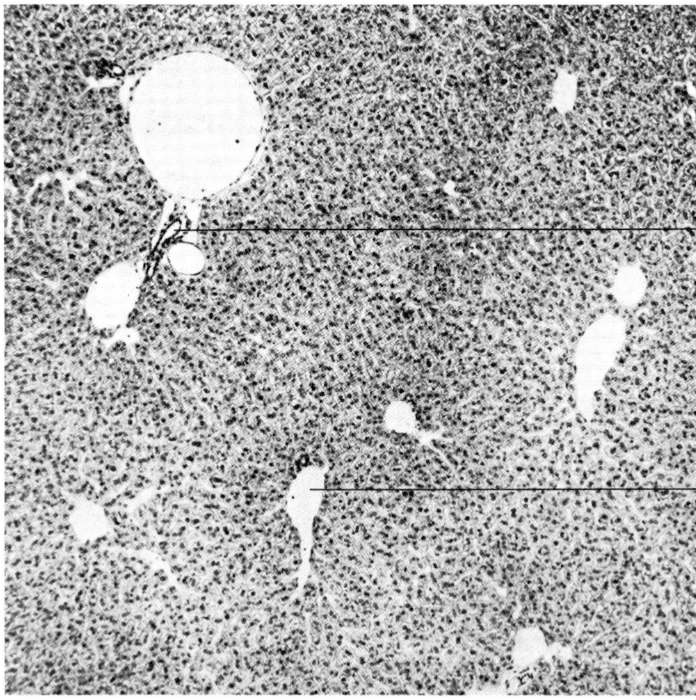

Fig. 13-38. Liver, section ($\times 100$).

two in a cell) and one or more nucleoli. The cell outline is often indistinct and the cytoplasm extremely variable in appearance. It may be granular, vacuolated, deep staining, or very pale. The cells are arranged two deep in cords separated by sinusoids. Bile capillaries are located between the adjoining faces of the cells, the opposite surfaces being in contact with sinusoids. Bile capillaries unite to form bile ducts lined by cuboidal epithelium (Fig. 13-38).

Megakaryocytes are present in the stroma of the liver during the first few weeks of postnatal life and strain differences in the time of their disappearance have been reported (Fortuyn, 1933). Extramedullary hematopoiesis is seen frequently in adult mice in the reticular tissue of the interlobular septa (Dunn, 1954).

The mucosa of the gall bladder is folded except when the sac is distended. The lining of cuboidal epithelium is surrounded by a fibroelastic lamina propria, a thin layer of interwoven smooth muscle fibers, and an outer coat which is in part adventitia and in part serosa (Fig. 13-39*A*). The common bile duct is lined with cuboidal cells changing to columnar where the duct enters the intestinal wall. The walls of the cystic duct and common bile ducts contain a scattering of alveolar glands opening directly into the lumen (Fig. 13-39*B*).

Pancreas

Gross anatomy. The pancreas is a diffuse pink gland suspended in the mesenteries between stomach, duodenum, and ascending and transverse colons. It extends posteriorly in the duodenal loop, lying close to the mesenteric attachment. It is divided into irregular lobes and lobules. There are several excretory ducts, some joining the bile duct where this traverses the pancreas before entering the duodenum (Fig. 13-40). Usually at least one duct enters the duodenum independently near the entrance of the bile duct.

Microscopic anatomy. The exocrine pancreas is a compound tubuloalveolar gland lacking a connective tissue capsule but surrounded by loose vascular connective tissue of the mesentery and separated into lobes and lobules by septa of loose fibroelastic tissue. The small secretory alveoli, resembling those of the parotid and lacrimal glands, are closely packed and have extremely small lumina (Fig. 13-40). The cells are pyramidal and polyhedral with large round basal nuclei. Around and below the nucleus the cytoplasm is deeply staining and basophilic, whereas above it the cyto-

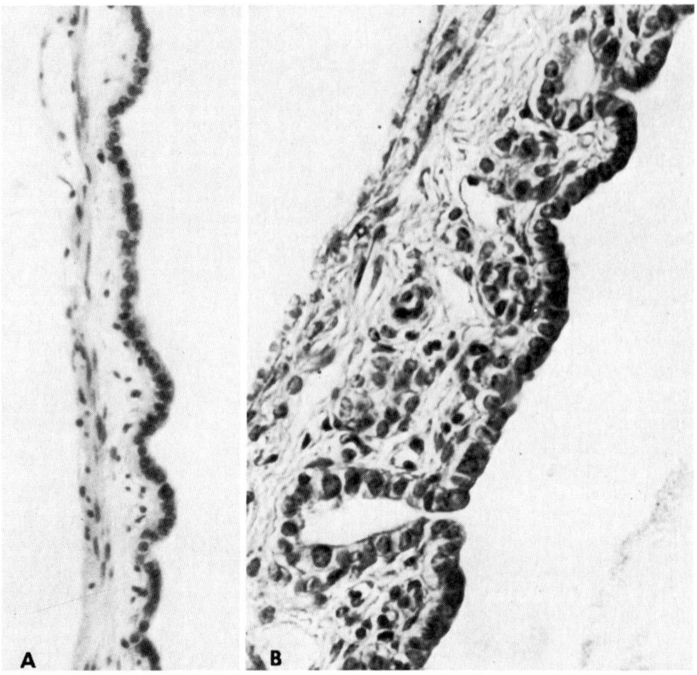

Fig. 13-39. *A*. Section of the gall bladder wall (×400). *B*. Section of the bile duct wall (×400).

plasm is lightly staining, acidophilic, and granular. Variations in cell height, granulation, position of nucleus, and density of stain occur with secretory phases.

The secretion is collected in minute intercalated ducts, lined by very flat cells with elongated nuclei. The interlobar and excretory ducts are lined with cuboidal epithelium and are surrounded by delicate connective tissue supporting vessels and nerves. The endocrine portions, the islets of Langerhans, are always closely associated with the septal ducts and blood vessels (Fig. 13-40).

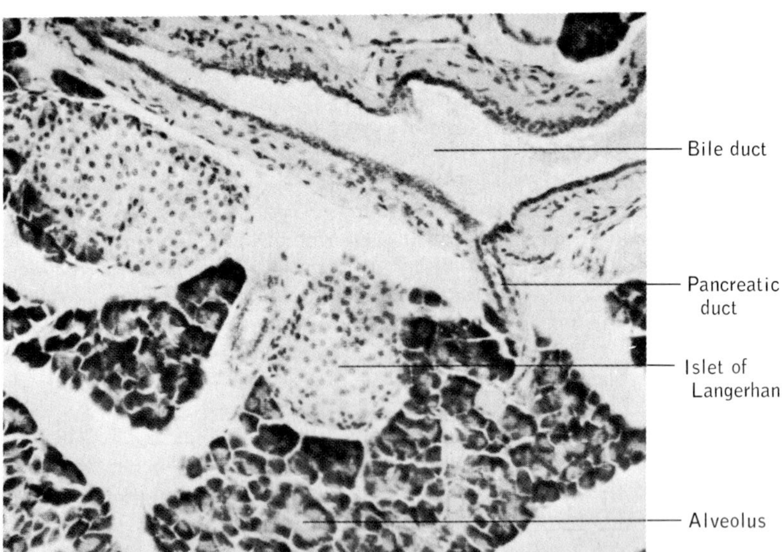

Fig. 13-40. Section through the pancreas showing several lobules of alveoli, two islets of Langerhans, and a pancreatic duct emptying into the bile duct (×150).

RESPIRATORY SYSTEM

The respiratory system consists of portions anterior and posterior to the pharynx where respiratory and digestive tracts cross. The air passageways of the anterior division are surrounded by bones of the skull giving the rigidity necessary to prevent collapse at inspiration. The air ducts of the posterior division are supported and given rigidity by cartilaginous plates and rings.

Anterior respiratory tract

This includes nostrils, nasal cavities, and pharyngeal duct (nasopharynx). In addition to functioning as air passageways to the lungs, the nasal cavities are modified for olfaction. For detailed descriptions of the anatomy and histology of the anterior respiratory tract of the mouse and rat see Kelemen (1953) and Kelemen and Sargent (1946).

Gross anatomy. The nostrils (anterior nares) are small openings on either side of the midline near the tip of the snout. Externally they are guarded by curved folds of thickened skin and internally they communicate through vestibules with the anterior nasal cavities, two narrow lateral chambers separated by a median septum. Two sets of bony ridges, the dorsal and lateral turbinals, decrease the size of each cavity. A long narrow channel from the vestibular region of each anterior cavity leads ventrally and posteriorly to open on the roof of the oral cavity. The median septum extends posteriorly for a short distance and at its end the anterior nasal cavities open through choanae (posterior nares) into the pharyngeal duct, an undivided tubelike structure. The anterior nasal cavities also extend posteriorly into two blindly ending cavities, the ethmoid sinuses, dorsal to the pharyngeal duct. These ethmoid sinuses are separated by a median septum resting on the roof of the pharyngeal duct and are highly developed olfactory organs each containing seven rows of turbinals. The right and left anterior nasal cavities and right and left ethmoid sinuses are in communication through a "window" in the nasal septum just anterior to the choanae (Kelemen, 1953).

The pharyngeal duct is a straight undivided tube from choanae to pharynx. Its posterior portion is the nasopharynx which is dorsal to the soft palate and communicates through a narrow oval opening with the oropharynx (see the digestive system). There is a narrow oblique slit on either lateral wall of the nasopharynx opening into the Eustachian tube.

Microscopic anatomy. The nostrils, vestibules, and channels to the oral cavity are lined by stratified squamous epithelium. This changes abruptly within the anterior nasal cavities to pseudostratified columnar ciliated with many goblet cells, a type of epithelium that lines most of the respiratory tract. The turbinals of nasal cavities and ethmoid sinuses are covered with olfactory epithelium of pseudostratified columnar type containing specialized bipolar sensory cells. At the posterior border of the soft palate, the epithelium becomes stratified squamous continuous with that of oral cavity and pharynx. The fibrous connective tissue underlying the epithelium merges with the periosteal connective tissue of the skull bones. There are numerous branched alveolar glands with short ducts in the lamina propria of the walls and floor of the anterior nasal cavities and on the roof of the pharynx dorsal and posterior to the rim of the soft palate. The dorsal and ventral portions of the nasal septum are composed of bone, the central portion of cartilage.

Posterior respiratory tract

This portion includes larynx, trachea, bronchi, and lungs. The larynx and trachea are embedded in the muscles of the neck region ventral to the esophagus, and the bronchi and lungs are in pleural sacs within the thoracic cavity.

Gross anatomy. The larynx is a chamber with walls partially composed of cartilaginous plates. The anterior end or top of the larynx is on the floor of the pharynx immediately posterior to the base of the tongue, and the opening into it, the glottis, is a slit between dorsoventral folds (the vocal cords) projecting from its lateral walls. A triangular flap of tissue, the epiglottis, projects from the anteroventral border into the pharynx. When the larynx is raised in the act of swallowing, its top and the epiglottis fit into the nasopharynx preventing the entrance of swallowed material into the air passages. The ventral wall contains the large shield-shaped thyroid cartilage to which the epiglottis is attached. The cricoid cartilage, which encircles the larynx posterior to the thyroid cartilage, is wider on the dorsal than ventral side. The paired arytenoid cartilages support the dorsal rim of the glottis and extend ventrally into the vocal cords.

The trachea extends from the narrowed end of the larynx into the thoracic cavity where it branches dorsal to the aortic arch into left and right bronchi. The tracheal walls are stiffened by cartilage rings that are incomplete dorsally, leaving a flexible strip in the wall adjacent to the esophagus. The rings, especially the most anterior ones, are uneven, branching and fusing with one another. The large primary bronchi, completely encircled by irregular cartilage plates, lead into the right and left lungs where further divisions take place. There are no cartilages in the walls of bronchi within the lungs.

The lungs are divided into lobes, the normal pattern being a single lobe on the left and four on the right: three in a row (anterior, middle, and posterior); and one (the median, cardiac, or infracardiac lobe) against the diaphragm to the left of the midline. This pattern of lobation was the most frequent of nine patterns observed in a study of eight strains (Browder, 1942). The lungs are covered by a serous membrane, the visceral or pulmonary pleura, and are suspended within pleural cavities lined by parietal pleura. The medial walls of the two pleural cavities meet anterior and posterior to the pericardial cavity to form the double-walled mediastinal septum. The left lung lobe lies in the left pleural cavity, three of the right lobes lie in the right cavity, and the median lobe lies in a fold of the mediastinal septum to the left of the inferior vena cava.

Microscopic anatomy. The lingual surface of the epiglottis and the anterior surface of the vocal cords are covered by stratified squamous epithelium. The remainder of the larynx, trachea, and larger bronchi are lined by pseudostratified columnar ciliated (respiratory) epithelium with many goblet cells. The lamina propria of all regions is fibroelastic connective tissue supporting numerous alveolar glands in the larynx and anterior end of the trachea. Some of the alveoli are composed of serous and some of mucous cells, and their short ducts are lined by ciliated cuboidal cells. There are no subepithelial glands on the epiglottis, vocal cords, posterior portion of the trachea, bronchi, or smaller respiratory tubes. The lamina propria merges with the dense fibrous perichondrium of the laryngeal, tracheal, and bronchial cartilages. The cartilages are hyalin except for those of the epiglottis and arytenoids in which there are elastic and reticular fibers. The cartilages of the larynx are connected with each other and to adjacent structures by strands of fibrous connective tissue and bundles of striated muscle. The open ends of the tracheal rings are joined by smooth muscles attached to the dorsal surface of the perichondrium. When the muscle is contracted, the mucous membrane in this region is thrown into longitudinal folds. There are no muscle layers surrounding either larynx or trachea; the connective tissue adventitia anchors them to surrounding muscles and bone. The primary bronchi resemble the posterior portion of the trachea in structure except for having cartilage in the dorsal wall, a layer of smooth muscle, and a serosa (visceral pleura).

The bronchi branch repeatedly after entering the lung, diminishing in size with each division. The secondary and tertiary bronchi resemble the primary bronchi except that their walls lack cartilage and serosa. The smaller bronchi are lined

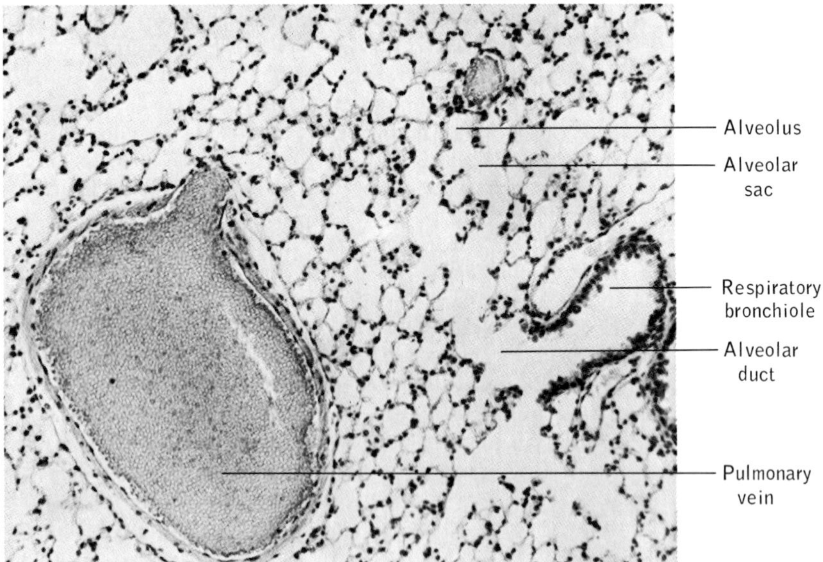

Fig. 13-41. Lung, section (×150).

by simple columnar ciliated cells and the bronchioles by low columnar epithelium lacking both cilia and goblet cells. As the tubes become smaller their walls become thinner and contain less connective tissue and smooth muscle. The terminal bronchioles give rise to respiratory bronchioles, each branching into several alveolar ducts and alveolar ducts lead into alveolar sacs, each composed of several alveoli (Fig. 13-41). Terminal and respiratory bronchioles are lined with cuboidal epithelial cells surrounded by thin connective tissue sheets containing scattered smooth muscle cells. The cuboidal epithelium ends abruptly at the junctions of respiratory bronchioles and alveolar ducts where it is replaced by squamous epithelium. Alveolar ducts, alveolar sacs, and alveoli have very thin walls invested with fine close-meshed networks of large thin-walled capillaries. Electron microscopic studies of the structure of alveoli and pulmonary capillaries have shown that both alveolar epithelium and capillary endothelium are continuous uninterrupted layers with a thin, structureless basement membrane between them (Low, 1953; Karrer, 1956). Two types of alveolar epithelial cells have been described: one with its thick nucleus-containing portion protruding into the alveolar lumen, and the other often in a niche of the alveolar wall (Karrer, 1956). Free macrophages, usually in contact with the alveolar epithelium adjacent to an alveolar septum, were seen in small numbers in all lungs studied (Karrer, 1958). Alveoli are separated from alveolar sacs by interlobular connective tissue supporting arteries and veins.

The lungs receive blood from two sources: aerated blood through bronchial arteries from the systemic circulation, and venous blood through the pulmonary arteries from the heart. The pulmonary arteries and veins follow the bronchi, and the very thin capillaries invest the alveolar walls. The walls of the pulmonary veins contain cardiac muscle (Fig. 13-41).

URINARY SYSTEM

The urinary system includes kidneys, ureters, urinary bladder, and urethra.

Kidney

Gross anatomy. The kidneys are paired bean-shaped organs lying retroperitoneally against the dorsal body wall on either side of the vertebral column. They are not attached to the body wall, but are held loosely in place by adipose tissue. The right kidney is larger, heavier, and located more anteriorly. The anterior pole of the right kidney is usually at the level of the 12th rib, that of the left kidney at the level of the 13th or just posterior to it. There is a sex difference, the male kidney being consistently heavier and larger throughout life. The shape and size of the kidney varies somewhat with strain of mice; in strain C58, 10 to 12 per cent of animals have one or both kidneys reduced in size or missing (Hummel, 1954).

The kidney is dorsoventrally flattened and has an extensive convex lateral and a short concave medial border. The concavity is the hilus where blood vessels and the ureter join the kidney. Two layers, cortex and medulla, can be seen without the aid of a lens if the kidney is bisected. The cortex follows the contours of the convex border, and the medulla is like a broad pyramid with its convex base fitted against the concave surface of the cortex. The apex of the pyramid is the papilla which is surrounded by the pelvis, the expanded funnel-like anterior end of the ureter.

Development of the kidney, ureter, and associated blood vessels has been described by Brown (1931) and Carter (1954).

Microscopic anatomy. The kidney is composed of units or nephrons held together by delicate, richly vascular connective tissue strands and enclosed in a thin connective tissue capsule. The ventral surface is covered by peritoneum. Each nephron is a tubule with a widened end, Bowman's capsule, enclosing a tuft of blood capillaries, the glomerulus. The tubule consists of proximal and distal convoluted portions with a straight segment, Henle's loop, between. The nephrons drain into straight collecting tubules that join others to form large tubules opening into the pelvis near the tip of the papilla. The nephron is structurally similar to that of man, as is the blood-vessel architecture. For histological details of each region and for the arrangement of arteries, veins, and capillaries see a histology text.

The outer zone of the cortex is composed predominantly of glomeruli and convoluted tubules (Fig. 13-42). The straight tubules of Henle's loops are grouped into bundles and give a rayed appearance to the inner zone of the cortex. The medulla has a striated appearance, being composed chiefly of straight collecting tubules converging toward the papilla.

Certain features of the mouse kidney are noteworthy. Compared to male animals of many species, the mouse has a glomerular volume (number and size) of about one-half of that predicted on the basis of kidney size and volume (Rytand,

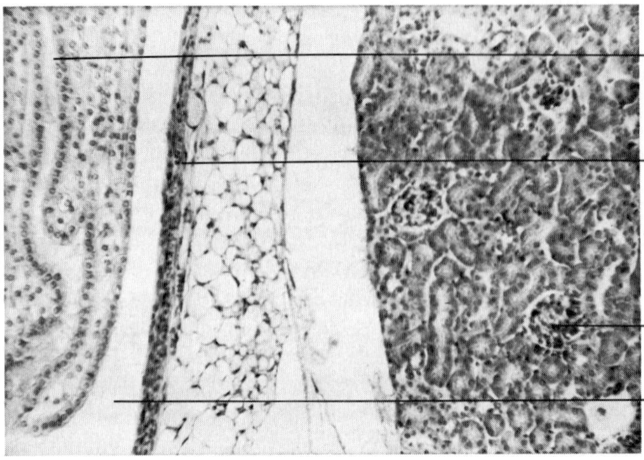

Fig. 13-42. Section through kidney hilus region showing portions of cortex, pelvis, and papilla (×65).

1938). The granular cells in the walls of the afferent glomerular arteries (the juxtaglomerular apparatus) are easily demonstrated in the mouse, in contrast to the difficulties of revealing them in man (Dunn, 1949*a*). There is a sex difference in the relative number of Bowman's capsules that have parietal linings of cuboidal rather than squamous cells. Most of the parietal epithelium in females and young and castrated males is the squamous type found in kidneys of most other animals, whereas many of the Bowman's capsules of the adult male are lined by cuboidal cells (Crabtree, 1941).

Ureter

Gross anatomy. The ureter extends from the kidney to the urinary bladder passing dorsal to uterine horn and vas deferens. The widened anterior end, the pelvis, is surrounded by kidney cortex. The ureters enter the dorsal wall of the

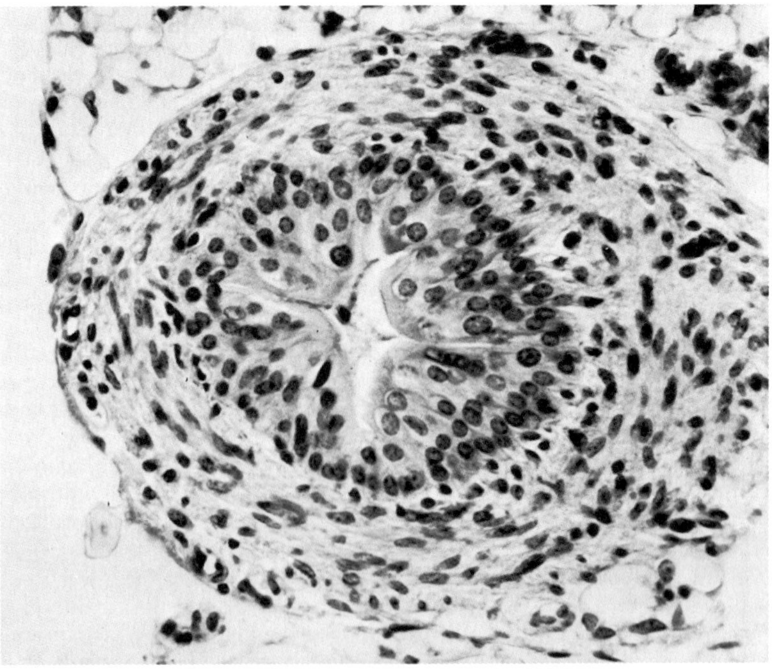

Fig. 13-43. Ureter, cross section (×400).

neck of the bladder separately, lateral to the entrances of the vasa deferentia in the male. An intramural part courses through the bladder muscles in a slightly oblique direction, serving as a barrier to reverse movements of urine.

Microscopic anatomy. The epithelium of the pelvis is continuous with that covering the papilla and lining the ducts of the larger collecting tubules, as well as with that lining the rest of the ureter. The wall of the pelvis is composed of a thin layer of transitional epithelium surrounded by layers of fibrous connective tissue and loose connective tissue containing adipose cells. The epithelium covering the papilla is transitional in deeper regions and usually simple low columnar to cuboidal over the tip and lining the collecting ducts (Fig. 13-42). The wall of the ureter is composed of transitional epithelium, a fibrous lamina propria, an inner circular and an outer longitudinal layer of smooth muscle fibers, and an adventitia of loose connective tissue and adipose cells (Fig. 13-43). The mucosa is in low longitudinal folds.

Urinary bladder

Gross anatomy. The pear-shaped bladder is in the posterior abdominal cavity in the midline of the body ventral to the colon. It varies in size with the amount of urine contained. Posteriorly it narrows into a neck continuous with the urethra. The bladder is attached to the ventral body wall by the ventral ligament.

Microscopic anatomy. The bladder is lined by transitional epithelium consisting of two to four layers of cells when the bladder is empty and one or two layers when distended. The fibrous lamina propria is richly vascular, occasionally containing aggregations of lymphocytes. In the empty bladder the mucosa is thrown into wide irregular folds (Fig. 13-44) which disappear in the distended organ. The muscularis is wide and consists of smooth muscle bundles of irregular size and direction separated by considerable amounts of connective tissue. Around the neck of the bladder, the muscle bundles are circularly arranged to form a sphincter. The adventitia is fibroelastic tissue covered in part by a serous membrane continuous with the ventral ligament and elsewhere merging with adjacent loose connective tissue.

Urethra (female)

Gross anatomy. The female urethra is a short tube, slightly flattened dorsoventrally, extending from the neck of the bladder to an external opening in the clitoral fossa, a depression near the tip of the clitoris just anterior to the vaginal orifice. The male urethra, modified to function as a duct for both urine and male sex cells, is described under the male genital system.

Microscopic anatomy. The mucous membrane is in longitudinal folds, the epithelium is transitional, and the lamina propria is dense fibrous connective tissue supporting numerous blood vessels. Alveolar urethral glands, with short ducts opening directly into the lumen, are present in the connective tissue posterior to the neck of the bladder on ventral and lateral sides only. At the neck of the bladder, the smooth muscle layers continuous with the musculature of the bladder

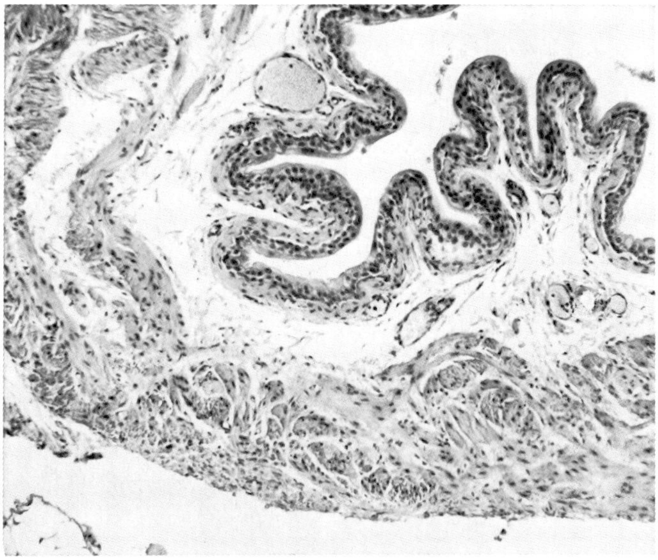

Fig. 13-44. Urinary bladder, section (×100).

are replaced by striated muscle forming an especially thick circular coat on the ventral side. A loose connective tissue adventitia surrounds the muscle layers and merges with adjacent structures. Near the external orifice, the urethra passes through the clitoris where the structure of the urethral wall is modified (Fig. 13-62). The clitoris is described with the female genital system.

MALE GENITAL SYSTEM

The male genital system consists of testes, excretory ducts, accessory glands, and urethra and penis (Fig. 13-45).

Testis

Gross anatomy. The paired testes are located in the posterior body cavity on either side of the urinary bladder or in the scrotal sacs. The scrotal sacs are extensions of the body cavity into the subcutaneous tissue on either side of the penial urethra, just anterior to the anus and base of the tail. Their cavities remain in communication with the body cavity through inguinal canals that remain open throughout life, and the testes are often in a retracted position. The fat body attached to the epididymis occupies the inguinal canal when the testis is in the scrotal sac.

The testis is an oval body, consisting of coiled

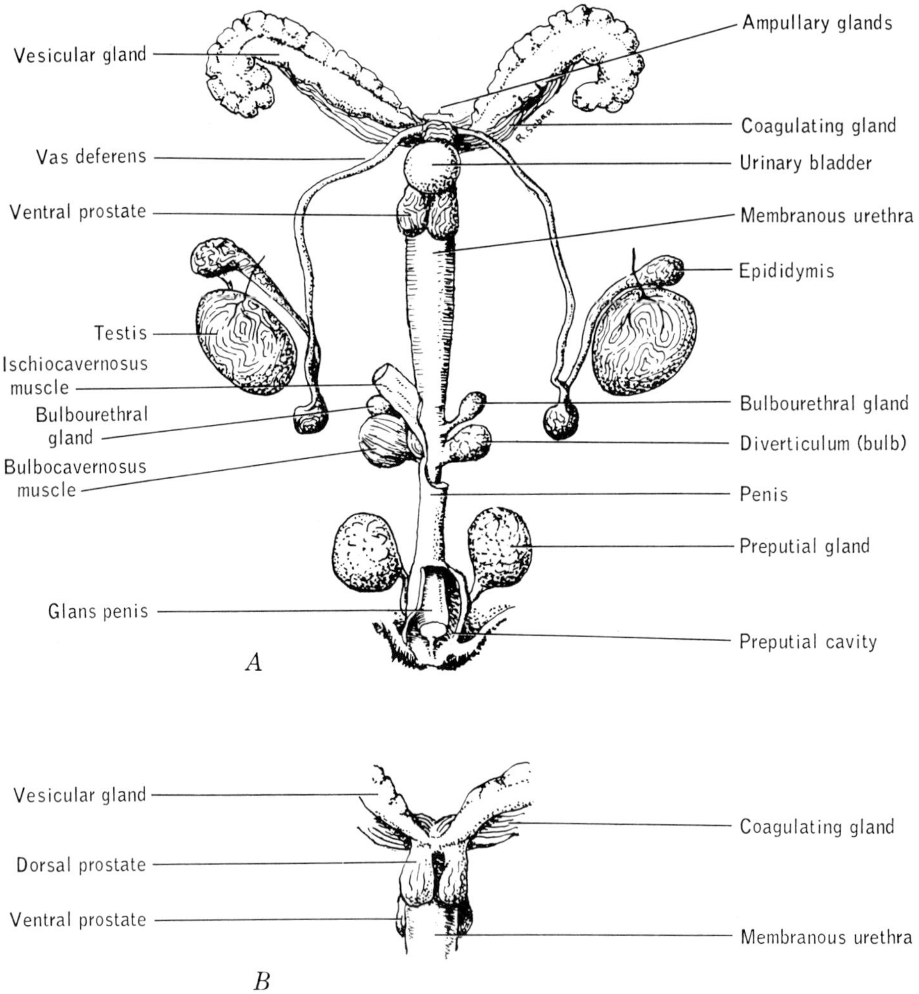

Fig. 13-45. Drawings of the male genital system ($\times 2\frac{3}{4}$) *A*. Ventral view with the penis straightened, with parts of left side dissected (bulbourethral gland and urethral diverticulum), and with part of prepuce cut away. *B*. Dorsal view of the anterior end of the urethra.

tubules held together by connective tissue and covered by a tough membrane, the tunica albuginea. In a mature male, the testis weighs about 85 mg, and measures about 8.5 × 5 × 5 mm. The total tubule length has been estimated at between 1.7 and 2 m (Bascom and Osterud, 1925). The testis has exocrine and endocrine functions; mature male sex cells, spermatozoa, are produced in the testis and discharged into the excretory ducts; and male hormones, the androgens, are excreted directly into the blood stream. The sex cells mature in the walls of the tubules, and the endocrine cells are scattered in the intertubular connective tissue.

Microscopic anatomy. The tunica albuginea or capsule of the testis is a thick layer of fibrous connective tissue with a covering of mesothelial cells. Thin septa with blood vessels project into the gland and divide it into irregular lobules. The lobules contain the convoluted seminiferous tubules and the richly vascular intertubular stroma with groups of interstitial (Leydig) cells. The tubules are made up of specialized seminiferous epithelium resting on a basement membrane and covered by thin connective tissue merging with the intertubular stroma. The seminiferous epithelium contains two kinds of cells: sustentacular Sertoli cells, and male sex cells undergoing spermatogenesis and spermiogenesis. For details of these processes, see Chap. 7, 11.

The Sertoli cells are spaced at fairly regular intervals against the basement membrane; they have indistinct outlines and large oval often indented nuclei with compound nucleoli consisting of one central acidophilic and two peripheral basophilic bodies. When a Sertoli cell is fulfilling its normal function of supporting spermatozoa, it is pyramidal in shape with the apex directed toward the lumen and the nucleus perpendicular to the tubule wall; when resting, the cell is polygonal and the nucleus is parallel to the wall. Under certain abnormal conditions, which result in degeneration or atrophy of the seminiferous cells, the tubules are lined with the more resistant Sertoli cells.

The spermatogenic cells are arranged in layers that vary in number with phase of spermatogenic activity; this process does not take place simultaneously in all tubules. The most primitive sex cells are the spermatogonia, resting on the basement membrane and interrupted at intervals by Sertoli cells. Cells of the other layers are those resulting from divisions of the spermatogonia: the spermatocytes, spermatids, and immature spermatozoa. Spermatids are small spherical cells that remain in close association with Sertoli cells during their transformation into spermatozoa. The heads of immature spermatozoa are buried in Sertoli cells, and the tails extend into the lumen. The fully mature spermatozoa become free of the sustentacular cells, pass into the lumen, and thence into excretory ducts.

A mature spermatozoon, found in the lumina of testis tubules and excretory ducts, is made up of head, middle piece, and tail. The head is sickle-shaped and flattened, and the tail long and flagellate. There are strain differences in the breadth and length of the head, the shape of the posterior end of the head, and the shape of the middle piece (Braden, 1958; Sharma, 1960).

The interstitial cells, which secrete androgenic hormones, are in irregularly sized groups in the intertubular loose connective tissue in close association with capillaries. The cells are small with indistinct cell outlines, acidophilic cytoplasm, and large round nuclei containing one or two nucleoli and coarse chromatin granules.

Excretory ducts

Gross anatomy. The excretory ducts include the rete, the efferent ducts, the epididymis, and the ductus (vas) deferens (Fig. 13-45, 13-46). Near the hilus of the testis within the tunica, the tubules become straight and are gathered into a network, the rete, from which three to five efferent ducts emerge. The efferent ducts pierce the tunica and become enclosed within the capsule of the epididymis where they unite to form a single duct, the duct of the epididymis. The epididymis consists of three regions, caput, corpus, and cauda (Fig. 13-46). In the caput and corpus, the duct is extremely convoluted, but less so in the cauda. The ductus deferens is a straight tubule extending from the cauda to the urethra, passing ventral to the ureter and widening into an ampulla before entering the dorsal wall of the urethra near the neck of the bladder. The ductus deferens and closely associated blood vessels and nerves make up the spermatic cord.

Microscopic anatomy. In the hilus region of the testis where the convoluted tubules transform into straight tubules, the epithelial lining is low columnar; in the irregular spaces of the rete it is low cuboidal or squamous. The efferent ducts are lined by alternating groups of tall and low columnar cells, giving the lumen a characteristic scalloped outline (Fig. 13-48). Basement membranes and a few circularly arranged smooth muscle fibers

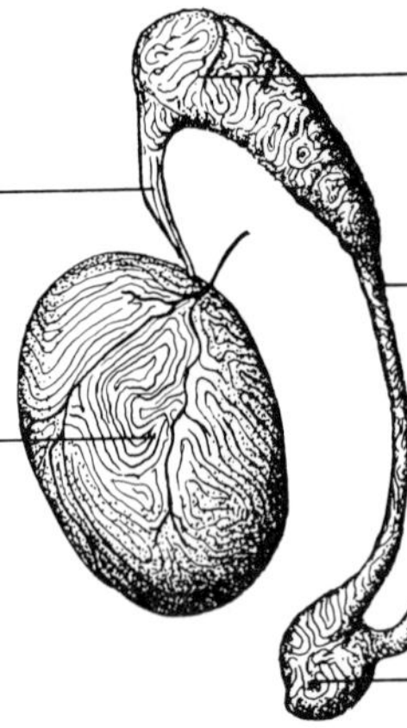

Fig. 13-46. Drawing of the testis, efferent ducts, epididymis, and vas deferens (×5.4).

complete the duct walls. Loose connective tissue fills the spaces between ducts and the composite is surrounded by a fibrous connective tissue capsule continuous with that of the epididymis.

The epididymis is covered throughout its lengtn by a fibrous connective tissue sheath which projects into the caput and divides it into seven or eight segments or lobules (Fig. 13-47). The first of these segments contains portions of the efferent duct and the others contain the very coiled duct of the epididymis. This duct is lined by columnar or cuboidal cells bearing tufts of large nonmotile stereocilia on their free surfaces (Fig. 13-48). Secretion granules are visible in the cytoplasm between the free margins and the basal nuclei, and excretion takes place in cytoplasmic extensions between stereocilia. The lumen is wide in the first and second segments, is narrowed in the third and subsequent segments, and again becomes wider toward the cauda. In the second segment the epithelial cells are very tall and not all of their nuclei are basal. In the third and subsequent segments of the caput and in the corpus and cauda, the epithelium is made up of a single layer of low columnar cells with centrally located nuclei. There is a layer of small round cells between the epithelium and the basement membrane throughout and, surrounding these, one or two layers of circularly arranged smooth muscle fibers.

The mucosa of the ductus deferens, consisting of columnar epithelium and a delicate fibrous lamina propria, projects into the lumen in tall, regularly spaced longitudinal folds and short, irregular circular elevations. The tall columnar cells have tufts of stereocilia on their free surfaces, oval centrally located nuclei, and light-staining cytoplasm with secretion granules (Fig. 13-49). Between the mucosal folds the epithelium appears

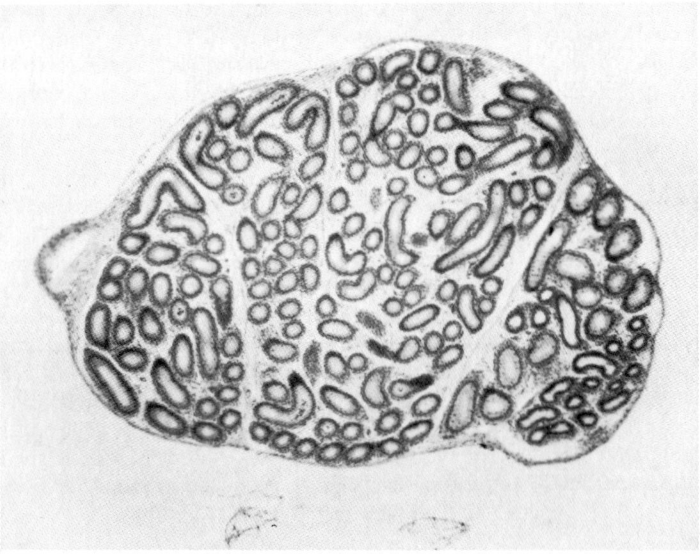

Fig. 13-47. Section of the caput of the epididymis, showing separation into lobules (×14).

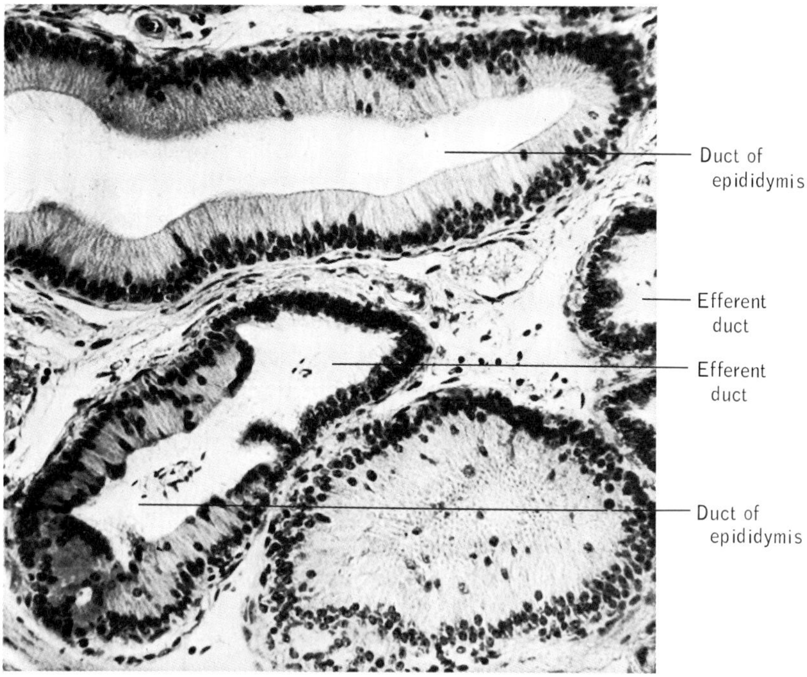

Fig. 13-48. Section of the first lobule of the caput to show the efferent duct entering the duct of the epididymis (×200).

to be pseudostratified and, as in the epididymis, flat cells with small densely staining nuclei lie between the epithelial cells and the basement membrane. In the ampulla, the epithelium changes abruptly to low columnar without stereocilia, and the mucosa is in tall irregular folds. The fibrous lamina propria is very thin and is surrounded by a wide layer of circularly arranged smooth muscle

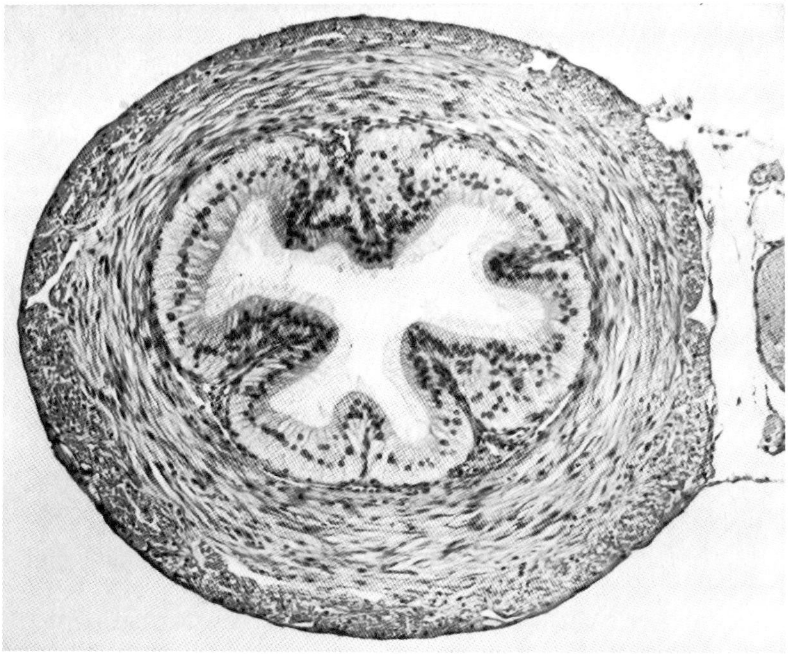

Fig. 13-49. Vas deferens, cross-section (×150).

fibers and a much thinner layer of longitudinal fibers (Fig. 13-49). The serosa is thin except at the point of mesotubarium attachment, where the fibrous tissue supports the vessels and nerves that accompany the duct in the spermatic cord.

Urethra and penis

Gross anatomy. The male urethra is a long duct extending from the urinary bladder to an opening on the tip of the penis. The portion of the urethra from the neck of the bladder to the pelvic girdle has a relatively thin wall and is referred to as the membranous urethra. The penial urethra is surrounded by the erectile, muscular, and fibrous tissues of the penis.

The urethra enters the penis at the posterior border of the pelvic girdle, where a bulbous diverticulum extends laterally and posteriorly. From this point the penis extends anteriorly along the ventral wall within the subcutaneous tissue to terminate on the ventrally elevated genital papilla. The external orifice of the urethra is at the tip of the glans, the club-shaped terminal end of the penis, which is covered by a reflected fold of skin, the prepuce or foreskin. A small bone, the os penis, projects slightly beyond the orifice.

The body of the penis is made up of three masses of erectile tissue (corpus cavernosum urethrae and two corpora cavernosa penis) within heavy connective tissue membranes that also serve to attach the penis to the pelvic girdle. The thin cavernosum urethrae, which is ventral to the urethra, expands proximally over the urethral diverticulum or bulb and extends distally a short distance between the heavier corpora cavernosa penis. The corpora cavernosa penis extend laterally and dorsally and almost encircle the urethra. They are separated proximally by individual coverings but are a single mass distally. At the region of the urethral diverticulum, the two corpora diverge and their tunicas merge with the connective tissues of the bulbocavernosus muscle that covers the urethral diverticulum and the ischiocavernosus muscle that inserts on the ischium. Distally the corpora cavernosa penis extend to the glans.

Microscopic anatomy. The membranous urethra is lined by transitional epithelium except over the colliculus seminalis on the dorsal wall at the neck of the bladder. This elevation is covered by low columnar epithelium, a continuation of that of the ducts of the ampullary and vesicular glands. The richly vascular lamina propria supports the urethral glands of Littré, which form a complete sheath around the urethra from the neck of the bladder to the junction with the penis (Fig. 13-51). These glands are groups of alveoli made up of secretory cells with oval basal nuclei and cytoplasm containing basophilic secretion granules. Their short ducts, lined by cuboidal epithelium, open separately into the lumen. A thick layer of striated muscle surrounds the gland layer, this in

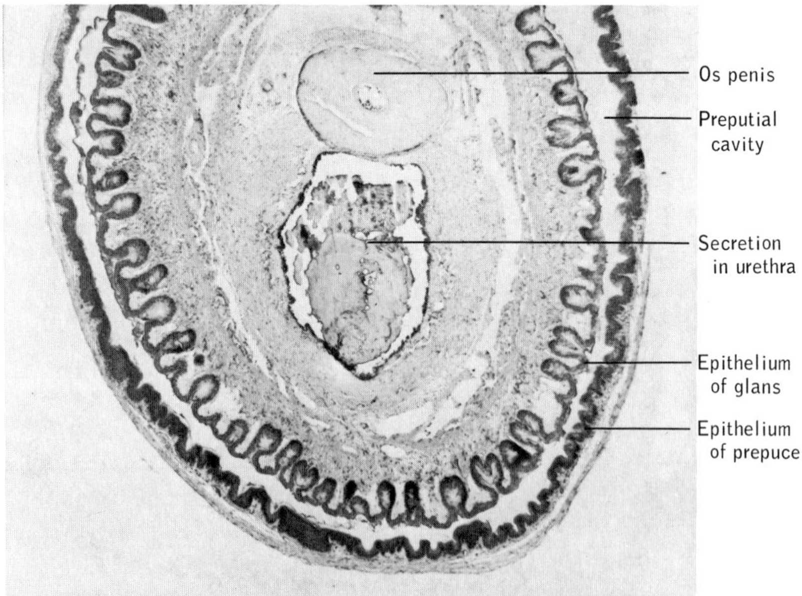

Fig. 13-50. Cross-section through the glans penis, preputial cavity, and prepuce (×40).

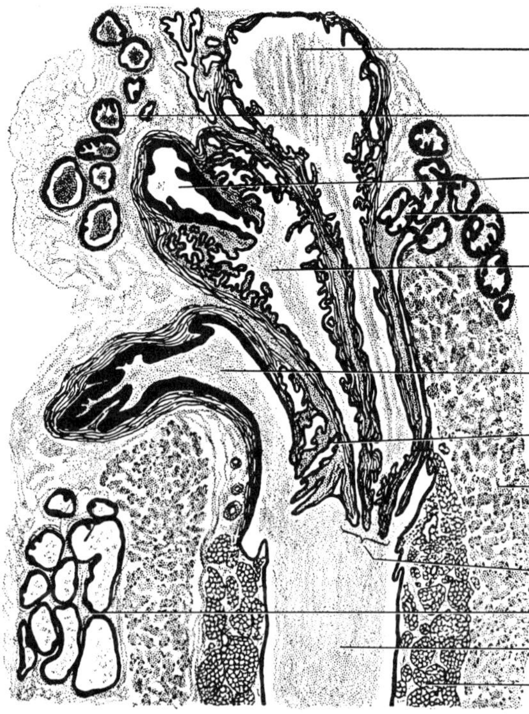

Fig. 13-51. Cephalic end of the male urethra. A composite drawing of three adjacent sagittal sections (×13).

turn being surrounded by connective tissue merging with that of adjacent structures.

The urethral diverticulum is lined also with transitional epithelium, its thickness depending on dilatation and the amount of secretion present. Glands similar to those in the membranous urethral wall open directly into the lumen by short ducts. The diverticulum is partially surrounded by an extension of the corpus cavernosus urethrae and its membrane composed of an inner circularly arranged smooth-muscle sheath and an outer sheath of fibrous connective tissue. Trabeculae from this muscular and fibrous membrane penetrate among the glands to form cavernous spaces lined with endothelium. When the spaces are distended with blood, groups of glands are widely separated; when the spaces are collapsed, the glandular tissue appears compact. The connective tissue of the outer sheath merges with the perimysium of the bulbocavernosus and ischiocavernosus muscles, which are composed of thick bundles of striated muscle.

Transitional epithelium lines the urethra from the diverticulum to near the tip of the penis, where it changes to stratified squamous at the orifice. Glands are absent and the fibrous lamina propria is continuous with the surrounding cavernous spaces of the corpora cavernosa penis. In the proximal penis, each corpus is surrounded by a tunica, but distally these disappear except for a narrow septum, and the cavernous spaces intercommunicate. The os penis is located in the connective tissue of the septum.

The glans is covered and the preputial cavity lined with hairless skin. Over the glans, the epidermis projects deeply into the dermis and cornified filiform papillae project from the pits thus formed (Fig. 13-50). A very vascular dermis fills the narrow spaces between the epidermal invaginations, and the subcutaneous connective tissue merges with that surrounding os penis and urethra. The epidermis of the prepuce has a thick surface layer of noncornified squamous cells and a width varied by the dermal papillae that project into it at frequent intervals (Fig. 13-50). The fibrous subcutaneous tissue contains smooth muscle fibers.

Accessory glands

Gross anatomy. The neck of the bladder and the anterior end of the urethra are surrounded by accessory glands and their ducts (Fig. 13-45, 13-51). The largest and most prominant are the paired, elongated, curved vesicular glands (the seminal vesicles of older literature), which are serrated medially and hooked at the narrowed tips. Each vesicular gland has a wide duct that enters the urethra with the ampulla of the vas deferens on an elevation, the colliculus seminalis, located on the dorsal wall of the urethra near the neck of the bladder (Fig. 13-51). Infrequently, the vas deferens and vesicular-gland duct join on one or both sides before entering the urethra.

The paired coagulating glands are attached to the lesser curvatures of the vesicular glands. They are less prominent and less opaque than the vesicular glands and their much folded mucosa is visible through their thin, somewhat transparent, walls. Each gland has two ducts entering the dorsal wall of the neck of the bladder anterior to the colliculus seminalis (Fig. 13-51).

The ampullary glands are groups of branched tubular glands with many short ducts that open directly into the wide vestibules of the ampullae (Fig. 13-51).

The dorsal and ventral prostates are the other glands of this region. The dorsal prostate has many ducts, some entering the urethra lateral to all other ducts. The ventral prostate has several ducts entering the ventral walls of the urethra.

The bulbourethral (Cowper's) glands are paired structures lateral to the junction of the membranous urethra and penis. The main body of the gland is at the side of the urethral diverticulum buried in the bulbocavernosus muscle, the tail is between the diverticulum and the ischiocavernosus muscle, and the duct enters the urethra immediately anterior to the diverticulum (Fig. 13-45).

The preputial glands are large dorsoventrally flattened leaf-shaped glands lying close together in the subcutaneous tissue near the end of the penis. Their ducts empty into the lateral wall of the preputial cavity.

Microscopic anatomy. The ampullary glands are groups of branched tubules lined by low columnar cells with large oval nuclei. The lamina propria is very thin and the mucous membrane is thrown into many deep longitudinal folds. The tubules are surrounded by a very thin layer of smooth muscle cells and are held together by a thin connective tissue membrane. In eosin-stained sections the secretion is deep red and homogeneous and tends to shrink away from the epithelial lining.

The vesicular gland has a large elongated internal cavity with medial alveolar outpocketings. The mucosa, especially on the medial side, is thrown into many fine intricate folds. The epithelium consists of tall columnar cells with distinct boundaries and large oval basal nuclei. The slightly basophilic cytoplasm contains dark secretion granules surrounded by lighter staining areas or halos. When the lumen is distended by secretion, the epithelial cells are low columnar and do not contain secretion granules. Smooth muscle fibers and a connective tissue sheath surround the gland. The secretion is intensely acidophilic in prepared sections and tends to appear as though cracked into parallel fissures.

The mucous membrane of the coagulating glands forms curved longitudinal folds, some projecting far into the lumen. The epithelium consists of a layer of columnar cells with round centrally located nuclei and acidophilic cytoplasm. Each gland usually has two ducts lined by low columnar cells with deeply staining nuclei and slightly basophilic cytoplasm. The ducts have a folded mucosa giving the lumen a wavy outline. The gland tubules are surrounded by a delicate layer of smooth muscle fibers, and the mass of tubules is contained within a connective tissue sheath which attaches it to the vesicular gland. The secretion is a homogeneous faintly acidophilic substance appearing cracked in stained sections.

The tubules of the dorsal prostate glands are structurally similar to those of the coagulating glands, although considerably narrower. The mucous membrane is folded in the secretory tubules but not in the ducts. The secretion also is similar to that of the coagulating gland except that "cracking" is rare. The gland tubules of the ventral prostate are lined by low columnar cells with spherical deeply staining nuclei and slightly basophilic cytoplasm. The mucosa is folded in inactive but not in distended tubules. The gland tubules have thin coats of smooth muscle and are held together and surrounded by a fibrous sheath. The secretion is slightly acidophilic and tends to mass into globules of varying size.

The bulbourethral gland is composed of body and tail. The body is partially covered by the bulbocavernosus muscle and the tail is surrounded by the ischiocavernosus muscle. A thin connective tissue membrane projects into the gland separating the alveoli and supporting the blood vessels. The tubules and alveoli of the body of the gland are lined by tall columnar cells of uneven height. The granular cytoplasm is slightly basophilic and very light staining, and the nuclei are small, darkly staining, and flattened against the cell bases. The basement membrane is well developed. The alveoli of the tail portion are lined by low columnar cells with granular dark-staining baso-

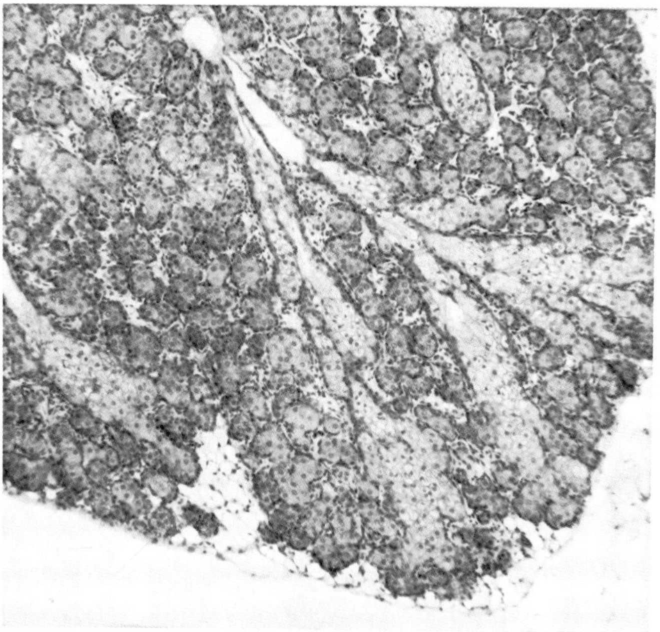

Fig. 13-52. Preputial gland, section (×100).

philic cytoplasm and round basal nuclei. Interspersed among these cells are small groups of more lightly staining cells similar to those of the body of the gland. The ducts of body and tail are lined by cuboidal epithelium.

The preputial glands are large sebaceous glands made up of groups of flat polyhedral cells with pale-staining nuclei, surrounded by a connective tissue capsule (Fig. 13-52). In an actively secreting cell, the major portion is filled with an oily secretion, and when the nucleus degenerates the dead secretion-packed cell is excreted. Each gland has a long duct lined with stratified squamous epithelium opening into the preputial cavity. Near the orifice the epithelial cells of the duct and the subcutaneous tissue around it usually contain some pigment (except in albino animals).

FEMALE GENITAL SYSTEM

The female genital system is composed of ovaries, oviducts, uterus, and vagina (Fig. 13-53). For a detailed description of the anatomy and development of ovary, oviduct, and uterus see Agduhr (1927).

Ovary

Gross anatomy. The ovaries are small paired spherical bodies functioning to produce mature female sex cells (ova) and sex hormones. They are located at the posterolateral poles of the kidneys, each attached by the mesovarium to the dorsal body wall and enclosed in a thin transparent elastic capsule or bursa. The periovarian space is shut off from the abdominal cavity except for a tiny tunnel-like channel through the bursa (Wimsatt and Waldo, 1945). Blood vessels and nerves supported in the mesovarium enter and leave the ovary through a stalk at the hilus where the periovarian space is interrupted. The surface of the ovary is smooth in prepuberal females but becomes nodular after sexual maturity because of the presence of follicles and corpora lutea. Conspicuous strain and age differences in ovarian size are largely the result of differences in numbers of follicles and corpora lutea (Fekete, 1946, 1953).

Microscopic anatomy. The capsule is a thin membrane composed of loose connective tissue, blood vessels, nerves, and a few smooth muscle fibers covered on inner and outer surfaces by mesothelium. Near the hilus of the ovary, adipose tissue is accumulated between the mesothelial layers to form the fat body (Fig. 13-54). The surface of the ovary is covered with a layer of cuboidal cells with large nuclei, small amounts of cytoplasm, and slightly rounded free surfaces. This "germinal" epithelium rests on a prominent basement membrane and follows the contours of the ovary, being thinner where follicles and corpora lutea bulge the surface (Fig. 13-56).

Two poorly defined areas, a central vascular medulla and a compact outer cortex, are visible in sections (Fig. 13-54). The large blood vessels that join the ovary at its hilus branch in the dense

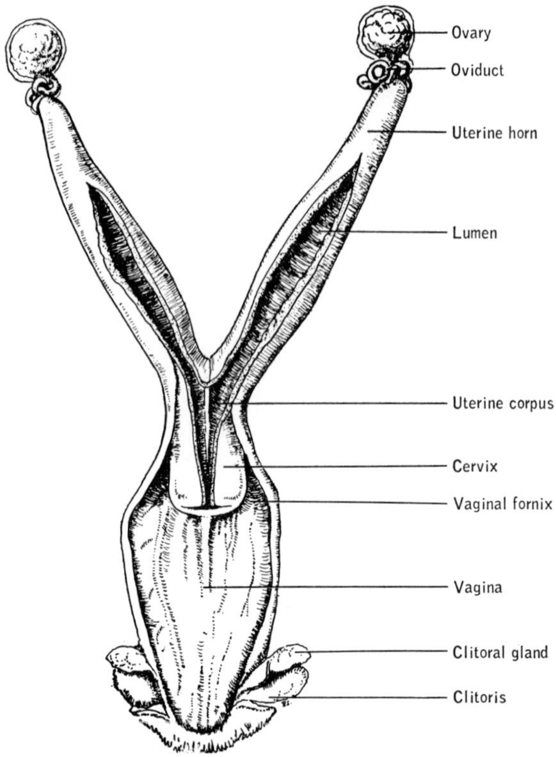

Fig. 13-53. Drawing of the female genital system (×2.7). Parts of ventral walls of the uterine horn, uterine body, and vagina have been cut away.

fibrous connective tissue stroma of the central medullary portion. Rudiments of the rete ovarii may persist in the medulla near the hilus as blind tubules or cords of epithelial cells, and cysts with this derivation are especially large and numerous in the ovaries of C57L mice (Fekete, 1953). Other mesonephric remnants may persist as the epoophoron, a small mass of cells in the meso-

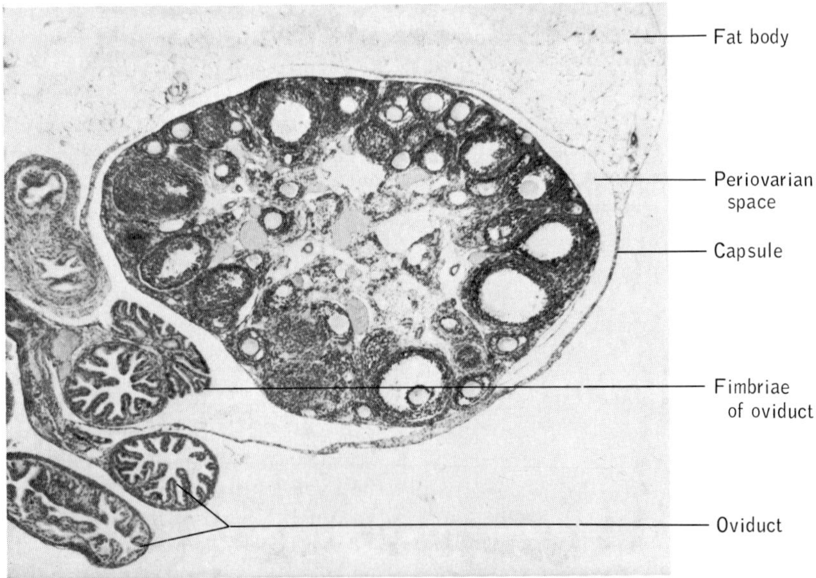

Fig. 13-54. Ovary, ovarian bursa, fat body, fimbriated end of the oviduct within the periovarian space, and coils of the oviduct, section (×40).

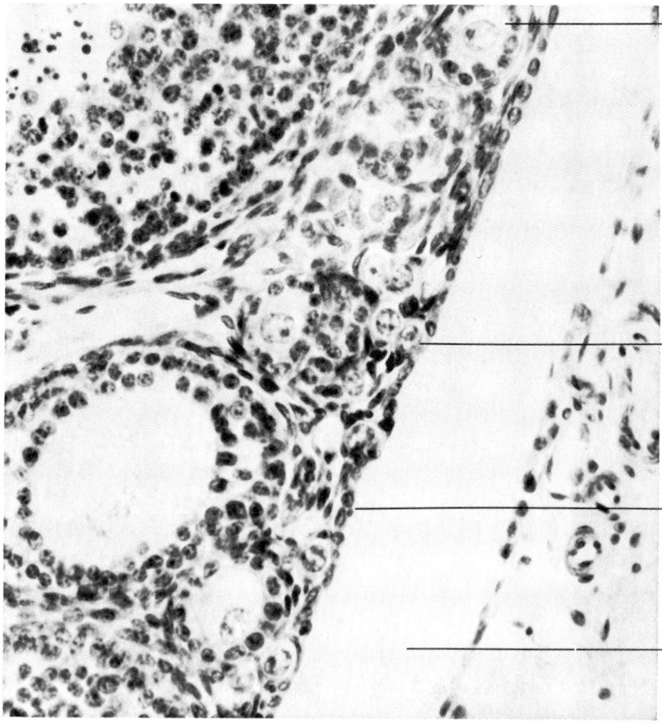

Fig. 13-55. Section of the ovary to show clusters of primordial follicles under the germinal epithelium (×400).

varium. The cortex is composed of strands of loose fibrous stroma separating developing follicles and corpora lutea and supporting groups of interstitial cells and blood vessels. The primordial follicles, consisting of single layers of flat cells surrounding oocytes, are massed beneath the germinal epithelium in the tunica albuginea (Fig. 13-55). The oocyte or ovum is a clear spherical cell with a vesicular nucleus containing small chromatin granules and a prominent nucleolus. Although a follicle usually contains a single ovum, polyovular follicles are not uncommon in young mice (Kent, 1960) and follicles with two to five ova are numerous in ovaries of strain C58 mice (Fekete, 1950). The number of primordial follicles is gradually depleted by ovulation and atresia, and the tunica albuginea in the senile ovary is dense and conspicuous. Depletion of follicles and oocytes proceeds at different rates in mice of different strains (Jones and Krohn, 1961).

During maturation, which occurs simultaneously in several follicles, both oocyte and follicle enlarge, and the single layer of flat cells is replaced by a many-layered stratum granulosum, made up of small basophilic follicular or granulosa cells. As the follicle grows larger it moves centrally and acquires a liquor-filled cavity or antrum and an encapsulating sheath derived from stromal cells. Later two layers, a vascular theca interna and a fibrous theca externa differentiate in the sheath. Blood and lymph vessels penetrate the externa to form a plexus in the interna, but the granulosa layer is avascular. The oocyte loses its direct connection with granulosa cells through deposition around it of a zone of acellular material, the zona pellucida. A mature (Graafian) follicle bulges into the periovarian space, separated from it by the stretched theca and germinal epithelium. The oocyte and first polar body surrounded by the zona pellucida and a cluster of follicle cells, the cumulus oophorus, retains a slim attachment to the thinning granulosa layers. The theca interna cells are hypertrophied and vacuolated. At ovulation, the germinal epithelium ruptures and the oocyte with its cluster of cells is released into the periovarian space. For a discussion of the processes of oogenesis and ovulation see Chap. 7, 11.

Immediately on release of ova, the follicles begin to change into corpora lutea (Fig. 13-56). The ruptured epithelium heals, granulosa cells hypertrophy, and connective tissue and capillaries

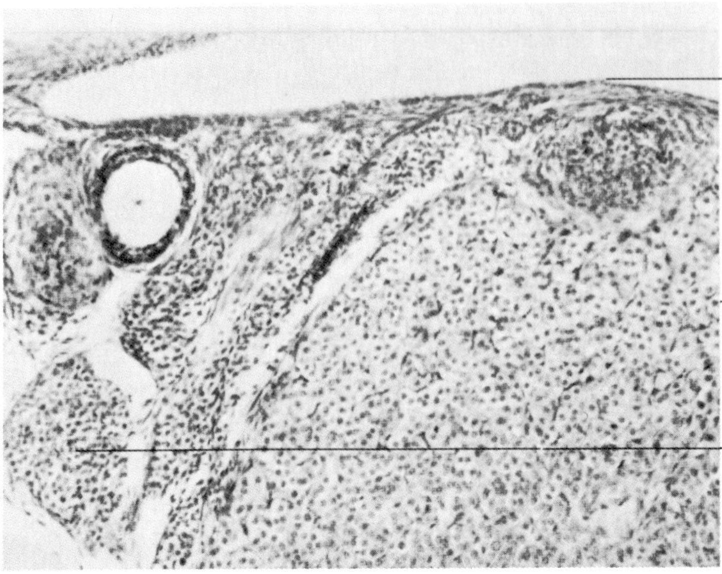

Fig. 13-56. Section of the ovary through a corpus luteum (×150).

grow inward from the theca interna, obliterating the antrum. The granulosa cell of the mature follicle is small and basophilic with a large oval nucleus, whereas the mature lutein cell that develops from it is large with clear vacuolated eosinophilic cytoplasm and a large vesicular nucleus. Several follicles mature at each estrus, so several generations of corpora lutea may be present in an active ovary. The cells of newly formed corpora are small and slightly basophilic, and those of older bodies are large and eosinophilic and are arranged in cords around branching sinusoids. Many of the ova and follicles do not mature but undergo atresia during one or another stage of development. Atretic follicles with pycnotic cells, fragmented ova, and remnants of zona pellucida are scattered throughout the cortical stroma.

The interstitial cells are similar morphologically to mature lutein cells and are found in irregular groups among the fibrous stromal cells. Their origin is in doubt and they may derive from

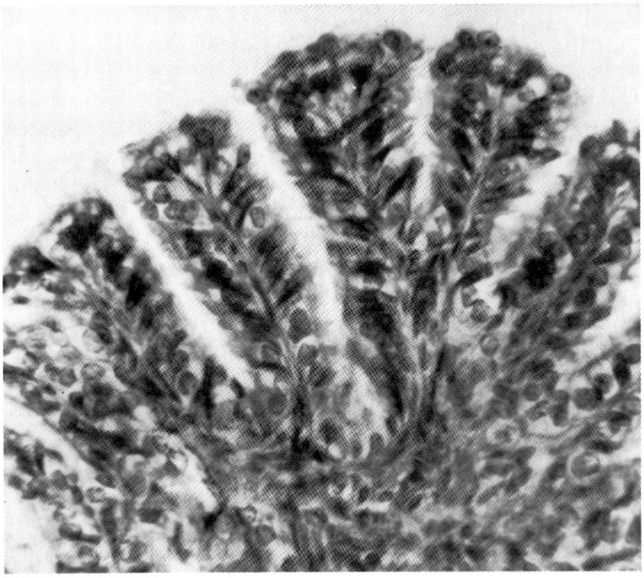

Fig. 13-57. Fimbriae of oviduct, section (×400).

several sources: hypertrophied stromal cells, remnants of corpora lutea, and theca interna cells (Fig. 13-56). The interstitial cells are believed to be a source of androgens as well as of estrogens; other secretory cells of the ovary are those of corpus luteum and theca interna and possibly granulosa cells. In ovaries of old mice large cells containing a brown pigment are frequent components, their number, size, and clumping patterns being strain characteristics (Fekete, 1946, 1953). Histochemical observations indicate that the pigment is identical to that occurring in cells of aged adrenals (Deane and Fawcett, 1952).

Oviduct

Gross anatomy. The oviduct (uterine or fallopian tube) is a long (1.8 cm) narrow coiled tube connecting the periovarian space with the uterine horn. It is suspended from the dorsal body wall by the mesotubarium, a double-walled membrane continuous with mesovarium, ovarian bursa, and the mesometrium of the uterus. There are three segments of oviduct: the widened ampulla near the ovarian bursa; the long, narrow, tightly coiled isthmus; and the intramural portion within the uterine wall. The ampulla opens through an infundibulum into the periovarian space. Fringe-like processes, the fimbriae, surround the opening and extend into the periovarian space (Fig. 13-54). The oviduct joins the uterus on its dorsolateral wall slightly posterior to its rounded anterior end, passes obliquely through the wall, and opens at the tip of a projection, the colliculus tubarius. Kuhlmann (1965, personal communication) describes the colliculus as a projection varying in length from 0.2 to 0.6 mm surrounded at the base by a furrow and with circular folds in its wall. He finds strain differences in the depth of the furrow, the number of folds, and the shape and length of the projection. In strain BALB/c females the colliculus is mound-shaped with a normal length of about 0.3 mm, whereas in strain 129 females, it is cylindrical or conical, has two or more encircling folds, and has a normal length of 0.4 mm; in abnormal situations the colliculus may exceed 1 mm in length.

Microscopic anatomy. The mucosa is elaborately folded in the ampulla of the oviduct and is in four to six longitudinal folds in the intramural and colliculus tubarius portions. The amount and direction of folding varies in different portions of the isthmus (Agduhr, 1927). The epithelium lining the ampulla and infundibulum and covering the fimbriae is ciliated columnar (Fig. 13-57). The cells are tall with oval centrally located nuclei, eosinophilic cytoplasm, and long motile cilia. Scattered among these cells are club-shaped nonciliated cells protruding into the lumen. The isthmus is lined with pseudostratified and low columnar epithelium containing an occasional ciliated cell (Fig. 13-58), and the intramural portion is lined with simple columnar epithelium. The lamina propria consists of a thin layer of connective tissue containing a few elastic and smooth muscle fibers. The muscularis, of circularly arranged smooth muscle fibers, is thin in the isthmus and becomes progressively thicker distally. In the intramural and colliculus tubarius portions, the muscularis of the oviduct merges with the circular muscle layer of the uterine wall (Fig. 13-59).

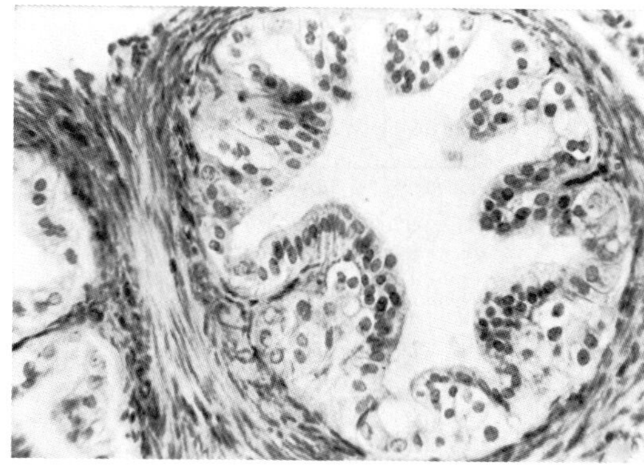

Fig. 13-58. Isthmus region of the oviduct, cross-section (×200).

Uterus

Gross anatomy. The uterus is a Y-shaped tubular structure divided into two lateral horns (cornua) and a single median body (corpus) (Fig. 13-53). The uterine horns extend posteromedially from the oviducts to a position dorsal to the urinary bladder where they unite to form the corpus. The horns are suspended from the dorsal body wall by the heavy broad ligaments or mesometria through which blood and lymph vessels and nerves course at regular intervals. The body of the uterus consists of a cranial portion, containing two cavities separated by a median septum, and a caudal undivided portion, the neck or cervix, projecting into the cavity of the vagina (Fig. 13-53). The walls of the cervix and vagina are continuous dorsally and ventrally but not laterally where the lumen of the vagina extends anteriorly into deep fornices. In the rat the median septum extends the length of the corpus and the two lateral chambers open separately into the vaginal lumen.

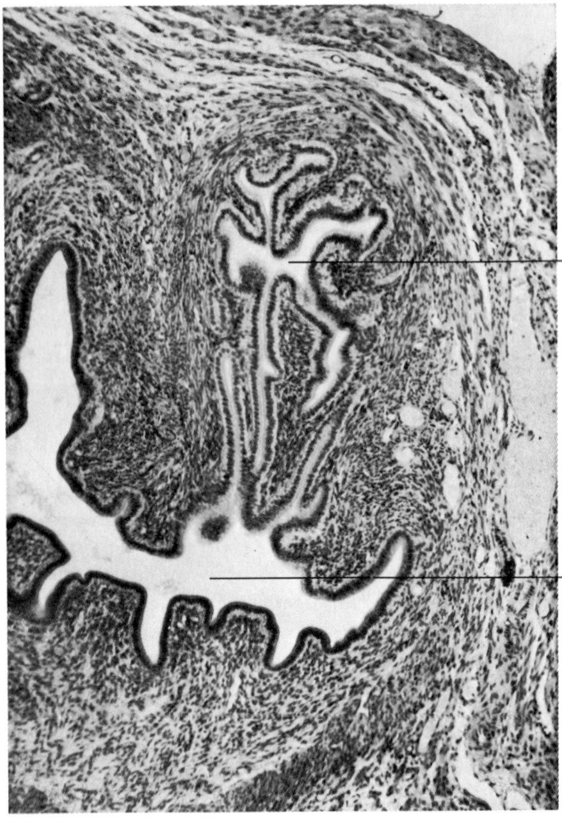

Fig. 13-59. Section through the intramural oviduct, colliculus tubarius, and uterus ($\times$100).

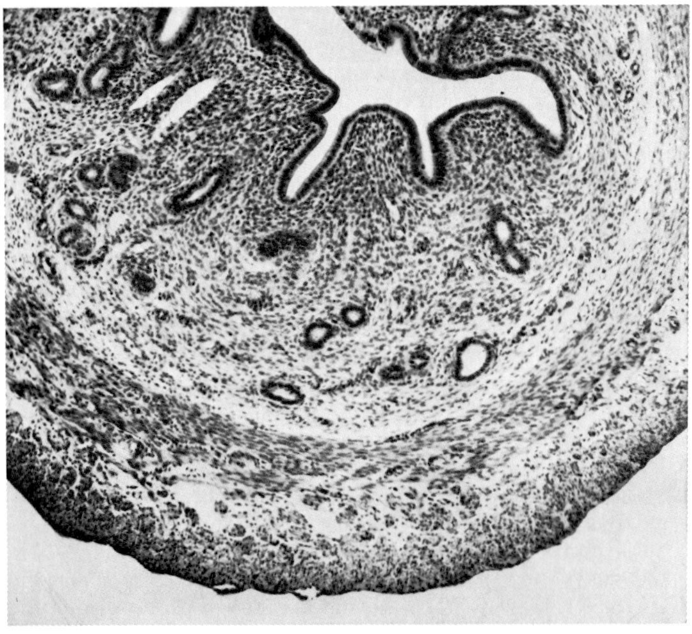

Fig. 13-60. Uterine horn, cross-section ($\times$100).

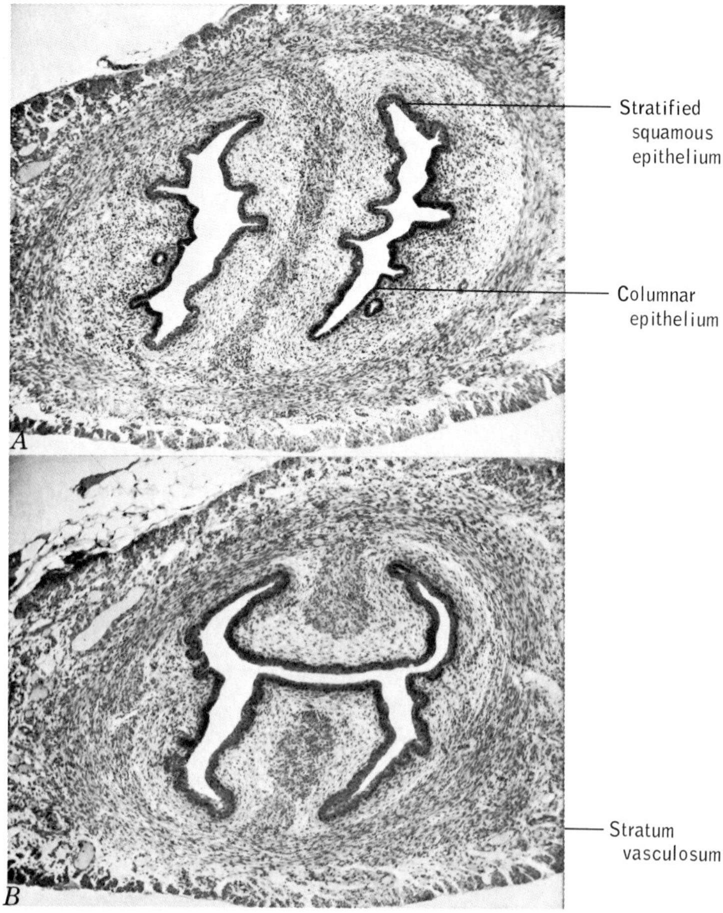

Fig. 13-61. Uterine corpus, cross-sections through two levels (×50). *A*. Cavity divided by a median partition. *B*. Undivided cavity caudal to level of *A*.

Microscopic anatomy. The mucosa of the uterine horns, called the endometrium in nonpregnant females, is elevated into transverse folds and is well supplied with blood vessels and nerves (Fig. 13-60). The epithelium is simple columnar extended into branched tubular glands projecting into the endometrial stroma, which is composed of reticular tissue containing many small polyhedral cells and many lymphocytes. The muscle layer, or myometrium, consists of inner circular and outer longitudinal layers of smooth muscle with a layer of very vascular loose connective tissue, the stratum vasculosum, between. The outer covering is serosa continuous with the mesometrium.

As the uterine horns come together in the midline, their medial walls lose serosa, stratum vasculosum, and some muscle fibers and fuse to form a partition that extends through the lumen almost to the level of the vaginal fornices. Some circular muscle fibers are retained in the center of the partition (Fig. 13-61*A*). The lateral cavities are lined with epithelium that changes gradually from simple columnar to stratified squamous. The most anterior portions of the corpus are lined entirely by epithelium similar to that of uterine horns, but elsewhere, especially on the medial walls, patches of stratified squamous epithelium are interspersed (Fig. 13-61*A*). The undivided cavity of the corpus (Fig. 13-61*B*) and the cervical canal are lined by stratified squamous epithelium continuous with that of the vagina. The lamina propria of the uterine corpus is less cellular and more fibrous than that of the horns

Circular and longitudinal smooth muscle fibers and serosa complete the wall. The wall of the cranial two-thirds of the cervix contains circularly arranged smooth muscle; the wall of the caudal one-third contains collagenous fiber bundles that become loose and widely separated during pregnancy (Leppi, 1964).

Vagina and clitoris

Gross anatomy. The short, thick, muscular vagina extends from the uterine corpus and cervix to an external opening anterior to the anus on the ventral body surface. The vagina is loosely attached to the rectum dorsally and to the urethra ventrally. On the anterior wall of the vaginal opening is a small ventrally extending elevation, the clitoris, covered by skin and hair on its anterior and lateral surfaces. The urethra opens near the tip of the clitoris, within a shallow depression, the clitoral fossa. Small pear-shaped clitoral glands, homologous with the preputial glands of the male, are embedded anterolaterally in the subcutaneous connective tissue; a single duct from each extends to an opening in the lateral wall of the clitoral fossa.

Microscopic anatomy. The wide dorsoventrally flattened lumen of the vagina is lined with stratified squamous epithelium, which undergoes marked changes in number of layers during the estrous cycle (Chap. 11). The mucosa is folded into longitudinal elevations and contains no glands. The lamina propria is fibrous and the muscularis thin, the inner circular and outer longitudinal layers being intermingled with considerable connective tissue. The outer covering is adventitia continuous with the connective tissues

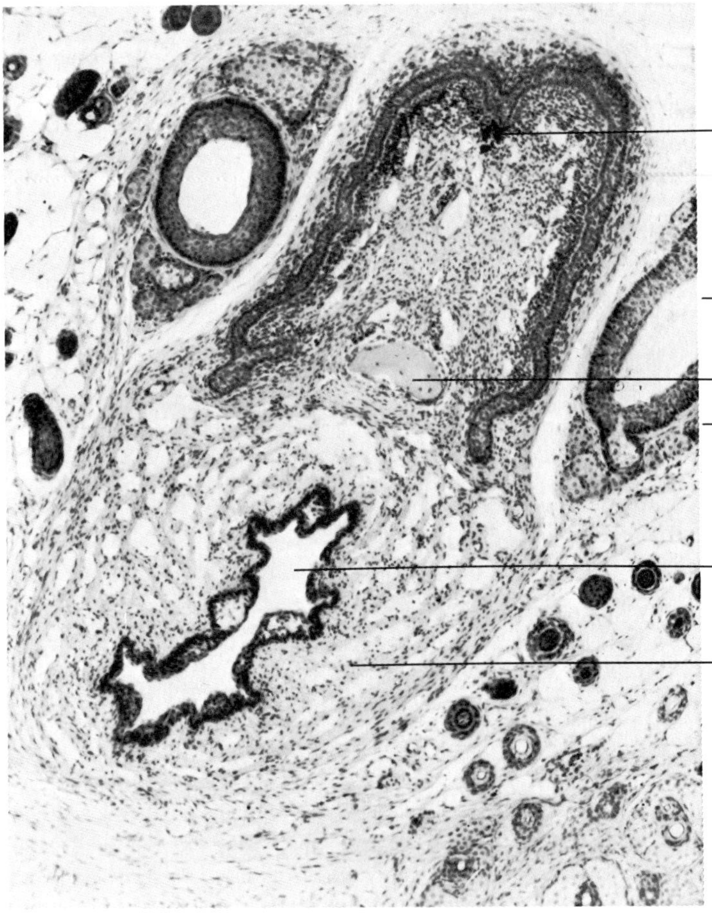

Fig. 13-62. Section through the clitoris to show erectile tissue surrounding the urethra, the clitoral gland and duct, and the small os clitoris ($\times 100$).

surrounding rectum and urethra. At the vaginal opening, the epithelium is stratified squamous cornified, and the muscularis contains some striated muscle fibers.

The posterior face of the clitoral elevation is covered with vaginal epithelium and the tip, sides, and anterior face are covered by skin with hair. The paired clitoral glands, one on either side at the base of the elevation, are sebaceous type consisting of groups of large, pale-staining, often-vacuolated cells, surrounded by a thin connective tissue capsule. A single large hair follicle occupies the center of each gland. The excretory ducts are lined with stratified squamous epithelium continuous with that of the clitoral fossa.

Within the clitoris, the lining of the urethra is transitional epithelium except at the orifice where it is stratified squamous. A small group of glands, different in structure from the urethral glands of more anterior regions, is located in the lamina propria anterior to the lumen. The alveoli are very small and composed of four or five pyramidal cells with large central nuclei and acidophilic cytoplasm. Their short ducts, which open directly into the urethral lumen, are lined by stratified columnar epithelium. Erectile tissue, comparable to that of the penis but with a finer fibrous network, surrounds the urethra near the tip of the clitoris, and the connective tissue on the anterior face of the clitoris encloses a small bone, homologous with the os penis (Fig. 13-62). There are no muscle layers surrounding the urethra within the clitoris. The connective tissue of the cavernous spaces merges with the surrounding subcutaneous tissue and dermis.

SUMMARY

The descriptions of gross and microscopic anatomy in this chapter are based on our observations as well as on those recorded in the literature. In many instances where information on development of organs or regions of the body was available, we included brief summaries or references to embryological and developmental studies. We also included some observations on strain, sex, and age differences in morphology. Under skeleton, bones were listed in regional groups, but no structural details were given. Under the circulatory system heading, only the heart, large arteries and veins, and their principal branches and tributaries were described and no attempt was made to trace or name the smaller arteries, veins, and capillary networks. Descriptions of nervous and muscular systems and of blood and blood-forming tissues were omitted entirely.

LITERATURE CITED

Agduhr, E. 1927. Studies on the structure and development of the bursa ovarica and the tuba uterina in the mouse. Acta Zool. 1:1–133.

Albert, S., and R. M. Johnson. 1960. Lymph node morphology and metabolism in mammary tumor-susceptible and -resistant mice. Cancer Res. 20:246–250.

Andrew, W., and N. V. Andrew. 1942. Senile involution of the thyroid gland. Amer. J. Pathol. 18:849–863.

Andrew, W., and N. V. Andrew. 1957. An age involution in the small intestine of the mouse. J. Gerontol. 12:136–149.

Arnesen, K. 1958. The secretory apparatus in the thymus of mice. Acta Pathol. Microbiol. Scand. 43:339–349.

Auerbach, R. 1960. Morphogenetic interactions in the development of the mouse thymus gland. Develop. Biol. 2:271–284.

Auerbach, R. 1964. Experimental analysis of mouse thymus and spleen morphogenesis, p. 95–111. *In* R. A. Good and A. E. Gabrielsen [ed.] The thymus in immunobiology. Harper & Row, New York.

Baer, P. N., and J. E. Lieberman. 1959. Observations on some genetic characteristics of the periodontium in three strains of inbred mice. Oral Surg. 12:820–829.

Bascom, K. F., and H. L. Osterud. 1925. Quantitative studies of the testicle. II. Pattern and total tubule length in the testicles of certain common mammals. Anat. Rec. 31:159–169.

Bateman, N. 1954. Bone growth: a study of the grey-lethal and microphthalmic mutants of the mouse. J. Anat. 88:212–262.

Billingham, R. E., and W. K. Silvers. 1960. The melanocytes of mammals. Quart. Rev. Biol. 35: 1–40.

Blumenthal, H. T. 1955. Aging processes in the endocrine glands of various strains of normal mice. J. Gerontol. 12:253–267.

Borghese, E. 1952. Foyer d'hématopöièse dans la glande surrénale foetale de *Mus musculus*. Acta Anat. 16:54–71.

Borodach, G. N., and W. Montagna. 1956. Fat in skin of the mouse during cycles of hair growth. J. Invest. Dermatol. 26:229–232.

Braden, A. W. H. 1958. Strain differences in the

morphology of the gametes of the mouse. Austral. J. Biol. Sci. 12:65–71.

Brookreson, A. D., and C. W. Turner. 1959. Normal growth of mammary gland in pregnant and lactating mice. Proc. Soc. Exp. Biol. Med. 102: 744–745.

Browder, S. 1942. Factors influencing lung lobation in the mouse. I. Genetic factors: a preliminary report. Anat. Rec. 83:31–39.

Brown, A. L. 1931. An analysis of the developing metanephros in mouse embryos with abnormal kidneys. Amer. J. Anat. 47:117–172.

Bunker, L. E., Jr. 1959. Hepatic fusion, a new gene in linkage group I of the mouse. J. Hered. 50:40–44.

Carter, T. C. 1951. The genetics of luxate mice. I. Morphological abnormalities of heterozygotes and homozygotes. J. Genet. 50:277–299.

Carter, T. C. 1954. The genetics of luxate mice. IV. Embryology. J. Genet. 52:1–35.

Chase, H. B. 1954. Growth of the hair. Physiol. Rev. 34:113–126.

Chase, H. B., and G. J. Eaton. 1959. The growth of hair follicles in waves. Ann. N.Y. Acad. Sci. 83:365–368.

Chase, H. B., H. Rauch, and V. W. Smith. 1951. Critical stages of hair development and pigmentation in the mouse. Physiol. Zool. 24: 1–8.

Chen, J. M. 1952. Studies on the morphogenesis of the mouse sternum. I. Normal embryonic development. J. Anat. 86:373–386.

Chiquoine, A. D. 1958. The identification and electron microscopy of myoepithelial cells in the Harderian gland. Anat. Rec. 132:569–584.

Clark, S. L., Jr. 1963. The thymus in mice of strain 129/J, studied with the electron microscope. Amer. J. Anat. 112:1–34.

Cohn, S. A. 1955. Histochemical observations on the Harderian gland of the albino mouse. J. Histochem. Cytochem. 3:342–353.

Cohn, S. A. 1957. Development of the molar teeth in the albino mouse. Amer. J. Anat. 101:295–319.

Cook, M. J. 1965. The anatomy of the laboratory mouse. Academic Press, London. 143 p.

Coupland, R. E. 1960. The post-natal distribution of the abdominal chromaffin tissue in the guinea-pig, mouse and white rat. J. Anat. 94:244–256.

Cowie, A. T., and S. J. Folley. 1961. The mammary gland and lactation, p. 590–642. *In* W. C. Young [ed.] Sex and internal secretions, 3rd ed. Vol. I. Williams and Wilkins, Baltimore.

Crabtree, C. 1941. The structure of Bowman's capsule as an index of age and sex variations in normal mice. Anat. Rec. 79:395–413.

Crelin, E. S. 1960. The development of bony pelvic sexual dimorphism in mice. Ann. N.Y. Acad. Sci. 84:479–512.

Danforth, C. H., and E. Center. 1953. Development and genetics of a sex-influenced trait in the livers of mice. Proc. Nat. Acad. Sci. 39:811–817.

Davidson, P., and M. H. Hardy. 1952. The development of mouse vibrissae *in vivo* and *in vitro*. J. Anat. 86:342–356.

Deane, H. W., and D. W. Fawcett. 1952. Pigmented interstitial cells showing "brown degeneration" in the ovaries of old mice. Anat. Rec. 113:239–245.

Delost, P., and P. Chirvan-Nia. 1958. Différences raciales dans l'involution de la zone X surrénalienne chez la souris adulte vierge. Compt. Rend. Soc. Biol. 152:453–455.

Dry, F. W. 1926. The coat of the mouse (*Mus musculus*). J. Genet. 16:287–340.

Dun, R. B. 1958. Growth of the mouse coat. VI. Distribution and number of vibrissae in the house mouse. Austral. J. Biol. Sci. 11:95–105.

Dunn, T. B. 1944. Ciliated cells of the thyroid of the mouse. J. Nat. Cancer Inst. 4:555–557.

Dunn, T. B. 1949*a*. Some observations on the normal and pathological anatomy of the kidney of the mouse. J. Nat. Cancer Inst. 9:285–301.

Dunn, T. B. 1949*b*. Melanoblasts in the stroma of the parathyroid glands of strain C58 mice. J. Nat. Cancer Inst. 10:725–733.

Dunn, T. B. 1954. Normal and pathologic anatomy of the reticular tissue in laboratory mice, with a classification and discussion of neoplasms. J. Nat. Cancer Inst. 14:1281–1434.

Dunn, T. B., and A. Kessel. 1945. Paneth cells in carcinomas of the small intestine in a mouse and in a rat. J. Nat. Cancer Inst. 6:113–118.

Engeset, A., and E. Tjötta. 1960. Lymphatic pathways from the tail in rats and mice. Cancer Res. 20:613–614.

Fawcett, D. W. 1952. A comparison of the histological organization and cytochemical reactions of brown and white adipose tissue. J. Morphol. 90:363–406.

Fekete, E. 1941. Histology, p. 89–167. *In* G. D. Snell [ed.] Biology of the laboratory mouse. Blakiston, Philadelphia.

Fekete, E. 1946. A comparative study of the ovaries of virgin mice of the dba and C57Black strains. Cancer Res. 6:263–269.

Fekete, E. 1950. Polyovular follicles in the C58 strain of mice. Anat. Rec. 108:699–707.

Fekete, E. 1953. A morphological study of the ovaries of virgin mice of eight inbred strains showing quantitative differences in their hormone producing components. Anat. Rec. 117:93–114.

Ferguson, D. J. 1956. Endocrine control of mammary glands in C_3H mice. Surgery 39:30–36.

Figge, F. H. J., and R. H. Davidheiser. 1957. Porphyrin synthesis by mouse Harderian gland extracts: sex, age, and strain variation. Proc. Soc. Exp. Biol. Med. 96:437–439.

Forsthoefel, P. F. 1959. The embryological development of the skeletal effects of the luxoid gene in the mouse, including its interactions with the luxate gene. J. Morphol. 104:89–142.

Forsthoefel, P. F. 1963. Observation on the sequence of blastemal condensations in the limbs of the mouse embryo. Anat. Rec. 147:129–137.

Fortuyn, A. B. D. 1933. On the age at which the megakaryocytes disappear in the liver of the mouse. Peking Nat. Hist. Bull. 7:227.

Foster, C. L. 1943. Studies on the parathyroid of the mouse. I. The cytology of the normal gland in relation to its secretory activity. J. Endocrinol. 3:244–253.

Froud, M. D. 1959. Studies on the arterial system of three inbred strains of mice. J. Morphol. 104:441–478.

Gaunt, W. A. 1956. The development of enamel and dentine on the molars of the mouse, with an account of the enamel-free areas. Acta Anat. 28:111–134.

Gaunt, W. A. 1961. The presence of apical pits on the lower cheek teeth of the mouse. Acta Anat. 44:146–158.

Gibbs, H. F. 1941. A study of the post-natal development of the skin and hair of the mouse. Anat. Rec. 80:61–81.

Gorbman, A. 1947. Functional and morphological properties in the thyroid gland, ultimobranchial body, and persisting ductus pharyngiobranchialis IV of an adult mouse. Anat. Rec. 98:93–101.

Gorbman, A., and H. A. Bern. 1962. A textbook of comparative endocrinology. Wiley, New York. 468 p.

Green, E. L. 1962. Quantitative genetics of skeletal variations in the mouse. II. Crosses between four inbred strains. Genetics 47:1085–1096.

Green, E. L., and C. W. McNutt. 1941. Bifurcated xiphisternum and its relationship with short ears in the house mouse. J. Hered. 32:94–96.

Greene, E. C. 1955. The anatomy of the rat. Hafner, New York. 370 p.

Grewal, M. S. 1962. The development of an inherited tooth defect in the mouse. J. Embryol. Exp. Morphol. 10:202–211.

Grüneberg, H. 1952. The genetics of the mouse, 2nd ed. Nijhoff, The Hague. 650 p.

Grüneberg, H. 1963. The pathology of development. Wiley, New York. 309 p.

Halmi, N. S., and W. D. Gude. 1954. The morphogenesis of pituitary tumors induced by radiothyroidectomy in the mouse and the effects of their transplantation on the pituitary body of the host. Amer. J. Pathol. 30:403–420.

Hardy, M. H. 1949. The development of mouse hair *in vitro* with some observations on pigmentation. J. Anat. 83:364–384.

Hummel, K. P. 1954. Aplasia of the kidney in mice of strain C58. Anat. Rec. 118:391. (Abstr.)

Hummel, K. P. 1958. Accessory adrenal cortical nodules in the mouse. Anat. Rec. 132:281–295.

Hunt, R. D. 1963. Aberrant thyroid tissue in the mouse. Science 141:1054–1055.

Jacobs, B. B. 1958. Variations in thyroid morphology of mice. Proc. Soc. Exp. Biol. Med. 97: 115–118.

Job, T. T. 1915. The adult anatomy of the lymphatic system in the common rat (*Epimys norvegicus*). Anat. Rec. 9:447–458.

Jones, E. C., and P. L. Krohn. 1961. The relationships between age, numbers of oocytes and fertility in virgin and multiparous mice. J. Endocrinol. 21:469–495.

Jones, I. C. 1950. The effect of hypophysectomy on the adrenal cortex of the immature mouse. Amer. J. Anat. 86:371–403.

Karrer, H. E. 1956. The ultrastructure of mouse lung. General architecture of capillary and alveolar walls. J. Biophys. Biochem. Cytol. 2:241–252.

Karrer, H. E. 1958. The ultrastructure of mouse lung: the alveolar macrophage. J. Biophys. Biochem. Cytol. 4:693–700.

Kelemen, G. 1953. Nonexperimental nasal and paranasal pathology in hereditarily obese mice. Arch. Otolaryngol. 57:143–151.

Kelemen, G., and F. Sargent. 1946. Nonexperimental pathologic nasal findings in laboratory rats. Arch. Otolaryngol. 44:24–42.

Kelsall, M. A. 1946. Number of Peyer's patches in mice belonging to high and low mammary tumor strains. Proc. Soc. Exp. Biol. Med. 61:423–424.

Kent, H. A., Jr. 1960. Polyovular follicles and multinucleate ova in the ovaries of young mice. Anat. Rec. 137:521–524.

Kerr, T. 1946. The development of the pituitary

of the laboratory mouse. Quart. J. Microscop. Sci. 87:3–29.

Kutuzov, H., and H. Sicher. 1953. Comparative anatomy of the mucosa of the tongue and the palate of the laboratory mouse. Anat. Rec. 116:409–425.

Lee, Y. B., H. Elias, and I. Davidsohn. 1960. Vascular pattern in the liver of the mouse. Proc. Anim. Care Panel 10:25–32.

Leppi, T. J. 1964. A study of the uterine cervix of the mouse. Anat. Rec. 150:51–66.

Liebelt, R. A. 1959. Postnatal development of two types of fat depots in the NH and CBA inbred strains of mice. Amer. J. Anat. 105:197–218.

Little, C. C., and H. MacDonald. 1945. Abnormalities of the mammae in the house mouse. J. Hered. 36:285–288.

Low, F. N. 1953. The pulmonary alveolar epithelium of laboratory mammals and man. Anat. Rec. 117:241–264.

Mann, S. J. 1962. Prenatal formation of hair follicle types. Anat. Rec. 144:135–142.

Markert, C. L., and W. K. Silvers. 1956. The effects of genotype and cell environment on melanoblast differentiation in the house mouse. Genetics 41:429–450.

McPhail, M. K., and H. C. Read. 1942. The mouse adrenal. I. Development, degeneration and regeneration of the X-zone. Anat. Rec. 84:51–73.

Melaragno, H. P., and W. Montagna. 1953. The tactile hair follicles in the mouse. Anat. Rec. 115:129–150.

Miller, R. A. 1950. Cytological phenomena associated with experimental alterations of secretory activity in the adrenal cortex of mice. Amer. J. Anat. 86:405–438.

Montagna, W., and H. B. Chase. 1950. Redifferentiation of sebaceous glands in the mouse after total extirpation with methylcholanthrene. Anat. Rec. 107:83–91.

Munger, B. L. 1958. A light and electron microscopic study of cellular differentiation in the pancreatic islets of the mouse. Amer. J. Anat. 103:275–312.

Nandi, S. 1958. Endocrine control of mammary-gland development and function in the C3H/HeCrgl mouse. J. Nat. Cancer Inst. 21:1039–1063.

Rauch, H. 1952. Strain differences in liver patterns in mice. Genetics 37:617. (Abstr.)

Raynaud, A. 1961. Recent studies on the morphogenesis of the mammary gland of the mouse, p. 30–42. *In* S. K. Kon and A. T. Cowie [ed.] Milk: the mammary gland and its secretion. Vol. I. Academic Press, London.

Richardson, F. L. 1951. Further studies on the mammary gland development in male mice at nine weeks of age. Anat. Rec. 111:669–694.

Richardson, F. L. 1953. The mammary gland development in normal and castrate male mice at nine weeks of age. Anat. Rec. 117:449–465.

Richardson, F. L., and G. Hall. 1960. Mammary tumors and mammary-gland development in hybrid mice treated with diethylstilbestrol for varying periods. J. Nat. Cancer Inst. 25:1023–1039.

Richardson, F. L., and K. P. Hummel. 1959. Mammary tumors and mammary-gland development in virgin mice of strains C3H, RIII, and their F_1 hybrids. J. Nat. Cancer Inst. 23:91–107.

Richardson, F. L., and B. Pearson. 1954. Alkaline phosphatase activity during carcinogenesis of mammary tumors in mice implanted with stilbestrol pellets. J. Nat. Cancer Inst. 14:1123–1135.

Russell, E. S. 1946. A quantitative histological study of the pigment found in the coat-color mutants of the house mouse. I. Variable attributes of the pigment granules. Genetics 31:327–346.

Rytand, D. A. 1938. The number and size of mammalian glomeruli as related to kidney and body weight, with methods for their enumeration and measurement. Amer. J. Anat. 62:507–520.

Santisteban, G. A. 1960. The growth and involution of lymphatic tissue and its interrelationships to aging and growth of the adrenal glands and sex organs in CBA mice. Anat. Rec. 136:117–126.

Sharma, K. N. 1960. Genetics of gametes. IV. The phenotype of mouse spermatozoa in four inbred strains and their F_1 crosses. Proc. Roy. Soc. Edinb. B 68:54–71.

Sidman, R. L., and D. W. Fawcett. 1954. The effect of peripheral nerve section on some metabolic responses of brown adipose tissue in mice. Anat. Rec. 118:487–507.

Smith, C. 1964. The microscopic anatomy of the thymus, p. 71–84. *In* R. A. Good and A. E. Gabrielsen [ed.] The thymus in immunobiology. Harper & Row, New York.

Smith, C., and C. P. Clifford. 1962. Histochemical study of aberrant parathyroid glands associated with the thymus of the mouse. Anat. Rec. 143:229–238.

Smith, C., and B. K. Hénon. 1959. Histological

and histochemical study of high endothelium of post-capillary veins of the lymph node. Anat. Rec. 135:207–214.

Smith, C., and L. N. Ireland. 1941. Studies on the thymus of the mammal. I. The distribution of argyrophil fibers from birth through old age in the thymus of the mouse. Anat. Rec. 79:133–154.

Smith, C., E. C. Thatcher, D. Z. Kraemer, and E. S. Holt. 1952. Studies on the thymus of the mammal. VI. The vascular pattern of the thymus of the mouse and its changes during aging. J. Morphol. 91:199–220.

Snook, T. 1950. A comparative study of the vascular arrangements in mammalian spleens. Amer. J. Anat. 87:31–78.

Stein, K. F. 1957. Genetical studies on the skeleton of the mouse. XXI. The girdles and the long limb bones. J. Genet. 55:313–324.

Straile, W. E. 1960. Sensory hair follicles in mammalian skin: the tylotrich follicle. Amer. J. Anat. 106:133–148.

Suntzeff, V., and P. Angeletti. 1961. Histological and histochemical changes in intestines of mice with aging. J. Gerontol. 16:225–229.

Van Ebbenhorst Tengbergen, W. J. P. R. 1955. The morphology of the mouse anterior pituitary during the oestrous cycle. Acta Endocrinol. 18: 213–218.

Van Heyningen, H. E. 1961. The initiation of thyroid function in the mouse. Endocrinology 69:720–727.

Von Bartheld, F., and J. Moll. 1954. The vascular system of the mouse epiphysis with remarks on the comparative anatomy of the venous trunks in the epiphyseal area. Acta Anat. 22:227–235.

Walker, B. E., and F. C. Fraser. 1956. Closure of the secondary palate in three strains of mice. J. Embryol. Exp. Morphol. 4:176–189.

Webster, S. H., and E. J. Liljegren. 1955. Organ body-weight ratios for certain organs of laboratory animals. III. White Swiss mouse. Amer. J. Anat. 97:129–153.

Wellings, S. R., K. B. DeOme, and D. R. Pitelka. 1960. Electron microscopy of milk secretion in the mammary gland of the C3H/Crgl mouse. I. Cytomorphology of the prelactating and the lactating gland. J. Nat. Cancer Inst. 25:393–421.

Wimsatt, W. A., and C. M. Waldo. 1945. The normal occurrence of a peritoneal opening in the bursa ovarii of the mouse. Anat. Rec. 93:47–53.

Wirtschafter, Z. T., and J. K. Tsujimura. 1961. The sesamoid bones in the C3H mouse. Anat. Rec. 139:399–408.

14

Teratogenesis[1]

Charles P. Dagg

Teratologists are concerned primarily with the causes, development, and anatomy of morphologically abnormal individuals. Most investigators have focused their attention upon gross structural defects produced either by mutant genes, by noxious environmental stimuli, or by a combination of genetic and environmental conditions. This chapter deals mainly with studies on environmentally induced skeletal abnormalities in mice. The coverage is restricted to a few papers selected to illustrate some of the major factors that must be considered in teratogenic experiments with these animals.

Abnormalities arising from the action of single or multiple genetic factors are discussed in Chap. 8 and 15. Reviews summarizing experimentation with other animals and covering aspects of teratogenesis not dealt with in this chapter have been published by European Society (1963), Fishbein (1960, 1963), Giroud and Tuchmann-Duplessis (1962), Kalter and Warkany (1959), Nishimura (1964), Wilson and Warkany (1964), and Wolstenholme and O'Connor 1960).

As with many other types of investigations, the advantages in using mice are found in their relatively low cost, small size with consequent reduction in space required for breeding and maintenance, in the availability of genetically uniform strains, and in the large number and diversity of mutants affecting the development of structures of interest to teratologists.

Malformations induced by teratogens are the end result of morphogenetic interactions between the genes of the embryo which confer the potentialities for normal or abnormal development and the environment of the embryo which includes the dam's genotype and environment, in addition to the teratogen. Specific effects of some of these factors are known. The ones that will be considered are: time of treatment, dose-response relationships, route of administration, interaction between agents, protective treatments, diet, seasonal effects, maternal influences, the maternal genotype, and the embryonal genotype.

ENVIRONMENTAL FACTORS

Time of treatment

A list of teratogenic treatments and their principal morphological effects in various strains of mice is presented in Table 14-1. Descriptions and photographs of abnormalities of the head and skull can be found in Kalter and Warkany (1957) and Kalter (1963). Abnormalities of the axial skeleton are illustrated in Kalter and Warkany (1957), Kalter (1963), and Murakami and Kameyama (1963), and photographs of various limb and foot deformities are presented in Kalter and Warkany (1957), Nishimura and Kuginuki (1958) and Kageyama (1961*a*).

In general, each organ of the embryo passes through a period of development during which it is particularly susceptible to teratogenic treatments. This developmental phase is commonly referred to as the critical period for that organ. In most cases the critical period corresponds to the time at which the organ is developing most rapidly, so that by the time organogenesis is com-

[1] The writing of this chapter was supported in part by Public Health Service Research Grant HD 00473 from the National Institute of Child Health and Human Development.

Table 14-1. PHYSICAL AND CHEMICAL TERATOGENIC AGENTS

Treatment	Strain[a]	Malformation or structure affected	Reference
X-radiation	C57BL ♀ × NB ♂	Cleft palate Forefeet and fore-limbs Hind feet and hind limbs Vertebrae and ribs Tail	Russell (1954)
	129/Rr	Exencephaly Vertebrae and ribs Tail	Runner and Dagg (1960)
	Outbred (ddN)	Exencephaly Cleft palate Forefeet and fore-limbs Hind feet and hind limbs Vertebrae and ribs	Murakami et al. (1961)
Hypoxia	Outbred	Exencephaly Cleft palate Vertebrae and ribs	Ingalls et al. (1952)
	C57BL/6J, C57BR/cdJ, BALB/cSc	Exencephaly	Ingalls et al. (1953)
	C57BL/6J, C57BR/cdJ, A/J, DBA/1J	Vertebrae and ribs	
	C57BL/6J, C57BR/cdJ, A/J, BALB/cSc, DBA/1J	Sternum	
	Outbred (ddN)	Exencephaly Cleft palate with cleft lip Cleft palate Hind feet and limbs Vertebrae and ribs Tail Omphalocele	Murakami and Kameyama (1963)
Fasting 24 hours	129/Rr	Exencephaly Vertebrae and ribs	Runner (1954)
	129/Rr, C57BL/6J	Exencephaly Vertebrae and ribs	Miller (1962)
Amniotic sac puncture	Not identified	Cleft palate	Trasler, et al. (1956)
Nitrogen mustard	Outbred (S)	Cleft palate	Nishimura and Takagaki (1959*a*)
	Outbred (S), Outbred (hybrid)	Forefeet Hind feet	
	Partially inbred (white)	Cleft palate Micrognathia Forefeet Hind feet	Thalhammer and Heller-Szöllösy (1955)

Table 14-1. PHYSICAL AND CHEMICAL TERATOGENIC AGENTS (*Continued*)

Treatment	Strain[a]	Malformation or structure affected	Reference
Triethylene melamine	Outbred (dd), Outbred (H)	Cleft palate Vertebrae and ribs	Kageyama (1961*b*)
	Outbred (dd), Outbred (H)	Cleft palate Forefeet Hind feet Tail	Kageyama and Nishimura (1961)
Trypan blue	CBA	Tail	Waddington and Carter (1952)
	BALB/c	Exencephaly Tail	Hamburgh (1952)
	129/Rr	Exencephaly Vertebrae and ribs Tail	Runner (1954)
Cortisone	A/J, C57BL/J, Black and tan/J, "Naked"	Cleft palate	Fraser and Fainstat (1951)
	C57BL/6J, CBA, C3H, DBA/1J, A/J	Cleft palate	Kalter (1964)
	A/St, C3H	Cleft palate	Loevy (1963)
Hydrocortisone	Outbred	Cleft palate	Ingalls and Curley (1957)
	A/J	Cleft palate	Pinsky and DiGeorge (1965)
ACTH	A/J	Cleft palate	Heiberg et al. (1959)
Dexamethasone	A/J	Cleft palate	Pinsky and DiGeorge (1965)
Prednisolone	A/J	Cleft palate	Pinsky and DiGeorge (1965)
Thalidomide	A/St	Hydrocephaly Vertebrae Forelimbs Hind limbs	DiPaolo (1963)
2,3-Dimercaptopropanol (BAL)	Outbred	Cleft palate Forefeet Hind feet	Nishimura and Takagaki (1959*a*)
p-Chlorodimethylaminoazobenzene	Outbred (dd)	Cleft palate Forefeet Hind feet	Sugiyama et al. (1960)
Dimethylaminoazobenzene	Outbred (dd)	Hind feet	Sugiyama et al. (1960)
Methylcholanthrene	C3H/J	Tail	Savkur et al. (1961)
Urethane	CBA ♀ × C57BL ♂, Outbred (albino)	Exencephaly	Sinclair (1950)

Table 14-1. PHYSICAL AND CHEMICAL TERATOGENIC AGENTS (*Continued*)

Treatment	Strain[a]	Malformation or structure affected	Reference
Ethylurethan	Outbred	Cleft palate Forefeet Hind feet Tail	Nishimura and Kuginuki (1958)
	C3H	Exencephaly	Kageyama (1961*a*)
	C3H, Outbred (dd), Outbred (H), Outbred (hybrid)	Cleft palate	
	Outbred (H), Outbred (hybrid)	Forefeet	
	C3H, Outbred (dd), Outbred (H), Outbred (hybrid)	Hind feet	
	C3H	Omphalocele	
	C3H, Outbred (dd)	Tail	
Mitomycin C	BALB/cGn	Cleft palate Forefeet Hind feet Tail	Nishimura (1964)
Methionine sulfoximine	Outbred (dd)	Hind feet Tail	Nishimura et al. (1962)
Methionine sulfoxide	Outbred (dd)	Hind feet	Nishimura et al. (1962)
Nicotine	Outbred (S)	Forefeet Hind feet	Nishimura and Nakai (1958)
Caffeine	Outbred (SMA)	Cleft palate Forefeet Hind feet	Nishimura and Nakai (1960)
5-Hydroxytryptamine	Outbred	Brain and skull Microphthalmia Forefeet Hind feet Omphalocele Tail	Poulson et al. (1963)
Iodoacetate	129/Rr	Exencephaly Vertebrae and ribs	Runner (1959)
Tolbutamide	129/Rr	Exencephaly	Smithberg and Runner (1963)
	129/Rr, BALB/cRr	Vertebrae and ribs	
Insulin	129/Rr	Exencephaly Vertebrae and ribs	Smithberg and Runner (1963)
Alloxan	Outbred	Cleft palate	Watanabe and Ingalls (1963)
Vitamin A excess	A/J, C57BL	Cleft palate	Walker and Crain (1960)
	Outbred	Exencephaly Cleft palate Foot deformities	Giroud and Martinet (1960)

Table 14-1. PHYSICAL AND CHEMICAL TERATOGENIC AGENTS (*Continued*)

Treatment	Strain[a]	Malformation or structure affected	Reference
Galactoflavin	DBA/1J, 129/J, C57BL/6J, A/J	Cleft palate	Kalter and Warkany (1957)
	DBA/1J, A/J, 129/J	Forefeet and forelimbs	Kalter (1964)
	DBA/1J, A/J, C57BL/6J, 129/J	Hind feet	
6-Aminonicotinamide	Strains not identified	Cleft palate Vertebrae Sternum Forefeet Hind feet and hind limbs	Pinsky and Fraser (1959)
	A/J, C57BL	Cleft palate Vertebrae	Goldstein et al. (1963)
	Outbred	Cleft palate	Ingalls et al. (1964)
9-Methyl folic acid	129/Rr	Exencephaly Vertebrae and ribs	Runner (1954)
5-Bromodeoxyuridine	A/St ♀ × C3H-MF ♂	Hind feet	DiPaolo (1964)
5-Chlorodeoxyuridine	C57BL/Ks, C57BL/10Gn, C57BL/6J, A/J, BALB/cJ	Cleft palate Forefeet Hind feet Tail	Nishimura (1964)
8-Azaguanine	Outbred (dd)	Cleft palate Forefeet Hind feet	Nishimura and Nimura (1958)
5-Fluorouracil	BALB/cGn, 129/Rr	Cleft palate Forefeet Hind feet Tail	Dagg (1960)
	C57BL/Ks, C57BL/6J, C57BL/10Gn, 129/Sv, SJL/J	Hind feet Tail	Dagg (1963)
5-Fluorodeoxyuridine	129/Rr	Cleft palate Hind feet Tail	Dagg and Kallio (1962)

[a] ddN, S, dd, H, and SMA are strain symbols.

plete the organ no longer responds to teratogens. Occasionally the organ may be malformed by treatment prior to the appearance of its earliest visible rudiment. As used by various investigators, the term *critical period* refers either to the time of maximum frequency of response or to the entire period during which a particular malformation can be produced. In the latter case the length of the critical period is often dependent upon the intensity of the stimulus. An increase in intensity frequently extends the critical period slightly to earlier and later developmental stages.

Critical periods for some of the teratogens that produce gross morphological changes are shown in Table 14-2. There is some confusion in the literature about specific critical periods because not all investigators have used the same method for counting the days of gestation. Some have

Table 14-2. CRITICAL PERIODS FOR VARIOUS MALFORMATIONS PRODUCED BY TERATOGENIC AGENTS IN MOUSE EMBRYOS

Malformation	Teratogen	Days of gestation: Treated	Critical period	Maximum response	Reference
Exencephaly	Vitamin A excess	6–14	6–9	7	Giroud and Martinet (1960)
	X-radiation	7–10	7–8	7	Murakami et al. (1961)
	Hypoxia	2–18	8	8	Ingalls et al. (1952)
Cleft palate	X-radiation	0–13	7–8, 10–13	8, 10–11	Russell (1954)
	X-radiation	7–17	9–13	9, 11–12	Heitz and Martinet (1961)
	Alloxan	8–13	8–12	8	Watanabe and Ingalls (1957)
	p-Chlorodimethyl-aminoazobenzene	8–15	9	9	Sugiyama et al. (1960)
	Ethylurethan	9–14	9–12	9–10	Nishimura and Kuginuki (1958)
	Vitamin A excess	6–14	9–13	9, 11–12	Giroud and Martinet (1960)
	Hypoxia	7–11	9–10	10	Murakami and Kameyama (1963)
	Hypoxia	2–18	13–14	14	Ingalls et al. (1952)
	5-Fluorouracil	9–13	10–13	10, 12–13	Dagg (1960)
	Caffeine	9–14	9–13	12–13	Nishimura and Nakai (1960)
	Hydrocortisone	8–16	9–14	11–12	Ingalls and Curley (1957)
	Triethylene melamine	9–15	9–13	11–12	Kageyama (1961*b*)
	6-Aminonicotinamide	8–14	8–14	13	Goldstein et al. (1963)
	6-Aminonicotinamide	8–15	9–14	13	Ingalls et al. (1964)
	Nitrogen mustard	9–14	11–13		Thalhammer and Heller-Szöllösy (1955)
Enlarged hind feet (Polydactyly and hyperphalangy)	p-Chlorodimethyl-aminoazobenzene	8–15	9, 10, 14	9	Sugiyama et al. (1960)
	Triethylene melamine	8–15	9–10	10	Kageyama and Nishimura (1961)
	5-Fluorouracil	9–13	10–11	10	Dagg (1960)
	X-radiation	0–13	6, 8–10	10	Russell (1954)
	Nitrogen mustard	9–14	11	11	Thalhammer and Heller-Szöllösy (1955)
	Dimethylamino-azobenzene	8–15	7, 10, 11, 13		Sugiyama et al. (1960)
	Ethylurethan	9–14	9–10		Nishimura and Kuginuki (1958)
Reduced hind feet (Ectro-, brachy- and syndactyly)	Triethylene melamine	9–15	10–12	11	Kageyama and Nishimura (1961)
	X-radiation	0–13	10–13	11	Russell (1954)
	5-Fluorouracil	9–13	10–13	13	Dagg (1960)
	Ethylurethan	9–14	11–12		Nishimura and Kuginuki (1958)
	Caffeine	9–14	11–14		Nishimura and Nakai (1960)
Reduced forefeet (Ectro-, brachy- and syndactyly)	Triethylene melamine	9–15	10–12	11	Kageyama and Nishimura (1961)
	X-radiation	0–13	10–12	11	Russell (1954)
	Ethylurethan	9–14	9–12	11–12	Nishimura and Kuginuki (1958)
	5-Fluorouracil	9–13	11–13	13	Dagg (1960)
	Caffeine	9–14	10–13		Nishimura and Nakai (1960)
Malformed ribs	X-radiation	0–13	6–12	7, 11	Russell (1954)
	Hypoxia	2–18	8–9	8–9	Ingalls et al. (1952)
	Hypoxia	7–11	9	9	Murakami and Kameyama (1963)
	Triethylene melamine	9–15	9–15	12–13	Kageyama (1961*b*)
Malformed vertebrae	X-radiation	0–13	6–8	7	Russell (1954)
	X-radiation	7–10	7–9	9	Murakami et al. (1961)
	Hypoxia	2–18	8–13	8–9	Ingalls et al. (1952)
	Hypoxia	7–11	7–11	9	Murakami and Kameyama (1963)
	6-Aminonicotinamide	8–14	8–9	9	Goldstein et al. (1963)

regarded the first 24 hours as day 0 while others have designated the same time as day 1. To facilitate comparisons of critical periods for the different teratogens listed in Table 14-2 the first 24 hours has been called day 1 in every case.

As shown in Table 14-2, the critical periods for various organs may overlap. Each organ, however, has a distinctive and relatively limited period of susceptibility to teratogens. Occasionally, a teratogenic treatment has shown two separate periods of maximum response for a particular malformation. For example, there are two distinct periods of high frequencies of cleft palate following treatment with either X-radiation, excess vitamin A, or 5-fluorouracil.

The over-all critical period and the time of maximum response for a particular malformation may be different for different agents. Cleft palate is a noteworthy example of a malformation for which the critical period is markedly dependent upon the nature of the teratogen (Table 14-2). These great differences in critical periods may be explained by assuming that different biochemical and, perhaps, developmental events were being interfered with in the separate cases.

The period during which a particular malformation can be induced in an organ is usually of much shorter duration than the critical period for the organ as a whole. For example, polydactylous hind feet were produced by treatment of 10-day embryos with low doses of fluorouracil. Treatment the following day resulted in both polydactyly and oligodactyly, and treatment of 12-day embryos caused oligodactyly only (Dagg, 1960, 1963). Furthermore, as expected, the location of defects in the axial skeleton has varied with the time of exposure. Similar malformations—fusions, deletions, and reductions of vertebral elements—were found in different and relatively localized areas of the spinal column depending on the age of the embryos when subjected to X-radiation or to hypoxia (Murakami et al., 1961; Murakami and Kameyama, 1963).

Dose-response relationships

The response to a teratogenic agent is dependent upon the intensity and the duration of the stimulus. Typically the range of doses can be divided into a subthreshold, a teratogenic, and a lethal range. The subthreshold range includes all doses not teratogenic according to criteria established by the observer. Obviously, the type of abnormalities that are looked for and the standards used to distinguish normal from abnormal development will vary with the investigator, and an embryo classified as normal in one laboratory might be regarded as defective in another. In the teratogenic range, an increase in intensity or duration of treatment usually is reflected in an increase in frequency and severity of the defect. With some agents, malformations can be produced without appreciably affecting the survival of the embryos, whereas other agents appear to be embryocidal over the entire teratogenic range. Generally, the frequency of embryonal deaths rises progressively as the intensity or duration of the stimulus is increased until all embryos are destroyed. At this point the pregnant dam may herself show signs of being affected.

In order for an agent to be teratogenic it must show some specificity in regard to the tissues affected. The damage produced must be differential so that certain structures can be malformed without, at the same time, destroying the entire conceptus. In some instances the teratogenic range is so narrow that it is difficult to demonstrate that the agent is teratogenic at sublethal doses.

Occasionally, dose-response curves have shown a plateau, so that after a given frequency of malformations was reached, further increases in dosage were not accompanied by higher proportions of deformed embryos. As an example, 200 mg/kg of 5-chlorodeoxyuridine given to A/J and BALB/cJ dams at 10 days after mating produced malformed hind feet in approximately 40 per cent of the fetuses (Nishimura, 1964). The incidence of malformed hind feet did not change significantly when 400 or 600 mg/kg was given, although the incidence of cleft palates did increase in strain A/J showing that the plateaued response curve was not characteristic of all organs. It is likely that the hind feet at this stage of development were capable of response in only 40 per cent of the A/J and BALB/cJ embryos. A plateaued response curve was not seen in three other strains of mice: C57BL/Ks, C57BL/10Gn, and C57BL/6J.

Similarly, in an experiment designed to test the interaction of four different factors on the cleft palate frequency, Warburton et al. (1962) used two dose levels of cortisone, two mouse strains, two maternal weight categories, and two commercial brands of food. With one brand of food and with the heavy mothers an increase in the dose of cortisone greatly increased the frequency of cleft palate in strain A/J (from 26 to 100 per cent), whereas there was little or no increase in strain C57BL (from 52 to 59 per cent).

The schedule of doses can be a factor in determining embryonal responses. Isaacson and

Chaudhry (1962) gave pregnant A/J mice intramuscular injections of cortisone commencing at $11\frac{1}{2}$ days of gestation. Four injections at 6-hour intervals of 0.625 mg for a total of 2.5 mg produced a higher incidence of cleft palates than a single dose of 2.5 mg.

The two most common methods of administering drugs are to give amounts in proportion to the mother's body weight and to give the same absolute amount to all animals. An inverse relationship between the maternal weight and the frequency of cleft palate induced by a given dose of cortisone was found by Kalter (1956, 1957), and a similar relationship for 6-aminonicotinamide was reported by Pinsky and Fraser (1959). Kalter (1964) found that when the dosage of cortisone was based on maternal body weight, the influence of weight was largely eliminated. This effect of body weight is not universal. Dagg (1963) gave fluorouracil on a weight basis and found that the fetuses in the larger females, receiving proportionately the larger amounts of the teratogen, had the higher frequencies of malformed hind feet. In this example the response appeared to be more closely related to the absolute amount of teratogen than to the mother's body weight.

The response to teratogenic treatments has been shown to vary quantitatively and qualitatively at different times of the year. Kalter (1959*a*) found a seasonal variation in the frequency of cortisone-induced cleft palate in the offspring of (C57BL/6J × A/J)F_1 females crossed to A/J males. When the treatment was given during winter months, November to April, 56 per cent of the offspring were malformed. A lower incidence, 36 per cent cleft palate, was obtained during the summer, May to October. Ingalls et al. (1953) observed a seasonal variation in anomalies produced by exposure of pregnant females to 5 hours of hypoxia on the 14th day of gestation. Strain C57BL/6J responded with a higher rate of rib and vertebral malformations during the winter than during the summer. In these examples the variation was quantitative, in contrast to a qualitative variation found by Kalter and Warkany (1961) in experiments with vitamin A. Excessive amounts of vitamin A were administered during the early stages of embryonic development to A/J, DBA/1J, and C3H/J strains. Treatment during winter months and during late March and April caused frequent malformations of the palate and pinna and mild and moderate microstomia. However, none of the defects of the central nervous system, anus, and tail, observed during winter months, appeared in the fetuses treated in late March and April. In all of these examples the teratogens were more effective during winter than during summer.

The nutritional status of the pregnant female would be expected to play an important role in teratogenesis. The influence of two diets, designated A and B, on the teratogenic action of cortisone has been reported by Warburton et al. (1962). On diet A the incidence of cleft palate was higher in strain A/J (72 per cent) than in C57BL/6 (6 per cent). On diet B the frequency was lower in strain A/J (26 per cent) than in C57BL/6 (52 per cent). It was concluded that both genetic and nutritional factors influenced the embryos' responses to cortisone, and genetic factors affected the influence of the nutritional factors.

Interaction between agents

Investigators studying the effects of simultaneously applied combinations of teratogenic agents have found four modes of interaction: addition, potentiation, noninteraction, and interference. Examples of additive interaction, in which the combined effects of two agents approximated the sum of the individual effects, were found by Runner and Dagg (1960). Strain 129/Rr pregnant females were fasted during the ninth day of gestation and 24 per cent of the embryos developed a syndrome of abnomalities of the ribs and axial skeleton. A similar syndrome was noted in 47 per cent of the embryos whose mothers were kept in a hypoxic atmosphere, in 17 per cent of the embryos from mothers injected with trypan blue, and in 13 per cent of those irradiated with 100 R of X-rays. Combination of fasting with each of the other three treatments had an additive effect: Fasting with hypoxia gave 75 per cent abnormals; with trypan blue, 44 per cent; and with 100 R of X-radiation, 33 per cent.

Other investigators have observed potentiation in which the combined effects of two agents exceeded the sum of their individual effects. Kalter (1960) fed a restricted diet, amounting to approximately 40 per cent of the normal intake during the middle third of pregnancy to (C57BL/6J × A/J)F_1 females mated to A/J males and found approximately 6 per cent cleft palates. Intramuscular cortisone injections at 0.5 and 1.0 mg for 4 consecutive days starting at $11\frac{1}{3}$ days of gestation produced 1 per cent and 12 per cent cleft palates, respectively. The combination of the restricted diet and the two separate doses of cortisone gave 37 per cent and 51 per cent cleft palates, clearly more than the expected frequencies of 7 per cent and 18 per cent had the effects been arithmetically additive. Smithberg and Runner (1963) also observed potentiation in interaction and discovered differences between strains in this regard. Tolbutamide (a sulfonamide) and nicotinamide together showed potentiated interaction in strain 129/Rr and in strain C57BL/6J but not in strain BALB/cRr. Insulin was potentiated by nicotinamide in strain BALB/cRr but not in strain 129/Rr (C57BL/6 was not tested).

The potentiation of a teratogenic agent by a second agent not itself teratogenic or presented in subteratogenic doses was seen by Woollam and Millen (1960*a*). Thyroxine, given on days 11 and 12 in doses of 0.1 mg, had no deleterious effect on the survival of young, but it considerably reduced the chances of survival of young also exposed to X-radiation. A comparable situation was reported by Dagg and Kallio (1962). A teratogen, 5-fluoro-2′-deoxyuridine (FUDR), caused cleft palate, malformed hind feet, and abnormal tails in strain 129/Rr mice when injected 10 days after mating. The natural metabolite, thymidine, was not teratogenic alone, but when it was injected at a dose of 160 mg/kg immediately after FUDR at 20 mg/kg the percentages of surviving embryos with cleft palates and malformed hind feet were markedly increased. The frequency of malformed tails, however, was decreased, indicating that the potentiating interaction had some degree of organ specificity.

Examples of noninteraction of two agents, in which the combined effects were approximately those produced by the single treatment having the larger effect, have been presented by Runner and Dagg (1960) who found that iodoacetate and 9-methyl folic acid were nonadditive with 24 hours of fasting in strain 129/Rr.

The type of interaction is not solely a function of the properties of the agents but is also dependent upon the genotype of the treated animals. As noted above, the experiments by Smithberg and Runner (1963) with tolbutamide and nicotinamide demonstrated potentiation in strains 129/Rr and C57BL/Ks. However, in strain BALB/cRr the teratogens were nonadditive.

The fourth type of interaction, interference, in which the combined effects are less than those produced by the single treatment having the larger effect has not been observed in mouse embryos. In this regard it is of interest to note that hypoxia will protect the embryo against X-radiation on day 11½ of pregnancy (Russell et al., 1951; Russell and Russell, 1954). The mice were kept in an atmosphere containing 5 per cent oxygen for 10 minutes before and during the radiation treatment. The hypoxia by itself had no effects on the embryo under these conditions. Combined with irradiation, however, hypoxia provided marked protection against the radiation effects on viability, birth weight, tail length and shape, and forefoot and hind foot structure.

Protective treatments

The damage produced by some teratogens can be prevented by the administration of other agents. Typically, investigations of this sort have either or both of two purposes: to gain information about the modes of action of teratogens and to determine whether different teratogens that produce similar malformations can be counteracted by the same supplemental treatment, thereby providing evidence for similar or dissimilar modes of teratogenic action.

By withholding all food for 24 hours on the ninth day of pregnancy Runner and Miller (1956) produced abnormalities in 22 per cent of the embryos of strain 129/Rr. The defects consisted of cranioschisis and vertebral and rib deformities. Certain supplements fed by stomach tube, such as glucose, casein, amino acids, corn oil, and acetoacetate, reduced the frequency to between 2 and 11 per cent (Runner, 1959).

The offspring of mice injected with 6-aminonicotinamide displayed a variety of malformations in the axial and appendicular skeletons (Pinsky and Fraser, 1959). In subsequent tests, matings were made between strains C57BL/6 and A/J and pregnant females were given a single intramuscular injection of 6-aminonicotinamide at approximately 9½ or 11½ days after the vaginal plug was observed (Pinsky and Fraser, 1960). If a standard dose of nicotinamide (depending upon the dam's body weight) was given simultaneously with 6-aminonicotinamide, there were no malformations and no increase in number of resorptions. However, when the same dose was given 2 hours after the teratogen, there was a marked increase in frequency of both malformations and resorptions. If twice the standard dose of nicotinamide was given after 2 hours, the frequency of malformations was reduced for treatments at 9½ days but not for those at 11½ days after mating. On the basis of these and other investigations, it was concluded that the teratogenicity of 6-aminonicotinamide arises from its ability to form a nicotinamide adenine dinucleotide (NAD) analogue that is inactive in some NAD-dependent enzymatic reactions. The differential effects at 9½ and 10½ days appeared to indicate that the requirements of the maternal-fetal system vary during embryogenesis.

Thymidine, in appropriate amounts, protected strain 129/Rr embryos against fluorodeoxyuridine (Dagg and Kallio, 1962). At 10 days after mating, 20 mg/kg of fluorodeoxyuridine caused 92 per cent of the fetuses to have malformed tails, 39 per cent malformed hind feet, and 18 per cent cleft palates. Thymidine was given immediately after the teratogen in a series of doses ranging from 2.5 to 640 mg/kg. Tail abnormalities decreased in frequency with 5 to 640 mg/kg of thymidine. The frequency of malformed hind feet dropped to a low level of 4 per cent at 20 mg/kg of thymidine and underwent a reversal, climbing to 88 per cent at 160 mg/kg. The palate was not protected at any dose; instead, in the range of 80 to 640 mg/kg of thymidine, the proportion of embryos with cleft palate was greater than in the controls. The results cannot be explained by assuming a single mode of action of either the drug or the supplement. Fluorodeoxyuridine is partially transformed to fluorouracil and therefore could inhibit both DNA and RNA synthesis in embryonic tissues. The synthesis of one or both of the nucleic acids may be the critical events in the affected organs, the level of teratogen required for effective inhibition varying between organs. Furthermore, low doses of thymidine may bypass the blockade of DNA synthesis created by fluorodeoxyuridine, and thereby protect some organs (tail and feet). At higher doses the amount of thymine formed from thymidine may be sufficient to block the degradation of the fluorouracil formed from fluorodeoxyuridine, thereby prolonging the maintenance of effective levels of the teratogen.

Peer et al. (1958) reported that folic acid, but not riboflavin, protected Swiss albino embryos against cortisone-induced cleft palate. Using a cross between C57BL/6J females and A/J males and giving vitamins in food rather than by injection, Kalter (1959*b*) was unable to confirm that folic acid protected against cortisone.

According to Runner (1964) X-radiation protected embryonic ribs against damage that would have resulted from fasting, hypoxia, or iodoacetate given alone. Since irradiation (130 R) by itself caused abnormal vertebrae, but practically no abnormal ribs in 27 per cent of the young, it it was concluded that fasting, hypoxia, and iodoacetate did not protect against irradiation, but the reverse must be the case. Furthermore, the data showed that in the dual treatments (X-radiation with fasting and X-radiation with iodoacetate) there were additive and potentiated interactions in the production of vertebral anomalies.

The teratogenic action of X-radiation is reduced or prevented altogether by a variety of supplementary treatments. As mentioned in the section on Interaction between Agents, Russell et al. (1951) and Russell and Russell (1954) observed a protective effect of hypoxia against radiation damage to the embryo. Woollam and Millen (1960*b*) administered cysteamine (β-mercaptoethylamine) 5 minutes before 300 R of X-radiation and noted a reduction of radiation-induced deformities of the brain and skull from 23 to 2 per cent. Cysteamine had but little protective effect if given 30 minutes after exposure. Rugh and Grupp (1960) studied 15 agents for effectiveness in protecting 8½ day CF1 mouse embryos against embryonic death and malformations caused by 200 R and found that only cysteamine, cystamine, and hypoxia were beneficial.

GENETIC FACTORS

Without question, genetic factors in both embryo and pregnant female play major roles in determining the type, frequency, and the severity of defects that develop spontaneously or through experimental intervention.

Spontaneous deformities

Whenever sufficiently large numbers of mice from either inbred or noninbred strains are examined, some individuals show such marked departures from morphological norms for their respective strains that they are classifiable as abnormal. Each strain of mice has characteristic, relatively uniform frequencies of one or more spontaneous defects. The contributors to *Inbred Strains of Mice,* No. 3 (1963) reported some of the characteristic spontaneous deviants and their frequencies. These include: 10 to 15 per cent cleft lip with or without cleft palate in sublines of the A strains, 8 to 20 per cent microphthalmia and anophthalmia in strains C57BL/6 and C57BL/10, 4 per cent open eyelids in A/JFr, 1 to 3 per cent hydrocephalus in C57BL/6J and C57BL/10J, and 5 per cent kinked or bent tails in LG/Ml.

Spontaneous defects may occur more frequently in one sex than the other. According to Dickie (*Inbred Strains of Mice,* No. 3) 20 per cent of the females and 3 per cent of the males of C57BL/10J have microphthalmia and anophthalmia. Spontaneous cleft lip–cleft palate was found slightly more often in females (20 per cent) than in males (13 per cent) of strain A/HeJ (Dagg, unpublished).

A few of these deviants are the consequence of mutations, but most are environmental rather than genetic in origin. The absence of genetic differences between normal and aberrant animals of the same strain can be established from the lack of parent-offspring correlations when normal and deviant phenotypes are mated. In such cases, crosses between normal individuals produce as many phenodeviants as do crosses between phenodeviants. Although genetic differences are not responsible for most morphological variations between individuals within a given strain, differences between strains in frequencies and types of phenodeviants are strongly influenced by genetic factors. The evidence supports the interpretation that sporadic variants are a result of interaction between numerous genetic loci and environmental stresses. The combined effect of the several loci serves to position the population close to a threshold for development of a particular abnormality, and a less-than-optimal environment is a sufficient stimulus to complete the conditions for atypical development. The nature of the environmental stimulus or stimuli are not known, nor is it known whether the stimuli are identical for all similarly affected individuals in a single litter or in different litters. The effective environmental stimulus may be external to the females in some cases: for example, background radiation from all sources. Viral or bacterial infections and improperly balanced or otherwise inadequate diets may be sufficient stimuli. Inconstant poorly defined physiological states, for example changing physiology associated with aging of the female, may be important. The environmental stimulus may be localized within the uterus. Differences in blood supply to embryos, the effects of crowding, the proximity of dead and resorbing embryos, temperature, and many other factors may determine that some embryos will fail to conform to the strain-typical morphological pattern. Some of these localized uterine differences may be secondary consequences of systemic physiological changes in the female.

The spontaneous cleft lip–cleft palate that develops with predictable regularity in various sublines of strain A mice is genetically determined but strongly influenced by environmental factors (Grüneberg, 1952). From a study of inbred lines in which there was approximate homozygosis of genetic factors including those for harelip, Reed (1936*b*) concluded that harelip resulted either when a small number of cumulative genes were present in homozygous condition and when environmental conditions were favorable, or when a single gene in homozygous condition was present in association with several modifying genes and appropriate environmental factors. Among the factors affecting the development of harelip were sex of the individual, litter size, and age of the mother. As much as 76 per cent of the total variation was attributed to unknown environmental factors (Reed, 1963*a*).

Trasler (1960) has found a possible clue to the nature of one of the uterine factors influencing development of spontaneous cleft lip in strain A/J. Embryos in the uterine site nearest the ovary developed cleft lip (with or without cleft palate) significantly more often than embryos in other positions in the uterus. The ovarian site was less favorable whether there were many or few embryos in the horn, and therefore the difference was probably due to an undetermined inherent quality of that area and not to crowding and competition for nutrients.

Woollam and Millen (1960*a*) reported a significant decrease in the incidence of cleft lip–cleft palate in the A/St strain following injection of the dam with thyroxine on days 11 and 12 of pregnancy. The average number of young per litter was not affected by the treatment. Cortisone injections, also, reduced the percentage of embryos with spontaneous cleft lip with cleft palate in the A/J strain (Walker and Crain, 1959). The rate of embryonal mortality increased in treated litters, and therefore the lowered frequency of cleft lip–cleft palate embryos was interpreted as a consequence of a differential lethal effect of cortisone on defective embryos rather than a healing effect. Hypervitaminosis A which, like cortisone, produced cleft palate without cleft lip did not alter the incidence of cleft lip–cleft palate, although excessive vitamin A reduced the litter size about as much as did cortisone. Similarly, a riboflavin-deficient galactoflavin-containing diet caused a doubling of the normal resorption rate without changing the frequency of spontaneous cleft lip–cleft palate in strain A/J (Kalter and Warkany, 1957). This diet did affect the incidence of four other deformities that occurred "spontaneously" in the controls. Increases were found in the frequencies of open eyes in strain A/J and cleft palate, miscellaneous eye defects, and brachygnathia in strain C57BL/6J. An increase in the percentage of harelip mice following treatment of the dam during the first half of pregnancy with an anterior pituitary preparation, Preloban, was reported by Steiniger (1940).

Hyperphalangy or polydactyly on the preaxial side of the hind foot occurs with a low incidence in all of the C57BL strains and sublines. In C57BL/10Dg the frequency is aproximately 0.05 per cent and in C57BL/Ks it is 1.5 to 3 per cent (Dagg, 1963). The difference between polydactylous and nonpolydactylous animals of strain C57BL/Ks is not genetic.

When 10-day embryos of strain C57BL/Ks were exposed to X-radiation, hyperphalangous hind feet were found in a large number of the fetuses (Dagg, 1964). The dose-response curve over the range of 75 to 150 R did not show evidence of a threshold dose for polydactyly. Definite thresholds were obtained for two other strains: 129/Rr and BALB/cGn. These results were explained by assuming that genetically determined predispositions and uterine environmental factors interacted on the 10th day of gestation to produce spontaneous hyperphalangy in 1.5 per cent of the controls. Furthermore it was assumed that the entire population of embryos was distributed continuously with regard to this tendency toward abnormality, and therefore the normal control embryos as a group lay just below a threshold for spontaneous hyperphalangy. It was also assumed that X-radiation acted directly upon the foot plate and thereby, even at very low doses, caused an additional small number of embryos to cross the threshold of abnormality. As a teratogen, X-radiation may be unique in not showing clearly defined threshold doses under certain circumstances, because it can act directly upon the fetus. Chemical agents would be expected to show thresholds because their actions are subject to inhibition by naturally occurring metabolites in the mother and fetus, and because they are subject to detoxification and excretion by the pregnant dam.

It follows from the assumptions on the mechanism of interaction of X-radiation and genetically determined tendencies toward spontaneous hyperphalangy that a crucial factor in the interaction is the relationship between the time of gene action on the affected organ and the critical period for experimentally produced malformations of the same type. How closely these two events should correspond is not known. However, it would be expected that an increase in frequency of spontaneous deformities would not occur in some circumstances because the teratogen was not effective at the proper time. For some deformities the genetic and the teratogenic events need not take place within a short span of time, especially if those deformities can be induced over a relatively long period of development. An example is cleft palate.

Differences between strains

In virtually all teratogenic experiments in which different strains were compared, the frequencies of particular defects and syndromes of defects, that is, the total arrays of defects and their frequencies, were found to vary between strains. These differences between strains are generally regarded as genetic in origin with multiple genetic factors, rather than single genes, as the basis. However, the set of multiple factors responsible for differences between strains in response to one teratogen may not be identical to the set responsible for differences in response to another. Evidence for this conclusion is found in the observations that the relative order of strains in terms of frequencies of a particular malformation and interstrain differences in syndromes depend upon the nature of the teratogens. Furthermore, the set of multiple factors involved in the production of malformations in one organ may not be identical with those that induce maldevelopment in other organs of the same embryo.

The effects on the axial skeleton produced by fasting pregnant females throughout the ninth day of pregnancy are not strain-specific (Miller, 1962). The same array of defects was found in strain 129/Rr, C57BL/6J, and LG/Rr. The frequency of malformations was higher in strain 129/Rr than in C57BL/6J, and the maximum frequency occurred in strain 129/Rr a day later than in C57BL/6J.

Smithberg and Runner (1963) determined the relative responses of three strains to the hypoglycemia-producing sulfonamide, tolbutamide. The axial skeleton was more susceptible in strain 129/Rr than in BALB/cRr, which in turn was more susceptible than C57BL/6J. Insulin was much more teratogenic in strain 129/Rr, with 62 per cent of the fetuses malformed, than in strain BALB/cRr, with 3 per cent malformed.

Dagg (1963) treated strains 129/Rr, C57BL/6J, and BALB/cRr with 5-fluorouracil. When strains were compared with respect to the frequency of malformed hind feet, strain C57BL/6J was the most susceptible, BALB/cRr embryos were the most resistant, and strain 129/Rr was intermediate. Thus, the relative order of susceptibility of strains 129/Rr and C57BL/6J was the same for fasting as for tolbutamide, but the relative order of the three strains—129/Rr, C57BL/6J, and

BALB/cRr—was not the same for tolbutamide and fluorouracil.

Ingalls et al. (1953) compared responses of several strains to temporary hypoxia on the ninth day of gestation and found the relative order of sensitivity to the induction of malformed vertebrae and ribs to be, from most to least susceptible: A/J (75 per cent), C57BL/6J, DBA/1J, C57BR/cdJ, and BALB/cSc (17 per cent). Sternal deformities were in a different order: DBA/1J (74 per cent), C57BR/cdJ, BALB/cSc, A/J, and C57BL/6J (23 per cent).

Kalter and Warkany (1957) subjected strains A/J, DBA/1J, 129/J and C57BL/6J to a riboflavin-deficient galactoflavin-containing diet for 4 days starting at 9⅓ or 10⅓ days after mating. Skeletal malformations were found in the skull, vertebrae, ribs, and appendages of fetuses. At a low dose of galactoflavin the DBA/1J and 129/J fetuses displayed the entire syndrome of abnormalities. At the same dose level C57BL/6J fetuses showed no deformities and the A/J fetuses only a few. A higher dose level did not produce more malformations in DBA/1J or 129/J, but C57BL/6J fetuses showed moderate frequencies of certain abnormalities, and the A/J fetuses developed the entire syndrome. Each strain presented its own pattern of defects. For example, strain 129/J had the highest frequency of digital defects, A/J had the highest incidence of open eye, and DBA/1J had proportionately more fetuses with cleft palate and micromelia.

Kalter and Warkany (1957) compared the frequencies of cleft palate produced by riboflavin deficiency with those produced by cortisone in the same strains (Kalter, 1954). Cortisone caused cleft palate in 100 per cent of the A/J fetuses, in 75 per cent of DBA/1J, and in 20 per cent of C57BL/6J. Riboflavin deficiency caused cleft palate in 3 per cent of A/J fetuses, in 41 per cent of DBA/1J, and in 13 per cent of C57BL/6J. The palates of strain A/J embryos were also more susceptible than those of C57BL to excessive vitamin A (Walker and Crain, 1960). Ninety-six per cent of the A/J fetuses and 57 per cent of C57BL fetuses from treated dams had cleft palate. Furthermore, according to a system of rating the stages of palatal closure (Walker and Fraser, 1956), the palates of A/J fetuses were the more seriously affected.

Since the incidence of a particular malformation is dependent upon the developmental age of the embryo when exposed to a teratogen, some interstrain differences in frequencies of malformations may be a consequence of dissimilarities in developmental ages in contrast to chronological (postconceptual) ages at the time of treatment. With some teratogens a change in time of treatment by 12 hours produces marked changes in frequencies of particular malformations (Dagg, 1963, 1964).

Walker and Crain (1960) devised a system of grading embryos on the basis of morphological criteria. Developmental stages of forefeet, hind feet, ears, hair follicles, and eyes are assigned numerical values. The morphological rating for an embryo is calculated by adding the numerical values for each of the five developmental stages. Photographs to assist in classifying embryos according to these developmental features have been published by Trasler (1964).

The crown-rump length of C57BL/Ks, 129/Rr, and BALB/cGn embryos has been used as a measure of relative developmental ages (Dagg, 1963, 1964). In these embryos, as in the series described by Grüneberg (1943), the crown-rump length is correlated with gross changes in external features of the embryos, such as foot pads, branchial arches, and length of tail. The C57BL/Ks embryos were found to be about 12½ hours more advanced than the BALB/cGn and 129/Rr embryos. The (C57BL/Ks ♀ × BALB/cGn ♂) hybrid embryos were equal in length to homozygous C57BL/Ks embryos, whereas the (BALB/cGn ♀ × C57BL/Ks ♂) hybrid embryos were intermediate in size relative to parental strain embryos.

There is considerable variation between and within litters in crown-rump lengths (Dagg, 1964). The largest embryos of strain BALB/cGn examined at 10 days after conception were equal in size to the smallest 11-day embryos. If the difference between average crown-rump lengths of 10- and 11-day embryos was used as an estimate of 1 day's development, then the largest and the smallest 10-day embryos and the largest and smallest 11-day embryos differed by 1 day of development, and in one litter the differences in crown-rump lengths corresponded to 1 day's development.

Inheritance of susceptibility

The first informative and conclusive investigation of the inheritance of interstrain differences in response to a teratogen was made by Kalter. The results have been described in detail (Kalter, 1954, 1957, 1964) and will be summarized briefly. Four daily intramuscular injections of cortisone beginning about the 11th day of pregnancy produced cleft palate in 100 per cent of

A/J embryos, in 19 per cent of C57BL, and in 12 per cent of CBA. Reciprocal crosses were made between CBA and A/J. The F_1 embryos, identical in genotype except for the sex chromosomes, had nearly equal frequencies of cleft palates which were intermediate relative to the parental frequencies. In the reciprocal crosses between C57BL and A/J the F_1 embryos had different frequencies of deformity: 44 per cent in the A/J mothers and 3 per cent in the C57BL mothers. Since, again, the embryos were genetically uniform, the differences between the two types of F_1 embryos must have been influenced by maternal factors. To determine whether this maternal effect was nuclear or cytoplasmic in origin, F_1 mothers from the mating of (A/J ♀ × C57BL ♂) or the reciprocal cross were mated to A/J males. The nuclear genotypes of these two types of F_1 mothers were identical, but the egg cytoplasm came exclusively from one or the other strain. These backcross embryos had equal repsonses in both types of mothers and, therefore, the maternal effect observed in the F_1 embryos was probably not due to cytoplasmic factors transmitted in the egg. Since the (A ♀ × A ♂) embryos and the (A ♀ × C57BL ♂) embryos were growing in the same type of mothers and since these embryos responded differently to cortisone, genetic factors operating in the embryo must also have played a role in determining the response to teratogens.

A study of strain differences in reaction to cortisone (Loevy, 1963) also demonstrated genetic influences on the response to a teratogen and, more interestingly, provided evidence of a patroclinous effect; that is, the responses of reciprocal F_1 embryos were different and each type of F_1 embryo tended to be more like the paternal rather than the maternal strain. Cortisone treatment induced cleft palate in all of the embryos of the A/St strain and 36 per cent of the C3H embryos. The hybrid embryos produced by mating C3H females with A/St males had a higher frequency of cleft palates than did the embryos of the reciprocal cross. Figures published by Kalter (1964), who used the same strains but a higher dose of cortisone, corroborate these findings.

Fluorouracil, given at 10 days after mating, caused hind foot deformities, primarily hyperphalangy, more often in strain 129/Rr than in BALB/cGn (Dagg, 1963). The amount of drug injected was based on maternal weight at the time of treatment. The frequency of malformed hind feet was significantly higher in (BALB/c × 129) F_1 embryos developing in BALB/c mothers than in those developing in strain 129/Rr mothers. By itself, this finding could have been interpreted as evidence of a patroclinous effect, but an inspection of the data revealed that the BALB/c females were heavier than 129/Rr females, and, therefore, received proportionately more of the drug. When comparisons were made between groups with the same ranges in weight, the patroclinous effect disappeared.

The work of Goldstein et al. (1963) supports the idea that embryonal genes influencing responses to teratogens may act by producing variations in response of specific organ systems rather than, or in addition to, altering the embryo's general susceptibility to a given teratogen. Vertebral fusions produced by 6-aminonicotinamide occurred more often in A/J embryos (89 per cent) than in C57BL/6J (56 per cent). The F_1 embryos sired by A/J fathers had a higher incidence of vertebral fusions than those sired by C57BL/6J (67 per cent vs. 45 per cent), a patroclinous reciprocal cross difference. Several alternative explanations for the patroclinous difference were offered: (1) random variation; (2) a factor for resistance transmitted in the sperm but not in the egg; (3) a maternal uterine or cytoplasmic factor for resistance transmitted in the A/J strain, interacting with an embryonal genetic factor for resistance transmitted in C57BL/6J, thereby resulting in F_1 embryos which are genetically intermediate but, because of maternal factors for resistance, the embryos developing in A/J mothers had a lower frequency of vertebral fusions than did embryos developing in C57BL/6J mothers; and (4) differences in rates of development of the two types of hybrid, so that F_1 embryos in C57BL/6J mothers were treated at the point of maximum vertebral sensitivity, whereas embryos in A/J mothers had developed faster and were treated at a point of lower sensitivity. In the latter case the rate of development would be maternally rather than paternally determined.

The frequency of cleft palate produced by 6-aminonicotinamide was higher in strain A/J embryos (76 per cent) than in C57BL/6J (11 percent). In the reciprocal crosses the response was greater in F_1 embryos developing in A/J mothers (36 per cent) than in embryos developing in C57BL/6J mothers (4 per cent), a matroclinous reciprocal cross difference. Since the reciprocal cross differences were in opposite directions for vertebral fusions and cleft palates, the strain differences were regarded as organ-specific. To determine whether the maternal effect was a reflection of maternally determined differences in the embryos' environments or was due to differences in factors transmitted through the egg cytoplasm, the two types of F_1 mothers (A/J ♀ × C57BL/6 ♂ and C57BL/6 ♀ × A/J ♂) were backcrossed to A/J ♂ and treated with 6-aminonicotinamide. The frequency of deformed embryos in (A/J ♀ × C57BL/6J ♂) F_1 mothers was higher than the frequency in F_1 mothers produced by the reciprocal cross (24 per cent vs. 6 per cent). These results were interpreted as suggesting that the teratogenic response to 6-aminonicotinamide is influenced by factors transmitted through the egg cytoplasm.

Hybrid embryos treated with X-radiation tend to resemble the more resistant parental strain in frequency of abnormalities. Rugh et al. (1961) irradiated embryos of strains C57BL/6, CF1, and the (CF1 ♀ × C57BL/6 ♂) F_1 hybrids. The percentages of total implantations that developed into normal fetuses were 5 for C57BL/6, 38 for CF1, and 44 for the hybrids. The exposure of 10-day embryos to 150 R of X-rays caused polydactylous hind feet in 38 per cent of strain C57BL/Ks and 5 to 8 per cent in strain BALB/cGn (Dagg, 1964). The F_1 embryos from reciprocal crosses were like the resistant strain, showing abnormalities in 3 to 9 per cent.

A provisional estimate has been made of the minimum number of gene pairs involved in differences between strains 129/Rr and BALB/cGn in response to fluorouracil, with the conclusion that at least two pairs of genes would be required to account for the observed differences (Dagg, 1963). The estimate was made from F_1 embryos and from two successive backcrosses to the BALB/c strain and was based on several assumptions: (1) Alleles determining susceptibility were dominant to those for resistance; (2) if more than one pair of genes were involved, they segregated independently and had nearly equal effects; and (3) the genes were effective only in the fetus or entire conceptus and the maternal genotype was of no consequence. It should be noted that the assumption of dominance by susceptibility-determining genes was based on a similar frequency of malformed hind feet in F_1 embryos and homozygous 129/Rr embryos. This assumption may not be valid because the F_1 embryos appeared to be more advanced in development than 129/Rr embryos, and because 129/Rr embryos were more susceptible at more advanced stages of development (10½ vs. 10 days of gestation). If dominance is absent and if the alleles have additive effects, the minimum number of loci would be greater than two.

Since responses to teratogens are at least partially determined by the genotype, as shown by the investigations of interstrain differences in susceptibility to teratogens, it would be expected that under certain circumstances teratogens would affect the phenotypic expression of single genes which by themselves produce morphogenetic changes.

A reduction in penetrance of a mutant gene following maternal treatment with a teratogenic agent has been reported by Watney and Miller (1964). The mutation, identified as a recurrence of lid gap (*lg*) in C3H/Ml, caused open eyelids at birth in 77 per cent of the offspring from *lg/lg* × *lg/lg* matings. Normally the eyelids close on approximately the 17th day of gestation and do not reopen until 2 weeks after birth. A single low dose of cortisone administered intramuscularly on the 15th day of gestation to *lg/lg* females, mated to *lg/lg* males, completely inhibited the penetrance of the gene so that all newborn mice had closed eyes. This low dose of cortisone did not increase neonatal mortality nor did it cause cleft palates, whereas higher doses or multiple doses produced both of these effects.

Barber (1957) reported an effect of trypan blue on embryos heterozygous for the eyeless (*ey*) genes. The number of gene pairs, one or two, involved in the inheritance of anophthalmia appears to be undecided (Chap. 8). Barber mated anophthalmic mice of one colony to normal-eyed mice of another and treated the pregnant dams with trypan blue on the seventh, eighth, and ninth days of gestation. Anophthalmia was found in 15 of 40 fetuses and newborn animals from these mothers. The same treatment regimen had no effect on fetuses from matings of normal-eyed parents, and none of the fetuses from untreated dams in matings between anophthalmic and normal mice had anophthalmia. It was concluded that maternal treatment with trypan blue resulted in the expression of a recessive congenital trait that would otherwise be suppressed. Subsequently, Barber et al. (1959) reported that injections of cortisone into pregnant dams on days 7, 8, and 9 of gestation had no effect on the size of the eyes in normal embryos nor in embryos heterozygous for the eyeless genes.

Beck (1963) confirmed the findings made by Barber (1957) on the effect of trypan blue on embryos carrying the eyeless gene and suggested that a single dose of the mutant gene acted to bring the embryo closer to a threshold for reduced eye size than is the case in a homozygous normal individual. Beck (1964) demonstrated that genes other than the eyeless gene also influence the frequency of anophthalmia in the offspring of mothers treated with trypan blue. In the untreated control groups, 2.5 per cent of the fetuses from C57BL/10 females mated to ZRDCT *ey/ey* males were microphthalmic or anophthalmic, and 7 per cent of the fetuses from C57BL/6 females mated to ZRDCT *ey/ey* males had these eye defects. The difference in incidence of defect in the two types of mothers was not significant. When C57BL/10 females, mated to ZRDCT *ey/ey* males, were treated with trypan blue, 5 per cent of fetuses had abnormally small eyes, indicating no increase in frequency over the control level. In contrast, 36 per cent of the fetuses from C57BL/6 females mated to ZRDCT *ey/ey* males and treated with the dye had anophthalmia or microphthalmia.

Teratogenic agents may increase the penetrance and expressivity of mutant genes that cause gross skeletal malformations of the limbs. The autosomal genes luxate (*lx*) and luxoid (*lu*) produce a variety of skeletal changes including preaxial hyperphalangy of the hind feet in heterozygotes and tibial hemimelia in homozygotes (Carter, 1951; Green, 1955). In strain C57BL/10 the penetrance of *lx*/+ is 0.68, and for *lu*/+ the penetrance is 0.85 (Forsthoefel, 1958). In the same strain, 0.25 mg of 5-fluorouracil injected into pregnant females 10 days after conception produced hyperphalangous hind feet in 42 per cent of the fetuses (Dagg, 1965). At a higher dose, 0.50 mg, the incidence of hyperphalangy increased and some of the fetuses showed tibial hemimelia. To test the combined effects of the mutants in the teratogen, C57BL/10-*lu*/+ and -*lx*/+ males were mated to C57BL/10-+/+ females and the females were injected with 0.25 mg of fluorouracil. The pregnant dams, therefore, were carrying either +/+ and *lu*/+ embryos or +/+ and *lx*/+ embryos. Hyperphalangous hind feet were found in 68 per cent of the fetuses sired by *lu*/+ males and in 77 per cent of those sired by *lx*/+ males. In addition, many fetuses had tibial hemimelia, a characteristic of homozygotes or embryos treated with the higher dosage. The results were interpreted as demonstrating that fluorouracil increased the penetrance of *lu* and *lx* in heterozygotes and, furthermore, caused some of the heterozygotes to manifest the homozygous phenotype.

Kobozieff and Pomriaskinsky-Kobozieff (1953, 1962) have described a mutant resembling luxoid (*lu*) in almost all essential features. The heterzygotes are polydactylous and the homozygotes show tibial hemimelia and luxation. The penetrance varies on different genetic backgrounds; for example, in crosses between strain MO females and heterozygous polydactylous males only 7 per cent of the offspring are polydactylous instead of the expected 50 per cent. The effect of trypan blue on the penetrance and expressivity of this mutant was studied by injecting the dye on the eighth day of gestation into MO females that had been mated to MO males or heterozygous polydactylous males (Kobozieff et al., 1959). Neither the penetrance nor the expressivity was changed by the treatment. Other malformations, including exencephaly, occurred with equal frequencies in both types of matings.

Pennycuik (1965) investigated the effects of acute exposure to high temperatures on prenatal development of mice carrying the sex-linked gene tabby (*Ta*). Normal females (+/+) of an outbred stock called TS were mated to males carrying tabby (*Ta*/Y). Pregnant females, therefore, contained tabby (*Ta*/+) female fetuses and normal (+/Y) males fetuses. At different gestational ages, from 1 to 18 days, the dams were exposed to 43°C for 1 hour. Foot malformations were produced by treatment at 11 days, but mice with deformed feet were distributed almost equally between the sexes, so that it was unlikely that the tabby gene influenced the sensitivity of the feet to heat damage. One of the effects of the tabby gene is to reduce the number of secondary vibrissae. Exposure to heat at 12 days of gestation also reduced the number of secondary vibrissae. The effect of heat on vibrissa development was greater in mice carrying the tabby gene than in normal sibs, and the effects of heat and the tabby gene were found to be additive.

SUMMARY

The production of developmental abnormalities by teratogenic agents is a complex event involving the interplay of multiple environmental and genetic factors. The most important experimental variables are the nature of the agent, the developmental stage at which it is applied, and the dose or intensity of treatment. Some of the factors that may modify the embryonal response are diet, season of the year, other teratogens, protective agents, maternal weight and age, position of the embryo in the uterus, and the physiological condition of the mother. In a few cases, cytoplasmic factors passed through the egg may exert an effect on the response to a teratogen. More commonly, genes acting in the embryo and the mother determine and modify the responses to teratogens.

Differences between strains of mice, whether they are reflections of maternal or embryonal genotypes or both, are polygenic in origin. Different sets of genes may be implicated in responses to different types of teratogens, and different sets of genes may be operating in the responses of different embryonal organs. In a few experimentally devised situations, the interaction of single genes and teratogens has been studied.

LITERATURE CITED

Barber, A. N. 1957. The effects of maternal hypoxia on inheritance of recessive blindness in mice. Amer. J. Ophthalmol. 44:94–101.

Barber, A. N., C. Afeman, and J. Willis. 1959. Inheritance of congenital anophthalmia in mice. II. Effects of cortisone and maternal immunization with brain. Amer. J. Ophthalmol. 48:763–769.

Beck, S. L. 1963. Frequencies of teratologies among homozygous normal mice compared with those heterozygous for anophthalmia. Nature 200:810–811.

Beck, S. L. 1964. Sub-line differences among C57 black mice in response to trypan blue and outcross. Nature 204:403–404.

Carter, T. C. 1951. The genetics of luxate mice. I. Morphological abnormalities of heterozygotes and homozygotes. J. Genet. 50:277–299.

Dagg, C. P. 1960. Sensitive stages for the production of developmental abnormalities in mice with 5-fluorouracil. Amer. J. Anat. 106:89–96.

Dagg, C. P. 1963. The interaction of environmental stimuli and inherited susceptibility to congenital deformity. Amer. Zool. 3:223–233.

Dagg, C. P. 1964. Some effects of X-irradiation on the development of inbred and hybrid mouse embryos, p. 91–102. *In* W. D. Carlson and F. X. Gassner [ed.] Effects of ionizing radiation on the reproductive system. Pergamon Press, New York.

Dagg, C. P. 1965. Effects of fluorouracil on the penetrance of two skeletal mutants in mice. Anat. Rec. 151:341. (Abstr.)

Dagg, C. P., and E. Kallio. 1962. Teratogenic interaction of fluorodeoxyuridine and thymidine. Anat. Rec. 142:301–302. (Abstr.)

DiPaolo, J. A. 1963. Congenital malformations in strain A mice. J. Amer. Med. Ass. 183:139–141.

DiPaolo, J. A. 1964. Polydactylism in the offspring of mice injected with 5-bromodeoxyuridine. Science 145:501–502.

European Society for the Study of Drug Toxicity. 1963. Effects of drugs on the foetus. Proc. Vol. I. Excerpta Medica Foundation, Amsterdam. 58 p.

Fishbein, M. [ed.] 1960. First international conference on congenital malformations. Lippincott, Philadelphia. 314 p.

Fishbein, M. [ed.] 1963. Second international conference on congenital malformations. International Medical Congresses, New York. 442 p.

Forsthoefel, P. F. 1958. The skeletal effects of the luxoid gene in the mouse, including its interactions with the luxate gene. J. Morphol. 102: 247–288.

Fraser, F. C., and T. D. Fainstat. 1951. Production of congenital defects in the offspring of mice treated with cortisone. Pediatrics 8:527–533.

Giroud, A., and M. Martinet. 1960. Action tératogène de l'hypervitaminose A chez la souris en fonction du stade embryonnaire. Compt. Rend. Soc. Biol. 154:1353–1355.

Giroud, A., and H. Tuchmann-Duplessis. 1962. Malformations congénitales. Role des facteurs exogènes. Pathol.-Biol. 10:119–151.

Goldstein, M., M. F. Pinsky, and F. C. Fraser. 1963. Genetically determined organ specific responses to the teratogenic action of 6-aminonicotinamide in the mouse. Genet. Res. 4:258–265.

Green, M. C. 1955. Luxoid—a new hereditary leg and foot abnormality in the house mouse. J. Hered. 46:91–99.

Grüneberg, H. 1943. The development of some external features in mouse embryos. J. Hered. 34:89–92.

Grüneberg, H. 1952. The genetics of the mouse, 2nd ed. Nijhoff, The Hague. 650 p.

Hamburgh, M. 1952. Malformations in mouse embryos induced by trypan blue. Nature 169:27.

Heiberg, K., H. Kalter, and F. C. Fraser. 1959. Production of cleft palates in the offspring of mice treated with ACTH during pregnancy. Biol. Neonat. 1:33–37.

Heitz, F., and M. Martinet. 1961. Dualité des stades sensibles dans le développement du palais chez la souris mise en évidence par les rayons X. Compt. Rend. Soc. Biol. 155:707–709.

Inbred Strains of Mice, No. 3. 1963. The Jackson Laboratory, Bar Harbor, Maine. 104 p.

Ingalls, T. H., F. R. Avis, F. J. Curley, and H. M. Temin. 1953. Genetic determinants of hypoxia-induced congenital anomalies. J. Hered. 44: 185–194.

Ingalls, T. H., and F. J. Curley. 1957. The relation of hydrocortisone injections to cleft palate in mice. New Engl. J. Med. 256:1035–1039.

Ingalls, T. H., F. J. Curley, and R. A. Prindle. 1952. Experimental production of congenital anomalies. New Engl. J. Med. 247:758–768.

Ingalls, T. H., E. F. Ingento, and F. J. Curley. 1964. Acquired chromosomal anomalies induced in mice by a known teratogen. J. Amer. Med. Ass. 187:836–838.

Isaacson, R. J., and A. P. Chaudhry. 1962. Cleft palate induction in strain A mice with cortisone. Anat. Rec. 142:479–484.

Kageyama, M. 1961*a*. Differences in susceptibility to induction of congenital malformations by ethylurethan among various strains of mice. Acta Anat. Nippon. 36:1–9.

Kageyama, M. 1961*b*. Disturbances in the skeletal development of mouse embryos induced by injection of triethylene melamine (TEM) during pregnancy. Acta Anat. Japon. 36:246–255.

Kageyama, M., and H. Nishimura. 1961. Developmental anomalies in mouse embryos induced by triethylene melamine (TEM). Acta Scholae Med. Univ. Kioto 37:318–327.

Kalter, H. 1954. The inheritance of susceptibility to the teratogenic action of cortisone in mice. Genetics 39:185–196.

Kalter, H. 1956. Modification of teratogenic action of cortisone in mice by maternal age, maternal weight, and litter size. Amer. J. Physiol. 185:65–68.

Kalter, H. 1957. Factors influencing the frequency of cortisone induced cleft palate in mice. J. Exp. Zool. 134:449–467.

Kalter, H. 1959*a*. Seasonal variation in frequency of cortisone-induced cleft palate in mice. Genetics 44:518–519. (Abstr.)

Kalter, H. 1959*b*. Attempts to modify the frequency of cortisone-induced cleft palate in mice by vitamin, carbohydrate, and protein supplementation. Plast. Reconstr. Surg. 24: 498–504.

Kalter, H. 1960. Teratogenic action of a hypocaloric diet and small doses of cortisone. Proc. Soc. Exp. Biol. Med. 104:518–520.

Kalter, H. 1963. Congenital malformations of the central nervous system. Amer. J. Clin. Nutrit. 12:264–274.

Kalter, H. 1964. Interplay of intrinsic and extrinsic factors, p. 57–80. *In* J. G. Wilson and J. Warkany [ed.] Teratology: principles and techniques. The University of Chicago Press, Chicago.

Kalter, H., and J. Warkany. 1957. Congenital malformations in inbred strains of mice induced by riboflavin-deficient, galactoflavin-containing diets. J. Exp. Zool. 136:531–566.

Kalter, H., and J. Warkany. 1959. Experimental production of congenital malformations in mammals by metabolic procedures. Physiol. Rev. 39:69–115.

Kalter, H., and J. Warkany. 1961. Experimental production of congenital malformations in strains of inbred mice by maternal treatment with hypervitaminosis A. Amer. J. Pathol. 38: 1–21.

Kobozieff, N., and N.-A. Pomriaskinsky-Kobozieff. 1953. Recherches sur la constitution génotypique des souris luxées et polydactyles. Compt. Rend. Soc. Biol. 147:196–199.

Kobozieff, N., and N.-A. Pomriaskinsky-Kobozieff. 1962. Hémimélie chez la souris. II. Étude morphologique des homozygotes atteints de différentes anomalies du squelette. C.–Membres posterieurs: polydactylie intégral et ceinture pelvienne. Rec. Méd. Vét. 138:485–505.

Kobozieff, N., H. Tuchmann-Duplessis, L. Mercier-Parot, and N.-A. Pomriaskinsky-Kobozieff. 1959. Influence du bleu trypan sur la fréquence d'apparition et la gravité des malformations chez des souris présentant une polydactylie héréditaire. Rec. Méd. Vét. 135:317–324.

Loevy, H. 1963. Genetic influences on induced cleft palate in different strains of mice. Anat. Rec. 145:117–122.

Miller, J. R. 1962. A strain difference in response to the teratogenic effect of maternal fasting in the house mouse. Can. J. Genet. Cytol. 4:69–78.

Murakami, U., and Y. Kameyama. 1963. Vertebral malformation in the mouse foetus caused by maternal hypoxia during early stages of pregnancy. J. Embryol. Exp. Morphol. 11:107–118.

Murakami, U., Y. Kameyama, A. Majima, and T. Sakurai. 1961. Patterns of radiation malformations of the mouse fetus and subjected stage of development. Annu. Rep. Res. Inst. Environ. Med., Nagoya Univ. 9:71–81.

Nishimura, H. 1964. Chemistry and prevention of congenital anomalies. Charles C Thomas, Springfield, Ill. 119 p.

Nishimura, H., M. Kageyama, and K. Hayashi. 1962. Teratogenic effect of the methionine derivatives upon the mouse embryos. Acta Scholae Med. Univ. Kioto 38:193–197.

Nishimura, H., and M. Kuginuki. 1958. Congenital malformations induced by ethyl-urethan in mouse embryos. Okajimas Folia Anat. Japon. 31:1–13.

Nishimura, H., and K. Nakai. 1958. Developmental anomalies in offspring of pregnant mice treated with nicotine. Science 127:877–878.

Nishimura, H., and K. Nakai. 1960. Congenital malformations in offspring of mice treated with caffeine. Proc. Soc. Exp. Biol. Med. 104:140–142.

Nishimura, H., and H. Nimura. 1958. Congenital malformations in mouse embryos induced by 8-azaguanine. J. Embryol. Exp. Morphol. 6: 593–596.

Nishimura, H., and S. Takagaki. 1959*a*. Developmental anomalies in mice induced by 2,3-dimercaptopropanol (BAL). Anat. Rec. 135: 261–267.

Nishimura, H., and S. Takagaki. 1959*b*. Congenital malformations in mice induced by nitrogen mustard. Acta Scholae Med. Univ. Kioto 36: 20–26.

Peer, L. A., W. H. Bryan, L. P. Strean, J. C. Walker, W. G. Bernhard, and G. C. Peck. 1958. Introduction of cleft palate in mice by cortisone and its reduction by vitamins. J. Int. Coll. Surg. 30:249–254.

Pennycuik, P. R. 1965. The effects of acute exposure to high temperatures on prenatal development in the mouse with particular reference to secondary vibrissae. Austral. J. Biol. Sci. 18:97–113.

Pinsky, L., and A. M. DiGeorge. 1965. Cleft palate in the mouse: a teratogenic index of glucocorticoid potency. Science 147:402–403.

Pinsky, L., and F. C. Fraser. 1959. Production of skeletal malformations in the offspring of pregnant mice treated with 6-aminonicotinamide. Biol. Neonat. 1:106–112.

Pinsky, L., and F. C. Fraser. 1960. Congenital malformations following a two-hour inactivation of nicotinamide by its analogue, 6-aminonicotinamide, in pregnant mice. Brit. Med. J. 2:195–197.

Poulson, E., J. M. Robson, and F. M. Sullivan. 1963. Teratogenic effect of 5-hydroxytryptamine in mice. Science 141:717–718.

Reed, S. C. 1936*a*. Harelip in the house mouse. I. Effects of the external and internal environments. Genetics 21:337–360.

Reed, S. C. 1936*b*. Harelip in the house mouse. II. Mendelian units concerned with harelip and application of the data to the human harelip problem. Genetics 21:361–374.

Rugh, R., and E. Grupp. 1960. Protection of the embryo against the congenital and lethal effects of X-irradiation (Part I and Part II). Atompraxis 6:209–217.

Rugh, R., E. Grupp, and M. Wohlfromm. 1961. Evidence of prenatal heterosis relating to X-ray induced congenital anomalies. Proc. Soc. Exp. Biol. Med. 106:219–221.

Runner, M. N. 1954. Inheritance of susceptibility to congenital deformity—embryonic instability. J. Nat. Cancer Inst. 151:637–649.

Runner, M. N. 1959. Inheritance of susceptibility to congenital deformity. Metabolic clues provided by experiments with teratogenic agents. Pediatrics 23:245–251.

Runner, M. N. 1964. General mechanisms of teratogenesis, p. 95–103. *In* J. G. Wilson and J. Warkany [ed.] Teratology: principles and techniques. The University of Chicago Press, Chicago.

Runner, M. N., and C. P. Dagg. 1960. Metabolic mechanisms of teratogenic agents during morphogenesis. Nat. Cancer Inst. Monogr. 2: 41–54.

Runner, M. N., and J. R. Miller. 1956. Congenital deformity in the mouse as a consequence of fasting. Anat. Rec. 124:437–438.

Russell, L. B. 1954. The effects of radiation on mammalian prenatal development, p. 861–918. *In* A. Hollaender [ed.] Radiation biology. Vol. I, part II. McGraw-Hill, New York.

Russell, L. B., and W. L. Russell. 1954. An analysis of the changing radiation response of the developing mouse embryo. J. Cell. Comp. Physiol. 43(Suppl. 1):103–147.

Russell, L. B., W. L. Russell, and M. H. Major. 1951. The effect of hypoxia on the radiation induction of developmental abnormalities in the mouse. Anat. Rec. 111:455. (Abstr.)

Savkur, L. D., B. K. Batra, and B. N. Sridharan. 1961. Effect of 20-methylcholanthrene on mouse embryos. II. Strain C3H (Jax). J. Reprod. Fertil. 2:374–380.

Sinclair, J. G. 1950. A specific transplacental effect of urethane in mice. Texas Rep. Biol. Med. 8:623–632.

Smithberg, M., and M. N. Runner. 1963. Teratogenic effects of hypoglycemic treatments in inbred strains of mice. Amer. J. Anat. 113:479–489.

Steiniger, F. 1940. Über die experimentelle Beeinflussung der Ausbildung erblicher Hasenscharten bei der Maus. Z. Menschl. Vererb. Konst. 24:1–12.

Sugiyama, T., H. Nishimura, and K. Fukui. 1960. Abnormalities in mouse embryos induced by several aminoazobenzene derivatives. Okajimas Folia Anat. Japon. 36:195–205.

Thalhammer, O., and E. Heller-Szöllösy. 1955.

Exogene Bildungsfehler ("Miszbildungen") durch Lostinjektion bei der graviden Maus. Z. Kinderheilk. 76:351–365.

Trasler, D. G. 1960. Influence of uterine site on occurrence of spontaneous cleft lip in mice. Science 132:420–421.

Trasler, D. G. 1964. Strain differences in susceptibility to teratogenesis: survey of spontaneously occurring malformation in mice, p. 38–56. *In* J. G. Wilson and J. Warkany [ed.] Teratology: principles and techniques. The University of Chicago Press, Chicago.

Trasler, D. G., B. E. Walker, and F. C. Fraser. 1956. Congenital malformations produced by amniotic-sac puncture. Science 124:439.

Waddington, C. H., and T. C. Carter. 1952. Malformations in mouse embryos induced by trypan blue. Nature 169:27–28.

Walker, B. E., and B. Crain, Jr. 1959. The lethal effect of cortisone on mouse embryos with spontaneous cleft lip-cleft palate. Texas Rep. Biol. Med. 17:637–644.

Walker, B. E., and B. Crain, Jr. 1960. Effects of hypervitaminosis A on palate development in two strains of mice. Amer. J. Anat. 107:49–58.

Walker, B. E., and F. C. Fraser. 1956. Closure of the secondary palate in three strains of mice. J. Embryol. Exp. Morphol. 4:176–189.

Warburton, D., D. G. Trasler, A. Naylor, J. R. Miller, and F. C. Fraser. 1962. Pitfalls in tests for teratogenicity. Lancet 2:1,116–1,117.

Watanabe, G., and T. H. Ingalls. 1963. Congenital malformations in offspring of alloxan-diabetic mice. Diabetes 12:66–72.

Watney, M. J., and J. R. Miller. 1964. Prevention of a genetically determined eye anomaly in the mouse by the administration of cortisone during pregnancy. Nature 202:1,029–1,031.

Wilson, J. G., and J. Warkany [ed.] 1964. Teratology: principles and techniques. The University of Chicago Press, Chicago. 279 p.

Wolstenholme, G. E. W., and C. M. O'Connor [ed.] 1960. Ciba Foundation symposium on congenital malformations. Little, Brown, Boston. 308 p.

Woollam, D. H. M., and J. W. Millen. 1960*a*. Influence of thyroxine on the incidence of harelip in the "Strong A" line of mice. Brit. Med. J. 1:1,253–1,254.

Woollam, D. H. M., and J. W. Millen. 1960*b*. The modification of the activity of certain agents exerting a deleterious effect on the development of the mammalian embryo, p. 158–177. *In* G. E. W. Wolstenholme and C. M. O'Connor [ed.] Ciba Foundation symposium on congenital malformations. Little, Brown, Boston.

15

Genes and Development[1]

Margaret C. Green

The mouse has been the subject of numerous studies of the effects of genes on development. All of the work prior to 1951 has been thoroughly reviewed by Grüneberg (1952), and his later book (Grüneberg, 1963) is a review of developmental studies on genes affecting the skeleton. Other chapters in this book deal with the effects of genes on the development of pigment (Chap. 21), blood (Chap. 17), metabolic and endocrine disorders (Chap. 19, 20), and behavior (Chap. 32, 33). In this chapter I shall not try to describe in any detail the results of studies on the effects of individual genes on development in the mouse. I shall rather try to summarize the kinds of contributions to understanding of development that have been made by these studies. It will be clear immediately that these investigations have raised more questions than they have answered.

There are two main reasons for studying the developmental effects of mutant genes in an animal like the mouse. One is to discover the causes of pathological development of conditions similar to inherited human diseases in the hope of devising cures and preventive measures. The other is to discover how genes control development by investigating how mutant genes change normal development. Studies on the mouse may be expected to contribute to both of these purposes since the mouse is subject to a number of known hereditary diseases similar to those of man, with many others probably yet to be discovered, and since the genetic constitution of the mouse is better known and more easily manipulated than that of any other mammal.

BIOCHEMICAL MUTANTS

In seeking answers to the general question of how genes control development, the first step is to frame more specific questions. In multicellular organisms many kinds of cells are formed and it is clear that although each kind contains the whole array of genes present in the zygote, only certain genes are active in each different kind of cell. In some way genes are called into action at particular times and promote changes which may then call certain other genes into action or repress the activity of others already active. Probably the important basic questions for the field of developmental genetics are: What determines when a particular gene will act, and what does it do when it does act?

The ultimate answers to these questions will have to be in biochemical terms. A start has been made toward answering them in work with inducible enzymes in bacteria, organisms without cellular differentiation but with various gene-controlled functions inducible by environmental changes. The important principles discovered in this work have been set forth by Jacob and Monod (1961). In bacteria there are "structural" genes which carry the information determining the structure of certain enzymes made by the cell and transmit the information to the cell by way of messenger ribonucleic acid. In addition, there are specific regulatory loci controlling the production of repressor substances which, by interaction with environmental factors, determine whether or not a particular structural gene will become active. The interaction between environmental factors and the

[1] The writing of this chapter was supported in part by research grants G15826 and G18485 from the National Science Foundation.

products of regulatory loci are relatively simple in bacteria. In higher organisms the controlling mechanisms are undoubtedly enormously more complicated and may involve chains of interactions between many genes. For some suggested schemes see Monod and Jacob (1961). Evidence is accumulating that many hormones produce their effects by "turning on" the process by which genes make proteins (Williams-Ashman, 1964) but it is not yet known why only certain genes in certain cells respond. There is evidence that the histones, proteins which are usually bound in some degree with the DNA of chromosomes, may act as repressors of gene action and play a role in the response of genes to hormonal stimulation as well as to other kinds of activation (Bonner and Ts'o, 1964). Elucidation of the details of these processes would constitute a large part of the answer to the question of how genes control development.

In the mouse a number of loci are known to determine variations in amount or kind of protein. By definition those controlling changes in amino acid composition of the protein chain are regarded as structural loci. Such differences have been demonstrated directly only for the hemoglobin loci (*Hba* and *Hbb*, Chap. 17) but are inferred for other loci on the basis of electrophoretic or physical differences. Some other such loci are the albino (*c*, Chap. 21) and β-glucuronidase (g, Chap. 19) loci, and probably also the transferrin (*Trf*), serum esterase-1 (*Es-1*), salivary amylase (*Amy-1*), pancreatic amylase (*Amy-2*), erythrocytic esterase-1 (*Ee-1*), isocitrate dehydrogenase (*Id-1*), and immunoglobulin (*Ig-1*, *Ig-2*) loci (Chap. 8). Other loci may be either structural or regulatory: serum esterase-2 (*Es-2*), kidney esterases (*Es-3*, *Es-4*), erythrocytic esterase-2 (*Ee-2*), and prealbumin component (*Pre*) (Chap. 8), which determine presence or absence of electrophoretically demonstrated proteins; hemolytic complement (*Hc*, Chap. 8) which determines the presence or absence of complement activity; and levulinate dehydratase (*Lv*, Chap. 19), and liver catalase (*Ce*, Chap. 8), which determine high and low levels of an enzyme. Pyrimidine degrading (*Pd*, Chap. 19) may be a regulatory locus of some kind, since it appears to govern the activity of three enzymes simultaneously.

Since these genes are concerned with proteins, the primary product (after messenger RNA) of the genes, the activity by which these loci are recognized is probably fairly close to the primary gene action. For the albino locus, for example, it is probable that the primary product of the locus is tyrosinase. It is also reasonably certain that tyrosinase is formed in the melanosomes of the melanocyte and produces melanin there (Moyer, 1963). At least part of the question, "What does the *c* locus do?" is thus answered, but nothing at all is known about what causes the *c* locus to go into action at this particular place. Most and probably all of the genes mentioned in the preceding paragraph are active in only one or, at most, a few kinds of cells and have no detectable activity in other cells. There is at present no clue as to the nature of this difference between cells or how such a difference is brought about.

MORPHOLOGICAL MUTANTS

Leaving the still largely unexplored area of gene action at the primary biochemical level in the mouse, we now consider what can be learned from a study of the more remote effects of genes, principally at the physiological or morphological level. The morphological mutants can be classified in many different ways, but for convenience they will be considered here under two headings which are not necessarily mutually exclusive: mutants providing natural experiments and pleiotropic mutants.

Natural experiments

To study the mechanisms of differentiation and morphogenesis and the interrelationships of various developmental processes, embryologists perform experiments which change the conditions at a particular stage and observe the consequences in later stages. Mammalian embryos are not easily accessible to the embryologist, and, although many ingenious experimental procedures have been devised for dealing with them, it is still difficult to interfere mechanically with a mouse embryo in its normal uterine environment without destroying that environment. Mutant genes can often serve the same purpose as the embryologist's experiment by interfering with a developmental process and allowing the later consequences to be observed under natural conditions. Some restraint is usually necessary, however, in interpreting the causative relationships in a sequence of gene effects, since it is not always certain whether a late effect of a mutant gene is the result of a particular earlier effect or an effect of the mutant acting through some other pathway. A few cases of natural experiments are described below.

The effect of alleles at the *W* locus (Chap. 17) has been used to settle an old dispute on the origin of the primary germ cells (Mintz and Russell, 1957). The gonads of *W/W* mice at birth are very deficient in germ cells. By an alkaline phosphatase staining technique which selectively stains primordial germ cells, Mintz and Russell showed that stained cells were present in the yolk sac of normal embryos at 8 days of gestation. Between 8 and 12 days they increased in number and moved along the wall of the hindgut to the germinal ridges. In *W/W* embryos the stained cells were present in normal numbers at 8 days, but this number increased very little between then and 12 days. The deficiency of migrating cells in embryos known to be defective in capacity to produce germ cells is very strong evidence for the extragonadal origin of the germ cells.

The apical ectodermal ridge of the limb buds has been shown to play an important role in the outgrowth of the limb buds (Saunders, 1948; Zwilling, 1956). The sequence of events in the development of the limbs of mice with syndactylism (*sm/sm*, Chap. 8) offers further evidence for the importance of the apical ectodermal ridge. The third and fourth digits of affected mice are fused. Grüneberg (1960) has shown that the first detectable effect of *sm* in homozygotes is a patchy, localized hyperplasia of the epidermis of the limb buds and tail accompanied by hyperplasia of the apical ectodermal ridge of the limbs. The footplates subsequently become enlarged and concave ventrally, presumably as a result of overgrowth caused by excess stimulation by the enlarged apical ectodermal ridge. The abnormal shape of the footplates pushes the developing digits together with resulting syndactyly.

Bateman (1954) used the mutant grey-lethal (*gl*, Chap. 8) to study the normal pattern of bone growth. The gene causes retardation of accretion and lack of erosion of bone. By comparing the size and shape of bones of normal mice with those of their mutant sibs at 3 weeks of age, Bateman was able to show the exact sites of both accretion and erosion in the growth of normal bone.

It is known from the results of experiments with amphibians and birds that differentiation of the otic vesicle into a normal ear is dependent on the inductive influence of the neural tube. Two mutants of the mouse, kreisler (*kr*) and dreher (*dr*), provide evidence supporting this relationship for mammals (Deol, 1964*a, b*). In kreisler, segmentation of the rhombencephalon and its associated neural crest is abnormal, so that the otic vesicle is prevented by intervening ganglion cells from coming into contact with the neural tube. Subsequent development of the labyrinth is abnormal and similar to that obtained in experiments with amphibia in which the relationship between neural tube and otic capsule is moderately disturbed (Deol, 1964*a*). Dreher mice also have an abnormally formed labyrinth, but the abnormality is not so severe as in kreisler. Deol (1964*b*) suspected that the neural tube might be abnormal in these mice also. His investigation confirmed this suspicion and showed that the abnormality of the neural tube precedes that of the otic vesicle by at least one day.

Mice bearing the mutant dominant hemimelia (*Dh*) lack a spleen, have a small stomach, short intestine, an abnormal urogenital system, preaxial polydactyly or oligodactyly of the hind feet, and tibial hemimelia (Searle, 1964). At 9 days of gestation, before any of these structures are present, the splanchnic mesoderm of the anterior and posterior parts of the coelom is abnormal in mutant embryos, more so in *Dh/Dh* than in *Dh/+*. At this stage normal splanchnic mesoderm has an epithelial-like structure which is less evident or absent in the mutants (M. C. Green, unpublished data). The occurrence of defective mesoderm at this early stage suggests that normal organization of the splanchnic mesoderm is necessary for normal development of the spleen, stomach, intestine, urogenital system, and hind limbs. By analogy with the known inductive effects of the epithelium of the apical ectodermal ridge of the limb bud, it can be postulated that the effect of the splanchnic mesoderm may be dependent on its epithelial-like structure and may be inductive in nature.

Pleiotropic mutants

Many mutants of the mouse are pleiotropic; that is, they have two or more apparently unrelated effects. Mutants of this kind provide evidence for developmental relationships that might not be revealed in any other way. The evidence is especially convincing if there is more than one locus with the same array of effects. Multiple effects of a mutant at a single locus may conceivably be due to the "mutant" being in fact a small deficiency affecting several loci. The occurrence of two or more independent loci with the same array of effects virtually rules out this explanation. Some examples of such cases are discussed at length in other chapters; here I will only mention them and draw attention to a few other cases.

An outstanding example of pleiotropism in the mouse is the multiple effect of mutants at the *W* and *Sl* loci (Chap. 17). Both loci affect development of erythrocytes, pigment cells, and germ cells. It has not been possible, in spite of a large amount of work by many investigators, to discover a common basis for the three defects, but the existence of these two loci, as recognized by their mutant alleles, makes it almost certain that the three types of affected cells, or their precursors, have some common gene-controlled property. This property must be specific to these cells and not shared by other kinds of cells.

Recessive mutants at two different loci (piebald, *s*, and lethal spotting, *ls*) both cause irregular white spotting, and both also cause megacolon in some or all of the animals homozygous for each mutant. The megacolon is associated with a deficiency of intrinsic ganglion cells in the rectum and lower part of the colon (Bielschowsky and Schofield, 1962; Lane, 1966).

It has long been a question among students of pigmentation whether white spotting is due to a defect in pigment cells themselves or to an inhospitable environment in the skin which prevents pigment cells from becoming established there (Markert and Silvers, 1956; Chap. 21). Ganglion cells and pigment cells are both known to be derived from neural crest. The simultaneous deficiency of pigment cells and ganglion cells in mutants at two loci makes it very likely that the defect in both cases is in the neural crest and that the spotting in these cases is indeed due to defective pigment cells. Experiments by Mayer and Maltby (1964) and Mayer (1965) on the development of pigment cells in *ls/ls* and *s/s* mice have added evidence in support of this hypothesis (Chap. 8, 21).

The circling mutants, of which there are nearly 20, are characterized by a tendency to run in circles, head-tossing, hyperactivity, and in many cases deafness (Chap. 8, 32). All have defective labyrinths, the vestibular part only being defective in those not deaf, the cochlea also in those deaf. Most of these mice cannot orient themselves properly when submerged in water and may have difficulty finding their way to the surface. The deafness is easily explained by abnormalities of the cochlea; the inability to orient themselves is explained by abnormalities of the semicircular canals, sacculus, and utriculus. The circling, head-tossing, and hyperactivity, however, are not obvious consequences of the observed pathological conditions. Their nearly universal association with abnormalities of the vestibular labyrinth suggests a close causative connection. A further interesting point in this case is the very large number of independent mutations with similar effects on behavior. At first glance the number seems inordinately large. However, the labyrinth is a complicated structure and many genes must contribute to its development. The striking symptoms and the fact that the abnormalities are quite compatible with life allow mutations at any of these loci to be easily detected. In contrast, mutations affecting the eye, for example, may be quite as numerous but not so easily detected.

A number of mutant genes causing pre- or postaxial abnormalities of the limbs are known. Most of them are listed in Table 15-1 which gives their effects on the limbs as well as on variation in number of presacral vertebrae and ribs (see Chap. 8 for more complete descriptions). Five of the 11 mutant genes affect the number of ribs or presacral vertebrae or both. Of the remaining six, two are known not to affect the number of pre-

Table 15-1. PLEIOTROPIC EFFECTS OF GENES CAUSING LIMB DEFECTS

	Effect on limbs				
Gene	Part affected	Limb most affected	Presacral vertebrae	Ribs	Reference
Dh	Preaxial	Hind only	Fewer	Fewer	Searle (1964)
lx	Preaxial	Hind only	Fewer	No effect	Carter (1951)
lu	Preaxial	Hind	More	More	Forsthoefel (1958)
lst	Preaxial	Hind	No effect	No effect	Forsthoefel (1962)
fi	Preaxial	Hind	No effect	?	Grüneberg (1955)
py	Preaxial	Hind	?	?	Holt (1945)
pc	Preaxial	Hind	?	?	Gluecksohn-Waelsch et al. (1956)
Xt	Preaxial	Hind and fore	?	?	Lyon et al. (1964)
ol	Postaxial	Fore	?	Fewer	Freye (1954)
px	Postaxial	Fore	Fewer	More	Searle (1964)
Po	Postaxial	Fore only	?	?	Nakamura et al. (1963)

sacral vertebrae, and four have apparently not been observed for this character. It should also be mentioned here that differences in number of ribs and presacral vertebrae are commonly observed between inbred strains (Green, 1951, 1954, 1962; Searle, 1954; McLaren and Michie, 1954, 1955), and are also a further effect of several mutant genes which cause short or kinked tails (*Ts, dm, tk, sc, ur, vt,* and *T;* Chap. 8). Carter (1954) has proposed a theory to explain the relationship of hind limb abnormalities and vertebral shifts in the case of *lx*. The theory supposes that the limb is formed by the interaction of an inductor, presumably resident in the ectoderm, with an underlying competent tissue, presumably the mesoderm. A forward shift of the inductor would place it out of line with the competent tissue and result in a limb that is deformed and displaced forward. The forward displacement would result in a forward shift of the pelvic girdle and therefore fewer presacral vertebrae. Under this scheme a backward shift of the competent tissue could be invoked to explain the abnormalities produced by *lu* (Forsthoefel, 1959). This scheme is almost certainly wrong in detail (see Zwilling and Ames, 1958; Forsthoefel, 1959; and Grüneberg, 1963, p. 235–237 for more detailed discussion), but it can surely be no coincidence that abnormal hind limbs are so frequently associated with a change in the anteroposterior position of the pelvic girdle.

During the ninth day of embryonic life in the mouse the hind limb begins to form. It forms at the level of the posterior end of the coelom. Prior to this time the coelom and tail gut extend nearly to the end of the tail bud. During the ninth day the tail bud and the tail gut grow rapidly and soon leave the coelom far behind. At about the same time segmentation, which has been proceeding along the midgut region, arrives at the hind limb level and the end of the coelom. Closure of the nerve cord takes place at the hind limb level at about this time also. It seems possible that variation in the timing of these various events might well produce variation in the position of the limb bud and hence of the pelvic girdle. Some variations might be compatible with normal limb formation and others not. These relationships have not been investigated. To do so by means of experimental procedures may be very difficult in mammals. Mutants which cause changes in limb level with and without limb abnormalities, as well as inbred strains with differences in limb level, provide the natural experiments by which these relationships may be investigated.

A further generalization can be made from the facts in Table 15-1. Preaxial limb defects tend to be confined to, or more severe in, the hind limbs; postaxial limb defects are more severe in the forelimbs. It seems worthwhile to draw attention to this correlation, even though no explanation is apparent. There seems to be no relationship between the direction of the vertebral shift and the pre-or postaxial nature of the limb defect.

Not included in Table 15-1 are other limb mutants which cause syndactyly (*Os, sm, sy*), fusion of bones of the forearm and shank (*ld*), or shortening of the limbs (*bp, sho*). These mutants tend to affect both fore and hind limbs, though not always equally, and no effect on number of ribs or presacral vertebrae has been reported for any of them.

SEX-LINKED MUTANTS

This chapter would not be complete without the mention of the special features associated with the action of sex-linked genes in development. Several lines of evidence led Lyon (1961) to propose the hypothesis that in female mice, and perhaps in other female mammals as well, one of the two X chromosomes is inactivated early in embryonic development. Either the maternal or the paternal X can be the one inactivated, but once the decision has been made, all cells descended from a particular cell have the same inactivated X. Evidence in favor of this hypothesis is as follows.

In many kinds of cells of female mice and other mammals, one chromosome of the proper size to be considered an X chromosome is heteropycnotic (Ohno and Hauschka, 1960; Tjio and Östergren, 1958; Chap. 7). The heteropycnotic chromosome is probably the one inactivated. The Barr body seen in the interphase nuclei of females of many species of mammals is probably the heteropycnotic X. (See McKusick, 1962, for summary of the evidence and for references.)

The Lyon hypothesis predicts that for sex-linked genes with localized gene action (as, for example, genes determining the color of pigment cells) heterozygous females will be mosaics. This is in fact the case in the mouse. All known sex-linked coat color mutants in the mouse cause a mosaic or mottled phenotype with patches of normal and mutant color in heterozygous females. This is true of *Mo*, Mo^{br}, *To*, Mo^{dp}, and *Blo* (Chap. 8, 21). The sex-linked mutants, tabby (*Ta*) and striated (*Str*), which affect hair structure, also have mosaic coats with transverse stripes of normal hair interspersed with stripes of hair of abnormal structure (Lyon, 1963). In addition, females heterozygous for translocations between the X chromosome and an autosome bearing a color mutant, with the wild-type allele on the translocated chromosome and the mutant allele on the normal chromosome, have a variegated phenotype with patches of normal and mutant fur (Russell and Bangham, 1961; Cattanach, 1961).

When the phenotype produced by sex-linked genes is not due to localized gene action, the genes may appear either semidominant or recessive, depending on the mode of gene action and the number of normal cells necessary to insure a normal phenotype. Thus, jimpy (*jp*), a lethal mutant which causes abnormal behavior and severe defects of myelination of the central nervous system (Sidman et al., 1964), is usually completely recessive in its effects on behavior, but Phillips (1954) has reported one presumably heterozygous female showing the abnormal behavior. Lyon (1961) suggests that this individual may represent an example of the rare instance when by chance all the cells responsible for the jimpy trait had the normal gene inactivated. The mutant bent-tail (*Bn*), on the other hand, has an intermediate effect in heterozygotes and fails to manifest itself in a proportion of them, suggesting that the effect may be proportional to the number of determining cells in which the mutant allele is active.

Mice with an X/O chromosome constitution are known to exist and to be normal viable fertile females (Welshons and Russell, 1959). This shows that only one X chromosome is necessary for normal development. The two X chromosomes in normal females appear to do no more than the single X of normal males, judging by the similarity of effect of sex-linked mutants in homozygous females and hemizygous males. A *Ta*/*Ta* female is no more severely affected than a *Ta*/Y male. Nothing is known about the specific proteins controlled by any sex-linked loci in mice, but in man there is no difference between males (X/Y) and females (X/X) in the level of several substances controlled by specific normal sex-linked alleles, for example, glucose 6-phosphate dehydrogenase and antihemophilic globulin (McKusick, 1962). The Lyon hypothesis offers a ready explanation for this phenomenon of "dosage compensation."

Translocations between the X and the autosomes involving linkage groups I and VIII produce variegated phenotypes in mice heterozygous for the translocation and bearing a color mutant on the intact autosome (Russell and Bangham, 1961; Cattanach, 1961; Russell, 1963). Russell (1963) has mapped the autosomal break point for a number of these translocations and has shown that the amount of variegation of the autosomal color mutants is different in the different translocations and not necessarily related to the closeness of the locus to the break. The results are consistent with the hypothesis that the capacity to inactivate translocated autosomes is not evenly distributed along the X. Mapping of the X-chromosome break points of these translocations should reveal the distribution of the inactivating capacity on the X.

If the Lyon hypothesis is indeed true, the tissue in which a sex-linked gene is active should be a mosaic of normal and abnormal cells in females heterozygous for a mutant allele of that gene. If the abnormality is recognizable in individual cells by some means, one may thus be able to identify the tissue in which the gene has its effect. Thus females heterozygous for jimpy (*jp*/+) might be expected to show patchy defects of myelination in the central nervous system if the *jp* locus is active in myelin-producing cells. If myelin is found to be completely normal in heterozygous females, the site of gene action is probably elsewhere. The site of action of autosomal loci may be studied in the same way if suitable translocations can be obtained between the X chromosome and the autosome bearing the locus. For example, one such translocation (Cattanach, 1961) has a piece of the chromosome bearing the wild-type allele of albino (*c*) in linkage group I transposed to the X chromosome. A mutant at the closely linked locus shaker-1 (*sh-1*) causes degeneration of parts of the membranous labyrinth. It is not known whether the *sh-1* locus is included in the transposed piece, but if it is, in females of the appropriate genotype carrying the translocation, the degeneration should be patchy if the site of gene action is in the degenerating cells themselves. This method of investigating gene action has not been explored in the mouse.

Russell et al. (1964) have used X-autosome translocations to investigate certain other properties of genes. If the autosomal loci translocated to the X chromosome are inactivated in approximately half of the cells, mice can be obtained that are mosaic for patches of cells hemizygous and homozygous for a mutant allele. For recessive color mutants, comparison of the two types of tissue may show whether the mutant is completely inactive or not. By this method, pink-eyed dilution (*p*) was shown to have some activity, since

p/O cells appear as light mottling on a p/p background. Brown (b) has no activity, since b/O is indistinguishable from b/b. The same system was used to determine whether lethal mutants at these loci were cell-lethal. Most were found not to be cell-lethal.

SUMMARY

Studies of the effects of genes on development in the mouse may be expected to contribute to knowledge of how genes control differentiation. Although the explanations will ultimately have to be in biochemical terms, very little is known about the primary biochemical action of most mutants in the mouse.

Some examples are given of mutants which produce altered proteins. It is not yet known how the action of these loci is controlled.

Mutants may sometimes serve the same purpose as embryologists' experiments by interfering with a developmental process and allowing the later consequences to be observed. Mutants with two or more apparently unrelated effects (pleiotropic mutants) may provide evidence for developmental relationships not otherwise easily detectable. Examples of both these kinds of mutants are briefly described with a more extended discussion of pleiotropic mutants affecting the limbs and axial skeleton.

The inactive X-chromosome hypothesis provides an explanation for the frequent mosaicism of females heterozygous for sex-linked mutants, and for "dosage compensation" in the action of sex-linked genes in males and females. It also provides a new approach to the analysis of developmental effects of sex-linked genes and of genes made effectively sex-linked by X-autosome translocations.

LITERATURE CITED

Bateman, N. 1954. Bone growth: a study of the grey-lethal and microphthalmic mutants of the mouse. J. Anat. 88:212–262.

Bielschowsky, M., and G. C. Schofield. 1962. Studies on megacolon in piebald mice. Austral. J. Exp. Biol. Med. Sci. 40:395–404.

Bonner, J., and P. O. P. Ts'o [ed.] 1964. The nucleohistones. Holden-Day, San Francisco.

Carter, T. C. 1951. The genetics of luxate mice. I. Morphological abnormalities of heterozygotes and homozygotes. J. Genet. 50:277–299.

Carter, T. C. 1954. The genetics of luxate mice. IV. Embryology. J. Genet. 52:1–35.

Cattanach, B. M. 1961. A chemically-induced variegated-type position effect in the mouse. Z. Vererb. 92:165–182.

Doel, M. S. 1964*a*. The abnormalities of the inner ear in *kreisler* mice. J. Embryol. Exp. Morph. 12:475–490.

Deol, M. S. 1964*b*. The origin of the abnormalities of the inner ear in dreher mice. J. Embryol. Exp. Morph. 12:727–733.

Forsthoefel, P. F. 1958. The skeletal effects of the luxoid gene in the mouse, including its interactions with the luxate gene. J. Morphol. 102:247–287.

Forsthoefel, P. F. 1959. The embryological development of the skeletal effects of the luxoid gene in the mouse, including its interactions with the luxate gene. J. Morphol. 104:89–141.

Forsthoefel, P. F. 1962. Genetics and manifold effects of Strong's luxoid gene in the mouse, including its interactions with Green's luxoid and Carter's luxate genes. J. Morphol. 110:391–420.

Freye, H. 1954. Anatomische und entwicklungsgeschlichtliche Untersuchungen am Skelett normaler und oligodactyler Mäuse. Wiss. Z. Martin-Luther-Univ. 3:801–824.

Gluecksohn-Waelsch, S., D. Hagedorn, and B. F. Sisken. 1956. Genetics and morphology of a recessive mutation in the house mouse affecting head and limb skeleton. J. Morphol. 99:465–479.

Green, E. L. 1951. The genetics of a difference in skeletal type between two inbred strains of mice (BalbC and C57blk). Genetics 36:391–409.

Green, E. L. 1954. Quantitative genetics of skeletal variations in the mouse. I. Crosses between three short-ear strains (P, NB, SEC/2). J. Nat. Cancer Inst. 15:609–624.

Green, E. L. 1962. Quantitative genetics of skeletal variations in the mouse. II. Crosses between four inbred strains (C3H, DBA, C57BL, BALB/c). Genetics 47:1,058–1,096.

Grüneberg, H. 1952. The genetics of the mouse, 2nd ed. Nijhoff, The Hague. 650 p.

Grüneberg, H. 1955. Genetical studies on the skeleton of the mouse. XV. Relations between major and minor variants. J. Genet. 53:515–535.

Grüneberg, H. 1960. Genetical studies on the skeleton of the mouse. XXV. The development of syndactylism. Genet. Res. 1:196–213.

Grüneberg, H. 1963. The pathology of development. Wiley, New York. 309 p.

Holt, S. B. 1945. A polydactyl gene in mice capable of nearly regular manifestation. Ann. Eugen. 12:220–249.

Jacob, F., and J. Monod. 1961. Genetic regulatory mechanisms in the synthesis of proteins. J. Mol. Biol. 3:318–356.

Lane, P. W. 1966. Association of megacolon with two recessive spotting genes in the mouse. J. Hered. 57:29–31.

Lyon, M. F. 1961. Gene action in the X-chromosome of the mouse. Nature 190:372–373.

Lyon, M. F. 1963. Attempts to test the inactive-X theory of dosage compensation in mammals. Genet. Res. 4:93–103.

Lyon, M. F., R. J. S. Phillips, and A. G. Searle. 1964. The overall rates of dominant and recessive lethal and visible mutation induced by spermatogonial X-irradiation of mice. Genet. Res. 5:448–467.

Markert, C. L., and W. K. Silvers. 1956. The effects of genotype and cell environment on melanoblast differentiation in the house mouse. Genetics 41:429–450.

Mayer, T. C. 1965. The development of piebald spotting in mice. Develop. Biol. 11:319–334.

Mayer, T. C., and E. Maltby. 1964. An experimental investigation of pattern development in lethal spotting and belted mouse embryos. Develop. Biol. 9:269–286.

McKusick, V. A. 1962. On the X chromosome of man. Quart. Rev. Biol. 37:70–175.

McLaren, A., and D. Michie. 1954. Factors affecting vertebral variation in mice. 1. Variation within an inbred strain. J. Embryol. Exp. Morphol. 2:149–160.

McLaren, A., and D. Michie. 1955. Factors affecting vertebral variation in mice. 2. Further evidence on intrastrain variation. J. Embryol. Exp. Morphol. 3:366–375.

Mintz, B., and E. S. Russell. 1957. Gene-induced embryological modifications of primordial germ cells in the mouse. J. Exp. Zool. 134:207–237.

Monod, J., and F. Jacob. 1961. Teleonomic mechanism in cellular metabolism, growth, and differentiation. Cold Spring Harbor Symp. Quant. Biol. 26:389–401.

Moyer, F. H. 1963. Genetic effects on melanosome fine structure and ontogeny in normal and malignant cells. Ann. N.Y. Acad. Sci. 100:584–606.

Nakamura, A., H. Sakamoto, and K. Moriwaki. 1963. Genetical studies of post-axial polydactyly in the house mouse. Annu. Rep. Nat. Inst. Genet. Jap. (1962) 13:31.

Ohno, S., and T. S. Hauschka. 1960. Allocycly of the X-chromosome in tumors and normal tissues. Cancer Res. 20:541–545.

Phillips, R. J. S. 1954. *Jimpy*, a new totally sex-linked gene in the house mouse. Z. Indukt. Abstamm. Vererb. 86:322–326.

Russell, L. B. 1963. Mammalian X-chromosome action: inactivation limited in spread and in region of origin. Science 140:976–978.

Russell, L. B., and J. W. Bangham. 1961. Variegated-type position effects in the mouse. Genetics 46:509–525.

Russell, L. B., J. W. Bangham, and C. S. Montgomery. 1964. The use of X-autosome translocations in the mouse for the study of properties of autosomal genes. Genetics 50:281–282. (Abstr.)

Saunders, J. W. Jr. 1948. The proximo-distal sequence of origin of the parts of the chick wing and the role of the ectoderm. J. Exp. Zool. 108:363–404.

Searle, A. G. 1954. Genetical studies on the skeleton of the mouse. X. Rarer variants in the A and C57BL pure lines. J. Genet. 52:103–110.

Searle, A. G. 1964. The genetics and morphology of two "luxoid" mutants in the house mouse. Genet. Res. 5:171–197.

Sidman, R. L., M. M. Dickie, and S. H. Appel. 1964. Mutant mice (*quaking* and *jimpy*) with deficient myelination in the central nervous system. Science 144:309–311.

Tjio, J. H., and G. Östergren. 1958. The chromosomes of primary mammary carcinomas in milk virus strains of the mouse. Hereditas 44:451–465.

Welshons, W. J., and L. B. Russell. 1959. The Y-chromosome as the bearer of male determining factors in the mouse. Proc. Nat. Acad. Sci. 45:560–566.

Williams-Ashman, H. G. 1964. Some experimental approaches to the molecular basis of sex hormone action, p. 103–123. *In* P. Emmelot and O. Mühlbock [ed.] Cellular control mechanisms and cancer. Elsevier, Amsterdam.

Zwilling, E. 1956. Reciprocal dependence of ectoderm and mesoderm during chick embryo limb development. Amer. Natur. 90:257–265.

Zwilling, E., and J. F. Ames. 1958. Polydactyly, related defects and axial shifts—a critique. Amer. Natur. 92:257–266.

16

Physiological Characteristics[1]

Seldon E. Bernstein

Maintenance of the internal environment within narrowly restricted limits is an essential characteristic of all mammalian systems. Control is exerted through highly refined and sensitive regulatory systems employing positive and negative feedback methods and functioning chemically or electrically in modifying both cellular and organic behavior. Many of the controls are exerted through, by, and upon the blood vascular system. Hence, an indication of the extent of homeostasis, permissible deviations, as well as significant disparities resulting from pathological processes, can be obtained by measuring the physical and chemical attributes of blood. This becomes even more apparent when it is recalled that most somatic cells have a direct dependency upon the fluid constituent of blood for their survival, since plasma is the vehicle for provision of nutrients, removal of excretory products, temperature regulation, and transmission of control information.

The cellular components of blood, though no less important in the maintenance of the dynamically balanced internal milieu, are considered in Chap. 17. The fluid portion containing dissolved solids will be emphasized in this chapter. We are concerned mainly then with the quantitative and qualitative evaluation of the materials carried in the aqueous vehicle, the vehicle itself, and the removal of solid, soluble, and gaseous products of cellular metabolism from plasma.

Inasmuch as significant deviations from normal may be of diagnostic value when well-defined baselines are established, the primary function of this chapter is to present available information on the quantitative physiological characteristics of the noncellular components of blood and urine and on the variables affecting these characteristics in the laboratory mouse. Circulation, excretion, and respiration are emphasized while metabolism, neurophysiology, muscle physiology, sensory perception, acclimatization, endocrine balance, and nutrition are discussed in Chap. 5, 11, 19, 20, 26, 32, and 33.

CIRCULATION

Composition of plasma

Mouse plasma is composed of water, proteins, salts, and other solids in solution. Its major constituent is water and this is reflected in its specific gravity and conversely in its content of total solids. Table 16-1 contains representative values.

Specific gravity. Historically, specific gravity has been determined by weighing a known volume of fluid and comparing its weight with an equal volume of water at a particular temperature. Since large volumes are necessary for accurate determinations, other methods have been relied upon for work with mice. Hydrometry has been employed but surface tension complicates this method of measurement. Refractometry is more convenient and, in general, the values obtained parallel those of other methods. Its usefulness, however, is limited to relatively clear fluids and its sensitivity is not great. The gradient tube technique of Lowry and Hunter (1945), the

[1] The writing of this chapter was supported in part by Contract AT(30-1)-1800 with the U.S. Atomic Energy Commission and by Public Health Service Research Grant HD 00254 of the National Institute of Child Health and Human Development.

Table 16-1. QUANTITATIVE SERUM CHARACTERISTICS

Strain		Genotype	No. obs.	Total serum protein, g/100 ml (mean and S.E.)	Specific gravity	Reference
Balb/Gw				6.22 ± 0.07		Thompson et al. (1954)
CBA/Strong					1.060	Dougherty and White (1944)
C3H				3.99		Gleason and Friedberg (1953)
C3H				6.1 to 6.7		Rask-Nielsen et al. (1960)
E				6.83 ± 0.07		Thompson et al. (1954)
K				6.19 ± 0.04		Thompson et al. (1954)
LGW				6.20 ± 0.05		Thompson et al. (1954)
pHH				4.94		Dolyak and Weir (1956)
pHL				5.05		Dolyak and Weir (1956)
RI				5.96 ± 0.07		Thompson et al. (1954)
S				6.20 ± 0.03		Thompson et al. (1954)
Z				5.81 ± 0.13		Thompson et al. (1954)
C57BL/6J	♀	+/+	24	5.0 ± 0.1	1.0211	Bernstein (original data)
	♂	+/+	25	5.2 ± 0.1	1.0219	Bernstein (original data)
	♀	*ja*/+	23	4.8 ± 0.1	1.0205	Bernstein (original data)
	♂	*ja*/+	17	5.4 ± 0.1	1.0222	Bernstein (original data)
	♀	*ha*/+	17	5.5 ± 0.1	1.0225	Bernstein (original data)
	♂	*ha*/+	15	5.3 ± 0.1	1.0221	Bernstein (original data)
DBA/2J	♀	+/+	22	5.4 ± 0.1	1.0223	Bernstein (original data)
	♂	*ha*/+	17	5.3 ± 0.1	1.0221	Bernstein (original data)
WC/Re	♀	+/+	24	5.7 ± 0.1	1.0233	Bernstein (original data)
	♂	+/+	19	6.0 ± 0.1	1.0239	Bernstein (original data)
	♀	W^v/+	20	4.8 ± 0.1	1.0205	Bernstein (original data)
	♀	*Sl*/+	23	5.5 ± 0.1	1.0225	Bernstein (original data)
	♂	*Sl*/+	22	6.0 ± 0.1	1.0239	Bernstein (original data)

falling drop method of Barbour and Hamilton (1926), or the copper sulfate technique of Phillips et al. (1943) all appear to be reliable and simple, and they are the ones usually employed in the determination of specific gravity of mouse serum and urine.

Electrolytes and trace minerals. Little information about electrolytes and trace minerals in mouse blood is available. Small amounts of data are available, however, on magnesium, phosphorus, and iodine. Eveleth (1937) found that whole blood contains 10.3 mg/100 ml and plasma contains 7.6 mg/100 ml of magnesium. Rapoport and Guest (1941) reported that inorganic phosphorus was present in the concentration of 7.4 to 7.9 mg/100 ml of whole blood, and organic phosphorus was distributed as ATP (12.1 to 16.0 mg/100 ml of cells), diphosphoglycerate (51.8 to 54.0 mg/100 ml of cells), and organic acid–soluble phosphorus (84.1 to 85.8 mg/100 ml of cells). Morris and Courtice (1955) found that phospholipid phosphorus was present in plasma in the amount of 6.97 ± 0.6 mg per cent. Iodine is primarily available in a protein-bound form (3.8 μg/100 ml of mouse plasma), while total plasma iodine was 4.5 μg/100 ml according to Taurog and Chaikoff (1946). Much work remains to be done in the determination of the mineral content of mouse plasma.

Plasma proteins. In man and mouse the plasma proteins have been separated on the basis of precipitation of proteins by salts of various bases into albumins and globulin components and on the basis of electrophoretic mobility into prealbumins, α-globulins, β-globulins, and γ-globullins. Biochemists have further subfractioned them into α_1-, α_2-, α_3-, β_1-, β_2-, and β_3-globulin groups, etc., and into categories defined on the basis of presumed function (transferrins, haptoglobins, etc.), whereas immunologists and geneticists have split them into groups based on their immunological specificity.

The distribution of albumins and of α-, β-, and γ-globulins in various mouse strains and in different genotypes has been investigated by Geinitz (1954), Dolyak and Weir (1956), White and

Dougherty (1946), Gleason and Friedberg (1953), Yoon (1961), and others, (Table 16-2). The paper by Dolyak and Weir (1956) also presents electrophoretic mobility values for each of the components mentioned. They obtained values for albumin near 6.4; for α-globulin 4.7; β-globulin 2.7; and γ-globulin 1.3×10^{-5} cm per volt per second.

Polymorphism of serum components appears to be the rule rather than the exception. In the mouse several electrophoretically detectable protein variant systems are known. These include transferrins (Shreffler, 1960; Cohen, 1960; Cohen and Shreffler, 1961); prealbumins (Shreffler, 1964*a*); and esterases (Popp, 1961; Petras and Church, 1962). Serological variants in many γ-globulins are recognizable by Ouchterlony immuno-diffusion techniques with isoimmune serum (Kelus and Moor-Jankowski, 1961; Wunderlich and Herzenberg, 1962). Ashton and Braden (1961) found serum β-globulin variants, and Cinader and Dubiski (1963) discovered α-globulin allotype variants. Schreffler and Owen (1963) presented information on quantitative differences in the concentration of a specific serum globulin not related to γ-globulin variants. Schreffler (1964*b*) later showed that the serum trait was controlled directly by the complex *H-2* locus. These hereditary variations are described in Chap. 8 under the gene symbols *Trf, Pre, Es-1, Es-2, Ig-1, He,* and *Ss*.

Free serum amino acids. Physiological differences are often associated with variations in free amino acid level of tissues and fluids. Hrubant (1958) found several differences in the free amino acid composition of serum attributable to the action of selected genes. Quoting from his paper:

> This research disclosed statistically significant differences between the genotypes in mice of the SEC/2Gn-*d* strain for glutathione (*DD* high), glycine (*Dd* high), alpha alanine (*DD* high), and valine-norvaline (measured together) (*dd* high). No differences were found between the genotypes in the CBA/Ca-*se* strain. Between the genotypes segregating in the Furless strain, there were significant differences found with respect to glutathione (*fsfs* high), aspartic acid (*fsfs* high), lysine (*Fsfs* high), and isoleucine-leucine (measured together) (*FsFs* high). Significant differences in the level of arginine, lysine, valine-norvaline, and glutamic acid were found between the sexes within a particular strain. Male and female SEC/2 mice differed in the proportion of arginine in their blood, the males being higher. CBA males differed from the females in their proportion of lysine (females higher than males), and valine-norvaline (males higher than females), while the proportion of glutamic acid in Furless males exceeded that in females by a statistically significant amount.

Steel et al. (1950) gave normal values of individual amino acids found in plasma. They reported that the concentration of amino acids in mouse blood was essentially unaffected by variations in dietary intake.

Table 16-2. PROTEIN CONSTITUENTS OF SERUM

Strain	Genotype	Albumin	α-globulin	β-globulin	γ-globulin	Reference
		Protein fractions as per cent of total serum protein				
pHH	♀	64.1 ± 1.5	14.5 ± 1.8	16.9 ± 1.6	4.5 ± 0.6	Dolyak and Weir (1956)
pHL	♀	65.7 ± 2.0	13.5 ± 1.4	16.5 ± 2.1	4.3 ± 0.2	Dolyak and Weir (1956)
"mice"		60.8 ± 0.7	12.1 ± 0.4	20.4 ± 0.6	6.7 ± 0.2	Thompson et al. (1954)
		42.9 ± 5.1	20.6 ± 2.3	20.2 ± 2.9	16.3 ± 1.6	Geinitz (1954)
Tm	+/? ♂ & ♀	66.95	10.59	14.96	7.50	Yoon (1961)
	tm/tm ♂ & ♀	74.76	6.22	12.69	6.33	Yoon (1961)
Qv	+/? ♂ & ♀	66.36	11.18	14.44	8.02	Yoon (1961)
	qv/qv ♂ & ♀	60.55	11.46	17.17	10.82	Yoon (1961)
Wd	+/? ♂ & ♀	70.45	9.00	12.89	7.66	Yoon (1961)
	wd/wd ♂	70.81	10.25	11.73	7.21	Yoon (1961)
	wd/wd ♀	73.95	5.89	11.34	8.82	Yoon (1961)
		Concentration in grams/100 ml				
C3H		3.38	0.092	0.463	0.058	Gleason and Friedberg (1953)

Table 16-3. TOTAL BLOOD VOLUMES

Strain	Genotype	Blood volume, ml/100 g body wt	Method	Reference
Akm	+/+	12.1 ± 0.8	Radio-iodoprotein exsanguination	Kaliss and Pressman (1950)
C57	+/+	4.9 ± 0.17	Exsanguination	Taylor (1945)
dba	+/+	5.23 ± 0.31	Exsanguination	Taylor (1945)
$WBB6F_1$	$W/+$	5.5	Fe^{59} labeled RBC	Keighley et al. (1962)
$WBB6F_1$	W/W^v	5.5	Fe^{59} labeled RBC	Keighley et al. (1962)
Not reported	+/+	5.59	Extraction of hemoglobin	Grüneberg (1941)
Not reported	+/+	6.3	Perfusion and exsanguination	Oakley and Warrack (1940)
$(Ak \times Rf)F_1$	+/+	5.2	Perfusion and exsanguination	Furth and Sobel (1946)
	+/+	10.9	Dye dilution with T1824	Furth and Sobel (1946)
	+/+	5.7	Dye T1824 and exsanguination	Furth and Sobel (1946)

Other serum constituents. Isolated reports on other serum constituents have been encountered. These deal with glucose levels, serum cholesterol, and vitamins. Hiestand et al. (1947) reviewed the situation with respect to blood sugar levels. They reported that unstarved Purdue Swiss mice had a mean blood sugar of 173.8 mg/100 ml of blood while those fasted for 24 hours averaged 108.9 mg. Ritchey et al. (1947) summarized the results obtained on determinations of vitamin concentrations, and Morris and Courtice (1955) indicated that mice have characteristic values of total fatty acids of 10.0 ± 0.4 mEq/liter, and total cholesterol of 97 ± 4.4 mg per cent. Inbred strains, however, do differ in serum cholesterol level (Bruell, 1963) and the action of the genes controlling this level is modified by sex.

Blood volume

The over-all size of the cell-filled fluid component of blood has been estimated by several means. Even though blood volume is influenced by many things, it is reasonably constant under normal laboratory conditions, being estimated at 5.5 ml/100 g of body weight for adults. It varies with age (Grüneberg, 1941), blood or fluid loss, vascular dilation due to heat or drugs, and salt imbalance. Hypoxia and altitude also have an influence. Short-term compensation is achieved by dilation or contraction of liver, spleen, and the vascular tree, so the amount of fluid available to the heart for distribution is remarkably constant.

Various means have been employed to determine blood or plasma volume. They often yield different results as seen in Table 16-3. An interesting comparison of methods with simultaneous determinations of hematocrit, plasma volume, red cell volume, and total blood volume has been presented by Wish et al. (1950).

Estimates of total blood volume may be obtained by exsanguination or by the injection of known amounts of material which are diluted by the blood but which escape slowly from the vascular system. Brilliant Vital Red T1824 and iodinated serum albumin (I^{131}) have been employed but yield larger blood volume values than do methods employing erythrocytes tagged with radioisotopes (P^{32}, Cr^{51}, Fe^{59}), since the former are soluble and expand into pericapillary spaces and lymphatics. It goes without saying that the tag must be distributed homogeneously, that the sample must be representative, and that the investigator must be able to measure the dilution easily if the isotope dilution method is to be of value.

Blood pressure and heart rate

It is well known that blood pressure depends upon the pumping action of the heart, the peripheral resistance, the quantity of blood in the arterial system, the viscosity of blood, and upon the elasticity of the vascular walls. The mouse has been little studied in this regard. In fact, there are only four reports dealing with determination of systolic pressure (Table 16-4).

Mean arterial pressures depend upon the elasticity of the arterial walls and the volume of blood within the arterial system. The volume in the system is dependent on cardiac output, which is a function of cardiac frequency and cardiac stroke volume. Cardiac frequency or heart rate in turn is coupled with the rate of metabolism, and any alteration in metabolism is reflected in alteration of the heart rate and cardiac output.

Table 16-4. BLOOD PRESSURE AND PULSE RATE

Strain	Age, months	No. obs.	Systolic pressure, mm Hg (mean ± S.E.)	Pulse count per minute (mean ± S.E.)	Reference
A	Adult	20	107.8 ± 1.8		Edwards and Reinecke (1953)
A/J ♂	9	800	83.9 ± 1.7	589 ± 11	Schlager (1965)[a]
BALB/cJ ♂	9	800	104.9 ± 1.7	494 ± 11	Schlager (1965)[a]
CBA/J ♂	7	800	97.4 ± 1.9	657 ± 8	Schlager (1965)[a]
CBA ♂	2	12	83 ± 5		Henry et al. (1963)
CBA ♂	14	11	103 ± 6		Henry et al. (1963)
CBA ♂	20	10	99 ± 4		Henry et al. (1963)
C3H	2–5		111 (95 to 138)		Wu and Visscher (1947)
C3H	13–14		136 (114 to 165)		Wu and Visscher (1947)
C3H	31–32		151 (138 to 164)		Wu and Visscher (1947)
C57BL/6J ♂	9	800	93.3 ± 2.2	633 ± 11	Schlager (1965)[a]
DBA/2J ♂	7	800	89.1 ± 2.0	614 ± 11	Schlager (1965)[a]
RF/J ♂	6	800	96.0 ± 1.8	595 ± 11	Schlager (1965)[a]
SJL/J ♂	9	800	96.0 ± 2.0	639 ± 14	Schlager (1965)[a]
129/J ♂	8	800	88.7 ± 2.0	630 ± 8	Schlager (1965)[a]

[a] Personal communication.

Richards et al. (1953) reported that "newborn mice" have a heart rate of 286 ± 56.8 (standard deviation) beats per minute and "adult mice" have a heart rate of 632 ± 51.3 (standard deviation). Other values appear in Table 16-4. There seems to be no literature on peripheral resistance or cardiac stroke volume.

Blood pH

The pH of the blood is related to the base-binding capacity of hemoglobin and the extent of oxygenation of hemoglobin. For practical purposes, hemoglobin and phosphate esters together constitute the major portion of the nondiffusible anions available for this purpose. The pH of red blood cells is on the alkaline side of the hemoglobin isoelectric point and as a result the alkali-binding capacity increases linearly with pH. Moreover, it has been shown that oxygenated hemoglobin binds more alkali than reduced hemoglobin. Thus, a decrease in carbon dioxide tension or an increase in oxygenation of hemoglobin results in an elevated blood pH. Wolfe's (1965) observations (Table 16-5) bear out these generalizations, for in every case the pH of arterial blood (left ventricular) is higher than venous blood obtained from the right ventricle, which in turn is generally more alkaline than sinus blood. Marked variations in strains have been observed by Wolfe (1961), Weir (1955), Weir and Clark (1955), Dolyak and Weir (1956), and others, but these workers report that neither sex nor age had a significant effect on blood pH. Heritable differences in pH were not necessarily associated with variations in total serum protein (Dolyak and Weir, 1956).

There is some indication that diet and environmental temperature may influence blood pH (Weir, 1949). Weir found that blood pH increased with temperature. His data indicated that the pH of blood from mice of strain RI rose from 7.302 to 7.336 when their environmental temperature was elevated from 24 to 32°C, and this was true for stain Ba as well, for pH's of 7.228 and 7.270 were observed in mice kept at 24 and 32°C, respectively. The pH elevation may have been due to rise in body temperature, but Weir found evidence against this since rectal temperatures of five mouse strains did not correlate with their blood pH's.

RESPIRATION

Oxygen consumption, respiratory rate, and basal metabolism

Surprisingly little information is available on respiratory rate, oxygen consumption, and basal metabolism in mice. Most of the data, presented in Tables 16-6, 16-7, and 16-8, were obtained from Spector's *Handbook of Biological Data* (1956). This material is supplemented by information obtained from other investigators. See Ta-

Table 16-5. MEANS AND STANDARD ERRORS OF MEANS OF BLOOD pH

Strain	No. of mice	*Venous blood* Orbital sinus	*Venous blood* Right ventricle	*Arterial blood* Left ventricle	Reference
A/HeJ	22	7.31 ± 0.008	7.30 ± 0.011	7.40 ± 0.009	Wolfe (1965)[a]
AKR/J	22	7.32 ± 0.010	7.32 ± 0.008	7.39 ± 0.007	Wolfe (1965)[a]
C3H/HeJ	30	7.29 ± 0.010	7.27 ± 0.010	7.38 ± 0.014	Wolfe (1965)[a]
CBA/J	24	7.29 ± 0.012	7.29 ± 0.009	7.37 ± 0.012	Wolfe (1965)[a]
C57BL/6J	21	7.30 ± 0.012	7.29 ± 0.008	7.36 ± 0.010	Wolfe (1965)[a]
DBA/1J	17	7.23 ± 0.020	7.26 ± 0.014	7.37 ± 0.011	Wolfe (1965)[a]
DBA/2J	30	7.29 ± 0.012	7.30 ± 0.007	7.41 ± 0.009	Wolfe (1965)[a]
SJL/J	14	7.25 ± 0.018	7.30 ± 0.007	7.40 ± 0.009	Wolfe (1965)[a]
SWR/J	12	7.28 ± 0.015	7.31 ± 0.011	7.43 ± 0.009	Wolfe (1965)[a]
129/ReJ	23	7.32 ± 0.006	7.30 ± 0.006	7.37 ± 0.014	Wolfe (1965)[a]
Ba	14		7.358	7.470	Weir (1949)
pHH	50		7.428 ± 0.0065	7.480 ± 0.0062	Wolfe (1959)
pHL	37		7.381 ± 0.0084	7.475 ± 0.0076	Wolfe (1959)
RI	16		7.439	7.540	Weir (1949)
S	12		7.417	7.538	Weir (1949)
Z	2		7.30	7.40	Weir (1949)

[a] Personal communication.

bles 16-6 and 16-7 for a summary of respiratory characteristics and energy metabolism, and 16-8 for consumption of oxygen by mouse tissues.

Body temperature

For the mouse, maintenance of a constant body temperature is a major physiological problem. This stems largely from the high surface-to-volume ratio characteristic of mice. The large surface area presents an excellent opportunity for heat loss through radiation and convection, and additional heat loss takes place through vaporization in the lungs. A high metabolic rate thus is necessary if a balance between heat production and

Table 16-6. RESPIRATORY CHARACTERISTICS OF MICE

Character	Mean value and range	Reference
Lung ventilation (Wt and type of mouse not specified)		
Respirator frequency	163 (84–230)/minute	Guyton (1947)
Tidal volume (ml)	0.15 (0.09–0.23)	Guyton (1947)
Minute volume (liters)	0.024 (0.011–0.036)	Guyton (1947)
Oxygen consumption (Wt and type of mouse not specified)		
Basal metabolism	1530 mm³ O_2/g per hr	Morrison (1948)
Resting metabolism	3500 mm³ O_2/g per hr	Pearson (1947)
Oxygen consumption		
"House mouse" 15.8 g, 24–28°C	1.53 ml/g per hr	Morrison (1948)
Swiss strain "White mice" ♂ 21.2 g, 28.5°C	1.59 ml/g per hr	Morrison (1948)
DBA/2J ♂ (35 to 42 days of age, 29°C)	1.24 ml/g per hr	Schlesinger and Mordkoff (1963)
C57BL/6J ♂ (35 to 42 days of age, 29°C)	1.47 ml/g per hr	Schlesinger and Mordkoff (1963)
B6D2F$_1$ ♂ (35 to 42 days of age, 29°C)	1.38 ml/g per hr	Schlesinger and Mordkoff (1963)
DBA/2J ♀ (35 to 42 days of age, 29°C)	1.47 ml/g per hr	Schlesinger and Mordkoff (1963)
C57BL/6J ♀ (35 to 42 days of age, 29°C)	1.41 ml/g per hr	Schlesinger and Mordkoff (1963)
B6D2F$_1$ ♀ (35 to 42 days of age, 29°C)	1.48 ml/g per hr	Schlesinger and Mordkoff (1963)

Table 16-7. BASAL AND RESTING ENERGY METABOLISM. (*Modified from Spector, 1956, with permission of the author and publisher.*)

Characteristic	Adult albino	Adult dwarf	Adult obese
Body weight (kg)	0.02	0.008	0.06
Body surface (m^2)	0.005	0.004	0.01
Basal metabolism			
kcal/kg per day	170	125	130
kcal/m^2 per day	525	280	550

heat loss is to be achieved. The requirement for heat production is accentuated at low environmental temperatures. See Table 16-9 for the relationship between environmental temperature and caloric expenditure necessary to maintain heat balance.

Mice, in general, have poor temperature regulatory mechanisms and, when young, are even poikilothermic (Barnett, 1956). External temperatures above 33°C usually produce in adults an elevation in body temperature proportional to the increase above this level. Mice exposed to temperatures below 18.5°C develop a hypothermia corresponding to the drop below this point (Mills et al. 1939). Between 18 and 33°C, only minor variations in rectal temperatures are encountered (Table 16-10).

Table 16-8. TISSUE OXYGEN CONSUMPTION. (*Modified from Spector, 1956, with permission of the author and publisher.*)

Tissue	mm^3 O_2/mg dry wt per hr	Medium[a]
Adrenal	6.0	S
Brain cortex	32.9	RP
Cerebral cortex	11.0	R
Embryo	10.4	RG
Kidney cortex	46.1	RP
Liver	8.8–13.8	RG
Liver	18.7	R
Liver	23.1	RP
Lung	7.3–8.0	RG
Lung	12.0	RP
Ovary	9.0	S
Pituitary	8.0	S
Newborn skin	6.1	RG
Spleen	16.9	RP

[a] S = Serum. RP = Calcium free Ringer's solution with added pyruvate, fumarate, glutamate, and gluclose. R = Standard Ringer's solution. RG = Ringer's solution plus glucose.

Table 16.9. RELATIONSHIP OF ENVIRONMENTAL TEMPERATURE TO HEAT PRODUCTION. (*Modified from Spector, 1956, with permission of the author and publisher.*)

Temperature, °C	Heat production, kcal/m^2 per day
14.6	1741
20.0	1037
24.9	953
29.9	879
35.3	1009

McLaren (1961), studying the effects of strain, sex, age, and sexual condition on body temperature, reported a highly significant decrease in temperature with age in strain C3H/Bi, but no change was observed in C57BL/How. There were no significant effects of pregnancy or lactation in the C57BL strain, whereas in mothers of the C3H strain there was an elevation of 1°C shortly after delivery. The mother's temperature returned gradually to control levels when the litter was 20 days old, a time when effective lactation had ceased.

With a drop in body temperature there is a decrease in heart rate. Richards et al. (1953) found that heart rate was lowered from 600 beats per minute to 200 or fewer when body temperature was lowered from 37 to 20°C.

EXCRETION

A fully functional integrated excretory system is essential in maintenance of the internal milieu. It not only provides for constancy of the internal environment but also provides for the removal of end products of metabolism from the blood and maintenance of proper osmotic pressure. Such a system regulates the outflow of gases, water, salts, and a wide variety of nitrogenous and nonnitrogenous organic compounds. It should be emphasized that although the kidney is one of the most important excretory organs it is not the only one. In fact, every mammalian surface is a potential excretory organ. Excretion of water and salts, for example, occurs through the skin. The liver is known to excrete bile and sterols; the lungs regulate blood pH by removing carbon dioxide, and viscosity of the blood is regulated in part by excretion of water by this organ. The intestine, though not ordinarily considered as an excretory

Table 16-10. BODY TEMPERATURE

Environmental temperature, °C	Genotype	Rectal temperature (mean ± S.E.)	Reference
20	+/+	36.97 ± 0.11	Weir (1947)
20	*wa-2/wa-2*	37.08 ± 0.16	Weir (1947)
20	+/+	37.12 ± 0.12	Weir (1947)
20	*pn/pn*	36.74 ± 0.13	Weir (1947)
20	*wa-2/wa-2 hr/hr*	36.53 ± 0.13	Weir (1947)
20	*sh/sh*	36.89 ± 0.08	Weir (1947)
20	+/+	37.11 ± 0.08	Weir (1947)
5	+/+	31.2	Congdon (1912)
16	+/+	34.2	Congdon (1912)
21	+/+	38.1	Larson et al. (1940)
25	+/+	35.3	Congdon (1912)
36.7	+/+	40.0	Larson et al. (1940)
	+/+	37.39 ± 0.39	Herter and Sgonina (1938)
	c/c	37.68 ± 0.26	Herter and Sgonina (1938)
	v/v	37.37 ± 0.41	Herter and Sgonina (1938)

organ, is involved in regulation of fluid volume, and the excretion of excess calcium and magnesium. Nevertheless, major emphasis will be placed here on the production of urine and upon its physical and chemical attributes.

Water balance

Since the main constituent of urine is water, it is appropriate to consider its metabolic turnover before turning to nonaqueous components. Barbour and Trace (1952) estimated that a 21-g albino mouse turns over 4.28 ml water per day. Water intake averages 2.12 ml and metabolic water amounts to 2.16 ml. Urine output averages 0.9 ml while the remainder (3.38 ml) is lost through respiration or in the feces or is incorporated into protoplasm. In an adult mouse it appears that the major portion of excreted water is vaporized and removed from the respiratory system by pulmonary ventilation.

Density of urine

Mouse urine generally is highly concentrated as indicated by an unusually high specific gravity and high content of total solids. These values depend to a considerable extent upon the intake of water and the excretion of urine, for as the water intake rises, and as the urine output increases, there is a progressive decrease in the total solids content of urine per unit volume. This is evident from the data presented in Tables 16-11 and 16-12.

In most strains there are few significant quantitative changes with age. However, Hummel (1964, personal communication) reported that strains SWR and MA deviate significantly in regard to water balance (see Chap. 20 and 29). For example, 4-month-old females of strain MA drank on the average 5.6 ml of water per day and excreted 2.9 ml urine per day (total solid concentration of 9.7 per 100 g). Values for water consumption and urine output increased gradually with age, while urine total solids decreased. At 17 and 19 months of age water consumption in this strain was up to 28.6 ml per day. Silverstein et al. (1961*a*) reported similar observations for strain STR/N. They found that the mean urine osmolarity of STR/N mice (298 milliosmols/kg) was approximately one-fifth of the control value, and sodium content (15.5 vs. 151 mEq/liter) and total ionic concentration (3780 vs. 43,800 ppm) in the urine was much below that found in controls. No protein or reducing substances were detected. The plasma of these polydipsic mice was normal in osmolarity (325 milliosmols/kg) and total ionic concentration (7960 ppm). Twenty-four-hour urine output equalled 20.8 ml and urine total solids were found to be 1.1 g per 100 g. Calculations of 24-hour urine total solids showed little deviation from normal.

Proteinuria

In contrast to most other mammals, proteinuria is a normal condition in mice. The high total-solids content of urine and high specific gravity reflect to some extent this passage of protein into the urine. It has been recognized as being normal for mice since 1933 (Parfentjev and Perlzweig) and is probably characteristic of all strains (Table 16-13). Finlayson and Baumann (1958) presented electrophoretic evidence indicating that the proteins excreted were the same regardless of the strain. Wicks (1941) showed that males usually excreted more protein than females. This apparently stemmed from the influence of testosterone on kidney function. Castration decreased the protein content of male urine, whereas testosterone increased the proteinuria. Oestrone and progesterone had no effect on protein content (Thung, 1962). Wicks (1941) also showed that proteinuria was influenced by diet, being higher on a 33 per cent casein diet than on a 5 per cent diet.

The protein excreted has bound to it both cholesterol and Δ^7-cholestenol. Finlayson and Baumann (1957) noted strain differences in the amount of sterols encountered. Rockland mice, for example, excreted 3.0 μg of cholesterol per day, and 0.3 μg of Δ^7-cholestenol, whereas Swiss Webster mice excreted 0.7 μg and 0.1 μg, respectively, of the two compounds.

Mice normally excrete a considerable amount of albumin and prealbumin but very little α_2-globulin. Mice with a minor to medium degree of experimental amyloidosis of liver and spleen (Chap. 29) showed no definite abnormal excretion of protein, but mice with kidney amyloidosis added β-globulins to the normal excretion of albumin and prealbumin.

Aminoaciduria

More attention has been devoted to amino acid excretion. Harris and Searle (1953) found that taurine was the only ninhydrin-positive material present on paper chromatograms of urine of mice fed on a standard diet. This was present in urine of nearly all mice whether they were from inbred strains (A, C57BL, or CBA), or were homozygous for mutant genes (A^w, a^t, *a*, *b*, c^{ch}, c^e, *c*, *d*, d^l, *f*, *fi*, *je*, Mi^{wh}, *p*, *ru*, *sh-2*, *se*, *un*, *Va*, W^v), or were heterozygous for others (A^y, *ch*, *gl*, *hy-3*, *mi*, *Sd*, or *W*). Differences in the concentration of ex-

Table 16-11. QUANTITATIVE CHARACTERISTICS OF URINE

Strain	Sex	Genotype	No. obs.	Mean specific gravity	Mean total solids, g/100 g	Mean urine osmolality, osmol/kg	Reference
A/HeN	♂		12	1.076		2.56	Silverstein (1961)
A/LN	♂		7	1.063		2.14	Silverstein (1961)
BALB/cAnN	♂		8	1.078		2.63	Silverstein (1961)
C57L/HeN	♂		13	1.050		1.72	Silverstein (1961)
DBA/2JN	♂		9	1.067		2.27	Silverstein (1961)
LAF_1			50	1.065			Bahn et al. (1957)
STR/N	♂		12	1.030		1.06	Silverstein (1961)
STR/1N	♂		8	1.037		1.29	Silverstein (1961)
STR/1N	♀		13	1.041		1.42	Silverstein (1961)
"mice"	♂			1.058			Parfentjev and Perlzweig (1933)
C57BL/6J	♀	+/+	21		12.5		Bernstein (original data)
	♂	+/+	25		14.5		Bernstein (original data)
	♀	*ja*/+	21		15.8		Bernstein (original data)
	♂	*ja*/+	17		15.1		Bernstein (original data)
	♀	*ha*/+	16		15.0		Bernstein (original data)
	♂	*ha*/+	15		14.8		Bernstein (original data)
DBA/2J	♀	+/+	18		14.3		Bernstein (original data)
	♀	*ha*/+	19		12.5		Bernstein (original data)
WC/Re	♀	+/+	21		12.8		Bernstein (original data)
	♂	+/+	17		16.1		Bernstein (original data)
	♀	W^v/+	17		12.1		Bernstein (original data)
	♀	*Sl*/+	18		12.5		Bernstein (original data)
	♂	*Sl*/+	17		13.8		Bernstein (original data)

Table 16-12. WATER BALANCE IN YOUNG ADULTS (3 TO 5 MONTHS OF AGE)

Strain	Sex	No. of obs.	Body wt, g	Average water intake, ml/24 hr	Average urine output, ml/24 hr	Average urine total solids, g/100 g	Reference
A/HeN	♂	12	29.7	4.8			Silverstein (1961)
A/LN	♂	7	24.1	5.7			Silverstein (1961)
BALB/cAnN	♂	8	25.6	4.3			Silverstein (1961)
BALB/cJ	♂	3		3.3	1.3	16.2	Hummel (1965)[a]
BALB/cJ	♀	3		4.3	1.3	13.8	Hummel (1965)[a]
C3HeB	♂	3		2.9	1.0	13.7	Hummel (1965)[a]
C3HeB	♀	3		3.0	0.9	8.9	Hummel (1965)[a]
C57L/HeN	♂	13	26.4	4.6			Silverstein (1961)
DBA/2JN	♂	9	26.5	5.4			Silverstein (1961)
$D1C3F_1$	♂	5		2.4	1.0	15.2	Hummel (1965)[a]
$D1C3F_1$	♀	6		2.5	1.0	13.3	Hummel (1965)[a]
IHB		9	29.8		1.7		McNutt and Dill (1963)
MA	♂	5		4.4	1.5	12.5	Hummel (1965)[a]
MA	♀	5		5.6	2.9	9.7	Hummel (1965)[a]
NH		10	25.3		1.7		McNutt and Dill (1963)
STR/N	♂	12	29.1	10.0			Silverstein (1961)
STR/1N	♂	8	33.7	10.0			Silverstein (1961)
STR/1N	♀	13	28.4	6.8			Silverstein (1961)
SWR	♂	4		4.0	1.7	15.2	Hummel (1965)[a]
SWR	♀	6		6.4	3.6	11.1	Hummel (1965)[a]

[a] Personal communication.

creted taurine were also observed, it being very much higher in C57BL (0.4 per cent) than in strain A mice (0.04 per cent), though there was no significant difference in serum concentration of taurine in the two strains.

Bennett (1961) found no qualitative effect of the brain hernia gene (*bh*) on excretion of amino acids in urine, but pointed out that the amount of amino acid excreted in *bh/bh* individuals was constantly aberrant. She was concerned because these mice developed polycystic kidney disease as part of their syndrome. Goodman (1956) ana-

Table 16-13. PROTEINURIA

Strain	Sex	*Milligrams urinary protein nitrogen per* Mouse/day (range)	100-g mouse/day (range)	Protein content, mg/ml (mean and range)	Reference
C3H	♂	0.6 –1.0	3.4 –3.8		Finlayson and Baumann (1958)
C3H	♀	0.03–0.05	0.15–0.25		Finlayson and Baumann (1958)
Rockland	♂	1.2 –3.1	4.2 –7.5		Finlayson and Baumann (1958)
Sutter stock	♂	1.0 –2.8	3.1 –7.5		Finlayson and Baumann (1958)
Sutter stock	♀	0.25–0.65	0.75–1.7		Finlayson and Baumann (1958)
Swiss Webster	♂	2.1 –2.2	5.3 –7.2		Finlayson and Baumann (1958)
CBA	♂			25.8 (19.8–32.2)	Thung (1962)
CBA	♀			10.6 (6.2–17.1)	Thung (1962)
C57BL	♂			15.3 (7.6–29.2)	Thung (1962)
C57BL	♀			6.8 (5.5– 7.9)	Thung (1962)
WLL	♀			9.6 (7.4–12.0)	Thung (1962)

lyzed urinary amino acids after feeding D, L, and DL forms of tyrosine, methionine, phenylaline, tryptophane, and histidine to mice of the W^v/W^v, $W^v/+$, and $+/+$ genotypes. She concluded that $+/+$ mice utilize both D and L isomers, whereas W^v/W^v use only L forms and excrete D forms. Heterozygotes are intermediate. The differential excretion of these amino acids was attributed to the action of the *W*-allele, since differences in genetic background (C57BL vs. BALB/c) did not influence excretion pattern.

In general, urinary excretion of amino acids by mice depends upon the types and amounts of amino acids ingested, although Steel et al. (1950) noted that significant alterations in the quantity of amino acids excreted had little relationship with the amounts encountered in the blood. The reader is referred to their paper for the influence of age and diet on urinary amino acid appearing in urine.

Creatinuria

Creatinine and creatine are, next to urea, the most abundant nitrogen-containing substances present in urine. Creatine does not normally appear in the urine of mammals but in the mouse it is a normal constituent. Specific reference is made to its excretion in mice with hereditary muscular dystrophy in Chap. 19. Many physiologists often assume that total creatine, creatine per gram of body weight, total creatinine, and creatinine per gram of body weight are relatively constant for a given mouse, and the data generally bear out the assumption that creatinine excretion is constant and related to body weight. Since creatinine in a 24-hour sample shows less variation than do other urinary constituents, many investigators employ a creatinine:creatine index. Madison (1952) observed that the index values depend upon the nutritional status of the mouse, since fasting supposedly increased the creatine output. She also found sex differences in creatinine excretion, females excreting larger amounts of creatinine than males. Her data also suggested that the short-ear gene (*se*) may affect creatine and possibly creatinine excretion, though the significance was not immediately apparent, since *se* appeared to reduce excretion of these substances in males and increase it in females.

Other organic compounds

Compounds other than those already considered appear in the urine. Among these are allantoin, chlorides, glucose, urea, uric acid, hippuric acid, purine bases, indican, 17-ketosteroids, and estrogens. The concentration of these compounds reported in the literature are summarized in Table 16-14, and the influences of age, strain, and sexual condition on excretion of these substances may be found in a paper by Karnofsky et al. (1944).

Table 16-14. COMPOSITION OF NORMAL MOUSE URINE

Constituent	*Weight,*[a] *mg/24 hr* ♀	♂	Per cent[b]	Per cent of total nitrogen[b]
Allantoin	8.38	8.11	0.92	8.7
Ammonia nitrogen	4.68	5.48	0.36	7.3
Chloride	5.79	5.75	0.62	
Chondromucoid protein			1.52	6.0
Creatine	0.855	1.024	0.09	
Creatinine	0.567	0.670	0.10	1.7
Glucose	1.98	3.09		
Hippuric acid			0.28	
Indican			0.005	
Inorganic phosphorus			0.43	
Inorganic sulfate			0.15	
Organic sulfate	1.58	1.64		
Purine bases			0.01	
Titerable acidity	4.68	5.67		
Total nitrogen	40.2	40.8	4.15	
Total phosphorus	2.77	3.23		
Total sulfur			0.27	
Urea nitrogen	24.3	29.8	5.70	65.0
Uric acid			0.04	

[a] Madison (1952).
[b] Parfentjev and Perlzweig (1933).

Inorganic ions

Strain differences in the excretion of sodium and potassium have been investigated by McNutt and Dill (1963). These authors indicate that strains IHB (salt-resistant), and NH (salt-sensitive) differ in excretion of potassium but do not differ perceptibly in the amount of excreted sodium (122 mEq/liter). Urinary potassium concentration was slightly but not significantly higher in IHB (103 ± 7.5 mEq/liter) than in NH (91 ± 7.8 mEq/liter) mice, though the 24-hour urine volumes and total amounts of sodium and potassium were identical. Sodium administered in the form of 4 per cent NaCl in the drinking water elevated urine concentration of sodium to 310 mEq/liter and potassium to 274 mEq/liter within 24 hours in the IHB strain and to 430 and 196 mEq/liter, respectively, in the NH strain. The NH strain had consistently higher mean values for sodium in urine compared with the IHB strain and lower mean potassium values. The differences between means, however, were not statistically significant.

SUMMARY

In this chapter I have limited the presentation to the physiological and biochemical features of respiration, circulation, and excretion, and excluded other areas of physiology such as sense perception, nutrition, and metabolism, which are covered in other chapters. Within this restricted area attention has been devoted to normal values and permissible deviations and emphasis has been placed on variations attributable to differences of strain, sex, or genotype, and to the influence of diet and environment.

Physiologically the mouse is like other mammals, differing from the larger members of the class by its small size, its consequent large surface-to-volume ratio, and its higher metabolic rate. Despite superficial differences, the mouse possesses the same complex physiological control systems characteristic of other terrestrial mammals.

The laboratory mouse has been little studied physiologically and much micromethodology needs to be developed before its full potential as an experimental animal is realized.

LITERATURE CITED

Ashton, G. C., and A. W. H. Braden. 1961. Serum β-globulin polymorphism in mice. Austral. J. Biol. Sci. 14:248–253.

Bahn, R., J. Furth, E. Anderson, and E. Gadsden. 1957. Morphologic and functional changes associated with transplantable ACTH-producing pituitary tumors in mice. Amer. J. Pathol. 33: 1,075–1,085.

Barbour, H. G., and W. F. Hamilton. 1926. The falling drop method for determining specific gravity. J. Biol. Chem. 69:625–640.

Barbour, H. G., and J. Trace. 1937. Standard metabolism in the white mouse. Amer. J. Physiol. 118:77–86.

Barnett, S. A. 1956. Endothermy and ectothermy in mice at −3°C. J. Exp. Biol. 33:124–133.

Bennett, D. 1961. A chromatographic study of abnormal urinary amino acid excretion in mutant mice. Ann. Hum. Genet. 25:1–6.

Bruell, J. H. 1963. Additive inheritance of serum cholesterol level in mice. Science 142:1,664–1,666.

Cinader, B., and S. Dubiski. 1963. An alpha-globulin allotype in the mouse. Nature 200:781.

Cohen, B. L. 1960. Genetics of plasma transferrins in the mouse. Genet. Res. 1:431–438.

Cohen, B. L., and D. C. Shreffler. 1961. A revised nomenclature for the mouse transferrin locus. Genet. Res. 2:306–308.

Congdon, E. D. 1912. The surroundings of the germ plasm. III. The internal temperature of warm-blooded animals in artificial climates. Arch. Entwickl. Mech. Organismen 33:703–715.

Dolyak, F., and J. A. Weir. 1956. Electrophoretic analyses of the serum proteins of high and low blood-pH line of mice. Trans. Kansas Acad. Sci. 59:346–350.

Dougherty, T. F., and A. White. 1944. Influence of hormones on lymphoid tissue structure and function. Endocrinology 35:1–14.

Edwards, C. C., and R. M. Reinecke. 1953. Effect of ischemia of the tail of the mouse on the subsequent local blood pressure. Amer. J. Physiol. 174:289–292.

Eveleth, D. F. 1937. Comparison of the distribution of magnesium in blood cells and plasma of animals. J. Biol. Chem. 119:289–292.

Finlayson, J. S., and C. A. Baumann. 1957. Protein-bound sterols in rodent urine. Amer. J. Physiol. 190:297–302.

Finlayson, J. S., and C. A. Baumann. 1958. Mouse proteinuria. Amer. J. Physiol. 192:69–72.

Furth, J., and H. Sobel. 1946. Hypervolemia secondary to grafted granulosa-cell tumor. J. Nat. Cancer Inst. 7:103–113.

Geinitz, W. 1954. Serum proteins of animals frequently used for experiments or serum production. Klin. Wochensch. 32:1,108–1,111.

Gleason, T. L., and F. Friedberg. 1953. Filter paper electrophoresis of serum proteins from small animals. Physiol. Zool. 26:95–100.

Goodman, R. M. 1956. The effect of the W^v allele in the mouse on the differential excretion of the optical isomers of several amino acids. J. Exp. Zool. 132:189–218.

Grüneberg, H. 1941. Growth of the blood of the sucking mouse. J. Pathol. Bacteriol. 52:323–339.

Guyton, A. C. 1947. Measurement of respiratory volumes of laboratory animals. Amer. J. Physiol. 150:70–77.

Harris, H., and A. G. Searle. 1953. Urinary amino acids in mice of different genotypes. Ann. Eugen. 17:165–167.

Henry, J. P., J. P. Meehan, G. Santisteban, and

P. Stevens. 1963. Age variation of the blood pressure of male CBA mice. Fed. Proc. 22:455. (Abstr.)

Herter, K., and K. Sgonina. 1938. Vorzugstemperatur und Haut beschaffenheit bei Mäusen. Z. Vergl. Physiol. 26:366–415.

Hiestand, W. A., M. F. Hadley, S. E. Mercer, and B. K. Sandock. 1947. Hemoglobin and glycemia levels of the adult white mouse. Proc. Soc. Exp. Biol. Med. 65:324–326.

Hrubant, H. E. 1959. A chromatographic analysis of the free amino acids in the blood plasma of three inbred strains of the house mouse. Genetics 44:591–608.

Kaliss, N., and D. Pressman. 1950. Plasma and blood volumes of mouse organs as determined with radioactive iodoproteins. Proc. Soc. Exp. Biol. Med. 75:16–20.

Karnofsky, D. A., I. T. Nathansen, and J. C. Aub. 1944. The urinary excretion of estrogens, 17-ketosteroids, creatine, and creatinine in high and low mammary tumor strains of mice. Cancer Res. 4:772–778.

Keighley, G., E. S. Russell, and P. H. Lowy. 1962. Response of normal and genetically anemic mice to erythropoietic stimuli. Brit. J. Haematol. 8:429–441.

Kelus, A., and J. K. Moor-Jankowski. 1961. An isoantigen BA of mouse gamma-globulin present in inbred strains. Nature 191:1,405–1,406.

Larson, W. P., M. Levine, R. B. Bieter, and W. F. McLimans. 1940. Study of mouse temperatures with reference to the effect of temperature on sulfanilamide therapy. J. Bacteriol. 39:45.

Lowry, O. H., and T. H. Hunter. 1945. The determination of serum protein concentration with a gradient tube. J. Biol. Chem. 159:465–474.

Madison, C. R. 1952. A search for quantitative differences in the normal constituents of the urine of short ear and normal mice. J. Exp. Zool. 120:457–468.

McLaren, A. 1961. Some causes of variations of body temperature in mice. Quart. J. Exp. Physiol. 46:38–45.

McNutt, W., and R. E. Dill. 1963. Strain differences in mice on a high salt diet. J. Hered. 54: 297–303.

Mills, C. A. 1939. Medical climatology. Charles C Thomas, Springfield, Ill. 296 p.

Morris, B., and F. C. Courtice. 1955. The protein and lipid composition of plasma of different animal species determined by zone electrophoresis and chemical analysis. Quart. J. Exp. Physiol. 40:127–137.

Morrison, P. R. 1948. Oxygen consumption in several mammals under basal conditions. J. Cell. Comp. Physiol. 31:281–291.

Oakley, C. L., and G. H. Warrack. 1940. The blood volume of the mouse. J. Pathol. Bacteriol. 50:372–377.

Parfentjev, I. A., and W. A. Perlzweig. 1933. The composition of the urine of white mice. J. Biol. Chem. 100:551–555.

Pearson, O. P. 1947. The rate of metabolism of some small mammals. Ecology 28:127–145.

Petras, M., and T. A. Church. 1962. The inheritance of a serum esterase component in *Mus musculus*. Genetics 47: 976. (Abstr.)

Phillips, R. A., D. D. Van Slyke, V. P. Dole, K. Emerson, P. B. Hamilton, and R. M. Archibald. 1943. Copper sulfate method for measuring specific gravities of whole blood plasma. U.S. Naval Research Unit, Rockefeller Institute, New York. 51 p.

Popp, R. A. 1961. Inheritance of different serum esterase patterns among inbred strains of mice. Genetics 46:890. (Abstr.)

Rabinowitz, J. L., J. Burns, G. D. Chase, and H. C. Allen, Jr. 1959. Studies on dystrophic mice and their littermates. I. Blood volume determinations. Atompraxis 5:487–488.

Rapoport, S., and G. M. Guest. 1941. Distribution of acid-soluble phosphorus in the blood cells of various vertebrates. J. Biol. Chem. 138: 269–282.

Rask-Nielsen, R., J. Clausen, and H. E. Christensen. 1960. Biochemical changes in experimental amyloidosis. II. Paper electrophoretic investigations on serum and urine from mice with experimental amyloidosis, p. 190–195. *In* Protides of the biological fluids. Proc. 2nd Colloq. Bruges, 1959. Elsevier, Amsterdam.

Richards, A. G., E. Simonson, and M. B. Visscher. 1953. Electrocardiogram and phonogram of adult and newborn mice in normal conditions and under the effect of cooling, hypoxia and potassium. Amer. J. Physiol. 174:293–298.

Ritchey, M. G., L. F. Wicks, and E. L. Tatum. 1947. Biotin, choline, inositol, p-aminobenzoic acid and vitamin B_6 in transplantable mouse carcinomas and in mouse blood. J. Biol. Chem. 171:51–59.

Schlesinger, K., and A. M. Mordkoff. 1963. Locomotor activity and oxygen consumption. J. Hered. 54:177–182.

Shreffler, D. C. 1960. Genetic control of serum transferrin type in mice. Proc. Nat. Acad. Sci. 46:1,378–1,384.

Shreffler, D. C. 1964*a*. Inheritance of a serum prealbumin variant in the mouse. Genetics 49:629–634.

Shreffler, D. C. 1964*b*. A serologically detected variant in mouse serum: further evidence for genetic control by the histocompatibility-2 locus. Genetics 49:973–978.

Shreffler, D. C., and R. D. Owen. 1963. A serologically detected variant in mouse serum: inheritance and association with the histocompatibility-2 locus. Genetics 48:9–26.

Silverstein, E. 1961. Urine specific gravity and osomolality in inbred strains of mice. J. Appl. Physiol. 16:194–196.

Silverstein, E., L. Sokoloff, O. Mickelsen, and G. E. Jay. 1961*a*. Primary polydipsia and hydronephrosis in an inbred strain. Amer. J. Pathol. 38:143–159.

Spector, W. S. 1956. Handbook of biological data. Wright Air Development Center Technical Report 56-273, ASTIA Document No. AD110501. Carpenter Lithographic & Printing Co., Springfield, Ohio. (*Also* publ. as Handbook of biological data, Nat. Acad. Sci. Saunders, Philadelphia, 1960.)

Steel, B. F., M. S. Reynolds, and C. A. Baumann. 1950. Effect of diet on amino acids in blood of mice of various ages. Arch. Biochem. 25:124–132.

Taurog, A., and I. L. Chaikoff. 1946. On the determination of plasma iodine. J. Biol. Chem. 163:313–322.

Taylor, A. 1945. Changes in hemoglobin concentration, total hemoglobin, and blood volume associated with tumor growth. U. Texas Publ. No. 4507:95–102.

Thompson, S., J. F. Foster, J. W. Gowen, and O. E. Tauber. 1954. Hereditary differences in serum proteins of normal mice. Proc. Soc. Exp. Biol. Med. 87:315–317.

Thung, P. J. 1962. Physiological proteinuria in mice. Acta Physiol. Pharmacol. Neer. 10:248–261.

Weir, J. A. 1947. The temperature of the mouse in health and disease. Proc. Iowa Acad. Sci. 54:383–388.

Weir, J. A. 1949. Blood-pH as a factor in genetic resistance to mouse typhoid. J. Infect. Dis. 84:252–274.

Weir, J. A. 1955. Male influence on sex ratio of offspring in high and low blood-pH lines of mice. J. Hered. 46:277–283.

Weir, J. A., and R. D. Clark. 1955. Production of high and low blood-pH lines of mice by selection and inbreeding. J. Hered. 46:125–132.

White, A., and T. F. Dougherty. 1946. The role of lymphocytes in normal and immune globulin production, and the mode of release of globulin from lymphocytes. Ann. N.Y. Acad. Sci. 46:859–883.

Wicks, L. F. 1941. Sex and proteinuria in mice. Proc. Soc. Exp. Biol. Med. 48:395–400.

Wish, L., J. Furth, and R. H. Storey. 1950. Direct determinations of plasma, cell, and organ-blood volumes in normal and hypervolemic mice. Proc. Soc. Exp. Biol. Med. 74:644–648.

Wolfe, H. G. 1959. Blood-pH differences in two inbred strains of mice. J. Hered. 50:155–158.

Wolfe, H. G. 1961. Selection for blood-pH in the house mouse. Genetics 46:55–75.

Wu, C. H., and M. B. Visscher. 1947. Measurement of blood pressure in the mouse with special reference to age. Fed. Proc. 6:231. (Abstr.)

Wunderlich, J. R., and L. A. Herzenberg. 1962. A second gamma-globulin isoantigen (allotype). Genetics 47:995. (Abstr.)

Yoon, C. H. 1961. Electrophoretic analysis of the serum proteins of neurological mutations in mice. Science 134:1,009–1,010.

17

Blood and Blood Formation[1]

Elizabeth S. Russell and Seldon E. Bernstein

Healthy adult mice maintain relatively constant numbers of each of the formed elements of the blood. Maintenance of typical blood cell numbers involves constant regulated replenishment from precursor cells in hematopoietic tissues. Evidences of the importance of genetic factors for hematopoietic balance are found in the significant quantitative differences observed between peripheral blood of mice from different inbred strains and in the various hereditary anemias. Genes are also responsible for differences in the chemical structure of the hemoglobin molecules found in erythrocytes of mice from different inbred strains.

In this chapter we attempt, first, to coordinate diverse information on mouse hematology, directing attention to genetic and developmental variations and considering characteristic values and quantitative variations for each of the formed elements in the blood of normal adult mice. Second, we discuss genetically controlled biochemical polymorphisms of adult mouse hemoglobins, and follow this with a description of blood formation in healthy adults, with discussion of stem cells, erythropoietic stimuli, and erythroid homeostatic mechanisms. Third, we describe changes in blood and blood formation during normal development. Fetal erythropoiesis is presented in some detail, followed by a brief summary of the extensive changes in peripheral blood values which occur during the period of rapid postnatal growth.

In the remainder of the chapter we describe the 12 inherited anemias in the mouse. There is much evidence of faulty erythroid differentiation in six of the single-gene induced anemias, of peripheral cell defects in five other anemias, and of an immune mechanism in one case.

The hereditary anemias of the mouse are intrinsically interesting as constitutional diseases and are also very useful tools for analyzing the function of blood-forming tissue. We indicate the manner of inheritance of each anemic condition and describe the peripheral blood and blood-forming tissue of mice of affected genotypes at all stages of life-history which have been investigated. We also summarize critical evidence on the nature of gene action in causation of certain extensively studied anemias. Several of the anemia-producing genes affect viability and have pleiotropic effects in other tissues, which are discussed in Chap. 29.

QUANTITATIVE VARIATIONS OF ERYTHROCYTES

From 2 months until late in life there is little variation in the number of circulating erythrocytes in the blood of individual mice or of different mice of the same inbred strain, but significant differences in erythrocyte number, hematocrit level, and hemoglobin content have been recorded

[1] The writing of this chapter was supported in part by Public Health Service Research Grant CA 01074 from the National Cancer Institute, in part by Public Health Service Research Grant HD 00254 from the National Institute of Child Health and Human Development, and in part by Contract AT(30-1)-1800 with the U.S. Atomic Energy Commission.

Table 17-1. MEANS AND STANDARD ERRORS OF MEANS FOR ERYTHROCYTE CHARACTERISTICS OF 18 INBRED STRAINS.[a] *(After Russell et al., 1951, with permission of the authors and of the Society for Experimental Biology and Medicine.)*

Strain	RBC[b] (mean and S.E.)	Hematocrit percentage (mean and S.E.)	Mean cell volume, μ^3 (mean and S.E.)	Hemoglobin, g/100 cc blood (mean and S.E.)	Hemoglobin, g/cc cells (mean)	Reticulocytes, per cent (mean)
A/J	9.42 ± 0.28	42.5 ± 0.4	45.1 ± 1.4	12.9 ± 0.2	0.30	3.5
A/HeJ	9.48 ± 0.18	42.5 ± 0.5	44.8 ± 1.0	12.7 ± 0.2	.30	2.9
AKR/J	9.38 ± 0.24	45.6 ± 1.0	48.5 ± 1.6	13.9 ± 0.2	.30	2.3
BALB/cAnJ	10.14 ± 0.15	46.5 ± 0.8	45.9 ± 1.1	14.5 ± 0.2	.31	3.3
BALB/cJ	10.51 ± 0.16	48.0 ± 0.7	45.7 ± 1.0	15.0 ± 0.2	.31	2.9
CBA/J	10.04 ± 0.27	45.0 ± 1.3	44.8 ± 1.8	13.5 ± 0.2	.31	2.6
C3H/J	8.79 ± 0.24	39.5 ± 0.7	44.9 ± 1.4	12.2 ± 0.4	.31	2.8
C3H/ScJ	9.63 ± 0.26	43.0 ± 1.0	44.7 ± 1.6	13.2 ± 0.3	.30	2.2
C57BL/6pJ	9.70 ± 0.15	43.4 ± 0.8	44.7 ± 1.0	13.0 ± 0.3	.30	2.5
C57BL/6J	9.66 ± 0.09	44.0 ± 0.4	45.5 ± 0.6	13.3 ± 0.2	.30	2.6
C57BR/cdJ	10.54 ± 0.17	50.0 ± 0.5	47.4 ± 0.9	14.6 ± 0.2	.29	2.4
C57L/HeJ	9.82* ± 0.20	50.6* ± 0.4	51.5* ± 1.1	14.9 ± 0.2	.29	2.6
DBA/1J	10.52* ± 0.27	43.8* ± 0.6	41.6* ± 1.2	13.2 ± 0.2	.30	1.5
DBA/WaJ	9.93 ± 0.27	43.0 ± 0.6	43.3 ± 1.1	12.5 ± 0.2	.29	2.6
DBA/2J	10.30 ± 0.25	42.6 ± 0.5	41.4 ± 1.1	12.7 ± 0.1	.30	3.1
I/J	10.27 ± 0.27	46.8 ± 0.7	45.6 ± 1.5	13.5 ± 0.1	.29	2.4
RIII/J	9.63 ± 0.25	44.5 ± 0.6	46.2 ± 1.3	13.7 ± 0.2	.31	2.8
ST/J	9.88* ± 0.19	44.1* ± 1.1	44.6* ± 1.4	12.1 ± 0.2	.31	2.1

[a] All means are based on 10 animals, except that those marked * are based on 20 animals.
[b] Red blood count in millions per cubic millimeter.

between mice of different inbred strains (summary and bibliography, Heinecke, 1962; Russell et al., 1951; Table 17-1).

The genetic constancy of inbred mice is demonstrated by the observation in independent strain surveys, long separated in time, of significantly higher erythrocyte numbers in DBA mice, intermediate numbers in C57BL mice, and significantly lower values in C3H mice (Law and Heston, 1941; Russell et al., 1951; Pujman et al., 1954; Storer, 1964, personal communication). Observed differences between inbred strains in hemoglobin content and hematocrit levels (ratio of packed red-cell volume to total blood volume) result largely from differences in number of erythrocytes, the mean erythrocyte volume and mean corpuscular volume varying little in healthy adults (Table 17-1). Russell (1963) has summarized information on adaptations of hematological methods for use with mice.

Erythrocyte numbers somewhat lower than those in Table 17-1 and larger mean cell volumes may be expected in normal mice younger than 2 months of age (Russell, 1949; Russell and Fondal, 1951), and lower erythrocyte numbers are often found in very old mice, particularly in certain inbred strains (Strong and Francis, 1937; Goulden and Warren, 1944; Ewing and Tauber, 1964). Growth of spontaneous or transplanted tumors also frequently causes a drop in circulating erythrocytes, through increased destruction rather than through failure of red cell formation (Davies et. al., 1962; Lockner et al., 1963; Wiadrowski and Metcalf, 1963; Yonehiro and Aust, 1963). Higher erythrocyte numbers and hematocrit values may be observed in mice kept at high altitudes (Lushbaugh and Russell, unpublished data) or exposed to lowered oxygen tension at normal pressure (Keighley et al., 1962).

In adult mice at sea level, reticulocytes typically make up 1 to 3 per cent of total erythrocytes (Table 17-1), a value which corresponds well with recent calculations of erythrocyte lifespan. Following exposure to erythropoietic stimuli, reticulocyte percentages are elevated in some circumstances to 20 to 50 per cent of circulating red cells. Using C^{14}-labeled glycine as hemoglobin precursor, von Ehrenstein (1958) observed that labeled erythrocytes had a mean lifespan of 40 to 43 days in TE and C3H mice. Almost identical values were found following incorporation in vivo of P^{32}-labeled diisopropylfluorophosphate into erythrocytes of (CBA × C57BL)F_1 mice (van Putten, 1958). The half-survival time for transfused isologous normal Cr^{51}-labeled erythrocytes had been estimated at 20 days (Goodman and Smith, 1961) and at 25.6 days (Bernstein, unpublished data).

QUANTITATIVE VARIATIONS OF LEUKOCYTES

Total and differential leukocyte counts of mice vary markedly (summary and bibliography, Heinecke, 1961), being influenced by genetic and environmental factors and methods of handling. Surveys demonstrate significant differences between inbred strains in total leukocyte counts (Gowen and Calhoun, 1943; Russell et al., 1951; Table 17-2, 17-3). A description of all types of mouse leukocytes and of their precursors is given by Dunn (1954).

Almost all of the types of leukocytes which have been observed in human blood are also seen in mouse blood. The proportion of lymphocytes to total leukocytes is much higher in the mouse than in man and it should be noted that basophilic leukocytes are almost never observed in adult mouse blood (Heinecke, 1961). A typical range of total and differential leukocyte counts in mice of six inbred strains differing in resistance to mouse typhoid demonstrated highly significant strain difference in all cell types (Gowen and Calhoun, 1943; Table 17-3). Statistically significant differences between strains in the percentage of granulocytes were also observed in the survey of 18 inbred strains recorded in Table 17-2 (Russell et al., 1951). Independent direct chamber counts of granulocytes and lymphocytes of mice from seven inbred strains demonstrated significant strain differences in both kinds of cells with particularly high lymphocyte levels in C57BL/6J (mean 8, 120 lymphocytes/mm^3) and particularly high granulocyte levels in C3H/J (mean 1,970 granulocytes/mm^3) (Budds et al.,

Table 17-2. MEANS AND STANDARD ERRORS OF MEANS FOR LEUKOCYTE CHARACTERISTICS OF 18 INBRED STRAINS. ALL MEANS ARE BASED ON 10 ANIMALS.

(After Russell et al., 1951, with permission of the authors and of the Society for Experimental Biology and Medicine.)

		Per cent granulocytes			
		♀♀		♂♂	
Strain	WBC[a] Mean and S.E.	Mean and S.E.	S.E./mean	Mean and S.E.	S.E./mean
A/J	8.71 ± 1.37	10.7 ± 1.1	0.10	14.5 ± 5.0	0.35
A/HeJ	6.14 ± 0.68	19.8 ± 2.3	.12	37.2 ± 7.5	.20
AKR/J	6.80 ± 0.65	21.6 ± 4.5	.21	23.8 ± 5.2	.22
BALB/cAnJ	8.70 ± 1.03	15.4 ± 2.3	.15	29.6 ± 5.0	.17
BALB/cJ	8.54 ± 1.04	14.9 ± 2.1	.14	15.3 ± 3.8	.25
CBA/J	6.44 ± 0.33	26.8 ± 4.5	.17	24.2 ± 2.5	.10
C3H/J	7.34 ± 0.80	21.4 ± 4.3	.20	27.5 ± 5.1	.19
C3H/ScJ	5.07 ± 0.30	19.0 ± 3.9	.21	22.0 ± 2.2	.10
C57BL/6pJ	10.61 ± 0.64	12.1 ± 3.8	.31	15.3 ± 5.1	.33
C57BL/6J	11.43 ± 1.04	8.2 ± 1.2	.15	10.4 ± 3.5	.34
C57BR/cdJ	9.43 ± 0.78	10.6 ± 1.1	.10	11.0 ± 2.4	.22
C57L/HeJ	10.82 ± 1.13	8.5 ± 1.3	.15	6.7 ± 1.4	.21
DBA/1J	8.60 ± 1.60	16.5 ± 1.5	.09	20.4 ± 4.2	.21
DBA/WaJ	8.32 ± 0.96	19.6 ± 0.9	.05	17.2 ± 1.5	.09
DBA/2J	9.28 ± 0.29	16.7 ± 2.3	.14	18.1 ± 1.3	.07
I/J	11.62 ± 1.41	13.2 ± 3.2	.24	18.1 ± 2.9	.16
RIII/J	5.87 ± 0.64	26.2 ± 2.8	.11	18.6 ± 3.7	.20
ST/J	7.74 ± 1.23	14.9 ± 2.4	.16	19.4 ± 3.7	.19

[a] White blood count in thousands per cubic millimeter.

Table 17-3. MEAN TOTAL NUMBERS AND DIFFERENTIAL DISTRIBUTIONS OF LEUKOCYTES IN BLOOD OF MICE FROM SIX INBRED STRAINS.

(After Gowen and Calhoun, 1943, with permission of the authors and of the Journal of Infectious Diseases.)

Strains	Number of mice	WBC[a]	Lymphocytes[b]	Large monocytes[b]	Neutrophils[b]	Eosinophils[b]	Degenerates[b]
Ba	38	11.1	65.1	1.15	18.2	2.05	13.6
L	41	11.6	74.1	1.31	14.2	0.93	9.4
E	45	14.6	69.4	2.58	14.0	3.76	10.3
Z	42	16.1	75.3	1.21	9.4	1.93	12.0
RI	41	22.5	67.3	0.67	13.6	2.10	16.4
S	45	19.6	72.8	0.71	15.0	1.98	10.4
F value		*31.3*[c]	*6.3*[c]	*5.0*[c]	*3.2*	*6.4*[c]	*3.4*

[a] White blood count in thousand per cubic millimeter.
[b] Mean per cent of total leukocytes.
[c] Significant beyond the 0.01 level.

1953). In some but not all inbred strains, higher granulocyte counts were observed in females than in males. Mice of one particular inbred strain, SM/J, are characterized by exceedingly low leukocyte counts(mean 2,320 WBC/mm^3) and the difference in leukocyte number between these and mice of the LG/Rr strain (mean, 8,380 WBC/mm^3) is controlled by a small number of genes (Chai, 1957). Further evidence of the heritability of differences in leukocyte count is the production of two stocks of mice, LCH, with a mean leukocyte count of 17,500 WBC/mm^3, and its counterpart, LCL, with a mean of 3,600 WBC/mm^3, by 13 successive generations of selection for high vs. low total leukocyte counts (Weir and Schlager, 1962); the difference in lymphocyte counts was much higher than the difference in granulocytes. Nongenetic factors also influence numbers of white cells. Striking diurnal variations have been reported in total leukocytes, lymphocytes, and eosinophils, all with peak values in the morning, approximately three times the minimal values observed in the evening (Halberg and Visscher, 1950; Panzenhagen and Speirs, 1953; Brown and Dougherty, 1956; Halberg et al., 1953). Repeated daily bleeding from the tail of mice from five inbred strains increased the lymphocyte count by 50 to 120 per cent and altered the differential distribution of cell types (Metcalf and Buffett, 1957). Nembutal anesthesia during blood-letting affected leukocyte values by lowering agranulocyte count in males only and granulocyte counts in both sexes (Budds et al., 1953).

NUMBER OF THROMBOCYTES

Few studies enumerating mouse blood platelets have been reported, but these agree on typical values of 0.7 to 1.2 × 10^6 platelets/mm^3 of whole blood, with a mean circulating lifespan of approximately 4 days (Petri, 1933; Jacobson, 1944; Odell and McDonald, 1960, 1961; Heinecke, 1963). Use of a phase-contrast microscope facilitates counting. The function of these units in blood-clotting is discussed in Chap. 18.

QUALITATIVE PHYSICOCHEMICAL DIFFERENCES IN HEMOGLOBINS

Each of the adult mouse hemoglobins whose structure has been examined has been found to be made up of four globin chains in two unlike pairs. Because of their striking similarities to the α- and β-chains of human hemoglobin, these mouse polypeptide chains are called α-chains and β-chains, or, in the case of certain minor hemoglobin components, β-like chains. Several differing qualities of adult mouse hemoglobins have been found to depend upon action of alleles at two independent genetic loci: *Hbb* (probably a complex locus), believed to control structure of β- and β-like chains; and *Hba*, believed to control structure of α-chains.

The first genetically controlled hemoglobin characteristic reported in mice was a difference in electrophoretic pattern. Several hemoglobins produced a homogeneous pattern, suggesting the presence of a single hemoglobin component, whereas others produced a diffuse pattern, suggesting the presence of two or more hemoglobin components (Ranney and Gluecksohn-Waelsch, 1955). Both of these electrophoretic patterns are widely distributed among inbred strains (Russell and Gerald, 1958; Popp, 1963*b*; Heinecke and Wagner, 1964; Table 17-4). Despite complexities to be discussed later, it is convenient to refer to these electrophoretic patterns as caused by single vs. diffuse hemoglobins. Different numbers of bands have been reported in electrophoretic and chromatographic patterns of diffuse hemoglobins (Welling and van Bekkum, 1958; Rosa et al.,

Table 17-4. KNOWN CHARACTERISTICS OF THE HEMOGLOBINS OF 48 INBRED STRAINS

Inbred strain	α-Chain solubility[a]	α-Chain fingerprint[a]	Electrophoretic pattern	β-Chain tryptophan-containing peptide
A/J, A/He, A/Cum	Like B6		Diffuse	Absent
AKR/J, AKR/Up	Like B6	Like B6	Diffuse	Absent
BALB/cJ	Like SEC	Like SEC	Diffuse	Absent
CBA/Cum	Like C3H		Diffuse	
C3H/Cum	Proto α^{C3H}		Diffuse	
C3H/J, C3HeB/J, C3H.K/Sn-*cc*, C3H.K/Sn-*CC*	Like C3H		Diffuse	
CFW/Rl, CFCW/Rl	Like B6		Diffuse	
C57BL/6J, C57BL/6J-*cc*	Proto α^{B6}	Proto α^{B6}	Homogeneous	Present
C57BL/6Cum	Like B6	Like B6	Homogeneous	Present
C57BR/cdJ			Homogeneous	
C57L/J, C57L/Rl	Like B6		Homogeneous	
C58/J			Homogeneous	
DBA/1J			Diffuse	
DBA/2J, DBA/2Cum	Like B6		Diffuse	
FLEX/1	Unlike B6, SEC		Homogeneous	
FLEX/2	Unlike B6, SEC		Diffuse	
FL/Re-*f*/*f*, FL/Re-+/+	Like B6	Proto α^{FL}	Diffuse	
FUS/Ls	Like B6		Homogeneous	
FU/Rl	Unlike B6, SEC		Diffuse	
HBS	Like B6		Homogeneous	Present
HBD	Like B6		Diffuse	Absent
MA/My			Diffuse	
NB	Proto α^{NB}	Proto α^{NB}	Homogeneous	Present
RF/J, RFM/Up	Like B6		Diffuse	
RUS/Rl	Like B6		Diffuse	
SEA/Rl	Unlike B6, SEC		Diffuse	
SEC/1Re-*Se*	Like SEC	Proto α^{SEC}	Homogeneous	Present
SEC/Rl	Proto α^{SEC}	Like SEC	Homogeneous	
SWR/J	Like B6	Like FL	Homogeneous	Present
WB/Re			Diffuse	
WC/Re	Unlike B6, SEC		Diffuse	
WH/Re			Diffuse	
WK/Re			Homogeneous	
11G/1Rl	Like B6		Homogeneous	
11G/2Rl	Unlike B6, SEC		Diffuse	
101/Cum	Like B6		Diffuse	
129/Rl	Like B6		Diffuse	

[a] *Proto* indicates prototype example of designated character.

SOURCES: Popp (1963*b*); Russell and Gerald (1958); Hutton et al., (1962*a*); Wolfe et al., (1963); Hutton et al., (1964).

1948; Ranney et al., 1960; Morton, 1962; Mehrotra and Cardinali, 1963; Rhinesmith et al., 1964; Riggs, 1965) and even in single hemoglobins (Rosa et al., 1958; Rhinesmith et al., 1964). These differences certainly arise at least in part from differences in experimental conditions. However, the electrophoretic difference observed by Morton (1965) appears to be genetically controlled. A rapid rate of alkali denaturation has been shown for a minor component of CBA hemoglobin, and a normal rate, similar to that of C57BL hemoglobin, for the major component (Welling and van Bekkum, 1958). Studies of fresh and stored hemoglobins from DBA and "Kink" mice showed that the "diffuse hemoglobin from DBA contains several components, at least one of which may undergo aggregation" shown as a heavy peak upon ultracentrifugation, whereas the "single hemoglobin from 'Kink' mice exhibits homogeneity similar to that of human A" (Ranney et al., 1960). Polymerization of certain mouse hemoglobins to eight-chain double molecules may account for part of the diffuse electrophoretic pattern (Riggs, 1965).

Diffuse hemoglobins are more soluble in buffered salt solutions, over a wider pH and molarity range, than are any of the single hemoglobins (Popp and Cosgrove, 1959; Popp and St. Amand, 1960; Popp, 1962*c*; Wolfe et al., 1963). Associated with the difference in solubility is a difference in crystal formation; the homogeneous hemoglobin of C57BL/6 precipitates from buffered salt solution as large hexagonal crystals, the diffuse hemoglobin of strain 101 as a flocculent amorphous precipitate (Popp and Cosgrove, 1959). Two single hemoglobins, from C57BL/6J and SEC/1Re, "showed a single major hemoglobin peak upon chromatography on Amberlite CG-50 resin columns," whereas four diffuse hemoglobins, from AKR/J, FL/Re, C3H/J, and DBA/2J, "showed a major peak in the same region as the single hemoglobins plus an additional minor component which was approximately 20 per cent of the total hemoglobin" (Hutton et al., 1962*a*). All of these qualities of single and diffuse hemoglobins fit well with the concept that there is one stable major hemoglobin in erythrocytes giving a homogeneous electrophoretic pattern, and unequal quantities of two other hemoglobins (at least one of which is somewhat unstable) in erythrocytes giving a diffuse electrophoretic pattern.

Total amino acid compositions have been reported for both α- and β-chains of C57BL/6J (Riggs, 1963) and DBA/2J (Riggs, 1965). Differences in structure of both α- and β-chains of mouse hemoglobins have been identified by two-dimensional chromatography-electrophoresis "fingerprints" (Hutton et al., 1962*a*; Popp, 1962*d*, *f*; Popp et al., 1963) and by counter-current analysis (Rifkin et al., 1965). The amino acid composition of each of the tryptic peptides of the α-chains of C57BL/Cum, BALB/cJ, and NB/Rl has been reported (Popp, 1965*a*), along with a nearly complete sequence of the entire α-chain of C57BL/Cum (Popp, 1965*b*). The amino acid compositions of tryptic peptides from SEC/1Re-*Se* are very similar to or identical with those of BALB/cJ (Rifkin et al., 1965). The amino acid compositions of tryptic peptides of the β-chain of C57BL/6J have also been established (Rifkin et al., 1965).

Peptide analysis of hemoglobins from AKR/J and FL/Re mice showed that the two chromatographically separated hemoglobins within a single mouse had the same α-chain but differing non-α-chains (Hutton et al., 1962*a*). Neither the β-chain of the major AKR/J and FL/Re hemoglobins, nor the β-like chain of their minor hemoglobins, was identical with the β-chain of C57BL/6 hemoglobin. Analysis of this situation is complicated by the tendency of diffuse hemoglobins to polymerize (Riggs, 1965). Popp (1963*a*) has reviewed the experimental methods used in the study of mammalian hemoglobins.

In F_1, F_2, and backcross generations from crosses between inbred mice with electrophoretically single and diffuse hemoglobins, the patterns segregated as if controlled by alleles at a single genetic locus called *Hb* when first discovered (Gluecksohn-Waelsch et al., 1957; Gluecksohn-Waelsch, 1960; Morton, 1965). Later Popp (1965*b*) used the symbol *Hbb* to distinguish this locus from that for the α-chain (*Hba*). The *Hbb* locus has been located in the first genetic linkage group near the albino locus (Popp and St. Amand, 1960; Popp, 1962*a*, *b*; Hutton et al., 1962*b*; Wolfe et al., 1963). The linkage tests involved many different sources of the single (Hbb^s) and diffuse (Hbb^d) hemoglobin electrophoretic types and different methods of identifying segregants. Linkage tests involving double identification of hemoglobin type by electrophoretic pattern and by presence or absence of a particular tryptophan-containing peptide established that an allele controlling the two β- and β-like chains of diffuse hemoglobins from AKR/J and FL/Re mice segregated as a unit from the allele for the single β-chain of C57BL/6J hemoglobin (Hutton et al., 1962*b*). Similar results were obtained in the F_2 generation from crosses between NB/Rl and BALB/cJ (Popp, 1962*f*). If the basis for the diffuse hemoglobin electrophoretic pattern is the presence of two unlike hemoglobins, then the allele at the *Hbb* locus carried by AKR/J, FL/Re, and BALB/c mice is compound, probably consisting of two adjacent cistrons, either or both of which may be homologous to the single cistron of the *Hbb* allele in C57BL/6J mice (Gluecksohn-Waelsch, 1960).

Different alleles at the *Hba* locus (formerly called *Sol*), controlling structure of the α-chain, have been detected through effects on solubility (Popp, 1962*b*, *c*), through differences in fingerprint pattern (Hutton et al., 1962*a*; Popp, 1962*e*) or through counter-current analysis (Rifkin et al., 1965), or by a combination of two techniques (Hutton et al., 1964). The *Hba* alleles of many inbred strains have been identified (Popp, 1963*b*; Table 17-4) and in certain cases the positions of amino acid substitutions have been established (Popp, 1965*a*, *b*; Rifkin et al., 1965). In crosses between inbred strains differing in both α-chain and β-chain structures, differences in the two classes of polypeptide chains assort independently (Popp, 1962*c*, 1963*b*; Hutton et al., 1962*b*; Wolfe et al., 1963). Numerous linkage tests, however, have failed to locate the position of the *Hba* locus on the genetic linkage map (Popp, 1962*c*, 1963*b*; Hutton et al., 1964). All of these genetically induced variations in hemoglobin electrophoretic type, solubility, and structure of α- and β-chains represent normal variations; none has been associated in any way with pathological conditions. Thus in the mouse no hemoglobinopathies have been discovered.

BLOOD FORMATION IN NORMAL ADULT MICE

Except in conditions of severe hematological stress, bone marrow is the principal postnatal site of erythropoiesis and myelopoiesis in the mouse. Most of the bony cavities contain red marrow throughout life, with 85 to 95 per cent of available marrow space filled with hematopoietic cells (Endicott and Gump, 1947; Brecher et al., 1948), and one may easily obtain approximately 5 to 10 million cells from one femur of an adult. Recognition of the differentiating cell types in the marrow of mice presents considerable difficulty, since staining qualities do not correspond well with human marrow cells. An excellent description of classifiable cell types, particularly in the granulocytic series, has been presented by Endicott and Gump (1947), who also gave information on relative numbers of each type (Table 17-5).

It is generally accepted that hematopoiesis starts from undifferentiated stem cells in the mar-

Table 17-5. MYELOGRAMS OF 2-MONTH-OLD FEMALE MICE.
(From Endicott and Gump, 1947, with permission of the authors and of Blood, the Journal of Hematology.)

	C57BR (20)			*CFW (14)*		
Cell classification	Index[a] (mean and S.D.[b])		Per cent in smear (mean)	Index (mean and S.D.[b])		Per cent in smear (mean)
Neutrophil segmenters and metamyelocytes	19.4	6.9	20.7	23.0	6.5	24.4
Neutrophil myelocytes and premyelocytes	11.6	2.8	12.4	15.0	4.4	15.9
Eosinophils (all types)	9.3	3.1	10.0	7.1	1.0	7.5
Total granulocytes	40.0	9.4	42.9	45.1	7.1	47.8
Normoblasts	20.4	6.5	21.9	21.3	6.7	22.6
Pronormoblasts	8.0	3.3	8.6	4.3	2.5	4.6
Total erythropoietic cells	28.4	7.5	30.4	25.6	6.7	27.1
Blast cells	0.3	0.25	0.3	0.8	0.5	0.8
Lymphocytes (all types)	22.6	5.9	24.2	21.4	6.3	22.7
Miscellaneous cells	1.8	0.6	1.9	1.5	0.5	1.6
Marrow cellularity	0.933 ± 0.029[c]			0.943 ± 0.021[c]		

[a] Per cent in smear multiplied by proportion of active cellular marrow.
[b] Standard deviation.
[c] Standard error.

row or spleen (Lajtha, 1962). However, very small numbers of stem cells may be present in the peripheral blood (Goodman and Hodgson, 1962).

Estimates of the number of such stem cells in different blood-forming tissues of normal mice have been obtained by transplanting known numbers of cells into lethally irradiated mice. The number of colonies of hematopoietic cells which develop in the spleens of recipients appears to be proportional to the number of effective stem cells in the donor tissue (Till and McCulloch, 1961; Siminovitch et al., 1963; Cudkowicz et al., 1964). Normal marrow contains approximately 10 colony-forming units (CFU) per 10^5 cells, and normal spleen approximately one CFU per 10^5 cells.

Erythropoiesis may be stimulated in normal mice by treatments which reduce the circulating red-cell mass such as phenylhydrazine-poisoning or bleeding; by exposure to hypoxia with or without reduced air pressure (Grüneberg, 1939; Prentice and Mirand, 1957; Brecher and Stohlman, 1959; Keighley et al., 1962); and by administration of the erythropoietic hormone, erythropoietin (Jacobson and Doyle, 1962). In contrast, erythropoiesis may be completely suppressed in mice by increasing the total red-cell mass either by hypertransfusion (Jacobson et al., 1957) or by hypoxia followed by return to normal oxygen tension (Cotes and Bangham, 1961). Reinitiation of blood formation in polycythemic mice, usually measured by Fe^{59} uptake, has provided a sensitive assay system in which very small amounts of erythropoietin can be recognized and measured (Cotes and Bangham, 1961; Gurney et al., 1961).

Erythropoiesis reinitiated in polycythemic mice by small doses of exogenous erythropoietin (Filmanowicz and Gurney, 1961; Gurney et al., 1961) provides some of the best information on the sequence of morphological events in erythroid maturation. On the first and second days after treatment with erythropoietin, spleens of polycythemic mice showed numerous early erythroid cells considered to be proerythroblast(s), oval to round cell(s) of about 20 to 30 μ diameter with a large pale nucleus that contained only a sparse amount of fine chromatin. On the second and third days these spleens contained very large numbers of later erythroid cells considered to be normoblast(s), smaller round cells of about 10 to 20 μ diameter with a more darkly staining chromatin-clumped nucleus, and cytoplasm varying in color. Not until days 3 and 4 did reticulocytes appear in the circulation (Filmanowicz and Gurney, 1961). This fits well with isotope-incorporation data which show for normal mice an interval of 3 to 4 days between administration of labeled heme precursors such as glycine and δ-aminolevulinic acid and the appearance in the circulating blood of erythrocytes with labeled protoporphyrin (Altman et al., 1953; Altman and Russell, 1964). It should be noted, however, that the time interval between administration of Fe^{59} and near-maximal labeling of erythrocytes is only 24 hours (Gurney et al., 1962), suggesting that Fe^{59} incorporation occurs largely during late stages in erythroid maturation.

BLOOD FORMATION BEFORE BIRTH

The blood-islands of the yolk sac are the only source of red cells in the mouse embryo from the eighth through the 11th day of development (deAberle, 1927). Products of yolk sac hematopoiesis, known as the primitive generation of red blood cells, are very large nucleated cells, (mean cell volume four to five times that of adult erythrocytes) comparable to proerythroblasts and normoblasts in bone marrow (Fig. 17-1). They are often seen in mitosis in blood of 11- and 12-day fetuses, but at 13 to 14 days their nuclei are pycnotic (Fig. 17-2). Cells of this primitive generation

Fig. 17-1 to 17-5. Reproductions of paintings by Marcia L. Craig of cells from mouse blood smears.

Fig. 17-1. Heart blood from 12-day C57BL/6 fetus, showing proerythroblast (*A*), nucleated erythrocytes with pachychromatic nucleus (*B*), and dividing hemoglobinized cell (*C*) (Wright-Lepehne stain).

Fig. 17-2. Heart blood from 14-day C57BL/6 fetus, showing well-hemoglobinized primitive erythrocytes with pycnotic nucleus (*A*), intermediate-generation erythrocytes in various stages of hemoglobinization (*B*), and stippled cells (*C*) (Wright-Lepehne stain).

Fig. 17-3. Erythroid precursor cells from the hematopoietic liver of a 14-day C57BL/6 fetus, arranged according to presumed order of maturation: proerythroblast (*A*), basophilic erythroblasts (*B*, *C*, *D*), polychromatophilic erythroblast (*E*), normoblast (*F*), and reticulocyte (*G*) (Wright-Lepehne stain).

Fig. 17-4. Peripheral blood cells of newborn flexed mouse. Note siderotic granules, microcytosis, and hypochromia. (Stained with potassium ferricyanide, counterstained with Biebrich scarlet.)

Fig. 17-5. Peripheral blood cells of newborn jaundiced mouse. Note marked anisocytosis, polychromatophilia, and presence of nucleated erythrocytes (modified Wright-Giemsa stain).

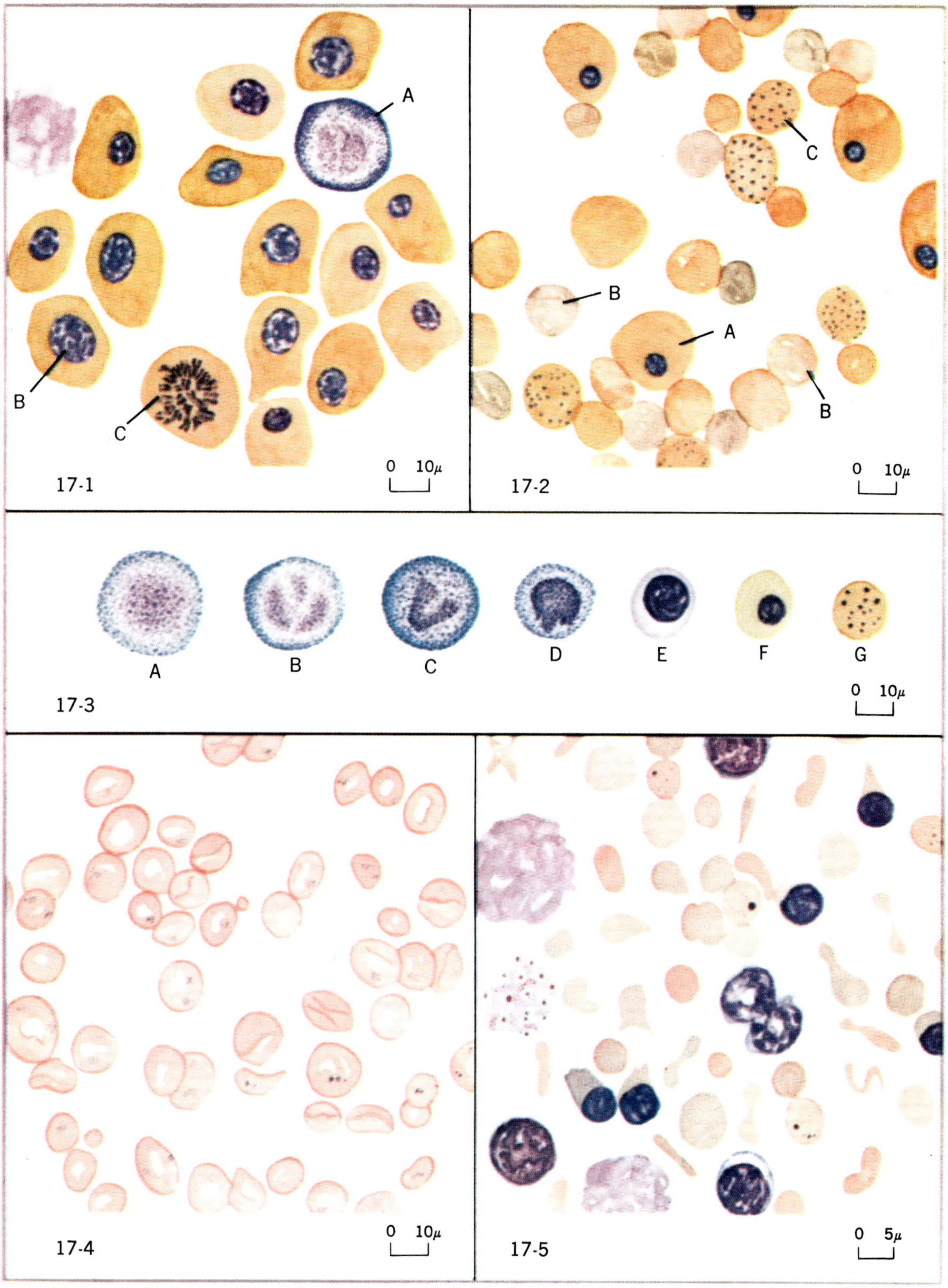

Fig. 17-1 through 17-5. See facing page for legends.

Table 17-6. FETAL ERYTHROCYTE POPULATIONS

Genotype	Mean count[a]	12 days (mean and S.E.)	13 days (mean and S.E.)	14 days (mean and S.E.)	15 days (mean and S.E.)	16 days (mean and S.E.)	Source
		Erythrocyte counts in fetuses aged					
C57BL/6-+/+	T	4.6 ± 0.4	5.4 ± 0.2	9.8 ± 0.4	16.7 ± 1.0	26.2 ± 0.6	b
	L	3.4 ± 0.2	3.0 ± 0.2	3.6 ± 0.1	3.6 ± 0.3	2.1 ± 0.2	
	N	97%	61%	32%	7%		
FL/Re-+/+	T	3.4 ± 0.2	3.4	6.4	13.9	21.4	c
	L	2.6	2.0	2.4	3.2	3.4	
	N	95%	71%	38%	6%	<1%	
FL/Re-*f*/*f*	T	2.7 ± 0.2	2.1 ± 0.1	4.5	7.9	14.3	c
	L	1.8	1.3	1.2	0.8	0.7	
	N	95%	86%	25%	8%	<1%	c
FL/Re-*W*/*W*	T	1.6	2.0	2.2	2.0	3.5	c
	L	1.0	1.3	1.5	1.3	1.9	
	N	95%	94%	72%	40%	4%	
FL/Re-*W*/*W f*/*f*	T	1.6	1.4	2.1	2.0	1.8	c
	L	1.0	0.6	1.0	0.6	0.5	
	N	94%	96%	66%	38%	9%	
+/+ (non-inbred)	T					26.4	d
W^v/W^v (non-inbred)	T					12.9	d
W/W^v (non-inbred)	T					9.8	d
W/*W* (non-inbred)	T					5.0	d
+/+ (non-inbred)	T			10.0–20.0		20.0–30.0	e
an/*an* (non-inbred)	T			5.0–6.0		5.0–10.0	e

[a] T = mean total erythrocytes in hundred thousands per cubic millimeter; L = mean large erythrocytes in hundred thousands per cubic millimeter; and N = per cent nucleated erythrocytes.
[b] Craig and Russell (1964).
[c] Russell and McFarland (1965).
[d] Russell and Fondal (1951).
[e] Kunze (1954).

contain hemoglobin components not found at any later stages (Craig and Russell, 1963, 1964). In C57BL/6J and FL/Re-+/+ fetuses the number of these cells in the blood increases slightly from the 12th to the 15th day of fetal life and diminishes rapidly thereafter (Table 17-6; Craig and Russell, 1964; Russell and McFarland, 1965).

From the 12th through the 16th day of fetal life the liver is the only hematopoietic organ and is an important source of new erythrocytes up to birth. It may contain hematopoietic foci even up to 1 week after birth (deAberle, 1927; Grüneberg, 1942*a*, *b*; Hertwig, 1956; Borghese, 1959). Borghese (1959) described hematopoietic cells in the fetal liver, spleen, and bone marrow. During its period of maximum erythropoietic activity the liver contains almost no myeloid precursors, though Borghese (1959) identified very small numbers of myoblasts and neutrophilic and basophilic leukocytes at 15 days. The erythroid precursor cells are of three distinct sizes (Fig. 17-3) and are quite similar to cell types seen later in bone marrow. Red blood cells produced in the liver are not nucleated (Fig. 17-2) and are smaller than the primitive nucleated cells (8 μ in diameter) (Grüneberg, 1942*a*) but larger than erythrocytes of the adult (6 μ in diameter). Cells of this "intermediate" generation (Grüneberg, 1942*a*) are released into the circulation in increasing numbers and the total blood count rises rapidly (Table 17-7). In the spleen the first regular occurrence of erythropoiesis, plus some myelopoiesis, is on the 15th day of development; later, at 17 days, the spleen becomes the first major site of myelopoiesis (Borghese, 1959). Borghese considered that bone-marrow hemopoiesis, which appears first on the 16th day of development, is largely myeloid up to birth.

The thymus has been identified as the first source of lymphocytes and its development described (Auerbach, 1961; Ball, 1963).

NORMAL BLOOD VALUES FROM BIRTH TO TWO MONTHS

The typical erythrocyte number for newborn normal mice ranges from 3.6 to 5.6 $\times 10^6$ RBC/mm^3, with differences between animal stocks and between observers (Table 17-7). Within 1 month this number rises to 6.5 to 8.6 $\times 10^6$ cells, while the body weight increases by a factor of 10 and the total blood volume increases nearly as much, so that the circulating red-cell mass has increased 15 to 20 times (Grüneberg, 1941; Russell and Fondal, 1951). The mean cell volume of erythrocytes decreases sharply during the same interval from 100 to 110 μ^3 at birth to 48 to 63 μ^3 at 28 days and continues to drop slightly until stable values for erythrocyte number and mean cell volume appear at the age of 8 to 10 weeks. Typical values for normal mice in various stocks at 0, 14, and 28 days are given in Table 17-7, where they are compared with abnormal values observed in littermate anemic mice. The change in erythrocyte mean cell volume comes from a combination of two factors. The "intermediate cells" formed in liver hematopoiesis gradually drop out of the circulation and are replaced by younger and smaller cells. The size of newly formed erythrocytes also decreases, at least during the first week after birth (Grüneberg, 1942*a*).

HEREDITARY ANEMIAS

Action of mutant alleles at 11 different loci in the mouse are known to produce anemias by altering either the process of hematopoiesis, the nature of the erythrocyte, or both (Table 17-7). In addition to brief reviews of published information on each of these anemias, we present original data on those which are under investigation at The Jackson Laboratory.

Effects of mutants at each of these loci can be recognized before birth of individuals of the

Table 17-7. POSTNATAL INCREASE IN MEAN NUMBER AND MEAN VOLUME OF ERYTHROCYTES IN NORMAL AND CERTAIN GENETICALLY ANEMIC MICE

		Birth to 1 day		*7 days*		*14 days*		*28 days*		*Adult*	
Strain	Genotype	RBC[a] (mean ± S.E.)	MCV, μ^3 (mean)	RBC (mean ± S.E.)	MCV, μ^3 (mean)	RBC (mean ± S.E.)	MCV, μ^3 (mean ± S.E.)	RBC (mean ± S.E.)	MCV, μ^3 (mean ± S.E.)	RBC (mean ± S.E.)	MCV, μ^3 (mean ± S.E.)
WBB6F$_1$[b]	+/+			4.2 ± 0.1	76	5.8 ± 0.2	58 ± 2	8.5 ± 0.2	52 ± 1	10.5 ± 0.2	47 ± 1
WBB6F$_1$[b]	*W*/+	4.6 ± 0.1	100	4.7 ± 0.1	69	6.2 ± 0.1	56 ± 1	8.8 ± 0.2	50 ± 1	10.9 ± 0.1	46 ± 1
WBB6F$_1$[b]	$W^v/+$			3.6 ± 0.1	86	5.3 ± 0.2	61 ± 3	7.7 ± 0.1	55 ± 1	9.7 ± 0.1	50 ± 1
WBB6F$_1$[b]	W/W^v	2.1 ± 0.1	129	1.6 ± 0.1	100	3.0 ± 0.1	87 ± 2	4.9 ± 0.1	70 ± 1	5.5 ± 0.1	69 ± 1
WB/Re[b]	*W*/*W*	1.4 ± 0.1	148	0.8 ± 0.1	98					3.5 ± 0.2[c]	83 ± 3
WC/Re[b]	*W*/*W*	1.7 ± 0.1				1.9 ± 0.6	110 ± 4			4.5 ± 0.2[c]	72 ± 2
WCB6F$_1$[b]	+/+			4.4 ± 0.1	77	5.4 ± 0.2	58 ± 3	8.5 ± 0.2	51 ± 1	10.0 ± 0.2	48 ± 1
WCB6F$_1$[b]	*Sl*/+	4.0 ± 0.1	112	3.7 ± 0.1	83	4.4 ± 0.2	68 ± 2	7.4 ± 0.2	58 ± 1	8.8 ± 0.1	52 ± 1
WCB6F$_1$[b]	$Sl^d/+$			3.6 ± 0.1	90	4.6 ± 0.1	65 ± 1	7.3 ± 0.2	58 ± 1	8.8 ± 0.1	52 ± 1
WCB6F$_1$[b]	Sl/Sl^d	1.4 ± 0.1	142	1.0 ± 0.1	90	1.9 ± 0.2	95 ± 3	3.3 ± 0.2	81 ± 1	4.1 ± 0.1	81 ± 2
B6D2F$_1$[b]	Sl^d/Sl^d	1.1								4.2	
BAN/Re[b]	+/+	4.4 ± 0.1	110	4.8 ± 0.1	77	5.3 ± 0.1	58 ± 1	9.3 ± 0.2	50 ± 1	9.9 ± 0.1	49 ± 1
BAN/Re[b]	*an*/*an*	2.7 ± 0.1	125	2.4 ± 0.1	129	3.4 ± 0.1	85 ± 3	5.0 ± 0.2	70 ± 1	6.1 ± 0.2	60 ± 1
FL/Re[b]	+/+	3.6 ± 0.1	103	4.0 ± 0.1	87	5.4 ± 0.1	69 ± 2	7.7 ± 0.1	58 ± 1	9.5 ± 0.1	51 ± 1
FL/Re[b]	*f*/+	3.7 ± 0.1	128	4.2 ± 0.1	89	5.4 ± 0.1	68 ± 1	8.2 ± 0.2	56 ± 1	9.7 ± 0.1	51 ± 1
FL/Re[b]	*f*/*f*	2.8 ± 0.1	117	3.9 ± 0.1	96	5.3 ± 0.1	66 ± 1	7.8 ± 0.1	57 ± 1	9.1 ± 0.1	51 ± 1
noninbred[d]	*dm*/+	4.2	107					9.5	50	10.8	48
	dm/*dm*	1.6	142					6.2	64	7.1	60
noninbred[e]	+/+	4.0	110	4.6	77	5.6	62	6.5	63	8.2	57
	sla/Y	3.1	110	3.2	69	4.4	62	5.0	53	6.2	50
noninbred[b]	−/+	3.9 ± 0.2		4.2 ± 0.2		5.9 ± 0.3	62	7.6 ± 0.4	60	9.3 ± 0.2	53
	mk/*mk*	3.7 ± 0.2		3.7 ± 0.4		6.5 ± 0.5	51	10.6 ± 0.6	38	13.5 ± 0.4	31

[a] Red blood count in millions per cubic millimeter.
[b] Original data, 1964–65.
[c] Means for rare surviving adults (see text).
[d] Stevens and Mackensen (1958).
[e] Grewal (1962).

severely affected genotypes. Particular allelic combinations for four of these loci (*W* series, *Sl* series, Hertwig's anemia, and diminutive) lead to macrocytic anemia (suggestive of defective hematopoiesis), which in animals of certain genotypes is sufficiently severe to cause perinatal death. Genes at two loci (flexed, tail-short) have effects on hematopoiesis limited to a particular developmental stage. Particular combinations of alleles at three loci (jaundice, hemolytic anemia, and spherocytosis) lead to severe hemolytic anemia. Mice of the NZB inbred strain also develop hemolytic anemia but many genes probably contribute to this disease. Mice with sex-linked anemia or microcytic anemia have abnormally shaped erythrocytes. In this chapter we limit discussion to the hematological effects of these gene substitutions, since pleiotropic effects in other tissues and secondary pathological effects are discussed in Chap. 29.

W series anemias

The most extensively investigated mouse hereditary anemias are those caused by action of alleles at the dominant spotting or *W* locus of linkage group XVII. Four dominant alleles (W, W^v, W^b, and W^j) with different effects on red-cell number and on pigmentation have been described (deAberle, 1927; Grüneberg, 1939, 1942*c;* Fekete et al., 1941; Russell et al., 1957; Ballantyne et al., 1961), as well as several remutations not distinguishable in effect from previously described *W* alleles (Strong and Hollander, 1953; Russell et al., 1957). Individuals of all genotypes with two dominant alleles (i.e., W/W, W/W^j, W^v/W^v) are sterile and have black eyes and white hair (Chap. 29).

Animals with two dominant *W* alleles are characterized by severe macrocytic anemia, with deficient hematopoiesis already apparent at the beginning of fetal liver hematopoiesis, at 12½ days in fetal life (Attfield, 1951; Russell and Fondal, 1951; Borghese, 1959). The erythrocyte number is slightly reduced and mean cell volume slightly increased in $W^v/+$ mice (Grüneberg, 1942*c;* Russell, 1949). Wherever erythrocyte numbers of mice with different *W* genotypes on the same genetically homogeneous background have been compared, the rank order was always the same: $+/+ = W/+ = W^j/+ > W^v/+ > W^v/W^v > W/W^v > W/W = W/W^j = W^j/W^j$. Mice of the W/W, W/W^j and W^j/W^j genotypes regularly die within a few days after birth, and some of the W/W^v, and W^v/W^v mice also die during the rapid growth stage between birth and weaning; those surviving to adulthood typically live more than 1 year. On a nonselected heterogeneous genetic background, mean erythrocyte counts were 2.2×10^6 RBC/mm³ for W^v/W^v, 1.4×10^6 RBC/mm³ for W/W^v, and 0.8×10^6 RBC/mm³ for W/W newborn mice (Russell and Fondal, 1951). On specially selected genetic backgrounds, typical mean erythrocyte numbers are higher (1.5 to 1.8×10^6 RBC/mm³ in W/W newborn mice), and postnatal survival is considerably longer (mean for W/W mice in strain WB/Re, 10 days) with an occasional individual surviving to adulthood (Russell and Lawson, 1959). Two F_1 hybrid genetic backgrounds (WB/Re × C57BL/6J and WC/Re × C57BL/6J) have proven particularly favorable for survival and vigor of severely anemic (W/W^v) mice and have been adopted as standard for many experiments on physiological, biochemical, and radiological reactions of *W* series anemic mice (Table 17-7).

In spite of exceedingly low erythrocyte counts in mice of the most severely anemic genotypes, they are afflicted with hypoplastic rather than aplastic anemia. The absolute number of erythrocytes shows the same relative increase from the 16th day of fetal life to birth in severely anemic (W/W) and normal ($+/+$) mice (Russell and Fondal, 1951). The proportion of reticulocytes in the blood of adult W/W^v mice (Niece et al., 1963) and of newborn W/W mice (deAberle, 1927) is higher than that in normal littermates. Fetal liver and newborn marrow of severely anemic (W/W, W/W^v, and W^v/W^v) mice are hypoplastic (Russell, 1955; Borghese, 1959), but near-normal marrow cellularity is characteristic of adult W/W^v and W^v/W^v mice (Russell et al., 1953).

There is much evidence of an altered time-course of hematopoiesis in mice of anemic genotypes. In 12-day normal ($+/+$) fetal livers, all stages of erythropoiesis were seen, but only early stages (proerythroblasts and basophilic erythroblasts) were seen in W/W fetal livers (Borghese, 1959). In adult W^v/W^v bone marrow, the ratio of total erythroid to total myeloid tissue is normal, but the ratio of early to late erythroid cells is still abnormally high, suggesting a delay in erythroid maturation (Russell et al., 1953). When isotopically labeled heme-precursors are injected into $+/+$ and W/W^v littermates, erythrocytes with labeled protoporphyrin appear in circulating blood of $+/+$ mice after 3 days but only after more than 10 days in the W/W^v mice (Altman and Russell, 1964). The great differential radiosensitivity of W/W^v mice, whose typical $LD_{50:30}$ is approximately 240 R, has been traced to delay in the regeneration of erythroid tissue (Russell et al., 1963; Chap. 22). These differences in hematopoiesis may result from deficiency in number or differentiating capacity of erythropoietic stem cells (McCulloch et al., 1964).

It is possible to cure the anemia of W/W, W/W^v, and W^v/W^v mice completely and permanently by implantation of isologous normal (+/+) blood-forming tissue from adult marrow or 15-day fetal hematopoietic livers (Russell et al., 1959; Russell, 1960; Bernstein, 1963). This happens readily without irradiation of the recipient, presumably because of more efficient functioning of the injected cells (Bernstein and Russell, 1959), and host irradiation does not enhance acceptance of histoincompatible fetal cells (Bernstein et al., 1959; Bernstein, 1963). This finding demonstrates that the W series defective hematopoiesis depends entirely upon genotype of the blood-forming tissue and is completely independent of influences from other parts of the body.

Hematopoietic balance of normal (+/+) and anemic (W/W^v) mice differs both in the size of circulating red-cell mass maintained under stable conditions and in the nature of response to erythropoietic stimuli. Following polycythemia induced by hypertransfusion, normal and anemic mice show the same increase in red-cell volume, decrease in reticulocytes and temporary cessation of blood formation; the hematocrit level at which new erythrocytes are again released into the circulation (46 to 48 per cent in +/+, 38 to 40 per cent in W/W^v) depends upon the genotype of functioning blood-forming tissue (Niece et al., 1963). Normal and W/W^v anemic mice respond to lowered oxygen tension by producing more erythrocytes, with proportional increase at least as great in W/W^v as in +/+ mice, but W/W^v mice are very deficient in response to erythropoietin (Keighley et al., 1962). A minimal Fe^{59} uptake response, approximately $\frac{1}{150}$th that in plethorized normal mice, was observed when erythropoietin was injected into plethorized W/W^v mice (Thompson et al., 1963; Keighley et al., 1966). The regulation of hematopoiesis in anemic (W/W^v) mice differs in some critical way, not yet understood, from that in normal (+/+) mice.

Steel series anemias

The steel (Sl) locus first described by Sarvella and Russell (1956) is rather mutable, having given rise to a number of alleles. In populations of mice exposed to X-irradiation at Oak Ridge and at Harwell more than 30 mutations at this locus have been encountered, and in nonirradiated populations of C57BL/6J and DBA/2J at The Jackson Laboratory no fewer than five mutations have appeared in a 3-year period. Of these perhaps steel (Sl) and steel-Dickie (Sl^d) have been most thoroughly investigated hematologically.

Under ordinary laboratory conditions Sl/Sl homozygotes die at about the 15th day of embryonic life according to Sarvella and Russell, whereas Sl/Sl^d heterozygotes are as viable as Sl^d/Sl^d homozygotes. Twenty per cent of Sl^d/Sl^d mice survive for 30 days postpartum, and the mean survival time of these is 79 days on the C57BL/6J background. The percentage weaned and the mean survival time, however, are dependent on genetic background, as evidenced by the fact that on a (C57BL/6J × DBA/2J)F_1 background these values are 25 per cent and 113 days respectively.

All mice carrying two dominant Sl alleles are black-eyed, white, infertile anemics, with a phenotype very similar to that of W series anemics. Evidence has accumulated, however, to show that the steel locus, linked with dystrophy (dy), Ames waltzer (av) and pigmy (pg), is located in linkage group IV and is not linked with W in linkage group XVII (Wolfe, 1963; Nash, 1964).

The anemia is observable grossly at 14 days of embryonic life, continues through birth, and is manifest in adults as a severe, persistent, macrocytic anemia in which the cells are larger than normal, but contain a normal amount of hemoglobin per unit cell volume. These and other quantitative hematological characteristics of +/+, $Sl/+$, $Sl^d/+$, Sl/Sl^d, and Sl^d/Sl^d genotypes appear in Table 17-7. There is a decreased number of nucleated cells per femur (5×10^6 vs. 1×10^7 for normal mice) but, in contrast with the W series, the colony-forming capacity of these cells is not diminished and their radiosensitivity as measured by colony-forming capacity after exposure to Co^{60} radiation in vitro is normal (McCulloch et al., 1965). Marrow cellularity, however, drops precipitously after exposure to 200 R of X-rays and Sl^d/Sl^d homozygotes have an $LD_{50:30}$ of less than 200 R (Bernstein, unpublished). Leukocyte levels of Sl/Sl^d mice appear normal (mean, 11,600 WBC/mm^3).

Sl^d/Sl^d erythrocytes labeled with radioactive sodium chromate (Cr^{51}) or with ferrous citrate (Fe^{59}) have normal half-survival values (26 days). Iron incorporation studies reveal a normal plasma iron clearance time (18 to 20 minutes) and a mean half-appearance time in erythrocytes (12 to 13 hours) which is not different from that of +/+ siblings. Both relative and absolute reticulocyte counts are unusually high, being 9.1 ± 1.4 per cent and 350,000 ± 46,000 per mm^3 respectively, suggesting that reticulocytes in these animals have an abnormally long circulation time.

Steel anemias are unresponsive to erythropoietin. Doses up to 5 A units per day (Keighley, 1962) per mouse are ineffective. Repetition of this experiment with 95 A units given in two doses on successive days also failed to elicit a response (Bernstein and Keighley, unpublished). This is several thousand times the amount of erythropoietin which produces a measurable response in nonanemic individuals. Serum from Sl/Sl^d mice contains roughly the equivalent of 5 A units of purified erythropoietin per milliliter of serum. The relationship between high concentration of serum erythropoietin and response to lowered oxygen tension is not clear, since these mice are poorly if at all responsive to continuous exposure to anoxia (Bernstein, unpublished data) but show a positive hematopoietic response when subjected to 10 per cent oxygen 8 hours per day for 3 weeks (Keighley, 1965, personal communication). Sl^d/Sl^d marrow cells transplanted to *W* series anemics cure the anemia of the latter and respond normally to erythropoietin when they are in this environment (Bernstein and Newburger, unpublished data).

Attempts at therapy of this anemia through implantation of normal isologous blood-forming tissues have been uniformly unsuccessful, as have been therapeutic attempts with normal serum and parenterally administered cobalamine, folinic acid, folic acid, pyridoxal, phosphate, Fe citrate, vitamin E, and ascorbic acid in normal serum, nor do these anemics respond to testosterone. These failures, however, do not rule out the possibility that the steel series anemias are the products of a metabolic defect in some other cells which secondarily affect the rate of erythrocyte production. This position is strengthened by results obtained from the parabiotic union of Sl/Sl^d and +/+, a condition under which both parabionts become normal in cell numbers and sizes (Bernstein, unpublished data).

Hertwig's anemia

Another inherited macrocytic anemia appeared first in offspring of a heavily irradiated male mouse (Hertwig, 1942). The mutant allele (*an*) appears to be completely recessive, and is closely linked to the *b* locus in linkage group VIII; stocks developed to maintain this anemia-producing gene usually employ this linkage for identification of affected animals and carriers (Chap. 8). Affected individuals are born in the expected proportion (23 per cent observed in F_2 generation from crosses between carriers in Hertwig's stock) but are smaller and weaker than normal littermates and not completely viable. Kunze (1954) reported a few cachectic newborn mice with mean erythrocyte counts of 0.9×10^6 RBC/mm^3 and viable newborn mice with a mean of 1.6×10^6 RBC/mm^3. Only 4 per cent survived more than 1 month and had adult erythrocyte counts of 3 to 5×10^6 RBC/mm^3.

Somewhat higher values have been observed in the BAN/Re inbred strain segregating for *an* and + (Table 17-7). In this stock mean survival time for all *an/an* anemics was 212 days and the median survival time was 63 days. Ten of 168 weaned BAN/Re-*an/an* mice survived more than 500 days (McFarland and Russell, unpublished data). It is clear that inbreeding with forced heterozygosis for *an*/+ has in this case resulted in the accumulation of favorable modifying genes (Chap. 2).

Faulty hematopoiesis may be recognized in *an/an* fetuses as early as the 12th or 13th day by a high proportion of proerythroblasts in the fetal liver (Kunze, 1954) and by the 14th day by a reduced proportion of nonnucleated intermediate-generation erythrocytes in the circulating blood (Hertwig, 1956). During the postnatal growth spurt the absolute and relative number of reticulocytes is lower in anemic (*an/an*) mice than in their normal littermates, but in adult survivors Kunze (1954) reported increased relative number of reticulocytes. The number of leukocytes also is markedly reduced in *an/an* mice during the first month of postnatal life, with values of 1.0 to 2.3×10^3 WBC/mm^3 in anemic mice in contrast to 2.5 to 4.1×10^3 WBC/mm^3 in normal littermates (Kunze, 1954).

Transitory siderocytic anemia

An anemia associated with the recessive gene "flexed" (*f*) seems to affect chiefly the intermediate generation erythrocytes, which stem largely from fetal liver hematopoiesis (Grüneberg, 1942*a*, *b*). During the 12th day of development. when blood of both normal and flexed (*f/f*) fetuses contains more than 90 per cent of large nucleated red blood cells (Table 17-6), the difference between genotypes is minimal; the nucleated blood cells are completely hemoglobinized in fetuses of both genotypes. On the 13th to 15th days, however, flexed fetuses form somewhat fewer nonnucleated erythrocytes than do comparable normal fetuses (Kamenoff, 1935; Grüneberg, 1942*a;* Russell and McFarland, 1965). Blood cells of the intermediate generation in *f/f* fetuses are extremely abnormal siderocytes, containing little hemoglobin and numerous siderotic granules (identified by Prussian blue stain) (Grüneberg, 1942*b;* Fig. 17-4. This defect is maximal at 13 to 14 days but by the 16th day partially hemoglobinized erythrocytes are frequently seen.

Studies of *f/f* mice combining this fetal anemia with *W* series anemia provide new insight into the causation of both conditions (Russell and McFarland, 1965). Concomitant determinations of erythrocyte number and size and proportion of cell types have demonstrated effects of the *f* and *W* gene substitutions on erythropoiesis in both primitive and intermediate cell generations in mouse fetuses congenic with inbred strain FL/Re. Blood of normal fetuses (+/+ +/+) contains approximately 3×10^5 large nucleated erythrocytes per cubic millimeter at 12 to 16 days gestation (Table 17-6). Flexed anemic fetuses (+/+ *f/f*) have similar numbers of nucleated erythrocytes at 12 days, but significantly fewer at 16 days. Dominant-spotted anemic fetuses (*W/W* +/+) have fewer nucleated erythrocytes at all ages. Normal fetuses show rapid increase in number of smaller well-hemoglobinized erythrocytes (Table 17-6), and flexed fetuses produce lesser numbers of abnormal nonnucleated erythrocytes. *W/W* +/+ fetuses produce very small numbers of well-hemoglobinized nonnucleated erythrocytes. The primary defect in *W/W* +/+ fetuses appears to be in cell formation, that in +/+ *f/f* fetuses a transitory deficiency of hemoglobinization. Doubly anemic fetuses (*W/W f/f*) combine both defects and die before birth.

The blood of *f/f* newborn mice contains only 4.64 g of Hb/100 cc, in contrast to 10.26 g/100 cc in normal mice (Grüneberg, 1942*a*). After birth the blood values of *f/f* mice improve rapidly, so that by 7 days the red-cell number is nearly normal (Table 17-7) and the hemoglobin content of blood greatly improved (6.38 g/100 cc, compared with 8.83 g/100 cc in normal mice) (Grüneberg, 1942*a*). The proportion of siderocytes drops sharply after birth, in exact correspondence to the drop in erythrocyte mean cell volume which accompanies replacement of intermediate generation by definitive generation erythrocytes (Grüneberg, 1942*b*).

Erythrocyte number and size are normal in adult *f/f* mice, although Grüneberg reported 3 per cent siderocytes in the blood of adult *f/f* mice in contrast to none in normals beyond 1 week postnatal (Grüneberg, 1942*b*).

Margolis and Russell (1965) found high levels of the enzyme δ-aminolevulinic dehydratase (ALD) in livers and spleens of FL/Re-+/+ adult mice, but low levels in both FL/Re-*f*/+ and FL/Re-*f/f* adults (Chap. 19). Following injection of phenylhydrazine to stimulate extramedullary hematopoiesis, the splenic ALD level increased sharply in FL/Re-+/+ mice but little if at all in *f/f* and *f*/+ mice. In the dehydratase effect *f* is dominant; in all other effects it seems to be recessive. We have no clue as to the relationship of this enzymatic defect to the transitory difficulty of hemoglobinization in *f/f* fetuses.

Anemia of "diminutive" mice

The recessive mutant gene "diminutive" (*dm*) affects body size, development of the entire axial skeleton, and erythrocyte formation (Stevens and Mackensen, 1958). At all ages the red blood cells of *dm/dm* mice are macrocytic. Newborn mice are severely anemic, having approximately one-third of the normal erythrocyte count (Table 17-7), and many die shortly after birth either from anemia or because of small size and complications associated with their skeletal abnormalities. The macrocytic anemia persists throughout life, but is less extreme in surviving adults, which are often fertile.

Transitory anemia of "tail-short" mice

Effects of a dominant gene, tail-short (*Ts*), first recognized in adults through its effect on the tail, have been traced back to reduction of the number of blood-islands in the yolk sac on the eighth day of embryonic life (Deol, 1961). Defects of the skeleton of *Ts*/+ fetuses are not seen until the 11th day of development. During the 14th to 16th days of development, *Ts*/+ fetuses have much less blood than their normal littermates, as judged by intensity of hemoglobin coloration in cleared specimens (Deol, 1961). This anemic condition, apparently stemming from deficiency of yolk sac hematopoiesis, is completely corrected before birth. It is highly probable that the effects of *Ts* on the skeleton are secondary to the primary transitory anemia of the primitive cell generation.

Sex-linked anemia

A sex-linked gene, *sla* (Falconer and Isaacson, 1962), causes a mild degree of anemia (erythrocytes 75 per cent of normal number, Table 17-7; Grewal, 1962). Both *sla/sla* females and *sla/Y* males are completely viable and fertile, though their growth is somewhat impaired. Erythrocytes of anemic individuals show some poikilocytosis and anisocytosis, tending to be thicker than normal (normal mean cell volume, reduced cell diameter). Their resistance to isotonic saline is normal. The percentage of reticulocytes is markedly higher in *sla/sla* anemics than in normal mice during the rapid growth phase and slightly higher in adults. Both liver and bone marrow of newborn *sla/sla* and *sla/Y* anemics contained few hematopoietic foci. At all stages the marrow is deficient in erythroid precursors.

Jaundice

Neonatal jaundice (*ja*) is inherited as a recessive lethal (Stevens et al., 1959). Early manifestations appear in utero; later developments give rise to hepatosplenomegaly, cardiac enlargement, microcytic anemia, jaundice, and death.

Affected mice are grossly anemic as early as 14 days postconception but are not jaundiced until several hours after birth. At this time skin, serum, and urine become highly pigmented. Death usually occurs within 24 hours, possibly as a result of bilirubin toxemia, kernicterus, or anoxia.

Hematological studies by Bernstein indicate that *ja/ja* mice have a severe microcytic anemia (Table 17-8). The majority of circulating erythrocytes are abnormal in size, shape, and staining capacity (Fig. 17-5). Target cells, schistocytes, stipple cells, and many immature erythroid and myeloid elements appear in the circulation. The presence of large numbers of nucleated cells (up to 450,000 per mm^3) distort the true extent of the microcythemia and complicate nearly all biochemical procedures.

Erythrocyte lifespan values obtained by transfusing Cr^{51}-labeled cells between mice with the same and different genotypes gave mean half-survival times (ST/2, corrected for elution) of 25.6 days for adult +/+ erythrocytes and 17.3 days for adult *ja*/+ erythrocytes in either +/+ or *ja*/+ recipients. When Cr^{51}-tagged erythrocytes from newborn mice were transfused to adult +/+ mice, ST/2 values of 1.6 days and 0.5 days, respectively, were obtained for erythrocytes of *ja*/+ and *ja/ja* genotypes. No alterations in ST/2 were observed when *ja*/+ and *ja/ja* mice were used as recipients. These results confirm the hemolytic nature of the disease and suggest the presence of an intracorpuscular defect. Splenectomy does not cure the anemia nor increase viability. Determination of marrow cellularity and colony-forming capability showed that jaundiced anemics have a hyperplastic marrow containing a normal proportion of colony-forming cells.

Jaundice hererozygotes (*ja*/+) have a mild but well-compensated form of the disease, with reduced erythrocyte lifespan (see above). Trans-

Table 17-8. HEMATOLOGICAL CHARACTERISTICS OF MICE BEARING ONE OR MORE MUTANT GENES CAUSING HEMOLYTIC ANEMIA OR JAUNDICE[a]

	Newborn					*Adult*				
Characteristic	+/+ Normal	*ja*/+ Carrier	*ja/ja* Jaundiced	*ha*/+ Carrier	*ha/ha* Hemolytic anemia	+/+ Normal	*ja*/+ Carrier	*ja/ja* Jaundiced	*ha*/+ Carrier	*ha/ha* Hemolytic anemia
RBC[b]	3.8	4.0	2.6	3.6	1.9	10.4	10.6	4.4	10.6	4.3
Hematocrit (per cent)	38.8	39.1	18.2	38.2	16.4	48.7	48.1	27.0	48.1	29.1
Mean cell volume (μ^3)	104	101	69	104	83	46.9	45.4	61.3	45.1	67.0
WBC[c]			400			10.5	12.5		10.7	42.5
Hemoglobin										
Concentration (g/100 ml)	15.3	12.9	4.1	15.8	6.7	16.2	16.3		16.2	6.4
MCH[d]	40.2	32.3	15.6	43.2	24.2	16.7	15.2		15.2	14.9
MCHC[e]	39.4	15.6	22.5	41.3	40.8	31.4	33.7		33.7	21.9

[a] All numbers in this table represent mean values determined from samples of at least 10 mice.
[b] Red blood count in millions per cubic millimeter.
[c] White blood count in thousands per cubic millimeter.
[d] Mean corpuscular hemoglobin in micromicrograms.
[e] Mean corpuscular hemoglobin concentration in per cent.

fusion of these cells into splenectomized recipients restores ST/2 to normal values. A moderate decrease in the agranulocyte-to-granulocyte ratio, coupled with an elevation in total white count, indicates that nonspecific marrow hyperplasia is a compensatory characteristic of the carrier condition.

Newborn mice homozygous for jaundice have a smaller amount of hemoglobin per cell (15.6 $\mu\mu$g) than do their +/+ siblings (32.9 $\mu\mu$g) and a diminished mean corpuscular hemoglobin concentration (22.5 vs. 39.4 per cent). Hypochromia associated with erythrocyte abnormalities of size and shape are frequently encountered in hemoglobinopathies and in maternal-fetal incompatibility reactions, but paper, starch, and agar electrophoretic analyses and studies of solubility and spectral absorption failed to detect abnormal hemoglobin molecules in these mice. Moreover, standard immunological techniques developed for the mouse failed to present evidence of maternal antibodies responsible for their jaundice and anemia. Additional evidence against the immunological causation of the disease was provided by an analysis of relation between litter size, litter seriation, and proportion of affected siblings. No sensitization was seen since litter size (average 7.2) of successive litters was not diminished, and the proportion of affected siblings in the first litter (22.5 per cent) was not smaller than that found in succeeding litters.

No metabolic defect has yet been detected in *ja/ja* or *ja/+* erythrocytes, although the analyses have been confounded by presence of large numbers of nucleated and reticulated cells which are extremely difficult to separate from older more differentiated cells.

Hemolytic anemia

Hemolytic anemia (gene symbol *ha*), the second hemolytic disease of the newborn to be considered here, is inherited as a simple Mendelian recessive. In most of its manifestations it resembles jaundice (*ja*), but they are not alleles. All information reported here comes from Bernstein's previously unpublished studies.

The major phenotypic difference between jaundice and hemolytic anemia is in viability. Some homozygous hemolytic anemics (*ha/ha*) become adults. Pathologically, developmentally, and hematologically the two conditions are strikingly similar. Spherocytes, schistocytes, target cells, anisocytotic and poikilocytotic erythrocytes, erythroblasts, and numerous myeloid elements are seen in the circulating blood and the total white-cell count is elevated from 10,700 to 45,500 per mm^3. The peripheral blood presents the classic picture of a severe persistent hemolytic disease with marrow hyperplasia and hyperbilirubinemia (Table 17-8).

Affected individuals are grossly identifiable as early as 14 days postconception. They are pale but not jaundiced until several hours after birth. A small percentage die in utero, but the majority come to term, and most die within a week of birth. Viability depends upon genetic background.

A maternal-fetal incompatibility has not been detected nor has a hemoglobinopathy been implicated (Bernstein and Wolfe, unpublished data). Red-cell glucose 6-phosphate dehydrogenase and phosphokinase activities and lactic acid content appear normal (Hutton, Schimke, and Bernstein, unpublished data). Red-cell survival, however, is drastically reduced (ST/2 = 0.55 days vs. 24 days for +/+ littermates). An intracorpuscular defect is implicated by the fact that Cr^{51}-labeled +/+ red blood cells transfused to *ha/ha* recipients have normal chrome half-survival times.

Heterozygous carriers of hemolytic anemia are afflicted with a mild, almost completely compensated, form of the disease. Circulating erythrocytes of carriers are nearly normal in size and hemoglobin concentration, but their erythrocyte survival time (ST/2) is diminished from 24 to 16 days. Moreover, it is clear from the analyses of total and differential leukocyte counts that *ha/+* heterozygotes show a significant increase over +/+ homozygotes in the number of cells in each of the granulocyte series, but that the total number of lymphocytic elements is unchanged. This finding suggests that compensatory hyperplasia of the marrow involves both erythroid and myeloid elements.

Spherocytosis

A third hemolytic disease of the newborn has been reported (Joe et al., 1962). It is a severe hemolytic anemia with jaundice, microcytosis, and hepatomegaly, and is inherited as an autosomal recessive lethal. Genetic tests implicated a third locus unrelated to *ja* and *ha*, and Joe et al. (1962) designated the new locus as spherocytosis (*sph*).

The circulating blood of homozygous spherocytic (*sph/sph*) mice contains various abnormal erythroid elements designated as elliptocytes, cells with Howell-Jolly bodies, nucleated red cells, poikilocytotic and polychromatophilic erythrocytes, numerous siderocytes and spherocytes, and huge numbers of reticulocytes. Mean cell diameters of affected red blood cells were small (4.68 vs.

8.69 μ) and the mean corpuscular hemoglobin concentration was nearly normal (25.0 vs. 26.9 per cent in +/+ littermates).

Afflicted mice were detected at and before birth by pale color, but during the first few hours after birth they gradually acquired a gross yellow coloration. A level of serum bilirubin (5.8 mg per 100 ml) considerably higher than that in nonaffected sibs (0.54 mg per 100 ml) was found. No *sph/sph* individuals survived longer than 24 hours. Attempts to distinguish between carrier (*sph*/+) and noncarrier (+/+) mice by hematological methods proved fruitless.

Microcytosis

Microcytic anemia or microcytosis, inherited as an autosomal recessive (*mk/mk*), shows certain characteristics not seen in other mouse anemias (Nash et al., 1964). These include extreme microcytosis, presence of target cells, and almost complete compensation of the anemia in adults through increase in numbers of erythrocytes.

The first anemic animals, observed in a newborn litter in a mixed animal stock involving C3H/J, C57BL/6J, and DBA/2 ancestors, were readily distinguishable from normal littermates by pale coloration. In two partially inbred stocks segregating for microcytosis one-third to one-half of the anemic offspring die before weaning, usually during the first postnatal week, but surviving anemics develop into fertile adults.

The erythrocyte number of newborn *mk/mk* mice is normal (Table 17-7) and their obvious anemia due solely to reduced erythrocyte volume. by the age of 2 weeks, *mk/mk* mice have surpassed their normal sibs in red-cell count and by 8 weeks their average count is over 13 million erythrocytes per mm^3, almost 50 per cent higher than that of normal littermates, with the difference between genotypes increasing during the postnatal period of rapid growth so that the mean cell volume of erythrocytes from adult *mk/mk* mice is less than two-thirds that of normal littermates. Smears of *mk/mk* erythrocytes show benzidine peroxidase hemoglobin stain restricted to the periphery of the cells. Mean corpuscular hemoglobin is decreased in proportion to the decrease in cell volume, but concentration of hemoglobin per unit volume of packed red cells remains almost normal. These findings suggest that *mk/mk* erythrocytes may be abnormally thin in cross-section. Lifespan of *mk/mk* erythrocytes has not been reported. Reticulocyte percentages are significantly elevated in microcytic mice (mean 13.2 per cent for adult females, 11.6 per cent for males, in contrast to 3.7 and 3.6 per cent for normal littermates) and methylene-blue–stained smears show the reticulum concentrated near the cell border in a characteristic "Christmas wreath."

Other hematological defects

A completely normocytic anemia found in some but not all of Strong's luxoid (*lst/lst*) newborn mice has been traced to hemorrhage from lacerations which result rather naturally from action of that gene (Kuharcik and Forsthoefel, 1963). A hemolytic anemia of varying severity has been reported as developing in a high proportion of adult mice of the NZB/Bl inbred strain (Helyer and Howie, 1963). Affected mice showed reticulocytosis, splenomegaly, jaundice, anemia, and elevated antibody titre. The condition, which seems to be a disease of autoimmunity, is discussed in Chap. 29.

SUMMARY

In adult mice certain genetically controlled quantitative and qualitative variations of peripheral blood are compatible with health. Information is presented on the range of normal values for each of the formed elements in the blood and significant strain differences are demonstrated in numbers of erythrocytes, granulocytes, and agranuloyctes. Hemoglobin polymorphisms characteristic of normal mice are caused by independently inherited differences in hemoglobin α- and β-chain structures. We have presented methods available for classifying hemoglobin phenotypes.

Blood formation is initiated in the marrow or spleen of normal adult mice through stimulation of stem cells and proceeds through a characteristic pattern of differentiation and release of reticulocytes. We have summarized critical experiments which employed mice to increase understanding of the initiation and regulation of hematopoiesis. Before birth, blood formation stems first from blood-islands in the yolk sac and later from the fetal liver. We have described quantitative and qualitative changes of the erythron during prenatal development and in the postnatal period of rapid growth.

We have reviewed published investigations concerning each of the 12 recognized hereditary anemias of the mouse and provided new information on certain of the anemias under investigation at The Jackson Laboratory. The anemias due to single-gene substitutions include four macrocytic anemias, two transitory anemias limited to early developmental stages, four anemias with hemolysis and jaundice, and two with other peripheral cell defects.

The processes of normal and abnormal blood formation in mice and the characteristics of their completed blood cells provide excellent material for the study of gene action in mammals. Many kinds of information are important to such studies, including counts and measurements of the blood cells in normal mice at all stages in their life history; biochemical studies of hemoglobin and of enzymes affecting hematopoiesis; studies of blood formation in the adult and in the fetus; and studies of mouse hereditary anemias.

LITERATURE CITED

Altman, K. I., and E. S. Russell. 1964. Heme synthesis in normal and genetically anemic mice. J. Cell. Comp. Physiol. 64:293–301.

Altman, K. I., E. S. Russell, K. Salomon, and J. K. Scott. 1953. Chemopathology of hemoglobin synthesis in mice with hereditary anemia. Fed. Proc. 12:168.

Attfield, M. 1951. Inherited macrocytic anaemias in the house mouse. III. Red blood cell diameters. J. Genet. 50:250–263.

Auerbach, R. 1961. Experimental analysis of the origin of cell types in the development of the mouse thymus. Develop. Biol. 3:336–354.

Ball, W. D. 1963. A quantitative assessment of mouse thymus differentiation. Exp. Cell Res. 31:82–88.

Ballantyne, J., F. G. Bock, L. C. Strong, and W. C. Quevedo, Jr. 1961. Another allele at the *W* locus of the mouse. J. Hered. 52:200–202.

Bernstein, S. E. 1963. Problems and potentialities of hematopoietic tissue transplants. Trans. New Engl. Surg. Soc. 44:76–83.

Bernstein, S. E., and E. S. Russell. 1959. Implantation of normal blood-forming tissue in genetically anemic mice, without X-irradiation of the host. Proc. Soc. Exp. Biol. Med. 101:769–773.

Bernstein, S. E., E. S. Russell, and F. A. Lawson. 1959. Limitations of whole-body irradiation for inducing acceptance of homografts in cases involving a genetic defect. Transpl. Bull 6(23):106–108.

Borghese, E. 1959. The present state of research on *WW* mice. Acta Anat. 36:185–220.

Brecher, G., K. M. Endicott, H. Gump, and H. P. Brawner. 1948. Effects of x-ray on lymphoid and hemopoietic tissues of albino mice. Blood 3:1,259–1,274.

Brecher, G., and F. Stohlman, Jr. 1959. Humoral factors in erythropoiesis, p. 110–132. *In* L. M. Tocantins [ed.] Progress in hematology. Vol. 2. Grune & Stratton, New York.

Brown, H. E., and T. F. Dougherty. 1956. The diurnal variation of blood leukocytes in normal and adrenalectomized mice. Endocrinology 58:365–375.

Budds, O. C., E. S. Russell, and G. E. Abrams. 1953. Effects of genetics and anesthesia upon granulocyte and agranulocyte levels in seven inbred mouse strains. Proc. Soc. Exp. Biol. Med. 84:176–178.

Chai, C. K. 1957. Leukopenia: an inherited character in mice. Science 126:125.

Cotes, P. M., and D. R. Bangham. 1961. Bio-assay of erythropoietin in mice made polycythaemic by exposure to air at a reduced pressure. Nature 191:1,065–1,067.

Craig, M. L., and E. S. Russell. 1963. Electrophoretic patterns of hemoglobin from fetal mice of different inbred strains. Science 142:398–399.

Craig, M. L., and E. S. Russell. 1964. A developmental change in hemoglobins correlated with an embryonic red cell population in the mouse. Develop. Biol. 10:191–201.

Cudkowicz, G., A. C. Upton, L. H. Smith, D. G. Gosslee, and W. L. Hughes. 1964. An approach to the characterization of stem cells in mouse bone marrow. Ann. N.Y. Acad. Sci. 114:571–582.

Davies, A. J. S., A. M. Cross, and K. Lapis. 1962. Anaemia associated with the NK/lymphoma in mice. Brit. J. Cancer 16:770–781.

DeAberle, S. B. 1927. A study of the hereditary anemia of mice. Amer. J. Anat. 40:219–247.

Deol, M. S. 1961. Genetical studies on the skeleton of the mouse. XXVIII. Tail-short. Proc. Roy. Soc. B 155:78–95.

Dunn, T. B. 1954. Normal and pathologic anatomy of the reticular tissue in laboratory mice, with a classification and discussion of neoplasms. J. Nat. Cancer Inst. 14:1,281–1,433.

Endicott, K. M., and H. Gump. 1947. Hemograms and myelograms of healthy female mice of C-57 brown and CFW strains. Blood (Spec. issue) 1:61–63.

Ewing, K. L., and O. E. Tauber. 1964. Hematological changes in aging male C57BL/6Jax mice. J. Gerontol. 19:165–167.

Falconer, D. S., and J. H. Isaacson. 1962. The genetics of sex-linked anaemia in the mouse. Genet. Res. 3:248–250.

Fekete, E., C. C. Little and A. M. Cloudman. 1941. Some effects of the gene W^v (dominant spotting) in mice. Proc. Nat. Acad. Sci. 27:114–117.

Filmanowicz, E., and C. W. Gurney. 1961. Studies on erythropoiesis. XVI. The response to a single dose of erythropoietin in the polycythemic mouse. J. Lab. Clin. Med. 57:65–72.

Gluecksohn-Waelsch, S. 1960. The inheritance of hemoglobin types and other biochemical traits in mammals. J. Cell. Comp. Physiol. 56(Suppl.): 89–101.

Gluecksohn-Waelsch, S., H. M. Ranney, and B. F. Sisken. 1957. The hereditary transmission of hemoglobin differences in mice. J. Clin. Invest. 36:753–756.

Goodman, J. W., and G. S. Hodgson. 1962. Evidence for stem cells in the peripheral blood of mice. Blood 19:702–714.

Goodman, J. W., and L. H. Smith. 1961. Erythrocyte life span in normal mice and in radiation bone marrow chimeras. Amer. J. Physiol. 200:764–700.

Goulden, F., and F. L. Warren. 1944. The hemoglobin content of the blood of mice of the RIII and CBA strains. Cancer Res. 4:421–424.

Gowen, J. W., and M. L. Calhoun. 1943. Factors affecting genetic resistance of mice to mouse typhoid. J. Infect. Dis. 72:40–56.

Grewal, M. S. 1962. A sex-linked anaemia in the mouse. Genet. Res. 3:238–247.

Grüneberg, H. 1939. Inherited macrocytic anemias in the house mouse. Genetics 24:777–810.

Grüneberg, H. 1941. The growth of the blood of the suckling mouse. J. Pathol. Bacteriol. 52:323–329.

Grüneberg, H. 1942*a*. The anaemia of flexed-tailed mice (*Mus musculus L.*). I. Static and dynamic hematology. J. Genet. 43:45–68.

Grüneberg, H. 1942*b*. The anaemia of flexed-tailed mice (*Mus musculus L.*). II. Siderocytes. J. Genet. 44:246–271.

Grüneberg, H. 1942*c*. Inherited macrocytic anaemias of the house mouse. II. Dominance relationships. J. Genet. 43:285–293.

Gurney, C. W., R. Degowin, D. Hofstra, and J. Byron. 1962. Applications of erythropoietin to biological investigation, p. 151–161. *In* L. O. Jacobson and M. Doyle [ed.] Erythropoiesis. Grune & Stratton, New York.

Gurney, C. W., N. Wackman, and E. Filmanowicz. 1961. Studies on erythropoiesis. XVII. Some quantitative aspects of the erythropoietic response to erythropoietin. Blood 17:531–546.

Halberg, F., and M. B. Visscher. 1950. Regular diurnal physiological variation in eosinophil levels in five stocks of mice. Proc. Soc. Exp. Biol. Med. 75:846–847.

Halberg, F., M. B. Visscher, and J. J. Bittner. 1953. Eosinophil rhythm in mice: range of occurrence; effects of illumination, feeding and adrenalectomy. Amer. J. Physiol. 174:109–122.

Heinecke, H. 1961. Das Blutbild der Maus (Eine Übersicht). I. Das normale quantitative und qualitative weisse Blutbild. Z. Versuchstierk. 1:16–37.

Heinecke, H. 1962. Das Blutbild der Maus (Eine Übersicht). II. Das normale rote Blutbild. Z. Versuchstierk. 1:141–159.

Heinecke, H. 1963. Das Blutbild der Maus. III. Thrombozyten-Normalwerte. Z. Versuchstierk. 3:77–80.

Heinecke, H., and M. Wagner. 1964. Zur Frage der Hämoglobintypen der Maus. Z. Versuchstierk. 5:88–97.

Helyer, B. J., and J. B. Howie. 1963. Spontaneous auto-immune disease in NZB/Bl mice. Brit. J. Haematol. 9:119–131.

Hertwig, P. 1942. Neue Mutationen und Koppelungsgruppen bei der Hausmaus. Z. Indukt. Abstamm. Vererb. 80:220–246.

Hertwig, P. 1956. Erbliche anemien bei Mäusen. Verhandl. Deut. Zool. Ges. Hamburg, p. 185–193.

Hutton, J. J., J. Bishop, R. Schweet, and E. S. Russell. 1962*a*. Hemoglobin inheritance in inbred mouse strains. I. Structural differences. Proc. Nat. Acad. Sci. 48:1,505–1,513.

Hutton, J. J., J. Bishop, R. Schweet, and E. S. Russell. 1962*b*. Hemoglobin inheritance in inbred mouse strains. II. Genetic studies. Proc. Nat. Acad. Sci. 48:1,718–1,724.

Hutton, J. J., R. S. Schweet, H. G. Wolfe, and E. S. Russell. 1964. Hemoglobin solubility and α-chain structure in crosses between two inbred mouse strains. Science 143:252–253.

Jacobson, L. O. 1944. The effect of estrogens on the peripheral blood and bone marrow of mice. Endocrinology 34:240–244.

Jacobson, L. O., and M. Doyle [ed.] 1962. Erythropoiesis. Grune & Stratton, New York. 339 p.

Jacobson, L. O., E. Goldwasser, L. F. Plzak, and W. Fried. 1957. Studies on erythropoiesis. Part IV. Reticulocyte response of hypophysectomized and polycythemic rodents to erythropoietin. Proc. Soc. Exp. Biol. Med. 95:243–249.

Joe, M., J. M. Teasdale, and J. R. Miller. 1962. A new mutation (*sph*) causing neonatal jaundice in the house mouse. Can. J. Genet. Cytol. 4:219–225.

Kamenoff, R. J. 1935. Effects of the flexed-tailed gene on the development of the house mouse. J. Morphol. 58:117–155.

Keighley, G. 1962. Experience with assays, units and standards of erythropoietin, p. 17–32. *In* L. O. Jacobson and M. Doyle [ed.] Erythropoiesis. Grune & Stratton, New York.

Keighley, G. H., P. Lowy, E. S. Russell, and M. W. Thompson. 1966. Analysis of erythroid homeostatic mechanisms in normal and genetically anemic mice. Brit. J. Haemat. (In press.)

Keighley, G. H., E. S. Russell, and P. Lowy. 1962. Response of normal and genetically anaemic mice to erythropoietic stimuli. Brit. J. Haematol. 8:429–441.

Kuharcik, A. M., and P. F. Forsthoefel. 1963. A study of the anemia in Strong's luxoid mutant. J. Morphol. 112:13–21.

Kunze, H. 1954. Die Erythropoese bei einer erblichen Anämie röntgenmutierte Mäuse. Folia Haematol. 72:392–436.

Lajtha, L. G. 1962. Stem cell kinetics and erythropoietin, p. 140–150. *In* L. O. Jacobson and M. Doyle [ed.] Erythropoiesis. Grune & Stratton, New York.

Law, L. W., and W. E. Heston. 1941. Total erythrocyte counts as quoted by E. Fekete, p. 93. *In* G. D. Snell [ed.] Biology of the laboratory mouse. Blakiston, Philadelphia.

Lockner, D., K. Sletten, and G. de Hevesy. 1963. Studies on cancer anaemia; organ weights, blood values, and iron metabolism in normal and tumour-bearing mice. Brit. J. Cancer 17:238–354.

Margolis, F. L., and E. S. Russell. 1965. Delta-aminolevulinate dehydratase activity in mice with hereditary anemia. Science 150:496–497.

McCulloch, E. A., L. Siminovitch, and J. E. Till. 1964. Spleen-colony formation in anemic mice of genotype WW^v. Science 144:844–846.

McCulloch, E. A., L. Siminovitch, J. E. Till, E. S. Russell, and S. E. Bernstein. 1965. The cellular basis of the genetically determined hemopoietic defect in anemic mice of genotype Sl/Sl^d. Blood. 26:399–410.

Mehrotra, T. N., and G. Cardinali. 1963. Study of mouse hemoglobins by starch-gel electrophoresis. Fed. Proc. 22:601.

Metcalf, D., and R. F. Buffett. 1957. Lymphocytosis response in mice and its relation to thymus and adrenal. Proc. Soc. Exp. Biol. Med. 95:576–579.

Morton, J. R. 1962. Starch-gel electrophoresis of mouse haemoglobins. Nature 194:383–384.

Morton, J. R. 1965. Mouse News Letter 32:20–21.

Nash, D. J. 1964. Mouse News Letter 30:53–54.

Nash, D. J., E. Kent, M. M. Dickie, and E. S. Russell. 1964. The inheritance of "mick," a new anemia in the house mouse. Amer. Zool. 14:404–405. (Abstr.)

Niece, R. L., E. C. McFarland, and E. S. Russell. 1963. Erythroid homeostasis in normal and genetically anemic mice: reaction to induced polycythemia. Science 142:1,468–1,469.

Odell, T. T., Jr., and T. P. McDonald. 1960. Peripheral counts and survival of blood platelets of mice. Fed. Proc. 19:63. (Abstr.)

Odell, T. T., Jr., and T. P. McDonald. 1961. Lifespan of mouse blood platelets. Proc. Soc. Exp. Biol. Med. 106:107–108.

Panzenhagen, H., and R. Speirs. 1953. Effect of horse serum, adrenal hormones, and histamine on the number of eosinophils in the blood and peritoneal fluid of mice. Blood 8:536–544.

Petri, S. 1953. Morphologic und Zahl der Blutkörperchen bei 7-ca. 30 g. Schweren normalen weissen laboratoriumsmäusen. Acta Pathol. Microbiol. Scand. 10:159–238.

Popp, R. A. 1962*a*. Studies on the mouse hemoglobin loci. II. Position of the hemoglobin locus with respect to albinism and shaker-1 loci. J. Hered. 53:73–75.

Popp, R. A. 1962*b*. Studies on the mouse hemoglobin loci. III. Heterogeneity of electrophoretically indistinguishable single-type hemoglobins. J. Hered. 53:75–77.

Popp, R. A. 1962*c*. Studies on the mouse hemoglobin loci. IV. Independent segregation of *Hb* and *Sol:* effect of the loci on the electrophoretic and solubility properties of hemoglobins. J. Hered. 53:77–80.

Popp, R. A. 1962*d*. Studies on the mouse hemoglobin loci. V. Differences among tryptic peptides of the β-chain governed by alleles at the *Hb* locus. J. Hered. 53:142–146.

Popp, R. A. 1962*e*. Studies on the mouse hemoglobin loci. VI. A third allele, Sol^3, at the *Sol* locus. J. Hered. 53:147–148.

Popp, R. A. 1962*f*. Studies on the mouse hemoglobin loci. VII. Differences among tryptic peptides of the α-chain governed by alleles at the *Sol* locus. J. Hered. 53:148–151.

Popp, R. A. 1963*a*. Mammalian hemoglobins, p. 299–322. *In* W. J. Burdette [ed.] Methodology in mammalian genetics. Holden-Day, San Francisco.

Popp, R. A. 1963*b*. Hemoglobin loci: mice classified for their *Hb* and *Sol* alleles. Science 140:893–894.

Popp, R. A. 1965*a*. The separation and amino acid composition of the tryptic peptides of the α-chain of hemoglobin from C57BL mice. J. Biol. Chem. 240:2,863–2,867.

Popp, R. A. 1965*b*. Hemoglobin variants in mice. Fed. Proc. 24:1,252–1,257.

Popp, R. A., and G. E. Cosgrove. 1959. Solubility of hemoglobin as red cell marker in irradiated mouse chimeras. Proc. Soc. Exp. Biol. Med. 101:754–758.

Popp, R. A., D. M. Popp, and B. C. Webb. 1963. Amino acid analyses of tryptic peptides of the α-chain of mouse hemoglobin. Amer. Zool. 3:490. (Abstr.)

Popp, R. A., and W. St. Amand. 1960. Studies on the mouse hemoglobin locus. I. Identification of hemoglobin types and linkage of hemoglobin with albinism. J. Hered. 51:141–144.

Prentice, T. C., and E. A. Mirand. 1957. Effect of acute liver damage plus hypoxia on plasma erythropoietin content. Proc. Soc. Exp. Biol. Med. 95:231–234.

Pujman, V., S. Prokopová, and R. Reichlova. 1954. Bemerkungen zum Blutbild der Maus. Acta Soc. Zool. Bohemoslov. 18:289–297.

Ranney, H. M., and S. Gluecksohn-Waelsch. 1955. Filter-paper electrophoresis of mouse haemoglobin: preliminary note. Ann. Hum. Genet. 19:269–272.

Ranney, H. M., G. M. Smith, and S. Gluecksohn-Waelsch. 1960. Haemoglobin differences in inbred strains of mice. Nature 188:212–214.

Rhinesmith, H. S., H. H. Li, B. L. Millikin, and L. C. Strong. 1964. Mouse hemoglobin. I. Chromatographic analysis. Anal. Biochem. 8:407–414.

Rifkin, D., M. M. Rifkin, and W. Konigsberg. 1965. Amino acid compositions of tryptic peptides of two strains of mouse hemoglobin. Fed. Proc. 24:532. (Abstr.)

Riggs, A. 1963. The amino acid composition of some mammalian hemoglobins: mouse, guinea pig and elephant. J. Biol. Chem. 238:2,983–2,987.

Riggs, A. 1965. Hemoglobin polymerization in mice. Science 147:621–623.

Rosa, J., G. Schapira, J. C. Dreyfus, J. de Grouchy, G. Mathé, and J. Bernard. 1958. Different heterogeneities of mouse haemoglobin according to strains. Nature 182:947–948.

Russell, E. S. 1949. Analysis of pleiotropism of the *W*-locus in the mouse: relationship between the effects of *W* and W^v substitution on hair pigmentation and on erythrocytes. Genetics 34:708–723.

Russell, E. S. 1955. Review of the pleiotropic effects of *W*-series genes on growth and differentiation, p. 113–126. *In* D. Rudnick [ed.] Aspects of synthesis and order in growth. Princeton University Press, Princeton, N.J.

Russell, E. S. 1960. Genetic aspects of implantation of blood-forming tissue. Fed. Proc. 19:573–578.

Russell, E. S. 1963. Techniques for the study of anemias in mice, p. 558–564. *In* W. J. Burdette [ed.] Methodology in mammalian genetics. Holden-Day, San Francisco.

Russell, E. S., S. E. Bernstein, F. A. Lawson, and L. J. Smith. 1959. Long-continued function of normal blood-forming tissue transplanted into genetically anemic hosts. J. Nat. Cancer Inst. 23:557–566.

Russell, E. S., S. E. Bernstein, E. C. McFarland, and W. R. Modeen. 1963. The cellular basis of differential radiosensitivity of normal and genetically anemic mice. Radiat. Res. 20:677–694.

Russell, E. S., and E. L. Fondal. 1951. Quantitative analysis of the normal and four alternative degrees of an inherited macrocytic anemia in the house mouse. I. Number and size of erythrocytes. Blood 6:892–905.

Russell, E. S., and P. S. Gerald. 1958. Inherited electrophoretic hemoglobin patterns among 20 inbred strains of mice. Science 128:1,569–1,570.

Russell, E. S., and F. A. Lawson. 1959. Selection and inbreeding for longevity of a lethal type. J. Hered. 50:19–25.

Russell, E. S., F. A. Lawson, and G. Schabtach. 1957. Evidence for a new allele at the *W*-locus of the mouse. J. Hered. 48:119–123.

Russell, E. S., and E. C. McFarland. 1965. Erythrocyte populations in fetal mice with and without two hereditary anemias. Fed. Proc. 24:240. (Abstr.)

Russell, E. S., E. F. Neufeld, and C. T. Higgins. 1951. Comparison of normal blood picture of young adults from 18 inbred strains of mice. Proc. Soc. Exp. Biol. Med. 78:761–766.

Russell, E. S., L. J. Smith, and F. A. Lawson. 1956. Implantation of normal blood-forming tissues in radiated genetically anemic hosts. Science 124:1,076–1,077.

Russell, E. S., C. M. Snow, L. M. Murray, and J. P. Cormier. 1953. The bone marrow in inherited macrocytic anemia in the house mouse. Acta Haematol. 10:247–259.

Sarvella, P. A., and L. B. Russell. 1956. Steel, a new dominant gene in the house mouse. J. Hered. 47:123–128.

Siminovitch, L., E. A. McCulloch, and J. E. Till. 1963. The distribution of colony-forming cells among spleen colonies. J. Cell. Comp. Physiol. 62:327–336.

Stevens, L. C., and J. A. Mackensen. 1958. The inheritance and expression of a mutation in the mouse affecting blood formation, the axial skeleton, and body size. J. Hered. 49:153–160.

Stevens, L. C., J. A. Mackensen, and S. E. Bernstein. 1959. A mutation causing neonatal jaundice in the house mouse. J. Hered. 50:35–39.

Strong, L. C., and L. D. Francis. 1937. The blood of female mice (breeders) of cancer-susceptible (A) and cancer-resistant (CBA) strains. Arch. Pathol. 23:202–206.

Strong, L. C., and W. F. Hollander. 1953. Two non-allelic mutants resembling "*W*" in the house mouse. J. Hered. 44:41–44.

Thompson, M. W., E. S. Russell, and E. C. McFarland. 1963. Response of polycythemic WW^v anemic mice to erythropoietin. Proc. XI Int. Congr. Genet. 1:185–186. (Abstr.)

Till, J. E., and E. A. McCulloch. 1961. A direct measurement of the radiation sensitivity of normal mouse bone marrow cells. Radiat. Res. 14:213–222.

Van Putten, L. M. 1958. The life span of red cells in the rat and the mouse as determined by labeling with DFP^{32} in vivo. Blood 13:789–794.

Von Ehrenstein, G. 1958. The life span of the erythrocytes of normal and of tumour-bearing mice as determined by glycine-2-^{14}C. Acta Physiol. Scand. 44:80–91.

Weir, J. A., and G. Schlager. 1962. Selection for leucocyte count in the house mouse and some physiological effects. Genetics 47:1,199–1,217.

Welling, W., and D. W. van Bekkum. 1958. Different types of haemoglobin in two strains of mice. Nature 182:946–947.

Wiadrowski, M., and D. Metcalf. 1963. Erythrocyte osmotic fragility in AKR mice with lymphoid leukaemia. Nature 198:1,103–1,104.

Wolfe, H. G. 1963. Mouse News Letter 29:40.

Wolfe, H. G., E. S. Russell, and S. O. Packer. 1963. Hemoglobins in mice; segregation of genetic factors affecting electrophoretic mobility and solubility. J. Hered. 54:107–112.

Yonehiro, E. G., and J. B. Aust. 1963. Studies on some factors influencing anemia in tumor-bearing animals. J. Amer. Med. Ass. 186:550–553.

18

Blood Coagulation[1]

Hans Meier and Warren G. Hoag

This chapter summarizes what is known about normal and abnormal coagulation of blood in inbred mice. Until 1960 mouse blood had not been studied with respect to the amounts of various clotting factors present normally in healthy mice and absent in a number of hemorrhagic syndromes.

NORMAL BLOOD CLOTTING

Methods of assaying normal mouse blood for clotting factors and for the amounts of each component necessary for normal clotting are given by Meier et al. (1960, 1962) and Meier (1963). Knowledge of clotting factors extracted from mouse blood is of value in (1) the diagnosis of abnormal clotting in man, (2) the study of the mechanism of clotting as a biological process, and (3) the assay of certain liver functions. Blood coagulation is a highly complicated biological phenomenon. A tentative and oversimplified scheme of the clotting mechanism is shown in Fig. 18-1. It is presented to familiarize the reader with the various synonyms in common use and to provide some understanding of the different factorial interactions. A separate assay is available for the three stages, each consisting of several linked reactions.

ABNORMAL BLOOD CLOTTING

Several disorders of the clotting mechanism including both excessive bleeding and hypercoagulability have been described in inbred mice.

Stuart-Prower factor deficiency

A spontaneous bleeding or hemorrhagic diathesis occurred in certain inbred strains of mice at The Jackson Laboratory in 1960 (Meier et al., 1962). About 80 per cent of males of the SWR/J strain and 20 per cent of males of the DBA/2J strain died with bilateral hemothorax and sometimes with multiple bleedings elsewhere in the body, either subcutaneous, perivesicular, or subdural. The mice were anemic and jaundiced. The major histopathological findings were myocarditis and various degrees of hepatomegaly. Elevated lactic dehydrogenase was associated with myocarditis. Anemia and icterus seemed to be consequences of extensive hemorrhage rather than of hepatic dysfunction since alkaline and acid serum phosphatase and glutamic-oxalic transaminase were, except for lowered serum glutamic-pyruvic transaminase, within normal range or only slightly altered.

The affected mice had single or multiple "prothrombin complex" deficiencies, including deficiencies of the Stuart-Prower factor or factor X, plasma thromboplastin component (PTC) or factor IX, serum prothrombin conversion factor or factor VIII, and prothrombin. Activities of the first three factors, determined by specific assays using both human and normal mouse blood were less than 3 per cent of normal and the last was less than 10 per cent. Accelerator globulin (Ac-G or factor V), Hageman factor, and fibrinogen were normal. The clotting abnormality which first appeared in the course of the disease was always Stuart-Prower factor deficiency.

[1] The writing of this chapter was supported in part by the National Hemophilia Foundation and in part by Public Health Service Research Grant CA 04691 from the National Cancer Institute.

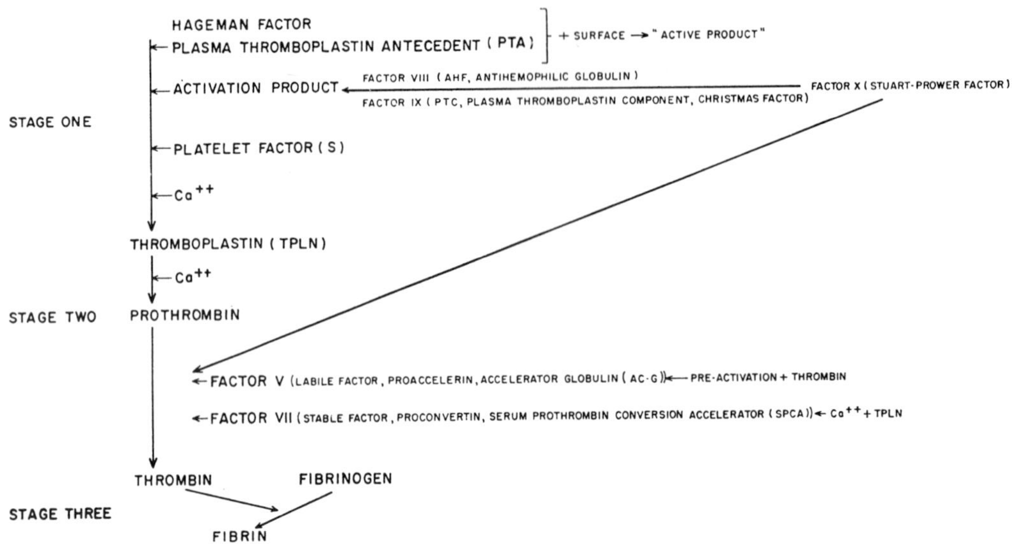

Fig. 18-1. Factors in blood coagulation.

Two features were particularly noteworthy: (1) occurrence only in males of two strains and (2) consistency of the lesions both microscopically and biochemically. Although myocarditis was the most severe histological alteration, liver dysfunction with respect to clotting factor(s) was the most important clinical consequence, leading eventually to death from loss of blood. The simultaneous occurrence of the condition in two strains suggested that the disease was due to an environmental agent rather than to a mutation and that these strains differed from the others in their genetic susceptibility. A search for possible agents implicated ethylene glycol and higher polymers produced by ethylene oxide sterilization of wood shavings used as mouse-cage bedding. The presence of one or more glycols in shavings gassed with ethylene oxide was shown chromatographically. Dilute solutions of ethylene glycol, administered by gavage, eliminated or greatly decreased the activity of factor X in less than 72 hours (Allen et al., 1962). In addition, we had observed the spontaneous factor X deficiency only in those bleeding mice that had prolonged contact with sterilized bedding. Mice not only touch shavings with parts of their bodies, but also continually chew them. Discontinuing gas sterilization eliminated the hemorrhagic disease.

No explanation is available for the peculiar strain and sex distribution, other than the obvious genetic differences in response to a toxic agent.

Factor VII deficiency

Clotting factor determinations in mice have been found to be extremely sensitive indicators of liver function. For example, clotting abnormalities associated with liver damage caused by a variety of bacterial diseases and toxicities have appeared before other functional deficiencies of the liver were detected. An example is factor X deficiency induced by ethylene glycol. However, liver damage usually affects several clotting factors. For instance, hypoprothrombinemia is usually most severe in chloroform poisoning, but the accelerator globulin (factor V) is also greatly diminished.

Of all clotting factors, a factor VII deficiency occurs most often in association with impaired liver function. It readily results from hypovitaminosis K in most animal species including man (Alexander et al., 1959). However, in mice, a diet without any vitamin K complex did not induce a factor VII deficiency. Coupled with other evidence this suggests that under ordinary conditions intestinal microorganisms are able to supply much, if not all, of the required vitamin K.

Signs of vitamin K deficiency in female mice may be induced by *dl*-α-tocopherol quinone, a vitamin K and E analogue (Woolley, 1945). Hemorrhages presumably due to factor VII deficiency occur only in the reproductive system of pregnant mice. The compound is without effect in males and nonpregnant females. However, pregnant females that received daily oral doses of 100 mg of the quinone had prothrombin times similar to the controls. Feeding of 3,3-methylene-bis-4-hydroxycoumarin, another antimetabolite of vitamin K, caused some signs of the vitamin K deficiency but did not produce resorption of fetuses and vaginal hemorrhages in pregnant mice (Woolley, 1945).

Left auricular thrombosis

Left auricular thrombosis occurs in high incidence (66 per cent) among older breeding females or inactive breeders of the BALB/cJ strain. The significance of repeated pregnancies rather than age alone is indicated by the fact that BALB/cJ males and virgin or ovariectomized females never develop the condition. Although repeated pregnancies in females of most strains do not affect coagulation or only irregularly cause some reduction in certain precoagulant factors, BALB/cJ mice suffer from severe deficiencies of PTC or factor IX (about 40 per cent of normal), antihemophilic factor (AHF) or factor XIII (about 60 per cent of normal), Stuart-Prower factor or factor X (about 50 per cent of normal), and prothrombin (about 33 per cent of normal). These deficiencies occur shortly before parturition. Hageman factor and Ac-G are always normal. These prothrombin-complex abnormalities disappear a few days after birth; plasma prothrombin rebound may be 20 to 25 per cent above normal within 2 days of parturition. The rebound phases are only moderately or not at all reflected in the "one-stage" prothrombin time. Attempts to measure fibrinolytic activity in blood from thrombotic mice fail because of the very low activity found normally in mouse blood (Meier et al., 1960).

We believe that the postpartum prothrombin-rebound, similar to that observed in man following Dicumarol and heparin therapy, may have clinical implications. The thrombus probably forms rather quickly and is primary, whereas endocardial damage is probably secondary. The clinical condition associated with advanced stages of auricular thrombosis consists of unthriftiness, labored breathing from lung congestion, and usually subcutaneous edema. At necropsy the left auricle is three to five times the normal size. Since the disease can be diagnosed readily or at least suspected in breeders pregnant more than six times, it will be useful in studies of thrombosis formation, lysis, and potential fibrinolytic compounds (Meier and Hoag, 1962).

A similar condition can be induced in young adult mice of a number of strains by feeding a hyperlipotropic diet containing 28 per cent fat as lard and 8 per cent protein as casein. After about 7 weeks on this diet the mice develop atrial necrosis, resulting in formation of mural thrombi and reaching critical or terminal dimensions in 10 to 12 weeks. Betaine (hydrochloride, 2 mg/100 g of diet) does not prevent the lesions (Ball, 1962).

SUMMARY

We have referred to the pertinent literature on aspects of normal blood clotting in mice. The papers cited give the methods for factorial assays and the normal values of all factors involved in coagulation. A simplified scheme of the factors concerned in blood coagulation has been presented. The examples of single or multiple factor deficiency syndromes discussed include the Stuart-Prower (factor X), factors VII and VIII, and the plasma thromboplastin component (factor IX). Each deficiency depends upon a particular genetic susceptibility whether or not it occurs spontaneously or is induced.

LITERATURE CITED

Alexander, B., A. Kilman, R. Coleman, E. Scholtz, and A. DiFrancesco. 1959. New "hemophiloid" defects: some clinico-laboratory and experimental abnormalities in thromboplastin generation, p. 137–157. *In* K. M. Brinkhous and P. DeNichola [ed.] Hemophilia and other hemorrhagic states. University of North Carolina Press, Chapel Hill.

Allen, R. C., H. Meier, and W. G. Hoag. 1962. Ethylene glycol produced by ethylene oxide sterilization and its effects in an inbred strain of mice. Nature 193:387–388.

Ball, C. T. 1962. Cardiac lesions in mice fed high fat, low protein diets. Anat. Rec. 142:212.

Meier, H. 1963. Experimental pharmacogenetics. Academic Press, New York. 213 p.

Meier, H., R. C. Allen, and W. G. Hoag. 1960. Normal blood clotting of inbred mice. Amer. J. Physiol. 201:375–378.

Meier, H., R. C. Allen, and W. G. Hoag. 1962. Spontaneous hemorrhagic diathesis in inbred mice due to single or multiple "prothrombin complex" deficiencies. Blood 19:501–514.

Meier, H., and W. G. Hoag. 1962. Studies on left auricular thrombosis in mice. Exp. Med. Surg. 19:317–322.

Woolley, D. W. 1945. Some biological effects produced by α-tocopherol quinone. J. Biol. Chem. 159:59–66.

19

Inherited Metabolic Variations[1]

Andrew A. Kandutsch and Douglas L. Coleman

Since every heritable difference between two mouse strains must involve a quantitative or qualitative difference in metabolism, a number of chapters in this book deal either directly or indirectly with metabolism. Especially pertinent are chapters dealing with pigmentation (Chap. 21), nutrition (Chap. 5), endocrinology (Chap. 20), physiology (Chap. 15, 16, 17, 18), and pathology (Chap. 29, 30). The subject matter discussed in this chapter is, therefore, not a complete survey of heritable metabolic characteristics but is a selective presentation of some established differences between inbred and mutant strains not discussed in detail in other chapters.

PROTEIN AND AMINO ACID METABOLISM

The existence of serum and tissue isoantigens (Chap. 24) and of strain-dependent differences in electrophoretic patterns of urinary and serum proteins (Thompson et al., 1954; Finlayson et al., 1963) imply variations in protein metabolism. Such differences in protein structure cannot yet be explained in metabolic terms. Hereditary differences in the excretion and utilization of amino acids have been reported. The amount of taurine excreted in the urine of C57BL strain mice is approximately 10 times that excreted by strains A and CBA (Harris and Searle, 1953). Although mice and rats are able to utilize both D and L forms of certain amino acids, mice bearing the W^v gene are unable to utilize the D forms (Goodman, 1955, 1958). Nutrition studies provide some evidence for the existence of differences in the rates of protein turnover in different strains (Fenton and Marsh, 1956; Fenton, 1957). Those strains found to exhibit the most marked differences in protein turnover also exhibit marked differences in carbohydrate metabolism and in their susceptibility to obesity.

METABOLIC PATTERNS ASSOCIATED WITH OBESITY

The major pathways of carbohydrate, protein, and lipid metabolism are intricately interrelated and, in mammals, are regulated by a complicated interplay of hormonal factors. It is, therefore, not unexpected to find that in every instance of extensive study an alteration in the metabolism of one class of compounds is accompanied by disturbances in the metabolism of one or both of the other classes and by evidence for altered rates of secretion of, or sensitivity to, endocrine hormones.

Obesity in inbred strains

An interest in the metabolic factors underlying the development of obesity has been a major stimulus for the investigation of metabolic differences between inbred strains of mice. Inbred strains range in their tendency toward obesity from extremely resistant to highly susceptible. Since the occurrence and degree of obesity in

[1] The writing of this chapter was supported in part by Public Health Service Research Grants CA 01329, CA 02758, CA 05873, and CA 07976 from the National Cancer Institute and AM 06871 from the National Institute for Arthritis and Metabolic Diseases.

mice of a particular strain is to a large extent dependent upon the nature of the diet (Fenton and Dowling, 1953), reports of mouse strains as resistant or susceptible to obesity are sometimes at variance. Genetic divergence of sublines of the same strain may also contribute to discrepancies between the observations of different workers. Thus, Fenton and Dowling (1953) classify strain A/Fn as severly obese, whereas Yamamoto et al. (1963) refer to DBA/2JN and A/LN (now designated AL/N) as lean and STR/1N as obese.

With the exception of the obese conditions determined by the single genes, obese (*ob*), adipose (*ad*), and the A^y allele at the agouti locus, little is known of the number or nature of genes that influence the development of obesity. On the basis of a single study, inconclusive in this respect, Yamamoto et al. (1963) suggested that inheritance of obesity in the STR/1N strain is polygenic with some evidence for dominance of greater body weight or heterosis.

Serum and plasma levels of total lipids, phospholipids, and cholesterol are elevated in obese mice but there is no simple relationship between the degree of obesity and plasma lipid levels (Yamamoto et al., 1963; Zomzely and Mayer, 1958). Although the number of genes regulating plasma levels of total lipid, phospholipids, and cholesterol has not been determined, the over-all behavior is additive (Yamamoto et al., 1963). Studies of serum cholesterol levels among six inbred strains selected at random indicated wide variation ranging from 128 mg per 100 ml for C57BL/6 to 208 mg per 100 ml for C3H. Levels in males were consistently higher than in females. Levels in F_1 hybrids were found to be a linear function of the cholesterol level of the dam, the cholesterol level of the sire, and the sex of the subject. The three factors do not interact (Bruell et al., 1962; Bruell, 1963).

Six strains of mice can be grouped into three categories on the basis of the tendency to become obese. Strain I/Fn is completely resistant to obesity; LT and C57BL/Fn are moderately obese; A/Fn, C3H/Fn, and BUB are severely obese (Fenton and Dowling, 1953; Lyon, 1957). Strain I has a relatively high rate of protein turnover (Fenton and Dowling, 1953; Fenton, 1957). Glycogen levels in the muscle of the lean strain I are elevated, whereas glycogen levels in the liver are depressed relative to levels found in more obese animals (Lyon, 1957). Results of studies aimed at providing an explanation for the abnormal distribution of carbohydrate in strain I appear to implicate muscle phosphorylase. Strain C57BL has a lower nutritional requirement for vitamin B_6 (a cofactor for phosphorylase) than does strain I (Lyon et al., 1958). Total concentrations of vitamin B_6 in the livers of the two strains are similar, but the level of pyridoxal 5′-phosphate is greater in the C57BL strain and the level of pyridoxamine 5′-phosphate is higher in the I strain (Lyon et al., 1962). Muscle phosphorylase levels in the two strains are similar, but there is a striking difference in the levels of the active (*a*) form of the enzyme. The amount of the active form found in I strain mice ranged from 0 to 5 per cent of the total phosphorylase, while in C57BL mice the amount of active form averaged 60 per cent of the total. The amount of the active form of phosphorylase in the muscle of the C57BL strain is elevated by dietary restrictions and by treatment with epinephrine, but these experimental procedures have no effect on the amount of active form present in I strain muscle. The failure of these procedures to induce an increase in the amount of active phosphorylase in strain I muscle appears to be explained by the finding that phosphorylase *b*-kinase is essentially absent from the muscle of I strain mice. Phosphorylase *b*-kinase is present in heart muscle from both strains and the increase in the amount of active phosphorylase in this organ following dietary restriction or treatment with epinephrine is similar in both strain I and C57BL. Although these observations are of much interest and seem to be involved in the development of the abnormal metabolic picture presented by strain I mice, their significance in relation to the observed differences in glycogen storage and utilization is not clear. Glycogenolysis in strain I proceeds with facility even when amounts of phosphorylase *a* are too low to detect (Lyon and Porter, 1962, 1963). A possible role of endocrine hormones in the development of the metabolic patterns characteristic of strain I has been suggested (Fenton, 1960; Fenton and Duguid, 1962).

The results of other studies concerned with elucidating the metabolic basis for obesity in inbred strains indicate that the obese C3H/Fn strain has a higher glucose tolerance than the relatively lean C57BL/Fn strain (Bloom and Fenton, 1956), and the obese A/Fn strain is able to mobilize fat to the liver more rapidly than the I strain (Fenton, 1960). An association of altered carbohydrate metabolism with the development of obesity appears to be a uniform finding. This relationship and the nature of other metabolic changes associated with obesity have been extensively studied in obese (*ob/ob*) mice and to a lesser extent in yellow ($A^y/-$) mice.

Obesity determined by single genes

Obesity in ob/ob mice. Obese (*ob/ob*) mice do not mate unless the diet is restricted and the lines are usually maintained by breeding animals heterozygous for the gene (Ingalls et al., 1950). The obesity is classified as "metabolic," indicating that the primary lesion is a defect in the metabolism of tissues. Characteristics of this class of obesities, as exemplified by the obese mouse, have been extensively reviewed and compared with those of the "regulatory" type in which the primary impairment is of the central mechanism regulating food intake (Mayer, 1953, 1960; Meier, 1963). Obese mice often attain weights of 80 g and approximately 90 per cent of the excess weight is due to fat. Obesity is accompanied by high levels of blood glucose and the syndrome is characterized as obese-hyperglycemic. Blood levels of total lipid and cholesterol are also elevated.

The energetics of the obese condition in mice bearing the *ob* gene is explained by an increased food intake, by a diminished rate of oxygen consumption, and by greatly diminished activity. Fasting reduces blood glucose levels to normal, but body composition remains obese when weight is reduced to normal or below normal. Both hormonal and metabolic disturbances have been found and it is still not clear whether the primary defect is hormonal or enzymatic (see Chap. 20). The size and number of islets of Langerhans and the ratio of beta to alpha cells are increased in obese-hyperglycemic mice (Mayer, 1960; Hellerstrom and Hellman, 1963). Despite the apparent hypersecretion of insulin, there is a progressive development of hyperglycemia and glucosuria. Blood levels of glucose are abnormally resistant to alteration by administered insulin, whereas the hyperglycemia is increased by treatment with growth hormone, ACTH, or glucagon. Hyperplasia of the cortical layers of the adrenals has been detected and corticosterone production is elevated in obese mice (Carstensen et al., 1961).

The major metabolic factors responsible for the development of obesity appear to be an increased rate of lipogenesis from acetate and glucose and an imbalance between rates of fatty acid esterification and lipolysis. Metabolism of acetate to carbon dioxide in vivo is diminished, whereas lipogenesis from acetate and glucose is greatly increased (Hughes and Tolbert, 1958; Shigeta and Shreeve, 1964). Studies of acetate metabolism by liver and adipose tissue in vitro are in general agreement with results obtained in vivo (Hollifield et al., 1960; Christophe et al., 1961*b;* Hellman et al., 1962*a*). Since citrate is the main precursor of acetate for the extramitochondrial synthesis of fatty acids, the observation that the specific activity of the citrate cleavage enzyme is elevated in obese mice may be correlated with the increased rate of lipogenesis from glucose and acetate in vivo (Kornacker and Lowenstein, 1964). For reasons that are not clear, studies of glucose metabolism by liver, adipose tissue, and diaphragm in vitro are not in agreement with findings in vivo. Metabolism of glucose to carbon dioxide, glycogen, glycerol-glyceride, and fatty acids is depressed in tissues of obese mice (Christophe et al., 1961*b;* Leboeuf et al., 1961; Lochaya et al., 1961; Hellman et al., 1961, 1962*b*). Unlike the response to insulin or growth hormone in vivo the effects of these hormones on the metabolism of acetate and glucose in vitro are normal.

Studies in vitro indicate the presence of a variety of other changes in metabolic rates and enzyme levels. In keeping with the hyperglycemic state of the animal, liver glycogen turnover is increased and levels of liver phosphorylase are elevated (Mayer, 1960). Although the rate of conversion of glucose to glycerol-glyceride by adipose tissue from obese mice is low, the rate at which glycerol is incorporated into glycerides is much increased (Lochaya et al., 1963). Levels of glycerol kinase in adipose tissue of obese mice are twice those in adipose tissue of non-obese littermates (Treble and Mayer, 1963). The rate at which free fatty acids are released from adipose tissue of obese mice in vitro is similar to that of normal controls, but there is a diminished ability of insulin to inhibit and epinephrine to augment the rate (Marshall and Engle, 1960; Leboeuf et al., 1961). The altered balance between fatty acid esterification and lipolysis indicated by these studies may be of primary importance to the development of the obese condition.

Obesity in other genotypes. Another gene, adipose (*ad*), determines an obese condition superficially similar to that determined by the *ob* gene. Like *ob/ob* mice, *ad/ad* mice do not breed successfully. A double homozygote for adipose and pituitary dwarf (*dw*) has been produced. The dwarfism does not interfere with the adiposity nor does the adiposity affect the skeletal growth of the dwarf (Falconer and Isaacson, 1959; Batt and Harrison, 1960). The metabolism of adipose *ad/ad* mice has not yet been compared with the metabolism of obese *ob/ob* mice.

A^y, a dominant gene determining yellow coat color, also results in the development of obesity. Animals homozygous for the gene die before birth and all yellows are heterozygous. Weights attained by yellow obese mice may be two to three times those of normal controls. The development of obesity is influenced by the genetic background and the amount of activity that the mice are permitted. Apparently some yellow mice do not become obese. There is a pronounced sex effect, females being more obese than males (Grüneberg, 1952; Dickie and Woolley, 1946). The metabolic characteristics of the obese condition determined by the A^y gene have not been so well studied as those of the obesity determined by the *ob* gene. In general, metabolic lesions identified appear to be similar in the two obesities, but are less pronounced in mice bearing the A^y gene (Zomzely and Mayer, 1959; Carpenter and Mayer, 1958; Mayer, 1960; Wolff, 1963).

The genetics of an obese condition present in the NZO strain of mice developed in New Zealand has not been studied sufficiently to determine whether the condition is determined by one or more genes, but the characteristics of the obesity have been to some extent compared with those of obesity determined by the *ob* gene. NZO obese mice attain weights of 50 to 70 g and are able to breed. As in *ob/ob* mice, the islets of Langerhans are enlarged and there is evidence for hypersecretion of insulin (Sneyd, 1964). NZO mice are moderately hyperglycemic and resistant to insulin (Bielschowsky and Bielschowsky, 1956; Chap. 20). Metabolic characteristics of the strain have not been studied extensively, but they appear to differ in some respects from those of the obese-hyperglycemic syndrome determined by the *ob* gene. The rate of oxygen consumption and the rate at which acetate is metabolized to carbon dioxide are normal. As in *ob/ob* mice, the rate at which glucose is converted to lipid is increased, suggesting increased lipogenesis from carbohydrate (Subrahmanyam, 1960).

CREATINE METABOLISM IN MICE WITH HEREDITARY MUSCULAR DYSTROPHY

A disease analogous to human muscular dystrophy arose by mutation in an inbred strain of mice (129/Re) and is inherited as an autosomal recessive (symbol *dy*) (Michelson et al., 1955; Chap. 29). The biochemistry of this mutant is complex and many seemingly unrelated biochemical abnormalities have been found associated with the *dy* locus. Although most aspects of this type of dystrophy have been reviewed (Harman et al., 1963; Meier, 1963), information dealing with creatine metabolism has been considered only briefly and is discussed more fully below.

Abnormalities in the pattern of storage and excretion of creatine invariably accompany muscular wasting and have been studied extensively in the dystrophic mouse. Although the amount of creatine in liver and kidney of homozygous dystrophic (*dy/dy*) mice is normal, decreased amounts are found in serum and muscle (Kandutsch and Russell, 1958). A creatinuria is also observed, the significance of which is difficult to evaluate since all mice normally excrete approximately equal amounts of creatine and creatinine in the urine (Madison, 1952). Most mammals excrete little or no creatine but only its anhydride, creatinine. If dystrophic mice are maintained on a commercial laboratory chow, both creatine and creatinine are excreted at nearly normal levels. When the animals are placed on a semisynthetic ration of more nearly optimal nutritional quality the creatine excretion increases markedly, whereas the creatinine excretion remains normal. Concurrent with the increased excretion of total creatine in dystrophic mice, the activity of glycine transamidinase increases from 50 per cent of normal on the chow ration to 150 per cent of normal on the semisynthetic ration (Coleman and Ashworth, 1960). Since this enzyme controls the production of guanidoacetic acid, the immediate precursor to creatine, this implies that a changed rate of creatine synthesis may be involved in the change from one ration to the other. A lessening of the muscle lesions and an improvement in over-all health of the dystrophic mice is also associated with rations that lead to increased transamidinase activity (Coleman and West, 1961).

Dietary creatine greatly represses the activity of glycine transamidinase in the kidneys of both mammals and birds (Walker, 1960, 1961) and this mechanism has been suggested as an effective and sensitive means of control of creatine biosynthesis. The addition of the enzyme substrate, glycine, to creatine-containing rations completely prevents the repression of this enzyme (Coleman, 1961; Van Pilsum, 1961; Fitch et al., 1961*a*). Coleman (1961) found that as little as 0.25 per cent of creatine in the diet was sufficient to lower the transamidinase activity significantly in dystrophic mice, whereas the enzyme in normal mice was much more stable, requiring at least 1 per cent of creatine before significant repression of the enzyme occurred. The addition of glycine to the diet at a level about 10 times that of creatine restored levels of transamidinase activity to those normally found in mice maintained on diets devoid of creatine. The marked fluctuations in kidney transamidinase activity in dystrophic mice when compared with those seen in normal mice demonstrate an unusual degree of metabolic instability. The elevated transamidinase activities present in dystrophic mice maintained on diets devoid of creatine indicate an important difference between the hereditary dystrophy and that caused by a deficiency of vitamin E, which is characterized by markedly decreased levels of kidney transamidinase. The decrease in vitamin E–deficient animals is thought to be due to the repression of kidney transamidinase caused by increased leakage of creatine from muscle into the blood stream (Fitch et al., 1961*b*). In the dystrophic mouse the increase in the total excreted creatine and creatinine is proportional to the increase in transamidinase activity and for some reason the increased level of endogenous creatine does not trigger the repression mechanism, even though these mice are more sensitive to dietary creatine. These findings suggest that the mechanism of transamidinase repression in dystrophic mice may be more complicated than that in vitamin E–deficient animals.

ENZYME ACTIVITIES CONTROLLED BY SINGLE GENES

A number of enzyme activities under the control of single genes have been identified in mice. As yet, abnormal levels of these activities have not been associated with any pathological or other abnormal conditions, suggesting that reduced levels of the enzymes are normally adequate and that any selective value of high or low levels may be in the response of the animal to unusual environmental conditions.

β-Glucuronidase

Interest in the genetic factors controlling tissue β-glucuronidase in inbred mice began with the studies of Morrow et al. (1950) who found that C3H mice exhibited markedly less β-glucuronidase activity than did those of the DBA, C57BL, and A strains. Subsequent investigations on the inheritance of low and high tissue β-glucuronidase (Law et al., 1952) demonstrated that the activity of this enzyme is controlled by alleles at a single locus. The allele (+) determining high β-glucuronidase activity was shown to be dominant to the allele (*g*) responsible for low enzyme activity. This locus affects the β-glucuronidase activity differently in various tissues. For example, spleens from low strains (*g/g*) contain much more β-glucuronidase activity than do the liver and kidney (1.6 vs. 0.36 and 0.34 units respectively), whereas in high strains (+/+) the spleen activity (3.6 units) is equal to the liver activity (3.5 units) and about twice as high as that in kidneys (1.7 units). Investigations reveal a marked sex difference in the activity of kidney β-glucuronidase in adults. Kidneys from males contain about twice the β-glucuronidase activity seen in females regardless of whether the genotype is +/+ or *g/g*. No sex differences are seen in immature mice (Morrow et al., 1951). Fishman and Farmelant (1953) have amplified these findings using mice of 129/J (+/+) and C3H (*g/g*) strains. Testosterone propionate produces marked increases in kidney β-glucuronidase in intact or castrated mice of both sexes in both high and low strains. In high strains (+/+) this androgen sometimes increases the activity of the liver and spleen enzyme slightly, but this increase was never seen in mutant (*g/g*) mice. Estrogen did not affect the kidney enzyme but rather increased the liver and spleen enzyme in both sexes. It is important to note that these treatments never increased the level of β-glucuronidase activity in *g/g* mice to that seen in untreated wild-type (+/+) mice, suggesting that the control exerted by the *g* locus is independent of that exerted by hormones.

Comparisons of certain properties (pH optima, Michaelis constants, and activation by DNA) of the β-glucuronidases isolated from two high strains (129/J and DBA) and two low strains (C3H and AKR) suggested that both the *g* locus and hormones affect only the amount and not the structure of the protein (Sie and Fishman, 1953). In probing this problem still further Paigen (1961*a*) found that the β-glucuronidase from mutant mice was more sensitive to heat and alkali and concluded that the enzyme structure is, indeed, altered in *g/g* mice. Also, he established that this enzyme is located in both the lysosomal and microsomal fraction of tissue homogenates. The relative amounts of the enzyme at the two sites were different in the two genotypes, and this shift in intracellular location was controlled by a single genetic factor which is identical to or closely linked with the *g* locus which controls the structure of the enzyme. Paigen concluded that the changes in structure and relative concentration of the enzyme are reflected in changes in the structure of the cytoplasmic particles carrying the enzyme, a situation which could arise, most simply, if the changes in particle structure were a consequence of the alteration in the structure of the enzyme protein. In a study of the development of enzyme levels in various tissues it was found that the decreased activity in the livers from *g/g* mice was due to two effects, lower initial activity at birth followed by a failure of subsequent normal development after 11 days (Paigen, 1961*b*). In contrast, β-glucuronidase activity in the spleen is only slightly lower at birth in *g/g* mice and subsequently increases normally until maturity. The thymus presents an intermediate case in that normal activity in both genotypes is seen at birth but that later development is slower in *g/g* than in normal mice. Genetic analysis again indicates that the gene controlling these developmental sequences in all these tissues is identical to or closely linked to the *g* locus. Thus the *g* locus controls the differentiation of tissues with respect to β-glucuronidase and determines the structure of the enzyme and its intracellular location. This developmental difference seems to be related to the responsiveness of the normal and mutant types of β-glucuronidase to some stimulus, either genetic or environmental, which normally regulates enzyme production. Critical studies such as these illustrate the complexities which develop when attempts are made to analyse the mechanism or mechanisms whereby genes control enzymes in mammalian systems.

Catalase

The low levels of catalase found in most tumors and in livers from mice bearing tumors prompted many investigations into the activity of this enzyme in various organs from different inbred strains of mice. Greenstein and Andervont (1942) found that catalase activity was nearly the same for all strains with the exception of C57BL which had about 50 per cent of the normal level of liver catalase activity while maintaining normal catalase activity in other tissues. Rechcigl and Heston (1963) found two closely related C57BL sublines that have liver catalase values similar to those seen in other strains, whereas all the other C57BL sublines studied have the low level of liver catalase originally reported. Genetic analysis revealed that the activity of liver catalase is controlled by a single gene (*Ce:* Heston, 1964, personal communication) and that the presence of high liver catalase in the C57BL/6He and C57BL/6An sublines represents a reverse mutation to the wild-type allele.

Catalase in most tissues exists both in particulate and soluble forms (Adams and Burgess, 1959), and the enzyme activity found in whole tissue homogenates varies greatly depending on the degree of disruption of the subcellular particles. Feinstein (1959) observed species differences in the amount of supernatant and particulate catalases. In pig liver 89 per cent of the catalase activity was found in the supernatant while in mouse liver only 27 per cent was in the supernatant with the other 73 per cent bound rather tightly to the particulate fraction. Feinstein found that the nonionic detergent Triton X 100 released all the catalase from the particulate material and suggested its routine use in catalase assays where total activity is desired. Thus the *Ce* locus could act either by causing the particulate catalase to be bound more tightly to the binding particles of mouse liver or by increasing the proportion of catalase in association with the particles, rather than decreasing the total amount of catalase actually present in the whole liver.

Feinstein et al. (1964) screened 12,000 progeny from irradiated male mice for blood catalase. One male and one female had blood catalase levels which were about one-half that observed for the rest of the group. Breeding of these two mice established that the trait was inherited, probably as a recessive. Further genetic and biochemical tests will be required to determine whether one or more genes are involved in producing this enzyme reduction and whether these genes also control the catalase levels of other tissues.

δ-Aminolevulinate dehydratase

Strain differences in the hepatic activity of δ-aminolevulinate dehydratase were found to be under the control of a single genetic locus (*Lv*) (Russell and Coleman, 1963). In all mice the enzyme activity is relatively high in fetal liver, falls to a low level shortly after birth, and then increases to the adult value. Mice with the low adult enzyme level (Lv^b/Lv^b) have activities predictably lower than those of mice with high adult activity (Lv^a/Lv^a) at all stages of development. Although this enzyme is involved in porphyrin synthesis and is extremely low in fetal livers from Lv^b/Lv^b homozygotes, no anemias or other blood dyscrasias are characteristic of these mice. A study on porphyrin synthesis by mouse harderian gland (Figge and Davidheiser, 1957) showed similar patterns of high and low enzyme activity in the various strains studied. Presumably the enzyme activity in this organ could also be controlled by this locus. Unpublished results (Coleman) suggest that the dehydratase activity in kidney and spleen is also under the control of the *Lv* locus.

Another gene, flexed (*f*), linked or allelic with the *Lv* locus, influences the amount of δ-aminolevulinate dehydratase in mouse tissues and is also responsible for a transitory anemia (Chap. 17).

Pyrimidine degrading sequence

Pyrimidine degradation in mice has been shown to be under genetic control (Dagg et al., 1963). Catabolism of uracil and thymine involves their breakdown in three steps to CO_2, ammonia, and β-alanine. Certain strains (SJL/J and RF/J) have rapid rates of degradation in vivo, whereas most other strains have relatively slow rates. Genetic analysis of the F_1, F_2, BC_1, and BC_2 generations indicated that this biochemical trait is inherited as a single factor (*Pd*) showing no dominance. Studies in vitro indicate that homozygous mice (Pd^b/Pd^b) having a high rate of over-all degradation have a high level of all three enzyme activities, whereas homozygous Pd^a/Pd^a mice having a low rate of degradation in vivo have much less of all three enzyme activities. The F_1 hybrids have intermediate activities for these enzymes. Since the activities of all three enzymes segregate together, it appears that the three enzymes are controlled by the *Pd* locus. A study of the development of the activity of the first enzyme in the sequence revealed that there was a finite but small amount of enzyme present at birth in both genotypes. The amount increased in both strains until normal adult values were reached (6 days for Pd^a/Pd^a and 21 days for Pd^b/Pd^b). The *Pd* gene does not appear to control the structure of this enzyme but rather acts as a regulator which at 6 days, in Pd^a/Pd^a mice, affects the rate of accumulation of the pyrimidine degrading enzymes, resulting in an equilibrium concentration much lower than is seen in the Pd^b/Pd^b genotype. One gene controlling the activity of three enzymes simultaneously is unusual but not impossible. It may be that *Pd* is an operator gene which controls the amount of synthesis determined by three linked structural genes in a manner similar to that proposed in bacterial systems (Jacob and Monod, 1961). *Pd* might also be a regulator gene which determines the production of a repressor substance which acts upon the operator to control amount of synthesis. There is, however, no evidence at this time to rule out the possibility that this gene affects an unknown metabolic reaction which controls the three enzyme systems in some indirect fashion.

Esterases

The development of various electrophoretic techniques, coupled with rapid histochemical techniques for enzyme analysis, has greatly facilitated studies of protein types and concentrations in various mouse tissues. Abnormalities in cholinesterases (Hamburgh, 1958; Yoon and Harris, 1962) have been observed in neurological mutants. Differences in other nonspecific esterases have been found in different inbred strains and in mutant mice within an inbred strain (Meier et al., 1962; Popp and Popp, 1962; Petras, 1963; Ruddle and Roderick, 1965). In some cases (Popp and Popp, 1962; Petras, 1963; Ruddle and Roderick, 1965), single gene differences have been found associated with these different electrophoretic profiles, whereas in other cases adequate genetic studies have not been carried out to determine whether one or many genes are involved. For a more detailed discussion of these changes in esterase activity, the reader is referred to discussions by Meier (1963) and Ruddle and Roderick (1965).

SUMMARY

Heritable characteristics such as obesity, leanness, high levels of cholesterol, lipid and carbohydrate in the serum, and muscular dystrophy are of major interest in relation to human health problems, and a relatively large effort has been devoted to their study. Elucidation of the basic metabolic factors that are involved in the development of these characteristics is, however, extremely difficult. The characteristic may be well defined but influenced by a number of genes, as are levels of serum lipid and cholesterol, implying that a number of metabolic changes are involved in the production of the phenotype, or the characteristic may be determined by a single gene (*ob*, A^y, *ad*, obesities; *dy*, muscular dystrophy). Determination by a single gene implies that the primary metabolic change is at the level of a single protein or reaction, but major metabolic pathways are interwoven and governed by hormonal and feedback mechanisms to such an extent that numerous secondary changes in metabolism appear, and the observed phenotype may result from the combined effects of a number of these. Studies of hereditary obesities, indicate the presence of alterations in carbohydrate and lipid metabolism and in the secretion of, and sensitivity to, hormones. The dystrophic mouse exhibits alteration in carbohydrate, lipid, nucleic acid, and creatine metabolism. Lean strain I mice differ biochemically from other strains in a number of ways, most strikingly, in the almost complete loss of phosphorylase *b*-kinase.

In contrast to studies aimed at defining the metabolic causes of complex heritable physical characteristics, identification of genes that alter single enzymes has, in general, resulted from a search for the gene after the alteration had been defined. Single genes that control the enzymes, β-glucuronidase, catalase, δ-aminolevulinate dehydratase, and pyrimidine-degrading enzymes have been identified. The observation that three enzymes functioning in pyrimidine degradation are controlled by a single gene is of especial interest. Further studies of this type should clarify the extent to which genetic models derived from studies with bacteria are applicable to mammals.

LITERATURE CITED

Adams, D. H., and E. A. Burgess. 1957. The effect of the degree of homogenization on the catalase activity of liver "homogenates." Brit. J. Cancer 11:310–325.

Batt, R. A. L., and G. A. Harrison. 1960. Features of the "adipose" mouse. Heredity 15:335.

Bielschowsky, M., and F. Bielschowsky. 1956. The New Zealand strain of obese mice: their response to stilboestrol and insulin. Austral. J. Exp. Biol. Med. Sci. 34:181–198.

Bloom, R. A., and P. F. Fenton. 1956. Glucose tolerance in relation to obesity and food intake. Amer. J. Physiol. 184:438–440.

Bruell, J. H. 1963. Additive inheritance of serum cholesterol level in mice. Science 142:1,664–1,666.

Bruell, J. H., A. F. Daroczy, and H. K. Hellerstein. 1962. Strain and sex differences in serum cholesterol levels of mice. Science 135:1,071–1,072.

Carpenter, K. J., and J. Mayer, 1958. Physiologic observations on yellow obesity in the mouse. Amer. J. Physiol. 193:499–504.

Carstensen, H., B. Hellman, and S. Larsson. 1961. Biosynthesis of steroids in the adrenals of normal and obese-hyperglycemic mice. Acta Soc. Med. Upsal. 64:139–151.

Christophe, J., B. Jeanrenaud, J. Mayer, and A. E. Renold. 1961*a*. Metabolism *in vitro* of adipose tissue in obese-hyperglycemic and goldthioglucose-treated mice. I. Metabolism of glucose. J. Biol. Chem. 236:642–647.

Christophe, J., B. Jeanrenaud, J. Mayer, and A. E. Renold. 1961*b*. Metabolism *in vitro* of adipose tissue in obese-hyperglycemic and goldthioglucose-treated mice. II. Metabolism of pyruvate and acetate. J. Biol. Chem. 236:648–652.

Coleman, D. L. 1961. Effects of dietary creatine and glycine on transamidinase activity in dystrophic mice. Arch. Biochem. Biophys. 94:183–186.

Coleman, D. L., and M. E. Ashworth. 1960. Influence of diet on transamidinase activity in dystrophic mice. Amer. J. Physiol. 199:927–930.

Coleman, D. L., and W. T. West. 1961. Effects of nutrition on growth, lifespan, and histopathology of mice with hereditary muscular dystrophy. J. Nutr. 73:273–281.

Dagg, C. P., D. L. Coleman, and G. M. Fraser. 1964. A gene affecting the rate of pyrimidine degradation in mice. Genetics 49:979–989.

Dickie, M. M., and G. W. Woolley. 1946. The age factor in weight of yellow mice. J. Hered. 37:365–368.

Falconer, D. S., and J. H. Isaacson. 1959. Adipose, a new inherited obesity of the mouse. J. Hered. 50:290–292.

Feinstein, R. N. 1959. Solubilization of particulate catalase with nonionic detergent. Arch. Biochem. Biophys. 79:399–400.

Feinstein, R. N., J. E. Seaholm, J. B. Howard, and W. L. Russell. 1964. Acalalasemic mice. Proc. Nat. Acad. Sci. 52:661–662.

Fenton, P. F. 1957. Hereditary factors in protein nutrition. Amer. J. Clin. Nutr. 5:663–665.

Fenton, P. F. 1960. Studies on genetically determined metabolic patterns. Amer. J. Clin. Nutr. 8:748–751.

Fenton, P. F., and M. T. Dowling. 1953. Studies on obesity. I. Nutritional obesity in mice. J. Nutr. 49:319–331.

Fenton, P. F., and J. R. Duguid. 1962. Growth hormone and cardiac glycogen: influence of environmental and genetic factors. Can. J. Biochem. Physiol. 40:337–341.

Fenton, P. F., and J. M. Marsh. 1956. Inherited metabolic patterns in mice. Caloric requirements for protein utilization and determination of protein minima. J. Nutr. 60:465–472.

Figge, F. H. J., and R. H. Davidheiser. 1957. Porphyrin synthesis by mouse harderian gland extract, sex, age, and strain variations. Proc. Soc. Exp. Biol. Med. 96:437–439.

Finlayson. J. S., M. Potter, and C. R. Runner. 1963. Electrophoretic variation and sex dimorphism of the major urinary protein complex in inbred mice: a new genetic marker. J. Nat. Cancer Inst. 31:91–107.

Fishman, W. H., and M. H. Farmelant. 1953. Effects of androgens and estrogens on β-glucuronidase in inbred mice. Endocrinology 52: 536–545.

Fitch, C. D., C. Hsu, and J. S. Dinning. 1961*a*. Partial reversal of creatine inhibition of transamidinase by dietary glycine. Biochim. Biophys. Acta 52:194–195.

Fitch, C. D., C. Hsu, and J. S. Dinning. 1961*b*. The mechanism of kidney transamidinase reduction in vitamin E-deficient rabbits. J. Biol. Chem. 236:490–492.

Goodman, R. M. 1955. Effect of W^v locus in the mouse on differential excretion of isomers of several amino acids. Proc. Soc. Exp. Biol. Med. 88:283–287.

Goodman, R. M. 1958. *In vitro* amino acid metabolism of tissues from a mouse mutant showing differential patterns of amino acid excretion. Fed. Proc. 17:57. (Abstr.)

Greenstein, J. P., and H. B. Andervont. 1942. The liver catalase activity of tumor-bearing mice and the effect of spontaneous regression and of removal of certain tumors. J. Nat. Cancer Inst. 2:345–355.

Grüneberg, H. 1952. The genetics of the mouse, 2nd ed., p. 40–42. Nijhoff, The Hague.

Hamburgh, M. 1958. The distribution of acetylcholinesterase in the mouse brain. Anat. Rec. 130:311. (Abstr.)

Harman, P. J., J. P. Tassoni, R. L. Curtis, and M. B. Hollinshead. 1963. Muscular dystrophy in the mouse, p. 407–456. *In* G. H. Bourne and M. N. Golarz [ed.] Muscular dystrophy in man and animals. Karger, New York.

Harris, H., and A. G. Searle. 1953. Urinary amino-acids in mice of different genotypes. Ann. Eugen. 17:165–167.

Hellman, B., S. Larsson, and S. Westman. 1961. Aspects of the glucose and amino acid metabolism in the liver and the diaphragm of normal and obese-hyperglycaemic mice. Acta Physiol. Scand. 53:330–338.

Hellman, B., S. Larsson, and S. Westman. 1962*a*. Acetate metabolism in isolated epididymal adipose tissue from obese-hyperglycaemic mice of different ages. Acta Physiol. Scand. 56:189–198.

Hellman, B., S. Larsson, and S. Westman. 1962*b*. The metabolism of variously labelled glucose in fatty livers from mice with congenital hyperglycaemia and obesitas. Acta Endocrinol. 39: 457–464.

Hellerström, C., and B. Hellman. 1963. Quantitative studies on isolated pancreatic islets of mammals. I. Peptidase activity in normal and obese-hyperglycaemic mice. Acta Endocrinol. 42:615–624.

Hollifield, G., W. Parson, and C. R. Ayers. 1960. *In vitro* synthesis of lipids from C-14 acetate by adipose tissue from four types of obese mice. Amer. J. Physiol. 198:37–38.

Hughes, A. M., and B. M. Tolbert. 1958. Oxidation of acetate, glucose, or glycine to carbon dioxide in mice exhibiting the hereditary obesity syndrome. J. Biol. Chem. 231:339–345.

Ingalls, A. M., M. M. Dickie, and G. D. Snell. 1950. Obese, a new mutation in the house mouse. J. Hered. 41:317–318.

Jacob, F., and J. Monod. 1961. Genetic regulatory mechanisms in the synthesis of proteins. J. Mol. Biol. 3:318–356.

Kandutsch, A. A., and A. E. Russell. 1958. Creatine and creatinine in tissues and urine of mice with hereditary muscular dystrophy. Amer. J. Physiol. 194:553–556.

Kornacker, M. S., and J. M. Lowenstein. 1964. Citrate cleavage enzyme in livers of obese and nonobese mice. Science 144:1,027–1,028.

Law, L. W., A. G. Morrow, and E. M. Greenspan. 1952. Inheritance of low liver glucuronidase activity in the mouse. J. Nat. Cancer Inst. 12:909–916.

Leboeuf, B., S. Lochaya, N. Leboeuf, F. C. Wood, Jr., J. Mayer, and G. F. Cahill, Jr. 1961. Glucose metabolism and mobilization of fatty acids by adipose tissue from obese mice. Amer. J. Physiol. 201:19–22.

Lochaya, S., J. C. Hamilton, and J. Mayer. 1963. Lipase and glycerokinase activities in the adipose tissue of obese-hyperglycaemic mice. Nature 197:182–183.

Lochaya, S., N. Leboeuf, J. Mayer, and B. Leboeuf. 1961. Adipose tissue metabolism of obese mice on standard and high fat diets. Amer. J. Physiol. 201:23–26.

Lyon, J. B., Jr. 1957. Muscle and liver glycogen levels in lean and obese strains of mice. Amer. J. Physiol. 190:434–438.

Lyon, J. B., Jr., J. A. Bain, and H. L. Williams. 1962. The distribution of vitamin B6 in the tissues of two inbred strains of mice fed complete and vitamin B6-deficient rations. J. Biol. Chem. 237:1,989–1,991.

Lyon, J. B., Jr., and J. Porter. 1962. The effect of pyridoxine deficiency on muscle and liver phosphorylase of two inbred strains of mice. Biochim. Biophys. Acta 58:248–254.

Lyon, J. B., Jr., and J. Porter. 1963. The relation of phosphorylase to glycogenolysis in skeletal muscle and heart of mice. J. Biol. Chem. 238: 1–11.

Lyon, J. B., Jr., H. L. Williams, and E. A. Arnold. 1958. The pyridoxine-deficient state in two strains of inbred mice. J. Nutr. 66:261–275.

Madison, C. R. 1952. A search for quantitative differences in the normal constituents of the urine of short ear and normal mice. J. Exp. Zool. 120:457–468.

Marshall, N. B., and F. L. Engel. 1960. The influence of epinephrine and fasting on adipose tissue content and release of free fatty acids in obese-hyperglycemic and lean mice. J. Lipid Res. 1:339–342.

Mayer, J. 1953. Genetic, traumatic and environmental factors in the etiology of obesity. Physiol. Rev. 33:472–508.

Mayer, J. 1960. The obese hyperglycemic syndrome of mice as an example of "metabolic" obesity. Amer. J. Clin. Nutr. 8:712–718.

Meier, H. 1963. Experimental pharmacogenetics, p. 11–78. Academic Press, New York.

Meier, H., E. Jordan, and W. G. Hoag. 1962. The zymogram technique as a tool for study of genetic differences. J. Histochem. Cytochem. 10:103–104.

Michelson, A. M., E. S. Russell, and P. J. Harman. 1955. Dystrophia muscularis: a hereditary primary myopathy in the house mouse. Proc. Nat. Acad. Sci. 41:1,079–1,084.

Morrow, A. G., D. M. Carroll, and E. M. Greenspan. 1951. A sex difference in the kidney glucuronidase activity of inbred mice. J. Nat. Cancer Inst. 11:663–669.

Morrow, A. G., E. M. Greenspan, and D. M. Carroll. 1950. Comparative studies of liver glucuronidase activity in inbred mice. J. Nat. Cancer Inst. 10:1,199–1,203.

Paigen, K. 1961*a*. The effect of mutation on the intracellular location of β-glucuronidase. Exp. Cell Res. 25:286–301.

Paigen, K. 1961*b*. The genetic control of enzyme activity during differentiation. Proc. Nat. Acad. Sci. 47:1,641–1,649.

Petras, M. L. 1963. Genetic control of a serum esterase component in *Mus musculus*. Proc. Nat. Acad. Sci. 50:112–116.

Popp, R. A., and D. M. Popp. 1962. Inheritance of serum esterases having different electrophoretic patterns. J. Hered. 53:111–114.

Rechcigl, M., Jr., and W. E. Heston. 1963. Tissue catalase activity in several C57BL substrains and in other strains of inbred mice. J. Nat. Cancer Inst. 30:855–864.

Ruddle, F. H., and T. H. Roderick. 1965. The genetic control of three kidney esterases in C57BL/6J and RF/J mice. Genetics 51:445–454.

Russell, R. L., and D. L. Coleman. 1963. Genetic control of hepatic δ-aminolevulinate dehydratase in mice. Genetics 48:1,033–1,039.

Shigeta, Y., and W. W. Shreeve. 1964. Fatty acid synthesis from glucose-l-H^3 and glucose-l-C^{14} in obese-hyperglycemic mice. Amer. J. Physiol. 206:1,085–1,090.

Sie, H., and W. H. Fishman. 1953. A comparative study of certain properties of liver β-glucuronidase of inbred mice. Cancer Res. 13: 590–592.

Sneyd, J. G. T. 1964. Pancreatic and serum insulin in the New Zealand strain of obese mice. J. Endocrinol. 28:163–172.

Subrahmanyam, K. 1960. Metabolism in the New Zealand strain of obese mice. Biochem. J. 76: 548–556.

Thompson, S., J. F. Foster, J. W. Gowen, and O. E. Tauber. 1954. Hereditary differences in serum proteins of normal mice. Proc. Soc. Exp. Biol. Med. 87:315–317.

Treble, D. H., and J. Mayer. 1963. Glycerolkinase activity in white adipose tissue of obese-hyperglycaemic mice. Nature 200:363–364.

Van Pilsum, J. F. 1961. Dietary control of kidney transamidinase. Fed. Proc. 20:225 (Abstr.)

Walker, J. B. 1960. Metabolic control of creatine biosynthesis. I. Effect of dietary creatine. J. Biol. Chem. 235:2,357–2,361.

Walker, J. B. 1961. Metabolic control of creatine biosynthesis. II. Restoration of transamidinase activity following creatine repression. J. Biol. Chem. 236:493–498.

Wolff, G. L. 1963. Growth of inbred yellow (A^ya) and non-yellow (aa) mice in parabiosis. Genetics 48:1,041–1,058.

Yamamoto, R. S., L. B. Crittenden, L. Sokoloff, and G. E. Jay, Jr. 1963. Genetic variations in plasma lipid content in mice. J. Lipid Res. 4: 413–418.

Yoon, C. H., and S. R. Harris. 1962. Cholinesterase studies of neurologic mutant in mice. I. Alterations in serum cholinesterase levels. Neurology 12:423–426.

Zomzely, C., and J. Mayer. 1958. Levels of serum cholesterol in obese mice. Nature 182:1,738–1,739.

Zomzely, C., and J. Mayer. 1959. Fat metabolism in experimental obesities. IX. Lipogenesis and cholesterogenesis in yellow obese mice. Amer. J. Physiol. 196:611–613.

20

Endocrine Variations[1]

Chen K. Chai and Margaret M. Dickie

There is abundant evidence documenting genetic variation in the endocrine physiology of mammals. In this chapter we will not attempt to describe the basic endocrinology of the mouse but will discuss some of the heritable differences of mice in hormone levels, responses to hormone treatment, results of organ ablation and transplantation, and hormonal imbalance.

Some endocrine variations are the result of single-gene mutations but most are due to multiple factors or polygenes. Systematic genetic approaches to the problem of endocrine variation are lacking but the many differences between genetically different groups of mice justify the statement that endocrine variations are under genetic control. For descriptions of the anatomy and pathology of endocrine glands and their target organs, see Chap. 13 and 27.

VARIATIONS IN STRAINS OR STOCKS

In this section we discuss some of the variations of endocrine glands and target organs among different types of mice, including age groups, and we describe the sexual dimorphism of some of the tissues that provide sensitive indices of sex steroidal activity. It should be noted that in comparing the effects of experimental endocrine manipulations with the untreated condition, the untreated baseline must be ascertained for each strain or stock. These differences are presumed to be polygenic in contrast to single-gene effects which are described in the next section.

Pituitary

The hypophysis shows few aging changes either grossly or microscopically, but cysts, cholesterol clefts, and chromophobic or basophilic tumors are found in some animals. The gland has a wet weight of about 1 to 2 mg and a dry weight of about 0.30 to 0.85 mg. The pituitary is consistently heavier in the female than in the male and there are also strain differences in the gland weights (Dickie, unpublished data) (Fig. 20-1).

The hormone content of the pituitary gland has been éstimated by using various pituitary-ovary-uterine relationships. Uterine weight shows a correlation with gonadotropin content (Runner, 1953). The feeble ovulatory response in prepuberal animals shows the low gonadotropin content of obese female pituitaries (Runner, 1952). The effect of hormone administration on organ weights in intact and hypophysectomized mice of five strains led Mühlbock et al. (1952) to conclude that strain DBA pituitaries have the highest gonadotropin content, C57BL and CBA an intermediate, and strains 020 and A the lowest content.

Thyroid

Few aging changes are observed in the thyroid, but thyroid hormone activity is relatively reduced by 16 months of age (Chai et al., 1964). Strains C57BL/6J and C57BR/cdJ and their hybrids ($B6BRF_1$) have a significantly higher turnover rate and lower I^{131} uptake than strains BALB/cJ and A/J and their CAF_1 hybrids (Chai et al.,

[1] The writing of this chapter was supported in part by Public Health Service Research Grant CA 03108 from the National Cancer Institute and by Contract AT(30-1)-3249 with the U.S. Atomic Energy Commission.

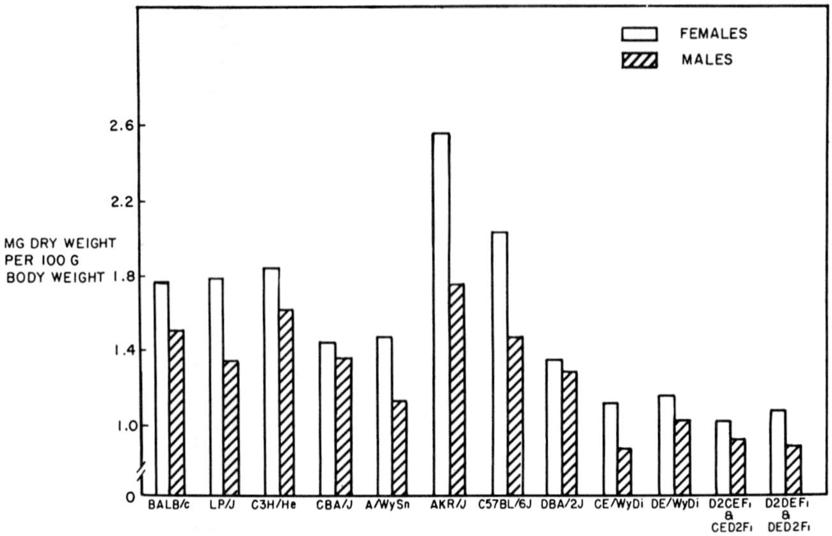

Fig. 20-1. Pituitary gland weights of some inbred strains and hybrid mice.

1957). Thyroid activity as measured by secretion rates is different in females and males. The average thyroid secretion rates, expressed as micrograms of L-thyroxine per 100 g of body weight per day indicate that strains BALB/cJ and A/J and the CAF_1 hybrid have a significantly lower secretion rate than strains C57BL/6J and C57BR/cdJ and the $B6BRF_1$ hybrid. Strain C57BL/6J has the highest secretion rate, and the $B6BRF_1$ hybrid has a rate intermediate between the parental strains. These results suggest that thyroid activity is under polygenic control and that different genetic factors may control the discharge of thyrotropic hormone and thyroid hormone (Amin et al., 1957). Thyroid-secretion rates can also be measured using thyroid weight and follicular cell height (Chai, 1958). Although values may be higher, perhaps because of the older ages of the animals, the order and magnitude of differences of secretion rate using this latter method generally agree with the estimates using I^{131} (Table 20-1), Chai (1960*a*) estimated heritability for I^{131} uptake to be about 10 per cent in a hybrid mouse population crossbred from six inbred strains.

Table 20-1. COMPARISON OF ESTIMATES OF THYROID SECRETION RATE, EXPRESSED AS μg L-THYROXINE/100 g BODY WEIGHT PER DAY, USING THYROID WEIGHTS, FOLLICULAR CELL HEIGHTS, AND THE THYROID COMPENSATION METHOD

Strain or hybrid[a]	Thyroid weight[b]	Cell height[b]	Compensation method[c]
A/J	3.77 μg	3.50 μg	2.13 μg
C57BL/6J	4.80	5.76	4.19
C57BR/cdJ			3.35
DBA/1J	6.31	6.31	
DBA/2J	6.73	6.45	
BALB/cJ			2.44
$B6AF_1$	6.53	5.76	
$B6D2F_1$	8.05	6.71	
$B6BRF_1$			3.76
CAF_1			2.35

[a] Strain and hybrid designations changed to conform with nomenclature rules (Staats, 1964). See Chap. 6 for strain abbreviations used in designations of hybrids.
[b] Chai (1958).
[c] Amin et al. (1957).

Silverstein et al. (1960 *a, b*) studied the ratio of I^{131} count in the thyroid to that in serum (T/S) and the amount of protein-bound iodine (PBI) as indices of thyroid activity in several inbred strains of mice and their F_1 hybrids from 7 to 16 months of age. They found that there is an inverse relationship between the T/S ratio and gland weight, that gland weight does not change significantly with a change in body weight, and that PBI is not associated with gland weight (Table 20-2).

Adrenal

Aging changes in the adrenal cortex include an increase in the number of subcapsular cells, rearrangement of fasciculata cells, degeneration of

Table 20-2. RATIO OF I^{131} COUNT IN THYROID TO THAT IN SERUM (T/S) AND PLASMA PROTEIN-BOUND IODINE CONCENTRATION (PBI) IN VARIOUS STRAINS OF MICE AND THEIR HYBRIDS[a]

Strain	*Mean T/S*[b]		*Mean PBI*[c]
	Males	Females	Males
STR/1N	96 ± 12.8	62 ± 5.6	0.9 ± 0.4
DBA/2JN	69 ± 5.0	53 ± 6.2	2.6 ± 0.4
C57L/HeN	112 ± 10.4	105 ± 35.4	2.5 ± 0.4
STR/N	43 ± 5.8		2.0
A/HeN	26 ± 7.2	25 ± 4.7	
A/LN	42 ± 4.9	40 ± 3.6	1.6
(STR/1N × A/LN)F_1	124 ± 12.5	70 ± 10.9	1.5 ± 0.5
(A/LN × STR/1N)F_1	76 ± 5.1	62 ± 6.3	1.7 ± 1.1
(DBA/2JN × A/LN)F_1	58 ± 3.4	67 ± 4.1	
(A/LN × DBA/2JN)F_1	68 ± 4.5	69 ± 5.0	1.0 ± 0.1
(C57L/HeN × A/HeN)F_1	36 ± 3.0	31 ± 3.3	1.0 ± 0
(A/HeN × C57L/HeN)F_1	49 ± 6.7	54 ± 4.8	1.0 ± 0.1

[a] From Silverstein et al. (1960*a*, *b*).

[b] $\frac{\text{Counts/min/g thyroid}}{\text{Counts/min/ml plasma}} \pm$ S.E. of the mean.

[c] Micrograms/100 ml plasma ± S.E. of the mean.

the X-zone, and appearance of foci of heterochromatic cells and brown fat or ceroid pigment in the juxtamedullary area (Woolley et al., 1953). In strain A the adrenals of old animals have amyloid deposits in the cortex (Heston et al., 1945). In strain NH the adrenals develop cortical hypertrophy as a result of early spontaneous gonadal atrophy (Kirschbaum et al., 1946). In old BALB/cHuDi females and males there are severe cortical changes which include some "A" cell type carcinomas (Dickie, unpublished data). Adrenals are always smaller and dark red in males in contrast to the light pinkish (more opaque) large adrenals of females, and there are strain differences in average weights of the glands (Dickie, unpublished data) (Fig. 20-2).

There are few reports concerning the adrenal medulla in mice, but some tumors such as pheochromocytomas have been observed (Smith et al., 1949; Dickie, 1954). Microscopic medullary tumors have been found in the adrenals of old D2CEF_1 mice and in offspring of these F_1 animals backcrossed to strain CE (Dickie, 1958).

Ovary

Strain differences are apparent in the morphology and in the weight of ovaries ranging from 2 to 3 mg in strain NH to 25 to 65 mg in strain DBA/2WyDi (Dickie, unpublished data). In strain C57BL/6J there is an increasing amount of ceroid pigment in the senile ovary but morphological changes follow an orderly pattern of development and aging (Fekete, 1953). In DBA, however, corpora lutea are retained and may persist over many estrous cycles. As aging advances the corpora lutea become hyalinized and calcified and there are many atretic follicles and follicular cysts (Fekete, 1953). Strain C58/J ovaries are distinguishable by the presence of many polyovular follicles (Fekete, 1953). Strains A and RIII ovaries have highest number of oocytes (15,000) postpartum, and CBA ovaries the lowest (10,000) among strains examined.

There is a gradual decline in the number of oocytes as age advances, CBA ovaries losing 28 per cent of their remaining oocytes every 20 days while strains A and RIII ovaries lose only 10 to 13 per cent in the same time. At 437 days of age there are few or no oocytes in the CBA ovaries, about 365 in A, and around 540 in RIII ovaries. The decline in number of oocytes is not affected by bearing young. The decline in fertility of the strains is probably due to defects in hormonal control of the ovary and uterine environment rather than in loss of oocytes (Jones and Krohn, 1961). Other strains such as C3HeB/FeJ (Hummel, 1965, personal communication) and CE/J develop granulosa cell tumors. In crosses of

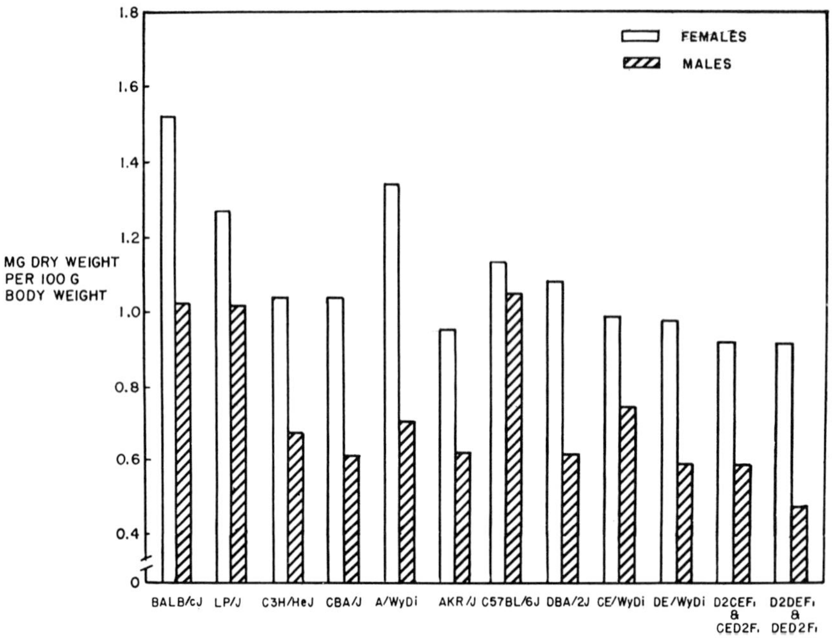

Fig. 20-2. Adrenal gland weights in some inbred strains and hybrid mice.

DBA/2 with strain CE, ovarian weight in the F_1 hybrids ($D2CEF_1$ and $CED2F_1$) is slightly more than that in the two parental strains, morphology resembles that of the DBA/2 ovary, and tumor incidence parallels that of strain CE. In backcrosses of these F_1 animals to strain CE, the ovarian tumor incidence is double that observed in strain CE. The morphology of the nontumorous ovaries is often similar to that of strain DBA/2 (Dickie, 1958).

Related target organs

The uteri of virgin $D2CEF_1$ and $CED2F_1$ hybrids spontaneously develop a hyperestrogenic syndrome characterized by cystic glandular hyperplasia, adenomyosis, and adenomatous hyperplasia of the endometrium. Neonatally gonadectomized hybrids develop the same syndrome (Christy et al., 1951). Investigations with these animals reveal that the uteri are extremely sensitive to exogenous estrogen and that adrenalectomy has no effect on the syndrome, but breeding or hypophysectomy prevents its development. After the syndrome is established at about 9 months of age, the animals remain in constant estrus. Few senile changes occur, there is no decrease in weight, and no tumors develop. The syndrome does not interfere with pregnancy, since all virgin females at 12, 15, and 18 months of age became pregnant when mated (Dickie et al., 1957). Study of the ovaries of these animals, with a morphology like that of strain DBA and a tumor incidence like that of strain CE, suggests that although the uterus appears to be under a hyperestrogenic influence the retention and maintenance of many generations of corpora lutea indicates a hyperprogestational condition (Atkinson et al., 1954).

The time of opening of the vagina and the initiation of estrous cycling vary among strains. Vaginal opening occurs at 28 days in BALB/cJ mice and at 37 days in C57BL/6J and A/J mice and $B6AF_1$ hybrids. Cycling begins about 10 days after the vagina opens (Liu and Chai, 1961). Environmental factors such as diet and number of animals per pen may also affect the time of vaginal opening and initiation of cycling.

The sexually dimorphic submandibular gland, dark in females and almost white in males, shows few changes with advancing age. The male glands are consistently heavier (Fig. 20-3). Raynaud's (1960) studies indicate that, although hormones of the testis and thyroid govern the development of the tubules in males, those of the thyroid and adrenal govern the development in females. The secretion of the tubules contains a ninhydrin-positive material suggesting the presence of amyloid, but does not contain glycogen, cholesterol, lipids, mucopolysaccharides, or mucoproteins. There may be a hitherto unidentified component in the tubular portion of the gland promoting growth and differentiation of the sympathetic ganglia. This factor is found in higher concentration in males and androgen injections increase the yield in females (Caramia et al., 1962).

The dimorphic nature of Bowman's capsule of the kidney was described by Crabtree (1940). Since then either the structure of Bowman's capsule or kidney weight has been used as an index of androgenic activity. Analyses of variance of kidney weights in backcrosses of $D2CEF_1$ and $CED2F_1$ hybrids to strain CE show that male kidneys are always heavier than those of females, and after 12 months the weight decreases slightly in males and increases in females (Dickie, 1958). The rate of change in alkaline phosphatase activity in the proximal and distal segments of the convoluted tubule between birth and 36 days of age varies among strains or stocks. C57 differentiates most rapidly, DBA is slowest, and C, NIH, C3H, and CFW are intermediate. The differentiation of these regions of intense alkaline phosphatase activity can be accelerated by either estrogen or testosterone and is retarded in males following castration (Longley and Fisher, 1956). The weights of both the submandibular gland and the kidney are sensitive indices of androgenic activity. The prostates reflect estrogenic as well as androgenic stimulation.

Female mice have five pairs of mammary glands and male mice have some gland rudiments with no nipples (Chap. 13). The presence of rudiments in males is a strain-controlled characteristic that responds to varying hormonal conditions (Richardson and Cloudman, 1947; Richardson, 1951, 1953). Extensive investigations of the mammae indicate that the genome, hormonal balance, and extrachromosomal agents all affect the etiology of mammary cancer in mice (Bittner, 1945; Bern, 1960).

VARIATIONS DUE TO SINGLE GENES

Many point mutations cause secondary endocrine changes, but few mutations appear to cause primary endocrine defects. Some examples of mutations with endocrine effects are given, and where possible they are contrasted with endocrinopathies under polygenic control.

Snell's dwarf (*dw/dw*)

The classic case of primary endocrine defects caused by a single-gene mutation is Snell's dwarf, discovered in 1929. Dwarf mice have juvenile body proportions, myxedema, some accumulation of fat, and attain a size only about one-fourth to one-third that of normal littermates (Fig. 20-4;

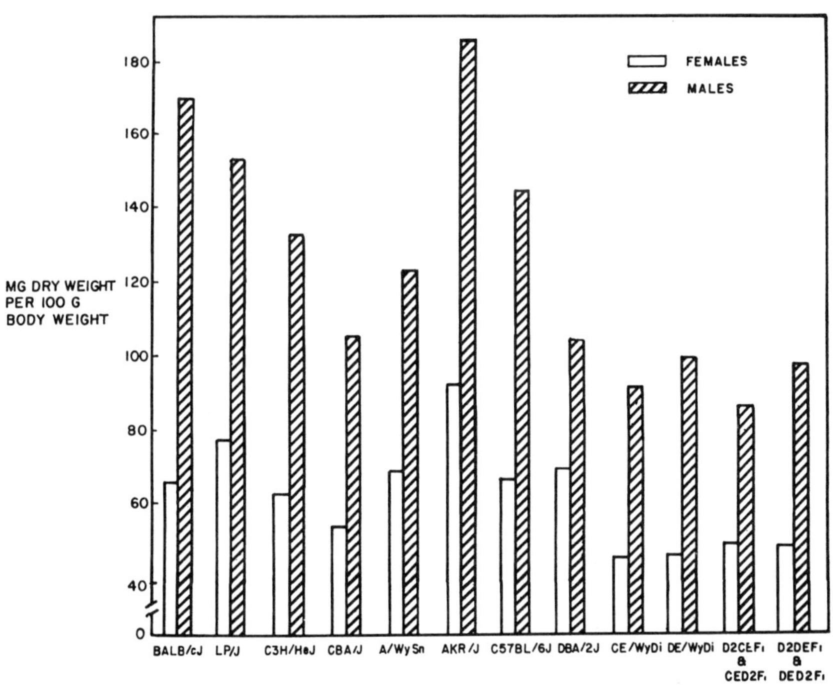

Fig. 20-3. Submandibular gland weights in some inbred strains and hybrid mice.

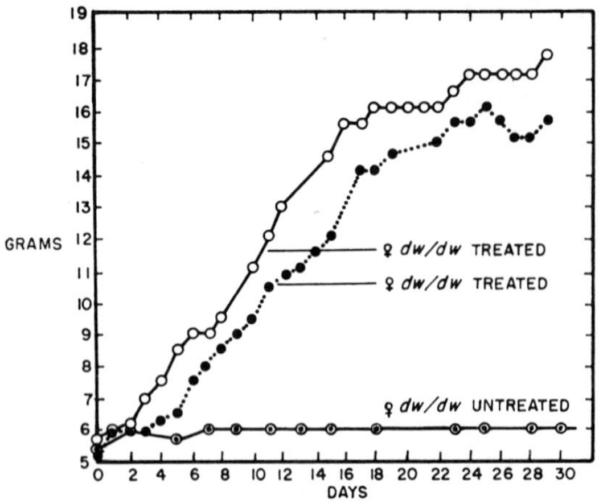

Fig. 20-4. Body weights of treated and untreated dwarf females. (*From Smith and MacDowell,* 1930, *with permission of Wistar Institute of Anatomy and Biology, publishers of the Anatomical Record.*)

Grüneberg, 1952). Elftman and Wegelius (1959) confirmed previous findings of the deficiency of eosinophiles and provided cytological evidence that there is also a deficiency of thyrotropic basophiles in the anterior lobe of the pituitary of dwarf mice.

The concentration of protein-bound iodine in the serum of dwarf mice is about 1.3 g/100 ml compared to 7.3 g/100 ml in serum of normal siblings. The thyroid glands of untreated dwarf mice do not accumulate radioactive iodine, but injection of thyrotropin significantly increases the iodine uptake in the glands (Wegelius, 1959). This confirms the idea that the defective thyroids and myxedema are secondary and suggests that the primary defect may be lack of thyrotropic hormone in the pituitary.

A combination of thyroxine and growth hormone exerts a pronounced growth-promoting effect and causes an increase in body length and weight (Nielsen, 1953). Growth hormone alone gives a lesser effect, and thyroxine alone the least effect.

Secondary effects of the *dw* gene have been observed in other organs including the thymus, adrenal cortex, and gonads. Both sexes are sterile but some males become fertile and increase in size when given daily injections of pituitary from a normal mouse the same age as the dwarf (Grüneberg, 1952).

Carbohydrate metabolism is atypical, resembling that of hypophysectomized normal mice. Dwarf mice have prolonged hypoglycemia after fasting and are extremely sensitive to insulin, reacting to a dose only 3 per cent of that needed to cause shock in normally sensitive mice. When dwarfs are given growth hormone, adrenocorticotropic hormone, and cortisone in combination, these compounds act as anti-insulin agents preventing convulsions and allowing the mice to withstand a dose four times greater than that causing shock in untreated dwarf mice (Mirand, 1953).

Ames dwarf (*df/df*)

Another mutant, Ames dwarf, phenotypically similar to Snell's dwarf, has been discovered in an irradiated stock (Schaible and Gowen, 1961). The body weight of Ames dwarf mice averages 12 g at 2 months. Treatment with growth hormone increases weight in males to 22 to 24 g and allows them to sire young, whereas females reach 18 to 21 g and remain infertile. Growth hormone alone appears to be more effective in Ames dwarf mice than in Snell's dwarf mice (Bartke, 1963).

Obese (*ob/ob*)

In this recessive mutant (Ingalls et al., 1950), the islets of Langerhans are greatly enlarged, the β-cells being hypertrophied and very active. Obese mice have attained a maximum weight of 128 g but most weigh between 80 and 100 g and have lifespans of about 14 to 16 months. There is increased glucose 6-phosphatase activity in the β-cells, and studies of distribution of enzymes in islet tissue suggest the existence of an active hexose monophosphate shunt (Hellman and Hellerström, 1962). Obese mice have high blood sugar levels (150 to 400 mg/100 ml of blood) and are insulin resistant. Obese mice of both sexes are infertile but under long-term diet restriction some males will mate (Lane and Dickie, 1954). They become sensitive to insulin, the islet cells become normal, and lifespan is greatly increased (Lane and Dickie, 1958).

Parabiosis using intact or castrated obese and nonobese siblings in all possible combinations showed that the pituitaries of obese mice can produce more gonadotropin than they normally do, but are less competent to do this than the nonobese pituitaries. Obese male pituitaries appeared to have a slightly higher gonadotropin content than obese female pituitaries (Lane, 1959). Small doses of estrogen produce constant estrus in obese females (Drasher et al., 1955).

The frequency of silver-positive A_1 cells in

obese (*ob/ob*) mice is one-half that found in normal mice, but in normal heterozygous (*ob*/+) or homozygous (+/+) littermates made obese with goldthioglucose no significant decrease in the frequency of the silver-positive (A_1) cells is observed, although there is a more than 100 per cent increase in islet volume (Hellman, 1961; Petersson and Hellman, 1962). Alloxan treatment of obese mice lowers the blood glucose level and causes an increase in granulation of the β-cells, but produces a degranulation in normal siblings (Solomon and Mayer, 1962). Methods of isolating pure islet material, devised by Hellerström (1964), show that the mean weight of islets from normal mice is 1.2 μg (range 0.2 to 3.7 μg) and from obese mice is 11.3 μg (range 2.8 to 79.7 μg).

Endocrinopathies similar to that in the obese mice, but not caused by mutations, are found in other types of mice. An obese strain has been developed by selection of obese animals from a mixed colony (Bielschowsky and Bielschowsky, 1956). Although there were some obese animals in the first 10 generations, they appeared regularly after the 12th. This syndrome is not due to a single gene but appears to have a polygenic basis. NZO (obese strain) females and males are fertile, attain a weight of 50 to 60 g, and have a blood glucose level over 200 mg/100 ml after 12 months. They have a greater oxygen consumption and carbon dioxide output per gram of body weight than normal (NZC strain), are resistant to insulin, and have hypertrophied islets of Langerhans comprised of enlarged β-cells and few α-cells. Adrenals of NZO mice are larger than those of NZC and have less lipid. NZO and NZC animals oxidize acetate-1-C^{14} and acetate-2-C^{14} at the same rate but the oxidation of both glucose-6-C^{14} and glucose-1-C^{14} is about 12 per cent lower in NZO (Subrahmanyam, 1960).

Similar islet hypertrophy and concomitant obesity are also observed in the following types of mice: those treated with corticosteroids; those bearing ACTH-secreting pituitary tumors; genetically yellow mice (carrying A^y); strain NH with early gonadal atrophy and cortical hypertrophy; and LAF_1 mice after castration. Adrenalectomy prevents such changes in these animals but is not effective in animals made obese with hypothalamic lesions or in the genetically obese (Hausberger and Ramsey, 1959; Hausberger, 1960). If total food intake is restricted in LAF_1 mice bearing ACTH-secreting tumors and in CBA mice made obese with goldthioglucose, there is no fat deposition or islet hypertrophy (Hausberger, 1961).

Grey-lethal (*gl/gl*)

A major defect in grey-lethal mice appears to be lack of secondary bone absorption but it is suggested that the primary defect involves the parathyroid gland. When parathyroids from grey-lethal mice are grown in tissue culture or in intracerebral grafts, they can initiate osteoclastic resorption of normal bone as well as grey-lethal bone (Hirsch, 1962). Grüneberg (1963) concluded that parathormone is inactivated or destroyed too rapidly in grey-lethal mice.

Polydypsic mutation

The adrenals of mice of strain DE/J become enlarged and opalescent after 12 to 14 months of age. Their wet weight increases from 6 mg in females and 3 mg in males up to 25 mg. Large areas of the cortex are replaced by a cellular homogeneous material. Coincident with these adrenal changes, water intake is increased from an average of 4 to 8 ml/day to 30 to 70 ml/day and excretion as urine is increased from 2 to 5 ml/day to 25 to 65 ml/day. Tolerance to water deprivation is greatly impaired in these animals. Specific gravity of the urine measures 1.055 (normal mice, 1.36 to 1.78) and total solids 1.6 per cent (normal mice, 18 to 20 per cent); tests of the urine for sugar, acetone, albumin, and heme were negative. The adrenals, liver, and testes have heavy deposits of an acellular material. Some kidney changes are found in older animals.

Measurement of water intake in (DE $\times$ DBA/2WyDi)F_1 and F_2 hybrids and in backcrosses of the F_1 hybrids to DE indicate that the polydypsia of DE mice is due to a recessive gene (Dickie, unpublished data).

VARIATIONS IN RESPONSE TO HORMONE ADMINISTRATION

Different organs of an individual, as well as individuals of different strains of mice, vary in their sensitivities to exogenous hormones. This sensitivity is an important factor in tumorigenesis.

Many hormones emanating from the pituitary gland are used to test the sensitivity and competence of tissues. These include growth hormone or somatotropin (GH or STH), adrenocorticotropic hormone (ACTH), thyrotropic or thyroid stimulating hormone (TSH), the gonadotropins (follicular stimulating hormone, FSH, and luteinizing or interstitial cell stimulating hormone, LH or ICSH), and lactogenic or mammotropic hormone (LTH or MH). Other compounds such as pregnant mare serum gonadotropin (PMS or PMSG) containing both FSH and LH and human chorionic gonadotropin (HCG) containing primarily LH, are also used.

Steinetz and Beach (1963) observed that, although injection of estradiol cyclopentylpropionate (ECP) and relaxin promote interpubic ligament formation in intact female mice, the hormones are less effective in hypophysectomized animals. Addition of powdered thyroid to the diet increased the response in the hypophysectomized animals, but only by addition of STH was a response obtained indistinguishable from ECP action in intact mice.

Response to growth hormones, as measured by body length and weight increment, is greater in C3H than in A and C57BL mice, and females of all three strains are more sensitive than males, but all responses are slight. Growth hormone produced no increase in tumor incidence in these strains (Moon et al., 1952).

Human chorionic gonadotropin (HCG) causes hemorrhagic follicular development in the ovaries of prepuberal mice. The ovaries of strain A mice are unresponsive to HCG, but other strains are sensitive to this hormone (Hummel, 1942). Carr (1949) confirmed this finding, noting that strain A mice are unresponsive and other strains, such as C57, are responsive. He suggested that this response is inherited as a single dominant gene with incomplete penetrance. Genetic studies of this response using strains BALB/cJ, A/J, and the hybrid CAF_1, indicate that BALB/c ovaries are sensitive, A/J ovaries are not, and the F_1 hybrid response falls between the parental responses but is closer to that of strain A/J (Chai, 1960*b*). There are slight differences in the ovarian responses of DBA/1J and DBA/2J, and both are sensitive.

PMS has been used to study the competence of the C57BL/6J ovary at various ages. Prior to 14 months of age the ovary has a fairly uniform weight increase during PMS treatment; after 14 months the response gradually decreases. After 17 months the ovary does not become heavier following PMS injections, suggesting that the ovary is then not competent to respond to exogenous gonadotropin (Green, 1957).

Lactogenesis occurs only when somatotropin and mammotropin are used in combination with cortisol or cortisone in strain BALB/c (Nandi and Bern, 1961). Rivera (1964, personal communication) indicated that many times more somatotropin is required to induce lactogenesis in strain A than in strain C3H mice.

Inbred strains vary in response to insulin injection. Zr stock mice can tolerate only minute doses of insulin, C57BL mice tolerate about four times as much insulin, and strain KL mice survive insulin dosages 310 times greater than that which kills Zr, or 91 times that which kills C57BL (Chase et al., 1948). Adrenalectomy increases sensitivity to insulin; all dosages are fatal to C57BL and strain KL mice convulse when given more than 8 units (Katsh, 1953). Beyer (1955) demonstrated an insulinase in the liver of KL mice that degrades insulin very rapidly. Using strains A2G, C57BR/cd, CBA, and DBA/1, five hybrids, and a randombred stock, Brown (1961) found that the strains with smaller body weight have greater sensitivity to insulin than strains with greater body weight.

More than 40 hormonal compounds have been isolated from adrenal glands. These include corticoids, aldosterone, and sex steroids (Grant, 1960; Short, 1960). Most of these compounds have been used in species other than the mouse, but a few examples are given of cortical hormone activity in the mouse.

One response of an animal to stress is a rapid decrease in the eosinophiles in the blood. Strain differences in response to cortical hormones can be shown using the eosinopenic response (Wragg and Speirs, 1952). C57BL/6J, C57BR/cdJ, $B6BRF_1$ and 129 BRF_1 have a maximum eosinopenic response after injection of 6 μg of cortisone, but 96 μg of cortisone are needed to produce a maximum eosinopenic response in strains 129/Rr, DBA/1J, and BALB/cJ. The response in males is more reliable than that in females.

Cortical hormones may potentiate leukemogenesis in some inbred strains, or suppress or delay development of spontaneous or carcinogen-induced leukemia, depending upon age and experimental conditions (Kirschbaum, 1957). Adrenal androgens injected into immature NMRI females resulted in absence of preovulatory follicles and corpora lutea, in follicular atresia, and in involution of the adrenal X-zone and made the mice less sensitive to superovulatory doses of PMS and HCG (Varon and Christian, 1963).

Prolonged estrogen treatment produces pituitary tumors in strain C57BL (Gardner and Strong, 1940). In neonatally castrated CE mice this treatment not only produces pituitary tumors but prevents postcastrational adrenal changes (Woolley and Little, 1946). Interstitial cell tumors of the testes are produced following estrogen treatment in strains A, BALB, and CE, but not in C3H, I, or C57BL (Shimkin and Grady, 1942). BALB/c males develop interstitial cell testicular tumors after 180 days of estrogen treatment, but in similarly treated C3H males the testes atrophy and no other changes are found (Shimkin et al., 1963). Implantation of estrogen pellets causes urinary bladder dilatation and increased water intake in C57BL, A, and CE, but not in C3H males. Regulation of water intake in the three sensitive strains reduces the severity of the dilatation and increases the survival rate (Thompson, 1955).

Leukemia can be induced in certain strains by treatment with estrogens, carcinogens, radiation, or combinations of these (Kirschbaum, 1956, 1957). In strains A, C3H, CBA, C12I, JK, C57, and PM, leukemia is induced as a result of estrogenic treatment. The amount of hormone injected and the length of treatment influence the incidence of the disease. Addition of testosterone reduces the incidence to control levels. Intact mice of strain DBA/2 are resistant to induction of leukemia by methylcholanthrene but are susceptible to the action of the carcinogen when castrated. Castrated animals become resistant when given testosterone. In C57BL mice, castration does not alter susceptibility to leukemogenesis unless the animals are X-irradiated. If the thymus is shielded during irradiation, estrogen will not produce leukemia.

In high mammary cancer lines estrogen treatment of males causes mammary tumor development, but large doses of estradiol benzoate (EB) may inhibit mammary gland growth. Smaller doses of EB and addition of corticoids give responses like a large dose of EB (Dierwichter, 1962). Female mice of the Riet strain need 10 times more estrogen than C57BL or DBA to stimulate lobule alveolar mammary development (Mühlbock, 1948).

The relative response of the uterus in females of various strains to short-term estrogen administration as measured by wet weights, DNA, RNA, and protein nitrogen content is: 129 > DBA/1J > C3HeB/FeJ > C57BL/6J > C3H/HeJ (Drasher, 1955). Uteri of females of strains C3H/HeJ, C3HeB/FeJ, and DBA/1J show a rapid regression rate, but those of strains 129 and C57BL/6J regress slowly. These results indicate that the intact animals of C3HeB/FeJ, C3H/HeJ, and DBA/1J either have a higher level of ovarian hormone secretion than 129 or C57BL/6J, or the uterine tissue is more sensitive to hormone withdrawal. Direct application of estrogen to the vaginal epithelium elicits a greater response than subcutaneous injection of the hormone (Mühlbock, 1957). The amount of estrogen required to produce vaginal cornification in 50 per cent of injected subjects varies approximately fivefold among the strains tested by Trentin (1950; Table 20-3). There is no increase in vaginal sensitivity to estrogen at puberty, contrary to findings in rats (Liu and Chai, 1961).

Table 20-3. MINIMUM AMOUNT OF ESTRADIOL BENZOATE GIVING POSITIVE VAGINAL SMEARS IN APPROXIMATELY 50 PER CENT OF THE MICE. *(From Trentin, 1950, reprinted from Cancer Research by permission of the University of Chicago Press.)*

Strain or hybrid[a]	Estradiol benzoate, μg
C57	0.06
JK	0.08
C3H	0.1
C3H without MTA[b]	0.08–0.1
(C57 × CBA)F_1	0.08–0.1
(CBA × C57)F_1	0.1
CBA	0.15
(CBA × A)F_1	0.2
N	0.25–0.27
A	0.25–0.27
A without MTA	0.27

[a] Strain and hybrid designations changed to conform with nomenclature rules (Staats, 1964).
[b] Mammary tumor agent.

In intravaginal tests (i.e., direct application of hormone to the vagina) the median effective dose of estrogen is the same in strains C57 and CBA but is less for the F_1 hybrid. However, when estrogen is given subcutaneously the median effective dose is less in C57 than in CBA and is lower in the F_1 hybrid than in either parental strain (Claringbold and Biggers, 1955). The vagina and endometrium of C57BL mice appear to be most sensitive to doses of α-estradiol, DBA less sensitive (Mühlbock, 1948) and A least sensitive, whereas the reverse is true for sensitivity of the mammary glands of these strains to estrogen (Silberberg and Silberberg, 1951). The results suggest that there are differences in sensitivity of the target organs of inbred strains to a given dose of hormone.

Testosterone injections in females cause an increase in submandibular gland weight and change the structure from female to male type (Chap. 13). Swigart et al. (1963) used the presence of amyloid in the tubules of the submandibular gland to measure the response of strain C57BL/6J mice to varying doses of estrogens and androgens and found that the gland is extremely sensitive to minute doses of androgens and to some doses of

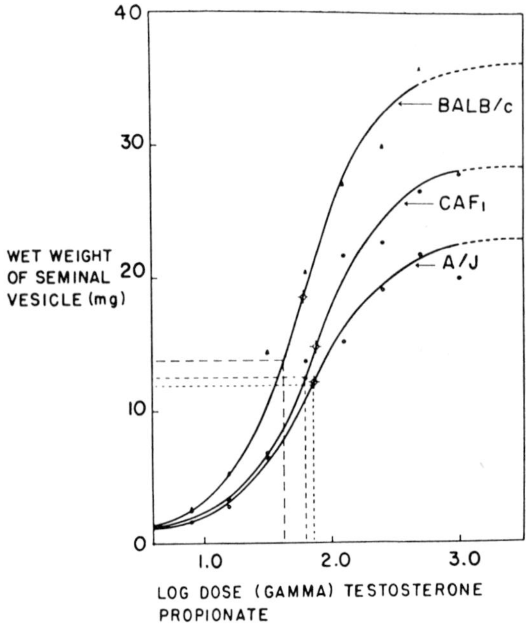

Fig. 20-5. Plots of seminal vesicle weight (wet) against log dose of testosterone propionate in A/J, BALB/cJ, and CAF_1 hybrid mice. (*From Chai, 1956, with permission of the author and of Wistar Institute of Anatomy and Biology, publishers of the Anatomical Record.*)

estrogens. Androgen administration causes a gain in both water content and dry tissue residue in both sexes (Atkinson et al., 1959).

Testosterone propionate injections cause a four- to fivefold increase in kidney β-glucuronidase concentration in strains 129 and DBA and a 95- to 125-fold increase in C3H, AKR, CAF_1, and $B6AF_1$ (Chap. 19; Fishman and Farmelant, 1953).

Increase in seminal vesicle weight was used as the index of response to testosterone propionate in two inbred strains of mice, BALB/cJ and A/J, and their hybrid CAF_1 (Chai, 1956). The mean responses of BALB/c are greater than the mean responses of A and the mean response of the hybrids falls between those of the parental strains. Typical sigmoid dose-response curves for the three kinds of mice are plotted in Fig. 20-5. On the basis of the dose-response relationship and the seminal vesicle weights of the controls, A/J mice appear to secrete about one-third more testicular androgen than BALB/c. The F_1 hybrids secrete more than BALB/c but less than A. These estimates are approximations since they are estimated graphically, but the consistency of the dose-response in the various groups is noticeable.

OTHER TYPES OF ENDOCRINE IMBALANCE

Another method of studying the physiology of the endocrine system and its target organs is to create artificial imbalances by ablation or addition of an endocrine gland or by its transplantation to another site. This section provides examples of some of the strain differences in response to various kinds of imbalance.

Hypophysectomy has been used to create an assay animal that can provide answers to many questions. The mortality rate for the operation in mice is high and depends upon the age, strain, and skeletal and vascular variations. A mouse that is completely hypophysectomized does not gain any weight in the postoperative months. When pituitary remnants remain, the body and organ weights become normal for the age of the mouse (Dickie, unpublished data; Table 20-4).

Thyroidectomy, by radiation, by surgical procedures, or by treatment with goitrogens, produces thyrotropic pituitary tumors in most strains investigated (Furth, 1955; Clifton, 1959), but in C57BL males such tumors develop only when the mice are castrated as well as X-irradiated (Edel-

Table 20-4. ASSOCIATION OF BODY AND ORGAN WEIGHTS WITH WEIGHTS OF PITUITARY GLAND REMNANTS IN D2CEF$_1$ HYBRID FEMALE MICE

Condition	Pituitary or remnant wt, mg	Body wt, g	Submandibular gland wt, mg	Ovary wt, mg	Adrenal gland wt, mg	Uterus wt, mg
Intact	2.04	27.5	101.0	34.00	5.26	128.0
Operated	1.00	26.0	74.4	33.54	5.98	192.0
Operated	0.64	24.0	77.0	18.10	6.04	156.4
Operated	0.60	26.0	68.5	22.34	6.58	263.4
Operated	0.48	25.5	67.4	15.88	5.22	206.4
Operated	0.10	16.5	36.4	7.64	4.00	101.0
Operated	0.00	13.0	16.1	1.60	1.78	3.8

mann and Gorbman, 1955). When thyrotropic pituitary tumors are transplanted into C57BL mice, the hosts have a blood serum concentration of thyroid stimulating hormone (TSH) of about 1 unit per ml, or about 2,000 times normal. The concentration decreases slightly in each transplant generation. There is a concomitant increase in ovarian weight suggesting that some gonadotropins may also be present, but there is no evidence of any adrenocorticotropic hormone (ACTH) secretion (Bates et al., 1957). Adrenotropic pituitary tumors (basophilic) arise as a result of ionizing radiation, and their growth is enhanced in adrenalectomized animals. Animals bearing these tumors are obese and have elevated liver glycogen and marked lymphocytopenia. Mammotropic pituitary tumors (eosinophilic), also produced after whole-body ionizing radiation, are similar to tumors found in C57BL after prolonged estrogen treatment (Furth and Clifton, 1957). Pituitary tumors, probably gonadotropic (Furth, 1957), arise in neonatally gonadectomized hybrids between strains CE and DBA/2 and between DE and DBA/2 subsequent to the development of adrenal cortical carcinomas (Dickie and Lane, 1956).

The endocrine alteration used in many experiments is gonadectomy. We have learned that the response to this operation is strain-controlled. Strain DBA responds to neonatal gonadectomy by development of adrenal cortical hyperplasia with concomitant feminizing stimulation of the accessory reproductive organs of both sexes. Strain CE develops adrenal cortical carcinoma and there is evidence of both androgenic and estrogenic stimulation of the accessory reproductive organs of both sexes (Woolley and Little, 1945). Certain strains such as C57BR (Woolley and Dickie, 1947), A/WyDi, LP/J, C57BL/10, and DE/J have no major postcastrational adrenal changes (Dickie, unpublished). Strain NH undergoes early spontaneous gonadal atrophy and develops cortical hypertrophy (Kirschbaum et al., 1946). The response of the adrenal must depend upon the genic complement of the adrenal cells themselves, because the response is not altered by transplantation to a host with a different adrenal postcastrational response (Huseby and Bittner, 1951).

The postcastrational response in F_1 reciprocal hybrid mice studied by Woolley et al. (1952, 1953) and by Dickie and Lane (1956) is always formation of adrenal cortical carcinoma rather than the nodular hyperplasia response or absence of response. The incidence of the changes in the F_1 generation and in the backcrosses to the parental strains provides additional evidence that postcastrational adrenal changes are under genetic control (Dickie, 1958).

The postcastrational response has been used to study endocrine tumors, effects of hormones on the adrenals, hormonal effects on other systems, and effects of gene substitutions in various stocks of mice. Thirty days after castration adrenals of female BALB/c mice decrease in size, those of strains C3H, A, and C57BL, and of $AC3F_1$ hybrids do not change, and strain CE adrenals increase in size. Administration of estradiol benzoate to castrated mice for 30 days causes an increase in adrenal size in strains BALB/c, C3H, and $AC3F_1$ hybrids, no change in C57BL and A, and a decrease in CE adrenal size (Martinez and Smith, 1958).

Implantation of biologically potent sex-steroid pellets into gonadectomized CE mice prevents postcastrational development of adrenal cortical carcinoma (Woolley, 1950). When neonatally gonadectomized $D2CEF_1$ and $CED2F_1$ hybrid mice are hypophysectomized and observed up to 10 months of age, long after adrenal cortical carcinomas appear in nonhypophysectomized castrated mice, no postcastrational changes are evident (Atkinson and Dickie, 1958). C3H mice unilaterally adrenalectomized at the time of castration have some hyperplastic changes in the remaining adrenal, but there is little evidence of hormonal stimulation of the accessory reproductive organs (Martinez and Bittner, 1955).

The lipid content of the adrenals of intact $D2CEF_1$ and $CED2F_1$ hybrid mice is about 50 per cent in females and about 20 per cent in males throughout life. The lipid content of the adrenals of gonadectomized hybrids decreases to about 25 to 30 per cent in females and increases to 35 to 40 per cent in males even though cortical carcinomas are present. When other strains and hybrids are castrated and the lipid content of the adrenals is measured 35 days later, the response observed in the $D2CEF_1$ and $CED2F_1$ hybrids, i.e., the decrease in adrenal lipid in females and an increase in males, occurs in all the other strains tested with the exception of AKR/J and C57BL/6J males and LP/J, $D2DEF_1$, and $DED2F_1$ females (Fig. 20-6). These observations suggest that the lipid concentration is not a factor of tumorigenesis but a response to castration itself (Swigart and Hilton, 1963, 1964, personal communication).

Histological changes in the terminal tubules of the submandibular gland due to androgenic secretion appear before any changes are evident in the adrenals of these castrated strains and hybrids

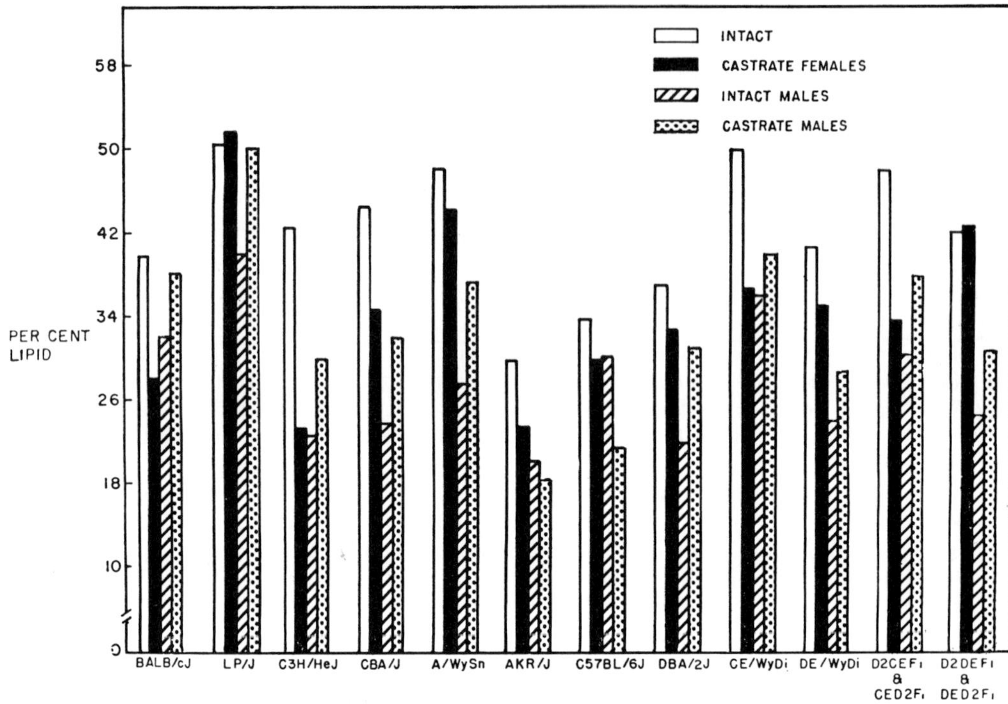

Fig. 20-6. Lipid concentration in adrenal glands of intact and castrated mice of some inbred strains and hybrids.

(Swigart and Hilton, 1964, personal communication).

Although abnormalities arising in the adrenal cortex as a result of castration appear to secrete sex hormones, those changes found in intact animals in old age result in no unusual stimulation of the target organs. In aged animals the capability to produce hormones has diminished and the accessory reproductive organs are probably no longer capable of response to hormonal stimulation.

Ovaries transplanted into the spleens of gonadectomized animals, removed before 2 months residence in the spleen and reimplanted into an ovarian capsule, will release fertilizable ova that produce live young. After 2 months of residence in the spleen the ovary is incapable of producing ova when reimplanted in an ovarian capsule. Tumorigenesis of intrasplenic ovarian grafts occurs at 3 months in strain A, 5 months in strains C57BL and BALB/c, and at 7 months in DBA (Hummel et al., 1953; Hummel, 1954). The rapid destruction by hepatic tissue of the gonadal hormones from these ovarian grafts entering into the portal circulation and the concomitant increase in gonadotropin level is believed to be the cause of tumorigenesis of these grafts (Gardner, 1953). However, transplantation per se is also an important factor in tumorigenesis of the grafts, because ovaries transplanted directly from one ovarian capsule to another ovarian capsule become tumorous 12 to 21 months after transplantation. Invariably the hosts develop adrenal tumors as well as the ovarian tumors, presumably because of the imbalance created by the presence of an abnormal ovary (Hummel et al., 1953; Hummel, 1954).

A comparison of intrasplenic grafts with subcutaneous and intratesticular grafts of ovarian tissue showed that only the intrasplenic grafts were tumorous in the strains used (Gardner, 1955). Similarly, a comparison of intrasplenic grafts with intrarenal and intrahepatic grafts of ovarian tissue showed that only the intrasplenic grafts became tumorous, but when grafts at other sites were made simultaneously with intrasplenic grafts, microscopic tumors were found in the ovarian tissue at both sites (Lipschutz and Cerisola, 1962). Jett et al. (1961) found that only 19 per cent of intrasplenic ovarian grafts in $CNHF_1$ weanling mice become tumorous, but 97 per cent become tumorous when placed in castrated weanling mice, and 74 per cent become tumorous in adults which had adrenal adenomas. Therefore, the site of transplantation, age, hormonal status, and the genetic constitution of the hosts are also important in the development of these tumors.

We think that reproductive performance, normal and pathological aging changes in the endocrine system and its related target organs, and responses to hormonal manipulation, i.e., the total life history, all contribute to a characterization of the capabilities or potentials of the endocrine system of an inbred strain. It is not possible here to develop a detailed analysis of any inbred strain, but we have seen that some strains have many tissues which are sensitive to hormonal changes and other strains have few such tissues which are sensitive to hormonal manipulation. The extremes of response to hormonal imbalance also contribute to the characterization of the strain. It appears that inbred strains vary not only in the sensitivity of the tissues within the strain, but the strains vary in their over-all capability to compensate for hormonal imbalance.

SUMMARY

In this chapter we have described the effects of five single-gene mutations on the endocrine system, Snell's dwarf, Ames dwarf, obese, grey-lethal, and a polydypsic mutation; the dwarfs have defects in the pituitary, the obese in the islets of Langerhans, grey-lethal probably in the parathyroid, and the polydypsic mutant probably in the adrenal. We have cited some of the characteristic hormone levels and aging phenomena in many inbred strains and hybrids, the responses of tissues of such animals to exogenous hormones, and to other hormonal manipulation such as ablation and transplantation. There are great variations in responses to endocrine alterations among genetically different groups of mice. It is evident that the differences between inbred strains indicate that the responses to hormones, rate of secretion, and tissue sensitivity, i.e., endocrine variations, are quantitative genetic characters affected by multiple factors or polygenes.

Inbred strains of mice, with genetically fixed characteristics, are perhaps the best research animals available for study of the genetics of endocrine variation and endocrine relationships.

LITERATURE CITED

Amin, A., C. K. Chai, and E. P. Reineke. 1957. Differences in thyroid activity of several strains of mice and F_1 hybrids. Amer. J. Physiol. 191:34–36.

Atkinson, W. B., and M. M. Dickie. 1958. Essential role of the hypophysis in hypercorticism and hyperovarianism in DBA × CE and reciprocal mice. Proc. Soc. Exp. Biol. Med. 99: 267–269.

Atkinson, W. B., M. M. Dickie, and E. Fekete. 1954. Effects of breeding on the development of ovarian, adrenal and uterine lesions in DBA × CE and reciprocal hybrid mice. Endocrinology 55:316–325.

Atkinson, W. B., F. Wilson, and S. Coates. 1959. The nature of the sexual dimorphism of the submandibular gland of the mouse. Endocrinology 65:114–117.

Bartke, A. 1963. The response of two genetically different types of dwarf mice to growth hormone. Genetics 48:882. (Abstr.)

Bates, R. W., E. Anderson, and J. Furth. 1957. Thyrotrophin potency of transplantable pituitary tumors of mice through four transfers. Endocrinology 61:549–554.

Bern, H. A. 1960. Nature of the hormonal influence in mouse mammary cancer. Science 131: 1,039–1,040.

Beyer, R. E. 1955. A study of insulin metabolism in an insulin tolerant strain of mice. Acta Endocrinol. 19:309–332.

Bielschowsky, M., and F. Bielschowsky. 1956. The New Zealand strain of obese mice; their response to stilbesterol and insulin. Austral. J. Exp. Biol. Med. Sci. 34:181–198.

Bittner, J. J. 1945. Inciting influences in the etiology of mammary cancer in mice, p. 63–98. *In* F. R. Moulton [ed.] A.A.A.S. Research Conference on Cancer. American Association for the Advancement of Science, Washington, D.C.

Brown, A. M. 1961. The pattern, sensitivity and precision of the response to insulin in random bred, inbred and hybrid strains of mice. J. Pharm. Pharmacol. 13:670–678.

Caramia, F., P. U. Angeletti, and R. Levi-Montalcini. 1962. Experimental analysis of the mouse submaxillary salivary gland in relationship to its nerve-growth factor content. Endocrinology 70:915–922.

Carr, J. G. 1949. Inheritance of a differential response to gonadotrophic hormone in mice. Heredity 3:262.

Chai, C. K. 1956. Comparison of two inbred strains of mice and their F_1 hybrids in response to androgen. Anat. Rec. 126:269–282.

Chai, C. K. 1958. Endocrine variation. J. Hered. 49:143–148.

Chai, C. K. 1960*a*. Endocrine variation: heritability of iodine metabolism in the thyroids of mice, p. 215–224. *In* O. Kempthorne [ed.] Biometrical genetics. Pergamon Press, New York and London.

Chai, C. K. 1960*b*. Response of inbred and F_1 hybrid mice to hormones. Nature 185:514–518.

Chai, C. K., A. Amin, and E. P. Reineke. 1957. Thyroidal iodine metabolism in inbred and F_1 hybrid mice. Amer. J. Physiol. 188:499–502.

Chai, C. K., J. L. Morrison, and J. L. Lenz. 1964. Changes in thyroid gland during lifespan of mice. J. Hered. 55:270–275.

Chase, H. B., M. S. Gunther, J. Miller, and D. Wolffson. 1948. High insulin tolerance in an inbred strain of mice. Science 107:297–299.

Christy, N. P., M. M. Dickie, W. B. Atkinson, and G. W. Woolley. 1951. The pathogenesis of uterine lesions in virgin mice and in gonadectomized mice bearing adrenal cortical and pituitary tumors. Cancer Res. 6:413–422.

Claringbold, P. J., and J. D. Biggers. 1955. The response of inbred mice to oestrogens. J. Endocrinol. 12:9–14.

Clifton, K. H. 1959. Problems in experimental tumorigenesis of the pituitary gland, gonads, adrenal cortices, and mammary glands: a review. Cancer Res. 19:2–22.

Crabtree, C. 1940. Sex difference in the structure of Bowman's capsule in the mouse. Science 91: 299.

Dickie, M. M. 1954. The use of F_1 hybrid and backcross generations to reveal new and/or uncommon tumor types. J. Nat. Cancer Inst. 15: 791–799.

Dickie, M. M. 1958. Adrenal tumors and other pathological changes in reciprocal backcross mice. I. Backcrosses to strain CE. Ph.D. Thesis, Brown University, Providence, R.I. 84 p.

Dickie, M. M., W. B. Atkinson, and E. Fekete. 1957. The ovary, estrous cycle and fecundity of DBA $\times$ CE and reciprocal hybrid mice in relation to age and the hyperovarian syndrome. Anat. Rec. 127:187–200.

Dickie, M. M., and P. W. Lane. 1956. Adrenal tumors, pituitary tumors and other pathological changes in F_1 hybrids of strain DE $\times$ strain DBA. Cancer Res. 16:48–52.

Dierwichter, R. A. 1962. Inhibition of mammary growth by high doses of estrogen. Yale J. Biol. Med. 34:608–619.

Drasher, M. L. 1955. Strain differences in the response of the mouse uterus to estrogens. J. Hered. 46:190–192.

Drasher, M. L., M. M. Dickie, and P. W. Lane. 1955. Physiological differences in uteri of obese stock mice. J. Hered. 46:209–212.

Edelmann, A., and A. Gorbman. 1955. Endocrine factors influencing the development of hypophyseal tumors in mice. Proc. Amer. Ass. Cancer Res. 2:13–14. (Abstr.)

Elftman, H., and O. Wegelius. 1959. Anterior pituitary cytology of the dwarf mouse. Anat. Rec. 135:43–49.

Fekete, E. 1953. A morphological study of the ovaries of virgin mice of eight inbred strains showing quantitative differences in their hormone producing components. Anat. Rec. 117: 93–114.

Fishman, W. H., and M. H. Farmelant. 1953. Effects of androgens and estrogens on β-glucuronidase in inbred mice. Endocrinology 52: 536–545.

Furth, J. 1955. Experimental pituitary tumors. Recent Progr. Hormone Res. 11:221–249.

Furth, J. 1957. Discussion of problems related to hormonal factors in initiating and maintaining tumor growth. Cancer Res. 17:454–463.

Furth, J., and K. H. Clifton. 1957. Experimental pituitary tumors, p. 3–17. *In* C. E. W. Wolstenholme and M. O'Connor [ed.] Ciba Foundation Colloquia on Endocrinology Vol. XII. Little, Brown, Boston.

Gardner, W. U. 1953. Hormonal aspects of experimental tumorigenesis. Adv. Cancer Res. 1: 173–232.

Gardner, W. U. 1955. Development and growth of tumors in ovaries transplanted into the spleen. Cancer Res. 15:109–117.

Gardner, W. U., and L. C. Strong. 1940. Strain-limited development of tumors of the pituitary gland in mice receiving estrogens. Yale J. Biol. Med. 12:543–548.

Grant, J. K. 1960. The biosynthesis of the adrenocortical steroids, p. 24–39. *In* F. Clark and J. K. Grant [ed.] The biosynthesis and secretion of adrenocortical steroids. Cambridge University Press, Cambridge.

Green, J. A. 1957. Some effects of advancing age on the histology and reactivity of the mouse ovary. Anat. Rec. 129:333–348.

Grüneberg, H. 1952. Endocrine organs, p. 122–129. *In* H. Grüneberg, The genetics of the mouse. Nijhoff, The Hague.

Grüneberg, H. 1963. Section on grey-lethal, p. 77–84. *In* H. Grüneberg, The pathology of development; a study of inherited skeletal disorders in animals. Wiley, New York.

Hausberger, F. X. 1960. Changes in adrenal and pancreatic histology related to obesity after castration. Anat. Rec. 136:208. (Abstr.)

Hausberger, F. X. 1961. Effect of food restriction on body composition and islet hypertrophy of mice bearing corticotrophin-secreting tumors. Acta Endocrinol. 37:336–342.

Hausberger, F. X., and A. J. Ramsay. 1959. Islet hypertrophy in obesity of mice bearing ACTH-secreting tumors. Endocrinology 65:165–171.

Hellerström, C. 1964. A method for microdissection of intact pancreatic islets of mammals. Acta Endocrinol. 45:122–132.

Hellman, B. 1961. The occurrence of argyrophil cells in the Islets of Langerhans of American obese-hyperglycemic mice. Acta Endocrinol. 36:596–602.

Hellman, B., and C. Hellerström. 1962. Oxidative enzymes in the pancreatic islets of normal and obese-hyperglycemic mice. Z. Zellforsch. 56: 97–106.

Heston, W. E., C. D. Larsen, and M. K. Deringer. 1945. Variations in occurrence of pathologic calcification, nephritis, and amyloidosis in mice fed control and modified diets. J. Nat. Cancer Inst. 6:41–47.

Hirsch, M. S. 1962. Studies on the response of osteopetrotic bone explants to parathyroid explants *in vitro*. Bull. Johns Hopkins Hosp. 110: 257–264.

Hummel, K. P. 1942. Differences in response of mice of different strains to human pregnancy urine. Endocrinology 30:74–76.

Hummel, K. P. 1954. Induced ovarian and adrenal tumors. J. Nat. Cancer Inst. 15:711–715.

Hummel, K. P., E. Fekete, and C. C. Little. 1953. Transplantation and ovarian tumors in mice. Anat. Rec. 117:628. (Abstr.)

Huseby, R. A., and J. J. Bittner. 1951. Differences in adrenal responsiveness to post-castrational alteration as evidenced by transplanted adrenal tissue. Cancer Res. 11:954–961.

Ingalls, A. M., M. M. Dickie, and G. D. Snell. 1950. Obese, a new mutation in the house mouse. J. Hered. 41:317–318.

Jett, J. D., H. S. Tullos, J. J. Trentin, and M. E. DeBakey. 1961. Adrenal versus ovarian estrogen in endocrine tumor suppression. Proc. Amer. Ass. Cancer Res. 3:238. (Abstr.)

Jones, E. C., and P. L. Krohn. 1961. The relationships between age, numbers of oocytes and fertility in virgin and multiparous mice. J. Endocrinol. 21:469–495.

Katsh, S. 1953. High insulin tolerance modified by adrenalectomy in the mouse. Anat. Rec. 117:624. (Abstr.)

Kirschbaum, A. 1956. Endocrine aspects of experimental neoplasia. Amer. J. Med. 21:659–670.

Kirschbaum, A. 1957. The role of hormones in cancer: laboratory animals. Cancer Res. 17: 432–453.

Kirschbaum, A., M. H. Frantz, and W. L. Williams. 1946. Neoplasms of the adrenal cortex in non-castrate mice. Cancer Res. 6:707–711.

Lane, P. W. 1959. The pituitary-gonad response of genetically obese mice in parabiosis with thin and obese siblings. Endocrinology 65:863–868.

Lane, P. W., and M. M. Dickie. 1954. Fertile, obese male mice. J. Hered. 45:56–58.

Lane, P. W., and M. M. Dickie. 1958. The effect of restricted food intake on the life span of genetically obese mice. J. Nutr. 64:549–554.

Lipschutz, A., and H. Cerisola. 1962. Ovarian tumors due to a functional imbalance of the hypophysis. Nature 193:145–147.

Liu, F. T. Y., and C. K. Chai. 1961. Sensitivity to estrogen of uteri of ovariectomized mice in relation to age. Proc. Soc. Exp. Biol. Med. 106: 521–522.

Longley, J. B., and E. R. Fisher. 1956. A histochemical basis for changes in renal tubular function in young mice. Quart. J. Microscop. Sci. 97:187–195.

Martinez, C., and J. J. Bittner. 1955. Postcastrational adrenal tumors in unilaterally adrenalectomized C3H mice. Cancer Res. 15:612–613.

Martinez, C., and J. M. Smith. 1958. Adrenal changes following gonadectomy in mice of different strains. Acta Physiol. Latinoamer. 8:84–90.

Mirand, E. A. 1953. The effect of growth hormone on carbohydrate metabolism in the hereditary hypopituitary dwarf mouse. Anat. Rec. 117: 620–621. (Abstr.)

Moon, H. D., M. E. Simpson, C. H. Li, and H. M. Evans. 1952. Effect of pituitary growth hormone in mice. Cancer Res. 12:448–450.

Mühlbock, O. 1947. On the susceptibility of different inbred strains of mice for oestrone. Acta Brev. Neer. 15:18–20.

Mühlbock, O. 1948. The oestrone-sensitivity of the mammary gland in female mice of various strains. Acta Brev. Neer. 16:22–27.

Mühlbock, O., W. J. P. R. van Ebbenhorst Tengbergen, and G. Maurik. 1952. Differences in the gonadotrophic hormone content and the cytology in the anterior lobe of the hypophysis of male mice of various strains. Acta Endocrinol. 9:48–58.

Nandi, S., and H. A. Bern. 1961. The hormones responsible for lactogenesis in BALB/cCrgl mice. Gen. Comp. Endocrinol. 1:195–210.

Nielsen, E. L. 1953. Studies on hereditary dwarfism in mice. XIV. Effect of thyroxin and growth hormone on growth. Acta Pathol. Microbiol. Scand. 32:316–334.

Petersson, B., and B. Hellman. 1962. The pancreatic islet tissue in mice with obesity induced by goldthioglucose. Acta Pathol. Microbiol. Scand. 55:401–406.

Raynaud, J. 1960. Controle hormonal de la glande sous-maxillaire de la souris. Bull. Biol. 94:399–523.

Richardson, F. L. 1951. Further studies on the mammary gland development in male mice at nine weeks of age. Anat. Rec. 111:669–693.

Richardson, F. L. 1953. The mammary gland development in normal and castrate male mice at nine weeks of age. Anat. Rec. 117:449–466.

Richardson, F. L., and A. Cloudman. 1947. The mammary gland development in male mice at 9 weeks of age. Anat. Rec. 97:223–237.

Runner, M. N. 1952. Study of ovarian and pituitary function of the obese mouse, p. 6–7. *In* Conference on the obese mouse. The Jackson Laboratory, Bar Harbor, Maine.

Runner, M. N. 1953. Hereditary differences in pituitary activity in inbred strains of mice. Anat. Rec. 117:547. (Abstr.)

Schaible, R., and J. W. Gowen. 1961. A new dwarf mouse. Genetics 46:896. (Abstr.)

Shimkin, M. B., and H. G. Grady. 1942. Interstitial cell tumors of testes induced with stilbesterol in mice. Cancer Res. 2:730. (Abstr.)

Shimkin, M. B., P. M. Shimkin, and H. B. Andervont. 1963. Effect of estrogens on kidney weight in mice. J. Nat. Cancer Inst. 30:135–141.

Short, R. V. 1960. The secretion of sex hormones by the adrenal gland, p. 59–84. *In* F. Clark and J. K. Grant [ed.] The biosynthesis and secretion of adrenocortical steroids. Cambridge University Press, Cambridge.

Silberberg, M., and R. Silberberg. 1951. Susceptibility to estrogen of breast, vagina and endometrium of various strains of mice. Proc. Soc. Exp. Biol. Med. 76:161—164.

Silverstein, E., L. Sokoloff, O. Mickelsen, and G. E. Jay, Jr. 1960*a*. Thyroid function in various strains of mice: T/S ratio, PBI and thyroid weight. Amer. J. Physiol. 199:203–208.

Silverstein, E., L. Sokoloff, G. E. Jay, Jr., and O. Mickelsen. 1960*b*. Thyroid function, osteoarthritis and epiphyseal closure in various strains of mice. Amer. J. Physiol. 199:209–211.

Smith, F. W., W. U. Gardner, M. H. Li, and H. Kaplan. 1949. Adrenal medullary tumors (pheochromocytomas) in mice. Cancer Res. 9: 193–198.

Smith, P. E., and E. C. MacDowell. 1930. An hereditary anterior-pituitary deficiency in the mouse. Anat. Rec. 46:249–257.

Solomon, J., and J. Mayer. 1962. Long-term hypoglycemia and regranulation of β-cells following alloxan administration in hereditarily obese-hyperglycemic mice. Fed. Proc. 21:396. (Abstr.)

Staats, J. 1964. Standardized nomenclature for inbred strains of mice, Third listing. Cancer Res. 24:147–168.

Steinetz, B. G., and V. L. Beach. 1963. Hormonal requirements for interpubic ligament formation in hypophysectomized mice. Endocrinology 72:771–776.

Subrahmanyam, K. 1960. Metabolism in the New Zealand strain of obese mice. Biochem. J. 76:548–556.

Swigart, R. H., F. K. Hilton, B. Foster, and M. M. Dickie. 1963. Amylase activity in the submandibular gland of mice: evidence of estrogen stimulation. Anat. Rec. 145:290. (Abstr.)

Thompson, J. S. 1955. An effect of estrogens on water intake. Can. J. Biochem. Physiol. 33:10–13.

Trentin, J. J. 1950. Vaginal sensitivity to estrogen as related to mammary tumor incidence in mice. Cancer Res. 10:580–583.

Varon, H. H., and J. J. Christian. 1963. Effects of adrenal androgens on immature female mice. Endocrinology 72:210–222.

Wegelius, O. 1959. The dwarf mouse—an animal with secondary myxedema. Proc. Soc. Exp. Biol. Med. 101:225–227.

Woolley, G. W. 1950. Effect of hormonal substances on adrenal cortical tumor formation in mice. Cancer Res. 10:250. (Abstr.)

Woolley, G. W., and M. M. Dickie. 1947. Genetic and endocrine factors in adrenal cortical tumor formation. Cancer Res. 7:722. (Abstr.)

Woolley, G. W., M. M. Dickie, and C. C. Little. 1952. Adrenal tumors and other pathological changes in reciprocal crosses in mice. I. Strain DBA × strain CE and the reciprocal. Cancer Res. 12:142–152.

Woolley, G. W., M. M. Dickie, and C. C. Little. 1953. Adrenal tumors and other pathological changes in reciprocal crosses in mice. II. An introduction to results of four reciprocal crosses. Cancer Res. 13:231–245.

Woolley, G. W., and C. C. Little. 1945. The incidence of adrenal cortical carcinoma in gonadectomized male mice of the extreme dilution strain over one year of age. Cancer Res. 5:506–509.

Woolley, G. W., and C. C. Little. 1946. Prevention of adrenal cortical carcinoma by diethylstilbestrol. Cancer Res. 6:491 (Abstr.)

Wragg, L. E., and R. S. Speirs. 1952. Strain and sex differences in response of inbred mice to adrenocortical hormones. Proc. Soc. Exp. Biol. Med. 80:680–684.

21

Pigmentation[1]

H. Glenn Wolfe and Douglas L. Coleman

Much has been written about phenogenesis of pigment in the mouse. Pigmentation of mice with part or major emphasis on genetic aspects is reviewed by Grüneberg (1952) and Deol (1963), of rodents and carnivores by Little (1958), and of mammals in general by Chase and Mann (1960), Billingham and Silvers (1960), and Silvers (1961).

Methodology in pigment cell research in mammals has been discussed by Silvers (1963). The approaches to the general problem of pigmentation aside from conventional microscopic methods have involved a detailed comparison of the various genetically influenced types of pigment with respect to (1) the actual amount and distribution of the pigment as measured colorimetrically and histologically, (2) the amount of pigment-forming enzymes present, e.g., tyrosinase as measured enzymatically, and (3) the size, shape, and distribution of the developing pigment granules as seen with the electron microscope. The same techniques have been applied to potentially pigmented tissue or pigmented tissues transplanted to mice of other genotypes or to chick coelom and more recently to skin explants cultured in vitro.

It is rather obvious that pigment differences were studied early because they are conspicuous and are convenient and easy to classify in genetic experiments. At the same time pigment differences in mice or other mammals provide an excellent way to study a single cell type that is more or less autonomous, being little affected by surrounding tissues. Investigators in the field of mammalian genetics will recognize this as a rare attribute.

DIFFERENCES IN PIGMENTATION UNDER SINGLE GENE CONTROL

For convenience and as an aid to later discussion, some of the "pigment genes" of the mouse and their principal effects are listed in Table 21-1. We have attempted to classify these genes as to probable site of gene action. For some the experimental evidence is good, whereas for others the classification is based largely on inference from observed effects.

Pigment genes may affect kind or intensity of coat color or affect the distribution of pigment over the integument of the mouse, i.e., cause white spotting or variegation (mottling). Several affect both color and spotting, e.g., white (Mi^{wh}) (Fig. 21-2*C*), and many affect or are affected by other physiological systems (are pleiotropic). Sterility, anemia, choreic behavior, anomalies of the axial skeleton, failure of secondary bone resorption, reduced eye size, megacolon, and lethality are frequently associated with reduced eye and coat color, white spotting, and mottling. The most pronounced of these related effects are mentioned in Table 21-1. The listing of a gene is not intended to mean that its primary effect is on pigment or pigment distribution. For example, restriction of pigment caused by *W*- or

[1] The writing of this chapter was supported in part by Public Health Service Research Grants CA 01074 and CA 05873 from the National Cancer Institute, by Contract AT(30-1)-1800 from the U.S. Atomic Energy Commission, and by Grant E-76 from the American Cancer Society.

Table 21-1. SOME GENES AFFECTING PIGMENT IN THE MOUSE

Symbol	Name	General effect on pigment	Probable site of gene action[a]	Related effects
A^y	Yellow	Yellow coat color, black eye and skin pigment	1	A^y/A^y prenatal lethal; $A^y/-$ becomes obese, susceptible to tumors
A^{vy}	Viable yellow	Coat color varies from agouti to yellow	1	A^{vy}/A^{vy} viable; some become obese; $A^{vy}/-$ sometimes become obese
A^w	White-bellied agouti	Agouti dorsum, white ventrum (some yellow pigment)	1	
A or +	Agouti	Hairs typically with black base, yellow subterminal band, black tip	1	
A^i	Intermediate agouti	Resembles agouti, but with darkened middorsum	1	
a^{td}	Tanoid	Black dorsum, white ventrum (some yellow pigment); some agouti hairs	1	
a^t	Black-and-tan	Black dorsum, tan (yellow) ventrum	1	
a	Nonagouti	Black-pigmented hairs except for some yellow hairs around ears, mammae, perineum	1	
a^e	Extreme nonagouti	All hairs black-pigmented	1	
a^x		Pseudoallele of agouti; a^x/a black, with pale belly		a^x/a^x prenatal lethal
As	Agouti-suppressor	Pseudoallele (?) of agouti; agouti but with darkened middorsum		
B^{lt}	Light	Black pigment reduced in hairs, especially proximally	3a	
b^c	Cordovan	Dark brown pigment	3a	
b	Brown	Brown pigment	3a	
bg	Beige	Skin and hair pigment diluted		
Blo	Blotchy	Variegated	4	*Blo/Blo* and *Blo*/Y viable, vibrissae kinked at birth, straight at weaning
bt	Belted	Pigment absent in a variable-sized belt, depending on genetic background	1	
c^{ch}	Chinchilla	Reduced melanins, particularly phaeomelanin; hairs have reduced or absent distal pigment with darker bases giving chinchilla effect	3c	
c^h	Himalayan	Reduced melanins, pigment intense at body extremities	3c	
c^e	Extreme dilution	Coat appears dirty white from near elimination of melanins	3c	
c	Albino	Absence of melanins	3c	
d	Dilute	Hair and skin pigments appear diluted	3d	Susceptible to audiogenic seizures
d^l	Dilute-lethal	Hair and skin pigments appear diluted	3d	d^l/d^l postnatal lethal, spontaneous convulsions, myelin degeneration
f	Flexed tail	Belly spot on some genetic backgrounds	2	Kinked tail, transitory siderocytic anemia

Fig. 21-1. Coat colors of mice. *A*. Yellow ($A^y/-$); phaeomelanin. *B*. Agouti ($+/+$); typical agouti banding. *C*. Agouti chinchilla (c^{ch}/c^{ch}); note drastic reduction in phaeomelanin. *D*. Nonagouti (a/a); pigment principally eumelanin except for few hairs around mammae, ears, and perineum. *E*. Black-and-tan (a^t/a); note regional differences in pigmentation, i.e., ventrum tan (yellow), dorsum black. *F*. Nonagouti brown ($a/a\ b/b$); pigment change from black to brown. *G*. Nonagouti himalayan ($a/a\ c^h/c^h$); thermolabile tyrosinase, pigmentation on extremities only. *H*. Nonagouti pink-eyed dilution ($a/a\ p/p$); drastic reduction of eumelanin.

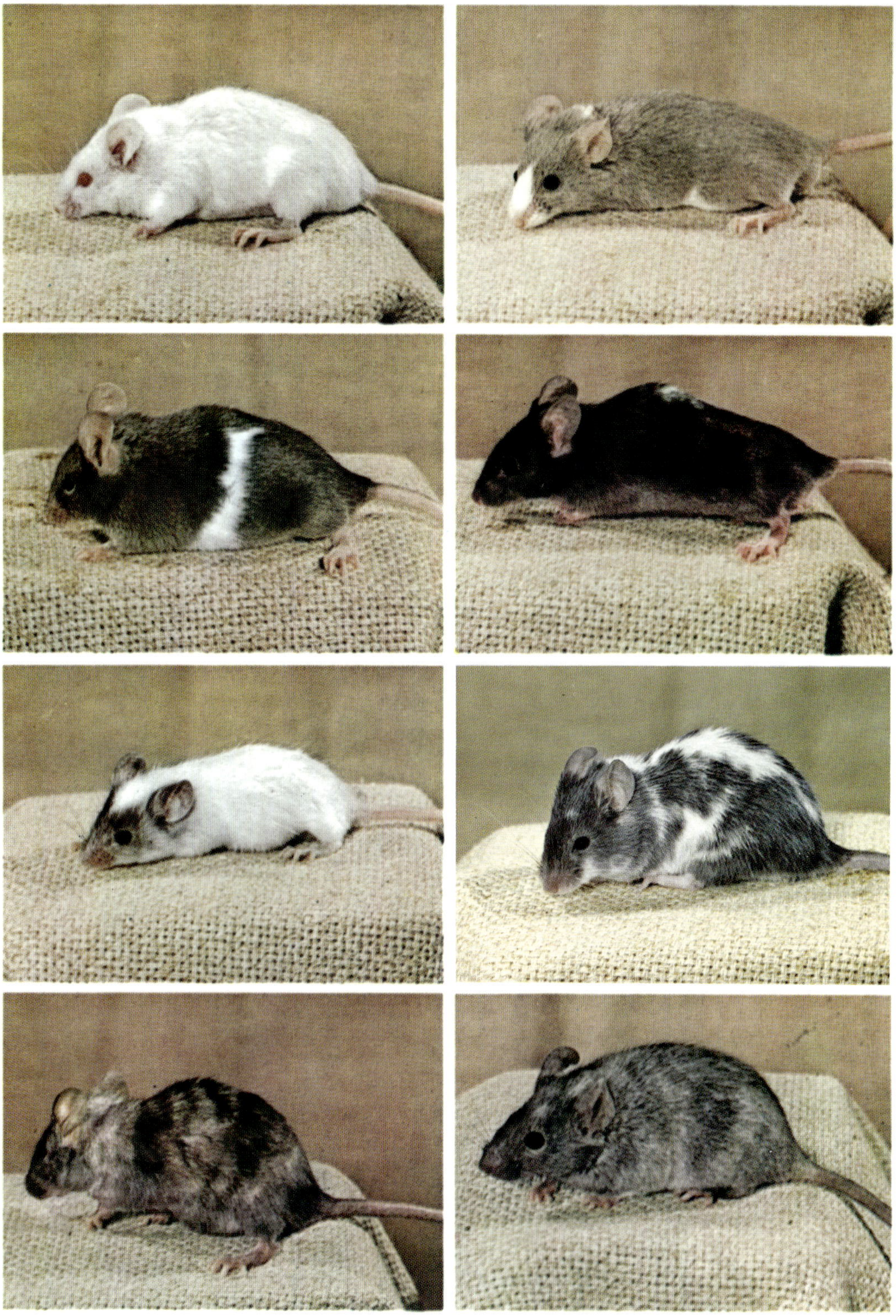

A B C D E F G H

Fig. 21-2. Coat colors of mice. *A.* Albino (*c/c*); no pigment. *B.* Nonagouti brown dilute piebald spotting (*a/a b/b d/d s/s*); recessive type of spotting due to *s* allele; pigmented areas diluted because of *d* allele. *C.* Nonagouti white (*a/a* Mi^{wh}/+); dominant spotting especially on ventrum, reduction of hair pigment in pigmented hairs. *D.* Nonagouti viable dominant spotting (*a/a* W^{v}/+) (Chap. 17, pleiotropism). *E.* Nonagouti patch viable dominant spotting (*a/a Ph* W^{v}/+ +); hyperadditive interaction between *Ph* and W^{v}. *F.* Varitint-waddler (*Va*/+); variegation (mosaicism) not due to X-chromosome inactivation. *G.* Nonagouti tortoise (*a/a To*/+); variegation attributed to X-chromosome inactivation. *H.* Cattanach translocation. Segment of chromosome I inserted in the X chromosome; variegation attributed to X-chromosome inactivation.

Table 21-1. SOME GENES AFFECTING PIGMENT IN THE MOUSE *(Continued)*

Symbol	Name	General effect on pigment	Probable site of gene action[a]	Related effects
gl	Grey-lethal	Yellow pigment missing from agouti hairs	1	*gl/gl* postnatal lethal, failure of secondary bone resorption
ln	Leaden	Diluted hair and skin pigment, no effect on retinal pigment	3d	
ls	Lethal spotting	Extensive white spotting of the piebald type	2	*ls/ls* postnatal lethal, megacolon
m	Misty	Diluted coat color, white tail tip, belly spot	2	Females breed poorly
Mi^{wh}	White	Mi^{wh}/Mi^{wh} nonpigmented except for retina; $Mi^{wh}/+$ has reduced eye, hair, and skin pigment	2	Mi^{wh}/Mi^{wh} has reduced eye size, viability, and fertility
mi	Microphthalmia	*mi/mi* nonpigmented; *mi/+* has variable spotting, hairs fully pigmented but pigment in choroid and iris reduced	2	*mi/mi* postnatal lethal, eyes drastically reduced in size or missing, failure of secondary bone resorption, incisors do not erupt
mi^{bw}	Black-eyed white	mi^{bw}/mi^{bw} all-white with black eyes	2	
mi^{sp}	Mi-spotted	mi^{sp}/mi^{sp} fully pigmented, black eyes; mi^{sp}/Mi^{wh} pale yellow with variable white spotting	2	
Mo	Mottled	Variegated	4	*Mo/Mo* or *Mo*/Y prenatal lethal
Mo^{br}	Brindle	Variegated	4	Mo^{br}/Mo^{br} or Mo^{br}/Y prenatal lethal
Mo^{dp}	Dappled	Variegated	4	Mo^{dp}/Mo^{dp} or Mo^{dp}/Y prenatal lethal
p	Pink-eyed dilution	Reduced eye, skin, and hair pigment, particularly eumelanin	3b	Greater response to induced ovulation
pa	Pallid	Reduced eye, skin, and hair pigment, particularly eumelanin		Abnormal behavior due to absence of otoliths
pe	Pearl	Reduced eye and hair pigment		Females breed poorly
Ph	Patch	White spotting on dorsum and ventrum, sometimes appearing as a belt	2	*Ph/Ph* prenatal lethal; *Ph/+* have slightly shortened heads
ru	Ruby eye	Reduced eye, skin, and hair pigment	3b	
s^l	Piebald lethal	Extensive white spotting	2	s^l/s^l postnatal lethal, megacolon
s	Piebald spotting	Variable white spotting depending on genetic background	2	Small percentage with megacolon
si	Silvered	Hairs having no pigment or reduced pigment interspersed among hairs having pigment	2	
Sl	Steel	Reduced skin and hair pigment in *Sl/+*, variable white spotting	2	*Sl/Sl* prenatal lethal; *Sl/+* have slight macrocytic anemia
Sl^d	Steel-Dickie	Sl^d/Sl^d black-eyed white; $Sl^d/+$ have reduced skin and hair pigment	2	Sl^d/Sl^d viable, but anemic and sterile
Sl^{gb}	Grizzle-belly	Sl^{gb}/Sl^{gb} black-eyed white; $Sl^{gb}/+$ have reduced skin and hair pigment	2	Sl^{gb}/Sl^{gb} postnatal lethal, macrocytic anemia
Sp	Splotch	*Sp/+* have variable white belly spotting	2	*Sp/Sp* prenatal lethal, cranial and caudal rachischisis

Table 21-1. SOME GENES AFFECTING PIGMENT IN THE MOUSE (*Continued*)

Symbol	Name	General effect on pigment	Probable site of gene action[a]	Related effects
Sp^d	Delayed splotch	$Sp^d/+$ have variable white belly spotting	2	Sp^d/Sp^d prenatal lethal, caudal rachischisis and later death than *Sp*
To	Tortoise	Variegated	4	*To/To* or *To*/Y prenatal lethal
tp	Taupe	Slightly diluted coat color	2	Females do not nurse their young
Va	Varitint-waddler	Variegated	2	Defective labyrinth in *Va/Va* and *Va*/+
W	Dominant spotting	*W/W* black-eyed white; *W*/+ has variable white spotting	2	*W/W* prenatal or early postnatal lethal, macrocytic anemia
W^a	Ames dominant spotting	W^a/W^a black-eyed white; $W^a/+$ has reduced skin and hair pigment, variable white spotting	2	W^a/W^a prenatal or early postnatal lethal; W^a/W^a and $W^a/+$ have macrocytic anemia
W^j	Jay's dominant spotting	W^j/W^j black-eyed white; $W^j/+$ has reduced skin and hair pigment, variable white spotting	2	W^j/W^j prenatal or early postnatal lethal; W^j/W^j and $W^j/+$ have macrocytic anemia
W^v	Viable dominant spotting	W^v/W^v black-eyed white; $W^v/+$ has reduced skin and hair pigment, variable white spotting	2	W^v/W^v has reduced viability, severe macrocytic anemia and is sterile; $W^v/+$ has less severe macrocytic anemia
	X-autosome translocations	Variegated	4	

[a] 1. Follicular environment.
2. Melanoblast differentiation. Time of gene action has not been determined nor have relative roles of melanoblast-tissue environment been fully evaluated.
3. Within melanoblast (melanocyte): (*a*) polymerization of melanin, as related to periodicity of binding sites on protein matrix of melanosome; (*b*) alteration of protein matrix of melanosome itself; (*c*) alteration of tyrosinase structure; (*d*) alteration of melanocyte morphology.
4. Variegation explained by X chromosome–inactivation hypothesis.

Sl-series alleles (Fig. 21-2*D*, *E*) is generally considered secondary to more fundamental metabolic disturbances (Chap. 17). The reader is referred to theoretical discussions of pleiotropism by Goldschmidt (1958), Hadorn (1961), and Grüneberg (1963).

The inheritance of coat color in mice, including spotting patterns, is not a new concern of geneticists. Indeed albinism (*c*) (Fig. 21-2*A*) in mice was the first to be analyzed in mammals by means of the newly rediscovered principles of Mendel (Cuénot, 1902; Castle and Allen, 1903). The acquisition and analysis of mutant genes yellow (A^y), white bellied agouti (A^w), nonagouti (*a*), dilute (*d*), brown (*b*), dominant spotting (*W*), pink-eyed dilution (*p*), and piebald spotting (*s*) soon followed until today some 70 genes are known affecting color or spotting at about 40 loci (see Fig. 21-1, 21-2 for coat colors of mice). Interesting accounts of early genetic studies of pigmentation in mice may be found in Little (1913) and Wright (1917).

PIGMENT CELLS: ORIGIN, LOCATION, AND FUNCTION IN ADULT STRUCTURES

All pigment in the mouse is in the form of granules (or modified granules) produced by a specialized cell, the melanocyte. This pigment may be either of two basic kinds, eumelanin (black or brown) and phaeomelanin (yellow). Melanocytes in a single hair follicle that are producing eumelanin and phaeomelanin (both kinds of pigment may be present within a single melanocyte) are illustrated in Fig. 21-3. All prospective pigment cells, or melanoblasts, except those of the retinal pigment epithelium (which come from the outer wall of the optic cup) arise from cells of the neural crest (Rawles, 1940,

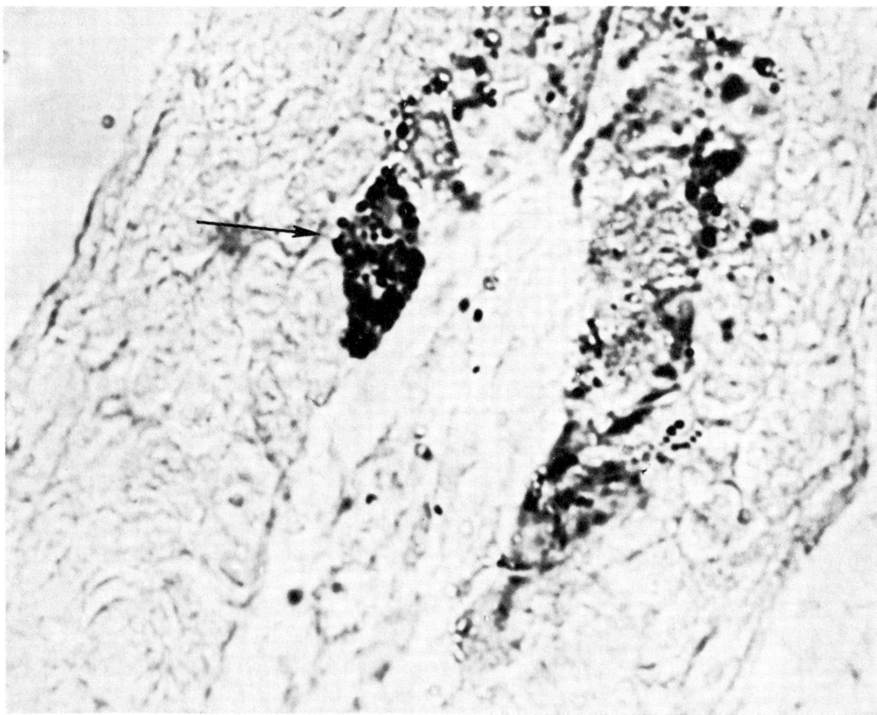

Fig. 21-3. Unstained tangential section (8μ) of transitional hair follicle showing black and yellow pigment in individual melanocytes (arrow). In the black and white photographs the yellow pigment appears either as areas of less density or as light ovoid granules (×1,212). (*From Fig. 2A, Galbraith,* 1964, *with permission of the author and Journal of Experimental Zoology.*)

1947 and reviews by Rawles, 1948, 1953). These melanoblasts migrate mediolaterally from the neural crest during the 8th to 12th day of fetal development to their defined locations, in the usual case to the hair bulb matrix of the receiving hair follicle (Rawles, 1940, 1947). In an undifferentiated state melanoblasts cannot be distinguished by any known criterion from the other embryonic cells with which they are associated. The experimental procedure used by Rawles (1947) to prove the neural crest origin of pigment cells in the mouse was to isolate tissues from various axial levels at timed developmental stages and transplant them to the embryonic coelom of White Leghorn chick hosts. Only those tissues containing presumptive neural crest, either histologically recognizable neural crest or cells migrating from the neural crest, produced melanocytes (Rawles, 1947).

Some investigators have used the silver nitrate staining reaction to chart the migration of melanoblasts in time fetuses of mice (Danneel and Cleffmann, 1954; Schumann, 1960; Danneel and Schumann, 1961). However, a serious limitation of the method is that melanoblasts apparently are not stainable upon arrival in a region. Schumann (1960) first detected pigment cells in the dermis near the ear primordium in 13-day fetuses of the C57 strain and in limb buds on day 17. Transplant data by Rawles (1947) indicate dispersal of melanoblasts by day 12 to all regions of the fetus with the possible exception of extremities. This is further corroborated by Mayer and Reams (1962), who traced the migration of melanoblasts into connective tissue of leg musculature by transplants of fetal leg muscle of PET/LSU mice of differing ages to chick coelom. Melanoblasts reach the dorsal surface of the leg by day 12 and migrate via the dermal-epidermal interface into the interior mesoderm by the end of day 14.

Indirect information about early behavior and number of melanoblasts has come from studies of X-ray–induced mosaics by Russell and Major (1957). By extrapolation from the size of mutant color spots in adults developed from fetuses irradiated at 10¼ days they estimated the modal number of prospective pigment cells to be from 150 to 200 at this stage of development. Melanocytes are not normally found in trunk-skin epidermis except in association with the hair bulb. They are, however, present in the basal layer of the epidermis and dermis of tail, feet, nose, ear, scrotum, genital papilla, and eyelids as well as meninges of the brain, particularly between the olfactory and cerebral hemispheres, in the parathyroid, thymus, and harderian gland, and in the nictitans and choroid (Markert and Silvers, 1956). Melanocytes are also found in ovary and spleen (Billingham and Silvers, 1960) and are nearly ubiquitous in distribution in connective tissue of PET/MCV mice (now extinct), being consistently absent only from the connective tissue of the gut mucosa (Nichols and Reams, 1960). Melanocytes were particularly abundant in connective tissue of lungs, kidney, rib cartilages, gonads, intercostal and extremity muscles, and semicircular canals of PET/MCV mice.

The dendritic or branched melanocytes of neural crest origin function in the epidermis as unicellular melanin-secreting glands (Billingham and Silvers, 1960). Because the function of melanocytes is so intimately connected with the hair follicle the growth and differentiation of hair follicles and hair will be briefly discussed.

Development of the hair follicle is initiated in the mouse embryo in the period extending from about 7 days before birth to 4 or more days after birth (Chase, 1954). Vibrissa follicles develop first, followed by the larger coat hairs, then the smaller ones. The hair follicle forms as a solid cylindrical downgrowth of cells that includes melanoblasts from the basal layer of the epidermis. The stages of hair growth are *anagen*, growing phase (six substages of 17 days duration); *catagen*, the transitional phase (2 days duration); and *telogen*, the resting phase (of variable duration until the next hair cycle) (Dry, 1926; Chase et al., 1951). Pigment cells normally first develop melanin during anagen III. Melanization and hair growth are maximal during anagen VI. Melanin granules are deposited, presumably through the end processes of dendritic melanocytes by a kind of cytocrine activity, into the cortical and medullary cells of the growing hair as they are pushed through the outer cone of pigment cells of the hair bulb matrix.

Each new hair passes through the same growth phases and acquires its own complement of melanocytes. The source of melanocytes for succeeding hair generations is not known but it could be from (1) mitotic descendants of melanocytes or reactivated melanocytes, (2) a "papilla reservoir" of melanoblasts or from melanoblasts that migrate into the papilla from dermal tissues, or (3) undifferentiated stem cells in the vicinity of the hair follicle. The hair follicle and its relation to pigment cells is discussed in reviews by Chase (1954, 1958) and Chase and Mann (1960).

The developing hair very early becomes keratinized and may become any one of four principal types making up the coat of the mouse: monotrich, awl, auchene, or zigzag (Dry, 1926; see Chap. 13 for anatomy of skin and its derivatives). A number of mutations in the mouse deter development of the hair follicle or alter the normal distribution or proportions of hair types which in turn indirectly affects coat color (Chap. 8). The mutant Tabby (*Ta*) is an example. A good discussion of some hair and pigment mutants, with emphasis on follicle development, may be found in Chase and Mann (1960).

The hair of the mouse is an accurate chronometer of events occurring within the melanocyte and of interaction between the melanocyte and the hair follicle. A history of pigment production by the melanocytes is contained within the hair bulb matrix. Russell (1946, 1948, 1949*a*, 1949*b*) utilized this fact in studies of hair pigment granules and pigment volume in mice of 36 different genotypes. Number, size, shape, clumping, and arrangement of granules within medullary cells and along the hair shaft and type of pigment produced were each found to contribute in an important way to the final coat color expressed (Russell, 1946). Some important generalizations emerging from the study are: (1) The alleles at the agouti locus (A^y, A^w, a) control a reversible trigger mechanism causing either yellow or black pigment granules to be deposited in the hair, (2) the principal effect of substituting brown (b) for black (+) is a qualitative change in pigment from black to brown with some reduction in pigment volume, (3) the alleles at the albino locus (c^{ch}, c^e, c) appear to control quantity of pigment only, (4) the dilute allele (d) causes irregularity of pigment deposition (granular clumping), reduced cortical pigment, and uneven pigment distribution within the cells of the hair, and (5) pink-eyed dilution (p) affects size of granules and level of pigmentation. These shred-like granules assume a kind of flocculent clumping and hairs have greatly reduced pigmentation distally (Russell, 1949*b*). A similar clumping of granules gives the "dilute" effect to hairs of d/d mice because of decreased light absorption although pigment volume is little affected.

Though the variable attributes of pigment granules in the hair are in many ways interdependent, four key pigmentation characteristics appear to be relatively independent of each other. These are (1) granule color, (2) granule size, (3) degree of pigmentation, and (4) granule clumping (Russell, 1949*a*).

It has become evident from electron microscope studies of melanocytes of the retinal pigment epithelium of the mouse that one may investigate not only the ontogeny and migratory behavior of the melanoblast, but also the ontogeny of pigment granules within the melanocyte (Moyer, 1961, 1963). The fine structure of melanosomes in neural crest-derived melanocytes of the choroid, hair follicles, and iris in the mouse are identical except for size and uniformity to those of the retinal pigment epithelium (Moyer, 1963). Those of the retina are irregular in size and shape and are generally much larger (Markert and Silvers, 1956; Moyer, 1963). The morphogenesis and fine structure of pigment granules is discussed later in this chapter.

TRANSPLANTATION AND RELATED EXPERIMENTS

Autonomy of the pigment cell is inferred from the discrete nature of patches of mutated somatic cells having different hair color persisting through successive hair molts. Experimentally this is demonstrated when the integrity of skin of mice of one color genotype is maintained following transplantation to a histocompatible host of another color genotype, and again when pigment develops from presumptive neural crest that is transplanted to the anterior chamber of the mouse eye, to mouse spleen, or to any suitable environment in another host species, e.g., chick coelom. Again, autonomy is demonstrated in vitro when melanocytes in the skin of young mice continue to produce melanin after explantation to a suitable culture medium.

In transplants of skin between fetal and newborn mice differing in color genotype the graft develops pigmentation like that of the donor except at the periphery of the graft (Reed and Sanders, 1937; Reed, 1938*b*). Here the hairs develop a color characteristic of homologous regions of the host except in the case of the agouti locus (to be discussed later). Reed (1938*b*), Reed and Henderson (1940), and Silvers and Russell (1955) have shown conclusively that melanoblast migration into alien hair follicles accounts for the "pigment spread." The migration involves hair pigment rather than skin pigment and does not occur after follicle differentiation (Reed and Henderson, 1940; Chase, 1949). Pigment spread is not observed when transplants of general trunk-skin epidermis are made between adult mice (Billingham and Medawar, 1948). The slight pigment spread that is observed when pigmented ear or tail skin is transplanted to white spotted or albino trunk areas or when nonpigmented ear skin is transplanted to a pigmented trunk area followed by treatment with a skin irritant known to stimulate melanogenesis is attributed to the passive transfer of pigment cells as a consequence of cell movements during wound healing (Silvers, 1956*b*, 1958*d*).

By transplanting fetal or neonatal skin to a neonatal host, it is possible by virtue of the migratory ability of melanoblasts to produce the effect of having melanocytes of a given color genotype function in the hair follicle of another color genotype. The method has been of great value in the study of gene action at the level of the melanocyte and hair follicle, but these studies have been possible only through the existence of the requisite inbred mouse strains, their F_1 hybrids, and by appropriate gene substitutions in congenic color stocks of mice of the type developed by E. S. Russell.

Agouti locus

The agouti alleles determine phaeomelanin (A^y, A^{vy}) or eumelanin (a, a^e), both phaeomelanin and eumelanin in the same hair at different levels (A^w, $+$), or a dorsal pigmentation different from ventral pigmentation (A^w, a^{td}, a^t) (Table 21-1). There is much overlapping between members of the series, especially as new mutant genes are discovered so that it is difficult to place them in discrete classes. Mice of genotype a/a (Fig. 21-1*D*), for example, have some yellow hairs in the region of the mammae, ears, and perineum, whereas mice of phenotype $A^{vy}/-$ and $a^{td}/-$ show some agouti coloration (Dickie, 1962; Loosli, 1963). For a description of the agouti pattern of hair the reader is referred to Dry (1928), Kaliss (1942), Russell (1949*b*), and Galbraith (1964).

Phaeomelanin is found only in hair and melanocytes of the hair bulb. Other pigment in yellow-haired phenotypes, for example, that of the eyes, tail, feet, and ear skin, is eumelanin. Vibrissae and monotrichs of agouti mice are not banded. The two kinds of pigment and pattern differences specified by alleles at the agouti locus make it particularly suitable experimental material for the type of study to be described.

Silvers and Russell (1955) and Silvers (1958*a*,

1958*b*) transplanted ventral or dorsal skin from fetal or newborn mice to the backs of newborn mice differing at the agouti locus (A^y, A^w, +, a^t, and *a*). Use was also made of genes *c*, c^e, c^{ch}, *p*, and Mi^{wh} in suitable combinations to produce a diluted or all-white donor skin to better visualize the pigmentation potentiality of the invading host melanoblasts. Control experiments and prior experiments of Reed (1938*b*) and Reed and Henderson (1940) established that hair follicles in the diluted or all-white donor graft could support differentiation and melanogenesis by invasive melanoblasts from the pigmented histocompatible host. Interpretations were based on a microscopic examination of hairs at the periphery of the graft.

In a typical experiment, dorsal skin from a fetal or newborn mouse of genotype $A^y/a\ c^e/c^e$ (near-white with black eyes) was transplanted to the dorsum of a newborn mouse of genotype $a/a\ c^e/+$ (black). The kind of pigment produced by melanoblasts of genotype $a/a\ c^e/+$ migrating into hair follicles of the donor of genotype $A^y/a\ c^e/c^e$ was always that dictated by the agouti locus genotype of the donor graft, in this case yellow. The one apparent exception, that of ventral $a/a\ c^e/c^e$ graft transplanted to the dorsum of a $A^w/a\ c^e/+$ host with consequent appearance of yellow hairs proved later to be due to the presence of normally occurring yellow hairs of *a/a* mice in the region of the nipple areolae (Silvers, 1958*a*, 1958*b*).

Results of transplantation experiments at the agouti locus are best summarized by a quotation from Silvers (1958*b*): ". . . the agouti locus genotype of the receiving hair follicle determines whether a melanocyte will produce eumelanin or phaeomelanin. The receiving hair follicle also determines the pattern of pigmentation. . . . ventrality and dorsality are not important *per se,* but together with their genetic constitution present different follicular environments which affect the expression of melanocytes."

The next logical question in the retrograde analysis of gene action at the agouti locus would be: In what way is the cellular environment of the follicle altered to produce a change in response of the melanocyte? A study of the action of the various agouti alleles in tissue culture by Cleffmann (1963) has furnished some important clues as to how this may be done (discussed later in this chapter).

The compound nature of the agouti locus, suspected by Wallace (1954) has been proved by observing crossing over within the locus (Russell et al., 1963). The mutation a^x, derived from irradiated spermatogonia, is recessive to A^y, A^w, +, and a^t and dominant to *a*, the a^x/a type being distinguishable from *a/a* by a slightly paler belly (Russell et al., 1963). The a^x homozygote is lethal. From the balanced lethal mating of $A^y/a^x \times A^y/a^x$, with or without genetic markers on either side, six (out of 2,247) exceptional agouti offspring presumed to be recombinants were recovered. Another pseudoallele (?), agouti suppressor (*As*), which causes a darkening of the middorsum has been reported (Phillips, 1961, 1963). The existence of dual genetic determiners for dorsal and ventral pigmentation is at once implied but in the absence of additional data it must remain a possibility only. Regional differences in pigmentation can be plausibly explained in terms of physiologically different cell environments (see comments of Loosli, 1963).

Brown, albino, dilute, leaden, and pink-eyed dilution loci

Transplantation experiments, of the type described for the agouti locus or as an adjunct to experiments at the agouti locus, have shown that gene action of each of the brown (*b*), albino (*c*), dilute (*d*), leaden (*ln*), and pink-eyed dilution (*p*) loci is localized within the melanoblast (melanocyte) as opposed to other cells of the hair follicle (Reed, 1938*b*; Reed and Henderson, 1940; Silvers and Russell, 1955; Silvers, 1957, 1958*a*, 1958*b*). Of these the *d* and *ln* loci appear to alter melanocyte morphology, which may in turn interfere with inoculation of pigment granules into the cells of the growing hair. In mice of genotypes *d/d* (Fig. 21-2B) and *ln/ln*, melanocytes have fewer and finer dendritic processes, and pigment granules are clumped around the nucleus (Markert and Silvers, 1956). This type of pigment cell has been called nucleopetal as opposed to the normal type of pigment cell called nucleofugal (Markert and Silvers, 1956). Because of the clumped nature of granules in the hair and the decrease in light absorption, the coat of such mice appears "dilute" or "leaden." Another mutant, "slate" (a remutation of beige, *bg*), shows clumping of granules much like *d* and *ln*, but melanocyte morphology is unaffected (Pierro and Chase, 1963; Pierro, 1963). This mutant has not been utilized in transplantation experiments.

Genic action within the melanoblast was demonstrated for the *d* and *ln* loci by transplants of neural crest–containing tissue of genotypes +/+ and *d/d* or +/+ and *ln/ln* to the anterior chamber of the eye of albino or pink-eyed dilution hosts (Markert and Silvers, 1959). Each of these environments supports melanoblast differentiation and function. Melanoblasts having a nucleofugal genetic potential differentiate into nucleofugal melanocytes only, whereas melanoblasts having a nucleopetal genetic potential differentiate into nucleopetal, nucleofugal, and intermediate forms. The results were interpreted to mean that host environment greatly influences number and size of dendritic extensions of melanocytes within limits established by the genotype. Those melanoblasts differentiating in the spleen had fewer and finer dendrites than did those in the less restrictive environment of the eye. Some *d/d* and *ln/ln* melanocytes in the eye were indistinguishable from nucleofugal melanocytes (Markert and Silvers, 1959).

White spotting vs. albinism

White spotting must be considered etiologically distinct from albinism. Whereas amelanotic melanocytes (clear cells) are present in albino hair follicles of mice (Fig. 21-4), hereditary white spotting is characterized by the absence of these and other pigment cells from the hair follicle (Chase et al., 1951; Silvers, 1953, 1956*a*, 1958*c*; Mayer and Maltby, 1964). That the large hyalinated cells found in hair bulbs of albino mice are amelanotic melanocytes is supported by different kinds of experimental evidence. Silvers (1953, 1956*a*) observed in histological sections that clear cells occupied locations comparable to those of melanocytes of lightly pigmented types such as $p/p\ c^{ch}/c^{ch}$ or d/d genotypes, or of the nearly nonpigmented $A^y/a\ c^e/c^e$ genotype. Melanocytes of black and yellow mice that are depigmented by biotin deficiency or by X-irradiation are similar in morphology to the follicular clear cells of albino (Chase and Rauch, 1950; Quevedo, 1956, 1957). Further, clear cells of albino hair follicles and melanocytes of pigmented types have comparable sensitivity to radiation; both are readily destroyed by X-irradiation of the resting hair follicles.

The most convincing evidence is by Silvers (1958*c*). Isografts of neural crest–free tissue (limb buds of 10- to 10½-day fetuses) and neural crest–containing tissue (somites and neural tube

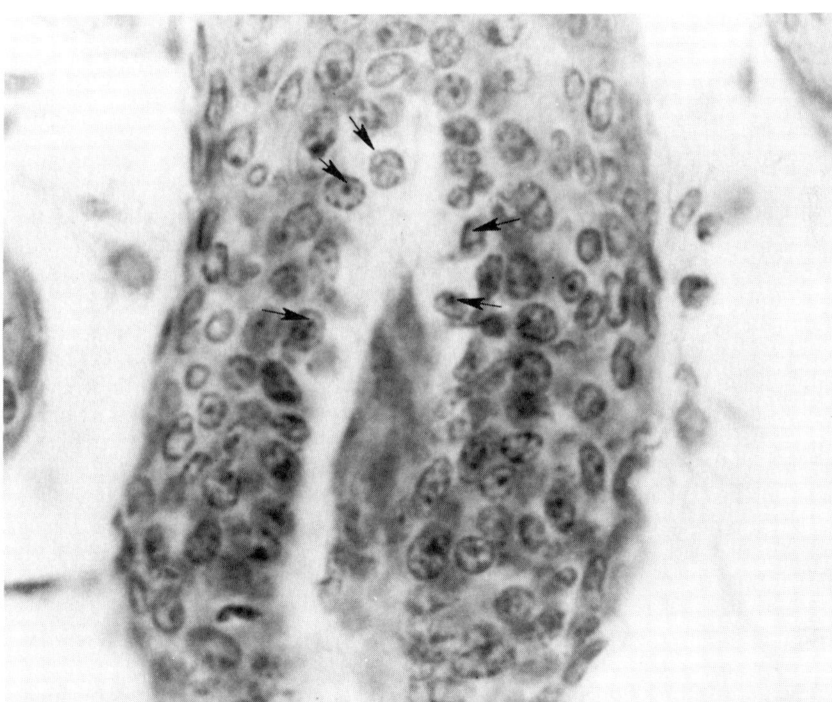

Fig. 21-4. Hair bulb of graft of intact somites plus the neural tube taken from 10½-day albino embryo. Note the position of the large clear cells (arrows) in the distal portion of the matrix (×750). (*From Fig. 1, Silvers, 1958c, with permission of the author and Anatomical Record.*)

from anterior trunk levels of 10- to 10½-day fetuses or skin plus adhering mesenchyme of limb buds of 12- to 12½-day fetuses) from mice of genotype *a/a c/c* (albino) were implanted in *a/a c/c* hosts. Compatible F_1 hybrid hosts of genotype *a/a b/+ d/+* received the same two kinds of grafts from *a/a b/b d/d* (dilute brown) donors. Grafts developing from tissues having no neural crest cells, whether from albino or pigmented donors, lacked clear cells and were characterized by hair matrices of regularly arranged cells of equal size; those containing neural crest cells from albino donors produced hair bulbs with clear cells, while those from potentially pigmented donors produced melanocytes.

Theoretically a mutant gene causing restriction of pigment on the integument (white spotting) may act at any stage from differentiation of the melanoblast from the neural crest to the time of final differentiation of melanoblast to melanocyte within the hair follicle or in the skin. It is generally agreed that melanocytes in white spotted areas are either absent, undifferentiated, or abnormally differentiated. Possible reasons for the absence of melanoblasts from a white spot are enumerated by Billingham and Silvers (1960) and Silvers (1961). These are (1) a neural crest defect, e.g., as might result from a disturbance of the region of the developing embryo which includes the neural crest and which therefore interferes with the differentiation of its cells, (2) a migratory defect which results in the melanoblast not reaching all or some areas of the embryo, or (3) a failure of potentially pigmented cells to survive in the "spotted environment."

Important alternatives to distinguish, which may not be mutually exclusive, are whether gene action is mediated via the cellular environment or resides within the melanoblast. Markert and Silvers (1956) believed, though did not establish, that cellular environment is responsible for failure of melanoblast differentiation at some early stage in development for mutant spotting genes steel (*Sl*), white (Mi^{wh}), dominant spotting (*W*), viable dominant spotting (W^v), varitint waddler (*Va*), taupe (*tp*), piebald spotting (*s*), tortoise (*To*), flexed (*f*), and belted (*bt*) (see Fig. 21 of Markert and Silvers, 1956). This was mostly inferred from the nonrandom distribution of melanocytes among different tissues (retina, choroid, hair follicle, ear skin, harderian gland, nictitans) which emphasized the importance of given tissue in supporting melanoblast differentiation. Evidence presented later in this chapter shows that *ls* (and therefore probably *s*, which has similar pleiotropic effects) acts through the neural crest. Splotch (*Sp*) and silver (*si*) are believed to act within the melanoblast at an early stage of melanoblast differentiation.

All-white mice of genotype Mi^{wh}/Mi^{wh} have frequently been referred to as having "one big spot" because of the demonstrable absence of melanocytes in skin and hair follicles (Silvers, 1953, 1956). Working with alleles at this locus, Markert and Silvers explanted mi^{bw}/mi^{bw} (black-eyed white) embryonic tissue containing neural crest into the anterior chamber of the eye of albino hosts. No pigment cells were ever obtained from these grafts, indicating failure of neural crest differentiation (Markert, 1960). Since control grafts from potentially pigmented neural crest frequently did not produce pigment, results must be interpreted with caution. Any effect of a "spotted environment" would have to be transitory or involve a certain critical period since white spotted areas do not inhibit melanogenesis. This has been shown both by transplants of neural crest cells from potentially pigmented donors (Silvers and Russell, 1955) and by pigment spread to white spotted regions (Silvers, 1956*b*).

Mayer and Maltby (1964) reexamined the problem of melanoblast differentiation in white spotted mice having the mutant genes belted (*bt*) and lethal spotting (*ls*). They demonstrated, by a series of grafts from different regions of fetal mice to chick coelom, that melanoblasts *were present* in prospective white spotted areas in the vicinity of hair follicles of 12½-day fetuses in *bt/bt* mice. However, no melanoblasts were demonstrated in prospective white spotted areas of presumed *ls/ls* mice of equivalent age or older. Thus white spotting in *bt/bt* mice is due either to a failure of melanoblasts to gain entrance into the developing hair follicles or to a failure to differentiate, whereas white spotting in *ls/ls* mice represents either a failure to migrate into the dermis of the white areas or failure to differentiate in this region even when transplanted to the suitable environment of the chick coelom. Neural crest disturbance is suspected in lethal spotting (from occurrence of megacolon) and in splotch (from occurrence of cranial and caudal rachischisis).

The presence of amelanotic melanocytes in a "white spot" is suggested in a study by Takeuchi (1964) who observed transfilter pigment spread between Mi^{wh}/Mi^{wh} skin and melanoma in tissue culture similar to, but less than, that observed between albino and melanoma under the same conditions. If amelanotic melanocytes are present in skin from mice of genotype Mi^{wh}/Mi^{wh} they exhibit little if any tyrosinase activity. Tyrosine-2-C^{14} incorporated in vitro into skin of 5-day-old mice was greatly reduced in all-white Mi^{wh}/Mi^{wh} or white spotted regions of Mi^{wh}/mi^{sp} genotypes (Wolfe and Coleman, 1964).

The degree of spotting is modified greatly by other genes. Some, such as those affecting *s* and *W*, are discussed later in this chapter. Other pigment genes frequently interact in a synergistic manner, as for example patch (*Ph*) and viable dominant spotting (W^v). The double heterozygote $Ph/+ \ W^v/+$ (Fig. 21-2*E*) usually has pigment restricted to the head (Grüneberg and Truslove, 1960). A similar restriction of pigment is found in double heterozygotes of *Ph* with *W*, *Sl*, or Sl^d (Wolfe, unpublished data). This suggests a common cause for *W*- and *Sl*-series mutants, each of which exhibits a triad of effects, i.e., varying degrees of macrocytic anemia, infertility, and reduced pigmentation (Chap. 17, 29). Yellow (A^y) generally causes reduction in size of white spots when in combination with spotting genes (Dunn et al., 1937).

Different kinds of white spotting at different loci need not have a common etiology. About 27 mutant alleles at 13 loci exhibiting various degrees of white spotting and diverse pleiotropic effects have been described (see Table 21-1). Thus spotting genes probably act in varied ways and at different times.

BIOSYNTHESIS OF MELANIN

Naturally occurring melanin is a polymer of indole-5,6-quinone which is copolymerized with protein to form melanin granules. The indole-5,6-quinone arises from the multistep oxidation of tyrosine by the enzyme tyrosinase (Lerner and Fitzpatrick, 1960; Mason, 1955, 1959; Swan, 1963). Whether this enzyme is required in all steps of this reaction is not known. The best evidence suggests that tyrosinase must be involved in the initial conversion of tyrosine to dihydroxyphenylalanine (DOPA) and probably in the conversion of DOPA to dopaquinone, although this reaction and all subsequent reactions in the series can proceed spontaneously but at a slower rate in the absence of tyrosinase. Although tyrosine is generally accepted to be the precursor of eumelanin (black or brown), tryptophan has been suggested as the natural precursor of phaeomelanin (yellow) and considerable effort has been made to demonstrate that neither tyrosine or tyrosinase are directly involved in the production of this pigment (Foster, 1951; Nachmias, 1959).

Methods of enzymatic assay

Assays for tyrosinase activity are complicated both because of the insolubility of the enzyme as it occurs in mouse skin and because of the variety of products which are produced by the enzyme before any true pigment (melanin) is formed. Since the conversion of tyrosine to DOPA is extremely slow, involving a long lag period even in the presence of large amounts of enzyme, many investigators have used DOPA as the preferred substrate in enzyme assays. This reaction proceeds rapidly and the rate can be conveniently assayed manometrically by measuring oxygen uptake or colorimetrically by measuring the formation of dopaquinone, a colored precursor of melanin. These methods have serious limitations (Fitzpatrick and Kukita, 1959) as DOPA is not the natural substrate (Kim and Tchen, 1962) and it is capable of being oxidized by a variety of nonspecific oxidases. Thus, unless pure solutions of tyrosinase are being studied it is difficult to assess how much oxygen is being consumed by spontaneous oxidation of DOPA or by nonspecific oxidases present in the crude tissues. If a colorimetric assay is used instead, the variety of colored products (all absorbing at different wavelengths), formed from DOPA spontaneously and enzymatically, make an assay based on color development alone quite difficult. Some of the same difficulties are involved using tyrosine as substrate, and it seems that in most experiments involving skin slices or tissues from mice a variety of methods using both substrates should be employed.

Fitzpatrick and Kukita (1959) developed a radioautographic assay system which provided a direct measurement of the amount of radioactive tyrosine incorporated into the newly synthesized melanin of melanocytes. Although this method was difficult to quantify, it was a considerable improvement over the previous histochemical technique using DOPA as substrate (Bloch, 1927). This method was modified in subsequent investigations by Coleman (1962) and Kim and Tchen (1962). In both modifications, after incubation with radioactive tyrosine the tissue was dissolved or otherwise treated in order to plate it out on planchets for actual counting of the amount of radioiosotope incorporated. These procedures were specific for rate of melanin formation from the natural substrate tyrosine and were extremely quantitative. It should be emphasized that such an assay system is not a direct measurement of the tyrosinase present in the tissue but only of that which is physiologically active under the assay conditions.

Effect of genic substitution on tyrosinase activity

Tyrosinase activity and its relationship to the amount and types of melanin found in various coat color mutations in mice has been studied in great detail primarily by geneticists (Russell and Russell, 1948; Foster, 1951, 1959, 1963), although later the problem of gene action in pigment mutants has interested workers in other disciplines (Fitzpatrick and Kukita, 1959; Coleman, 1960, 1962; Moyer, 1961, 1963). In general, the results of all these investigations are in good agreement, but in some cases it is hard to make valid comparisons between the work of different investigators because of the varieties of mouse genotypes and the diverse assay systems employed.

Tyrosinase activity in skin was found to vary with age of the mouse (Foster, 1951; Coleman, 1962), more specifically with the stage of the hair cycle (Fitzpatrick and Kukita, 1959). The latter authors using the radioautographic method and studying the changes in tyrosinase activity during the hair cycle found no detectable tyrosine incorporation until the appearance of pigmented melanocytes in the hair bulb. This occurred about 4 days after the hair was plucked. Tyrosinase activity increased to a maximum between the sixth and 14th days and by the 24th day no activity could be demonstrated. Foster (1951) found that skins of baby mice between the ages of 4 and 9 days had the highest tyrosinase activity. Also skins from mice of this age are particularly amenable for study since they can be pulverized easily in a mortar and pestle after freezing with dry ice. Coleman (1962) established that in mice of at least two strains (C57BL/6J and C57BR/cdJ) the maximum tyrosinase activity (without resorting to plucking) occurred at $5 \pm \frac{1}{2}$ days.

Albino locus. The alleles at the albino (c) locus have been studied extensively by the histochemical DOPA oxidase technique (Russell and Russell, 1948). This method demonstrated that the DOPA oxidase activity corresponded in the series of alleles at this locus to what would be expected from visual examination of the genotypes, i.e., $+/+ > c^{ch}/c^{ch} > c^e/c^e > c/c$. No activity was detected in the albino (Fig. 21-2*A*). Similar results were obtained by Coleman (1962) using the radioactive assay for tyrosinase. It was of interest that the genotype $c/+$ was distinguishable from $+/+$ biochemically, whereas no visual differences can be observed. In every case, mice heterozygous for the alleles at this locus had intermediate and predictable activities when compared with the homozygous genotypes.

The allele himalayan (c^h) was of particular interest in these studies since the tyrosinase in the skins from mice homozygous for this allele (Fig. 21-1*G*) was heat labile. This thermolability of the tyrosinase suggests that this allele and thus, by inference, all alleles at the c locus control the structure and not the quantity of the enzyme produced. Further evidence (Coleman, unpublished data) comes from an electrophoretic study of tyrosinases isolated from the skins of mice carrying various alleles at the c locus. Whereas the tyrosinase from wild-type mice $(+/+)$ produced two distinct tyrosinase bands on electrophoresis, similar preparations from chinchilla mice (c^{ch}/c^{ch}) (Fig. 21-1*C*) produced only one band, which was faster moving than either of the two wild-type tyrosinases. Again when enzyme from himalayan mice was similarly studied, two tyrosinases were produced, the fastest-moving of which corresponded to the fastest-moving wild-type component. The slower-moving component from the himalayan mice was however much slower moving than its counterpart from wild-type mice.

Brown locus. The action of genic substitution at the b locus on tyrosinase activity in mice has been studied by several investigators (Russell and Russell, 1948; Foster, 1951, 1959; Fitzpatrick and Kukita, 1959; Coleman, 1962). All these investigations established that the brown mice (Fig. 21-1*F*) surprisingly had equally as much tyrosinase as black mice (Fig. 21-1*D*). In fact in most studies it appeared the brown mice actually had twice the amount of tyrosinase. Numerous explanations have been advanced to explain this apparently anomalous result. Foster (1959) suggested that under conditions in vivo the substrate, tyrosine, is limiting, a situation which is not the case in vitro. Fitzpatrick and Kukita (1959) thought that the brown granules developed more slowly and at the time of assay, being not so fully developed, retained more active tyrosinase sites on their surface. Coleman (1962) showed that limitation of substrate in vivo is not involved, since injecting trace amounts of tyrosine-C^{14} into baby mice and letting the incubation run in vivo gave identical results to those seen in vitro. Also the time course of development of black and brown granules appeared to be the same (see earlier section) in both genotypes of mice. The wild-type allele in these studies was seen to be dominant and no heterozygous effect could be noticed in these enzyme assays. The gene appears to have a marked effect on the size and shape of the pigment granules as well as influencing tyrosinase activity. This suggests that the primary function of this gene is the production of the basic matrix of the granule which provides binding sites for structural tyrosinase as controlled by the *c* locus.

Agouti locus. The series of alleles at the agouti locus is responsible for the presence or absence of yellow banding in hair (see Fig. 21-1*B* for typical agouti pattern). The control exerted by this locus is sensitive enough to change the type of pigment from eumelanin to phaeomelanin within the space of one or two medullary cells. Much confusion has developed concerning the mode of action of this locus on pigmentation. Foster (1951) found that oxygen uptake in pulverized yellow (A^y/a) skin was greatly stimulated when tryptophan was substituted for tyrosine as substrate. Concurrent with this was a production of a yellow pigment rather than the abnormal black pigment formed by these genetically yellow skins from tyrosine. Little tyrosinase activity was seen in these preparations, and when nonagouti (a/a) skin was mixed with yellow (A^y/a) skin an actual inhibition of the tyrosinase activity normally present in the a/a skin was seen. These data suggested that tryptophan was the precursor of yellow pigment and that the yellow pigment was produced in spite of the residual tyrosinase activity present because of the tyrosinase inhibitor found in yellow skin (see also Nachmias 1959). Similarly Fitzpatrick and Kukita (1959), although demonstrating the incorporation of tyrosine-C^{14} in vitro into pigment, suggested that tyrosine was not really the natural precursor because this pigment formed in vitro from tyrosine was black rather than the expected yellow. Coleman (1962) showed that tryptophan injected into baby mice was not incorporated into yellow, black, or brown pigment in vivo, whereas tyrosine-C^{14} was incorporated into all types of pigment, although at a somewhat reduced rate in yellow mice when compared with their nonagouti (a/a) black or brown counterparts. Skin slices also failed to incorporate tryptophan-C^{14} but did incorporate tyrosine-C^{14} at about one-third the normal rate seen for nonagouti mice. These observations strongly suggest that tyrosine, not tryptophan, is the natural precursor to both eumelanin and phaeomelanin. The reduced rate at which tyrosine was incorporated into phaeomelanin suggests that the normal sequence of events leading to eumelanin formation is interrupted or diverted by the presence of the yellow allele, possibly resulting in a smaller polymer with the changed physical properties.

An inhibitor, as observed by Foster (1951), is implicated in this interruption of melanogenesis. Tissue culture studies have demonstrated that all hair melanocytes in skin explants from mice homozygous for the + allele at the *b* locus when grown in vitro produce black pigment regardless of whether the skin came from nonagouti (a/a), yellow (A^y/a), or agouti ($+/+$) mice (Cleffman, 1963). On addition of sulfhydryl (SH) compounds such as reduced glutathione to the medium, all skin explants could be made to produce yellow pigment which was indistinguishable from naturally occurring yellow pigment. Although all genotypes could be made to produce yellow pigment in vitro, the maximum concentration of SH required to change the developing pigment from black to yellow varied from genotype to genotype. Thus A^y/a melanocytes required less SH to produce yellow pigment than did a/a melanocytes. The amount required to produce this change, once established, remained constant throughout the growth period in vitro. In contrast skins from mice homozygous for the agouti alleles ($+/+$, A^w/A^w, and a^t/a^t) exhibited a rhythmic cyclic requirement in the amount of SH required to keep the pigment yellow. Thus during the first days of growth in vitro (period when black tips of hair are being made) a high level of SH was required equal to that seen in a/a mice. During the period when yellow pigment would be produced less SH was required and after this the requirement increased again as the cells attempted to synthesize black pigment.

Pink-eyed dilution locus. Mice homozygous for pink-eyed dilution (p/p), have markedly reduced pigmentation in the hair and eyes (Fig. 21-1*H*). This gene reduces the activity of tyrosinase as measured by incorporation studies (Fitzpatrick and Kukita, 1959; Coleman, 1962) and histochemical studies (Russell and Russell, 1948). In contrast Foster (1959, 1963), using the manometric method with both tyrosine and DOPA as substrates, found that this allele greatly enhanced the enzyme activity over that seen in wild-type ($+/+$) mice. This increased oxygen uptake does not measure the amount of pigment formed as do the other methods and may represent nonspecific tyrosine or DOPA oxidases which are not involved in pigment formation per se but which in fact compete for substrates and thus prevent normal pigment synthesis. Studies on the fine structure (see later section) suggest that pink-eye prevents the normal cross-linking seen between parallel fibers within pigment granules. Just how this will correlate with the biochemical action is at present unknown.

Other loci. Other genes affecting pigmentation, such as ruby-eye (*ru*), leaden (*ln*), and dilute (*d*), have also been studied with respect to tyrosinase activity. Briefly, mice homozygous for ruby-eye (*ru/ru*) have slightly decreased tyrosinase activity. This may be related to the delay in development of the granule as seen morphologically. Leaden mice (*ln/ln*) have normal tyrosinase activities as do dilute mice (*d/d*). Both these genes have the effect of diluting visible pigmentation by causing an abnormal clumping of a normal number of granules.

Effect of pigment genes on other enzymes

Although the dilute locus has no direct effect on tyrosinase it has been shown to reduce the activity of phenylalanine hydroxylase, which converts phenylalanine to tyrosine (Coleman, 1960). In mice homozygous for the dilute allele (*d/d*) the phenylalanine hydroxylase activity is reduced 30 to 50 per cent when compared with congenic nondilute (+/+) mice. The activity of this enzyme is reduced still further (85 per cent) in mice homozygous for the dilute lethal allele (d^l/d^l). This decrease is not caused by an enzyme deficiency but instead by an abnormal production of an inhibitor of phenylalanine hydroxylase associated with the microsomal fraction of liver homogenates. Neither the inhibitor nor the enzyme is present in newborn mice (Rauch and Yost, 1963) but both develop in the first 3 weeks of postnatal life in both dilute and nondilute genotypes. However, in normal mice (+/+) the production of the inhibitor stops between the second and third week and the amount of inhibitor decreases to adult levels. This disappearance in inhibitor permits an increase in the effective concentration in vivo of phenylalanine hydroxylase. In dilute mice (*d/d*) or dilute lethal mice (d^l/d^l) the inhibitor concentrations remain high, preventing normal phenylalanine hydroxylase activity in vivo in these dilute genotypes.

This inhibition of phenylalanine hydroxylase activity would be expected to decrease tyrosine levels in adult mice, and thus lack of substrate for the tyrosinase reaction could cause the diluted pigmentation. However, this is not the case, since Russell (1948) established that dilute hairs contain as many pigment granules as do nondilute. Instead the dilution in pigment is related to the clumping of the individual granules. These clumps originate in the follicular melanocytes which are morphologically abnormal in that they lack dendritic processes. The transplantation experiments of Markert and Silvers (1956) indicate that melanocytic differentiation in dilute genotypes begins as early as the ninth day of gestation, prior to the development of either phenylalanine hydroxylase or the inhibitor. This would not necessarily rule out an abnormal concentration of phenylalanine or of its metabolites at some critical period of development as the cause of this abnormal pigmentation. In fact, Wilde (1955) has shown that phenylalanine plays an important role in the differentiation of neural crest derivatives including melanocytes.

Dilute mice, although not actually lacking in phenylalanine hydroxylase, do exhibit some symptoms of the human condition phenylketonuria. Dilute (*d/d*) mice are subject to audiogenic seizures, whereas dilute lethal mice (d^l/d^l) have spontaneous convulsive seizures. Both dilute genotypes produce and excrete abnormal amounts of phenylketones and phenylacetic acid which could be toxic to the central nervous system. Kelton and Rauch (1962) have demonstrated myelin degeneration in dilute lethal mice, a condition often seen in human phenylketonurics. Thus we see striking phenotypical similarities between human phenylketonuria and dilution in mice even though the genetic defect is quite different.

For the effects of the agouti locus, particularly the A^y allele on mouse metabolism, the reader is referred to Chap. 5, 19, and 26.

ULTRASTRUCTURE OF PIGMENT GRANULES

The fine structure of the developing pigment granule in the mouse eye has been studied extensively by Moyer (1959, 1960, 1961, 1963). Electron microscopy has revealed that melanogenesis progresses in a definite sequence which can be interrupted or mediated by the action of genes (Seiji et al., 1963; Moyer, 1961, 1963). The same general pattern of development of the melanosome was observed for all genotypes studied. In the earliest stage (stage 1), thin unit fibers aggregate to form compound fibers within a membranous boundary. The shape of the melanosome develops as these fibers cross-link and become oriented parallel to each other (stage 2). When this orientation is complete, melanin is deposited on this matrix at definite sites along these fibers (stage 3). As this deposition continues the details of the matrix are obscured by the electron-dense melanin (stage 4). This last stage represents the typical mature melanin granule as seen through the light microscope.

Closer examination revealed that the compound fibers within the melanosome (stage 2) show a primary or first-order periodicity of approximately 65 to 85 A. This measurement varies slightly with the genotype. As the melanin is deposited at discrete sites along these fibers, a second-order periodicity is conferred on the fiber which is controlled by the *b* locus.

Genetic variation in retinal melanosomes

Of the six genetic loci studied by Moyer, only the albino (*c*), brown (*b*), and pink-eyed dilution (*p*) loci were found to affect the fine structure of the developing melanosome. The dilute (*d*) and leaden (*ln*) loci although not altering the fine structure or tyrosinase activity of the melanosome do affect the clumping of granules. Although these two genes closely mimic each other in most respects, no clumping of retinal melanosomes was observed in mice homozygous for *ln* whereas clumping was seen for dilute (*d/d*) homozygotes. Similarly the dilute locus affects phenylalanine hydroxylase whereas leaden does not, suggesting that these two loci effect their action by different means. The gene ruby-eye (*ru*) was also without effect on the size, shape, or fine structure of the melanosome, but instead seemed to delay the onset of pigmentation.

Mice homozygous for the recessive allele brown (*b/b*) when compared with the black (+/+) mice revealed differences in size, shape, and fine structure of the melanin granule. The melanin in mature brown granules as seen with the electron microscope was flocculent, coarsely granular, and often the underlying matrix could still be distinguished. In contrast, the melanin of black granules was finely granular and completely obscured the underlying matrix in mature granules. The shape of the brown granules usually was oval to spherical, whereas black granules varied from long rods to spheres but most were oval or rod-shaped. Brown granules were in general smaller but considerable variation was seen in these retinal preparations. In stage-3 granules, when melanin was just beginning to be deposited, second-order periodicity was observed which in black granules was in the order of 200 A whereas in brown granules it was only 113 A. The first-order periodicity seen in all granules remained unchanged. This second-order periodicity was interpreted to represent the active site of the enzyme tyrosinase. Thus the shorter distance between sites in brown granules would predict increased tyrosinase activity in brown tissues compared with those from black. This indeed is the case (see earlier section).

The albino (*c*) locus affects the amount of melanin within the granule as well as the size and shape of the granule. Biochemical evidence suggests that it controls the structure of the enzyme tyrosinase. In albino mice (*c/c*) the initial sequence of melanosome development is unaltered but no pigment is ever deposited on the thickened (stage-2) fibers, suggesting nonfunctional tyrosinase. Since there is no melanization of these fibers no second order periodicity was observed so no judgment could be made regarding whether the nonfunctional tyrosinase was present on these fibers. In mice homozygous for the alleles which do allow pigmentation (+/+, c^{ch}/c^{ch}, and c^h/c^h) normal second-order periodicity was seen and its magnitude was related only to the allele present at the *b* locus. In all cases the resulting mature granules were normal with respect to size and number. However, in mice homozygous for the allele extreme dilution (c^e/c^e) the granules were smaller in size and fewer in number.

A study of melanosomes from pink-eyed dilution homozygotes (*p/p*) revealed that early in development the fibers of melanosomes do not become arranged in the orderly parallel fashion. Also much less cross-linking is seen. This leads to a mature granule smaller in size and extremely irregular in shape. Although the lack of orientation of the fiber matrix makes it difficult to assess the granularity of the melanin deposited, occasional melanosomes are found in which at least part of the matrix is fairly well oriented. A study of melanin in these areas shows that the *p* locus does not affect the granularity of the melanin as determined by the *b* locus nor is the second-order periodicity altered from what would be predicted.

COLOR MOSAICISM

Aside from nongenetic changes in coat color, a coat having phenotypically distinct pigment areas or variegated (mottled) coat is indicative of (1) genetic change within a genome, (2) chimerism from polar body retention or fusion of zygotes, or (3) activity of a chromosome, part of a chromosome, or gene alternating with that of its homologue in given parts of the integument. The possible genetic causes of mosaicism in mice are reviewed by Grüneberg (1952), and Russell (1964), in mammals by Robinson (1957), and in somatic cells of mammals by Klein (1963). Color mosaics are theorized by Robinson (1957) to be a result of (1) somatic gene mutation, (2) gross chromosome change, or (3) cellular dynamics. Gross chromosome changes include nondisjunction, deletion, inversion, duplication, and trisomy; cellular dynamics include somatic crossing over, somatic reduction, and redistribution of chromosomes. Mosaics may be somatic, somatic-gonadal, or gonadal type but are most often somatic. Robinson (1957) could find 32 cases of color mosaic mice reported in the literature. Of these, 10 were known to involve the gonad.

Wolfe (1963) described an unusual type of mosaic mouse. An apparent allele of pink-eyed dilution spontaneously reverts with relatively high frequency (0.2 to 0.3 per cent in homozygotes) to wild type. Mosaic mice varied from those having a few dark hairs at only one place on the body to heavily mottled animals, the latter usually somatic-gonadal type. Eye color in mottled mice ranged from pink eye to full color, and often showed bilateral asymmetry. Russell and Major (1956) reported a similar type of mosaicism involving the mutant gene pearl (*pe*), which is unstable on strain 101 background but stable on C3H background.

Mice having wild-type patches of pigmented hairs are often observed in color mutants. Schaible and Gowen (1960) for example reported an incidence of mosaics of 6 per cent for heterozygotes of white (Mi^{wh}), 10 per cent for heterozygotes of Ames dominant spotting (W^a), and 96 per cent for heterozygotes of varitint-waddler (*Va*) mice in their colony. These ordinarily involve somatic tissue only.

Color mosaicism has been induced by chemical mutagens (Strong, 1947, 1948) and radiation (Russell and Major, 1957; Schaible, 1963). Russell and Major (1957) X-irradiated 10¼ day fetuses of genotype $b/+\ c^{ch}\ p/++\ d\ se/++$ from the cross C57BL × NB with 100 or 150 R. Controls consisted of irradiated fetuses of C57BL × C57BL matings, which were homozygous wild-type for the four color loci under study, and non-irradiated offspring of both types of mating. After correcting for mosaic animals observed in the three types of controls, the frequency of expression of recessive at one or another of the four color loci was calculated to be 10.1 per cent for the 100 R group (235 mice irradiated). Spots were randomly distributed and a single mutational event was postulated in each instance. None had mosaic eyes. The most conservative explanation is somatic mutation or deletion at the loci studied (Russell and Major, 1957). The spotting observed in progeny of C57BL × NB matings differed from the small white spots which were always mid-ventral and which were found only in irradiated fetuses of C57BL × C57BL matings. The significant increase in midventral spotting was interpreted to mean that irradiation killed some of the prospective pigment cells. Schaible (1963) X-rayed mouse embryos of genotype $Mi^{wh}/+\ bt/+$ ranging in age from ½ to 11½ days in 1-day increments. There was a significant increase in number of mosaics in 2½- and 10½-day stages.

One type of mosaicism can be explained on the basis of the X-chromosome inactivation hypothesis (Lyon, 1961). According to this hypothesis only one X chromosome or part of the X chromosome (Russell, 1963) is active in any one somatic cell. Inactivation is random, occurs early in embryonic development, and gives rise to clones of cells in alternate states of genic activity. Thus females heterozygous for X chromosome–linked pigment genes (see Fig. 21-2*G*) or for pigment genes located in autosomal segments translocated to the X chromosome (see Fig. 21-2*H*) exhibit varied degrees of mosaicism. Eight different translocations to the X chromosome have been reported to date. Seven of these have been translocations of linkage group I and the other involves linkage group VIII (Russell, 1963). Details concerning chromosome structure and behavior of X-chromosome inactivation mosaics can be found in Chap. 7 and 15. The mechanism appears to be the same as that for sex-linked semidominant genes (usually lethal in the homozygote). In these, for example tortoise (*To*) (Fig. 21-2*G*), the heterozygous female has alternate patches of black and yellow hairs (Dickie, 1954), the particular color presumably depending on which X chromosome is active in a given part of the integument.

A type of variegation that resembles the X-chromosome inactivation mosaic but which is not X-linked is varitint-waddler (*Va*, linkage group XVI). The coat of varitint-waddler mice of genotype *a/a* (Fig. 21-2*F*) has alternate areas of black, dilute (grey), and white (Cloudman and Bunker, 1945). The cause of this mosaicism is not known.

QUANTITATIVE INHERITANCE

In addition to named pigment genes there are many individually unidentified genes that modify or have slight effects on color or spotting. One has only to witness the expression of a given "spotting gene" on different inbred strain backgrounds to be convinced of this. The most thoroughly studied of the so-called modifying genes are the "*k* complex" and the "*m* (*W*) complex," affecting pattern and amount of white in piebald spotting (*s*) and dominant spotting (*W*) respectively, investigated extensively by L. C. Dunn and his associates (Dunn, 1937, 1942; Dunn and Charles, 1937; Dunn et al., 1937; Charles, 1938). An excellent account of their work is given by Grüneberg (1952). The modifying genes, some of which in the "k complex" are spotting genes in their own right, have small cumulative effects on amount of white and appear to exhibit a threshold effect. Divergent sublines obtained by selection and homozygous for *s* were nearly all-white but with black eyes or nearly full-colored at the other extreme.

Essentially the same phenomenon has been observed during selection for increased amounts of white hair starting from a heterogeneous stock of nearly full-colored mice in experiments initiated in 1931 by Goodale, discontinued, then begun again by Kyle. Approximately two-thirds of the coat of mice of the selected population is white and some individuals reached the phenotypic limit of black-eyed white in the eighth generation of continued selection (Kyle and Goodale, 1963).

Though studies of genetic modifiers and of polygenic nature of small variations in coat color have added to our understanding of pigmentation, the great progress made in the genetics and physiology of mammalian pigmentation is directly attributable to the occurrence of different kinds of coat color mutations in mice and to their general availability in inbred strains. Thus gene action may be studied in a single cell type that is nearly autonomous. Also as Russell (1946) pointed out the path between primary gene action and character must be relatively short in pigment formation.

SUMMARY

Most of what is known of genetic and physiological control of pigmentation in mammals has come from studies of named pigment genes in the mouse. With a variety of techniques the biochemistry and morphogenesis of a single cell type, the pigment cell, and its unique product, melanin, have been studied.

The source of all pigment in mice is a specialized pigment cell, the melanocyte. Two basic types of melanin are elaborated, eumelanin (black or brown) and phaeomelanin (yellow). Melanin forms as granules and remains within the melanocyte or is incorporated into developing hair cells. Melanocytes have two embryonic origins, the neural crest and wall of the optic cup. Melanocytes differentiating from neural crest (dendritic type) usually become associated with skin and hair follicle and less frequently with connective tissue of a great variety of organs. Melanocytes differentiating from the optic cup (nondendritic type) are found only in the retinal epithelium.

It has been possible with many color mutants to distinguish between two alternative modes of gene action by transplantation of embryonic tissue or undifferentiated skin to hosts of unlike color genotype, i.e., whether the site of gene action is within, or outside, the melanoblast. The production of phaeomelanin in hair follicles is dependent on follicular environment of the melanocytes under the control of agouti-series alleles. Gene action at other loci studied resides within the melanoblast and fine-structure analysis of some of these have revealed differences in periodicity of synthetic sites of melanin or organization of the protein matrix in the developing melanocyte (melanosome). It is more difficult to trace the pathway between primary gene product and effect in those color mutants whose phenotype is a function of cells surrounding the melanocyte. Melanocytes of hair follicles that normally produce eumelanin when cultured in vitro can experimentally be made to produce phaeomelanin on addition of sulfhydryl compounds to the culture medium. The type of pigment produced *in situ* is associated with mitotic rate in the hair bulb. These findings have contributed significantly to our understanding of, but do not yet define gene action at, the agouti locus.

It is generally agreed that melanocytes in white spotted areas of mutant mice are either absent, undifferentiated, or abnormally differentiated. Since genes causing white spotting may act at any stage from differentiation of melanoblasts from neural crest, through migration, to their differentiation into melanocytes at defined sites, the problem becomes one of placing time of gene action and of assessing relative importance of melanoblast and melanoblast-environment. In some types of white spotting, neural crest disturbance probably interferes with the differentiation of melanoblasts from crest cells or the migration of melanoblasts. There is experimental evidence for the presence of amelanotic melanocytes in white spotted areas of some mutants and equal evidence for their absence in white spotted areas of other mutants. In one belted mutant melanocytes have been demonstrated by tissue transplants in the vicinity of hair follicles in prospective white areas of fetal skin. Either the melanoblasts fail to survive and differentiate or are unable to gain entrance into developing hair follicles.

Variegated spotting (mottling) is related in some cases to an active state of one X chromosome (or part of the X chromosome) and to inactivity of its homologue in given parts of the integument. Thus a female mouse heterozygous for a pigment gene located on the X chromosome may show at least two alternative types of pigment.

Biochemical studies suggest that the *c* locus controls the structure of tyrosinase and that the *b* locus alters the amount of tyrosinase deposited on the fibers in a pigment granule. In the process the *b* locus must also alter the shape of the mature granule. Possibly subunits of the protein controlled by the *b* locus make up the parallel fibers seen in early melanosomes and bind the product of the *c* locus (tyrosinase) in a certain fixed ratio. Since brown skins have more tyrosinase activity than black skins and approximately the same number of granules, it appears that the protein produced by the *b* locus, as well as providing a structural framework for the attachment of the tyrosinase molecules, also influences the activity of these molecules, and this in turn must alter the final molecular structure of the melanin synthesized. The *p* locus appears to control the protein which provides the cross-linkages seen in the early melanosomes. This cross-linkage is basic to the structure of the final granule and probably influences tyrosinase activity in many ways.

LITERATURE CITED

Billingham, R. E., and P. B. Medawar. 1948. Pigment spread and cell heredity in guinea pigs' skin. Heredity 2:29–47.

Billingham, R. E., and W. K. Silvers. 1960. The melanocytes of mammals. Quart. Rev. Biol. 35:1–40.

Bloch, B. 1947. Das Pigment. Handbuch der Haut- und Geschlechtskrankheiten 1 (1):434–541.

Castle, W. E., and G. M. Allen. 1903. The heredity of albinism. Proc. Amer. Acad. Arts Sci. 38:603–622.

Cattanach, B. M. 1961. A chemically-induced variegated-type position effect in the mouse. Z. Vererb. 92:165–182.

Charles, D. R. 1938. Studies on spotting patterns. IV. Pattern variation and its developmental significance. Genetics 23:523–547.

Chase, H. B. 1949. Greying of hair. I. Effects produced by single doses of X-rays on mice. J. Morphol. 84:57–80.

Chase, H. B. 1954. Growth of the hair. Physiol. Rev. 34:113–126.

Chase, H. B. 1958. The behavior of pigment cells and epithelial cells in the hair follicle, p. 229–237. *In* W. Montagna and R. A. Ellis [ed.] The biology of hair growth. Academic Press, New York.

Chase, H. B., and S. J. Mann. 1960. Phenogenetic aspects of some hair and pigment mutants. J. Cell. Comp. Physiol. 56:103–111.

Chase, H. B., and H. Rauch. 1950. Greying of hair. II. Response of individual hairs in mice to variations in X-radiation. J. Morphol. 87:381–391.

Chase, H. B., H. Rauch, and V. W. Smith. 1951. Critical stages of hair development and pigmentation in the mouse. Physiol. Zool. 24:1–10.

Cleffmann, G. 1963. Agouti pigment cells *in situ* and *in vitro*. Ann. N.Y. Acad. Sci. 100:749–760.

Cloudman, A. M., and L. E. Bunker. 1945. The varitint-waddler mouse. J. Hered. 36:259–263.

Coleman, D. L. 1960. Phenylalanine hydroxylase activity in dilute and non-dilute strains of mice. Arch. Biochem. Biophys. 91:300–306.

Coleman, D. L. 1962. Effect of genic substitution on the incorporation of tyrosine into the melanin of mouse skin. Arch. Biochem. Biophys. 96: 562–568.

Cuénot, L. 1902. La loi de Mendel et l'hérédité de la pigmentation chez les souris. Arch. Zool. Exp. Gén., 3e Sér. 10:27–30.

Danneel, R., and G. Cleffmann. 1954. Die Einwanderung per Pigmentzellen in die Haut und die Haare bei Nagetieren. Biol. Zentrabl. 73: 414–428.

Danneel, R., and H. Schumann. 1961. Über die Entstehung und Vererbung des Blesse bei Mäusen. Z. Vererb. 92:69–73.

Deol, M. S. 1963. Inheritance of coat color in laboratory rodents, p. 177–196. *In* Lane-Petter [ed.] Animals for research. Principles of breeding and management. Academic Press, London.

Dickie, M. M. 1954. The tortoise shell house mouse. J. Hered. 45:158–159.

Dickie, M. M. 1962. A new viable yellow mutation in the house mouse. J. Hered. 53:84–88.

Dry, F. W. 1926. The coat of the mouse (*Mus musculus*). J. Genet. 16:287–340.

Dry, F. W. 1928. The agouti coloration of the

mouse (*Mus musculus*) and the rat (*Mus norvegicus*). J. Genet. 20:131–144.

Dunn, L. C. 1937. Studies on spotting patterns. II. Genetic analysis of variegated spotting in the house mouse. Genetics 22:43–64.

Dunn, L. C. 1942. Studies of spotting patterns. V. Further analysis of minor spotting genes in the house mouse. Genetics 27:258–267.

Dunn, L. C., and D. R. Charles. 1937. Studies on spotting patterns. I. Analysis of quantitive variations in the pied spotting of the house mouse. Genetics 22:14–42.

Dunn, L. C., E. C. MacDowell, and G. A. Lebedeff. 1937. Studies on spotting patterns. III. Interaction between genes affecting white spotting and those affecting color in the house mouse. Genetics 22:307–318.

Fitzpatrick, T. B., and A. Kukita. 1959. Tyrosinase activity in vertebrate melanocytes, p. 489–524. *In* M. Gordon [ed.] Pigment cell biology. Proc. Fourth Conf. Biol. Normal and Atypical Pigment Cell Growth. Academic Press, New York.

Foster, M. 1951. Enzymatic studies of pigment-forming abilities in mouse skin. J. Exp. Zool. 117:211–246.

Foster, M. 1959. Physiological studies of melanogenesis, p. 301–314. *In* M. Gordon [ed.] Pigment cell biology. Proc. Fourth Conf. Biol. Normal and Atypical Pigment Cell Growth. Academic Press, New York.

Foster, M. 1963. Discussion of paper, tactics in pigment-cell research, p. 336–343. *In* W. J. Burdette [ed.] Methodology in mammalian genetics. Holden-Day, San Francisco.

Galbraith, D. B. 1964. The agouti pattern of the mouse: a quantitative and experimental study. J. Exp. Zool. 155:71–89.

Goldschmidt, R. B. 1958. Theoretical genetics, p. 393–400. University of California Press, Berkeley.

Grüneberg, H. 1952. The genetics of the mouse. Nijhoff, The Hague. 300 p.

Grüneberg, H. 1963. The pathology of development, p. 2–4, 17. Wiley, New York.

Grüneberg, H., and G. M. Truslove. 1960. Two closely linked genes in the mouse. Genet. Res. 1:69–90.

Hadorn, E. 1961. Developmental genetics and lethal factors, p. 50, 182–271, 259–270. Wiley, New York.

Kaliss, N. 1942. The morphogenesis of pigment in the hair follicles of the house mouse. J. Morphol. 70:209–218.

Kelton, D. E., and H. Rauch. 1962. Myelination and myelin degeneration in the central nervous system of dilute-lethal mice. Exp. Neurol. 6: 252–262.

Kim, K., and T. T. Tchen. 1962. Tyrosinase of the goldfish *Carassius auratus* L. I. Radio-assay and properties of the enzyme. Biochim. Biophys. Acta 59:569–576.

Klein, G. 1963. Genetics of somatic cells, p. 407–468. *In* W. J. Burdette [ed.] Methodology in mammalian genetics. Holden-Day, San Francisco.

Kyle, W. H., and H. D. Goodale. 1963. Selection progress toward an absolute limit for amount of white hair on mice. p. 154–155. *In* S. J. Geerts [ed.] Proc. XI Int. Congr. Genet. Vol. 1. Pergamon Press, New York. (Abstr.)

Lerner, A. B., and T. B. Fitzpatrick. 1950. Biochemistry of melanin formation. Physiol. Rev. 30:91–126.

Little, C. C. 1913. Experimental studies of the inheritance of color in mice. Publ. Carnegie Inst. No. 179:13–102.

Little, C. C. 1958. Coat color genes in rodents and carnivores. Quart. Rev. Biol. 33:103–137.

Loosli, R. 1963. Tanoid—a new agouti mutant in the house mouse. J. Hered. 54:26–29.

Lyon, M. F. 1961. Gene action in the X-chromosome of the mouse (*Mus musculus* L.). *Nature* 190:372–373.

Markert, C. L. 1960. Biochemical embryology and genetics. Nat. Cancer Inst. Monogr. 2:3–12.

Markert, C. L., and W. K. Silvers. 1956. The effects of genotype and hair environment on melanoblast differentiation in the house mouse. Genetics 41: 429–450.

Markert, C. L., and W. K. Silvers. 1959. Effects of genotype and cellular environment on melanocyte morphology, p. 241–248. *In* M. Gordon [ed.] Pigment cell biology. Proc. Fourth Conf. Biol. Normal and Atypical Pigment Cell Growth. Academic Press, New York.

Mason, H. S. 1955. Comparative biochemistry of the phenolase complex. Adv. Enzymol. 16:105–184.

Mason, H. S. 1959. Structure of melanins, p. 563–582. *In* M. Gordon [ed.] Pigment cell biology. Proc. Fourth Conf. Biol. of Normal and Atypical Pigment Cell Growth. Academic Press, New York.

Mayer, T. C., and E. L. Maltby. 1964. An experimental investigation of pattern development in lethal spotting and belted mouse embryos. Develop. Biol. 9:269–286.

Mayer, T. C., and W. M. Reams, Jr. 1962. An experimental analysis and description of the melanocytes in the leg musculature of the PET strain of mice. Anat. Rec. 142:431–441.

Moyer, F. H. 1959. The fine structure of the developing melanin granule. Anat. Rec. 134:612. (Abstr.)

Moyer, F. H. 1960. Some effect of pigment mutations on the fine structure of mouse melanin granules. Anat. Rec. 138:372. (Abstr.)

Moyer, F. H. 1961. Electron microscope studies on the origin, development, and genetic control of melanin granules in the mouse eye, p. 469–486. *In* G. K. Smelser [ed.] The structure of the eye. Academic Press, New York.

Moyer, F. H. 1963. Genetic effects on melanosome fine structure and ontogeny in normal and malignant cells. Ann. N.Y. Acad. Sci. 100:584–606.

Nachmias, V. T. 1959. Tryptophan oxidation by yellow mouse skin. Proc. Soc. Exp. Biol. Med. 101:247–250.

Nichols, S. E., Jr., and W. M. Reams, Jr. 1960. The occurrence and morphogenesis of melanocytes in the connective tissues of the PET/MCV mouse strain. J. Embryol. Exp. Morphol. 8:24–32.

Phillips, R. J. S. 1961. Mouse News Letter 24:34.

Phillips, R. J. S. 1963. Mouse News Letter 29:37–38.

Pierro, L. J. 1963. Pigment granule formation in slate, a coat color mutant in the mouse. Anat. Rec. 146:365–372.

Pierro, L. J., and H. B. Chase. 1963. Slate—a new coat color mutant in the mouse. J. Hered. 54: 46–50.

Quevedo, W. C., Jr. 1956. Effect of biotin deficiency on follicular melanocytes of mice. Proc. Soc. Exp. Biol. Med. 93:260–263.

Quevedo, W. C., Jr. 1957. Loss of clear cells in the hair follicles of X-irradiated albino mice. Anat. Rec. 127:725–734.

Rauch, H., and M. T. Yost. 1963. Phenylalanine metabolism in dilute-lethal mice. Genetics 48: 1,487–1,495.

Rawles, M. E. 1940. The development of melanophores from embryonic mouse tissues grown in the coelom of chick embryos. Proc. Nat. Acad. Sci. 26:673–680.

Rawles, M. E. 1947. Origin of pigment cells from the neural crest in the mouse embryo. Physiol. Zool. 20:248–266.

Rawles, M. E. 1948. Origin of melanophores and their role in development of color patterns in vertebrates. Physiol. Rev. 28:383–408.

Rawles, M. E. 1953. Origin of the mammalian pigment cell and its role in the pigmentation of hair, p. 1–15. *In* M. Gordon [ed.] Pigment cell growth. Academic Press, New York.

Reed, S. C. 1938. Determination of hair pigments. III. Proof that expression of the black-and-tan gene is dependent upon tissue organization. J. Exp. Zool. 79:337–346.

Reed, S. C., and J. M. Henderson. 1940. Pigment cell migration in mouse epidermis. J. Exp. Zool. 85:409–418.

Reed, S. C., and G. Sanders. 1937. Time of determination of hair pigments in the mouse. Growth 1:194–200.

Robinson, R. 1957. Mosaicism in mammals. Genetica 29:120–145.

Russell, E. S. 1946. A quantitative histological study of the pigment found in the coat-color of the house mouse. I. Variable attributes of the pigment granules. Genetics 31:327–346.

Russell, E. S. 1948. A quantitative histological study of the pigment found in the coat-color mutants of the house mouse. II. Estimates of the total volume of pigment. Genetics 33:228–236.

Russell, E. S. 1949*a*. A quantitative histological study of the pigment found in the coat-color mutants of the house mouse. III. Interdependence among the variable granule attributes. Genetics 34:133–145.

Russell, E. S. 1949*b*. A quantitative histological study of the pigment found in the coat-color mutants of the house mouse. IV. The nature of the effects of genic substitution in five major allelic series. Genetics 34:146–166.

Russell, L. B. 1963. Mammalian X-chromosome action: inactivation limited in spread and in region of origin. Science 140:976–978.

Russell, L. B. 1964. Genetic and functional mosaicism in the mouse, p. 153–181. *In* M. Locke [ed.] The role of chromosomes in development. 23rd Symp. Soc. for the Study of Development and Growth.

Russell, L. B., and M. H. Major. 1957. Radiation-induced presumed somatic mutations in the house mouse. Genetics 42:161–175.

Russell, L. B., M. N. C. McDaniel, and F. N. Woodiel. 1963. Crossing-over within the *a* "locus" of the mouse. Genetics 48:907. (Abstr.)

Russell, L. B., and W. L. Russell. 1948. A study of the physiological genetics of coat color in the mouse by means of the dopa reaction in frozen sections of skin. Genetics 33:237–262.

Schaible, R. H. 1963. Developmental genetics of spotting patterns in the mouse. Ph.D. Thesis. Iowa State University, Ames.

Schaible, R. H., and J. W. Gowen. 1960. Delimitation of coat pigment areas in mosaic and piebald mice. Genetics 45:1,010. (Abstr.)

Schumann, H. 1960. Die Entstehung der Scheckung bei Mäusen mit weisser Blesse. Develop. Biol. 2:501–515.

Seiji, J., K. Shimao, M. S. C. Birbeck, and T. B. Fitzpatrick. 1963. Subcellular localization of melanin biosynthesis. Ann. N.Y. Acad. Sci. 100: 497–533.

Silvers, W. K. 1953. Histological distinction between hair follicles of albino and spotted genotypes of the mouse. Genetics 38:691–692.

Silvers, W. K. 1956*a*. Pigment cells: occurrence in hair follicles. J. Morphol. 99:41–56.

Silvers, W. K. 1956*b*. Pigment cell migration following transplantation. J. Exp. Zool. 132:539–550.

Silvers, W. K. 1957. Melanoblast differentiation secured from different mouse genotypes after transplantation to adult mouse spleen or chick embryo coelom. J. Exp. Zool. 135:221–238.

Silvers, W. K. 1958*a*. An experimental approach to action of genes at the agouti locus in the mouse. II. Transplants of newborn *aa* ventral skin to $a^t a$, $A^w a$, and *aa* hosts. J. Exp. Zool. 137:181–188.

Silvers, W. K. 1958*b*. An experimental approach to action of genes at the agouti locus in the mouse. III. Transplants of newborn A^w–, A–, and a^t–skin to A^y–, A^w–, A–, and *aa* hosts. J. Exp. Zool. 137:189–196.

Silvers, W. K. 1958*c*. Origin and identity of clear cells found in hair bulbs of albino mice. Anat. Rec. 130:135–144.

Silvers, W. K. 1958*d*. Pigment spread in mouse skin. J. Exp. Zool. 139:443–457.

Silvers, W. K. 1961. Genes and the pigment cells of mammals. Science 134:368–373.

Silvers, W. K. 1963. Tactics in pigment cell research, p. 323–343. *In* W. J. Burdette [ed.] Methodology in mammalian genetics. Holden-Day, San Francisco.

Silvers, W. K., and E. S. Russell. 1955. An experimental approach to action of genes at the agouti locus in the mouse. J. Exp. Zool. 130: 199–220.

Strong, L. C. 1947. The induction of germinal mutations by chemical means. Amer. Natur. 81:50–59.

Strong, L. C. 1948. The induction of melanotic tumors and pigmented hair, p. 358–367. *In* R. W. Miner and M. Gordon [ed.] The biology of melanomas. Spec. Pub. N.Y. Acad. Sci. Vol. 4, New York.

Swan, G. A. 1963. Chemical structure of melanins. Ann. N.Y. Acad. Sci. 100:1,005–1,019.

Takeuchi, T. 1964. Pigment spread from mouse melanoma into albino skin. Amer. Zool. 4:322–323.

Wallace, M. E. 1954. A mutation or a crossover in the house mouse? Heredity 8:89–105.

Wilde, C. E., Jr. 1955. The role of phenylalanine in the differentiation of neural crest cells. Ann. N.Y. Acad. Sci. 60:1,015–1,025.

Wolfe, H. G. 1963. Two unusual mutations affecting pigmentation in the mouse, p. 251. *In* S. J. Geerts [ed.] Proc. XI Int. Congr. Genet. Vol. 1. Pergamon Press, New York. (Abstr.)

Wolfe, H. G., and D. L. Coleman. 1964. Mi-spotted, a mutation in the mouse. Genet. Res. 5:432–440.

Wright, S. 1917. Color inheritance in mammals. II. The mouse. J. Hered. 8:373–378.

22

Acute Responses to Ionizing Radiation[1]

John B. Storer

Ionizing radiation is probably the most extensively studied of all environmental hazards and of the various test organisms used to study responses to radiation, the laboratory mouse has been the most widely utilized. Consequently, there is a voluminous literature on the radiation biology of the mouse that cannot be adequately summarized here. Rather, certain basic principles relating to total-body exposure will be considered. For more detailed and specialized information the reader is referred to a number of review articles and books on the subject, for example, Bloom (1948), Nickson (1952), Hollaender (1954), Bacq and Alexander (1955), Mitchell et al. (1956), Cronkite and Bond (1956), Bond and Robertson (1957), Lushbaugh (1957), *Proceedings of the 2nd UN International Conference on Peaceful Uses of Atomic Energy* (1958), Lacassagne and Gricouroff (1958), Brues (1959), Mole (1959), Storer and Grahn (1960), Haley and Snider (1962), *Report of the UN Scientific Committee on the Effects of Atomic Radiation* (1962), Loutit (1962), and Thomson (1962).

The deleterious effects of X-rays were discovered soon after Roentgen's original report on this type of radiation, although it required some years before the potentially harmful long-term effects were fully appreciated. Since the value of X-rays in clinical therapy was quickly apparent, it is natural that early animal experimentation was oriented toward the solution of problems encountered in radiation therapy. Although the older literature contains excellent reports on the effects produced by total-body radiation exposure, this aspect of radiation biology was not greatly emphasized until the atomic age when it became clear that individual human beings and even entire populations might receive such exposures. Since then major emphasis in animal experimentation has been on the effects of total-body exposure and for this reason the following discussion will be confined largely to such effects. Further, because of limitations of space, only acute responses to irradiation will be considered.

PHYSICAL CONSIDERATIONS AND DOSAGE MEASUREMENTS

The term *ionizing radiation* includes all electromagnetic radiations and high energy particles capable of ionizing atoms in matter on being absorbed. Although the types of electromagnetic radiations (X-rays and γ-rays) capable of producing ionization are not increasing, advancing technology is providing an increasing range of energies (wavelengths) and both the types and energies of ionizing particles available from accelerators will continue to increase. The distinction between electromagnetic and corpuscular (particulate) radiation is useful for a number of reasons. The electromagnetic radiations (X-rays and γ-rays) have been the more thoroughly studied and are the ones most commonly available to biologists. X-ray generators and γ-emitting sources such as Co^{60} or Ce^{137} are relatively inexpensive and are standard equipment in most hospitals and medical schools. Consequently they are accessible to a majority of biologists. Methods of measuring

[1] The writing of this chapter was supported in part by Contracts AT(30-1)-2313 and AT(30-1)-3314 with the U.S. Atomic Energy Commission.

the radiation output and, in turn, the radiation dose to biological objects have been well standardized. Because of the ease with which these radiations penetrate tissues it is possible to obtain relatively uniform dosage throughout an object the size of a mouse. Corpuscular radiations include all charged particles of sufficient energy to produce ionization. Examples are: electrons and positrons (β-particles), hydrogen nuclei (protons), deuterium nuclei (deuterons), tritium nuclei (tritons), helium nuclei (α-particles), and the "stripped" nuclei of additional light elements. As particle accelerators continue to get bigger and better, the list of available charged particles (atomic nuclei) will increase. Even though some of these ionizing particles (specifically α- and β-particles) are freely available from the radioactive decay of naturally occurring and artificially produced isotopes, most of them require large and expensive accelerators. Consequently they are far less commonly used than X-rays or γ-rays. An additional disadvantage resides in the fact that because of their mass they are relatively poorly penetrating at low energies, and uniform dosage is difficult to achieve. Dosage measurement (dosimetry) is complex and is a specialty in itself, usually lying outside the training and competence of biologists.

For completeness, one additional type of corpuscular radiation should be mentioned, namely, neutrons. These are unlike the previous particles in that they are uncharged and do not produce ionizations directly. Rather they react with atomic nuclei to produce charged corpuscular or electromagnetic radiations which secondarily produce ionizations. For example, they may collide with the hydrogen nucleus (proton) and impart to it sufficient energy to produce ionizations along its path. An example of induced electromagnetic radiation is the capture of a very low energy (thermal) neutron by hydrogen, with the emission of a γ-ray [the $H^1(n, \gamma)H^2$ reaction]. Since the radiation dosage from neutrons is a complex function of neutron energy and elemental composition of tissue, dosimetry is extraordinarily complicated and neutron radiation biology is nearly a subspecialty within the field of radiation biology.

In view of the foregoing considerations it is clear why the majority of biologists interested in the response of animals to ionizing radiation have worked with X-ray or γ-ray sources.

When tissues are exposed to ionizing radiation, the energy absorbed causes excitation of some of the atoms. In many cases (though not all) the excitational energy is sufficiently high that orbital electrons are ejected (ionization occurs). It follows from this fact that rearrangement of chemical bonds may occur and thus the chemical structure of molecules may be altered. When such derangements occur in molecules of biological importance, impairment of the integrity of structure or function may result. Additionally, the ionization of water, a major constituent of tissue, may lead to the production of highly reactive radicals that can react with biologically important molecules. The surprising fact is not, then, that ionizing radiation can produce profound biological effects, but that so little energy absorption is needed for these effects. Calculations of the percentage of atoms ionized in a gram of tissue by a biologically effective dose indicate that an extremely small fraction is affected. Nevertheless, the effects amplify to produce dramatic changes. This, along with the unusual latency for the appearance of effects, make ionizing radiations unique among environmental agents of biological importance. The high penetrating power of X- and γ-rays, the relatively uniform energy deposition throughout the biological specimen, and the brief exposure times required combine to make these radiations invaluable as tools in biological research.

When radiation is used as a research tool or when its effects are investigated, it is essential to know the amount of energy absorbed during an exposure (the radiation dose). It has been recommended by the International Commission on Radiological Units and Measurements (1956) that all radiation doses be expressed in terms of rads. This basic dose unit corresponds to the amount of radiation which delivers 100 ergs per gram. Despite this recommendation, the unit "roentgen" or "R" is still in widespread use. The reasons for continued use arise from historical considerations and convenience. There is no serious objection to the use of the "roentgen" for electromagnetic radiation (X- and γ-rays) of moderate energies. Further, since I R equals approximately 0.95 rads for commonly used sources, little confusion arises from the concurrent use of both units. In the majority of cases, calibrations are made in terms of roentgens. These may be converted to rads by multiplication by the appropriate factor. Because the two units are related by a constant and because rads are calculated rather than measured units, many investigators see little point in using the recommended rad unit. To relate doses of corpuscular radiations to doses of electromagnetic radiation the use of rads is mandatory, however, since corpuscular radiations cannot be measured in terms of roentgens.

A wide variety of methods for measuring X- and γ-ray doses are available. These methods include the use of ionization chambers (usually thimble chambers), calorimeters, chemical dosimeters, photographic films, and various glass and plastic products which undergo dose-dependent changes. The nearly universally employed method in biological studies is the use of commercially available ionization chambers. There are certain potential pitfalls associated with all these dosimeters, however, and the experimental biologist is wise to consult a qualified radiation physicist on the proper use of the equipment. In cases where such consultation is not feasible the investigator himself should be familiar with the physical principles on which the method is based as well as the limitations of the specific equipment under unusual exposure conditions such as very low energy radiation, unfiltered beam, etc. The limits of accuracy and in some cases the necessary corrections to be applied should also be known.

Because of the general excellence of dosimetry equipment for the more conventional energies of X- and γ-rays, the commonest errors in dosimetry are not the fault of the equipment but arise from the physical arrangements employed in making the radiation exposures. Common errors in measuring radiation dose to biological objects result from: failure to consider the lack of uniformity of an X-ray beam, failure to provide proper filtration of the beam, inadequate source (target) to specimen distance, interspersed shielding, failure to take into account scattered radiation, and temporal variations in the voltage and filament current in the X-ray generator.

In the past, comparisons of results of biological experiments between two or more laboratories have been made difficult or even impossible by the lack of even minimum standardization of exposure methods and dose measurement. To correct this unfortunate situation the International Commission on Radiological Units and Measurements has prepared a report on radiobiological dosimetry. This report, issued as *National Bureau of Standards Handbook 88*, may be obtained from the Superintendent of Documents, U.S. Government Printing Office, Washington 25, D.C. It is strongly urged that all biologists working with radiation utilize this handbook as a guide for making and for reporting radiation exposures. Good practices are stressed and common errors are pointed out. Adoption of their recommendations would go far in eliminating some of the existing confusion and sources of disagreement concerning the biological effects of radiation.

No mention has been made of the personnel hazard attendant to the use of sources of ionizing radiation. Compliance with state and federal regulations to keep personnel exposure at an irreducible minimum has been assumed.

For a review of physical principles and methods of dosimetry see the chapters by Fano, Marinelli and Taylor, and Franck and Platzman in *Radiation Biology* edited by Hollaender (1954).

ACUTE LETHAL RESPONSE

Causes of death

If a group of mice is exposed to 600 to 700 rads of X-rays delivered in a brief period (the exact dose for the response described depends on strain, age, and other variables described below) and observed over several days, one is struck by the apparent lack of any gross evidence of any effect for several days. The animals appear entirely normal. Although progressive weight loss is occurring, the blood count is declining, and various internal tissues are becoming atrophic, no clue to these derangements is given by the external appearance of the mice. About the fifth or sixth day following exposure a few of the mice may begin to appear lethargic with ruffled fur and a generally unkempt appearance. By the eighth day a few of the mice may be dead. Daily mortality increases to a peak between the 10th and 14th day then gradually subsides. After the 20th day deaths are infrequent and the surviving mice show evidence of recovery from their injury. Thirty days after exposure it is impossible to tell an irradiated survivor from a normal mouse by inspection. Because of this characteristic slow time-course of injury and recovery a longer observation period is used in radiation studies than is commonly used in studies of response to drugs. A 30-day observation period for evaluating the acute lethal response of mice to radiation exposure is widely used.

If the radiation dose is increased to 1,000 to 1,500 rads, a somewhat different pattern of illness and death emerges. With mice from our colony, it is obvious by the third or fourth day that the animals have been severely injured. They show evidence of diarrhea and an accompanying dehydration. Loss of weight is pronounced and the animals tend to sit quietly in the cage. On the fifth day some deaths occur and on the sixth day there is a sharp peak of mortality. Those surviving beyond 6 days may show a secondary mortality peak in the 8- to 12-day period. None survive beyond about 14 days.

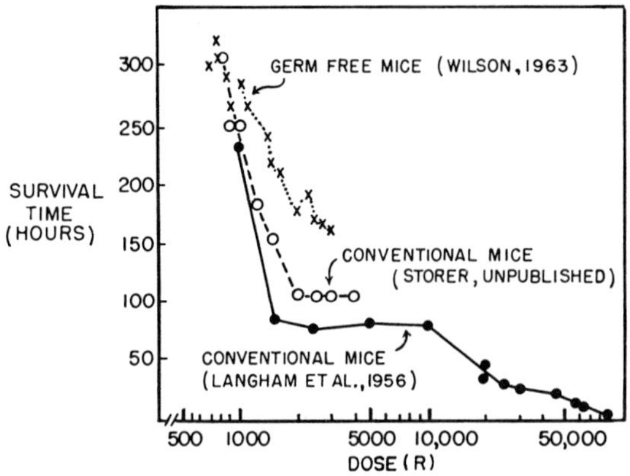

Fig. 22-1. Effect of radiation dose on survival time.

As the radiation dose is further increased up to about 10,000 rads there is little, if any, effect on survival time. All the mice die in the 5- to 7-day period to produce a curious plateau in the curve relating dose and mean survival time. Other workers have found that this stable death time occurs somewhat earlier than 5 to 7 days and many reports of a 3½-day survival time have been published. We believe the basic phenomenon to be the same but that the exact survival time depends strongly on environmental factors, particularly the intestinal bacterial flora.

With radiation doses higher than about 10,000 rads the survival time again becomes dose-dependent, i.e., the higher the dose the shorter the mean survival. Signs of neurological damage appear at these very high doses with convulsive seizures a prominent part of the syndrome. With extremely high exposures delivered at high dose rates the survival time decreases to a few hours or less and, in fact, it is possible to kill animals under the beam.

Because of these dose-dependent peaks in mortality and changing clinical signs of injury, it has been suggested that there are various modes of death in mice exposed to total-body irradiation and further that each modality has its own characteristic death time. At least four such modes of death have been established. Deaths at the lowest doses of radiation, where the peak in mortality occurs at 10 to 14 days, are due primarily to bone marrow injury. The evidence for this is as follows. Death at these dosages can be prevented by shielding a portion of the bone marrow such as that contained in one femur or by inoculating the exposed animals with bone marrow cells. In the dose range between about 1,000 and 2,000 rads the small peak in mortality at 8 or 9 days (in those animals surviving for this period) is apparently due to damage to oral structures (Quastler et al., 1956; Goepp and Fitch, 1962). Doses of about this magnitude to the head only or slightly higher doses to the oral structures only will reproduce this survival time with adequate evidence of histological injury to the tongue and pharynx to explain the cause of death. The survival time of 5 to 6 days (3½ days in some laboratories) at doses up to about 10,000 rads is due to intestinal injury (Quastler et al., 1951; Bond et al., 1950; Rajewsky, 1955; Quastler, 1956). These survival times can be produced by irradiation of the abdomen or by irradiation of the exteriorized intestine with the remainder of the animal shielded. The short survival times resulting from massive exposures can be duplicated by irradiation restricted to the head. Direct damage to the brain is responsible for death.

There is a tendency to oversimplify "causes of death" in irradiated mice in view of the identified modalities discussed above. It should not be assumed, for example, that marrow damage alone is responsible for death in the low dose range. It is of major importance but there is a complex interplay of damage to a number of tissues and tissue functions leading to death. Neither should it be assumed that the tissues listed are the only ones damaged by radiation or that these are the only modes of death. There is undoubtedly a hierarchy of causes of death. For example, radiation to the kidneys only or the heart only can also be fatal if the radiation dose is sufficiently high to compromise proper function.

The remarkable effect of environment on survival time, particularly in the gut-death range, deserves further consideration. Figure 22-1 shows survival time as a function of dose for mice maintained in three different environments. Maximum survival times were found in germ-free mice (Wilson, 1963). Minimum times were obtained in an early study (Langham et al., 1956). In this particular investigation a very great range of doses was utilized and the short survival from massive doses (central nervous system death) is illustrated. Later studies (Storer, unpublished data) with mice rigidly controlled in a closed colony but not maintained either germ-free or pathogen-free showed survival times intermediate between the other two studies. There were, of course strain differences as well as differences in radiation sources in these three investigations. Nevertheless, it is considered highly probable that environmental differences, and specifically differences in bacterial flora, account for the somewhat dissimilar results. As animal stocks and animal-colony management improve it may be anticipated that the survival curve will shift upward toward the line representing the germ-free mice.

Median lethal dose ($LD_{50:30}$)

The most widely used quantitative estimation of radiation sensitivity of a population of experimental animals is the median lethal dose or LD_{50}. For reasons discussed earlier, a 30-day observation period is the standard interval used in tabulating mortality and the dose necessary to kill 50 per cent of the animals within this period is called the $LD_{50:30}$. In practice, the LD_{50} is an interpolated value based on the response of subsamples to graded doses of radiation. Justification for this methodology is based on the following. If it were possible to measure precisely the minimum dose necessary to kill each individual mouse, a plot of the frequency with which each dose is lethal would result in a normal curve as shown in Fig. 22-2*A*. Note that in this figure the dose scale is shown in terms of standard deviations from the mean minimum lethal dose for the population. Thus if the mean is 700 R and the standard deviation is 50 R the dose points corresponding to −3, −2, −1, . . . , +3 are 550, 600, 650, . . . , 850 R. If such measurements could be made in practice, the precision of the assay would be greatly improved, since each mouse would yield a quantitative item of measurement data which could be handled by conventional statistical methods (calculation of mean, variance, standard deviation, etc.). Unfortunately this cannot be done. If a mouse is exposed to a dose of radiation and observed for 30 days it will either survive or die in the interval. If it survives, the minimum lethal dose was not attained. If it dies, the minimum has been reached or exceeded and, if exceeded, there is no way of knowing how much less radiation would have been lethal. If no recovery from radiation injury occurred it would be possible to keep adding small increments of dose until death resulted. But recovery does occur and this method cannot be used. To get around these difficulties, a curve of the cumulative frequency with which graded radiation doses equal or exceed the minimum lethal dose is employed. From Fig. 22-2*A* it is apparent that as the dose increases the minimum lethal dose will be reached or exceeded for more and more animals. The shaded area in the figure indicates the fraction of the total population that will die following a dose which is −1 S.D. from the mean. In our example this dose corresponds to 650 R. Figure 22-2*B* shows the cumulative frequency of

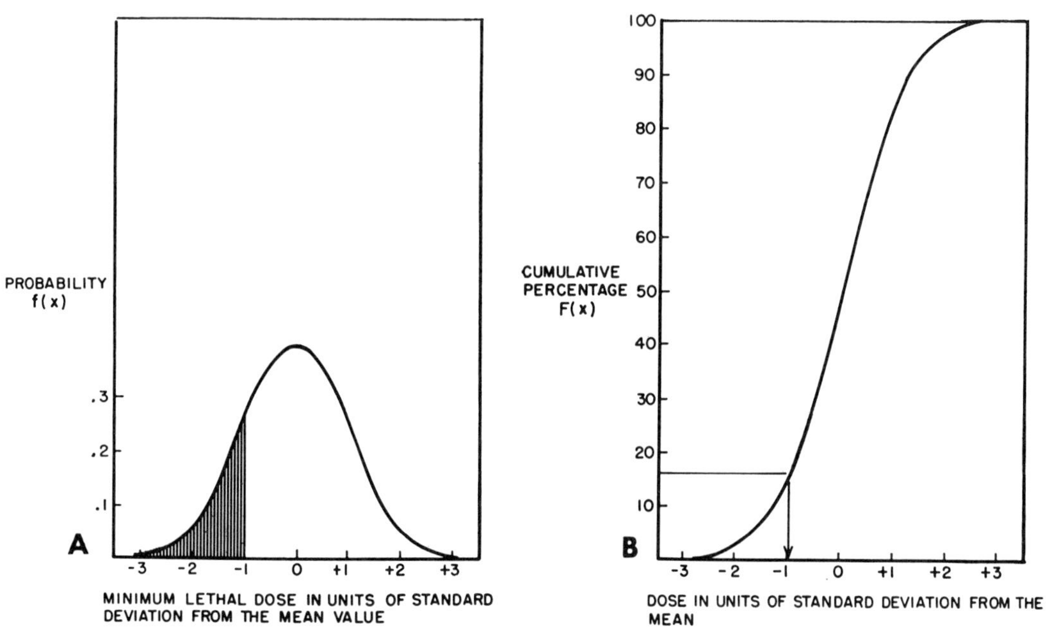

Fig. 22-2. *A*. The probability density function $f(x)$ of the normal distribution. *B*. The cumulative distribution function $F(x)$ of the normal distribution. Note that the percentage of the total area of curve *A* which is shaded corresponds exactly to the cumulative percentage shown on curve *B* for an equal dosage value.

death with increasing dose. Note that at a standard deviation of −1 (indicated by arrows in the figure) the expected percentage of deaths is 15.9. Basically the cumulative frequency curve converts areas under the normal curve (Fig. 22-2A) to percentages.

Figure 22-3 shows an experimentally obtained dose-response curve with percentage dead plotted as a function of radiation dose. This method utilizes a quantal or "all-or-none" response in that the only biological measurement is whether the mouse lives or dies. The dose necessary to kill 50 per cent of the animals could be estimated very roughly from a graphical plot of the data. This method is occasionally used but is imprecise. A polynominal regression line could be fitted to the data and the LD_{50} calculated, but the computation is too complex for routine use. Empirical transformations of scale (usually the ordinate), such as angular (arc-sine) transformations, have been used to convert the data to a straight line suitable for a least-squares fit. The most widely used method for fitting these regression lines, however, is based on the probit transformation (Gaddum, 1933; Bliss, 1938; Finney, 1947). Basically this method converts cumulative percentage mortality to the corresponding number of standard deviations from the mean associated with the particular cumulative frequency (percentage). Thus at 0 S.D. the minimum lethal dose is exceeded for 50 per cent of the animals. At −1.0 S.D. the minimum is exceeded for 15.9 per cent, and at +1.0 S.D. the corresponding cumulative percentage is 84.1. To avoid the problem of signs, 5 is added to all these standard deviations and the resulting values called *probits*. Appropriate weighting factors taking into account the number of mice in each dose group and the deviation from 50 per cent mortality are used in the least-squares computation with the transformed variable. (Points close to 50 per cent mortality are properly weighted more heavily than points toward the tails of the distribution.) With this method the median lethal dose, standard error, confidence intervals, etc., are easily computed.

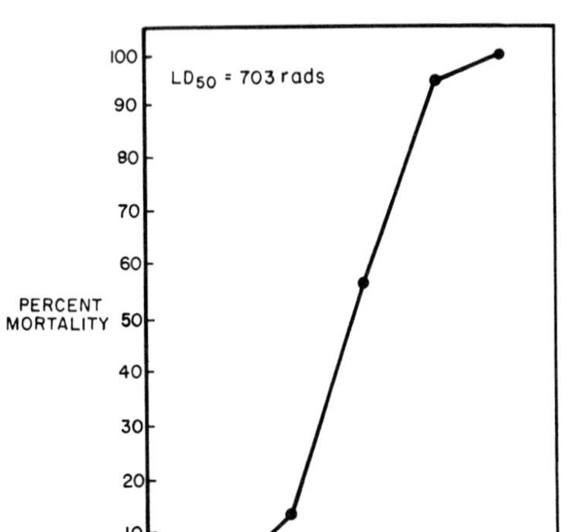

Fig. 22-3. Example of dose-response curve. Percentage dead vs. dose. Data for 129/J female mice. (*Storer, unpublished data*)

The frequency distribution curves shown in Fig. 22-2 and 22-3 are of the same shape as those encounterd with a very great variety of toxic agents. Radiation differs from most toxic agents, however, in the very remarkable steepness of the slope of the cumulative curve (Fig. 22-3) or the narrowness of the distribution (Fig. 22-2). Usually an increase in dosage of 30 per cent or less (increase from 600 to less than 800 R) will change survival from 100 per cent to zero. Because of this fact it makes little difference whether arithmetic or log dose is used in computations. For other toxic agents the distribution is nearly always normal with log dose and the transformation to log dose is required.

One final point of interest concerns the normal distribution of radiation sensitivity in mouse populations. When such a distribution is found in genetically heterogeneous populations, it can be concluded that a portion of the variability is attributable to inherent genetic differences in sensitivity. Such distributions are also seen, however, with inbred strains in which there is high genetic uniformity. Environmental differences, perhaps extending back to the early life of the mice, probably account for some of this variability. It also seems highly likely that temporal fluctuations in the resistance of an individual animal occur, in view of what is known of the day-to-day fluctuations in other physiological processes. Thus a mouse might fall on the low end of the distribution 1 day (highly sensitive) and shift to the other end of the distribution a few days later. Sacher (1956) has discussed the application of this concept to the question of natural mortality and to radiation-induced mortality.

Factors influencing the $\text{LD}_{50:30}$

Strain. Genetic constitution is one of the major factors influencing radiation resistance in mice. It is well known that there are major strain differences in sensitivity (Grahn, 1958*a*, 1958*b*,

Table 22-1. RADIATION SENSITIVITY ($LD_{50:30}$) OF TEN INBRED MOUSE STRAINS[a]

Strain	$LD_{50:30}$ ± S.E., rads	
	Males	Females
BALB/cJ	<570	585 ± 12
A/J	590 ± 22	642 ± 8
RF/J	628 ± 20	713 ± 15
SWR/J	629 ± 10	614 ± 6
C57BL/6J	647 ± 15	670 ± 6
CBA/J	656 ± 9	689 ± 8
C3HeB/J	676 ± 11	673 ± 7
SJL/J	713 ± 11	774 ± 13
C57BR/J	729 ± 9	738 ± 8
129/J	734 ± 10	774 ± 13

[a] Storer (unpublished data). All mice 3 to 4 months of age at time of irradiation.

1960; Grahn and Hamilton, 1957; Kohn and Kallman, 1956, 1957; Reinhard et al., 1954; Stadler and Gowen, 1957; Frölén et al., 1961). Differences are found not only in the single-dose LD_{50} but also in the length of survival under repeated radiation exposures. For example, Roderick (1963*a*) has determined the survival times of 27 inbred mouse strains under daily exposures to 100 R. An array of mean survival times was obtained with the most resistant strain (129/J) living twice as long as the most sensitive strain (CBA/J). Since the survival times obtained by this method are known to be correlated with the single-dose LD_{50} (Grahn, 1958*a*), it seems safe to assume that a similar ordering of sensitivity in these strains would have been obtained for LD_{50} determinations. Table 22-1 shows the single-dose LD_{50} values for 10 strains of mice in our laboratory (Storer, unpublished data). It can be seen that there are major differences in these values. Because of these genetically controlled variations in sensitivity, inbred mouse strains provide an ideal experimental tool for determining the nature of radiation resistance. The correlated characteristics so far identified have been discussed by Roderick (1963*a, b*). Generally, those characters considered to be indices of vigor are positively associated with resistance. Thus, radiation resistance may be a nonspecific measure of vigor. Nevertheless, continued study to identify genetically controlled physiological factors responsible for resistance seems worthwhile.

Hybrids are usually more resistant than either parental strain except in those cases in which parental strains of very widely divergent sensitivity are used. In this case the hybrids may fall between parental strains though closer to the more resistant parent (Frölén et al., 1961). Single-gene substitutions in an inbred line produce little or no effect on sensitivity unless the gene also produces major physiological disturbances (Doolittle, 1961). For example, the genes for short ear (*se*) and for obesity (*ob*) produce detectable shifts in sensitivity (Storer, unpublished data; Roderick, 1963, personal communication) as well as other gross disturbances. Genes such as those responsible for severe anemia produce a drastic shift in sensitivity as might be expected (Chap. 17). Bernstein (1962, 1963) has reported an LD_{50} of about 200 R for congenitally anemic mice which makes these animals at least three times as sensitive as their normal siblings. It appears, then, that given a reasonably normal phenotype, genetic control of resistance probably represents a complex interaction of a large number of gene loci as appears to be the case with other characteristics of vigor.

Age. There are a number of other variables besides genetic constitution that result in differences in radiation resistance. Within a strain, major changes in resistance are found with increasing age. Both Abrams (1951) and Lindop and Rotblat (1959) found higher resistance at birth than at 30 days of age. After minimum resistance at 30 days, resistance increases to a maximum in young adulthood, remains on a plateau for a variable length of time, and then declines in old age (Lindop and Rotblat, 1959; Sacher, 1957; Spalding and Trujillo, 1962). There is not complete agreement when the plateau of maximum resistance is attained and the time might well be a function of strain. In general, resistance increases from the time of weaning to about 3 or 4 months of age and misleading results in comparisons of LD_{50}'s can be obtained if the animals are exposed during this period of rapidly changing resistance. All too often, however, investigations are conducted with mice in this age range. We recommend the routine use of animals at least 3 and preferably 4 months of age.

Sex. The sex of mice has relatively little effect on resistance to single doses of radiation. Males tend to be slightly less resistant than females but many exceptions are found. Frequently there is so little difference in survival that data can be pooled for analysis. When mice are given daily exposures to radiation at levels resulting in mean survival times of 20 to 40 days, however, there is a marked sex difference in survival time with the females dying significantly earlier (Sacher and Grahn, 1964). This difference can be abolished by ovariectomy (Hamilton et al., 1963), so it is clear that the sensitivity is related to endocrine function rather than chromosome complement.

Environment. As might be anticipated, environmental variables are of extreme importance in influencing LD_{50} estimations. Radiation doses in this range result in severely damaged mice whose lives hang precariously in balance and relatively minor deleterious environmental factors can result in a lethal outcome.

Diets deficient in essential constituents resulting in malnourished animals produce significantly lowered radiation resistance. The LD_{50} may differ, however, with changes in standard diets. Storer (unpublished data) was able to show a difference of about 50 rads in the LD_{50} of mice fed two standard commercially available laboratory mouse foods. No gross evidence of malnutrition was apparent in the more sensitive population but apparently a subclinical deficiency existed.

Seasonal variations in sensitivity have also been noted. Roderick (1963*a*) found that survival under daily exposure was shorter in the summer months than through the rest of the year. He attributed the difference to variations in temperature since his animal rooms were not air-conditioned. Presumably a reversal of this effect (higher sensitivity in the winter) might occur in temperature-controlled quarters if recirculation of air with the attendant hazard of airborne bacterial contamination is practiced in colder weather.

Significant "caging effects" have been reported for mice by Raventos (1955) and for rats by Hahn and Howland (1963). Raventos exposed mice to an approximate LD_{50} dose and calculated the expected number of cages in which 1, 2, . . . , 10 mice should die. The observed results deviated significantly from expected with an excess of cages with low mortality and high mortality. The cage environment, then (presumably the bacterial flora of the cage), contributed significantly to the results. Hahn and Howland tested the effects of crowding on radiation resistance in rats. Group-caged rats were found to be more sensitive than singly caged rats. Caution is therefore required in interpreting minor differences in LD_{50} values as a result of experimental manipulation unless equal numbers of mice are present in each cage and unless treatments are randomly distributed within cages.

Radiation exposure in the moderate-to-high dose range causes an extreme depression in the immunological competence of mice. For this reason it is obvious that endemic infections in a mouse population may reach epidemic proportions following irradiation. Carriers of pathogenic bacteria such as *Salmonella typhimurium* are virtually worthless for radiation sensitivity studies because of the severe infection that may result after irradiation. Even mice carrying such normally nonpathogenic forms as *Pseudomonas aeruginosa* are of limited usefulness since an overwhelming bacteremia may occur and confuse the interpretation of the mortality data. Ideally, radiation studies should be performed with pathogen-free mice (with specific listings of forms considered pathogenic), with mice contaminated only with known and specified bacteria such as *Escherichia coli*, or with germ-free mice. Improvements in methods for raising and maintaining mice may ultimately make animals of these types widely available. Until such time as they come into general use, rigid control and sanitation procedures, which are advisable in any mouse colony, are mandatory in stocks used in radiation experiments.

One final environmental stress commonly encountered is that associated with the shipment of mice from the supplier to the user. It is generally recognized that shipment is stressful and most users allow a reasonably long period of recuperation and acclimatization to the new animal quarters before use. This procedure combined with rigid culling and, if feasible, bacteriological testing prevents misleading results arising from combined shipping and radiation stresses.

Radiation quality and intensity. In the foregoing discussion of variables affecting the LD_{50} of mice, the quality (type and energy) or intensity (dose rate) of the incident radiation was not discussed. If mice are exposed to the same type and energy of radiation at roughly the same dose rate, then all the factors mentioned can influence the LD_{50}. Variations in quality or intensity also lead to marked differences in LD_{50} values. In general, with X- or γ-rays, a reduction in dose rate leads to an increased LD_{50}. Fortunately, in the range of dose rates commonly used (15 or 20 rads per minute to several hundred rads per minute) the differences in duration of exposure are not sufficient to cause significant differences in the LD_{50}. As exposure times are lengthened (dose rate lowered), however, there is a progressive increase in the LD_{50} until a plateau in LD_{50} is approximated with exposure times of greater than 12 hours and up to about 36 hours (Brown et al., 1960, 1962; Stearner and Tyler, 1963; Tyler and Stearner, 1964). If the dose rate becomes too low, a lethal dose may not be accumulated in 30 days and the $\text{LD}_{50:30}$ becomes the wrong end point. The mean survival time is commonly used at these very low dose rates. At some point the mice cease to die from the acute deleterious effects of radiation and die instead from the chronic deleterious effects of the earlier phases of the exposure. In other words there is a shortening of natural longevity (Chap. 26) which often appears to be nonspecific. Failure to die from acute effects, even though doses well in excess of the usual $LD_{50:30}$ are accumulated, is presumably due to the intervention of recovery mechanisms which reverse the acute injury at a rate equal to its production. Thus a new quasi-steady state is attained (Sacher, 1955). The shift from the acute to chronic mechanism of death is, of course, not clear-cut but may occur in mice at dose rates of approximately 20 to 40 R per day. It should be noted that the transition apparently does not occur at all in some species (Sacher, 1955).

Fractionation of radiation exposures similarly increases the total dose required for an LD_{50}. The extent of the increase is a complex function of number of fractions and time between fractions. Additional information on fractionation can be found in articles by Brown et al. (1960, 1962), Tyler and Stearner (1964), Sacher (1958), and Kereiakes et al. (1957).

With more densely ionizing radiations (more ion pairs per unit of path in tissue) such as fission-energy neutrons, the LD_{50} is less responsive to variations in dose rate or, in other words, each increment of dose is more nearly additive with previous increments. For example, Vogel et al. (1957*a*) have shown that the LD_{50} in mice exposed to fission neutrons was approximately the same when the dose was delivered in 1.5 hours or over a 24-hour period. With γ-rays delivered over the same time intervals the LD_{50} increased from 929 to 1,324 R. This lesser rate-dependence for neutrons arises from the fact that recovery is slower in neutron-irradiated mice (Vogel et al., 1957*b*, 1959; Storer, 1959).

Relative biological effectiveness. When mice are exposed to different types and energies of ionizing radiation, the dose required to produce a specified level of biological effect is found to vary. These differences may be found even in those cases in which the dosage to the object is uniformly distributed throughout the tissue and the dose rates are identical. Thus the potency of various radiations (X-rays, neutrons, protons, α-particles, etc.) differs. The characteristic potency for producing a specific biological effect is known as the relative biological effectiveness (RBE) of the radiation. Since RBE is a relative value or ratio, it is obvious that the baseline radiation on which the comparison is based must be specified. The International Commission on Radiological Units and Measurements (1956) and the RBE Committee to the International Commissions on Radiological Protection and on Radiological Units and Measurements (1963) have defined the standard as X-rays producing an average of 100 ion pairs per micron of path or an average linear energy transfer (LET) of about 3 kev per micron. These units (ion pairs per micron and LET) refer to energy absorption in water. As previously explained, when radiations are absorbed some of the energy is dissipated in the ionization of atoms. For each ionizing event a pair of ions is formed, since the ejected electron becomes attached to another atom to produce a second ion of opposite sign. Filtered X-rays of energies in the range of 200 to 250 KVP produce an average of about 100 such ionizations per micron of path through water. Such radiation is said to produce a "specific ionization" (SI) of 100 ion pairs per micron. This older terminology (SI) is no longer in general use since it erroneously implies that all the energy is dissipated in the production of ionizations. The preferred is LET, for which the unit is 1,000 electron volts (kev) per micron of path. This does not specify how the energy is dissipated but refers strictly to energy absorption. On the average, one ion pair is formed for about every 33 electron volts of energy absorbed. It follows that an SI of 100 ion pairs per micron equals approximately 3 kev per micron, since roughly 3,000 electron volts of energy are absorbed for every 100 ion pairs formed.

To return to the question of baseline radiation it should be pointed out that the choice of the International Commission was probably partly a matter of expediency since much of the early work on RBE used X-rays for a baseline. A strong case can be made for the use of high energy γ-rays such as those from Co^{60} because of the wide availability of this radiation, because the average LET is close to the theoretical minimum, and because for nearly all biological responses radiations in this LET range are minimally effective. In practice most authors now use high energy γ-rays as the baseline.

RBE values are calculated by dividing the dose of baseline radiation (X- or γ-rays) required to produce a specific level of biological effect by the dose of another radiation required to produce an identical level of effect. Doses of both radiations must be expressed in the same units (preferably rads). Ideally the term RBE should be

Table 22-2. RELATIVE BIOLOGICAL EFFECTIVENESS (RBE) OF VARIOUS IONIZING RADIATIONS FOR THE PRODUCTION OF ACUTE LETHALITY IN MICE

Radiation	Approx. mean LET, kev/μ	RBE[a]	References
X-rays	3	1.4	Upton et al. (1955)
	3	1.3	Storer et al. (1957)
Fast neutrons	9	1.6	Stone (1948)
	14	1.2	Riley et al. (1955)
	14	1.6	Carter et al. (1956)
	14	4.0	Hagen and Zirkle (1950)
	20	1.5	Lawrence (1937), Lawrence and Tennant (1937)
	20	3.2	Evans (1948)
	24	2.3	Evans and Leinfelder (1953)
	24	3.2	Evans (1949)
Fission neutrons	45	2.3	Storer et al. (1957)
	45	2.0	Upton et al. (1955)
	45	3.6	Henshaw et al. (1947)
	45	4.4	Vogel et al. (1954)
	45	1.7	Delihas and Curtis (1958)
0.6-mev protons from thermal-neutron capture in tissue	65	2.4	Brennan et al. (1954), Storer et al. (1957)
1-mev neutrons	70	2.8	Riley et al. (1955)
α-Particles	110	1.4	Hollcroft and Lorenz (1951)
α-Particles and Li^7 recoils	190	1.3	Storer et al. (1957)

[a] Co^{60} or radium γ-rays were used as the baseline for RBE of X-rays. Baselines for the other radiations were either X- or γ-rays.

reserved for those studies in which all other exposure variables are held constant and only the effect of differences in LET studied. In practice this ideal is rarely attained. With animals the size of a mouse there are often gross differences in dose to various tissues and in depth doses. Terms such as "equal effect dose ratio" or "relative potency" might be preferable for these latter types of study, but by usage it appears that the term RBE will continue to be applied except in those cases where there is such a gross difference in variables other than LET that the effect of LET is completely overshadowed.

Table 22-2 summarizes some examples of RBE values for a number of types of radiation for the production of 30-day mortality in mice. It can be seen from this table that the reported RBE values range up to about 4, with the more densely ionizing radiations (high LET) being the more effective. It should be pointed out that the LD_{50} is not a particularly good choice of biological end-point because of a curious propensity for high LET radiations to cause intestinal death rather than the marrow death typical of the baseline radiation. Although the end point is the same, the mechanism for its production is different. Other test systems in the mouse, however, give roughly comparable RBE values except for the lens of the eye where neutrons are much more effective (RBE of 5 to 10) than X-rays in the production of opacities or cataracts. For reviews of RBE see the chapter by Zirkle in Hollaender (1954) and articles by Bora (1959), Storer et al. (1957), and Cronkite and Bond (1956).

An excellent example of a major effect on the LD_{50} of differences in the pattern of dose deposition in the presence of roughly equal LET values is provided by the work of Grahn et al. (1956). These investigators found LD_{50} values in mice of 767 R, 1,022 R, and 1,633 R (mean tissue dose) with X-rays generated at 250, 100, and 80 KVP, respectively. Exposure to the lower energy X-rays apparently resulted in sufficient heterogeneity of tissue dose that some tissues (marrow) received considerably less than the average dose. This resulted in the increased LD_{50} values at lower energies.

In summary, physical differences in the radiations employed can result in major differences in biological response due either to inherent differences in rates of loss of energy along the path of the radiation (differences in LET) or to macroscopic variations in energy deposition.

Modification of the lethal response

Pretreatment with chemical agents. It has long been recognized that certain environmental manipulations will decrease the radiation sensitivity of experimental animals. For example, the occlusion of blood vessels by ligation or pressure will increase the resistance of the tissue rendered hypoxic (Jolly, 1924; Mottram, 1924). Tight taping of the chest to interfere with respiration (Evans et al., 1942) or severe chilling of newborn animals (Lacassagne, 1942; Storer and Hempelmann, 1952) lead to increased radioresistance. The idea of using chemical agents to increase resistance, however, is of later origin. Because of the potential practical importance of antiradiation compounds, a considerable effort has been made to identify effective nontoxic agents. The search, however, has not been entirely successful. Patt et al. (1949) seem to have provided the major impetus to this line of research with their report that the amino acid, cysteine, when administered to rats prior to 800 R of X-rays, significantly increased survival. Since that time, literally thousands of agents have been tested and dozens have been reported to offer slight to moderate protection. Among the most effective compounds reported are para-aminopropiophenone (Storer and Coon, 1950), 2-mercaptoethylamine (Bacq and Herve, 1952), S-(2-aminoethyl) isothiouronium (Doherty and Burnett, 1955) and 5-hydroxytryptamine (Gray et al., 1952). All these compounds will increase the LD_{50} of mice by 60 to 80 per cent. Thus if the LD_{50} for control mice is 600 R, the LD_{50} in the pretreated animals might be about 1,000 R. A wide array of somewhat less effective compounds is available and these have been tabulated and discussed by Thomson (1962).

The basic difficulty with the use of chemical protectants is that they appear to be effective only at close to toxic levels. This fact, together with the relatively small degree of protection afforded (less than a factor of two), seems to preclude any immediate practical application to human beings.

All these agents apparently exert their "antiradiation" effect by influencing the indirect action of radiation, i.e., the production of reactive radicals. It is known from radiation chemistry that irradiation of water leads to the production of short-lived highly reactive radicals. If oxygen is present in the water the variety and yield of these radicals is increased. It is further known that the physical exclusion of oxygen from biological test systems increases radiation resistance and the presence of oxygen decreases resistance (the so-called "oxygen effect"). It seems reasonable to suppose that any chemical agent that reduces oxygenation in vital tissues by interference with oxygen transport, vasoconstriction, severe hypotension, or any other method, should be radioprotective. This seems to be the case and one class of agents can be so characterized. It is not clear, however, that the thiols (cysteine, glutathione, 2-mercaptoethylamine (MEA), etc.) exert their effect by this mechanism. Instead, these agents may act as radical traps or scavengers. According to this concept the thiols and certain other agents would successfully react with the free radicals to inactivate them and thus protect vital cellular functions. Objections have been raised to this theory as an oversimplification and as inconsistent with certain data. For a discussion of theories of mechanism of action see Thomson (1962). In any event it is likely that chemical agents protect by their effects on the indirect action of radiation and for this reason there may be a theoretical limit to their effectiveness by virtue of their inability to influence the direct action.

As indicated earlier, the agents producing hypoxia and certain of the thiols are the most effective protective agents. A wide variety of pharmacological agents has been tested with generally disappointing results with the exception of 5-hydroxytryptamine (serotonin). Certain sulfur-containing compounds other than thiols are slightly effective but generally do not compare with the thiols. Some amines are also slightly effective. None of these agents has been shown to be of benefit when given after radiation exposure.

Some common pitfalls in evaluating "antiradiation" agents should be pointed out. It is, unfortunately, a common practice to inject a group of mice with an agent and then expose these mice along with a control group to a single normally lethal or near-lethal dose of radiation and tabulate 30-day survival. If survival is significantly increased in the treated group, there is a temptation to assume that the protectant is highly effective in changing sensitivity. Because of the steepness of the slope relating per cent mortality in probits to dose, a minor shift in population sensitivity results in a major improvement in survival at a single-dose level. While there is no objection to the single-dose method as a screening procedure to identify promising agents, the proper method for evaluating effectiveness is to run the entire radiation dose-response curve and compare LD_{50}

values. This consideration takes on additional importance in view of the fact that pretreatment may cause a significant widening of the radiation sensitivity distribution (increased variance). In this case the dose-response curve will be much flatter in the treated group than in the control group and it is possible for the treated group to show a relatively decreased mortality at high doses, no change at the LD_{50} level, and increased mortality at low doses.

Curiously enough, radioprotective agents are relatively ineffective in protecting against high LET radiations such as neutrons (Vogel et al., 1959). This ineffectiveness may result from differences in radical production by these radiations.

Therapeutic procedures. Of the various attempts at therapy of radiation illness in mice, the best results by far have been obtained from the inoculation of hematopoietic tissue from normal donors. Under optimal circumstances the LD_{50} can be increased by about a factor of two by this method. Rekers et al. (1950) appear to have been the first to attempt this form of therapy using irradiated dogs. Their results, however, were equivocal probably for technical reasons and interest lagged until Jacobson and co-workers (1951) and Lorenz et al. (1951) independently, but working in close association, were able to demonstrate that implanted spleen (which is normally hematopoietic in rodents) or bone marrow cells would increase survival in irradiated mice. Because the results were so striking and because of the fundamental importance and possible practical usefulness of this technique, widespread interest was aroused and many researchers began work on the problem. Initially considerable time and effort were spent in attempts to identify the mechanism by which the effect was mediated. Basically the two conflicting hypotheses were (1) that a humoral substance was acting as a specific stimulus to hematopoietic regeneration or (2) that the intact implanted cells "reseeded" the depleted marrow and proliferated to repopulate the marrow spaces. Proponents of the humoral hypothesis leaned heavily on the observation that xenogeneic marrow (for example, from a guinea pig or rabbit) would protect mice. It was argued that foreign cells such as these could not survive and proliferate in the mouse. Considerable effort was therefore expended in unsuccessful attempts to extract the humoral material or to protect with cell-free preparations. Finally Main and Prehn (1955) and Lindsley et al. (1955) proved conclusively that the regenerated marrow was of donor type. In retrospect it should have been realized, perhaps, that xenogeneic cells could survive in a heavily irradiated host. As early as 1914 Murphy showed that heteroplastic transplants were feasible by growing mouse tumors in heavily irradiated rats.

The technique of protection by marrow inoculation is beautiful in its simplicity. A suspension of bone marrow (or spleen or fetal liver) cells is made in a suitable isotonic medium and inoculated intravenously or intraperitoneally into irradiated recipients. The number of nucleated cells injected can be estimated by counting cells in a sample of the suspension with a standard hemocytometer chamber. The outcome of the procedure is predictable in terms of the genetic theory of histocompatibility. Autogeneic (same animal) or isogeneic (same strain) implants are curative with no late sequelae other than those associated with nonspecific life shortening. Allogeneic (different strain and specifically one with dissimilar histocompatibility loci) or xenogeneic (different species) transplants will often enable survival for 30 days but serious delayed sequelae result from the grafting procedure. With allogeneic grafts a peculiar "secondary disease" frequently occurs as a result of the implanted cells becoming immune to the cells of the host in which they reside. This immunological conflict results in delayed death of a high proportion of the mice. For example, Davis et al. (1963) reported that 65 per cent of allogeneically grafted mice died within 20 weeks. Some animals, however, did not succumb and lived about as long as isogeneically grafted mice. It should be noted that during the period of survival of either allogeneic or xenogeneic grafts, the host animals have immunological characteristics typical of the donor. Skin grafts, for example, from the donor strain are not rejected. Mice injected with cells from a different species may also show secondary disease but in addition show a high incidence of delayed death associated with marrow failure. In this case the graft was established temporarily to enable short-term survival but subsequently rejected (graft failure) leaving the host vulnerable to death from inadequate hematopoietic function (de Vries and Vos, 1959).

The only other therapeutic procedure that is even moderately effective in raising the LD_{50} is the administration of antibiotics. It has long been known that heavy irradiation may lead to a generalized bacteremia in experimental animals (see reviews by Miller, 1956; Bond et al., 1954). With the development of broad-spectrum antibiotics it seemed logical to determine whether these agents would influence survival in irradiated mice. Some slight beneficial effect has been reported with a number of antibiotics. Streptomycin is probably the most effective of those tested. Generally the results have been disappointing with the increase in survival significant but not particularly remarkable (Hammond, 1954).

OTHER QUANTITATIVE RESPONSES

In the preceding sections considerable space has been devoted to the $LD_{50:30}$ and to factors which modify it. This was done because the acute lethal response shows a quantitative relationship between dose and incidence of mortality. Also the simplicity of the method makes it the most widely used end point in mammalian radiation biology. There is, of course, a large number of other quantitative responses of mice which can also be used and factors influencing the LD_{50} will also influence these other responses.

Quantitative relationships between radiation dose and time-dependent changes in the body weight or the weight of various tissues have been established (reviewed by Storer et al., 1957). Any tissue showing a loss of cellularity as a result of radiation exposure is a potential candidate for use as a quantitative indicator. Those commonly used include the thymus, spleen, intestine, and testis. For optimum accuracy in results the same factors must be controlled as for the LD_{50}. Homogeneous populations of relatively young age should be used. These methods offer advantages over the lethal response in that a much wider range of radiation dosages can be employed, the waiting period between exposure and measurement is generally shorter, and the end points are less complex in terms of interactions of a great number of effects.

Other measurement data such as peripheral blood cell counts, hematocrits, incidence of mitotic figures, or incidence of mitotic abnormalities in various tissues can also be quantified with radiation dose. These methods are technically more complex and it is often difficult to obtain observations on sufficient numbers of animals to establish a high degree of precision in the estimates.

The magnitudes of various biochemical and physiological responses are also dose-dependent. These are rarely used to evaluate either strain differences in or the experimental modification of radiation sensitivity because of the variety of more easily measured end points available. Histopathological changes are sometimes necessarily used as semiquantitative end points. Such changes, however, require a subjective evaluation or grading by the observer and for this reason are less precise than objective measurement or enumeration data. For measuring dose-dependent late effects the shortening of longevity provides an excellent index in large samples.

HISTOPATHOLOGY

A detailed summarization of the massive amount of knowledge of the histopathological effects of radiation exposure is beyond the scope of this section. The reader is referred to the monograph by Lacassagne and Gricouroff (1958) and to extensive reviews by Furth and Upton (1953), Lushbaugh (1957), and Bloom and Bloom (1954).

There appears to be general agreement that ionizing radiation is deleterious in its effects on tissue and that the initial reaction is an interference with cell function that may lead to cell death. The sporadic reports of a stimulating effect of radiation on cellular proliferation can be explained as secondary reactions to an initial depression. Tissues and organs vary tremendously in their radiation resistance as judged from histological changes. The causes of this variability are poorly understood. It is likely that there is a greater uniformity of functional impairment than of altered histological appearance. With improved methods of measuring function it has been necessary to revise the concepts of resistance and to classify tissues previously considered radioresistant as relatively sensitive.

The problem of classification of organs in terms of radiation sensitivity is further complicated by the fact that usually not all cell types in the organ are equally sensitive. The resistance of the organ then is limited by the resistance of the most sensitive essential cell type. Thus, if the cells lining the vascular supply are severely damaged, the vessels may become sclerotic and secondarily damage other cell types that were relatively little injured by the direct radiation exposure. Not all tissues show maximal evidence of histological injury at the same time. An ordering of relative sensitivity at 3 days after exposure would be entirely different from an ordering at 10 days, 30 days, or 100 days. Despite these problems, the rule-of-thumb propounded by Bergonié and Tribondeau in 1906 for predicting tissue sensitivity is still useful. This rule states that (1) the most rapidly proliferating cells, (2) cells retaining a capability for division the longest, and (3) cells that are the least differentiated are the most radiosensitive. Many exceptions, particularly with tumors, have been noted but the rule remains generally valid. Finally, it should be noted that all tissues of mice including bone can be damaged severely if a sufficiently large dose of radiation is delivered and a sufficiently long observation period is employed.

The specific cytological appearance of injury in various tissues is more a function of the cell type than of the injuring agent. Cells and tissues have only a limited number of ways of reacting to injury and radiation elicits a cytological response similar to that seen with many other agents. Thus, radiation produces no unique cytological or histopathological changes and the diagnosis of radiation damage depends on the clinical history or on the over-all changes produced in a number of different tissues. This latter method of diagnosis is far from foolproof, since a number of chemical toxins produce changes so similar to radiation that they are classified as "radiomimetics."

Obvious histological evidence of injury is most easily produced (smallest radiation doses are required) in hematopoietic cells, the mucosa of the small intestine, lymphoid tissue, and the germinal epithelium of the testes. A few hours after exposure of mice to even sublethal radiation doses the bone marrow cells show striking evidence of cellular destruction. Degenerating cell nuclei and nuclear debris are common. Phagocytosis of damaged cells is seen. In a day or two the cellular debris has been largely cleaned up and a marked loss of cellularity is apparent. The missing cells may be replaced by a gelatinous intercellular substance or by hemorrhage. All cell types are affected although erythrocyte and leukocyte precursors seem the most sensitive. Primitive reticulum cells are resistant. Gradually foci of hematopoiesis reappear and by the 10th to 14th day the marrow may actually appear hyperplastic. A similar sequence occurs in the lymphoid tissue (thymus, splenic follicles, lymph nodes, etc.). There is an initial intense breakdown of cells followed by atrophy of the organs from cell depletion and a gradual recovery.

The events in the testes follow a somewhat slower time-course. Here the cells principally affected are the spermatogonia, some of which are extremely sensitive and may be killed by as little as 5 rads. Initial rhexis and pycnosis occur in these cells but the effect is not striking because of the continued presence of great numbers of normal cells in later stages of maturation. With many of the progenitor cells dead, however, there is little replacement of the later stages as they proceed through their normal course of maturation and differentiation into spermatozoa. Consequently, by 4 weeks the germinal epithelium is largely absent in mice exposed to a few hundred rads and the testis is extremely atrophic. Regeneration even after high doses occurs by proliferation of the few surviving stem cells to repopulate the tubule. Such regeneration does not necessarily occur in all tubules, however, and at long times after exposure completely denuded tubules can be found. The interstitial cells of the testis are not noticeably damaged by even large doses of X-rays (Lacassagne and Gricouroff, 1958).

There is not complete agreement on the sequence of damage in the small intestine. A few hours after a moderate dose of radiation there is considerable cellular debris in the intestinal crypts. Undoubtedly many of the dead cells represent lymphocytes, although germinative mucosal cells may also be injured. The debris persists for several days before being gradually removed. The intestinal villi become shortened with fewer cells covering a villus. This effect presumably results from a prolonged mitotic delay with little or no replacement of cells that normally migrate up the villus and shed from the tip into the intestinal lumen. Concurrent with the shortening of the villus the mucosal cells which are normally columnar become flattened and almost squamous in appearance. With high radiation doses the mucosa may ulcerate and the barrier against bacterial invasion may be broken. Even in the absence of ulceration it is apparent that the functional integrity of the barrier is severely compromised. Gradually mitoses reappear in the base of the crypts and with continued proliferation the normal appearance is restored. The stomach and large intestine respond similarly but are more resistant and higher radiation doses are required to produce severe effects.

The mouse ovary is also extremely sensitive to radiation, particularly in terms of functional damage. The primitive oocytes are especially sensitive and doses of 50 rads or less may destroy many of them. This, of course, results in a reduced reproductive performance with sterility occurring at an earlier age than normally. More mature follicles are relatively resistant and single pregnancies are common after even high doses of radiation. Mice and rats are apparently unique among mammals in their ability to continue through estrus cycles even though all ova have been destroyed by radiation (Lacassagne and Gricouroff, 1958). They are probably not in hormonal balance, however, and with progressive atrophy and scarring of the ovaries the estrogen levels decrease to a point where secondary sex characteristics are affected. This contrasts with the males where secondary sex characteristics are maintained by continued function of the interstitial cells.

A number of other tissues show significant histological changes after relatively small doses of radiation, but the changes are generally either much slower in developing or much less spectacular than those seen in the tissues described above. Mitotic arrest following low radiation doses can be detected in most tissues that are normally dividing. Careful cytological evaluation of the mitotic figures at the time they reappear reveals a high incidence of chromosomal aberrations such as broken, lagging, or "sticky" chromosomes. Unequal division of the chromosomal material and an abnormal complement in the daughter cells may result. Presumably these cells may be sterile or may produce multinucleated cells. Replacement of the abnormal cells by proliferation of grossly unaffected cells occurs rapidly, however, and chromosomal abnormalities usually decrease or disappear after a few cell generations to yield apparently normal tissues. Aberrations also are produced in nondividing tissues but their presence is undetected until the tissue is stimulated to divide. For example, a few hundred rads greatly increase the incidence of these abnormalities in liver cells and these abnormalities can be detected by inducing the liver to proliferate in response to partial hepatectomy. Repeated partial hepatectomies result in a loss of aberrations presumably because of the inability of abnormal cells to continue division (Curtis et al., 1964).

Some tissues such as the kidney are often considered radioresistant because of the lack of evidence of damage in the early period after exposure. This erroneous notion arises because of the slow rate of evolution of kidney damage. Intercapillary glomerulosclerosis is a common late sequela in mice after sublethal radiation exposure. A number of other tissues similarly show damage only after long latent periods.

For detailed descriptions of the histopathological effects the reader is referred to the bibliographies appended in standard texts on pathology or radiation biology.

HEMATOLOGY

Changes in the peripheral cell counts following irradiation are largely predictable from the above described histopathological changes in lymphoid and hematopoietic tissues. Significant depression in the lymphocyte count can be demonstrated following very small (20 to 30 rads or less) doses of X-rays. Following higher doses the lymphocyte count drops precipitously, reaching a minimum in 36 to 48 hours. The duration of the depression is a function of the dose size with recovery beginning at 10 to 15 days after 300 to 400 rads and nearly normal levels are usually restored by 25 to 30 days.

The granulocytes are slightly more resistant but also show a marked drop after moderate doses. Frequently there is an initial granulocytosis preceding the decline which presumably represents an accelerated release from the marrow. After the first postirradiation day the granulocyte levels decline rapidly to a minimum at about 4 to 5 days. The duration of minimal counts is again dose-dependent with recovery occurring at about the same time as with the lymphocytes.

Probably because of the longer normal survival time of erythrocytes, severe anemia is rarely produced by irradiation. A gradual decline in the hematocrit and red cell count occurs following moderate doses with a minimum reached at 10 to 15 days. Recovery occurs equally gradually and by 30 days the values are usually near normal.

Platelets, like the granulocytes, also may show an initial increase followed by a gradual decline to very low levels. Minimum values occur at 10 to 15 days followed by fairly rapid recovery to near normal levels at 25 to 30 days. It is interesting to note that despite very low platelet levels, mice rarely show the tendency to hemorrhage that is seen in irradiated guinea pigs, dogs, or men.

Different strains of mice are known to differ with respect to: normal peripheral blood cell counts (Russell et al., 1951); radiation resistance as judged by the LD_{50} (Grahn 1958*a*, *b*, 1960; Grahn and Hamilton, 1957; Kohn and Kallman, 1956, 1957; Reinhard et al., 1954; Stadler and Gowen, 1957; Frölén et al., 1961) or survival under daily exposure (Roderick 1963*a*); and rate of recovery from radiation injury (Kohn and Kallman, 1957). In view of the fact that the extent of fall of the granulocytes may be prognostic of mortality (Brues and Rietz, 1948; Cronkite et al., 1956) or may be an index of radiation sensitivity, it would seem logical to exploit these strain differences to determine whether the initial peripheral count or radiation-induced changes in peripheral counts are correlated either with radiation sensitivity or recovery rate. Curiously, this does not appear to have been done.

Detailed descriptions of hematological effects of ionizing radiation are found in the reviews by

Jacobson et al. (1949), and Cronkite and Brecher (1955).

IMMUNE RESPONSE

Exposure to ionizing radiation interferes seriously with the ability of mice to respond immunologically to the administration of various antigenic substances. The precise effects observed are a complex function of type of antigenic stimulus, dose of radiation, and time of exposure. While the literature on the subject is conflicting and sometimes confusing, certain general principles have emerged. The primary response is more sensitive than the secondary response and less than 100 rads may be sufficient for significant depression of the primary response. The initial stages of the reaction leading to antibody formation are more sensitive than the later stages. This conclusion follows from the observation that a greater depression is obtained when radiation is given before antigen than when antigen is given before radiation. All mechanisms of response to antigenic stimuli are depressed in that cellular immunity as well as circulating antibodies are affected.

Because of the depression in immune response it is obvious that heavily irradiated mice are more susceptible to both endogenous and exogenous bacterial and viral infections. This point was discussed briefly previously. Ability to reject incompatible tissue grafts is also impaired. This explains the successful transplantation of incompatible bone marrow. Grafting of other tissues may also result in temporary "takes." Completely alien tumors may grow rapidly in irradiated mice and may even kill the recipients before immune mechanisms recover sufficiently to cause rejection. For these reasons the heavily irradiated mouse is, in a sense, a self-contained, self-operating tissue-culture system at least temporarily. Irradiation, therefore, provides a powerful tool for studies on the nature of the immune process and for studies of tissue and organ transplantation.

Extensive reviews on the subject are provided by Leone (1962), Talmage (1955), and Taliaferro (1957).

REPRODUCTION

Basically there are two mechanisms by which ionizing radiation can influence reproductive performance. Transmissible genetic defects (chromosomal derangements or point mutations) may be induced which lead to failure of implantation, early death and resorption of the fetus, or stillbirths. In totally irradiated mice this mechanism is of relatively minor importance compared with the somatic effects on the breeding animals which may markedly reduce the total numbers of viable offspring. The genetic effects are covered in Chap. 10 and will not be reviewed here. From a consideration of the histopathological effects on the testis and ovary and the relative ease with which sterility can be induced, particularly in females, it is apparent that direct somatic effects may sharply reduce numbers of offspring during the reproductive life of mice. Males are relatively more resistant than females. Moderate to large doses of radiation may produce a period of sterility by the killing of large numbers of spermatogonia which are the stem cells from which later stages in spermatogenesis are derived. This sterility is only temporary at less than lethal doses and most of the tubules regenerate and spermatozoa are again formed. The sperm count may remain slightly lower than normal since not all tubules regenerate, but apparently there is normally a great excess of spermatozoa and a modest depression in numbers does not interfere seriously with the ability to inseminate successfully.

Because of the high sensitivity of primitive ova and because ova are not continuously supplied from germinal cells capable of indefinite proliferation, female mice are easily sterilized with even modest doses of radiation. With 100 rads or less the female may become permanently sterile after bearing one or two litters. (These litters early after radiation can occur presumably because the more mature ova are most resistant than the more primitive ones.) With higher doses the incidence of complete sterility is greatly increased.

Impaired reproductive performance in the female could also occur if disturbances in the endocrine balance or in the ability to provide a suitable uterine environment were induced. While such changes presumably could be induced by high doses of radiation to localized regions, they are not a factor with total-body radiation because the extreme sensitivity of the mouse ovary is limiting. It might be noted that the corpora lutea and the uterine mucosa are relatively resistant structures.

SUMMARY

I have attempted to describe briefly the physical basis by which radiation damage is produced. The importance of good practices in dosimetry is stressed. Clinical signs and symptoms of radiation injury and the presumed modes of death from this injury are discussed. I have placed particular emphasis on methods for the quantitative estimation of biological injury and factors which influence the accuracy of the estimation. Variables affecting sensitivity to irradiation are included along with a brief description of the histopathology of radiation injury.

Radiation provides an extremely valuable tool for studies in hematology, immunology, aging, cancerogenesis, and response to stress. Radiation doses are easily quantified and the responses are unusually uniform if the extraneous variables are carefully controlled.

LITERATURE CITED

Abrams, H. L. 1951. Influence of age, body weight, and sex on susceptibility of mice to the lethal effects of X-radiation. Proc. Soc. Exp. Biol. Med. 76:729–732.

Bacq, Z. M., and P. Alexander. 1955. Fundamentals of radiobiology. Butterworth, London. 389 p.

Bacq, Z. M., and A. Herve. 1952. Protection chimique contre le rayonnement X. Bull. Acad. Roy. Med. Belg. 18:13–58.

Bergonié, J., and L. Tribondeau. 1906. Interprétation de quelques résultats de la radiothérapie et essai de fixation d'une technique rationelle. Compt. Rend. Acad. Sci. 143:983–985. Cited by Lacassagne and Gricouroff (1958).

Bernstein, S. E. 1962. Acute radiosensitivity in mice of differing *W* genotype. Science 137: 428–429.

Bernstein, S. E. 1963. Modification of radiosensitivity of genetically anemic mice by implantation of blood-forming tissue. Radiat. Res. 20: 695–702.

Bliss, C. I. 1938. The determination of dosage-mortality curves from small numbers. Quart. J. Pharmacol. 11:192–216.

Bloom, W. [ed.] 1948. Histopathology of irradiation from external and internal sources. National Nuclear Energy Series IV–221. McGraw-Hill, New York. 808 p.

Bloom, W., and M. A. Bloom. 1954. Histological changes after irradiation, p. 1,091–1,144. *In* A. Hollaender [ed.] Radiation biology. Vol. I, Part 2. McGraw-Hill, New York.

Bond, V. P., and J. S. Robertson. 1957. Vertebrate radiobiology (lethal actions and associated effects). Annu. Rev. Nucl. Sci. 7:135–162.

Bond, V. P., M. S. Silverman, and E. P. Cronkite. 1954. Pathogenesis and pathology of post-irradiation infection. Radiat. Res. 1:389–400.

Bond, V. P., M. N. Swift, A. C. Allen, and M. C. Fishler. 1950. Sensitivity of abdomen of rat to X-irradiation. Amer. J. Physiol. 161:323–330.

Bora, K. C. 1959. Factors affecting the relative biological efficiencies of ionizing radiations, p. 278–299. *In* J. G. Bugher, J. Coursaget, and J. F. Loutit [ed.] Progress in nuclear energy. Ser. VI. Biological Sciences. Vol. 2. Pergamon Press, New York.

Brennan, J. T., P. S. Harris, R. E. Carter, and W. H. Langham. 1954. The biological effectiveness of thermal neutrons on mice. Nucleonics 12(2):48–56; 12(4):31–35.

Brown, J. A. H., M. J. Corp, and R. H. Mole. 1962. The effect of dose-rate and fractionation on acute mortality in X-irradiated mice. Int. J. Radiat. Biol. 5:369–377.

Brown, J. A. H., M. J. Corp, and D. R. Westgarth. 1960. Effect of dose-rate and fractionation of X-ray dose on acute lethality in mice. Int. J. Radiat. Biol. 2:371–381.

Brues, A. M. [ed.] 1959. Low-level irradiation. American Association for the Advancement of Science, Washington, D.C. 148 p.

Brues, A. M., and L. Rietz. 1948. Acute hematologic radiation response of the guinea-pig as a function of lethality, p. 183–187. *In* Biological and Medical Divisions quarterly report for period Aug. to Oct. 1948. ANL-4227 (Argonne National Laboratory, Chicago.)

Carter, R. E., V. P. Bond, and P. H. Seymour. 1956. The relative biological effectiveness of fast neutrons in mice. Radiat. Res. 4:413–423.

Cronkite, E. P., and V. P. Bond. 1956. Effects of radiation on mammals. Annu. Rev. Physiol. 18: 483–526.

Cronkite, E. P., V. P. Bond, and C. L. Dunham. 1956. Some effects of ionizing radiation on human beings. Document TID-5358, U.S. Government Printing Office, Washington, D.C. 106 p.

Cronkite, E. P., and G. Brecher. 1955. The protective effect of granulocytes in radiation injury. Ann. N.Y. Acad. Sci. 59:815–833.

Curtis, H. J., J. Tilley, and C. Crowley. 1964. The elimination of chromosome aberrations in liver cells by cell division. Radiat. Res. 22: 730–734.

Davis, W. E., Jr., L. C. Cole, W. A. Foley, and V. J. Rosen, Jr. 1963. Leukemia incidence and longevity in radiation-induced homologous mouse chimeras. Radiat. Res. 20:43–52.

Delihas, N., and H. J. Curtis. 1958. The relative biological effectiveness of fission neutrons for the production of acute mortality in mice. Radiat. Res. 8:166–180.

De Vries, M. J., and O. Vos. 1959. Delayed

mortality in radiation chimeras: a pathological and hematological study. J. Nat. Cancer Inst. 23:1403–1440.

Doherty, D. G., and W. T. Burnett, Jr. 1955. Protective effect of S,β-Aminoethylisothiouronium·Br·HBr and related compounds against X-radiation death in mice. Proc. Soc. Exp. Biol. Med. 89:312–314.

Doolittle, D. P. 1961. The effect of single gene substitutions on resistance to radiation in mice. Genetics 46:1501–1509.

Evans, T. C. 1948. Effects of small daily doses of neutrons on mice. Radiology 50:811–833.

Evans, T. C. 1949. The fast neutron hazard. Nucleonics 4(3):2–8.

Evans, T. C., J. P. Goodrich, and J. C. Slaughter. 1942. Temperature and radiosensitivity of skin of newborn rats: effects of decreased circulation and breathing during irradiation. Radiology 38:201–206.

Evans, T. C., and P. J. Leinfelder. 1953. The relative effectiveness of fast neutrons and X-radiation in the production of lens damage in mice. Radiat. Res. 1:130. (Abstr.)

Finney, D. J. 1947. Probit analysis. Cambridge University Press, New York. 256 p.

Frölén, H., K. G. Lüning, and C. Rönnbäck. 1961. The effect of X-irradiation on various mouse strains due to their genetic background. I. Lethality after acute irradiation. Radiat. Res. 14: 381–393.

Furth, J., and A. C. Upton. 1953. Vertebrate radiobiology: histopathology and carcinogenesis. Annu. Rev. Nucl. Sci. 3:303–338.

Gaddum, J. H. 1933. Reports on biological standards. III. Methods of biological assay depending on a quantal response. Spec. Rep. Ser. Med. Res. Counc., London, No. 183. Cited by Finney (1947).

Goepp, R., and F. Fitch. 1962. Pathological study of oral radiation death in mice. Radiat. Res. 16:833–845.

Grahn, D. 1958*a*. The genetic factor in acute and chronic radiation toxicity. Proc. 2nd Int. Conf. Peaceful Uses Atomic Energy (Geneva) 22: 394–399.

Grahn, D. 1958*b*. Acute radiation response of mice from a cross between radiosensitive and radioresistant strains. Genetics 43:835–843.

Grahn, D. 1960. The genetics of radiation toxicity in animals, p. 181–200. *In* Caldecott and Snyder [ed.] Radioisotopes in the biosphere. Center for Continuation Study, The University of Minnesota Press, Minneapolis.

Grahn, D., and K. F. Hamilton. 1957. Genetic variation in the acute lethal response of four inbred mouse strains to whole body X-irradiation. Genetics 42:189–198.

Grahn, D., G. A. Sacher, and H. Walton, Jr. 1956. Comparative effectiveness of several X-ray qualities for acute lethality in mice and rabbits. Radiat. Res. 4:228–242.

Gray, J. L., J. T. Tew, and H. Jensen. 1952. Protective effect of serotonin and of para-aminopropiophenone against lethal doses of X-radiation. Proc. Soc. Exp. Biol. Med. 80:604–607.

Hagen, C. W., and R. E. Zirkle. 1950. Comparative biological actions of cyclotron fast neutrons and X rays. I. Lethal action on mice and rabbits. Univ. Chicago Rep. CH-3903.

Hahn, E. W., and J. W. Howland. 1963. Modification of irradiation response of female rats by population density. Radiat. Res. 19:676–681.

Haley, T. J., and R. S. Snider. [ed.] 1962. Response of the nervous system to ionizing radiation. Academic Press, New York. 783 p.

Hamilton, K. F., G. A. Sacher, and D. Grahn. 1963. A sex difference in mouse survival under daily gamma irradiation and its modification by gonadectomy. Radiat. Res. 18:12–16.

Hammond, C. W. 1954. The treatment of post-irradiation infection. Radiat. Res. 1:448–458.

Henshaw, P. S., E. F. Riley, and G. E. Stapleton. 1947. Plutonium project; biologic effects of pile radiations. Radiology 49:349–359.

Hollaender, A. [ed.] 1954. Radiation biology. Vol. I. High energy radiation. McGraw-Hill, New York. 1,265 p.

Hollcroft, J., and E. Lorenz. 1951. The 30-day LD_{50} of two radiations of different ion density. J. Nat. Cancer Inst. 12:533–544.

International Commission on Radiological Units and Measurements. 1956. National bureau of standards handbook 62. U.S. Government Printing Office, Washington, D.C. 48 p.

International Commissions on Radiological Protection and on Radiological Units and Measurements. Report of the RBE Subcommittee. 1963. Health Phys. 9:357–384.

Jacobson, L. O., E. K. Marks, and E. Lorenz. 1949. The hematological effects of ionizing radiations. Radiology 52:371–395.

Jacobson, L. O., E. L. Simmons, E. K. Marks, E. O. Gaston, M. J. Robson, and J. H. Eldredge. 1951. Further studies on recovery from radiation injury. J. Lab. Clin. Med. 37:683–697.

Jolly, J. 1924. Actions des rayons X sur les cellules. Modifications de la radiosensibilité par ligature des connexions vasculaires. Compt. Rend. Soc. Biol. 91:351–354.

Kereiakes, J. G., W. H. Parr, and A. T. Krebs. 1957. Fractionated dose effects on survival and organ weights in X-irradiated mice. Amer. J. Physiol. 191:131–134.

Kohn, H. I., and R. F. Kallmann. 1956. The influence of strain on acute X-ray lethality in the mouse. I. LD_{50} and death rate studies. Radiat. Res. 5:309–317.

Kohn, H. I., and R. F. Kallman. 1957. The influence of strain on acute X-ray lethality in the mouse. II. Recovery rate studies. Radiat. Res. 6:329–338.

Lacassagne, A. 1942. Chute de la sensibilité aux rayons X chez la souris nouveau-née en état d'asphyxie. Compt. Rend. Acad. Sci. 215:231–232.

Lacassagne, A., and G. Gricouroff. 1958. Action of radiation on tissues. Grune & Stratton, New York. 199 p.

Langham, W. H., K. T. Woodward, S. M. Rothermel, P. S. Harris, C. C. Lushbaugh, and J. B. Storer. 1956. Studies of the effect of rapidly delivered, massive doses of gamma rays on mammals. Radiat. Res. 5:404–432.

Lawrence, E. O. 1937. The biological action of neutron rays. Radiology 29:313–322.

Lawrence, J. H., and R. Tennant. 1937. The comparative effects of neutrons and X-rays on the whole body. J. Exp. Med. 66:667–688.

Leone, C. A. [ed.] 1962. Effects of ionizing radiations on immune processes. Gordon and Breach, New York. 518 p.

Lindop, P. J., and J. Rotblat. 1959. Shortening of lifespan of mice as a function of age at irradiation. Gerontologia 3:122–127.

Lindsley, D. L., T. T. Odell, Jr., and F. G. Tousche. 1955. Implantation of functional erythropoietic elements following total-body irradiation. Proc. Soc. Exp. Biol. Med. 90:512–515.

Lorenz, E., D. Uphoff, T. R. Reid, and E. Shelton. 1951. Modification of irradiation injury in mice and guinea pigs by bone marrow injection. J. Nat. Cancer Inst. 12:197–201.

Loutit, J. F. 1962. Irradiation of mice and men. The University of Chicago Press, Chicago. 154 p.

Lushbaugh, C. C. 1957. Vertebrate radiobiology (the pathology of radiation exposure). Annu. Rev. Nucl. Sci. 7:163–184.

Main, J. M., and R. T. Prehn. 1955. Successful skin homografts after the administration of high dosage X irradiation and homologous bone marrow. J. Nat. Cancer Inst. 15:1023–1030.

Miller, C. P. 1956. The effect of irradiation on natural resistance to infection. Ann. N.Y. Acad. Sci. 66:280–291.

Mitchell, J. S., B. E. Holmes, and C. L. Smith [ed.] 1956. Progress in radiobiology. Charles C. Thomas, Springfield, Ill. 557 p.

Mole, R. H. 1959. Some aspects of mammalian radiobiology. Radiat. Res. Suppl. 1:124–148.

Mottram, J. C. 1924. On the skin reactions to radiation exposure and their avoidance in therapy; an experimental investigation. Brit. J. Radiol. 29:174–180.

Murphy, J. B. 1914. Heteroplastic tissue grafting affected through Roentgen-ray lymphoid destruction. J. Amer. Med. Ass. 62:1459. Cited by Murphy (1926).

Murphy, J. B. 1926. The lymphocyte in resistance to tissue grafting, malignant disease, and tuberculous infection. Monogr. Rockefeller Inst. Med. Res. No. 21. 168 p.

Nickson, J. J. [ed.] 1952. Symposium on radiobiology: the basic aspects of radiation effects on living systems. Wiley, New York. 465 p.

Patt, H. M., E. B. Tyree, R. L. Straube, and D. E. Smith. 1949. Cysteine protection against X irradiation. Science 110:213–214.

Quastler, H. 1956. The nature of intestinal radiation death. Radiat. Res. 4:303–320.

Quastler, H., M. K. Austin, and M. Miller. 1956. Oral radiation death. Radiat. Res. 5:338–353.

Quastler, H., E. F. Lanzl, M. E. Keller, and S. W. Osborne. 1951. Acute intestinal radiation death. Amer. J. Physiol. 164:546–556.

Rajewsky, B. 1955. Radiation death in mammals, p. 81–92. *In* Z. M. Bacq and P. Alexander [ed.] Radiobiology symposium 1954 (Liege). Academic Press, New York.

Raventos, A. 1955. A factor influencing the significance of radiation mortality experiments. Brit. J. Radiol. 28:410–414.

Reinhard, M. C., E. A. Morand, H. L. Galtz, and J. G. Hoffman. 1954. Mouse-strain differences in response to radiation. Proc. Soc. Exp. Biol. Med. 85:367–370.

Rekers, P. E., M. P. Coulter, and S. L. Warren. 1950. Effect of transplantation of bone marrow into irradiated animals. Arch. Surg. 60:635–637.

Riley, E. F., P. S. Leinfelder, T. C. Evans, and P. B. Rhody. 1955. Relative cataractogenic effectiveness of fast neutron radiation from different sources. Radiat. Res. 3:342.

Roderick, T. H. 1963*a*. The response of twenty-seven inbred strains of mice to daily doses of whole-body X-irradiation. Radiat. Res. 20:631–639.

Roderick, T. H. 1963*b*. Selection for radiation resistance in mice. Genetics 48:205–216.

Russell, E. S., E. F. Neufeld, and C. T. Higgins. 1951. Comparison of normal blood picture of young adults from 18 inbred strains of mice. Proc. Soc. Exp. Biol. Med. 78:761–766.

Sacher, G. A. 1955. A comparative analysis of radiation lethality in mammals exposed at constant average intensity for the duration of life. J. Nat. Cancer Inst. 15:1125–1144.

Sacher, G. A. 1956. On the statistical nature of mortality, with especial reference to chronic radiation mortality. Radiology 67:250–257.

Sacher, G. A. 1957. Dependence of acute radiosensitivity on age in adult female mouse. Science 125:1039–1040.

Sacher, G. A. 1958. Reparable and irreparable injury: a survey of the position in experiment and theory, p. 283–313. *In* W. D. Claus [ed.] Radiation biology and medicine. Addison-Wesley, Reading, Mass.

Sacher, G. A., and D. Grahn. 1964. Survival of mice under duration-of-life exposure to gamma rays. I. The dosage-survival relation and the lethality function. J. Nat. Cancer Inst. 32:277–321.

Spalding, J. F., and T. T. Trujillo. 1962. Radiosensitivity of mice as a function of age. Radiat. Res. 16:125–129.

Stadler, J., and J. W. Gowen. 1957. Contributions to survival made by body cells of genetically differentiated strains of mice following X-irradiations. Biol. Bull. 112:400–421.

Stearner, S. P., and S. A. Tyler. 1963. Radiation mortality in the mouse: model of the kinetics of injury accumulation. Radiat. Res. 20:619–630.

Stone, R. S. 1948. Neutron therapy and specific ionization. Amer. J. Roentgenol. 59:771–785.

Storer, J. B. 1959. Rate of recovery from radiation damage and its possible relationship to life shortening in mice. Radiat. Res. 10:180–196.

Storer, J. B., and J. M. Coon. 1950. Protective effect of para-aminopropiophenone against lethal doses of X-radiation. Proc. Soc. Exp. Biol. Med. 74:202–204.

Storer, J. B., and D. Grahn. 1960. Vertebrate radiobiology: late effects. Annu. Rev. Nucl. Sci. 10:561–582.

Storer, J. B., P. S. Harris, J. E. Furchner, and W. H. Langham. 1957. The relative biological effectiveness of various ionizing radiations in mammalian systems. Radiat. Res. 6:188–288.

Storer, J. B., and L. H. Hempelmann. 1952. Hypothermia and increased survival rate of infant mice irradiated with X-rays. Amer. J. Physiol. 171:341–348.

Taliaferro, W. M. 1957. Modification of the immune response by radiation and cortisone. Ann. N.Y. Acad. Sci. 69:745–764.

Talmage, D. W. 1955. Effect on ionizing radiation on resistance and infection. Annu. Rev. Microbiol. 9:335–346.

Thomson, J. F. 1962. Radiation protection in mammals. Reinhold, New York. 212 p.

Tyler, S. A., and S. P. Stearner. 1964. Accumulation of acute injury in the mouse subjected to split or fractionated dose of X-radiation or gamma radiation. Radiat. Res. 21:16–25.

United Nations. 1958. Biological effects of radiation. Proc. 2nd Int. Conf. Peaceful Uses Atomic Energy (Geneva). Vol. 22. United Nations, Geneva. 552 p.

United Nations. 1962. Report of the Scientific Committee on the Effects of Atomic Radiation. General Assembly. Official Records: 17th Session, Suppl. No. 16 (A/5216) United Nations, New York. 442 p.

Upton, A. C., F. P. Conte, G. S. Hurst, and W. A. Mills. 1955. The relative biological effectiveness of fast neutrons, X-rays and γ-rays for acute lethality in mice. Radiat. Res. 3:355. (Abstr.)

Vogel, H. H., Jr., J. W. Clark, and D. L. Jordan. 1954. The relative biological effectiveness of fast neutrons and cobalt^{60} γ-radiation. Radiat. Res. 1:233. (Abstr.)

Vogel, H. H., Jr., J. W. Clark, and D. L. Jordan. 1957*a*. Comparative mortality after 24 hour, whole-body, exposures of mice to fission neutrons and cobalt-60 gamma-rays. Radiat. Res. 6:460–468.

Vogel, H. H., Jr., J. W. Clark, and D. L. Jordan. 1957*b*. "Recovery" after fission neutrons: paired equal doses with varying time intervals. Radiat. Res. 7:458–459.

Vogel, H. H., Jr., D. L. Jordan, and S. Lesher. 1959. Injury and recovery in neutron irradiated animals, p. 221–235. *In* J. H. Martin [ed.] Radiation Biology, Proc. 2nd Australasian Conf. Radiation Biol. (Melbourne, 1958). Butterworth, London.

Wilson, N. R. 1963. Survival studies of whole-body X-irradiated germfree (axenic) mice. Radiat. Res. 20:477–483.

23

Responses to Drugs[1]

Hans Meier and John L. Fuller

Mice of randombred stocks are widely used in the pharmaceutical industry for the testing of drugs. This chapter, however, concentrates on genetic variation in responses to drugs as observed in inbred mice, hybrids, and mutant stocks. Differences of this kind may affect drug assays and, more importantly, shed light on the mechanism of drug action. Some mutant stocks may be useful for assessing drugs for their curative, ameliorative, or prophylactic effects. A survey of this one species illustrates the role of genotype in determining responses to drugs and also exemplifies some of the unsolved problems of pharmacogenetics.

GENETICS OF RESPONSES TO DRUGS

The existence of strain differences in response to drugs has been amply documented. The mode of transmission of response patterns has been less investigated and few of the differences observed have been related to a single locus with major effects. The quantitative genetics of pharmacological reactions is necessarily complex because of the multiplicity of possible phenotypes for study (Green and Meier, 1965). One cannot simply assume that the intensity of all observed responses to a particular drug will be determined by the same hereditary makeup.

Most of the human pharmacogenetic literature (Kalow, 1962) concerns single-locus effects on sensitivity related to sex-linked differences in glucose-6-phosphate dehydrogenase (Price-Evans and Clarke, 1961). In contrast, studies of strain differences have predominated in the mouse, though a few single-locus systems have been reported. For example, Dagg et al. (1964) found single-locus control of the rate of metabolism of the pyrimidines, uracil, dihydrouracil, and thymine.

The analysis of the responses of inbred, hybrid, and randombred mice to insulin (Brown, 1961*a*) and to sodium pentobarbitone (Brown, 1961*b*) illustrates the quantitative approach to strain differences. A randombred strain, LAC, was generally less sensitive to insulin than were inbred strains. Hybrids between several of the inbred strains were compared with respect to ED_{50} and mean slope of the dose-response regression line (Table 23-1). It is apparent from the table that the outcomes of the two crosses, one of them made reciprocally, follow no simple rule. The hybrid may resemble one parent in ED_{50} and neither in average slope. Similar results were obtained with pentobarbitone sodium (Brown, 1961*b*). Brown states, "The ability to forecast the usefulness of any F_1 hybrid for a particular pharmacological assay will depend on prior knowledge of the response of its parents, but the confirmatory experimental results may be disappointing."

When sensitivity in assay can be correlated with other strain characteristics, the factors responsible for variation in pharmacological response may be indicated. There is a significant relation-

[1] The writing of this chapter was supported in part by Public Health Service Research Grants CA 04691 from the National Cancer Institute and MH 01775 from the National Institute of Mental Health.

Table 23-1. CONVULSIVE THRESHOLD OF INBRED AND HYBRID MICE TREATED WITH INSULIN.
(*After Brown,* 1961*a, with permission of the author and the Journal of Pharmacy and Pharmacology.*)

Parents						Hybrids	
Female			*Male*				
Stock	ED50[a]	Slope	Stock	ED50	Slope	ED50	Slope
A2G	953	1.929	CBA	1,125	0.905	1,190	1.815
CBA	1,125	0.905	A2G	953	1.929	1,100	0.985
A2G	885	1.962	DBA/1	875	2.842	890	1.283

[a] Milliunits/kg mouse.

ship between average body weight and ED$_{50}$ for insulin (Fig. 23-1*A*), but the correlation between weight and slope is less clear (Fig. 23-1*B*).

GENETICS AND BIOASSAYS

The choice of animals for bioassay procedures has been widely debated (Becker, 1962). A part of the argument has centered upon the relative merits of inbred mice and their F$_1$ hybrids. There is some evidence that hybrids are phenotypically more uniform than inbreds (Chap. 9). Thus, given equal sample size, one should expect greater reproducibility from hybrids. On the other hand, hybrids may possess better homeostatic mechanisms and thus be less affected by the administration of a drug or hormone (Chai, 1960). It is not unlikely that the desiderata of uniformity and sensitivity may be negatively associated. Thus the choice of animals for bioassay is a matter for empirical determination. Likewise one has a choice of statistics for judging the results of a bioassay. Chai (1960) has used $\lambda = s/b$ as a measure of efficiency, where s is the standard deviation within groups and b is the regression coefficient of the dose-response curve. He found no superiority of hybrids over inbreds in a number of hormone assays. Becker (1962) compared drug responses by an F-statistic, related to λ, based on analysis of variance.

Apparently, then, there is no general rule for selecting the most sensitive assay animal in advance. Randombred mice are more widely used than either inbreds or hybrids in commercial assay work, where not only the sensitivity but the cost of subjects is important. Careful cost studies

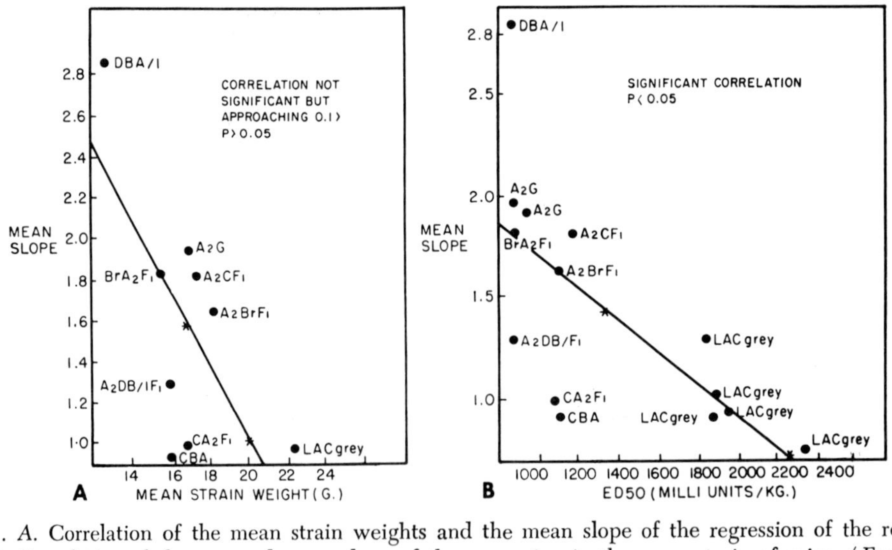

Fig. 23-1. *A*. Correlation of the mean strain weights and the mean slope of the regression of the response to insulin. *B*. Correlation of the ED$_{50}$ and mean slope of the regression in the same strain of mice. (*From Brown,* 1961*a, reproduced by permission of the author and the Journal of Pharmacy and Pharmacology.*)

might show that the greater uniformity of inbred and F_1 hybrid mice would more than compensate for the added expense of their production. Becker and Chai have both suggested that stocks of mice selected on the basis of specific drug sensitivity might be valuable for assay purposes. At least one such strain—a stock of Swiss mice with a high incidence of audiogenic seizures—has proved valuable in testing anticonvulsant and tranquilizing drugs (Swinyard et al., 1963). The further development of such stocks is a matter of economics. It is almost certain that selection can produce strains with increased or decreased specific drug sensitivities as needed.

It is sometimes argued that heterogeneous mice are more desirable for toxicity testing, since the range of sensitivities provides a better chance of detecting idiosyncratic responses which may be serious. The merit of this point of view is questionable. The occurrence or nonoccurrence of exceptional toxic responses in heterogeneous mice is not conclusive with respect to idiosyncratic toxic responses in man. The advantages of genetic uniformity and the consequently increased reliability of toxicity measurement seem to be greater than the advantages of heterogeneity.

VARIATION IN PHARMACOLOGICAL REACTIONS

Variations in pharmacological response have been found using a wide variety of drugs and strains of mice. It seems likely that any broad survey of mouse strains will uncover significant differences. The following sections describe examples of differential responses grouped according to the pharmacological or chemical characteristics of the substances used. Additional cases of variation in response to chemical agents are found in Chap. 9, 32, and 33.

Central nervous system stimulants

The need and importance of specifying the strain of mice used in studies on drug effects is emphasized by Weaver and Kerley's (1962) investigations of the response of several strains to *d*-amphetamine. Confirming the reports of other investigators Weaver and Kerley showed that aggregated mice are more susceptible than isolated mice to the lethal effects of amphetamine.

Mice tested were the Swiss-Webster, C57BL/6, DBA/2 strains and B6D2F_1 hybrids. The median lethal dose (LD_{50}) of amphetamine was different in each. Aggregated Swiss-Webster mice were five- to 10-fold more susceptible than isolated mice, but no evidence for increased lethality was found in aggregated C57BL/6 mice. Piperacetazine (2-acetyl-10-(3-4-beta-hydroxyethyl piperidino) phenothiazine), phenobarbital, and metaglycodol (2-m-chlorophenyl-3-methyl-2, 3-butanediol) reduced the lethal effects of amphetamine in aggregated Swiss-Webster mice and not in aggregated B6D2F_1 mice. Also, B6D2F_1 mice appeared more susceptible than Swiss-Webster mice to metaglycodol, piperacetazine, and the convulsant effects of pentylenetetrazol, but less susceptible to strychnine.

Insulin

Chase et al. (1948) first reported a high tolerance of mice to insulin; mice of the KL strain survive about 300 times the dose lethal for LT/Ch mice. Apparently the cause of the difference is the presence of an insulin-destroying enzyme "insulinase" in the livers of KL mice which protects them from extreme hypoglycemia (Beyer, 1955). Quantitative differences in insulin tolerance are also found in other strains (Brown, 1961*a*). Among nine strains tested C57BR/cd was by far the most sensitive (650 milliunits/kg); DBA/2, BALB/c, and A2G were alike in sensitivity (900 milliunits/kg). Sensitivity was measured by the approximate ED_{50} convulsive responses.

Serum from obese mice (*ob/ob*) has greater insulin-like activity than serum from normal littermates, yet the glucose tolerance of obese mice is normal (Mayer et al., 1953). In contrast, NZO (New Zealand Obese) mice reveal a typical diabetic type of glucose intolerance in about two-thirds of the individuals (Crofford and Davis, 1963). The mice having normal glucose tolerance tests were consistently normal; those that showed impaired glucose tolerance showed a consistent impairment. The average blood sugar concentration of NZO mice is 22mg/100 ml higher than that in mice of the control strain. Insulin sensitivity tests revealed them to be four to five times more resistant to insulin-induced convulsions than the controls.

Hormones other than insulin

Clear evidence for a genetic basis of hormone response in mice has been presented by Chai (1960) and by others (Chap. 20). The response of C57BL/6J, C57BR/cdJ, and their F_1 hybrids (B6BRF_1) to androgen (testosterone proprionate) was measured by the weights of seminal vesicles in 4-week-old castrated males. There appeared to be no difference in the slope of the three dose-response curves. Results of chorionic gonadotropin assays in 1-month-old females of strains DBA/1J, DBA/2J, and A/J were as follows: the regression of the dose-response curve was largest for BALB/cJ and least in A/J; that of the CAF_1 hybrid fell between that of the parental strains, but close to the A. The difference in slope between the two DBA strains, though not large, seemed to show genetic divergence as a consequence of separation for more than 20 generations.

The thyroid provides an additional example of heritable variations in hormonal activity. Iodine metabolism differs between inbred strains of mice and their F_1 hybrids as determined by several physiological measures (Chai, 1958; Amin et al., 1957).

In the course of the development of an assay for adrenal cortical hormones it was noted that adrenalectomized mice (maintained on salt solution) responded to epinephrine in two different ways, some showing a decrease in eosinophils and others an increase during a period of 3 to 4 hours following subcutaneous injection (Speirs and Meyer, 1949). Those that responded with eosinophilia turned out to have accessory adrenals and could survive after removal of the extra salt. After one dose of epinephrine the eosinopenic response became refractory to ACTH, histamine, etc., but not to 11-oxy-corticosteroids. Thus Speirs (1953) used epinephrine pretreatment to prepare mice as assay subjects for cortisone. It is well known that accessory adrenal cortical tissue is present in many mammals. Although knowledge of presence or absence, incidence, etc. of accessory nodules should be required by those using animals for experiments involving adrenalectomy, pertinent information is available only for inbred mice and certain of their hybrids (Hummel, 1958; Chap. 13).

The finding that growth rate, protein turnover, and nitrogen requirement are heritable suggests that strains of mice differ in their ability to secrete pituitary growth hormone or in their sensitivity to this hormone (Russell and Bloom, 1956). There is no direct chemical method for the determination of growth hormone, but cardiac glycogen is, within limits, an indicator of growth hormone activity (Russell and Bloom, 1956). At room temperature the cardiac glycogen of food-deprived mice was lowered. Large doses of growth hormone were required to maintain a normal level in fasting mice. At 30°C the cardiac glycogen of A strain mice increases during a 24-hour fast and administration of growth hormone increases the glycogen level still further. At this elevated temperature fasting lowers the cardiac glycogen of the I strain mice, and growth hormone raises the glycogen above the level of fed animals. It was suggested that a fasting stimulus, requiring the mediation of the hypophysis, evokes a greater secretion of growth hormone in A strain mice (Adrouny and Russell, 1956) or that the tissues of these mice are more sensitive to the hormone (Fenton and Duguid, 1962).

The availability of hairless mice has recently given impetus to cosmetological hormone research. Two types of hairless mice, one suitable for study up to the age of about 4 months and the other suitable up to about 2 months of age were used by Homburger et al. (1961). Several materials successfully smoothed the skin by hydration and densification of dermal connective tissue. Among them were the estrogens, estrogen and progesterone, ethisterone, and pregnenolone. Testosterone made the skin look rougher. Microscopically, the beneficial effects consisted of densification and hydration of the dermis with widening of its papillae and consequently stretching and flattening of the skin folds. Interestingly, a commercial cream base produced a desirable effect through hydration of epidermal cells. When hormonal ingredients were given in solvents rather than cream base there was complete absence of epidermal swelling.

Clearly then, this study not only holds considerable promise for assay studies, but also has opened up avenues for more basic investigations in cosmetic research dealing with mechanism of action rather than simply clinical effectiveness of compounds.

Psychotropic drugs

After treatment with iproniazid, high mortality and hepatic injury occurred in AKR mice, whereas at the same dosage there was no evidence of toxicity in C57BL/6 and DBA/2 mice (Rosen, 1951). Mice carrying the dilute genes (d, d^l), such as the P strain and the sublines of DBA (DBA/1 and DBA/2), are both phenylketonuric and subject to audiogenic seizures (Chap. 19). The alleles d and d^l have a depressant effect upon phenylalanine hydroxylase activity; d causes about a 50 per cent and d^l about an 85 per cent inhibition (Coleman, 1960). Meier (1963*a*, *b*) reported large differences in iproniazid toxicity between mice of the following genotypes, $+/+$, $d^l/+$, and d^l/d^l. For example, after treatment with iproniazid phosphate LD_{50}'s were 1,010, 976, and 715 mg per kilogram of body weight, respectively, for the three genotypes. Iproniazid did not prevent convulsions, but enhanced them both in rapidity of onset and severity. In contrast dl-α-ethyl-tryptamine (Monase) prevented the seizures. The cause of the failure of iproniazid is still hypothetical but the possibility exists that it may elevate brain catecholamines rather than serotonin. It may be pertinent that medmain, an antiserotonin compound, has been shown to induce seizures in mice (Woolley, 1959).

Protection against sound-induced convulsions (audiogenic seizures) was obtained in noninbred Swiss mice by various phenothiazine ataractics (Plotnikoff and Green, 1957; Plotnikoff, 1958, 1960), whereas chlorpromazine was entirely inactive in an inbred strain of Swiss mice (Plotnikoff, 1960). Changes in drug resistance were observed during the course of inbreeding (Plotnikoff, 1961). Chlorpromazine exerted diminishing protection from parental to each succeeding generation; a similar trend was observed with other phenothiazines, promazine, perphenazine, prochlorperazine, and trifluoperazine. In contrast, sodium phenobarbital uniformly gave protection against convulsions in all inbred generations tested. The lowered response to protective effects of chlorpromazine in each succeeding generation was very probably the result of selection or to chance fixation of recessives during the inbreeding.

Considerable variation in sensitivity to chlorpromazine has been found among the widely used inbred strains. In one study subjects were adult mice of both sexes from C57BL/6J, DBA/2J, A/HeJ, C3HeB/FeJ, and all their possible hybrids (Huff, 1962). Differential drug sensitivities were revealed by use of tests of activity after administration of chlorpromazine, 1 to 4 mg/kg intraperitoneally. With 4 mg/kg, 90 per cent of strain C57BL/6J mice, but only 7 per cent of strain C3HeB/FeJ were inactive in an open-field test. A/HeJ and DBA/2J mice showed complete depression in 62 per cent and 67 per cent of the individuals, respectively. Results from study of hybrids indicate that a simple genetic mechanism, possibly involving no more than two loci, could be responsible. Weight differences were not correlated with differential suppression of activity nor was the suppression of exploratory activity correlated with appearance of ataxia (Huff, 1962).

A mutant with spastic symptoms (gene symbol *spa*) that lends itself for use in assessment of sedative drugs was discovered by Chai (1961). Chai et al. (1962) found that aminooxyacetic acid abolished the symptoms whereas dilantin was only moderately effective. Hydroxylamine and trimethadione were totally ineffective.

Catecholamines

Few data are available concerning the normal organ content of adrenaline and noradrenaline in mice. Owing to fundamental differences in the techniques of measurement, comparison of the results of different studies is extremely difficult. However, De Schaepdryver and Preziosi's (1959) investigation of the effect of drugs on the adrenaline and noradrenaline content of adrenal glands, heart, liver, and spleen of normal adult white inbred mice will be mentioned. In some extracts of heart and spleen no adrenaline could be detected, suggesting that it may not be a regular constituent of these organs in mice and that its occurrence may depend on the presence of chromaffin cell groups irregularly scattered throughout the organism. Reserpine, insulin, nicotine, and histamine (in order of decreasing activity) lower the catecholamine content of the mammalian adrenal gland. After depletion of adrenaline and noradrenaline the time required for restoration of the catecholamines may be remarkably long. The rate of resynthesis apparently depends on the substance used and not on the degree of depletion. Of the two reserpine-like drugs used in this study only the one with sedative action, desmethoxyreserpine, caused complete depletion of noradrenaline in the adrenal gland. The amount of this catecholamine was only transiently lowered after reserpiline, which is supposedly devoid of tranquilizing properties. Sedative doses of chlorpromazine, mepazine, perphenazine, and promazine provoked only a slight depression of adrenal catecholamines, as did sedative doses of meprobamate. This implies that sedation can occur without depletion of adrenal noradrenaline. Furthermore, the data presented proof of the preferential depletion of adrenaline and noradrenaline after pharmacological stimulation. Iproniazid, which largely prevents catecholamine depletion after reserpine by blockade of monoamine oxidases, also protected against adrenal gland catechol depletion provoked by other pharmacological agents.

Lack or decreased amounts of catecholamines in adipose tissue from genetically obese mice may be responsible for one of their metabolic defects, decreased lipolysis, since mobilization of free fatty acids from adipose tissue is regulated in part by catecholamines and ACTH. However, although fat depots of normal mice have been measured, there are as yet no measurements for obese mice. Sidman et al. (1962) found that noradrenaline is present in the epididymal white fat and especially in the interscapular brown fat of the mouse. They reported values of 0.05 and 0.49 μg per gram of wet tissue, respectively. Adrenaline levels were only about 10 per cent of those obtained for noradrenaline.

Determination of adrenal epinephrine and norepinephrine levels in normal and dilute-lethal (d^l/d^l) mice disclosed differences with respect to both total amounts and rate of increase. During the period from birth to 3 weeks of age there is a rapid rise of both amines, but at a greater rate and to higher levels in the mutant. By 21 days dilute-lethal mice have about 25 per cent more epinephrine and over twice as much norepinephrine as do normal mice (Doolittle and Rauch 1965).

Antioxidants

Antioxidants have been found to prolong lifespan of mice. The half-survival time of C3H mice was prolonged by 2-mercapto-ethylamine hydrochloride (1 per cent by weight, incorporated into pellets) from 14.5 to 18.3 months, an increase of 26 per cent, while hydroxylamine hydrochloride (1 per cent by weight) produced a slight prolongation, 7 per cent. Cysteine hydrochloride (1 per cent by weight) and hydroxylamine hydrochloride (2 per cent by weight) increased the half-survival time of AKR male mice from 9.6 to 11.0 and 11.2 months, respectively, an average prolongation of about 15 per cent. Ascorbic acid (2.0 per cent) and 2-mercapto-ethanol (0.5 per cent) had no significant effect (Harman, 1961).

None of the antioxidants studied, 2-mercapto-ethylamine hydrochloride (1 per cent by weight), 2,2′ diaminodiethyl disulfide (1 per cent by weight), and hydroxylamine hydrochloride (1 and 2 per cent by weight), prolonged the life of Swiss male mice. Hydroxylamine hydrochloride (1 and 2 per cent by weight) produced a marked decrease in the tumor incidence of C3H female mice. This latter finding suggests the possibility of prophylactic cancer chemotherapy by this or other anticancer agents. Similarly, encouraging results with reducing agents against Ehrlich's ascites tumor have been reported. In addition to the compounds already mentioned, N-methyl-formamide and potassium arsenite were found to produce tumor inhibition. The use of antioxidants tested the hypothesis that endogenously produced free radicals such as HO· and HO_2·, contribute both to aging (Harman, 1956*a*, *b*, 1957) and to the incidence of spontaneous tumors (Harman, 1961), since reducing substances could act as free radical inhibitors.

Purines and pyrimidines

Investigations comparing the susceptibilities of various strains of inbred mice to teratogenic stimuli are numerous (Chap. 14). For example, 12-day embryos of two strains, BALB/c and 129, were approximately equally sensitive to 5-fluorouracil (30 to 40 mg/kg). However, strain 129 was more sensitive at all ages to a dose of 20 mg/kg. At this dosage effective teratogenic concentrations of fluorouracil apparently do not reach the BALB/c embryos or, alternatively, these embryos readily inactivate the drug. If the blood concentration is the critical factor, then the resistance of the BALB/c mice may be due to a slow rate of uptake from the peritoneal fluid or to rapid rates of catabolism or excretion.

Purines and pyrimidines apparently influence lipolysis. Dole (1961) found that they (as well as caffeine and pyrophosphate) increased the lipolytic action of epinephrine, adrenocortical-stimulating hormone (ACTH), thyroid-stimulating hormone (TSH), and glucagon. Indeed obese mice treated with a combination of purines, caffeine, and epinephrine gained less weight than untreated controls (Meier, 1963*a*, *b*).

Carcinogens

Urethan (ethyl carbamate) has been reported to augment the induction of lymphoid leukemias by X-rays, estrogen, and cholanthrene (Doell and Carnes, 1962). Since urethan alone did not augment the incidence of tumors in low-leukemia strains, it may be classified as a co-leukemogen. The drug has been found not to affect the incidence of leukemia in the high-leukemia strains, AKR and C58. In Swiss albino mice injected at birth or given toxic doses in the drinking water as adults, urethan augmented the leukemia incidence and shortened the latent period. Later it was shown that administration of urethan to newborn C57BL/6 mice induced thymic lymphomas with a high frequency approached only by that following divided doses of whole-body irradiation. Urethan also induces pulmonary tumors in certain strains of mice, among them strain BALB/c. Reciprocal lung grafts from BALB/c and DBA (1 day old) to their F_1 hybrids treated with urethan revealed that susceptibility to the carcinogenic action of urethan is determined by the intrinsic properties of the lung graft rather than the host (Shapiro and Kirschbaum, 1951). Young rapidly growing Swiss mice are more sensitive to the tumorigenic effects of urethan than mice just arriving at maturity. It has been suggested that urethan brings about the adenomatous state by acting upon the nucleus of alveolar lining cells (Rogers, 1951).

Considerable progress has been made in the study of the inheritance of spontaneous and induced experimental tumors. It has long been known that the inheritance of susceptibility to lung tumors is controlled by many genes. Tumors appear when the combined effects of genetic and nongenetic factors surpass a physiological threshold (Heston, 1942*a*, *b*). Chemical carcinogens are potent factors which greatly increase the probability of the occurrence of the tumors above that characteristic of a particular strain when untreated. Agents studied included dibenz (a, h) anthracene, 3-methylcholanthrene, urethan, nitrogen, sulfur mustard (Heston, 1950), and radiation (Lorenz et al., 1946). High concentration of inhaled oxygen has been found to increase the number of pulmonary tumors in the susceptible strain A mice injected with dibenz (a, h) anthracene over that in mice injected likewise but kept in air (reviewed by Heston, 1956). Bloom and Falconer (1964) confirmed the multifactoral inheritance of susceptibility to lung-tumor development in mice. However, they clearly demonstrated that about three-fourths of the difference in susceptibility between the C57BL/Fa and A/Fa strains is attributable to a single gene, *ptr* (pulmonary tumor resistance). The presence of a single recessive major gene conferring low susceptibility was established in analyses of all possible F_1 hybrids between six inbred strains, A/Fa, C57BL/Fa, RIII/Fa, JU/Fa, and KL. Strain C57BL which had the lowest susceptibility behaved differently from the other strains in crosses; it did not give F_1's with intermediate susceptibilities, but behaved as a "recessive" in all the crosses (Chap. 9).

Miscellaneous chemicals

Jay (1955) determined the mean sleeping time produced by hexobarbital (125 mg/kg) in 12 inbred strains. Values ranged from 18 minutes in SWR/HeN to over 48 minutes for A/LN. Ambrus et al. (1955) found that Swiss ICR mice are 6.6 times more sensitive to the effects of histamine than C3H/J. Strain and sex differences in response to serotonin (5-HT) were found for a number of strains. For example, a dose of 135 mg/kg of 5-HT creatinine sulfate intraperitoneally killed half of the animals in a group of C3H/HeJ in 24 minutes, whereas 98.5 mg/kg killed half of the C57BL/6J in 34 minutes; these animals had been adrenalectomized prior to the treatments (Meier, 1963*a*, *b*). Exposure of mice of several strains to minute amounts of chloroform in the air results in potentially fatal kidney lesions in males but less so in females. Death may follow exposure by as little as 1 hour, but there may be a delay of several weeks. In such cases the lesions are of the same type as those in early death. Greatest susceptibility to chloroform is found in strains C3H, C3Hf, A, HR, and DBA. Male mice of C57BL/6, C57BR/cd, C57L, and ST are resistant to amounts of chloroform that are lethal to the mice of the strains listed above (Deringer et al., 1953; Shubik and Ritchie, 1953).

SUMMARY

Responses to drugs are in part influenced by hereditary factors. In some of the examples of hereditary determination of reactions to drugs, the differences between strains are probably due to many pairs of genes; others are apparently due to single pairs of genes. The over 200 inbred strains of mice with the large number of known genes and the many mutant stocks provide suitable materials to study the genetic basis for differential drug responses.

LITERATURE CITED

Adrouny, G. A., and J. A. Russell. 1956. Effects of growth hormone and nutritional status on cardiac glycogen in the rat. Endocrinology 59: 241–251.

Ambrus, J. L., P. S. Guth, S. Goldstein, M. E. Goldberg, and J. W. E. Harrison. 1955. Toxicity of histamine and antagonism between histamine and anti-histamines in various strains of mice. Proc. Soc. Exp. Biol. Med. 88:457–459.

Amin, A., C. K. Chai, and E. P. Reineke. 1957. Differences in thyroid activity of several strains of mice and F_1 hybrids. Amer. J. Physiol. 191: 34–36.

Becker, W. A. 1962. Choice of animals and sensitivity of experiments. Nature 193:1264–1266.

Beyer, R. E. 1955. A study of insulin metabolism in an insulin tolerant strain of mice. Acta Endocrinol. 19:309–332.

Bloom, J. L., and D. S. Falconer. 1964. A gene with major effects on susceptibility to induced lung tumors in mice. J. Nat. Cancer Inst. 33: 607–618.

Brown, A. M. 1961*a*. The pattern, sensitivity and precision of the response to insulin in randombred, inbred, and hybrid strains of mice. J. Pharm. Pharmacol. 13:670–678.

Brown, A. M. 1961*b*. Sleeping time responses of mice—randombred, inbred and F_1 hybrids—to pentobarbitone sodium. J. Pharm. Pharmacol. 13:679–687.

Chai, C. K. 1958. Endocrine variation, thyroid function in inbred and F_1-hybrid mice. J. Heredity 49:143–148.

Chai, C. K. 1960. Response of inbred and F_1 hybrid mice to hormone. Nature 185:514–518.

Chai, C. K. 1961. Hereditary spasticity in mice. J. Heredity 52:241–243.

Chai, C. K., E. Roberts, and R. L. Sidman. 1962. Influence of aminooxyacetic acid, alpha-aminobutyrate transaminase inhibitor, on hereditary spastic defects in the mouse. Proc. Soc. Exp. Biol. Med. 109:491–495.

Chase, H. B., M. S. Gunther, J. Miller, and D. Wolffson. 1948. High insulin tolerance in an inbred strain of mice. Science 107:297–299.

Coleman, D. L. 1960. Phenylalanine hydroxylase activity in dilute and non-dilute strains of mice. Arch. Biochem. Biophys. 91:300–306.

Crofford, O. B., and C. K. Davis, Jr. 1963. Glucose intolerance, insulin resistance and growth characteristics of New Zealand Obese mice. Fed. Proc. 22:387. (Abstr.)

Dagg, C. P., D. L. Coleman, and G. M. Fraser. 1964. A gene affecting the rate of pyrimidine degradation in mice. Genetics 49:979–989.

Deringer, M. K., T. B. Dunn, and W. E. Heston. 1953. Results of exposure of strain C3H mice to chloroform. Proc. Soc. Exp. Biol. Med. 83: 474–479.

De Schaepdryver, A. F., and P. Preziosi. 1959. Pharmacological depletion of adrenaline and noradrenaline in various organs of mice. Arch. Int. Pharmacodyn. 111:177–221.

Doell, R. G., and W. H. Carnes. 1962. Urethan induction of thymic lymphoma in C57BL mice. Nature 194:588–589.

Dole, V. P. 1961. Effect of nucleic acid metabolites on lipolysis in adipose tissue. J. Biol. Chem. 236:3125–3130.

Doolittle, C. H., and H. Rauch. 1965. Epinephrin and norepinephrin levels in dilute-lethal mice. Biochem. Biophys. Res. Commun. 18:43–47.

Fenton, P. F., and T. R. Duguid, Jr. 1962. Growth hormone and cardiac glycogen: influence of environmental and genetic factors. Can. J. Biochem. Physiol. 40:337–341.

Green, E. L., and H. Meier. 1965. Use of laboratory animals for the analysis of genetic influences upon drug toxicity. Ann. N.Y. Acad. Sci. 123:295–304.

Harman, D. 1956*a*. Reducing agents as chemotherapeutic agents in cancer. Clin. Res. 4:54–55.

Harman, D. 1956*b*. Aging: a theory based on free radical and radiation chemistry. J. Gerontol. 11:298–300.

Harman, D. 1957. Prolongation of the normal life span by radiation protection chemicals. J. Gerontol. 12:257–263.

Harman, D. 1961. Prolongation of the normal life span and inhibition of spontaneous cancer by antioxidants. J. Gerontol. 16:247–254.

Heston, W. E. 1942*a*. Genetic analysis of susceptibility to induced pulmonary tumors in mice. J. Nat. Cancer Inst. 3:69–78.

Heston, W. E. 1942*b*. Inheritance of susceptibility to spontaneous pulmonary tumors in mice. J. Nat. Cancer Inst. 3:79–82.

Heston, W. E. 1950. Carcinogenic action of the mustards. J. Nat. Cancer Inst. 11:415–423.

Heston, W. E. 1956. Effects of genes located in chromosomes III, V, VII, IX and XIV on the occurrence of pulmonary tumors in the mouse. Cytologia Suppl. p. 219–224.

Homburger, F., A. Treger, J. R. Baker, and C. M. Crooker. 1961. The use of hairless mice for study of cosmetics. Proc. Sci. Sec. Toilet Goods Assoc. 35:6–11.

Huff, S. D. 1962. A genetically controlled response to the drug chlorpromazine, p. 50. 34th Ann. Rep. 1962/63. The Jackson Laboratory, Bar Harbor, Maine.

Hummel, K. P. 1958. Accessory adrenal cortical nodules in the mouse. Anat. Rec. 132:281–291.

Jay, G. E., Jr. 1955. Variation in response of various mouse strains to hexobarbital (Evipal). Proc. Soc. Exp. Biol. Med. 90:378–380.

Kalow, W. 1962. Pharmacogenetics. Saunders, Philadelphia.

Lorenz, E., W. E. Heston, M. K. Deringer, and A. B. Eschenbrenner. 1946. Increase in incidence of lung tumors in strain A mice following long continued irradiation with gamma-rays. J. Nat. Cancer Inst. 6:349–353.

Mayer, J., R. E. Russell, M. W. Bates, and M. M. Dickie. 1953. Metabolic, nutritional and endocrine studies of the hereditary obesity—diabetes syndrome of mice and mechanism of its development. Metabolism 2:9–21.

Meier, H. 1963*a*. Experimental pharmacogenetics. Academic Press, New York, 213 p.

Meier, H. 1963*b*. Potentialities for and present status of pharmacological research in genetically controlled mice. Adv. Pharmacol. 2:161–209.

Plotnikoff, N. P. 1958. Bioassay of potential tranquilizers and sedative agents against audiogenic seizures in mice. Arch. Int. Pharmacodyn. 64:130–135.

Plotnikoff, N. P. 1960. Ataractics and strain differences in audiogenic seizures in mice. Pharmacologia 1:429–432.

Plotnikoff, N. P. 1961. Drug resistance due to inbreeding. Science 134:1881–1882.

Plotnikoff, N. P., and D. M. Green. 1957. Bioassay of potential ataraxic agents against audiogenic seizures in mice. J. Pharmacol. Exp. Therap. 119:294–298.

Price-Evans, D. A., and C. A. Clarke. 1961. Pharmacogenetics. Brit. Med. Bull. 17:234–240.

Rogers, S. 1951. Age of host and other factors affecting the production with urethan of pulmonary adenomas in mice. J. Exp. Med. 93: 427–499.

Rosen, F. 1951. The relationship of certain vitamin deficiencies to the toxicity of iproniazid. Ann. N.Y. Acad. Sci. 80:885–897.

Russell, J. A., and W. Bloom. 1956. Hormonal control of glycogen in the heart and other tissues in rats. Endocrinology 58:83–94.

Shapiro, J. R., and A. Kirschbaum. 1951. Intrinsic tissue response to induction of pulmonary tumors. Cancer Res. 11:644–647.

Shubik, P., and A. C. Ritchie. 1953. Sensitivity of male dba mice to the toxicity of chloroform as a laboratory hazard. Science 117:285.

Sidman, R. L., M. Perkins, and N. Winer. 1962. Noradrenaline and adrenaline content of adipose tissue. Nature 193:36–37.

Speirs, R. S. 1953. Eosinopenic activity of epinephrine in adrenalectomized mice. Amer. J. Physiol. 172:520–526.

Speirs, R. S., and R. K. Meyer. 1949. The effects of stress, adrenal and adrenocorticotropic hormones on the circulating eosinophils of mice. Endocrinology 45:403–429.

Swinyard, E. A., A. W. Castellion, G. B. Fink, and L. S. Goodman. 1963. Some neurophysiological and neuropharmacological characteristics of audiogenic-seizure-susceptible mice. J. Pharmacol. Exp. Therap. 140:375–384.

Weaver, L. C., and T. L. Kerley. 1962. Strain difference in response of mice to d-Amphetamine. J. Pharmacol. Exp. Therap. 135:240–244.

Woolley, D. W. 1959. Antimetabolites. Science 129:615–621.

24

Genetics of Tissue Transplantation[1]

George D. Snell and Jack H. Stimpfling

Intensive study of the phenomena associated with the acceptance and rejection of tissue transplants has led to the establishment of two principles. First, rejection depends on genetic disparity of donor and host. Second, the processes involved in rejection are immunological in nature. We shall be concerned here only with the genetic factors in transplant rejection. For information concerning the immunological phenomena the reader may consult Snell (1963), Billingham and Silvers (1963), or Russell and Monaco (1965).

DEFINITIONS

The vocabulary of transplantation includes certain terms that may not be generally familiar. The problem of unfamiliarity is compounded by a diversity of usages, including some that are notably inconsistent. To smooth the path of the reader we shall employ a vocabulary recently suggested (Snell, 1964*b*) which essentially eliminates the inconsistencies and contradictions. The key terms are defined herewith.

Histocompatibility is an adjective used to indicate relevance to the growth or failure to grow of tissue transplants. Thus *histocompatibility genes* are the genes that determine susceptibility and resistance to transplants.

Isogenic is an adjective used to indicate genetic identity of animals or tissues. The members of one inbred strain are isogenic. *Isogeneic* is a variant used, usually as a contrasting term to allogeneic, to indicate similarity of origin.

Coisogenic strains are strains genetically identical except for a difference at a single genetic locus. Since true coisogenicity is a theoretical ideal seldom if ever attained in practice, it is convenient to have a term applicable to those strains actually available which provide an imperfect but useful approximation to the coisogenic state. Such strains may be called *congenic*.

An *isogenic graft* or *isograft* is a graft between genetically identical individuals. Typically, isografts are either grafts between animals of a single highly inbred strain, between the F_1 hybrids produced by crossing inbred strains, or between identical twins.

An *allogeneic graft* or *allograft* is a graft between genetically disparate individuals of the same species. More specifically, it is a graft in which the grafted tissue carries a histocompatibility allele or alleles, and hence presumably an alloantigen or alloantigens, foreign to the recipient. Homograft is also used in this sense, but because its use has led also to the use of homologous with a meaning exactly opposite from the traditional one, it is not recommended.

An *alloantigen* or *isoantigen* is the heteromorphic product of a heteromorphic locus (a locus existing in two or more allelic forms), and a product such that alternative forms are antigenic in individuals lacking them. It incites an immune response when transferred *within* the species. Al-

[1] The writing of this chapter was supported in part by Public Health Service Research Grant CA 01329 from the National Cancer Institute and by a Public Health Service Research Career Award K6 CA 21784.

loantigen is a new and unfamiliar term in this context. We shall use it in preference to isoantigen because it leads to a more consistent vocabulary. The type of antigenic stimulus here referred to is provided by allografts, not by isografts; by grafts between individuals that are nonisogenic, not between individuals that are isogenic. It is therefore more appropriate to speak of the active substances as alloantigens than as isoantigens.

An *alloantibody* or *isoantibody* is the antibody incited by an alloantigen when transferred within the species. We shall again use alloantibody rather than isoantibody for the sake of consistency. It is allografts and not isografts that lead to the formation of alloantibody.

Codominant genes are genes whose end products are individually demonstrable in the heterozygote. The classical example is the A and B blood group alleles in man, where both the A and the B properties are demonstrable on heterozygous, AB red cells. Codominance should be distinguished from partial dominance, which is a blending condition leading to a heterozygote intermediate between the two homozygotes.

TYPES OF TRANSPLANTS

Both transplantable tumors and various normal tissues have been used in studies of the genetics of transplantation. Nearly all early work was done with tumors, and tumors are still the instrument of choice for many studies, but increasing use is being made of normal-tissue transplants.

The great advantage of tumors is the rapidity with which transplants can be performed. One experienced person can graft 500 or more animals in a day. If either leukemias, which are easily prepared as suspensions of single cells (Snell, 1953), or ascites tumors, which grow as suspensions of single cells in the fluid-filled peritoneal cavity, are used, an accurately measured dose of cells can be administered. In tests of histocompatibility, the tumors, whether of the ascites type or in the solid form, are typically implanted subcutaneously on the flank. Intramuscular implants have also been used. Death or survival of the host is taken as the end point.

If properly employed, tumors can be sensitive indicators of histocompatibility differences. Prior immunization of the recipient by one or more injections of tissue from the donor strain is frequently used to increase the reaction against the ultimate graft. Leukemias are particularly sensitive indicators of histocompatibility differences, probably because of their especial susceptibility to antibody. Tumors are also easily stored in the frozen state (see, e.g., Hauschka et al., 1959). On the other hand tumors, since they are living cell populations, may change with successive passages from host to host and hence lose desirable characteristics; they may become infected, either as the result of careless transplantation or because of an unsuspected infection in their host; and they may not be available in a particular inbred strain or hybrid which it is desired to use as a donor. (For further information on the characteristics of transplantable tumors, see Chap. 28 or Snell, 1959.)

The normal tissue most commonly used in studies of the genetics of histocompatibility is skin. Techniques for skin grafting in mice are well established (Billingham and Silvers, 1961), and this tissue is among the more sensitive indicators of histocompatibility differences. Transplants of ovaries into ovariectomized hosts (see, e.g., Linder, 1961) have also been used. With this tissue, growth or rejection may be determined either by the continuation of the estrus cycle, or, if ovaries are placed in the ovarian capsule, by the production of young. For purposes other than genetic studies, a great deal of use has been made of suspensions of bone marrow and the lymphoid tissues. Because the cells of such transplants become widely dispersed and are not easily distinguished from host cells, there is usually no clear end point of rejection. Cudkowicz (Cudkowicz and Stimpfling, 1964*a*), however, has developed an ingenious method of measuring the growth in irradiated hosts of transplanted marrow cells by the use of a radioactive DNA precursor. This method has proved suitable for genetic studies.

FIVE LAWS OF TRANSPLANTATION

Studies which established the basic principles of the genetics of tissue transplantation were started by Little and Tyzzer (1916) and continued by Little, Bittner, Cloudman, and Strong. All these studies were based on the use of inbred strains of mice. These early studies were summarized in some detail in the first edition of this book (Little, 1941). Because recent data are more broadly informative, most of the early data will be omitted here and attention concentrated instead on the conclusions drawn from them. It should also be noted that when we do discuss early data, designations of substrains will be omitted. The major substrains (e.g., DBA/1 and DBA/2) were either not established or not recognized at that time.

The results of Little and co-workers can be summarized as five laws or dicta of transplantation. There are exceptions to most of these laws, especially to the second and third. These will be discussed later. Despite the exceptions, the laws have a very general validity. The laws may be stated as follows:

1. Grafts within inbred strains are successful; or stated in more general terms, isografts succeed.

2. Grafts between inbred strains are not successful; or stated in more general terms, allografts fail.

The remaining laws apply to the results which are obtained when inbred strains are crossed to produce F_1, F_2, and backcross generations, and grafts exchanged between the inbred parents and animals of these generations or between animals of the hybrid generations themselves.

3. Grafts from either inbred parent strain to the F_1 hybrid succeed, but grafts in the reverse direction fail.

4. Grafts from F_2 or subsequent generations also grow in all F_1 mice.

5. Grafts from either inbred parent strain are accepted by some members of an F_2 generation, but rejected by others. Usually the rejections are much more frequent than the acceptances. The same is true of grafts made from one inbred parent strain to a backcross produced by crossing the F_1 to the opposite parent strain. The proportion of rejections in this generation is, for any one given cross, donor, and type of tissue, usually higher than in the F_2.

REPRESENTATIVE DATA

The key to the understanding of the genetics of tissue transplantation is law 5, which concerns the results of transplants made to F_2 and backcross generations. We shall, therefore, examine in some detail representative results obtained in these two generations.

Of the many tests made with transplantable tumors, two are summarized in Table 24-1. Cloudman (1932) crossed strains A and DBA, raised F_1, F_2, and backcross (BC) generations, and implanted them with a transplantable mammary tumor indigenous to strain A. This tumor grew in all animals of strain A but, except for occasional temporary growths, failed to grow in strain DBA. It grew in all of 92 F_1 mice and in all of 69 mice produced by crossing the F_1 to the susceptible (strain A) parent. It grew, however, in only 10 of 116 mice of the opposite backcross and 60 of 219 F_2 mice.

Amos et al. (1955*a*) ran a similar test using strains C57BL and BALB/c, and a C57BL transplantable leukemia. An interesting variation in their method was the use, in some of the mice, of an "immunizing" injection of C57BL blood 6 to 10 days prior to tumor implantation. When BC mice were unimmunized, 23 out of 28 succumbed (results not shown in table). When they were immunized, all of 39 survived. In the immunized F_2, 27 out of 29 survived.

Although there have been many experiments with mice in which tumors have been transplanted from an inbred strain to its F_1, F_2, and BC descendants, only two comparable experiments using the more tedious techniques of skin grafting have been reported. These are of considerable

Table 24-1. DEATHS PRODUCED BY TUMORS INDIGENOUS TO INBRED STRAINS WHEN TRANSPLANTED TO F_1, F_2, AND BC MICE

		Recipients			*Per cent mice dying*		
Cross	Tumor donor	Generation	Number	Number dying	Observed	Expected	No. loci assumed
A × DBA[a]	A	F_1	92	92	100	100	
		F_2	219	60	27.4	31.6	4
		BC	116	10	8.6	6.2	4
C57BL × BALB/*c*[b]	C57BL	F_2	29	2	6.9	7.5	9
		BC	39	0	0	0.2	9

[a] Cloudman (1932).

[b] Amos et al. (1955*a*). The F_2 and BC mice in this experiment were preimmunized with 0.1 ml of strain C57BL blood 6 to 10 days prior to implantation of a C57BL leukemia.

Table 24-2. SURVIVAL OF SKIN GRAFTS FROM INBRED MICE WHEN TRANSPLANTED TO F_1, F_2, AND BC GENERATIONS

Cross	Skin donor	Recipients: Generation	Recipients: No. of mice	No. of grafts surviving for various intervals: >60 days	>120 days	>200 days	Number expected (no. of loci assumed)
BALB/c × DBA/2[a]	BALB/c ♂♂	F_2 ♂♀	120	10	4	3	2.81 (13)
	BALB/c ♂♂	BC ♂♀	99	3	0		0.01 (13)
A × CBA[b]	A ♀♀	F_1	11			11	
	CBA ♂♂	F_1	10			10	
	A ♂♀	F_2 ♂♀	120	3	2	0	1.60 (15)
	CBA ♂♀	F_2 ♂♀	154	2	1	0	2.05 (15)

[a] Prehn and Main (1958).
[b] Barnes and Krohn (1957).

interest, and both are summarized in Table 24-2. In the first, Prehn and Main (1958) crossed strains BALB/c and DBA/2 and grafted BALB/c skin to F_2 and BC generations. Of 120 grafts made to the F_2, three were still healthy at 200 days and were regarded as definite acceptances. All other grafts were rejected, although a few persisted for 60 or more days before sloughing. Of 99 grafts to backcross mice, none survived permanently, though again there were a few late rejections. The other experiment was carried out by Barnes and Krohn (1957). These authors crossed strains A and CBA, and grafted skin from both parent strains to F_1 and F_2 mice. As expected, all grafts to the F_1 were accepted. About one-fourth of all grafts to the F_2 had been rejected by 12 or 15 days. The survival of these grafts was approximately that expected when grafts are exchanged between inbred strains. Many other grafts looked healthy at this time. Thereafter rejections of grafts continued to occur, and by 180 days all of 274 grafts, with one possible exception, had been rejected. For purposes of discussion, the authors treated three grafts (one of strain A skin and two of CBA skin) which showed "an autograft-like condition 100 days after grafting" as acceptances. It should be noted, however, that control autografts and isografts were still healthy after 6 months or a year.

These four experiments all showed a combination of rejections and acceptances (or at least very long survivals) in the segregating generations receiving grafts from the parent strains, but the proportion of acceptances varied considerably. Specifically, the percentage of acceptances in the four F_2's were 27.4, 6.9, 2.5, and 0 (or 1.7 if 100-day survival is taken as an acceptance), and for the corresponding backcross generations, 8.6, 0 (out of 39 mice), and 0 (out of 99 mice). The last experiment did not include a backcross. The consistently lower percentage of graft survival in the backcross is notable.

GENETIC THEORY OF TRANSPLANTATION

The genetic theory of transplantation in essentially its present form was adumbrated in an early paper by Little (1914) and was fully described in a number of subsequent papers (see, e.g., Little, 1941). Although a considerable wealth of detail has been added, and certain minor modifications have been made, the theory developed by Little and co-workers is still fully valid in its essentials.

The key to the genetics of susceptibility and resistance to allografts must be found in the F_1, F_2, and BC generations. The absence of a 3:1 ratio in the F_2 and of a 1:1 ratio in the backcross suggests multiple-factor rather than single-factor inheritance. In most multiple-factor inheritance the condition which is dominant in the F_1 tends to be frequently repeated in the F_2, yet the exact opposite occurs here. In the transplantation studies, susceptibility is dominant in F_1, but uncommon in F_2. The theory of transplantation, designed to accommodate these facts, may be stated as follows:

The growth of transplants is determined by multiple genes which may be called histocompatibility genes or by the abbreviated designation *H* genes. Permanent or progressive growth occurs if, and only if, all histocompatibility alleles present in the graft are also present in the host. Modern research provides an intelligible explanation of this rule in terms of the end products of histocompatibility genes. These may be presumed to be alloantigens which, when transferred to a host lacking them, have the special property of inciting an immune response. Each gene is presumed to determine a single alloantigen, a situation which has led to the dictum, "one gene, one antigen." The fact that alloantigens must be *foreign* to the recipient to evoke an immune response accounts for the *acceptance* of grafts which carry no histocompatibility alleles foreign to the host, and the *rejection* of grafts which do carry foreign alleles. Although this extension of the genetic theory of transplantation helps us to understand it and can be supported by numerous facts, it would take us too far afield to examine it in detail. We shall for the most part examine the theory merely as a formal exercise in genetics.

We shall represent the different histocompatibility loci by the symbols *H-1*, *H-2*, *H-3*, etc. We shall assume that histocompatibility genes are codominant (individually expressed in the heterozygote), and shall therefore use only the capital form for the initial letter, rather than the capital and small letters which are sometimes used to indicate the dominant and recessive states. Different alleles at one locus will be indicated by superscript small letters. Thus alleles at *H-1* are represented by $H\text{-}1^a$, $H\text{-}1^b$, $H\text{-}1^c$, etc., alleles at *H-2* by $H\text{-}2^a$, $H\text{-}2^b$, $H\text{-}2^c$, etc.

Suppose, then, that two strains, A and B, with a single histocompatibility difference are crossed.

Table 24-3. THE EXPECTED OUTCOME OF TRANSPLANTS MADE FROM STRAIN A TO STRAIN B, AND TO F_1, F_2, AND BC GENERATIONS PRODUCED BY CROSSING STRAIN A AND STRAIN B[a]

Generations	*Genotypes and outcomes of transplants*		
	Strain A		Strain B
P	$H\text{-}2^a/H\text{-}2^a$ donor	×	$H\text{-}2^b/H\text{-}2^b$ all −
F_1		$H\text{-}2^a/H\text{-}2^b$ all +	
F_2	$H\text{-}2^a/H\text{-}2^a$ 1/4 +	$H\text{-}2^a/H\text{-}2^b$ 2/4 +	$H\text{-}2^b/H\text{-}2^b$ 1/4 −
BC to strain B		$H\text{-}2^a/H\text{-}2^b$ 1/2 +	$H\text{-}2^b/\text{-}H2^b$ 1/2 −

[a] A plus (+) sign indicates graft acceptance, a minus (−) sign graft rejection.

Let us suppose that the difference is at the *H-2* locus. The parents, the different hybrid generations, and the expected outcome of transplants from the A parent may then be represented as shown in Table 24-3. The only animals represented in this table which lack the $H\text{-}2^a$ of the graft donor are the $H\text{-}2^b/H\text{-}2^b$ parent, and the mice of this same homozygous $H\text{-}2^b$ genotype which comprise one-fourth of the F_2 generation and one-half of the BC generation. These mice will resist the $H\text{-}2^a/H\text{-}2^a$ graft; all other mice will accept it. The expected results are:

F_1 all susceptible
F_2 $\frac{3}{4}$ susceptible
BC $\frac{1}{2}$ susceptible

The expected results in F_2 when the parents differ at two histocompatibility loci are shown in Fig. 24-1. The expected results for a backcross are similar, except that the ratio of the four classes is 1:1:1:1 instead of 9:3:3:1. The expected proportions of susceptible and resistant mice for two factors are thus:

F_1 all susceptible
F_2 $(\frac{3}{4})^2 = \frac{9}{16}$ susceptible
BC $(\frac{1}{2})^2 = \frac{1}{4}$ susceptible

Generalizing from these two examples, we may say that if inbred strains differing at n histocompatibility loci are crossed, the expected proportion of susceptible animals in subsequent generations will be:

F_1 all susceptible
F_2 $(\frac{3}{4})^n$ susceptible
BC $(\frac{1}{2})^n$ susceptible

To get a picture of what this means in terms of the actual proportion of susceptible mice for different numbers of histocompatibility factors, we may appropriately select those values of n which happen to give the best fit for the four crosses shown in Table 24-1 and 24-2. These values are, in the order in which the crosses appear in the tables, $n = 4$, $n = 9$, $n = 13$, and $n = 15$. The corresponding expected percentages of susceptible mice in F_2 and backcross generations are shown in Table 24-4.

It is apparent from these values that in a backcross, even with no more than nine histocompatibility loci segregating, large numbers of mice may have to be raised to get even a single mouse that will accept a graft from the parental strain. We refer here of course to the parental strain not used in producing the backcross. For nine factors,

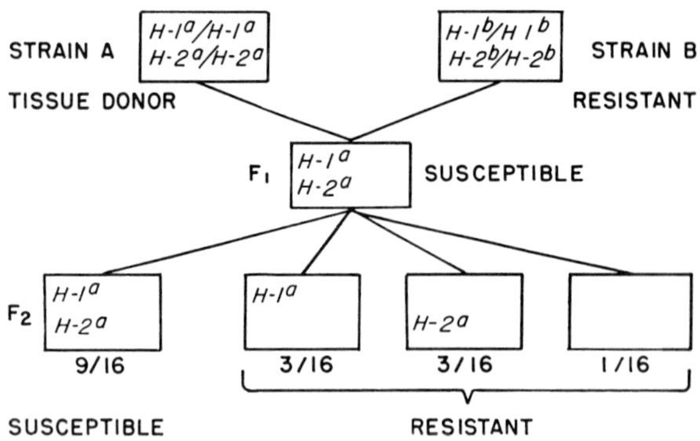

Fig. 24-1. Diagram showing the expected portion of susceptible and resistant mice in an F_2 generation when susceptibility and resistance are determined by two loci both showing codominance. Grafts from strain A will grow in those, and only those, recipients carrying both $H\text{-}1^a$ and $H\text{-}2^a$. In like manner, grafts from strain B require the presence of both $H\text{-}1^b$ and $H\text{-}2^b$.

in fact, only about one mouse in 500 will, on the average, accept a parental graft. For an F_2 with nine factors, approximately one mouse in 13 will accept a parental graft. It is not surprising therefore that in two of the three crosses in which both F_2 and BC generations were raised, there were positive grafts in the F_2 only.

The concordance, in any one individual experiment, between observed results for F_2 and BC generations and the predictions for these same generations provides an important test of the validity of the theory. An inspection of the tables will show that the concordance is satisfactory for the four crosses summarized. We may also say that it was satisfactory for most of the early experiments in which transplantable tumors were implanted in both F_2 and BC generations.

There is, however, a considerable lack of concordance when the four experiments cited are compared with respect to the indicated number of histocompatibility loci. The number ranges all the way from four to not fewer than 15. Sometimes the same cross challenged with different tumors has given widely differing ratios. This diversity of results requires explanation and raises questions concerning the validity of estimates as to the number of histocompatibility loci. The experiments cited have also shown curious differences in the survival of skin grafts in segregating generations. Even though many grafts were rejected within the first 2 weeks, others persisted for months before being sloughed. This raises interesting questions about the properties of the histocompatibility loci concerned. Moreover, a comparison of the results of this type of experiment with possible theories of graft rejection will show that the assumption of simple dominance is as admissible as the assumption of codominance. Dominance rather than codominance was, in fact, the assumption made in the original theory (Little, 1941). We must then ask whether the assumption of codominance is necessary. Although the facts so far presented can be taken as evidence that the genetic theory of transplantation as set forth above has essential validity, they leave a number of questions unanswered. We shall defer answers to these questions until additional evidence has been assembled.

Table 24-4. EXPECTED PERCENTAGE OF SUSCEPTIBLE MICE IN F_2 AND BC GENERATIONS ON THE ASSUMPTION THAT SUSCEPTIBILITY AND RESISTANCE IS DETERMINED BY 4, 9, 13, OR 15 HISTOCOMPATIBILITY LOCI

	Per cent susceptible	
n	F_2	BC
4	31.6	6.2
9	7.5	0.2
13	2.4	0.01
15	1.3	0.003

CONGENIC RESISTANT LINES

Grafts made to F_2 and backcross generations provide evidence for the existence of multiple histocompatibility loci, but do not establish the individual identity of the loci. They do not permit any separation of the loci one from another. Such separation has been accomplished by two methods, the first primarily serological, the second primarily genetic. Even though historically the serological method was the first to be employed, the one we shall examine first is the genetic.

Method of production

Coisogenic and congenic strains have already been defined. Congenic strains in which the significant difference is at a single histocompatibility locus can be produced by an appropriate series of crosses, with implantation of tumor to select resistant animals in every second or third generation. Such strains may be called *congenic resistant* or CR strains, because they resist grafts from their congenic partner. The theory of the production of such lines is discussed in Snell (1948), Snell and Bunker (1965), and Chap. 2. The cross-intercross system which has usually been employed is illustrated in Fig. 24-2. The end result of matings made according to this system is a line, A.B, or a group of lines A.B(1), A.B(2), A.B(3), etc., each of which may be presumed to differ from strain A at a single histocompatibility locus. Perfect coisogenicity is never attained (discussion in Chap. 2), but it is possible to attain an approximation thereto adequate for the purposes for which the lines are designed.

Symbols for designating congenic lines

Symbols for designating inbred strains of mice are discussed in Chap. 6. Because congenic strains of mice present certain special problems not covered by the usual nomenclature rules, some further details are given here.

Congenic resistant (CR) strains are usually designated by a symbol consisting of the symbols of the two strains used in the initial cross, separated by a period. The strain to which all subsequent crosses are made, and which therefore provides the congenic partner to the CR line, appears first. Symbols are usually abbreviated, using standard abbreviations where these are available. Thus a line congenic with C57BL/10 in which the gene for resistance came from BALB/c would be called B10.C. Where several lines are derived from the same cross, these are distinguished by appending a number, or a number and letter, in parentheses. We have used the letter M to mean production by the cross-intercross system, N to mean production by the cross-backcross-intercross system, and NX to mean production by backcrossing a gene with a visible dominant effect, and known to be linked with a histocompatibility gene, onto an inbred background. Thus B10.C(41N) and B10.C(47N) are CR lines produced by cross-backcross-intercross matings in which a histocompatibility allele from BALB/c has been introduced onto a C57BL/10ScSn background. Once the foreign allele in the CR line has been identified, an alternative symbol in the form C57BL/10-H-7^b (= B10.C(47N)), in accordance with rule 5 of Strain Rules, Chap. 6, becomes permissible.

Analysis by linkage

Once a group of lines on one or more inbred backgrounds is established, the problem remains of identifying the histocompatibility locus by which each is distinguished. Two CR lines from the same initial cross may be different, but they may also be identical. The process by which each line is produced merely selects an allele at *some* foreign histocompatibility locus for introduction onto the chosen background; it provides no foreknowledge of what locus this will be.

Two methods of analyzing CR lines once they are established are available. The first of these, described in this section, is analysis by linkage. If a histocompatibility locus can be shown to occupy a particular position on the linkage map (Chap. 8), it is of necessity different from a histocompatibility locus occupying any other position on the map.

The first step in analysis by linkage may be essentially a fortuitous one. In several instances, at first entirely by accident but in later studies partly by choice, the parents of congenic lines have differed by genes with visible effects closely linked with histocompatibility genes. Such genes serve as "markers." The introduction of the histo-

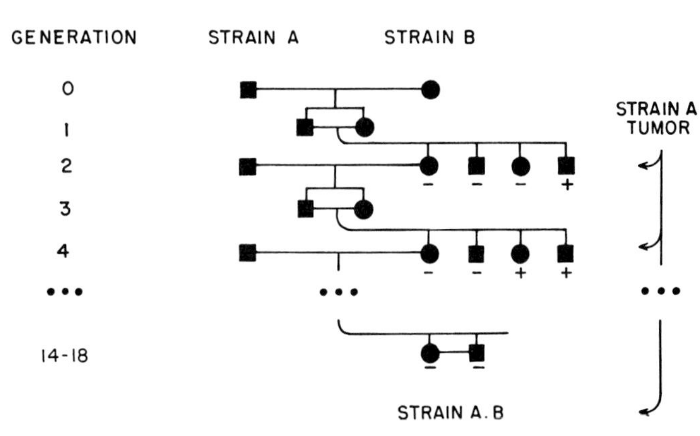

Fig. 24-2. Diagram showing the cross-intercross method of producing congenic resistant lines. Graft acceptance = +, graft rejection = —.

Table 24-5. LINKAGE OF ALBINISM (*c*) WITH RESISTANCE (−) IN THE 12TH (INTERCROSS) GENERATION OF CR LINE B10.C(41N)[a]

	++	+−	*c*+	*c*−
Observed	43	2	3	11
Expected, no linkage	33.3	11.1	11.1	3.7
Expected, 8.5% crossing over in ♀♀, 5.4% in ♂♂[b]	42.2	2.0	2.0	12.8

[a] Snell, unpublished data.

[b] These crossover values were established by other tests (Table 24-7).

compatibility gene onto the inbred background introduces the marker gene also, and in the later intercross generations the linkage becomes apparent through an association of the marker gene with resistance. This has happened a number of times. An example is given in Table 24-5, which shows a clear association of albinism (*c*) and resistance found in the 12th generation (an intercross generation) in the production of CR line B10.C(41N). This line was derived from an initial cross between C57BL/10 and the albino strain BALB/c (or actually C.B6, a congenic partner of BALB/c with the *H-2* allele of C57BL/6 substituted for the *H-2* allele of BALB/c).

Three histocompatibility loci, *H-1*, *H-3*, and *H-4* have been identified by this means. *H-1* and *H-4* are in linkage group I, but, as subsequent studies showed, in different positions in this linkage group, and *H-3* is in linkage group V. Histocompatibility-2 in linkage group IX would have been identified by this means had it not already been identified by serological tests and by linkage tests of another type. The pertinent congenic resistant strains are shown in Table 24-6.

Once linkages are identified by a chance association of resistance with a marker gene, it is possible to set up crosses specifically designed to give accurate information on the crossover percentage. This has been done for *H-1*, *H-2*, *H-3*, and *H-4*. The details will not be described here, but some of the results are summarized in Table 24-7 and Fig. 8-3. Further information is given under the discussion of the separate loci.

Analysis by the F_1 test

More often than not, congenic resistance lines are, so far as all visible traits are concerned, identical with their congenic partners. It is only the exceptional line that carries a distinguishing marker gene. Some other method than linkage must therefore be used for the analysis of most lines.

An effective alternative is provided by the F_1 test (Fig. 24-3). In this test, two CR lines on the same inbred background are crossed, and the F_1 hybrid is challenged with a transplant from the inbred partner. If the two lines are identical (as in the case of lines 2 and 3 in the figure), the F_1 of necessity reduplicates their genotype, and likewise their resistance. If the two lines are different (as in the case of lines 1 and 2 in the figure), the two genotypes complement each other and produce a susceptible hybrid. A special case arises when the two lines come from different initial crosses (strains 3 and 4 of the figure). Here the two lines may differ from the common partner at the same locus, but by different alleles. The hybrid, $H\text{-}2^b/H\text{-}2^c$ will, in this case, usually be resistant to $H\text{-}2^a/H\text{-}2^a$ donor tissues, but not always. The alleles may complement each other.

Table 24-6. FOUR HISTOCOMPATIBILITY LOCI IDENTIFIABLE BY THEIR LINKAGE RELATIONS

Inbred strain	Congenic resistant partner	Marker gene introduced with resistance	Linkage group of marker gene	Histocompatibility locus identified
C3H	C3H.K	*c*	I	*H-1*
C57BL/10	B10.129(5M)	*c*	I	*H-1*
	B10.C(41N)	*c*	I	*H-1*
	B10.LP	A^w	V	*H-3*
	B10.129(21M)	*p*	I[a]	*H-4*
A	A.CA	*Fu*	IX	*H-2*

[a] Further studies have shown that, though *H-1* and *H-4* are both in the first linkage group, they are at different positions in this group.

Fig. 24-3. Diagram showing how the F_1 test can be used to determine whether two CR strains on the same inbred background differ from the background strain at the same (lines 2 and 3, and 3 and 4) or different (lines 1 and 2) histocompatibility loci.

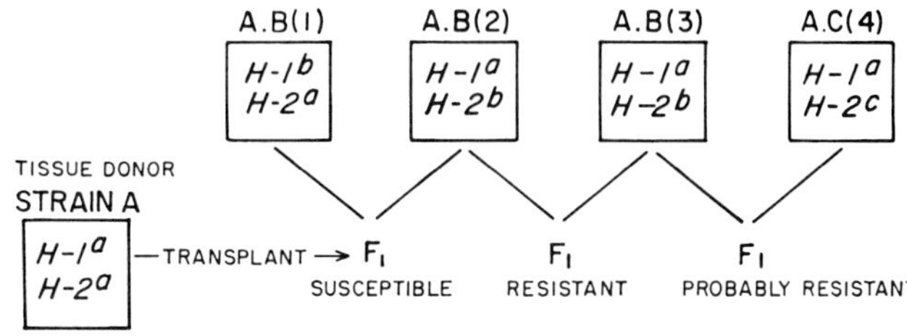

Table 24-7. CROSSOVER DATA FOR *H-1*, *H-2*, *H-3*, AND *H-4*

Crossing over between	*Per cent crossing over when heterozygous parent is* ♀	♂	References
H-1–c	8.5	5.4	Snell and Stevens (1961)
H-2–Fu	6.8	4.1	Allen (1955*a*)
H-2–T	15.4	8.2	Allen (1955*b*)
H-3–we	0.6–3.6[a]		Snell and Bunker (1964)
H-3–a	19.6–22.6[a]		Snell and Bunker (1964)
H-4–p	No proven crossovers		Snell and Stevens (1961)

[a] The lower value is based on crossovers proven genetically, the higher value on animals which responded as crossovers to the tumor inoculation test, but which were not all proven genetically.

At least one case is known where the complementation is complete (Snell et al., 1953). How often complementation occurs is unknown, but it is likely that, at loci where there are multiple alleles, partial complementation is rather common, but complete complementation rare.

If then, two CR lines on the same genetic background give a susceptible hybrid, it may be inferred with considerable confidence that they differ from this common partner at *different* loci. If one identifies a locus *H-1*, the other must identify a distinct locus which can be called *H-2*. If three lines give susceptible hybrids in all three possible F_1's, then three loci are identified. A necessary proviso is that the lines be on a common background, though the method can, in theory, be extended to lines on different backgrounds provided alleles are shared in common.

Table 24-8 shows the application of the F_1 test to two lines, B10.C(41N) and B10.C(47N), derived from an initial cross between C57BL/10 and BALB/c. As already mentioned, B10.C(41N) is an albino line, and on the basis of the linkage of resistance with albinism is presumed to differ from C57BL/10 at the *H-1* locus. Both lines were

Table 24-8. ANALYSIS OF CR LINES B10.C(41N) AND B10.C(47N) by the F_1 TEST[a]

Test F_1's				*Simultaneous controls*		
Known parent	Difference from C57BL/10	Unknown parent	Fraction dying	Strain	Fraction dying	Conclusion
				C57BL/10	50/50	
				B10.C(47N)	0/27	
B10.LP	*H-3*	B10.C(41N)	10/10	B10.LP	0/10	B10.C(41N) is not *H-3*
B10.BY	*H-1*	B10.C(41N)	0/10			B10.C(41N) is *H-1*
B10.129(5M)	*H-1*	B10.C(41N)	0/10	B10.129(5M)	0/10	B10.C(41N) is *H-1*
B10.129(5M)	*H-1*	B10.C(47N)	10/10			B10.C(47N) is not *H-1*
B10.C(41N)	*H-1*	B10.C(47N)	10/10	B10.C(41N)	0/10	B10.C(47N) is not *H-1*
B10.D2	*H-2*	B10.C(47N)	10/10			B10.C(47N) is not *H-2*
B10.LP	*H-3*	B10.C(47N)	10/10	B10.LP	0/19	B10.C(47N) is not *H-3*
B10.129(21M)	*H-4*	B10.C(47N)	10/10	B10.129(21M)	0/19	B10.C(47N) is not *H-4*

[a] Snell and Bunker (1965 and unpublished data). The F_1's included in this table were run in several separate groups. Strains run as controls with each group always included the susceptible congenic partner, C57BL/10, and usually one or both of the CR strains used as parents in any given cross. All mice were preimmunized with three injections of C57BL/10 thymus and challenged with a C57BL/10 transplantable leukemia.

Table 24-9. USE OF THE F_1 TEST TO TYPE VARIOUS INBRED STRAINS FOR ALLELE *H-7*^{a} OF STRAIN C57BL/10 [a]

Test F_1's: Known parent	Unknown parent	Fraction dying	Conclusions
C57BL/10		10/10	Susceptible control, *H-7*^{a}
B10.C(47N)		0/10	Resistant control, *H-7*^{b}
B10.C(47N)	AKR	10/10	AKR is *H-7*^{a}
B10.C(47N)	C57BR/cd	10/10	C57BR/cd is *H-7*^{a}
B10.C(47N)	C57L	10/10	C57L is *H-7*^{a}
B10.C(47N)	C58	10/10	C58 is *H-7*^{a}
B10.C(47N)	DBA/1	10/10	DBA/1 is *H-7*^{a}
B10.C(47N)	DBA/2	7/7	DBA/2 is *H-7*^{a}
B10.C(47N)	129	10/10	129 is *H-7*^{a}
B10.C(47N)	A	0/10	A is not *H-7*^{a}
B10.C(47N)	BALB/c	0/8	BALB/c is not *H-7*^{a}
B10.C(47N)	C3H	0/10	C3H is not *H-7*^{a}

[a] Snell and Bunker (unpublished data). All mice preimmunized with three injections of C57BL/10 thymus and challenged with a C57BL/10 transplantable leukemia.

crossed with a panel of CR tester stocks of known histocompatibility genotypes. Three of the tester stocks, B10.129(5M), B10.LP, and B10.129 (21M), are listed in Table 24-6. These stocks test for *H-1, H-3,* and *H-4,* respectively. The other stocks used were B10.BY which provides an alternative test for *H-1* and B10.D2 which tests for *H-2*. As expected, from prior evidence derived from linkage, line B10.C(41N) gave resistant hybrids with *H-1* tester stocks, but positive hybrids with other stocks. The conclusion that it differs from C57BL/10 at *H-1*, or, as we may say for convenience, that it is an "*H-1* line," is confirmed. Line B10.C(47N), on the contrary, gave susceptible hybrids with all stocks. The inference is that it does not differ from C57BL/10 at *H-1, H-2, H-3,* or *H-4,* and therefore must differ from it at some previously unidentified histocompatibility locus. Since Amos et al. (1963) have assigned the symbols *H-5* and *H-6* to loci identified by serological methods (see below), the locus identified by B10.C(47N) may be assigned the symbol *H-7*. The evidence that *H-7* is different from *H-5* and *H-6* will be discussed later.

ANALYSIS OF HISTOCOMPATIBILITY ALLELES

Multiple allelic systems are common in genetics, and it is not likely that histocompatibility loci present an exception. What methods are available for the analysis of histocompatibility alleles? Five approaches have been used; we shall discuss only the three most useful ones. One of these involves an extension of the F_1 test we have just described. We shall examine this first.

As applied to the identification of alleles, the F_1 test employs a CR pair and an unrelated inbred strain, the unknown. The three strains may be represented as follows:

Tissue donor	Known parent	Unknown parent
A	A.B	U
(*A*) *H-1*^{a}	(*A*) *H-1*^{b}	(*U*) *H-1*^{x}

Relevant information as to the genotypes of the three lines is given in the formulae (*A*) *H-1*^{a}, (A) *H-1*^{b}, and (*U*) *H-1*^{x}, where *H-1* is the locus at which strains A and A.B differ, (*A*) is the total genotype of strain A exclusive of *H-1*, (*U*) the same for strain U (the unknown), *H-1*^{a} and *H-1*^{b}, the known *H-1* alleles of strain A and A.B respectively, and *H-1*^{x}, the unknown *H-1* allele of strain U. If strains A.B and U are crossed, the formula of the F_1 hybrid can then be written (*A*)/(*U*) *H-1*^{b}/*H-1*^{x}. It is apparent that if *H-1*^{x} is, in fact, *H-1*^{a}, this hybrid will be susceptible to strain A transplants since every gene found in strain A can also be found in the F_1. When *H-1*^{x} is not *H-1*^{a} the situation is more complex. Usually the transplant will be rejected, but there is also the possibility that *H-1*^{b} and *H-1*^{x} will show a complementary relationship to *H-1*^{a}. Thus if each allele determines antigenic specificities as follows:

$H\text{-}1^a$	$H\text{-}1^b$	$H\text{-}1^x$
1,2,3	1,-2,-3	-1,2,3

the hybrid $H\text{-}1^b/H\text{-}1^x$ will be susceptible even though $H\text{-}1^x$ is not identical with $H\text{-}1^a$. The conclusions to be drawn from the F_1 test are then as follows:

F_1 resistant: unknown is not $H\text{-}1^a$.

F_1 susceptible: unknown probably is $H\text{-}1^a$, but may instead only share some of the specificities of $H\text{-}1^a$.

The F_1 test can also be reversed, using A.B as the tissue donor and A as the known parent. The allele tested for is then $H\text{-}1^b$. It is also possible to treat the tissue donor as the unknown and the second parent as the known.

An example of the typing of 10 inbred strains (the unknowns) by the F_1 test is shown in Table 24-9. The tissue donor is strain C57BL/10 and the known parent CR strain B10.C(47N). Since the donor and the known parent differ at *H-7*, with the allele of the donor being $H\text{-}7^a$, this is a test for $H\text{-}7^a$. Of the 10 hybrid combinations, seven proved susceptible, all mice succumbing to the tumor, and three resistant. The parents of the resistant F_1's were strains A, BALB/c, and C3H. Since the gene for resistance in strain B10.C(47N) came from strain BALB/c there would have been an inconsistency had BALB/c not produced a resistant F_1. Presumably the *H-7* alleles in BALB/c and B10.C(47N) are the same, and both may be designated $H\text{-}7^b$. All that has been proved formally, however, is that strains A, BALB/c and C3H do not carry $H\text{-}7^a$. A considerable presumption is also established that the other seven strains are $H\text{-}7^a$, though there is a possibility that one or more of them carries an *H-7* allele different from both $H\text{-}7^a$ and $H\text{-}7^b$, but sufficiently like $H\text{-}7^a$ so that any missing components of $H\text{-}7^a$ are supplied by $H\text{-}7^b$.

Table 24-10 gives information on the distribu-

Table 24-10. DISTRIBUTION OF HISTOCOMPATIBILITY ALLELES AND ALLOANTIGENIC SPECIFICITIES AMONG INBRED STRAINS AND CONGENIC RESISTANT SUBLINES

	Alleles[a]			*Alloantigenic specificities*[b]						
Strain	*H-1*	*H-3*	*H-7*	*H-5*	*H-6*	Kappa	Iota	TL	θ-AKR	θ-C3H
A	not-*c*	not-*a*	not-*a*	+	+	−	+	+	−	+
AKR	not-*c*	not-*a*	*a*					−	+	−
BALB/c	not-*c*	not-*a*	not-*a*					−	−	+
C3H/He	*a*	not-*a*	not-*a*	−	+	−	+	−	−	+
C3H.K	*b*	not-*a*	not-*a*							
C3H/St				+	−	−	−			
C57BL	*c*	*a*	*a*	−	+	−	+	−	−	+
B10.BY	*b*?	*a*	*a*							
B10.C(47N)	*c*	*a*	not-*a*							
B10.129(5M)	*b*	*a*	*a*	−						
B10.129(13M)	*c*	not-*a*	*a*							
B10.129(21M)	*c*	*a*	*a*	−	+					
C57BR/cd	*c*	*a*	*a*							
C57L	*c*	*a*	*a*						−	+
C58	*c*	not-*a*	*a*	−	+	−	−	+	−	+
DBA/1		not-*a*	*a*	−	−	−	+		−	+
DBA/2	*d*	not-*a*	*a*	−	−	−	+		−	+
F/St						−	−			
RF									+	−
SWR									−	+
YBR/HeHa						+	+			
129	*b*?	not-*a*	*a*	+	+	−	+		−	+

[a] Snell et al. (1953), Snell (1958*b*), Snell and Stevens (1961), Snell (unpublished data), Snell and Graff (unpublished data).

[b] Data for *H-5* and *H-6* from Amos et al. (1963), data for Kappa and Iota from Spencer et al. (1964), data for TL from Old et al. (1963), data for θ-AKR and θ-C3H from Reif and Allen (1964).

tion, within a number of inbred and CR strains, of known alleles of *H-1*, *H-3*, and *H-7*. (For comparable information on *H-2* see Table 24-13.) It should be remembered that evidence that a particular strain does *not* carry a certain allele is more conclusive than evidence that it does carry it. The information in Table 24-10 on alloantigenic specificities will be discussed later. An interesting sidelight provided by this table is the tendency of lines known to be related to have similar histocompatibility genotypes (e.g., C57BL/10, C57BR/cd, C57L, and C58; A and BALB/c; DBA/1 and DBA/2).

The primary value of the F_1 test in studies of allelic systems is in typing a variety of strains for already identified alleles. By showing that a given strain lacks the alleles of both members of a congenic pair, a third allele can be proven, but beyond this the method cannot go. Other methods are necessary for establishing longer series of alleles.

A useful method may be available when multiple congenic resistant lines on a common background have been produced by more than one initial cross. Sometimes these lines will differ from the background line at the same locus but by different alleles. This happened with *H-1*, partly by design in this instance since albinism could be used as a marker. The relevant lines are B10.BY, B10.129(5M), and B10.C(41N). Evidence from the F_1 test and from linkage of resistance with the albino locus prove that these all differ from C57BL/10 at *H-1*. One method of demonstrating different alleles in such lines is to exchange skin grafts between them. Strains with different alleles should be histoincompatible. This method is straightforward but requires a high degree of coisogenicity in the lines if the results are to be trustworthy. This method has not been used extensively to date but undoubtedly will be used more in the future.

A third method of analyzing alleles uses alloantisera and is essentially identical with the method used for the study of blood groups in man. This method is described in connection with the discussion of *H-2*.

Information on two other methods of studying histocompatibility alleles not described here will be found in Snell et al. (1953) and Snell et al. (1957).

Evidence concerning the multiplicity of alleles is summarized in Table 24-10 and 24-13 and discussed in connection with the descriptions of the separate loci.

In concluding this section on histocompatibility alleles, it should be reemphasized that, whatever the method used, evidence of dissimilarity of alleles is more conclusive than evidence of identity.

DIFFERENCES IN "STRENGTH" OF HISTOCOMPATIBILITY LOCI

Once pairs of congenic resistant lines with single histocompatibility differences were available, it became possible to characterize individual histocompatibility loci in a way that had been quite impossible when all information was based on comparisons of inbred strains with multiple differences or the segregating generations derived therefrom. Not surprisingly, the different loci showed marked individuality.

One of the first differences that became apparent was in the magnitude of the barriers which different loci oppose to transplants. Some of the barriers are easily transgressed, others are transgressed with great difficulty. This has been referred to as a difference in the "strength" of the loci (Counce et al., 1956). This is not an altogether happy expression, but the usage is now well established, and in the absence of any better alternative we shall follow it here.

Some results with tumor transplants made from strain C57BL/10 to 15 lines congenic with C57BL/10 and with skin grafts exchanged in both directions between C57BL/10 and the same lines are shown in Table 24-11. It should be noted that all tumors were transplanted to hosts immunized three times with C57BL/10 tissue and that the tumors were selected for their specificity. Without prior immunization or with other tumors more deaths would have resulted. Mice receiving skin grafts were unimmunized.

It will be seen that there is a great difference in the behavior of the transplants in the different lines. In several lines all animals survived the tumor transplants; in other lines deaths ranged up to 78 per cent. There were similar differences with skin grafts. Grafts from C57BL/10 to strain B10.D2 showed a median survival of 9 days, grafts from C57BL/10 to B10.129(21M) a median survival of 127 days. In the weaker combinations there was great variability in the interval to rejection. In these combinations also chronic rejections, characterized by apparently incipient rejection followed by partial recovery, were common. In general there was a close correlation between the percentage of deaths from tumor grafts and the length of survival of skin grafts.

The data from skin grafts bear on the question of whether histocompatibility loci show dominance

Table 24-11. DIFFERENCES IN THE "STRENGTH" OF VARIOUS HISTOCOMPATIBILITY LOCI AS MEASURED BY GRAFTS OF SKIN AND C57BL/10 TRANSPLANTABLE LEUKEMIAS MADE BETWEEN CONGENIC STRAIN PAIRS

		Per cent survival of leukemic allografts in immunized hosts[a]		*Median survival (and range), in days, of skin allografts in unimmunized hosts*[b]	
Locus	CR Strains	♂♂	♀♀	From C57BL/10	To C57BL/10
H-2	B10.D2	0	0	9	9
H-1	B10.BY	0	0	15 (14–20)	>60 (34–>60)
	B10.129(5M)	0	0	28 (21–36)	>100 (24–>100)
	B10.C3H(40NX)	4	0	17 (15–23)	42 (22–58)
	B10.C(41N)	0	0	25 (21–58)	>60 (26–>60)
	B10.D2(58N)	17	0	33 (18–61)	>60 (27–>60)
H-3	B10.LP	2	2	21 (16–27)	35 (28–>78)
	B10.LP-*a*	10	2	24 (21–28)	38 (24–49)
H-7	B10.C(47N)	2	1	22 (16–58)	44 (21–93)
H-8	B10.D2(57N)	30	16	45 (26–>56)	34 (23–41)
H-4	B10.129(21M)	39	44	127 (43–>261)	29 (29–34)
H-9	B10.C(45N)	67	6	>200 (17–>225)	>225 (24–>225)
H-10	B10.129(9M)	70	61	>75 (21–>250)	>225 (46–>225)
H-11	B10.129(10M)	78	59	124 (24–224)	>200 (19–>250)
	B10.D2(55N)	55	45	123 (22–>200)	138 (30–>175)
Y-linked ♂ to ♀				26 (16–65)	

[a] Snell and Bunker (1965 and unpublished data). The number of animals per group ranged from 10 to 231.
[b] Graff et al. (in press and unpublished data), except that results for B10.D2 and B10.LP are from Berrian and McKhann (1960*a*, *b*), results with B10.129(5M) grafted to C57BL/10 from Billingham and Silvers (unpublished data), and results for Y-linked ♂ to ♀ grafts from Eichwald et al. (1958). The 26 days survival reported by the last named authors is a mean rather than a median.

or codominance. In every instance where skin grafts were made reciprocally within a strain pair they were rejected in both directions. There was a great difference in the two directions in some instances, but never 100 per cent survival in either direction. Thus with the strain pair B10. 129(5M) and C57BL/10, most of the grafts survived beyond 100 days when strain C57BL/10 was the host, though only 28 days in the reciprocal direction. It follows that none of the alleles tested can be regarded as determining the absence of an antigen (though some may approximate this). It also follows, since grafts from any pair of strains grow in the F_1 (one exception will be discussed later), that the alloantigens determined by each strain are expressed in the F_1. This constitutes codominance by definition.

Other aspects of Table 24-11 will be considered when the separate loci are discussed.

THE *H-1* LOCUS

The *H-1* locus is defined by a group of congenic resistant strain pairs (Table 24-6 and 24-11) which have been shown to share a common difference and by its linkage with albinism. At least four alleles have been identified (Table 24-10) and it seems likely that there are others as yet undetected. Skin allograft rejection in *H-1* CR pairs ranges from a median of 15 days for grafts made from C57BL/10 to B10.BY to more than 100 days for grafts made from B10.129(5M) to C57BL/10 (Table 24-11). As already noted the last named pair shows a marked reciprocal difference.

Ovarian grafts exchanged between *H-1* pairs survive much longer than skin grafts; probably they survive permanently in at least some instances (Snell and Stevens, 1961). This is probably a peculiarity of the tissue, not of the locus, since long survival of ovarian grafts across other non-*H-2* barriers has been noted (Linder, 1961). The *H-1* locus is a blood group locus as well as a histocompatibility locus. Appropriate antisera will distinguish between the erythrocytes of the congenic partners C57BL/10 and B10.129(5M) (Stimpfling, unpublished data; Snell and Graff, unpublished data).

THE *H-2* LOCUS

History

It is generally accepted that of the dozen or more histocompatibility systems known in the mouse, the histocompatibility-2 or *H-2* system of linkage group IX is essentially unique in both its complexity and in the immunogenic complexity and potency of its end product. The term *H-2* is derived from antigen II, a cellular antigen initially identified by Peter A. Gorer with the aid of a rabbit antimouse serum. Gorer also demonstrated that the rejection of a tumor allograft resulted in the formation of humoral antibodies specific for antigen II present on the erythrocytes of the tumor-donating strain. Further, it was shown by tumor grafting and blood typing mice of the F_2 and the appropriate backcross generations that antigen II was probably determined by a dominant gene and that compatibility of the tumor and host with respect to antigen II was a necessary but not entirely sufficient condition for the progressive growth of the graft (Gorer, 1937, 1938, 1942). The observations of Gorer provided direct evidence of an immune reaction to allotransplants and demonstrated the existence of at least one antigen system shared by a tumor and the normal cells of the strain to which the tumor was indigenous.

Genetic tests made by Gorer et al. (1948) showed that there is a linkage between *H-2* and the gene *Fu* located in linkage group IX, and further tests by Allen (1955*a*, *b*) served to locate *H-2* with respect to other genes in this linkage group (Table 24-7 and Fig. 24-4).

The linkage of *H-2* with *Fu* and *T* has been used to detect alleles at the *H-2* locus. The method will not be discussed here as it has been fully described elsewhere (Snell et al., 1953). Originally three variants were identified and assigned the symbols *H-2*, $H\text{-}2^d$, and *h-2*. Strain A was classified as *H-2*, strain DBA/2 as $H\text{-}2^d$, and strain C57BL as *h-2*. The symbol *h-2* with a small letter was used because in the early studies no alloantigenic

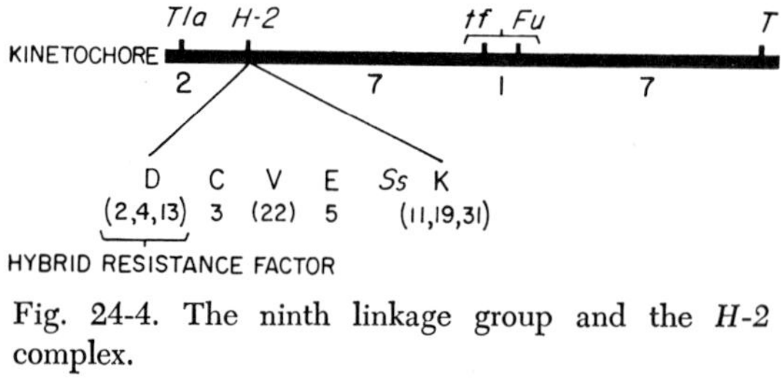

Fig. 24-4. The ninth linkage group and the *H-2* complex.

activity was detected in the red cells of mice with this allele. Several other strains were tentatively put in this category. Subsequently all alleles were found to determine effective alloantigens and to show codominance. The symbol *h-2* was therefore dropped and replaced in all cases with *H-2*, with a superscript small letter added to indicate the allele. The number of identified alleles was also greatly increased. As of 1965, 20 alleles are positively known. However, tests of a randombred strain showed it to be segregating for a number of alleles as yet unidentified (Rubinstein and Ferrebee, 1964). Undoubtedly the ultimate number of alleles will greatly exceed 20.

Early studies, both serological and genetic (Gorer, 1938; Gorer et al., 1948; Snell et al., 1953), demonstrated that the *H-2* alloantigen(s) is characterized by multiple components or specificities. Thus Gorer showed by absorbing a C57BL anti-A serum with CBA cells that the serum contained two specificities. Probably both specificities were determined by *H-2*, though this was not proved at the time. Subsequently it was shown in transplantation experiments that all F_1 hybrids of the genotype $H\text{-}2^d/H\text{-}2^k$, whatever the source of these two alleles, were susceptible to transplantable strain A tumors of the genotype $H\text{-}2^a/H\text{-}2^a$. This led to the suggestion that the alleles $H\text{-}2^d$, $H\text{-}2^k$, and $H\text{-}2^a$ determined, respectively, the "components" D, K, and DK (Snell et al., 1953). D was probably identical with Gorer's original antigen II. The presence of K in strains typed genetically as $H\text{-}2^k$ or $H\text{-}2^a$ was later demonstrated by serological methods also.

Serological studies

Much of our information concerning the *H-2* locus has been derived from serological studies. One of the surprises of histocompatibility in the mouse is the ease with which alloantibodies reactive with *H-2* specificities are produced and demonstrated, and the relative difficulty of producing, or at least of demonstrating, alloantibodies reactive with the specificities determined by other loci. Red-cell agglutination has been by all odds the most useful serological technique for studying *H-2*. A limited use has also been made of the cytotoxic test which employs lymphocytes as the target cells. These tests have shown that the alloantigen determined by *H-2* is remarkably complex. Because they reveal individual specificities, or at least can reveal such specificities when appropriate antisera are used, they nicely complement the methods of tissue transplantation which respond to groups of specificities more

Table 24-12. THE MAXIMUM NUMBER OF ALLOANTIGENIC SPECIFICITIES DETECTABLE WITH FOUR GENETICALLY UNRELATED STRAINS OF MICE

Mouse strains	*Categories of combinations of specificities*													
	$n-1$				$n-2$						$n-3$			
A	—	2	3	4	—	—	7	—	9	10	11	—	—	—
B	1	—	3	4	—	6	7	8	—	—	—	12	—	—
C	1	2	—	4	5	6	—	—	9	—	—	—	13	—
D	1	2	3	—	5	—	—	8	—	10	—	—	—	14

nearly representative of the totality of a given allele. While Gorer first demonstrated the *H-2* antigen by the use of rabbit antisera, all studies are now carried out with alloantisera produced in one strain of mice by the injection or transplantation of the tissue of another strain.

It was observed early in the investigation of *H-2* alloantigens that strong hemagglutination reactions in saline media could be obtained routinely only with cells of certain strains while the results with cells from other strains often gave irregular reactions. Further, some sera of high titer became inactive following storage for a few hours while others retained their activity after many weeks of storage. Subsequently, it was found that some apparently inactive antisera agglutinated erythrocytes if the cells were suspended in human serum. A further improvement in the test consisted of diluting antisera in an appropriate dextran solution and suspending red blood cells in normal human serum previously absorbed with mouse tissue (Gorer and Mikulska, 1954). A modification of this test was described in which cells are suspended in saline and antibody diluted in polyvinylpyrrolidone or PVP (Stimpfling, 1961). The use of developing or conglutinating agents such as dextran and PVP in mouse hemagglutination tests is necessitated by what appears to be the "incomplete" character of mouse alloantibodies.

Besides the hemagglutination test, several other serological techniques have been used to study one or another aspect of histocompatibility phenomena. Papers describing leukocyte agglutination tests, cytotoxic tests, hemolytic tests, the fluorescent antibody technique, and various other methods relevant to the serological study of histocompatibility genes and antigens will be found in the Bibliography of Techniques at the end of this book.

A serological technique useful with mice is absorption in vivo. Animals of the appropriate genotype are injected intraperitoneally with 0.2 to 0.3 ml of antiserum and bled 2 or more hours later. The method is much simpler than absorption in vitro with liver or red cells, and with a good antiserum the lowered titer presents no difficulties.

The immunization of mice with allogeneic tissues or purified tissue derivatives usually results in the formation of antibodies detectable by one or more of the above techniques. Usually these antibodies consist of a mixture of specificities. Since much of the value of the serological method lies in its ability to reveal single antigenic components or specificities, it is important to examine the method by which individual specificities are detected.

Gorer and Mikulska (1959) have shown that there is a simple mathematical relationship between the number of genetically different inbred strains available for testing and the maximum number of antigenic specificities, other than specificities related to sex, that can be detected. The requirements for the detection of different specificities are: (1) that with respect to available strains they be present in at least one and absent in at least one, and (2) that with respect to the available strains their distributions show at least one difference. With two strains not more than two specificities are demonstrable. The situation for four strains is illustrated in Table 24-12. It will be seen that in this situation the theoretical total of demonstrable specificities is 14. More generally, the maximum number of specificities that can be revealed with n inbred strains is the sum of the number of combinations of $(n-1)$ in n, plus the number of combinations of $(n-2)$ in n, plus the number of combinations of $(n-3)$ in n, etc., or

$$\binom{n}{n-1} + \binom{n}{n-2} + \binom{n}{n-3} + \cdots$$

$$+ \binom{n}{n-n+1}, \quad \text{where} \quad \binom{n}{x} = \frac{n!}{x!(n-x)!}$$

The terms of this formula developed by Gorer

and Mikulska, as it so happens, are the coefficients of the binomial expansion, exclusive of the first and last which are both 1. Since the sum of these coefficients is 2^n, the above formula is equivalent to $2^n - 2$. If we apply these two formulas to the situation illustrated in Table 24-12, where $n = 4$, we find that the maximum number of demonstrable specificities equals $4!/(3!1!) + 4!/(2!2!) + 4!/(1!3!) = 2^4 - 2 = 14$. It will be seen that as n increases the number of detectable specificities increases rapidly. With 10 strains it becomes 1,022. It should be emphasized that this is the theoretical maximum. Because the distribution of specificities in any group of strains will virtually never conform to the optimum pattern, and because antibodies are produced less easily to some specificities than to others, the actual values will almost always fall below that theoretically obtainable. It should also be emphasized that the mere detection of specificities tells nothing about the locus by which they are determined. Actually most of the easily demonstrable specificities in the mouse are directed against *H-2*, but added sources of information are necessary if specificities are to be related positively to a given locus. The easiest of all methods for accomplishing this is to use as donor and recipient a congenic strain pair with a single known histocompatibility difference. Such pairs have been used extensively to produce antisera whose specificities are limited to *H-2*.

Most antisera prepared against allogeneic tissues contain multiple specificities. The problem of producing monospecific antisera can be illustrated by further recourse to Table 24-12.

An antiserum prepared in strain A against strain B may contain antibodies specific for components 1, 6, 8, and 12. The specificity of this antiserum can be increased by appropriate absorption, either in vivo or in vitro, with the tissues of a third, crossreactive strain. For instance, absorption of strain A anti strain B serum with tissues from strain C will selectively remove antibodies specific for components 1 and 6, but leave behind antibodies specific for components 8 and 12. A further refinement in the preparation of serotyping reagents can be attained by producing antisera in hybrids. If, for example, an antiserum is made in hybrids between strain A and strain C against tissues of mice from strain B, the antibodies will be specific for components 8 and 12. An anti-12 serum can be obtained by absorption in vitro or in vivo with strain D tissue. Actually this antiserum may still not be monospecific, but, in addition to its reactivity with 12, may be capable of reaction with one or more specificities not indicated in Table 24-12 and capable of absorption only by recourse to additional strains.

By use of procedures outlined above, 30 or more serologically distinct specificities have been detected on mouse red cells. The majority of these are determined by the *H-2* locus.

The specificities which characterize the *H-2* alloantigen were originally designated by capital letters. Ultimately, however, the number of reported specificities outran the letters of the alphabet, and it became necessary to assign compound symbols such as A^1 and B^1 (Stimpfling and Pizarro, 1961). Such symbols are awkward and confusing. We are therefore adopting a numerical system of nomenclature proposed later (Snell et al., 1964). This parallels and is largely based on a similar system proposed for the Rh blood group of man (Rosenfield et al., 1962). Each specificity is designated by an Arabic numeral, its absence by a minus sign (—), or, where the minus sign alone would not be clear, by the appropriate numeral preceded by a minus sign, e.g., −7, −8, etc.

The *H-2* chart

Much of our knowledge of the *H-2* locus can be summarized in the form of a chart which lists (1) the known *H-2* alleles, (2) the antigenic specificities which have been found associated with the different alleles, and (3) the inbred strains carrying each allele. Such a chart is presented in Table 24-13.

It will be seen in the table that there are several gaps in the listing of specificities. Numbers 15, 18, 20, 21, 23, 24, and 26 are vacant. Most of these gaps are due to current uncertainty about the specificities to which these numbers (or the corresponding letters) were assigned. Two of the missing specificities, 18 (R), identified by an RIII anti-C3H antiserum, and 26 (Z), identified by a C3H anti-RIII antiserum, are determined by a gene or genes believed to be in linkage group IX but distinct from *H-2* (Hoecker and Pizarro, 1962). They, therefore, do not belong in the *H-2* table. Shreffler (1965, personal communication) believes that specificities 6 and 28 may be the same. Contrary to previous reports, he finds that strain AKR.M reacts with (will absorb) an anti-6, whereas strain P does not react with it.

The *H-2* table provides a convenient summary of available knowledge concerning *H-2*, but it fails to reveal a number of known complexities of *H-2* immunogenetics. These complexities constitute pitfalls concerning which anyone contemplating experimental use of *H-2* should be fore-

Table 24-13. THE KNOWN ALLELES AND ALLOANTIGENIC SPECIFICITIES OF THE *H-2* SYSTEM AND THE INBRED STRAINS THAT CARRY THEM

H-2 alleles	*Alloantigenic specificities; old symbols (letters) shown at top*[a]																									Inbred strains
	A	D^b	C	D	E	F	G	H	I	J	K	M	N	P	Q	S	V	Y	A^1	B^1	C^1	D^1	E^d	D^k	K^b	
a	1	–	3	4	5	6	–	8	–	10	11	13	14	–	–	–	–	25	27	28	29	–	–	–	–	A, AKR.K, B10.A
b	–	2	–	–	5	6	–	–	–	–	–	–	14	–	–	–	22	–	27	28	29	–	–	–	33	A.BY, C3H.SW, C57BL/6, C57BL/10, C57L, CC57BR, CC57W, D1.LP, LP/J, ST/a, 129
c	–	–	3	4	?	*	–	8	–	*	–	13	*	–	–	–	*	*	27	28	29	–	31	*	*	D1.C
d	–	–	3	4	–	6	–	8	–	10	–	13	14	–	–	–	–	–	27	28	29	–	31	–	–	BALB/c, C57BL/Ks, B10.D2, DBA/2, ST.T6, WH, YBL/Rr, YBR/Wi
e	–	–	3	–	5	6	–	–	–	–	–	13	*	*	*	–	*	25	27	28	29	30	*	*	*	STOLI
f	–	–	–	–	–	?	7	8	9	*	–	–	*	–	–	–	*	–	27	–	–	–	–	*	*	A.CA, B10.M, RFM/Un
g	–	2	–	–	–	6	*	?	*	*	–	–	14	*	–	–	22	*	*	*	*	*	31	–	–	HTG
h	1	2	3	–	5	6	*	8	*	*	11	–	?	*	–	–	–	*	*	*	*	*	–	–	–	HTH, B10.A(2R)
i	–	–	3	4	5	6	*	*	*	*	–	13	*	*	–	–	22	*	*	*	*	*	–	–	33	HTI, B10.A(5R)
j	–	–	–	–	–	6	?	–	*	–	–	–	–	*	–	–	22	?	?	*	*	*	*	*	*	JK/St
k	1	–	3	–	5	–	–	8	–	–	11	–	–	–	–	–	–	25	–	–	–	–	–	32	–	AKR, B10.BR, CBA, CE, CHI, C3H, C57BR/a, C57BR/cd, C58, D1.ST, MA/J, RF/J, ST/bJ, 101
l	–	–	–	–	–	6	?	–	*	10	–	–	–	*	–	–	22	–	?	?	*	*	*	*	*	I/St, N/St(?)
m	1	–	3	–	5	?	–	8	–	*	11	13	*	–	–	–	*	*	27	28	29	30	–	*	*	AKR.M
n	1	–	–	–	5	6	*	8	*	10	–	–	14	*	–	–	–	*	?	?	*	*	*	*	*	F/St
o	1	*	3	–	5	*	*	*	*	*	–	–	*	*	*	*	*	*	*	*	*	*	31	*	*	HTO/Sf
p	–	–	?	–	5	?	–	–	–	*	–	–	*	16	–	–	*	*	–	–	–	–	–	*	*	P/Sn, C3H.NB
q	–	–	3	–	5	6	–	8	–	–	11	13	–	–	17	–	–	*	27	28	29	30	–	*	*	DBA/1, C/St, BUB
r	–	–	3	–	5	–	–	8	–	*	11	–	*	–	–	–	*	25	–	–	–	–	?	*	*	RIII/J, RIII/Wy, LP.RIII
s	–	–	3	–	5	6	7	–	–	*	–	–	*	–	–	19	*	–	–	28	–	–	–	–	–	A.SW, SJL
w	–	2	*	–	*	*	*	–	–	*	–	–	*	*	*	–	*	*	*	*	*	*	–	*	*	WB/Re, WC/Re

[a] This table is taken from Snell et al. (1964) with a few modifications and additions suggested by later work. An asterisk (*) in the body of the table indicates that no test has been made for the particular component; a negative (–) symbol denotes the absence of the component; a question mark (?) indicates that the presence or absence of the component is uncertain. The references on which the table is based and the alleles and specificities which they identify or the strains which they type are: Allen (1955*a*), allele *H-2[f]*; Amos et al. (1955*b*), specificities 2, 5, 6, 11, 31; Amos (1958), strains C/St, CHI; Amos (1959), alleles *H-2[e]*, *H-2[j]*, *H-2[l]*, *H-2[n]*, specificities 1, 10, 14, 22, strains STOLI, JK/St, I/St, N/St, F/St; Brondz and Yegorov (1964), strains CC57BR, CC57W; Gorer et al. (1948), alleles *H-2[a]*, *H-2[b]*, *H-2[d]*, specificity 4, strains A, C57BL, DBA/2; Gorer (1956), strain YBR/Wi; Gorer (1959), specificities 32, 33, strains ST/a, ST/bJ; Gorer and Mikulska (1959), alleles *H-2[g]*, *H-2[h]*, *H-2[i]*, strains C58, HTG, HTH, HTI; Hoecker et al. (1954), allele *H-2[s]*, specificities 2, 3, 5, 11, 16, 17, 19, strain A.SW; Hoecker et al. (1959), allele *H-2[e]*, specificities 7, 8, 9, 25, strain STOLI; Popp and Amos (1965), strain RFM/Un; Popp et al. (1958), strain 101; Rubin (1960), strain CE; Shreffler (1965), allele *H-2[o]*, strain H-2O/Sf; Snell and Higgins (1951), alleles *H-2[b]*, *H-2[p]*, strains P and BALB/c; Snell (1951), allele *H-2[k]*, strain CBA; Snell et al. (1953), alleles *H-2[q]*, *H-2[r]*, specificities 4, 11, strains C57L, LP, 129, C57BL/Ks, YBL/Rr, AKR, C3H, C57BR/a, C57BR/cd, ST/bJ, DBA/1, RIII/Wy; Snell (1958*b*), alleles *H-2[c]*, *H-2[m]*, strains AKR.K, A.BY, C3H.SW, D1.LP, D1.C, B10.D2, ST.T6, A.CA, B10.M, D1.ST, AKR.M, LP. RIII; Snell (unpublished data), strains WH, C58, MA/J, RF/J, C3H.NB, BUB, RIII/J; Stimpfling and Pizarro (1961), specificities 13, 27, 28, 29, 30; Stimpfling and Richardson (1965), strains B10.A, B10.A(2R), B10.A(5R), B10.BR; Stimpfling and Snell (unpublished data), allele *H-2[w]*, specificities 1, 8, 31, strains SJL, WB/Re, WC/Re.

warned. Not all of these can be described here, but a few salient facts must be mentioned.

The alloantigenic specificities determined by *H-2* differ in their "strength" as indicated by the ease with which alloantibodies are induced, by the titer of the resulting antiserum, and by the ease and consistency with which agglutination is obtained when presumably active sera are titered against presumably reactive red cells. While there is no sharp division of the specificities into "strong" and "weak" groups, specificities, 1, 2, 3, 4, 5, 9, 11, 16, 17, and 19 tend to be easily demonstrable. In donor-recipient combinations differing at both strong and weak specificities, antibodies to the latter may appear late or not at all. Sometimes the same specificity behaves differently on different genetic backgrounds. Often specificities which cannot be demonstrated by agglutination can be demonstrated because antibodies are removed by absorption.

Curiously, two *H-2* specificities, 32 and 33, have so far been demonstrated only by the cytotoxic test (Gorer, 1959). They are, therefore, present on lymphocytes, but are absent from, or at least are difficult to demonstrate on, erythrocytes.

There are probably minor variants of some *H-2* alleles. By their very nature there is more uncertainty about these variants than about the major differences. None are included in the table. One of them, $H\text{-}2^{d'}$, has been detected serologically (Gorer, 1956) as well as by the methods of tissue transplantation (Snell et al., 1953). This variant occurs in strains YBL and YBR. The distinguishing specificity, called D′, is also omitted from the table. There are probably minor variants of $H\text{-}2^k$, but how these are distributed among the different strains typed as $H\text{-}2^k$ is undetermined (Snell et al., 1953).

It has already been noted that there is an unusual relationship between the alleles, $H\text{-}2^d$, $H\text{-}2^k$, and $H\text{-}2^a$, in that the hybrid $H\text{-}2^d/H\text{-}2^k$, whatever may be the source from which these alleles are derived, is susceptible to strain A ($H\text{-}2^a/H\text{-}2^a$) transplantable tumors. This complementarity of $H\text{-}2^d$ and $H\text{-}2^k$ is also revealed in the specificities listed in Table 24-13. Allele $H\text{-}2^a$ has no specificities not present in alleles $H\text{-}2^d$ and $H\text{-}2^k$, and, with the exception of specificities 31 and 32 which occur in $H\text{-}2^d$ and $H\text{-}2^k$, respectively, but not in $H\text{-}2^a$, has all the specificities present in $H\text{-}2^d$ and $H\text{-}2^k$. If we assume that 31 and 32 are not represented in $H\text{-}2^a$ by some alternative specificity but rather by a true absence of antigenicity, all the requirements of the compatibility relationships of these strains are met. Of course hybrids carry incompatibilities at other loci, but these are overridden by the rather virulent transplantable tumors which have been used in these tests.

It has been suggested that the unusual relationship of these three alleles can be explained by the assumption that $H\text{-}2^a$ was derived from a crossover within the *H-2* locus occurring in a mouse heterozygous for $H\text{-}2^d$ and $H\text{-}2^k$ (Gorer, 1959).

One of the interesting questions about *H-2* specificities is whether any of them are "allelic." The concept of allelism was originated in connection with alternative or mutually exclusive genes occupying the same locus. It could reasonably be applied to alloantigenic specificities if it could be shown that these sometimes occupy the same site on the antigen molecule and are therefore mutually exclusive. The extension of the term would gain added justification if proof were also forthcoming that mutually exclusive specificities were determined by mutually exclusive or allelic subunits of a complex gene. No firm information of this sort is available concerning the specificities determined by *H-2*. However, it will be seen from the table that some specificities do appear to show a mutually exclusive distribution. Thus specificities 5 and 31 have never been found in combination.[1] Partly because of this 5 and 31 were originally assigned the symbols E and E^d, implying an allelic relationship. While allelism of specificities may ultimately be proved, the evidence now available is of a very uncertain nature, and we have followed Snell et al. (1964) in treating all specificities as nonallelic.

A serologically detected serum variant determined by a locus designated *Ss*, although without known effects on histocompatibility, is of interest in connection with *H-2* because of its linkage relations. Two alleles are known. Ss^h determines a high level of a specific serum protein, Ss^l a low level. Numerous inbred strains have been typed for *Ss*. All strains which are $H\text{-}2^k$, and AKR.M, the only known $H\text{-}2^m$ strain, are Ss^l; all other strains are Ss^h (Shreffler and Owen, 1963; Shreffler, 1964, 1965). The unusual linkage relations of *H-2* and *Ss* are described below.

Crossing over within *H-2*

Crossing over within the *H-2* locus was first observed by Allen (1955*b*) and by Amos et al. (1955*b*). Allen, using progeny tests and tumor transplants to type animals from a cross $H\text{-}2^a + T/H\text{-}2^f\ Fu + \times H\text{-}2^b + +/H\text{-}2^b + +$, found one proven and one doubtful crossover between "components" 4 and 11 in 284 mice. Amos et al., using red cell agglutination, absorption in vivo, and progeny testing to type the animals from a cross $H\text{-}2^b/H\text{-}2^a \times H\text{-}2^b/H\text{-}2^b$, found one crossover between specificities 2 and 5 in 32 mice. Subsequently the initial report of Amos et al. was extended by additional publications (Gorer, 1959; Gorer and Mikulska, 1959; Amos, 1962; see also Table 24-13 for the complete serotype information on the three crossovers reported in these studies), and three other crossover studies were carried out (Pizarro et al., 1961; Shreffler, 1964, 1965; Stimpfling and Richardson, 1965).

Fourteen crossover alleles obtained in these studies, each of which has been typed for five or more specificities, are shown in Table 24-14. Four of the crossovers were also typed for *Ss* (Shreffler, 1965). It will be seen that these crossovers serve to identify five regions within the *H-2* locus or "complex," and that they place *Ss* within *H-2*. The regions and the specificities assigned to each are D (2,4,13), C (3), V (22), E (5), and K (11,19,31). *Ss* lies between E and K. Informa-

[1] Subsequent to writing this chapter, 5 and 31 were found in combination in allele $H\text{-}2^o$ (Shreffler, unpublished data) and possibly in $H\text{-}2^r$ (Snell, unpublished data).

Table 24-14. CROSSOVERS SEPARATING *H-2* INTO FIVE REGIONS AND LOCATING THE POSITION OF *Ss*

Genotype of heterozygous parent	Crossover allele							
	Symbol[a]	No. of occurrences	*Specificities grouped according to region*[b]					
			D	C	V	E	*Ss*	K
$H\text{-}2^b/H\text{-}2^d$	$H\text{-}2^{g-Go}$	1	2, −4, −13	−3	22 ×	−5, −33		
$H\text{-}2^a/H\text{-}2^b$	$H\text{-}2^{h-Go}$	1	2, −4, −13 ×	3	−22, −33			11
	$H\text{-}2^{h-Sg}$	5	−4, −13 ×	3				11
	$H\text{-}2^{i-Go}$	1	−2, 4, 13	3 ×	22, 33			−11
	$H\text{-}2^{i-Sg}$	2	4, 13	3	. . . . . .	×	. . . . . .	−11
$H\text{-}2^dSs^h/H\text{-}2^kSs^l$	$H\text{-}2^{o-Sf}Ss^h$	1	−4, −13			5 ×	Ss^h	−11, 31
	$H\text{-}2^{o-2Sf}Ss^l$	1	−4, −13			5	Ss^l ×	−11, 31
	$H\text{-}2^{a-Sf}Ss^l$	1	4, 13	. . . . . ×	. . . . . .	 5	Ss^l	11, −31
$H\text{-}2^aSs^l/H\text{-}2^sSs^h$	$H\text{-}2^{?-Sf}Ss^l$	1	4, 13			5	Ss^l ×	−11, 19, −31

[a] The *Go*, *Sg*, and *Sf* in the superscript of the allele symbol indicate crossovers obtained by Gorer and co-workers, Stimpfling and Richardson (1965), and Shreffler (1965), respectively.
[b] The location of the crossover for each allele is shown by a times (×) sign.

tion as to specificities not shown in the table also locates 1 and 8 in C, V, or E. So far as is known, the *Ss* protein is neither antigenically similar to nor structurally associated with the *H-2* antigen. If this lack of relationship is confirmed by further studies, it will be necessary to regard the K region of *H-2* as a separate locus.

It was suggested above that allele $H\text{-}2^a$ was derived by crossing over between alleles $H\text{-}2^d$ and $H\text{-}2^k$. If this is indeed the origin of $H\text{-}2^a$, strains which are $H\text{-}2^a$ should also be Ss^l whereas actually they are Ss^h. The meaning of this discrepancy is not clear at the present time.

The percentage of crossing over between three of the regions, based on the study of Stimpfling and Richardson (1965), is shown in Table 24-15. The values range from 0.55 for the D–K interval to 0.05 for the C–K interval, both in heterozygous females. In accordance with results usually obtained in the mouse, crossing over, for two of the intervals, is substantially more frequent in heterozygous females than in heterozygous males,

Table 24-15. PER CENT CROSSING OVER BETWEEN DIFFERENT REGIONS OF THE *H-2* LOCUS[a]

Crossing over between regions	*Per cent*	
	In heterozygous ♀♀	In heterozygous ♂♂
D and K	0.55 ± 0.22	0.19 ± 0.14
D and C	0.25 ± 0.17	0.09 ± 0.19
C and K	0.05 ± 0.05	0.06 ± 0.02

[a] Data of Stimpfling and Richardson (1965).

but the differences are not statistically significant. Shreffler (1965) has reported similar values. Somewhat higher values have been reported by other investigators but these were based on many fewer animals.

The crossover obtained by Allen (1955*b*) indicated that the order of the *H-2* regions in relation to the genes *Fu* and *T* of linkage group IX is D K *Fu T*. This order has subsequently been confirmed (Stimpfling and Richardson, 1965; Shreffler, 1965). Information in regard to *H-2* and its relation to other identified loci in linkage group IX is shown in Fig. 24-4. Other factors shown in the figure in close association with *H-2* will be discussed later.

Transplantable tumors induced in hybrids between two inbred lines usually fail to grow in either parental strain. If the *H-2* antigens of one parent are lost from the hybrid tumor, the tumor is then able to grow in the other parental strain. This provides the basis for an assay system used by several investigators to select variants lacking all or some part of one of the parental *H-2* complexes (Mitchison, 1956; Hellström, 1961; Dhaliwal, 1964; Klein and Klein, 1964). On the basis of serological and transplantation criteria, it has been shown with several types of tumors, including sarcomas, carcinomas, and leukemias, that the loss of *H-2* alloantigens was persistent, irreversible, and specific. Hellström (1961) tested 12 variants of two $(A \times A.SW)F_1$ lymphomas and found that 11 had lost specificity 11 and one had lost both specificities 4 and 11. The loss of 4 without 11 has not been observed. The loss of alloantigens from hybrid tumors is compatible with the interpretation that the variants arose by mitotic crossing over. However, other genetic mechanisms cannot be excluded. In any case, these observations are consistent with the existence of functionally independent regions of the *H-2* locus. They also suggest, since 11 can be lost without 4 but not 4 without 11, that the kinetochore or spindle-fiber attachment is adjacent to the D (4) rather than the K (11) region of *H-2*. This is the basis for the position assigned to the kinetochore in Fig. 24-4.

The *H-2* system has intrinsic interest, apart from the role it plays in tissue transplantation, as a model for the study of complex genes in mammals. Studies on the serological and genetic characteristics of *H-2* may provide some insight into the immunogenetic properties of similar types of blood-group systems in other vertebrates.

Chemistry of the *H-2* alloantigen

Because of the unique role which *H-2* plays in graft rejection, most efforts to isolate an alloantigen from mouse tissues have centered on the products of this locus. Several methods of extraction have been employed to obtain immunologically active, cell-free preparations from both normal and tumor tissues (Castermann and Oth, 1959; Herzenberg and Herzenberg, 1961; Brent et al., 1961; Manson et al., 1963; Kandutsch and Stimpfling, 1963, 1965). Assays of the various preparations for biological activity have included tests of their ability to induce accelerated rejection of skin allografts, elicit the formation of hemagglutinins, absorb hemagglutinins from appropriate antisera, and abrogate resistance to tumor allografts (immunological enhancement).

A variety of tissues has been used as a source of alloantigens, including spleen, lymph nodes, liver, and transplantable tumors. Irrespective of the source of starting material, procedure of extraction, or assay system used, the available information indicates that the major part of *H-2* activity is associated with the membranous structures of cells.

The earliest studies on the properties of transplantation antigens were carried out by Snell (1952). At that time the relation of the immunologically active fractions to *H-2* was unknown. Kandutsch undertook a study of the chemical properties of the enhancing or *H-2* substance in 1954 and has succeeded in preparing a soluble alloantigenic lipoprotein in a relatively homogeneous form, according to electrophoretic and ultracentrifugal criteria (Kandutsch and Reinert-Wenke, 1957; Kandutsch and Stimpfling, 1963, 1965). The preparation described by Kandutsch was obtained from a particulate fraction of a strain A tumor by extraction with the detergent, Triton. In the presence of Triton, its sedimentation coefficient suggested a relatively low molecular weight. In the absence of Triton and at a pH in the region of neutrality, the material was almost completely insoluble in water. Digestion with phospholipase A extracted from snake venom rendered the Triton-soluble lipoprotein soluble in water with no gross change in lipid or amino acid composition. Alloantigenic activity of both the Triton-soluble lipoprotein and the product obtained by venom-digestion was demonstrated by the hemagglutinin production, hemagglutination-inhibition, accelerated skin graft rejection, and enhancement tests. The preparations were shown to have specific activity corresponding to the *H-2* specificities 4, 8, and 11. The activity appeared to be independent of the lipid content of different preparations, suggesting that it resided largely or entirely in the protein component.

The investigation of the chemical properties of histocompatibility antigens is of considerable interest, not only because of their role in tissue transplantation but also because some or all of these antigens are components of cell membranes and probably contribute in varying degrees to maintaining the structural and functional integrity of living cells.

THE *H-3* COMPLEX

The *H-3* locus was originally defined by the congenic lines C57BL/10 and B10.LP, nonagouti (*a*) and white-bellied agouti (A^w), respectively, in their coat colors, and by the association, demonstrated by crosses of these lines, of white-bellied agouti with resistance. Crossing over between the *a* locus and *H-3* was estimated at about 10 per cent (Snell, 1958*b*).

This rather close linkage between a dominant gene affecting coat color and a gene for resistance provided a favorable situation for further analysis of the *H-3* locus. +, A^w, or a^t from various sources was introduced onto a C57BL/10 background, and the resulting lines subjected to appropriate tests. In most instances the resulting lines resisted transplants from C57BL/10, and linkage tests showed that the resistance was due to a gene or genes associated with the introduced "marker" gene. In two instances, the introduced marker was lost by crossing over, but resistance remained and was shown still to be due to a gene close to the *a* locus. Two of the resulting lines B10.129 (13M), a nonagouti line, and B10.129(14M) a white-bellied agouti line, came from the same initial cross. Both these lines resist transplants of C57BL/10 tissues, and the resistance has been shown to be due to a gene in linkage group V, but the F_1 hybrid between them is susceptible. Both give resistant F_1 hybrids when crossed to strain B10.LP. Somewhat similar results have been obtained with other lines (Snell and Bunker, 1964; Snell, unpublished data).

Further tests will be necessary before the meaning of these results is finally known, but the probable interpretation is that there are two rather closely linked histocompatibility loci in linkage group V. Line B10.129(14M) differs from C57BL/10 at the *H* locus closer to *a*, lines B10.129(13M) and B10.LP-*a* at the locus farther from *a*. Line B10.LP differs from C57BL/10 at both loci.

By the use of strain B10.UW which carries the linked genes a^t, *un*, and *we* on a C57BL/10 background, one locus of the *H-3* complex has now been located definitely close to *we*, and on the side of *we* away from *a* (Snell and Bunker, 1964; see also the linkage map in Fig. 8-3). This locus will retain the symbol *H-3*. No symbol has been assigned to the other postulated locus.

Until the possible multiplicity of *H* loci in linkage group V is resolved, the question of possible multiple alleles at *H-3* cannot be tested. However, a group of standard inbred stocks has been typed for $H\text{-}3^a$, using as the "known parent" a line which should provide a test for *H-3* only, with the results shown in Table 24-10. The only standard inbred strains sharing $H\text{-}3^a$ with C57BL/10 are C57L and C57BR/cd; all other lines tested carry some other allele.

Skin grafts made from C57BL/10 to either B10.LP or B10.LP-*a* survive about 3 weeks, grafts made in the opposite direction about 5 weeks (Table 24-11). The similarity of results in the two combinations is interesting since, as indicated above, strain B10.LP apparently has an *H* difference from C57BL/10 not shared with B10.LP-*a* as well as one shared with this strain. Curiously, also, skin grafts exchanged between B10.LP and B10.LP-*a* survive for very long periods (Berrian and McKhann, 1960*b;* Graff and Snell, unpublished data) despite the proven difference between the strains.

Attempts to find red-cell agglutinins associated with *H-3* have been unsuccessful (Amos, Stimpfling, and Snell, unpublished data).

THE *H-4* LOCUS

H-4 is identified by the congenic pair C57BL/10 and B10.129(21M) and by its linkage with pink-eye (*p*). No proven crossovers between *H-4* and *p* have occurred (Snell and Stevens, 1961). Skin grafts exchanged between members of this strain pair showed a median survival time of 127 days when C57BL/10 was the donor, 29 days when B10.129(21M) was the donor. This is one of the most striking differences between reciprocal grafts that has yet been observed (Table 24-11).

THE *H-5* AND *H-6* LOCI

H-5 has been established by Amos et al. (1963) by serological techniques. It is defined by antisera, produced in strain C57BL against transplantable lymphoma 6C3HED, which are reactive with the red cells of certain inbred strains but not of others. Reactivity was tested either by red-cell agglutination or by the usually more reliable method of removal of agglutinating activity by absorption. When absorption was used, the preferred red cells for testing residual activity were those of strain 129. When red cells of certain other strains were used, they apparently reacted not only with *H-5* antibodies but with antibodies of other specificities which these antisera contained, thereby confusing the results.

Table 24-10 summarizes some of the results. In this table the symbol + means that the red cells of the indicated strain were reactive, the symbol − that they were nonreactive. Amos et al. regard reactivity as indicative of the presence of a specific allele, $H\text{-}5^a$, but this conclusion rests on the assumption that *H-5* is very unlike *H-2*, determining only one specificity per allele rather than the multiplicity of specificities determined by *H-2*. It is not clear that this assumption is justified by the evidence so far available. This is the reason for using + and − in Table 24-10 to summarize the evidence concerning *H-5* and *H-6* which have been established by serological methods, rather than the letters, indicative of alleles, which are used to summarize the evidence from loci studied by transplantation methods.

It will be seen from the Table that the *H-5* antisera reacted with the cells of strain 129, but did not react with the cells of strains C57BL, B10.129(5M), or B10.129(21M). Since the *H-1* and *H-4* alleles, respectively, of the two last named strains came from strain 129, this distribution of reactivities proves the nonidentity of *H-5* with *H-1* or *H-4*. Were the antisera in question reactive, for example, with *H-1*, B10.129 (5M) would have to react like 129. Nonidentity of *H-5* with *H-3* was proven by showing absence of linkage between *H-5* specificity and the *a* locus. Similarly nonidentity with *H-2* was proven by independent segregation of *H-5* and *H-2* specificities. In these linkage tests, the *H-5* specificity segregated in approximately the ratios expected of a unit mendelian factor. These data adequately establish *H-5* as a new locus.

It should be noted, however, that the tests of *H-5* characterize it as a blood group locus, not necessarily as a histocompatibility locus. Amos et al. believe that it is a histocompatibility locus because they have found the *H-5* alloantigen on numerous tissues in addition to the erythrocytes. This is strong presumptive evidence, but final proof must probably await the establishment of congenic lines with an *H-5* difference and tests with actual transplants.

Evidence for *H-6* is of essentially the same nature as evidence for *H-5*, though proof that it is distinct from previously identified loci is somewhat less satisfactory (Amos et al., 1963). The preferred sera for tests for *H-6* specificity are made in strains C3H/St against the C3H/He ascites sarcoma MC2M. Evidence that *H-6* is not *H-2* or *H-3* comes from linkage data and should be rather conclusive. Evidence with respect to *H-1* and *H-4* comes from the data summarized in Table 24-10 on the strain distribution of alleles and specificities. To evaluate this with respect to *H-6*, we need to take stock of just what these data can tell us.

The serological methods used to study *H-5* and *H-6* reveal specificities, not alleles. Specificities and alleles would, in effect, be the same thing if single alleles determined single specificities. Whether this ever occurs and, if so, with what frequency, we do not know. Certainly it is very far from being the case at *H-2*. To further complicate matters, it is difficult to be sure that antisera are monospecific. An antiserum with two specificities may react with three different alleles. Thus anti-(1,2) will react with (1,2), (1,—2), and (—1,2). The F_1 transplantation test also does not type for alleles, but rather for a *combination* of specificities. Evidence from the F_1 test that two alleles are different should be conclusive. Evidence that they are the same is not conclusive, but proves that at least one and sometimes a considerable number of specificities are shared in common. With these two distinct typing systems, each with its own deficiencies, how can information be interconverted between them?

Unfortunately the answer is that this can usually only be done with assurance under certain rather special circumstances. The key situation is as follows. If two strains are known positively to carry *identical* alleles at a certain locus, and if these strains react *differently* with an antiserum, then the antiserum cannot be testing for the locus in question. An examination of the data in Table 24-10 will show that *H-5* probably does meet this requirement with respect to *H-1* and *H-4*, but that *H-6* does not. In fact the "*H-6*" specificity could very well be one of the specificities determined by *H-1*. At the time that Amos et al. published their report there was some evidence to the contrary. C3H/He which is plus with respect to *H-6*, and DBA/1 and DBA/2 which are minus, were all originally reported as carrying allele $H\text{-}1^a$ (Snell and Stevens, 1961). As noted above, however, evidence that alleles are the same is less conclusive than evidence that they are different, and subsequent tests have classified DBA/2 as $H\text{-}1^d$ (Snell and Graff, unpublished). DBA/1 has not yet been retyped. The distinctness of *H-1* and *H-6* is therefore uncertain. As Amos et al. point out, it is not even entirely certain that the *H-5* and *H-6* specificities are not themselves the product of one locus, though differences in the tissue distribution of corresponding alloantigens (see below) make this unlikely.

The evidence that *H-6* is a histocompatibility locus as well as a blood-group locus is the same as in the case of *H-5*.

OTHER BLOOD GROUP LOCI

Loci *H-5* and *H-6* are blood group loci believed to be concerned also with histocompatibility because their antigens occur in tissues other than blood. Evidence has been presented which may identify four other blood group loci. The degree of overlap between blood group and histocompatibility loci is very uncertain, but we describe these additional loci briefly because a relation to histocompatibility may ultimately be demonstrated. It is already known that the antigens determined by some of them are in tumors or in normal tissues other than blood.

Spencer et al. (1946) have reported two alloantigenic specificities, Kappa and Iota, identified by alloantisera against certain strains of mouse cells grown in vitro. The antibodies were demonstrated by red cell agglutination, but the cells that induced them were not of hematogenous origin, so the antigens seem not to be confined to blood. The distribution of the specificities in different inbred strains is shown in Table 28-10. Kappa occurs only in strain YBR/HeHa, Iota in all strains tested except C3H/St, C58, and F/St. Both specificities segregate independently of *H-2*.

Popp and Popp (1964), using an antiserum produced in C3H mice against RFM tissues and subsequently absorbed with tissues of strain A.CA, have identified a specificity which occurs only in strain RFM of 13 strains tested. The specificity is not related to *H-2*. It is not confined to erythrocytes, and hence may be of significance for histocompatibility.

Singer et al. (1964) have reported a blood group locus *Ea-1*, with alleles *Ea-1*^{a}, *Ea-1*^{b}, and *Ea-1*^{o}, detected in wild house mice. Antibodies are induced by the injection of erythrocytes, and unlike other alloantibodies in mice behave as saline or complete agglutinins. Four phenotypes are demonstrable, A, B, AB, and O. All of 13 inbred strains tested are O. The locus is distinct from *H-2*.

H-7, H-8, H-9, H-10, H-11

Besides the histocompatibility loci so far described, five others have been identified by the application of the F_1 test to lines congenic with C57BL/10 (Snell and Bunker, 1965; Table 24-8). These loci have tentatively been assigned the symbols *H-7, H-8, H-9, H-10,* and *H-11*. The distribution of allele *H-7*^{a} in various inbred strains has been determined by the F_1 test (Table 24-10). The distribution suggests a difference between *H-7* and the blood group loci *H-5* and *H-6* (compare, e.g., strains DBA/1 and 129), but this evidence is subject to the same qualifications as are other comparisons with these two loci. Amos et al. (1963) tested, with their *H-5* typing antiserum, the congenic lines that identify *H-8* and *H-11* and found that they react like C57BL/10. This establishes a presumption that *H-8* and *H-11* are distinct from *H-5,* the presumption being particularly strong in the case of *H-11* because of the key role of strain 129 in the identification of both *H-5* and *H-11*. Aside from the results of these few tests, the nonidentity of these presumed new loci with *H-5* and *H-6* is unproven.

The loci, the CR strains that identify them, and some results with tumor and skin allografts are given in Table 24-11. The table gives similar data for *H-1, H-2, H-3,* and *H-4*. The principal point of interest is the marked differences in the strength of the barrier to allografts imposed by the congenic strain pairs. This was noted in an earlier discussion of the table. Presumably this points to comparable differences in the "strength" of the different loci, but there is a distinct possibility that different allelic combinations within the same locus would alter the results. (See for example the different *H-1* combinations in Table 24-11.) On the basis of evidence at present available, the comparative immunogenic strength of the nine loci listed in the table is, in descending order: *H-2, H-1, H-3, H-7, H-8, H-4* (but this is relatively "strong" in the reciprocal direction), *H-9, H-10, H-11* (but the *H-11* allele present in strain B10.D2(55N) would place this locus in a higher position).

TWO THYMIC ALLOANTIGENS

Two alloantigens, probably genetically distinct, have been identified in the thymus of mice.

The TL antigen is determined by a locus *Tla* in linkage group IX about two crossover units from *H-2* (Fig. 24-4). It is present in thymus, as demonstrated by the capacity of thymus to absorb appropriate antisera, and in certain leukemias as demonstrated by a cytotoxic test. (Hence the designation TL = thymus leukemia.) It is not present in liver, spleen, lymph nodes, or red cells. The strain distribution is shown in Table 24-10. It is present in the thymus of strains A and C58, lacking in the thymus of strains AKR, BALB/c, C3H/He and C57BL/6. Curiously, it is present in some leukemias arising in strains that lack it. It has been possible to produce reactive antisera only when leukemias possessing the antigen are used as the immunizing tissue, not when thymus is so employed. The TL antigen may not affect histocompatibility; C57BL/6 mice with a high titer of TL alloantibody did not reject leukemias possessing the antigen (Old et al., 1963; Boyse et al., 1963, 1964; Boyse et al., 1965).

Reif and Allen (1964) have discovered a second and probably distinct thymic antigen which occurs in two antithetical forms. θ-AKR is present in strains AKR and RF, lacking in all other strains tested. θ-C3H is lacking in AKR and RF, present in all other strains tested (Table 24-10). The antigens are demonstrated by the cytotoxic action on thymic lymphocytes of alloantisera prepared in *H-2*-compatible strains AKR and C3H. The antigens are present in brain as well as thymus, lacking or present in very low concentrations in other tissues tested. Genetic determination has not been demonstrated, but the distinct strain

Table 24-16. EVIDENCE FROM GRAFTS TO C57BL × BALB/c RECIPROCAL HYBRIDS FOR AN X–LINKED HISTOCOMPATIBILITY FACTOR. (*Bailey*, 1963)

Genotype of host[a]	Genotype of first donor[a]	Outcome[b]	Genotype of second donor[a]	Outcome[b]	Foreign element
X^CX^B F_1 ♀	X^BX^C F_1 ♀	+			none
X^BY^C F_1 ♂	X^BY^C F_1 ♂	+			none
X^CY^B F_1 ♂	X^BY^C F_1 ♂	−	X^BY^C F_1 ♂	−	X^B and Y^C
			X^BY^B inbred ♂	−	X^B
			X^BX^B inbred ♀	−	X^B
			X^CY^C inbred ♂	+	Y^C
			X^CX^C inbred ♀	+	none

[a] X^B and Y^B = X and Y derived from strain C57BL, X^C and Y^C = X and Y derived from strain BALB/c.
[b] + indicates acceptance, − rejection. Second grafts showed accelerated rejection.

distribution certainly points to determination by a single locus. There is no evidence as to whether or not the antigens affect histocompatibility.

THE X–LINKED HISTOCOMPATIBILITY FACTOR

Bailey (1963) has demonstrated the presence of a histocompatibility factor on the X chromosome of mice. The evidence is summarized in Table 24-16. Male F_1 hybrids between strains C57BL and BALB/c bearing an X^B (X from strain C57BL) rejected grafts with an X^C (X from strain BALB/c) regardless of the sex of the donor and those bearing an X^C rejected grafts with an X^B. Grafts of all other genotypes were accepted. (Male to female grafts were not made.) Rejection was slow, many grafts surviving beyond 35 days, and in the combination CBF_1 to BCF_1, 14 of 36 grafts had not been rejected at 63 days when the test was terminated. The X locus thus determines a weak immunogen. Perhaps it is especially weak, since the grafts used were tail skin placed on the tail, and there is some reason to believe that grafts of this type are particularly responsive to allogeneic differences (Bailey, personal communication). There is no means at present of telling whether the effect is due to one locus or several, and no evidence as to the number and distribution of alleles at this locus or these loci beyond the fact that strains C57BL and BALB/c must have different alleles.

THE Y–LINKED HISTOCOMPATIBILITY FACTOR

In 1955 Eichwald and Silmser reported that grafts of skin made from male to female within an inbred strain are rejected, whereas grafts from male to male, female to female, and female to male are accepted. This sex-associated incompatibility has been widely studied. Since the Y chromosome of the X/Y male donor is foreign to the X/X female recipient in male-to-female grafts, and in this one combination only, the phenomenon has been generally interpreted as due to a Y-linked histocompatibility gene (or genes).

The rejection due to the Y factor is weak and shows pronounced strain differences. The rejection is entirely consistent in strain C57BL, but mean rejection time has varied from 26 to 45 days in the hands of different investigators. Table 24-11 includes the results of one typical experiment. In other strains, e.g., A and BALB/c, only a fraction of the grafts are rejected, while in strain CBA two extensive experiments failed to produce any rejections at all (Eichwald et al., 1958; Klein and Linder, 1961; Zaalberg, 1959). (But at least one author has reported some rejections in this combination. See Hauschka et al., 1961, for a more complete summary of published results.)

In strain C57BL, in which rejections occur regularly, a second graft is rejected more rapidly than a primary one. Thus 13 C57BL females that had rejected an initial male graft in an average of 26 days, rejected a second graft in an average of 13 days. Also testosterone treatment or castration of the female hosts failed to alter rejection time (Eichwald et al., 1958). The sex-determined rejection thus behaves, as do other rejections of allografts, like an immune phenomenon incited by alloantigenic differences between donor and host.

If the rejection of male-to-female skin grafts is, in fact, due to an antigenic difference between the sexes, the obvious place to look for the genetic determinant of the difference is on the Y chromosome. A Y-linked histocompatibility gene would explain the results, at least so far as strain C57BL is concerned. A Y-linked gene, however, as Fox (1959) has pointed out, is not the only possible explanation. Important modifications of the phenotype, including, apparently, differences in the presence or absence of antigens, can be brought about by modifications in the proportions or balance of the various chromosomes. Studies with *Drosophila,* for example, provide precedents for postulating that the male antigen is determined, not by a Y-linked factor, but by the presence of one rather than two X's.

Celada and Welshons (1963) have carried out experiments which discriminate between these alternatives. Their results confirm the presence of a histocompatibility factor on the Y chromosome. The experiments made use of an exceptional stock of mice in which some animals have the genetic constitution of X/0 (one X and no Y) (Welshons and Russell, 1959) and of a single exceptional mouse of the genotype XXY. The X/0 animals are fertile females; the XXY mouse was a male. Since these animals were not from inbred stocks, the role of the Y could not be tested by direct male-to-female skin grafts. Instead, the presence or absence of the male antigen was demonstrated indirectly by the capacity of tissues bearing it to immunize. Females were immunized with spleen cells and then challenged with an isogenic male graft. Another unusual feature of the system was the use, not of skin, but of male spleen cells from mice presensitized to rat erythrocytes as the challenge. Reduced production of anti-erythrocyte antibody was taken as evidence of preimmunization. The method involves other interesting details, but these need not concern us here. The point of genetic significance is that XXY tissue did immunize and X/0 tissue did not. Evidently the male antigen can safely be attributed to the presence of the Y rather than the absence of a second X. It is truly a Y-linked histocompatibility factor (or factors). These experiments also show that the Y-determined antigen is carried by spleen cells as well as by skin.

While the existence of a Y-linked factor may be regarded as established, the absence of male-to-female graft rejection in strain CBA suggests that the factor may be absent in this strain. But other explanations of this curious strain difference have been proposed. At least four alternatives are possible: (1) The Y of strain CBA lacks the male factor; (2) the male factor is on the X as well as the Y of this strain; (3) CBA's have the Y factor, but they also have an autosomal factor which produces an identical antigen (Michie and McLaren, 1958); (4) the CBA's are less effective than the C57BL's in mounting an immune response against weak antigens or at least against this particular antigen. Tests have now been reported which discriminate quite effectively between these alternatives.

Zaalberg (1959) tested C57BL, CBA, and F_1 hybrid male skin in F_1 male and female hybrids. His most critical results are given in Table 24-17. The rejection of CBA male skin proves that the CBA Y chromosome carries the male factor, and that the CBA X lacks it. This result also proves that CBA does not carry an autosomal replica of the Y factor or at least not a replica that acts as a dominant. Since all known histocompatibility factors in the mouse are dominant (or more precisely codominant), including the Y factor, the postulate of a recessive autosomal replica of the Y factor is improbable. This one simple experiment thus rules out the first three of the four proposed explanations. Klein and Linder (1961) obtained the same sort of result in the critical combination of grafts from CBA males to (CBA × C57BL)F_1 females. It must be added that three out of 23 of Zaalberg's CBA male to F_1 female grafts were accepted, a problem we shall return to later. Linder and Klein obtained 100 per cent rejection.

Zaalberg's tests establish one other point. Forty of 40 C57BL male to (C57BL × CBA)F_1 male grafts were accepted. Since donor and recipient in this combination have Y chromosomes from strains C57BL and CBA, respectively, the histo-

Table 24-17. EVIDENCE THAT THE Y-LINKED HISTOCOMPATIBILITY FACTOR IS PRESENT IN THE Y AND ABSENT IN THE X OF STRAIN CBA. (*Zaalberg,* 1959)

Donor[a]	Recipient[a]	Fraction of grafts accepted	Conclusions
♂ $X^B\mathbf{Y^B}$	♀ X^BX^{CBA}	2/21	The male factor is on $\mathbf{Y^B}$; both X^B and X^{CBA} lack it
♂ $X^{CBA}\mathbf{Y^{CBA}}$	♀ X^BX^{CBA}	3/23	The male factor is on $\mathbf{Y^{CBA}}$; both X^B and X^{CBA} lack it
♂ $X^B\mathbf{Y^B}$	♂ $X^B\mathbf{Y^{CBA}}$	40/40	The male factors on $\mathbf{Y^B}$ and $\mathbf{Y^{CBA}}$ are identical

[a] X^B and X^{CBA} designate the X chromosomes of strain C57BL and CBA, respectively, and $\mathbf{Y^B}$ and $\mathbf{Y^{CBA}}$ their respective Y chromosomes. Boldface type indicates the presence of the male histocompatibility factor.

compatibility antigens determined by the two Y chromosomes must be identical. The genetic formulas used to designate donor and recipient in Table 24-17 should help one to visualize this situation.

Experiments by Billingham and Silvers (1960), utilizing the capacity of neonatally injected marrow to induce tolerance to subsequent grafts of the same genotype, confirm the presence of the male factor in strain CBA. C57BL females made tolerant to CBA male tissue accepted C57BL male skin. CBA males must carry the antigen that normally causes C57BL male skin to be rejected. Similar results were obtained with strains A, AU, and C3H. Controls injected neonatally with female tissue were not rendered tolerant.

We are left then, as the most likely explanation of the absence of male-to-female graft rejection in strain CBA, with the postulate that CBA does not mount an effective immunological response against the male antigen. Differences in immunological response in different strains of mice are well established, and Klein and Linder (1961) have summarized data which suggest a correlation between the immunological capacity of several strains and their tendency to show male-to-female graft rejection.

These authors have also tested CBA male skin in females from a backcross produced by crossing (CBA ♀ × C57BL ♂)F_1 males to CBA females. In a test thus designed, all sex chromosomes of both donor and recipient are of CBA origin. Of 110 backcross mice grafted, 85 rejected their grafts. The rejection times of the sloughed grafts were more varied than in grafts from CBA males to F_1 females. These results fit the postulate that the CBA Y chromosome carries the same male factor as the Y of strain C57BL and that the difference between the strains resides in autosomal modifying factors, presumably factors that affect the organs responsible for the immune response. The authors tentatively suggest that two dominant autosomal genes are involved, but interpretation of multiple-factor situations is difficult, and the number of factors should be regarded as uncertain at the present time. The results do not fit the possibility discussed above that the exceptional response of strain CBA is due to a recessive autosomal replica of the Y factor. This would require 50 per cent rejected grafts in the backcross instead of the 77 per cent observed.

It remains to consider the explanation of those cases, exemplified by strain A, in which some male-to-female isografts are accepted and others rejected. This situation is examined in the next section.

PENETRANCE OF HISTOCOMPATIBILITY GENES

In a number of the studies cited, donor and recipient have differed one from the other at a weakly immunogenic locus. The results have been ambiguous in the sense that some but not all of the grafts have been rejected. Thus only 23 of 39 grafts of strain A skin made from males to females were rejected at the end of 5 months (Eichwald et al., 1958). Similar results are common in the case of tumor allografts; some animals of a given genotype are killed and others survive (Table 24-11). Since these grafts are made between inbred animals, the results may be puzzling to persons to whom inbreeding has become synonymous with uniformity. Actually, results of this sort are no novelty. Geneticists have known for years of examples of variation in inbred strains, sometimes of a very striking nature (see Wright, 1934, for the classic example). They also know a great variety of mutant genes which, both on inbred and noninbred backgrounds, may fail to show 100 per cent expression. To describe this last phenomenon they have coined the term "penetrance," defined as the percentage of animals in a given population in which a given gene is expressed, or in which phenotype corresponds to genotype (e.g., the penetrance of the Y chromosome factor in A male to A female skin grafts is 59 per cent).

While results of this sort may at first seem puzzling, a little consideration will show that they have an obvious explanation. All traits, even those most directly related to gene action, are determined at least in part by environment. Imperfect penetrance is thus merely a reflection of the interaction of environment with genotype. In noninbred animals, an important part of the "environment" of any given gene is the rest of the gene complex. In inbred animals, the gene complex is uniform, but its expression may be modified by any or all of a great variety of forces to which it is exposed, forces which begin at conception and do not terminate till death. Variation of a given trait within an inbred strain thus becomes a direct measure of the role of environment in determining that trait.

Our knowledge of the environmental forces influencing tissue allografts is far from complete, but the role of certain factors is well established.

Table 24-18. RELATION OF THE AGE OF DONOR AND HOST TO THE PER CENT SUCCESSFUL SKIN GRAFTS MADE FROM MALE TO FEMALE IN STRAIN A. (*Mariani et al.*, 1959)

Approximate age of donor, weeks	Approximate age of host, weeks	Per cent successful grafts (survival >150 days)
6	5	75
40	5	65
7	8	40
6	33	47
31	39	0

One of the more important is age of the host, with age of the donor in some cases also playing a role. A typical experiment with skin grafts, reported by Mariani et al. (1959) is summarized in Table 24-18. The percentage of success of male-to-female skin grafts made in strain A was 75 when donors and hosts were both young (6 and 5 weeks of age, respectively), but was 0 per cent when both were fully mature (31 and 39 weeks). Other age combinations gave intermediate results. The role of the age of the host is well established. Maximum resistance is reached at about 12 weeks; presumably this coincides with the attainment of full immunological maturity (see Snell, 1958*a*, for summary). The effect of age of donor is less striking, but the relatively prolonged survival of immature skin is well documented (see e.g., Cannon et al., 1954; Billingham and Silvers, 1964).

There are a number of other known environmental factors that influence the outcome of grafts, e.g., the tissue employed, the site in which it is placed, and prior immunization of the recipient. Even if all these known factors are controlled, some variation persists in the outcome of allografts, especially allografts where the genetic disparity between donor and host is marginal. A translocation of the Y factor to an autosome has been invoked to explain results of this type obtained with male-to-female grafts (Hauschka and Holdridge, 1962). Possibly such explanations are justified by the facts in a few instances. However, imperfect penetrance of genes on an inbred background due to shifts in minor and largely obscure environmental factors is well known for a variety of genes and there is no reason to doubt that it applies to histocompatibility genes also. A weak level of resistance, a delicate balance between the mechanisms of immunity and tolerance, and fluctuations in minor environmental factors are quite sufficient to account for these cases of imperfect penetrance. What the minor environmental factors are we can only guess, but the state of the hair-growth cycle in the graft, the amount of blood in the graft, the site of the graft relative to host blood vessels, minor technical fluctuations, the state of the estrus cycle of the host, and the health and infective state of the host are some of the possibilities.

Of course, if one were able to measure directly the presence of the histocompatibility alloantigens and use this as the measure of gene penetrance, the penetrance of histocompatibility genes would doubtless be found to be almost, if not quite, 100 per cent.

EXCEPTIONS TO THE LAWS OF TRANSPLANTATION

That there are exceptions to some of the five laws of transplantation described in an earlier section should now be apparent. The commonest exceptions concern law 2 which asserts the incompatibility of allografts. There is some reason to suppose that allografts are, in fact, almost always *resisted*, except perhaps in the case of certain tissues (e.g., cartilage), or tissues placed in certain "privileged sites" (e.g., the anterior chamber of the eye), but they certainly are not always rejected. Exceptions are very common with tumors, especially across non-*H-2* barriers. Generally speaking, the tendency of tumors to grow progressively in foreign hosts increases with the number of transplant generations through which they have been carried. "Antigenic simplification" and increased capacity to survive the immunological attack of the host have been suggested as explanations. Probably the latter is the more important. Some tissues, e.g., ovary, grow quite easily across weak histocompatibility barriers (Linder, 1961; Snell and Stevens, 1961). Even skin, probably the most sensitive of all histocompatibility indicators, at least in the absence of prior immunization sometimes transgresses the weaker barriers.

Law 3, which asserts the compatibility of tissues grafted from an inbred parent to its F_1 hybrid, was long regarded as universally valid, but a major exception has been found (Snell, 1958*a*; Snell and Stevens, 1961) and analyzed in a series of ingenious experiments (Cudkowicz and Stimpfling, 1964*a*, *b*; 1965*a*, *b*). The phenomenon is noted when lymphoid tissues of the genotype $H\text{-}2^b/H\text{-}2^b$ are grafted to recipients heterozygous for $H\text{-}2^b$ and some other $H\text{-}2$ allele. By the use of alleles derived by crossing over within the $H\text{-}2$ locus, it has been shown that the effect is controlled by a factor which is either a part of or close to the D region of the $H\text{-}2$ locus (Fig. 24-4). It has been suggested that an antigenic specificity normally produced by allele $H\text{-}2^b$ is not produced when $H\text{-}2^b$ is combined with another $H\text{-}2$ allele. Heterozygotes can therefore react against $H\text{-}2^b$ homozygotes. Since, however, the phenomenon is noted only with grafts of lymphoid tissues and not, for example, with skin, the postulated specificity must be formed only in such tissues. Also there is some reason to doubt that rejection of $H\text{-}2^b/H\text{-}2^b$ marrow by F_1 hybrids is an immune phenomenon. The test for successful marrow transplantation employed by Cudkowicz and Stimpfling is carried out in irradiated hosts. C3H hosts irradiated with 1,000 R show no resistance to C57BL/10 marrow by this test despite the "strong" $H\text{-}2$ difference, yet in (C57BL/10 × C3H)F_1 hosts receiving the same radiation dose the hybrid resistance is fully manifested (Cudkowicz and Stimpfling, 1964*a*). On the other hand hybrids pretreated with marrow show an increased resistance suggestive of immunity in genetic combinations expected to show the hybrid effect, and hybrids pretreated with spleen show a reduced resistance suggestive of tolerance (Cudkowicz and Stimpfling, 1964*c*).

The rejection of 5×10^5 cells of $H\text{-}2^b/H\text{-}2^b$ marrow by heterozygous recipients noted by Cudkowicz and Stimpfling occurs within 5 days. When 10^7 nucleated marrow cells are used, repopulation of the recipient usually though not always occurs. Popp and Cudkowicz (1965) have shown that the failures under these conditions are not related to $H\text{-}2^b$, though they are dependent on the use of heterozygous recipients.

Other but much less striking examples of hybrid resistance have been reported with transplantable sarcomas and carcinomas in genotypes not confined to $H\text{-}2^b$ (Hellström, 1964). Studies of this sort have led to the detection of the interesting phenomenon of syngeneic preference in which cells in tissue culture are damaged by contact with histoincompatible cells or even histoincompatible alloantigens (Hellström et al., 1964; Möller, 1965). If hybrid resistance, at least in some instances, is indeed a manifestation of syngeneic preference, then the role of the F_1 hybrid hosts is to provide an environment for transplants in which intimate contact can occur between cells which are genetically disparate and probably disparate with respect to the alloantigens of their cell membranes, and yet in which no ordinary immunological reaction can mask the resulting cell damage.

ARE THERE "HYBRID ANTIGENS" IN THE MOUSE?

In rabbits and some other species, animals heterozygous at certain blood group loci have been shown to possess an alloantigen or alloantigens not determined by the alleles of these loci acting alone. A search for similar "hybrid antigens" in mice gave negative results. Martinez et al. (1959) rendered strain A mice tolerant to C3H alloantigens, and then grafted them simultaneously with C3H and (A × C3H)F_1 skin. Both grafts were accepted. The same results were obtained in the reciprocal combination. Had heterozygosity at any of the histocompatibility loci by which strains A and C3H differ led to the formation of alloantigens not present in the homozygotes, the hybrid grafts should have been rejected. The test of course gives no information about loci at which the strains used are alike or about alloantigens confined to tissues other than skin.

NUMBER OF HISTOCOMPATIBILITY LOCI

In an earlier section we discussed the formulas by which the percentage of susceptible animals in the segregating generations of a cross is related to the number of segregating histocompatibility loci, but we deferred actual attempts to estimate the number of loci. We are now in a better position to evaluate such data as are available. There are at least three and perhaps four factors that would lead to underestimation of the number of loci: (1) Incomplete penetrance would increase the proportion of survivors and therefore lower the estimate. This is particularly true of tumor transplants but we have seen that it can apply to skin grafts also. (2) Linked loci would tend to behave like a single locus and would add less to the estimate of gene number than would the same loci if unlinked. Linked histocompatibility loci are already known and more will certainly be discovered. (3) If the strains crossed carry the same allele at any locus, this locus would make no contribution to the estimate of gene number. How often this is likely to occur depends on the number of alleles at different loci. At present this is largely an immeasurable factor. (4) It may well be that, because of an unequal distribution of alloantigens in the different tissues, some types of grafts fail to detect loci that would be demonstrable were other tissues employed.

There are no comparable factors that would tend to lead to overestimation of the number of loci. However, in the two tests with skin grafts summarized in Table 24-2 some use was made of male donors with female recipients. This could lead to rejections due to the Y chromosome factor and to an elevation of the estimate of locus number. An extenuating circumstance is that the strains used happen to be ones in which the Y chromosome effect is particularly weak. Strains CBA and DBA/2 practically do not show it at all. Perhaps the results can be taken at their face value.

These tests led to estimates of approximately 13 and 15 loci, respectively. In the light of the sources of error mentioned above, these are clearly minimum figures. It should also be remembered that in Barnes and Krohn's tests, *all* grafts were sloughed by 200 days. While this implies that some of the loci acting are very weakly immunogenic, it leads to a necessary but indeterminate elevation of estimated locus number.

Chai and Chiang (1963) devised a mathematical method for estimating histocompatibility gene number from the percentage acceptance of grafts exchanged between sibs in a partially inbred strain. Using data from the 14th to 19th generations of brother-sister inbreeding of strain LG, they arrived at an estimate of 18 histocompatibility loci. If there are selective forces favoring heterozygosity at histocompatibility loci, this would tend to make this figure an overestimate. But the other factors already enumerated would tend to make it an underestimate. The figure would not include the X and Y chromosome loci.

Bailey and Mobraaten (1964), using the same genetic device but grafts of tail skin to the tail rather than the more familiar thoracic grafts, arrived at a figure of 32 loci. In a second test using similar grafts but animals of an N5 generation, the same authors found evidence for 29 loci. The high estimates from these last studies possibly reflect an unusual responsiveness of the tail skin method to allogeneic differences (Bailey, personal communication).

Quite evidently estimates of about 14 or 15 histocompatibility loci which have frequently appeared in the literature are substantial underestimates. Quite evidently also the number of loci which we call histocompatibility loci is a function of the methods we use to detect them.

MUTATIONS AT HISTOCOMPATIBILITY LOCI

There is considerable evidence that, with the passage of time, occasional minor or "weak" histoincompatibilities arise within inbred strains, presumably due to mutation. Linder (1963) performed 600 skin grafts between sublines of CBA separated 0 to 18 generations. (The number of generations was taken as the sum of the generations separating donor and recipient from their common ancestral brother-sister pair.) There were no rejections in grafts between 280 pairs separated zero to eight generations, three delayed rejections (82 to 100 days) in 320 similar grafts in generations 10 to 18. Four additional grafts in the latter group showed signs of incipient rejection but were nevertheless retained. Linder was cautious in interpreting the histoincompatibility thus revealed as due to mutation, but since strain CBA had passed through some 120 generations of brother-sister inbreeding prior to the beginning of the tests, mutation seems a more likely explanation than continued segregation. Billingham et al. (1954) had previously reported signs of incipient rejection in grafts between sublines of A and CBA, and Snell and Stevens (1961) found that sublines 6 and 10 of strain C57BL, separated some 25 years after prolonged inbreeding, could be easily differentiated with a transplantable tumor if the hosts were given prior immunization with donor tissue. Godfrey and Searle (1963) noted no rejections in 82 grafts of tail skin made between sublines of C3H which had been subjected to prolonged gamma irradiation. Donor and host were separated by an average of about 34 generations.

Major differences at the *H-2* locus have also been reported as appearing within inbred strains, but the circumstances of discovery of these variants were such as to make it difficult to rule out contamination. In one case a variety of other differences has been shown, by several methods, to accompany the *H-2* difference (see, e.g., Snell, 1958*b*).

DISTRIBUTION OF HISTOCOMPATIBILITY ALLOANTIGENS IN CELLS AND TISSUES

The distribution of the histocompatibility alloantigens in cells and tissues is a complex problem about which little is known. Yet it is an important problem in transplantation studies, since the potential for histoincompatibility of a particular

Table 24-19. RELATIVE CONCENTRATION IN DIFFERENT TISSUES OF THE *H-2*, *H-5*, AND *H-6* TRANSPLANTATION ALLOANTIGENS

	H-2		*H-5*	*H-6*
Tissue	Amos et al. (1963)	Basch and Stetson (1962)[a]	Amos et al. (1963)	Amos et al. (1963)
Spleen	4	4	1	1–3
Liver	3	3	0–2	2
Thymus		2		
Lung	2	2	2	2
Adrenal		2		
Gut	2			4
Kidney	1	1	4	1
Red cell	1	0	4	4
Testis	0		3	3
Heart		½		
Muscle	0	0	1	1
Brain	0	0	1	3

[a] The values given by Basch and Stetson are transformed to a numerical system approximating that used by Amos et al. As thus expressed, only the relative concentrations are indicated, with 4 indicating the highest concentration, 0 the lowest.

tissue obviously resides in the alloantigens which it carries.

Much of the information so far gathered on tissue distribution concerns three loci, *H-2*, *H-5*, and *H-6*, and is based on the capacity of tissue homogenates to absorb the appropriate antibodies. Basch and Stetson (1962) studied the tissue distribution of the *H-2* alloantigen and Amos et al. (1963) of *H-2*, *H-5*, and *H-6*. Some of the results of these authors are summarized in Table 24-19. The figures in the table indicate the relative, not the absolute, concentration of the different antigens. It will be seen that the two different studies gave essentially identical results concerning the tissue distribution of the *H-2* substance, but that when different alloantigens were compared, the distributions showed striking differences. Thus high concentrations of *H-2* are found in the spleen and liver; of *H-5* in the kidney, red cells, and testis; of *H-6* in the gut, red cells, testis, and brain. While both spleen and liver are relatively rich in *H-2*, the properties of *H-2* appear to differ in these two organs. The *H-2* substance in liver is more easily sedimented than that in spleen, and it is less effective in inducing antibody formation, particularly following a single injection of lyophilized tissue (Pizarro et al., 1963).

The Y-determined antigen, like *H-2*, *H-5*, and *H-6*, probably occurs in a variety of tissues, since male-to-female interstrain grafts of a number of tissues besides skin are rejected (see, for example, Gittes and Russell, 1961). Katsh et al. (1964) have shown that there is accelerated rejection of male skin in isogeneic females injected with 1 million sperm and delayed rejection (tolerance) in females injected with 8 million sperm. Apparently the Y antigen is present on sperm, in fact a comparison of the effectiveness of sperm as compared with a preparation of spleen cells in producing immunity and tolerance suggests that it is present on sperm in particular abundance.

As Billingham et al. (1956) point out, the capacity of neonatally injected spleen to induce tolerance to a subsequent skin graft from the same donor strain proves that all the transplantation alloantigens of the latter tissue must be present on the former. It does not necessarily follow, however, that the relative concentration of the antigens is the same in both tissues.

As to the intracellular distribution of the histocompatibility alloantigens, the best available evidence points to an association with the cell membrane. The evidence is quite firm in the case of *H-2* and of those other alloantigens (*H-1*, *H-5*, *H-6*) which are demonstrable by red-cell agglutination, but is of an uncertain nature in all other cases. The *H-2* alloantigen, at least, is probably associated with the membranes of the endoplasmic reticulum as well as with the surface membrane. For references and further details, see Snell (1962, 1964*a*).

SUMMARY AND CONCLUSIONS

In summary, the compatibility of tissue grafts is governed by 15 or more loci called histocompatibility loci. Fifteen is a minimum estimate; the actual number may be substantially more than this. These loci are widely distributed throughout the chromosomes. Histocompatibility genes have so far been identified on the X and Y, and in linkage groups I, V, and IX.

Histocompatibility genes probably act as such because their end products are alloantigens, i.e., possess the property of inciting an immune response when transferred via a graft from an individual possessing them to an individual lacking them. Even though all histocompatibility genes may determine alloantigens, the reverse proposition, that all alloantigens play a role in histocompatibility, is probably not true. For example, a gene whose alloantigenic end product is confined to the fluid components of the blood will presumably not affect histocompatibility.

Histocompatibility loci differ in their immunogenicity or "strength." Some impose a much more powerful barrier to transplants than do others. The *H-2* locus is seldom transgressed even by transplants of most tumors; the Y-chromosome locus is essentially without effect on some genetic backgrounds even when tested by such a sensitive agency as skin grafts. These properties of the loci presumably relate to the character, quantity, or location of the alloantigen which each determines.

For a few of the histocompatibility loci of the mouse, there is rather conclusive evidence that their end product is a component of the cell membrane. It is tempting to speculate that this is true of most or all of the 15 or more such loci, excepting, perhaps, some of those with particularly weak effects. The histocompatibility loci would then be the loci whose normal function is to produce the chemical building blocks from which the cell membrane is formed. While such evidence as is available indicates a wide distribution in the different tissues of the various histocompatibility alloantigens, the relative concentrations of these substances may vary greatly. As to the developmental factors which determine these variations in concentration we have no knowledge, but in some way the different *H* loci contribute more or less to the cell—more specifically, perhaps, to the cell membrane—according to the tissue in which they operate. They thereby contribute to the individuality of tissues as well as to the individuality of the organism as a whole.

LITERATURE CITED

Allen, S. L. 1955*a*. *H-2*^{f}, a tenth allele at the histocompatibility-2 locus in the mouse as determined by tumor transplantation. Cancer Res. 15:315–319.

Allen, S. L. 1955*b*. Linkage relations of the genes histocompatibility-2 and fused tail, brachyury and kinky tail in the mouse as determined by tumor transplantation. Genetics 40:627–650.

Amos, D. B. 1958. Genetic studies on tumor immunity: an analysis of C3H sublines and their tumors. Ann. N.Y. Acad. Sci. 71:1009–1021.

Amos, D. B. 1959. Some iso-antigenic systems of the mouse. Proc. 3rd Can. Cancer Conf. 3:241–258.

Amos, D. B. 1962. Isoantigens of mouse red cells. Ann. N.Y. Acad. Sci. 97:69–82.

Amos, D. B., P. A. Gorer, and Z. B. Mikulska. 1955*a*. The antigenic structure and genetic behavior of a transplanted leukosis. Brit. J. Cancer 9:209–215.

Amos, D. B., P. A. Gorer, and Z. B. Mikulska. 1955*b*. An analysis of an antigenic system in the mouse (the *H-2* system). Proc. Roy. Soc. B 144:369–380.

Amos, D. B., M. Zumpft, and P. Armstrong. 1963. H-5.A and H-6.A, two mouse isoantigens on red cells and tissues detected serologically. Transplantation 1:270–283.

Bailey, D. W. 1963. Histoincompatibility associated with the X chromosome in mice. Transplantation 1:70–74.

Bailey, D. W., and L. E. Mobraaten. 1964. Estimates of the number of histocompatibility loci at which the BALB/c and C57BL/6 strains of mice differ. Genetics 50:233. (Abstr.)

Barnes, A. D., and P. L. Krohn. 1957. The estimation of the number of histocompatibility genes controlling the successful transplantation of normal skin in mice. Proc. Roy. Soc. B 146:505–526.

Basch, R. S., and C. A. Stetson. 1962. The relationship between hemagglutinogens and histocompatibility antigens in the mouse. Ann. N.Y. Acad. Sci. 97:83–94.

Berrian, J. H., and C. F. McKhann. 1960*a*. Strength of histocompatibility genes. Ann. N.Y. Acad. Sci. 87:106–111.

Berrian, J. H., and C.F. McKhann. 1960*b*. Transplantation immunity involving the *H-3* locus: graft survival times. J. Nat. Cancer Inst. 25: 111–123.

Billingham, R. E., L. Brent, and P. B. Medawar. 1956. Quantitative studies on tissue transplantation immunity. III. Actively acquired tolerance. Phil. Trans. Roy. Soc. B 239:357–414.

Billingham, R. E., L. Brent, P. B. Medawar, and E. M. Sparrow. 1954. Quantitative studies on tissue transplantation immunity. I. The survival times of skin homografts exchanged between members of different inbred strains of mice. Proc. Roy. Soc. B 143:43–58.

Billingham, R. E., and W. K. Silvers. 1960. Studies on tolerance of the Y chromosome antigen in mice. J. Immunol. 85:14–26.

Billingham, R. E., and W. K. Silvers. [ed.] 1961. The transplantation of tissues and cells. Wistar Institute Press, Philadelphia. 149 p.

Billingham, R. E., and W. K. Silvers. 1963. Sensitivity to homografts of normal tissues and cells. Annu. Rev. Microbiol. 17:531–564.

Billingham, R. E., and W. K. Silvers. 1964. Studies on homografts of foetal and infant skin and further observations on the anomalous properties of pouch skin grafts in hamsters. Proc. Roy. Soc. B 161:168–190.

Boyse, E. A., L. J. Old, and S. Luell. 1963. Antigenic properties of experimental leukemias. II. Immunological studies *in vivo* with C57BL/6 radiation-induced leukemias. J. Nat. Cancer Inst. 31:987–995.

Boyse, E. A., L. J. Old, and S. Luell. 1964. Genetic determination of the *TL* (thymus-leukaemia) antigen in the mouse. Nature 201:779.

Boyse, E. A., L. J. Old, and E. Stockert. 1965. TL: a leukemia-specific antigen determined by the genome of the leukemic cell. Symp. Immunopathol. (Monte-Carlo). (In press.) (Abstr.)

Brent, L., P. B. Medawar, and M. Ruszkiewicz. 1961. Serological methods in the study of transplantation antigens. Brit. J. Exp. Pathol. 42: 464–477.

Brondz, B. D., and I. K. Yegorov. 1964. Antigenic structure of locus *H-2* of the mouse strains CC57BR and CC57W. Folia Biol. (Prague) 10:90–93.

Cannon, J. A., R. A. Weber, and W. P. Longmire, Jr. 1954. Factors influencing the survival of successful skin homografts in the chicken. I. Effects of varying age of donor and recipient. Ann. Surg. 139:468–472.

Castermans, A., and A. Oth. 1959. Transplantation immunity: separation of antigenic components from isolated nuclei. Nature 184:1,224–1,225.

Celada, F., and W. J. Welshons. 1963. An immunogenetic analysis of the male antigen in mice utilizing animals with an exceptional chromosome constitution. Genetics 48:139–151.

Chai, C. K., and M. S. M. Chiang. 1963. A method of estimating the number of histocompatibility loci in a sib-mating mouse population. Genetics 48:1,153–1,161.

Cloudman, A. M. 1932. A comparative study of transplantability of eight mammary gland tumors arising in inbred mice. Amer. J. Cancer 16:568–630.

Counce, S., P. Smith, R. Barth, and G. D. Snell. 1956. Strong and weak histocompatibility differences in mice and their role in the rejection of homografts of tumors and skin. Ann. Surg. 144:198–204.

Cudkowicz, G., and J. H. Stimpfling. 1964*a*. Deficient growth of C57BL marrow cells transplanted in F_1 hybrid mice: association with the histocompatibility-2 locus. Immunology 7:291–305.

Cudkowicz, G., and J. H. Stimpfling. 1964*b*. Hybrid resistance to parental marrow grafts: association with the K region of *H-2*. Science 144: 1,339–1,340.

Cudkowicz, G., and J. H. Stimpfling. 1964*c*. Induction of immunity and of unresponsiveness to parental marrow grafts in adult F_1 hybrid mice. Nature 204:450–453.

Cudkowicz, G., and J. H. Stimpfling. 1965*a*. Hybrid resistance controlled by *H-2* region: correction of data. Science 147:1,056.

Cudkowicz, G., and J. H. Stimpfling. 1965*b*. Lack of expression of parental isoantigen(s) in F_1 hybrid mice, p. 144–148. *In* Proc. X Cong. Int. Soc. Blood Transfusion. S. Karger, Basel.

Dhaliwal, S. S. 1964. Histocompatibility variations in mouse tumor cells and embryonic tissue with the use of C57BL/10Sn strain and its isogenic resistant strains. J. Nat. Cancer Inst. 32:1,001–1,022.

Eichwald, E. J., and C. R. Silmser. 1955. (Note without title.) Transplant. Bull. 2:148–149.

Eichwald, E. J., C. R. Silmser, and I. Weissman. 1958. Sex-linked rejection of normal and neoplastic tissue. I. Distribution and specificity. J. Nat. Cancer Inst. 20:563–575.

Fox, A. S. 1959. Genetic determination of sex-specific antigens. J. Nat. Cancer Inst. 23:1,297–1,311.

Gittes, R. F., and P. S. Russell. 1961. Male histocompatibility antigens in mouse endocrine tissues: functional and histologic evidence. J. Nat. Cancer Inst. 26:283–303.

Godfrey, J., and A. G. Searle. 1963. A search for histocompatibility differences between irradiated sublines of inbred mice. Genet. Res. 4:21–29.

Gorer, P. A. 1937. The genetic and antigenic basis of tumour transplantation. J. Pathol. Bacteriol. 44:691–697.

Gorer, P. A. 1938. The antigenic basis of tumor transplantation. J. Pathol. Bacteriol. 47:231–252.

Gorer, P. A. 1942. The role of antibodies in immunity to transplanted leukaemia in mice. J. Pathol. Bacteriol. 54:51–65.

Gorer, P. A. 1956. Some recent work in tumor immunity. Adv. Cancer Res. 4:149–186.

Gorer, P. A. 1959. Some new information on *H-2* antigens in mice, p. 25–30. *In* F. Albert and P. B. Medawar [ed.] Biological problems of grafting. Blackwell Scientific Publications, Ltd., Oxford.

Gorer, P. A., S. Lyman, and G. D. Snell. 1948. Studies on the genetic and antigenic basis of tumor transplantation. Linkage between a histocompatibility gene and 'fused' in mice. Proc. Roy. Soc. B 135:499–505.

Gorer, P. A., and Z. B. Mikulska. 1954. The antibody response to tumor inoculation. Improved methods of antibody detection. Cancer Res. 14: 651–655.

Gorer, P. A., and Z. B. Mikulska. 1959. Some further data on the *H-2* system of antigens. Proc. Roy. Soc. B 151:57–69.

Graff, R. J., W. H. Hildemann, and G. D. Snell. 1965. Skin and tumor allografts in immunized and unimmunzied mice congenic at various non-*H-2* histocompatibility loci. Transplantation. (In press.)

Hauschka, T. S., S. T. Grinnell, M. Meagher, and D. B. Amos. 1961. Sex-linked incompatibility of male skin and primary tumors transplanted to isologous female mice, p. 271–294. *In* Genetics and cancer. 13th Annu. Symp. Fundamental Cancer Res. University of Texas Press, Austin.

Hauschka, T. S., and B. A. Holdridge. 1962. A cytogenetic approach to the Y-linked histocompatibility antigen of mice. Ann. N.Y. Acad. Sci. 101:12–22.

Hauschka, T. S., J. T. Mitchell, and D. J. Niederpruem. 1959. A reliable frozen tissue bank: viability and stability of 82 neoplastic and normal cell types after prolonged storage at −78°C. Cancer Res. 19:643–653.

Hellström, K. E. 1961. Studies on the mechanism of isoantigenic variant formation in heterozygous mouse tumors. II. Behavior of *H-2* antigens D and K: cytotoxic tests on mouse lymphomas. J. Nat. Cancer Inst. 27:1,095–1,105.

Hellström, K. E. 1964. Growth inhibition of sarcoma and carcinoma cells of homozygous origin. Science 143:477–478.

Hellström, K. E., I. Hellström, and G. Haughton. 1964. Demonstration of syngeneic preference *in vitro*. Nature 204:661–664.

Herzenberg, L. A., and L. A. Herzenberg. 1961. Association of *H-2* antigens with the cell membrane fraction of mouse liver. Proc. Nat. Acad. Sci. 47:762–767.

Hoecker, G., S. Counce, and P. Smith. 1954. The antigens determined by the *H-2* locus: a Rhesus-like system in the mouse. Proc. Nat. Acad. Sci. 40:1,040–1,051.

Hoecker, G., and O. Pizarro. 1962. The histocompatibility antigens, p. 54–71. *In* A. P. Cristoffanini and G. Hoecker [ed.] Proceedings International Symposium Tissue Transplantation. Universidad Chile, Santiago.

Hoecker, G., O. Pizarro, and A. Ramos. 1959. Some new antigens and histocompatibility factors in the mouse. Transplant. Bull. 6:407–411.

Kandutsch, A. A., and U. Reinert-Wenck. 1957. Studies on a substance that promotes tumor homograft survival (the "enhancing substance"). J. Exp. Med. 105:125–139.

Kandutsch, A. A., and J. H. Stimpfling. 1963. Partial purification of tissue isoantigens from a mouse sarcoma. Transplantation 1:201–216.

Kandutsch, A. A., and J. H. Stimpfling. 1965. Properties of purified isoantigenic preparations from a mouse sarcoma, p. 1–12. *In* Counc. Int. Organ. Med. Sci. Symp. Immuropathol. Schwabe, Basel.

Katsh, G. F., D. W. Talmage, and S. Katsh. 1964. Acceptance or rejection of male skins by isologous female mice; effect of injection of sperm. Science 143:41–42.

Klein, E., and G. Klein. 1964. Studies on the mechanisms of isoantigenic variant formation in heterozygous mouse tumors. III. Behavior of the *H-2* antigens D and K when located in the *trans* position. J. Nat. Cancer Inst. 32:569–578.

Klein, E., and O. Linder. 1961. Factorial analysis of the reactivity of C57BL females against isologous male skin grafts. Transplant. Bull. 27:457–460.

Linder, O. E. A. 1961. Comparisons between survival of grafted skin, ovaries and tumors in mice across histocompatibility barriers of different strength. J. Nat. Cancer Inst. 27:351–373.

Linder, O. E. A. 1963. Skin compatibility of different CBA sublines separated from each other in the course of a varying number of generations. Transplantation 1:58–60.

Little, C. C. 1914. A possible Mendelian explanation for a type of inheritance apparently non-Mendelian in nature. Science 40:904–906.

Little, C. C. 1941. The genetics of tumor transplantation, p. 279–309. *In* G. D. Snell [ed.] Biology of the laboratory mouse. Blakiston, Philadelphia.

Little, C. C., and E. E. Tyzzer. 1916. Further studies on inheritance of susceptibility to a transplantable tumor of Japanese waltzing mice. J. Med. Res. 33:393–425.

Manson, L. A., G. V. Foschi, and J. Palm. 1963. An association of transplantation antigens with microsomal lipoproteins of normal and malignant mouse tissues. J. Cell. Comp. Physiol. 61:109–118.

Mariani, T., C. Martinez, J. M. Smith, and R. A. Good. 1959. Age of donor and host in sex histo-incompatibility to skin isografts. Proc. Soc. Exp. Biol. Med. 102:751–755.

Martinez, C., F. Shapiro, and R. A. Good. 1959. Absence of gene interaction in mouse hybrids, revealed by studies of immunological tolerance and homotransplants. Proc. Soc. Exp. Biol. Med. 101:658–660.

Michie, D., and A. McLaren. 1958. A proposed genetic analysis of the Eichwald-Silmser effect. Transplant. Bull. 5:17–18.

Mitchison, N. A. 1956. Antigens of heterozygous tumors as material for the study of cell heredity. Proc. Roy. Phys. Soc. 25(2):45–48.

Möller, E. 1965. Contact-induced cytotoxicity by lymphoid cells containing foreign isoantigens. Science 147:873–879.

Old, L. J., E. A. Boyse, and E. Stockert. 1963. Antigenic properties of experimental leukemias. I. Serological studies *in vitro* with spontaneous and radiation-induced leukemias. J. Nat. Cancer Inst. 31:977–986.

Pizarro, O., G. Hoecker, P. Rubinstein, and A. Ramos. 1961. The distribution in the tissues and the development of *H-2* antigens of the mouse. Proc. Nat. Acad. Sci. 47:1,900–1,906.

Pizarro, O., P. Rubinstein, and G. Hoecker. 1963. Properties of histocompatibility-2 antigens from different tissues. Guy's Hosp. Rep. 112: 392–401.

Popp, D. M., and D. B. Amos. 1965. An *H-2* analysis of strain RFM/U mice. Transplantation 3:501–508.

Popp, R. A., G. E. Cosgrove, and R. D. Owen. 1958. Genetic differences in hemoglobin as markers for bone marrow transplantation in mice. Proc. Soc. Exp. Biol. Med. 99:692–694.

Popp, R. A., and G. Cudkowicz. 1965. Independence of deficient early growth and later regression of $(C57BL \times 101)F_2$ marrow grafts in $(C57BL \times 101)F_1$ hybrid mice. Transplantation 3:155–160.

Popp, D. M., and R. A. Popp. 1964. A non-*H-2* erythrocyte isoantigen on RFM mice. Genetics 50:276–277. (Abstr.)

Prehn, R. T., and J. M. Main. 1958. Number of mouse histocompatibility genes involved in skin grafting from strain BALB/cAn to strain DBA/2. J. Nat. Cancer Inst. 20:207–209.

Reif, A. E., and J. M. V. Allen. 1964. The AKR thymic antigen and its distribution in leukemias and nervous tissues. J. Exp. Med. 120:413–433.

Rosenfield, R. E., F. H. Allen, S. N. Swisher, and S. Kochwa. 1962. A review of Rh serology and presentation of a new terminology. Transfusion 2:287–312.

Rubin, B. A. 1960. The determination of the *H-2* genotype of CE mice. Transplant. Bull. 26:153–154.

Rubinstein, P., and J. W. Ferrebee. 1964. The *H-2* phenotypes of random-bred Swiss-Webster mice. Transplantation 2:715–721.

Russell, P. S., and A. P. Monaco. 1965. The biology of tissue transplantation. Little, Brown, Boston. 207 p.

Shreffler, D. C. 1964. A serologically detected variant in mouse serum: further evidence for genetic control by the histocompatibility-2 locus. Genetics 49:973–978.

Shreffler, D. C. 1965. The Ss system of the mouse —a quantitative serum protein difference genetically controlled by the *H-2* region. p. 11–19. *In* J. Palm (ed.) Isoantigens and cell interactions. Wistar Institute Press, Philadelphia.

Shreffler, D. C., and R. D. Owen. 1963. A serologically detected variant in mouse serum: inheritance and association with the histocompatibility-2 locus. Genetics 48:9–25.

Singer, M. F., M. Foster, M. L. Petras, P. Tomlin, and R. W. Sloan. 1964. A new case of blood group inheritance in the house mouse, *Mus musculus*. Genetics 50:285–286. (Abstr.)

Snell, G. D. 1948. Methods for the study of histocompatibility genes. J. Genet. 49:87–108.

Snell, G. D. 1951. A fifth allele at the histocompatibility-2 locus of the mouse as determined by tumor transplantation. J. Nat. Cancer Inst. 11:1,299–1,305.

Snell, G. D. 1952. Enhancement and inhibition of the growth of tumor homoiotransplants by pretreatment of the hosts with various preparations of normal and tumor tissue. J. Nat. Cancer Inst. 13:719–729.

Snell, G. D. 1953. A cytosieve permitting sterile preparation of suspensions of tumor cells for transplantation. J. Nat. Cancer Inst. 13:1,511–1,515.

Snell, G. D. 1958*a*. Histocompatibility genes of the mouse. I. Demonstration of weak histocompatibility differences by immunization and controlled tumor dosage. J. Nat. Cancer Inst. 20:787–824.

Snell, G. D. 1958*b*. Histocompatibility genes of

the mouse. II. Production and analysis of isogenic resistant lines. J. Nat. Cancer Inst. 21:843–877.

Snell, G. D. 1959. Transplantable tumors, p. 293–345. *In* F. Homburger [ed.] The physiopathology of cancer, 2nd ed. Hoeber-Harper, New York.

Snell, G. D. 1963. The immunology of tissue transplantation, p. 323–352. *In* Conceptual advances in immunology and oncology. 16th Annu. Symp. Fundamental Cancer Res. (Univ. Texas, Houston). Hoeber-Harper, New York.

Snell, G. D. 1964*a*. Methods for the study of histocompatibility genes and isoantigens. Introduction. Meth. Med. Res. 10:1–7.

Snell, G. D. 1964*b*. The terminology of tissue transplantation. Transplantation 2:655–657.

Snell, G. D., and H. P. Bunker. 1964. Histocompatibility genes of mice. IV. The position of *H-3* in the fifth linkage group. Transplantation 2:743–751.

Snell, G. D., and H. P. Bunker. 1965. Histocompatibility genes of mice. V. Five new histocompatibility loci identified by congenic resistant lines on a C57BL/10 background. Transplantation 3:235–252.

Snell, G. D., and G. F. Higgins. 1951. Alleles at the histocompatibility-2 locus in the mouse as determined by tumor transplantation. Genetics 36:306–310.

Snell, G. D., G. Hoecker, D. B. Amos, and J. H. Stimpfling. 1964. A revised nomenclature for the histocompatibility-2 locus of the mouse. Transplantation 2:777–784.

Snell, G. D., P. Smith, and F. Gabrielson. 1953. Analysis of the histocompatibility-2 locus in the mouse. J. Nat. Cancer Inst. 14:457–480.

Snell, G. D., and L. C. Stevens. 1961. Histocompatibility genes of mice. III. *H-1* and *H-4*, two histocompatibility loci in the first linkage group. Immunology 4:366–379.

Snell, G. D., N. Wheeler, and M. Aaron. 1957. A new method of typing inbred strains of mice for histocompatibility antigens. Transplant. Bull. 4:18–21.

Spencer, R. A., T. S. Hauschka, D. B. Amos, and B. Ephrussi. 1964. Co-dominance of isoantigens in somatic hybrids of murine cells grown *in vitro*. J. Nat. Cancer Inst. 33:893–903.

Stimpfling, J. H. 1961. The use of PVP as a developing agent in mouse hemagglutination tests. Transplant. Bull. 27:109–111.

Stimpfling, J. H., and O. Pizarro. 1961. On the antigenic products of the $H\text{-}2^m$ allele in the laboratory mouse. Transplant. Bull. 28:102–106.

Stimpfling, J. H., and A. Richardson. 1965. Recombination within the histocompatibility-2 locus of the mouse. Genetics 51:831–846.

Welshons, W. J., and L. B. Russell. 1959. The Y-chromosome as the bearer of male determining factors in the mouse. Proc. Nat. Acad. Sci. 45:560–566.

Wright, S. 1934. An analysis of variability in number of digits in an inbred strain of guinea pigs. Genetics 19:506–536.

Zaalberg, O. B. 1959. An analysis of the Eichwald-Silmser effect. Transplant. Bull. 6:433–435.

25

Cell, Tissue, and Organ Culture[1]

Charity Waymouth

The principal purpose of cell, tissue, and organ culture is to isolate, at each level of organization, the parts from the whole organism for study in experimentally controlled environments. It is characteristic of intact organisms that a high degree of interrelationship exists and interaction occurs between the component parts. Cultivation in vitro places cells beyond the effects of the organism as a whole and of the products of all cells other than those introduced into the culture. Artificial environments may be designed to imitate the natural physiological one, or varied at will by the deliberate introduction of particular variables and stresses.

Virtually all types of cells or aggregates of cells may be studied in culture. Living cells can be examined by cinephotomicrography, and by direct, phase-contrast, interference, fluorescence, or ultraviolet microscopy. Fixed cells from culture are suitable for cytological, cytochemical, histological, histochemical, and electron microscopical study. Populations of cells from monolayers or suspension cultures are used for nutritional, biochemical, and immunological work.

Organ culture, the cultivation of whole organs or parts thereof, is particularly suitable for studies of development, of inductive interactions, and of the effects of chemical and physical agents upon the physiological functions of specific organs.

Both cell and organ culture have applications in pathology, e.g., for comparative, developmental, and diagnostic studies of tissues from normal and diseased donors, for investigations on carcinogenesis, somatic cell genetic variation, viral susceptibility, etc. Cell cultures are widely used in microbiological studies, for investigations of the effects of radiation, and for screening drugs, especially carcinogenic, mutagenic, and radiomimetic agents.

Cell nutrition has been generally excluded from material selected for this chapter, since this topic has been fully reviewed elsewhere (Waymouth, 1954, 1960, 1965). The researches to be referred to here have been chosen because they (1) contribute significantly to our understanding of the biology of the mouse, as distinguished from observations of general interest in cell biology (e.g., the use of mouse cells for testing or screening drugs or carcinogens), (2) refer to cells obtained from inbred strains of mice or from mutants, or (3) suggest application to such cells.

CELL AND TISSUE CULTURES

Techniques

Tissues were first cultured before the turn of the century and since that time a multitude of techniques, each designed for the solution of a particular problem, has been devised. Basic information on technical procedures and their numerous applications in mammalian biology may be found in the textbooks of Parker (1961), Paul (1961), White (1963), and Merchant et al. (1964). Other reviews include those edited by White (1957) and Stevenson (1962); a number particularly emphasizing cell nutrition by Stewart and Kirk (1954), Waymouth (1954, 1960, 1965), Hanks (1955), Biggers et al. (1957), Geyer (1958), Morgan (1958), Swim (1959), Paul (1960); and one dealing principally with cell

[1] The writing of this chapter was supported in part by a grant from The John A. Hartford Foundation.

biochemistry by Levintow and Eagle (1961). The *Bibliography of the Research in Tissue Culture, 1884–1950* (Murray and Kopech, 1953) covers the literature of that period very completely, and keys to papers in tissue culture for the years 1961 onwards are to be published (Murray and Kopech, 1965, 1966).

Principal cell types

Although the chick embryo was the most generally used source of material for cultivation in the period 1910 to 1940, tissues of the embryonic and adult mouse have gained favor spectacularly since 1940, particularly since the introduction of antibiotics, which have reduced the need for strictly aseptic techniques in some types of experiment. The extensive cultivation of mouse cells owes much to the pioneering work of Earle, Evans, Sanford, and their colleagues at the National Cancer Institute. Among the major contributions of this group have been the improvement and standardization of techniques, including mass-culture methods and nutrition by chemically defined media, the development of cell lines and clones, the comparative study of the cytological and biochemical properties of long-established cell lines, and the investigation of malignant transformation in vitro.

Each of the basic cell types (fibrocytes or mechanocytes, epitheliocytes, and amebocytes (Willmer, 1960*a*, *b*)) has its special characteristics in vitro. Experience has shown that it is important, when describing cells grown in vitro, to give details of the species, age, and sex of the source of the cells, the tissue of origin, and to state whether they are normal or neoplastic. Cells and tissues freshly isolated from the animal are designated "primary cultures" (Fedoroff, 1966). A "primary cell line" refers to a population of cells derived by direct isolation from an animal and is not necessarily capable of serial proliferation indefinitely. An established cell line refers to a population of cells which has been serially transplanted at least 60 times in vitro. Primary or established cell lines or cell strains should receive designations according to the principles recommended by the 1957 International Tissue Culture Meeting (Anon., 1958; Paul, 1958; Fedoroff, 1966).

Applications of cell culture

Only a few representative examples of the use of cell cultures in mouse biology can be cited.

The mitotic process and its modification by stimulants or suppressors have been studied in many cell types (Fell and Hughes, 1949). Exact chromosome counts, to establish the degree of divergence from diploidy, have been made on 10-day fetal mouse cells in culture by Hungerford (1955). The chromosome complements of newborn and adult mice can be counted, without killing the animals, in primary tissue cultures of tail tips or ear fragments (Edwards, 1961). The mitotic cycle has been analysed by Defendi and Manson (1963), actinomycin D-resistant and -sensitive systems of RNA synthesis identified by Paul and Struthers (1963), and the duration of the DNA-synthetic period of mouse somatic cells shown to be probably constant (Cameron, 1964).

Visible light (Earle, 1928*a*, *b*, *c;* Frédéric, 1954) has some inhibitory effects upon living cells. The lethal effects of X-irradiation can be quantified on mouse cells (Reid and Gifford, 1952), and the effects of radiation upon cell constituents (Whitmore et al., 1958) and upon DNA and RNA synthesis (Whitfield and Rixon, 1959; Till, 1961*a*) can be studied. Also methods of chemical protection of irradiated cells (Whitfield et al., 1962) can be applied. X-ray–induced chromosome aberrations can be analysed (Chu and Monesi, 1960). Ultraviolet light, which inhibits cell division in strain L cells, does not significantly affect DNA synthesis (Whitfield et al., 1961). The survival of irradiated cells can be compared in vivo and in vitro, as has been done by McCulloch and Till (1962) for mouse bone marrow exposed to Co^{60} γ-rays.

Differentiation at the cellular level has mostly been studied in organ, rather than cell, cultures. However, Ginsburg (1963) has seen the differentiation of mouse thymus lymphoid cells into mast cells, which can be produced in large numbers and grown in suspension (Ginsburg and Sachs, 1963). Muscle differentiation can also be followed in tissue culture, and the tissue culture technique has been applied by Pearce (1963) to the study of muscular dystrophy.

The uses of tissue culture in the study of cancer have been reviewed by Murray (1959). More recent work includes that of Mitsutani et al. (1960) on the development of a near-diploid cell strain from a smoke-condensate–induced leiomyosarcoma in a male C3H mouse and that of Fernandes and Koprowska (1963) on cell lines from normal cervix uteri of C3H mice and on cells from uteri treated for varying lengths of time with benzpyrene.

Comparison of enzyme activities in cells in culture with those from the mouse have been made (e.g., of β-galactosidase by Maio and Rickenberg,

1960). Estimations of β-glucuronidase in cell lines from C3H mice (which have a lower activity in respect of this enzyme than most inbred mouse strains, especially in their livers) have demonstrated that most cell lines have activities many times higher than those of the highest activity mouse livers (Kuff and Evans, 1961). On the other hand, mouse cell strains after long cultivation in vitro seem uniformly to have a very low catalase activity in comparison with freshly isolated mouse tissues (Peppers et al., 1960). Long-term cultures of fibroblasts seem to undergo greater variations in enzyme content than, for example, liver cells. Westfall et al. (1958) found that the arginase and rhodanese activities in liver cell lines were high, as in the tissue of origin, but that fibroblast lines varied widely in the activities of one or both of these enzymes. This variation may be related to Klein's (1960, 1961) experience that induction of arginase depends upon other factors than the substrate. In most cases arginase induction requires the presence of RNA as well as arginase. The enzyme patterns of established cell strains are among the most useful traits for characterizing cells in culture (Westfall, 1962; Conklin et al., 1962).

Stable and unstable characters of cell cultures

Cloning. In certain respects cells cultivated in vitro exhibit considerable stability; in others they undergo extensive alterations from the parent cells. Cloning—the process of deriving a population of cells from a single cell—has enabled cell lines of common origin to be followed and has given many new insights into the potentialities of cells and into the circumstances under which they retain or lose characteristics derived from their parents (Sanford et al., 1948; Hobbs et al., 1957; Sanford et al., 1961*b*).

Variations within clones. The many cell lines and clones established from C3H mice by Sanford et al. (1961*e*) usually underwent morphological, biochemical, and chromosomal changes quite early in their culture history. But, once established, many of the characteristics were of remarkable stability and persisted over many years of serial cultivation. A clone of sarcoma-producing cells (originating from normal C3H connective tissue) gave rise to "high" and "low" sarcoma-producing lines (Sanford et al., 1954; Sanford et al., 1958), and sublines of these were characterized by widely differing patterns of chromosome number and character (Chu et al., 1958), of several enzyme activities (Sanford, 1958; Westfall et al., 1958; Sanford et al., 1959; Scott et al., 1960; Peppers et al., 1960; Sanford et al., 1961*e*), and of glycolytic activity (Woods et al., 1959).

Stable characteristics of cell lines. Some functions persist over many transplant generations in culture. These include the production of melanin by a mouse melanoma (Sanford et al., 1952) and of steroids by an adrenal tumor (Sato and Buonassisi, 1961) and the synthesis of 5-hydroxytryptamine, histamine, and heparin by the mouse mast-cell tumor P-815 (Dunn and Potter, 1957; Schindler et al., 1959; Green and Day, 1960; Day and Green, 1962). The polynucleotide sequences of DNA characteristic of the mouse are retained in L cells after more than 20 years in vitro in spite of the very abnormal karyotype exhibited (McCarthy and Hoyer, 1964).

Immunological characteristics. Immunological specificity persists over very long periods of cultivation in homologous, heterologous, or chemically defined media, particularly mouse-strain specificity in tumors cultivated in vitro. Immunological methods have been used for species identification of cells in culture (Coombs et al., 1961; Coombs, 1962; Fedoroff, 1962; Brand and Syverton, 1962), for studying antigenic differences between cell lines (Coriell et al., 1958; Kite and Merchant, 1961; McKenna and Blakemore, 1962), and for identification of particular antigens, such as the *H-2* transplantation antigen (Manson et al., 1962*a, b;* Cann and Herzenberg, 1963*a, b*).

Neoplastic transformations. It has been repeatedly observed that cells of normal origin undergo malignant change after more or less prolonged cultivation in vitro (Earle and Nettleship, 1943; Sanford et al., 1950; Evans et al., 1958; Sanford et al., 1961*d, e*). Neoplastic transformation of cells originating from normal tissues and the maintenance or loss of the capacity to produce tumors on transplantation into suitable hosts have been intriguing problems illustrating the range of capacities of the cell. Many of these transformations have been "spontaneous" (i.e., unexplained) (Earle and Nettleship, 1943; Sanford, 1958, 1962; Evans et al., 1958; Sanford et al., 1959; Shelton et al., 1963). Others have been deliberately induced by chemical carcinogens (Earle, 1943; Earle et al., 1950; Sanford et al., 1950; Shelton and Earle, 1951; Berwald and Sachs, 1963) or by viruses (Dawe and Law, 1959; Dulbecco and Vogt, 1960; Vogt and Dulbecco, 1960; Sanford et al., 1961*c;* Sachs and Medina, 1961; Pearson, 1962).

Tumors retain their capacity to grow in histocompatible hosts and in general do not acquire the ability to transcend histocompatibility barriers, except after being stored at −70°C (Morgan et al., 1956), though some degree of immunological incompatibility can develop in long-term cultures (Sanford et al., 1954, 1956). Strains of cells, originating from normal cells and acquiring within one or two years the ability to produce tumors, may undergo a progressive reduction in tumor-producing capacity after several more years in vitro (Earle et al., 1950). The cells with reduced tumor-producing ability could, however, grow in animals which had received X-irradiation (Sanford et al., 1956). The strain L, for example, after 10 years in vitro could produce tumors in 15 per cent of unirradiated, and in 64 per cent of irradiated C3H hosts. This raises the question whether the progressive immunological incompatibility is due to changes in the cell line or to changes in the inbred strain over the period of 10 or more years since the cell isolation. After 13 years the L cells retained their C3H specificity, i.e., would not grow in any of several other strains of mice. The differences in the "high" and "low" cancer-producing lines, both of which produce some immunity in C3H mice, and the fact that the "low" line will grow in irradiated C3H hosts, led to the conclusion (Sanford et al., 1958) that the faster-growing "high" tumor line can establish a tumor before resistance develops in the host. Cell strains and derived single-cell clones, established in culture from C3H carcinomas carrying mammary tumor agent, were found to vary in the persistence of the agent. In some cell strains the agent was demonstrable after 6 to 12 months of rapid cell proliferation in vitro; in others the agent disappeared (Sanford et al., 1961*a*).

Experimental control of malignant change. Attempts have been made to place the "spontaneous" transformation of normal to malignant cells under experimental control. Evans et al. (1964) have followed the progressive changes in cultures initiated from minced C3H embryos by testing their ability to produce tumors on intraocular implantation. The longer the cells have been maintained in vitro, the more quickly tumors appeared after implantation. No tumors resulted from a limited number of cultures grown for up to 211 days in a chemically defined medium. Cells grown in a medium supplemented with 10 per cent horse serum were able to produce tumors from about 120 days.

Barski and Cassigena (1963) aimed to produce parallel malignant and nonmalignant cell lines from adult female C57BL lung, analogous to the spontaneously derived parallel C3H lines of Sanford et al. (1950), for use in their studies of cell hybridization (see below). Such pairs of lines were derived, one from a culture frequently subcultured with the aid of trypsin, the other from a less frequently transferred culture, subcultured by mechanical dispersion and not exposed to trypsin. Both lines early became aneuploid (mean chromosome number in both cell lines at the 10th passage of the trypsinized line and the sixth passage of the nontrypsinized line was 68). The trypsinized line (PT) was highly malignant from the 16th passage (184 days), whereas the nontrypsinized line (PG) was not malignant up to the 37th passage (436 days). Further evidence is needed to determine whether the differences in morphology and malignancy are causatively related to the trypsin treatment. Todaro and Green (1963) suggested that the process of establishment of cell lines may require a reduction in the "leakiness" of cells to small molecules, and trypsin treatment increases "leakiness" (Phillips and Terryberry, 1957; Magee et al., 1958).

Sarcomatous change in carcinomas. Tissue cultures have contributed to the elucidation of the well-known "sarcomatous change" frequently observed in transplanted carcinomas. In an extensive study Sanford et al. (1961*d*) examined 18 different cell strains derived from C3H mammary gland tumors. In general, tumors maintained in culture for up to 25 weeks grew as differentiated mammary carcinomas on retransplantation into mice. Cells transplanted after this time grew as sarcoma-like tumors. It appeared that tumors which had been carried in mouse serial passage were morphologically more stable than primary tumors put into culture. The "sarcomatous change" was apparently not a unitary process. In some instances, with hepatomas, melanomas, and thyroid tumors (Sanford et al., 1952), as well as with mammary tumors, the stroma may undergo malignant change. In other cases, the carcinoma cells themselves change morphologically and assume a fibroblastic appearance. The opposite change (from fibroblastic to epithelioid cell types) has been studied in human cells of both normal and malignant origin by Ludovici et al. (1962*a*, *b*), who produced a significantly higher (64 per cent) proportion of cultures showing this alteration by treatment with a trypsin-antibiotic mixture, than was seen in controls not so treated (7 per cent).

Chromosomal variation and somatic cell genetics. Nutrient-dependent and nutrient-independent, drug-resistant and drug-sensitive, radiation-resistant and radiation-sensitive clones (Hauschka, 1957; Fedoroff and Cook, 1959;

Fischer, 1959; Hsu and Kellogg, 1959; Biesele et al., 1959; Roosa and Herzenberg, 1959; Hsu, 1961; Cann and Herzenberg, 1963*a*, *b*) are among the tools making possible the study of the genetics of mammalian cell and neoplastic cell populations (Merchant and Neel, 1962; Harris, 1964; Krooth, 1964).

Rothfels and Parker (1959) reported, what is now a rather common experience, that freshly explanted tissues (in their case from CF_1 mice) grow rapidly at first, then pass into a long (6 months or more) period of survival without growth, and finally, in some instances, enter a new phase of proliferation from which cell lines may be established. The chromosomes of such cell lines are usually heteroploid and heterotypic (i.e., contain chromosomes differing markedly from the normal 40 telocentrics of the mouse). Bimodal chromosome distributions are not uncommon, as in the case of Rothfels and Parker's culture 23855-8 from CF_1 kidney, which contained approximately equal numbers of cells with around 38 and 70 chromosomes respectively, a pattern which persisted through at least 14 subcultures (12 months). Hsu (1961) and Chu (1962) commented upon the rapid departure from diploidy observed even in primary cultures with mouse cells and contrast this with the greater karyotypic stability of man, rat, and many other mammals. Todaro and Green (1963) developed established cell lines from trypsin-disaggregated 17- to 19-day mouse embryos using trypsin at each transfer. The usual decline in growth rate during early passages was encountered but, at from 15 to 30 generations in vitro, the growth rate rose. At the beginning of the third phase which, under their conditions, was less than 3 months, the cells responsible for the upturn in growth rate were diploid, but they shifted (often rapidly) to the tetraploid range. Marker chromosomes appeared later.

Long-established cell strains, like most cancers (which Hauschka (1958) describes as "multiclonal mosaics of altered karyotypes"), have characteristic and identifiable karyotypes, even though within a strain there may be considerable variation of numbers and types of chromosomes. It is rare to find in cultures of mouse cells that the chromosomes are all, or even sometimes predominantly, telocentric. However, an analysis by Levan and Hsu (1960) of NCTC 2940—a cell line which originated from a C3H mammary carcinoma—after being carried for about 2 years in vitro, showed only telocentrics in a stem line number of $s = 84$ chromosomes. Another mammary tumor cell line, NCTC 2777, of hypotetraploid number ($s = 73$), contained only telocentrics, with the marked exception of one large, bizarre, and multiform heterometacentric chromosome.

Five established strains of mouse cell (three of them sublines of NCTC clone 929, strain L) were found by Hsu and Klatt (1958) to exhibit karyotypic polymorphism and to contain "marker" chromosomes highly characteristic of individual cell strains and wholly distinct from normal mouse chromosomes. In another study Hsu (1959) observed modal chromosome numbers of 67 to 73 in 12 mouse cell strains. Highly polyploid cells are not uncommon. Levan and Biesele (1958) observed a gradual increase in the number of polyploid cells in cultures of mouse embryo cells. Polyploid cells can usually be found early in the life of cultures, and their proportion in the population can be increased by treatment with colchicine (Hsu and Kellogg, 1960). One subline (L-P59) of NCTC clone 929 strain L, studied by Hsu (1960), contained 63 to 65 chromosomes, including a very conspicuous long subtelocentric (chromosome D). Derived subline Amy from L-P59 had an average chromosome number of 128 (with two D chromosomes), and subline Barbara had 58 to 59 without the D marker but with a large metacentric chromosome known as Victoria. In mixed cultures the stem line L-P59 rapidly overgrew Amy or Barbara. Moreover, the proportion of D chromosomes in L-P59 cultures was found to be variable according to the frequency of subculture. Old cultures, or cultures subdivided only every 2 weeks, contained on an average more than 1.5 D chromosomes per cell. In cultures subdivided twice a week, the population changed to one with less than one D chromosome per cell. The D chromosome and a probable isochromosome T were lost from one subline (L-M) (Hsu and Merchant, 1961) and new distinctive markers were reported 2 years after the first study. Hsu (1961) has reviewed the topic of chromosomal evolution in cell populations.

Occasionally mouse cell lines of diploid mode have been observed. Billen and Debrunner (1960) had cells from normal mouse bone marrow which remained diploid for more than 1 year. A line (H_2), started from cells from the peritoneum of a C3H mouse, retained its diploid character for at least 5 months before becoming predominantly tetraploid with a minority of hypodiploid (38) cells, which gradually diminished (Hsu et al., 1961). Another series of intriguing hypodiploid cell lines are the MB III lymphoblasts which originated in 1935 from a spontaneous lymphosarcoma

T86157 in a 286-day-old female mouse (De Bruyn et al., 1949). The primary line (MB I) of lymphosarcoma cells contained a mixed population of tumor-producing lymphoblasts ($s = 40$ or 41) and tumor-negative fibroblasts ($s = 56$). The MB III lines are sublines of lymphoblasts, free from fibroblasts, which have become tumor-negative and hypodiploid ($s = 30$ to 32). In contrast to many normal cells which undergo "spontaneous" transformation in vitro into tumor-producing cells, neither MB III (lymphoblasts) nor MB II (fibroblasts), in spite of great morphological variability and frequent mitotic disturbances, produces tumors in vivo after about 27 years of life in vitro (De Bruyn and Hansen-Melander, 1962).

The radiosensitivity of sublines of strain L mouse cells, as measured by their ability to form macroscopic colonies, was found to be independent of chromosome number in cell lines with mean chromosome numbers between 53 and 109 (Till, 1961*b*). Chromosomal anomalies acquired in vivo, by injecting a teratogen during pregnancy, persisted in the fetal tissues during cell culture, the treated cells showing 50 per cent polyploidy, and the controls only 2 per cent (Ingalls et al., 1963).

Cell hybridization. By making cultures of populations of two cell types, each containing conspicuous marker chromosomes, cells can be produced containing both sets of chromosomes. Sorieul and Ephrussi (1960), Barski (1961), and Barski et al. (1961*a*) found such "hybrid" cells in mixed cultures of cells from NCTC 2472 (a high-cancer line) and NCTC 2555 (a low-cancer line). Both of these lines took their ultimate origin from a single clone of normal cells, which produced the original "high" and "low" lines (Sanford et al., 1954) later designated NCTC 1742 and NCTC 2049 respectively. NCTC 2472 was derived from NCTC 1742 (Sanford et al., 1961*e*) and NCTC 2555 from NCTC 2049 (Woods et al., 1959). The high-cancer line NCTC 2472 (N1) has a modal chromosome number of 55 telocentrics, one being very long (Barski et al., 1961*b*). The low-cancer line NCTC 2555 (N2) has a modal chromosome number of 62, with from 9 to 19 two-armed chromosomes. By 104 days in mixed culture M type cells began to appear, with 115 to 116 chromosomes, of which 9 to 15 were metacentric, and in which the extra-long telocentric chromosome could usually be identified. Cells of the M type were never found in cultures of N1 or N2 cells alone (Barski and Cornefert, 1962) but M cells could be produced in vivo as well as in vitro, in tumors produced by inoculating C3H mice with mixed-cell populations. The hybrid characteristics of cloned M lines remained stable for at least 1 year. Barski and Belehradek (1963) have demonstrated cinephotomicrographically that nuclear transfer can take place in mixed cultures of N1 cells with normal mouse embryo cells. This may occur repeatedly in mixed cultures, or, as Ephrussi and Sorieul (1962) point out, it is not impossible that hybrid populations "arose from a single mating event involving modal cells of the two parental lines, followed by rapid segregation."

ORGAN CULTURES

Techniques

Techniques for culturing organs are described in the textbooks of Parker (1961), Paul (1961), White (1963), and Merchant et al. (1964), and are referred to in the major papers and review articles of the principal practitioners of the method, e.g., Wolff (1952), Fell (1953, 1954, 1955, 1958, 1964), Gaillard (1942, 1948, 1953), Borghese (1958), Kahn (1958), Lasnitzki (1958, 1965), Trowell (1959, 1961*b*), and Grobstein (1962).

Applications

Organ culture is used principally for (1) the maintenance of structural organization in tissues which are to be subjected to experimentally varied environments (e.g., to hormones, drugs, or radiation); (2) the study of morphogenesis, differentiation, and function in excised organs or presumptive organs; and (3) for comparison of the growth and behavior of explanted organs with the growth and behavior of similar organs in situ.

Almost every organ of the mouse has been cultivated in vitro. Some of the principal references to the cultivation of mouse organs are listed in Table 25-1.

The environmental variables studied by means of organ cultures include: radiation (Trowell, 1961*a;* Lasnitzki, 1961*a, c, d;* Borghese, 1961), vitamins, mainly vitamin A (Fell and Mellanby, 1952; Lasnitzki 1958, 1961*b, c,* 1962; New 1962), and carcinogens (Lasnitzki, 1958; Lasnitzki and Lucy, 1961). Organ culture is peculiarly suitable for the study of hormones (Fell, 1964) and provides an excellent way of distinguishing the effects of individual, or combinations of several, hormones on particular structures (Table 25-2).

The mammary gland has been one of the most commonly grown organs of the mouse and, besides its use for investigation of responses to hormones, has been studied for its secretory activity (Lasfargues, 1957*b;* Lasfargues and Feldman,

Table 25-1. ORGANS OF THE MOUSE CULTIVATED IN VITRO

Organ	References
Adipose tissue, brown	Sidman (1956*a*, *b*)
Adipose tissue, white	Trowell (1959)
Adrenal	Trowell (1959)
Bone marrow[a]	Berman and Kaplan (1960), Trowell (1959)
Brain[a]	Trowell (1959)
Buccal epithelium	New (1962)
Calvarium	Goldhaber (1960)
Ductus deferens	Trowell (1959)
Ear epidermis	Gelfant (1960)
Egg	Mintz (1962*a*, *b*, *c*, 1963), Tarkowski (1959*a*, *b*, 1961, 1963), Whitten (1956, 1957)
Embryonic shield	Grobstein (1950)
Embryos, developing	Bennett (1958), Biggers et al. (1962), Grobstein and Holtzer (1955), Grobstein and Parker (1954), McLaren and Biggers (1958), New and Stein (1963)
Eye	Lucas (1958), Lucas and Trowell (1958)
Fibula	Fell and Mellanby (1950, 1952)
Gonads, primordial	Asayama and Furusawa (1961), Borghese and Venini (1956), Wolff (1952), Wolff and Weniger (1954)
Hair and hair follicles	Cleffmann (1963), Hardy (1949, 1951)
Kidney	Ansevin and Buchsbaum (1962), Gluecksohn-Waelsch and Rota (1962), Trowell (1959)
Liver[a]	Ingram (1962), Trowell (1959)
Lung	Alescio (1960*a*, *b*), Borghese (1961), Trowell (1959)
Lymph nodes	Trowell (1959)
Mammary glands	Elias (1957, 1962), Elias and Rivera (1959), Hardy (1950), Koziorowska (1962), Lasfargues (1957*a*, *b*, *c*, 1960, 1962), Lasfargues and Feldman (1963), Lasfargues and Murray (1959), Lasfargues et al. (1958), Moretti and DeOme (1962), Prop (1959, 1960, 1961), Rivera (1963), Rivera and Bern (1961), Trowell (1959)
Metanephros	Auerbach and Grobstein (1958), Grobstein (1955*a*, 1956, 1957, 1959), Grobstein and Dalton (1957)
Ovary	Francke (1946), Guthrie (1953), Jacobs (1963), Lostroh (1959), Martinovitch (1938, 1939), Salzgeber (1962), Trowell (1959), van de Kerckhove (1958)
Palate	New (1962)
Pancreas[a]	Grobstein (1962), Trowell (1959), Wells and Borghese (1961, 1963)
Parathyroid	Gaillard (1961*b*), Trowell (1959)
Pineal	Trowell (1959)
Pituitary	Trowell (1959)
Prostate	Franks (1959, 1961), Franks and Barton (1960), Lasnitzki (1954, 1955*a*, *b*, 1962), Lasnitzki and Lucy (1961), Seaman (1956), Trowell (1959)
Radius	Fell (1956), Fell and Mellanby (1950, 1952), Gaillard (1961*a*, *b*, 1962, 1963),
Retina	Lucas (1958), Lucas and Trowell (1958), Sidman (1961, 1963), Sidman and Mottla (1961)
Seminal vesicle	Gaillard (1942), Trowell (1959)
Skeletal tissues	Fell (1958)
Skin	Cleffmann (1963), Gillette et al. (1961), Hardy (1949, 1951), New (1962), Trowell (1959)
Spinal ganglia	Trowell (1959)
Spleen[a]	Auerbach (1963), Pinkel (1963), Trowell (1959)
Sternum	Chen (1952*a*, *b*, 1953)
Stomach	Davenport et al. (1948)
Sublingual glands	Borghese (1950*a*)
Submandibular glands	Borghese (1950*a*, *b*), Grobstein (1953*a*, *b*)
Submaxillary glands	Dawe and Law (1959)
Sympathetic ganglia	Trowell (1959)
Teeth and tooth germs	Szabó (1954), Yoshioka (1960)
Testis[a]	Lostroh (1960), Trowell (1959)

Table 25-1. ORGANS OF THE MOUSE CULTIVATED IN VITRO (*Continued*)

Organ	References
Thymus[a]	Auerbach (1960, 1961*a*, *b*, 1963), Lieberman (1958), Trowell (1959)
Thyroid	Seaman and Stahl (1956), Trowell (1959)
Tibia	Biggers et al. (1961), Fell and Mellanby (1950, 1952)
Tongue	New (1962)
Trachea	Trowell (1959)
Ulna	Fell and Mellanby (1950, 1952)
Ureter	Trowell (1959)
Uterus	Gaillard (1942), Lostroh (1963), Trowell (1959)
Vagina	Biggers et al. (1956), Hardy et al. (1953), Lasnitzki (1961*a*, *b*, *c*), Martin (1959)
Vibrissae	Davidson and Hardy (1952)

[a] Of the large numbers of organs Trowell (1959) grew in culture, these survived poorly under his conditions.

1963) and as a vehicle for the mammary tumor agent (Lasfargues et al., 1958). The toxic effects of steroid hormones on mammary adenocarcinomas of C3H mice in organ culture have been examined (Rivera et al., 1963).

The cultivation of mouse ova (Whitten, 1956, 1957; Tarkowski, 1959*a*, *b*) has made possible the production in vitro of genotypically mosaic embryos from fused eggs (Tarkowski, 1961, 1963; Mintz, 1962*c*, 1963).

Organ culture has contributed significantly to our understanding of embryonic induction and of the control of morphogenesis by the juxtaposition of specific cell types. The effects of specific mesenchymal elements upon epithelial structures has been elucidated, e.g., by Borghese (1950),

Table 25-2. EFFECTS OF HORMONES ON MOUSE ORGAN CULTURES

Organ	Hormone(s)	References
Bone	Parathormone	Gaillard (1961*a*, *b*)
	Thyroid and parathyroid hormones	Gaillard (1962, 1963)
Brown fat	Insulin	Sidman (1956*a*, *b*)
Mammary gland	Estrone, progesterone, cortisol, growth hormone, mammatropic hormone	Elias (1957)
	Insulin	Elias (1962), Lasfargues (1962)
	Prolactin	Prop (1961)
	Combination of several hormones	Lasfargues and Murray (1959), Elias and Rivera (1959), Lasfargues (1960), Rivera and Bern (1961), Koziorowska (1962), Moretti and DeOme (1962), Lasfargues and Feldman (1963), Rivera (1963)
Ova	Progesterone	Whitten (1957)
Ovaries	Gonadotropin	Lostroh (1959)
	Gonadotropin and prolactin	Jacobs (1963)
Prostate	Estrone	Lasnitzki (1954)
	Testosterone	Lasnitzki (1955*a*)
		Franks and Barton (1960)
	Insulin and gonadotropins	Franks (1961)
Skin	Cortisone	Gillette et al. (1961)
Seminal vesicles	Estrone	Gaillard (1942)
Testis	Gonadotropin	Lostroh (1963)
Uterus	Estrone	Gaillard (1942)
	Insulin and aldosterone	Lostroh (1963)
Vagina	Estrogens	Biggers et al. (1956)
	Estrogens, testosterone, stilbestrol, cortisone	Martin (1959)
	Estrone	Lasnitzki (1961*a*, *b*, *c*)

Grobstein (1953*a*, *b*, *c*, 1955*a*, *b*, 1956, 1957, 1959, 1962), Grobstein and Dalton (1957), Auerbach and Grobstein (1958), and Auerbach (1960, 1961*a*, *b*). Cartilage induction has been studied in BALB/c × C3H embryos by Grobstein and Parker (1954) and Grobstein and Holtzer (1955) and in *T/T* embryos by Bennett (1958).

The pioneer work of Hardy (1949, 1951) in the growth of hair and hair follicles has been followed up by Cleffmann (1963) with studies of pigment formation in the hair-follicle melanocytes of agouti mice.

Developmental anomalies and inherited diseases and defects have been less studied by means of organ culture than might be expected. Elegant examples of the possibilities are shown in the growth of normal retinas of CBA (Lucas and Trowell, 1958) and BALB/c (Sidman, 1961) mice, in the analysis of the changes occurring in inherited retinal dystrophy in C3H animals by Lucas (1958) and by Sidman (1961, 1963), and in the examinations by organ culture of the potentialities of the kidney rudiments of prospectively kidneyless (*Sd/Sd*) mice (Gluecksohn-Waelsch and Rota, 1963).

SUMMARY

The mouse has become one of the species of choice for furnishing tissue for cell and organ cultures. A fund of basic information on mouse tissue culture is growing, much of it concerning work with tissues from inbred strains of mice. In so far as the genetic history of the cells and organs cultivated, as well as their subsequent history under cultivation, is pertinent to their observed behavior, this information will be valuable for future work on the characterization, function, and variation of somatic cells.

LITERATURE CITED

Alescio, T. 1960*a*. Osservazioni su culture organotipiche di polmone embrionale di topo. Arch. Ital. Anat. Embriol. 65:323–363.

Alescio, T. 1960*b*. La culture du poumon embryonnaire de souris. Anat. Anz. 109:144–149.

Anon. 1958. Nomenclature of cell strains. J. Nat. Cancer Inst. 20:439.

Ansevin, K. D., and R. Buchsbaum. 1962. Capacity for histological reconstruction by mouse kidney cells in relation to age. J. Gerontol. 17:130–137.

Asayama, S. I., and M. Furusawa. 1961. Sex differentiation of primordial gonads of the mouse embryo after cultivation in vitro. Jap. J. Exp. Morphol. 15:34–47.

Auerbach, R. 1960. Morphogenetic interactions in the development of the mouse thymus glands. Develop. Biol. 2:271–284.

Auerbach, R. 1961*a*. Genetic control of thymus lymphoid differentiation. Proc. Nat. Acad. Sci. 47:1,175–1,181.

Auerbach, R. 1961*b*. Experimental analysis of the origin of cell types in the development of the mouse thymus. Develop. Biol. 3:336–354.

Auerbach, R. 1963. Developmental studies of mouse thymus and spleen. Nat. Cancer Inst. Monogr. 11:23–33.

Auerbach, R., and C. Grobstein. 1958. Inductive interaction of embryonic tissues after dissociation and reaggregation. Exp. Cell Res. 15:384–397.

Barski, G. 1961. Clones cellulaires "hybrids" isolés à partir de cultures cellulaires mixtes. Compt. Rend. Acad. Sci. 253:1,186–1,188.

Barski, G., and J. Belehradek. 1963. Transfert nucléaire intercellulaire en cultures mixtes. Exp. Cell Res. 29:102–111.

Barski, G., J. L. Biedler, and F. Cornefert. 1961*a*. Modification of characteristics of an *in vitro* mouse cell line after an increase of its tumor-producing capacity. J. Nat. Cancer Inst. 26:865–889.

Barski, G., and R. Cassigena. 1963. Malignant transformation *in vitro* of cells from C57BL mouse normal pulmonary tissue. J. Nat. Cancer Inst. 30:865–883.

Barski, G., and F. Cornefert. 1962. Characteristics of "hybrid" type cloned cell lines obtained from mixed cultures *in vitro*. J. Nat. Cancer Inst. 28:801–821.

Barski, G., S. Sorieul, and F. Cornefert. 1961*b*. "Hybrid" type cells in combined cultures of two different mammalian cell strains. J. Nat. Cancer Inst. 26:1,269–1,291.

Bennett, D. 1958. *In vitro* study of cartilage induction in T/T mice. Nature 181:1,286.

Berman, I., and H. S. Kaplan. 1960. The functional capacity of mouse bone marrow cells growing in diffusion chambers. Exp. Cell Res. 20:238–239.

Berwald, Y., and L. Sachs. 1963. In vitro cell transformation with chemical carcinogens. Nature 200:1,182–1,184.

Biesele, J. J., J. L. Biedler, and D. J. Hutchison. 1959. Chromosomal status of drug resistant sublines of mouse leukemia L1210, p. 295–307. *In* Genetics and cancer. 13th Symp. Fundamental Cancer Res. University of Texas Press, Austin.

Biggers, J. D., P. J. Claringbold, and M. H. Hardy. 1956. The action of oestrogens on the vagina of the mouse in tissue culture. J. Physiol. 131:497–515.

Biggers, J. D., R. B. L. Gwatkin, and R. L. Brinster. 1962. Development of mouse embryos in organ cultures of Fallopian tubes on a chemically defined medium. Nature 194:747–749.

Biggers, J. D., R. B. L. Gwatkin, and S. Heyner. 1961. Growth of embryonic avian and mammalian tibiae on a relatively simple chemically defined medium. Exp. Cell Res. 25:41–58.

Biggers, J. D., L. M. Rinaldini, and M. Webb. 1957. The study of growth factors in tissue culture. Symp. Soc. Exp. Biol. 11:264–297.

Billen, D., and G. A. Debrunner. 1960. Continuously propagating cells derived from normal mouse bone marrow. J. Nat. Cancer Inst. 25:1,127–1,139.

Borghese, E. 1950*a*. The development *in vitro* of the submandibular and sublingual glands of *Mus musculus*. J. Anat. 84:287–302.

Borghese, E. 1950*b*. Explantation experiments on the influence of the connective tissue capsule on the development of the epithelial part of the submandibular gland of *Mus musculus*. J. Anat. 84:303–318.

Borghese, E. 1958. Organ differentiation in culture, p. 704–773. *In* W. D. McElroy and B. Glass [ed.] A symposium on the chemical basis of development. Johns Hopkins Press, Baltimore.

Borghese, E. 1961. The effect of ionizing radiations on mouse embryonic lungs developing *in vitro*. Ann. N.Y. Acad. Sci. 95:866–872.

Borghese, E., and M. A. Venini. 1956. Culture *in vitro* di gonadi embrionale di *Mus musculus*. Symp. Genet. (Pavia) 5:69–83.

Brand, K. G., and J. Syverton. 1962. Results of species-specific hemagglutination tests on "transformed," nontransformed, and primary cell cultures. J. Nat. Cancer Inst. 28:147–157.

Cameron, I. L. 1964. Is the duration of DNA synythesis in somatic cells of mammals and birds a constant? J. Cell Biol. 20:185–188.

Cann, H. M., and L. A. Herzenberg. 1963*a*. In vitro studies of mammalian somatic cell variation. I. Detection of H-2 phenotype in cultured mouse cell lines. J. Exp. Med. 117:259–265.

Cann, H. M., and L. A. Herzenberg, 1963*b*. In vitro studies of mammalian somatic cell variations. II. Iso-immune cytotoxicity with a cultured mouse lymphoma and selection of resistant variants. J. Exp. Med. 117:267–283.

Chen, J. M. 1952*a*. Studies on the morphogenesis of the mouse sternum. I. Normal embryonic development. J. Anat. 86:373–387.

Chen, J. M. 1952*b*. Studies on the morphogenesis of the mouse sternum. II. Experiments on the origin of the sternum and its capacity for self-differentiation *in vitro*. J. Anat. 86:387–401.

Chen, J. M. 1953. Studies on the morphogenesis of the mouse sternum. III. Experiments on the closure and segmentation of the sternal bands. J. Anat. 87:130–149.

Chu, E. H. Y. 1962. Chromosomal stabilization of cell strains. Nat. Cancer Inst. Monogr. 7:55–62.

Chu, E. H. Y., and V. Monesi. 1960. Analysis of X-ray induced chromosome aberrations in mouse somatic cells *in vitro*. Genetics 45:981.

Chu, E. H. Y., K. K. Sanford, and W. R. Earle. 1958. Comparative chromosomal studies on mammalian cells in culture. II. Mouse sarcoma-producing cell strains and their derivatives. J. Nat. Cancer Inst. 21:729–751.

Cleffmann, G. 1963. Agouti pigment cells *in situ* and *in vitro*. Ann. N.Y. Acad. Sci. 100:749–761.

Conklin, J. L., M. M. Dewey, and R. H. Kahn. 1962. Cytochemical localization of certain oxidative enzymes. Amer. J. Anat. 110:19–27.

Coombs, R. R. A. 1962. Identification and characterization of cells by immunologic analysis, with special reference to mixed agglutination. Nat. Cancer Inst. Monogr. 7:91–98.

Coombs, R. R. A., M. R. Daniel, B. W. Gurner, and A. Kelus. 1961. Species-characterizing antigens of "L" and "ERK" cells. Nature 189:503–504.

Coriell, L. L., M. G. Tall, and H. Gaskill. 1958. Common antigens in tissue culture cell lines. Science 128:198–199.

Davenport, H. W., V. Jensen, and L. A. Woodbury. 1948. The secretion of acid by the mouse stomach *in vitro*. Gastroenterology 11:227–239.

Davidson, P., and M. H. Hardy. 1952. The development of mouse vibrissae *in vivo* and *in vitro*. J. Anat. 86:342–356.

Dawe, C. J., and L. W. Law. 1959. Morphologic changes in salivary-gland tissue of the newborn mouse exposed to parotid-tumor agent *in vitro*. J. Nat. Cancer Inst. 23:1,157–1,177.

Day, M., and J. P. Green. 1962. The uptake of amino acids and the synthesis of amines by neoplastic mast cells in culture. J. Physiol. 164:210–226.

De Bruyn, W. M., and E. Hansen-Melander. 1962. Chromosome studies in the MB mouse lymphosarcoma. J. Nat. Cancer Inst. 28:1,333–1,354.

De Bruyn, W. M., R. Korteweg, and E. Kits van Waveren. 1949. Transplantable mouse lymphosarcoma T86157(MB) studied *in vivo, in vitro* and at autopsy. Cancer Res. 9:282–293.

Defendi, V., and L. A. Manson. 1963. Analysis of the life-cycle in mammalian cells. Nature 198:359–361.

Dulbecco, R., and M. Vogt. 1960. Significance of continued virus production in tissue cultures rendered neoplastic by polyoma virus. Proc. Nat. Acad. Sci. 46:1,617–1,623.

Dunn, T. B., and M. Potter. 1957. A transplantable mast-cell neoplasm in the mouse. J. Nat. Cancer Inst. 18:587–596.

Earle, W. R. 1928*a*. Studies upon the effect of light on blood and tissue cells. I. The action of light on white blood cells *in vitro*. J. Exp. Med. 48:457–474.

Earle, W. R. 1928*b*. Studies upon the effect of light on blood and tissue cells. II. The action of light on erythrocytes *in vitro*. J. Exp. Med. 48:667–681.

Earle, W. R. 1928*c*. Studies upon the effect of light on blood and tissue cells. III. The action of light on fibroblasts *in vitro*. J. Exp. Med. 48:683–694.

Earle, W. R. 1943. Production of malignancy *in vitro*. IV. The mouse fibroblast cultures and changes seen in the living cells. J. Nat. Cancer Inst. 4:165–212.

Earle, W. R., and A. Nettleship. 1943. Production of malignancy in vitro. V. Results of injections of cultures into mice. J. Nat. Cancer Inst. 4:213–227.

Earle, W. R., E. Shelton, and E. L. Schilling. 1950. Production of malignancy *in vitro*. XI. Further results from reinjection of *in vitro* cell strains into strain C3H mice. J. Nat. Cancer Inst. 10:1,105–1,113.

Edwards, R. G. 1961. Identification of the chromosome complements of new-born and adult living mice. Nature 192:1,316–1,317.

Elias, J. J. 1957. Cultivation of adult mouse mammary gland in hormone-enriched synthetic medium. Science 126:842–844.

Elias, J. J. 1962. Response of mouse mammary duct end-buds to insulin in organ culture. Exp. Cell Res. 27:601–604.

Elias, J. J., and E. Rivera. 1959. Comparison of the responses of normal, precancerous and neoplastic mouse mammary tissues to hormones in vitro. Cancer Res. 19:505–511.

Ephrussi, B., and S. Sorieul. 1962. Mating of somatic cells *in vitro*, p. 81–97. *In* D. J. Merchant and J. V. Neel [ed.] Approaches to the genetic analysis of mammalian cells. The University of Michigan Press, Ann Arbor.

Evans, V. J., N. M. Hawkins, B. B. Westfall, and W. R. Earle. 1958. Studies on culture lines derived from mouse liver parenchymatous cells grown in long-term culture. Cancer Res. 18:261–266.

Evans, V. J., G. A. Parker, and T. B. Dunn. 1964. Neoplastic transformations in C3H mouse embryonic tissue *in vitro* determined by intraocular growth. I. Cells from chemically defined medium with and without serum supplement. J. Nat. Cancer Inst. 32:89–121.

Fedoroff, S. 1962. Method for distinguishing between human and mouse cells in tissue culture. Nature 196:394–395.

Fedoroff, S. 1966. Report on animal tissue culture nomenclature. Tissue Culture Assoc. Comm. on nomenclature. To be published.

Fedoroff, S., and B. Cook. 1959. Effects of human blood serum on tissue culture. II. Development of resistance to toxic human serum in fibroblast-like cells (Earle's L strain) obtained from a C3H mouse. J. Exp. Med. 109:615–632.

Fell, H. B. 1953. Recent advances in organ culture. Sci. Progr. 162:212–231.

Fell, H. B. 1954. The effect of hormones and vitamin A on organ cultures. Ann. N.Y. Acad. Sci. 58:1,183–1,187.

Fell, H. B. 1955. The effect of hormones on differentiated tissues in culture, p. 138–148. *In* R. W. Smith, O. H. Gaebler, and C. N. H. Long [ed.] The hypophyseal growth hormone, nature and actions. McGraw-Hill, New York.

Fell, H. B. 1956. Effect of excess vitamin A on organized tissues cultivated *in vitro*. Brit. Med. Bull. 12:35–37.

Fell, H. B. 1958. The physiology of skeletal tissue in culture. Lect. Sci. Basis Med. 6:28–45.

Fell, H. B. 1964. The role of organ cultures in the study of vitamins and hormones. Vit. Horm. 22:81–127.

Fell, H. B., and A. F. W. Hughes. 1949. Mitosis in the mouse: A study of living and fixed cells in tissue cultures. Quart. J. Microscop. Sci. 90:355–380.

Fell, H. B., and E. Mellanby. 1950. Effects of hypervitaminosis A on foetal mouse bones cultivated *in vitro*. Brit. Med. J. 2:535–539.

Fell, H. B., and E. Mellanby. 1952. The effect of hypervitaminosis A on embryonic limb-bones cultivated *in vitro*. J. Physiol. 116:320–349.

Fernandes, M. A. R., and I. Koprowska. 1963. Tissue culture studies on cells from mouse cervix subjected to carcinogenic treatment. Acta Cytol. 7:215–223.

Fischer, G. A. 1959. Nutritional and amethopterin-resistant characteristics of leukemic clones. Cancer Res. 19:372–376.

Francke, C. 1946. The effect of gonadotrophins on explanted fragments of ovaries from immature mice. Endocrinology 39:430–431.

Franks, L. M. 1959. A factor in normal human serum that inhibits epithelial growth in organ cultures. Exp. Cell Res. 17:579–581.

Franks, L. M. 1961. The growth of mouse prostate during culture *in vitro* in chemically defined and natural media, and after transplantation *in vivo*. Exp. Cell Res. 22:56–72.

Franks, L. M., and A. A. Barton. 1960. The effects of testosterone on the ultrastructure of mouse prostate *in vivo* and in organ cultures. Exp. Cell Res. 19:35–50.

Frédéric, J. 1954. Effets de différentes longueurs d'onde du spectre visible sur des cellules vivantes cultivées in vitro. Compt. Rend. Soc. Biol. 148:1,678–1,682.

Gaillard, P. J. 1942. Hormones regulating growth and differentiation in embryonic explants. Masson, Paris. 82 p.

Gaillard, P. J. 1948. Growth, differentiation and function of explants of some endocrine glands. Symp. Soc. Exp. Biol. 2:139–144.

Gaillard, P. J. 1953. Growth and differentiation of explanted tissues. Int. Rev. Cytol. 2:331–401.

Gaillard, P. J. 1961*a*. The influence of parathyroid extract on the explanted radius of albino mouse embryos. II. Proc. Koninkl. Ned. Akad. Wetensch. Ser. C64:119–128.

Gaillard, P. J. 1961*b*. Parathyroid and bone in tissue culture, p. 20–45. *In* R. O. Greep and R. V. Talmage [ed.] The parathyroids. Charles C Thomas, Springfield, Ill.

Gaillard, P. J. 1962. A comparative study of the influence of thyroxine and of parathyroid extract on the histological structure of the explanted embryonic radius rudiment. Acta Morphol. Neer.-Scand. 5:21–36.

Gaillard, P. J. 1963. Observations on the effect of thyroid and parathyroid secretions on explanted mouse radius rudiments. Develop. Biol. 7:103–116.

Gelfant, S. 1960. A study of mitosis in mouse ear-epidermis *in vitro*. III. Effects of glucolytic and Krebs' cycle intermediates. IV. Effects of metabolic inhibitors. Exp. Cell Res. 19:65–82.

Geyer, R. P. 1958. Nutrition of mammalian cells in tissue culture. Nutr. Rev. 16:321–323.

Gillette, R. W., A. Findley, and H. Conway. 1961. Effect of cortisone on skin maintained in organ culture. J. Nat. Cancer Inst. 27:1,285–1,309.

Ginsburg, H. 1963. The in vitro differentiation and culture of normal mast cells from mouse thymus. Ann. N.Y. Acad. Sci. 103:20–39.

Ginsburg, H., and L. Sachs. 1963. Formation of pure suspensions of mast cells in tissue culture by differentiation of lymphoid cells from the mouse thymus. J. Nat. Cancer Inst. 31:1–39.

Gluecksohn-Waelsch, S., and T. R. Rota. 1963. Development in organ tissue culture of kidney rudiments from mutant mouse embryos. Develop. Biol. 7:432–444.

Goldhaber, P. 1960. Behavior of bone in tissue culture. Pub. Amer. Ass. Adv. Sci. 64:349–372.

Green, J. P., and M. Day. 1960. Heparin, 5-hydroxytryptamine and histamine in neoplastic mast cells. Biochem. Pharmacol. 3:190–205.

Grobstein, C. 1950. Behavior of the mouse embryonic shield in plasma clot cultures. J. Exp. Zool. 115:297–314.

Grobstein, C. 1953*a*. Epithelio-mesenchymal specificity in the morphogenesis of mouse submandibular rudiments *in vitro*. J. Exp. Zool. 124:383–413.

Grobstein, C. 1953*b*. Analysis in vitro of the early organization of the rudiments of the mouse submandibular gland. J. Morphol. 93:19–44.

Grobstein, C. 1953*c*. Inductive epithelio-mesenchymal interaction in cultured organ rudiments of the mouse. Science 118:52–55.

Grobstein, C. 1955*a*. Inductive interaction in the development of the mouse metanephros. J. Exp. Zool. 130:319–339.

Grobstein, C. 1955*b*. Tissue interaction in the morphogenesis of mouse embryonic rudiments in vitro, p. 233–256. *In* D. Rudnick [ed.] Aspects of synthesis and order in growth. Princeton University Press, Princeton, N.J.

Grobstein, C. 1956. Trans-filter induction of tubules in mouse metanephrogenic mesenchyme. Exp. Cell Res. 10:424–440.

Grobstein, C. 1957. Some transmission characteristics of the tubule-inducing influence of mouse metanephrogenic mesenchyme. Exp. Cell Res. 13:575–587.

Grobstein, C. 1959. Autoradiography of the interzone between tissue in inductive interaction. J. Exp. Zool. 142:203–213.

Grobstein, C. 1962. Interactive processes in cytodifferentiation. J. Cell. Comp. Physiol. 60 (Suppl. 1):35–48.

Grobstein, C., and A. J. Dalton. 1957. Kidney tubule induction in mouse metanephrogenic mesenchyme without cytoplasmic contact. J. Exp. Zool. 135:57–73.

Grobstein, C., and H. Holtzer. 1955. In vitro studies of cartilage induction in mouse somite mesoderm. J. Exp. Zool. 128:333–358.

Grobstein, C., and G. Parker. 1954. *In vitro* induction of cartilage in mouse somite mesoderm by embryonic spinal cord. Proc. Soc. Exp. Biol. Med. 85:477–481.

Guthrie, M. J. 1953. Enhancement of growth in the explanted ovary of the newborn mouse by inorganic phosphate and adenine. Anat. Rec. 115:314–315.

Hanks, J. H. 1955. Nutrition of cells *in vitro,* p. 74–82. *In* An introduction to cell and tissue culture. Burgess, Minneapolis.

Hardy, M. H. 1949. The development of mouse hair *in vitro* with some observations on pigmentation. J. Anat. 83:364–384.

Hardy, M. H. 1950. The development *in vitro* of the mammary glands of the mouse. J. Anat. 84: 388–393.

Hardy, M. H. 1951. The development of pelage hairs and vibrissae from skin in tissue culture. Ann. N.Y. Acad. Sci. 53:546–561.

Hardy, M. H., J. D. Biggers, and P. J. Claringbold. 1953. Vaginal cornification of the mouse produced by oestrogens *in vitro*. Nature 172: 1,196–1,197.

Harris, M. 1964. Cell culture and somatic variation. Holt, New York, 547 p.

Hauschka, T. S. 1957. Tissue genetics of neoplastic cell populations. Can. Cancer Res. Conf. 2 (1956):305–345.

Hauschka, T. S. 1958. Correlation of chromosomal and physiologic changes in tumors. J. Cell. Comp. Physiol. 52:197–267.

Hobbs, G. L., K. K. Sanford, V. J. Evans, and W. R. Earle. 1957. Establishment of a clone of mouse liver cells from a single isolated cell. J. Nat. Cancer Inst. 18:701–708.

Hsu, T. C. 1959. Mammalian chromosomes *in vitro*. XI. Variability among progenies of a single cell, p. 129–134. *In* Biol. Contrib. No. 5914, Univ. Texas, Austin.

Hsu, T. C. 1960. Mammalian chromosomes *in vitro*. XIII. Cyclic and directional changes of population structure. J. Nat. Cancer Inst. 25: 1,339–1,353.

Hsu, T. C. 1961. Chromosomal evolution in cell populations. Int. Rev. Cytol. 12:69–161.

Hsu, T. C., D. Billen, and A. Levan. 1961. Mammalian chromosomes *in vitro*. XV. Patterns of transformation. J. Nat. Cancer Inst. 27:515–541.

Hsu, T. C., and D. S. Kellogg. 1959. Genetics of *in vitro* cells, p. 183–204. *In* Genetics and cancer. 13th Symp. Fundamental Cancer Res. University of Texas Press, Austin.

Hsu, T. C., and D. S. Kellogg. 1960. Mammalian chromosomes *in vitro*. XII. Experimental evolution of cell populations. J. Nat. Cancer Inst. 24: 1,067–1,093.

Hsu, T. C., and O. Klatt. 1958. Mammalian chromosomes *in vitro*. IX. On genetic polymorphism in cell populations. J. Nat. Cancer Inst. 21:437–473.

Hsu, T. C., and D. J. Merchant. 1961. Mammalian chromosomes *in vitro*. XIV. Genotypic replacement in cell populations. J. Nat. Cancer Inst. 26:1,075–1,083.

Hungerford, D. A. 1955. Chromosome numbers of ten day fetal mouse cells. J. Morphol. 97: 497–510.

Ingalls, T. H., E. F. Ingenito, and F. J. Curley. 1963. Acquired chromosomal anomalies induced in mice by injection of a teratogen in pregnancy. Science 141:810–812.

Ingram, R. L. 1962. Maintenance of organized adult liver tissue *in vitro*. Exp. Cell Res. 28: 370–380.

Jacobs, B. B. 1963. In vivo assay of function of mouse ovaries following culture in hormone enriched medium. Exp. Cell Res. 32:431–441.

Kahn, R. H. 1958. Organ culture in experimental biology. Univ. Michigan Med. Bull. 24:242–252.

Kite, J. H., and D. J. Merchant. 1961. Studies of some antigens of the L-strain mouse fibroblast. J. Nat. Cancer Inst. 26:419–434.

Klein, E. 1960. On substrate-induced enzyme formation in animal cells cultured *in vitro*. Exp. Cell Res. 21:421–429.

Klein, E. 1961. Studies on the substrate-induced arginase synthesis in animal cell strains cultured *in vitro*. Exp. Cell Res. 22:226–232.

Koziorowska, J. 1962. The influence of ovarian hormones and insulin on the mouse mammary glands cultivated in vitro. Acta Med. Pol. 3: 237–245.

Krooth, R. S. [ed.] 1964. Somatic cell genetics. The University of Michigan Press, Ann Arbor. 301 p.

Kuff, E. L., and V. J. Evans. 1961. β-Glucuronidase activities of cultured cells derived from

C3H mouse liver. J. Nat. Cancer Inst. 27:667–678.

Lasfargues, E. Y. 1957*a*. Cultivation and behavior *in vitro* of the normal mammary epithelium of the adult mouse. Anat. Rec. 127: 117–129.

Lasfargues, E. Y. 1957*b*. Cultivation and behavior *in vitro* of the normal mammary epithelium of the adult mouse. II. Observations on the secretory activity. Exp. Cell Res. 13:553–562.

Lasfargues, E. Y. 1957*c*. Comparative behavior *in vitro* of the normal and neoplastic mammary epithelium of the mouse. Proc. Amer. Ass. Cancer Res. 2:224.

Lasfargues, E. Y. 1960. Action de l'oestradiol et de la progestérone sur des cultures de glandes mammaires de jeunes souris. Compt. Rend. Soc. Biol. 154:1,720–1,722.

Lasfargues, E. Y. 1962. Concerning the role of insulin in the differentiation and functional activity of mouse mammary tissues. Exp. Cell Res. 28:531–542.

Lasfargues, E. Y., and D. G. Feldman. 1963. Hormonal and physiological background in the production of B particles by the mouse mammary epithelium in organ culture. Cancer Res. 23:191–196.

Lasfargues, E. Y., D. H. Moore, and M. R. Murray. 1958. Maintenance of the milk factor in cultures of mouse mammary epithelium. Cancer Res. 18:1,281–1,285.

Lasfargues, E. Y., and M. R. Murray. 1959. Hormonal influences on the differentiation and growth of embryonic mouse mammary gland in organ culture. Develop. Biol. 1:413–435.

Lasnitzki, I. 1954. The effect of oestrone alone and combined with 20-methylcholanthrene on mouse prostate glands grown *in vitro*. Cancer Res. 14:632–639.

Lasnitzki, I. 1955*a*. The effect of testosterone propionate on organ cultures of the mouse prostate. J. Endocrinol. 12:236–240.

Lasnitzki, I. 1955*b*. The influence of A hypervitaminosis on the effect of 20-methylcholanthrene on mouse prostate glands grown in vitro. Brit. J. Cancer 9:434–441.

Lasnitzki, I. 1958. The effect of carcinogens, hormones and vitamins on organ cultures. Int. Rev. Cytol. 7:79–121.

Lasnitzki, I. 1961*a*. The effect of radiation on the normal and oestrone-treated mouse vagina grown *in vitro*. Brit. J. Radiol. 34:356–361.

Lasnitzki, I. 1961*b*. Effect of excess vitamin A on the normal and oestrone-treated mouse vagina grown in chemically defined media. Exp. Cell Res. 24:37–45.

Lasnitzki, I. 1961*c*. Action and interaction of radiation, oestrogen and vitamin A on the mouse vagina grown in vitro. Colloq. Int. Centre Nat. Res. Sci. 101:73–84.

Lasnitzki, I. 1961*d*. The effect of X rays on cellular differentiation in organ culture. Ann. N.Y. Acad. Sci. 95:873–881.

Lasnitzki, I. 1962. Hypovitaminosis-A in the mouse prostate gland cultured in chemically defined medium. Exp. Cell Res. 28:40–51.

Lasnitzki, I. 1965. The action of hormones on cell and organ cultures, p. 591–658. *In* E. N. Willmer [ed.] Cells and tissues in culture. Vol. 1. Academic Press, New York.

Lasnitzki, I., and J. A. Lucy. 1961. Amino acid metabolism and arginase activity in mouse prostate glands grown *in vitro* with and without 20-methylcholanthrene. Exp. Cell Res. 24:379–392.

Levan, A., and J. J. Biesele. 1958. Role of chromosomes in cancerogenesis, as studied in serial tissue culture of mammalian cells. Ann. N.Y. Acad. Sci. 71:1,022–1,053.

Levan, A., and T. C. Hsu. 1960. The chromosomes of two cell strains from mammary carcinomas of the mouse. Hereditas 46:231–240.

Levintow, L., and H. Eagle. 1961. Biochemistry of cultured mammalian cells. Annu. Rev. Biochem. 30:605–640.

Lieberman, M. 1958. Cultivation of mouse thymus in diffusion chambers *in vivo*. Proc. Amer. Ass. Cancer Res. 2:321.

Lostroh, A. J. 1959. The response of ovarian explants from post-natal mice to gonadotropins. Endocrinology 65:124–132.

Lostroh, A. J. 1960. *In vitro* response of mouse testis to human chorionic gonadotropin. Proc. Soc. Exp. Biol. Med. 103:25–27.

Lostroh, A. J. 1963. Effects of insulin and of DL-aldosterone on protein synthesis by mouse uteri in organ culture. Exp. Cell Res. 32:327–332.

Lucas, D. R. 1958. Inherited retinal dystrophy in the mouse: its appearance in eyes and retinae cultured *in vitro*. J. Embryol. Exp. Morphol. 6:589–592.

Lucas, D. R., and O. A. Trowell. 1958. *In vitro* culture of the eye and the retina of the mouse and rat. J. Embryol. Exp. Morphol. 6:178–182.

Ludovici, P. P., C. Ashford, and N. F. Miller. 1962*a*. Studies on chemically induced and "spontaneous" alterations of morphology and growth potential in human cell culture. Cancer Res. 22:788–796.

Ludovici, P. P., C. Ashford, and N. F. Miller. 1962*b*. Studies on the chemicals required to induce alterations of morphology and growth potential in human cell culture. Cancer Res. 22: 797–803.

Magee, W. E., M. R. Sheek, and B. P. Sagik. 1958. Methods of harvesting mammalian cells grown in tissue culture. Proc. Soc. Exp. Biol. Med. 99:390–392.

Maio, J. J., and H. V. Rickenberg. 1960. The β-galactosidase of mouse strain L-cells and mouse organs. Biochim. Biophys. Acta 37:101–106.

Manson, L. A., G. V. Foschi, J. F. Duplan, and O. B. Zaalberg. 1962*a*. Isolation of transplantation antigens from a leukemic cell grown "in vitro." Ann. N.Y. Acad. Sci. 101:121–130.

Manson, L. A., G. V. Foschi, and J. Palm. 1962*b*. *In vivo* and *in vitro* studies of histocompatibility antigens isolated from a cultured mouse cell line. Proc. Nat. Acad. Sci. 48:1,816–1,821.

Martin, L. 1959. Growth of the vaginal epithelium of the mouse in tissue culture. J. Endocrinol. 18:334–342.

Martinovitch, P. N. 1938. The development *in vitro* of the mammalian gonad. Ovary and ovogenesis. Proc. Roy. Soc. B 125:232–249.

Martinovitch, P. N. 1939. The effect of subnormal temperature on the differentiation and survival of cultivated *in vitro* embryonic and infantile rat and mouse ovaries. Proc. Roy. Soc. B 128: 138–143.

McCarthy, B. J., and B. H. Hoyer. 1964. Identity of DNA and diversity of messenger RNA molecules in normal mouse tissues. Proc. Nat. Acad. Sci. 52:915–922.

McCulloch, E. A., and J. E. Till. 1962. The sensitivity of cells from normal mouse bone marrow to gamma radiation "in vitro" and "in vivo." Radiat. Res. 16:822–832.

McKenna, J. M., and W. S. Blakemore. 1962. Serological comparisons among tissue culture cell lines. Nature 195:1,009–1,010.

McLaren, A., and J. D. Biggers. 1958. Successful development and birth of mice cultivated *in vitro* as early embryos. Nature 182:877–878.

Merchant, D. J., R. H. Kahn, and W. H. Murphy. 1964. Handbook of cell and organ culture, 2nd ed. Burgess, Minneapolis. 263 p.

Merchant, D. J., and J. V. Neel [ed.] 1962. Approaches to the genetic analysis of mammalian cells. The University of Michigan Press, Ann Arbor. 97 p.

Mintz, B. 1962*a*. Experimental study of the developing mammalian egg: removal of the zona pellucida. Science 138:594–595.

Mintz, B. 1962*b*. Incorporation of nucleic acid and protein precursors by developing mouse eggs. Amer. Zool. 2:432.

Mintz, B. 1962*c*. Formation of genotypically mosaic mouse embryos. Amer. Zool. 2:432.

Mintz, B. 1963. Growth *in vitro* of t^{12}/t^{12} lethal mutant mouse eggs. Amer. Zool. 3:550.

Mitsutani, M., Y. Ohnuki, Y. H. Nakanishi, and C. M. Pomerat. 1960. The development of a near-diploid *in vitro* strain from a smoke-condensate induced mouse tumor. Texas Rep. Biol. Med. 18:455–469.

Moretti, R. L., and K. B. DeOme. 1962. Effect of insulin on glucose uptake by normal and neoplastic mouse mammary tissues in organ culture. J. Nat. Cancer Inst. 29:321–329.

Morgan, J. F. 1958. Tissue culture nutrition. Bacteriol. Rev. 22:20–45.

Morgan, J. F., L. F. Guerin, and H. J. Morton. 1956. The effect of low temperature and storage on the viability and mouse strain specificity of ascitic tumor cells. Cancer Res. 16:907–912.

Murray, M. R. 1959. Uses of tissue culture in the study of malignancy, p. 469–516. *In* F. Homburger [ed.] The physiopathology of cancer, 2nd ed. Hoeber-Harper, New York.

Murray, M. R., and G. Kopech. 1953. A bibliography of the research in tissue culture, 1884–1950. Academic Press, New York. 2 vol.

Murray, M. R., and G. Kopech. 1965. Current tissue culture literature. Vol. 5. October House, New York.

Murray, M. R., and G. Kopech. 1966. Current tissue culture literature. [1961–64]. Vols. 1–4. October House, New York.

New, D. A. T. 1962. Action of vitamin A on rodent skin and buccal epithelium in organ culture, p. 20. Annu. Rep. Strangeways Res. Lab., Cambridge, England.

New, D. A. T., and K. F. Stein. 1963. Cultivation of mouse embryos in vitro. Nature 199:297–299.

Parker, R. C. 1961. Methods of tissue culture, 3rd ed. Hoeber-Harper, New York. 358 p.

Paul, J. 1958. A note on nomenclature of cell strains. Virology 5:175.

Paul, J. 1960. Environmental influences on the metabolism and composition of cultured cells. J. Exp. Zool. 142:475–505.

Paul, J. 1961. Cell and tissue culture, 2nd ed. Livingstone, Edinburgh and London. 312 p.

Paul, J., and M. G. Struthers. 1963. Actinomycin D-resistant RNA synthesis in animal cells. Biochem. Biophys. Res. Commun. 11:135–139.

Pearce, G. W. 1963. Tissue culture in the study of muscular dystrophy, p. 178–191. *In* G. H. Bourne and M. N. Golarz [ed.] Muscular dystrophy in man and animals. Hafner, New York.

Pearson, H. E. 1962. Reaction of certain mouse and hamster tumor tissue cultures to polyoma virus. Proc. Soc. Exp. Biol. Med. 111:332–334.

Peppers, E. V., B. B. Westfall, H. A. Kerr, and W. R. Earle. 1960. Note on the catalase activity of several mammalian cell strains after long cultivation in vitro. J. Nat. Cancer Inst. 25:1,065–1,068.

Phillips, H. J., and J. E. Terryberry. 1957. Counting actively metabolizing tissue cultured cells. Exp. Cell Res. 13:341–347.

Pinkel, D. 1963. Successful cultivation of spleen fragments in organ culture. Proc. Soc. Exp. Biol. Med. 112:242–245.

Prop, F. J. A. 1959. Organ culture of total mammary gland of the mouse. Nature 184:379–380.

Prop, F. J. A. 1960. Development of alveoli in organ cultures of total mammary glands of six weeks old virgin mice. Exp. Cell Res. 20:256–258.

Prop, F. J. A. 1961. Sensitivity to prolactin of mouse mammary glands in vitro. Exp. Cell Res. 24:629–631.

Reid, T. R., and M. P. Gifford. 1952. A quantitative study of the effects of X-radiation on cells in vitro. J. Nat. Cancer Inst. 13:431–439.

Rivera, E. M. 1963. Hormonal requirements for survival and growth of mouse primary mammary ducts in organ culture. Proc. Soc. Exp. Biol. Med. 114:735–738.

Rivera, E. M., and H. A. Bern. 1961. Influence of insulin on maintenance and secretory stimulation of mouse mammary tissues by hormones in organ culture. Endocrinology 69:340–353.

Rivera, E. M., J. J. Elias, H. A. Bern, N. P. Napalkov, and D. Pitelka. 1963. Toxic effects of steroid hormones on organ cultures of mouse mammary tumors, with a comment on the occurrence of viral inclusion bodies. J. Nat. Cancer Inst. 31:671–687.

Roosa, R. A., and L. A. Herzenberg. 1959. The selection of variants resistant to folic acid antagonists and 5-fluorouridine in cell cultures of a lymphocytic neoplasm. Proc. Amer. Ass. Cancer Res. 3:58.

Rothfels, K. H., and R. C. Parker. 1959. The karyotypes of cell lines recently established from normal mouse tissues. J. Exp. Zool. 142:507–520.

Sachs, L., and D. Medina. 1961. In vitro transformation of normal cells by polyoma virus. Nature 189:457–460.

Salzgeber, B. 1962. Étude du développement de l'ovaire de souris greffé dans l'embryon de poulet après culture *in vitro*. Arch. Anat. Microscop. Morphol. Exp. 51:1–10.

Sanford, K. K. 1958. Clonal studies on normal cells and their neoplastic transformation *in vitro*. Cancer Res. 18:747–752.

Sanford, K. K. 1962. Transformations in relation to malignancy. Acta Cytol. 6:400–401.

Sanford, K. K., H. B. Andervont, G. L. Hobbs, and W. R. Earle. 1961*a*. Maintenance of the mammary tumor agent in long term cultures of mouse mammary carcinoma. J. Nat. Cancer Inst. 26:1,185–1,191.

Sanford, K. K., A. B. Covalesky, L. T. Dupree, and W. R. Earle. 1961*b*. Cloning of mammalian cells by a simplified capillary technique. Exp. Cell Res. 23:361–372.

Sanford, K. K., T. B. Dunn, A. B. Covalesky, L. T. Dupree, and W. R. Earle. 1961*c*. Polyoma virus and production of malignancy in vitro. J. Nat. Cancer Inst. 26:331–357.

Sanford, K. K., T. B. Dunn, B. B. Westfall, A. B. Covalesky, L. T. Dupree, and W. R. Earle. 1961*d*. Sarcomatous change and maintenance of differentiation in long term cultures of mouse mammary carcinoma. J. Nat. Cancer Inst. 26:1,139–1,183.

Sanford, K. K., W. R. Earle, and G. D. Likely. 1948. The growth *in vitro* of single isolated tissue cells. J. Nat. Cancer Inst. 9:229–246.

Sanford, K. K., W. R. Earle, E. Shelton, E. L. Schilling, E. M. Duchesne, G. D. Likely, and M. M. Becker. 1950. Production of malignancy *in vitro*. XII. Further transformations of mouse fibroblasts to sarcomatous cells. J. Nat. Cancer Inst. 11:351–375.

Sanford, K. K., G. L. Hobbs, and W. R. Earle. 1956. The tumor-producing capacity of strain L mouse cells after 10 years *in vitro*. Cancer Res. 15:162–166.

Sanford, K. K., G. D. Likely, and W. R. Earle. 1954. The development of variations in transplantability and morphology within a clone of mouse fibroblasts transformed to sarcoma-producing cells *in vitro*. J. Nat. Cancer Inst. 15:215–237.

Sanford, K. K., G. D. Likely, V. J. Evans, C. J. Mackey, and W. R. Earle. 1952. The production of sarcomas from cultured tissues of hepatoma, melanoma and thyroid tumors. J. Nat. Cancer Inst. 12:1,057–1,077.

Sanford, K. K., R. M. Merwin, G. L. Hobbs, M. C. Fioramonti, and W. R. Earle. 1958. Studies on the difference in sarcoma-producing capacity of two lines of mouse cells derived *in vitro* from one cell. J. Nat. Cancer Inst. 20: 121–145.

Sanford, K. K., R. M. Merwin, G. L. Hobbs, J. M. Young, and W. R. Earle. 1959. Clonal analysis of variant cell lines transformed to malignant cells in tissue culture. J. Nat. Cancer Inst. 23:1,035–1,059.

Sanford, K. K., B. B. Westfall, E. H. Y. Chu, E. L. Kuff, A. B. Covalesky, L. T. Dupree, G. L. Hobbs, and W. R. Earle. 1961*e*. Alterations in morphology, arginase and β-glucuronidase within a clone of mouse tumor cells *in vitro*. J. Nat. Cancer Inst. 26:1,193–1,219.

Sato, G., and V. Buonassisi. 1961. The culture of differentiated tumors. Abstr. 1st meeting Amer. Soc. Cell Biol. p. 189.

Schindler, R., M. Day, and G. A. Fischer. 1959. Culture of neoplastic mast cells and their synthesis of 5-hydroxytryptamine and histamine *in vitro*. Cancer Res. 19:47–51.

Scott, D. B. M., A. M. Pakoskey, and K. K. Sanford. 1960. Analysis of enzymatic activities of clones derived from variant cell lines transformed to malignant cells in tissue culture. J. Nat. Cancer Inst. 25:1,365–1,379.

Seaman, A. R. 1956. The *in vitro* cultivation of the prostate gland of the adult mouse in alkaline fluid medium. Exp. Cell. Res. 11:283–288.

Seaman, A. R., and S. Stahl. 1956. The uptake of I^{131} by the thyroid gland of the adult mouse after *in vitro* cultivation. Exp. Cell Res. 11:220–221.

Shelton, E., and W. R. Earle. 1951. Production of malignancy in vitro. XIII. Behavior of recovery cultures. J. Nat. Cancer Inst. 11:817–837.

Shelton, E., V. J. Evans, and G. A. Parker. 1963. Malignant transformation of mouse connective tissue grown in diffusion chambers. J. Nat. Cancer Inst. 30:377–392.

Sidman, R. L. 1956*a*. Histogenesis of brown adipose tissue *in vivo* and in organ culture. Anat. Rec. 124:581–595.

Sidman, R. L. 1956*b*. The direct effect of insulin on organ culture of brown fat. Anat. Rec. 124:723–734.

Sidman, R. L. 1961. Tissue culture studies of inherited retinal dystrophy. Dis. Nerv. Syst. (Monogr. Suppl.) 22(4):14–20.

Sidman, R. L. 1963. Organ culture analysis of inherited retinal dystrophy in rodents. Nat. Cancer Inst. Monogr. 11:227–246.

Sidman, R. L., and P. Mottla. 1961. Cell division and migration in organ cultures of mouse retina. Excerpta Med. (Sect. I) 15:558.

Sorieul, S., and B. Ephrussi. 1961. Karyological demonstration of hybridization of mammalian cells in vitro. Nature 190:653–654.

Stevenson, R. E. [ed.] 1962. Analytic cell culture. Nat. Cancer Inst. Monogr. 7:290 p.

Stewart, D. C., and P. L. Kirk. 1954. The liquid medium in tissue culture. Biol. Rev. 29:119–153.

Swim, H. E. 1959. Microbiological aspects of tissue culture. Annu. Rev. Microbiol. 13:141–176.

Szabó, G. 1954. Studies on the cultivation of teeth *in vitro*. J. Anat. 88:31–44.

Tarkowski, A. K. 1959*a*. Experiments on the development of isolated blastomeres of mouse eggs. Nature 184:1,286–1,287.

Tarkowski, A. K. 1959*b*. Experimental studies on regulation in the development of isolated blastomeres of mouse eggs. Acta Theriol. 3:191–267.

Tarkowski, A. K. 1961. Mouse chimaeras developed from fused eggs. Nature 190:857–860.

Tarkowski, A. K. 1963. Studies on mouse chimaeras developed from eggs fused *in vitro*. Nat. Cancer Inst. Monogr. 11:51–71.

Till, J. E. 1961*a*. Radiation effects on the division cycle of mammalian cells *in vitro*. Ann. N.Y. Acad. Sci. 95:911–919.

Till, J. E. 1961*b*. Radiosensitivity and chromosome number in strain L mouse cells in tissue culture. Radiat. Res. 15:400–409.

Todaro, G. J., and H. Green. 1963. Quantitative studies of the growth of mouse embryo cells in culture and their development into established cell lines. J. Cell Biol. 17:299–313.

Trowell, O. A. 1959. The culture of mature organs in a synthetic medium. Exp. Cell Res. 16:118–147.

Trowell, O. A. 1961*a*. Cytocidal effects of radiation on organ cultures. Ann. N.Y. Acad. Sci. 95:849–865.

Trowell, O. A. 1961*b*. Problems in the maintenance of mature organs in vitro. Colloq. Int. Centre Nat. Res. Sci. 101:237–254.

Van de Kerckhove, D. 1958. L'ovaire périnatal de la souris blanche en culture organotypique. Compt. Rend. Ass. Anat. 104:754–759.

Vogt, M., and R. Dulbecco. 1960. Virus-cell interaction with a tumor-producing virus. Proc. Nat. Acad. Sci. 46:365–370.

Waymouth, C. 1954. The nutrition of animal cells. Int. Rev. Cytol. 3:1–68.

Waymouth, C. 1960. Growth in tissue culture, p. 546–587. *In* W. W. Nowinski [ed.] Fundamental aspects of normal and malignant growth. Elsevier, Amsterdam.

Waymouth, C. 1965. Construction and use of synthetic media. *In* E. N. Willmer [ed.] Cells and tissues in culture, p. 99–142. Vol. 1. Academic Press, New York.

Wells, L. J., and E. Borghese. 1961. Sviluppo embrionale *in vitro* del pancreas di topo. Boll. Zool. 28:235–239.

Wells, L. J., and E. Borghese. 1963. Development of the pancreas of the mouse embryo *in vitro:* acini and islets. Gen. Comp. Endocrinol. 3:265–273.

Westfall, B. B. 1962. Characterization of cells in tissue culture. Nat. Cancer Inst. Monogr. 7:147–157.

Westfall, B. B., E. V. Peppers, V. J. Evans, K. K. Sanford, N. M. Hawkins, M. C. Fioramonti, H. A. Kerr, G. L. Hobbs, and W. R. Earle. 1958. The arginase and rhodanese activities of certain cell strains after long cultivation *in vitro.* J. Biophys. Biochem. Cytol. 4:567–570.

White, P. R. [ed.] 1957. Proceedings of the decennial review conference on tissue culture. J. Nat. Cancer Inst. 19:467–843.

White, P. R. 1963. The cultivation of animal and plant cells, 2nd ed. Ronald, New York. 228 p.

Whitfield, J. F., and R. H. Rixon. 1959. Effects of X radiation on multiplication and nucleic acid synthesis in cultures of L strain mouse cells. Exp. Cell Res. 18:126–137.

Whitfield, J. F., R. H. Rixon, and T. Youdale. 1961. Effects of ultraviolet light on multiplication and deoxyribonucleic acid synthesis in cultures of L strain mouse cells. Exp. Cell Res. 22:450–454.

Whitfield, J. F., R. H. Rixon, and T. Youdale. 1962. Prevention of mitotic delay in irradiated suspension cultures of L mouse cells by agmatine. Exp. Cell Res. 27:143–147.

Whitmore, G. F., J. E. Till, R. B. L. Gwatkin, L. Siminovitch, and A. F. Graham. 1958. Increase of cellular constituents in X-irradiated mammalian cells. Biochim. Biophys. Acta 30:583–590.

Whitten, W. K. 1956. Culture of tubal mouse ova. Nature 177:96.

Whitten, W. K. 1957. The effect of progesterone on the development of mouse eggs *in vitro.* J. Endocrinol. 16:80–85.

Willmer, E. N. 1960*a.* Tissues in culture and in the body. Symp. Soc. Exp. Biol. 14:28–40.

Willmer, E. N. 1960*b.* Cytology and evolution. Academic Press, New York. 430 p.

Wolff, E. 1952. Sur la différenciation sexuelle des gonades de souris explantées in vitro. Compt. Rend. Acad. Sci. 234:1,712–1,714.

Wolff, E., and J. P. Weniger. 1954. Recherches préliminaires sur les chimères d'organes embryonnaires d'oiseaux et de mammifères en culture in vitro. J. Embryol. Exp. Morphol. 2:161–171.

Woods, M. W., K. K. Sanford, D. Burk, and W. R. Earle. 1959. Glycolytic properties of high and low sarcoma-producing lines and clones of mouse tissue culture cells. J. Nat. Cancer Inst. 23:1,079–1,088.

Yoshioka, W. 1960. Studies on the isologous transplantation of tooth germs in mice. II. Effects of hormones on the isologous transplantation of tooth germs previously cultured *in vitro.* Arch. Histol. Jap. 20:19–34.

26

Lifespan and Aging Patterns[1]

Elizabeth S. Russell

This chapter contains a brief presentation of knowledge up to 1964 concerning the longevity of inbred mice. Mice have long been favored subjects in investigations of the nature of aging. Much of the available information on their lifespan comes from isolated experiments each of which involved one or a very few genotypes. Some of the best material for comparison of the relative longevity of different genetic types comes from life-histories of mice of many inbred strains maintained together under a single regimen, as in the pedigreed expansion stocks of a supply colony.

The first purpose of this chapter is to provide information about lifespans of mice from a variety of inbred strains and F_1 hybrids to assist investigators in the design of experiments. Although the data summarized here provide a reliable comparison of the potentialities of many genotypes, readers who wish to put this information to practical use are cautioned against expecting exact correspondence between these data and observations in their own laboratories. In any experiment involving lifespan, concomitant controls are of course essential. The second purpose of this chapter, particularly pertinent to mouse pathology, is to present a discussion of the aging patterns of mice and of variation between genetic groups of mice in the incidence of particular pathological conditions. Many of the research uses of inbred mice have developed from their reliable and predictable development of specific types of tumors, described in detail in Chap. 27. Differences between genotypes are also known in the incidence of particular types of nonneoplastic pathological conditions. To the extent that these differences are genetically controlled, such conditions may be termed constitutional disease (Chap. 29). In this chapter I will deal with relationships between lifespan and the incidence of pathological conditions and will discuss genetic and environmental influences on longevity of mice.

LIFESPANS OF INBRED AND F_1 HYBRID MICE

Continuity of a particular inbred strain implies that mice must live long enough to produce offspring, but beyond that minimal requirement there are great differences between strains in characteristic lifespan. Comfort (1959) stated that the typical "life span (reckoned as the last decile survival) ranges from 1.3 to 3 years in various strains of mice." Mice from strains with the shortest lifespans (i.e., AKR/J and C3H/J females) are usually extremely susceptible to a specific kind of neoplasm. Certain long-lived strains and F_1 hybrids have been much favored in radiation experiments. Since 1959, CBA (Neary, 1960; Comfort, 1959; Alexander and Connell, 1960, 1963), C3HeB (Comfort, 1959; Grahn and Hamilton, 1964), RF and RFM (Upton, 1960; Storer, 1962; Spalding and Strang, 1962), A (Duhig and Warren, 1960), LAF_1 (Upton et al., 1960; Upton, 1960; Neary, 1960; Lesher et al., 1960; Cole et al., 1960; Sacher and

[1] The preparation of this chapter and the collection of much of the information reported have been supported in part by Public Health Service Research Grant CA 01074 from the National Cancer Institute.

Grahn, 1964), (101 × C3H)F_1 (Cosgrove et al., 1964), and noninbred CF#1 (Finkel et al., 1959; Curtis and Gebhard, 1959) have been so used. Mean lifespans reported for nonbred controls in these radiation experiments ranged from 460 days (strain A ♀♀, Duhig and Warren, 1960) to 866 days (CBA ♀♀, Neary et al., 1962), with some difference between means for the same strain in different laboratories. In many of these control groups, the oldest survivors lived more than 1,000 days. Differences in lifespan have been found between lines of mice selected for large and small body size, with large mice surviving longer in one experiment (Chai, 1959), and selected small mice in another (Roberts, 1961). In crosses between these selected lines, as in all hybrid crosses which have been reported, F_1 hybrid offspring survived significantly longer than did mice of either parental type. However, one male in a line (NS) selected for low body weight survived 1,330 days, which possibly is the longest recorded lifespan for the laboratory mouse (Roberts, 1961). A detailed study of the effects of dietary differences on lifespans of mice gives comparative data on nonbred C57BL/6J, A/J, and DBA/2J (Silberberg et al., 1961, 1962*a*, *b*). Similarly, Mühlbock's (1959) study of factors influencing lifespan of inbred mice gives comparative data for DBA_f, O20, CBA, C3H, and $C3H_f$, and Murray's (1965) analysis of factors influencing rate of mammary tumor incidence gives comparative data for C3H, C3HeB, DBA, A, and MA/My.

Data collected at The Jackson Laboratory provide a basis for comparing larger numbers of inbred strains and F_1 hybrids. This material, previously unpublished, may be suitable for demonstrating: (1) range of lifespans characteristic of mice in single genetically homogeneous populations, (2) differences between inbred strains in means and range of lifespan, (3) effects of sex, breeding, environmental conditions, and hybrid vigor on lifespan.

Pedigreed expansion stocks

Large colonies of mice from many different inbred strains, all descended from mice in the Foundation Stocks of The Jackson Laboratory, are maintained on a common regimen in the Pedigreed Expansion Stocks (PES) of The Jackson Laboratory (Green, 1964). Inbred but nonpedigreed offspring of PES mice constitute the Production Stocks (PS) of The Jackson Laboratory.

Table 26-1. MEAN LIFESPANS (m) AND STANDARD ERRORS OF MEANS (S.E.) IN DAYS IN SAMPLES OF n INBRED MICE IN THE PEDIGREED EXPANSION STOCKS

Inbred strain	Mating[a]	(1) *Bred* ♀♀ (1948–56) m ± S.E.	n	(2) *Virgin* ♀♀ (1952–57) m ± S.E.	n	(3) ♂♂ (1948–56) m ± S.E.	n	(4) *Bred* ♀♀ (1960–62)[b] m ± S.E.	n
A/HeJ	1	400 ± 5	380	520 ± 9	265	478 ± 7	300	512 ± 11	62
A/J	1	405 ± 7	299	481 ± 8	310	503 ± 11	159	512 ± 9	59
AKR/J	1	269 ± 3	563	256 ± 3	396	272 ± 5	194		
BALB/cJ	1	462 ± 7	377	532 ± 15	210	485 ± 9	266		
C3H/J	1	280 ± 1	1447	398 ± 6	352	407 ± 7	421		
C3HeB/J	3	492 ± 9	386	583 ± 11	200			571 ± 11	66
CBA/J	1	359 ± 11	84					523 ± 8	68
C57BL/6J	1	561 ± 8	525	653 ± 13	193	539 ± 7	446	695 ± 9	99
C57BR/cdJ	1	512 ± 10	175	588 ± 14	175	475 ± 13	136		
C57L/J	1	492 ± 13	126	577 ± 19	96	532 ± 13	127		
DBA/1J	1	410 ± 5	514	582 ± 11	174	438 ± 9	208	602 ± 13	52
DBA/2J	1	407 ± 6	478	547 ± 9	196	415 ± 9	243	661[c] ± 11	64
LP/J	3	517 ± 7	257	706 ± 11	131				
MA/MyJ	3	486 ± 7	367	583 ± 8	197				
P/J	3	369 ± 9	180	450 ± 9	175				
129/J	3	504 ± 11	179	621 ± 18	68				

[a] 1 = Single pair mating, 3 = three females with one male.
[b] Improved husbandry conditions.
[c] In 1960–1962, DBA/2J had much lower mammary tumor incidence (Hoag, 1963).

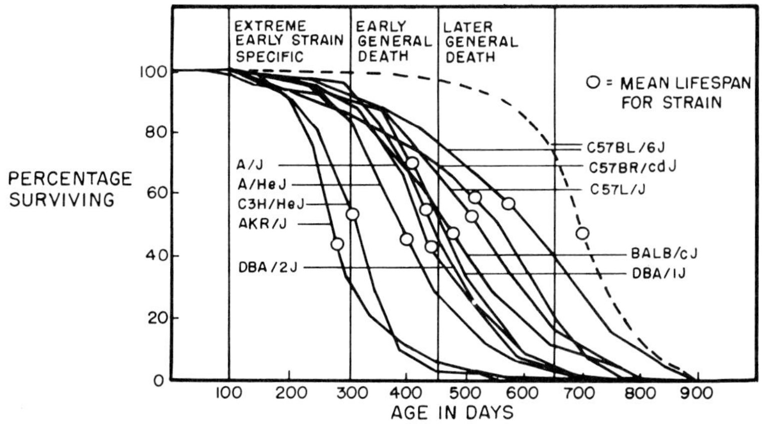

Fig. 26-1. Survival curves for bred female mice from 10 inbred strains in the Pedigreed Expansion Stocks, The Jackson Laboratory. Solid line, 1948–1956. Dashed line, 1960–1962.

In order to provide data for comparison of the characteristics of mice from 16 different inbred strains, total lifespans were determined for all breeding females and a great majority of the males used as breeders during the first 5 years of the PES colony from 1948 to 1952, plus a representative sample of breeders used from 1952 to 1956. In the sense of uniformity of timespan, place, and most of the environmental conditions, the mice belonged to a single population, but genetically they consisted of 16 different subpopulations of varying sizes. All mice were fed Purina Fox Chow or Purina Laboratory Chow, kept in wooden boxes until 11 months of age, and then retired in groups of two to five. Males killed by fighting shortly after retirement have been eliminated from the data. The physical condition of all mice was checked at least once per week for their entire lifetime; the majority of the animals were killed and autopsied when moribund; time of death was recorded for the remainder.

The mean lifespans (Table 26-1, col. 1, 3) and curves for per cent survival (Fig. 26-1, 26-2) varied considerably among the component inbred strains. In AKR/J, all were dead by 550 days; in C57BL/6J, the mean longevity of females was 561 days, that of males 539 days. Losses prior to 300 days were restricted almost exclusively to males and females of the AKR/J and to females of the C3H/J inbred strains. Between 300 and 649 days, there were many deaths in all strains (except those mentioned above), but there were still significant differences between strains in mean lifespan. Beyond 650 days, the proportion remaining alive was less than 20 per cent in all strains except the three related ones C57BL/6J,

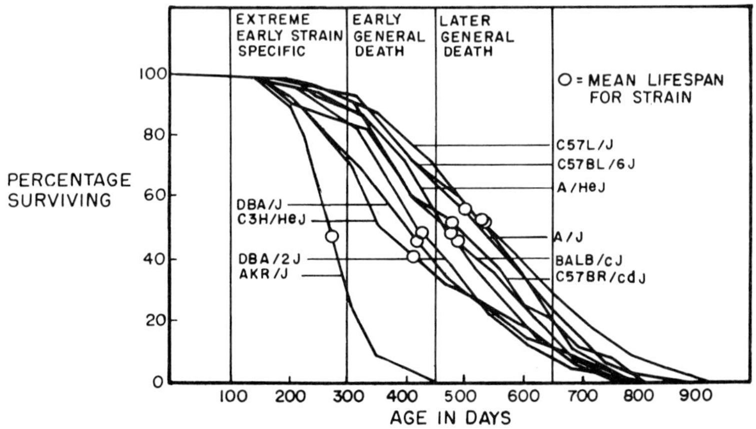

Fig. 26-2. Survival curves for male mice from 10 inbred strains in the Pedigreed Expansion Stocks, The Jackson Laboratory.

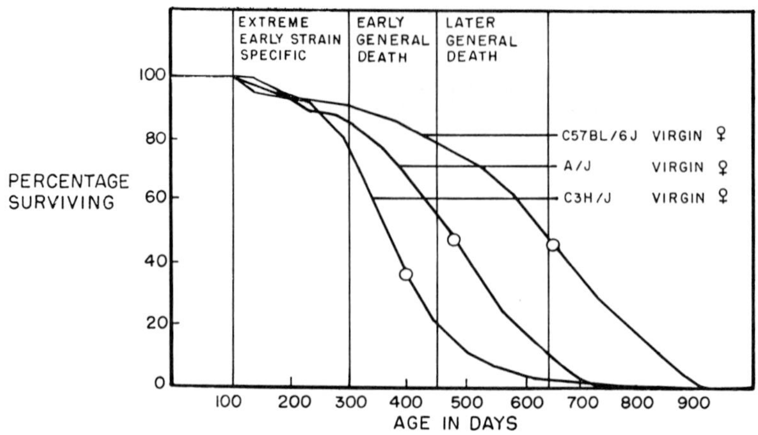

Fig. 26-3. Survival curves for virgin female mice from three inbred strains in the Pedigreed Expansion Stocks, The Jackson Laboratory.

C57L/J, and C57BR/cdJ, and no mice survived beyond 925 days. From 1952 to 1957, lifespans of virgin females of these same inbred strains in the PES were determined (Table 26-1, col. 2). For 14 of 15 tested inbred strains, virgins survived significantly longer than did their bred sisters. Only in C57BL/6J did as many as one-half of the mice live more than 650 days (Fig. 26-3).

Following transfer of the Pedigreed Expansion Stocks in 1959 to new improved quarters, with change in diet, housing, cleanliness, and pathogen control, significantly increased lifespans were observed in all of the seven inbred strains for which more than 50 new life-histories of bred females were followed (Table 26-1, col. 4). Animals were maintained *Salmonella*-free and housed as single pairs in stainless-steel pens until retired at 8 to 10 months of age. They were fed Old Guilford pellets (11 per cent fat, 19 per cent protein). Observation regimen was the same as in the original study. In each case increased longevity was observed, associated with delay in onset of mortality, with no increase in maximum lifespan. Fifty-five per cent of the C57BL/6J bred females survived more than 650 days (Fig. 26-1).

Hybrid mice from production stocks

F_1 hybrid mice from crosses between two inbred strains, combining genetic homogeneity with hybrid vigor, are favored animals for many types of research which do not require production of genetically homogeneous offspring. To provide information on characteristics of six different types of F_1 hybrids, total lifespans were determined for virgin animals born in the Production Stocks (PS) between 1954 and 1958, maintained in groups of five like-sexed animals per pen, and observed according to the PES regimen (Table 26-2). For each F_1 hybrid, the survival curves of males (Fig. 26-4) and of virgin females were very similar, and the mean and maximum lifespans exceeded those for males and virgin females of either parent strain. In five of the six F_1 hybrids, more than half of the mice of both sexes lived more than 650 days.

CHARACTERISTIC AUTOPSY FINDINGS

When inbred mice are autopsied at advanced ages, the incidence of specific pathological lesions varies according to strain, sex, and breeding history (Dunn, 1954*a*, *b*; Hoag, 1963; Chap. 20, 27, 29). The official listing of inbred strains of mice (Staats, 1964) gives information on characteristic

Table 26-2. MEAN LIFESPANS (m) AND STANDARD ERRORS OF MEANS (S.E.) IN DAYS IN SAMPLES OF n NONBRED F_1 HYBRID MICE (1954–1958)[a]

	Virgin ♀♀		♂♂	
Hybrid	m ± S.E.	n	m ± S.E.	n
AKD2F_1	544 ± 10	198	507 ± 11	187
B6AF_1	714 ± 14	194	673 ± 16	195
BRB6F_1	744 ± 17	122	692 ± 20	135
B6D2F_1	729 ± 12	171	723 ± 12	160
CAF_1	678 ± 11	208	661 ± 11	213
LAF_1	725 ± 13	186	704 ± 14	196

[a] Husbandry conditions correspond to those in Table 26-1, col. 1.

tumor incidences for many inbred strains and sublines. Incidences observed for different colonies of the same inbred strain tend to be very similar, but significant differences have been noted (Law et al., 1955; Hoag, 1963). The longevity studies of bred animals from PES of The Jackson Laboratory provide good material for comparison of the aging patterns of mice from many different inbred strains, since a large proportion of these mice were autopsied when moribund, with abnormal-appearing tissues diagnosed by a pathologist (E. A. Fekete, P. Borges, or E. D. Murphy of The Jackson Laboratory staff). All strain differences reported here were derived from comparisons of animals dying within the same age interval. The marked differences in pathological findings between strains compared at equivalent ages demonstrate a genetic contribution to the causation of many disease processes. High percentages of mammary tumors were observed in C3H/J, A/HeJ, DBA/1J, and DBA/2J females, plus a moderate number in A/J females, in contrast to low percentages in all other strains. In all of these high mammary tumor strains, time of appearance of mammary tumors (influenced by genes, mammary tumor inciter, and hormonal stimulation) was the major determinant of female lifespan (Hummel and Little, 1963; Mühlbock, 1959; Murray, 1965; Pilgrim and Dowd, 1963). C3H/J females showed the greatest susceptibility to mammary cancer (95 to 99 per cent of all deaths), with a highly significant tendency to die extremely early. Incidence of reticular neoplasia differed greatly among the 10 tested inbred strains. The incidence of lymphatic leukemia was significantly higher ($P < 0.0001$) and time of appearance earlier in the AKR/J than in all other strains observed; because of the early onset, there is little doubt that this is a specific AKR/J characteristic. A low incidence of reticular neoplasia was also observed in very old mice of several inbred strains, and the question arose as to whether this was a nonspecific result of the aging process. This is particularly pertinent because the incidence was highest in C57BL/6, which had the largest proportion of long survivors. However, proof of genetic contribution to the causation of late-appearing neoplasms of reticular tissue was provided by analysis of frequencies in the late general death period. C57BL/6J mice of both sexes dying between 450 and 649 days showed a significantly higher ($P < 0.001$) incidence of reticular neoplasia (frequently reticulum cell sarcoma) than did mice of all other inbred strains dying in the same age interval.

The appearance of primary lung tumors was also restricted to a small number of inbred strains. The incidence was especially high (23 to 59 per cent) in A/HeJ and A/J mice of both sexes. Statistical analysis demonstrated that at each age interval these mice were significantly more susceptible to lung tumors than were mice of all other inbred strains observed. The incidence of lung tumors increased slightly with advancing age in A/HeJ males. The lower incidence observed in A/HeJ females almost certainly resulted from early mammary tumor deaths among these females. Arrangement of autopsy records by age groups disclosed an age-dependent susceptibility to primary lung tumors in BALB/c mice of both sexes. Before 450 days these mice showed a low incidence of lung tumors (10 to

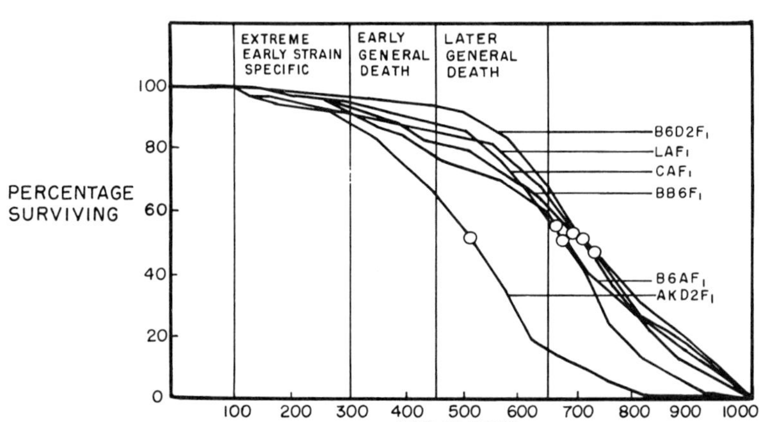

Fig. 26-4. Survival curves for male mice produced in six different F_1 hybrid crosses in the Production Stocks, The Jackson Laboratory.

13 per cent), but the frequency increased markedly in later autopsies (34 to 45 per cent). Between 450 and 649 days BALB/c mice, though having significantly fewer lung tumors than mice of either A strain, still had a significantly higher incidence than was observed in the combined populations of the other seven inbred strains.

Hepatomas were observed only in the older mice and were limited almost entirely to males. Among mice autopsied between 450 and 649 days, the incidence of hepatoma was high in C3H, C57BR/cd, and C57L males. The incidence was significantly higher in C3H than in the other two susceptible strains, whose incidence was, in turn, significantly greater than that in the other seven strains observed.

Certain nonneoplastic lesions also appeared to be restricted to particular inbred strains. Calcified areas on the heart were frequently observed in DBA/2 and C3H females, very rarely in other mice. Mice of the A/HeJ and A/J strains showed high incidence of papillonephritis, increasing with advancing age to nearly 100 per cent. This condition is believed to be related to amyloidosis, also characteristic of these strains (Dunn, 1954*b;* West and Murphy, 1965). Kidney lesions were also observed frequently in C57BL/6, C57BR/cd, and C57L mice, increasing with advancing age so that approximately 50 per cent of animals over 450 days were affected.

Aging C57BL/6 mice show hematological changes with gradual decrease in the number of erythrocytes and consequent drop in the ratio of red-cell mass to total blood volume (Ewing and Tauber, 1964). This effect of age on blood values is probably not limited to this strain. Decline in erythrocyte number apparently does not have a strong direct effect on lifespan, since the same type of change was observed in short-lived and long-lived individuals.

ENVIRONMENTAL INFLUENCES

Broad distribution of age at death within inbred strains and differences in mean survival of mice from the same inbred strains under altered conditions demonstrate that environmental factors have significant effects on longevity. Intercurrent infections, parasitism, fighting, and occasional malocclusion account for sporadic deaths. Three specific nongenetic factors (mammary tumor virus, breeding history, and diet) affecting lifespan have been identified in controlled experiments. Individual variations in these factors may contribute to observed differences in lifespan within a single colony of a particular inbred strain. Other identifiable within-strain life-history variables, such as season of birth, age of parents at birth, or lifespan of parents, have no discernible effects upon mouse lifespan (Schlager, 1965, personal communication).

Effects on mean lifespan of certain identifiable variables have been demonstrated in a few carefully controlled experiments. High-fat diets throughout life shorten the lifespan of C57BL/6J, DBA/2J, and YBR mice (Silberberg and Silberberg, 1954; Silberberg et al., 1955). Caloric restriction lowered the mean lifespan of C57BL/6J, A/J, and DBA/2 mice by increasing juvenile mortality, whereas lard-enriched diets during growth increased the mean lifespans of C57BL/6 males and DBA/2 females, but not of other groups (Silberberg et al., 1961, 1962*a, b*). Genetically obese (*ob/ob*) mice fed ad libitum had shorter lifespans than normal littermates, but restricting their food intake cancelled the difference in survival (Lane and Dickie, 1958). Dystrophic mice have been shown to survive longer on a good than on a poor laboratory diet (Coleman and West, 1961). Crowding has been shown to reduce mean lifespan of $C3H_f$ mice (Mühlbock, 1959), and type of cage to influence both lifespan and, secondarily, tumor incidence in CF#1 mice (Finkel and Scribner, 1955). The frequently observed differences in mean lifespan between breeding and virgin females indicate that producing and rearing offspring reduces life expectancy. Forced breeding, with reduced intervals between successive litters, accentuates this life-shortening effect (Mühlbock, 1959; Murray, 1965). Lifespan differences attributable to breeding history are particularly noticeable in strains with a high susceptibility to mammary tumors. In these strains the presence of the tumor-inducing virus greatly decreases lifespan of females (Hummel and Little, 1963; Murray, 1965). Evidence of this is also seen in the data given earlier in this chapter in the difference in mean lifespan between C3H/HeJ (carrying the virus) and C3HeB/J (virus-free) females. The extreme increase in mean lifespan of DBA/2J females between 1956 and 1960 can also be explained in part by loss of the mammary tumor virus during the intervening period (Hoag, 1963).

GENETIC INFLUENCES

Considerable variations in lifespan have been observed in mice, both within and between genetically homogeneous groups. The differences observed between mice of the same inbred strain in the Pedigreed Expansion Stocks cannot be attributed to residual variability within the strain for genes affecting lifespan, since there was no correlation between lifespan of parent and of progeny (Schlager, 1965, personal communication). However the variations in means and distributions of lifespan between inbred strains of mice, each representing the replication of the same or nearly the same homozygous genotype, provide valuable material for the detection of polygenic influences on lifespan and for the identification of certain environmental factors with significant effects on longevity. The highly significant differences in lifespan observed when mice from many inbred strains are maintained together under a common regimen clearly demonstrate a large genetic contribution to the determination of total lifespan, but give no indication of the number or nature of the effective genes. Comparisons of mean lifespans and mean litter sizes of productive breeding females in 12 inbred strains showed highly significant correlations, indicating that the genetic factors which favor longevity also lead to increased fertility (Roderick and Storer, 1961). Differences between inbred strains in aging pattern or incidence of particular pathological lesions indicate differences between strains in tissues susceptible to early degeneration. The uniform observation that F_1 hybrid animals survive longer than those of either parental inbred strain indicates that the inbred strains are homozygous for different recessive genes which affect lifespan. Further analysis of the formal genetics of aging would be very difficult. The few studies known are limited to demonstration of effects of certain deleterious genes such as lethal yellow (A^y) (Silberberg et al., 1955; Heston and Vlahakis, 1963; Hrubant, 1964), the viable black-and-tan allele (a^t) (Hrubant, 1964), obese (*ob/ob*) (Lane and Dickie, 1958), and dystrophia muscularis (*dy/dy*) (Coleman and West, 1961). In genetically controlled stocks where these specific mutant genes are maintained congenic with particular inbred strains, the lifespans of mice carrying mutant alleles were significantly shorter than those of homozygous normal littermates.

SUMMARY

Information on lifespans of mice from numerous inbred strains and hybrids at The Jackson Laboratory has been presented. These data reveal highly significant strain and sex differences and longer lives for virgin than for bred females. Under comparable environmental conditions the mean and maximum lifespans of F_1 hybrid mice always exceeded those observed for either parental strain. The incidence of certain kinds of neoplasia, including mammary gland tumors, reticular neoplasia, primary lung tumors, and hepatomas, differed markedly from strain to strain.

Differences between strains demonstrated the importance of genetic factors in determination of lifespan. Life-shortening effects have also been demonstrated for four specific mutant genes. Broad distribution within strains of age at death showed that environmental factors influence lifespan. Improvement of husbandry conditions increased the mean lifespans of mice from many inbred strains in the PES stocks of The Jackson Laboratory. Significant effects have been demonstrated for diet, crowding, differences in breeding history, and presence vs. absence of the mammary tumor virus.

In the design of experiments involving lifespan determination or in other studies involving old mice, both environmental and genetic controls are important. The investigator should be aware of strain differences in total lifespan and in aging pattern and should select mice of appropriate genotypes. However, because of effects of environmental variables, it is important that concomitant controls be included in all studies.

LITERATURE CITED

Alexander, P., and D. I. Connell. 1960. Shortening of the life span of mice by irradiation with x-rays and treatment with radiomimetic chemicals. Radiat. Res. 12:38–48.

Alexander, P., and D. I. Connell. 1963. The failure of the potent mutagenic chemical, ethyl methane sulphonate, to shorten the life-span of mice, p. 259–265. *In* R. J. C. Harris [ed.] Cellular basis and aetiology of late somatic effects of ionizing radiations. Academic Press, London.

Chai, C. K. 1959. Life span in inbred and hybrid mice. J. Hered. 50:203–208.

Cole, L. J., P. C. Nowell, and J. S. Arnold. 1960. Late effects of x-radiation. The influence of dose fractionation on life span, leukemia, and nephrosclerosis incidence in mice. Radiat. Res. 12:173–185.

Coleman, D. L., and W. T. West. 1961. Effects of nutrition on growth, lifespan and histopathology of mice with hereditary muscular dystrophy. J. Nutrit. 73:273–281.

Comfort, A. 1959. Natural aging and the effects of radiation. Radiat. Res. (Suppl. 1):216–234.

Cosgrove, G. E., A. C. Upton, C. C. Congdon, D. G. Doherty, K. W. Christenberry, and D. G. Gosslee. 1964. Late somatic effects of x-radiation in mice treated with AET and isologous bone marrow. Radiat. Res. 21:550–574.

Curtis, H. J., and K. L. Gebhard. 1959. Radiation induced ageing in mice, p. 210–216. *In* J. C. Bugher, J. Coursaget, and J. F. Loutit [ed.] Progress in nuclear energy. Biological Sciences. Vol. 2. Pergamon Press, New York.

Duhig, J. T., and S. Warren. 1960. Intermittent radiation and the life span of mice. Arch. Pathol. 70:486–496.

Dunn, T. B. 1954*a*. Normal and pathologic anatomy of the reticular tissue in laboratory mice, with a classification and discussion of neoplasms. J. Nat. Cancer Inst. 14:1,281–1,433.

Dunn, T. B. 1954*b*. The importance of differences in morphology in inbred strains. J. Nat. Cancer Inst. 15:573–589.

Ewing, K. L., and O. E. Tauber. 1964. Hematological changes in aging male C57BL/6Jax mice. J. Gerontol. 19:165–167.

Finkel, M. P., B. O. Biskis, and G. M. Scribner. 1959. The influence of strontium-90 upon life span and neoplasms of mice, p. 199–209. *In* J. C. Bugher, J. Coursaget, and J. F. Loutit [ed.] Progress in nuclear energy. Biological Sciences. Vol. 2, Pergamon Press, New York.

Finkel, M. P., and G. M. Scribner. 1955. Mouse cages and spontaneous tumors. Brit. J. Cancer 9:464–472.

Grahn, D., and K. F. Hamilton. 1964. Influence of sex, environment, and radiation factors on life shortening and tumor incidence in $C3H_f$ mice. Radiat. Res. 22:191. (Abstr.)

Green, E. L. [ed.] 1964. Handbook on genetically standarized Jax mice. The Jackson Laboratory, Bar Harbor, Maine. 82 p.

Heston, W. E., and G. Vlahakis. 1963. Influence of the A^y gene on mammary-gland tumors, hepatomas, and normal growth in mice. J. Nat. Cancer Inst. 26:969–983.

Hoag, W. G. 1963. Spontaneous cancer in mice. Ann. N.Y. Acad. Sci. 108:805–831.

Hrubant, H. E. 1964. Specific genetic control of life span. J. Gerontol. 19:451–452.

Hummel, K. P., and C. C. Little. 1963. Comparative virulence of the mammary tumor agent from different sources; qualitative and quantitative differences. J. Nat. Cancer Inst. 30:593–604.

Lane, P. W., and M. M. Dickie. 1958. The effect of restricted food intake on the life span of genetically obese mice. J. Nutr. 64:549–554.

Law, L. W., T. B. Dunn, and P. Boyle. 1955. Neoplasms in the C3H strain and in F_1 hybrid mice of two crosses following introduction of extracts and filtrates of leukemic tissues. J. Nat. Cancer Inst. 16:495–540.

Lesher, S., K. Hamilton, D. Grahn, and G. Sacher. 1960. The causes of death for LAF_1 mice exposed to low dose, daily, Co^{60} gamma irradiation. Radiat. Res. 12:451.

Mühlbock, O. 1959. Factors influencing the life-span of inbred mice. Gerontologia 3:177–183.

Murray, W. S. 1965. Biological significance of factors influencing the incidence of mammary cancer in mice. J. Nat. Cancer Inst. 34:21–41.

Neary, G. J. 1960. Ageing and radiation. Nature 187:10–18.

Neary, G. J., E. V. Hulse, and R. H. Mole. 1962. The relative biological efficiency of fast neutrons and gamma-rays for life-shortening in chronically irradiated CBA mice. Int. J. Radiat. Biol. 4:239–248.

Pilgrim, H. I., and J. E. Dowd. 1963. Correcting for extraneous death in the evaluation of morbidity or mortality from tumor. Cancer Res. 23:45–48.

Roberts, R. C. 1961. The lifetime growth and reproduction of selected strains of mice. Heredity 16:369–381.

Roderick, T. H., and J. B. Storer. 1961. Correlation between mean litter size and mean life span among 12 inbred strains of mice. Science 134:48–49.

Sacher, G. A., and D. Grahn. 1964. Survival of mice under duration-of-life exposure to gamma rays. I. The dosage-survival relation and the lethality function. J. Nat. Cancer Inst. 32:277–321.

Silberberg, M., and R. Silberberg. 1954. Factors modifying the lifespan of mice. Amer. J. Physiol. 177:23–26.

Silberberg, R., S. R. Jarrett, and M. Silberberg. 1961. Life span of mice fed enriched or restricted diets during growth. Amer. J. Physiol. 200:332–334.

Silberberg, R., S. R. Jarrett, and M. Silberberg. 1962*a*. Longevity of female mice kept on various dietary regimens during growth. J. Gerontol. 17:239–244.

Silberberg, R., M. Silberberg, and S. R. Jarrett. 1962*b*. Effects of diet during growth; studies in male mice of various strains. Pathol. Microbiol. 25:56–66.

Silberberg, R., M. Silberberg, and S. Riley. 1955. Life span of 'yellow' mice fed enriched diets. Amer. J. Physiol. 181:128–130.

Spalding, J. F., and V. G. Strang. 1962. Reduced survival time in descendants of five generations of x-irradiated sires. Radiat. Res. 16:159–164.

Staats, J. 1964. Standardized nomenclature for inbred strains of mice, Third listing. Cancer Res. 24:147–168.

Storer, J. B. 1962. Evaluation of radiation response as an index of aging in mice. Radiat. Res. 17:878–902.

Upton, A. C. 1960. Ionizing radiation and aging. Gerontologia 4:162–176.

Upton, A. C., A. W. Kimball, J. Furth, K. W. Christenberry, and W. H. Benedict. 1960. Some delayed effects of atom-bomb radiations in mice. Cancer Res. 20 (No. 8, Part 2):1–60.

West, W. T., and E. D. Murphy. 1965. Sequence of deposition of amyloid in strain A mice and relationship to renal disease. J. Nat. Cancer Inst. 35:167–174.

27

Characteristic Tumors[1]

Edwin D. Murphy

The tumors of the mouse described in this chapter have been selected primarily on the basis of frequency of occurrence among the available inbred strains and the amount of research interest shown in them. Many less frequent types repeatedly appear as incidental findings in tabulations of tumors of untreated mice of inbred strains. Some rarer types that have occurred spontaneously in mice at The Jackson Laboratory have been included for completeness. Selected tumors that rarely occur spontaneously but are readily induced have been included, particularly tumors that can be induced by hormonal imbalance. Emphasis has been given to the induced tumors of types important in human pathology. More references have been cited for the less well known tumors than for the common types which have been extensively reviewed.

DEFINITION AND PROPERTIES OF TUMORS

Willis (1960) proposed a workable definition for distinguishing true tumors from inflammatory and reparative proliferations, hyperplasias, and malformations with excess of tissue. A tumor "is an abnormal mass of tissue, the growth of which exceeds and is uncoordinated with that of the normal tissues, and persists in the same excessive manner after cessation of the stimuli which evoked the change." Every pathologist can think of exceptions but these do not invalidate the general applicability of the definition.

Classification of tumors

The commonly used and most useful classification of tumors is histogenetic, that is, the tumors are named according to the tissues from which they arise and of which they consist. In most tumors the neoplastic tissue consists of cells of a single type and, with experience, one can readily classify them. The types of histological differentiation found in tumors appear to be inherent in the parent tissues. Foulds (1940) concluded that most adult normal cells have a greater capacity for divergent differentiation than was formerly supposed and that it is unlikely that tumor cells acquire new capacities. The few kinds of tumors in which there is uncertainty regarding the precise tissue of origin require further histopathological research. Meanwhile in these cases we must settle for noncommittal identifying names.

Willis (1960) illustrated the application of histogenetic classification to both human and animal tumors. Cloudman (1941) presented a histological classification of mouse tumors. Dunham and Stewart (1953) gave a classification of transplantable and transmissible animal tumors.

Benign or malignant

In addition to histogenetic classification, it is of practical value in human oncology to attempt to predict the behavior of a tumor from its morphology. The tumors of any given cell type may show a wide range of difference in structure, mode of growth, rate of growth, and danger to the host. Some, called *benign*, are well differen-

[1] The writing of this chapter was supported in part by Public Health Service Research Grant CA 05985 from the National Cancer Institute, by a Fulbright award, and by a guest professorship at Universität Frankfurt, Institut für Humangenetik und Vergleichende Erbpathologie.

tiated, grow only by expansion with the formation of a capsule, grow slowly, and are dangerous only in terms of position, accidental complication, or excessive hormone production. Others, often less well differentiated, grow rapidly, invade adjacent tissues, spread by metastasis, and unless extirpated at an early stage will kill the host; these are *malignant* tumors. Between these extremes there may be tumors of intermediate behavior. Thus, the terms benign and malignant are relative and arbitrary.

Criteria for diagnosing malignancy include imperfect differentiation and variation in the size, shape, and staining quality of the cell and the nucleus, invasion of adjacent tissues, and metastasis. In general the degree of malignancy is roughly proportional to the degree to which tumors fail to attain histological differentiation; the most anaplastic tumors are the most malignant. Metastasis depends on the invasion of blood vessels, lymphatics, or serous and other cavities, with the detachment of tumor cells or cell clusters and the establishment of distant secondary deposits. Lymphatic metastasis is less common in mice than in man. When tumors occur in inbred animals, there is available an additional criterion of malignancy: successful transplantation with progressive growth (Chap. 28).

Stages of development

The fact that tumors develop in a series of stages was first clearly recognized by Rous and Beard (1935), in studies on virus-induced skin tumors in rabbits, and by Greene (1940), in studies on the development of spontaneous mammary and uterine tumors in rabbits. Foulds (1954, 1958) has generalized under the term *progression* the concept of the development of a tumor by irreversible, qualitative changes in one or more of its characters. The concept includes the early "precancerous" and neoplastic changes and the extended development of neoplastic characters that occurs during serial transplantation. A tumor may change, often abruptly, in growth rate, histological structure, invasiveness, or responsiveness to extraneous stimuli such as hormones or chemotherapeutic agents. These characters tend to be independently variable and subject to independent progression. Progression is independent of the duration or size of a tumor. Some tumors pass through, or bypass, all the theoretically possible developmental stages before they are grossly or even histologically recognizable. Others may become stabilized at any stage for the lifetime of the host. Continued serial transplantation, however, almost inevitably results in continued progression toward an endpoint of rapid growth rate, loss of functional and histological differentiation, and loss of responsiveness to extraneous stimuli.

In mice, Berenblum and Shubik (1947) demonstrated stages of initiation and promotion in the induction of skin tumors by a carcinogenic hydrocarbon. Foulds (1956) carried out detailed studies on mammary tumors developing from plaques which grow in response to pregnancy and regress after parturition. Many tumors of endocrine glands and their target organs go through a dependent or conditional phase in which removal of the causative stimulus is followed by regression (Furth, 1953). In such tumors, some cells ultimately become altered if not during residence in the primary host then after transplantation and give rise to autonomous growth, which continues even when the causative stimulus is eliminated.

Genetic factors

Heston (1963) stated that the development of inbred strains has constituted probably the greatest advance in all cancer research. Genetic factors are immediately apparent in the different incidences of spontaneous tumor types among inbred strains and in the differing susceptibilities of these strains to the effects of carcinogens. The process of inbreeding fixes genes concerned with susceptibility to tumor formation (Chap. 2). At the same time, transmitted oncogenic viruses may be carried along as in the case of the mammary tumor agent. Additional environmental factors, bacterial, viral, nutritional, and other, may be perpetuated by laboratory conditions (Chap. 4, 30). In general, the susceptibility of inbred strains and their hybrids to specific tumor induction is related to the frequency of spontaneous tumors of the same type. Potent carcinogens in high doses can induce subcutaneous sarcomas in mice of any strain, but graded doses reveal strain differences. Susceptibility to the development of most tumors appears to be inherited as a multiple-factor character with alternative expression (Heston, 1963). The character appears when the combined action of the genetic and nongenetic factors surpasses a threshold (Chap. 9). The nongenetic factors include physical and chemical carcinogens, hormones, nutritional factors, and viruses.

CHARACTERISTIC TUMORS

The most common tumors of inbred strains of mice are mammary tumors (in breeding females), lymphocytic leukemia, primary lung tumor, hepatoma (in males), and reticulum cell sarcoma (in older animals). Andervont and Dunn (1962) showed that wild mice maintained under laboratory conditions have a similar predominance of these tumor types, except for lymphocytic leukemia. Mammary tumors occur in high incidence in breeding females of strains C3H, DBA, A, and DD. In virgin females the incidence is high in C3H and DD, intermediate in DBA, and low in

Table 27-1. CHARACTERISTIC TUMORS OF INBRED STRAINS, SEX AND AGE INCIDENCE

Tumor type	Strain	Incidence, per cent	Type mouse	Average age, months	Reference
Mammary gland tumor	C3H	99	Breeding ♀♀	7.2	a
		100	Virgin ♀♀	8.8	a
	DBA/2	77	Breeding ♀♀	15	b
		Lower	Virgin ♀♀		c
	A	84	Breeding ♀♀		c
		5	Virgin ♀♀		c
	DD	84	Breeding ♀♀	7.7	a
		75	Virgin ♀♀	10.2	a
Lymphocytic leukemia	AKR	92		8.0	d
	C58	90		10.0	d
Primary lung tumor	A	90		After 18	e
	SWR	80		After 18	f
Hepatoma	C3HeB	91	Breeding ♂♂	21.4	b
		58	Virgin ♀♀	24.1	b
		30	Breeding ♀♀	21.0	b
Reticulum cell neoplasm, Type B	SJL	91	Virgin ♀♀	13.3	g
		High	Breeding ♀♀, ♂♂		g
Reticulum cell neoplasm, Type A	C57BL	15	Breeding ♀♀	22.3	h
Ovarian tumor	C3HeB/De	37	Breeding ♀♀	21.5	b
		47	Virgin ♀♀	24.3	b
	C3HeB/Fe	64	Breeding ♀♀	After 19	i
		22	Virgin ♀♀	After 19	i
	RIII	60	Breeding ♀♀	After 17	i
		50	Virgin ♀♀	After 17	i
	CE	34	♀♀	After 20	j
Pituitary tumor	C57L	33	Breeding ♀♀	Old	k
	C57BR/cd	33	Breeding ♀♀	Old	k
Hemangioendothelioma	HR/De	24	♂♂ and ♀♀	22	l
Adrenal cortical tumor	CE	100	Gonadectomized ♀♀	After 6	m
		79	Gonadectomized ♂♂	After 7	n
Testicular teratoma	129	1	♂♂	Congenital	o
		82	(Male gonadal ridges transplanted to adult testes)		p
Myoepithelioma	BALB/c	4			d
	A	Similar			d
Skin papilloma	HR/De	9	Hairless	22	l
Skin carcinoma	HR/De	3	Hairless	18	l
Subcutaneous sarcoma	C3H/J	3	♀♀	25	q
Harderian gland tumor	C3H	1		20.7	r

[a] Heston et al. (1964). [b] Deringer (1959). [c] Little (1941). [d] Law et al. (1955). [e] Heston (1942). [f] Law (1948). [g] Murphy (1963). [h] Heston (1964). [i] Staats (1964), Hummel (1965, personal communication). [j] Dickie (1954). [k] Russell (1965, personal communication). [l] Deringer (1962*b*). [m] Woolley and Little (1945*a*). [n] Woolley and Little (1945*b*). [o] Stevens (1959). [p] Stevens (1964). [q] Dunn et al. (1956). [r] Heston et al. (1960*b*).

Table 27-2. PERCENTAGE INCIDENCE OF TUMORS OF SOME LONG-LIVED INBRED STRAINS AND WILD MICE

	C3HeB/De[a]				C3Hf/He[b,c]		C57BL/He[d]	DBA/2eBDe[e]		HR/De[f]			Wild mice[g]		
	V♀♀[h]	B♀♀	FB♀♀	♂♂	V♀♀	B♀♀	B♀♀	V♀♀	B♀♀	V♀♀	B♀♀	♂♂	V♀♀	B♀♀	♂♂
Number of mice	99	155	82	21	100	471	314	76	51	48	63	52	72	99	54
Average age, months	23	19	19	21	20	19	21	22	21	22	16	21	24	27	21
Mammary carcinomas	4	55	74		2	40		3	6	4	30			6	
Type A		17	26			18								3	
Type B	4	23	28			19			6					2	
Type C								3							
Adeno-canthoma		17	28			10									
Miscellaneous		5	4											1	
Mixed		<1	5												
Reticulum cell neoplasms	2	3			3	3	25	20	27	12	10	8	11	12	5
Type A	2	<1					15	12	10	6	3	6	<1		
Type B		2					11	8	18	6	6	2	11	12	5
Primary lung tumor	14	5	4	24	10	6	1	4	4	17	3	8	15	28	16
Hepatoma	59	30	38	90	17	25	<1	5		6	2	12	3	4	9
Ovarian tumors	47	37	29		8	6	<1			4	6		3	6	
Granulosa cell tumor	20	23	13				<1			2	5		3	6	
Tubular adenoma	25	14	15												
Luteoma	2	<1	1												
Papillary cystadenoma										2	2				
Hemangioendo-thelioma	2	5	1		2	3	2	12	16	19	14	29	3	8	
Lymphocytic neoplasm	1					<1	3	3	8	6	8	6		2	
Granulocytic leukemia								1	2						
Plasma cell neoplasm			1					1							
Mast cell neoplasm														1	
Subcutaneous sarcoma	7	2	1		3	3			4				1	2	
Osteogenic sarcoma					1	<1							1		
Uterine tumors															
Sarcoma	3				2		<1			2	3				
Leiomyoma								1							
Adenocarci-noma											2				
Endometrial sarcoma									2						
Adrenal cortical tumor				5				1							
Pheochromo-cytoma			1										3		2
Harderian gland tumor				5					2	2					
Myoepithelioma			1								2				
Skin tumors															
Papilloma		<1				<1					2	4			
Squamous cell carcinoma		<1				<1	<1				2	6			
Mammary gland sarcoma											2				
Teratoma															2
Islet cell tumor											2				
Granular myoblastoma					1										
Kidney carcinoma						<1									
Bladder papilloma					1										
Clitoral gland tumor							<1								

[a] Deringer (1959). [b] Heston and Deringer (1952). [c] Heston et al. (1964). [d] Heston (1964). [e] Deringer (1962a). [f] HR/De (haired), Deringer (1951, 1956). [g] Andervont and Dunn (1962). [h] V = virgin, B = bred, FB = force-bred.

A. Tumors of high incidence tend to occur during the latter half of the first year of life and in the early part of the second year. Tumors of low incidence tend to occur during the latter half of the second year and the first half of the third year. Characteristic tumors of inbred strains are listed in Table 27-1. The spectrum of tumors observed in some longer-lived sublines and in wild mice is shown in Table 27-2.

MAMMARY TUMORS

The "mammary tumor of the mouse has probably been the most completely studied of all tumors" (Dunn, 1959). Accessibility to palpation, predictable frequency in a number of inbred strains, and ready transplantability have made the mammary tumor an invaluable tool for investigations in genetic, viral, hormonal, chemotherapeutic, nutritional, and other facets of cancer research. The discovery of the milk agent (Staff, Jackson Memorial Laboratory, 1933) greatly stimulated research on mammary tumors. The agent has the characteristics of a virus and is transmitted by the milk of high mammary tumor strain females to the young. It appears to modify the responsiveness of the mammary gland tissue so that with a favorable genetic constitution and the proper hormonal stimulation, tumors develop at a comparatively early age. A usually much lower incidence of mammary tumors of variable morphology occurs in older mice in the absence of the milk agent (Table 27-2). Mammary tumors have been induced by estrogenic hormones (Lacassagne, 1932; Gardner et al., 1959), implantation of pituitary glands (Mühlbock and Boot, 1959; Heston, 1964), chemical carcinogens (Andervont and Dunn, 1953), and by polyoma virus (Dawe et al., 1959).

Classification

Many classifications of mammary tumors have been proposed. Apolant (1906) made the first detailed histological study of mammary tumors of the mouse. His classification was standard for many years. The histological classification of Dunn (1959) has the advantage of simplicity and has been extensively applied to the analysis of histological types in a number of inbred strains, with and without the milk agent. Dunn's classification appears in Table 27-3 and an application in Table 27-2.

Table 27-3. CLASSIFICATION OF MAMMARY TUMORS

Histological type	Synonyms
Adenocarcinoma Type A	Typical mammary tumor, mammary adenoma, mammary tumor of basic acinar structure, alveolar carcinoma, small tubular carcinoma, microtubular carcinoma
Type B	Variable tumor, papillary cystadenocarcinoma, carcinoma simplex, intratubular carcinoma
Type C	Fibroadenoma, adenofibroma
Adenoacanthoma	Keratinized mammary tumor, adenosquamous carcinoma, adenocancroid, mammary tumor with squamous metaplasia
Carcinosarcoma	Carcinoma with spindle cell formation, anaplastic carcinoma, mixed tumor
Sarcoma	Fibrosarcoma
Miscellaneous	

Gross appearance

Because of the extent of the mammary tissue in the female mouse, from the cervical region to the vulva on the ventral surface and almost to the midline in the back, the tumors may be found at almost any subcutaneous site in the body. In gross appearance, tumors may be round, oval, or coarsely nodular, and well circumscribed. The tumor tissue is usually grayish white and soft and often contains blood-filled cysts and central areas of necrosis.

Adenocarcinoma, Type A

The tumor is composed of uniform small acini, or tubules lined by a single layer of small cuboidal epithelial cells (Fig. 27-1). The tissue appears well differentiated and may show foci of secretory activity. Type A is the characteristic tumor of strain C3H with the mammary tumor agent and together with Type B represents the "typical" adenocarcinomas in mice with the agent.

Adenocarcinoma, Type B

The category represents a diversified group of glandular epithelial patterns, several of which may appear in a single tumor. Areas similar to Type A may be found as well as cysts filled with blood or clear fluid, intracystic papillary projections, irregular cords and tubes, and solid sheets of cells (Fig. 27-2). The amount of stroma may vary.

Adenocarcinoma, Type C

The tumor is composed of multiple cysts of varying size, lined by a single layer of cuboidal

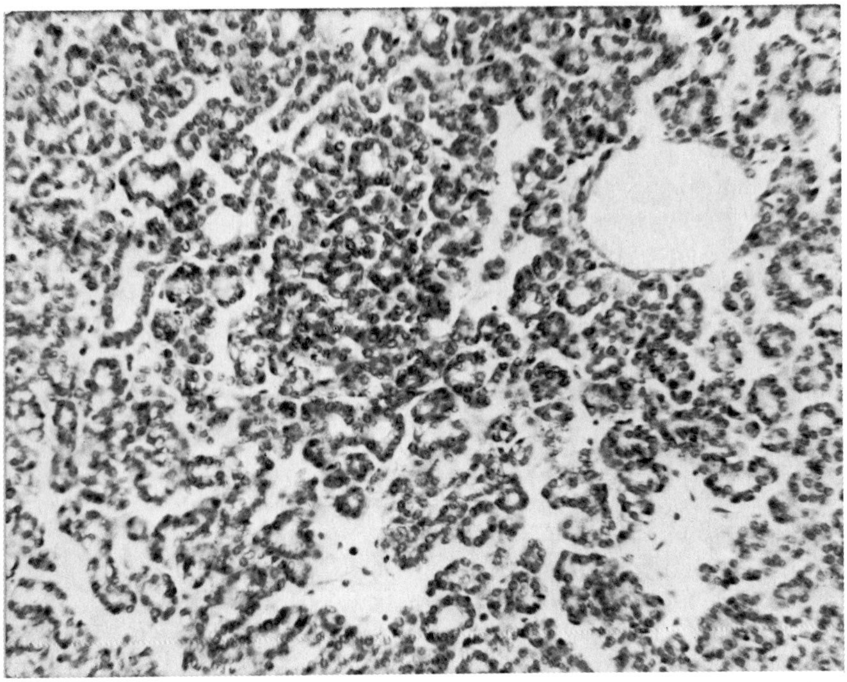

Fig. 27-1. Mammary adenocarcinoma, Type A, with characteristic acinar structure (×250).

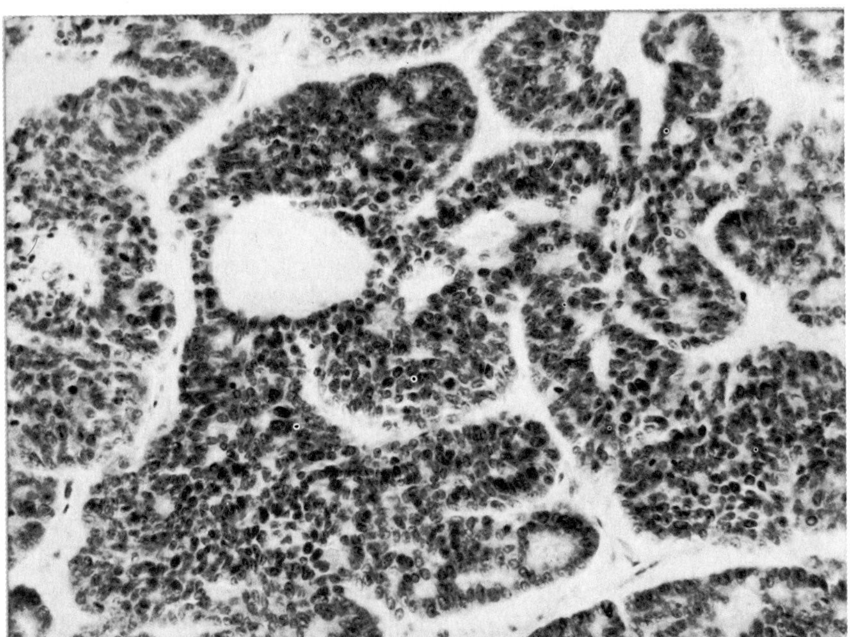

Fig. 27-2. Mammary adenocarcinoma, Type B, with varying epithelial patterns (×250).

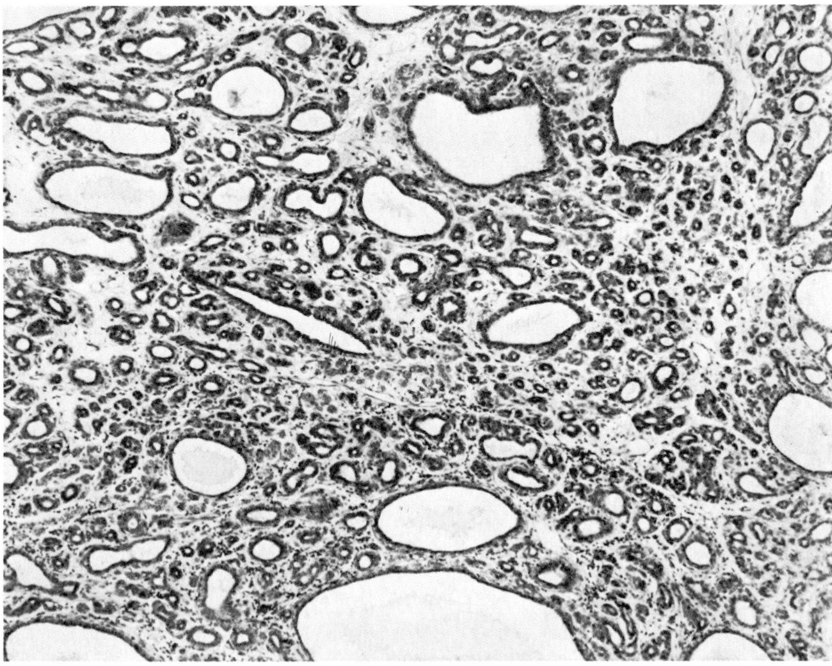

Fig. 27-3. Mammary adenocarcinoma, Type C, with cysts of varying size and prominent stroma (×80).

epithelial cells, which are closely surrounded by a spindle cell layer (Fig. 27-3). The connective tissue stroma usually appears edematous. Type C has nearly always been found in very old mice that lacked the agent.

Adenoacanthoma

Although foci of stratified squamous epithelium may occur in any type of mammary tumor, the term *adenoacanthoma* is restricted to tumors in which at least one-fourth of the section shows epidermoid differentiation (Fig. 27-4). The glandular elements of the tumors resemble Types A and B.

Carcinosarcoma

In these tumors there are irregular nests of epithelial cells closely intermixed with spindle cells resembling fibroblasts. Both elements may show numerous mitotic figures. The type occurs frequently among tumors induced by carcinogenic hydrocarbons. On repeated transplantation, purely glandular tumors may undergo a stage of carcinosarcoma and finally become pure sarcomas.

Sarcomas of the subcutaneous tissue must be distinguished from mammary tumors. However, some sarcomas may be derived from the stroma of the mammary gland and may appear in increased number after experimental procedures that increase the incidence of epithelial tumors. Cloudman (1941) presented a list of tumors and other lesions, arising in the area of the mammary gland, that may be mistaken for mammary tumors. For the sake of completeness, salivary gland tumors, lymphocytic and reticulum cell neoplasms, and squamous cell carcinoma of the skin should be added. These lesions can be distinguished by histological study.

Miscellaneous mammary tumors

These tumors that fit none of the other categories include a peculiar giant-cell type, epithelial tumors bearing no resemblance to the structure of mammary tissue, and tumors with abundant fibrous stroma resembling the characteristic human "scirrhous" carcinoma. Organoid tumors composed of ducts and acini radiating from a central area have been described (Dunn, 1959). The central area is keratinized in the molluscoid type.

Preneoplastic and early neoplastic change

The preneoplastic and early neoplastic changes in the mammary gland of high-tumor strains have often been studied. The most important precancerous change in high-tumor strains with the milk agent appears to be the "hyperplastic nodule," composed of a localized proliferation of acini. These areas are well demonstrated in whole

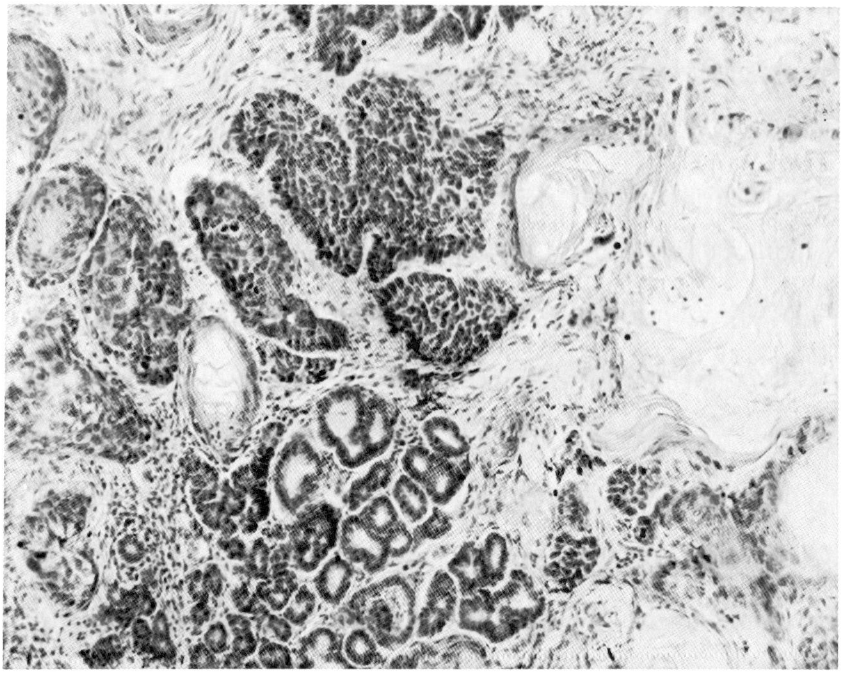

Fig. 27-4. Mammary adenocanthoma, with glandular tumor tissue showing prominent epidermoid differentiation (×150).

mounts of the mammary glands (Chap. 13). The nodules also occur in old female mice without the agent. Foulds (1956) has described an early neoplastic "plaque" which grew only during pregnancy and regressed after parturition. The plaques consisted of radially arranged branching tubules. These growths were dependent on the hormonal stimulation of pregnancy.

TUMORS OF THE HEMATOPOIETIC SYSTEM

The classification of tumors of the hematopoietic system in the mouse is adapted from Dunn (1954*a*) and appears in Table 27-4. Only the more common types will be described here. See Dunn's monograph for a description of variants and rarer types.

Lymphocytic neoplasms

The generalized form of lymphocytic neoplasm, lymphocytic leukemia, is the most frequent and most extensively studied form of leukemia in mice. The incidence is high in relatively young mice of high leukemia strains, such as AKR and C58, but the tumor appears sporadically in mice of other strains, usually at a more advanced age. The incidence was low in noninbred mice reviewed by Horn and Stewart (1952). Dunn and Andervont (1963) found only two lymphocytic neoplasms

Table 27-4. CLASSIFICATION OF TUMORS OF THE HEMATOPOIETIC SYSTEM OF THE MOUSE

Cell of origin	Tumor type
Undifferentiated	Stem-cell leukemia
Lymphocyte	Lymphocytic neoplasm—localized (lymphosarcoma)
	Lymphocytic neoplasm—generalized (lymphocytic leukemia)
Granulocyte	Granulocytic leukemia
Reticulum cell, Type A	Reticulum cell sarcoma—localized
	Monocytic leukemia—generalized
Reticulum cell, Type B	Hodgkin's-like lesion
Plasmacyte	Plasmacytoma—localized
	Plasmacytic leukemia—generalized
Tissue mast cell	Mastocytoma—localized
	Mast-cell leukemia—generalized
Unclassified	

among 225 wild mice maintained under laboratory conditions. Lymphocytic neoplasms have been induced by a number of viruses (Moloney, 1960; Lieberman and Kaplan, 1959), by X-irradiation (Furth and Furth, 1936; Kaplan, 1964), by carcinogenic hydrocarbons (Law, 1941), and by estrogenic hormones (Gardner et al., 1959).

A characteristic case shows general enlargement of the lymph nodes, thymus, and spleen. The involved organs are soft and white and often show hemorrhagic areas. The kidney and the liver may be enlarged and pale and contain nodules of soft white tissue. Ascites may occur in advanced cases. Microscopically (Fig. 27-5), the leukemic cell infiltrates the internal organs and may be found in blood smears. Two types of leukemic cells occur. One closely resembles a normal lymphocyte, with a deeply basophilic small round nucleus and a thin rim of clear basophilic cytoplasm. In other cases, the cell is larger than the normal adult lymphocyte, with a more vesicular nucleus which may be round or slightly indented and with more abundant cytoplasm.

A typical leukemic blood picture with lymphocytosis may not occur until late in the course of the disease. Cases are found in which the process appears localized and may be diagnosed as lymphosarcoma.

Reticulum cell neoplasms

The solitary fixed cells of the reticular framework of lymphatic and hematopoietic organs give rise to several distinctive tumors in the mouse. These reticulum cell sarcomas occur in older mice of a variety of inbred strains, particularly in their long-lived hybrids, and are less well known than the tumors of younger mice. Since they involve lymph nodes, spleen, and other organs, in many studies they tend to be lumped grossly with the leukemias. However, they are readily separated histologically.

Reticulum cell neoplasm, Type A. This tumor of usually well-differentiated reticulum cells has been described in the literature as histiocytoma, monocytoma, reticuloendothelioma, and reticuloendotheliosis. The process may be localized or generalized and tumor cells may be found in the peripheral blood. Gorer (1946) reported an incidence of 15 to 20 per cent in strain C57BL mice over 18 months of age (see also Table 27-2). The tumors occur sporadically in old animals of other inbred strains.

At autopsy, enlargement of the liver and ascites are usually found. Involvement of the uterus is common in females. Spleen, lungs, kidneys, thymus, mesenteric, and other lymph nodes may be involved. The tumor tissue is firm, usually white,

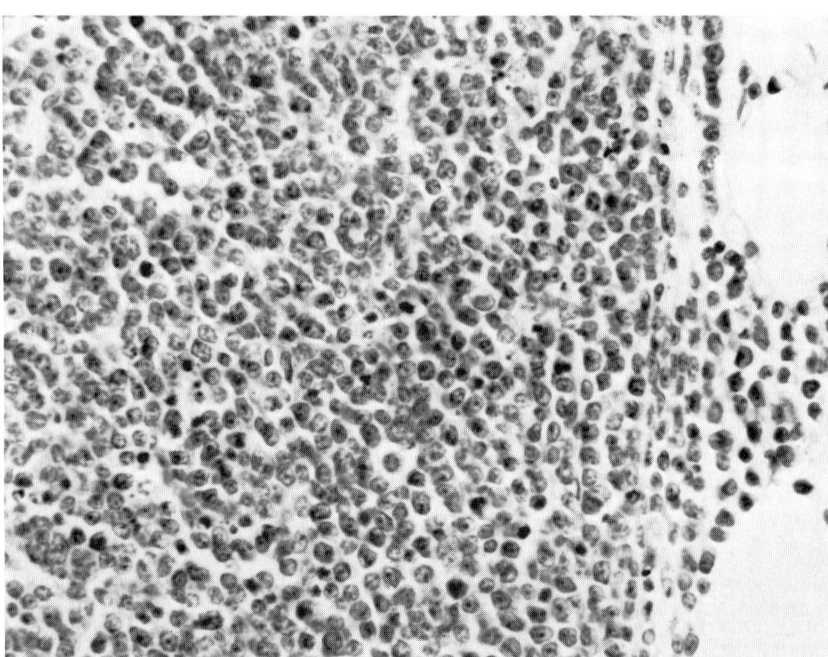

Fig. 27-5. Lymphocytic leukemia involving lymph node. Note capsular invasion (×400).

often shows hemorrhagic foci, and may have a distinctive orange color due to hematoidin. The tumor cell has eosinophilic cytoplasm and a heavily stained basophilic nucleus, and shows great variation in the size and shape of the cell and the nucleus (Fig. 27-6). Differentiation is indicated by erythrophagia and hemosiderin within the tumor cells. Multinucleated cells are frequent and may resemble Langhans-type giant cells. A common variant contains sheets of small spindle cells, with heavily basophilic ovoid nuclei and scanty cytoplasm, resembling fibrosarcoma. Both patterns may appear in the same animal, either intermixed or in separate deposits. An angiomatous pattern is occasionally seen in the tumor in the liver.

Transplantation to mice of the same strain is usually successful but growth is slow and may extend for 10 to 12 months. The Type A neoplasm may be compared with the more differentiated types of reticulum cell sarcoma in man, such as the clasmatocytic lymphoma described by Gall and Mallory (1942). The tumor induced by the Friend virus (Friend, 1957) has been derived by Buffett and Furth (1959) from the reticulum cell. It resembles some of the leukemic forms of Type A.

Reticulum cell neoplasm, Type B. This multicellular tumor is more common than Type A and may be the most common tumor in older mice of otherwise low-tumor strains and of long-lived hybrids. Jobling (1910) and other early investigators used the term *Hodgkin's disease;* later investigators, *Hodgkin's-like*. The tumor occurs in 25 per cent of strain C57L/He mice, 18 months of age (Heston, 1963). It is a characteristic tumor of old age in mice of many inbred strains and in wild mice maintained under laboratory conditions (Dunn and Andervont, 1963). Murphy (1963) has reported an incidence of more than 90 per cent at an average age of 13 months in a new inbred strain, SJL/J. Stansly and Soule (1962) presented evidence for a filtrable agent that can induce Type B reticulum cell neoplasms.

At autopsy the mesenteric node is usually greatly enlarged. Peyer's patches are frequently involved and the tumor appears to spread to the pancreatic and renal nodes. Nodular involvement of the white pulp of the spleen is common and discrete nodules may be found in the liver. Advanced cases may involve kidney, lungs, mediastinal, and peripheral nodes. A minority of cases may be primary in some other node or the spleen. The tumor tissue is firm and white, with little hemorrhage. The process develops slowly; therefore localized cases may readily be found. Transplantation has been successful in a limited percentage of trials.

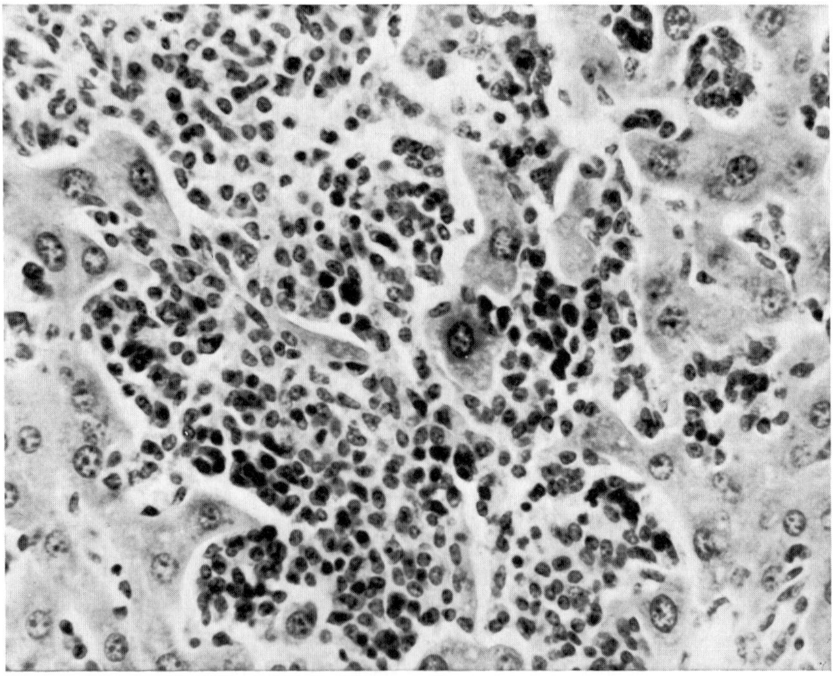

Fig. 27-6. Reticulum cell neoplasm, Type A, involving liver (×400).

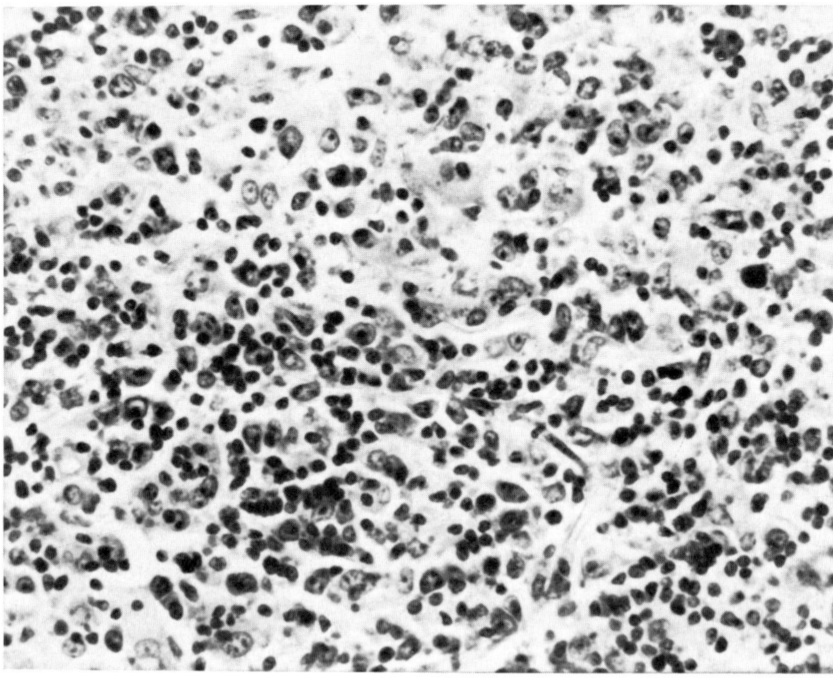

Fig. 27-7. Reticulum cell neoplasm, Type B, with intermixture of large, pale reticulum cells and lymphocytes (×400).

Microscopically, there is a background of large, pale reticulum cells intermixed with lymphocytes and plasma cells (Fig. 27-7). Tumor giant cells and multinucleated cells resembling foreign-body giant cells and Langhans' giant cells may be found. In strain SJL/J eosinophils may be prominent along with fibrosis (Fig. 27-8), and in several cases tumor giant cells were observed which duplicate the classic features of the Sternberg-Reed cells described in human Hodgkin's disease (Murphy, 1963).

Granulocytic leukemia

Granulocytic leukemia is rare in comparison with lymphocytic leukemia in the mouse. It can be induced in strain RF mice by ionizing radiation (Upton, 1961). Graffi (1957) described granulocytic leukemias in mice inoculated with viruses isolated from five different tumors. Grossly, the distribution and appearance of the tumors are usually indistinguishable from those of lymphocytic leukemia, unless the green color of chloroleukemia is present. A high granulocyte count with many undifferentiated cells is found in the peripheral blood. In the tissues, collections of relatively immature granulocytic cells are found (Fig. 27-9). Invasion of the capsule of lymph nodes and infiltration of fatty tissue are helpful in distinguishing this leukemia from extramedullary hematopoiesis.

Plasma cell neoplasm

Rask-Nielsen and Gormsen (1956) have reported a low incidence of plasma cell leukemia in several inbred strains. Plasma cell neoplasms have been induced by Plexiglas fragments (Merwin and Redmon, 1963) and mineral oil (Potter and Boyce, 1962) introduced into the peritoneal cavity of BALB/c mice. The inducing agents appear to exert their effect by their physical rather than their chemical properties (Potter and MacCardle, 1964). Transplantable plasma cell tumors secrete a variety of proteins related to γ- and β-immunoglobulins and their subunits in the form of Bence-Jones proteins (Fahey, 1961).

Dunn (1954*a*) has described both localized and generalized forms of plasma cell neoplasm. The localized type begins in the ileocecal area of old strain C3H mice, and the neoplastic cells extend through all coats of the intestine, through the mesenteric fat, and involve the medullary sinuses of the mesenteric node. One of these tumors has been successfully transplanted and shows heavy infiltration of the gonads and the kidneys.

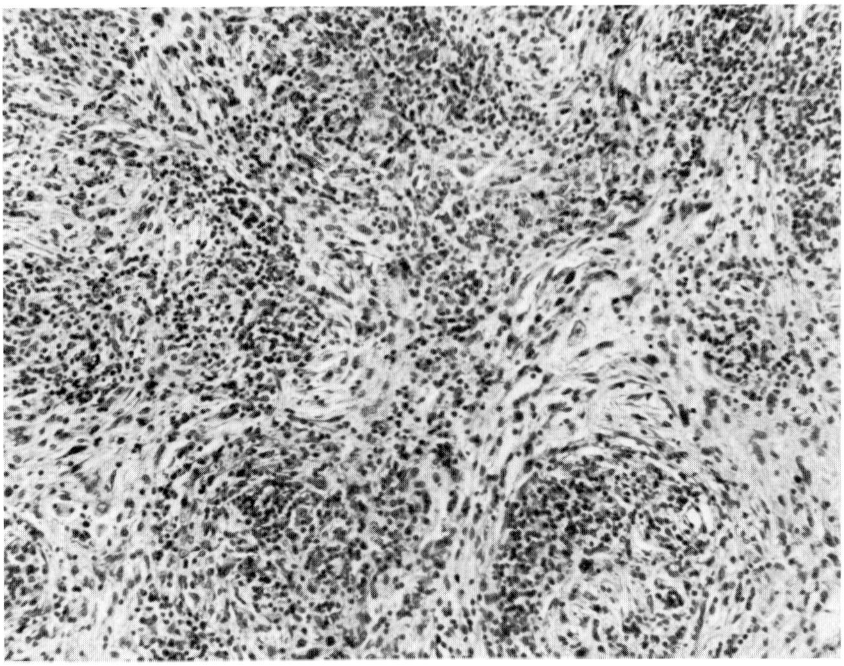

Fig. 27-8. A reticulum cell neoplasm, Type B, in strain SJL/J, with intermixture of reticulum cells, lymphocytes, tumor giant cells, and prominent fibrosis ($\times$150).

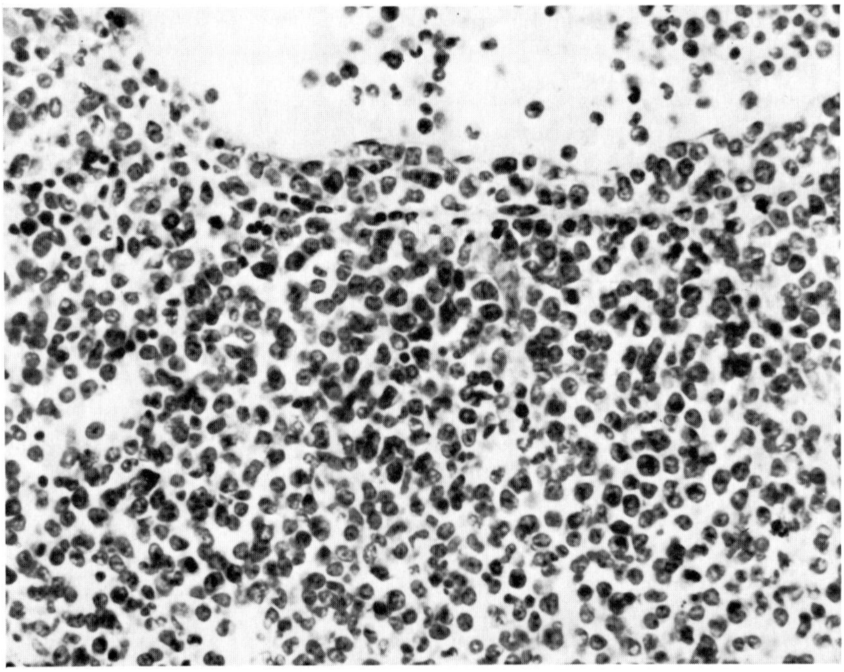

Fig. 27-9. Granulocytic leukemia involving liver, with immature cells in tissue and more differentiated cells in blood vessel ($\times$400).

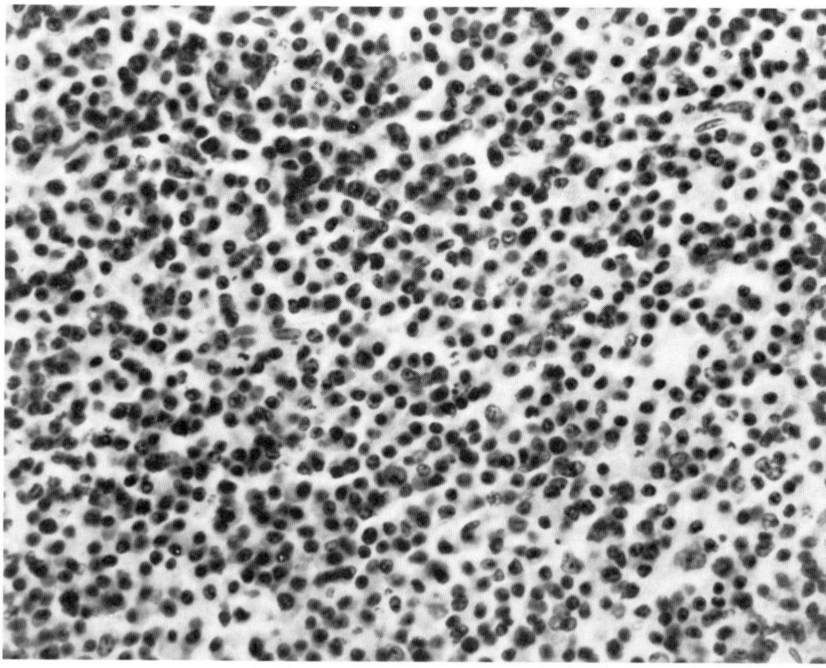

Fig. 27-10. Plasma cell neoplasm (×400).

The generalized type (Fig. 27-10) involves lymph nodes, spleen, and perivascular tissue in the kidneys and lungs. The tumor cell has an eccentric nucleus, with a clear area in the cytoplasm, and the Giemsa stain shows the violet-colored cytoplasm of the plasma cell. Russell bodies may occur in the tumor cells. These tumors must be distinguished from plasma cell hyperplasia of lymph nodes (especially frequent in old mice), from inflammatory plasma cell infiltrations, and from other neoplasms.

Mast cell neoplasm

True neoplasms of mast cells are rare in mice. They may be localized and termed *mastocytoma* or more generalized and termed *mast cell leukemia.* Dunn (1954*a*) observed several of these tumors at autopsy. They were usually localized masses in lymphatic or connective tissue. Rarely, the neoplasm was widely distributed, involving lymph nodes, spleen, liver, lungs, and kidneys. Microscopically, the neoplastic mast cell is larger than the normal, it is less heavily and uniformly granulated, and the granules generally do not stain so intensely. Transplantable mast cell tumors have been reported (Dunn and Potter, 1957; Rask-Nielsen and Christensen, 1963). These tumors are of particular interest because they may secrete heparin, histamine, and serotonin.

Stem cell leukemia

In human pathology this term refers to cases of acute leukemia in which the neoplastic cell is undifferentiated. Such tumors are not commonly distinguished in the mouse, but are probably included with the lymphocytic leukemias. Because of the preponderance of lymphocytic over granulocytic and other leukemias in the mouse, the undifferentiated forms are probably related to lymphocytic leukemia.

Misleading nonneoplastic lesions

Pathologists trained in human pathology are frequently misled by interstitial infiltrates of lymphocytes which occur with increasing frequency in the kidney, liver, and other organs of aging mice. The question of leukemic infiltration is often raised. The cell collections are usually perivascular, and on closer examination usually show a mixed population of reticulum cells, lymphocytes, and plasma cells.

A possible error is the misdiagnosis of extramedullary hematopoiesis as granulocytic leukemia. Extramedullary hematopoiesis in the spleen is physiological in the mouse. It requires little stimulus for the liver, lymph nodes, and other organs to respond with tumorlike infiltrates, with predominance of the granulocytic series. The spleen may be as large as in many cases of leukemia, but the numerous intermixed megakaryocytes usually rule out granulocytic leukemia. Barnes and Sisman (1939) have described and tabulated the important points for differentiating extramedullary hematopoiesis from the rare granulocytic leukemia in the mouse.

Simonds (1925) described an enlargement of the mesenteric lymph node in five mice of the Slye stock. The node was enlarged by wide blood-filled spaces, which disrupted the normal histology of the node. The process is not neoplastic but appears to be due to venous congestion. It is frequent in strain C3H and its hybrids and occurs in old C57BL mice.

PULMONARY TUMORS

Spontaneous pulmonary tumors are known to be frequent in only two species, man and mouse (Stewart, 1959*b*). Although the characteristic human tumor is bronchogenic in origin, the characteristic tumor of the mouse is alveologenic.

Alveologenic tumors

Alveologenic tumors occur spontaneously in high incidence in strains A, SWR, and BALB/c and can be induced in these and other susceptible strains by a variety of agents including urethan, carcinogenic hydrocarbons, nitrogen mustard, and γ-radiation (Shimkin, 1955). Even though these tumors have been variously diagnosed as adenoma, papillary cystadenoma, and adenocarcinoma, they may all be morphological variants of a single malignant neoplastic process. The pulmonary tumor of the mouse is a malignant neoplasm, as judged from its lack of encapsulation, local invasiveness, transplantability, and ability to metastasize.

In gross appearance, the tumors are rounded, pearly white nodules, often situated just below the pleura, and projecting slightly. The spontaneous tumors are frequently solitary and usually do not exceed two to four per animal. The induced tumors are almost invariably multiple. Microscopically, most of the tumors present a uniform pattern of closely packed columns of cuboidal or columnar cells (Fig. 27-11). The cells are rather uniform in size and shape, with acidophilic cytoplasm and round or oval nuclei. The sparse stroma is composed of mature fibrous tissue. Papillary

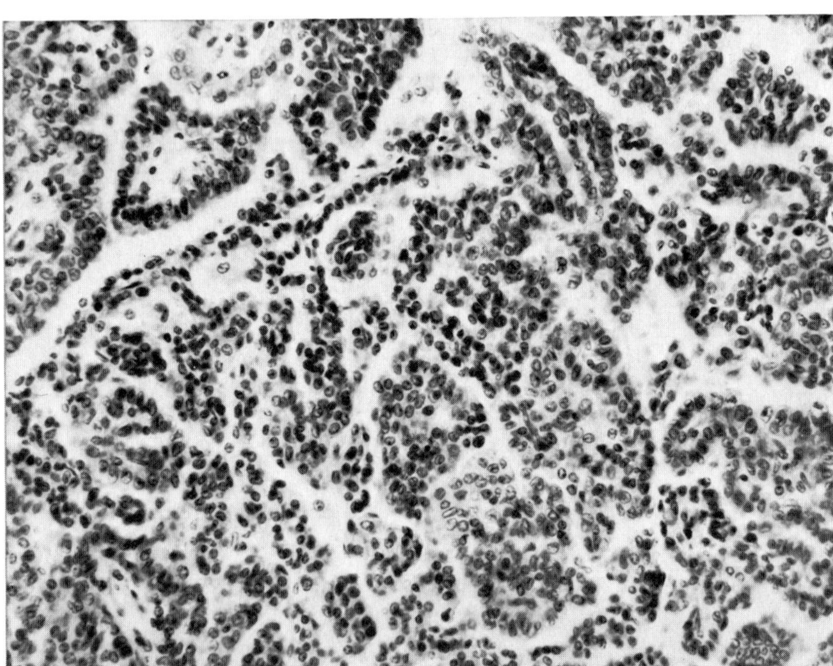

Fig. 27-11. Primary lung tumor, with columns of cuboidal cells and papillary formations (×250).

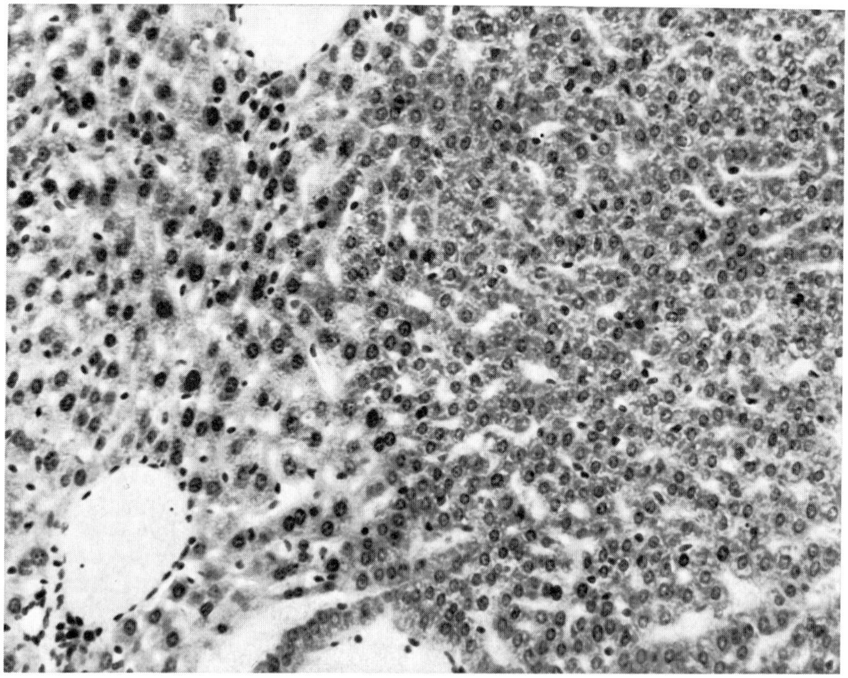

Fig. 27-12. Hepatoma, with cords of well differentiated liver cells separated by sinusoids (×250).

formation is frequent in larger tumors. Spontaneous and induced tumors are indistinguishable microscopically. The common metastatic tumors in the lungs of mice can usually be suspected from the presence of a primary tumor in another site and distinguished microscopically.

Bronchogenic tumors

Squamous cell carcinomas arising in bronchi have been induced by local application of radioactive substances (Gates and Warren, 1960) and by a combination of influenza virus and aerosols of hydrocarbons (Kotin and Wiseley, 1963).

HEPATIC TUMORS

Hepatomas (liver cell carcinoma, hepatocellular carcinoma) occur spontaneously in low incidence in a number of inbred strains, but are common in older males of strains C3H and CBA (Burns and Schenken, 1940; Gorer, 1940; Andervont, 1950*b*). Heston et al. (1960*a*) found an incidence of 85 per cent in C3H/He males. Deringer (1959) reported 91 per cent in C3HeB males and 58 per cent in virgin females. The tumors are usually solitary but may be multiple. They are usually elevated round or ovoid masses on the surface or margins of the lobes; some may be pedunculated. They may be gray or yellow or the same color as the liver. The histology is usually remarkably uniform and rather closely resembles normal liver (Fig. 27-12). Cords of cells are separated by sinusoids lined by flattened endothelial cells. However, a true lobular architecture is absent and bile ducts occur only at the periphery of the tumor. The size of the hepatoma cells and their nuclei shows a wide range of variation (Miyagi, 1952). Cytoplasmic hyaline inclusion bodies are frequently observed. The nontumorous portions of the liver usually appear normal, without evidence of cirrhosis or inflammatory processes. Cholangioma has not been observed in untreated mice.

Metastasis of hepatomas has been described and, although not all spontaneous tumors were successfully transplanted, Andervont and Dunn (1952) could not demonstrate a consistent histological difference between those that grew and those that failed to grow. Therefore, the spontaneous tumors of the liver must be considered malignant. It is unlikely that an attempt to distinguish between adenoma and carcinoma would be useful. The incidence of spontaneous hepatomas has been influenced by diet (Tannenbaum and Silverstone, 1949) and by castration (Andervont,

1950*b*). Hepatomas have been induced in the mouse by azo dyes, other chemical carcinogens, radioactive compounds, carbon tetrachloride, chloroform, and urethan (Heston et al., 1960*a*). Andervont and Dunn (1952) could find no identifiable qualitative histological difference between spontaneous and induced hepatomas.

TUMORS OF THE FEMALE REPRODUCTIVE TRACT

Ovarian tumors

Ovarian tumors occur only sporadically in most inbred strains, although nonneoplastic cysts are common. However, incidences of 34 per cent have been reported in CE females (Dickie, 1954) and 47 per cent in C3HeB/De virgin females (Deringer, 1959). The incidence in C3HeB/FeJ is 64 per cent after 19 months of age, and in RIII/J, 60 per cent after 17 months of age (Hummel, 1965, personal communication).

The most common types are granulosa cell tumors and tubular adenomas. A simple classification of spontaneous and induced ovarian tumors includes tubular adenoma, granulosa cell tumor, luteoma, papillary cystadenocarcinoma, and teratoma. Except for the latter two, these types have been induced by X-irradiation (Furth and Butterworth, 1936), by transplantation of ovaries to the spleen (Li and Gardner, 1949), by transplantation of ovaries to other sites in strain DBA mice (Hummel, 1954*a*), by remotely applied chemical carcinogens (Howell et al., 1954), and by genic deletion of ova (Murphy and Russell, 1963).

Tubular adenoma. Downgrowth of the so-called "germinal" epithelium is a common aging change in the ovaries of mice (Thung, 1961). It may be so extensive that the ovary is replaced by interlacing clefts and tubules lined by cuboidal to columnar epithelium resembling the germinal epithelium (Fig. 27-13). Only arbitrary morphological distinctions can be made between tubular adenomatous change and tubular adenoma, perhaps best at the point where the entire ovary is involved. Bali and Furth (1949) found that tubular adenomas could be transplanted and grew slowly, particularly in gonadectomized hosts.

Tubular adenoma is usually a prominent precursor in the formation of granulosa cell tumors and luteomas following genic deletion of ova, X-irradiation, and transplantation to the spleen. Gardner (1955) postulated that the germinal epithelium is the source of granulosa cell tumors. Bali and Furth (1949) have observed tubules in a tubular adenoma lined partly by germinal epithelium and partly by granulosa cells. At an intermediate stage in ovarian tumorigenesis there is frequently a proliferation of interstitial cells, which

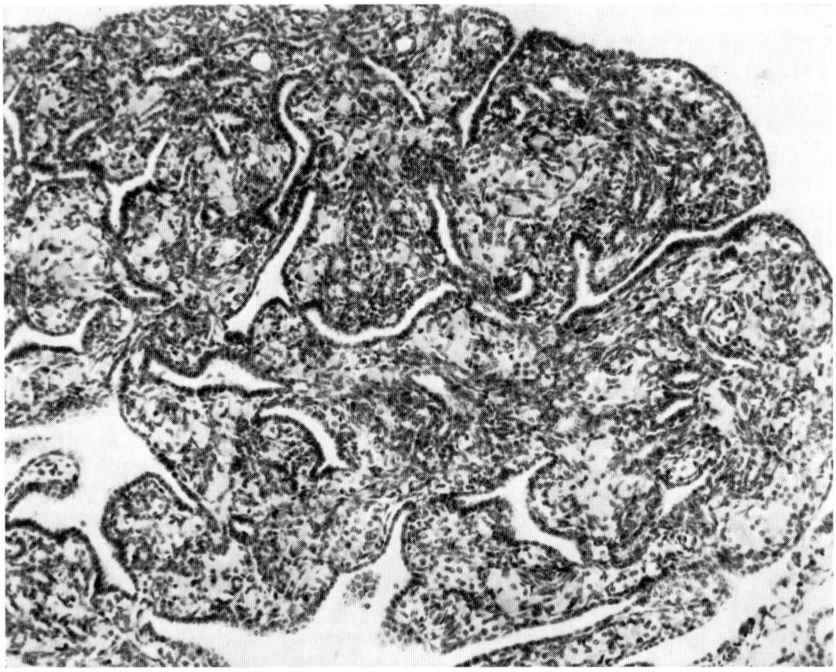

Fig. 27-13. Tubular adenoma of ovary, with clefts and tubules derived from germinal epithelium (×150).

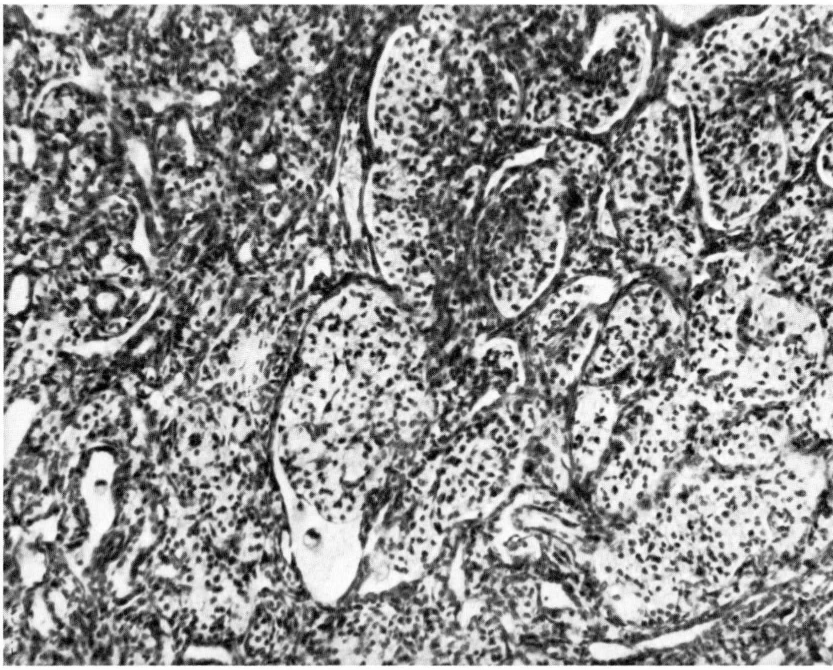

Fig. 27-14. Complex tubular adenoma of ovary, with proliferation of interstitial cells (×150).

can undergo luteinization (Fig. 27-14). The term *complex tubular adenoma* has been applied to these tumors (Bali and Furth, 1949). It is possible that some of these lipid-containing tumors have been called *luteomas* in the literature. Current opinion derives granulosa cell tumors and luteomas from the "interstitial" cells, that in turn are derived from theca cells (Guthrie, 1957; Mody, 1960). Thung (1959) has stressed the tremendous plasticity of the cells of the ovary and has suggested that the various tumors may arise from more than one original cell type.

Granulosa cell tumor. Estrogen-secreting tumors composed of cells resembling the characteristic cells of the membrana granulosa constitute the most common induced type of ovarian tumor in mice. The cells frequently show an elongated nucleus, with densely stippled chromatin especially around the nuclear membrane, and may be organized into sheets, cords, or pseudo-follicles (Fig. 27-15). As in the case of human granulosa cell tumors, there can be so many patterns that the diagnosis is not always certain on morphological grounds alone. Besides estrogen secretion the tumors may also produce hypervolemia by secretion of a substance termed "plethorin" (Furth and Sobel, 1946). Many of the cells of granulosa cell tumors may undergo a fatty change which is not true luteinization. However, true luteinization may occur and mixed tumors have been described (Bali and Furth, 1949). The spontaneous tumors of strain CE and its hybrids are frequently granulosa cell in type. They appear to develop from proliferations of peripheral stromal cells, forming cortical plaques and grossly observable mushroom-like caps (Fig. 27-16). Sertoli cell differentiation is common (Fig. 27-17).

Luteoma. Progesterone-secreting tumors composed of cells resembling those of the corpus luteum occur rarely spontaneously but have been induced by X-irradiation, transplantation to the spleen, and genic deletion of ova. The tumors are yellow, and the cells are polygonal, have abundant acidophilic cytoplasm, and are arranged in nodules separated by thin strands of reticular and collagenous fibers (Fig. 27-18). Furth and Sobel (1947) have described studies on a transplantable luteoma.

Teratoma or teratocarcinoma. Ovarian teratomas, derived from pluripotent cells, occur very rarely in mice. However, several have been observed in strain C3H mice at The Jackson Laboratory, and Fekete and Ferrigno (1952) have reported a transplantable ovarian teratoma that maintained its pleomorphic character through nine transplant generations. A further discussion of this important tumor type is given under the heading Testicular Teratoma.

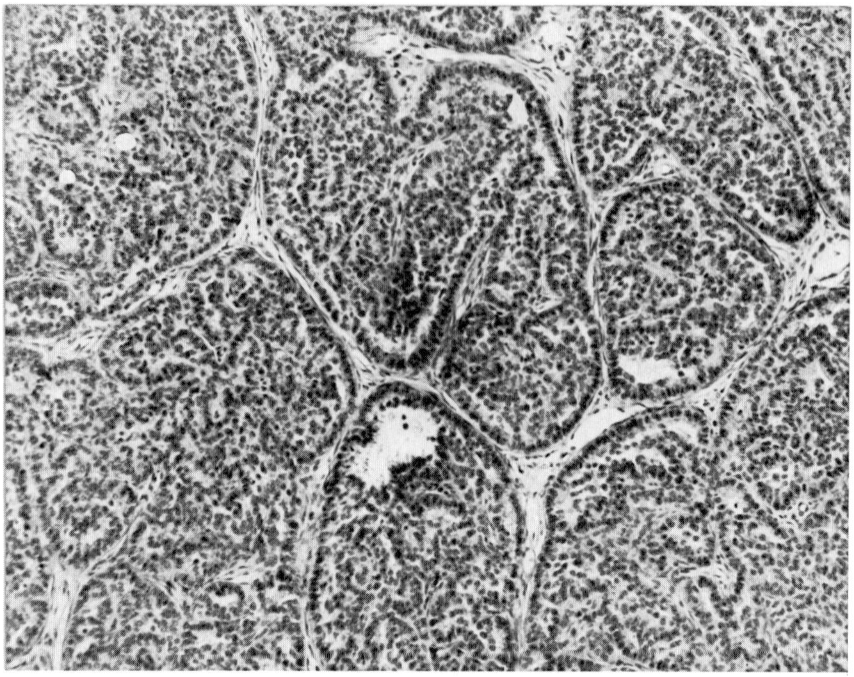

Fig. 27-15. Granulosa cell tumor, showing one of many possible glandular patterns (×150).

Other ovarian tumors. Papillary cystadenocarcinomas have been reported (Cloudman, 1941; Dunn, 1954*b*). A transplantable mucin-producing tumor has been described by Dunn (1954*b*). Nonspecific tumors of the ovary such as leukemias, reticulum cell sarcomas, hemangioendotheliomas, and fibrosarcomas have been described.

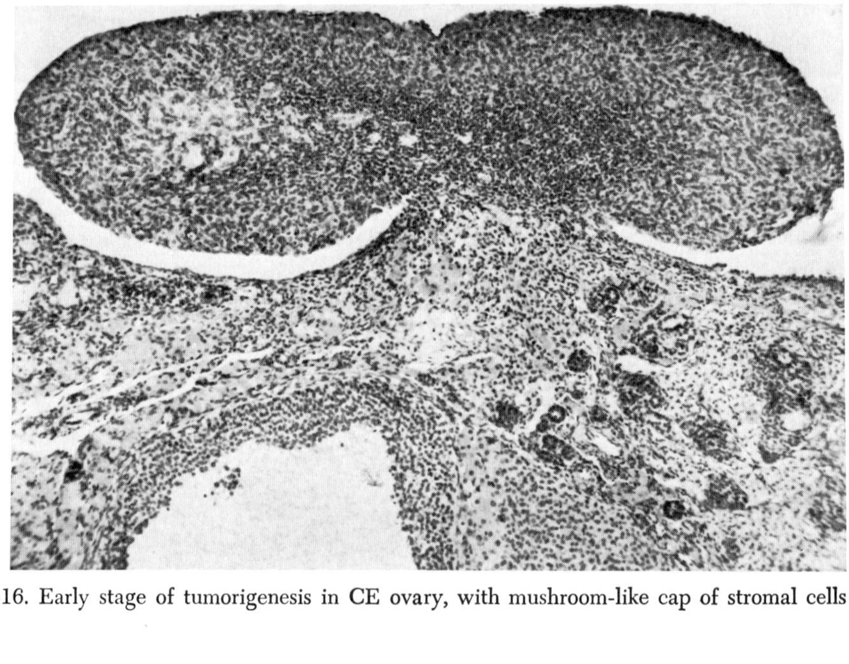

Fig. 27-16. Early stage of tumorigenesis in CE ovary, with mushroom-like cap of stromal cells (×96).

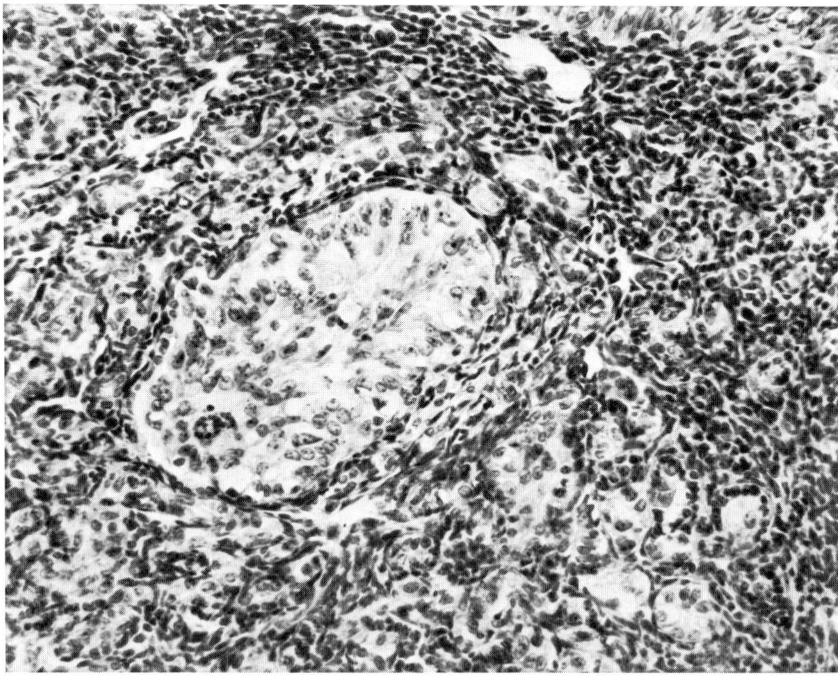

Fig. 27-17. Sertoli cell differentiation in an ovarian tumor in strain CE (×250).

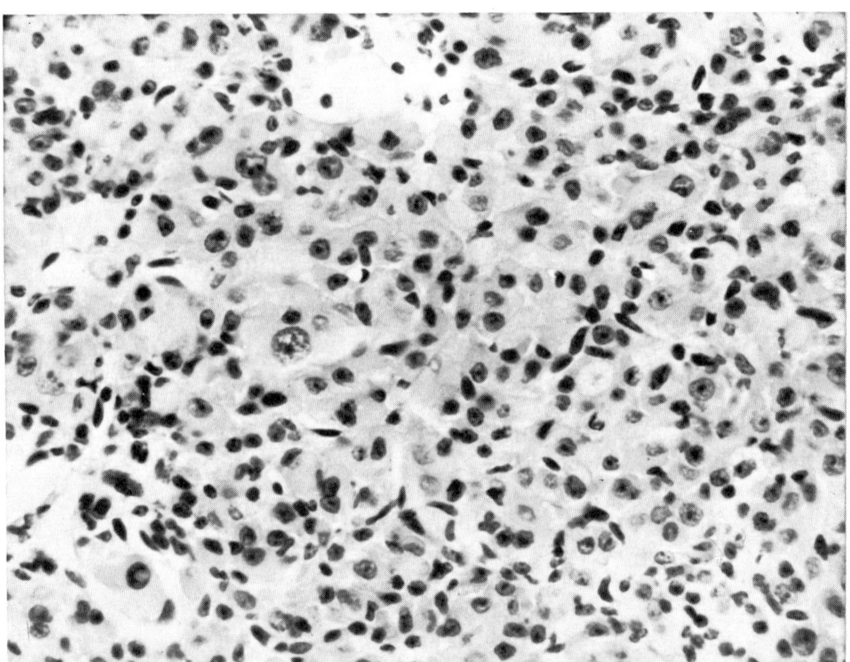

Fig. 27-18. Luteoma, composed of polygonal cells with abundant eosinophilic cytoplasm (×400).

Tumors of the uterine horns

Spontaneous tumors of the uterine horns are infrequent. The sporadic tumors are usually sarcomas, either fibrosarcomas or leiomyosarcomas. Spontaneous adenocarcinomas are rare in most inbred strains, however Dunn (1965, personal communication) has observed a number of transplantable adenocarcinomas in strain BALB/c and some similar tumors in C3H × C57BL hybrids (Dunn, 1954*b;* Heston, 1963). Glandular tumors appear to be less readily induced by carcinogens than sarcomas. Fibrosarcomas and leiomyosarcomas have been induced by methylcholanthrene-coated threads (Murphy, 1961). Rare leiomyomas resembling the common benign human uterine tumor have been reported (Table 27-2). Endometrial stromal sarcomas occur.

Tumors of the cervix and vagina

Spontaneous carcinomas of the cervix and vagina are extremely rare in mice, as in other laboratory animals, and in fact in all animals studied other than man. However, the mouse has become the laboratory animal of choice in the induction of these tumors by estrogenic hormones (Gardner et al., 1959; Dunn and Green, 1963), chemical carcinogens, and the combination of these agents (Murphy, 1961). There is one report of a strain of mice in which a high spontaneous incidence of cervical and vaginal carcinoma was observed (Gardner and Pan, 1948). Unfortunately, the strain was lost because of associated sterility (Gardner, et al., 1959).

Both the rare spontaneous carcinomas and the induced tumors range from well-differentiated squamous cell (epidermoid) carcinomas with extensive keratinization to anaplastic carcinomas with little or no evidence of differentiation. Most of the induced tumors, however, tend to be well differentiated and show much more keratin formation than occurs in human cervical carcinomas. A type of differentiation that occurs in rodents is the formation of mucin by the stratified squamous epithelium of the vagina and cervix. Mucin formation has been demonstrated in some of the tumors induced by methylcholanthrene (Murphy, 1961). The tumors metastasize to the lungs and are transplantable.

Tumors of the vulva

Papillomas and squamous cell carcinomas of the vulva have been observed in strain 129 females at The Jackson Laboratory.

TUMORS OF THE MALE REPRODUCTIVE TRACT

Interstitial cell tumors of the testis

Sporadically occurring spontaneous interstitial cell tumors have been reported in hybrids of strain A (Gardner, 1943), and in strains C (Hooker et al., 1946), BALB/c (Hummel, 1954*b*), and RF (Clifton et al., 1956). Andervont et al. (1960) have established the incidence of spontaneous hyperplasias and tumors in BALB/c as probably under 1 per cent. Interstitial cell hypertrophy, hyperplasia, and tumor formation have been reported in a high mammary tumor strain, designated strain H, in which the males also develop mammary tumors (Athias, 1945; Furtado Dias, 1958). Interstitial cell tumors have been induced by the administration of estrogens, particularly in strains A and BALB/c. Wide differences in susceptibility of inbred strains have been reported by Bonser (1944), Gardner et al. (1959), and Andervont et al. (1960). Experimentally produced cryptorchid testes of BALB/c mice developed a high incidence of interstitial cell hyperplasia and tumor formation (Huseby, 1958).

The tumors are yellowish brown and consist of masses of large polygonal or irregularly shaped cells with granular or vacuolated cytoplasm and nuclei varying in size (Fig. 27-19). The cells may contain a light brown pigment. Lymphatic metastasis is common. The induced tumors can be transplanted to animals of the same strain if estrogen is present. After serial transplantation the tumors lose their dependency on estrogen. The tumors produce androgen, losing this ability as they become more autonomous (Gardner et al., 1959).

Testicular teratoma

These tumors, derived from pluripotent cells, are extremely rare in male mice except for strain 129 in which approximately 1 per cent develop congenital tumors spontaneously (Stevens, 1959). It has been possible to raise the incidence as high as 10 per cent by introducing the steel gene (*Sl*) into strain 129 and by selecting males from second and later litters (Stevens and Mackensen, 1961). The transplantation of male gonadal ridges from strain 129 fetuses to adult testes has resulted in an 82 per cent incidence of teratoma (Stevens, 1964).

Grossly, larger tumors are hemorrhagic and the smaller ones appear as solid masses occupying one-fourth to three-fourths of the testicular volume. Usually the masses contain cysts filled with

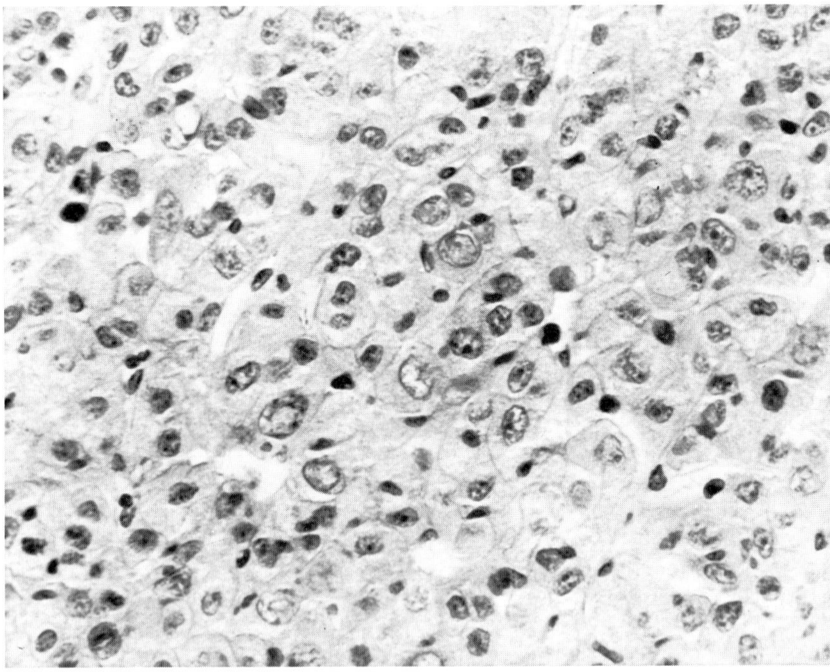

Fig. 27-19. Interstitial cell tumor of testis (×400).

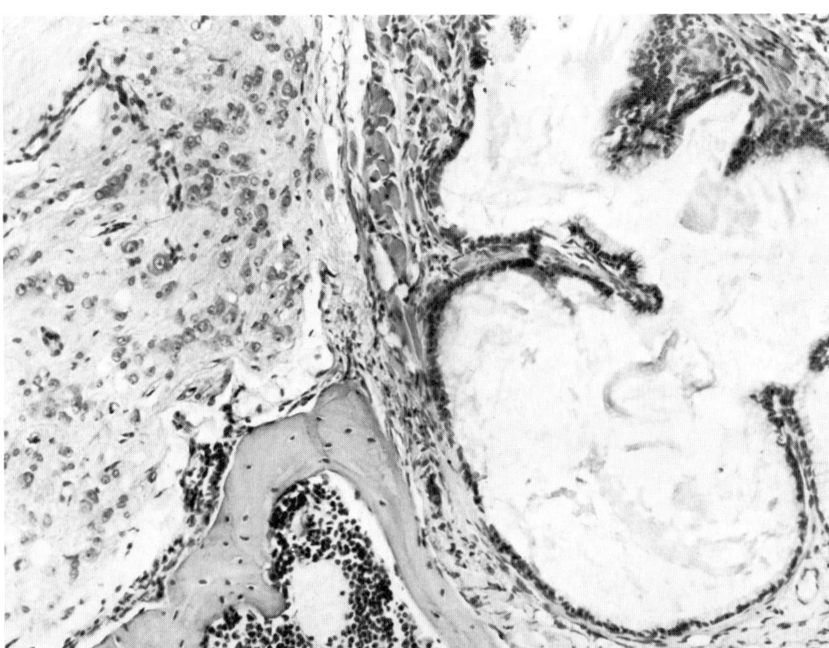

Fig. 27-20. Teratoma of testis, showing nervous tissue, bone, bone marrow, muscle, and glandular epithelium (×150).

clear or bloody fluid. Many of the teratomas contain palpable nodules of bone and cartilage. Histologically (Fig. 27-20), the most common components are nervous tissue, epithelia of various types, cartilage, bone with well-differentiated marrow, muscle, fat, and glandular tissue (Stevens and Little, 1954). Undifferentiated embryonic cells may be observed. The tumors are transplantable, but only a minority grow progressively. The grafts that merely survive are composed entirely of adult-type tissues. Those that grow progressively may consist purely of embryonal cells, mainly of undifferentiated cells, or principally of differentiated tissues (Stevens, 1958). Serial section study of testes in 15- to 19-day fetuses has demonstrated the origin of the teratomas within the seminiferous tubules and provided evidence of origin from the primordial germ cells (Stevens, 1962).

TUMORS OF OTHER ENDOCRINE GLANDS

Adrenal cortical tumors

Adrenal cortical tumors are rare in noninbred mice (Slye et al., 1921). They are readily induced by gonadectomy in strain CE (Woolley and Little, 1945*a*) and in (DBA × CE)F_1 hybrids (Woolley et al., 1952). They have been similarly induced in strains BALB/c, NH, CBA, C3H, and A (Frantz and Kirschbaum, 1949). Strain differences were noted in the secretion of androgen, estrogen, or both hormones by the tumors. The response of the adrenal gland to castration is influenced by genetic factors. Strain DBA responds with nodular hyperplasia (Fekete et al., 1941), while strains C57BR and C57BL show only slight increase in width of the cortex.

Fekete and Little (1945) have described the tumors arising in the adrenal cortex of gonadectomized mice of the CE strain. In most of the large tumors the predominant cells were polygonal and diffusely arranged (Fig. 27-21). Rows and cords of cuboidal cells tended to be more prominent at the periphery of the tumors. In some cases a syncytial type of cell resembling Sertoli cells of the testis formed a component of the tumor. Giant cells with yellow pigmented cytoplasm and often with multiple nuclei were found in many tumors. Blood vessels were numerous and consisted mainly of capillaries and sinusoids between groups of tumor cells. The larger tumors showed capsular invasion. Stewart et al. (1959) presented excellent illustrations of such a tumor and its transplants.

Adrenal medullary tumors

Tumors of the adrenal medulla in mice have been described by Smith et al. (1949) and as occasional findings by other investigators. Jones and

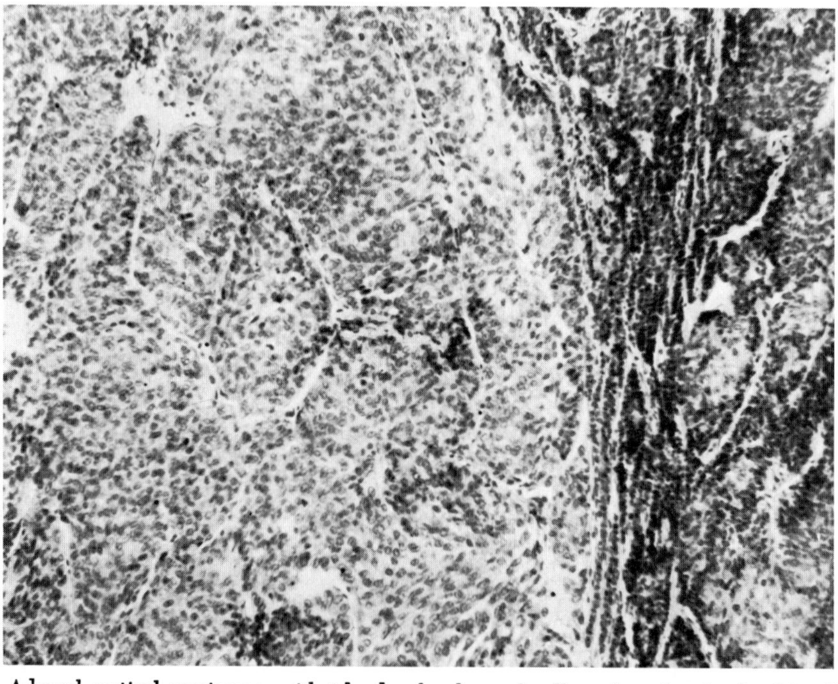

Fig. 27-21. Adrenal cortical carcinoma, with islands of polygonal cells and cords of cuboidal cells (×150).

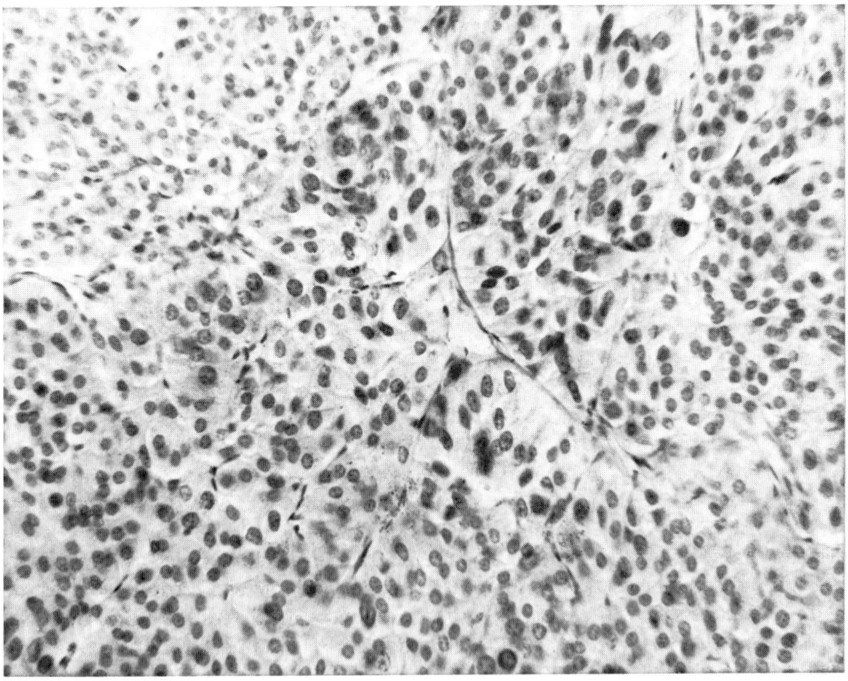

Fig. 27-22. Adrenal medullary tumor. Normal medullary cells are included in upper left (×225).

Woodward (1954) reported these tumors in untreated $(C3H \times I)F_1$ virgin females. The tumors closely reproduce the cell type and arrangement of the normal adrenal medulla of the mouse (Fig. 27-22). It is for this reason that the tumors are commonly termed "pheochromocytomas," not because of demonstrated chromaffin staining or endocrine properties.

Pituitary tumors

Spontaneous pituitary tumors have been considered rare in mice. Slye et al. (1931) and Gardner et al. (1936) have described single cases. Cloudman (1941) reported two adenocarcinomas in hybrids between C57BL and C57BR. Furth et al. (1960) found an incidence of 3.5 per cent in C57L and 1 per cent in LAF_1 hybrids. Pituitary tumors have been observed at The Jackson Laboratory in approximately one-third of retired female breeders of strains C57L/J and C57BR/cdJ (Russell, 1965, personal communication). These tumors are associated with mammary duct hyperplasia and, in more differentiated specimens, the cell type is acidophilic (Fig. 27-23).

Pituitary tumors have been induced in mice by chronic administration of estrogen, ionizing irradiation, goitrogenic drugs, surgical thyroidectomy, radiothyroidectomy, and gonadectomy (Gorbman, 1956). The tumors induced by gonadectomy in strain CE and its hybrids are associated with adrenal cortical tumors (Dickie and Woolley, 1949). Subcutaneous isografts of pituitaries develop into chromophobe adenomas (Mühlbock and Boot, 1959).

The classic cellular classification of pituitary tumors as acidophilic, basophilic, or chromophobic has been of limited value in mice. Most of the tumors have been described as "chromophobe," or lacking in stainable granules. In human pathology this designation usually signifies lack of hormonal function. In mice these tumors show a number of hormonal effects. They are best classified as mammotropic, thyrotropic, adrenotropic, somatotropic, or gonadotropic (Clifton, 1959). Functional characterization usually requires the study of transplants, since the inductive processes often produce masking hormonal responses in the diagnostic target organs.

Mammotropic activity has been demonstrated in spontaneous tumors and those induced by estrogens and by irradiation. Transplanted mammotropic tumors cause hyperplasia of all elements of the mammary glands with milk secretion, body growth, and disproportionate increase in weight of the viscera (Furth et al., 1956). The cell type, although usually described as chromophobic, may show acidophilic granules.

Thyrotropic tumors, induced by procedures

which eliminate or suppress thyroid function, arise from the aldehyde fuchsin–positive beta basophils (Halmi and Gude, 1954). Transplanted tumors cause massive thyroid hyperplasia with formation of dependent thyroid adenomas (Furth and Clifton, 1957).

Adrenotropic tumors were the first type to be described in irradiated mice (Furth et al., 1952). They were "chromophobic" by the usual staining methods. The tumors induced by gonadectomy have been described as "basophilic" and postulated to be gonadotropic (Dickie and Woolley, 1949).

The cranial cavity is often not routinely examined at autopsy; therefore pituitary tumors can be overlooked. Careful removal of the calvarium and brain may be tedious, but it is a simple matter to slice off the upper part of the skull and brain with a razor blade in order to examine the pituitary gland and the brain for tumors.

Thyroid tumors

Spontaneous tumors of the thyroid gland are rare in mice. Slye et al. (1926) reported several malignant epithelial and mesodermal tumors. Adenomas and adenocarcinomas are readily induced by goitrogens (Morris, 1955) and by thyrotropin-secreting pituitary tumors (Furth, 1954). Figure 27-24 illustrates such a tumor. Adenomatous tumors have been induced by polyoma virus (Stanton et al., 1959; Dawe et al., 1959).

TUMORS OF SKIN AND SKIN APPENDAGES

Spontaneous tumors of the skin have rarely been reported since creosoted wooden cages have been discarded. A low percentage of papillomas and squamous cell carcinomas occurs in both haired and hairless genotypes of strain HR/De (Deringer, 1951, 1956; Table 27-2). Extensive studies have been carried out on the induction of skin tumors in mice by painting with tars and pure carcinogenic hydrocarbons and by exposure to ultraviolet radiation (Stewart, 1959*a;* Boutwell, 1964). Hair-follicle tumors have been induced by polyoma virus (Dawe et al., 1959).

Papilloma and squamous cell carcinoma

The initial sequence of events during the induction of skin tumors in mice by painting with methylcholanthrene includes: epilation, hyperemia, appearance of areas of ulceration which may heal, and swelling due to edema of the dermis (Stewart, 1959*a*). Hyperplasia of the hair follicles in the marginal areas and epithelial hyperplasia with hyperkeratosis occur. Keratinized cysts, which

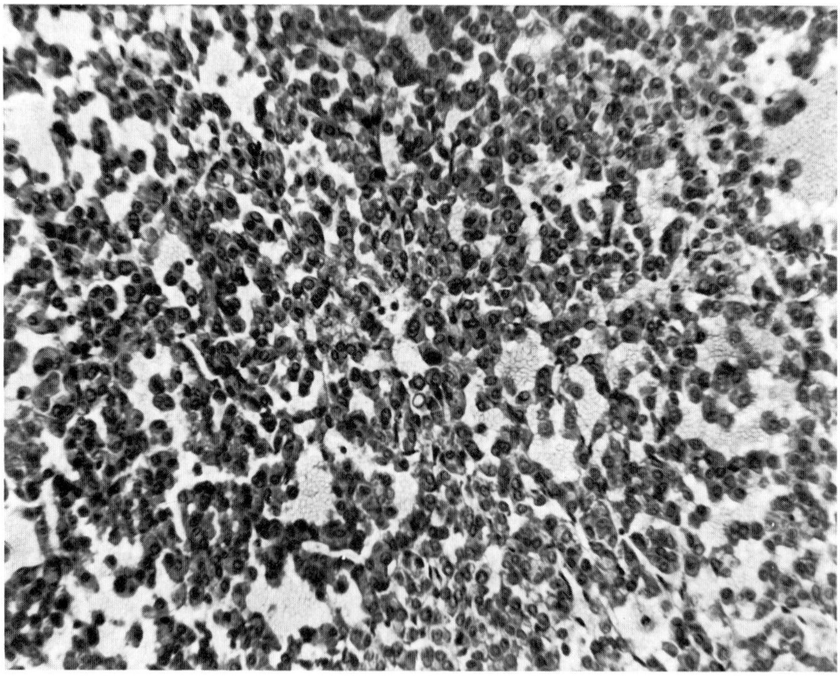

Fig. 27-23. Pituitary tumor occurring in strain C57BL/6 (×225).

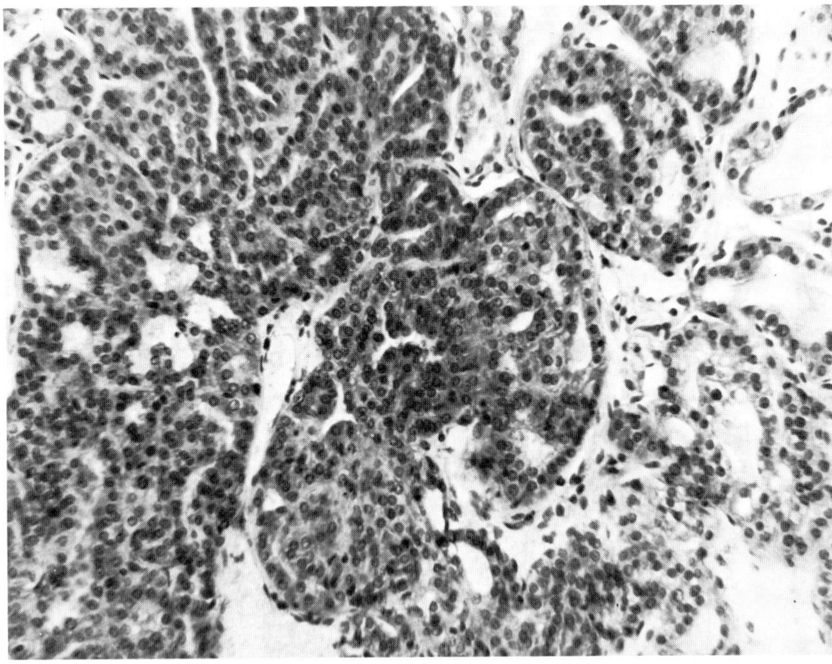

Fig. 27-24. Thyroid tumor induced by pituitary tumor which was induced by X-irradiation (×225).

may open on the surface, occur, and broad-based and pedunculated papillomas follow. The epithelial component of the papillomas is hyperplastic with frequent downgrowths of papillae into the stroma, but the cells are well oriented. Foci of carcinomatous change may develop anywhere in the hyperplastic epithelium, not necessarily in the papilloma. The pedunculated papillomas may even regress. In the carcinomatous foci the epithelial cells show increased variability in size, shape, and staining quality, show loss of orientation, frequently form epithelial pearls and keratin, and finally invade the stroma (Fig. 27-25). As the tumors grow larger and invade more extensively, they eventually ulcerate. They may metastasize to the regional lymph nodes and the lungs. Extremely anaplastic tumors with little evidence of differentiation occur. Forms with spindle-shaped epithelial cells are seen. Spontaneous basal cell tumors have rarely been reported. They have been induced by repeated applications of polyoxyethylene sorbitan monostearate (Tween 60) alone or after a single dose of a carcinogenic hydrocarbon (Della Porta et al., 1960).

Sebaceous gland tumors

Large round cells with pale foamy cytoplasm and relatively small, pale, oval central nuclei may occur in induced and in spontaneous tumors. The most rapidly growing parts of the tumor may contain small, round, deeply staining cells resembling the small undifferentiated cells of epidermoid carcinomas. Stratified squamous cells may also be found. In the mouse, the preputial and clitoral glands are specialized large sebaceous glands with large alveoli lined by pale sebaceous cells, and ducts lined by stratified squamous epithelium. Tumors of the clitoral and preputial glands have been observed. They may be remarkably organoid in structure. At The Jackson Laboratory a transplantable preputial gland tumor, ESR586, has been established and has maintained a highly differentiated form through over 150 transplant generations (Fig. 27-26). It has been found to contain large amounts of provitamin D and vitamin D. The tumor is a convenient sterol "factory" and has enabled Kandutsch and Russell (1960) to find a new pathway in intermediate sterol metabolism.

Melanoma

These highly pigmented tumors have occurred infrequently in strain DBA mice at The Jackson Laboratory (Cloudman, 1941). The commonest primary site has been on or near the tail, but the ear and other sites have been involved. The tu-

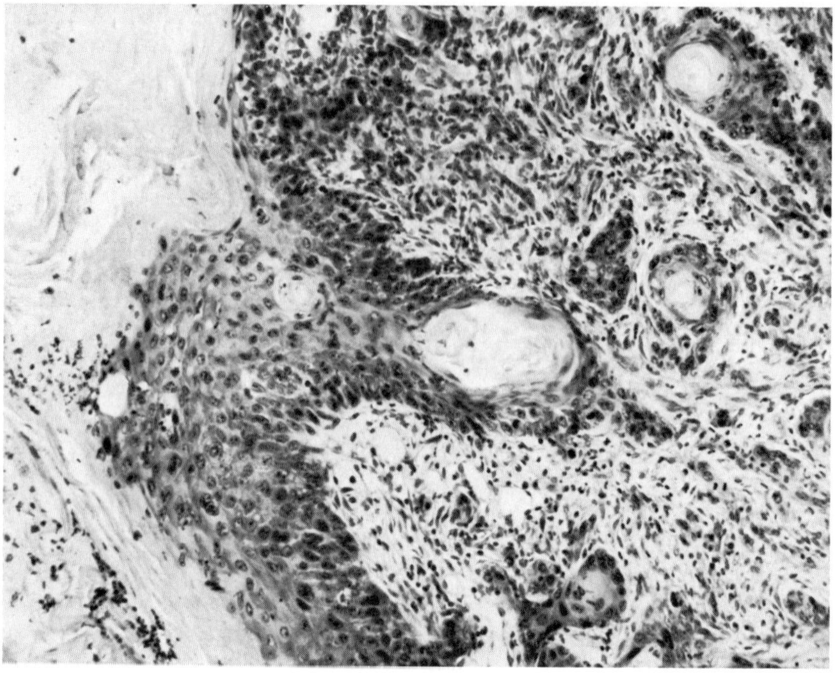

Fig. 27-25. Squamous cell carcinoma of skin, invading dermis and subcutaneous tissue (×180).

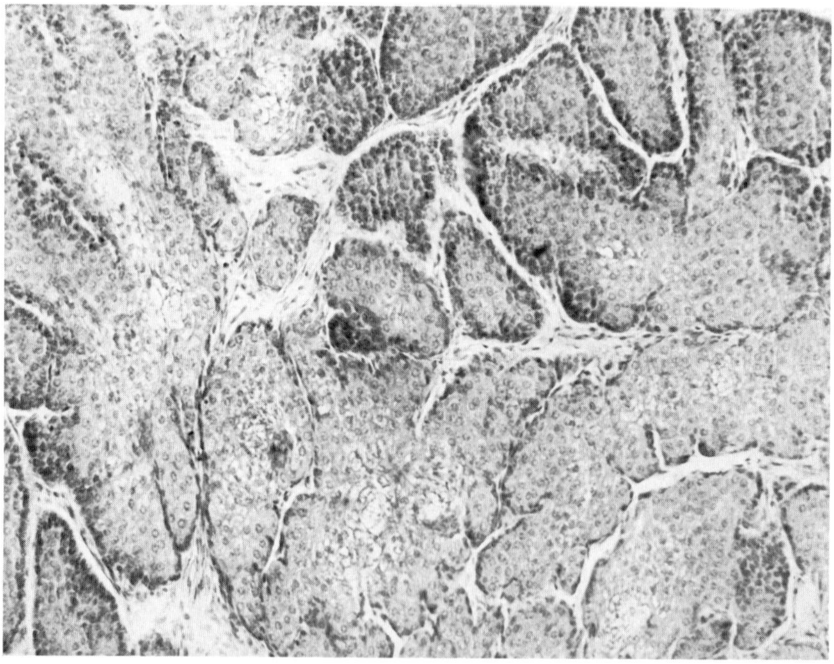

Fig. 27-26. Preputial gland tumor. Well-differentiated transplant (×150).

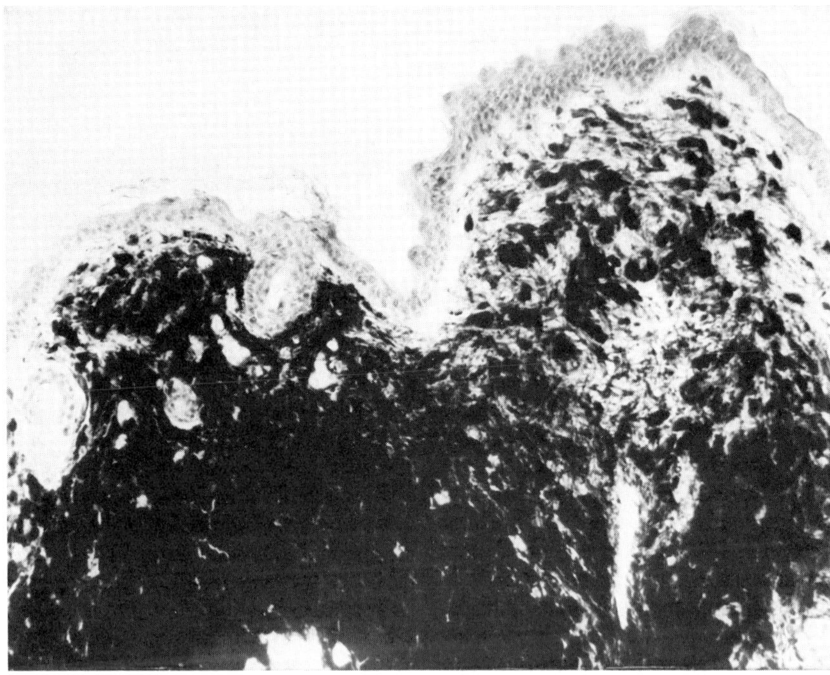

Fig. 27-27. Melanoma ($\times 150$).

mors are brown to black, smooth and rounded, and visible through the skin of the living animals. The microscopic architecture and cellular detail are heavily masked by the intense pigmentation (Fig. 27-27). The tumor cells may be spindle-shaped or large and oval, and they may be arranged in sheets, whorls, and interlacing bundles. The cytoplasm may be filled with small brown granules of melanin pigment. The tumors metastasize widely, particularly to lymph nodes and lungs. The transplantable Cloudman melanoma, S91, which arose in a DBA mouse, has been extensively studied. An amelanotic variant has been obtained by selection of less pigmented portions during the course of transplantation in albino strain BALB/c mice (Loustalot et al., 1952).

TUMORS OF THE ALIMENTARY TRACT

Induced tumors

The high incidence of tumors of the alimentary tract in man contrasts with the rarity of such tumors in mice. Spontaneous adenocarcinomas of the stomach, rectum, and colon and squamous cell carcinomas of the upper alimentary tract have only rarely been reported in mice. Adenocarcinomas of the small intestine and squamous cell carcinomas of the forestomach are readily induced by feeding chemical carcinogens (Stewart, 1953*a*). Adenocarcinomas of the stomach have been induced by intramural injection of methylcholanthrene (Stewart et al., 1953). Hyperplastic adenomatous gastritis, which occurs in strains I and DBA, has been mistaken for a malignant process (Stewart, 1953*a*).

Salivary gland tumors

Myoepithelioma. A spontaneous tumor of the parotid gland, which closely resembles the myoepithelioma of the parotid gland in man, has been observed repeatedly in strains A and BALB/c and twice in C58 (Law et al., 1955). The tumors characteristically form large central cysts containing a glairy mucoid substance. The microscopic pattern is that of sheets and cords of cells indistinctly separated into small alveoli by bands of connective tissue (Fig. 27-28). The cells are pleomorphic, a basal rounded cell lying adjacent to the connective tissue stroma. The cells become progressively more flattened and fusiform. Myoglia and fibroglia have been demonstrated in association with the fusiform cells (Lippincott et al., 1942). A pseudoglandular pattern suggesting acinar structures may appear. Focal keratinization occurs. The tumors, both primary and transplanted, may be associated with a granulocytic leukemoid reaction in the host (Bateman, 1951). Myoepitheliomas have been found rarely in subcutaneous areas where they may originate from mammary gland elements (Andervont and Dunn, 1950). Adenoacanthomas of the salivary glands have been induced by carcinogenic hydrocarbons (Bauer and Byrne, 1950).

Pleomorphic tumors. Gross (1953) and Stewart (1955) described the induction of a new type of salivary gland tumor following inoculation of newborn mice with cell-free filtrates of leukemic mouse tissues. The tumors are most prominent in the parotid and exorbital lachrymal glands (Law et al., 1955). They are usually bilateral. They are multinodular, the nodules varying from grayish white to pearly white, often resembling a bunch of grapes. When they become cystic, the contents are serous rather than mucoid. Histologically, both an epithelial and a mesenchymal or fibroblastic component can usually be distinguished (Fig. 27-29). The epithelial component consists of small tubules or minute cysts lined by a cuboidal epithelium. The connective tissue component can range from a poorly differentiated mesenchymal type of tissue, with basophilic fusiform cells separated by a loose network of fibers, to a well-differentiated tissue composed of fibroblasts with eosinophilic cytoplasm and intercellular collagen. The tumors are clearly multicentric in origin, undoubtedly reflecting their induction by the polyoma virus. On transplantation the mesenchymal component frequently outgrows the epithelial component (Law et al., 1955).

Similar tumors occur in the submaxillary, sublingual, and accessory salivary glands of the oropharynx, and the submucosal glands of the nasal passages and trachea (Dawe et al., 1959). Polyoma virus also induces tumors of thymus, bone, thyroid gland, hair follicles, mammary glands, subcutaneous connective tissue, renal medulla, adrenal medulla, and other sites (Dawe et al., 1959). Almost all of these tumors differ in many respects from the characteristic tumors of these tissues in the mouse (Dawe, 1960), except for the subcutaneous sarcomas (Law et al., 1955).

Intramandibular tumors

Carcinomas of the alveolar socket associated with exogenous hairs have been reported in nearly 1 per cent of old mice related to strain O20 (Van Rijssel and Mühlbock, 1955). Experimental in-

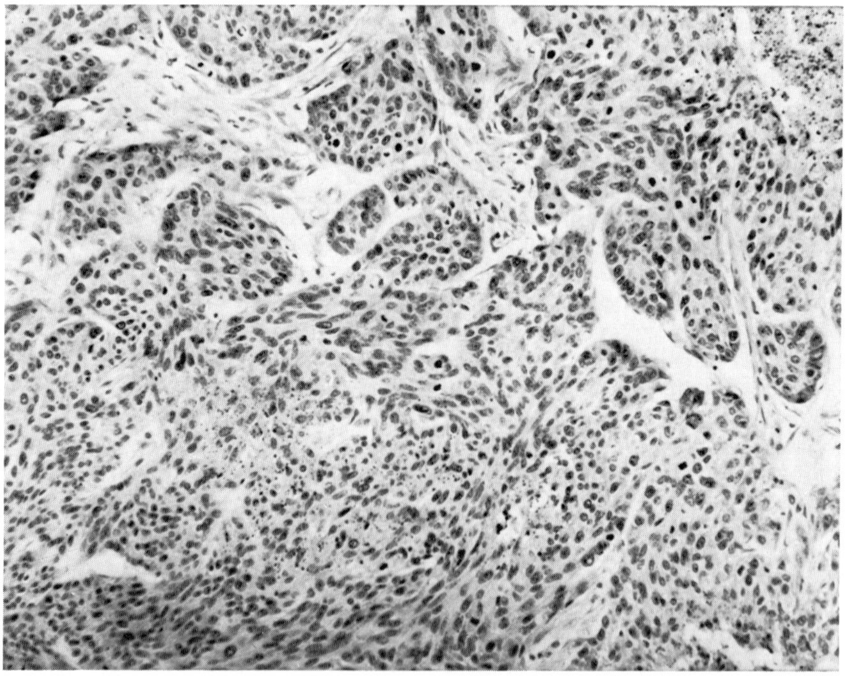

Fig. 27-28. Myoepithelioma of parotid gland, showing sheets and cords of fusiform cells surrounded by more rounded basal cells ($\times 150$).

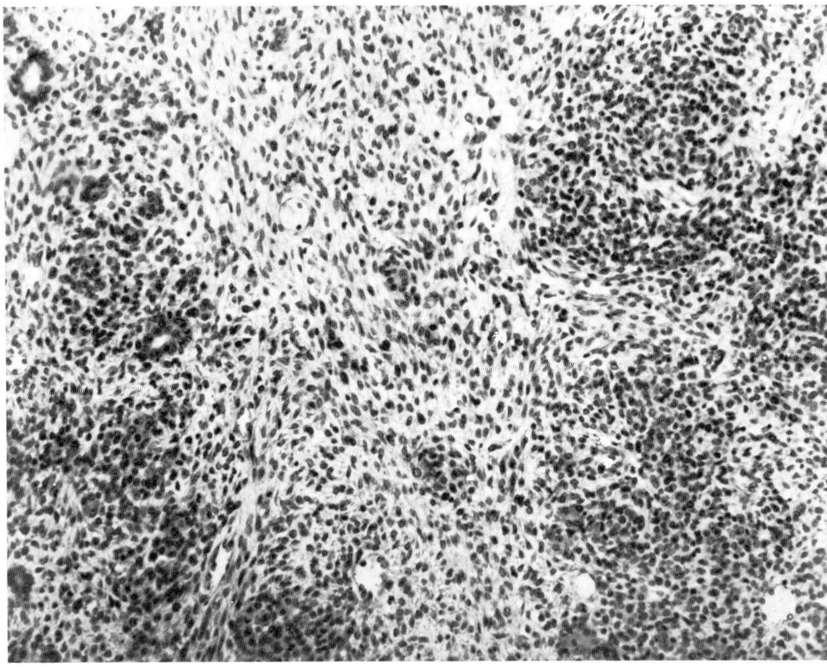

Fig. 27-29. Pleomorphic tumor of parotid gland induced by polyoma virus. Note glandular and mesenchymal elements (×150).

troduction of nylon threads, whisker hairs, and stainless steel wire into the alveolar socket induced many more of these tumors (Hollander and van Rijssel, 1963).

TUMORS OF THE MESODERMAL TISSUES

"A sarcoma is a malignant tumour arising from any nonepithelial mesodermal tissue—fibrous, mucoid, fatty, osseous, cartilaginous, synovial, lymphoid, haemopoietic, vascular, muscular or meningeal. The simplest nomenclature specifies each form of sarcoma by an appropriate prefix, fibro-, myxo-, lipo-, osteo-, etc." (Willis, 1960). There are benign forms for each of these tumors, but they are rarely reported for the mouse. The classification of some of these tumors requires special stains such as Mallory's phosphotungstic acid hematoxylin.

Fibrosarcoma

Fibrosarcoma is often used as a term for any tumor composed of spindle-shaped cells whose properties have not been further characterized. Under these circumstances the noncommittal term *sarcoma* or the descriptive term *spindle cell sarcoma* should be used. Statistically, the guess is more often right than wrong, but a number of more specialized mesodermal tumors are missed.

Fibrosarcomas occur spontaneously in the subcutaneous connective tissue and also in the internal organs. Dunn et al. (1956) have reported 106 subcutaneous sarcomas, probably derived from fibroblasts, among 4,049 female mice of strains C3H, C3Hf, C57BL, and their F_1 and backcross hybrids. None of the tumors occurred in C57BL or C57BL backcross females. Fibrosarcomas have been induced by a number of subcutaneously injected carcinogens (Stewart, 1953*b*), by long-term tissue culture (Sanford et al., 1950), and by subcutaneously implanted plastic films (Oppenheimer et al., 1959).

Fibrosarcomas are usually smooth, rounded, white, and often firm in texture. Microscopically, they are composed of elongated spindle-shaped cells arranged in bundles running in different directions (Fig. 27-30). The cytoplasm is pale, acidophilic, and usually scanty. Collagen fibers are present and fine, branching fibroglial fibrils can be demonstrated by the phosphotungstic acid hematoxylin stain. Reticulum fibers usually form a network embracing single cells. In poorly differentiated tumors, the spindle-shaped cells may be in the minority. The predominant cell is large and polyhedral. Multinucleated tumor giant cells may be formed.

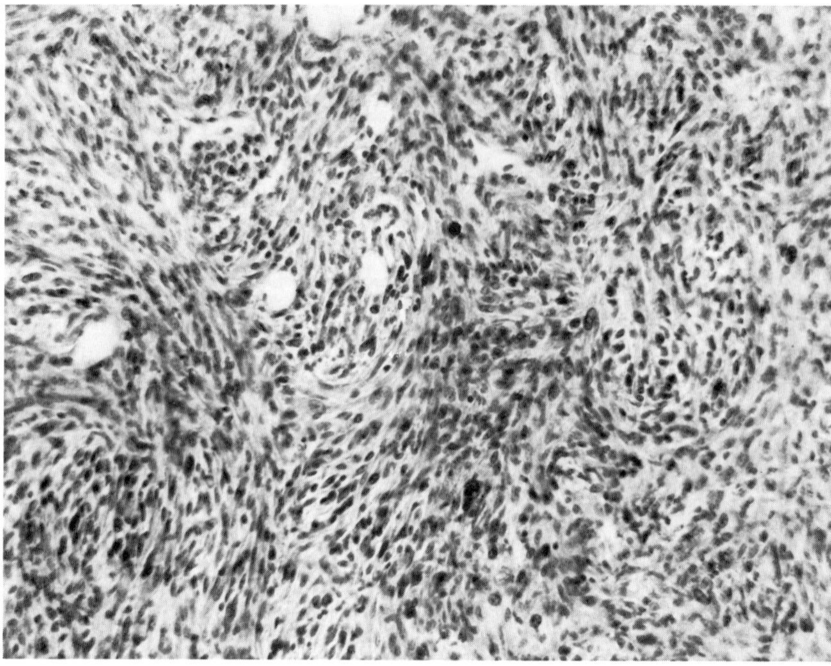

Fig. 27-30. Fibrosarcoma, with interlacing bundles of spindle-shaped cells forming collagen (×225).

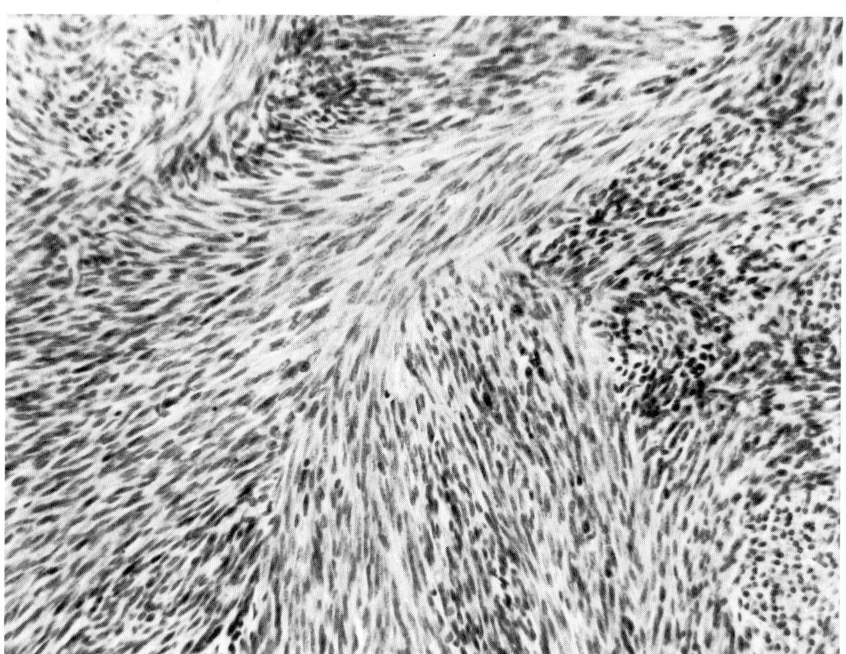

Fig. 27-31. Leiomyosarcoma of uterus, with interlacing bundles of spindle-shaped cells closely resembling smooth muscle cells (×225).

Leiomyosarcoma

Tumors of smooth muscle cells occur spontaneously in the uterus and have been induced by chemical carcinogens in this organ (Murphy, 1961) and in the alimentary tract (Saxén and Stewart, 1952). The tumors are composed of interlacing bundles of rather large spindle-shaped cells with abundant acidophilic cytoplasm (Fig. 27-31). Stroma is scanty and collagenous material minimal. Characteristic coarse, short, myoglial fibrils can be demonstrated by the phosphotungstic acid hematoxylin stain. Giant cells with one to several large nuclei may be present.

Rhabdomyosarcoma

Tumors of striated muscle have been observed in several strains of mice at The Jackson Laboratory. Stewart et al. (1959) gave a well-illustrated account of transplantable rhabdomyosarcoma H6668 that arose spontaneously in a BALB/c mouse. The most striking cell type is a large cell which may be round, sometimes oval, racquet-shaped, or straplike (Fig. 27-32). The nuclei are large, generally round, and centrally located. Giant cells with multiple nuclei are common. The cytoplasm is abundant and acidophilic. A number of the cells may contain longitudinal myofibrils which, in a few cells, may be arranged to show cross striations. Cross striations are not demonstrable in all tumors. There is also a small tumor cell which may be round, oval, or spindle-shaped.

Granular myoblastoma

This tumor of disputed histogenesis has been considered peculiar to man. It has been induced in the uterine cervix of mice by estrogen treatment (Murphy, 1961; Dunn and Green, 1963). The tumors are composed of large round cells packed with faintly eosinophilic granules of varying size. Dunn and Green (1965) have reported a transplantable tumor in strain C3H.

Liposarcoma

Malignant tumors of fatty tissue are extremely rare in the mouse. A transplantable liposarcoma has been established at The Jackson Laboratory in strain WB/Re mice.

Osteogenic sarcoma

Malignant tumors usually originating in bone and forming bone occur sporadically in mice. Most of the spontaneous osteogenic sarcomas re-

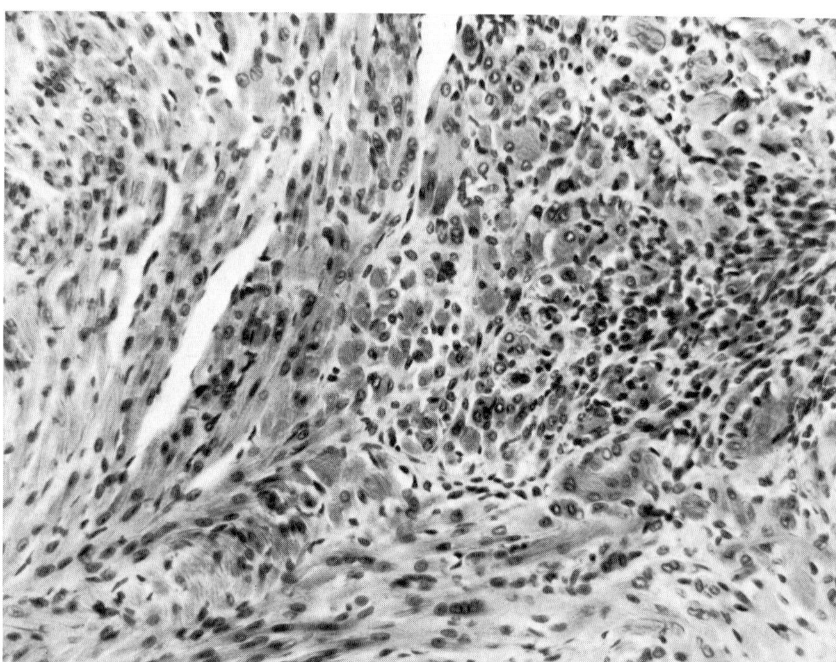

Fig. 27-32. Rhabdomyosarcoma with large straplike, multinucleated cells, appearing round in cross-section, and smaller spindle-shaped cells (×225).

ported for inbred strains have occurred in sublines of C3H and its hybrids (Dunn and Andervont, 1963; Hilberg, 1954). The incidence is below 1 per cent. Females are affected more frequently than males. Pybus and Miller (1938) derived sublines of the Simpson stock that developed a high incidence of bone tumors. The group of tumors showed the wide range of differentiation of the osteoblast: osseous, fibrous, cartilaginous, and osteoclastic. The most common type consisted principally of cancellous bone and osteoid tissue and ranged in structure from benign-appearing osteomas to osteogenic sarcomas with no sharp dividing line (Pybus and Miller, 1940). Next most common were fibrosarcomatous tumors. Several tumors were classified as giant-celled tumor, chondro-osteosarcoma, and osteoma of the compact type. Unfortunately, descendants of these mice no longer develop bone tumors.

Spontaneous osteogenic sarcomas have been transplanted but rarely maintain their bone-forming property for more than several generations (Stewart et al., 1959; Hilberg, 1956). Transplantable chondrosarcomas have been described (Ehrlich, 1906; Swarm, 1963). Osteogenic sarcomas have been induced in mice by X-irradiation and by bone-seeking radioactive substances (Glucksmann et al., 1957; Finkel et al., 1964), chemical carcinogens (Brunschwig and Bissell, 1938), and by polyoma virus. The majority of bone tumors induced by polyoma virus lacked the usual histological and cytological criteria of malignancy (Dawe et al., 1959), but metastasizing tumors have been described (Sjögren and Ringertz, 1962; Stanton et al., 1959).

An osteogenic sarcoma may be bony-hard or may be composed of softer tissue that is gritty when cut. Microscopically, the tumor can usually be identified by trabeculae of osteoid or partially ossified tissue (Fig. 27-33). More cellular portions are composed of interlacing bundles of spindle-shaped cells, resembling fibrosarcoma. In the trabeculae, cells are isolated in a matrix of hyaline material and may be rounded, resembling osteocytes, or may be spindle-shaped. The tumor cells may be palisaded along the borders of osteoid tissue, in the pattern of osteoblasts. There may be multinucleated giant cells having the appearance of osteoclasts. Foci of cartilage may be found.

Hemangioendothelioma

Malignant tumors of the vascular endothelium occur in low frequency in many inbred strains and in wild mice (Table 27-2). Deringer (1962*b*) found an incidence of 24 per cent in strain HR/De. Hemangioendotheliomas have been induced by carcinogenic hydrocarbons, ultraviolet radiation, 4-*o*-tolylazo-*o*-toluidine (Andervont, 1950*a*), and urethan (Deringer, 1962*b*). The spontaneous tumors occur in various sites, such as

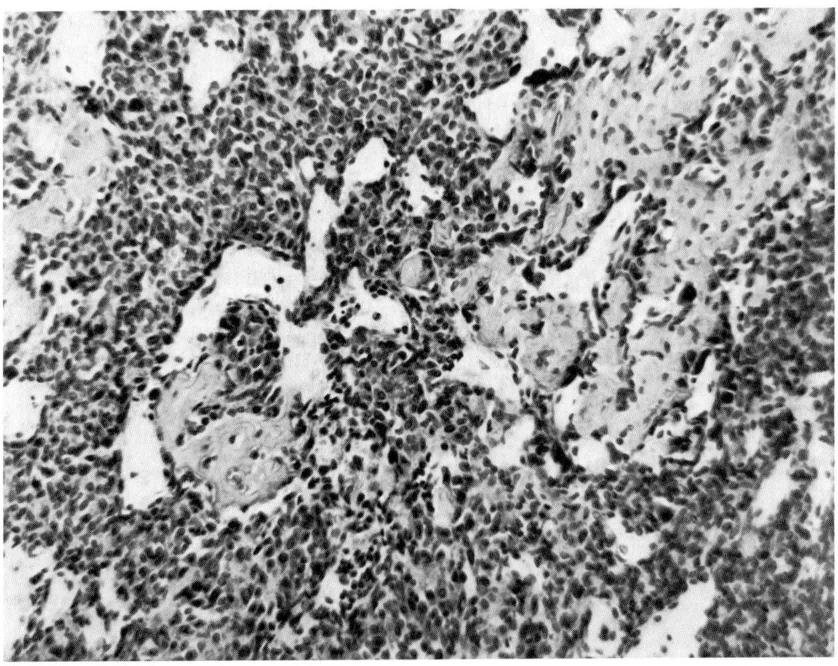

Fig. 27-33. Osteogenic sarcoma with trabeculae of osteoid tissue (×225).

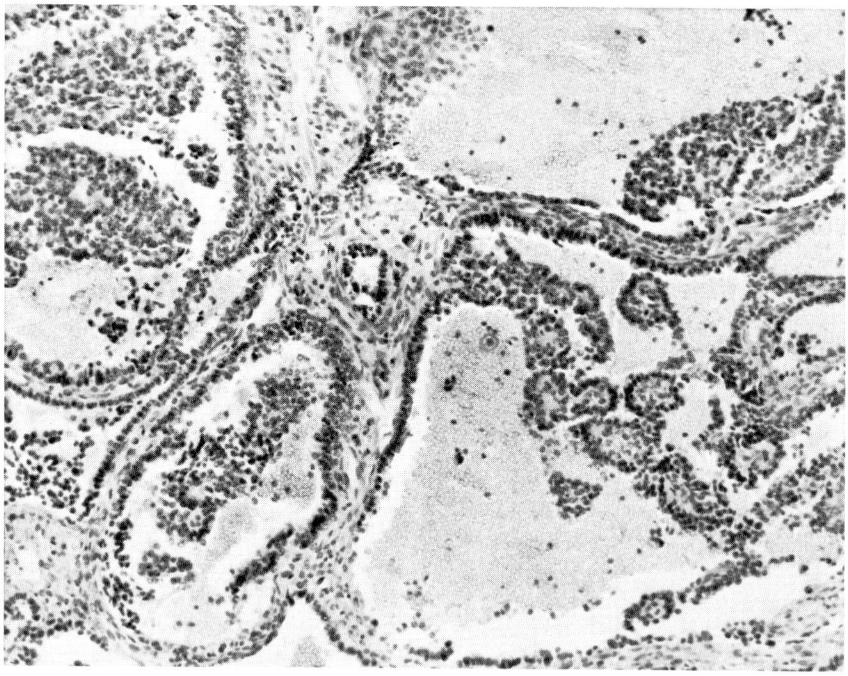

Fig. 27-34. Hemangioendothelioma, with vascular channels lined by neoplastic endothelial cells forming papillary structures (×150).

subcutaneous tissues, liver, spleen, ovaries, and mesentery.

The tumors form extremely vascular soft red masses. A collagenous fiber capsule and its extensions separate the masses into coarse and fine nodules. These fibers also form the supporting stroma for the neoplastic cells and blood vessels. The tumors are composed predominantly of blood vascular channels and sheets of neoplastic cells (Fig. 27-34). The tumor cells may be flat, round, polygonal, or spindle-shaped. Many of the tumor cells lining blood vessels resemble hyperplastic endothelial cells. More benign-appearing forms resembling cavernous hemangioma occur in the liver. Stewart et al. (1959) described and illustrated transplantable hemangioendothelioma H6221, which arose in the epididymis of a BALB/c mouse at The Jackson Laboratory.

LESS COMMON SITES OF SPONTANEOUS TUMORS

Brain and spinal cord

Spontaneous tumors of the brain and spinal cord are extremely rare in mice. The widely used transplantable tumor C1300 arose in the region of the spinal cord in an A/J mouse. Gorer (1947) identified it as a "round cell tumor" and added that it might possibly be a neuroblastoma. In noninbred mice, Horn and Stewart (1952) found reports of an ependymoma and an endothelioma of the brain and a "spindle cell" sarcoma of the spinal cord. Cloudman (1941) reported a medulloblastoma and a glioma in C57BL females. Stewart et al. (1950) reported two cases of primary tumors involving spinal nerve roots and meninges in strain NHO mice. Dickie (1965, personal communication) found a transplantable meningeal sarcoma in backcross DED2F_1 × DBA/2WyDi. A glioblastoma multiforme has been reported by Andervont et al. (1958) in a (BALB/c × C3H)F_1 mouse.

Direct implantation of carcinogenic hydrocarbons in the brain has induced glioblastoma multiforme, medulloblastoma, medulloepithelioma, astrocytoma, oligodendroglioma, spongioblastoma polare, ependymoma, pinealoma, and meningeal sarcoma (Stewart, 1953*c;* Peers, 1940; Zimmerman and Arnold, 1941).

Kidney

Rare adenomas and adenocarcinomas of the renal cortex have been reported by Tyzzer (1909), Haaland (1911), Slye et al. (1921), and Cloudman (1941). Figure 27-35 illustrates an adenocarcinoma. Claude (1958) reported the occurrence of bilateral renal adenocarcinomas in over 40 per cent of adults of a subline of BALB/c.

Papillary cystadenomas have been induced by X-irradiation (Berdjis, 1959; Rosen and Cole, 1962). Berdjis (1959) illustrated a clear cell adenocarcinoma (hypernephroma). Stevenson and von Haam (1962) reported that methylcholanthrene induced a renal cell adenocarcinoma and a number of tumors derived from the transitional epithelium of the pelvis. Transitional cell papillomas and carcinomas of the renal pelvis have been reported by Cloudman (1941).

Bladder

Spontaneous tumors of the urinary bladder are extremely rare. Cloudman (1941) mentioned papillomas and one transitional cell carcinoma. Heston and Deringer (1952) recorded a papilloma in strain C3Hf. Papillomas and transitional cell carcinomas have been induced by 2-acetylaminofluorene (Armstrong and Bonser, 1944), other aromatic amines (Bonser et al., 1956), and directly applied methylcholanthrene (Jull, 1951). Bonser and Jull (1956) described the histogenesis of the induced tumors.

Pancreas

Cloudman (1941) reported three adenocarcinomas and two islet cell tumors. Hueper (1936) reported a case of islet adenoma. Additional islet cell tumors have been recorded (Table 27-2). A high incidence of β-cell hyperplasias and tumors has been reported in $(C3Hf \times I)F_1$ hybrids (Jones, 1964).

Harderian gland

The Harderian gland is a retro-orbital lachrymal gland (Chap. 13). Spontaneous tumors are commonly reported as incidental findings in various inbred strains and hybrids (Table 27-2). Harderian gland tumors have been induced by X-irradiation (Furth et al., 1960) and by urethan (Tannenbaum and Silverstone, 1958). Large tumors cause protrusion of the eye. Microscopically, they are usually papillary and may be cystic or have solid adenomatous areas. Many of the cells closely resemble those typical of the normal gland, with foamy cytoplasm and basally located nuclei (Fig. 27-36). Harderian gland tumors may be invasive, metastasize to regional lymph nodes and lung, and be transplantable (Upton et al., 1960).

DEVELOPMENT OF NEW TUMOR TYPES

Manipulation of environmental factors

By refinement of techniques, most human tumor types can be duplicated in the mouse by direct application of chemical carcinogens to the com-

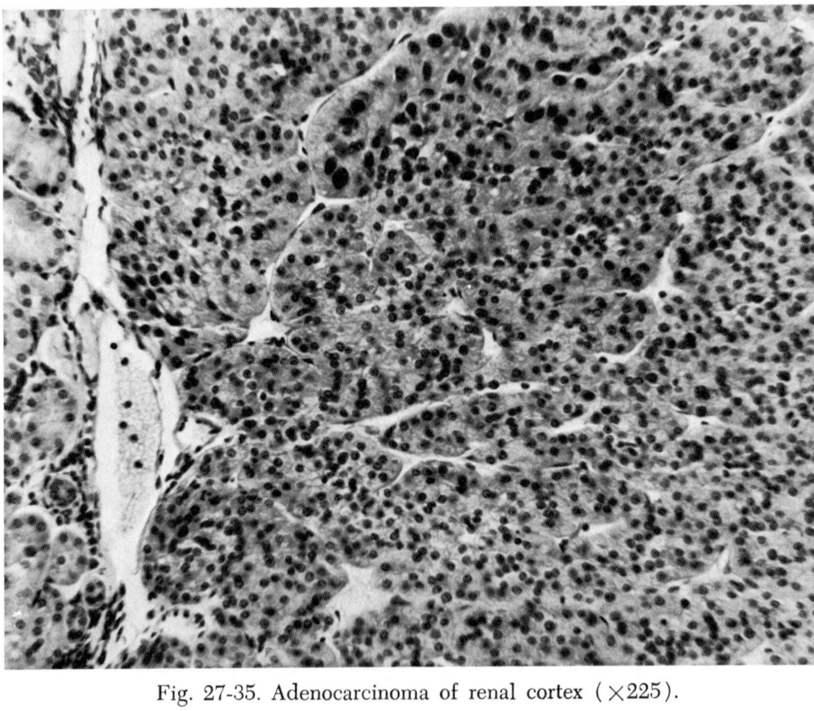

Fig. 27-35. Adenocarcinoma of renal cortex (×225).

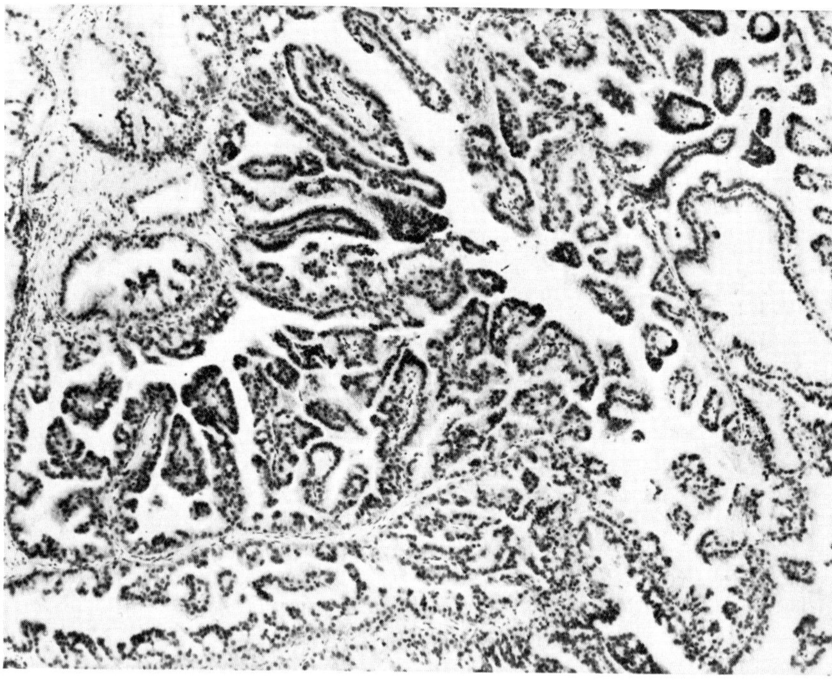

Fig. 27-36. Harderian gland tumor, with papillary structure (×80).

parable tissues. The small size of the mouse is no great hindrance. An increasing variety of tumor types is being produced by hormonal imbalance, irradiation, and oncogenic viruses. Combinations of genetic and environmental factors can be highly selective in the production of specific tumors.

Manipulation of genetic factors

More physiological models of human tumors can be expected from the development of new inbred strains and their hybrids and from the introduction of mutant genes into existing strains. The development of the SJL/J strain revealed a high incidence of reticulum cell sarcoma including close replicas of human Hodgkin's disease (Murphy, 1963). It has been shown that alleles at the *W* locus which limit the migration of primordial germ cells produce ovarian tubular adenomas in C57BL/6J mice (Russell and Fekete, 1958). By placing these genes on a hybrid background, it was possible to induce the characteristic range of ovarian tumors (Murphy and Russell, 1963). Dickie (1954) reported a wide variety of tumors in F_1 hybrids and backcross generations, involving strains CE, DBA, and DE, that were not characteristic of the parent strains. Adenocarcinomas of the uterus closely resembling the common human tumor have been found in C3H × C57BL hybrids (Heston, 1963; Dunn, 1954*b*). Spontaneous carcinomas of the cervix occurred in high incidence in the PM stock (Gardner and Pan, 1948), which has been lost because of sterility factors. These animals were derived from the stock in which Pybus and Miller (1938, 1940) described a high incidence of osteogenic sarcomas.

It is likely that many other analogues of important human tumor types can be developed by hybridizing our present strains. Selection of underlying genetic factors by the use of proper doses of carcinogens applied to segregating generations may be as effective as Snell's use of transplanted tumors to isolate the genetic factors concerned with tissue transplantation (Chap. 24).

SUMMARY

The characteristic spontaneous tumors of the mouse are described and illustrated. The five common tumors of the major inbred strains are mammary tumors, lymphocytic leukemias, primary lung tumors, hepatomas, and reticulum cell sarcomas. Ovarian tumors are frequent in several inbred strains, and hemangioendotheliomas in one. Additional spontaneous and some induced tumors, characteristic of the mouse, are described. The definition, classification, and biological properties of tumors are discussed briefly. Possibilities of manipulation of environmental and genetic factors in the development of experimental models for human tumor types are presented.

LITERATURE CITED

Andervont, H. B. 1950*a*. Induction of hemangioendothelioma and sarcomas in mice with *o*-aminoazotoluene. J. Nat. Cancer Inst. 10:927–941.

Andervont, H. B. 1950*b*. Studies on the occurrence of spontaneous hepatomas in mice of strains C3H and CBA. J. Nat. Cancer Inst. 11:581–592.

Andervont, H. B., and T. B. Dunn. 1950. Attempt to detect a mammary tumor-agent in strain C mice by X-radiation. J. Nat. Cancer Inst. 10:1,157–1,189.

Andervont, H. B., and T. B. Dunn. 1952. Transplantation of spontaneous and induced hepatomas in inbred mice. J. Nat. Cancer Inst. 13:455–503.

Andervont, H. B., and T. B. Dunn. 1953. Responses of strain DBAf/2 mice, without the mammary tumor agent, to oral administration of methylcholanthrene. J. Nat. Cancer Inst. 14:329–339.

Andervont, H. B., and T. B. Dunn. 1962. Occurrence of tumors in wild house mice. J. Nat. Cancer Inst. 28:1,153–1,163.

Andervont, H. B., T. B. Dunn, and H. Y. Canter. 1958. Susceptibility of agent-free inbred mice and their F_1 hybrids to estrogen-induced mammary tumors. J. Nat. Cancer Inst. 21:783–811.

Andervont, H. B., M. B. Shimkin, and H. Y. Canter. 1960. Susceptibilty of seven inbred strains and the F_1 hybrids to estrogen-induced testicular tumors and occurrence of spontaneous testicular tumors in strain BALB/c mice. J. Nat. Cancer Inst. 25:1,069–1,081.

Apolant, H. 1906. Die epithelialen Geschwülste der Maus. Arb. Königl. Inst. Exp. Therap. (Frankfurt a. M.) 1:7–62.

Armstrong, E. C., and G. M. Bonser. 1944. Epithelial tumours of the urinary bladder in mice induced by 2-acetyl-amino-fluorene. J. Pathol. Bacteriol. 56:507–512.

Athias, M. 1945. Lésions testiculaires chez des souris non cancéreuses, appartenant à une lignée très sujette au cancer de la glande mammaire. Arq. Patol. 17:397–417.

Bali, T., and J. Furth. 1949. Morphological and biological characteristics of x-ray induced transplantable ovarian tumors. Cancer Res. 9:449–472.

Barnes, W. A., I. E. Sisman. 1939. Myeloid leukemia and non-malignant extramedullary myelopoiesis in mice. Amer. J. Cancer 37:1–35.

Bateman, J. C. 1951. Leukemoid reactions to transplanted mouse tumors. J. Nat. Cancer Inst. 11:671–687.

Bauer, W. H., and J. J. Byrne. 1950. Induced tumors of the parotid gland. Cancer Res. 10:755–761.

Berdjis, C. C. 1959. Irradiation and kidney tumors. Histopathogenesis of kidney tumors in irradiated mice. Oncologia 12:193–202.

Berenblum, I., and P. Shubik. 1947. A new, quantitative, approach to the study of the stages of chemical carcinogenesis in the mouse's skin. Brit. J. Cancer 1:383–391.

Bonser, G. M. 1944. Mammary and testicular tumors in male mice of various strains following oestrogen treatment. J. Pathol. Bacteriol. 56:15–26.

Bonser, G. M., L. Bradshaw, D. B. Clayson, and J. W. Jull. 1956. A further study of the carcinogenic properties of ortho hydroxy-amines and related compounds by bladder implantation in the mouse. Brit. J. Cancer 10:539–546.

Bonser, G. M., and J. W. Jull. 1956. The histological changes in the mouse bladder following surgical implantation of paraffin wax pellets containing various chemicals. J. Pathol. Bacteriol. 72:489–503.

Boutwell, R. K. 1964. Some biological aspects of skin carcinogenesis. Progr. Exp. Tumor Res. 4:207–250.

Brunschwig, A., and A. D. Bissell. 1938. Production of osteosarcoma in a mouse by the intramedullary injection of 1,2-benzpyrene. Arch. Surg. 36:53–60.

Buffett, R. F., and J. Furth. 1959. A transplantable reticulum-cell sarcoma variant of Friend's viral leukemia. Cancer Res. 19:1,063–1,069.

Burns, E. L., and J. R. Schenken. 1940. Spontaneous primary hepatomas in mice of strain C3H: a study of incidence, sex distribution and morbid anatomy. Amer. J. Cancer 39:25–35.

Claude, A. 1958. Adénocarcinome rénal endémique chez une souche pure de souris. Rev. Franc. Etud. Clin. Biol. 3:261–262.

Clifton, K. H. 1959. Problems in experimental tumorigenesis of the pituitary gland, gonads, adrenal cortices, and mammary glands: a review. Cancer Res. 19:2–22.

Clifton, K. H., E. Bloch, A. C. Upton, and

J. Furth. 1956. Transplantable Leydig-cell tumors in mice. Arch. Pathol. 62:354–368.

Cloudman, A. M. 1941. Spontaneous neoplasms in mice, p. 168–233. *In* G. D. Snell [ed.] Biology of the laboratory mouse. Blakiston, Philadelphia.

Dawe, C. J. 1960. Cell sensitivity and specificity of response to polyoma virus. Nat. Cancer Inst. Monogr. 4:67–125.

Dawe, C. J., L. W. Law, and T. B. Dunn. 1959. Studies of parotid-tumor agent in cultures of leukemic tissues of mice. J. Nat. Cancer Inst. 23:717–797.

Della Porta, G., B. Terracini, K. Dammert, and P. Shubik. 1960. Histopathology of tumors induced in mice treated with polyoxyethylene sorbitan monostearate. J. Nat. Cancer Inst. 25:573–605.

Deringer, M. K. 1951. Spontaneous and induced tumors in haired and hairless strain HR mice. J. Nat. Cancer Inst. 12:437–445.

Deringer, M. K. 1956. The effect of subcutaneous inoculation of 4-*o*-tolylazo-*o*-toluidine in strain HR mice. J. Nat. Cancer Inst. 17:533–539.

Deringer, M. K. 1959. Occurrence of tumors, particularly mammary tumors, in agent-free strain C3HeB mice. J. Nat. Cancer Inst. 22:995–1,002.

Deringer, M. K. 1962*a*. Development of tumors, especially mammary tumors, in agent-free strain DBA/2eB mice. J. Nat. Cancer Inst. 28:203–210.

Deringer, M. K. 1962*b*. Response of strain HR/De mice to painting with urethan. J. Nat. Cancer Inst. 29:1,107–1,121.

Dickie, M. M. 1954. The use of F_1 hybrid and backcross generations to reveal new and/or uncommon tumor types. J. Nat. Cancer Inst. 15: 791–799.

Dickie, M. M., and G. W. Woolley. 1949. Spontaneous basophilic tumors of the pituitary glands in gonadectomized mice. Cancer Res. 9:372–384.

Dunham, L. J., and H. L. Stewart. 1953. A survey of transplantable and transmissible animal tumors. J. Nat. Cancer Inst. 13:1,299–1,377.

Dunn, T. B. 1954*a*. Normal and pathologic anatomy of the reticular tissue in laboratory mice, with a classification and discussion of neoplasms. J. Nat. Cancer Inst. 14:1,281–1,433.

Dunn, T. B. 1954*b*. The importance of differences in morphology in inbred strains. J. Nat. Cancer Inst. 15:573–589.

Dunn, T. B. 1959. Morphology of mammary tumors in mice, p. 38–84. *In* F. Homburger [ed.] The physiopathology of cancer, 2nd ed. Hoeber-Harper, New York.

Dunn, T. B., and H. B. Andervont. 1963. Histology of some neoplasms and non-neoplastic lesions found in wild mice maintained under laboratory conditions. J. Nat. Cancer Inst. 31:873–901.

Dunn, T. B., and A. W. Green. 1963. Cysts of the epididymis, cancer of the cervix, granular cell myoblastoma, and other lesions after estrogen injection in newborn mice. J. Nat. Cancer Inst. 31:425–455.

Dunn, T. B., and A. W. Green. 1965. A transplantable granular cell myoblastoma in strain C3H mice. J. Nat. Cancer Inst. 34:389–402.

Dunn, T. B., W. E. Heston, and M. K. Deringer. 1956. Subcutaneous fibrosarcomas in strains C3H and C57BL female mice, and F_1 and backcross hybrids of these strains. J. Nat. Cancer Inst. 17:639–655.

Dunn, T. B., and M. Potter. 1957. A transplantable mast-cell neoplasm in the mouse. J. Nat. Cancer Inst. 18:587–601.

Erhlich, P. 1906. Über ein transplantables Chondrom der Maus. Arb. Königl. Inst. Exp. Therap. (Frankfurt a. M.) 1:65–76.

Fahey, J. L. 1961. Physicochemical characterization of mouse myeloma proteins: demonstration of heterogeneity for each myeloma globulin. J. Exp. Med. 114:399–413.

Fekete, E., and M. A. Ferrigno. 1952. Studies on a transplantable teratoma of the mouse. Cancer Res. 12:438–440.

Fekete, E., and C. C. Little. 1945. Histological study of adrenal cortical tumors in gonadectomized mice of the ce strain. Cancer Res. 5:220–226.

Fekete, E., G. Woolley, and C. C. Little. 1941. Histological changes following ovariectomy in mice. I. dba high tumor strain. J. Exp. Med. 74:1–8.

Finkel, M. P., P. B. Jinkins, and B. O. Biskis. 1964. Parameters of radiation dosage that influence production of osteogenic sarcomas in mice. Nat. Cancer Inst. Monogr. 14:243–270.

Foulds, L. 1940. The histological analysis of tumours: a critical review. Amer. J. Cancer 39:1–24.

Foulds, L. 1954. The experimental study of tumour progression: a review. Cancer Res. 14:327–339.

Foulds, L. 1956. The histologic analysis of mammary tumors of mice. II. The histology of responsiveness and progression. The origins of tumors. J. Nat. Cancer Inst. 17:713–753.

Foulds, L. 1958. The biological characteristics of neoplasia, p. 27–44. *In* R. W. Raven [ed.] Cancer. Vol. 2. Butterworth, London.

Frantz, M. J., and A. Kirschbaum. 1949. Sex hormone secretion by tumors of the adrenal cortex of mice. Cancer Res. 9:257–266.

Friend, C. 1957. Cell-free transmisison in adult Swiss mice of a disease having the character of a leukemia. J. Exp. Med. 105:307–318.

Furtado Dias, T. 1958. Spontaneous testicular tumours in mice of the H strain. *In* L. Severi [ed.] II International symposium on mammary cancer. Division of Cancer Research, University of Perugia, Perugia, Italy.

Furth, J. 1953. Conditioned and autonomous neoplasms: a review. Cancer Res. 13:477–492.

Furth, J. 1954. Morphologic changes associated with thyrotrophin-secreting pituitary tumors. Amer. J. Pathol. 30:421–463.

Furth, J., R. F. Buffett, and N. Haran-Ghera. 1960. Pathogenesis and character of radiation-induced pituitary tumors. Acta Un. Int. Cancr. 16:138–142.

Furth, J., and J. S. Butterworth. 1936. Neoplastic diseases occurring among mice subjected to general irradiation with x-rays. II. Ovarian tumors and associated lesions. Amer. J. Cancer 28:66–95.

Furth, J., and K. H. Clifton. 1957. Experimental pituitary tumors and the role of pituitary hormones in tumorigenesis of the breast and thyroid. Cancer 10:842–853.

Furth, J., and O. B. Furth. 1936. Neoplastic diseases produced in mice by general irradiation with x-rays. I. Incidence and types of neoplasms. Amer. J. Cancer 28:54–65.

Furth, J., E. L. Gadsden, K. H. Clifton, and E. Anderson. 1956. Autonomous mammotropic pituitary tumors in mice: their somatotropic features and responsiveness to estrogens. Cancer Res. 16:600–607.

Furth, J., E. L. Gadsden, and A. C. Upton. 1952. ACTH secreting transplantable pituitary tumors. Proc. Soc. Exp. Biol. Med. 84:253–254.

Furth, J., and H. Sobel. 1946. Hypervolemia secondary to grafted granulosa cell tumors. J. Nat. Cancer Inst. 7:103–113.

Furth, J., and H. Sobel. 1947. Transplantable luteoma in mice and associated secondary changes. Cancer Res. 7:246–262.

Gall, E. A., and T. B. Mallory. 1942. Malignant lymphoma: a clinico-pathologic survey of 618 cases. Amer. J. Pathol. 18:381–429.

Gardner, W. U. 1943. Spontaneous testicular tumors in mice. Cancer Res. 3:757–761.

Gardner, W. U. 1955. Development and growth of tumors in ovaries transplanted into the spleen. Cancer Res. 15:109–117.

Gardner, W. U., and S. C. Pan. 1948. Malignant tumors of the uterus and vagina in untreated mice of the PM stock. Cancer Res. 8:241–256.

Gardner, W. U., C. A. Pfeiffer, and J. J. Trentin. 1959. Hormonal factors in experimental carcinogenesis, p. 152–237. *In* F. Homburger [ed.] The physiopathology of cancer, 2nd ed. Hoeber-Harper, New York.

Gardner, W. U., L. C. Strong, and G. M. Smith. 1936. An observation of primary tumors of the pituitary, ovaries, and mammary glands in a mouse. Amer. J. Cancer 26:541–546.

Gates, O., and S. Warren. 1960. The production of bronchial carcinomas in mice. Amer. J. Pathol. 36:653–671.

Glucksmann, A., L. F. Lamerton, and W. V. Mayneord. 1957. Carcinogenic effects of radiation, p. 497–539. *In* R. W. Raven [ed.] Cancer. Vol. 1. Butterworth, London.

Gorbman, A. 1956. Pituitary tumors in rodents following changes in thyroid function: a review. Cancer Res. 16:99–105.

Gorer, P. A. 1940. The incidence of tumours of the liver and other organs in a pure line of mice (Strong's CBA strain). J. Pathol. Bacteriol. 50:17–24.

Gorer, P. A. 1946. The pathology of malignant histiocytoma (reticuloendothelioma) of the liver in mice. Cancer Res. 6:470–482.

Gorer, P. A. 1947. Antibody response to tumor inoculation in mice with special reference to partial antibodies. Cancer Res. 7:634–641.

Graffi, A. 1957. Chloroleukemia of mice. Ann. N.Y. Acad. Sci. 68:540–558.

Greene, H. S. N. 1940. Familial mammary tumors in the rabbit. IV. The evolution of autonomy in the course of tumor development as indicated by transplantation experiments. J. Exp. Med. 71:305–324.

Gross, L. 1953. A filtrable agent recovered from Ak leukemic extracts, causing salivary gland carcinomas in C3H mice. Proc. Soc. Exp. Biol. Med. 83:414–421.

Guthrie, M. J. 1957. Tumorigenesis in intrasplenic ovaries in mice. Cancer 10:190–203.

Haaland, M. 1911. Spontaneous cancer in mice. Proc. Roy. Soc. B 83:532–540.

Halmi, N. S., and W. D. Gude. 1954. The morphogenesis of pituitary tumors induced by radiothyroidectomy in the mouse and the effects of their transplantation on the pituitary body of the host. Amer. J. Pathol. 30:403–419.

Heston, W. E. 1942. Inheritance of susceptibility to spontaneous pulmonary tumors in mice. J. Nat. Cancer Inst. 3:79–82.

Heston, W. E. 1963. Genetics of neoplasia, p. 247–268. *In* W. J. Burdette [ed.] Methodology in mammalian genetics. Holden-Day, San Francisco.

Heston, W. E. 1964. Induction of mammary gland tumors in strain C57BL/He mice by isografts of hypophyses. J. Nat. Cancer Inst. 32:947–955.

Heston, W. E., and M. K. Deringer. 1952. Test for a maternal influence in the development of mammary gland tumors in agent-free strain C3Hb mice. J. Nat. Cancer Inst. 13:167–175.

Heston, W. E., G. Vlahakis, and M. K. Deringer. 1960*a*. High incidence of spontaneous hepatomas and the increase of this incidence with urethan in C3H, C3Hf, and C3He male mice. J. Nat. Cancer Inst. 24:425–435.

Heston, W. E., G. Vlahakis, and M. K. Deringer. 1960*b*. Delayed effect of genetic segregation on the transmission of the mammary tumor agent in mice. J. Nat. Cancer Inst. 24:721–731.

Heston, W. E., G. Vlahakis, and Y. Tsubura. 1964. Strain DD, a new high mammary tumor strain, and comparison of DD with strain C3H. J. Nat. Cancer Inst. 32:237–251.

Hilberg, A. W. 1954. Osteogenic sarcoma of mice. Proc. Amer. Assoc. Cancer Res. 1:20. (Abstr.)

Hilberg, A. W. 1956. Morphologic variations in an osteogenic sarcoma of the mouse when transplanted to the kidney. J. Nat. Cancer Inst. 16:951–959.

Hollander, C. F., and T. G. van Rijssel. 1963. Experimental production of intramandibular carcinoma in mice by mechanical damage. J. Nat. Cancer Inst. 30:337–359.

Hooker, C. W., L. C. Strong, and C. A. Pfeiffer. 1946. A spontaneous transplantable testicular tumor in a mouse. Cancer Res. 6:503. (Abstr.)

Horn, H. A., and H. L. Stewart. 1952. A review of some spontaneous tumors in noninbred mice. J. Nat. Cancer Inst. 13:591–603.

Howell, J. S., J. Marchant, and J. W. Orr. 1954. The induction of ovarian tumours in mice with 9:10-dimethyl-1:2-benzanthracene. Brit. J. Cancer 8:635–646.

Hueper, W. C. 1936. Islet adenoma in the pancreas of a mouse. Arch. Pathol. 22:220–221.

Hummel, K. P. 1954*a*. Induced ovarian and adrenal tumors. J. Nat. Cancer Inst. 15:711–715.

Hummel, K. P. 1954*b*. A spontaneous transplantable testis tumor in a mouse. Proc. Amer. Assoc. Cancer Res. 1:21. (Abstr.)

Huseby, R. A. 1958. Interstitial cell tumours of the mouse testis: studies of tumorigenesis, dependency and hormone production, p. 216–230. *In* G. E. W. Wolstenholme and M. O'Connor [ed.] Hormone production in endocrine tumors. Ciba Found. Colloq. Endocrinol. Vol. 12. Little, Brown, Boston.

Jobling, J. W. 1910. Spontaneous tumors of the mouse. Monogr. Rockefeller Inst. Med. Res. 1:81–119.

Jones, E. E. 1964. Spontaneous hyperplasia of the pancreatic islets associated with glucosuria in hybrid mice, p. 189–191. *In* The structure and metabolism of the pancreatic islets. Pergamon Press, New York.

Jones, E. E., and L. J. Woodward. 1954. Spontaneous adrenal medullary tumors in hybrid mice. J. Nat. Cancer Inst. 15:449–461.

Jull, J. W. 1951. The induction of tumours of the bladder epithelium in mice by the direct application of a carcinogen. Brit. J. Cancer 5:328–330.

Kandutsch, A. A., and A. E. Russell. 1960. Preputial gland tumor sterols. III. A metabolic pathway from lanosterol to cholesterol. J. Biol. Chem. 235:2,256–2,261.

Kaplan, H. S. 1964. The role of radiation in experimental leukemogenesis. Nat. Cancer Inst. Monogr. 14:207–217.

Kotin, P., and D. V. Wiseley. 1963. Production of lung cancer in mice by inhalation exposure to influenza virus and aerosols of hydrocarbons. Progr. Exp. Tumor Res. 3:187–215.

Lacassagne, A. 1932. Apparition de cancers de la mamelle chez la souris mâle, soumis à des injections de folliculine. Compt. Rend. Acad. Sci. 195:630–632.

Law, L. W. 1941. The induction of leukemia in mice following percutaneous application of 9,10-dimethyl-,1,2-benzanthracene. Cancer Res. 1:567–571.

Law, L. W. 1948. Mouse genetics news, No. 2. J. Hered. 39:300–308.

Law, L. W., T. B. Dunn, and P. J. Boyle. 1955. Neoplasms in the C3H strain and in F_1 hybrid mice of two crosses following introduction of extracts and filtrates of leukemic tissues. J. Nat. Cancer Inst. 16:495–539.

Li, M. H., and W. U. Gardner. 1949. Further studies on the pathogenesis of ovarian tumors in mice. Cancer Res. 9:35–41.

Lieberman, M., and H. S. Kaplan. 1959. Leukemogenic activity of filtrates from radiation-induced lymphoid tumors of mice. Science 130: 387–388.

Lippincott, S. W., J. E. Edwards, H. G. Grady, and H. L. Stewart. 1942. A review of some spontaneous neoplasms in mice. J. Nat. Cancer Inst. 3:199–210.

Little, C. C. 1941. The genetics of spontaneous tumor formation, p. 248–278. *In* G. D. Snell [ed.] Biology of the laboratory mouse. Blakiston, Philadelphia.

Loustalot, P., G. H. Algire, F. Y. Legallais, and B. F. Anderson. 1952. Growth and histopathology of melanotic and amelanotic derivatives of the Cloudman melanoma S91. J. Nat. Cancer Inst. 12:1,079–1,117.

Merwin, R. M., and L. W. Redmon. 1963. Induction of plasma cell tumors and sarcomas in mice by diffusion chambers placed in the peritoneal cavity. J. Nat. Cancer Inst. 31:997–1,017.

Miyagi, T. 1952. Some observations on the volume of the nucleus of spontaneous hepatomas in mice. J. Nat. Cancer Inst. 13:627–645.

Mody, J. K. 1960. The action of four carcinogenic hydrocarbons on the ovaries of IF mice and the histogenesis of induced tumors. Brit. J. Cancer 14:256–266.

Moloney, J. B. 1960. Properties of a leukemia virus. Nat. Cancer Inst. Monogr. 4:7–33.

Morris, H. P. 1955. The experimental development and metabolism of thyroid gland tumors. Adv. Cancer Res. 3:51–115.

Mühlbock, O., and L. M. Boot. 1959. Induction of mammary cancer in mice without the mammary tumor agent by isografts of hypophyses. Cancer Res. 19:402–412.

Murphy, E. D. 1961. Carcinogenesis of the uterine cervix in mice: effect of diethylstilbestrol after limited application of 3-methylcholanthrene. J. Nat. Cancer Inst. 27:611–653.

Murphy, E. D. 1963. SJL/J, a new inbred strain of mouse with a high, early incidence of reticulum-cell neoplasms. Proc. Amer. Assoc. Cancer Res. 4:46. (Abstr.)

Murphy, E. D., and E. S. Russell. 1963. Ovarian tumorigenesis following genic deletion of germ cells in hybrid mice. Acta Un. Int. Cancr. 19: 779–782.

Oppenheimer, B. S., E. T. Oppenheimer, A. P. Stout, I. Danischefsky, and M. Willhite. 1959. Studies on the mechanism of carcinogenesis by plastic films. Acta Un. Int. Cancr. 15:659–663.

Peers, J. H. 1940. The response of the central nervous system to the application of carinogenic hydrocarbons. II. Methylcholanthrene. Amer. J. Pathol. 16:799–816.

Potter, M., and C. R. Boyce. 1962. Induction of plasma-cell neoplasms in strain BALB/c mice with mineral oil and mineral oil adjuvants. Nature 193:1,086–1,087.

Potter, M., and R. C. MacCardle. 1964. Histology of developing plasma cell neoplasia induced by mineral oil in BALB/c mice. J. Nat. Cancer Inst. 33:497–515.

Pybus, F. C., and E. W. Miller. 1938. Spontaneous bone tumours of mice. Amer. J. Cancer 33:98–111.

Pybus, F. C., and E. W. Miller. 1940. The histology of spontaneous bone tumours in mice. Amer. J. Cancer 40:54–61.

Rask-Nielsen, R., and H. E. Christensen. 1963. Studies on a transplantable mastocytoma in mice. I. Origin and general morphology. J. Nat. Cancer Inst. 30:743–761.

Rask-Nielsen, R., and H. Gormsen. 1956. On the occurrence of plasma-cell leukemia in various strains of mice. J. Nat. Cancer Inst. 16:1,137–1,147.

Rosen, V. J., Jr., and L. J. Cole. 1962. Accelerated induction of kidney neoplasms in mice after X radiation (690 rad) and unilateral nephrectomy J. Nat. Cancer Inst. 28:1,031–1,041.

Rous, P., and J. W. Beard. 1935. The progression to carcinoma of virus-induced rabbit papillomas (Shope). J. Exp. Med. 62:523–548.

Russell, E. S., and E. Fekete. 1958. Analysis of *W*-series pleiotropism in the mouse: effect of W^vW^v substitution on definitive germ cells and on ovarian tumorigenesis. J. Nat. Cancer Inst. 21:365–381.

Sanford, K. K., W. R. Earle, E. Shelton, E. L. Schilling, E. M. Duchesne, G. D. Likely, and M. M. Becker. 1950. Production of malignancy *in vitro*. XII. Further transformations of mouse fibroblasts to sarcomatous cells. J. Nat. Cancer Inst. 11:351–375.

Saxén, E. A., and H. L. Stewart. 1952. Histogenetic classification of induced gastric sarcomas in mice. J. Nat. Cancer Inst. 13:657–679.

Shimkin, M. B. 1955. Pulmonary tumors in experimental animals. Adv. Cancer Res. 3:233–267.

Simonds, J. P. 1925. Leukemia, pseudoleukemia, and related conditions in the Slye stock of mice. J. Cancer Res. 9:329–373.

Sjögren, H. O., and N. Ringertz. 1962. Histopathology and transplantability of polyoma-induced tumors in strain A/Sn and three coisogenic resistant (IR) substrains. J. Nat. Cancer Inst. 28:859–895.

Slye, M., H. F. Holmes, and H. G. Wells. 1921.

Primary spontaneous tumors in the kidney and adrenal of mice. J. Cancer Res. 6:305–336.

Slye, M., H. Holmes, and H. G. Wells. 1926. The comparative pathology of cancer of the thyroid, with report of primary spontaneous tumors of the thyroid in mice and in a rat. J. Cancer Res. 10:175–194.

Slye, M., H. F. Holmes, and H. G. Wells. 1931. Intracranial neoplasms in lower animals. Studies in the incidence and inheritability of spontaneous tumors in mice. Amer. J. Cancer 15:1,387–1,400.

Smith, F. W., W. U. Gardner, M. H. Li, and H. Kaplan. 1949. Adrenal medullary tumors (pheochromocytomas) in mice. Cancer Res. 9:193–198.

Staats, J. 1964. Standardized nomenclature for inbred strains of mice, Third listing. Cancer Res. 24:147–168.

Staff, Jackson Memorial Laboratory. 1933. The existence of non-chromosomal influence in the incidence of mammary tumors in mice. Science 78:465–466.

Stansly, P. G., and H. D. Soule. 1962. Transplantation and cell-free transmission of a reticulum-cell sarcoma in BALB/c mice. J. Nat. Cancer Inst. 29:1,083–1,105.

Stanton, M. F., S. E. Stewart, B. E. Eddy, and R. H. Blackwell. 1959. The oncogenic effect of tissue-culture preparations of polyoma virus on fetal mice. J. Nat Cancer Inst. 23:1,441–1,475.

Stevens, L. C. 1958. Studies on transplantable testicular teratomas of strain 129 mice. J. Nat. Cancer Inst. 20:1,257–1,275.

Stevens, L. C. 1959. Embryology of testicular teratomas in strain 129 mice. J. Nat. Cancer Inst. 23:1,249–1,295.

Stevens, L. C. 1962. Testicular teratomas in fetal mice. J. Nat. Cancer Inst. 28:247–267.

Stevens, L. C. 1964. Experimental production of testicular teratomas in mice. Proc. Nat. Acad. Sci. 52:654–661.

Stevens, L. C., Jr., and C. C. Little. 1954. Spontaneous testicular teratomas in an inbred strain of mice. Proc. Nat. Acad. Sci. 40:1,080–1,087.

Stevens, L. C., and J. A. Mackensen. 1961. Genetic and environmental influences on teratocarcinogenesis in mice. J. Nat. Cancer Inst. 27: 443–453.

Stevenson, J. L., and E. von Haam. 1962. Induction of kidney tumors in mice by the use of 20-methylcholanthrene-impregnated strings. Cancer Res. 22:1,177–1,179.

Stewart, H. L. 1953*a*. Experimental cancer of the alimentary tract, p. 3–45. *In* F. Homburger and W. H. Fishman [ed.] The physiopathology of cancer. Hoeber-Harper, New York.

Stewart, H. L. 1953*b*. Tumors induced by subcutaneously injected carcinogens, p. 46–61. *In* F. Homburger and W. H. Fishman [ed.] The physiopathology of cancer. Hoeber-Harper, New York.

Stewart, H. L. 1953*c*. Experimental brain tumors, p. 84–92. *In* F. Homburger and W. H. Fishman [ed.] The physiopathology of cancer. Hoeber-Harper, New York.

Stewart, H. L. 1959*a*. Experimental cutaneous carcinoma, p. 3–17. *In* F. Homburger [ed.] The physiopathology of cancer, 2nd ed. Hoeber-Harper, New York.

Stewart, H. L. 1959*b*. Pulmonary tumors in mice, p. 18–37. *In* F. Homburger [ed.] The physiopathology of cancer, 2nd ed. Hoeber-Harper, New York.

Stewart, H. L., W. V. Hare, and J. G. Bennett. 1953. Tumors of the glandular stomach induced in mice of six strains by intramural injection of 20-methylcholanthrene. J. Nat. Cancer Inst. 14:105–125.

Stewart, H. L., H. S. Kaplan, and J. G. Bennett. 1950. Report of two cases of identical primary tumors involving spinal nerve roots and meninges in strain NHO mice. J. Nat. Cancer Inst. 11:177–197.

Stewart, H. L., K. C. Snell, L. J. Dunham, and S. M. Schlyen. 1959. Transplantable and transmissible tumors of animals. Armed Forces Institute of Pathology, Washington, D.C. 378 p.

Stewart, S. E. 1955. Neoplasms in mice inoculated with cell-free extracts or filtrates of leukemic mouse tissues. I. Neoplasms of the parotid and adrenal glands. J. Nat. Cancer Inst. 15:1,391–1,415.

Swarm, R. L. 1963. Transplantation of a murine chondrosarcoma in mice of different inbred strains. J. Nat. Cancer Inst. 31:953–975.

Tannenbaum, A., and H. Silverstone. 1949. The genesis and growth of tumors. IV. Effects of varying the proportion of protein (casein) in the diet. Cancer Res. 9:162–173.

Tannenbaum, A., and H. Silverstone. 1958. Urethan (ethyl carbamate) as a multipotential carcinogen. Cancer Res. 18:1,225–1,231.

Thung, P. J. 1959. De histogenese van ovariumgezwellen bij mens en muis. Jaarb. Kankeronderzoek Kankerbestrijding Ned. p. 101–118.

Thung, P. J. 1961. Ageing changes in the ovary, p. 109–142. *In* G. H. Bourne [ed.] Structural aspects of ageing. Hafner, New York.

Tyzzer, E. E. 1909. A series of spontaneous tumors

in mice, with observations on the influence of heredity on the frequency of their occurrence. J. Med. Res. 21:479–518.

Upton, A. C. 1961. The dose-response relation in radiation-induced cancer. Cancer Res. 21:717–729.

Upton, A. C., A. W. Kimball, J. Furth, K. W. Christenberry, and W. H. Benedict. 1960. Some delayed effects of atom-bomb radiations in mice. Cancer Res. 20 (No. 8, Part 2):1–60.

Van Rijssel, T. G., and O. Mühlbock. 1955. Intramandibular tumors in mice. J. Nat. Cancer Inst. 16:659–689.

Willis, R. A. 1960. Pathology of tumours, 3rd ed. Butterworth, Washington, D.C. 1,002 p.

Woolley, G. W., M. M. Dickie, and C. C. Little. 1952. Adrenal tumors and other pathological changes in reciprocal crosses in mice. I. Strain DBA × strain CE and the reciprocal. Cancer Res. 12:142–152.

Woolley, G. W., and C. C. Little. 1945*a*. The incidence of adrenal cortical carcinoma in gonadectomized female mice of the extreme dilution strain. I. Observations on adrenal cortex. Cancer Res. 5:193–202.

Woolley, G. W., and C. C. Little. 1945*b*. The incidence of adrenal cortical carcinoma in gonadectomized male mice of the extreme dilution strain. Cancer Res. 5:211–219.

Zimmerman, H. M., and H. Arnold. 1941. Experimental brain tumors. I. Tumors produced with methylcholanthrene. Cancer Res. 1:919–938.

28

Transplanted Tumors[1]

Nathan Kaliss

Comprehensive treatments of the research uses of transplanted tumors and reviews of the literature will be found in Hauschka (1952), Kaliss (1961, 1965), Klein (1959), Snell (1958), and Woglom (1929). Bibliographic compendia appear from time to time (Handler, 1954 et seq.; 1963 et seq.), and occasional review articles in various areas of experimental cancer research appear in serials (*Advances in Cancer Research,* 1953 et seq.; Homburger, 1960 et seq.) and in the numerous journals dealing with cancer research and tissue grafting. The genetic relationships governing the fate of transplants apply with equal force to grafts of normal and cancer tissues and are detailed in Chap. 24.

Cancer research, from the beginnings of defined experimental approaches in the 19th century, was naturally concerned primarily with the initiating factors in cancerogenesis. With Pasteur's demonstration of the infectious etiology of disease, the search for similar causative agents for cancer was taken up with particular vigor at the turn of the century. Numerous attempts were made to transmit cancers within and between species by inoculating live tumors or tumor extracts. Some positive results were recorded, particularly in rodents, but the conditions for success were undefinable and the results unpredictable. What did become clear was that live cancer cells were required in the inoculum and that the resultant cancer was truly a graft and not a new growth generated by an infectious "cancer agent." Historical reviews of this exciting and vital period in cancer research will be found in Oberling (1952), Triolo (1964), and Woglom (1913, 1929).

Probably the first breakthrough to an understanding of the causes for the vagaries attending cancer grafting came with Jensen's demonstration that a mouse cancer could be consistently grafted to other mice but that the strain of mice used as host was a deciding factor in success (Triolo, 1964). Final clarification came first with Tyzzer's demonstration in the mouse that there was a genetic basis for the acceptance or rejection of cancer grafts, and Little's experimental verification of his hypothesis that Mendelian inheritance underlay graft-host compatibility (Little, 1921). Little advocated the use of inbred mice to provide genetically uniform research animals (Heston, 1949, 1959; Little, 1921, 1947), and inbred mice have since been the chief contributors to the development of experimental cancer research. Most of the important strains still in widespread use derive from those first developed by C. C. Little and L. C. Strong (Green, 1964; Heston, 1949, 1959; Staats, 1964). They have furnished a wide variety of cancers, indispensable to investigations in cancer biology and to an understanding of transplantation genetics and immunology.

It was suspected very early that the rejection of tumor grafts was accompanied by an immunological reaction (Triolo, 1964; Woglom, 1913, 1929) and this gave impetus to the hope for realizing an immunotherapy for cancer. In

[1] The writing of this chapter was supported in part by Public Health Service Research Grant CA 01594 from the National Cancer Institute, and in part by grant E-25 from the American Cancer Society.

attempts to equate the reactions to grafts with responses to infectious disease, extensive searches were made for specific antibodies in animals with regressing grafts, but without success. More rewarding were studies on the histopathology of the graft site, particularly those of Murphy (1926) and his colleagues, who demonstrated involvement of the lymphoid tissues and the lymphocyte in particular in the rejection of grafts. They noted the cellular responses as operative in grafts of both normal tissues and cancers and in infections such as tuberculosis. Though Murphy did not recognize the lymphocytic reaction as immunologically specific and allied to the evocation of antibodies, he foreshadowed the concept that these responses are dual manifestations of immune activity.

The first indication that immunologically specific mechanisms are involved in graft rejection was given by Gorer's (1942) demonstration that hemagglutinating antibodies appeared concomitantly with the rejection of tumor homografts in mice. Specificity of the lymphoid response to tumor grafts was demonstrated by Mitchison (1955), and Medawar (1957) and his colleagues extended this finding to normal tissue grafts. It is now generally recognized that the immunological defenses mobilized against infectious agents are identical with those evoked by tissue isoantigens (equals "alloantigens," Chap. 24), whether resident in normal tissues or cancers (Gorer, 1961; Chap. 31). This understanding puts in proper perspective the experimental conditions that must be satisfied in a search for cancer-specific antigens and points up the strictures attending the use of transplanted tumors for this purpose (Hauschka, 1952; Kaliss, 1961, 1965; Klein, 1959; Snell, 1958). Woglom (1929) has given an incisive critique of the flood of earlier unsuccessful attempts with tumor transplants and much of his analysis of the reasons for failure still holds true in the light of our present understanding. The paper will repay reading by anyone contemplating immunological researches with transplanted tumors.

TRANSPLANTATION IN CANCER RESEARCH

Beyond the initial motivations in developing tumor grafting outlined above, transplanted tumors have made major contributions to cancer biology which would otherwise have been unrealizable, or achievable with great difficulty (Furth, 1959). Accounts of some of the research areas that have been enriched follow.

Progression

The term "progression" is broadly inclusive of the changes in characteristics exhibited by tumors during their "lifetime" (Foulds, 1954, 1958; Furth, 1959; Furth et al., 1960; Klein and Klein, 1957). Examples are the alteration in endocrine tumors from dependence upon hormones for their survival to their eventual autonomy, the development of metastases, variations in growth rate and invasiveness, and the morphological evolution of chemically induced cancers (Chap. 27). The changes, seemingly sporadic and unpredictable, are often unobserved in autochthonous ("spontaneous") tumors, since there may be insufficient time for them to develop during the curtailed life of the afflicted animal. They do, however, occur, and the transplanted tumor, whose existence is prolonged beyond that of the original donor, is the essential tool for their detection and study. The information gained is of immediate clinical interest, since the choice of therapeutic measures may depend upon the peculiarities of a given cancer as exemplified in its life-history.

Cancerogenic viruses

The search for infectious cancerogenic agents operative in mammals has received its largest impetus with successes achieved within the last three decades (American Cancer Society, 1960; Dalton and Hagenau, 1962; Gross, 1961; Chap. 27, 30). The chief research instrument has been the variety of inbred strains of mice with well-characterized life-histories detailing the autochthonous tumors peculiar to each strain (Heston, 1949, 1959, 1963; Staats, 1964). Crucial proof that tumors are induced *de novo* by agents present in ultrafiltrates of tumor homogenates has been provided by the strain (genetic) specificity of such tumors when transplanted. Thus, filtrates prepared from leukemias arising in strain AKR mice produced leukemias when injected into newborn C3H mice. On transplantation, the latter leukemias were accepted by C3H mice but were rejected by AKR mice (Gross, 1961). This finding and the fact that the C3H strain normally has a very low incidence of leukemia was positive proof that some agent in the filtrate, and not live AKR leukemia cells, was responsible for the initiation of leukemias in the C3H mice injected at birth. We have here an instructive example of how an understanding of the genetics of transplantation, provided in the first place by studies in inbred mice, in turn supplied the essential tool for demonstrating an infectious origin for a mammalian cancer.

Chromosome morphology in relation to cancer

The suggestion was made early in the history of cancer research that the induction of chromosomal abnormalities (Chap. 7)—whether in the form of deviations from the normal diploid complement or of aberrant chromosome types—initiated the cancerous transformation (Bayreuther, 1960; Biedler et al., 1961; Ford et al., 1958; Hauschka, 1961; Hellström et al., 1963; Koller, 1963; Stevens and Bunker, 1964; Wakonig, 1960). There is still disagreement as to whether the departure from mitotic normality observed in many cancers is a consequence of cancerogenesis or is a necessary precondition. Studies of primary and transplanted tumors have shown that the former in many instances are diploid (Bayreuther, 1960; Hellström et al., 1961; Stevens and Bunker, 1964; Wakonig, 1960) and heteroploidy appears on repeated transplantation. One school of thought holds that heteroploidy is imposed on the cancer cell by its abnormal biology, with selective advantages thereby conferred assuring the continued growth of the cancer (Koller, 1963). In this respect, transplanted tumors may be considered as examples of the "progression" one might find if a given tumor were examined cytologically from its genesis and at intervals throughout its period of growth. A model study in this manner has been made with testicular teratomas in mice (Stevens and Bunker, 1964). These tumors permit histological detection at an unusually early stage of their inception. Diploidy was characteristic of the primary tumors, whereas aneuploidy and aberrant chromosomes appeared on transplantation.

Chemotherapy

Transplanted tumors—rather than "spontaneous" cancers—are most widely used for assaying the therapeutic effectiveness of a large and bewildering variety of compounds. The ready availability of grafts in large numbers and under presumably standard biological conditions has made transplanted tumors test objects of choice. An introduction to methodology and statements of the logic of various approaches to a rationale of chemotherapy will be found in a symposium of the American Cancer Society (1963). In this country, testing standards and procedures have been set up under the guidance of the National Institutes of Health's Cancer Chemotherapy National Service Center (Cancer Chemotherapy National Service Center, 1962). Periodic reports on the relative effectiveness of compounds are to be found in supplements to *Cancer Research* (Leiter, 1958 et seq.).

An interesting ancillary development has been the discovery that some radiomimetic drugs will suppress the immune response to grafts of normal or cancer tissues (Glynn et al., 1963; Hitchings and Elion, 1963; Humphreys et al., 1961, 1963; Uphoff, 1961). An extensive research field has developed in an attempt to use such drugs clinically to ensure survival of homografts of normal tissues and organs. This finding stresses the necessity for caution in interpreting the data of chemotherapeutic assays, since there may be a balance between the effects of a given drug on the cancer graft and on the immune response of the host. It emphasizes the fact that the host-graft relationship for a given assay combination must be clearly defined (Teller et al., 1964), for it cannot be taken for granted that the host may be simply acting as a passive provider of the conditions necessary for the graft's growth.

Biochemistry

Fundamental to the development of prophylaxis and therapy is an understanding of the biological properties peculiar to cancers, and biochemical characterization is essential for this understanding. Biochemical investigations concern the nature of both cancerous tissues (American Cancer Society, 1956; Bergel, 1961; Busch and Starbuck, 1964; Greenstein, 1954; Kit, 1960; LePage and Henderson, 1960; Potter, 1962) and the cancer-bearing animal (Furth, 1959; Greenstein, 1954). Transplanted tumors, as a putative source of large quantities of tissue of uniform character, seem to offer unique material for study. The choice of tumor, however, must be governed by the validity of its being specifically comparable with the normal tissue from which the tumor presumably arose, and this is not always unequivocably achievable, particularly with long-transplanted tumors. Potter (1962) has stated the problem and outlined the criteria to be met in choosing a model tumor. Statements of the main theories of the biochemistry of cancerogenesis will be found in a symposium of the American Cancer Society (1956) and in Bergel

(1961), Busch and Starbuck (1964), and Potter (1962).

CHARACTERISTICS OF TRANSPLANTED TUMORS

Types of tumors

The spectrum of transplanted tumors carried in mice reflects the range of autochthonous and experimentally produced tumors found in the laboratory (Chap. 27). Dunham and Stewart (1953) listed and briefly described 159 tumors, most of which were maintained by United States investigators. Stewart et al. (1959) have given detailed descriptions of the history, histology, and growth characteristics of 30 tumors selected as prototypes of the variety to be found in the mouse. Many more tumors have since been reported from a number of countries and new ones continually appear in the literature. It would be much beyond the scope of this chapter to attempt to list them.

Tumors originally of similar histological type may exhibit considerable changes on continued transplantation (Chap. 27; Stewart et al., 1959). Some tumors, for example, may lose their identity and become undifferentiated. It is, therefore, advisable to fix pieces (perhaps at every fifth transfer) for microscopic evaluation. Growth rates may vary considerably for tumors of the same histological type (Stewart et al., 1959), and some tumors characteristically metastasize to the lymph nodes or to organs distant from the site of tumor inoculation, while others do not (Stewart et al., 1959). The growth patterns of tumors may vary with the histological type, or may vary for tumors of the same type (Chap. 27; Stewart et al., 1959). Thus, certain lymphatic leukemias regularly disseminate when inoculated subcutaneously, while others produce only localized growths. Some endocrine tumors have special hormonal requirements for their continued growth on transplantation. Of particular note are the so-called dependent pituitary tumors which arise and will grow only in mice radiothyroidectomized by I^{131} (Furth et al., 1960). The tumors may become "autonomous" on continued transplantation and will then grow in normal hosts. These and other transplanted endocrine tumors have been of particular value in studying the interrelationships underlying the hormonal genesis of cancers (Chap. 20, 27).

The designation (coding) of transplanted tumor lines constitutes an esoteric and bewildering language of its own. This reflects the fact that there is no generally accepted convention for coding transplanted tumors, and each investigator devises his own system. Tumors reported for the first time should be precisely identified by a statement of their origin, their histological characterization, and their peculiarities on initial and continued transplantation.

Ascites tumors

Some tumors will grow as a dispersal of cells, free of a supporting stroma or vasculature, when inoculated intraperitoneally (Klein and Klein, 1960). A large volume of peritoneal fluid develops as the tumor cells multiply. This form of growth offers a particularly convenient research tool, since it permits rather precise cellular quantitation of an inoculum. (With solid tumors, quantitation is more difficult and inaccurate at best, though methods have been devised to disperse the growths into discrete cells [Boyse, 1960; Snell, 1953].) Ascites growths provide a relatively "pure culture" of tumor cells, though the fluid will contain some lymphoid cells and it may eventually become suffused with red blood cells. An account of the different research uses of ascites tumors will be found in Miner (1956).

Nongenetic factors affecting growth of tumor transplants

Even when there is genetic identity between tumor graft and host—and the graft is therefore expected to grow uniformly and without hindrance—extrinsic factors may alter the rate of growth and sometimes completely prevent it. Infectious contaminants—viral or bacterial—which transplanted tumors often acquire are a common source of trouble. The ambient temperature may alter the growth rate of subcutaneous grafts. Discussion of a variety of nongenetic conditions affecting tumor grafting will be found in Kaliss (1961), Klein (1959), and Snell (1958).

It is not an unusual experience that the primary transplant of a tumor originating in an inbred mouse will not grow in 100 per cent of hosts of the same inbred strain. With continued transfers the grafts will grow in all hosts and often at a much increased rate, so that transfers have to be made at shorter intervals to avoid losing the tumor line. The bases for these changes are not understood, and they have been variously attributed to "increased virulence" or to "antigenic simplification." "Virulence" is a descriptive general term for the unrestrained growth that distinguishes cancer as a disease and which is still the chief puzzle of cancer biology. "Antigenic simplification" implies that an autochthonous cancer may be immunogenic to its host and, by extension, to inbred mice when grafted to all individuals of the same strain, a possibility which is increasingly being demonstrated as a reality (Kaliss, 1965; Old and Boyse, 1965; Prehn, 1965).

Transplantation sites

A variety of sites has been used including the eye, brain, ovarian capsule, and splenic capsule, but the most usual are the subcutaneous (actually subpannicular in the mouse, in which the panniculus carnosus muscle layer is intimately associated with the overlying dermis), intraperitoneal, and intramuscular. Subcutaneous inoculations are usually placed in the dorsum or the flanks. For routine maintenance of most solid tumors, the inoculum is placed subcutaneously; the intraperitoneal site is used for ascites growths. As much as is feasible, the sterile procedures attending surgery should be observed during transplantation (Cancer Chemotherapy National Service Center, December, 1962). This is not so much to protect the host as to prevent infection of the graft.

STORING TUMORS

Tumors can be kept viable in the frozen state for prolonged periods (Hauschka et al., 1959; Kisileva, 1961; Klein et al., 1957; Morgan et al., 1956; Schmidt and Tessenow, 1960; Sugiura, 1961). Successful storage for as long as 2 years has been reported (Hauschka et al., 1959). The earlier methods used mechanical freezers or dry-ice chests maintaining a temperature of about −78°C. Preservation is better at even lower temperatures, and containers of simple design are now available for storage in liquid nitrogen (−147°C). Mechanical freezers are also made that can maintain temperatures much below that of dry ice. The usual method for preparing solid tumors for freezing is to immerse small pieces in various media (glycerol-glucose, for example) in sealed sterile ampoules. Ascitic tumors are likewise mixed in various proportions with a conditioning medium such as glycerol-glucose and sealed in sterile ampoules. The consensus is that the tumor-medium mixture should be slowly cooled to the final storage temperature and rapidly thawed (at 37°C) on removal. The tumors should be immediately inoculated after thawing. Not all tumors can be successfully stored.

Some authors (Hauschka et al., 1959; Kisileva, 1961; Klein et al., 1957) found no change in chromosome number or morphology, H-2 antigenic specificity, strain specificity, or response to chemotherapeutic drugs after storage, whereas, to the contrary, others have reported changes in transplantation specificity (Morgan et al., 1956). A slight increase in "latent period" sometimes occurs with the first transplant, but subsequent transplants directly from an animal donor will revert to the growth characteristic of a given tumor prior to freezing.

SUMMARY

Transplanted tumors were used from the beginning of experimental cancer research in parallel attempts to show that cancers were transmitted by infectious agents and that immunization against such agents could be achieved. These objectives were not realized, but it was demonstrated that transplanted cancers derived from live cells in the graft inoculum and that genetic identity between graft and host is the necessary condition for successful transplantation. It was also shown that graft rejection has an immunological basis whose operation is identical for grafts of cancers and normal tissues. With clarification of the conditions attending successful cancer transplantation, it has become possible to show that some cancers are generated by filtrable agents, and that cancer-specific antigens are realities. These findings have been made possible mainly by the establishment of a number of highly inbred strains of mice.

Transplanted cancers have contributed to several aspects of cancer biology: "progression," host-cancer interactions, cancer viruses, chromosome morphology, biochemistry. They are the chief test objects for possible therapeutic agents. Methods for maintaining transplanted tumors are described.

LITERATURE CITED

Advances in Cancer Research. 1953 et seq. Academic Press, New York.

American Cancer Society. 1956. Symposium on a critical appraisal of the biochemical characteristics of morphologically separable cancers. Cancer Res. 16:639–724.

American Cancer Society. 1960. Symposium on the possible role of viruses in cancer. Cancer Res. 20:669–830.

American Cancer Society. 1963. Symposium on problems basic to cancer chemotherapy. Cancer Res. 23:1,181–1,497.

Bayreuther, K. 1960. Chromosomes in primary neoplastic growth. Nature 186:6–9.

Bergel, F. 1961. Chemistry of enzymes in cancer. Charles C Thomas, Springfield, Ill. 122 p.

Biedler, J. F., L. J. Old, and D. A. Clarke. 1961. Chromosomal lesion associated with carcinogen-induced tumours in mice. Nature 192:286–288.

Boyse, E. A. 1960. A method for the preparation of viable cell suspensions from solid tumors. Transpl. Bull. 7(25):100–104.

Busch, H., and W. C. Starbuck. 1964. Biochemistry of cancer. Ann. Rev. Biochem. 33:519–570.

Cancer Chemotherapy National Service Center. April, 1962. A manual on quantitative drug evaluation in experimental tumor systems. Cancer Chemotherapy Reports No. 17. Public Health Service, Washington, D.C. 178 p.

Cancer Chemotherapy National Service Center. December, 1962. Protocols for screening. Cancer Chemotherapy Reports No. 25. Public Health Service, Washington, D.C., 184 p.

Dalton, A. J., and F. Hagenau [ed.] 1962. Tumors induced by viruses: ultrastructural studies. Academic Press, New York. 229 p.

Dunham, L. J., and H. L. Stewart. 1953. A survey of transplantable and transmissible animal tumors. J. Nat. Cancer Inst. 13:1,299–1,377.

Ford, C. E., J. L. Hamerton, and R. H. Mole. 1958. Chormosomal changes in primary and transplanted reticular neoplasms of the mouse. J. Cell. Comp. Physiol. 52 (Suppl. 1):235–269.

Foulds, L. 1954. The experimental study of tumor progression: a review. Cancer Res. 14:327–339.

Foulds, L. 1958. The natural history of cancer. J. Chron. Dis. 8:2–37.

Furth, J. 1959. A meeting of ways in cancer research. Thoughts on the evolution and nature of neoplasms. Cancer Res. 19:241–258.

Furth, J., U. Kim, and K. H. Clifton. 1960. On evolution of the neoplastic state: progression from dependence to autonomy. National Cancer Institute Monograph 2:149–178.

Glynn, J. P., S. R. Humphreys, G. Trivers, A. R. Bianco, and A. Goldin. 1963. Studies on immunity to leukemia L1210 in mice. Cancer Res. 23:1,008–1,015.

Gorer, P. A. 1942. The role of antibodies in immunity to transplanted leukemia in mice. J. Pathol. Bacteriol. 54:51–65.

Gorer, P. A. 1961. The antigenic structure of tumors. Adv. Immunol. 1:345–393.

Green, E. L. [ed.] 1964. Handbook on genetically standardized Jax mice. The Jackson Laboratory, Bar Harbor, Maine. 82 p.

Greenstein, J. P. 1954. Biochemistry of cancer, 2nd ed. Academic Press, New York. 653 p.

Gross, L. 1961. Oncogenic viruses. Pergamon Press, New York. 393 p.

Handler, A. H. Bibliography of tumor transplantation. Transplant. Bull. 1953–1962; Transplantation, 1963 et seq. (Issued from time to time.)

Hauschka, T. S. 1952. Immunologic aspects of cancer: a review. Cancer Res. 12:615–633.

Hauschka, T. S. 1961. The chromosomes in ontogeny and oncogeny. Cancer Res. 21:957–974.

Hauschka, T. S., J. T. Mitchell, and D. J. Niederpruem. 1959. A reliable frozen tissue bank: viability and stability of 82 neoplastic and normal cell types after prolonged storage at −78°C. Cancer Res. 19:643–653.

Hellström, K. E., I. Hellström, and H. O. Sjögren. 1963. Further studies on karyotypes of a variety of primary and transplanted mouse polyoma tumors. J. Nat. Cancer Inst. 31:1,239–1,253.

Heston, W. E. 1949. Development of inbred strains in the mouse and their use in cancer research. p. 9–31. *In* Genetics, cancer, growth, and social behavior. Roscoe B. Jackson Memorial Laboratory 20th Commemoration, Bar Harbor, Maine.

Heston, W. E. 1959. The impact of genetics upon cancer research. Roswell Park Mem. Inst. Bull. 4:105–122.

Heston, W. E. 1963. Genetics of neoplasia. p. 247–264. *In* W. J. Burdette [ed.] Methodology in mammalian genetics. Holden-Day, San Francisco.

Hitchings, G. H., and G. B. Elion. 1963. Chemical suppression of the immune response. Pharmacol. Rev. 15:365–405.

Homburger, F. [ed.] 1960 et seq. Progress in experimental tumor research. Lippincott, Philadelphia, and Hafner, New York.

Humphreys, S. R., M. A. Chirigos, K. L. Milstead, N. Mantel, and A. Goldin. 1961. Studies on the suppression of the homograft response with folic acid antagonists. J. Nat. Cancer Inst. 27:259–276.

Humphreys, S. R., J. P. Glynn, and A. Goldin. 1963. Suppression of the homograft response by pretreatment with antitumor agents. Transplantation 1:65–69.

Kaliss, N. 1961. The transplanted tumor as a research tool in cancer immunology. Cancer Res. 21:1,203–1,208.

Kaliss, N. 1965. Immunological enhancement and inhibition of tumor growth: relationship to various immunological mechanisms. Fed. Proc. 24:1,024–1,029.

Kisileva, N. S. 1961. The effect of prolonged storage in frozen state of transplantable tumors on the growth, metastatic spreading and strain specificity. Probl. Oncol. 7(2):22–27 [translated from Russian; Vop. Onkol. 7:179–187].

Kit, S. 1960. Nucleic acid synthesis in the neoplastic cell and impact of nuclear changes in the biochemistry of tumor tissue: a review. Cancer Res. 20:1,121–1,148.

Klein, G. 1959. The usefulness and limitations of tumor transplantation in cancer research: a review. Cancer Res. 19:343–358.

Klein, G., and E. Klein. 1957. The evolution of independence from specific growth stimulation and inhibition in mammalian tumour-cell populations. Symp. Soc. Exp. Biol. 11:305–328.

Klein, G., and E. Klein. 1960. Conversion of solid neoplasms into ascites tumors. Ann. N.Y. Acad. Sci. 63:640–661.

Klein, G., L. Révész, and E. Klein. 1957. Experiences with a frozen tumor bank. Transplant. Bull. 4:31–33.

Koller, P. C. 1963. The role of chromosome anomalies in cancer. Ciba Symp. 11:54–63.

Leiter, J. [ed.] 1958 et seq. Cancer chemotherapy screening data. Cancer Res. (Supplements). (Issued from time to time.)

LePage, G. A., and J. F. Henderson. 1960. Biochemistry of tumors. Progr. Exp. Tumor Res. 1:441–476.

Little, C. C. 1921. The relations of genetics to the problems of cancer research. Harvey Lectures, Series 17:65–88.

Little, C. C. 1947. The genetics of cancer in mice. Biol. Rev. 22:315–343.

Medawar, P. B. 1957. The immunology of transplantation. Harvey Lectures, Series 52:144–176.

Miner, R. W. [ed.] 1956. Ascites tumors as tools in quantitative oncology. Ann. N.Y. Acad. Sci. 63:637–1,030.

Mitchison, N. A. 1955. Studies on the immunological response to foreign tumor transplants in the mouse. I. The role of lymph node cells in conferring immunity by adoptive transfer. J. Exp. Med. 102:157–177.

Morgan, J. F., L. F. Guerin, and H. J. Morgan. 1956. The effect of low temperature and storage on the viability and mouse strain specificity of ascitic tumor cells. Cancer Res. 16: 907–911.

Murphy, J. B. 1926. The lymphocyte in resistance to tissue grafting, malignant disease, and tuberculous infection. Monogr. Rockefeller Inst. Med. Res. No. 21. 168 p.

Oberling, C. 1952. The riddle of cancer (translated by W. H. Woglom). Yale, New Haven. 238 p.

Old, L. J., and E. A. Boyse. 1965. Antigens of tumors and leukemias induced by viruses. Fed. Proc. 24:1,009–1,017.

Potter, V. R. 1962. Enzyme studies on the deletion hypothesis of carcinogenesis. p. 367–399. *In* The molecular basis of neoplasia. 15th Ann. Sympos. Fundamental Cancer Res. University of Texas Press, Austin.

Prehn, R. T. 1965. Cancer antigens in tumors induced by chemicals. Fed. Proc. 24:1,018–1,022.

Schmidt, F., and W. Tessenow. 1960. Methodische Versuche zur Einrichtung einer Tumorbank. Z. Krebsforsch. 63:284–293.

Snell, G. D. 1953. A cytosieve permitting sterile preparation of suspensions of tumor cells for transplantation. J. Nat. Cancer Inst. 13:1,511–1,515.

Snell, G. D. 1958. Transplantable tumors, p. 293–345. *In* F. Homburger [ed.] The physiopathology of cancer, 2nd ed. Hoeber-Harper, New York.

Staats, J. 1964. Standardized nomenclature for inbred strains of mice, Third listing. Cancer Res. 24:147–168.

Stevens, L. C., and M. C. Bunker. 1964. Karyotype and sex of primary testicular teratomas in mice. J. Nat. Cancer Inst. 33:65–78.

Stewart, H. L., K. C. Snell, L. J. Dunham, and S. M. Schlyen. 1959. Transplantable and transmissible tumors of animals. Atlas of tumor pathology, Sect. XII, Fasc. 40. Armed Forces Institute of Pathology, Washington, D.C. 378 p.

Sugiura, K. 1961. Frozen storage of 34 various solid and ascites tumors. Cancer Res. 21:496–501.

Teller, M. N., R. Wolff, and S. F. Wagshul. 1964. Host-tumor-drug relationships in experimental chemotherapy systems with allogeneic and xenogeneic host-tumor combinations. Cancer Res. 24:114–119.

Triolo, V. A. 1964. Nineteenth century foundations of cancer research. Origins of experimental research. Cancer Res. 24:4–27.

Uphoff, D. E. 1961. Drug-induced immunological "tolerance" for homotransplantation. Transplant. Bull. 28:12–16.

Wakonig, R. 1960. Aneuploidy in neoplasia—cause or result. Can. J. Genet. Cytol. 2:344–356.

Woglom, W. H. 1913. The study of experimental cancer. A review. Columbia University Press, New York. 288 p.

Woglom, W. H. 1929. Immunity to transplantable tumors. Cancer Rev. 4:129–214.

29

Constitutional Diseases[1]

Elizabeth S. Russell and Hans Meier

Constitutional diseases are here defined as malfunctions or pathological lesions whose etiology depends to a significant degree upon the action of genetic factors. If a disease occurs sporadically among genetically heterogeneous individuals, it may be impossible to distinguish between hereditary and environmental influences. The importance of genotype is then not determinable. In genetically controlled laboratory mice, however, it has been possible to identify a wide variety of constitutional diseases. These fall into two general classes, those resulting from action of identified single mutant genes and those with polygenic inheritance (Chap. 9). These latter conditions appear with different but characteristic frequencies in mice of different inbred strains. Environmental as well as hereditary factors may influence the incidence of conditions in this second genetic class.

We have chosen to omit discussion of prenatal lethal conditions caused by unit gene substitutions and of congenital malformations associated with particular inbred strains (Chap. 14). We have also omitted all discussion of cancer incidence (Chap. 26, 27), although strain differences in incidences of specific types of neoplasia clearly demonstrate the importance of genetic influences. The abnormal conditions discussed in this chapter are nonneoplastic constitutional diseases present through some part or all of postnatal life. Certain gross structural anomalies are included even though the malfunction responsible for their appearance may be limited to prenatal development.

A clear picture of the spectrum of pathological conditions induced by single genes may be obtained from the concise descriptions of effects associated with each recognized genetic locus (Chap. 8). Mutants at approximately one-half of the more than 250 named genetic loci in the mouse have effects sufficiently deleterious to be classified as constitutional diseases, many of them congenital. Thirty mutants invariably cause death between birth and maturity, and many others severely reduce viability. Neuromuscular syndromes, severe anemias, gross structural defects, and functional defects of ossification are frequent causes of early postnatal death. Polygenically inherited constitutional diseases are usually but not exclusively degenerative diseases appearing later in life.

Since pathologists are more interested in diseases than in the manner of their inheritance, the conditions to be described, whether dependent on one or many genes, have been combined into groups of related syndromes. These groups are presented in alphabetical order, and within each group all affected mutant genotypes of inbred strains are identified with a brief description of their pathology. The constitutional diseases associated with single-gene substitutions are also presented in Table 29-1. It should be stressed that the condition in each genotype listed is an independent disease, and each may represent a

[1] The writing of this chapter was supported in part by Public Health Service Research Grants CA 01074-14 and CA 04691-06 from the National Cancer Institute and by a grant from the Muscular Dystrophy Associations of America, Inc.

Table 29-1. CONSTITUTIONAL DISEASES ASSOCIATED WITH SINGLE-GENE SUBSTITUTIONS

Hereditary disease			
Class	Subclass	Mutant genes	Discussed in
Anemias	Macrocytic	*an, dm, Sl* series, *W* series	Chap. 17
	Transitory siderocytic	*f*	Chap. 17
	Yolk sac	*Ts*	Chap. 17
	Sex-linked	*sla*	Chap. 17
	Hemolytic	*ha, ja, mk, sph*	Chap. 17
Eye defects	Gross structural	*Bld, eb, Eo, ey-1, ey-2, gp, lg, lo, mi, my, oe, or*	Chap. 8
	Lens abnormality	*Cat, ec, lr*	Chap. 8, 32
	Retinal defect	*r, rd*	Chap. 8, 32
Labyrinth defects	Circling, due to gross structural abnormality	*Dc, dr, fi, kr, sy, Tw, Wt*	Chap. 8, 32
	Circling, due to cell degeneration	*av, je, pi, sh-1, sh-2, sr, sv, v, Va, wi*	Chap. 8, 32
	Cell degeneration, deafness only	*dn, su,* v^{df}	Chap. 8, 32
	Absence of otoliths	*pa*	Chap. 8
	Incompletely analyzed	*Gy, Kw, Q*	Chap. 8
Megacolon		*ls,* s^l, *s*	Chap. 15
Myopathies (primary)	Muscular dystrophy	*dy*	Chap. 29
	Muscular dysgenesis	*mdg*	Chap. 29
Neuromuscular syndromes	Proven myelin degeneration	d^l, *jp, qk, wl*	Chap. 8, 32
	Proven cerebellar defect	*rl, sq, wv*	Chap. 32
	Peripheral nerve degeneration	*dt*	Chap. 32
	Convulsive	*spa, tg, tn, Tr*	Chap. 32
	Uncoordinated	*ag, ax, du, fa, ji, la, Lc, lh, ti, tm, vc, wd*	Chap. 32
Obesity		A^y, A^{vy}, *ad, ob*	Chap. 19, 20, 32
Polydipsia		*Os*	Chap. 29
Pituitary dwarfism		*df, dw*	Chap. 20
Skeletal disorders	Systemic, membranous	*ch, de, pc, sc, se, sho, sy*	Grüneberg (1963)
	Systemic, cartilaginous	*cn, dw, mn, pg*	Grüneberg (1963)
	Systemic, osseous	*gl, mi*	Grüneberg (1963) Chap. 29
	Other systemic	*dm, Ds, my, Ph, sm, Ts*	Grüneberg (1963)
	Notochord	*Pt, Sd, tc, T* series	Grüneberg (1963)
	Unsegmented paraxial mesoderm	*Bn, vt*	Grüneberg (1963)
	Segmentation	*Cd, Fu,* Fu^{ki}, *Lp, pu, Rf, ur*	Grüneberg (1963)
	Sclerotome	*f, Hk, tk, un*	Grüneberg (1963)
	Appendicular skeleton	Many	Grüneberg (1963)
Urogenital system diseases	Abnormal kidney shape or size	*Dh, eb, lx, my, ol, sd*	Chap. 8
	Kidney malfunction	*bh, kd, ur*	Chap. 29

fundamentally different way of producing a particular pathological condition. If the condition or some aspect of its expression is discussed fully in another chapter of this book, we have referred to that source. The amount of information available varies greatly for different conditions and different genotypes. Some of the descriptions in this chapter must therefore be very brief. For two conditions, muscular dystrophy and the whole group of skeletal disorders controlled by single genes, we have referred to review articles.

AMYLOIDOSIS

Autopsies of certain inbred mice and their hybrids frequently reveal the presence of amyloid deposits, often associated with papillonephritis (the most common kidney disease of the mouse) (Chap. 26). Amyloidosis develops in a very high percentage of mice in some strains, notably the various sublines of the A strain (Dunn, 1954) and in SM/J. The incidence of amyloidosis drops in outcrosses of A/He to other strains (Heston and Deringer, 1948).

West and Murphy (1965) analyzed the distribution of amyloid in various organs of A/Sn males at different ages, the relationship of amyloidosis to kidney disease, the histochemistry of amyloid, and its possible association with lymphoid tissue reactions. No amyloid was noted in 1- to 3-month-old A/Sn males, but traces were found in the lungs of 4- to 5-month-old animals. In mice 6 months old and older there was a gradual increase in the number of organs (tongue, heart, skin, liver, testes, urinary bladder, and spleen) involved, as well as in the amount of amyloid present. By the 16th month all of the 15 organs surveyed contained amyloid. Amyloid was always present in the kidney before papillonephritis developed. Except for a consistently negative iodine test, the staining characteristics of amyloid in the mouse are similar to those in man (West and Murphy, 1965). The pathogenesis of amyloidosis is not entirely known although experimental procedures designed to elucidate it have been numerous (injections of casein, induction of various kinds of immunities, tumor transplantation, parabiosis). Electron microscopic evidence suggests that reticuloendothelial cells are of primary pathogenetic importance (Heefner and Sorenson, 1962). Histochemical tests demonstrated an increase in the acid mucopolysaccharide content of amyloid deposits of older animals (Christensen and Rask-Nielsen, 1962). Electrophoretic analysis of sera showed that progressive amyloid development is associated with hypergammaglobulinemia, later combined with increases of both β- and α_2-globulins.

ANEMIAS

Several different types of hereditary anemias have been reported in the mouse. Hematological features of each anemia are described in Chap. 17. In this chapter we describe the relationship between anemia and viability and the pathological conditions in nonhematopoietic tissues associated with certain of the hereditary anemias.

Reduction of viability is related both to the severity and to the nature of anemia. Mice with extreme hemolytic anemia (*ja/ja*, *sph/sph*) are usually born alive but die within 1 to 3 days. During their brief independent existence they develop severe jaundice and may also show cardiac hypertrophy, splenomegaly, and hepatomegaly (Stevens et al., 1959; Joe et al., 1962). If hemolytic anemia is slightly less severe, a small proportion of the mice (*ha/ha*) or larger proportion of the mice (*mk/mk*) may survive to adulthood.

The survival potentiality of mice with macrocytic anemia depends both on effects of the major anemia-producing genes and upon the genetic background. The period of rapid growth (birth to 5 weeks) is critical, since mice with defects in erythroid maturation have difficulty producing enough erythrocytes to keep pace with their increasing body size. In some stocks all *W/W* and *an/an* mice are born dead or die within 1 to 3 days (Russell and Fondal, 1951; Kunze, 1954), but in other specially selected stocks the survival of *W/W* or *an/an* mice is considerably enhanced (Russell and Lawson, 1959; Chap. 17). A higher proportion of mice of less severely affected genotypes (W/W^v, W^v/W^v, Sl/Sl^d, Sl^d/Sl^d, *dm/dm*) (Chap. 17) survive to adulthood, and these have near-normal life expectancy. In mice of certain viable genotypes growth is often impaired, possibly as a nonspecific result of the anemia (*an/an*, *dm/dm*, *f/f*, *sla/sla*) (Chap. 17).

Some of the pathological lesions associated with particular anemias, but appearing in nonhematopoietic tissues, arise as secondary consequences of specific anemias, but others appear to be independent pleiotropic effects of the action in other tissues of the same genes which affect hematopoiesis. Severe macrocytic anemia is characteristic of mice of all genotypes carrying two dominant *W* alleles (W^v/W^v, W/W^v) or two dominant *Sl* alleles (Sl/Sl^d, Sl^d/Sl^d). These mice are also sterile and have black eyes and white hair. For both genic series the sterility has been traced to failure of multiplication of primordial germ cells between the 8th and 12th days of fetal life (Bennett, 1956; Mintz, 1957; Mintz and Russell, 1957). The lack of hair pigment has been traced to failure of melanoblasts to arrive at the hair follicle sites during fetal development (Markert and Silvers, 1956) (Chap. 21). Each of these three characteristic effects of *W*-gene and *Sl*-gene substitutions appear to result from independent gene action in primarily affected tissues. In 100 per cent of W^v/W^v, W/W^v, and Sl/Sl^d females, absence of primordial germ cells leads eventually to development of ovarian tumors (Chap. 27).

Anemic mice of several different genotypes show skeletal abnormalities. Flexed (*f/f*) mice usually show tail flexure through fusion of vertebrae due to anomalies of intravertebral disks, and also show ventral white spotting. These pleiotropic effects may be secondary to fetal anemia (Kamenoff, 1935). Anemic diminutive mice (*dm/dm*) show abnormalities of the entire axial skeleton (Stevens and Mackensen, 1958). Skeletal defects of tail-short (*Ts*/+) mice are considered by Deol (1961) to be secondary effects of a transitory anemia of the primitive cell generation.

Surviving Hertwig's anemic mice (*an/an*) show several abnormalities. Adult *an/an* mice of both sexes have fewer definitive germ cells than do normal littermates and seldom produce offspring. The etiology of their sterility is, however, very different from that found in W/W^v and W^v/W^v mice. Testicular tubules of newborn *an/an* males contain primordial germ cells, but almost no spermatogenesis is seen in adult testes (Menner, 1957). In one breeding test, three of six 2-month-old *an/an* females became pregnant. One which was allowed to come to term produced six offspring, but died shortly after (Wolfe, 1964, personal communication). These findings suggest that in *an/an* mice, sterility is a secondary effect of anemia, rather than an independent effect upon primordial germ cells. Older *an/an* mice (beyond 6 to 8 months) frequently develop eye opacities, lose their hair completely, and become emaciated and kyphotic (McFarland, 1965, personal communication).

A hemolytic anemia of varying severity has been reported as developing in a high proportion of adult mice of the NZB/Bl inbred strain. It appears earlier (6 to 9 months) and more dramatically in males than in females (12 months). Affected mice show reticulocytosis, splenomegaly, jaundice, anemia, and elevated antibody titer (Helyer and Howie, 1963*a*). The condition seems to be an autoimmune disease.

The condition did not appear in mice of four other inbred strains in the same colony, nor in the F_1 generation of outcrosses, but did appear with considerable frequency in the backcrosses to the NZB/Bl inbred strain, suggesting but not proving a relatively simple inheritance pattern. However, details of expression vary according to genetic background (Helyer and Howie, 1963*b*).

The disease regularly has four manifestations: Coombs-positive hemolytic anemia with a variety of hematological, serological, and pathological manifestations; germinal centers in thymic medulla "with variety of secondary changes"; "undue frequency of severe (often lethal) membranous glomerulonephritis"; and positive serological tests for antinuclear factor in a few LE-positive tests (Burnet and Holmes, 1964; Holmes and Burnet, 1964). Fertility of NZB/Bl females is impaired in certain specific mating combinations, suggesting an autoimmune reaction (Bielschowsky and Bielschowsky, 1964). The disease shows significant similarities to human autoimmune conditions (Holmes and Burnet, 1963).

CARDIOVASCULAR DISEASES

In the following presentation, three syndromes will be discussed: dystrophic calcification, which occurs usually in the heart muscle, myocardial fibroplasia of dystrophic mice, and strain-specific arteriosclerosis. A fourth and frequent condition of certain strains, left auricular thrombosis, is considered elsewhere (Chap. 18).

Dystrophic calcification

Calcareous deposits, mainly involving the pericardium and myocardium, have been observed in a number of inbred strains. A survey revealed heart lesions in almost all retired breeders (more than 10 months old) of strains C3H/J, C3HeB/J, DBA/1J, and DBA/2J; mice of certain other strains, BALB/cJ and A/J, showed less involvement, and mice of the C57BL/6J, C57BL/10J, C57L/J, RIII/J, 129/J, MA/J, and SWR/J inbred strains had relatively few lesions (Hummel and Chapman, 1962). These calcium salt deposits are of the type previously reported by Clements (1956) as "spontaneous heart disease in DBA/2J mice" and as "calcareous pericarditis" of DBA mice by Hare and Stewart (1956). DiPaolo et al. (1964) described the lesions as occurring in mice of several related strains, C3H/St, DBA/St, and CHI/St. Although they found similar lesions also in noninbred Swiss mice, the distribution pattern was different in that the deposits also occurred along the epicardium. Importance of hereditary factors to the etiology of calcareous deposits on the heart is thus demonstrated by strain differences, including high incidence in mice from a group of related inbred strains.

Though a genetic susceptibility has been demonstrated, environmental factors also seem to play a role. DiPaolo et al. (1962) found that dietary restriction increased the incidence of calcium deposits. Hare and Stewart (1956) also thought nutritional deficiency might be important, since in affected animals they invariably observed gastritis related to the amount of roughage contained in the diet. Hummel and Chapman (1962) found more extensive muscle fiber damage at younger ages in breeding females of susceptible strains, except in DBA/2J where males also were severely affected. Virgin females were relatively free of lesions.

The pathogenesis of the calcifications is not known. They occur either as multifocal deposits in the myocardium or as diffuse pericardial plaques. In most cases both pericardium and myocardium are involved. These plaques are grossly visible as white specks on the surface of the pericardium or in cross-sections of the myocardium. Microscopically, fiber hyalinization and fibrosis (granulation tissue, scarring) are usually seen, but never an acute or chronic inflammatory reaction. Bacterial cultures have consistently been negative. The calcification is of the dystrophic type, with normal blood levels of both calcium and phosphorus. It is not possible to conclude whether these lesions are secondary to a metabolic (perhaps localized) aberration (Dunn, 1954).

Dystrophic calcification may also occur in organs other than the heart; the most frequent sites are the tunica albuginea of the testes and adrenals.

Myocardial fibroplasia

Occasionally cardiac lesions are found in mice of the 129/Re strain afflicted with muscular dystrophy. Involvement of the heart probably occurs in the final stages of the dystrophic process, which has a predilection for the skeletal musculature. The microscopic lesions include edema, vacuolar or hydropic fiber degeneration, pycnosis and karyolysis, sclerosis, myocytolysis, and fibrosis (fibroplasia). The predominant site is the septal region (Jasmin and Bajusz, 1962).

Vascular diseases

While blood-vascular diseases, both degenerative and inflammatory, are frequent in laboratory animals such as rabbits, rats, and guinea pigs, they are decidedly rare in mice. The absence of vascular lipid deposits and glomerular nodular changes in genetically obese (*ob/ob*) mice is particularly noteworthy, since this mutant is characterized by both hyperglycemia and vastly elevated serum cholesterol levels.

Despite the fact that mice may be subject to a large number of infectious agents, inflammatory lesions are uncommon. Periarteritis nodosa, thought to be the result of a chronic hypersensitivity reaction to an infectious organism, is not observed in mice. This is especially significant because many mouse colonies are chronically infected with bacterial organisms such as *Salmonella, Pasturella,* etc.

Three types of lesions involving the vascular system have been reported in mice. Two concern single reports. No publications, other than the original description, have recorded similar findings and no definite causes were ascertained. The first syndrome was described as a necrotizing arteritis in strain BL/De mice (Deringer, 1959) and consisted of amorphous subintimal and medial deposits in arteries and arterioles of a variety of tissues including ovaries (hilum), perirenal adipose tissue, adrenal cortex, uterus, and mesentery. Associated with the deposits were infiltration of the adventitia with lymphocytes and plasma cells. The second condition was a pulmonary phlebitis (and myocarditis) associated with rickettsia-like or coccobacillary-like bodies and polymorphonuclear leukocytes. This disease occurred during blind passages of tissue emulsions from a moribund mouse (Pappenheimer and Daniels, 1953).

The third type of lesion, arteriosclerosis, is seen quite commonly and consistently in males and females of strain BALB/cJ and SWR/J (Hummel and Chapman, 1962). Usually it is restricted to the coronary arteries but may involve also the vasa deferentia and mesenteric arteries. It consists of diffuse calcium deposits in the media; neither intima nor adventitia are affected. The cause is unknown.

EYE DEFECTS

Many inherited eye defects are recognized in mice (Table 29-1; Chap. 8). Several of them are gross structural defects resulting from gene action limited to embryonic development and thus are not pertinent to this chapter. Five mutant genes have been described, however, with postnatal effects on specific tissues of the eye. Dominant cataract (*Cat*) leads to liquefaction of the subcapsular zone of the lens. The lens epithelium ruptures in both lens-rupture (*lr/lr*) and ectopic (*ec/ec*) mice. The retinas of adult mice homozygous for rodless (*r/r*) or retinal degeneration (*rd/rd*) contain no differentiated rods. More extensive descriptions of these anomalies and accounts of experiments analyzing behavior differences associated with them will be found in Chap. 32.

LABYRINTH DEFECTS

Circling behavior, often accompanied by deafness, is one of the most frequently observed inherited disease syndromes in mice; 24 different mutant genes have been described with severe deleterious effects on gross structure or function of the inner ear (Table 29-1; Chap. 8). Seven of these have been shown to affect the gross structure of the labyrinth, each one interfering in a specific way with inductive relationships. Ten others produce circling and deafness, and three more deafness alone, through degeneration of cells in particular regions of the vestibular apparatus. One gene (*pa/pa*) leads to absence of otoliths. Analysis of behavior differences associated with these syndromes is given in Chap. 32.

LIPID STORAGE DISEASE

An inherited lipid storage disease, foam-cell reticulosis, appeared as a result of spontaneous mutation in CBA/H mice. "In affected homozygotes lipid-containing foam-cells replace the lymphoid tissue of the thymus and Peyer's patches and occur in smaller numbers in other tissues. Investigations suggest that the lipid may be a complex of lysolecithin and cholesterol and the disease is similar to Niemann-Pick disease in man. It is first detectable in mice of 2–3 months" (Hulse et al., 1965).

MEGACOLON

Innervation of the distal portion of the colon is defective in mice homozygous for either of two nonallelic mutant spotting genes (*ls/ls* and s^l/s^l) (Chap. 8). The absence of ganglia leads to development of megacolon and to death before 6 months. In one piebald (*s/s*) inbred strain (NZY), 10 per cent of the mice developed megacolon at ages ranging from 2 months to 2 years. The association between spotting and incidence of megacolon was even higher in the F_2 generation from outcrosses between mice of this inbred strain and solid-colored mice of two other strains (NZB and NZC). Although none of the solid-colored (+/−) offspring developed megacolon, the condition was found in 70 per cent of the spotted (*s/s*) offspring at earlier ages than in the NZY parental stock (Bielschowsky and Schofield, 1962). A tendency to megacolon has also been observed in a multicolored noninbred stock (Derrick and St. George-Grambauer, 1957). In all of these spotted mice with megacolon, segments of the gut were aganglionic or at least hypoganglionic. The accumulation of impacted feces apparently results from failure of peristalsis. Thus there is clear evidence of a defect in gut innervation associated with a number of different spotting mutants, and further evidence that this defect leads to megacolon.

MUSCLE DISEASE

Two inherited muscle diseases are known in the mouse: dystrophia muscularis and muscular dysgenesis. Dystrophia muscularis (*dy/dy*), the first hereditarily determined primary myopathy to be identified in an experimental animal, appeared as a deviant in the strain 129/Re (Michelson et al., 1955). Mice of the genotype *dy/dy*, which show pronounced muscle defect with no evidence of central or peripheral neural defect, have been extensively used for many types of research (see Staats, 1965, for references).

Affected animals may be recognized as early as 2 weeks after birth by a characteristic behavior syndrome (Loosli et al., 1961), including muscular weakness, periodic dragging of the rear feet, clasping of hind limbs when the animal is suspended by the tail, and spasmodic gaping or nodding of the head (Russell, 1963). In certain stocks segregating for the dystrophy gene, affected animals often show tetanic seizures, but these are not usually fatal. If dystrophic animals live long enough, they develop kyphosis and permanent paralysis of the hind limbs, often with contractures. Dystrophics almost never reproduce naturally. The dystrophic mouse research colony at The Jackson Laboratory is maintained by repeating ovarian transplantations in each generation (Stevens et al., 1957). Very old dystrophics often develop pneumonia, and renal calculi are common, especially in 129/Re-*dy/dy* mice. The lifespan of dystrophics is greatly reduced but varies according to genetic background and diet (Russell et al., 1962; Coleman and West, 1961).

Tests of the expression of the *dy/dy* genotype on a great variety of genetic backgrounds, including five different large F_2 populations in addition to a variety of linkage crosses, showed a proportion of recognized dystrophics in each population close to the expected 25 per cent (Loosli et al., 1961). The clinical symptoms of the dystrophic heterogeneous hybrids resembled those seen in 129/Re-*dy/dy*, except for increased vigor and longer lifespans. The dystrophy gene has been transferred to a different homogeneous genetic background (C57BL/6J) by use of the cross-intercross mating system (Loosli et al., 1961). Lifespans of C57BL/6J-*dy/dy* individuals, however, are even shorter than those of 129/Re-*dy/dy* individuals on the same regimen (Russell et al., 1962). For many kinds of experimentation the best animals appear in the F_1 hybrid (129/Re-*dy*/+ × C57BL/6J-*dy*/+) population, which segregates for especially long-lived, vigorous dystrophic animals and their normal littermates, essentially identical except for genes at the *dy* locus (Russell et al., 1962). These 129B6F_1-*dy/dy* dystrophics are easily identifiable by behavior at 2 weeks, are less impaired in their growth than are 129/Re-*dy/dy* dystrophics, and have longer lifespans (Russell et al., 1962).

For many research purposes it is extremely desirable to work with preclinical stages of the dystrophy syndrome, but affected individuals cannot be identified with certainty in these stages in segregating litters. The production of all-dystrophic litters has been achieved by artificial insemination of dystrophic juvenile or young adult females with sperm collected from dystrophic males (Wolfe and Southard, 1962). Sperm with normal histology and morphology were obtained from the ductus deferens of 129B6F_1-*dy/dy* males. Immature 129B6F_1-*dy/dy* females were subjected to priming doses of pregnant mare serum (PMS) and human chorionic gonadotropin (HCG). After 10 hours they were mated to vasectomized males and received dystrophic sperm by intrauterine injection. The os cervix route has been used since 1964 for insemination (Southard et al., 1965). This process has been used repeatedly to produce 100 per cent dystrophic litters, with success in approximately one-third of the trials, although it is sometimes necessary to foster the offspring on normal females or even to deliver them by caesarean section.

The histopathology of mouse hereditary muscular dystrophy shows many striking resemblances to that of human muscular dystrophy (Michelson et al., 1955; Banker and Denny-Brown, 1959; West and Murphy, 1960; Kitiyakara, 1961; Pearce and Walton, 1963; Harman et al., 1963, review and summary), although specific differences have been noted (Golarz and Bourne, 1960). At 2 weeks when dystrophic mice can first be identified by their behavior, *dy/dy* skeletal muscle already shows excessive variation in fiber size, with evidence of both coagulation necrosis and regeneration, nuclear rowing, and increase in connective tissue around the muscle fibers; these defects become more pronounced with advancing age (Russell, 1963). Electron microscope observations (Ross et al., 1960) showed swollen mitochondria, swollen vacuolated reticulum, and fragmented myofibrils. Activity and intracellular location of several enzymes in normal and dystrophic muscle have been studied by histochemical methods (Harman et al., 1963). Studies of the innervation pattern of dystrophic muscle by several methods showed only the "peripheral reactions of intact and healthy nerve fibers to an abnormal and predominantly fragmenting muscle periphery" (Harman et al., 1963). However, altered miniature end-plate potentials suggest functional and morphological denervation at the myoneural junction (Conrad and Glaser, 1964).

Although the histological lesions in hereditary muscular dystrophy are similar to those in vitamin E–deficiency dystrophy, no vitamin therapy, including treatment with vitamin E (Tubis et al., 1959) has had any effect on dystrophic mice. There is evidence that administration of certain anabolic androgenic steroids retards progress of the disease, but no suggestion of return of previously lost function as a result of steroid treatment (Dowben et al., 1964). Dietary improvement greatly increases lifespan of dystrophic mice (Coleman and West, 1961). At The Jackson Laboratory nutritional muscular dystrophy has been produced in 129/Re-+/+ normal mice by prolonged vitamin E deprivation in order to compare the two diseases on a common genetic background (Loosli, 1965). Histological lesions were similar in the two diseases, although 129/Re nutritional dystrophy was characterized by late onset, slow progress, and diffuse distribution of lesions. The nutritional disease was rapidly cured by treatment with α-tocopherol. An extremely interesting feature of this experiment was the difference in clinical manifestation of the two conditions. Mice with nutritional dystrophy showed a decrease in muscular strength in both fore- and hind limbs and were unable to hang on to a vertical wire-mesh food hopper. Although *dy/dy* mice had little strength in their hind limbs, they could cling to the hopper for some time with their forefeet. The mice with nutritional dystrophy never showed foot-dragging, clasping of the hind limbs when suspended, or spasmodic head-jerking.

Parabiosis experiments, with conjoined normal-dystrophic pairs, indicate clearly there is no circulating factor responsible for muscle breakdown in *dy/dy* mice (Pope and Murphy, 1960; Pope et al., 1964). The stress of parabiosis shortened the life of both normal and dystrophic partners, but there was some suggestion that the progress of dystrophic lesions might be slightly delayed in parabiosed dystrophics, possibly as a result of shared nutrition. Both normal and dystrophic muscle grow successfully in Algire diffusion chambers implanted in histocompatible normal and dystrophic hosts (O'Steen, 1962). Myoblasts and myotubes appear in implants of normal muscle, and only myoblasts appear, somewhat earlier than normal, in implants of dystrophic muscle. The dystrophy genotype of the recipient has little or no effect on development of the implanted muscle.

Alterations have been found in the physiology of dystrophic muscle that are similar to those in other genetic or induced dystrophies; these include disturbed K-to-Na balance (Baker et al., 1958; Young et al., 1959; Zierler, 1961), muscular weakness and altered twitch and tetanus tensions (Brust, 1964). Because of the higher connective tissue content of dystrophic muscle, it is difficult to assess strictly quantitative functional differences. However, qualitative differences have been reported, including heightened excitability and tendency to fibrillation (Harman et al., 1963). Contraction time of dystrophic muscle is within normal limits, but relaxation is much delayed. In a prolonged series of repeated stimulations, developed tension dropped much more slowly in dystrophic than in normal response. The qualitatively distinct fatigue pattern of dystrophic muscle seems to depend on some basic dissimilarity inherent in the dystrophy myopathy (Sandow and Brust, 1962; Conrad and Glaser, 1962; McComas and Mossawy, 1965).

The metabolism of dystrophic muscle is disturbed in many ways, including abnormal creatine-to-creatinine balance, deranged lipid, carbohydrate, and protein metabolism, and abnormal levels of a great many enzymes (Russell, 1963; Harman et al., 1963). Probably many of these alterations are results rather than causes of the basic defect. One promising approach to analysis of the interrelationships of changes is the grouping of enzymes with similar function. In a comparative study of levels in 20 enzymes in dystrophic muscle and in denervated muscle, TPN-requiring enzymes were elevated, DPN-requiring depressed, suggesting ". . . a shift from normal glycolytic and oxidative pathways to the pentose shunt" (McCaman, 1963). Hydrolytic and transferring enzymes tend to be elevated. The findings in dystrophic muscle were confirmed and extended by histochemical studies (Fennell and West, 1963).

In order to determine primary effects a useful approach is that of retrograde analysis, progressing toward earlier stages in an animal's life history, since the cause(s) of dystrophy must be shown to occur prior to the onset of clinical and histological manifestations. Alterations in muscle fibers, including breakdown and regeneration, have been observed in dystrophic mice as early as 3 days after birth (Laird and Walker, 1964; Platzer and Chase, 1964). Coagulation necrosis has been observed in the tongue, masseter, and psoas musculature of 20- to 21-day 129B6F$_1$-*dy/dy* fetuses (Meier et al., 1965). Elevated adenosinemonophosphatase activity, accumulation of triose phosphates, and higher kidney glycine transamidinase activity which already appear at 2 weeks may be more closely related to the original gene action than other lesions occurring at later stages (Gould and Coleman, 1961; Chap. 19). The fact that glycine transamidinase levels in *dy*/+ heterozygotes are intermediate between those observed in +/+ and *dy/dy* homozygotes also suggests close relation to primary gene action (Coleman and Ashworth, 1960).

Gould and Coleman (1961) have shown that a ketone body, measured by a method used to determine acetoacetate, accumulates when muscle homogenates from dystrophic mice of all ages and from normal mice under 2 weeks of age are incubated in phosphate buffer fortified with ATP and Mg ions. No such accumulation occurs when muscle homogenates from adult normal mice are incubated under similar conditions. In fact when normal muscle homogenates or ammonium sulfate fractions thereof were added to dystrophic muscle homogenates, accumulation of the ketone body was prevented. A decreased production of pyruvate was also observed in dystrophic muscle homogenates which accumulated this ketone body. This suggested that the ketone body may in fact not be acetoacetate as originally proposed but rather some intermediate in the glycolysis pathway which is a direct precursor to pyruvate (Coleman, 1965*a,b*). Reinvestigation of the identity of the ketone established that it consisted mainly of methylglyoxal. The methylglyoxal did not accumulate in dystrophic muscle homogenates but rather resulted from hydrolysis of other compounds, most probably triose phosphates, in the reaction mixture. These triose phosphates accumulate because of decreased activity of the enzymes glyceraldehyde-3-phosphate dehydrogenase and α-glycerophosphate dehydrogenase, which are directly involved in triose utilization and the subsequent production of pyruvate. Glyceraldehyde-3-phosphate dehydrogenase in normal muscle increased rapidly after birth, reaching an adult level at 3 to 4 weeks of age. In contrast, dystrophic muscle exhibits a deficiency as early as 10 days of age at which time further increase in activity stops, causing the deficiency to become progressively more severe. A similar situation was observed in the activity of α-glycerophosphate dehydrogenase. The failure of development of both of these enzymes at the age when dystrophic symptoms become manifest is of particular interest. The decreased activities of these enzymes in dystrophic muscle would effectively block glycolysis thus drastically reducing the energy available to the muscle cell. Also the deficiency of α-glycerophosphate which should result from the decreased activity of α-glycerophosphate dehydrogenase is of special significance in view of the proposed role of this compound as a major source of intramitochondrial energy.

A second and even more severe muscle disease is muscular dysgenesis (Gluecksohn-Waelsch, 1963) inherited as a unit recessive (*mdg*). Homozygous *mdg/mdg* mice, which die at birth, are very edematous and show "a severe, general deficiency of skeletal musculature; cardiac and smooth muscles appear normal" (Pai, 1965*a*). "The myopathy results from a genetically determined specific interference with skeletal muscle cell differentiation." Abnormalities were first seen in *mdg/mdg* fetuses at 13¼ days and appeared in any particular muscle "at the time when (its) myoblasts differentiate into striated myotubes, regardless of the embryonic age at which this occurs" (Pai, 1965*b*). Mutant skeletal muscle cells never develop into normal fully striated muscle fibers, show no acetylcholinesterase activity at the myoneural junction, and many of them degenerate before birth.

NEUROLOGICAL SYNDROMES

In Table 29-1 we have chosen to separate the mutants often classified as "behavioral" into two groups. The labyrinth mutants which cause circling behavior have already been discussed in this chapter. In this section we will consider mutants with effects in other parts of the nervous system, here called neurological syndromes. Mutations of this second group, leading to convulsive or uncoordinated behavior, have frequently been observed. Some mutants are so severe as to lead to death of the affected homozygotes before weaning age. We have listed first those conditions in which histological studies have demonstrated either myelin degeneration, gross defect of the cerebellum, or degeneration of the dorsal root ganglia of spinal nerves. Certain neuropathological and behavioral anomalies are discussed more extensively in Chap. 32, but it is worth remarking here that degeneration of myelin sheaths may lead both to convulsions (d^l/d^l) and to uncoordinated gait and tremor (*qk/qk, jp/jp, wl/wl*). All of the mutants for which cerebellar degeneration has been demonstrated have an abnormal gait and greatly reduced viability. The abnormality of mice with dystonia musculorum (*dt/dt*) is very different, involving a defect of spinal nerves. Histological study of one further condition, trembler (*Tr*/+), has failed to disclose pathological lesions in the nervous system (Chap. 8), and the condition has been listed simply as causing convulsive behavior. The histological basis of the other conditions has not been extensively investigated.

GENETIC OBESITY

In the mouse, mutant genes at three different loci (*ob, ad,* and two agouti alleles, A^y and A^{vy}) (Table 29-1) are known to produce excessive obesity, usually resulting in sterility. In addition, inbred strains of mice differ markedly in their response to the fat content of their dietary regimen. Mice of the inbred strain NZO regularly become extremely obese but remain fertile. Metabolic patterns associated with each of these genetic obesities are described elsewhere (Chap. 19), and changes in the islets of Langerhans associated with genetic obesity are discussed in Chap. 20. Physiological effects of obesity have been shown to alter incidence and time of appearance of a number of types of neoplasia. Obese mice also have an increased tendency to certain bone diseases (Sokoloff et al., 1962). Behavioral differences associated with *ob/ob* genotype are discussed in Chap. 33.

PITUITARY DWARFISM

The first known case in the mouse of an endocrine abnormality caused by a single-gene mutation was Snell's (1929) dwarf. Extensive research using a variety of approaches has demonstrated clearly that all of the defects of *dw/dw* mice result ultimately from an anterior pituitary defect (absence of eosinophils and thyrotropes) and lack of growth and thyrotropic hormones. Research on this condition is reviewed in the section on pituitary function (Chap. 20).

Another type of pituitary dwarf (*df/df*) (Schaible and Gowen, 1961) shows many similarities to Snell's dwarf, but also some differences in response to hormone therapy (Chap. 20). Further investigation of this syndrome may provide valuable insight into the interrelationship of cell types in the anterior pituitary.

POLYDIPSIA

Mice from two different inbred strains develop polydipsia-polyuria syndromes for quite different reasons. In 12- to 14-month-old DE mice, changes in the adrenal, including amyloid deposit and reduction of cortex, lead to a 5- to 10-fold increase in water intake and excretion (Chap. 20). The pituitaries of these animals are normal, and kidney changes come only after establishment of the polydipsic syndrome. Mice of the MA/J and MA/MyJ inbred strains also show a polydipsia-polyuria syndrome, but this is more probably associated with cysts pressing on the posterior lobe of the pituitary (Chap. 20). Urine secretion was measured in the "heavy drinkers" and was found to be about equal in amount to water consumed; the urine was light in color, free of sugar, and had a low specific gravity. Since kidneys and adrenals are grossly and histologically normal, abnormal secretion of antidiuretic hormone is suspected (Hummel, 1960).

Mild diabetes insipidus appears as a pleiotropic effect of the oligosyndactyly (*Os*) mutant gene. Certain modifier genes, themselves capable of producing a mild diabetes insipidus, also enhance the manifestation in *Os* animals. The severe condition associated with *Os* in the presence of modifying genes is probably renal in origin (Falconer et al., 1964).

SKELETAL DISORDERS

Inherited skeletal disorders, including both congenital and degenerative diseases, are very common in mice.

Initial malfunction

In a classic presentation of pathological skeletal development, Grüneberg (1963) described developmental effects of 39 different mutations in the mouse. Eleven of these are systemic disorders originating in membranous, cartilaginous, or osseous skeleton; eight are systemic disorders imposed upon the skeleton; eight stem from notochord disorder; eight from abnormality of the paraxial mesoderm; eight from abnormal segmentation; and eight as effects limited to the appendicular skeleton (Table 29-1). Most of these gene actions appear to be limited to embryonic life. In keeping with our emphasis on postnatal disease processes, we will limit discussion to four conditions. The reader is referred to Grüneberg's (1963) book for further analysis.

Although the earliest effects of the short-ear gene (*se*) can be traced back to defects in precartilaginous condensations, cartilage formation is still defective in young adult *se/se* mice, as shown by slow mending of fractured ribs (Green, 1958). A smaller callus forms, and a smaller quantity of cartilage is produced more slowly.

The disproportionate dwarfism associated with the gene for achondroplasia (*cn*) is a true chondrodystrophy, arising from a continuing abnormality in growth of cartilage (Dickie, 1965, personal communication). The shape of almost all bones is changed, and some bones are missing. Histological lesions in *cn/cn* cartilage present a "prima facie case of chondrodystrophy" (Grüneberg, personal communication).

The grey-lethal mouse is regarded by Grüneberg (1963) as "the prototype of systemic anomaly of the osseous skeleton." Although membranous and cartilaginous skeletons are completely normal, every element of the osseous skeleton is affected through failure of secondary bone resorption (Bateman, 1954). The ultimate cause of failure of resorption is not known. Grey-lethals have reduced numbers of osteoclasts (Barnicot, 1947). Transplantation experiments have shown *gl/gl* bone spicules capable of being absorbed in a normal host (Barnicot, 1948). This finding suggests deficiency of parathyroid hormone in *gl/gl* mice, but several experiments (Barnicot, 1948) have demonstrated parathormone activity in *gl/gl* parathyroids (Chap. 20). Grüneberg (1963) suggested that parathormone may be inactivated especially rapidly in *gl/gl* mice. Grey-lethals invariably die before 1 month of age, possibly because of malnutrition due to failure of tooth eruption (Grüneberg, 1935).

A similar but slightly less extreme failure of secondary bone resorption is seen in microphthalmic mice (*mi/mi*), which also usually die before weaning age. Reported investigations of this skeletal defect have been limited to analysis of reshaping of the bone (Bateman, 1954).

Skeletal abnormalities can be caused by interactions between mother and fetus. Females homozygous for the recessive gene hair-loss (*hl/hl*), when mated to normal (*hl/+* or *+/+*) males, produce stunted or short-lived *hl/+* offspring with bone fractures; their hairless offspring (*hl/hl*) grow normally (Hollander, 1960). Some demand of *hl/+* fetuses must be greater than that of *hl/hl* offspring. The mutant (*hl/hl*) mother's metabolism cannot meet this demand.

Osteoarthropathy

Osteoarthropathy of the knee joints occurs in old mice of the STR/1N inbred strain. It may result from an increased growth and vulnerability of cartilage, compounded by mechanical trauma, observed in young STR/1N mice (Silberberg and Silberberg, 1962). It is genetically controlled, apparently polygenic, and does not occur in progeny of outcrosses to other inbred strains. Genetic studies clearly showed independent inheritance of obesity, which occurs with high frequency in STR/1N, and the susceptibility to degenerative joint disease. Matings between STR/1N and lean AL/N produced F_1 hybrids almost as heavy as the STR/1N, with very low incidence of osteoarthropathy (Sokoloff et al., 1962). Although within the STR/1N strain there was an apparent correlation between plasma cholesterol levels, levels of other lipids, and the predisposition for osteoarthropathy, association between these characteristics disappeared with hybridization.

Experimental induction of obesity by feeding high-fat diets may lead to increased osteoarthritis (Sokoloff et al., 1960). These changes are secondary to obesity, the increased load borne by the joints being responsible for the development of the disease.

Degenerative joint disease has been recognized in mice of several strains. The knee joint is most commonly involved, demonstrating importance of mechanical influences in pathogenesis. However, different portions of the knee are affected in different strains, and related strains show lesions of similar structures. For example, in related C57L/HeN and C57BL/6JN the lateral condyle is involved, whereas in BRSUNT/N, BALB/cAnN and BL/HeN erosion of the patella and intercondylar fossa appeared most commonly. These findings indicate differences in structure of the knee joint among strains. Degenerative joint disease in older age groups is associated with progressive diminution in the number of chondrocytes. Since changes in articular cartilage may or may not be associated with osteoarthritis, the two conditions are not related. Although chondrodystrophy is a generalized degenerative age-induced lesion, osteoarthritis is a localized traumatic or destructive lesion (Sokoloff, 1956).

Osteoporosis

Osteoporosis occurs in mature obese (*ob/ob*) mice (Sokoloff et al., 1960; Sokoloff, 1960). Apparently it follows hypercorticism (Carstensen et al., 1961); similarly, the hyperglycemia (obese-hyperglycemic syndrome) is caused by increased production of cortico (gluco-) steroids.

Senile mice of strains C57BL/6, DBA/2, and A/J, 2 years of age or older, suffer from both osteoporosis and osteoarthrosis (Silberberg and Silberberg, 1962). In general the severity varies among strains, among animals within a strain, and between the sexes. Osteoporosis occurs much more frequently in mice of the DBA strain than in either C57BL/6J or A/J. Osteoarthrosis is more severe and more common in males than in females, whereas the reverse holds for osteoporosis. These sex differences may be related to the action of sex hormones. Indeed, an arthrosis-promoting effect of male sex hormones has been demonstrated in growing male and female C57BL/6 mice. Administration of estrogens during early adulthood, but not later in life, inhibited development of osteoarthrosis. Estrogen may tend to counteract endogenous testosterone and may also have a direct effect on the articular cartilage (Silberberg and Silberberg, 1963*a*, *b*).

Idiopathic necrosis

During a survey of spontaneous skeletal disorders in mice, aseptic necrosis of bone was encountered with some frequency. Although mice of a number of strains had lesions, the incidence was especially high (22 per cent) in females of the BL/HeN strains. Certain strains, e.g., AL/N, C57BL/6J, etc., were free of bone lesions. Pathologically, aseptic necrosis resembles such disorders of man as osteochondritis dessicans, aseptic necrosis of epiphyses, and primary types of bone infarcts. The lesions, affecting a number of skeletal structures, often resulted in fracture of the femoral neck, epiphyseal collapse, thoracolumbar kyphosis, or localized osteosclerosis. Because of these complications the animals were severely crippled. On occasion, however, healing took place (Sokoloff and Habermann, 1958).

The cause is unknown but the condition may be due to arteritis. Although by casual sectioning it was not possible to demonstrate involvement of femoral nutrient vessels, various branches of the iliac artery were frequently affected. In a few serial sections, segments of minute geniculate arteries were found to be completely occluded by fibrous tissue. It may be, therefore, that arterial occlusion resulting from necrotizing arteritis is the cause of the epiphyseal infarction (Sokoloff, 1959).

DISEASES OF THE UROGENITAL SYSTEM

Abnormalities and malfunctions of the urogenital system occur in mice either as results of the action of unit genes or as polygenically inherited characteristics of particular inbred strains.

Kidney abnormalities induced by single genes

At least six mutant genes (Table 29-1) in the mouse are known to produce defects in kidney shape or size (Chap. 8). One example is Short-Danforth (*Sd*); homozygotes (*Sd/Sd*) and some heterozygotes (*Sd/+*) die shortly after birth because they have no functioning kidney. This defect has been traced to reduced growth of the ureteric buds and failure of induction of the metanephros (Gluecksohn-Schoenheimer, 1945). Although normal tubules were formed whenever a renal pelvis was induced, recent tissue culture experiments suggest defects in both ureteric and metanephric tissue (Gluecksohn-Waelsch and Rota, 1963).

Defective kidney function has been demonstrated in mice with urogenital syndrome (*ur/ur*). Gross abnormalities are seen in this genotype on some but not all genetic backgrounds (Gluecksohn-Waelsch and Kamell, 1955). Even when kidney size is normal almost all *ur/ur* individuals die within 24 hours after birth. Their kidneys show much lower alkaline phosphatase activity than those of normal littermates. The plasma protein content is lower than that of normal littermates after but not before birth, indicating abnormal water retention. The few surviving mice develop hydronephrosis and polycystic degeneration.

Mice homozygous for kidney disease (*kd*) develop nephrosis at the age of 2 to 3 months (Hulse et al., 1965).

Mice with brain hernia (*bh/bh*) also show abnormal kidney histology and function. Kidneys appear histologically normal at birth, but later develop a polycystic condition (Bennett, 1959). Chromatographic studies of urine from *bh/bh* mice, 0 days to 3 weeks of age, showed pronounced proteinuria and generalized aminoaciduria. No elevation of amino acid content was observed in blood serum, suggesting that the defect is renal in origin. Histological abnormality appears after rather than before abnormal amino acid excretion. Aminoaciduria disappears in polycystic *bh/bh* adults, whose urine is unusually dilute.

Urinary calculi

Mice of the 129 strain have a high incidence of urinary calculi. They consist of glycolipoprotein that is different from the organic matrix of crystallized human or bovine calculi. The protein is similar in a general way to the Bence-Jones protein in content of phospholipid, cholesterol, and carbohydrate as well as in high serine level. The 129 strain of mice may be useful for the study of calculus formation (McGaughey, 1961). The serine content of the total amino acid residue on a weight basis is approximately 15 per cent; the glutamic acid content is also high (about 20 per cent) compared to that of many proteins, and aspartic acid content is low (approximately 5 per cent).

Obstructive uropathy has been a fairly constant observation at necropsy in male STR/1N mice before 16 months of age. Three forms have been observed: cystolithiasis, hemorrhagic occlusion of the urethral sinus, or suppurative vesicourethritis. The lesions develop about hard plugs of altered seminal material impacted in the urethral bulb. The urinary sediment is of mixed type and apparently does not represent a single organic metabolite. Although microorganisms are consistently found in association with the cystolithiasis, it is uncertain whether they are the cause or result of the urinary stasis. The lesions are genetically determined, do not occur in females, and their development in males is prevented by castration (Sokoloff and Barile, 1962).

Papillonephritis

Papillonephritis is the most common kidney disease of mice. It consists of multifocal necroses involving the most distal portions of the descending tubules. Nothing is known about its pathogenesis although it occurs in association with amyloidosis (West and Murphy, 1965). In strain A/Sn mice, amyloid deposits always precede papillonephritis. Papillonephritis represents a primary injury to the renal papilla (Dunn, 1944).

Glomerulosclerosis

Spontaneous glomerular hyalinization, or glomerulosclerosis, develops in RF mice about 8 to 20 months old (Gude and Upton, 1960). Its severity progresses with age. Histologically the lesions resemble those observed in mice following irradiation and those associated in man with hypertension, lupus erythematosus, and early diabetes. The glomerular changes are more diffuse than nodular in type and differ therefore from the typical Wilson-Kimmelstiel lesions of diabetes. Earliest lesions are thickening of the capillary wall and luminal dilation; these changes are followed by mesangial cell pycnosis, mesangial hyalinization, and eventually complete fibrosis of the capillary tufts. Originally, amyloidosis was implicated in the pathogenesis of the sclerotic lesions, but extensive histochemical studies suggested the presence of glycolipid, carbohydrate with a 1,2-glycol linkage, collagen, and fibrin. Acid mucopolysaccharides in both ground substance and basement membranes decrease as the disease progresses. Amyloid is entirely lacking (Gude and Upton, 1962).

Tendency to nephritis in old C57BL/6 mice

In studies of lifespan and incidence of particular types of pathoses in mice from many inbred strains (Chap. 26), nephritis was observed in 50 to 60 per cent of individuals surviving more than 450 days, in contrast to much lower percentages in most of the other inbred strains. This condition has been described by Dunn (1954): ". . . kidney damage secondary to amyloidosis apparently depends upon the site of deposition in the kidney. . . . In strain C57BL, the amyloid was deposited in the glomerulus, the tubules were remarkably well preserved, and the papilla was intact."

SUMMARY

This chapter serves as a guide to the recognized nonneoplastic constitutional diseases of the mouse. These are presented in a classified list arranged alphabetically according to the nature of the disease. The conditions described include those induced by action of specific mutant genes as well as those appearing with characteristic frequency in mice of particular inbred strains. We state the manner of inheritance of each condition when known and discuss certain of the diseases extensively. Cross-references guide the reader to further information about specific diseases.

LITERATURE CITED

Baker, N., W. H. Blahd, and P. Hart. 1958. Concentration of K and Na in skeletal muscle of mice with a hereditary myopathy (*Dystrophia muscularis*). Amer. J. Physiol. 193:530–533.

Banker, B., and D. Denny-Brown. 1959. A study of denervated muscle in normal and dystrophic mice. J. Neuropath. Exp. Neurol. 18:517–530.

Barnicot, N. A. 1947. The supravital staining of osteoclasts with neutral red: their distribution on the parietal bone of normal growing mice, a comparison with the mutants grey-lethal and hydrocephalus-3. Proc. Roy. Soc. B 134:467–485.

Barnicot, N. A. 1948. The local action of the parathyroid and other tissues on bone in intracerebral grafts. J. Anat. 82:233–248.

Bateman, N. 1954. Bone growth: a study of the grey-lethal and microphthalmic mutants in the mouse. J. Anat. 88:212–262.

Bennett, D. 1956. Developmental analysis of a mutation with pleiotropic effects in the mouse. J. Morphol. 98:199–229.

Bennett, D. 1959. Brain hernia, a new recessive mutation in the mouse. J. Hered. 50:264–268.

Bielschowsky, M., and F. Bielschowsky. 1964. Observations on NZB/BL mice; differential fertility in reciprocal crosses and the transmission of the auto-immune haemolytic anemia to NZB/BL × NZC/BL hybrids. Austral. J. Exp. Biol. Med. Sci. 42:561–568.

Bielschowsky, M., and G. C. Schofield. 1962. Studies on megacolon in piebald mice. Austral. J. Exp. Biol. Med. Sci. 40:395–404.

Brust, M. 1964. Effects of inhibitors on contractions of normal and dystrophic mouse muscles. Amer. J. Physiol. 206:1,036–1,042.

Burnet, F. M., and M. C. Holmes. 1964. Thymic changes in the mouse strain NZB in relation to the auto-immune state. J. Pathol. Bacteriol. 88: 229–241.

Carstensen, H., B. Hellman, and S. Larsson. 1961. Biosynthesis of steroids in the adrenals of normal and obese-hyperglycemic mice. Acta Soc. Med. Upsal. 66:139–151.

Christensen, H. E., and R. Rask-Nielsen. 1962. Comparative morphologic, histochemical, and serologic studies on the pathogenesis of casein-induced and reticulosarcoma-induced amyloidosis in mice. J. Nat. Cancer Inst. 28:1–33.

Clements, G. R. 1956. Spontaneous heart disease in DBA/2Jax mice. Fed. Proc. 15:511. (Abstr.)

Coleman, D. L. 1965*a*. Studies on the acetoacetate-like compound found in dystrophic mouse muscle homogenates. Arch. Biochem. Biophys. 111:489–493.

Coleman, D. L. 1965*b*. Accumulation of triose phosphates in dystrophic mouse muscle homogenates. Arch. Biochem. Biophys. 111: 494–498.

Coleman, D. L., and M. E. Ashworth. 1960. Influence of diet on transamidinase activity in dystrophic mice. Amer. J. Physiol. 199:927–930.

Coleman, D. L., and W. T. West. 1961. Effects of nutrition on growth, lifespan and histopathology of mice with hereditary muscular dystrophy. J. Nutr. 73:273–281.

Conrad, J. T., and G. H. Glaser. 1962. Neuromuscular fatigue in dystrophic muscle. Nature 196: 997–998.

Conrad, J. T., and G. H. Glaser, 1964. Spontaneous activity at myoneural junction in dystrophic mice. Arch. Neurol. 11:310–316.

Deol, M. S. 1961. Genetical studies on the skeleton of the mouse. XXVIII. Tail-short. Proc. Roy. Soc. B 155:78–95.

Deringer, M. K. 1959. Necrotizing arteritis in strain BL/De mice. Lab. Invest. 8:1,461–1,465.

Derrick, E. H., and B. M. St. George-Grambauer. 1957. Megacolon in mice. J. Pathol. Bacteriol. 73:569–571.

DiPaolo, J. A. 1962. Effect of tobacco diets on rodents. Nature 195:1,316.

DiPaolo, J. A., L. C. Strong, and G. E. Moore. 1964. Calcareous pericarditis in mice of several genetically related strains. Proc. Soc. Exp. Biol. Med. 115:496–497.

Dowben, R. M., L. Zuckerman, P. Gordon, and P. Sniderman. 1964. Effects of steroids on the course of hereditary muscular dystrophy in mice. Amer. J. Physiol. 206:1,049–1,056.

Dunn, T. B. 1944. Relationship of amyloid infiltration and renal disease in mice. J. Nat Cancer Inst. 5:17–28.

Dunn, T. B. 1954. The importance of differences in morphology in inbred strains. J. Nat. Cancer Inst. 15:573–585.

Falconer, D. S., M. Latyszewski, and J. H. Isaacson. 1964. Diabetes insipidus associated with oligosyndactyly in the mouse. Genet. Res. 5:473–488.

Fennell, R. A., and W. T. West. 1963. Oxidative and hydrolytic enzymes of homozygous dystrophic and heterozygous muscle of the house mouse. J. Histochem. Cytochem. 11:374–382.

Gluecksohn-Schoenheimer, S. 1945. The embryonic development of mutants of the *Sd*-strain in mice. Genetics 30:29–38.

Gluecksohn-Waelsch, S. 1963. Lethal genes and analysis of differentiation. Science 142:1,269–1,276.

Gluecksohn-Waelsch, S., and S. A. Kamell. 1955. Physiological investigations of a mutation in mice with pleiotropic effects. Physiol. Zool. 28: 68–73.

Gluecksohn-Waelsch, S., and T. R. Rota. 1963. Development in organ tissue culture of kidney rudiments from mutant mouse embryos. Develop. Biol. 7:432–444.

Golarz, M. N., and G. H. Bourne. 1960. Some histochemical observations on the muscles of mice with hereditary muscular dystrophy. Acta Anat. 43:193–203.

Gould, A., and D. L. Coleman. 1961. Accumulation of acetoacetate in muscle homogenates from dystrophic mice. Biochim. Biophys. Acta 47:422–423.

Gould, A., and D. L. Coleman. 1962. Acetoacetate metabolism in muscle homogenates from normal and dystrophic mice. Arch. Biochem. Biophys. 96:408–411.

Green, M. C. 1958. Effects of the short ear gene in the mouse on cartilage formation in healing bone fractures. J. Exp. Zool. 137:75–88.

Grüneberg, H. 1935. A new sublethal colour mutation in the house mouse. Proc. Roy. Soc. B 118:321–342.

Grüneberg, H. 1963. The pathology of development; a study of inherited skeletal disorders in animals. Wiley, New York. 309 p.

Gude, W. D., and A. C. Upton. 1960. Spontaneous glomerulosclerosis in aging RF mice. J. Gerontol. 15:373–376.

Gude, W. D., and A. C. Upton. 1962. A histologic study of spontaneous glomerular lesions in aging RF mice. Amer. J. Pathol. 40:699–709.

Hare, W. V., and H. L. Stewart. 1956. Chronic gastritis of the glandular stomach, adenomatous polyps of the duodenum and calcareous pericarditis in strain DBA mice. J. Nat. Cancer Inst. 16:889–911.

Harman, P. J., J. P. Tassoni, R. L. Curtis, and M. B. Hollinshead. 1963. Muscular dystrophy in the mouse, p. 407–456. *In* G. H. Bourne and M. N. Golarz [ed.] Muscular dystrophy in man and animals. Hafner, New York.

Heefner, W. A., and G. D. Sorenson. 1962. Electron microscopic observations on experimental amyloidosis in the spleen and lymph nodes of the mouse. Fed. Proc. 21:20. (Abstr.)

Helyer, B. J., and J. B. Howie. 1963*a*. Spontaneous auto-immune disease in NZB/Bl mice. Brit. J. Haematol. 9:119–131.

Helyer, B. J., and J. B. Howie. 1963*b*. Renal disease associated with positive lupus erythematosus tests in a crossbred strain of mice. Nature 197:197.

Heston, W. E., and M. K. Deringer. 1948. Hereditary renal disease and amyloidosis in mice. Arch. Pathol. 46:49–58.

Hollander, W. F. 1960. Genetics in relation to reproductive physiology in mammals. J. Cell. Comp. Physiol. 56 (Suppl. 1):61–72.

Holmes, M. C., and F. M. Burnet. 1963. The natural history of autoimmune disease in NZB mice; a comparison with the pattern of human autoimmune manifestations. Ann. Intern. Med. 59: 265–276.

Holmes, M. C., and F. M. Burnet. 1964. The inheritance of autoimmune disease in mice: a study of hybrids of the strains NZB and C3H. Heredity 19:419–434.

Hulse, E. V., M. F. Lyon, C. E. Rowe, and R. Meredith. 1965. Mouse News Letter 32:38.

Hummel, K. P. 1960. Pituitary lesions in mice of the Marsh strains. Anat. Rec. 137:366 (Abstr.)

Hummel, K. P., and D. C. Chapman. 1962. Heart lesions, p. 31. *In* Annual report for 1961–62, Roscoe B. Jackson Memorial Laboratory, Bar Harbor, Maine.

Jasmin, G., and E. Bajusz. 1962. Myocardial lesions in strain 129 dystrophic mice. Nature 193:181–182.

Joe, M., J. M. Teasdale, and J. R. Miller. 1962. A new mutation (*sph*) causing neonatal jaundice in the house mouse. Can. J. Genet. Cytol. 4: 219–225.

Kamenoff, R. J. 1935. Effects of the flexed-tailed gene on the development of the house mouse. J. Morphol. 58:117–155.

Kitiyakara, A. 1961. Cytologic study of dystrophia muscularis mouse muscles. Arch. Pathol. 71: 579–593.

Kunze, H. 1954. Die Erythropoese bei einer erblichen Anämie röntgenmutierter Mäuse. Folia Haematol. 72:391–436.

Laird, J. L., and B. E. Walker. 1964. Muscle regeneration in normal and dystrophic mice. Arch. Pathol. 77:64–72.

Loosli, R. 1965. Hereditary and nutritional muscular dystrophy. J. Hered. 56:75–81.

Loosli, R., E. S. Russell, W. K. Silvers, and J. L. Southard. 1961. Variability of incidence and clinical manifestation of mouse hereditary muscular dystrophy on heterogeneous genetic backgrounds. Genetics 46:347–355.

Markert, C. L., and W. K. Silvers. 1956. The effects of genotype and cell environment on the melanoblast differentiation in the house mouse. Genetics 41:429–450.

McCaman, M. W. 1963. Enzyme studies of skeletal muscle in mice with hereditary muscular dystrophy. Amer. J. Physiol. 205:897–901.

McComas, A. J., and S. J. Mossawy. 1965. Electrophysiological investigation of normal and dystrophic muscles in mice. *In* Third symposium on research in muscular dystrophy. Pitman's Medical Publishing Co., Ltd., London.

McGaughey, C. 1961. Excretion of uncrystallized urinary calculi composed of glycolipo-protein by normal and muscular dystrophic mice. Nature 192:1,267–1,269.

Meier, H., W. T. West, and W. G. Hoag. 1965. Preclinical histopathology of mouse muscular dystrophy. Arch. Pathol. 80:165–170.

Menner, K. 1957. Die postnatale Gonadenentwicklung bei Mäusen, die an einer angeborenen Anemie leiden. Wiss. Z. Martin-Luther-Univ. 6: 335–344.

Michelson, A. M., E. S. Russell, and P. J. Harman. 1955. Dystrophia muscularis: a hereditary primary myopathy in the house mouse. Proc. Nat. Acad. Sci. 41:1,079–1,084.

Mintz, B. 1957. Embryological development of primordial germ cells in the mouse: influence of a new mutation, W^j. J. Embryol. Exp. Morphol. 5:396–403.

Mintz, B., and E. S. Russell. 1957. Gene-induced embryological modifications of primordial germ cells in the mouse. J. Exp. Zool. 134:207–230.

O'Steen, W. K. 1962. Growth activity of normal and dystrophic muscle implants in normal and dystrophic hosts. Lab. Invest. 11:412–419.

Pai, A. C. 1965*a*. Developmental genetics of a lethal mutation, muscular dysgenesis (*mdg*), in the mouse. I. Genetic analysis and gross morphology. Develop. Biol. 11:82–92.

Pai, A. C. 1965*b*. Developmental genetics of a lethal mutation, muscular dysgenesis (*mdg*), in the mouse. II. Developmental analysis. Develop. Biol. 11:93–109.

Pappenheimer, A. M., and J. B. Daniels. 1953. Myocarditis and pulmonary arteritis in mice associated with the presence of rickettsia-like bodies in polymorphonuclear leucocytes. J. Exp. Med. 98:667–678.

Pearce, G. W., and J. N. Walton. 1963. A histological study of muscle from the Bar Harbor strain of dystrophic mice. J. Pathol. Bacteriol. 86:25–33.

Platzer, A. C., and W. H. Chase. 1964. Histologic alterations in preclinical mouse muscular dystrophy. Amer. J. Pathol. 44:931–946.

Pope, R. S., and E. D. Murphy. 1960. Survival of strain 129 dystrophic mice in parabiosis. Amer. J. Physiol. 199:1,097–1,100.

Pope, R. S., E. D. Murphy, and W. T. West. 1964. Decreased longevity of strain 129 dystrophic mice parabiosed at weaning. Amer. J. Physiol. 207:449–451.

Ross, M. H., G. D. Pappas, and P. J. Harman. 1960. Alterations in muscle fine structure in hereditary muscular dystrophy of mice. Lab. Invest. 9:388–403.

Russell, E. S. 1963. Genetic studies of muscular dystrophy in the house mouse. Proc. 2nd Int. Conf. Human Genet., Rome, 1961, p. 1,602–1,611.

Russell, E. S., and E. M. Fondal. 1951. Quantitative analysis of the normal and four alternative degrees of an inherited macrocytic anemia in the house mouse. Blood 6:892–905.

Russell, E. S., and F. A. Lawson. 1959. Selection and inbreeding for longevity of a lethal type. J. Hered. 50:19–25.

Russell, E. S., and E. C. McFarland. 1965. Erythrocyte populations in fetal mice with and without two different hereditary anemias. Fed. Proc. 24:240. (Abstr.)

Russell, E. S., W. K. Silvers, R. Loosli, H. G. Wolfe, and J. L. Southard. 1962. New genetically homogeneous background for dystrophic mice and their normal counterparts. Science 135:1,061–1,062.

Sandow, A., and M. Brust. 1962. Effects of activity on contractions of normal and dystrophic mouse muscles. Amer. J. Physiol. 202:815–820.

Schaible, R., and J. W. Gowen. 1961. A new dwarf mouse. Genetics 46:896. (Abstr.)

Silberberg, M., and R. Silberberg, 1962. Osteoarthrosis and osteoporosis in senile mice. Gerontologia 6:91–101.

Silberberg, M., and R. Silberberg. 1963*a*. Modifying action of estrogen on the evolution of osteoarthrosis in mice of different ages. Endocrinology 72:449–451.

Silberberg, M. and R. Silberberg. 1963*b*. Role of sex hormone in the pathogenesis of osteoarthrosis of mice. Lab. Invest. 12:285–289.

Silberberg, M., and R. Silberberg. 1964. Dyschondrogenesis and osteoarthrosis in mice. Arch. Pathol. 77:519–528.

Snell, G. D. 1929. Dwarf, a new Mendelian recessive character of the house mouse. Proc. Nat. Acad. Sci. 15:733–734.

Sokoloff, L. 1956. Natural history of degenerative joint disease in small laboratory animals. 1. Pathologic anatomy of degenerative joint disease in mice. Arch. Pathol. 62:118–128.

Sokoloff, L. 1959. Discussion of paper by Deringer. Lab. Invest. 8:1,465–1,466.

Sokoloff, L. 1960. Comparative pathology of arthritis. Adv. Vet. Sci. 6:193–250.

Sokoloff, L., and M. F. Barile. 1962. Obstructive genitourinary disease in male STR/1N mice. Amer. J. Pathol. 41:233–246.

Sokoloff, L., L. B. Crittenden, R. S. Yamamoto, and G. E. Jay, Jr. 1962. The genetics of degenerative joint disease in mice. Arthritis Rheum. 5:531–546.

Sokoloff, L., and R. T. Habermann. 1958. Idiopathic necrosis of bone in small animals. Arch. Pathol. 65:323–330.

Sokoloff, L., O. Mickelsen, E. Silverstein, G. E. Jay, Jr., and R. S. Yamamoto. 1960. Experimental obesity and osteoarthritis. Amer. J. Physiol. 198:765–770.

Southard, J. L., H. G. Wolfe, and E. S. Russell. 1965. Artificial insemination of dystrophic mice with mixtures of spermatozoa. Nature 208: 1,126–1,127.

Staats, J. 1965. *Dystrophia muscularis* in the house mouse: a bibliography. Z. Versuchstierk. 6:56–68.

Stevens, L. C., and J. A. Mackensen. 1958. The inheritance and expression of a mutation in the mouse affecting blood formation, the axial skeleton, and body size. J. Hered. 49:153–160.

Stevens, L. C., J. A. Mackensen, and S. E. Bernstein. 1959. A mutation causing neonatal jaundice in the house mouse. J. Hered. 50:35–39.

Stevens, L. C., E. S. Russell, and J. L. Southard. 1957. Evidence on inheritance of muscular dystrophy in an inbred strain of mice using ovarian transplantation. Proc. Soc. Exp. Biol. Med. 95: 161–164.

Tubis, M., N. Baker, and W. H. Blahd. 1959. Vitamin therapy in mice with an hereditary myopathy (Dystrophia muscularis). J. Nutr. 68: 595–601.

West, W. T., and E. D. Murphy. 1960. Histopathology of hereditary, progressive muscular dystrophy in inbred strain 129 mice. Anat. Rec. 137:279–295.

West, W. T., and E. D. Murphy. 1965. Sequence of deposition of amyloid in strain A mice and relationship to renal disease. J. Nat. Cancer Inst. 35:167–174.

Wolfe, H. G., and J. L. Southard. 1962. Production of all-dystrophic litters of mice by artificial insemination. Proc. Soc. Exp. Biol. Med. 109: 630–633.

Young, H. L., W. Young, and I. S. Edelman. 1959. Electrolyte and lipid composition of skeletal and cardiac muscle in mice with hereditary muscular dystrophy. Amer. J. Physiol. 197:487–490.

Zierler, K. L. 1961. Potassium flux and further observations on aldolase flux in dystrophic mouse muscle. Bull. Johns Hopkins Hosp. 108:208–215.

30

Infectious Diseases[1]

Warren G. Hoag and Hans Meier

In this chapter we will briefly describe pertinent features of some infectious diseases of inbred mice. We will not attempt to cover all aspects of the diagnosis, treatment, and prevention of mouse diseases.

VIRAL DISEASES

A number of viruses are indigenous to both wild and laboratory mice. Almost all produce no evidence of illness in nature and their presence is usually revealed only by a deliberate search. In some apparently healthy laboratory mice experimental stress may activate viruses and thus cause obvious clinical disease. A few viruses such as mouse pox (infectious ectromelia) cause spontaneous disease.

Of the viruses described in the following section some are sufficiently antigenic so that their presence or absence may be determined directly by serological testing or indirectly by measuring the antibody response in suckling mice after inoculation of suspect biological materials. These viruses include ectromelia, polyoma, K virus, mouse adenovirus, Theiler's GD VII, reoviruses, Manaker-Nelson hepatitis, and pneumonia viruses. For other poorly antigenic viruses, such as lymphocytic choriomeningitis (LCM), mouse salivary gland virus, thymic agent, and perhaps the Gledhill-type hepatitis virus, isolation procedures are required. Serial passages of biological material often bring to light viruses carried as latent infections. Some of the viruses can exist in a mouse colony whose members appear clinically healthy, but such mice may readily initiate a serious epidemic when brought into contact with healthy susceptible stocks.

Control measures for these viral diseases include prevention of virus introduction, destruction of all exposed and infected animals, disinfection, and reintroduction of healthy stocks. Another important measure is the establishment of specific-pathogen–free or germ-free colonies; the freedom from virus infection in such colonies has been well demonstrated (Rowe et al., 1962; Gledhill, 1962).

Mouse pox (infectious ectromelia)

Mouse pox is a highly infectious disease to which all strains of mice are susceptible, some more than others. Ectromelia is an excellent example of a virus carried and spread by apparently healthy mice. There may be no signs of illness, yet active virus may be readily recovered from viscera by blind passage. The virus obviously not only persists and multiplies despite a certain suppression by antibody, but also is shed through the feces. The virus was first described in England, but has since been reported from continental Europe as well as the United States (Saunders, 1958*a;* Trentin, 1953). It is from 100 to 150 μ in diameter, is destroyed by exposure to 55°C for 30 minutes, but withstands 50 per cent glycerol and 0.5 per cent phenol for several months at refrigerator temperature. Clinical aspects of the disease have been described in detail by Fenner (1949) and Trentin (1953). The disease is generalized, the virus multiplying in cells of the

[1] The writing of this chapter was supported in part by Public Health Service Research Grant CA 04691 from the National Cancer Institute.

skin, lymph nodes, liver, spleen, and other organs. In all these tissues intracellular eosinophilic inclusion bodies develop. The acute form, occurring in previously unexposed mice, is characterized by visceral lesions, usually hepatic necroses, the animals dying within days without external signs of illness. Sometimes inconspicuous external primary lesions are present such as swollen eyelids or pocked noses. The disease usually spreads rapidly throughout a susceptible colony, killing 50 to 95 per cent of the mice within weeks. The surviving mice develop circulating antibodies and often show a chronic form of the disease characterized by necrosis of the extremities, gangrene, crusting and scarring of skin, and regenerative lesions of internal organs. The foot lesions must be distinguished from those caused by *Streptobacillus moniliformis.* The most reliable diagnostic test for ectromelia infection is hemagglutination inhibition (HI). Either vaccinia or ectromelia virus is used as a red blood cell–agglutinating antigen. Serum from mice recovered from acute mouse pox will inhibit such agglutination of blood cells (Burnet and Boake, 1946; Briody, 1959). The test can be quickly and conveniently carried out with as little as 0.05 ml of serum. The virus is demonstrated by intraperitoneal inoculation of visceral suspensions into susceptible mice. These animals ordinarily die 4 to 6 days after inoculation and show the visible lesions of multiple focal necroses of the livers.

Mouse colonies can be protected from ectromelia by strict isolation to preclude contamination by infected mice and other materials. Once the disease is recognized, however, infected colonies should be destroyed and all presumably infected animals incinerated. Another preventive measure, applicable to valuable inbred stocks, is vaccination, either by scarification or intranasal instillation with vaccinia virus or with formolized liver suspensions from infected mice (Shope, 1954). Although both Levaditi and IHD-T strains of vaccinia virus have been successfully employed, IHD-T is preferred because it is more immunogenic, does not induce HI antibodies, and will not interfere with the HI test for diagnosis of infection. It is therefore possible to distinguish between antibodies formed as a result of infection and those formed as a result of vaccination. Vaccinated mice may transfer the infection to unvaccinated cage mates but not to mice in other cages of the same room. DBA mice seem to be more resistant to contact infection than other stocks (Briody, 1959).

Hepatoencephalitis group of viruses

Some viruses are capable of causing hepatitis, encephalitis, or both in suckling mice. These are designated as MHV, JHM, EHF 120, and H747 and are related by degrees of cross-immunity (complement-fixation, neutralization tests). All produce qualitatively similar histological lesions affecting mesothelial and endothelial surfaces. These viruses, particularly the neurotropic ones, have similar susceptible host species ranges and occur in latent form in mouse stocks. All are heat-labile and ether-sensitive and range in size from 80 to 120μ. The MHV or mouse hepatitis virus gives rise to intranuclear inclusion bodies, but to produce active disease requires the synergistic action of *Eperythrozoön coccoides,* a red blood cell parasite. The link between the two agents is complex. References may be found in Gledhill and Niven (1955) and Gledhill (1962).

Neurotropic viruses causing spontaneous encephalitis

There are three groups of viruses causing spontaneous encephalitis with distinctions based on the type of pathological lesion. Theiler's FV, FA, FO, and GD-VII affect the gray matter primarily, producing lesions comparable to those found in poliomyelitis (Theiler and Gard, 1940; Thompson et al., 1951). The second group includes the JHM virus which causes demyelinization (Olitsky and Lee, 1955; Pappenheimer, 1958*a*). The third group, which does not cause demyelinization, consists of SK, the Columbia strain of Jungeblut, LCM or lymphocytic choriomeningitis (Traub, 1939; Haas, 1954), EK virus, herpes simplex, and eastern equine encephalitis.

Theiler's mouse encephalomyelitis. Spontaneous infections of the central nervous system due to this virus are considerably more prevalent in younger mice. Unless the disease is rapidly fatal, gradual flaccid paralysis, usually of the hind limbs, develops. Microscopically, anterior-horn lesions quite similar to those of human poliomyelitis (ganglionic cell destruction) are observed. However, there is no serological relationship between Theiler's virus and that of poliomyelitis. Theiler's virus is recoverable from the intestinal tract oı central nervous system. Artificial infections are possible by virtually all routes, although the level of infective dose differs considerably. For a more detailed review see Maurer (1958*a*).

Lymphocytic choriomeningitis. Although mice of laboratory colonies may harbor this virus, the disease is carried in latent form with few animals showing clinical signs or dying. Clinical illness occurs only in young animals which may show photophobia, conjunctivitis, and convulsive leg movements. Tremors, spastic convulsions, and paralysis are observed in artificial infections. Microscopically, leukocytic infiltration of the meninges, pleural exudation, and splenomegaly are found. Virus is detected by intracerebral inoculation of mice with sterile suspensions of brain, blood, or spleen from suspect animals. In view of the wide host range (mice, monkeys, dogs, guinea pigs, and man) LCM represents a hazard to both experimental animals and man. Aside from direct contact transmission, as by the urine of infected mice, the virus may also be spread by ectoparasites and mosquitoes. For a review see Maurer (1958*b*).

Pneumotropic viruses

A number of viruses have been implicated as causes of respiratory disease in mice. Among these are Nigg's pneumonitis (Nigg and Eaton, 1944) and PVM or pneumonia virus of mice (Mirick et al., 1952; Volkert and Horsfall, 1947).

K virus

K virus infection is also a respiratory disease. In suckling mice the clinical signs are labored breathing and early death. Large basophilic inclusion bodies are observed in greatly hypertrophied alveolar lining cells. On occasion, focal disseminated fatty liver dystrophy is found. Since the first description of the disease and isolation of the virus, the K virus has been isolated in several parts of the world, notably Australia (Fisher and Kilham, 1953; Holt, 1959; Derrick and Pope, 1960). The virus presumably spreads by way of the urine and saliva, yet serological surveys of both laboratory and wild mouse colonies indicate that antibodies occur in only a small proportion of mice from infected colonies.

Livers of moribund suckling mice provide a source for a suitable complement-fixing antigen so that the virus may be detected by the presence of specific antibody, although intracerebral inoculation of newborn mice with suspect tissues is a more sensitive test. It has been suggested that the K virus may be generically related to polyoma in view of certain common biological properties such as size, stability, etc. (Kilham, cited by Rowe et al., 1962).

Salivary gland virus

This virus is ubiquitous in wild mouse populations, but has been observed in only a few laboratory colonies. Apparently the virus does not spread with ease (Rowe et al., 1962) since even in infected colonies its incidence is 3 per cent or less. Virus isolation techniques must be used for laboratory diagnosis. Serological techniques are of little value, and inclusion bodies are rarely observed despite continual excretion of virus in saliva (Rowe et al., 1962; Brodsky and Rowe, 1958). Two procedures (utilizing mouth swabs as a convenient source of virus) are generally employed: (1) isolation of the virus in mouse embryo tissue-culture, and (2) inoculation of newborn mice. The disease may be suspected in suckling mice by their malnourished appearance and at necropsy by the gross yellow discoloration of the edges of the liver.

Thymic virus

The thymic agent has been discovered in as high as one-half the mice of one laboratory colony and is highly prevalent in wild mice. It is pathogenic only for newborn mice and is recognizable by the production of gross thymic necroses, visible about 2 weeks after inoculation. Direct virus isolation is the most reliable diagnostic means, although infected mice produce neutralizing antibodies in low titer. The salivary glands represent the best source of virus since naturally or artificially infected mice excrete the virus in saliva for periods of more than one year. In contrast to the mouse salivary gland virus, the thymic agent does not propagate in tissue culture (Rowe and Capps, 1961).

Adenovirus

This virus regularly infects recently weaned mice after prolonged exposure to urine of carrier animals. Infection induces a good complement-fixing and neutralizing-antibody response. Spontaneous clinical disease rarely occurs. However, virus inoculation into suckling mice induces a fatal disease consisting in inflammation and necrosis of the heart, adrenals, and brown fat. Acidophilic intranuclear inclusion bodies occur in these and other tissues (Hartley and Rowe, 1960). The virus produces cytopathic effects in mouse kidney tissue cultures. Fluids from such infected cultures contain an excellent complement-fixing antigen which reacts strongly with adenoviral antisera prepared in guinea pigs (but not with human, monkey, or dog antisera). Apparently the virus does not cross the placental barrier, since it has not been encountered in caesarean-derived colonies (Rowe et al., 1962).

Reoviruses

The reovirus group is composed of three serotypes (Sabin, 1959). Antibodies to reoviruses are assayed by hemagglutination inhibition, neutralization, or complement-fixation tests. Reo 3 virus can be detected by tissue-culture cytopathogenicity, inoculation of suckling mice, or mouse antibody production tests. The sera of some mice contain hemagglutination inhibitor which is undoubtedly nonspecific since it is commonly encountered even in specific-pathogen-free and germ-free mouse colonies.

Only Reo 3 is indigenous to mice and has been isloated from tissue suspensions containing Moloney virus. The oncolytic virus of Nelson, obtained from a transplanted ascites tumor (Nelson and Tarnowski, 1960), has been identified as Reo 3. Reo 2 has been encountered as a focal infection in several wild mouse populations which may have acquired infection by contact with other species, such as man and cattle (Rowe et al., 1962).

Oncogenic and tumor-associated viruses

Gross (1961) has discussed the status of virus-induced neoplasms, including mouse leukemias as produced by cell-free leukemic filtrates, the induction of leukemia by filtrates of solid mouse tumors, and the filtrable agent causing mouse mammary tumors (Bittner's milk factor). Sinkovics (1962) has reviewed viral leukemias in mice and oncogenic viruses are discussed in Chap. 27 and 28. We refer here only to mouse polyoma infection and the lactic dehydrogenase-stimulating viruses (Gross, 1953; Stewart, 1953; Riley et al., 1960).

A detailed epidemiology of mouse polyoma infection has been developed after intensive studies. The classic techniques of virology for detection of antibodies reveal that mouse polyoma virus is widely disseminated among laboratory colonies and wild mouse populations. Focal reservoirs of infection are maintained by routes of transmission involving virus excretion in urine, feces, and saliva. Apparently the virus is not transferred transplacentally since mice derived by caesarean section for specific-pathogen-free or germ-free colonies have been negative to antibody tests. Polyoma virus (e.g., infected tissue culture fluid) induces multiple tumors when inoculated into newborn mice. Such tumors are of multicentric and mutiple histological origins, with mixed tumors of the salivary glands most frequent. The virus is capable of producing tumors in other species as well.

From the epidemiology of polyoma virus infection several points are notable: (1) Resistance of infant mice is perfectly correlated with presence of maternal antibody titers; no tumors develop in offspring of positive mothers; (2) spontaneous parotid tumors are rarely found in naturally infected mouse colonies; (3) high antibody titers are present in milk of infected mice; and (4) maternally derived antibodies are presumably present in fetuses, since it has been shown that mice born of resistant mothers but not having access to immune mouse milk are resistant to polyoma infection.

Certain transmissible agents have been found to be associated with many transplanted and induced spontaneous tumors of mice (Riley et al., 1960). These are manifested by biochemical response. Susceptible animals respond with five- to tenfold increases in serum lactic dehydrogenase (LDH) and by induction of other glycolytic enzymes after inoculation with plasma or organ extracts from tumor-bearing hosts. Natural transmission of the *lactic dehydrogenase-stimulating agent(s)* (LDH) occurs when normal mice are placed in the same cage with agent-infected mice or tumor-bearing mice. In most infected animals moderate splenomegaly occurs. Serum LDH elevation has been induced by inoculation with preparations of cells from infected mice (Riley et al., 1961). Although there is a close correlation between LDH levels and the growth of several transplanted tumors, elevation is not directly related to the neoplastic process. There is evidence to indicate that the LDH factor is a virus, specifically of mouse origin, and that changes in the LDH level of a tumor-bearing animal depend on whether or not the LDH factor has become associated with that tumor (Notkins et al., 1962). An increase in lactic dehydrogenase levels can also be induced by injection of several of the mouse hepatitis viruses. LDH virus may be eliminated as a contaminant of transplantable tumors by passage of such material through tissue culture. However, it has been reported that the agent can be propagated and maintained by serial passage on primary mouse embryo tissue-culture (Yaffee, 1962). It can be inactivated by radiation and is unstable in the presence of ether (Notkins et al., 1962).

Infantile diarrhea

Diarrheal disease of unweaned mice is a complex syndrome caused by a number of agents

(Pappenheimer, 1958*b;* Runner and Palm, 1953; Kraft, 1962). As is true for diarrheal disease in human infants, there is considerable evidence that no single pathogen is the causative agent (Thomlinson, 1964; Erlandson et al., 1964; Altman, 1964; Payne, 1960; Barua et al., 1962). Rather, under certain environmental conditions (crowding, poor sanitation, improper nutrition, variable temperatures, too low or too high humidity, inadequate ventilation, etc.) ubiquitous viral or bacterial organisms may be incriminated as producing diarrheal or septicemic signs in unweaned mice. We find that host factors, such as strain of mouse and parity play a determining role in morbidity and degree of clinical signs.

The clinical signs of the syndrome are: slight-to-severe diarrhea in which fecal material ranges from bright yellow to light brown in color. Nursing mothers will sometimes very efficiently clean up the anal region of the affected animal so that only a slight "pasting up" condition is noted. Animals affected by diarrhea (usually 4 to 10 days of age) usually recover, but many are discarded as runts at weaning time. The diarrhea is often followed by constipation contributing to later mortality. Mortality more often occurs in animals not showing signs of diarrhea. Such deaths occur early, 1 to 2 days after onset of the disease in a litter. Survivors in these litters show varying degrees of diarrhea.

At necropsy, the only common finding is the presence of light colored and watery fecal material in the intestinal tract, with occasional bubbles of gas. Histologically the tissues demonstrate a mild catarrhal inflammation. Some workers have reported inclusion bodies in epithelial cells from liver and intestinal tract.

Viral agents such as those described by Kraft (1962) are found in the intestinal contents and reportedly can be transmitted directly to susceptible animals through this medium. Various types of bacteria, coliforms, *Proteus* sp., and pseudomonads have been found in separate outbreaks of the disease.

Epidemiologically the disease usually occurs in cycles. In certain affected mouse colonies, diarrheal disease is rarely seen until the fall and winter months. Our studies suggest that lower humidity and less fresh outside air are involved in seasonal recurrence rather than length of day and seasonal variation in quality of diet.

Colonies in which diarrheal disease is enzootic may show longer periods between epizootic outbreaks when moved to new quarters. During an outbreak the disease can be controlled by the use of broad-spectrum antibiotics such as the oxytetracyclines or tetracyclines administered in the drinking water. Therapeutic doses are administered over a 2-week period to all animals in a breeding colony. The preferable medication period is during the last third of pregnancy. This treatment is repeated in 2 weeks. Such therapy results in marked increase in the numbers of healthy-looking mice weaned and a decrease in the incidence of diarrhea. Similar effects are produced by the use of air-filtering material placed about each breeding cage or by the use of aluminum foil covering 50 per cent or more of open cage surfaces. The effects shown by cage modifications (air filtering, foil covers) may be due to decreased aerosol dose levels. The use of certain types of highly absorbent bedding material also results in a higher recovery rate and a decrease in diarrheal disease. All of these control measures singly or in combination result initially in marked improvement of colony health, but usually over a period of months the disease reappears. Also, the effects of combining such control measures are not additive but only as good as the best measure used alone. Other effective control and prevention measures entail a variety of steps to isolate each breeding pen completely.

Much remains to be learned of the causes of infantile diarrheal disease, but the diarrhea should be considered as an accompaniment of certain diseases of suckling mice and not as a disease entity in itself.

BACTERIAL DISEASES

The spontaneously occurring or latently present bacterial diseases of mice are numerous, and the mouse is also susceptible to experimental infection by many pathogens from other species. Because of the small size of the animal, its habits, and the complexity of husbandry methods applied, it is extremely difficult to evaluate the health of a single animal by the usual clinical methods. Instead, an epidemiological approach is necessary. The population or a percentage thereof is observed for common factors which can be related to the various signs of morbidity and to mortality rates (Shope, 1964).

Salmonellosis

Infection by *Salmonella* sp. is one of the more common types of bacteriological infections of mice (Habermann and Williams, 1958*a;* Lane-

Petter, 1963). The most commonly found strain is *Salmonella typhimurium*. However, mice are highly susceptible to infection by most of the *Salmonella* sp. organisms and other strains have been reported as responsible for epizootics or as latently infecting a few mice in a colony (Wetmore and Hoag, 1960; Hoag et al., 1964*a*). The degree of virulence is dependent largely upon the dose and route of infection and the strain of host mouse as well as upon the inherent virulence of the organism itself. Salmonellosis is characterized initially by diarrhea or soft stools, sudden deaths, anorexia, and cachexia. Large numbers of animals are often lost not by death, but by the culling of underweight and poor-looking mice. The initial outbreak of salmonellosis soon subsides into a chronic form of the disease wherein the organisms are shed in fecal material. The adult carriers often appear healthy, but suckling and growing mice show great variation in body weight and rates of gain. Occasional breeding pairs are affected by attacks of mild diarrhea evidenced by soft, light-colored fecal pellets. Cull rates are high chiefly because of variations in weight and the poor appearance of weanling mice. At necropsy only a few animals show the classic "white spotted livers." Many will show enlarged spleens, but no gross necropsy findings are pathognomonic. Spleen, liver, and both large and small intestine are the tissues of choice for bacteriological examination. Tissues should be separately cultured for the presence of the organisms. In moribund or sick-looking animals the liver and spleen will yield *Salmonella*, whereas the chronically infected cases will usually yield organisms from the intestinal tract only.

For the isolation of *Salmonella* sp. enrichment media, preferably Tetrathionate Broth (Difco Laboratories), should be inoculated with a quantity of minced or ground tissues and the media incubated at 37°C for 72 hours (Hoag and Rogers, 1961). During this period subcultures to Brilliant Green Agar (Difco) are incubated at 37°C for 48 hours before being discarded as negative for salmonella growths. Serological identification of suspect bacterial colonies is then carried out with polyvalent or group-specific antisera. Specific identification is dependent upon further serological techniques.

The disease is not controllable by any known therapeutic measures. Broad-spectrum antibiotics such as tetracycline or oxytetracycline seem to be of some use in epizootics in increasing the numbers of survivors, but they will not cure chronically infected animals. Vaccines of various types have been found to be highly effective against clinical signs and mortality from the disease but do not prevent chronic infection and carrier states from developing (Hoag et al., 1964*b*). We have demonstrated that three differently prepared *S. typhimurium* vaccines were effective against as many as 100 LD_{50} challenge doses of *S. typhimurium* but all of the surviving animals, although appearing perfectly healthy, were shedding organisms in fecal material and had infected livers and spleens as long as 6 weeks after challenge.

The disease is controlled by elimination of infected or carrier stocks. Several (at least two and preferably six) weekly fecal tests should be performed before concluding that an animal is free of infection, since intermittent and low-level shedders are commonly encountered in a chronically infected colony. Sanitation is an important part of preventing spread of the disease, since an important mode of transmission is by contact with contaminated objects. Food supplies and bedding material should be treated as potential sources of infection. Occasional paratyphoid carriers are found among mouse handlers, but are not important sources of animal infection.

Pasteurellosis

Pasteurellosis due to *Pasteurella pseudotuberculosis*, *P. septica*, or *P. pneumotropica* may occur spontaneously in laboratory mice (Sellers et al., 1961; Tuffery, 1958; Hoag et al., 1962). The latter two organisms may be found in various tissues of apparently normal mice. Latently infected animals subjected to various types of stress (radiation, abrupt temperature changes) will often produce signs of acute disease: anorexia, lassitude, sudden deaths. The livers of animals infected with *P. pseudotuberculosis* present multiple focal abscesses. *P. pneumotropica* has been reported in outbreaks of pneumonia in mice and has also been found in normal-appearing lung tissue (Jawetz, 1948; Jawetz and Baker, 1948). In other instances this organism has been isolated from the brain, uterus, testes, liver, or spleen of normal-appearing mice. In rare instances septicemic deaths have been attributed to *P. pneumotropica*. Pasteurellosis is probably a stress-induced disease and the causative organism most commonly a latent infective one.

The pasteurella organism is susceptible to the tetracycline or oxytetracycline types of broad-spectrum antibiotics but is usually resistant to penicillin. Outbreaks in which signs of disease are thought to be caused by pasteurella organisms (on the basis of isolations from organ tissue) may be treated by administering the effective antibiotics in the drinking water. Response to the drugs is immediate and favorable.

Pseudomonas infections

Infection with *Pseudomonas aeruginosa* is common in laboratory mouse colonies (Flynn, 1963*a*). No signs of infection are manifest unless animals are subjected to radiation or other types of stress (cortisone, etc.) (Flynn, 1963*a;* Wensinck et al., 1957; Verder and Rosenthal, 1961). Such treatment results in rapid onset of septicemic disease (anorexia, listlessness) and high mortality. At necropsy the organisms are easily cultured from all organs, particularly liver and spleen. Although not important as a spontaneous disease-causing organism, *P. aeruginosa* is of major concern to investigators using mice in experiments employing radiation. The organism is shed in feces and urine from chronically infected animals. Nearly 100 per cent of the animals in a colony may become rapidly infected by contamination and recontamination of drinking water, caging equipment, and feeds. The infection can be controlled by hyperchlorination and acidification of water used for drinking and cleaning (Hoag et al., 1965). Sanitation procedures must be of the highest efficiency and detectable carriers eliminated. Carrier detection should not serve as a criterion for culling until *after* control procedures such as water treatment and intensified sanitation have been inaugurated, since it has been shown that most fecal shedders of the organisms are only transiently infected and will seemingly spontaneously "recover" if not exposed to further doses of the organisms in contaminated water. After control procedures have been in effect for several weeks, a test-and-slaughter program should be started. Pens containing mice shedding the organism should be removed from the colony.

Antibiotics and other drugs are of no value in treating the infection. In one instance when oxytetracycline was being administered in drinking water to inbred mice, it was found that the organism grew more rapidly in water containing hypertherapeutic levels of the drug (Hoag et al., 1965).

Klebsiella infections

Latent infections with capsulated lactose-fermenting bacilli, often loosely identified as *Klebsiella* group organisms, are not uncommon (Wilson and Miles, 1964). Such organisms are occasionally isolated from lung tissue of normal-appearing mice. Sometimes the organism is found in the nasopharyngeal region or in various body wastes. In certain outbreaks of pneumonia in mice, *Klebsiella* sp. can be recovered from 100 per cent of the lungs of sick or moribund mice. The organisms recovered vary greatly in virulence. Rarely, the bacterium can be recovered from abscesses of lung or from other internal organs and cutaneous areas. Improved sanitation and environmental constancy are recommended control measures.

Tyzzer's disease

Disease caused by *Bacillus piliformis* has been described by several workers (Fujiwara et al., 1963; Fujiwara et al., 1964; Saunders, 1958*b*). Most agree that the organism occurs in an infected colony as an intestinal saprophyte and that clinical disease occurs only after animals have been stressed, as by adverse nutrition or by the effects of stressors introduced during the course of an experiment. Onset of disease signs is abrupt with death occurring 2 to 3 days afterward. Diarrhea may occur as the major sign or may affect only a few of the sick animals. Anorexia and other signs of septicemia are manifest. Some immunity develops in recovered animals.

The condition may occur in mice of all ages, but the highest mortality rate is observed in 3- to 7-week-old animals. The most striking necropsy finding is the occurrence of numerous grayish-white spots (up to 2.5 mm in diameter) on the outer and cut surfaces of the liver. Mesenteric lymph nodes are usually enlarged and sometimes abscessed. Microscopically the organisms (long, thin, Gram-negative rods) are found intercellularly, surrounding the necrotic tissue areas, but are also seen in intact liver cells as well. The organisms have been reported as cultivatable in tissue culture, but cannot be grown on defined bacteriological media. Diagnosis is by histological sectioning and staining of tissues with subsequent demonstration of the stained organisms.

Differences between mouse strains in susceptibility have been demonstrated, but it is emphasized that any incidence in a mouse colony represents a disease potential and infected colonies must therefore be considered as hazardous for other research. Little work has been done to demonstrate the effectiveness of therapy, although oxytetracycline has been of value (Tuffery, 1958).

PPLO (Pleuropneumonia-like organisms)

Infections with organisms described as PPLO have been chiefly reported as causing catarrhal disease in mice (Nelson, 1958). Some outbreaks of disease of the upper respiratory tract are characterized by sudden onset and involvement of a large perecntage of a colony usually after chilling or overheating. Chattering and sniffling with variable nasal discharge are the observable signs of infection. Mortality rates are low, and recovery from the exudative catarrh is spontaneous. A few animals succumb to bronchopneumonia or may develop a chronic bronchiectatic pneumonia. In other types of upper respiratory disease caused by PPLO the onset is gradual and the outbreak appears enzootic in character. Because of the chronic nature and long course of the disease, the eventual involvement of large numbers of animals may give the appearance of a suddenly occurring epizootic disease, particularly if the mild symptoms in the early stages of the condition are unnoted. Labyrinthitis (manifested by disequilibrium) may occur in a few of the mice.

PPLO organisms are often found in animals from an apparently healthy colony and may be recovered from various tissues other than lung material (Freundt, 1959). The pathogenic potential of latent infections is always manifest but has not been quantitatively evaluated.

Streptobacillus moniliformis

Infections with *Streptobacillus moniliformis* are usually latent. In certain outbreaks where clinical signs include lameness and swelling of joints and extremities (due to edema), the organism is readily isolated from blood and organs, including affected joints (Wilson and Miles, 1964; Freundt, 1956). In other instances the organism is not found in the bacterial cell state, but is isolated with extreme difficulty in L form by PPLO culture techniques (Freundt, 1959). Clinical signs are often confused with those seen in chronic ectromelia, and for this reason an early differential diagnosis is indicated (Lane-Petter, 1963). Stress plays an important role in the activation of latent infection. In many outbreaks mice will arrive in apparently good health after exposure to adverse shipping conditions, but within 3 to 4 days 10 to 50 per cent will develop signs of the disease—swellings and ulcerations of feet and tail after the appearance of a sharp band of demarcation proximal to the swelling. Recovery is usually spontaneous, although severely affected animals are deformed because of deranged circulation to extremities with subsequent sloughing of the entire area distal to the necrotic demarcation band. In our experience antibiotics such as tetracycline and oxytetracycline seem to be useful in controlling outbreaks. In clinically affected animals these drugs serve to mitigate the severity of developing lesions.

PARASITIC DISEASES

Helminth infections

Mice may become infected with helminths from other species (Habermann and Williams, 1958*b*). Outbreaks of infection with *Taenia taeniformis* can occur where mice are housed in the same area with cats. Infections with the various helminths rarely produce clinical signs and are only potentially important as producing unpredictable variables in animals used in research. Heavily infected animals are below norms in weight and may be anemic. Some mouse colonies may be infected with *Hymenolopis nana* (the dwarf tapeworm). The life cycle of this parasite does not involve any secondary hosts, with infection taking place directly from eggs excreted in feces. Upon necropsy, the tapeworm is easily observed through the walls of the unopened intestinal tract.

Mouse colonies are often infected with the oxyurids *Syphacia obvelata* and *Aspiculuris tetraptera*. These mouse pinworms have a direct life cycle and spread through a colony rapidly because of the large numbers of eggs excreted and the ease with which the eggs are airborne. There are no clinical signs of oxyuriasis. There may be some weight and growth variation between infected and noninfected animals, but other signs such as poor hair coat, etc., are not clear-cut. There is some evidence that oxyuriasis may contribute to rectal prolapse. Diagnosis can be made by examination of fecal material or the anal region for oxyurid eggs by the various flotation or contact-tape techniques. Pinworm infection may be eliminated by treatment with various drugs such as the piperazine compounds, usually most efficacious when administered via drinking water (Hoag, 1961). Treatment must be accompanied by thorough cleaning of the room and caging equipment to remove the possibility of reinfection from egg-contaminated dust.

Arthropod infestations

Lice (*Polyplax serrata*) and mites (*Myobia musculi, Myocoptes musculinus, Myocoptes romboutsi, Radfordia affinis,* and *Psoregates simplex*) are the chief ectoparasites of laboratory mice (Flynn, 1963*b*). Their presence is manifested by hair loss and scratching, which may give rise to bacterially infected ulcerative wounds. It is not unusual for mice to die in a heavily infested colony as a result of these infected sores.

Control of ectoparasites is difficult. Various dusting powders containing DDT, methoxychlor, rotenone, or other parasiticides, and dips containing aramite or other materials have been used, but must be regularly or periodically applied. Bedding materials either treated with aromatic hydrocarbons (such as crude cedar wood oil) or naturally containing such materials (cedar wood shavings) are successful in suppressing infestations.

PROTOZOAN INFECTIONS

Several types of protozoan diseases have been reported in mice. Most are inapparent infections, noted only after synergistic activity with other agents such as viruses or after interference with immunological body defenses as by splenectomy. Control is usually by improving hygiene after elimination of infected animals or whole colonies.

Eperythrozoön coccoides occurs as a blood parasite in the form of disc-shaped structures on the surface of red blood cells. Splenectomy of infected mice results in a marked increase in the numbers of parasitized cells and onset of transient mild anemia. The presence of this parasite increases susceptibility to such infectious agents as the mouse hepatitis virus, lymphocytic choriomeningitis virus, and lactate dehydrogenase-elevating virus (Seamer et al., 1961; Riley, 1964).

Hemobartonella muris is another red blood cell parasite activated in infected mice after splenectomy (Griesemer, 1958). Infectivity is very low. The anemia produced is mild and transitory.

Encephalitozoön cuniculi causes mild febrile disease in mice, infects epithelial cells of the kidney papillae, and produces granulomatous lesions in brain tissue (Yost, 1958). These organisms occur as 1.5- to 2.0-μ Gram-positive rods. Transmission is apparently by way of urine, which may contain large numbers of organisms.

Klosiella muris has been described as causing an infection of mouse kidney epithelial cells (Dunn, 1949). Signs of disease are inapparent although, on necropsy, kidneys of infected animals may be surface-marked with varying numbers of tiny grayish-white foci. Transmission is by way of spore cysts released into the urine.

MYCOTIC INFECTIONS

Mycotic infections of mice usually produce a dermatitis. Such infections, which can be caused by any one of several *Trichophyton* or *Microsporum* sp., result in circumscribed encrusted areas with hair loss and are empirically called ringworm. The fungus from the lesions may be easily demonstrated by microscopic examination of skin scrapings (La Touche, 1957). Control of the colony infection is difficult. Outbreaks are chiefly important as sources of human infection as most of these fungi cause similar lesions in man.

SUMMARY

The mouse is susceptible to an array of infectious agents. Many of these agents produce signs of disease and must be eliminated or controlled if only to maintain sizable mouse populations. On the other hand, because the mouse is so important as a biological yardstick for the measurement of varied procedures, it becomes important as well to eliminate or control those infectious agents which do not produce observable disease, but which may affect the outcome of an experiment. The development of mouse colonies from caesarean-derived stock is an effective panacea for eliminating those disease agents which cannot penetrate the placental barriers. In another respect, it is important to study the synergistic or antagonistic effect of various disease agents either in combination with each other or with other factors so that information so derived may contribute toward man's understanding of his own disease problems. It would be unfortunate to eliminate completely all disease of mice before obtaining more complete knowledge of the etiology, epidemiology, and pathology of such naturally occurring conditions.

LITERATURE CITED

Altman, R. 1964. Clinical aspects of enterovirus infection. Postgrad. Med. 35:451–453.

Barua, D., P. L. Banerjee, B. K. Aikat, and C. L. Mukherjee. 1962. Infantile gastroenteritis outbreak due to *Escherichia coli*. J. Indian Med. Ass. 38:383–386.

Briody, B. A. 1959. Response of mice to ectromelia and vaccinia viruses. Bacteriol. Rev. 23: 61-95.

Brodsky, I., and W. P. Rowe. 1958. Chronic subclinical infection with mouse salivary gland virus. Proc. Soc. Exp. Biol. Med. 99:654–655.

Burnet, F. M., and W. C. Boake. 1946. The relationship between the virus of infectious ectromelia of mice and vaccinia virus. J. Immunol. 53:1,013.

Derrick, E. H., and J. H. Pope. 1960. Murine typhus in mice, rats, and fleas on the Darling Downs. Med. J. Austral. 11:924–928.

Dunn, T. B. 1949. Some observations on the normal and pathologic anatomy of the kidney of the mouse. J. Nat. Cancer Inst. 9:285–301.

Erlandson, A. L., Jr., M. A. Nemer, and I. A. Pearson. 1964. Characteristics of various experimental human adult virulent *Escherichia coli* infections in mice. J. Infect. Dis. 114:163–168.

Fenner, F. 1949. Mouse-pox (infectious ectromelia of mice). A review. J. Immunol. 63:341–373.

Fisher, E. R., and L. Kilham. 1953. Pathology of a pneumotropic virus recovered from C3H mice carrying the Bittner agent. Arch. Pathol. 55:14–19.

Flynn, R. J. 1963*a*. *Pseudomonas aeruginosa* infection and radiobiological research at the Argonne National Laboratory: effects, diagnosis, epizootiology and control. Lab. Anim. Care 13:25–35.

Flynn, R. J. 1963*b*. The diagnosis of some forms of ectoparasitism of mice. Lab. Anim. Care 13: 111–125.

Flynn, R. J. 1963*c*. The diagnosis of *Pseudomonas aeruginosa* infection in mice. Lab. Anim. Care 13:126–129.

Freundt, E. A. 1956. *Streptobacillus moniliformis* infection in mice. Acta Pathol. Microbiol. Scand. 38:231–245.

Freundt, E. A. 1959. Arthritis caused by *Streptobacillus moniliformis* and pleuropneumonia-like organisms in small rodents. Lab. Invest. 8: 1,358–1,366.

Fujiwara, K., Y. Takagaki, K. Maejima, K. Kato, M. Naiki, and Y. Tajima. 1963. Tyzzer's disease in mice: pathologic studies on experimentally infected animals. Jap. J. Exp. Med. 33:183–202.

Fujiwara, K., Y. Takagaki, M. Naiki, K. Maejima, and Y. Tajima. 1964. Tyzzer's disease in mice. Jap. J. Exp. Med. 34:59–75.

Gledhill, A. W. 1962. Viral diseases in laboratory animals, p. 99–112. *In* R. J. C. Harris [ed.] The problems of laboratory animal disease. Academic Press, New York.

Gledhill, A. W., and J. S. F. Niven. 1955. Latent virus infection as exemplified by mouse hepatitis virus (MHV). Vet. Rev. Annot. 1:82–90.

Griesemer, R. A. 1958. Bartonellosis. J. Nat. Cancer Inst. 20:949–956.

Gross, L. 1953. A filterable agent recovered from AK leukemic extracts, causing salivary gland carcinomas in C3H mice. Proc. Soc. Exp. Biol. Med. 83:414–416.

Gross, L. 1961. Oncogenic viruses. Pergamon Press, New York. 393 p.

Haas, V. H. 1954. Some relationships between choriomeningitis (LCM) virus and mice. J. Infect. Dis. 94:187–198.

Habermann, R. T., and F. P. Williams, Jr. 1958*a*. Salmonellosis in laboratory animals. J. Nat. Cancer Inst. 20:933–947.

Habermann, R. T., and F. P. Williams, Jr. 1958*b*. The identification and control of helminths in laboratory animals. J. Nat. Cancer Inst. 20:979–1,009.

Hartley, J. W., and W. P. Rowe. 1960. A new mouse virus apparently related to the adeno virus group. Virology 11:645–647.

Hoag, W. G. 1961. Oxyuriasis in laboratory mouse colonies. Amer. J. Vet. Res. 22:150–153.

Hoag, W. G., and J. Rogers. 1961. Techniques for the isolation of *Salmonella typhimurium* from laboratory mice. J. Bacteriol. 82:153–154.

Hoag, W. G., J. Strout, and H. Meier. 1964*a*. Isolation of *Salmonella* spp. from laboratory mice and from diet supplements. J. Bacteriol. 88: 534–536.

Hoag, W. G., J. Strout, and H. Meier. 1964*b*. Attempts to immunize mice against chronic and acute states of *Salmonella typhimurium* infection. Z. Versuchstierk. 5:121–129.

Hoag, W. G., J. Strout, and H. Meier. 1965. Epidemiological aspects of the control of *Pseudomonas* infection in mouse colonies. Lab. Anim. Care. 15:217–225.

Hoag, W. G., P. W. Wetmore, J. Rogers, and H. Meier. 1962. A study of latent *Pasteurella* infection in a mouse colony. J. Infect. Dis. 111:135–140.

Holt, D. 1959. Presence of K virus in wild mice in Australia. Austral. J. Exp. Biol. Med. Sci. 37: 183–191.

Jawetz, E. 1948. A pneumotropic *Pasteurella* of laboratory animals. I. Bacteriological and serological characteristics of the organism. J. Infect. Dis. 85:172–183.

Jawetz, E., and W. H. Baker. 1948. A pneumotropic *Pasteurella* of laboratory animals. II. Pathological and immunological studies with the organism. J. Infect. Dis. 86:184–196.

Kraft, L. M. 1962. Two viruses causing diarrhoea in infant mice, p. 115–130. *In* R. J. C. Harris [ed.] The problems of laboratory animal disease. Academic Press, New York.

Lane-Petter, W. [ed.] 1963. Animals for research. Academic Press, New York.

La Touche, C. J. 1957. Ringworm in laboratory mice, p. 526–532. *In* A. N. Worden and W. Lane-Petter [ed.] The Universities Federation for Animal Welfare handbook on the care and management of laboratory animals. Univ. Fed. Anim. Welfare, London.

Maurer, F. D. 1958*a*. Mouse poliomyelitis or Theiler's mouse encephalomyelitis, J. Nat. Cancer Inst. 20:871–874.

Maurer, F. D. 1958*b*. Lymphocytic choriomeningitis. J. Nat. Cancer Inst. 20:867–870.

Mirick, G. S., J. M. Smith, C. I. Leffwich, Jr., and W. V. Leffwich. 1952. The enhancing effect of urethane on the severity of infection with pneumonia virus of mice (PVM). J. Exp. Med. 95: 147–160.

Nelson, J. B. 1958. Infection in laboratory animals with pleuropneumonia-like organisms. J. Nat. Cancer Inst. 20:911–915.

Nelson, J. B., and G. S. Tarnowski. 1960. An oncolytic virus recovered from Swiss mice during passage of an ascites tumor. Nature 188: 866–867.

Nigg, C., and N. D. Eaton. 1944. Isolation from normal mice of pneumotropic virus which forms elementary bodies. J. Exp. Med. 79:497–509.

Notkins, A. L., R. L. Berry, J. B. Moloney, and R. E. Greenfield. 1962. Relationship of the lactic dehydrogenase factor to certain murine tumours. Nature 193:79–80.

Olitsky, P. K., and J. M. Lee. 1955. The immunologic response of mice from stocks resistant and susceptible to acute disseminated encephalomyelitis. J. Lab. Clin. Med. 45:81–86.

Pappenheimer, A. M. 1958*a*. Pathology of infection with the JHM virus. J. Nat. Cancer Inst. 20:879–891.

Pappenheimer, A. M. 1958*b*. Epidemic diarrheal diseases of suckling mice. J. Nat. Cancer Inst. 20:861–865.

Payne, F. J. 1960. *Escherichia coli* strains as etiological agents of diarrheal disease. Public Health Rep. 75:534–536.

Riley, V., E. Huerto, F. Lilly, D. Bardell, J. D. Loveless, and M. A. Fitzmaurice. 1961. Some characteristics of virus-like entities associated with 50 varieties of experimental tumors. Proc. Amer. Ass. Cancer Res. 3:262.

Riley, V., F. Lilly, E. Huerto, and D. Bardell. 1960. Transmissible agent associated with 26 types of mouse neoplasms. Science 132:545–547.

Rowe, W. P., and W. I. Capps. 1961. A new mouse virus causing necrosis of the thymus in newborn mice. J. Exp. Med. 113:831–844.

Rowe, W. P., J. W. Hartley, and R. J. Huebner. 1962. Polyoma and other indigenous mouse viruses. p. 131–142. *In* R. J. C. Harris [ed.] The problems of laboratory animal disease. Academic Press, New York.

Runner, M. N., and J. Palm. 1953. Factors associated with the incidence of infantile disease in mice. Proc. Soc. Exp. Biol. Med. 82:147–150.

Sabin, A. B. 1959. Reoviruses. A new group of respiratory and enteric viruses formerly classified as ECHO type 10 is described. Science 130:1,387–1,389.

Saunders, L. Z. 1958*a*. Mouse pox (infectious ectromelia). J. Nat. Cancer Inst. 20:875–877.

Saunders, L. Z. 1958*b*. Tyzzer's disease. J. Nat. Cancer Inst. 20:893–897.

Seamer, J., A. W. Gledhill, J. L. Barlow, and J. Hotchin. 1961. Effect of *Eperythrozoön coccoides* upon lymphocytic choriomeningitis in mice. J. Immunol. 86:512–515.

Sellers, T. F., J. J. Schulman, C. Bouvier, R. McCune, and E. D. Kilbourne. 1961. The influence of influenza virus infection on exogenous staphyloccal and endogenous murine bacterial infection of the bronchopulmonary tissues of mice. J. Exp. Med. 115:234–256.

Shope, R. E. 1954. Report of committee on infectious ectromelia of mice (mouse-pox). J. Nat. Cancer Inst. 15:405–408.

Shope, R. E. 1964. The epidemiology of the origin and perpetuation of a new disease. Perspect. Biol. Med. 7:263–278.

Sinkovics, J. G. 1962. Abnormal immune and other complicating phenomena associated with a viral mouse leukemia. J. Infect. Dis. 110:282–296.

Stewart, S. E. 1953. Leukemia in mice produced by a filterable agent present in AKR leukemic tissues with notes on a sarcoma produced by the same agent. Anat. Rec. 117:532. (Abstr.)

Theiler, M., and S. Gard. 1940. Encephalomyelitis of mice. I. Characteristics and pathogenesis of the virus. J. Exp. Med. 72:49–67.

Thomlinson, J. R. 1964. The pathogenesis of gastro-enteritis associated with *Escherichia coli.* Vet. Rec. 76:237–238.

Thompson, R., V. M. Harrison, and F. P. Myers. 1951. A spontaneous epizootic of mouse encephalomyelitis. Proc. Soc. Exp. Biol. Med. 77:262–266.

Traub, E. 1939. Epidemiology of lymphocytic choriomeningitis in a mouse stock observed for 4 years. J. Exp. Med. 69:801–817.

Trentin, J. J. 1953. An outbreak of mouse-pox (infectious ectromelia) in the United States; a presumptive diagnosis. Science 117:226–227.

Tuffery, A. A. 1958. The mouse, p. 243–278. *In* A. M. Worden and W. Lane-Petter [ed.] Universities Federation for Animal Welfare handbook on the care and management of laboratory animals. Univ. Fed. Anim. Welfare, London.

Verder, E., and S. M. Rosenthal. 1961. Role of infection in the delayed deaths of mice following extensive burn injury. Proc. Soc. Exp. Biol. Med. 108:501–505.

Volkert, M., and F. L. Horsfall. 1947. Studies on a lung tissue component which combines with pneumonia virus of mice (PVM). J. Exp. Med. 86:393–407.

Wensinck, F., D. W. Van Bekkum, and H. Renaud. 1957. A prevention of *Pseudomonas aeruginosa* infections in irradiated mice and rats. Radiat. Res. 7:491–499.

Wetmore, P. W., and W. G. Hoag. 1960. *Salmonella binza* and *Salmonella bredeney* from laboratory mice. J. Bacteriol. 80:283.

Wilson, F. S., and A. A. Miles. 1964. Topley and Wilson's principles of bacteriology and immunity, 5th ed. Williams & Wilkins, Baltimore. 2 vol.

Yaffee, D. 1962. The distribution and in vitro propagation of an agent causing high plasma lactic dehydrogenase activity. Cancer Res. 22: 573–580.

Yost, D. H. 1958. Encephalitozoön infection in laboratory animals. J. Nat. Cancer Inst. 20:957–963.

31

Immune Functions[1]

Henry J. Winn

Immunology in a broad sense encompasses a wide range of investigations bearing on the complex problems of resistance and susceptibility of higher forms of life to infectious diseases, and it draws heavily on a variety of medical, biological, and chemical disciplines. Experimental immunology has, however, focused largely on the antibody response, which in mammals tends to play a pivotal role in acquired immunity. This preoccupation with antibody, originally inspired by the dramatic demonstrations of its association with specifically increased resistance to some diseases and subsequently attributable to an increasing awareness of the practical and conceptual implications of specific immune reactions for many areas of biology, has nurtured the development of a conception of immunology that is much narrower in scope than that originally proposed. It is immunology in this restricted sense that is to be considered here. The basic principles of the field are presented in detail in a number of texts (Boyd, 1956; Raffel, 1961; Humphrey and White, 1963). The present discussion will be by no means exhaustive, but is intended to direct interested readers to published studies which provide basic information on immune functions in mice or in which some unique biological trait of these animals has been exploited in immunological investigations.

ANTIBODIES

Antibodies are serum globulins that are capable of combining specifically with substances (antigens) which provoke their formation by cells of the lymphoid tissues. They belong to a class of plasma proteins, designated γ-globulins or immunoglobulins, which shows a high degree of physical and chemical heterogeneity. Indeed the term γ-*globulin* may be somewhat misleading since antibody activity may be found in the β-globulins or α_2-globulins. However, despite this lack of true homogeneity, there is much justification for treating antibodies and similar proteins elaborated by plasmacytes and lymphoid cells as a family of closely related proteins under a single inclusive designation. In addition to their common tissue origin they have many overlapping physical and chemical properties and they have strikingly similar immunochemical characteristics (review by Fahey, 1962). The term *immunoglobulins* will be used here but it must not be interpreted to mean that all of the protein molecules so designated have known antibody activity, and the use of a collective designation should not obscure the fact that relatively small differences in molecular structure of antibodies can be responsible for marked differences in the biological activities of the molecules.

It has been possible to establish defined subfractions of the immunoglobulins and with respect to the mouse Fahey et al. (1964*a*) have described four major subdivisions. Two classes of 7 S γ-globulins differing in electrophoretic mobility and in antigenic structure were designated 7 S γ_1-globulins and 7 S γ_2-globulins. A third fraction designated γ_{1A}-globulin is found chiefly in the β-globulin fraction. It is heterogenous with respect to molecular size and antigenically distinct from

[1] The writing of this chapter was supported in part by Public Health Service Research Grant CA 01329 from the National Cancer Institute.

the other immunoglobulins. The fourth class, γ_1-globulin, is a high molecular weight fraction which migrates on electrophoresis with the mid γ-globulins, and contains characteristic antigenic determinants. An additional component, designated Xγ, was also demonstrated in the γ-globulins of mice but was considered to be devoid of immunological function and, accordingly, was not classified as an immunoglobulin. Each of the four major classes was shown to contain antibody activity. In a subsequent study Fahey et al. (1964*b*) described two subclasses of mouse 7 S γ_2-globulins which were designated γ_{2A}-globulins and γ_{2B}-globulins. Enzymatically produced subunits of mouse immunoglobulins have been studied by Fahey and Askonas (1962*a, b*). Classification of immunoglobulins on the basis of genetically determined isoantigens ("alloantigens," Chap. 24) is discussed in Chap. 8.

Purified antibodies obtained from the serum of mice immunized with protein conjugates were fractionated and some biological properties of the fractions were studied by Nussenzweig et al. (1964). Two types of precipitating antibody were found and designated γ_1 and γ_2. Both types were found in the γ-region on electrophoresis in agar gel, the γ_1 migrating slightly more rapidly than the γ_2. Mouse γ_1 antibodies were capable of sensitizing mice but not guinea pigs for passive cutaneous anaphylaxis, but they did not cause detectable lysis, in the presence of mouse complement, of sheep erythrocytes coated with the protein conjugates used as antigens. The γ_2-antibodies, in conjunction with mouse complement, lysed the coated red blood cells and were capable of transferring passive cutaneous anaphylaxis to guinea pigs but not to mice. These interesting differences may be related to the differences between mouse and guinea pig complement systems which are described in the next section.

The kinds and amounts of immunoglobulins elaborated in response to antigenic stimuli may be influenced by: (1) the chemical and physical properties of the antigen and the vehicle in which it is dissolved or suspended, (2) the route of injection, (3) the intensity and duration of immunization, (4) the age of the recipient animal, and (5) genetic factors. All of these factors interact and the roles they play in directing the course of immune responses are poorly understood.

Fahey and Humphrey (1962) studied the antibody responses of C3H/He mice to single intravenous injections of pneumococcal polysaccharide, hemocyanin, or sheep erythrocytes. Antisera were examiend 6 days after administration of antigen. Antibodies to both of the soluble antigens had the properties of low molecular weight (6.5 S) γ-globulins, whereas the bulk of the anti-erythrocyte antibodies were γ-macroglobulins (19 S) with a small proportion found among the lower molecular weight globulins. Subsequently, Fahey and Lawrence (1962) compared the physicochemical and immunochemical properties of antibodies that had been produced against hemocyanin in eight inbred strains of mice. Most of the antibodies from all of the mice were 6.6 S γ-globulins but there were marked differences in the electrophoretic properties of the antibodies from different strains. Winn (1965*a*) found that the response of a large number of inbred strains of mice to a single injection of sheep erythrocytes was confined almost exclusively to the elaboration of higher molecular weight antibodies. However, after hyperimmunization there were marked differences in the responses of various strains. C57BL/10J mice continued to produce large amounts of antibodies of the macroglobulin type with barely detectable amounts of low molecular weight antibodies. In A/J mice the production of high molecular weight antibodies was no longer detected, though their sera contained high titers of low molecular weight antibodies. Mice of other strains produced mixtures of the two types of antibodies in various proportions. The importance of genetic factors is also indicated in a study showing quantitative differences in the immune response to tetanus toxoid between four inbred lines of mice (Ipsen, 1959).

The immune responses of very young animals have been found to differ appreciably from those of mature animals. This aspect of immunization has not been investigated intensively in mice, but Boraker and Hildemann (1965) have examined the sera of young A/Jax and C57BL/6J mice that had received reciprocal grafts of skin and lymphoid tissues. They found that antibodies detected prior to 20 and 27 days of age in C57BL/6J and A/Jax mice, respectively, were of the macroglobulin type. Older animals of either strain produced 7 S antibodies as well as macroglobulins.

The interactions of mouse antibodies with specific antigens become manifest in a variety of ways that are characteristic of similar interactions of antibodies from other species. Most of these manifestations are alluded to in various parts of this discussion and require no further comment. It is appropriate, however, to mention the study of Anacker and Munoz (1961) on the precipitin reaction involving ovalbumin and the corresponding antibody contained in mouse peritoneal fluid. Mice repeatedly injected by the intraperitoneal route with antigen in Freund's adjuvant developed large volumes of ascites fluid containing high concentrations of antibody. The antibody was qualitatively indistinguishable from that found in serum and behaved very similarly to rabbit antibody in quantitative precipitin tests. Insoluble precipitates were found throughout the region of antibody excess and soluble complexes were found in the region of antigen excess. The amounts of antibody precipitated at equivalence reached, in some cases, the unusually high level of 2 mg of protein nitrogen per milliliter.

COMPLEMENT

Although it is widely recognized that complement may play an important role in many diverse immunological reactions, the detection and assay of this complex group of plasma proteins are generally based on its ability to lyse sheep erythrocytes sensitized with rabbit anti-sheep cell antibodies. Measurements based on this assay system have invariably revealed low or undetectable levels of activity in mouse sera. This has been attributed to deficiencies in various individual components of complement (Borsos and Cooper, 1961; Herzenberg et al., 1963). Borsos and Cooper showed that pools of serum from CFW mice contain all of the recognized factors of the complement system, that the C′2 and C′3 activities are quite low, and that mouse serum contains a potent inhibitor of lytic activity. These findings provide a suitable explanation for the failures of a number of workers to detect complement activity in mice. However, Herzenberg et al. reported that a deficiency of hemolytic complement in a highly inbred strain of mice was genetically determined. This deficiency was shown to be controlled by a single genetic locus (*Hc*) and subsequent studies have revealed a similar or identical deficiency in a number of inbred lines.

Erickson et al. (1963) showed that genes at this locus controlled the production of a serum protein identified by immunochemical techniques. Their evidence suggests that the allele carried by mice of deficient strains has no recognizable product. Subsequent studies by Terry et al. (1964) have suggested that deficient strains lack one of the components designated collectively as C′3. There is no indication that the health or immune functions of the deficient mice are impaired.

The hemolytic activity of mouse complement has been shown to be more efficiently activated by 7 S hemolysins than by high molecular weight antibodies (Winn, 1965*a*). This is in striking contrast to the activation of guinea pig complement which has served as a prototype in this area of research. Winn showed that this difference between complement from mice and guinea pigs could be attributed to differences in the C′1 macromolecules of the two species. Additional data presented in this report show that in hemolytic systems involving 19 S hemolysins, mouse serum markedly inhibits the activity of guinea pig complement but supplements that of rabbit serum. The inhibitory properties of mouse serum lead to some difficulties when mouse antisera are employed in complement-fixation tests. Problems of this nature are discussed by Winn (1965*b*), who has suggested ways of eliminating or compensating for this type of interference.

HYPERSENSITIVITY

The development of the immune state is often accompanied by a dramatic increase in reactivity to further contact with the same antigen. This increased reactivity, referred to as allergy or hypersensitivity, has diverse manifestations which lead to tissue destruction varying in degree of severity from evanescent wheal and flare reactions to chronic inflammation which may proceed to intense suppuration and necrosis. In some cases severe generalized illness or death of a sensitive animal may occur a few minutes after contact with antigen. The substances responsible for the development and expression of hypersensitivity are of widely varying chemical constitution, and they may or may not have intrinsic toxic properties. Microorganisms (dead as well as living), microbial metabolites, foreign proteins, pollen extracts, and some low molecular weight organic compounds provide only a sample of the known allergens.

Hypersensitive reactions are conveniently divided into immediate and delayed types on the basis of the temporal sequence of reactions following contact with antigen. Immediate-type reactions can be detected within seconds after contact and are often, though not necessarily, relatively short-lived. Delayed-type reactions are detectable by ordinary means only several hours after application of the incitant, and they continue to develop in intensity for periods of 18 to 72 hours. These differences in the tempo of response reflect basic differences in the mechanisms responsible for the development of the reactions. Reactions of the immediate type are invariably associated with serum antibodies, whereas delayed-type reactions involve cell-bound or cell-associated antibodies. A clear distinction between the two types is provided by demonstrations that a single antigen may induce either type of reactivity alone. Subdivision of the allergic reactions, particularly those of the immediate type, is useful and will be indicated below in discussing these reactions in mice.

The occurrence of delayed-type responses in mice has been reported (review by Crowle, 1959*a*), but they are not elicited so regularly as in other species and typical cutaneous reactions are rarely observed. Apparently, mice have the capacity to elaborate the immune substances that mediate delayed hypersensitive responses, but they seem to be constitutionally incapable of displaying reactions of the type observed in other animals. By employing a variety of special techniques, Crowle (1959*b*) has been able to demonstrate specifically induced inflammatory responses that are evidently manifestations of delayed reactivity.

Additional information on this point comes from studies on experimental allergic encephalomyelitis. There is a good deal of evidence that the production of this condition depends on the establishment of delayed-type reactivity to brain tissue. The ability to produce the disease regularly in mice supports the conclusion that these animals can develop reactivity of the delayed type (Lee and Olitsky, 1955).

Immediate-type allergy may be divided into two classes. Anaphylactic reactions result from the rapid release of short-lived pharmacologically active agents immediately following interaction of antigen and antibody in vivo. The symptoms and severity of these reactions vary considerably in different species principally because of differences in the amounts and kinds of active substances released and in the sensitivities of various animals to the action of these compounds. Arthus reactions are local responses initiated directly by the formation of antigen-antibody precipitates. Apparently, these precipitates have an irritant effect on the walls of small blood vessels, leading to the development of intense inflammatory responses. The reaction begins very shortly after application of a test antigen and continues to develop in intensity for several hours.

With respect to systemic anaphylaxis, older views that mice are highly resistant to the induction of allergic reactivity have been completely reversed. Burdon (1937) and Weiser et al. (1941) showed conclusively that mice can be sensitized and shocked, though the course of the reaction is not nearly so violent as that observed in other species. Quite commonly death does not occur until several hours after challenge. During this interval the mice usually remain quiet and move only if disturbed. The fur is ruffled, there may be scratching of the eyes and muzzle, and there may be some evidence of respiratory difficulty. As death approaches, breathing becomes more difficult, cyanosis may be marked, and the animals display evidence of paralysis or convulsive kicking. Occasionally, death occurs as soon as 15 minutes after injection of antigen.

Cameron (1956) established arbitrary classes of nonfatal shock on the basis of the severity and duration of symptoms in *W*-series mice. His report includes a study of the relative sensitizing abilities of various proteins, the efficacy of various routes for sensitizing and challenging, the effect of varying the number of sensitizing doses, and the duration of sensitization. Quantitative estimation of anaphylactic sensitivity in mice has been based on observations that shock is accompanied by a fall in body temperature (Kind, 1955) and a rise in hematocrit (Harris and Fulton, 1958). McMaster and Kruse (1955) have devised an extremely sensitive technique based on the detection of vasospasm in the ears of intravenously injected mice that had been previously sensitized with minute quantities of antigen.

Sobey and Adams (1959) have demonstrated in mice a high heritability of anaphylactic sensitivity to bovine plasma albumin, and Fink and Rothlauf (1954) reported variations in sensitivity to shock in five inbred strains of mice injected with ovalbumin. It is not clear whether these variations are attributable primarily to differences in the kinds and amounts of antibody formed in response to these antigens or to differences in sensitivity to the pharmacologically active compounds released as a result of the combination of antigen and antibody.

Pharmacological aspects of anaphylaxis in mice have been discussed by Austen and Humphrey (1963). The subject is too complex and controversial to treat here in any detail, but much experimental data suggest that histamine does not play an important role, and there is some evidence implicating serotonin as a mediator of shock.

A peculiar feature of systemic anaphylaxis in mice is brought to light by studies carried out in animals that have received injections of *Hemophilus pertussis* vaccine (review by Kind, 1958). This treatment often leads to hyperreactivity to histamine. When vaccine is administered simultaneously with a sensitizing antigen, the subsequent challenge with antigen results in a much higher mortality than in mice sensitized with antigen alone. The mechanism of this potentiating effect is not understood, but there seems to be no clear-cut relationship between the increased sensitivity to shock and the enhanced reactivity to histamine. Marked differences in the reactions of various inbred strains of mice are apparent but no comprehensive studies have been reported.

The combination of small amounts of antigen and antibody in cutaneous sites leads to an increased permeability of small vessels permitting local concentration of intravenously injected dyes. The reaction has been used principally for the detection of minute quantities of specific antigens or antibodies, though some characterization of the biological activities of different classes of antibodies may also be achieved. This cutaneous form of anaphylaxis has been described in the mouse by Ovary (1958) and immunochemical aspects of such reactions were reported by Nussenzweig et al. (1964).

A great deal of our information on the mechanism of anaphylaxis has come from studies carried out in vitro with isolated tissues from actively or passively sensitized animals. Experiments have been carried out with a variety of tissues from different animals (review by Austen and Humphrey, 1963), but the use of mice has been restricted to the study of isolated strips of uterine muscle (Fink and Rothlauf, 1955). Tissue placed in an organ bath with nutrient solution showed characteristic contractions after addition of the antigen to which the mice had been sensitized. The procedure is often referred to as the Schultz-Dale technique in recognition of the workers who first demonstrated the phenomenon with guinea pig tissue. These experiments with mouse tissue support the suggestion that serotonin rather than histamine plays an important role in anaphylaxis in mice.

Typical Arthus reactions are difficult to produce in mice and the inflammatory responses accompanying these reactions are not so intense as those observed in rabbits. Reactions have been elicited in the foot pads of immunized mice but the lip is a more convenient site of injection. Moreover, Arthus reactions are more regularly obtained by intralabial injection and in this area they are more readily distinguished from nonspecific inflammatory responses (Freund and Stone, 1956).

TRANSPLANTATION IMMUNOLOGY

The biology of tissue transplantation has been reviewed comprehensively by Russell and Monaco (1964) and many aspects of the problems of transplantation as they relate to mice are discussed in Chap. 24. The conception of homograft rejection as a manifestation of the immunization of the host against isoantigens of the graft is universally accepted, though the nature of the immune substances responsible for destruction of grafted tissues is not known precisely. There are striking differences in the sensitivity of mouse cells derived from various tissues to the destructive effects of isoimmune sera and complement (reviews by Winn, 1962; Stetson, 1963; Möller, 1963), and in some cases it has been shown that the presence of serum antibodies in the host may lead to prolonged survival of homografts (Kaliss, 1962; Snell et al., 1960). A substantial body of data indicates the participation of cellular or cell-bound antibodies in graft destruction (reviewed by Snell, 1963). Comprehensive reviews of the relative importance of humoral and cellular factors in graft rejection have been published by Gorer (1961), Brent (1958), and Amos (1962).

SUMMARY

This discussion has dealt briefly with a description of the kinds of immune substances that have been found in mice and with some aspects of their biological activities. The points which serve to distinguish immunological reactivities of mice from those of other species have been emphasized, and differences in reactivity between inbred strains of mice have been described. Important areas of immunology have been omitted because little definitive information with respect to mice is available or because the subject is such that no special treatment is required beyond that found in textbooks on general immunology.

LITERATURE CITED

Amos, D. B. 1962. The use of simplified systems as an aid to the interpretation of mechanisms of graft rejection. Progr. Allergy 6:468–538.

Anacker, R. L., and J. Munoz. 1961. Mouse antibody. I. Characterization of antibody in mouse peritoneal fluid. J. Immunol. 87:426–433.

Austen, K. F., and J. H. Humphrey. 1963. In vitro studies of the mechanism of anaphylaxis. Adv. Immunol. 3:1–96.

Boraker, D. K., and W. H. Hildemann. 1965. Maturation of alloimmune responsiveness in mice. Transplantation 3:202–223.

Borsos, T., and M. Cooper. 1961. On the hemolytic

activity of mouse complement. Proc. Soc. Exp. Biol. Med. 107:227–232.

Boyd, W. C. 1956. Fundamentals of immunology, 3rd ed. Interscience, New York. 776 p.

Brent, L. 1958. Tissue transplantation immunity. Progr. Allergy 5:271–348.

Burdon, K. L. 1937. Active sensitization of white mice. Proc. Soc. Exp. Biol. Med. 36:340–342.

Cameron, J. 1956. Anaphylactic shock in mice. Brit. J. Exp. Path. 37:470–476.

Crowle, A. J. 1959*a*. Delayed hypersensitivity in mice. J. Allergy 30:151–164.

Crowle, A. J. 1959*b*. Delayed hypersensitivity in several strains of mice studied with six different tests. J. Allergy 30:442–459.

Erickson, R. P., D. K. Tachibana, L. A. Herzenberg, and L. T. Rosenberg. 1963. A single gene controlling hemolytic complement and a β-mobility serum antigen in the mouse. Fed. Proc. 22:612. (Abstr.)

Fahey, J. L. 1962. Heterogeneity of gamma globulins. Adv. Immunol. 2:41–109.

Fahey, J. L., and B. A. Askonas. 1962*a*. Enzymatically produced subunits of proteins formed by plasma cells in mice. I. γ-globulin and γ-myeloma proteins. J. Exp. Med. 115:623–639.

Fahey, J. L., and B. A. Askonas. 1962*b*. Enzymatically produced subunits of proteins formed by plasma cells in mice. II. β-2A-myeloma proteins and Bence Jones proteins. J. Exp. Med. 115:641–653.

Fahey, J. L., and J. H. Humphrey. 1962. Antibodies with differing molecular size in mice. Immunology 5:104–109.

Fahey, J. L., and M. E. Lawrence. 1962. Antibody differences in several strains of mice. Fed. Proc. 21:19. (Abstr.)

Fahey, J. L., J. Wunderlich, and R. Mishell. 1964*a*. The immunoglobulins of mice. I. Four major classes of immunoglobulins: 7 S γ_2-, 7 S γ_1-, γ_A-, and 18 S γ_{1M}-globulins. J. Exp. Med. 120:223–242.

Fahey, J. L., J. Wunderlich, and R. Mishell. 1964*b*. The immunoglobulins of mice. II. Two subclasses of mouse 7 S γ_2-globulins: γ_{2a}- and γ_{2a}-globulins. J. Exp. Med. 120:243–251.

Fink, M. A., and M. V. Rothlauf. 1954. Variations in sensitivity to anaphylaxis and to histamine in inbred strains of mice. Proc. Soc. Exp. Biol. Med. 85:336–338.

Fink, M. A., and M. V. Rothlauf. 1955. In vitro anaphylaxis in the sensitized mouse uterus. Proc. Soc. Exp. Biol. Med. 90:477–480.

Freund, J., and S. H. Stone. 1956. Arthus reaction in the mouse and the rat after intralabial injections of antigens. J. Immunol. 76:138–145.

Fulton, J. D., W. E. Harris, and C. E. Craft. 1957. Hematocrit change as indication of anaphylactic shock in the mouse. Proc. Soc. Exp. Biol. Med. 95:625–627.

Gorer, P. A. 1961. The antigenic structure of tumors. Adv. Immunol. 1:345–393.

Harris, W. E., and J. D. Fulton. 1958. Quantification of anaphylaxis in mice. Proc. Soc. Exp. Biol. Med. 97:14–17.

Herzenberg, L. A., D. K. Tachibana, L. A. Herzenberg, and L. T. Rosenberg. 1963. A gene locus concerned with hemolytic complement in *Mus musculus*. Genetics 48:711–715.

Humphrey, J. H., and R. G. White. 1963. Immunology for students of medicine. Davis, Philadelphia. 450 p.

Ipsen, J. 1959. Differences in primary and secondary immunizability of inbred mice strains. J. Immunol. 83:448–457.

Kaliss, N. 1962. Elements of immunological enhancement: consideration of mechanisms. Ann. N.Y. Acad. Sci. 101:67–79.

Kind, L. S. 1955. Fall in rectal temperature as an indication of anaphylactic shock in the mouse. J. Immunol. 74:387–390.

Kind, L. S. 1958. The altered reactivity of mice after inoculation with *Bordetella pertussis* vaccine. Bact. Rev. 22:173–182.

Lee, J. M., and P. K. Olitsky. 1955. Simple method for enhancing development of acute disseminated encephalomyelitis in mice. Proc. Soc. Exp. Biol. Med. 89:263–266.

McMaster, P. D., and H. Kruse. 1949. Peripheral vascular reactions in anaphylaxis of the mouse. J. Exp. Med. 89:583–596.

Möller, G. 1963. On the role of isoimmune reactions in tissue transplantation. Balder, Stockholm. 56 p.

Nussenzweig, R. S., C. Merryman, and B. Benecerraf. 1964. Electrophoretic separation and properties of mouse antihapten antibodies involved in passive cutaneous anaphylaxis and passive hemolysis. J. Exp. Med. 120:315–328.

Ovary, Z. 1958. Passive cutaneous anaphylaxis in the mouse. J. Immunol. 81:355–357.

Raffel, S. 1961. Immunity, 2nd ed. Appleton-Century-Crofts, New York, 646 p.

Russell, P. S., and A. P. Monaco. 1964. The biology of tissue transplantation. New Eng. J. Med. 271:502–783 (six separate articles in No. 10 to 15).

Snell, G. D. 1963. The immunology of tissue transplantation, p. 323–352. *In* Conceptual advances in immunology and oncology: a collection of papers. Harper & Row, New York.

Snell, G. D., H. J. Winn, J. H. Stimpfling, and S. J. Parker. 1960. Depression of antibody of the immune response to homografts and its role in immunological enhancement. J. Exp. Med. 112:293–314.

Sobey, W. R., and K. M. Adams. 1959. The inheritance of anaphylactic sensitivity in mice using bovine plasma albumin as an antigen. Immunology 2:93–95.

Stetson, C. A. 1963. Role of humoral antibody in homograft reactions. Adv. Immunol. 3:97–130.

Terry, W. D., T. Borsos, and H. J. Rapp. 1964. Differences in serum complement activity among inbred strains of mice. J. Immunol. 92: 576–578.

Weiser, R. S., J. O. Golub, and D. M. Hamre. 1941. Studies on anaphylaxis in the mouse. J. Infect. Dis. 68:97–112.

Winn, H. J. 1962. The participation of complement in isoimmune reactions. Ann. N.Y. Acad. Sci. 101:23–45.

Winn, H. J. 1965*a*. Effects of complement on sensitized nucleated cells, p. 1. *In* G. E. W. Wolstenholme [ed.] Ciba Foundation symposium on complement. Little, Brown, Boston.

Winn, H. J. 1965*b*. Complement fixation in typing histocompatibility systems, p. 61–70. *In* P. S. Russell, H. J. Winn, and D. B. Amos [ed.] Proceedings of conference and workshop on tissue histocompatibility testing. Nat. Acad. Sci.–Nat. Res. Counc. Publ. 1,229, Washington, D. C.

32

Neural, Sensory, and Motor Functions[1]

John L. Fuller and Richard E. Wimer

The existence of numerous inbred strains and mutant types has made the mouse particularly useful to research workers interested in behavior genetics. Inbred strains of mice are also suitable for research on environmental sources of behavioral differences, since genetic variance within such strains can be safely neglected. A complete research design may vary genotype and environmental treatment simultaneously in order to study heredity-environment interactions.

This chapter deals with the biological systems most directly involved in behavior. Following discussion of the normal nervous, sensory, and motor systems, consideration is given to genetic anomalies of behavior and to audiogenic seizures which are characteristic of many strains of laboratory mice.

NERVOUS SYSTEM

Brain

A stereotaxic atlas of the mouse brain has been reported in abstract form only (Slotnick and Essman, 1964). The brain of the mouse is in many respects a smaller edition of the widely studied rat brain. Guides to the rat brain (Zeman and Innes, 1963; König and Klippel, 1963) can be used for general orientation with appropriate changes in scale, but these are not accurate for stereotaxic placement of lesions or electrodes. The histological regions of the mouse cortex have been described (Rose, 1929) and detailed accounts have been given of the cellular arrangement in the entorhinal cortex and the hippocampus (Lorente de Nó, 1933, 1934).

Although all normal mouse brains look similar to the naked eye and even under the microscope, it is possible that some of the observed variation in behavior is directly associated with characteristics of the nervous system. A truly quantitative neurology suitable for studying the genetics of neural characteristics has not been developed. Brains of inbred mice differ greatly in size, as is demonstrated in the next section, but the psychological significance of these variations is unknown. Inherited major anomalies of the central nervous system exist in great variety, however, and some are described in a section of this chapter on Inherited Neurological Defects.

Brain size

From birth to about 15 days, brain and body weight increase proportionally (Kobayashi, 1963). After 15 days body weight continues to rise, but brain growth is abruptly slowed down and brain weight may actually drop slightly between 16 and 23 days, presumably because of loss of water (Uzman and Rumley, 1958). Figure 32-1 shows the relationship of brain and body weights over a major portion of the lifespan in Swiss albino mice. The region of inflection of the curve of brain growth corresponds to the time of attainment of an adult electrocorticogram and full differentiation of dendrites of cerebral neurons.

Brain size varies widely among inbred strains as shown in Fig. 32-2. Two major principles are

[1] The writing of this chapter was supported in part by Public Health Service Research Grant MH 01775 from the National Institute of Mental Health.

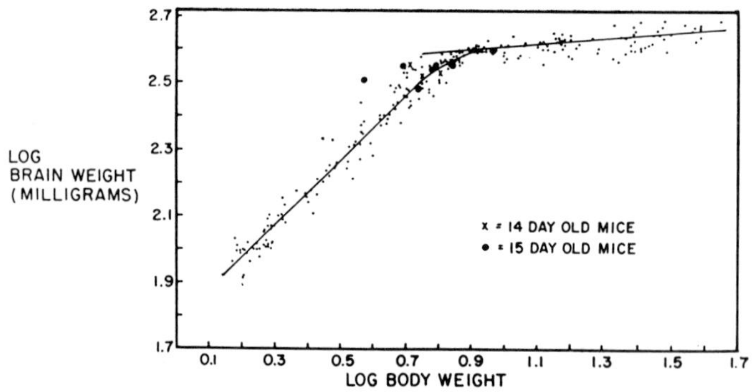

Fig. 32-1. Logarithmic values of brain weight (in milligrams) as a function of body weight (in grams) from birth until 271 days of age. The relationship changes abruptly at 14 to 15 days of age. (*From Kobayashi,* 1963, *with permission of the author and the American Journal of Physiology.*)

apparent: (1) Mean brain weights and mean body weights of strains are uncorrelated; (2) within most strains females have heavier brains in proportion to body weight (and often in absolute units). The ratio of between-strain variance to total variance of brain weight in 16 strains of diverse origins was 0.70 (Schwartzkroin, 1964, personal communication). This ratio is evidence for high heritability of brain size. The behavioral significance of these marked variations in brain size is not known.

Chemical constituents

Mouse brain like others is primarily water, protein, and lipids. Analyses of whole brain at several ages are set forth in Table 32-1. Proteins and strandin, a type of ganglioside, accumulate rather steadily up to 15 days. Phosphatides and cholesterol increase at a slightly slower rate up to about 25 days. Proteolipid proteins and cerebrosides are first found in significant amounts at 7 days and continue to increase slowly into maturity. These two classes of compounds appear to be associated with the process of myelinization.

Gamma-amino-butyric acid, believed to exert inhibitory effect upon neural transmission, is present at birth in amounts of 10 mg per 100 g of brain and rises to an adult level of 35 mg per 100 g at 30 days. The rate of increment is similar in the Goodale small, large, and CBA strains (Roberts et al., 1951).

Enzyme activity

Alkaline phosphatase, almost absent in adult brain, is present in large amounts in young fetuses (Chiquoine, 1954). Peaks of enzyme activity correspond to periods of morphological change as observed under the microscope. By 14 days of fetal age, alkaline phosphatase is low except in the telencephalon. Glutamic acid decarboxylase is present at birth and increases steadily up to 30 days with slow changes thereafter (Roberts et al., 1951). Thus maturation involves shifts in enzyme activity which presumably are related to morphological and functional changes.

The kinetic constants of the known enzymes of the glycolytic pathway have been determined in 10-day-old and adult mice (Lowry et al., 1964;

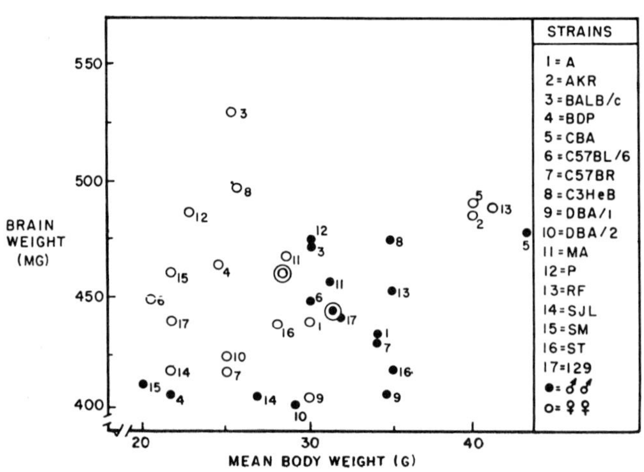

Fig. 32-2. Variation in the relationship between brain weight (in milligrams) and body weight (in grams) among adult males and females of 17 inbred strains. (*Data from Storer,* 1965, *personal communication.*)

Table 32-1. PERCENTAGE COMPOSITION OF WHOLE MOUSE BRAIN AT DIFFERENT AGES. *(From Folch-Pi, 1955, with permission of the author and Academic Press, Inc.)*

Age, days	Water	Protein	Strandin	Other lipids	Cholesterol	Phosphatides	Cerebrosides	Proteolipide protein
1	86.8	8.70	0.33	2.36	0.42	2.15	trace	trace
4	87.5	7.45	0.35	3.18	0.52	2.48	trace	trace
7	86.4	7.92	0.53	3.57	0.49	2.53	trace	trace
10	84.4	9.00		4.60	0.82	3.34	0.15	0.09
16	81.5	10.40	0.60	5.68	1.02	4.38	0.25	0.15
25	78.9	11.40		7.55	1.55	4.90	0.77	0.20
75	77.2	12.00	0.56	8.04	1.73	5.40	1.08	0.37
180	76.2	12.10	0.53	8.84	2.00	5.20	1.31	0.49

Lowry and Passoneau, 1964). The initial rates of change of the substrates in this pathway were used to compute the metabolic turnover of high energy phosphorus for adults (25 mmoles per kg of brain per minute) and for 10-day-old mice (10 mmoles per kg of brain per minute).

The chemical changes in mouse brain during development are like those described for the rat and similar species. The blood-brain barrier is less well developed in young mice than in adults. For example, injected glutamic acid enters the brain freely until about 21 days after birth (Himwich, 1962). Taking both chemical and structural features into consideration, it appears that the "critical period" for neuronal maturation falls within 10 to 15 days postpartum.

Biochemistry and behavior

To behaviorists the most interesting aspects of brain chemistry are the neural transmitters and modulators, for differences between strains in these substances might be rather closely related to variations in behavior. Maas (1962) found that the serotonin content of the brain stem (diencephalon, mesencephalon, and pons) was higher in BALB/c than in C57BL/10 mice and suggested that this substance might be related to behavioral differences between the strains. When brain serotonin in both strains was depleted by reserpine, their behavior became much more similar. Experiments with drugs led to the conclusion that the neurochemical difference was a function of binding mechanisms for serotonin rather than of its production (Maas, 1963).

Associations between behavior and specific biochemical events in two or three strains do not by themselves prove any functional genetic relationship between the two kinds of phenotypes. Surveys of wide scope involving many strains or correlations based upon data from segregating generations are necessary to ensure that the association stems from common dependence upon particular genes, and not to the fixation of unrelated genes within a strain.

Electrical activity

The electrocorticogram (ECoG) recorded with implanted electrodes from unanesthetized healthy adult mice shows four distinct patterns (Fig. 32-3). During sleep high voltage waves (200 to 300 μv) at 3 to 5 cycles per second are characteristic. The ECoG of drowsy mice shows intermittent spindles of 15 μv at 12 to 14 cycles per second and irregular variations in frequency and amplitude. During wakefulness the ECoG consists of 20 to 34 cycles per second waves up to 100 μv. Seizure patterns are seen following pentylenetetrazol administration and presumably occur in other convulsive states.

Changes of ECoG with age correspond to other indices of maturation. Before 5 to 6 days no clear records of coordinated electrical activity have been obtained. From this age on activity is found, at first irregular, but gradually assuming essentially adult form at about 16 days of age (Kobayashi et al., 1963).

EYE AND VISUAL CAPACITY

Anatomy and function

A cross-section of the mouse eye is presented in Fig. 32-4. Note that the lens nearly fills the globe and is almost spherical; the optical effect is to project a small, relatively undetailed, bright image upon the retina. Brückner (1951) reported that he could not detect the image of a 1-cm^2 figure at a distance of 30 cm when viewing with a microscope focused on the retina from behind

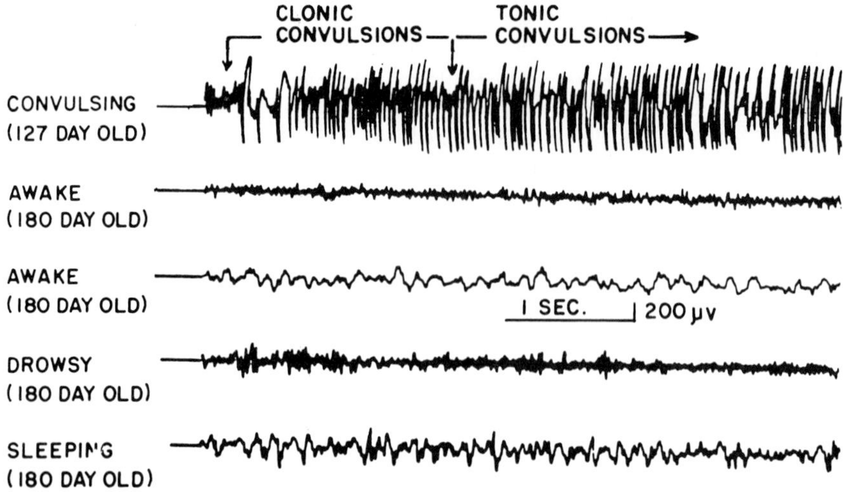

Fig. 32-3. Four typical patterns of the electrocorticogram in the mature mouse. (*From Kobayashi et al.*, 1963, *with permission of the authors and Plenum Press, Division of Consultants Bureau Enterprises, Inc.*)

the eyeball. The cornea covers about half the surface of the eyeball. The head of the optic nerve is extremely small. Indeed, Lashley (1932) reported estimates of ratio of receptor cells to ganglion cells in the mouse retina varying between 63 and 89. There is disagreement as to whether the mouse's receptors are all rods, or whether there are some cones as well. Karli (1952) reported having found a considerable number of cones and discussed observations by others both favoring and opposing his.

According to Walls (1942) the angle between the optic axis and body axis is about 60°, and the binocular field is probably about 40°. Bonaventure (1961) studied the spectral sensitivity curve of the mouse retina and found maximum sensitivity to be around 500 mμ.

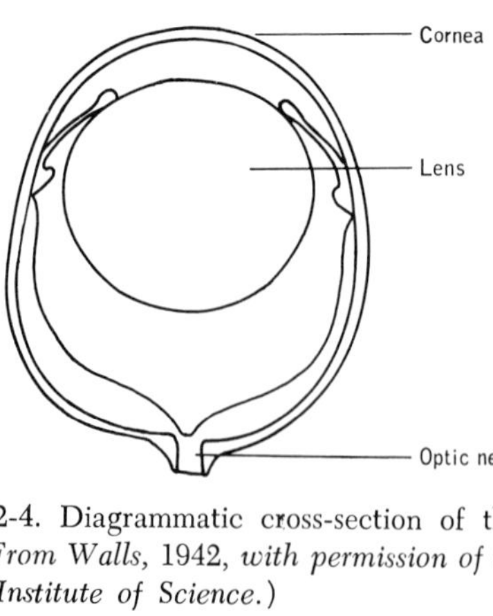

Fig. 32-4. Diagrammatic cross-section of the mouse eye. (*From Walls*, 1942, *with permission of the Cranbrook Institute of Science.*)

At birth the retina is composed of a ganglion cell layer, an inner fiber layer, and a thick layer of undifferentiated cells. Between 7 and 10 days the inner and outer limbs of the rods appear (Tansley, 1951). The general structure of the adult retina is established by the 12th day, and the eye typically opens about the 14th day (Sorsby et al., 1954). The eye of the young mouse is not fully mature at this time, for the hyaloid artery which supplies the embryonic lens is present until about 3 weeks after birth (Grüneberg, 1952), and the internal organization of the rods is not complete until about 4 weeks of age (Sorsby et al., 1954).

Visual capacity. It is possible to create a situation for inferring the visual capacity of an animal from its behavior in the presence of visual clues of brightness, hue, or form. Though simple in principle, the technique is difficult to apply, for possible extraneous clues must be rigidly controlled. In addition, failure to perform satisfactorily may be due to excessive difficulty of the task rather than limited receptor capacity of the organism.

This section will, with few exceptions, be limited to a review of the results obtained with a device, shown in Fig. 32-5, originally devised by Yerkes (1907) for use with the dancing mouse. The mouse leaves its nest box (A) and enters the choice area (B) through gate I. It must now choose between entry of the left and right chambers on the basis of visual cues provided by the experimenter. If the animal chooses correctly, it may return to its home cage and food by passing

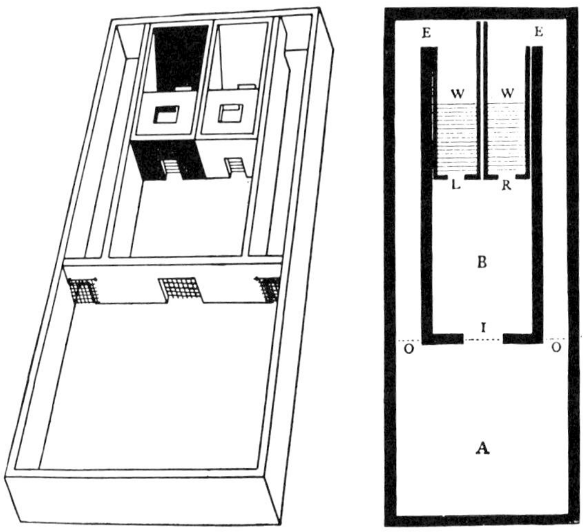

Fig. 32-5. Drawing and floor plans of Yerkes' discrimination box. A, nest box; B, entrance chamber; W, W, electric boxes; L, doorway of left electric box; R, doorway of right electric box; E, exit from electric box to alley; I, swinging door between A and B; O, swinging door between alley and A. (*From Yerkes, 1907, with permission of The Macmillan Company.*)

through (E) and (O). On the incorrect side, the exit (E) is blocked by glass and the animal receives a shock to its feet. The animal must then retrace and leave by the correct side. A cardboard is used in chamber (B) on occasion to force laggardly animals to choose rapidly. Location of the correct side should be varied from trial to trial by use of a table of random numbers or specially devised schedules (Hilgard, 1951). Common practice has been to give the animals 10 trials per day. Mice typically show ability to discriminate brightness differences after 100 to 200 trials. Consult Munn (1950) and Sutherland (1962) for an evaluation and discussion of techniques in this area.

Brightness discrimination. There is ample evidence that the mouse has sufficient visual capacity to learn brightness discrimination, i.e., to select the brighter or darker of two sides. For example, Bonaventure (1961) used brightness discrimination to determine the spectral sensitivity curve for the mouse eye under photopic and scotopic conditions of illumination. Threshold for light at each wavelength tested was defined as the lowest level of illumination producing choice behavior above chance levels. The resulting photopic and scotopic curves are shown in Fig. 32-6. Maximum sensitivity ocurs at 505 mμ, which is very close to the maximum absorption point for rhodopsin (500 mμ). Photopic and scotopic curves are very similar. This failure to obtain the Purkinje phenomenon—a shift in spectral sensitivity associated with change from cone to rod vision as illumination decreases—suggests that cones cannot play a very important part in mouse vision.

Color vision. Yerkes (1907) attempted to train dancing mice to discriminate green from blue, red from green, and blue from red lights. Physical brightness was varied over an extremely wide range to eliminate the factor of differential spectral sensitivity. He found no evidence for successful discrimination between green and blue, but some for discrimination between red and green and between red and blue. Waugh (1910) also tested for color vision and obtained some evidence for successful discrimination of red from green.

Hopkins (1927*a, b*), reasoning that mice apparently discriminating between hues might actually be responding to differences in brightness, attempted to find a gray paper of appropriate brightness which could not be discriminated from red or blue papers. Animals trained to go to blue also went to a specific light gray, but correctly discriminated between the blue and both darker and lighter grays. Mice trained to go to red confused it with a very dark gray. In a second experiment using light sources as stimuli, he found one animal which appeared able to discriminate red.

It is suspicious that red has always been the hue which mice have been reported to discriminate, for rods are particularly insensitive to red. The extreme variations in brightness used by experimenters to avoid this difficulty may not have been sufficient. Thus, neither the behavioral nor

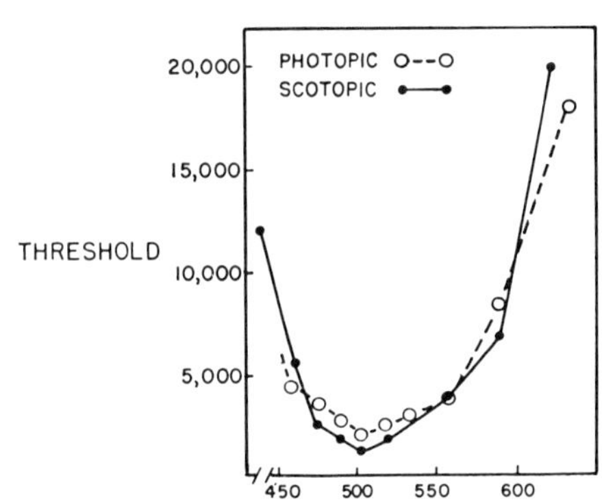

Fig. 32-6. Curves showing absolute thresholds in photopic and scotopic vision. Wavelength is shown on the abscissa, and light intensity in arbitrary units is shown on the ordinate. (*From Bonaventure, 1961, with permission of the author and Masson et Cie.*)

the anatomical literature on color vision is unequivocal.

Pattern discrimination. In studies on pattern vision, as on other visual capacities, Yerkes (1907) was the pioneer. He attempted to train mice to discriminate between a circle and a cross-shaped figure equated in brightness. Finding no evidence of ability to discriminate, he concluded that mice do not see very clearly and do not have very accurate perception of form. Waugh (1910), using similar stimuli and procedures, found some evidence for ability to discriminate.

Failure or poor performance may not necessarily indicate lack of capacity, for the task may be unnecessarily difficult. Boxberger (cited by Karli, 1954) has suggested a distance of 10 cm as the maximum for clear vision in the mouse. Both Yerkes and Waugh probably placed the forms to be discriminated considerably farther than that distance from the choice point. Karli (1954) circumvented this difficulty by forcing animals to each side 50 per cent of the time and allowing the animals to retrace. He found that mice discriminated between two rectangles differing only in orientation, one horizontal and the other vertical.

Other experimenters have successfully studied rather difficult discriminations using devices which permitted the animal to approach the stimuli closely. Rowley and Bolles (1935) placed the different stimuli to be discriminated on the choice doors and were able to study figure discrimination and transfer. Zimmerling (1934) placed mice in a circular enclosure having four openings located about its periphery. The animals were presented on each trial with four differently shaped openings such as circles, stars, triangles, or diamonds. Mice deprived of food for 12 to 15 hours learned to choose the correct openings on the basis of its visual characteristics and could then follow a U-shaped course to food. Runways behind incorrect openings were electrified to punish errors.

Wörner (1936) used the Zimmerling multiple-choice apparatus to study discrimination of form. Initially subjects were trained to select a circle presented with three triangles. When this problem was mastered the triangles were modified either by bulging the edges or rounding the corners to approximate the circle. Some mice were capable of rather fine discriminations. A similar series of experiments used ellipses which were made closer and closer to circles. Judging from the figures shown, it appears that mice can discriminate circles from ellipses with axes in a ratio of about 5:6. Some mice could transfer from discrimination of the cut-out openings to discrimination of painted figures surrounding uniformly shaped holes.

Boxburger (1953) compared visual discrimination in 10 white rats with that of 10 white mice. Stimulus cards were placed on swinging doors. The positive door opened easily, whereas pushing the negative door made an electrical contact and a bulb glowed as the locked door was encountered. Rats attained the criterion sooner on the initial three single-discrimination problems, but mice required fewer trials or the last three. There was little saving from problem to problem in the rats, but striking interproblem improvement was found in the mice. Multiple discrimination was also studied. After passing through one discrimination, a second was added, and so forth, until all six problems were included in one trial run. The arrangements of the correct stimuli varied from trial to trial for the 10 trials given per day. Over the 20 days mice consistently performed better than the rats on the six-stimulus test. Reetz (1957) also compared the learning performance of rats and mice. Using the Lashley jumping stand (Munn, 1950), he found rats could be more easily trained on all tasks. However, differences in performance were relatively small, and the better performing mice were not notably inferior to the better rats. The results clearly failed to justify the frequent assumption that mice are much poorer subjects for learning experiments than rats.

Genetic anomalies of vision

A number of genetic anomalies of the eye occur in mice, and some have become fixed in widely used inbred strains. It is important that an experimenter observing behavior be aware of these anomalies as sources of variation. They also may have special value for investigators interested in receptor function. Genes known to affect the eye are ectopic (*ec*), eyeless-1 (*ey-1*), eyeless-2 (*ey-2*), fidget (*fi*), lens rupture (*lr*), microphthalmia (*mi*), white (Mi^{wh}), ocular retardation (*or*), rodless retina (*r*), retinal degeneration (*rd*), and cataract (*Cat*). For further information about these genes see Chap. 8. The findings on stocks examined for retinal anomalies are presented in Table 32-2.

Rodless retina. Keeler (1924) was the first to find and investigate behaviorally a retinal defect which he called "rodless." The retina of the adult rodless mouse, according to Keeler (1927), was characterized histologically by (1) complete or nearly complete absence of the rod layer, (2) great reduction of the number of rows of nuclei in the adjacent external layer, and (3) lesser

Table 32-2. STOCKS AND INBRED STRAINS OF MICE WITH RECESSIVE INHERITED RETINAL DISEASE. (*Adapted from Sidman and Green, 1965, with permission of the authors and The American Genetic Association.*)

A. Strains with disease proved identical by 100% affected F_1 and F_2 offspring: CBA/J × C3H/FeJ (present report); Keeler rodless × Berlin-Dahlem stock.

B. Strains with disease proved identical by 100% affected F_1 offspring:

CBA/J × Brückner[a]	C3H/HeJ × P/J	(Sidman, 1965, personal communication)
BDP/J × Brückner[b]	C3H/HeJ × PL/J	(Sidman, 1965, personal communication)
P/J × Brückner[c]	C3H/HeJ × SJL/J	(Sidman, 1965, personal communication)
C3H/CaH × Brückner[d]	C3H/HeJ × ST/J	(Sidman, 1965, personal communication)
C3H/Ca × P/J[d]	C3H/HeJ × SWR/J	(Sidman, 1965, personal communication)

C. Stocks with histological changes indistinguishable from *rd*:

BUB/Bn	C3HeB/FeJ	DA/Hu	Albino-Bluhm[g]
C3H/An[e]	C3HeB/Hu	FL/Re	Basle waltzing stock[h,i]
C3H/Di	C3HfB/Hu	WB/Re	Basle and Zurich wild mice[i]
C3H/Ha[f]	C3H/St[f]	WC/Re	NIH wild stock[j]
C3H/HeHu-Sl^J	CFW[f]	WH/Re	Swiss albino stocks[k]
			Vienna stock[l]

D. Stocks with histologically normal retinas:

A[f]	BALB/cJ	C57BL/10Gn	DE/WyDi	LaA[f]	2-Prunt[f]
A/G[c]	BSL/Di	C57BL/10J	F[f]	LG/Rr[f]	5-Prunt[f]
A/HeJ	BrS[f]	C57BR/cd	FZ/Di	LOW[f]	RF/J
A/J	C/St[f]	C57L/J	H[f]	LP/J	RIII/AnJ
A/WySn	CBA/Ca	C58/J	HALB/Hu	MA/J	SM/J
A2G[d]	CBA/St[f]	DBA/1J	HD/Hu	MA/MyHu	WK/Re
AK[f]	CE/WyDi	DBA/1fHu	HG/Hu	MY/Hu	129/J
AKR/J	CHI[f]	DBA/1oHu	I[f]	MYA/Hu	129/Rr
B1+[f]	C57[f]	DBA/2[f]	IPBR[f]	N[f]	
AU/Ss	C57BL/Ks	DBA/2DeJ	JB/Di	PBR[f]	
BALB/cGn	C57BL/6J	DBA/2WyDi	JK[f]	PIN[f]	

NOTE: Sources are indicated by the following references. Wherever specific references are not given, the data were obtained at The Jackson Laboratory (Staats, 1964, personal communication).

[a] Theiler and Cagianut (1963).
[b] Sorsby et al. (1954).
[c] Tansley (1954).
[d] Lucas (1958).
[e] DiPaolo and Noell (1962).
[f] Paigen and Noell (1961).
[g] Fischer (1959).
[h] Bein (1947).
[i] Brückner (1951).
[j] Dunn and Andervont (1963).
[k] Noell (1958).
[l] Baumgartner and Paget (1955).

changes in other parts of the retina. Visual purple was said to be absent. The optic nerve, mitochondria of pigment epithelium, vascularization of the retina, and the ganglion cells and internal nuclei were all reported normal (Keeler, 1930). Differences between normal and rodless animals were not apparent until about 6 days after birth. Keeler (1927) believed that the condition was due to failure of the rods to develop rather than to their degeneration.

Since superficial observation of behavior gave no indication of blindness (an experience shared by others who have watched mice with retinal anomalies), Keeler (1927) tested rodless and control mice both in the Yerkes discrimination apparatus and in one of his own devising described in detail in Chap. 33. Neither these nor other tasks indicated that rodless mice could see (Keeler, 1928). Keeler et al. (1928) recorded the electroretinogram following photic stimulation in normal and rodless mice. They reported that rodless mice showed no electrical response to stimulation, concluding that if this were a necessary concomitant of vision, rodless mice were clearly blind. One disturbing note in Keeler's (1930) observations was that the pupil of the rodless eye exhibits "quasinormal" contractions in response to light. Hopkins (1927*c*) tested the ability of mice anatomically very similar to Keeler's rodless to discriminate red and gray papers in a

Yerkes discrimination apparatus. The mice showed clear evidence of ability to do so. One eye was removed prior to training in order to classify animals. Subsequently the other eye was removed. Performance then dropped to chance and remained there over a very long series of trials.

Retinal degeneration. An abnormality phenotypically quite similar to that produced by rodless but probably caused by a different gene (DiPaolo and Noell, 1962; Sidman and Green, 1965) has recently been investigated very extensively both anatomically and behaviorally. Discovered by Brückner (1951) in Germany, its effects have been described in detail by Tansley (1951, 1954) and by Noell (1958). According to Tansley, the retina appears to develop normally until 13 to 14 days of age. By day 16, most of the cells in the outer nuclear layer are dead, and nothing remains of the rods or their nuclei by 28 days. Later, the outer layers of the retina are lost and there is degeneration of the pigment epithelium. A cross of P/J with Tansley's animals is reported to have resulted in progeny all having affected retinas. A similar report has been made by Sorsby et al. (1954).

Karli (1952) examined mice supplied by Brückner and found that in many animals the degenerative process was interrupted after the rods disappeared, leaving a retina with bipolar cells and ganglion cells apparently normal. Further degeneration of these elements occurred between the ages of 3 and 8 months. Karli tested the residual visual capacity of animals without rods or cones, but with the remainder of their retinas intact.

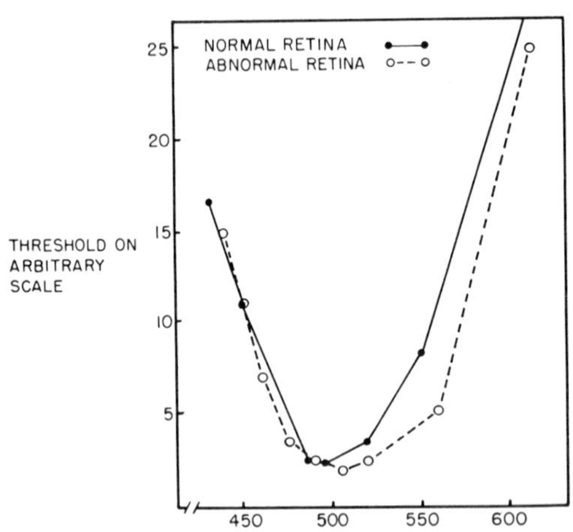

Fig. 32-7. Absolute threshold curves as a function of the wavelength of the light. Threshold units on the ordinate are arbitrary. Though not apparent from the figure, thresholds for *rd/rd* mice were estimated to be 10^5 greater than those of normal mice. (*From Bonaventure and Karli,* 1961, *with permission of the authors and Masson et Cie.*)

Karli (1952) found quasinormal pupillar contraction, similar to that described by Keeler (1927), and failed to find an electrical response from the retina of the Brückner mice. However, failure to obtain an ERG may not be crucial evidence for blindness. For according to Karli (1954) the ERG may be absent in both dogs and humans with satisfactory vision.

Using the Yerkes discrimination apparatus, Karli (1952) found no evidence for discrimination with a brightness difference of 50 meter-candles between the alternatives (a discrimination easily made by normal mice). However, he found clear evidence of discrimination learning when a difference of 500 meter-candles was used. Turning out the lights disrupted discriminative behavior. Controls for infrared emission appear adequate.

To demonstrate that the retina is somehow involved in successful performance of the task, Karli (1954) anesthetized the cornea with Buteline and the iris with homatropine and found that discrimination was unaffected. Surprisingly, he also showed that animals with the retinal anomaly could be trained to discriminate between rectangles in different orientations, an ability which was lost when the eyes were removed.

Bonaventure and Karli (1961) compared the spectral sensitivity curves for his presumptive *rd/rd* mice with one previously obtained for normal mice (Bonaventure, 1961). Both are presented in Fig. 32-7. A major difference, not apparent in the figure, is that the threshold for the abnormal mice was 10^5 higher than that for normal mice.

Where previously the total evidence favored Keeler's conclusion that mice without rods or cones cannot see, it now seems to vindicate Hopkins' assertion that some vision is possible.

OTHER SENSES

Ear and hearing

The Yerkes discrimination device has been used to test for deafness as well as blindness (Yerkes, 1907). However, simpler and more rapid techniques have been devised which work quite well. Yerkes himself also tested for deafness by producing loud noises: clapping his hands, shouting, whistling, exploding pistol caps, striking tuning forks, ringing an electric bell, using Galton whistles, making another mouse squeak. Reportedly, normal mice started violently and froze, and small wonder! Similarly, though less spectacularly, Deol and Kocher (1958) and Alford and Ruben (1963) used the "Pryer reflex," a twitching of the pinna to a sound stimulus, in making their assessments. It is claimed that this reflex is quite unambiguous in normal adult animals. Its occurrence has been shown to correlate reliably with the existence of cochlear potentials and acoustic nerve activity (Alford and Ruben, 1963).

Several investigators have studied the normal development of the ear and hearing in the mouse, and the picture is moderately detailed and quite consistent. According to Deol (1954, 1956), the vestibular part of the labyrinth is well developed at birth with generally adult proportions, though the cristae and maculae are still definitely immature. The cochlea, on the other hand, is still in initial stages of development, and though the hair cells are easily recognizable, there is as yet no trace of the tunnel of Corti. The cochlea is not completely developed until about the 14th day, at which time the normal mouse begins to hear. Alford and Ruben (1963), using both the Pryer reflex and electrical recording from the round window, reported that the modal time for appearance of the Pryer reflex is 12 days and the mean time for appearance of the cochlear potential is 11.6 days in a rather large sample of CBA/J mice.

Berlin (1963) reported acuity thresholds for hearing over a wide range of sound frequencies using the conditioned galvanic skin response as an index of hearing. He attached shocking electrodes to the forepaws and recording electrodes to the hind paws of CBA/J mice immobilized with bulbocapnine. Tones varying in frequency from 1 to 40 kc were presented prior to shock onset and change in body resistance was recorded. A modified method of limits was used to determine the minimum sound intensity required at each frequency to elicit three successive resistance changes to the tone. Berlin and Finck (1964, personal communication) obtained curves comparing frequency thresholds determined by both the GSR technique (solid line) and by microelectrode recording from auditory units (dotted line, Fig. 32-8). Commenting on the GSR technique, they state that simply presenting tone without shock and measuring resistance changes produced still more orderly and reproducible responses. Later work (Berlin, 1964, personal communication) continued to show maximum sensitivity to be between 10 and 20 kc, but peak frequency response appeared to be closer to 10 kc. Grüneberg et al. (1940) presented less

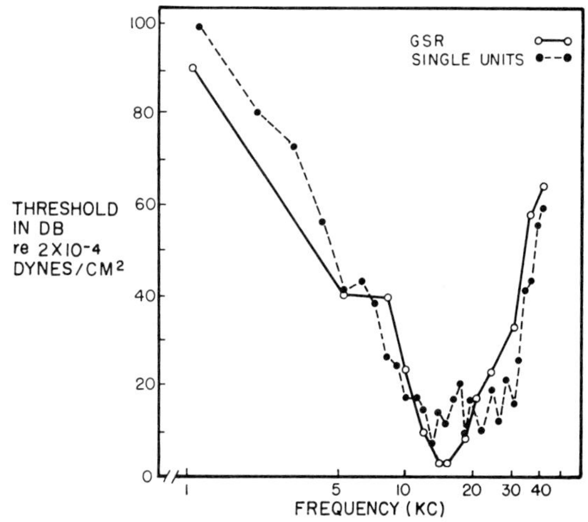

Fig. 32-8. Threshold changes with sound frequency as determined by galvanic skin response (GSR, solid line) and eighth nerve recordings (dotted line). Abscissa is sound frequency in thousands of cycles (KC), and ordinate is sound intensity in decibels. (*From Berlin and Finck,* 1964, *personal communication.*)

extensive data on the frequency response of the mouse ear.

Genes which are known to affect hearing are Ames' waltzer (*av*), deaf (v^{df}), deafness (*dn*), dreher (*dr*), jerker (*je*), kreisler (*kr*), pirouette (*pi*), shaker-1 (*sh-1*), shaker-2 (*sh-2*), spinner (*sr*), surdescens (*su*), Snell's waltzer (*sv*), shaker-with-syndactylism (*sy*), waltzer (*v*), varitint-waddler (*Va*), and whirler (*wi*). Mikaelian and Ruben (1964) compared developmental changes in eighth nerve activity and cochlear potentials of shaker-1 mice with those of normal CBA/J mice. For further information on these genes see Chap. 8 and Inherited Neurological Defects.

Chemical and tactile senses

It is generally thought that the primary receptors for the mouse are tactile and olfactory. The olfactory system has not been studied, though there is a literature on the role of olfactory stimuli in producing pregnancy blocking (Chap. 11).

The paucity of experiments on the chemical senses reflects the relatively inadequate state of knowledge in this area, a deficiency attributable to technical difficulties. Several studies on taste preference described in Chap. 33 may involve differential functioning of taste receptors, but the relative importance of central and peripheral factors in determining preferences has not been established. The mouse is known to have one taste bud in each of 50 to 70 fungiform papillae found in irregular rows just lateral to the median sulcus of the tongue (Kutuzov and Sicher, 1953).

Schaff (1933) observed what appears to be tactile discrimination. Mice lived in long runways with food at one end and housing at the other. Following a habituation period, a dividing wall of plaster of paris with slits 2 cm high and 1 to 7 mm wide was inserted. Mice always chose the largest slit to break through—distinguishing differences of 0.2 mm. Discrimination was as good in blinded mice as in sighted ones. Furthermore, the sinus hairs played no part, or a very small part, for dewhiskered mice were as competent as intact ones. Schaff believes that tactile receptors on the snout were involved.

Vestibular functions of the mouse have been studied largely in connection with inherited anomalies of this organ. As in the case of the eye, lesions produced by genes have resulted in natural experiments which contribute to physiology.

SENSORIMOTOR DEVELOPMENT

The behavior of mice changes rapidly during the first 2 weeks of life while the nervous system is maturing. Almost every day sees a new pattern appearing or an infantile one disappearing. The exact timing differs among individuals; the heavier infants may be a day or so ahead of more poorly nourished ones. However, no important or consistent differences between strains were reported by Williams and Scott (1953), Hertz (1964, personal communication), or Fox (1965), whose essentially similar accounts are the basis of the following section.

Locomotion at birth is limited to crawling sideways; by day 3 or 4 the young mouse can pivot. At day 6 or 7 it can stand and walk, at 9 days run unsteadily, and at 15 days move essentially as an adult. Up to 12 days the members of a litter usually remain in contact even when out of the nest, and they tend to stay close together up to about 20 days of age when some dispersion is seen. Young mice characteristically take long hops during the third week of life, earning them the name of "popcorn mice."

The usual mammalian trunk and limb reflexes can be demonstrated in infant mice. Crossed extension (a pinch on one hind foot leads to extension of the opposite limb) is strong at birth, weak by day 9, and gone by day 12. Forelimb and hind limb placing responses are weak or absent at birth but clearly seen from day 3 or 4 when the pup is brought in contact with a surface. A mouse is first able to right itself when placed on its back at 3 days; righting is accomplished with ease by 7 days. Upward pointing on a slope (negative geotropism) is seen when walking begins and is strongest at about 15 days. Baby mice will fall off an elevated surface before day 7, but between this age and 13 days they gradually develop an aversion to edges and avoid falling.

At birth, receptor capacities as judged by reflex responses appear limited to noxious and thermal stimuli. A generalized response to foot pinch in the neonate increases in strength and duration during the first week. In the second week the response becomes more discrete and may be limited to the stimulated limb if the stimulus is moderate. Auditory startle is reported to begin from the 10th to the 12th days and is strong at day 15. Eyes open between the 12th and 14th days. Mice nurse up to 16 or 24 days if not artificially weaned. A rooting reflex (forward pushing in response to contact on the side of the head) can be demonstrated during the early nursing period, but it weakens and usually disappears by day 15. Chewing movements are observable by 14 days and solid food is usually taken about day 16.

Grooming develops apparently out of rudimentary scratch reflexes by day 7 and is conspicuous by the 12th day. Other mice are groomed from about day 21 and the pattern then becomes a part of social behavior. Thus by 3 weeks, when less than half the eventual body weight is attained, the fundamental patterns of behavior have appeared except for fighting and sexual behavior. Deviations from the schedule of development are indicative of illness, trauma, or inherited defects.

MOTOR CAPACITY

Quantitative measurements of motor coordination and endurance are useful in studies of development, genetic differences, and drug effects. The rotarod, widely used in pharmacological laboratories, is an excellent device for evaluating coordination. Mice are placed on a slightly roughened rod about 2.5 cm in diameter which is rotated at constant speed by a motor. A speed of 1 rpm is used in our laboratory to detect gross aberrations of control. Adult mice after a few training trials can remain on the rod indefinitely at this speed and falling off indicates immaturity, structural defect, or intoxication. The task is more difficult at 6 rpm and individual variation is seen even among "normal" mice. At the faster speed one can detect motor effects of genes not generally considered to produce neuromuscular incoordination.

Swimming is frequently used as a test of vestibular function. Many of the waltzer-shaker mutants cannot maintain orientation to gravity when out of contact with a solid surface. Forced swimming has also been used as a test of endurance or as a stressor (Soltan, 1962). The walls of Soltan's apparatus were made of smooth material to prevent the animal from supporting its weight by resting its forefeet on the edges of the swimming tank, and air was bubbled through the water to keep the surface agitated. At 35°C laboratory mice can swim continuously for an hour or more. Endurance falls off in colder water.

There are many tests of "spontaneous" activity. The most commonly used are an open field, in which the amount of locomotion is measured by observation or by sensing devices, and rotating wheels which turn as the animal runs along the inner circumference. Vibration sensors can also be used to measure activity semiquantitatively in home cages or in special activity chambers. Since the animal is not forced to respond in these devices, they provide measures of motivation rather than of motor capacity alone and are treated in Chap. 33.

INHERITED NEUROLOGICAL DEFECTS

A large number of inherited defects of motor function have been described, many attributable to specific mutations. Brief descriptions of the mutant phenotypes with neurological effects are found in Chap. 8. The early literature was critically summarized by Grüneberg (1952). In this section we are concerned primarily with processes intervening between the genes and the behavioral phenotypes, more especially with those deviations compatible with a lifespan of at least several months. The treatment is not exhaustive but we have included examples of each of the major classes of phenotypes.

Genetics and classification

The most common finding in the mouse when single loci are involved is that the defect is dependent upon homozygosity for the mutant allele; that is, the mutant is recessive. A few mutations, for example waltzer-type (*Wt*) and trembler (*Tr*), produce marked effects when heterozygous and are called dominants. This appellation applies in the classic sense to trembler since *Tr*/+ and *Tr*/*Tr* individuals are practically identical (Grüneberg, 1952), but it is less precisely applied to waltzer-type since *Wt*/*Wt* mice die at the 11th day of gestation.

A few syndromes phenotypically very similar to mutant types have been shown by breeding tests to be multifactorial (e.g., careener, Chai and Chiang, 1962; zigzag, Lyon, 1960). These conditions seem to involve developmental thresholds. The genotype of the affected lines exerts precarious control over the differentiation of certain parts of the nervous system, resulting in a proportion of defective individuals.

Various attempts have been made to classify the neuromuscular mutants. Searle (1961) recognized three groups: (1) a waltzer-shaker group characterized by turning in circles, head shaking, and deafness; (2) a convulsive group characterized typically by spasticity and motor seizures under stress; and (3) an incoordinated group extending in more severe states to ataxia. A classification based upon observable behavior may not always correlate with one based upon neuropathological or biochemical findings. We may learn eventually that genes producing circling have very diverse metabolic effects or, conversely, that genes affecting the chemistry of myelin have different effects depending upon the period of development during which they are most active. For the purposes of this chapter, Searle's three-way classification is adequate, though it should be remembered that detailed investigations of individual genes will surely change the groupings, particularly if classification is based upon genetic control of chemical processes rather than the behavior phenotype (see Chap. 29 for further discussion).

Waltzer-shaker mutants

Mutants of this class have been known for centuries (Grüneberg, 1952, p. 179). Their prime triad of symptoms—circling, head shaking, and deafness—can all be caused by anomalies of the inner ear. Kocher (1960) studied eight mutants (deafness, *dn;* shaker-1, *sh-1;* shaker-2, *sh-2;* pirouette, *pi;* jerker, *je;* waltzer, *v;* deaf, v^{df}; and varitint-waddler, *Va*). In all the homozygotes (and in *Va* heterozygotes) he found regressive degeneration of the stria vascularis, neuroepithelium, and spiral ganglion, though the morphology of the bony labyrinth was unaffected. Damage is restricted primarily to the organ of Corti in two of the labyrinthine mutants v^{df} and *dn;* and these mice, though deaf, do not circle or shake their heads.

Table 32-3. BEHAVIORAL AND MORPHOLOGICAL EFFECTS IN SOME MUTANTS INVOLVING THE INNER EAR. (*From Lyon,* 1958, *with permission of the author and The Company of Biologists, Limited.*)

	Ear defect			*Responses to*		
Mutant	Otoliths	Horizontal canals	Vertical canals	Position changes	Horizontal movements	Vertical movements
Pallid (*pa*)	A[a]	N[a]	N	A	N	N
Zigzag[b]	N	A	N	N	A	N
Fidget (*fi*)	N	A	A	N	A	N
Twirler (*Tw*)	N/A	A	N/A	N/A	A	N/A
Dreher (*dr*)	N/A	A	A	N/A	A	A
Kreisler (*kr*)	A	A	A	A	A	A

[a] A = abnormal; N = normal.
[b] Not a simple mutant—multiple-factor determination.

Each mutant has a distinctive pattern of neuroepithelial degeneration. Some of them have temporary hearing, and the onset of deafness occurs among homozygotes as follows: *sh-1, sh-2,* 5 to 6 weeks; *pi,* 8 to 17 weeks, *v,* under 2 weeks (Kocher, 1960). A number of mutants show anomalies of the semicircular canals which produce waltzing movements, but they retain hearing because the cochlea does not degenerate. Examples are fidget (*fi*) (Truslove, 1956); twirler (*Tw*) (Lyon, 1958); and waltzer-type (*Wt*) (Stein and Huber, 1960). Zigzag is a polygenic character characterized by a reduction or absence of the horizontal canals, the remainder of the inner ear being essentially normal (Lyon, 1960). In some mutants, notably pallid (*pa*), absence of the otoliths in the vestibules impairs position responses (Lyon, 1953).

In summary, the labyrinthine mutants provide a relatively straightforward example of structural anomalies mediating particular behavioral defects. Table 32-3 illustrates this point. The unsolved problems in this group lie in the area of developmental rather than behavioral genetics. There is as yet no good explanation for the individualistic patterns of degenerative changes and failures of induction shown by this large group of genetically independent factors.

Convulsive syndromes

Mutants included here are a heterogeneous group with varying locomotor symptoms associated with exaggerated reflex responses and the occurrence of seizures following relatively mild stimulation. The lack of mutual exclusiveness between groups is illustrated by varitint-waddler (*Va*) (Cloudman and Bunker, 1945). Included in Kocher's (1960) list of labyrinthine mutants, it could equally well be classed as a convulsive mutant on the basis of its susceptibility to seizures during early life.

Trembler (*Tr*) is a dominant mutation manifested in heterozygotes by a triad of symptoms: spastic paralysis, action tremor, and frequent convulsions, particularly in young mice (Falconer, 1951). The electrocorticogram of trembler mice is normal (Braverman, 1953). The symptoms resemble myotonia gravis, but anticholinesterases, found helpful in this disease, are ineffective in restoring trembler mice to normality. Spastic mice (*spa/spa*) have similar symptoms but are more viable. Amino-oxyacetic acid, a γ-aminobutyrate transaminase inhibitor, has ameliorative effects on spastic mice (Chai et al., 1962). The authors suggest that spasticity results from an imbalance of excitatory and inhibitory factors in the central nervous system, which is rectified when γ-aminobutyric acid is allowed to accumulate. Tottering (*tg*) is another recessive mutation which causes convulsions in homozygotes (Green and Sidman, 1962). It could equally well be classed as a defect of coordination, since the gait is abnormal. No histological anomalies were detected in *tg/tg* mice by routine observation.

The dilute lethal gene (d^l) produces major effects on the nervous system as well as on pigment and phenylalanine metabolism. The biochemical aspects of the action of the *d* locus alleles are discussed in Chap. 21. At about 10 days of age d^l/d^l mice fall over while walking; clonic-tonic seizures with opisthotonus follow and death usually occurs by 18 to 20 days. Myelin degeneration in spinal tracts is the major neuropathological finding. Myelin appears on schedule but disintegrates within a few days (Kelton and Rauch, 1962).

Incoordination syndromes

Myelin is generally believed to play a major role in the nervous system because functional activation of tracts is associated with the process of myelinization. The severe symptoms and the eventual lethality of the d^l/d^l genotype are readily attributable to the loss of function in tracts connecting integrative centers in the brain stem with spinal motor neurons. Some mutants, however, with greatly reduced myelin display relatively mild clinical signs. Quaking mice (*qk*/*qk*) eat, swim, breed, and nurse well even though their entire central nervous system is very low in myelin by both chemical and histological criteria (Sidman et al., 1964). The defect appears to lie in synthesis since no signs of degeneration are seen. A marked tremor of the hind quarters is present but is compatible with life and reproduction under laboratory conditions. Jimpy (*jp*), a sex-linked recessive, produces similar but more severe symptoms and is usually lethal by 30 days of age (Sidman et al., 1964).

A number of inherited syndromes with major effects on the cerebellum are known. Reeler (*rl*) has been most thoroughly studied. Falconer (1951) hypothesized that the homozygotes were "mentally deficient" because of their inept behavior. Myelin formation is apparently normal, but there are cytoarchitectonic anomalies in cerebral cortex, hippocampus, and particularly in the cerebellum, which is reduced in size (Hamburgh, 1960, 1963; Meier and Hoag, 1962). Agitans (*ag*) (Hoecker et al., 1954) and staggerer (*sg*) mice (Sidman et al., 1962) also have defective cerebellums. Still another mechanism producing faulty locomotion is found in dystonia musculorum (*dt*), a neuropathy which mainly affects the dorsal root ganglia of spinal nerves (Duchen et al., 1962; Duchen and Strich, 1964). The number of identified neuromuscular mutants increases steadily and the neuropathological descriptions of many are yet to be published. This account can only suggest the variety of conditions subsumed under this heading. The final word on their classification must await the completion of thorough studies of each genetic entity. At least four kinds of analysis are possible: genetic, behavioral, neuropathological, and biochemical. For no one condition, save possibly dilute lethal (d^l), have all these levels been integrated. It seems particularly desirable to search for biochemical differences between the mutants as possible keys to the nature of genetic control over growth and differentiation. Behavioral methods of greater sophistication may also prove useful. While there may be little psychological interest in testing the learning of the more severely handicapped individuals, it would be interesting to know if quaking, for example, has peculiarities of learning associated with its deficient supply of myelin.

AUDIOGENIC SEIZURES

Laboratory mice of many strains convulse when subjected to intense sound stimulation. A susceptible subject ordinarily exhibits the following sequence of responses: startle, brief freezing or agitated walking, running at a more and more rapid pace, and finally a convulsion, which may be as mild as a standing spasm. More typically the mouse falls on its side and kicks rhythmically (clonic phase). Many mice go into tonic extension with fixation of trunk muscles in the inspiratory position. Death frequently results from such seizures unless the animal is given artificial respiration. Maximal audiogenic seizures have components similar to pentylenetetrazol or electroshock seizures, but the patterning is different (Swinyard et al., 1963).

Effects of age

Susceptibility varies with age. Seizures are rare before 15 days of age in DBA/2 mice, reach a peak of severity and incidence at 20 to 35 days, and become very infrequent after 70 days of age (Vicari, 1951*b*). Strain A mice, in contrast, are moderately susceptible (incidence 10 to 30 per cent) over a long period up to 245 days of age (Vicari, 1957). The O'Grady susceptible selected strain maintains nearly 100 per cent susceptibility between 20 and 60 days, but the proportion of maximal seizures is much less at older ages (Swinyard et al., 1963). The interstrain variations in age distribution suggest that susceptibility is a function of developmental patterns.

Factors affecting susceptibility

Prestimulation by subconvulsive doses of sound enhances or reduces susceptibility depending upon the temporal parameters of the treatment (Fuller and Williams, 1951; Fuller and Smith, 1953; Ginsburg and Fuller, 1954). Chilled mice are less susceptible (Fuller and Rappaport, 1952). In fact, seizure susceptibility is affected by such a variety of factors that it is difficult to replicate experiments quantitatively. This leads to some doubt regarding the hypothesis of Miller (1962) that very low radiation levels (0.2 mR per hour) were responsible for susceptibility changes in her colony. Indeed, Tacker and Furchtgott (1962) were unable to obtain an effect with 10 R of whole-body radiation.

The most effective sound frequencies for inducing seizures lie between 12,000 and 25,000 cycles per second (Frings and Frings, 1952). Sound pressures between 86 and 104 db above 0.002 dynes per cm^2 have been employed by most investigators. Although pure tone oscillators and white noise generators have been used, an ordinary doorbell suspended over a washtub or placed near a cage is perhaps the most common source of sound. Provided the stimulus is intense enough and in the proper frequency range, its physical dimensions do not appear to be critical determinants of variations in susceptibility.

Strain variation

Seizure susceptibility is apparently found in randombred Swiss albino stocks, in which selection for high and low susceptibility has been successful (Frings and Frings, 1953). Several inbred strains and hybrids have been investigated, but in only a few have the populations been large enough for accurate estimate of seizure risk (Hall, 1947; Fuller and Williams, 1951; Ginsburg, 1954; Vicari, 1951*b*, 1957).

Another source of difficulty in comparing susceptibility arises from differences in statistical treatment of data. Several workers (for example, Witt and Hall, 1949; Ginsburg, 1954) have characterized genetic groups by their proportion of convulsers, defined as animals which have a seizure on one or more of a fixed number of trials (usually four or five). Fuller et al. (1950) used seizure risk on a specified trial (usually the first) as an index of susceptibility. The methods yield quite different values. For example, if the one-trial risk is as low as 0.6, the risk of convulsing on one or more of five trials is 0.99 ($= 1 - 0.4^5$).

The death rate following maximal audiogenic convulsions can be as high as 80 per cent. Resuscitation is not always effective. Hybrids of strain DBA/2J with either strains AKR/J or A/J have been found highly susceptible to audiogenic seizures, but survive better than the parent strains.

Physiological basis

Swinyard et al. (1963) have adopted the view that the occurrence of a seizure is evidence of sustained activity in an "oscillator" group of neurons. They cite their results with low frequency electrical induction of seizures in resistant (CF#1) and susceptible (O'Grady) mice as evidence that maximal-seizure mice are more prone to oscillator discharge and spread than are minimal-seizure and resistant mice. Fuller and Smith (1953) hypothesized that occurrence of convulsions depended upon the relative rates of recruitment and blocking of motor circuits, these parameters presumably varying from strain to strain. They pointed out that seizure latencies were bimodally distributed and suggested either that different groups of neurons were involved in the fast and slow seizures, or that some humoral factor requiring about 30 seconds to become effective potentiated the late seizures. Direct confirmation of these theories by electrophysiological or neurochemical means is lacking.

Audiogenic seizures have served as phenotypes for a number of physiological investigations directed primarily at defining the gene-behavior relationship in biochemical terms. Ginsburg (1954) reported that glutamic acid and a number of other substances involved in the tricarboxylic acid energy cycle reduced seizure susceptibility; other substances increased it. He was unable to find any single principle for predicting enhancement or inhibitory effects. Particular genetic interest attaches to glutamic acid, which protected strain DBA/1 but not strain HS. Thus, similar behavioral phenotypes may have different physiological substrates and be genotypically unrelated. The glutamic acid effect is apparently not mediated through the adrenal glands, a mechanism once thought likely (Fuller and Ginsburg, 1954).

The metabolism of brains of audiogenic-seizure-susceptible DBA/1 and nonsusceptible C57BL/6 mice was studied by Abood and Gerard (1955). The brains did not differ in rate of glycolysis, activity of cytochrome oxidase, malic dehydrogenase, DPN cytochrome reductase, succinic dehydrogenase, and alkaline phosphatase. At 30 days, the age of maximum susceptibility to seizures, DBA mice were significantly lower in adenosine triphosphatase activity, a difference which disappeared at 45 to 50 days when the mice became resistant to convulsions. The rate of oxidative phosphorylation fell in DBA mice during the period of susceptibility. The authors concluded that vulnerability was related to a defect in the phosphorylating system.

The endocrine system is also involved in seizure susceptibility. The thyroid inhibitor 6n-propythiouracil protected DBA/2 effectively against convulsive death and decreased the risk of seizures (Vicari, 1951*a*). Insulin provided protection but castration increased susceptibility in

Table 32-4. RELATIONSHIP OF DRUGS AND ELECTROSHOCK TO AUDIOGENIC SEIZURE SUSCEPTIBILITY

Agent	Effect[a]	Reference
Caffeine	Caffeine produces more seizures in S than in R	Busnel and Lehmann (1961)
	No enhancement in low S	Moffatt et al. (1960)
Chlorpromazine	PD_{50} in DBA/1[b] = 4.8 mg/kg	Plotnikoff (1960)
	PD_{50} in SW[b] > 160 mg/kg	
Electroshock	S more susceptible to electroshock than R	Goodsell (1955)
		Hamburgh and Vicari (1960)
		Swinyard et al. (1963)
Meprobamate	PD_{50} in DBA/1 = 4.5 mg/kg	Plotnikoff (1960)
	PD_{50} in SW = 90 mg/kg	
	PD_{50} in S = 50 mg/kg	Busnel et al. (1958)
Nicotine	LD_{50} lower in S than in R	Frings and Kivert (1953)
	Seizures facilitated by subconvulsive drug	
Pentylenetetrazol	ED_{100} for S = 55 mg/kg	Busnel and Lehmann (1961)
	ED_{100} for R = 70 mg/kg	
	No difference R and S	Swinyard et al. (1963)
Reserpine	PD_{50} in DBA/1 = 5.0 mg/kg	Plotnikoff (1960)
	PD_{50} in SW = 9.0 mg/kg	
	Potentiation in 62%, protection in 38% of S	Bielec (1959)
	Enhancement of seizure in S	Busnel et al. (1958)
Strychnine	ED_{100} for S = 1.35 mg/kg	Busnel and Lehmann (1961)
	ED_{100} for R = 1.75 mg/kg	

[a] R and S signify resistant or susceptible lines, respectively; PD = protective dose; LD = lethal dose; ED = effective dose.

[b] Both DBA/1 and SW are susceptible strains, but they differ markedly in antiseizure response to three common tranquilizers.

DBA/1 and DBA/2 (Miller and Potas, 1956). Triiodothyronine administered to young mice did not alter seizure incidence except to advance the age at which seizures could be elicited in DBA/2 mice (Hamburgh and Vicari, 1960).

Pharmacology of seizures

Plotnikoff (1958) and Plotnikoff and Green (1957) demonstrated that tranquilizers such as chlorpromazine, meprobamate, and reserpine decrease seizure susceptibility. These and other pertinent studies are summarized in Table 32-4. In general, audiogenic-seizure–susceptible strains have a lower electroshock threshold and require less of a convulsant drug to produce a seizure, though Metrazol appears to be an exception. However, strain differences in effective drug levels are not simple functions of seizure susceptibility as shown by the tabulated data on anticonvulsant action of chlorpromazine, reserpine, and meprobamate.

SUMMARY

Contrary to popular belief as well as to our own when we started to assemble materials for this chapter, there *is* a substantial body of behaviorally relevant information on the biology of the mouse. Something is known of the mouse brain, its anatomy, chemical constituents, electrical activity, and development. The structure and visual capacity of the mouse eye have been studied in some depth. A good beginning has been made on hearing. Limited normative data on sensorimotor development and capacity exist as well. Knowledge is uneven, as is shown by the scanty information on such subjects as olfaction, taste, and electrophysiological characteristics of neurological mutants. No atlas of the mouse brain is generally available. These are a few of the gaps which must be filled.

The nervous and sensory systems of the laboratory mouse have attracted scientific interest primarily because it is the mammal whose genetics is best known and best controlled. A major part of the past research concerned mutant animals possessing defects of the brain, eye, and ear. The number of genes known already and the variety of their effects is impressive. We have mentioned 11 genes affecting the eye, 16 affecting hearing, and 18 affecting motor function. Among the variety of genes producing a gross syndrome there is a diversity of specific effect. For example, the incoordination syndrome can result from deficiencies in myelinization (quaking and jimpy), anatomical anomalies of the cerebellum (agitans and staggerer), and anomalies of dorsal root ganglia (dystonia musculorum). These gene-produced lesions have resulted in natural experiments which can contribute substantially to psychology as well as to physiology and biochemistry.

Genetical studies need not be restricted to single-gene effects. Quantitative traits, the effect of many genes of small individual effect acting together, are also very important. For example, variation in the brain to body weight ratio among inbred lines is great, and so also may be the variability in size of specific brain structures. The genetic diversity of the mouse makes it excellent material for research in this area. There is much to be done both with continuous and discrete traits. Both physiological and psychological differences between inbred strains and stocks segregating at a single locus need to be traced backward toward their origins.

Audiogenic seizures have been widely studied in mice as a model of maladaptive behavior in which response to an environmental trigger is modified by genetic background. Research has concentrated on elucidation of the mechanisms by which genes might control susceptibility and on chemical modification of susceptibility.

LITERATURE CITED

Abood, L. G., and R. W. Gerard. 1955. Phosphorylation defect in the brains of mice susceptible to audiogenic seizure, p. 467–472. *In* H. Waelsch [ed.] Biochemistry of the developing nervous system. Academic Press, New York.

Alford, B. R., and R. J. Ruben. 1963. Physiological, behavioral, and anatomical correlates of the development of hearing in the mouse. Ann. Otol. Rhinol. Laryngol. 72:237–247.

Baumgärtner, M., and O. E. Paget. 1955. Histologische Untersuchung eines rezessive erblichen Retinamerkmals bei der Hausmaus. Österreich. Zool. Z. 6:7–10.

Bein, H. J. 1947. Über vererbliche Aplasie des Sehnerven bei der Maus. Ophthalmologia 113: 12–37.

Berlin, C. I. 1963. Hearing in mice via GSR audiometry. J. Speech Hearing Res. 6:359–368.

Bielec, S. 1959. Influence of reserpine on the behavior of mice susceptible to audiogenic seizures. Arch. Int. Pharmacodyn. 119:352–358.

Bonaventure, N. 1961. Sur la sensibilité spectrale de l'appareil visuel chez la souris. Compt. Rend. Soc. Biol. 155:918–921.

Bonaventure, N., and P. Karli. 1961. Sensibilité visuelle spectrale chez des souris à rétine entièrement dépourvue de cellules visuelles photoréceptrices. Compt. Rend. Soc. Biol. 155: 2,015–2,018.

Boxberger, F. von. 1953. Vergleichende Untersuchungen über das visuelle Lernvermögen bei weissen Ratten und weissen Mäusen. Z. Tierpsychol. 9:433–451.

Braverman, I. M. 1953. Neurological actions caused by the mutant gene "trembler" in the house mouse (*Mus musculus*, L.). An investigation. J. Neuropathol. Exp. Neurol. 12:64–72.

Bruckner, R. 1951. Spaltlampenmikroskopie und Ophthalmoskopie am Auge von Ratte und Maus. Doc. Ophthalmol. 5–6:452–554.

Busnel, R. G., and A. Lehmann. 1961. Action de convulsivants chimiques sur les souris de lignеés sensible et résistante à la crise audiogène. Part III. Cafeine. J. Physiol. 53:285–286.

Busnel, R. G., A. Lehmann, and M. C. Busnel. 1958. Étude de la crise audiogène de la souris comme test psycho-pharmacologique: Son application aux substances de type "tranquilliseur." Pathol. Biol. 34:749–762.

Chai, C. K., and M. S. M. Chiang. 1962. The inheritance of careener, unbalanced locomotion in mice. Genetics 47:435–441.

Chai, C. K., E. Roberts, and R. L. Sidman. 1962. Influence of aminooxyacetic acid, a γ-aminobutyrate transaminase inhibitor, on hereditary spastic defect in the mouse. Proc. Soc. Exp. Biol. Med. 109:491–495.

Chiquoine, A. D. 1954. Distribution of alkaline phosphomonesterase in the central nervous system of the mouse embryo. J. Comp. Neurol. 100:415–439.

Cloudman, A. M., and L. E. Bunker, Jr. 1945. The varitint-waddler mouse. J. Hered. 36:259–263.

Deol, M. S. 1954. The anomalies of the labyrinth of the mutants varitint-waddler, shaker-2, and jerker in the mouse. J. Genet. 52:562–588.

Deol, M. S. 1956. A gene for uncomplicated deafness in the mouse. J. Embryol. Exp. Morphol. 4:190–195.

Deol, M. S., and W. Kocher. 1958. A new gene for deafness in the mouse. Heredity 12:463–466.

DiPaolo, J. A., and W. K. Noell. 1962. Some genetic aspects of visual cell degeneration in mice. Exp. Eye Res. 1:215–220.

Duchen, L. W., D. S. Falconer, and S. J. Strich. 1962. Dystonia musculorum. A hereditary neuropathy of mice affecting mainly sensory pathways. J. Physiol. 165:7P–9P.

Duchen, L. W., and S. J. Strich. 1964. Clinical and pathological studies of an hereditary neuropathy in mice (dystonia musculorum). Brain 87:367–378.

Dunn, T. B., and H. B. Andervont. 1963. Histology of some neoplasms and non-neoplastic lesions found in wild mice maintained under laboratory conditions. J. Nat. Cancer Inst. 31: 873–901.

Falconer, D. S. 1951. Two new mutants, "trembler" and "reeler," with neurological actions in the house mouse (*Mus musculus* L.) J. Genet. 50:192–201.

Fischer, H. 1959. Mikroskopische Untersuchungen an der Retina von Mäusen mit erblichen Augenaffektionen. Acta Biol. Med. Ger. 2:231–251.

Folch-Pi, J. 1955. Composition of the brain in relation to maturation, p. 121–133. *In* H. Waelsch [ed.] Biochemistry of the developing nervous system. Academic Press, New York.

Fox, M. W. 1965. Reflex ontogeny and behavioral development of the mouse. Anim. Behav. 8: 234–241.

Frings, H., and M. Frings. 1952. Acoustical determinants of audiogenic seizures in laboratory mice. J. Acoust. Soc. Amer. 24:163–169.

Frings, H., and M. Frings. 1953. The production of stocks of albino mice with predictable susceptibilities to audiogenic seizures. Behaviour 5:305–319.

Frings, H., and A. Kivert. 1953. Nicotine facilitation of audiogenic seizures in laboratory mice. J. Mammal. 34:391–393.

Fuller, J. L., C. Easler, and M. E. Smith. 1950. Inheritance of audiogenic seizure susceptibility in the mouse. Genetics 35:622–632.

Fuller, J. L., and B. E. Ginsburg. 1954. Effect of adrenalectomy on the anticonvulsant action of glutamic acid in mice. Amer. J. Physiol. 176: 367–370.

Fuller, J. L., and A. Rappaport. 1952. The effect of wetting on sound-induced convulsions in mice. J. Comp. Physiol. Psychol. 45:246–249.

Fuller, J. L., and M. E. Smith. 1953. Kinetics of sound induced convulsions in some inbred mouse strains. Amer. J. Physiol. 172:661–670.

Fuller, J. L., and E. Williams. 1951. Gene-controlled time constants in convulsive behavior. Proc. Nat. Acad. Sci. 37:349–356.

Ginsburg, B. E. 1954. Genetics and the physiology of the nervous system. Proc. Ass. Nerv. Ment. Dis. 33:39–56.

Ginsburg, B. E., and J. L. Fuller. 1954. A comparison of chemical and mechanical alterations of seizure patterns in mice. J. Comp. Physiol. Psychol. 47:344–348.

Goodsell, J. S. 1955. Properties of audiogenic seizures in mice and the effect of anticonvulsant drugs. Fed. Proc. 14:345. (Abstr.)

Green, M. C., and R. L. Sidman. 1962. Tottering —a neuromuscular mutation in the mouse. J. Hered. 53:233–237.

Grüneberg, H. 1952. The genetics of the mouse, 2nd ed. Nijhoff, The Hague. 650 p.

Grüneberg, H., C. S. Hallpike, and A. Ledoux. 1940. Observations on the structure, development, and electrical reactions of the internal ear of the shaker-1 mouse (*Mus musculus*). Proc. Roy. Soc. B 129:154–173.

Hall, C. S. 1947. Genetic difference in fatal audiogenic seizures. J. Hered. 38:3–6.

Hamburgh, M. 1960. Observations on the neuropathology of "reeler," a neurological mutation in mice. Experientia 16:460.

Hamburgh, M. 1963. Analysis of the postnatal developmental effects of "reeler," a neurological mutation in mice. A study in developmental genetics. Develop. Biol. 8:165–185.

Hamburgh, M., and E. Vicari. 1960. A study of some physiological mechanisms underlying susceptibility to audiogenic seizures in mice. J. Neuropathol. Exp. Neurol. 19:461–472.

Hilgard, E. R. 1951. Methods and procedures in the study of learning, p. 517–567. *In* S. S. Stevens [ed.] Handbook of experimental psychology. Wiley, New York.

Himwich, W. A. 1962. Biochemical and neurophysiological development of the brain in the neonatal period. Int. Rev. Neurobiol. 4:117–158.

Hoecker, G., S. Martinez, A. Markovic, and O.

Pizzaro. 1954. Agitans, a new mutation in the house mouse with neurological effects. J. Hered. 45:10–14.

Hopkins, A. E. 1927*a*. Vision and retinal structure in mice. Proc. Nat. Acad. Sci. 13:488–492.

Hopkins, A. E. 1927*b*. Experiments on color vision in mice in relation to the duplicity theory. Z. Vergl. Physiol. 6:300–344.

Hopkins, A. E. 1927*c*. Vision in mice with "rodless" retinae. Z. Vergl. Physiol. 6:345–360.

Karli, P. 1952. Rétines sans cellules visuelles—recherches morphologiques, physiologiques, et physiopathologiques chez les rongeurs. Arch. Anat. Histol. Embryol. 35:1–76.

Karli, P. 1954. Étude de la valeur fonctionnelle d'une rétine dépourvue de cellules visuelles photo-réceptrices. Arch. Sci. Physiol. 8:305–328.

Keeler, C. E. 1924. The inheritance of a retinal abnormality in white mice. Proc. Nat. Acad. Sci. 10:329–333.

Keeler, C. E. 1927. Rodless retina, an ophthalmic mutation in the house mouse, *Mus musculus*. J. Exp. Zool. 46:355–407.

Keeler, C. E. 1928. Blind mice. J. Exp. Zool. 51: 495–508.

Keeler, C. E. 1930. Hereditary blindness in the house mouse with special reference to its linkage relationships. *In* Bulletin No. 3, Howe Laboratory of Ophthalmology, Harvard Medical School, January. 11 p.

Keeler, C. E., E. Sutcliffe, and E. L. Chaffee. 1928. Normal and "rodless" retinae of the house mouse with respect to the electromotive force generated through stimulation by light. Proc. Nat. Acad. Sci. 14:477–484.

Kelton, D. E., and H. Rauch. 1962. Myelination and myelin degeneration in the central nervous system of dilute-lethal mice. Exp. Neurol. 6: 252–262.

Kobayashi, T. 1963. Brain-to-body ratios and time of maturation of the mouse brain. Amer. J. Physiol. 204:343–346.

Kobayashi, T., O. Inman, W. Buño, and H. E. Himwich. 1963. A multidisciplinary study of changes in mouse brain with age. Recent Adv. Biol. Psychiat. 5:293–308.

Kocher, W. 1960. Untersuchungen zur Genetik und Pathologie der Entwicklung von 8 Labyrinthmutanten (deaf-waltzer-shaker Mutanten) der Maus (*Mus musculus*). Z. Vererb. 91: 114–140.

König, J. F. R., and R. A. Klippel. 1963. The rat brain. Williams & Wilkins, Baltimore. 168 p.

Kutuzov, H., and H. Sicher. 1953. Comparative anatomy of the mucosa of the tongue and the palate of the laboratory mouse. Anat. Rec. 116: 409–425.

Lashley, K. S. 1932. The mechanism of vision. V. The structure and image-forming power of the rat's eye. J. Comp. Psychol. 13:173–200.

Lorente de Nó, R. 1933. Studies on the structure of the cerebral cortex. I. The area entorhinalis. J. Psychol. Neurol. 45:381–438.

Lorente de Nó, R. 1934. Studies on the structure of the cerebral cortex. II. Continuation of the study of the ammonic system. J. Psychol. Neurol. 46:113–177.

Lowry, O. H., and J. V. Passoneau. 1964. The relationships between substrates and enzymes of glycolysis in brain. J. Biol. Chem. 239:31–42.

Lowry, O. H., J. V. Passoneau, F. X. Hasselberger, and D. W. Schulz. 1964. Effect of ischemia on known substrates and cofactors of the glycolytic pathway in brain. J. Biol. Chem. 239:18–30.

Lucas, D. R. 1958. Retinal dystrophy strains. Mouse News Letter 19:43.

Lyon, M. F. 1953. Absence of otoliths in the mouse: an effect of the pallid mutant. J. Genet. 51:638–650.

Lyon, M. F. 1958. Twirler: a mutant affecting the inner ear of the house mouse. J. Embryol. Exp. Morphol. 6:105–116.

Lyon, M. F. 1960. Zigzag: a genetic defect of the horizontal canals in the mouse. Genet. Res. 1: 189–195.

Maas, J. W. 1962. Neurochemical difference between two strains of mice. Science 137:621–622.

Maas, J. W. 1963. Neurochemical difference between two strains of mice. Nature 197:255–257.

Meier, H., and W. G. Hoag. 1962. The neuropathology of "reeler," a neuro-muscular mutation in mice. J. Neuropathol. Exp. Neurol. 21: 649–654.

Meier, H., E. Jordan, and W. G. Hoag. 1962. The zymogram technique as a tool for the study of gentotypic difference. J. Histochem. Cytochem. 10:103–104.

Mikaelian, D. O., and R. J. Ruben. 1964. Hearing degeneration in shaker-1 mouse. Arch. Otolaryngol. 80:418–430.

Miller, D. S. 1962. Effects of low level radiation on audiogenic convulsive seizures in mice, p. 513–531. *In* T. J. Haley and R. S. Snider [ed.] Response of the nervous system to ionizing radiation. Academic Press, New York.

Miller, D. S., and M. Z. Potas. 1956. The influence of castration on susceptibility to audiogenic seizures in DBA mice. Anat. Rec. 124:336. (Abstr.)

Moffat, E., R. F. Krueger, R. O. Pfeiffer, and D. M. Green. 1960. Effects of analgesics on audiogenic seizures in mice. Fed. Proc. 19: 271. (Abstr.)

Munn, N. L. 1950. Handbook of psychological research on the rat. Houghton Mifflin, Boston. 598 p.

Noell, W. K. 1958. Studies on visual cell viability and differentiation. Ann. N.Y. Acad. Sci. 74: 337–361.

Paigen, K., and W. K. Noell. 1961. Two linked genes showing a similar timing of expression in mice. Nature 190:148–150.

Plotnikoff, N. P. 1958. Bioassay of potential tranquilizers and sedatives against audiogenic seizures in mice. Arch. Int. Pharmacodyn. 116: 130–135.

Plotnikoff, N. P. 1960. Ataractics and strain differences in audiogenic seizures in mice. Psychopharmacologia 1:429–432.

Plotnikoff, N. P., and D. M. Green. 1957. Bioassay of potential ataraxic agents against audiogenic seizures in mice. J. Pharmacol. Exp. Therap. 119:294–298.

Reetz, W. 1957. Unterschiedliches visuelles Lernvermögen von Ratten und Mäusen. Z. Tierpsychol. 14:347–361.

Roberts, E., P. J. Harman, and S. Frankel. 1951. Γ-aminobutyric acid content and glutamic decarboxylase activity in developing mouse brain. Proc. Soc. Exp. Biol. Med. 78:799–803.

Rose, M. 1929. Cytoarchitektonischer Atlas der Grosshirnrinde der Maus. J. Psychol. Neurol. 41:1–51.

Rowley, J. B., and M. M. Bolles. 1935. Form discrimination in white mice. J. Comp. Physiol. Psychol. 20:205–210.

Schaff, W. 1933. Raum- und Materialunterscheidung bei der grauen Hausmaus. Z. Vergl. Physiol. 18:622–653.

Searle, A. G. 1961. Tipsy, a new mutant in linkage group VII of the mouse. Genet. Res. 2:122–126.

Sidman, R. L., M. M. Dickie, and S. H. Appel. 1964. Mutant mice, *quaking* and *jimpy*, with deficient myelination in the central nervous system. Science 144:309–311.

Sidman, R. L., and M. C. Green. 1965. Retinal degeneration in the mouse; location of the rd locus in linkage group XVII. J. Hered. 56:23–29.

Sidman, R. L., P. W. Lane, and M. M. Dickie. 1962. Staggerer, a new mutation in the mouse affecting the cerebellum. Science 137:610–612.

Slotnick, B. M., and W. B. Essman. 1964. A stereotaxic atlas of the mouse brain. Amer. Zool. 4: 344. (Abstr.)

Soltan, H. C. 1962. Swimming stress and adaptation by dystrophic and normal mice. Amer. J. Physiol. 203:91–94.

Sorsby, A., P. C. Koller, M. Attfield, J. B. Davey, and D. R. Lucas. 1954. Retinal dystrophy in the mouse: histological and genetic aspects. J. Exp. Zool. 125:171–198.

Stein, K. F., and S. A. Huber. 1960. Morphology and behavior of waltzer-type mice. J. Morphol. 106:197–203.

Sutherland, N. S. 1962. The methods and findings of experiments on the visual discrimination of shape by animals. Experimental Psychology Society Monograph No. 1. Heffer, Cambridge. 68 p.

Swinyard, E. A., A. W. Castellion, G. B. Fink, and L. S. Goodman. 1963. Some neurophysiological and neuropharmacological characteristics of audiogenic-seizure-susceptible mice. J. Pharmacol. Exp. Therap. 140:375–384.

Tacker, R., S. Furchtgott, and E. Furchtgott. 1962. Low-level gamma irradiation and audiogenic seizures. Radiat. Res. 17:614–618.

Tansley, K. 1951. Hereditary degeneration of the mouse retina. Brit. J. Ophthalmol. 35:573–582.

Tansley, K. 1954. An inherited retinal degeneration in the mouse. J. Hered. 45:123–127.

Theiler, K., and B. Cagianut. 1963. Zur erblichen Netzhautdegeneration der Maus. Graefes Arch. Ophthalmol. 166:387–396.

Truslove, G. M. 1956. The anatomy and development of the fidget mouse. J. Genet. 54:64–86.

Uzman, L. L., and M. K. Rumley. 1958. Changes in the composition of the developing mouse brain during early myelinization. J. Neurochem. 3:170–184.

Vicari, E. M. 1951*a*. Effect of 6n-propylthiouracil on lethal seizures in mice. Proc. Soc. Exp. Biol. Med. 78:744–746.

Vicari, E. M. 1951*b*. Fatal convulsive seizures in the DBA mouse strain. J. Psychol. 32:79–97.

Vicari, E. M. 1957. Audiogenic seizures and the A/Jax mouse. J. Psychol. 43:111–116.

Walls, G. L. 1942. The vertebrate eye and its adaptive radiation. Cranbrook Press, Bloomfield Hills, Mich. 785 p.

Waugh, K. T. 1910. The role of vision in the mental life of the mouse. J. Comp. Neurol. Psychol. 20:549–599.

Williams, E., and J. P. Scott. 1953. The development of social behaviour patterns in the mouse, in relation to natural periods. Behaviour 6:35–65.

Witt, G., and C. S. Hall. 1949. The genetics of audiogenic seizures in the house mouse. J. Comp. Physiol. Psychol. 42:58–63.

Wörner, R. 1936. Über die Leistungsgrenze beim Auffassen figuraler Gestalten durch Mäuse. Biol. Zentralbl. 56:2–27.

Yerkes, R. M. 1907. The dancing mouse. Macmillan, New York. 290 p.

Zeman, W., and J. R. M. Innes. 1963. Craigie's neuroanatomy of the rat. Revised and expanded ed. Academic Press, New York. 230 p.

Zimmerling, H. 1934. Die Orientierung von Mäusen durch unselbständige transponierte Teilinhalte des optischen Wahrnehmungsfeldes. Biol. Zentralbl. 54:226–250.

33

Patterns of Behavior[1]

Richard E. Wimer and John L. Fuller

This chapter is divided into three major sections. The first surveys a wide range of studies on mouse behavior. It is also concerned with apparatus and testing procedure and with behavioral differences which have been established between genetic stocks. The second section deals with studies of environmental factors and their contribution to behavioral phenotypes. The final section deals with techniques for the genetic analysis of behavior.

SURVEY

Maintenance behavior

Eating and hoarding. That eating is cyclic is well known to all mouse breeders, for the sound of gnawing on pellets becomes audible in the mouse quarters late in the day. More precise records of daily variations in feeding activity can be obtained by the cumulative recorders used in operant conditioning studies (Anliker and Mayer, 1956). Normal mice show a strong 24-hour periodicity, the maximum rate occurring during the night. Genetically obese (*ob/ob*) mice and mice made obese by hypothalamic lesion or gold thioglucose injection eat at a uniform rate with only short irregular respites.

Hoarding of food is not seen in the laboratory unless conditions are arranged to favor it. Smith and Ross (1953*a*, *b*) found that C3H mice would carry food pellets to their cage from a supply connected temporarily by an alley. The effects of food deprivation were complex, and satiated animals sometimes hoarded more than deprived ones. The relevance of hoarding to eating behavior was questioned, since they found that wet and dry cotton packs were retrieved more readily than food pellets. A comparison of three strains on the hoarding test ranked BALB/c, C3H, and C57BL/10 in order from highest to lowest (Smith and Powell, 1955).

Drinking. Drinking, like eating, is cyclic and occurs mostly during the night. Mice given water ad libitum typically consume 4 to 6 ml each 24 hours. A number of strains manifest polydipsia. Parous MA/J and MA/MyJ females usually consume 10 to 50 ml per day; virgin females and males show a less extreme increase in water intake (Hummel, 1960). The polydipsia is probably caused by a reduction in antidiuretic hormone brought on by cystic degeneration in the posterior pituitary. An apparently similar condition, not inherited in a simple Mendelian fashion, has been reported in STR/N mice (Silverstein, 1961).

The intake of water by mice is considerably above minimum requirements. The effects of restricting time of access to water were studied by Hudson (1964, personal communication). Access was permitted for either one or two periods per day, with a total duration of 2, 4, 6, 8, or 10 minutes. The results, shown in Fig. 33-1, indicated a rather rapid loss of weight (10 to 17 per cent of original body weight) which leveled off at about 5 days. Partial recovery of the loss occurred in all except the 1 × 2-minute and 2 × 1-minute groups. Weight recovery in general paralleled ingestion curves. More water was drunk when access times were scheduled 12 hours rather than 24

[1] The writing of this chapter was supported in part by Public Health Service Research Grant MH 01775 from National Institute of Mental Health.

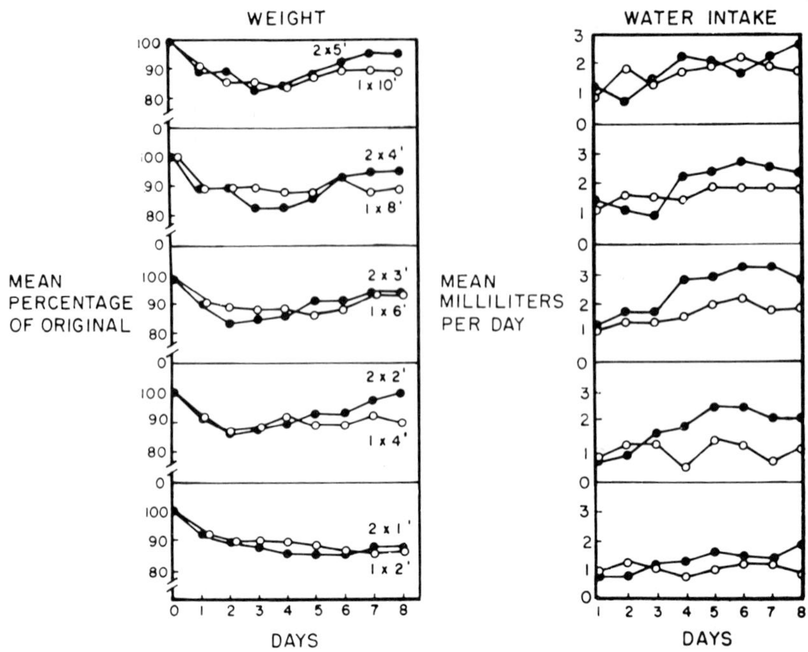

Fig. 33-1. Water intake and body weight of mice under varying conditions of scheduled drinking. The first number indicates the number of sessions per day (solid circle for two, open circle for one), and the second indicates their individual duration in minutes. (*Hudson*, 1965, *personal communication.*)

hours apart. It is apparent that water was being physiologically conserved, since weight increased while intake was far below the normal level with 24-hour access. The results suggest that 3 days of adaptation to scheduled drinking should precede experiments using water deprivation as a means of controlling drive.

Preference. A well-stocked grocery store is an excellent cafeteria for the determination of food preferences in mice. Chitty and Southern (1954) reported that the first supplies to be attacked were the cereal foods such as rolled oats, oatmeal, whole and ground rice, macaroni, and vermicelli. In a less preferred class were substances containing fats and proteins such as meats, butter, wax, etc. Finally, in the lowest class were sugar, chocolate, and a variety of dried fruits and preserves. Mice were said to avoid legumes even when other food was scarce. The earliest systematic experiment on food preferences was reported in the same source. Southern and his associates found rolled oats mixed with 20 per cent olive oil to be particularly attractive to mice.

Hoshishima et al. (1962) studied differences in preference for NaCl, saccharin, acetic acid, and phenylthiocarbamide (PTC) in strains NA-II, aa, SM, 0-61, C57, and C3H. Mice had access to distilled water and a test solution. Concentration of test solution was increased until threshold, defined as the lowest concentration of test solution necessary to produce significant differences between average 24-hour intake of the test solution and distilled water, was reached. Inspection of their data suggests comparatively little strain variation in threshold for NaCl and acetic acid, but larger variation for saccharin and PTC. Other preference studies have employed sodium sucaryl (Smith and Ross, 1960), sucrose octaacetate (Warren, 1963), propylthiouracil (Jacobs, 1962), and sucrose (Rodgers and McClearn, 1962).

The largest body of experimental research on preference concerns alcohol (Rodgers and McClearn, 1962*b*). McClearn and Rodgers (1959) measured intake of 10 per cent alcohol and water using strains C57BL/Crgl, A/2Crgl, DBA/2NCrgl, BALB/cCrgl, C3H/2Crgl, and AKR/Crgl. Strain C57BL/Crgl consumed significantly greater proportions of alcohol than the other strains, which did not differ. In a later study, Rodgers and McClearn (1962) reported preference among water and six concentrations of alcohol. Strain C57BL showed highest preference for a 12.5 per cent solution, and C3H preferred 10 per cent. Strains BALB/c and A avoided all concentrations of alcohol. Mirone (1958) investigated the effects of parental or earlier direct contact with alcohol on voluntary consumption of alcohol.

Biochemical differences related to alcohol metabolism have been found between C57BL/Crgl and DBA/2Crgl mice. The high-preference strain (C57BL) metabolizes ethanol more rapidly than the low-preference strain (Schlesinger, 1964). The results are consistent with the hypothesis that high-performance animals drink more alcohol because less acetaldehyde, a toxic intermediary in alcohol metabolism, accumulates in their blood. The content of alcohol dehydrogenase in the liver of C57BL/6 mice increased significantly after forced consumption of 10 per cent ethanol, but other liver enzymes remained unchanged. It is possible that the behavioral phenotype can be explained by a combination of biochemistry and reinforcement theory.

Social behavior

Laboratory mice live in groups sharing small cages, except when isolated for experimental purposes. Wild mice may also attain high population densities in grain ricks or other unusually favorable conditions (Elton, 1942). Though they commonly live in close proximity to one another, mice are not regarded as highly social mammals because group organization is minimal. Nonetheless, social behavior as shown in fighting, mating, and care of young has been investigated for many purposes, ranging from theoretically oriented investigations of the causes of fighting (Scott and Fredericson, 1951) to a search for drugs to control aggressiveness (Scriabine and Blake, 1962). Although some strain comparisons have been made, the genetics of social behavior has not been widely studied.

Fighting. Mouse breeders are well aware of the propensity for fighting shown by many strains, particularly by males. Severe wounds are inflicted in combat, and battles to the death are not uncommon. Several investigators have described the fighting of mice (reviewed in Scott and Fredericson, 1951). According to van Abeelen (1963*a*), the elements of male-male interaction include fixing (staring at rival), dancing, boxing, kicking, nosing, wrestling, biting, chasing or fleeing, and submissive posture. Tail rattling and fur fluffing are often seen when mice are paired and may represent emotional responses (Scott, 1947).

Laboratory mice housed in cages typically develop a social organization based upon exclusive dominance of one male (Uhrich, 1938). The despot ordinarily retains his position for some months. Linear dominance is occasionally observed in which an α-male dominates a β-male, who in turn dominates others below him in the social order. Equal or unsettled dominance relationships are seen in newly assembled groups. Only when mature animals were matched against juveniles was dominance not correlated with weight in Uhrich's colony of heterogeneous albino mice. A strong "home-cage" effect was seen in the usual advantage of residents over an introduced stranger, regardless of his previous high social status in another situation. Success in fighting was not the basis of success in mating.

Methods of studying fighting quantitatively can be divided according to whether subjects fight to a decision or are separated when fighting commences. The tests may also be separated into noncompetitive ones in which animals fight in bare cages and competitive ones in which animals have restricted access either to food or to shelter from electric shock (Scott and Fredericson, 1951).

Fighting tests are usually conducted in special pens with a removable center partition, thus minimizing handling. Male mice will fight readily in a bare arena, but somewhat more consistent results are reported when hungry mice compete for a loose pellet of food (Fredericson, 1950; Fredericson et al., 1950). Females (C57BL/10) which seldom fight spontaneously will fight regularly in a competitive situation (Fredericson, 1952). Success in competition to escape noxious stimuli has been used to evaluate aggressiveness of different strains (Doner et al., 1952) and effects of hormone treatments (Bevan et al., 1960). A somewhat different means of inducing fighting is to give foot shocks to mice confined within a narrow space. Most pairs attack each other promptly (Tedeschi et al., 1959).

Fighting until decision results in training for dominance or submission. Some investigators purposely pretrain for one or the other. Fighters are readily trained by the "dangler" method in which other mice, held by forceps so they cannot defend themselves, are presented to the potential fighter (Scott, 1946; Bauer, 1956). Mice may be trained for submission by exposure to such a trained fighter (Kahn, 1951). The round-robin method of matching members of a set of subjects in all possible pairings involves training of a less directed type (Beeman, 1947), but most subjects become habitually dominant or submissive during the course of testing.

Nondecision procedures usually involve isola-

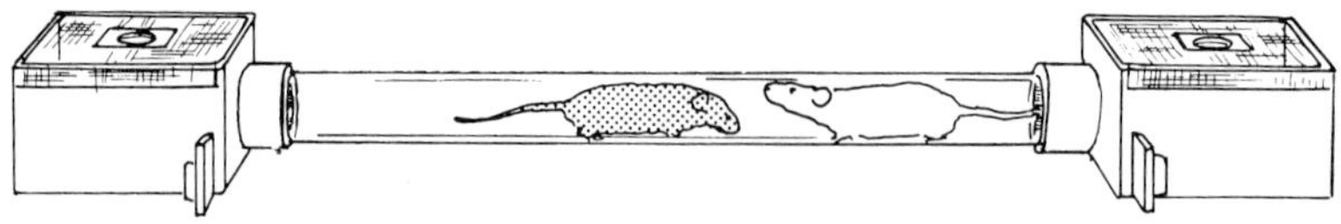

Fig. 33-2. A tube-dominance test apparatus. (*Drawing based on a photograph provided by Professor E. Howarth, University of Alberta.*)

tion after weaning (King, 1957*a*; Tollman and King, 1956), though shorter isolation periods have been employed in some drug assays (Scriabine and Blake, 1962). In such methods a *pair* of mice is taken as the unit, and aggressiveness is measured by the latent period until the beginning of a fight or the accumulated attacking time over a fixed period (Catlett, 1961). The advantage of the latency measure is that subjects can be separated before fighting is inhibited by injury or reinforced by victory. A great variety of other measures have been used ranging from simple occurrence or nonoccurrence of fighting in a pair to rating scales (Lagerspetz, 1964) and elaborate counts of specific acts (Banks, 1962).

Some methods of measuring social dominance do not involve actual fighting. In the tube-dominance test, mice deprived of food are trained to go down a narrow tube to a chamber where food is provided (Lindzey et al., 1961). On the dominance trial, two animals are started simultaneously at opposite ends, and the animal which retreats is scored as submissive (Fig. 33-2).

The amount of fighting between males varies according to previous experience (Scott, 1944). New males (C57BL/10) were accepted peacefully by mated males living in a large multiple-escape pen. After a period of isolation, during which residents were trained to fight, the pairs fought vigorously when they were again brought together.

Castration reduces male aggressive behavior as it does in many species. Beeman (1947) found that aggressive behavior in male C57BL/10 and BALB/c mice was eliminated if 25 or more days elapsed between extirpation of the testes and the initial encounter. She reported that implants of testosterone propionate promptly restored aggressive behavior to the level characteristic of the strain.

Bevan and his collaborators reported somewhat different findings. Replacement androgen therapy only partially restored aggressive behavior of castrated SWR mice, and doses of 600 μg per day of testosterone propionate even depressed it (Bevan et al., 1958). When pretraining, castration, and testosterone therapy were varied independently, the effects of the hormone on fighting success appeared to be mediated chiefly through an effect on body weight. Effects of pretraining were more striking than those of androgen (Bevan et al., 1960). An additional complication in the androgen-aggression relationship is that testosterone induces spontaneous aggression in prepuberally castrated males but not in females (Tollman and King, 1956). Some of the discrepancies between these studies are certainly byproducts of differences in techniques of measurement and in the handling of the subjects. It appears likely, however, that some genetic variation exists in the association between androgen and fighting and that more is involved than the sensitization of an "aggression center" by hormones.

Reduction of aggression is the purpose of much drug treatment of behavioral disorders in man. A comparison of various tranquilizers showed that meprobamate was very effective in reducing fighting in mice, though it had no effect on spontaneous motor activity (Tedeschi et al., 1959). Janssen et al. (1960), who used spontaneous fighting rather than foot shock–induced fighting as an index, found striking differences in the antifighting potency of drugs relative to their motor effects. It appears that tests of fighting behavior in mice have merit for evaluation of potential tranquilizing agents (Scriabine and Blake, 1962). One may conjecture that an ideal drug would reduce spontaneous fighting but leave competitive fighting intact, since success in obtaining food must have survival value.

Studies on the genetics of fighting in mice have been restricted, with one exception, to strain comparisons. Ginsburg and Allee (1942) found the descending order of fighting ability in three strains to be C57BL, C3H, and BALB/c. Several researchers have used BALB/c as a relatively nonaggressive strain, but little effort seems to have been made to survey the large number of available strains. Early experience in victory or defeat can produce great variation in fighting within a fixed genotype (Scott, 1946; Kahn, 1951). Furthermore, the type of test used is important. Pairs of BALB/c males usually share food pellets, but BALB/c males paired with C57BL/10 males become aggressive (Fredericson and Birnbaum, 1954). Lindzey et al. (1961) compared social dominance in pairings between three inbred strains using the tube-dominance apparatus. The outcomes in terms of percentage of animals winning a majority of contests for each pairing were A/alb (100) and DBA/8 (0); A/alb (86) and C3H (14); C3H (86) and DBA/8 (14).

Lagerspetz (1964) obtained good separation of aggressive and nonaggressive lines within two generations of selection from a heterogeneous population of albinos. Increased divergence was obtained in later generations, though the lines appeared to be fairly stable by the fourth generation. Only males could be tested because fighting was infrequent among females. The results suggested that a few genes make major contributions to the variation in fighting behavior observed in the original stock. Losing of fights reduced the aggressiveness score of both lines; winning enhanced fighting of initially aggressive animals but not of nonaggressive ones. Animals aroused by an immediately preceding fight would cross an electrified grid to attack an opponent or would choose an opportunity to fight rather than remain peaceful.

Mating. Courtship follows a pattern similar to that of other laboratory rodents but has species-specific characteristics. The basic sequence consists of elements described as sniffing, following, mounting, mounting-with-intromission, and postcopulatory grooming (van Abeelen, 1963*a*; Grant and Mackintosh, 1963; Lipkow, 1960; McGill, 1962). The sexually aroused male often crawls in front of the female or even under her ("rooting," McGill, 1962). There may be one to more than 100 intromissions before ejaculation, which is marked by the male rolling over on his side, often carrying the female with him. At ejaculation the male accessory glands produce a secretion which hardens to form the vaginal plug. Only one ejaculation usually occurs per day.

Inbred strains differ quantitatively in many aspects of mating behavior. Males of C57BL/6 typically gain intromission rapidly and ejaculate in 20 minutes. DBA/2 males are slow to achieve intromission, but ejaculate quickly thereafter. Slowest to achieve ejaculation are BALB/c males (average latency about 1 hour), largely because of a long period of courtship (McGill, 1962). Even more striking genetic effects are found in recovery time, the interval between successful copulations (McGill and Blight, 1963*a*). Recovery time in C57BL/6 males averages 4 days but only 1 hour in DBA/2 males. The F_1 hybrids resemble DBA/2 parents, and backcrosses to C57BL/6 yield intermediate latencies. The physiological and psychological bases of various components of mating behavior are complex (McGill and Blight, 1963*b*). Levine (1958) found that albino males (strain ST) sired 76 out of 88 litters from ST females when competing against pigmented males (CBA). Success in mating was not correlated with social dominance in a test of aggression. Although ST males had a slight advantage in initial encounters, all CBA males achieved dominance in continued competition (Levine, 1963). Solution of this apparent contradiction must be based upon direct observation of mating behavior in a competitive situation.

Caretaking. The parturient female generally constructs a hollow nest of any available material. She spends much time in the nest huddled over the young. The young are licked at frequent intervals, most often in the perineal region. Retrieving of young removed from the nest, amount of huddling, and nest building have been used as the basis of a quantitative rating scale for maternal behavior (Leblond and Nelson, 1937; Leblond, 1940).

Young mice elicit retrieving much more effectively than older ones. The percentage of successful retrievals with 1-day-old young was 83; 5-day-old, 78; 10-day-old, 54; and 15-day-old, 11. Leblond (1940) found that the expression of maternal behavior in mice was little influenced by hormones of parturition and lactation. Maternal behavior could be evoked in intact and hypophysectomized males and virgin females by leaving young in the cage, a process called "sensitization." Males were found which retrieved dislocated young mice as readily as did females.

The extensive investigations of Beniest-Noirot (1958) have confirmed the independence of so-called maternal behavior from hormonal control. Postparturient and virgin females and males performed equally well on her maternal behavior tests, which included retrieving, nest building, nursing care, and assumption of a nursing position. Similarly, adult mice of both sexes with and without breeding histories ate placentas, gnawed umbilical cords, and cleaned young when confronted with newborn mice. These activities are not maternal in a strict sense but are responses made to appropriate stimuli. The only probable hormonal effect noted by Beniest-Noirot was increased defense of the nest by postparturient females. Caretaking responses, though not dependent upon prior experience with infant mice, were more frequent in mice which had spent 3 days with their dam's second litter.

Whisker-eating, mutilation, and cannibalism of young are disturbances of caretaking behavior which vary in incidence among inbred strains. Their genetics has been reported only in a preliminary fashion (Hauschka, 1952*a*, *b*). Cannibalism is often a serious problem to the animal breeder and experimentalist, but no thoroughly satisfactory explanation has been advanced for its occurrence in certain lines and individuals. Disturbing the mother may be an inciting factor. Beniest-Noirot (1958) found more cannibalism in mothers selected for strong defense of their nests against intruders. Some experimenters working with neonatal mice routinely remove the mother and anesthetize her lightly with ether, while the young are being handled, to reduce the incidence of cannibalism.

Emotionality

Willingham (1956) measured defecation, urination, activity, freezing, emergence into open and enclosed areas, and squeaking in several different situations. The matrix of intercorrelations was factor-analyzed, and six factors were extracted. The first, called "elimination," had significant loadings on all the defecation and urination measures. The second factor was named "freezing." Willingham concluded that there are a variety of nearly independent types of emotional behavior and that the concept of general emotionality is an oversimplification.

Thompson (1953) scored 15 mouse strains on incidence of defecation in the open field (Table 33-1). Several other investigators have used elimination as a measure of timidity (Lindzey, 1951; Fuller et al., 1956).

An important challenge to the validity of elimination as an index of emotionality has recently come from Bruell (1963), who observed defecation in 25 different genotypes. Males defecated more than females, and hybrid males defecated more than their inbred sires. However, hybrid females defecated less than their inbred dams. Bruell's interpretation was that defecation of male mice in a strange environment is a territory-marking response. Thus, increased defecation in hybrids was interpreted as restoration of adaptive responses depressed by inbreeding.

Antalfi (1963) has shown that freezing (Willingham's second factor) in the presence of stress may be less strong in laboratory mice than in wild house mice. Freezing also has possible adaptive significance.

Table 33-1. PERCENTAGE OF MICE IN EACH OF 15 STRAINS DEFECATING DURING SIX 10-MINUTE TESTS. (*From Thompson, 1953, with permission of the author and the Canadian Journal of Psychology, published by the Canadian Psychological Association and the University of Toronto Press.*)

Strain	Per cent	Strain	Per cent	Strain	Per cent
AK/e	96	A/C1	78	C3H	52
BDP	85	DBA/2	70	C57BL/6	40
Obese	83	C57BR/a	66	AKR	34
LP	82	C57BL/10	58	MA/My	27
BALB/c	82	DBA/1	57	C3HeB	11

Learning

Conditioned responses. The earliest attempt to establish conditioned responses in mice came, most appropriately, from Pavlov (1923):

> The latest experiments (which are not yet finished) show that the conditioned reflexes, *i.e.*, the highest nervous activity, are inherited. At present some experiments on white mice have been completed. Conditioned reflexes to electric bells are formed, so that animals are trained to run to their feeding place on the ringing of the bell. The following results have been obtained:
>
> The first generation of white mice required 300 lessons. Three hundred times was it necessary to combine the feeding of the mice with the ringing of the bell in order to accustom them to run to the feeding place on hearing the bell ring. The second generation required, for the same result, only 100 lessons. The third generation learned to do it after 30 lessons. The fourth generation required only 10 lessons. The last generation which I saw before leaving Petrograd learned the lesson after 5 repetitions. The sixth generation will be tested after my return. I think it very probable that after some time a new generation of mice will run to the feeding place on hearing the bell with no previous lesson.

We do not know what Pavlov found on his return to Petrograd.

Denenberg (1958; 1959*a*, *b*; 1960) studied the formation of conditioned emotional responses in C57BL/10J and C57BL/10Sc mice. In one version of his procedure, the buzzer was sounded for 3 seconds and then followed by a short pause, after which the shock was presented for 1 second (Denenberg, 1958). Typical responses to shock were described as yelping, walking, running, climbing, and rearing. The occurrence of any of these responses to the buzzer was considered evidence of conditioning.

Operant conditioning. Studies on operant conditioning in the mouse have used sensory reinforcers—lights, primarily—and lever contact rather than pressing. Indeed, one of the earliest reports of successful utilization of light as a reinforcer used mice (Kish, 1955). Various studies have used males of strains C57BL/10J (Kish, 1955), C57BL/6J (Barnes and Baron, 1961; Kish and Baron, 1962; Baron and Kish, 1962), and DBA/1J (Barnes et al., 1959), with age range of 8 to 17 weeks. Pretraining in darkness is important (Kish and Baron, 1962) and the light source should be weak (Barnes et al., 1959). Barnes and Baron (1961) studied the effects of luminous line drawings as reinforcers, concluding that their effectiveness increased as they became more complex.

Barnes and Kish (1957) reported that termination of high intensity sound acts as a negative reinforcer, i.e., termination can be used to strengthen a response. Use of sound as a positive reinforcer for lever contact has also been studied. The definitive study is that of Barnes and Kish (1961), who reported the effects of 10 frequencies at five different intensity levels. Reinforcing effects were quite weak. In a later study comparing directly the effectiveness of light and sound, Baron and Kish (1962) reported that low intensity sound may have aversive properties. Clearly, light and luminous drawings are better choices as reinforcers.

Escape and avoidance. Essman and Jarvik (1961*a*) devised an apparatus consisting of a large container of water with an escape ramp fastened on one corner. The animal was placed in the water on the side opposite the escape ramp and time to reach and climb the escape ladder were recorded. Learning is extremely rapid (Fig. 33-3). The technique has been used for strain comparisons by Winston (1963) and for comparison of mutants by Denenberg et al. (1963).

The essence of avoidance conditioning is that the organism is trained to make a response which postpones the impending occurrence of noxious stimulation. A number of methods have been devised for the study of avoidance learning in mice. One technique employs a box with a grid floor partitioned in the middle by a hurdle. A warning stimulus—light or buzzer—begins a few seconds before shock is delivered to the floor grids on the side on which the animal is located. The animal is able to avoid shock by jumping the hurdle. The

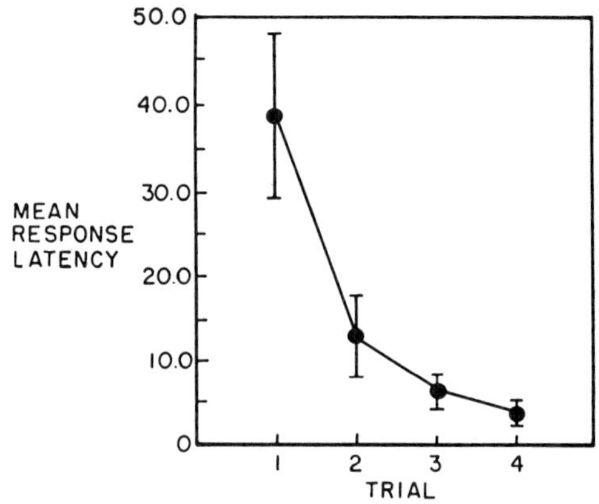

Fig. 33-3. Mean response latencies by trial on the water-escape test. (*From Essman and Jarvik, 1961a, with permission of the senior author and Psychological Reports.*)

warning stimulus goes on again a few seconds or minutes later, and the animal must quickly jump back if it is again to avoid the shock. Excellent descriptions of the use of such procedures with mice are given by Caldwell (1962) and by Royce and Covington (1960). Caldwell's technique worked moderately well with the female CF#1 mice he used.

Royce and Covington (1960) compared nine inbred strains maintained at The Jackson Laboratory. Nearly one-third of the animals could not be conditioned in 700 trials—half or more of strains C57BR/cd, AKR, C57BL/6, and A. Table 33-2 presents mean number of trials to a criterion of five successive avoidances for the strains in which conditioning was easiest. Subsequent work (Carran et al., 1964) showed that strain differ-

Table 33-2. STRAIN COMPARISONS FOR MEAN NUMBER OF TRIALS TO REACH CONDITIONING CRITERION OF FIVE SUCCESSIVE AVOIDANCES. (*From Royce and Covington, 1960, with permission of the senior author and the American Psychological Association.*)

	Mean no.	*Interstrain* t *values*				
Strains	trials	SWR	C3H	DBA	BALB/c	C58
SWR	51		0.67	2.14[a]	2.21[a]	2.52[a]
C3H	73			2.36[a]	2.26[a]	2.39[a]
DBA	125				0.39	1.46
BALB/c	137					1.20
C58	209					

[a] $P < 0.05$.

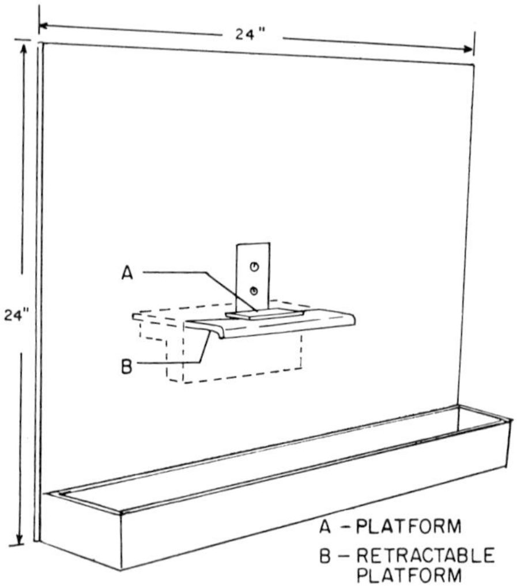

Fig. 33-4. Schematic view of Essman and Jarvik passive avoidance conditioning apparatus. (*From Essman and Jarvik,* 1961*a, with permission of the senior author and Springer-Verlag, Heidelberg.*)

ences in conditioning rate are dependent in part upon the shock voltage used. Whether the basis of the voltage dependence is associated with strain differences in shock sensitivity was not made clear. Two strains differing in conditionability were shown to diverge in measured skin resistance during training. King and Mavromatis (1956) and Stanley and Monkman (1956) also studied avoidance conditioning of mice.

A different kind of device has also been widely used. As described by Denenberg et al. (1958), it consists of a narrow rectangular box with partitions. The floor is an electrifiable grid. Near one end is a starting chamber in which the animal is restrained by a guillotine door. A safe chamber, its grid floor covered with wood, is located at the other end. The door of the starting chamber is opened and a 100-watt bulb is turned on. Five seconds later, the floor is electrified. The animal may escape or avoid the electric shock by opening the door of the safe chamber and entering. Use of the device was reported in several articles (Denenberg and Bell, 1959; Denenberg et al., 1963; Bell and Denenberg, 1963).

Ekström and Sandberg (1962) and Jarvik and Essman (1960) have also reported avoidance conditioning techniques. Ekström and Sandberg trained mice to avoid shock by jumping to a wire net hung on the wall. A rather large number of training trials was required for only moderate success.

The Jarvik and Essman device produced extremely rapid learning (Fig. 33-4). Subjects were placed on platform A and the shelf B was then pulled out. A step down from the platform to the shelf resulted in shock to the animal. Learning was measured by placing the animal back on the platform and measuring the amount of time it delayed before stepping down again. Comparison of shocked and nonshocked animals 24 hours after a single training trial resulted in 10 per cent of the previously shocked and 90 per cent of the unshocked animals leaving the platform in 10 seconds or less. This and similar devices have been used in three other studies (Essman and Alpern, 1964; Essman and Jarvik, 1961).

Maze learning. Early work on maze learning in mice was done by Yerkes (1907) and Vicari (1923, 1924, 1929). More recently, Lindzey and Winston (1962) compared the learning performance of several strains in a six-unit T maze. Twenty animals each of strains A/alb, C3H/Bi, DBA/8, and C57BL/1 were given one trial per day for 14 days in a study in which all animals received prior handling. Strain C3H/Bi was poorer than the other three strains. A second experiment indicated differential strain responsiveness to handling. Winston (1963) used a similar maze and procedures with strains A/alb, C3H/Bi, and DBA/8, and also showed strain C3H/Bi to be poorer. Enclosed mazes have also been used by Hall (1959) and Smith and Bevan (1957).

Morgan (1963) compared the effectiveness of food and a dark hiding place as reinforcers for mice of strains SW and C57BL. Animals were given 20 trials on the first day and 10 trials per day thereafter. All groups attained a level of 9 out of 10 correct by the end of 60 trials. Groups running to a dark hiding place learned faster. No

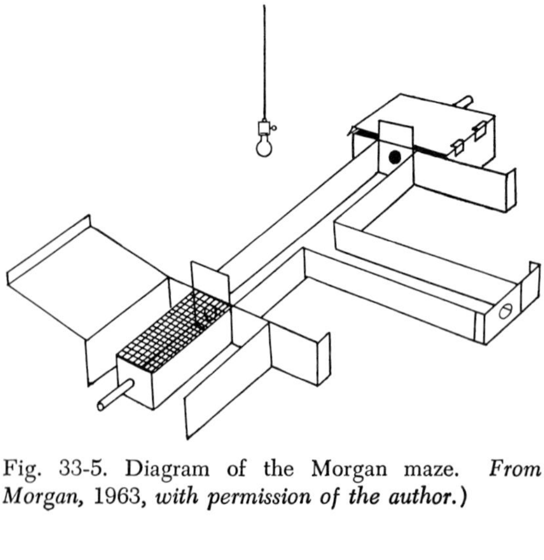

Fig. 33-5. Diagram of the Morgan maze. *From Morgan,* 1963, *with permission of the author.*)

strain differences or interaction with reinforcer were found. A diagram of the apparatus is presented in Fig. 33-5.

McClearn (1958, 1965) compared the performance of strains C3H/NT, C57BL, and BALB/c on an elevated maze. In decreasing order of performance, the strains were BALB/c, C57BL, and C3H/NT. Lindzey and Winston (1962) and Winston (1963) also found C3H mice to be poor learners, possibly because they are blind or nearly so (Chap. 32).

Meier and Foshee (1963) compared performances of strains AKR, BALB/c, C3H/He, C57BL/6, DBA/2, and noninbred CF#1 in a three-choice-point water maze in which escape from water was used as a reinforcer. The strains divided into two groups on the basis of time to reach an escape ramp; slower strains were BALB/c, AKR, and C3H/He. No error score analyses were included. This and a second experiment (Meier, 1964) demonstrate that level of performance for some strains depends in part upon the age at which the tests are given.

Discrimination learning. Much of the literature on discrimination learning is reviewed in Chap. 32, where information on sensory capacity of mice is given. However, two techniques of special interest are better described here.

What appears to be an extremely promising apparatus for the study of discrimination learning was devised by Keeler (1927) and used but once (Fig. 33-6). The device was made of two cylinders of sheet metal. The floors of the central cyclinder and those of the external chambers were independently wired for shock. There were six doorways from the central cylinder. A light was turned on in one of the external chambers and the floor of that compartment was isolated from the circuit, all others being electrified. The mouse was placed in the center cylinder, the center floor was electrified, and the mouse had to escape to the lighted compartment. Learning curves of normal and rodless (Chap. 32) mice tested in the device are presented in Fig. 33-7. It is evident that learning was extremely rapid.

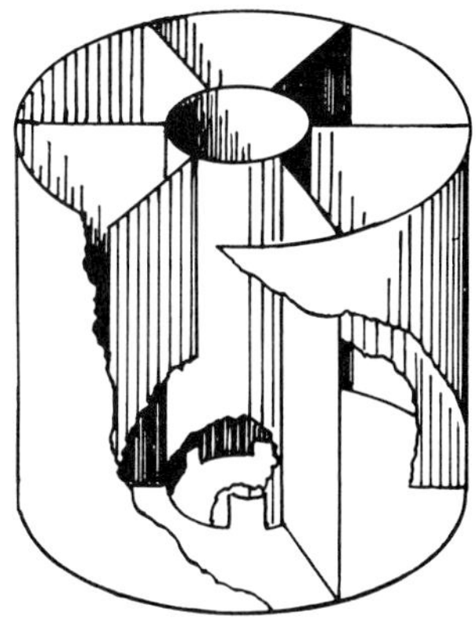

Fig. 33-6. The Keeler multiple choice box. (*From Fig. 4, Keeler,* 1927.)

A discrimination device with which we have had personal experience is the water maze devised by Waller et al. (1960; Fig. 33-8). Animals were

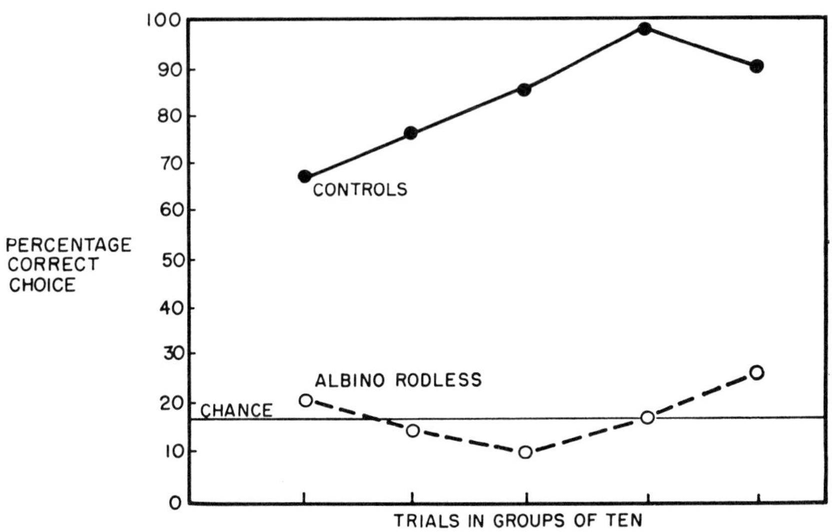

Fig. 33-7. Learning curves for mice in the Keeler multiple choice box. Trials are in groups of 10 along the abscissa. Percentage of correct choices is shown on the ordinate. (*From Fig. 2, Keeler,* 1928.)

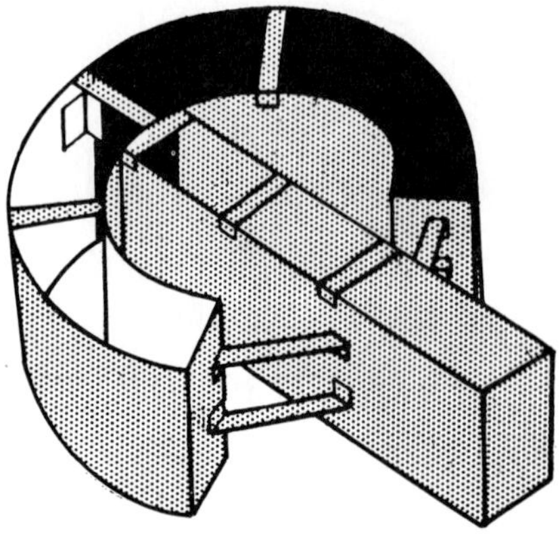

Fig. 33-8. Water maze for studying discrimination learning. (*From Waller et al.,* 1960, *with permission of the senior author and Psychological Reports.*)

trained to swim to either the white or the black side. The entire bottomless maze could be inverted in the water to reverse the orientation of the correct alternative. C57BL/6J males were given five trials per day for 12 consecutive days at one of three water temperatures. Waller et al. found that animals swam faster at lower temperatures, but that there was no difference in error performance among the three water temperatures used.

We have successfully used the device for black-white discrimination learning with six strains (Wimer and Weller, 1965). All but one strain performed significantly above chance by the fourth day. A total time investment of about 40 minutes per animal was required to achieve nearly errorless performance in some strains. In decreasing order of performance, the strains were C57BL/6J, DBA/2J, RF/J, AKR/J, and A/HeJ. Except for strains AKR/J and A/HeJ, the difference in performance between adjacent strains was not significant. However, differences between all strains separated by one other were significant. Our experience suggests that water temperature may affect accuracy of performance in some strains. DBA/2J and C57BL/6J discriminate horizontal from vertical striations with approximately equal success and at a level of performance only very slightly inferior to that for black-white discrimination (Wimer, unpublished data).

Exploratory activity

Exploratory activity—activity in an unfamiliar environment—is a comparatively well-studied mouse behavior. Thompson (1953) measured activity levels of 15 inbred strains in an enclosed arena (Table 33-3). In a later study he measured arena and Y-maze activity in larger samples of five strains (Thompson, 1956). Thompson's two studies showed that there is considerable variability between strains and that the ordering of the strains is invariant in the two testing devices, with the exceptions of strains C57BL/6J and AKR/J. McClearn (1959) attempted to establish the generality of Thompson's results with strains C57BR/cd, C57BL/10, LP, AKR, BALB/c, and A/J by observing their behavior in other situations in which one might reasonably expect the same ordering to be maintained. McClearn's findings indicated that Thompson's characterizations of strains has wide generality. The basis for these strain differences is not known. Thiessen (1961) attempted to relate body weight to activity, but his results were inconclusive.

McClearn's (1960) study of the effects of varying level of illumination suggests that there are limits to the conditions under which ordering of strains on activity is maintained. He measured activity levels of strains C57BL/Crgl and A/Crgl in an open field with barriers, in bright white light

Table 33-3. MEAN AMOUNT OF EXPLORATORY ACTIVITY SHOWN BY EACH OF 15 MOUSE STRAINS, AND THE PROBABILITY OF DIFFERENCES AMONG THEM. (*From Thompson,* 1953, *with permission of the author and the Canadian Journal of Psychology, published by the Canadian Psychological Association and the University of Toronto Press.*)

No.	Strain	Mean score	*Probability of differences*[a]	
			.05–.01	.01–.001
1	C57BR/a	459	3–4	5–15
2	C57BL/6	361	7–9	10–15
3	C57BL/10	359	7–9	10–15
4	DBA/1	334	8–10	11–15
5	MA/My	308	8–11	12–15
6	BDP	286	10–12	13–15
7	DBA/2	253	11–13	14–15
8	LP	194	13–15	
9	AKR	188	13–15	
10	C3H	177	13–15	
11	Obese	149	15	
12	C3HeB	117		
13	BALB/c	74		
14	AK/e	60		
15	A	20		

[a] Read, for example: No. 1, strain C57BR/a, has a mean of 459, significantly greater than No. 3 and 4 (P = .05–.01) and No. 5 through 15 (P = .01–.001).

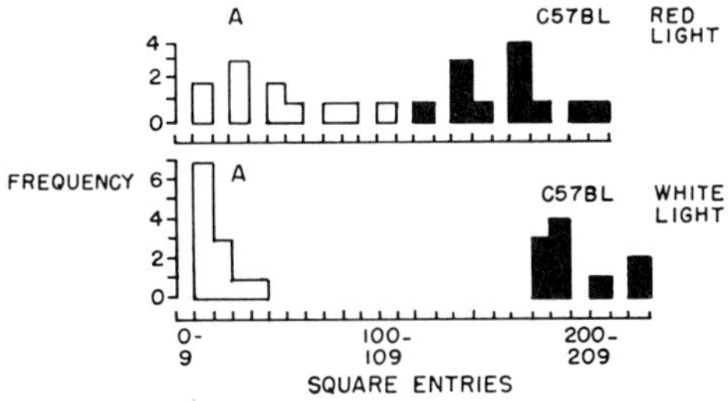

Fig. 33-9. Activity scores of C57BL and A mice under red and white illumination conditions. (*From McClearn, 1960, with permission of the author and the American Psychological Association.*)

and in dim red light. Note in Fig. 33-9 that strain A showed higher activity when tested under dim red light, whereas strain C57BL showed less. Although variation in level of illumination did not reverse the ordering of the strains, there was a significant interaction of strain and illumination. Difference in eye pigmentation between the strains might be partially responsible for the observed phenomenon. However, McClearn (1959) found that hybrids of strain C57BL/10 and A/J were intermediate in activity. Thus, more than eye pigmentation was involved, for the hybrids are dark-eyed like the C57BL/10 parents.

There are two other studies of exploratory activity using mice (Wimer and Sterns, 1964; Wimer and Fuller, 1965).

ENVIRONMENT AND PHENOTYPE

The purpose of this section is to present information on the effects of maintenance conditions and early environment on behavioral phenotypes.

Maintenance conditions

Light cycle. Cycles of approximately 24 hours (circadian rhythms) in laboratory mice have been so well demonstrated that artificially controlled light-dark cycles in the animal room must be regarded as essential for quantitative behavioral and physiological studies.

Strain I mice were much more susceptible to audiogenic seizures during the night (Halberg et al., 1955), and the peak hours of susceptibility could be shifted by artificially reversing the lighting cycle in the animal quarters (Halberg et al., 1958). Halberg et al. (1959) have shown also that variation in seizure susceptibility is merely one of many biological processes changing rapidly during the day (e.g., such diverse phenomena as blood corticosterone, number of mitoses in adrenal cortex, and total body activity). The importance of controlling time-of-day effects in psychological studies is clearly indicated. Higher paroxysmal activity in electrocorticograms (ECoG) has been found during the evening hours and may relate to increased susceptibility to audiogenic seizure (Harner, 1961). Since ECoG's were recorded under pentobarbital anesthesia, it is equally possible that differences were related primarily to a cycle in drug susceptibility. The relationship of the diurnal cycle to drug sensitivity is complex. Under constant illumination, variation in pentobarbital sleeping time with time of day was absent; under cyclic lighting, grouped subjects slept longer when tested in the light period but not in the dark (Davis, 1962) (Table 33-4). Periodicity in activity in waltzing mice was maintained in constant darkness, and even occurred in mice born and reared in the dark, though here the cycles were not correlated with the natural day-night cycle (Wolf, 1930).

Table 33-4. PENTOBARBITAL ANESTHESIA (60 mg/kg ip) UNDER VARYING CONDITIONS OF LIGHTING AND HOUSING. A AND B ARE DUPLICATE EXPERIMENTS. (*After Davis, 1962, with permission of the author and Experientia.*)

Prior lighting	Maintenance housing	Duration of anesthesia after drug injection, minutes: *In light period* A	*In light period* B	*In dark period* A	*In dark period* B
12-hour on-off cycles	Grouped	110	81	62	44
	Isolated	80	67	61	44
Continuous light	Grouped	86	68	94	60
	Isolated	75	59	78	59

Temperature. Temperature control is also widely regarded as essential in the animal room. Mice, however, have adapted quite successfully to ambient temperatures as low as −3°C, provided that ample nesting material was supplied (Barnett, 1956, 1959). Improved nest building was a major means of adaptation in both A and C57BL mice. Increased heat production of mice reared in the cold enabled them to withstand low-temperature stress much better than animals reared in heated laboratories, showing that physiological as well as behavioral adaptation had occurred. Improved body insulation played a minor role in adaptation, which must therefore have been based on increased heat production.

Population density. The relation between crowding and the endocrine system has been widely studied. Increased adrenal size and decreased size of male sex structures in crowded animals led Christian (1955) to postulate such endocrine effects as a mechanism for regulation of population density. Some studies have failed to confirm all of Christian's findings and have thus cast doubt on this hypothesis of density control. Southwick and Bland (1959) found no difference in the adrenal weights of CFW male mice reared in isolation or in groups of 2, 4, 8, and 16 per cage. However, they did find adrenal enlargement in wounded mice (presumably socially subordinate animals subjected to chronic stress). The males in groups of four or eight actually sired more young than isolates when females were introduced to their cages, a result contrary to Christian's hypothesis.

The effects of population density upon adrenal activity seem to be largely dependent upon the type of social organization in the group. A number of investigators have used eosinopenia and other measures as indicators of the stressfulness of various types of social grouping imposed by the experimenter (Southwick, 1959; Vandenbergh, 1960; Thiessen et al., 1962; Thiessen and Nealey, 1962; Bronson and Eleftheriou, 1963; Bronson, 1963). Enhanced adrenal function parallels social tension (grouping of strange males, subjection to defeat by fighters, as examples) rather than physical crowding alone. These findings are pertinent to research in ecological pathology, which is concerned with the effects of housing conditions upon cancer and other diseases.

C3H mice kept in isolation in a long-term experiment on nutrition showed a high incidence of convulsion (12 in 20) contrasted with subjects housed in one large cage (3 in 20). The mean lifespan of the isolates was 434 days; of the group-reared, 500 days (J. T. King et al., 1955). Group rearing also favored survival of C57BL/10 mice weaned early, apparently as a result of better heat conservation (J. A. King and Cannon, 1955). Prior social experience raises the lethal threshold for amphetamine and counteracts the enhanced susceptibility found when mice reared in isolation are aggregated just prior to toxicity testing (Mast and Heimstra, 1962).

It must appear from these latter studies that the relationship between the behavioral and physiological levels of integration are complex. The laboratory mouse is one of the best subjects for investigation of these interactions and has the further advantage of permitting an evaluation of genetic effects.

Early environment

Environment during both prenatal and postnatal developmental periods has a marked effect on behavioral phenotypes.

Prenatal. Lieberman (1963) provided an excellent demonstration of the effects of the prenatal environment. Pregnant females of strain C57BL/6 were given (1) saline injection, (2) epinephrine injection, (3) norepinephrine injection, (4) hydrocortisone injection, or (5) were crowded in a cage with 10 aggressive males. All treatments were given during the second trimester of pregnancy. Offspring were tested at 35 days of age in an open field test. In comparison with the saline controls, offspring of crowded and epinephrine-injected dams showed increased activity and decreased defecation. Offspring of hydrocortisone- and norepinephrine-injected dams showed decreased activity and increased defecation. Some, but not all, differences attained acceptable standards of statistical significance.

Weir and DeFries (1964) subjected pregnant females of strains BALB/cJ and C57BL/6J to a daily regimen of swimming in a water tank, exposure to intermittent loud tones in a tilt box, and placement in a brightly illuminated open field. Later test of the activity of offspring in an open field showed complex interactions of the treatment effect with genotype.

Postnatal. It is generally believed that certain kinds of experience may have the most profound and long-lasting effects if encountered early in postnatal life. That such experience may affect phenotype was dramatically demonstrated by Denenberg et al. (1964), who reared C57BL/10 young mice under a variety of conditions. For example, some mice (controls) were reared with other mice both before and after weaning, while other animals were transferred to a lactating female rat and reared with two male and two female offspring of her litter. Mice reared with rats from a very early age tended to be less active in an open field and less aggressive than mice reared with other mice. Preference for own species was very strongly affected by the treatment: Mice reared with rats preferred them to other mice.

A recent study by Ressler (1962) provides an excellent example in which strain of both parent and offspring of C57BL/10 and BALB/c mice contributed to an environmental effect. All offspring were foster-reared by parents of the same or different strain. Ressler found that BALB/c parents handled young of both strains more than C57BL/10 parents did. In addition, young BALB/c mice received more handling from both strains of parents than did C57BL/10 young.

Since variations of handling in infancy by humans have been shown to influence a variety of behavioral characteristics in adulthood, the superficially slight difference in the early environment of the young mice might be important. Ressler (1963) reported on differences in operant conditioning. Pressure on a wire mesh door was used as the response. For the first 15 minutes of testing, each press merely actuated a counter, while in the second half it briefly illuminated the dark box in which the animal was kept. The strain of the young mice had a significant effect on number of presses during the dark phase, while the strain of foster parents had a significant effect on the visual exploration score (number of presses during the phase in which the light was operative minus presses during the dark phase). Young of both strains engaged in more visual exploration if reared by BALB/c parents. Results are presented in Table 33-5.

King and his associates have produced differences in aggressiveness within strains by varying size of social group. King and Gurney (1954) treated C57BL/10 males in one of three ways. Some were housed individually from 20 to 100 days of age. Others were reared with sires and male sibs until 45 days and then isolated until over 100 days. A third group was reared with female sibs until 45 days and then isolated. Latency of fighting was measured after bringing together individuals reared similarly. Animals isolated at 20 days were found to be much less aggressive. King (1957*b*) used strains C57BL/10J and BALB/cJ and a variety of treatments incompletely replicated in both strains. King stated that his results showed C57BL/10J males isolated at 20 days to be slower to fight, whereas BALB/cJ males were unaffected by this treatment. Similar early treatment produced no major effects on sexual behavior (King, 1956).

The effect of early human handling on emotionality and fighting behavior in C57BL/10 mice has been studied by Levine (1959). Handling reduced the freezing component of emotionality and made animals more aggressive. Hall and Whiteman (1951) and Lindzey et al. (1960) also studied the effects of early experience on adult emotionality. Effects on other behavior have been reported by Baron et al. (1962), Bell and Denenberg (1963), Denenberg (1958, 1959, 1960), Denenberg and Bell (1960), and Stanley and Monkman (1956).

Table 33-5. MEAN MANIPULATION AND VISUAL EXPLORATION SCORES. (*From Ressler, 1963, with permission of the author and the American Psychological Association.*)

	C57BL/10 *Foster parents*		BALB/c *Foster parents*	
	C57BL/10 Young	BALB/c Young	C57BL/10 Young	BALB/c Young
Manipulation	38	61	38	78
Visual exploration	62	73	85	87

GENETIC ANALYSIS OF BEHAVIOR

The diversity of mouse stocks affords a corresponding diversity of possible analytic techniques for exploiting the genetic material, which varies from lines segregating at a single locus to those differing at many loci. The purpose of this section is to present examples of various techniques which have been applied to the genetic analysis of mouse behavior.

Stocks segregating at a single locus

The establishment of a behavioral difference between genotypes of a strain segregating at only a single locus is highly informative, for all possible pathways between gene and phenotype must converge on one biochemical process or on one regulator controlling a few related processes. A small number of studies on behavioral differences have been reported.

Studies of obesity (*ob*) illustrate the possibilities for physiological and biochemical analyses after establishment of a genetic effect. Reared under typical laboratory conditions, *ob/ob* mice attain a weight twice that of littermate non-obese sibs.

They characteristically eat heartily and are inactive and infertile. Both excessive weight and infertility may be eliminated by restriction of food intake (Runner and Gates, 1954; Lane and Dickie, 1954). Increased food consumption could be due either (1) to a defect in central mechanism regulating caloric intake or (2) to differences in the hedonic attributes of food ingestion per se. A series of experiments were performed to identify the underlying mechanisms.

Fuller and Jacoby (1955) decreased the caloric value of the regular laboratory food by diluting it with cellulose or increased it by adding fat. Normal mice responded rapidly to decreased caloric value by increasing consumption, but obese mice increased intake by a smaller amount. Both groups ate less of the fat-enriched diet, but the adjustment was again much greater in the non-obese mice. When bitter substances—quinine or caramelized sugar—were added to food, both obese mice and controls sharply reduced their consumption. However, controls slowly increased their ingestion over a few days, whereas obese mice continued to eat reduced amounts. Results, then, favored a defect in central mechanisms producing adaptive regulation of food intake.

The behavior of obese mice appears very similar to that produced in rats by surgical intervention in the hypothalamus. However, the hypothalamus of obese mice appears to be normal histologically (Maren, 1952). This does not, of course, eliminate the possibility either of subtle structural differences or of a biochemical disorder (Chap. 19). Injection of gold thioglucose may result in obesity, and because of this similarity of effect to hypothalamic lesions it has been assumed that the hypothalamus is the site of action. Hollifield et al. (1955) compared *ob/ob* mice with those made obese by injection of gold thioglucose. Gold thioglucose produced obese mice which did not increase activity during food deprivation nor show decreased activity following refeeding. Hereditarily obese mice, however, responded in a manner similar to controls. We may conclude (1) that the mechanism of hereditary obesity differs from that produced by gold thioglucose and (2) that hereditarily obese mice do show some of the behavioral effects of food deprivation normally associated with drive.

Known genes may share control of a phenotype with other genes which are not separately identifiable. An example is the dilute gene (*d*), which has attracted attention since Coleman (1960) demonstrated that homozygous *d/d* mice have a phenylketonuria-like condition. Since phenylketonurics often have convulsions, it was considered possible that the high incidence of audiogenic seizures in DBA/2 mice (*d/d*) might be caused in part by their deficiency in phenylalanine hydroxylase activity.

Direct evidence on this point is conflicting. No correlation between dilute phenotype and seizure susceptibility was found in (DBA/2J × C57BL/6J)F_1 hybrids (Fuller et al., 1950), but Huff and Huff (1962) found an excess of seizures in dilute animals from another source. Huff and Fuller (1962) failed to find an effect on susceptibility of *d*-locus substitution in a stock of DBA/1J mice segregating for + and *d*. They suspected that a mutation might have abolished the enzyme-inhibiting effects of the *d/d* genotype and thus interfered with its facilitation of seizures. Comparison between this and earlier studies was rendered difficult by an intervening change in diet of the mice. An interaction between the dilute locus and diet was found by Coleman and Schlesinger (1965). Mice from homozygous dilute strains (DBA/2J, BDP/J, and P/J) showed progressive increase in seizure susceptibility during 7 weeks on a low-pyridoxine (vitamin B_6) diet; nondilute C57BL/6J mice showed no such increase. Heterozygotes (D2B6F_1) showed a small increase.

The experiments using gene substitutions at the *ob* and *d* loci have concentrated on the analysis of the behavioral effects in physiological terms. Other studies of mutants have been directed toward finding an association between a gene and behavior.

Ashman (1957, personal communication) used strains SEC/1Gn-*se* and SEC/2Gn-*d* to study the effects of the alleles normal-ear (+) and short-ear (*se*), and dense coat color (+) and dilute coat color (*d*). Constitution of the stocks made it possible to compare +/+ *se*/+ with +/+ *se/se*, and *d*/+ *se/se* with *d/d se/se*. Ashman found no differences in activity in a tilt cage and open field, in conditionability, or in susceptibility to audiogenic seizure.

Les (1958) compared mutant and wild-type mice in stocks congenic at other loci. The mutants were furless (*fs*), short-ear (*se*), dilute (*d*), yellow (A^y), albino (*c*), black-and-tan (a^t), misty (*m*), and hairless (*hr*). He obtained measures of locomotor activity, transporting shavings, nibbling through a cardboard partition, defecation and urination in an open field, and adult body weight. More fortunate than Ashman, he established several behavioral differences. For example, animals homozygous for furless (*fs/fs*) and hairless (*hr/hr*) were less active than their normal sibs (*fs*/+ and *hr*/+), and yellow mice (A^y/+) were less active than their normal sibs (+/+). Differences in nibbling and adult body weight were also found.

Though not based upon stocks segregating at a single locus, results obtained by Winston and

Lindzey (1964) suggested a behavioral effect associated with the albino locus. Time to swim to an escape ramp was measured for pigmented (+/+) strains C3H/Bi, DBA/8, and JK and albino (*c/c*) strains A and BALB/c. The albino strains had longer time-scores. Of greater interest here were the performances of segregating stocks: An intercross of F_1 hybrids (A × DBA/8) × (A × DBA/8) resulted in F_2 offspring with coat color phenotypes albino (*c/c*) and pigmented (*c*/+ and +/+). Albinos had higher swimming scores. Backcross (A × DBA/8) × A also resulted in both albino (*c/c*) and pigmented offspring (*c*/+), and again albino mice were slower (had higher scores). The authors suggested that the recessive gene for albinism is also a recessive gene for slow water-escape behavior. Since the mating systems used did not minimize the likelihood of contribution of other genes in the A stock linked with the *c* allele and other genes in the DBA/8 stock linked with the + allele, no definite conclusions can be reached (Meier et al., 1965).

Stocks differing at many loci

Since most behavioral variation appears to be determined by many pairs of genes with small individual effect acting together (Broadhurst and Jinks, 1963), it may often not be possible to detect effects of a single-gene substitution. Several investigators using genetically more complex materials with discernible behavioral variation have applied the techniques of quantitative genetics to behavior. (For details of methods and assumptions see Chap. 9.)

Scaling. For a simple biometric analysis, the phenotypes of quantitative behavior genetics should have certain measurement characteristics. First, the effects of heredity and environment should act independently of each other, for interactions always make conceptualizations more difficult, and they frequently strain the statistical assumptions made in performing analyses. Specifically, variances in P_1, P_2, and F_1, all presumably environmental in origin, should be homogeneous for stocks reared under similar controlled conditions. Second, and for the same reasons, the genetic effects should be additive. Once the phenotypic values for P_1, P_2, and F_1 have been measured, the observed values for the F_2 and the two backcrosses should equal their expected values on some additive genetic model. Change of scale may eliminate interactions of genetic and environmental factors and produce genetic additivity. Such transformations should be considered whenever needed, for momentary awkwardness of working on an unfamiliar scale may be repaid by the long-term reduction in analytic complexity.

Bruell (1962) analyzed patterns of behavior according to Mather's (1949) two scaling criteria paraphrased above. Bruell's description of rationale and procedures for testing appropriateness of scale merit careful study. He measured the exploratory activity of strains A, C57BL/10, F_1's, F_2's, and backcrosses. The raw score was found adequate for the additivity criterion, and the variances of P_1, P_2, and F_1 were homogeneous. Figure 33-10 shows the discrepancies between observed and predicted values for F_2 and the two back-

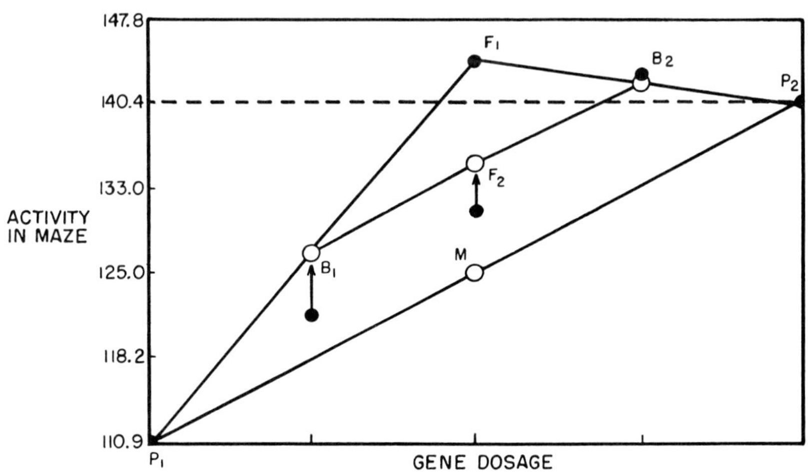

Fig. 33-10. Exploratory behavior test. P1 and P2 are strains A and C57BL/10. B_1, B_2, and F_2 are the open circles. (*From Bruell, 1962, with permission of the author and the Hoeber Medical Division of Harper and Row, Publ.*)

crosses. The apparent overdominance of the F_1 was not significant. In an analysis of activity-wheel data, Bruell (1962) compared raw scores, square root scores, and logarithmic scores for additivity, finding only the log scale to be adequate. He was unable to achieve homogeneity of variance among P_1, P_2, and F_1. Bruell also analyzed time to climb down a pole in a similar manner.

Physiological studies have sometimes indicated that inbred strains may be more variable than their F_1 hybrids. Mordkoff and Fuller (1959) studied variability in activity in stocks C57BL/6J, DBA/2J, B6D2F_1, and DX (a 4-way cross of strains C57BL/6J, DBA/2J, BALB/cJ, and C3HeB/FeJ; Green, 1964). A reduced variability in hybrids was or was not shown, depending upon the measure used. Ranked from highest to lowest in variability between animals within a group, the order was C57BL/6J, DX, B6D2F_1, and DBA/2J. However, the coefficient of variation (s/x) being used, they ranked C57BL/6J, DBA/2J, B6D2F_1, and DX. Reduced variability in hybrids has been hypothesized to be due to better developmental homeostasis—higher stability resulting from more biochemical and physiological versatility in the developing young heterozygous animal. Mordkoff and Fuller raised the question of whether developmental homeostasis would have the same meaning for behavior, suggesting that behavioral variability might produce better homeostasis. There is clearly need for careful distinction between homeostatic mechanisms and their consequences. A later study by Schlesinger and Mordkoff (1963) measured both locomotor activity and oxygen consumption in strains C57BL/6J, DBA/2J, and the F_1. They interpreted their results as showing less variability among F_1 animals on both measures.

Observations on the inheritance of seizure susceptibility in mice illustrate how polygenic systems, environmental effects, and choice of scale may combine to affect the results of an analysis. Fuller et al. (1950) selected strain DBA/2J (extremely susceptible to convulsions) and strain C57BL/6J (highly resistant) as parental stocks. Also tested were the F_1, F_2, and backcrosses. The measure was risk of convulsion during a 2-minute period of bell ringing. Particularly striking was the difference in seizure incidence between the original (32 per cent) and replicated (74 per cent) F_1 groups.

Fuller et al. (1950) assumed that a threshold model would account for their results. Two things are important in understanding the model: (1) the general nature of audiogenic seizures and (2) the measure used. Seizures result in some manner from massive discharges of central nervous system neurons produced by the cumulative effects of impulses entering auditory nerves. Differences in seizure susceptibility presumably reflect a continuum of differences in the rate of accumulation and dissipation of the effects of stimulation. Some mice convulse quickly during a test, others slowly; some severely, others lightly; some show great excitement without convulsing, others are overtly unresponsive. Because of problems of reliability

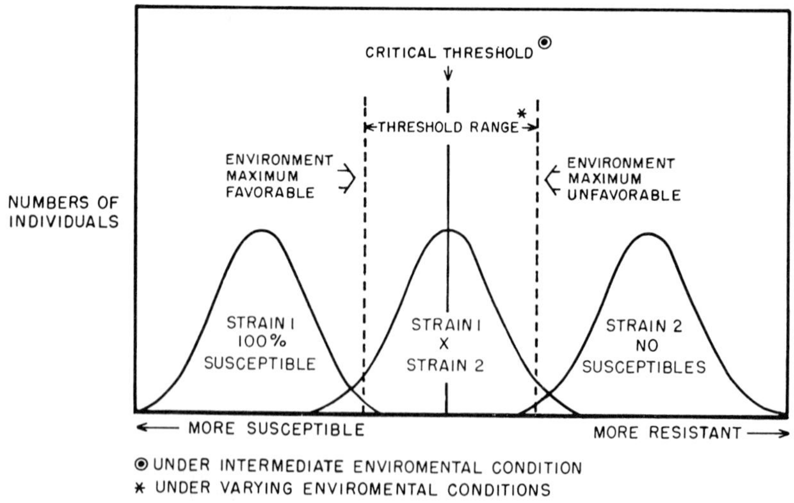

Fig. 33-11. Degree of audiogenic seizure susceptibility—controlled by many genes and by the environment. (*Adapted from Fuller et al.,* 1950.)

and scaling, the phenotypic measure generally used is dichotomous. That is, animals either do or do not convulse.

Fuller et al. (1950) proposed that the genotype of DBA/2 led to a phenotype almost always seizure susceptible and that of C57BL/6 to one almost always resistant. The F_1 genotype produced an intermediate phenotype leading to a seizure only with an environmental push. Figure 33-11 illustrates what might have happened if the two F_1 groups were reared or tested under somewhat different environmental conditions. It is important to remember that a phenotype characterized as being in one of two classes may be polygenically determined and that results of a genetic analysis are determined in part by the scale used.

Crosses, intercrosses, and backcrosses. As we have seen, the simplest quantitative model for any behavioral phenotype is that it is an additive function of heredity and environment. By use of this model and some basic statistical concepts of multiple-gene action, it is possible (1) to estimate the relative determinations of phenotype by genotype and (2) to make generalizations concerning the average contributions of genes from one parental stock as contrasted with those issuing from others.

McClearn (1961) crossed strains C57BL and A, intercrossed the resulting F_1 to produce an F_2, and backcrossed the F_1 to both parental stocks. The phenotype measured was number of squares traversed in an open field in a 3-minute period. Means of square root transforms, required to obtain homogeneity of variance among parental and hybrid stocks, are presented in Table 33-6. Deviation of the F_1 mean from the midparent value of 7.45 indicated dominance of C57BL genes over those from strain A for this phenotype. Note that the ranking of groups corresponded generally to the percentage of C57BL genes present. The relative contribution of the genetic factor—heritability in the broad sense—was estimated by measuring the variability in the parental and F_1 groups (in all of which variability between animals within stocks was presumably environmental in origin) and subtracting that value form the observed variability of the F_2's (in which both genetic and environmental sources of difference between individuals were operating). The ratio of residual genetic variability of the F_2's to their total observed variability was 0.69, indicating a rather high degree of genetic determination. A variety of related analyses were performed by McClearn and Rodgers (1961) in their studies of alcohol preference, by McClearn (1959) in his study of exploratory behavior, and by Fuller and Thompson (1960, p. 268) for activity. Heritabilities computed from crosses between inbred lines should not be used to estimate heritabilities in natural populations of unknown genetic diversity. Their greatest value is to guide experimenters to model systems which are promising for the analysis of genetic processes by a variety of methods.

Table 33-6. GENETIC CONSTITUTION OF GROUPS WITH MEANS AND VARIANCES OF ACTIVITY SCORES. (*From McClearn, 1961, with permission of the author and the American Psychological Association.*)

Group	Female parent	Male parent	*Square root transformation of offspring scores* Mean	Variance
C57	C57	C57	12.0	1.3
B(C57)	C57 × A	C57	10.6	2.9
F_1	C57	A	9.6	5.4
F_2	C57 × A	C57 × A	7.6	9.2
B(A)	C57 × A	A	4.7	5.1
A	A	A	2.9	1.6

Another quantitative technique is the diallel cross (Chap. 2, 9). In this technique, k strains are crossed in all possible combinations to produce $k(k-1)$ hybrids, and phenotypic measures are taken on members of all groups. From the resulting $k \times k$ matrix of phenotypic values classified by genotype, it is possible to assess (1) the general contribution (general combining ability) of genes from a specific parental strain when combined in hybrids with genes from all other strains, and (2) detect deviations (specific combining ability) from this general contribution occurring in specific crosses. Fuller (1964) applied this technique to the study of genetic factors in alcohol preference. The phenotypic values of parental stocks and hybrids are shown in Fig. 33-12. The parental strains differed significantly among themselves, and the analysis to evaluate differences in general and specific combining ability as measured in the hybrids showed both factors to be significant. Genes from the two extreme strains, C57BL/6J and DBA/2J, had consistent effects of raising or lowering alcohol preference scores, whereas genes from strains A/J and C3HeB/FeJ had negligible general combining ability. For example, hybrids of A/J resulting from a cross with

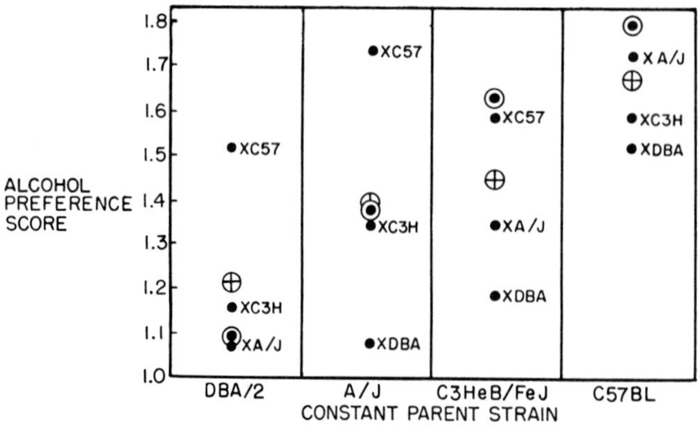

Fig. 33-12. Mean alcohol preference scores for parental strains (bull's-eye) and hybrids (solid circles). Average alcohol preference scores for a strain and crosses involving it are indicated by targets. (*Adapted from Fuller*, 1964.)

low-preference DBA/2J mice were like DBA/2J, hybrids from a cross with high-preference C57BL/6J mice were like C57BL/6J, and those from a cross with C3HeB/FeJ were almost identical with A/J. The dangers in generalizing from one specific cross are evident.

Bruell has related degree of heterosis or dominance observed in the F_1 to closeness of relation of parental strains. His stocks consisted of 31 F_1 hybrid groups produced from 13 parent strains, some of which were unrelated, moderately related, or related as sublines of a single strain. Bruell (1964*a*, *b*) measured exploration of a strange environment and running in activity wheels. Hybrids typically explored more and were more active in activity wheels. Furthermore, the degree of superiority of the hybrids over the parental strains *increased* as relatedness of the parental strains *decreased* (Fig. 33-13). Bruell (1965) also measured latency of descent from a pole and emergence from a dark tunnel into an open field. Mode of inheritance for these activities tended to be intermediate. Barnett and Scott (1964) suggested that mode of inheritance in the F_1 hybrid depends upon the adaptive significance of the behavior, heterotic inheritance being observed most often in behavior having homeostatic function. Bruell's (1964*a*) position is similar.

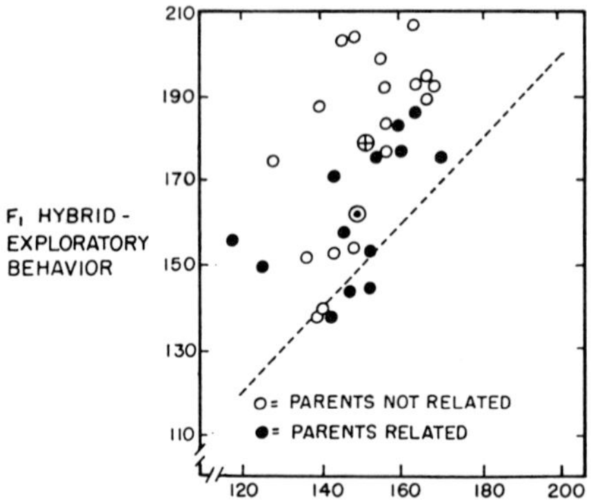

Fig. 33-13. Inheritance of exploratory behavior in F_1 hybrid mice. The diagonal is the "midparent line," i.e., a line based on the mean scores for parental stocks plotted on both axes. Scores of F_1 hybrids are arranged along the abscissa over the appropriate midparent value. The mean score for the hybrid is indicated along the ordinate. Note that most hybrids scored above the midparent line, and that the mean of the 18 hybrids whose parents were related (bull's-eye) is less than that of the 13 hybrids whose parents were unrelated (cross). (*From Bruell, 1964a, with permission of the author and the American Zoologist.*)

Selection

Selection is an important but neglected technique for behavior geneticists. Parent-offspring correlations obtained during very early phases of selection provide an estimate of heritability, and speed of response to selection provides a clue to the number of loci involved (Chap. 9). Examination of correlated responses to selection—changes in brain size or cell density, size of adrenals, etc.—could provide important insights into the mechanisms underlying a behavior pattern. Once phenotypically extreme stocks have been produced, they may be of value for a variety of purposes.

The power of selection as a tool for understanding the mechanisms underlying a behavior pattern is amply demonstrated in a study concerned with audiogenic seizures—a response to intense auditory stimulation characterized by running, convulsion, and death in highly susceptible mice. Frings and Frings (1953) selected separate lines for (1) high incidence of clonic-tonic seizures but low death rate, (2) high clonic seizure incidence, (3) low seizure susceptibility, and (4) susceptibility over a restricted age range. Results of selection indicated separate genetic control for age of susceptibility, severity of seizure, and resistance to death during seizure.

Two other behavioral selection studies have used the mouse: Dawson (1932) selected for running speed in a runway, and Lagerspetz (1964) developed strains differing in aggressiveness.

Multivariate phenotype

It is likely that genetically diverse stocks such as different inbred strains of mice will be behaviorally diverse for a variety of activities. Vanderpool and Davis (1962) employed a time-sampling technique to study the frequencies and durations of seven classes of orienting behavior in strains C57BL/6J, BALB/cJ, and DBA/2J. A variety of strain differences and similarities emerged.

Survey procedures have also been applied to identify behavioral differences associated with specific genetic loci. Thus, van Abeelen (1963*a*, *b*, *c*) applied ethological techniques, using an inventory of behavioral elements, including staring at the observer, hair fluffing, tail rattling, sniffing, reconnoitering, eating, digging, grooming, shaking the fur, wrestling, submissive posture, mounting, and many others. He reported, for example, that stocks carrying pink-eyed dilution (*p*, I) stared at the observer less and groomed their fur more. Stocks carrying jerker (*je*, XII) displayed clearly disturbed exploratory and aggressive behavior. Frequency of eating was not affected, but the eating posture was different.

Survey techniques have considerable potential value. First, they provide a profile of differences on a variety of activities. Second, they may identify specific activities and patterns of activities with higher heritabilities than those typically studied. A variety of multivariate statistical techniques useful in analysis are now available and practical (Cooley and Lohnes, 1962; Kendall, 1961).

SUMMARY

Decades of labor by geneticists have made available an extremely wide variety of genetically controlled mouse stocks. No such diversity exists in any other animal species except the fruit flies. There is a large body of detailed and behaviorally relevant genetic, anatomical, physiological, and biochemical knowledge concerning these mouse stocks. Further, as shown in early sections of this chapter, experience with apparatus and techniques for behavioral studies is substantial.

Though the presentation was expressly intended for behavior geneticists, the house mouse should be of interest to many different kinds of behaviorists. General experimental psychologists who seek behavioral laws applying across organisms should find value in both the genetic uniformity within inbred strains and the genetic variability between them. Uniformity within strains has the special advantage that observed differences among individuals are presumably environmental in origin, so that one has a genetically "noiseless" system in which to evaluate the consequences of experiential treatments. Genetic variability between strains offers the possibility of repeating treatments across genotypes as a reasonable first step to estimating their generality. Physiological psychologists will find a wide range of anatomical characteristics among standard inbred strains. Some effects produced by mutant genes—for example, retinal degeneration (*rd*)—can probably not be copied by any experimental treatment (Chap. 32).

Interest in the behavior of mice is not restricted to behaviorists. Biological problems in such fields as reproductive physiology and possibly resistance to cancer have been found to have psychological aspects. The mouse has been exploited rather extensively by neurochemists, though little advantage has been taken of the genetic diversity within the species. If this is remedied, psychochemistry may come to depend upon the cooperation of behaviorists, chemists, and laboratory mice. It should be noted that psychotropic drug assay and screening procedures make use of large numbers of mice, generally of unrecorded ancestry.

A final comment is required to place the genetic and environmental determinants of behavioral phenotype in proper perspective. Our writing has, as a matter of convenience and clarity of exposition and as a reflection of our own interests and those of the anticipated readers, placed great and oversimplified emphasis on genetic determinants of behavior. Let us set the record straight: Just as it is impossible to conceive of an organism without a genotype, so is it also impossible to conceive of one which does not develop and is not maintained and tested in some environment. Some of the genetically associated differences in behavior reported may originate in differences in the pre- and postnatal environments which the different stocks afford (Ressler, 1962, 1963). Both direction and degree of other genetically associated differences may depend upon more general experimenter-controlled conditions of maintenance (Cooper and Zubek, 1958; Lieberman, 1963) and test (McClearn, 1960).

We firmly believe that conceptions of behavior not jointly determined by heredity *and* environment are absurd (Moltz, 1965). More-traditional psychologists have been criticized for failure to pay adequate attention to genetic factors in behavior. Behavior geneticists potentially run the opposite risk. Intelligently united, both approaches will increase our understanding of behavior.

LITERATURE CITED

Anliker, J., and J. Mayer. 1956. An operant conditioning technique for studying feeding-fasting patterns in normal and obese mice. J. Appl. Physiol. 8:667–670.

Antalfi, S. 1963. Biological determination of the intensity of the alarm reaction in house mice. J. Comp. Physiol. Psychol. 56:889–891.

Banks, E. M. 1962. A time and motion study of prefighting behavior in mice. J. Genet. Psychol. 101:165–183.

Barnes, G. W., and A. Baron. 1961. The effect of sensory reinforcement on extinction behavior. J. Comp. Physiol. Psychol. 54:461–465.

Barnes, G. W., and G. B. Kish. 1957. Reinforcing properties of the termination of intense auditory stimulation. J. Comp. Physiol. Psychol. 50:40–43.

Barnes, G. W., and G. B. Kish. 1961. Reinforcing properties of the onset of auditory stimulation. J. Exp. Psychol. 62:164–170.

Barnes, G. W., G. B. Kish, and W. O. Wood. 1959. The effect of light intensity when onset or termination of illumination is used as reinforcing stimulus. Psychol. Rec. 9:53–60.

Barnett, S. A. 1956. Endothermy and ectothermy in mice at −3°C. J. Exp. Biol. 33:124–133.

Barnett, S. A. 1959. The skin and hair of mice living at a low environmental temperature. Quart. J. Exp. Physiol. 44:35–42.

Barnett, S. A., and S. G. Scott. 1964. Behavioral "vigour" in inbred and hybrid mice. Anim. Behav. 12:325–337.

Baron, A., J. J. Antonitis, and S. F. Schell. 1962. Effects of early restriction and facilitation of climbing on later climbing behavior of mice. J. Comp. Physiol. Psychol. 55:808–812.

Baron, A., and G. B. Kish. 1962. Low-intensity auditory and visual stimuli as reinforcers for the mouse. J. Comp. Physiol. Psychol. 55: 1,011–1,013.

Bauer, F. J. 1956. Genetic and experiential factors affecting social reactions in male mice. J. Comp. Physiol. Psychol. 49:359–364.

Beeman, E. A. 1947. The effect of male hormone on aggressive behavior in mice. Physiol. Zool. 20:373–405.

Bell, R. W., and V. H. Denenberg. 1963. The interrelationships of shock and critical periods in infancy as they affect adult learning and activity. Anim. Behav. 11:21–27.

Beniest-Noirot, E. 1958. Analyse du comportement dit maternel chez la souris. Cent. Nat. Rech. Sci. Monogr. Franc. Psychol. No. 1.

Bevan, J. M., W. Bevan, and B. F. Williams. 1958. Spontaneous aggressiveness in young castrate C3H male mice treated with three dose levels of testosterone. Physiol. Zool. 31:284–288.

Bevan, W., W. F. Daves, and G. W. Levy. 1960. The relation of castration, androgen therapy and pretest fighting experience to competitive aggression in male C57BL/10 mice. Anim. Behav. 8:6–12.

Broadhurst, P. L., and J. L. Jinks. 1963. The inheritance of mammalian behavior re-examined. J. Hered. 54:170–176.

Bronson, F. H. 1963. Density, subordination and social timidity in Peromyscus and C57BL/10J mice. Anim. Behav. 11:475–479.

Bronson, F. H., and B. E. Eleftheriou. 1963. Adrenal responses to crowding in Peromyscus and C57BL/10J mice. Physiol. Zool. 36:161–166.

Bruell, J. H. 1962. Dominance and segregation in the inheritance of quantitative behavior in mice, p. 48–67. *In* E. L. Bliss [ed.] Roots of behavior. Hoeber-Harper, New York.

Bruell, J. H. 1963. Emotional defecation in mice, a territory marking response. Amer. Psychol. 17:445. (Abstr.)

Bruell, J. H. 1964*a*. Inheritance of behavioral and physiological characters of mice and the problem of heterosis. Amer. Zool. 4:125–138.

Bruell, J. H. 1964*b*. Heterotic inheritance of wheel-running in mice. J. Comp. Physiol. Psychol. 58:159–163.

Bruell, J. H. 1965. Mode of inheritance of response time in mice. J. Comp. Physiol. Psychol. 60:147–148.

Caldwell, D. F. 1962. Effects of adrenal demedullation on retention of a conditioned avoidance response in the mouse. J. Comp. Physiol. Psychol. 55:1,079–1,081.

Carran, A. B., L. T. Yendall, and J. R. Royce. 1964. Voltage level and skin resistance in avoidance conditioning of inbred strains of mice. J. Comp. Physiol. Psychol. 58:427–430.

Catlett, R. H. 1961. An evaluation of methods for measuring fighting behaviour with special reference to *Mus musculus*. Anim. Behav. 9:8–10.

Chitty, D., and H. N. Southern [ed.] 1954. Control of rats and mice. Vol. 3. House mice. Oxford, London. 240 p.

Christian, J. J. 1955. Effect of population size on the adrenal glands and reproductive organs of male mice in populations of fixed size. Amer. J. Physiol. 182:292–301.

Coleman, D. L. 1960. Phenylalanine hydroxylase activity in dilute and nondilute strains of mice. Arch. Biochem. Biophys. 91:300–306.

Coleman, D. L., and K. Schlesinger. 1965. The effects of pyridoxine deficiency on audiogenic seizure susceptibility in inbred mice. Proc. Soc. Exp. Biol. Med. 119:264–266.

Cooley, W. W., and P. R. Lohnes. 1962. Multivariate procedures for the behavioral sciences. Wiley, New York. 211 p.

Cooper, R. M., and J. P. Zubek. 1958. Effects of enriched and restricted early environments on the learning ability of bright and dull rats. Can. J. Psychol. 12:159–164.

Davis, W. M. 1962. Day-night periodicity in pentobarbital response of mice and the influence of socio-psychological conditions. Experientia 18:235–237.

Dawson, W. M. 1932. Inheritance of wildness and tameness in mice. Genetics 17:296–326.

Denenberg, V. H. 1958. Effects of age and early experience upon conditioning in the C57BL/10 mouse. J. Psychol. 46:211–226.

Denenberg, V. H. 1959*a*. Interactive effects of infantile and adult shock levels upon learning. Psychol. Rep. 5:357–364.

Denenberg, V. H. 1959*b*. Learning differences in two separated lines of mice. Science 130:451–452.

Denenberg, V. H. 1960. A test of the critical period hypothesis and a further study of the relationship between age and conditioning in the C57BL/10 mouse. J. Genet. Psychol. 97:379–384.

Denenberg, V. H., and R. W. Bell. 1959. Relationships between social reactions and avoidance conditioning. Physiol. Zool. 32:51–56.

Denenberg, V. H., and R. W. Bell. 1960. Critical periods for the effects of infantile experience on adult learning. Science 131:227–228.

Denenberg, V. H., G. A. Hudgens, and M. X. Zarrow. 1964. Mice reared with rats: modification of behavior by early experience with another species. Science 143:380–381.

Denenberg, V. H., J. A. King, and D. W. Ehrenfeld. 1958. Effects of testosterone propionate on the avoidance learning of male C57BL/6 mice. Physiol. Zool. 31:244–247.

Denenberg, V. H., S. Ross, and M. Blumenfield. 1963. Behavioral differences between mutant and non-mutant mice. J. Comp. Physiol. Psychol. 56:290–293.

Doner, R. D., C. Inman, and R. T. Davis. 1952. Competitive behavior between inbred strains of mice. Proc. S.D. Acad. Sci. 31:172–176.

Ekström, N. A., and F. Sandberg. 1962. A method for quantitative determination of the inhibitory effect on the conditioned avoidance response in mice. Drug Res. 12:1,208–1,209.

Elton, C. 1942. Voles, mice and lemmings. Oxford, London. 496 p.

Essman, W. B., and H. Alpern. 1964. Single trial conditioning: methodology and results with mice. Psychol. Rep. 14:731–740.

Essman, W. B., and M. E. Jarvik. 1961*a*. A water-escape test for mice. Psychol. Rep. 8:58.

Essman, W. B., and M. E. Jarvik. 1961*b*. Impairment of retention for a conditioned response by ether anesthesia in mice. Psychopharmacologia 2:172–176.

Fredericson, E. 1950. The effects of food deprivation upon competitive and spontaneous combat in C57 black mice. J. Psychol. 29:89–100.

Fredericson, E. 1952. Aggressiveness in female mice. J. Comp. Physiol. Psychol. 45:254–257.

Fredericson, E., and E. A. Birnbaum. 1954. Competitive fighting between mice with different hereditary backgrounds. J. Genet. Psychol. 85: 271–280.

Fredericson, E., C. D. Fink, and J. R. Parker. 1955. Elicitation and inhibition of competitive fighting in food deprived mice. J. Genet. Psychol. 86:131–141.

Frings, H., and M. Frings. 1953. The production of stocks of albino mice with predictable susceptibilities to audiogenic seizures. Behaviour 5:305–319.

Fuller, J. L. 1964. Measurement of alcohol preference in genetic experiments. J. Comp. Physiol. Psychol. 57:85–88.

Fuller, J. L., R. M. Chambers, and R. P. Fuller. 1956. Effects of cortisone and of adrenalectomy on activity and emotional behavior of mice. Psychosomat. Med. 18:234–242.

Fuller, J. L., C. Easler, and M. E. Smith. 1950. Inheritance of audiogenic seizure susceptibility in the mouse. Genetics 35:622–632.

Fuller, J. L., and G. A. Jacoby, Jr. 1955. Central and sensory control of food intake in genetically obese mice. Amer. J. Physiol. 183:279–283.

Fuller, J. L., and W. R. Thompson. 1960. Behavior genetics. Wiley, New York. 396 p.

Ginsburg, B., and W. C. Allee. 1942. Some effects

of conditioning on social dominance and subordination in inbred strains of mice. Physiol. Zool. 15:485–506.

Grant, E. C., and J. H. Mackintosh. 1963. A comparison of the social postures of some common laboratory rodents. Behaviour 21:246–259.

Green, E. L. 1964. Fitness of populations of irradiated mice: plan of experiments. Genetics 50:417–421.

Halberg, F., J. J. Bittner, R. J. Gully, P. G. Albrecht, and E. L. Brackney. 1955. 24-Hour periodicity and audiogenic convulsions in I mice of various ages. Proc. Soc. Exp. Biol. Med. 88:169–173.

Halberg, F., E. Jacobsen, G. Wadsworth, and J. J. Bittner. 1958. Audiogenic abnormality spectra, twenty-four hour periodicity, and lighting. Science 128:657–658.

Halberg, R., R. E. Peterson, and R. H. Silber. 1959. Phase relations of 24-hour periodicities in blood corticosterone, mitoses in cortical adrenal parenchyma, and total body activity. Endocrinology 64:222–230.

Hall, C. S., and P. H. Whiteman. 1951. The effects of infantile stimulation upon later emotional stability in the mouse. J. Comp. Physiol. Psychol. 44:61–66.

Hall, M. J. 1959. The effects of early infantile experiences on maze learning in the mouse. Proc. W.Va. Acad. Sci. 31:79–82.

Harner, R. N. 1961. Electrocorticography and frequency analysis in mice; circadian periodicity in electrocerebral activity. Electroenceph. Clin. Neurophysiol. 13:752–761.

Hauschka, T. S. 1952*a*. Whisker-eating mice. J. Hered. 43:77–80.

Hauschka, T. S. 1952*b*. Mutilation patterns and hereditary cannibalism in mice. J. Hered. 43:117–123.

Hollifield, G., W. Parson, and K. R. Crispell. 1955. Studies of food drive and satiety in mice with gold thioglucose–induced obesity and the hereditary obesity-diabetes syndrome. Metabolism 4:537–541.

Hoshishima, K., S. Yokoyama, and K. Seto. 1962. Taste sensitivity in various strains of mice. Amer. J. Physiol. 202:1,200–1,204.

Huff, S. D., and J. L. Fuller. 1964. Audiogenic seizures, the dilute locus, and phenylalanine hydroxylase in DBA/1 mice. Science 144:304–305.

Huff, S. D., and R. L. Huff. 1962. Dilute locus and audiogenic seizures in mice. Science 136:318–319.

Hummel, K. P. 1960. Pituitary lesions in mice of the Marsh strains. Anat. Rec. 137:366. (Abstr.)

Jacobs, B. B. 1962. Propylthiouracil taste inheritance in mice. J. Hered. 53:183–186.

Janssen, P. A. J., A. H. Jageneau, and C. J. E. Niemegeers. 1960. Effects of various drugs on isolation-induced fighting behavior of male mice. J. Pharmacol. Exp. Therap. 129:471–475.

Jarvik, M. E., and W. B. Essman. 1960. A simple one-trial learning situation for mice. Psychol. Rep. 6:290.

Kahn, M. W. 1951. The effects of severe defeat at various age levels on the aggressive behavior of mice. J. Genet. Psychol. 79:117–130.

Keeler, C. E. 1927. Rodless retina, an opththalmic mutation in the house mouse, *Mus musculus*. J. Exp. Zool. 46:355–407.

Keeler, C. E. 1928. Blind mice. J. Exp. Zool. 51:495–508.

Kendall, M. G. 1961. Course in multivariate analysis, 2nd ed. Hafner, New York. 185 p.

King, J. A. 1956. Sexual behavior of C57BL/10 mice and its relation to early social experience. J. Genet. Psychol. 88:223–229.

King, J. A. 1957*a*. Intra- and interspecific conflict of *Mus* and *Peromyscus*. Ecology 38:355–357.

King, J. A. 1957*b*. Relationships between early social experience and adult aggressive behavior in inbred mice. J. Genet. Psychol. 90:151–166.

King, J. A., and H. Cannon. 1955. Effects of social relationships upon mortality in C57BL/10 mice. Physiol. Zool. 28:233–239.

King, J. A., and N. L. Gurney. 1954. Effect of early social experience on adult aggressive behavior in C57BL/10 mice. J. Comp. Physiol. Psychol. 47:326–330.

King, J. A., and A. Mavromatis. 1956. The effect of a conflict situation on learning ability in two strains of inbred mice. J. Comp. Physiol. Psychol. 49:465–468.

King, J. T., Y. C. P. Lee, and M. B. Visscher. 1955. Single versus multiple cage occupancy and convulsion frequency in C3H mice. Proc. Soc. Exp. Biol. Med. 88:661–663.

Kish, G. B. 1955. Learning when the onset of illumination is used as a reinforcing stimulus. J. Comp. Physiol. Psychol. 48:261–264.

Kish, G. B., and A. Baron. 1962. Satiation of sensory reinforcement. J. Comp. Physiol. Psychol. 55:1,007–1,010.

Lagerspetz, K. 1964. Studies on the aggressive behaviour of mice. Ann. Acad. Sci. Fenn. B 131:1–13.

Lane, P. W., and M. M. Dickie. 1954. Fertile, obese male mice. J. Hered. 45:56–58.

Leblond, C. P. 1940. Nervous and hormonal factors in the maternal behavior of the mouse. J. Genet. Psychol. 57:327–344.

Leblond, C. P., and W. O. Nelson. 1937. Maternal behavior in hypophysectomized male and female mice. Amer. J. Physiol. 120:167–172.

Les, E. 1958. A study of the effects of single locus heterozygosity on traits which may have survival value in eight stocks of laboratory mice. Ph.D. Diss. Ohio State Univ.

Levine, L. 1958. Studies on sexual selection in mice. I. Reproductive competition between albino and black-agouti males. Amer. Natur. 92:21–26.

Levine, L. 1963. Factors affecting mating competition in mice, p. 249–250. *In* S. J. Geerts [ed.] Proc. XI Int. Congr. Genet. (The Hague, 1963) Vol. 1. Pergamon Press, New York. (Abstr.)

Levine, S. 1959. Emotionality and aggressive behavior in the mouse as a function of infantile experience. J. Genet. Psychol. 94:77–83.

Lieberman, M. W. 1963. Early developmental stress and later behavior. Science 141:824–825.

Lindzey, G. 1951. Emotionality and audiogenic seizure susceptibility in five inbred strains of mice. J. Comp. Physiol. Psychol. 44:389–393.

Lindzey, G., D. T. Lykken, and H. D. Winston. 1960. Infantile trauma, genetic factors, and adult temperament. J. Abnorm. Soc. Psychol. 61:7–14.

Lindzey, G., and H. Winston. 1962. Maze learning and effects of pretraining in inbred strains of mice. J. Comp. Physiol. Psychol. 55:748–752.

Lindzey, G., H. Winston, and M. Manosevitz. 1961. Social dominance in inbred mouse strains. Nature 191:474–476.

Lipkow, J. 1960. Die Begattung bei der weissen Maus. Z. Tierpsychol. 17:182–187.

Maren, T. 1952. A summary of preliminary observations on genetically and chemically obese mice, p. 12–13. Conference on the obese mouse. Roscoe B. Jackson Memorial Laboratory, Bar Harbor, Maine.

Martin, G. M., D. Manring, and E. P. Benditt. 1960. Experimental turpentine abscesses in aggressive and non-aggressive inbred strains of mice. Fed. Proc. 19:20. (Abstr.)

Martin, P. G., and H. G. Andrewartha. 1962. Success in fighting of two varieties of mice. Amer. Natur. 96:375–376.

Mast, T. M., and N. W. Heimstra. 1962. Prior social experience and amphetamine toxicity in mice. Psychol. Rep. 11:809–812.

Mather, K. 1949. Biometrical genetics. Dover, New York. 158 p.

McClearn, G. E. 1958. Performance differences among mouse strains in a learning situation. Amer. Psychol. 13:405. (Abstr.)

McClearn, G. E. 1959. The genetics of mouse behavior in novel situations. J. Comp. Physiol. Psychol. 52:62–67.

McClearn, G. E. 1960. Strain differences in activity of mice: influence of illumination. J. Comp. Physiol. Psychol. 53:142–143.

McClearn, G. E. 1961. Genotype and mouse activity. J. Comp. Physiol. Psychol. 54:674–676.

McClearn, G. E. 1965. Genotype and mouse behaviour, p. 795–805. *In* S. J. Geerts [ed.] Proc. XI Int. Congr. Genet. (The Hague, 1963) Vol. 3. Pergamon Press, New York.

McClearn, G. E., and D. A. Rodgers. 1959. Differences in alcohol preference among inbred strains of mice. Quart. J. Stud. Alc. 20:691–695.

McClearn, G. E., and D. A. Rodgers. 1961. Genetic factors in alcohol preference of laboratory mice. J. Comp. Physiol. Psychol. 54:116–119.

McGill, T. E. 1962. Sexual behavior in three inbred strains of mice. Behaviour 19:341–350.

McGill, T. E., and W. C. Blight. 1963*a*. Effects of genotype on the recovery of sex drive in the male mouse. J. Comp. Physiol. Psychol. 56:887–888.

McGill, T. E., and W. C. Blight. 1963*b*. The sexual behaviour of hybrid male mice compared with the sexual behaviour of males of the inbred parent strains. Anim. Behav. 11:480–483.

Meier, G. W. 1964. Differences in maze performances as a function of age and strain of house mice. J. Comp. Physiol. Psychol. 58:418–422.

Meier, G. W., and D. P. Foshee. 1963. Genetics, age, and the variability of learning performances. J. Genet. Psychol. 102:267–275.

Meier, G. W., and D. P. Foshee; H. D. Winston and G. Lindzey. 1965. Albinism and water escape performance in mice. Science 147:307–308.

Mirone, L. 1958. The effect of ethyl alcohol on growth and voluntary consumption of alcohol by successive generations of mice. Quart. J. Stud. Alc. 19:388–393.

Moltz, H. 1965. Contemporary instinct theory and the fixed action pattern. Psychol. Rev. 72:27–47.

Mordkoff, A. M., and J. L. Fuller. 1959. Variability in activity within inbred and crossbred mice. J. Hered. 50:6–8.

Morgan, H. S. 1963. The effects of food and darkness on learning in two inbred strains of mice. M. A. Thesis. Cornell Univ.

Pavlov, I. P. 1923. New researches on conditioned reflexes. Science 58:359–361.

Ressler, R. H. 1962. Parental handling in two strains of mice reared by foster parents. Science 137:129–130.

Ressler, R. H. 1963. Genotype-correlated parental influences in two strains of mice. J. Comp. Physiol. Psychol. 56:882–886.

Rodgers, D. A., and G. E. McClearn. 1962*a*. Mouse strain differences in preference for various concentrations of alcohol. Quart. J. Stud. Alc. 23:26–33.

Rodgers, D. A., and G. E. McClearn. 1962*b*. Alcohol preference of mice, p. 68–95. *In* E. L. Bliss [ed.] Roots of behavior. Hoeber-Harper, New York.

Ross, S., and W. Smith. 1953. The hoarding behavior of the mouse. II. The role of deprivation, satiation, and stress. J. Genet. Psychol. 82:299–307.

Royce, J. R., and M. Covington. 1960. Genetic differences in the avoidance conditioning of mice. J. Comp. Physiol. Psychol. 53:197–200.

Runner, M. N., and A. Gates. 1954. Sterile, obese mothers. J. Hered. 45:51–55.

Schlesinger, K. 1964. Genetic and biochemical determinants of alcohol preference and alcohol metabolism in mice. Ph.D. Diss. Univ. California, Berkeley. 98 p.

Schlesinger, K., and A. M. Mordkoff. 1963. Locomotor activity and oxygen consumption: variability in two inbred strains of mice and their F_1 hybrids. J. Hered. 54:177.

Scott, J. P. 1944. An experimental test of the theory that social behavior determines social organization. Science 99:42–43.

Scott, J. P. 1946. Incomplete adjustment caused by frustration of untrained fighting mice. J. Comp. Physiol. Psychol. 39:379–390.

Scott, J. P. 1947. "Emotional" behavior of fighting mice caused by conflict between weak stimulatory and weak inhibitory training. J. Comp. Physiol. Psychol. 40:275–282.

Scott, J. P., and E. Fredericson. 1951. The causes of fighting in mice and rats. Physiol. Zool. 24:273–309.

Scriabine, A., and M. Blake. 1962. Evaluation of centrally acting drugs in mice with fighting behavior induced by isolation. Psychopharmacologia 3:224–226.

Silverstein, E. 1961. Effect of hybridization on the primary polydipsic trait of an inbred strain of mice. Nature 191:523.

Sirlin, J. L. 1956. Vacillans, a neurological mutant in the house mouse linked with brown. J. Genet. 54:42–48.

Smith, M. P., and S. Ross. 1960. Acceptance of sodium sucaryl by C57 black mice. J. Genet. Psychol. 96:101–104.

Smith, R. P., and W. Bevan. 1957. Maze performance, emotionality, audiogenic seizure susceptibility in rats and mice treated with 2-dimethylaminoethanol. Proc. Soc. Exp. Biol. Med. 96:382–385.

Smith, W., and E. K. Powell. 1955. The role of emotionality in hoarding. Behaviour 8:57–62.

Smith, W., and S. Ross. 1953*a*. The hoarding behavior of the mouse. I. The role of previous feeding experience. J. Genet. Psychol. 82:279–297.

Smith, W., and S. Ross. 1953*b*. The hoarding behavior of the mouse. III. The storing of "nonrelevant" material. J. Genet. Psychol. 82:309–316.

Southwick, C. H. 1959. Eosinophil responses of C57BR mice to behavioral disturbance. Ecology 40:156–157.

Southwick, C. H., and V. P. Bland. 1959. Effect of population density on adrenal glands and reproductive organs of CFW mice. Amer. J. Physiol. 197:111–114.

Stanley, W. C., and J. A. Monkman. 1956. A test for specific and general behavioral effects of infantile stimulation with shock in the mouse. J. Abnorm. Soc. Psychol. 53:19–22.

Tedeschi, R. E., D. H. Tedeschi, A. Mucha, L. Cook, P. A. Mattis, and E. J. Fellows. 1959. Effects of various centrally acting drugs on fighting behavior of mice. J. Pharmacol. Exp. Therap. 125:28–34.

Thiessen, D. D. 1961. Mouse exploratory behavior and body weight. Psychol. Rec. 11:299–304.

Thiessen, D. D., and V. G. Nealey. 1962. Adrenocortical activity, stress response and behavioral reactivity of five inbred mouse strains. Endocrinology 71:267–270.

Thiessen, D. D., J. F. Zolman, and D. A. Rodgers. 1962. Relation between adrenal weight, brain cholinesterase activity, and hole-in-wall behavior of mice under different living conditions. J. Comp. Physiol. Psychol. 55:186–190.

Thompson, W. R. 1953. The inheritance of behaviour: behavioural differences in fifteen mouse strains. Can. J. Psychol. 7:145–155.

Thompson, W. R. 1956. The inheritance of behavior; activity differences in five inbred mouse strains. J. Hered. 47:147–148.

Tollman, J., and J. A. King. 1956. The effects of

testosterone propionate on aggression in male and female C57BL/10 mice. Brit. J. Anim. Behav. 4:147–149.

Uhrich, J. 1938. The social hierarchy in albino mice. J. Comp. Psychol. 25:373–413.

Van Abeelen, J. H. F. 1963*a*. Mouse mutants studied by means of ethological methods. I. Ethogram. Genetica 34:79–94.

Van Abeelen, J. H. F. 1963*b*. Mouse mutants studied by means of ethological methods. II. Mutants and methods. Genetica 34:95–101.

Van Abeelen, J. H. F. 1963*c*. Mouse mutants studied by means of ethological methods. III. Results with yellow, pink-eyed dilution, brown and jerker. Genetica 34:270–286.

Vandenbergh, J. G. 1960. Eosinophil response to aggressive behavior in CFW albino mice. Anim. Behav. 8:13–18.

Vanderpool, D. L., and R. T. Davis. 1962. Differences in spontaneous behavior among inbred strains of mice. Psychol. Rep. 10:123–130.

Vicari, E. 1923. Hybridization and behavior. Eugen. Race State 2:75–77.

Vicari, E. 1924. Non-inheritance of the effects of training. Science 59:303.

Vicari, E. M. 1929. Mode of inheritance of reaction time and degrees of learning in mice. J. Exp. Zool. 54:31–88.

Waller, M. B., P. F. Waller, and L. A. Brewster. 1960. A water maze for use in studies of drive and learning. Psychol. Rep. 7:99–102.

Warren, R. P. 1963. Preference aversion in mice to bitter substance. Science 140:808–809.

Weir, M. W., and J. C. De Fries. 1964. Prenatal maternal influence on behavior in mice: evidence of a genetic basis. J. Comp. Physiol. Psychol. 58:412–417.

Willingham, W. W. 1956. The organization of emotional behavior in mice. J. Comp. Physiol. Psychol. 49:345–348.

Wimer, R. E., and J. L. Fuller. 1965. The effects of *d*-amphetamine sulphate on three exploratory behaviors. Can. J. Psychol. 19:94–103.

Wimer, R. E., and H. Sterns. 1964. Controlled visual input and exploratory activity in C57BL/6J mice. Percept. Mot. Skills 18:299–307.

Wimer, R., and S. Weller. 1965. Evaluation of a visual discrimination task for the analysis of the genetics of a mouse behavior. Percept. Mot. Skills 20:203–208.

Winston, H. D. 1963. Influence of genotype and infantile trauma on adult learning in the mouse. J. Comp. Physiol. Psychol. 56:630–635.

Winston, H. D., and G. Lindzey. 1964. Albinism and water escape performance in the mouse. Science 144:189–191.

Wolf, E. 1930. Die Aktivität der japanischen Tanzmaus und ihre rhythmische Verteilung. Z. Vergl. Physiol. 11:321–344.

Yerkes, R. M. 1907. The dancing mouse. Macmillan, New York. 290 p.

Bibliography of Techniques

George D. Snell and Katharine P. Hummel

This bibliography covers only material particularly pertinent to the mouse not covered in the main portion of this book. Only the more recent and important references are given for any one technique. General References includes those which describe many techniques or give bibliographies.

GENERAL REFERENCES

Avis, F. R. 1957. About mice and man. An introduction to mammalian biology. Walch, Portland, Maine. 194 p.

Billingham, R. E., and W. K. Silvers. 1961. Transplantation of tissues and cells. Wistar Inst. Press, Philadelphia. 149 p.

Burdette, W. J. [ed.] 1963. Methodology in mammalian genetics. Holden-Day, San Francisco. 646 p.

Cass, J. S., I. R. Campbell, and L. Lange. 1960. Laboratory animals, an annotated bibliography. Fed. Proc. 19 (No. 4, Part 3, Suppl. 6). 196 p.

Cass, J. S., I. R. Campbell, and L. Lange. 1963. Laboratory animals, an annotated bibliography. Fed. Proc. 22 (No. 2, Part 3, Suppl. 13). 250 p.

Gay, W. I. 1965. Methods of animal experimentation. Academic Press, New York. Vol. I, 382 p. Vol. II, 608 p.

BREEDING AND CARE

These subjects are discussed in Chap. 2 through 5. Certain subjects not dealt with in these chapters are covered below.

Ectoparasites: diagnosis and treatment

Flynn, R. J. 1954. Mouse mange. Proc. Anim. Care Panel 5:96–105.

Flynn, R. J. 1955. Ectoparasites of mice. Proc. Anim. Care Panel 6:75–91.

Flynn, R. J. 1963. The diagnosis of some forms of ectoparasitism of mice. Lab. Anim. Care 13: 111–125.

Heston, W. E. 1941. Parasites, p. 349–379. *In* G. D. Snell [ed.] Biology of the laboratory mouse. Blakiston, Philadelphia.

Keefe, T. J., J. E. Scanlon, and L. D. Wetherald. 1964. *Ornithonyssus bacoti* (*Hirst*) infestation in mouse and hamster colonies. Lab. Anim. Care 14:366–369.

Endoparasites: diagnosis and treatment

Chandler, A. C. 1952. Introduction to the worms, p. 236–252. *In* A. C. Chandler, Introduction to parasitology, 8th ed. Wiley, New York.

Fratta, I. D., and C. A. Slanetz. 1958. The treatment of oxyurid infested mice with piperazine citrate, stylomycin, caricide, phenothiazine, phthalofyne and gibberellic acid. Proc. Anim. Care Panel 8:141–146.

Habermann, R. T. 1951. Gastrointestinal parasites of laboratory animals. Proc. Anim. Care Panel 2:36–46.

Habermann, R. T., and F. P. Williams, Jr. 1957. The efficacy of some piperazine compounds and stylomycin in drinking water for the removal of oxyurids from mice and rats and a method of critical testing of anthelmintics. Proc. Anim. Care Panel 7:89–97.

Habermann, R. T., and F. P. Williams, Jr. 1963. Treatment of female mice and their litters with piperazine adipate in the drinking water. Lab. Anim. Care 13:41–45.

Habermann, R. T., F. P. Williams, Jr., and W. T. S. Thorp. 1954. Identification of some internal parasites of laboratory animals. Public

Health Service Publication No. 343, U.S. Government Printing Office, Washington, D.C. 29 p.
Herrlein, H. G. 1959. Elimination of oxyurids from the laboratory mouse. Proc. Anim. Care Panel 9:165–166.
Hoag, W. G. 1961. Oxyuriasis in laboratory mouse colonies. Amer. J. Vet. Res. 22:150–153.
King, V. M., and G. E. Cosgrove. 1963. Intestinal helminths in various strains of laboratory mice. Lab. Anim. Care 13:46–48.

GENETIC AND STATISTICAL TECHNIQUES

Rules for strain and gene nomenclature are given in Chap. 6; genetic, statistical, and breeding techniques are covered in Chap. 2 and 9; references on methods for testing linkage will be found in Chap. 8; and methods for the study of histocompatibility genes are described in Chap. 24. Useful general references on genetic and statistical techniques will be found in Chap. 2, 8, and 9.

CHROMOSOME CYTOLOGY

Bunker, M. C. 1961. A technique for staining chromosomes of the mouse with Sudan Black B. Can. J. Genet. Cytol. 3:355–360.
Bunker, M. C. 1965. Chromosome preparations from solid tumors of the mouse: a direct method. Can. J. Genet. Cytol. 7:78–83.
Evans, E. P., G. Breckon, and C. E. Ford. 1964. An air-drying method for meiotic preparations from mammalian testes. Cytogenetics 3:289–294.
Fechheimer, N. S. 1960. Mammalian chromosome counts: a simple method for making preparations. Nature 188:247–248.
Ford, C. E., and J. L. Hamerton. 1956. A colchicine, hypotonic citrate, squash sequence for mammalian chromosomes. Stain Technol. 31: 247–251.
Ford, E. H. R., and D. H. M. Woollam. 1963. A colchicine, hypotonic citrate, air drying sequence for foetal mammalian chromosomes. Stain Technol. 38:271–274.
Fox, M., and I. M. Zeiss. 1961. Chromosome preparations from fresh and cultured tissues using a modification of the drying technique. Nature 192:1,213–1,214.
Stich, H. F., and T. C. Hsu. 1960. Cytological identification of male and female somatic cells in the mouse. Exp. Cell Res. 20:248–249.
Tsuchida, R., and M. A. Rich. 1964. Chromosomal aberrations in viral leukemogenesis. I. Friend and Rauscher leukemia. J. Nat. Cancer Inst. 33: 33–47.
Welshons, W. J., B. H. Gibson, and B. J. Scandlyn. 1962. Slide processing for the examination of male mammalian meiotic chromosomes. Stain Technol. 37:1–5.

TECHNIQUES RELATED TO REPRODUCTION

Reproduction is discussed in Chap. 11. Certain techniques not covered in that chapter are dealt with below.

Vaginal smears for determining estrus

Allen, E. 1922. The oestrous cycle in the mouse. Amer. J. Anat. 30:297–371.
Fowler, R. E., and R. G. Edwards. 1957. Induction of superovulation and pregnancy in mature mice by gonadotrophins. J. Endocrinol. 15:374–384.
Snell, G. D. 1941. Reproduction, p. 55–88. *In* G. D. Snell [ed.] Biology of the laboratory mouse. Blakiston, Philadelphia.
Snell, G. D., K. P. Hummel, and W. H. Abelmann. 1944. A technique for the artificial insemination of mice. Anat. Rec. 90:243–253.

Induction of ovulation

Fowler, R. E., and R. G. Edwards. 1957. Induction of superovulation and pregnancy in mature mice by gonadotrophins. J. Endocrinol. 15: 374–384.
Lamond, D. R. 1960. Induction of ovulation in mice with placental gonadotrophins. J. Endocrinol. 20:277–287.
Runner, M. N., and A. Gates. 1954. Conception in prepuberal mice following artificially induced ovulation and mating. Nature 174:222–223.
Wilson, E. D., and M. X. Zarrow. 1962. Comparison of superovulation in the immature mouse and rat. J. Reprod. Fertil. 3:148–158.

Artificial insemination

Dziuk, P. J., and M. N. Runner. 1960. Recovery of blastocysts and induction of implantation following artificial insemination of immature mice. J. Reprod. Fertil. 1:321–331.
Kile, J. C., Jr. 1951. An improved method for the artificial insemination of mice. Anat. Rec. 109: 109–117.
Snell, G. D., K. P. Hummel, and W. H. Abelmann. 1944. A technique for the artificial insemination of mice. Anat. Rec. 90:243–253.
Wolfe, H. G., and J. L. Southard. 1962. Produc-

tion of all-dystrophic litters of mice by artificial insemination. Proc. Soc. Exp. Biol. Med. 109: 630–633.

Transplanting ova: unfertilized ova

Runner, M. N., and J. Palm. 1953. Transplantation and survival of unfertilized ova of the mouse in relation to postovulatory age. J. Exp. Zool. 124:303–316.

Transplanting ova: fertilized ova

Deringer, M. K. 1963. Technique for the transfer of fertilized ova, p. 563–564. *In* W. J. Burdette [ed.] Methodology in mammalian genetics. Holden-Day, San Francisco.

Fekete, E. 1947. Differences in the effect of uterine environment upon development in the dba and C57 black strains of mice. Anat. Rec. 98:409–415.

Fekete, E., and C. C. Little. 1942. Observations on the mammary tumor incidence of mice born from transferred ova. Cancer Res. 2:525–530.

McLaren, A., and D. Michie. 1956. Studies on the transfer of fertilized mouse eggs to uterine foster mothers. I. Factors affecting the implantation and survival of native and transferred eggs. J. Exp. Biol. 33:394–416.

Transplanting ovaries

Jones, E. C., and P. L. Krohn. 1960. Orthotopic ovarian transplantation in mice. J. Endocrinol. 20:135–146.

Palm, J. 1961. Transplantation of ovarian tissue, p. 49–56. *In* R. E. Billingham and W. K. Silvers [ed.] Transplantation of tissues and cells. Wistar Inst. Press, Philadelphia.

Parrott, D. M. V. 1960. The fertility of mice with orthotopic ovarian grafts derived from frozen tissue. J. Reprod. Fertil. 1:230–241.

Robertson, G. G. 1940. Ovarian transplantation in the house mouse. Proc. Soc. Exp. Biol. Med. 44: 302–304.

Russell, W. L., and J. G. Hurst. 1945. Pure strain mice born to hybrid mothers following ovarian transplantation. Proc. Nat. Acad. Sci. 31:267–273.

Stevens, L. C. 1957. A modification of Robertson's technique of homoiotopic ovarian transplantation in mice. Transplant. Bull. 4:106–107.

Transplanting mammary glands

Prehn, R. T. 1953. Tumors and hyperplastic nodules in transplanted mammary glands. J. Nat. Cancer Inst. 13:859–872.

Thompson, J. S. 1963. Transplantation of whole mammary glands in mice. Transplantation 1: 526–534.

EMBRYOLOGICAL PREPARATIONS AND TECHNIQUES

Information and references concerning embryological preparations and techniques will be found in Chap. 12 and 25. A few additional references are given below.

Carter, T. C. 1954. The genetics of luxate mice. IV. Embryology. J. Genet. 52:1–35. (A good description of method of collecting embryos, fixing, staining, etc., and making graphic reconstructions.)

Mintz, B. 1964. Formation of genetically mosaic mouse embryos, and early development of "lethal (t^{12}/t^{12})-normal" mosaics. J. Exp. Zool. 157: 273–292.

New, D. A. T., and K. F. Stein. 1963. Cultivation of mouse embryos *in vitro*. Nature 199:297–299.

Rawles, M. E. 1940. The development of melanophores from embryonic mouse tissues grown in the coelom of chick embryos. Proc. Nat. Acad. Sci. 26:673–680.

Orsini, M. W. 1962. Study of ova-implantation in the hamster, rat, mouse, guinea-pig, and rabbit in cleared uterine tracts. J. Reprod. Fertil. 3: 288–293.

ANATOMICAL PREPARATIONS

Injection of arterial system

West, W. T., and L. W. Gorham. 1962. Injection of the arterial system of the mouse. Stain Technol. 37:99–103.

Whole mounts of mammary glands

Dalton, A. J. 1945. Cytology of mammary tumors of the mouse, p. 7–12. *In* F. R. Moulton [ed.] A symposium on mammary tumors in mice. Science Press, Lancaster, Pa.

Richardson, F. L., and A. M. Cloudman. 1947. The mammary gland development in male mice at nine weeks of age. Anat. Rec. 97:223–237.

Preparing skins

Miller, D. F., and G. Blaydes. 1962. Methods and materials for teaching biological sciences, 2nd ed. McGraw-Hill, New York. 442 p.

Preparations for demonstrating bone and cartilage

Bechtol, C. O. 1948. Differential in toto staining of bone, cartilage and soft tissues. Stain Technol. 23:3–8.

Crary, D. D. 1962. Modified benzyl alcohol clearing of alizarin-stained specimens without loss of flexibility. Stain Technol. 37:124–125.

Green, M. C. 1952. A rapid method for clearing and staining specimens for the demonstration of bone. Ohio J. Sci. 52:31–33.

Grüneberg, H. 1953. Genetical studies on the skeleton of the mouse. VII. Congenital hydrocephalus. J. Genet. 51:327–358.

Noback, G. J. 1916. The use of the Van Wijhe method for the staining of the cartilaginous skeleton. Anat. Rec. 11:292–294.

Searle, A. G. 1954. Genetical studies on the skeleton of the mouse. IX. Causes of skeletal variation within pure lines. J. Genet. 52:68–102. (This gives a method for papain digestion of soft tissues.)

Williams, T. W., Jr. 1941. Alizarin red S and toluidin blue for differentiating adult or embryonic bone and cartilage. Stain Technol. 16: 23–25.

ANESTHESIA

General

Bolz, W. 1961. General anesthesia in domesticated, wild, zoo, and laboratory animals. Ferdinand Enke, Stuttgart. 256 p.

Croft, P. G. 1960. An introduction to the anesthesia of laboratory animals. Universities Federation for Animal Welfare, London. 32 p.

Lumb, W. V. 1963. Small animal anesthesia. Lea & Febiger, Philadelphia. 420 p.

Cooling

East, J., and D. M. V. Parrott. 1962. Operative techniques for newborn mice using anesthesia by cooling. J. Endocrinol. 24:249–250.

Ether

Buchsbaum, M., and R. Buchsbaum. 1962. Age and ether anesthesia in mice. Proc. Soc. Exp. Biol. Med. 109:68–70.

Avertin

Jones, E. C., and P. L. Krohn. 1960. Orthotopic ovarian transplantation in mice. J. Endocrinol. 20:135–146.

Pentobarbital

Billingham, R. E. 1961. Free skin grafting in mammals, p. 1–34. *In* R. E. Billingham and W. K. Silvers [ed.] Transplantation of tissues and cells. Wistar Inst. Press, Philadelphia.

Pilgrim, H. I., and K. B. DeOme. 1955. Intraperitoneal pentobarbital anesthesia in mice. Exp. Med. Surg. 13:401–403.

Methoxyflurane

Hagen, E. O., and J. M. Hagen. 1964. A method of inhalation anesthesia for laboratory mice. Lab. Anim. Care 14:13–15.

OPERATIVE TECHNIQUES

General

See also General References, especially Avis (1957) and Billingham and Silvers (1961).

Farris, E. J., and J. Q. Griffith [ed.] 1949. The rat in laboratory investigation, 2nd ed. Lippincott, Philadelphia. 542 p.

Siegler, R., and M. A. Rich. 1963. Artificial respiration in mice during thoracic surgery: a simple, inexpensive technic. Proc. Soc. Exp. Biol. Med. 114:511–513.

Adrenalectomy

Grollman, A. 1941. Biological assay of adrenal cortical activity. Endocrinology 29:855–861.

Llaurado, J. G. 1958. A method for rapid adrenalectomy in the rat. J. Anim. Tech. Ass. 8:75–78.

Gonadectomy

Avis, F. R. 1957. About mice and man. An introduction to mammalian biology. Walch, Portland, Maine. 194 p.

East, J., and D. M. V. Parrott. 1962. Operative techniques for newborn mice using anaesthesia by cooling. J. Endocrinol. 24:249–250.

Russell, W. L., and J. G. Hurst. 1945. Pure strain mice born to hybrid mothers following ovarian transplantation. Proc. Nat. Acad. Sci. 31:267–273.

Hypophysectomy

el-Bolkainy, M. N. 1963. Technique for hypophysectomy of the mouse. J. Nat. Cancer Inst. 30:1,077–1,089.

Falconi, G., and G. L. Rossi. 1964. Transauricular hypophysectomy in rats and mice. Endocrinology 74:301–303.

Jensen, E. 1963. An improved technique for

hypophysectomy of young mice. Acta Pathol. Microbiol. Scand. 59:301–303.

Lostroh, A. J., and C. W. Jordan, Jr. 1955. Improved procedure for hypophysectomy of the mouse. Proc. Soc. Exp. Biol. Med. 90:267–269.

Thomas, Frédéric. 1938. A technic for hypophysectomy of the mouse. Endocrinology 23:99–103.

Mammectomy

Dux, A. 1962. Total mammectomy in female mice and rats. Nature 196:287–288.

Staff of the Cancer Research Genetics Laboratory, University of California. 1963. Current applications of a method of transplantation of tissues into gland-free mammary fat pads of mice, p. 565–569. *In* W. J. Burdette [ed.] Methodology in mammalian genetics. Holden-Day, San Francisco.

Partial hepatectomy

Feigelson, M., P. Feigelson, and P. R. Gross. 1957. Xanthine oxidase activity in regenerating liver. J. Gen. Physiol. 41:233–242.

Radiothyroidectomy

Gorbman, A. 1950. Functional and structural changes consequent to high dosages of radioactive iodine. J. Clin. Endocrinol. 10:1,177–1,191.

Splenectomy

Haller, J. A., Jr. 1964. The effect of neonatal splenectomy on mortality from runt disease in mice. Transplantation 2:287–291.

Searle, A. G. 1959. Hereditary absence of spleen in the mouse. Nature 184:1,419–1,420. (Description of a "genetically splenectomized" mouse.)

Thymectomy

East, J., and D. M. V. Parrott. 1962. Operative techniques for newborn mice using anaesthesia by cooling. J. Endocrinol. 24:249–250.

Kaplan, H. S. 1950. Influence of thymectomy, splenectomy, and gonadectomy on incidence of radiation-induced lymphoid tumors in strain C57 black mice. J. Nat. Cancer Inst. 11:83–90.

Miller, J. F. A. P. 1962. Effect of neonatal thymectomy on the immunological responsiveness of the mouse. Proc. Roy. Soc. B 156:415–428.

O'Gara, R. W., and J. Ards. 1961. Intrasplenic transplantation of neonatal thymus. J. Nat. Cancer Inst. 27:277–297.

Sjodin, K., A. P. Dalmasso, J. M. Smith, and C. Martinez. 1963. Thymectomy in newborn and adult mice. Transplantation 1:521–525.

Parabiosis

Bunster, E., and R. K. Meyer. 1933. An improved method of parabiosis. Anat. Rec. 57:339–343.

Wilson, D. B. 1961. Parabiosis, p. 57–59. *In* R. E. Billingham and W. K. Silvers [ed.] Transplantation of tissues and cells. Wistar Inst. Press, Philadelphia.

Skin grafting

Bailey, D. W. 1963. Histoincompatibility associated with the X chromosome in mice. Transplantation 1:70–74.

Bailey, D. W., and B. Usama. 1960. A rapid method of grafting skin on tails of mice. Transplant. Bull. 7:424–425.

Billingham, R. E. 1961. Free skin grafting in mammals, p. 1–26. *In* R. E. Billingham and W. K. Silvers [ed.] Transplantation of tissues and cells. Wistar Inst. Press, Philadelphia.

Billingham, R. E., and P. B. Medawar. 1951. The technique of free skin grafting in mammals. J. Exp. Biol. 28:385–402.

INJECTIONS

Intravenous

See also Bleeding Methods.

Anderson, N. F., E. J. Delorme, M. F. A. Woodruff, and D. C. Simpson. 1959. An improved technique for intravenous injection of newborn rats and mice. Nature 184:1,952–1,953.

Barnes, D. W. H., C. E. Ford, and J. E. Harris. 1963. Intravenous injection of young mice. Transplantation 1:574.

Billingham, R. E. 1961. The induction of tolerance of homologous tissue grafts, p. 87–106. *In* R. E. Billingham and W. K. Silvers [ed.] Transplantation of tissues and cells. Wistar Inst. Press, Philadelphia.

Billingham, R. E., and L. Brent. 1956. Acquired tolerance of foreign cells in newborn animals. Proc. Roy. Soc. B 146:78–90.

Crispens, C. G., and N. Kaliss. 1961. A simple device for facilitating injections in the tail veins of mice. Amer. J. Clin. Pathol. 35:387–388.

Nickson, J. J., and S. S. Barkulis. 1948. An apparatus to facilitate intravenous injections in the mouse. Science 107:229–230.

Pinkerton, W., and M. Webber. 1964. A method of injecting small laboratory animals by the ophthalmic plexus route. Proc. Soc. Exp. Biol. Med. 116:959–961.

Intracardiac

Grazer, F. M. 1958. Technic for intravascular injection and bleeding of newborn rats and mice. Proc. Soc. Exp. Biol. Med. 99:407–409.

Postnikova, Z. A. 1960. A method of intracardiac injection in newborn rats and mice. Folia Biol. (Prague) 6:59–60.

TECHNIQUES RELATED TO BLOOD

Various techniques relating to hematology and blood clotting are described in Chap. 17 and 18. A few other techniques related to blood are cited.

Bleeding methods

See also Injections.

Ambrus, J. L., M. Ambrus, J. W. E. Harrisson, C. A. Leonard, C. E. Moser, and H. Cravitz. 1951. Comparison of methods for obtaining blood from mice. Amer. J. Pharm. 123:100–104.

Fuson, R. B., and O. N. Rambo, Jr. 1957. An improved method for the bleeding of mice. Transplant. Bull. 4:147–148.

Grice, H. C. 1964. Methods for obtaining blood and for intravenous injections in laboratory animals. Lab. Anim. Care 14:483–493.

Riley, V. 1960. Adaptation of orbital bleeding technic to rapid serial blood studies. Proc. Soc. Exp. Biol. Med. 104:751–754.

Russell, E. S. 1963. Techniques for the study of anemias in mice, p. 558–562. *In* W. J. Burdette [ed.] Methodology in mammalian genetics. Holden-Day, San Francisco.

Stimpfling, J. H., E. A. Boyse, and R. Mishell. 1964. Preparation of isoantisera in laboratory mice. Method. Med. Res. 10:18–21. (A description of bleeding from the tail. A very simple method if sterility is not important.)

Stone, S. H. 1954. Method for obtaining venous blood from the orbital sinus of the rat or mouse. Science 119:100.

Perfusion and infusion

Avis, F. R. 1957. The perfusion of organs in vitro, p. 169–182. *In* F. R. Avis, About mice and man. An introduction to mammalian biology. Walch, Portland, Maine.

Hint, H. C., and A. W. Richter. 1958. A simple intravenous infusion technique for mice. Method and some applications. Acta Pharmacol. Toxicol. 14:153–157.

Laszlo, J., S. R. Humphreys, and A. Goldin. 1960. Effects of glucose analogues (2-deoxy-D-glucose, 2-deoxy-D-galactose) on experimental tumors. J. Nat. Cancer Inst. 24:267–279.

Techniques for the study of anemias

Russell, E. S. 1963. Techniques for the study of anemias in mice, p. 558–562. *In* W. J. Burdette [ed.] Methodology in mammalian genetics. Holden-Day, San Francisco.

Blood pressure determinations

Chevillard, L., M. C. Arnal, and H. Giono. 1957. Pression artérielle et fréquence cardiaque de la souris. Compt. Rend Soc. Biol. 151:276–279.

Griffith, J. Q. 1934. Indirect method for determining blood pressure in small animals. Proc. Soc. Exp. Biol. Med. 32:394–396.

Henry, J. P., J. P. Meehan, P. Stevens, and G. A. Santisteban. 1965. Arterial pressure in CBA mice as related to age. J. Gerontol. 20:239–243.

Knudsen, P. J. 1957. Indirect determination of peripheral blood pressure in mice during light anaesthesia. Acta Physiol. Scand. 39:137–140.

McMaster, P. D. 1941. A method to determine the peripheral arterial blood pressure in the mouse. J. Exp. Med. 74:29–39.

Wu, C. H., and M. B. Visscher. 1948. Adaptation of the tail plethysmograph to blood pressure measurement in the mouse with some observations on the effects of temperature. Amer. J. Physiol. 153:330–335.

IMMUNOLOGICAL TECHNIQUES

Information on the use of alloantisera (isoantisera) in histocompatibility typing and blood group typing will be found in Chap. 24. Additional references dealing with immunological techniques are given in Chap. 31. Some other references are listed below.

Preparation of alloantisera (isoantisera)

Stimpfling, J. H., E. A. Boyse, and R. Mishell. 1964. Preparation of isoantisera in laboratory mice. Meth. Med. Res. 10:18–21.

Production of sperm autoantibodies

Snell, G. D., and H. Poucher. 1943. Relation of number of injections to the titer of sperm isoagglutinins in mice. Proc. Soc. Exp. Biol. Med. 54:261–263.

Red-cell agglutination techniques

Gorer, P. A., and Z. B. Mikulska. 1954. The antibody response to tumor inoculation: improved methods of antibody detection. Cancer Res. 14:651–655.

Stimpfling, J. H. 1961. The use of PVP as a developing agent in mouse hemagglutination tests. Transplant. Bull. 27:109–111.

Stimpfling, J. H. 1964. Methods for detection of hemagglutinins in mouse. Meth. Med. Res. 10:22–26.

Leukocyte agglutination techniques

Mishell, R. 1964. Leukocyte agglutination in mice. Meth. Med. Res. 10:35–38.

Cytotoxic technique

Boyse, E. A., L. J. Old, and I. Chouroulinkov. 1964. Cytotoxic test for demonstration of mouse antibody. Meth. Med. Res. 10:39–47.

Sanderson, A. R. 1964. Cytotoxic reactions of mouse iso-antisera: preliminary considerations. Brit. J. Exp. Pathol. 45:398–408.

Hemolysis and complement fixation

Winn, H. J. 1964. Hemolysis and complement fixation. Meth. Med. Res. 10:48–57.

Fluorescent antibody technique

Möller, G. 1964. Fluorescent antibody technique for demonstration of isoantigens in mice. Meth. Med. Res. 10:58–69.

Preparation of the *H-2* alloantigen

Kandutsch, A. A. 1964. Isolation of transplantation isoantigens of mice. Meth. Med. Res. 10:70–75.

TUMOR TRANSPLANTATION

Various techniques for low temperature storage of tumors are described in Chap. 28. References related to some other techniques are given below.

Transplantation of tumors

Cancer Chemotherapy National Service Center. December, 1962. Protocols for screening. Cancer Chemotherapy Reports No. 25. Public Health Service, Washington, D.C., 184 p.

Transformation of solid to ascites tumors

Goldie, H. 1956. Growth characteristics of free tumor cells in various body fluids and tissues of the mouse. Ann. N.Y. Acad. Sci. 63:711–719.

Klein, E. 1955. Transformation of solid into ascites tumors. Almqvist and Wiksells, Uppsala, Sweden. 40 p.

Klein, G., and E. Klein. 1956. Conversion of solid neoplasms into ascites tumors. Ann. N.Y. Acad. Sci. 63:640–661.

Yolk sac cultivation of tumors

Taylor, A., R. E. Hungate, and D. R. Taylor. 1943. Yolk sac cultivation of tumors. Cancer Res. 3:537–541.

MISCELLANEOUS

Determination of basal oxygen consumption

Mayer, J., R. E. Russell, M. W. Bates, and M. M. Dickie. 1952. Basal oxygen consumption of hereditarily obese and diabetic mice. Endocrinology 50:318–323.

Preparation of cell suspensions

Boyse, E. A. 1960. A method for the production of viable cell suspensions from solid tumors. Transplant. Bull. 7(25):100–104.

Cowan, W. K. 1962. The use of oxidised ascorbic acid for cell dispersal. J. Pathol. Bacteriol. 84:439–441.

Gwatkin, R. B. L., and J. L. Thomson. 1964. A new method for dispersing the cells of mammalian tissues. Nature 201:1,242–1,243.

Snell, G. D. 1953. A cytosieve permitting sterile preparation of suspensions of tumor cells for transplantation. J. Nat. Cancer Inst. 13:1,511–1,515.

Collection of thoracic duct lymph

Boak, J. L., and M. F. A. Woodruff. 1965. A modified technique for collecting mouse thoracic duct lymph. Nature 205:396–397.

Determination of viability of cells in suspension

Novelli, A. 1962. Amethyst violet as a strain for distinguishing cells with a damaged membrane from normal cells. Experientia 18:295–296.

Tennant, J. R. 1964. Evaluation of the trypan blue technique for determination of cell viability. Transplantation 2:685–694.

Index